THE BRITISH ISLANDS
AND THEIR
VEGETATION

CAMBRIDGE
UNIVERSITY PRESS
LONDON: BENTLEY HOUSE
NEW YORK, TORONTO, BOMBAY
CALCUTTA, MADRAS: MACMILLAN
TOKYO: MARUZEN COMPANY LTD

All rights reserved

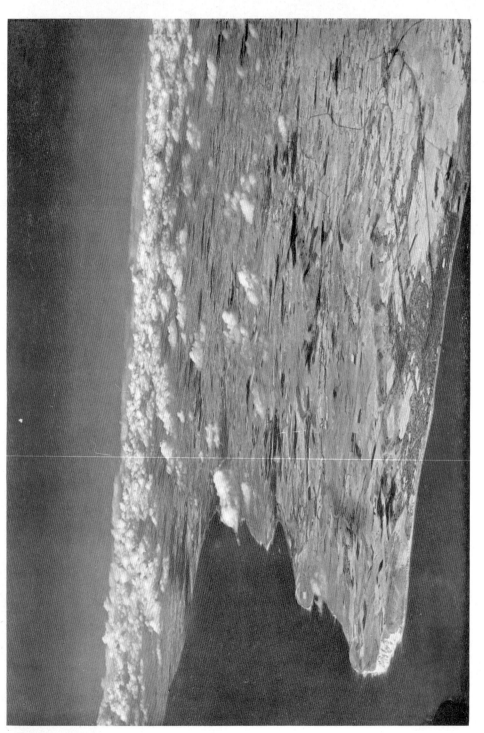

The country behind Beachy Head, East Sussex, seen from the air, looking WNW. The chalk cliffs of Beachy Head are seen in the left-hand bottom corner. Thence the escarpment of the South Downs, bearing chalk grassland, runs across the picture to the right, behind the town of Eastbourne in the foreground, and turning WNW can be traced in a sinuous line broken by two river valleys (Cuckmere and Ouse). Beyond Lewes on the Ouse (right centre) the escarpment runs westwards but is partly obscured by cumulus clouds which also cover the distant landscape at the top of the picture. Except for the Downs almost the whole of the country, which is here very poorly wooded, is seen to be cultivated.

Photograph from *The Times*.

THE BRITISH ISLANDS
AND THEIR
VEGETATION

BY

A. G. TANSLEY

M.A., F.R.S.

Lately Sherardian Professor of Botany, Oxford
President of the British Ecological Society
Editor and Part Author of Types of British Vegetation, 1911
Editor of The Journal of Ecology, 1916–37

With 162 Plates containing 418 photographs
and 179 Figures in the text

CAMBRIDGE
AT THE UNIVERSITY PRESS
1939

PREFACE

This book is intended to replace *Types of British Vegetation*, which was published in 1911, has been out of print for more than twenty years, and has long fetched fantastic prices in the secondhand market.

Meanwhile, the study of British vegetation in the field, and to some extent also in the laboratory, has made great progress. The knowledge of our natural and semi-natural plant communities is much wider and especially much deeper than it was in 1911. On the other hand, various problems that we then envisaged as relatively simple have shown themselves upon further study to be very complex indeed and not yet susceptible of satisfactory solution. Thus the writing of this book has been a very different task from the writing of *Types*. In 1911 we wrote practically all we knew and a good deal that we guessed; and though many of our guesses were not far from the truth others have not unnaturally turned out to be wide of the mark. A new generation of workers has grown up, with deeper knowledge and better training than those of the small band of pioneers in the early years of the century, and with no less enthusiasm and belief in the subject. Without the results of their work and without their generous, unstinted help this book could not have covered the ground with any adequacy.

Of necessity this is largely a compilative work, a great deal of the material being taken straight from published papers; but I have tried to weld the whole into a continuous story which can be read, and not merely "consulted", by the student—British or foreign—who wants trustworthy information about the British plant communities, but who is not an ecological specialist. For this reason I have tried to make the exposition self-explanatory, and many passages contain matter which is common knowledge to serious students of the subject. This treatment will also, I hope, help to relieve the unreadableness of long lists of species which have to be included if concreteness is to be given to the descriptions of the various communities.

In Part I (Environment) and Part II (History) what may be called the "background" of British vegetation is treated with considerable amplitude, more indeed than is strictly relevant to the vegetation itself. This wide treatment is expressed in the wording of the title and will, it is hoped, prove useful to students of British geography and others whose interests are not narrowly confined to plant ecology, as well as to foreign readers. For the data used in this part of the book I have relied mainly on good secondary sources (quoted in the Lists of References appended to each chapter), since it was not practicable to consult all the original authorities over so wide a field.

It is to be regretted that climate, as the paramount ecological factor-complex, cannot yet be brought into closer and more detailed relation with the vegetation of different parts of these islands; but it has to be recognised that the serious study of these relations has hardly been begun. We still have to be content with general and rather vague correspondences, lacking anything like a real quantitative basis. The available meteorological records are quite inadequate for any precise correlation between climatic factors and vegetation. It will only be when sufficient and suitable climatic and micro-climatic data are available for attempts at such correlation that we shall be in a position to appreciate the interaction of climatic with other factors.

In Part III I have tried to present what I believe to be the valid essentials of the modern theory of vegetation; and some supplementary explanation here, addressed rather to those interested in the theory of vegetation than to the general student, will avoid overloading the exposition in the text with theoretical discussion. My exposition is based of course on what has been called the "dynamic" point of view, which is simply the explicit recognition that natural and semi-natural vegetation is constantly changing, that certain uniformities in the direction, methods, and causes of change can be detected, and that positions of relative equilibrium are reached in which the conditions and composition of the vegetation remain approximately constant for a longer or shorter time. Such a position of relative stability is here called a "climax", whether it represents a major community corresponding with a climatic region and primarily determined by climate, or a minor community determined by some other factor or combination of factors. This usage departs from that of Clements and others who restrict the term climax to the climatic climax. My own view is that the position of relative equilibrium, corresponding with what I have called the mature "eco-system", is the fundamental ecological concept, and that relatively stable communities should receive the same general name, suitably qualified in each case, whatever the factors involved.

Of course these "positions of equilibrium" are seldom if ever really "stable". On the contrary, they contain many elements of instability and are very vulnerable to apparently small changes in the factor-complex. Recognition of "positions of stability" is a necessary first step in the understanding of vegetation. The more important sequel is study of the factors which maintain or disturb and often upset them.

The term *formation* I use, correspondingly, for a mature eco-system dominated by distinctive life forms. Since eco-systems developed under essentially similar (though not of course identical) conditions, and dominated by the same life forms, are often widely separated geographically and may possess entirely different floras, it is convenient to consider them as separate formations, but belonging to the same *formation-type*. Thus I regard not only a climatic type such as deciduous summer forest, but also an edaphic type such as reedswamp, as a formation-type. Each of these formation-types consists of different formations composed of various dominants of the same general and highly distinctive life form in different parts of the world. Each is a climax in the sense adopted because it represents a well-marked position of *relative* vegetational stability, which may persist indefinitely in equilibrium with relatively stable environmental conditions. In the case of deciduous summer forest the stabilising condition is a complex of climatic factors, in the case of reedswamp a stable water-level at a certain height above the soil. It is true that reedswamp quite normally gives way to land vegetation and ultimately to the climatic climax when the soil-level rises relatively to the water-level by silting or by accumulation of plant debris and the resulting humus. Nevertheless, reedswamp represents a position of *relative* stability, which may be maintained indefinitely under appropriate conditions and is marked by distinctive life form. Similarly, I regard pastured grassland as a formation-type, again marked by distinctive life forms, though here the stabilising factor is zoogenic and usually anthropogenic.

In this way the old conception of a plant formation as a natural unit of vegetation characterised by definite dominant life forms may be combined with the newer conception which recognises development as fundamental; for positions of relative stability in development correspond well enough with establishment of the dominance of specific life forms, and can thus be taken as fixed points for the schematisation of the various successional series which we see in vegetation.

Schematisation is always necessary when we are trying to bring observed facts together and arrive at a comprehensive view. But the conceptions which we form and the terms we employ or invent in the process should never be mistaken for facts of nature. They are simply creations of the human mind which assist in the understanding and correlation of the facts of nature. As such they are inevitably subject to differences of opinion, for different minds will always tend to prefer different criteria in constructing classifications. Ultimately the classifications that do least violence to the facts will survive and the beds of Procrustes will be discarded.

Meanwhile controversial heat is engendered only if the disputants confuse their concepts and schemes with the natural phenomena they refer to. Thus the characterisation of certain plant communities by specific dominant life forms in harmony with the whole complex of conditions in which the community lives is a fact, and so is the continuous change of vegetation between one state of relative equilibrium and another. But whether it is best to use the term "formation" for all communities of the same dominant life form (as many of the older plant geographers did), or for a major position of stability determined by a particular climatic complex (as

Clements does), or to apply the term to all positions of relative stability which are characterised by distinctive life forms (as I have done), or, finally, to discard the term altogether (as others would wish to do), is and must be a matter of opinion. It is not really significant for the progress of our knowledge of vegetation how the term "formation" is used (or whether it is used at all), provided we know the sense in which a particular author understands it, though it would doubtless be convenient if a particular use were universally accepted. What is important is that *both* the sets of facts—relating to life form and relating to positions of stability—should be clearly recognised.

The term *association* I use in the older European sense of a relatively major plant community dominated not only by distinctive life forms but by distinctive species. The very large number of small communities formerly (1920–30) described as "associations" by Du Rietz and some of his colleagues are now known by them as "sociations" (see p. 232n.). Following Clements I use *consociation* for a community of the rank of an association but with a single dominant species, *society* for a local dominance within an association or consociation. *Associes*, *consocies* and *socies* (again following Clements) are used for seral, i.e. transitory, communities which are undergoing change. Thus the same community may be an association *or* an associes. When we are dealing with a complete prisere we have, according to the usage of this book, succession from one formation to another, and thus associations and consociations become associes and consocies.

We must be on our guard against imagining that the element of rigidity unavoidably introduced by the application of such terms to the phenomena of nature is thereby imparted to the phenomena themselves. Vegetation is essentially kinetic, but may be arrested in various positions of relative stability through the constancy of incidence of particular factors. With the modification or removal of such factors change begins again. Further, a local dominant may become a general dominant with a shifting of the incidence of the factor complex. Thus an associes or a consocies may become stabilised as an association or a consociation, and these latter may lose their stability and become seral again; while the dominant of a society in one place may become the dominant of a consocies in another.

It is therefore not surprising that differences of opinion should arise as to the status to be assigned to a particular plant community. It is necessary to recognise that status may change and is never rigidly determined. But we cannot for that reason give up the attempt to determine status, which is a necessary procedure in attaining increased clarity.

Clements's method of avoiding these difficulties is to restrict the terms "formation" and "climax" to the major communities determined by the climatic complex, and the term "association" to the primary divisions of his (climatic) formations. If this is done it becomes necessary to invent a nomenclature for the numerous communities that do not belong to, but have to be interpreted in terms of, the climatic climax, which is given an absolute pre-eminence. Thus we get such terms as "proclimax", "dysclimax", "serclimax", and so on. As a practical terminology these are just as inconvenient as some of the awkward and cumbersome systems that are used on the continent, though they are more interesting because they represent attempts at dynamic interpretation.

If we take a middle course, and while recognising the fundamental nature of succession, identify climax with a condition of relative stability, we can retain formation and association in their old and most convenient meanings; but at the same time we must admit that there are no hard and fast limits between the categories denoted by the terms we employ, or, in other words, that communities may change their status.

I am unable to form an opinion as to the validity or usefulness of the terminology of plant communities invented by Dr J. Braun-Blanquet, and have therefore had to forgo any attempt to consider its possible application to British vegetation.

In the remainder of the book (Parts IV–IX) the vegetation itself is described. First the woodlands (Part IV)—the remaining fragments, nearly everywhere much modified, of the vegetation at one time dominant— the general *climatic* climax over most of the country; then the grasslands (Part V), occupying by far the largest area of any British formation, for the most part an anthropogenic formation-type replacing forest as the result of grazing. Next come the communities belonging to the hydroseres (Part VI) in their priseral sequence. Here we are able to follow the natural succession, because much more of the natural aquatic and semi-aquatic vegetation remains relatively undisturbed. Closely related to bog or moss, the climatic climax of the acidic hydrosere in the wettest climates, we have the heath formation (Part VII) which also appears in the xerosere on acidic sands. Upland heath or moor leads on naturally, with increasing altitude, to mountain vegetation (Part VIII). The last section (Part IX) deals with maritime vegetation, which belongs to priseres distinct from those of inland vegetation. I have not hesitated to devote to shingle beaches and their plant communities much more space than is warranted by their extent or importance in relation to other vegetation, because our knowledge of them has been so largely built up in this country.

Different plant communities have been treated somewhat differently according to the state of our knowledge. Thus general accounts are attempted of pedunculate oakwood, different types of beechwood, chalk grassland, salt marsh, and a few others of the better known types; while in dealing with some communities descriptions and lists of species are given from different localities where the community has been examined, because our knowledge is not yet comprehensive enough to make a general account possible. In other cases again a compromise between these two methods is adopted.

Lack of adequate knowledge of the vegetation over wide areas has inevitably led to unequal consideration of different parts of the country: some contribute but little to the general picture, others are almost entirely neglected. Data from the south-east of England (in a wide sense) are relatively over-prominent in the perspective because a very large proportion of the more thorough modern studies of vegetation have been made in that region.

It should be realised that the book is essentially a *description* of British vegetation and does not attempt to be a manual of British field ecology, though experimental as well as observational work is referred to where it has thrown light on the structure and development of our plant communities. Even purely descriptive work is very far from complete, and it is still true, as I wrote in the preface to *Types of British Vegetation* (p. xi) in 1911, that "the field of analytical and experimental ecology lies widely open to workers able and willing to devote themselves to the laborious tasks involved in the attack on the various problems underlying the phenomena of vegetation". We may, however, justly congratulate ourselves that there are now many young men of ability who turn naturally to ecological work as soon as they have taken their degrees, and that they are already contributing substantially to the advancement of the subject. As soon as the various "powers that be" have realised more fully that field ecology is the best of all trainings for forestry, for pastoral science, and for all careers in which a knowledge of the relation of vegetation to the land is of prime importance, the opportunities and facilities for such work will increase and multiply. That realisation has notably increased since the War, and it is still growing, though not so fast as we could wish. Botanists themselves have been slow to understand the nature and potentialities of the scientific study of vegetation. "Plants, yes; but vegetation, what is it but aggregates of plants?" The state of mind that is expressed in such a question shows a failure to recognise that because plants not only live in aggregates but have been developed and moulded by the gregarious life their behaviour cannot be understood unless these aggregates are studied for their own sakes, just as we must study human society if we would understand the behaviour of man. And this involves the detailed and laborious though fascinating tasks of the student of vegetation and ecology.

The study of vegetation, or synecology, always drives us to investigation of the behaviour of the individual species, or autecology, and as our knowledge of this deepens so does our understanding of the plant community and of the changes which it may undergo. Among the more pressing of the tasks referred to is therefore thorough investigation of the autecology of important and especially of dominant species, combined with what I have elsewhere called "the monographic method". By this I mean the following of a community throughout its geographical range and the study by observation and experiment of the behaviour of the dominant under the incidence of the different factors at work in different regions, of its relations to actual and potential competitors, and of its behaviour outside its range of dominance. This is the key to really fundamental understanding of the ecology of any community.

Determination of the actual modes and causes of the passage of one community into another under various conditions is the most vital point open to synecological attack, for it is always the behaviour of organisms under changing conditions that reveals their potentialities most clearly and throws the most light on their behaviour under more static and balanced conditions. That is why the study of succession is the most vital part of synecology. The few successions recorded in this book that have been studied with any approach to adequacy show clearly enough what can be gained through this method of attack; but they are still all too few.

The separation and study of ecotypes is another line of investigation that is urgently needed to elucidate the behaviour of important species. We very often suspect that it is the existence of different ecotypes of a species which accounts for its various behaviour in different habitats, but there can be no certainty and no real advance in knowledge till each case has been properly investigated; and this means a great deal of laborious cultural and genetical work. The fruitful investigations of Professor Stapledon's staff at Aberystwyth on genetically distinct ecotypes of pasture grasses and clovers are examples of what I mean, and valuable work of the same kind is being done at Kew and at Potterne by Dr Turrill, Mr Marsden Jones and others. What is wanted now is a joining of hands between the students of ecology proper, who come mainly from the universities, and the genetical taxonomists. The field is enormous and scarcely more than its fringes have been touched.

In regard to nomenclature of species there is no authoritative list of the names of British species which can safely be followed by one who is not a specialist. I therefore applied to the Kew Herbarium for assistance and (by kind permission of the Director and of the Curator of the Herbarium) Dr T. A. Sprague and his colleagues corrected my names in accordance with the most recent knowledge. I am very much indebted to them for the time they gave and the trouble they took over this task. It must, however, be distinctly understood that the names used are not to be taken as finally valid on the authority of Kew: they only represent the best that could be done at the moment, or rather in the spring of 1937. In spite of what I owe to Dr Sprague's kindly helpfulness I have decapitalised the initial letters of all the trivial names, contrary to his desire, because I am convinced that this is a reform (long since made by zoologists) which must ultimately be adopted by botanists. Into the arguments for and against the change this is not the place to enter.

Finally, I must discharge the pleasant duty of acknowledging the very substantial help I owe to friends, colleagues, former pupils and various specialists.

Mr E. G. Bilham's recently published work, *The Climate of the British Isles*, has been of great use in the revision of Chapters II and III, and the author has kindly given me supplementary information on various points, besides permitting the use of three of his figures.

My friend Mr C. G. T. Morison, Reader in Soil Science at Oxford, very kindly read and criticised the text of Chapters IV and V. The paragraph on p. 81 about the "weathering complex" I owe directly to him. Mr Charles Elton, Director of the Bureau of Animal Population at Oxford, willingly performed a similar service with Chapter VI. Mr V. S. Summerhayes, of the Royal Botanic Gardens, Kew, generously supplied me with unpublished information relating to his experiments on the effects of voles on grassland (p. 141).

Chapter VII (Pre-history) owes a great deal to the kindness of my friend Dr Godwin, University Lecturer in Botany at Cambridge, who has read the text more than once and has suggested important emendations. He has also given me the use of Fig. 42 from his recently published pollen analysis of Tregaron Bog peat. Professor C. Daryll Forde of Aberystwyth has been good enough to see that the archaeological statements are reasonably correct. I must also mention the special debt I owe to that admirable composite work, *An Historical Geography of England before* A.D. 1800, edited by Dr H. C. Darby (Cambridge, 1936), on which Chapter VIII is almost entirely based.

In Chapter XVI I am indebted to Dr J. S. Turner (now Professor of Botany at Sydney, N.S.W.), Dr A. S. Watt and Dr Paul Richards, all of the Cambridge Botany School, for detailed unpublished information on the Killarney woods, and to Dr Turner for permission to copy his woodland profile drawings. This contribution has added greatly to the value of the chapter. Dr Watt has kindly read Chapters XVIII and XIX, which are based almost entirely on his published work, and has cleared up some difficult points in the exposition. In Chapter XXI I owe the accounts of three high-level ashwoods to my former pupil, Dr J. L. Harley, and to Mr E. Price Evans, who most kindly allowed me to use their unpublished observations. To Dr Harley also, and to Dr W. H. Wilkins of the Department of Botany at Oxford, I owe information about the larger fungi of various woodland types. Mr R. Ross, of the British Museum (Natural History), has very generously allowed me to use, in Chapter XXIV, information contained in his unpublished thesis on the ecology of hawthorn scrub in south-west Cambridgeshire, and also to copy part of a diagram (Fig. 88) and to reproduce two photographs (Pl. 93, phots. 225–6).

Sir Reginald Stapledon, Director of the Welsh Plant Breeding Station Aberystwyth, has incurred my gratitude by reading and criticising the whole of Part V (Grasslands). I am happy to know that he approves generally of my treatment. Mr William Davies, of the same Station, has also kindly read and criticised Chapters xxv and xxvi. My pupil, Mr J. F. Hope Simpson, who has been continuing and extending my own work on chalk grasslands, has read the part of Chapter xxvii dealing with these, and has corrected and amplified some of the statements in accordance with his own (as yet unpublished) knowledge.

Dr W. H. Pearsall, of the University of Leeds (now Professor of Botany at Sheffield), has read Chapters xxx and xxxii, which are almost entirely based on his work, and as a result of correspondence and conversations with him several doubtful points have been cleared up and the chapters greatly improved. He has also very generously allowed the copying of unpublished maps of Esthwaite vegetation (Figs. 114, 115—reprinted as Figs. 124, 125). Dr Pearsall's help has been most valuable and I am very grateful to him.

Dr Godwin has read and criticised Chapters xxxii–xxxiv, and he and Miss Verona Conway have kindly allowed the use of their unpublished work on Tregaron Bog (p. 682) and the reproduction of Figs. 136 and 137. The chapters on the bog or moss formation were not easy to write, because it was impossible to ignore entirely the older work, which was carried out before the progress of knowledge had enabled us to form anything like a clear conception of the nature of bog communities and bog structure; until a wider range of British bogs has been reinvestigated it is difficult to be sure how far the older and the recent workers are talking about the same things. I am greatly indebted to Dr Godwin and especially to Professor Hugo Osvald of Stockholm, with both of whom I visited several of the Irish bogs in 1935, for help and information which have been the making of certain sections of these chapters.

To my friend and former colleague at Oxford, Dr A. R. Clapham, with whom I visited a wide range of Irish bogs in 1936, and a few of the Scottish Highland bogs as well as birch and pinewoods in 1937, I owe a deep debt of gratitude for constant help in the field and for lists of species collected or noted. Dr Clapham has also assisted me in many other ways, for example with the Highland woods described in Chapters xvi and xxii, and in the discussion of various theoretical problems which have arisen in the course of writing the book. To Mr H. Baker, also, another Oxford colleague, I owe thanks for putting at my disposal unpublished information about the vegetation of the upper Thames.

Of others who have rendered general assistance, Dr W. Watson of Taunton, the well-known lichenologist and bryologist, has contributed a great deal of useful information in correspondence. It is to him especially that the book owes a much fuller treatment of bryophytes and lichens than would have been possible without his published work. I regret that I have not been able to use his data so fully as I could have wished (for example, in the chapters on English woodlands) without overloading my accounts; but in several places the effect of his work will be apparent, and in the chapters on arctic-alpine vegetation I have not scrupled to reproduce several of his full lists of species in order to emphasise the predominance of these lower plants in the arctic-alpine communities. Mr H. N. Dixon of Northampton and Mr W. E. Nicholson of Eastbourne have most readily and courteously replied to questions about mosses and liverworts respectively.

Of the text-figures a certain number are reproduced directly from published figures, of which several are taken from the *Journal of Ecology*, but the majority have been drawn by my daughter, Mrs Tomlinson, who has taken very great pains to obtain the best results, and whose skilful work and excellent judgment in helping to design figures have contributed much to the value and appearance of the book. Many of the photographic plates are reprints from the half-tone blocks used in *Types of British Vegetation*, and many others are from the *Journal of Ecology*. Of the remaining photographs those marked with the initials R.J.L. have been specially taken for me by my son-in-law, Dr R. J. Lythgoe, to whom I owe sincere thanks for the time he has spent and the patience and skill he has shown in the difficult art of photographing vegetation. The photographer's or draughtsman's name, whenever known, is attached to each reproduction, except those from my own negatives, and permission to reproduce has always been willingly accorded. To Messrs A. Bourne & Co. Ltd., of 73 Ludgate Hill, who have made the whole of the blocks, I am much indebted for their careful and excellent work.

A. G. T.

GRANTCHESTER
CAMBRIDGE
March, 1939

CONTENTS

Preface — PAGE v

List of Figures in the Text — xxxi

List of Plates — xxxv

PART I

THE BRITISH ISLANDS AS ENVIRONMENT OF VEGETATION, pp. 1–146

Chapter I

PHYSICAL FEATURES AND GEOLOGICAL HISTORY, pp. 3–28

The continental shelf, 3. Endemic species, 4. Palaeogenic and Neogenic, 4. THE PALAEOGENIC REGION, 8. Scotland and northern England, 8. Wales, 10. South-western peninsula, 10. Ireland, 10. The great tectonic folds, 11. Caledonian folds, 11. Armorican folds, 13. Existing sculpture, 13. Palaeogenic soils, 14.

THE NEOGENIC REGION, 15. Permian and Trias, 16. The later Secondary rocks, 16. Jurassic, 18. Cretaceous, 18. Wealden, 18. Lower Greensand, 19. The Upper Cretaceous sea, 21. The Chalk, 21. Clay-with-Flints, 23. Eocene, 23. Oligocene, 23. Pliocene, 25. The Pleistocene ice age, 25. Glacial deposits, 25.

References, 28.

Chapter II

CLIMATE, pp. 29–54

Climate and the distribution of vegetation, 29. Microclimates, 29. Heat and moisture, 30. Temperature and rainfall, 30. Ratio of precipitation to evaporation, 31. Saturation deficit, 33. The "Meyer ratio", 33. Diurnal variation of saturation deficit, 34. Seasonal variation, 35. Mist and fog, 35. Sunshine, 36. Frost, 39. Snow, 40. Wind, 41. Direction, 41. Velocity, 41. Wind at high altitudes, 43. "Arctic-alpine" climate, 43.

General characters of the climate, 43. Distribution of barometric pressures, 43. Passage of depressions, 46. Cyclonic rainfall, 46. Orographical rainfall, 47. Convectional rainfall, 47. Effects of the westerly winds, 47. Highest precipitations, 53. Lowest precipitations, 54.

References, 54.

Chapter III

REGIONAL CLIMATES, pp. 55–77

Maritime climates, 55. Extreme Atlantic climate, 57. West coast of England and Wales, 59. South coast climate, 60. East coast climate, 61. Northern inland climate, 63. Scottish mountain climate, 63. English midland climate, 66. Irish inland stations, 67.

Seasonal distribution of rainfall, 67. Deviations from monthly and annual means, 68.

CLIMATE AND VEGETATION, 68. Climatic regions, 68. (1) South-eastern region, 69. Breckland climate, 70. (2) Western region, 71. (3) Northern region, 73.

CLIMATE AND AGRICULTURE, 74. Grassland, 74. Arable land, 75. Wheatland, 75. Barley, 76. Oats, 76. Sugar beet, 76.

References, 77.

CHAPTER IV

SOIL, pp. 78-100

Nature of soil, 78. Stratification, 78. Climatic soil types (world groups), 79. Action of water and heat, 79. Mineral materials, 80. Chemical weathering, 81. Function of calcium, 81. Action of hydrogen ions, 81. pH value, 81. Organic materials, 83. Humus, 83. Mull (mild humus), 83. Mor (raw humus), 83. Peat, 84. Functions of humus, 84. Formation of mature soil, 85.

British soil types, 85. Brown earths, 85. Meadow soils, 88. Fen peat, 88. Podsols, 88. Acid peat soils, 92. Rendzinas, 93.

Relations of climate, soil and vegetation, 95. Soil texture, 96. Gravel, 97. Sand, 97. Silt, 97. Clay, 97. Loam, 98. Calcareous soils, 98. Saline soils, 99. Organic soils, 99.

References, 100.

CHAPTER V

DISTRIBUTION OF ROCKS AND THE SOILS THEY PRODUCE, pp. 101-126

Distribution of outcrops and soils, 101. Relation to agriculture, 101. Order of treatment, 102.

NEOGENIC ROCKS. *Quaternary deposits*: Recent maritime soils, 102. River alluvium, 103. Peat soils: (1) fen peat, 104. (2) Bog or moss peat, 104. *Glacial drift*, 105. Irish drift, 105. Scottish drift, 105. English Midland drift, 106. Chalky Boulder Clay, 106. Breckland, 107. Clay-with-Flints, 107. *Tertiary deposits*: The south-eastern basins, 107. East Anglia, 107. The London and Hampshire basins, 111. *Secondary rocks*, 111. The Chalk, 112. Gault and Greensand, 113. Wealden beds, 113. Jurassic rocks, 115. Oolites, 115. Lias, 116. Triassic rocks, 116. Permian: Magnesian limestone, 117.

PALAEOGENIC ROCKS: General characters, 117. Carboniferous system: Mountain Limestone, Millstone Grit and Coal Measures, 119. Culm Measures, 119. Calciferous Sandstone, 120. Central Irish plain, 120. Limestone pavements, 120. Devonian system, 120. Tertiary granite intrusions, 121. Old Red Sandstone, 121. The older Palaeozoic rocks, 122. Archaean rocks, 123. Retrospect, 124.

Distribution of the main lithological types, 124.

References, 126.

CHAPTER VI

THE BIOTIC FACTOR, pp. 127-146

The ecosystem, 127. Animal and plant ecology, 127. The "biotic factor", 127. The "virgin" ecosystem, 128. Effect of human activity, 128. Anthropogenic factors, 129. Importance of the biotic factor, 129. Creation of grassland by grazing, 129. Cattle and sheep, 130. Variations in grazing intensity, 130. Sheep grazing and heather, 130. Effect of sheep grazing on bracken and mat-grass, 131. Waste grassland and grass verges, 132. Biotic zonation on common land, 132. *Rabbits*, 133. History of the rabbit in Britain, 133. Present distribution, 134. Effect of rabbits on vegetation, 134. Species not eaten by rabbits, 135. Effect on competition between species, 136. Degradation of vegetation, 139. Rabbits and maritime vegetation, 139. *Other animals*, 140. Effect of rodents on woodlands, 140. Mice and voles, 140. Effect of voles on grassland, 141. Interference with "the balance of nature", 142. Enemies of rodents, 142. Hares, 142. Squirrels, 142. Deer, 142. Red deer, 143. Roe deer, 144. Fallow deer, 144. Birds, 144. "Bird cliffs", 144. Invertebrates, 144.

Effects of fire, 145.

References, 146.

PART II

HISTORY AND EXISTING DISTRIBUTION OF VEGETATION, pp. 147–210

Chapter VII

PRE-HISTORY, pp. 149–170

Post-glacial history, 149. Pollen analysis, 149. Geo-chronology, 149. Effect of the Pleistocene glaciation, 151. Interglacial periods, 152. The last glaciation, 153. Relict species, 153. Late glacial (Sub-Arctic) period, 154. Pre-Boreal period, 154. Birches, 154. Pine, 156. Boreal period, 157. Hazel maximum, 158. Dominance of pine, 158. The "Lower Forestian", 159. Invasion of xerophytes, 159. Atlantic period, 159. Increase of alder, 162. Deciduous summer forest, 162. Oak, elm and lime, 162. Increase of *Sphagnum* bogs, 163. Sub-Boreal period, 163. Neolithic culture, 163. Distribution of Neolithic settlement, 164. Forest v. grassland on chalk and oolite, 164. Spread of grassland, 165. "Beaker" settlements, 165. Distribution of Bronze Age cultures, 166. Spread of pine and birch, 166. Sub-Atlantic period, 166. Iron Age settlements, 169.

References, 169.

Chapter VIII

THE HISTORICAL PERIOD, pp. 171–193

Belgic settlements, 171. The Roman occupation, 171. Romano-British agriculture, 171. Sheep grazing, 172. Forest, 172. Upper limits of forest, 173. Effects of the Saxon settlement, 174. Distribution of Saxon settlement, 174. The Danes and Norsemen, 176. Effects of the Norman conquest, 176. The Royal Forests, 176. Disafforestation and extension of cultivation, 177. Pasturage in forest, 178. The three kinds of land utilisation, 178. Regional variations in rural economy, 179. Corn growing and export, 180. Sheep raising and wool production, 180. Timber and fuel, 181. Origin of coppice-with-standards, 181. Depletion of forest resources, 181. The first suggestion of planting, 182. General and regional developments, 182. Dearth of timber in Scotland, 182. The seventeenth century, 183. Early Scottish planting, 184. The influence of Evelyn, 185. Land utilisation, 185. The draining of the Fens, 186. The eighteenth century, 187. Enclosure, 188. Plantations, 188. Destruction of the Highland forests, 188. Eighteenth century planting in Scotland, 189. Nineteenth century planting in Scotland, 190. The nineteenth and twentieth centuries in England, 190. Modern enclosure and building, 191. The Forestry Commission, 191. Efforts to preserve the countryside, 192. Country planning, 192.

References, 193.

Chapter IX

DISTRIBUTION OF THE FORMS OF VEGETATION, pp. 194–210

Natural and semi-natural vegetation, 194. "Permanent grass", 195. Forest and woodland, 196. Census of woodlands, 197. Status of deciduous woodland, 197.

Deciduous summer forest, 199. Northern coniferous forest, 199. Heath, 199. Moss or bog, 200. Arctic-alpine formation, 202. Distribution of the climatic formations, 202.

Grassland, 204. Climatic requirements, 204. Relation to grazing, 205. Origin of British grassland, 205. Arable and grassland, 205. Permanent grassland, 206. Chalk and limestone grassland, 206. Siliceous grassland, 207. Alluvial grassland, 208. Maritime and submaritime grassland, 208.

Fen and marsh, 208. Maritime formations, 209. Sea-cliff vegetation, 210. References, 210.

PART III

THE NATURE AND CLASSIFICATION OF VEGETATION, pp. 211–240

Chapter X

THE NATURE OF VEGETATION, pp. 213–233

Plant communities, 213. Life form, 213. Open and closed communities, 213. Dominance, 214. Layering, 214. Ecotones, 215. Habitat, 215. Historical factors, 216. Succession, 216. Climax vegetation, 218. The priseres, 219. Xeroseres, 219. Hydroseres, 220. Subseres, 220. Conditions in England, 221. Subclimax vegetation, 222. Pasture, 223. Heathland, 224. Climaxes determined by different factors, 224. Deflected seres (plagioseres), 225. Plagioclimaxes, 225.

THE NATURE OF A PLANT COMMUNITY: interrelations of the members, 227. The role of competition, 227. The ecosystem, 228.

THE NAMING OF PLANT COMMUNITIES: the hierarchy of communities, 228. Formation and formation-type, 229. Climax, formation and stability, 229. Association, 230. Consociation, 230. Society, 230. Seral communities, 232.

References, 233.

Chapter XI

DIFFERENT METHODS OF CLASSIFYING VEGETATION, pp. 234–240

RAUNKIAER, MOSS, CRAMPTON

Importance of life form, 234. Raunkiaer's life form classes, 234. Phytoclimates, 235. Phytoclimate and climax dominants, 235. Moss's scheme, 236. Edaphic formations, 236. "Progressive" and "retrogressive" succession, 236. Crampton's stable and migratory types, 238.

References, 240.

PART IV

THE WOODLANDS, pp. 241-484

Chapter XII

NATURE AND STATUS OF THE BRITISH WOODLANDS, pp. 243-266

DOMINANT AND OTHER TREES. THE MORE IMPORTANT SHRUBS

Natural, semi-natural and artificial woods, 243. Effects of enclosure by agricultural land, 244. Lack of seed parents, 244. Destruction of carnivorous animals, 244. Effects of methods of exploitation, 245.

THE DOMINANT TREES OF THE BRITISH WOODLANDS: The two oaks, 246. The pedunculate oak (*Quercus robur*), 246. The sessile or durmast oak (*Q. petraea* or *sessiliflora*), 247. The beech (*Fagus sylvatica*), 248. Present distribution of beech, 248. Soil preferences, 249. Relations of beech and oak, 249. The two birches (*Betula pendula* and *B. pubescens*), 250. The ash (*Fraxinus excelsior*), 252. The common pine (*Pinus sylvestris*), 253. Distribution and history, 253. Habitats, 254. The common alder (*Alnus glutinosa*), 255.

OTHER TREES: Yew (*Taxus baccata*), 256. Hornbeam (*Carpinus betulus*), 256. Holly (*Ilex aquifolium*), 257. The maples (*Acer pseudo-platanus* and *A. campestre*), 257. Wych elm (*Ulmus glabra*), 258. Other elms, 258. Rosaceous trees (*Sorbus aucuparia, S. aria, Pyrus malus, Prunus avium, P. padus*), 258. The limes (*Tilia cordata* and *T. platyphyllos*), 258. Willows (*Salix fragilis, S. alba*), 259. Poplars (*Populus tremula, P. nigra, P. canescens*), 259.

THE MORE IMPORTANT SHRUBS, 259. Hazel (*Corylus avellana*), 259. Hawthorn (*Crataegus monogyna*), 260. Blackthorn (*Prunus spinosa*), 261. Gorse (*Ulex*), 261. Juniper (*Juniperus communis*), 263. Roses (*Rosa*) and brambles (*Rubus* sect. *Eubatus*), 263. Sallows and osiers (*Salix caprea, atrocinerea, repens, viminalis, purpurea*, etc.), 264. OTHER SHRUBS: elder (*Sambucus nigra*), 264. CALCICOLOUS SHRUBS, 264: dogwood (*Cornus sanguinea*), privet (*Ligustrum vulgare*), wayfaring tree (*Viburnum lantana*), spindle-tree (*Euonymus europaeus*), buckthorn (*Rhamnus catharticus*), 264. Alder buckthorn (*Frangula alnus*), guelder rose (*Viburnum opulus*), box (*Buxus sempervirens*), sea buckthorn (*Hippophaë rhamnoides*), 265.

WOODY CLIMBERS, 266: ivy (*Hedera helix*), honeysuckle (*Lonicera periclymenum*), woody nightshade (*Solanum dulcamara*), traveller's joy (*Clematis vitalba*), 266.

References, 266.

Chapter XIII

OAKWOOD. INTRODUCTORY, pp. 267-290

PEDUNCULATE OAKWOOD (QUERCETUM ROBORIS)

Oak forest in the British Isles, 267. Semi-natural oakwood, 267. The two British oaks, 268. PEDUNCULATE OAKWOOD (Quercetum roboris): general distribution, 269. Former extent of oak forest, 270. Origin of coppice-with-standards, 270. "Damp" and "Dry" oakwood, 271. Ash-oak wood, 272. Soils, 273. Water content, 274. Humus content, 274. Light, 274.

Structure of the Quercetum: layering, 276. Lateral spacing, 277. Structure of coppice-with-standards, 277. Periodicity of field layer, 277. Composition of field layer in coppice, 278. The dominant tree layer, 278. Shrub layer, 279. Woody climbers, 279.

CONTENTS

FIELD LAYER SOCIETIES, 279. Societies of lighter and drier soils: Pteridietum, 280. *Pteridium-Holcus-Scilla*, 281. *Rubus-Pteridium*, 281. Society of *Anemone nemorosa*, 281. Society of *Scilla non-scripta*, 282. Societies of heavier and damper soils: *Mercurialis perennis*, 282. *Sanicula europaea*, 282. *Primula vulgaris, P. elatior*, 282. More local societies, 283. Societies of very damp or wet soils: *Ranunculus repens*, 283. *Filipendula ulmaria*, 283. *Urtica dioica*, 283. Societies of fully illuminated areas: *Melandrium dioicum, Galeobdolon luteum, Euphorbia amygdaloides*, 283. *Epilobium angustifolium*, 285. *Digitalis purpurea*, 285.
SPECIES OF THE FIELD LAYER, 285. Dry facies, 286. Medium to damp facies, 287. Damp to wet facies, 288. Calcareous facies—ash-oakwood 288.
References, 289.

CHAPTER XIV

PEDUNCULATE OAKWOOD (*continued*), pp. 291–302

General failure to regenerate, 291. Causes of failure, 291.
SUCCESSION, 293. On abandoned arable land—"Broadbalk Wilderness", 293. "Geescroft Wilderness", 294. Succession from arable and heath land, 294. Abandoned pasture, 295. Grass verges, 295. Developing oakwood on clay-with-flints, 296. Deflected succession, 297.
QUERCETUM ROBORIS UNDER EXTREME CONDITIONS: Wistman's Wood, 298. Situation and nature, 299. Soil, 300. Habit of trees, 300. Flora, 300. Vascular epiphytes, 301. "Lammas shoots", 302. Regeneration, 302.
References, 302.

CHAPTER XV

SESSILE OR DURMAST OAKWOOD (QUERCETUM PETRAEAE OR SESSILIFLORAE), pp. 303–324

ENGLISH WOODS

Soil conditions (introductory), 303.
HERTFORDSHIRE SESSILE OAKWOODS, 304. Soils, 305. Light intensities, 305. Composition and structure, 305. *Rubus* society, 307. *Pteridium* society, 307. Smaller societies, 308. "Path society", 308. "Marginal society", 309. Bryophytes, 309. Algae, 310. Effect of coppicing, 310.
Succession in Hertfordshire, 311. In Essex, 311. The three subseres, 311.
THE PENNINE OAKWOODS, 313. Structure and composition, 314. Heathy woods, 314. Societies of drier soils, 315. Damp soil with mild humus, 315. Stream sides and flushes, 316. Abundance of ash under high rainfall, 317.
Malvern oakwoods, 318. Other sessile oakwoods in the West of England, 319. Sessile oakwoods on Mountain Limestone, 320. Sessile oakwoods at the highest altitudes, 320.
References, 324.

CHAPTER XVI

SESSILE OAKWOODS (*continued*), pp. 325–349
WELSH, IRISH AND SCOTTISH WOODS

WELSH OAKWOODS, 325. At the Devil's Bridge, Aberystwyth, 326.
IRISH OAKWOODS: The Killarney woods, 327. The oak-holly wood, 327. Arbutus, 328. The "laurel character", 328. Soils and physiography, 329. Structure of the oakwoods: Type 1, 329. Bryophytic communities on boulders, 331. On fallen logs, 331. Epiphytic bryophytes, 332. Type 2, 334. Type 3, 336. Lichen flora, 337. The upper woods, 338. Local yew wood, 338. Status

of *Arbutus*, 338. Status of *Rhododendron ponticum*, 339. Succession, 340. Vascular epiphytes, 340. Glengariff woods, 340. Woods at Pontoon, 340. Donegal woodland and scrub, 342. Connemara and Mayo, 343. Island scrub, 343.

HIGHLAND OAKWOODS, 343. On better soils, 344. On intermediate soils, 346. On poor soils, 347. Disappearance of oak in the far north, 348. Oaks and hybridism in the Highlands, 348.

References, 349.

CHAPTER XVII

MIXED OAKWOOD ON SANDY SOILS, *pp.* 350–357

(QUERCETUM ROBORIS ET SESSILIFLORAE OR QUERCETUM ERICETOSUM)

Mixture of the two oaks, 350.

Quercetum ericetosum, 351. Soils, 352. Sherwood Forest oakwood, 352. Structure, 352. Regeneration and Succession, 353. "Oak-birch heath", 354. Cyclical alternation of forest and heath, 355. Nature of "oak-birch heath", 355.

References, 357.

CHAPTER XVIII

BEECHWOOD, *pp.* 358–385

INTRODUCTORY. BEECHWOOD ON CALCAREOUS SOIL (FAGETUM CALCICOLUM). CHALK SCRUB AND YEW WOOD. SERAL ASHWOOD

INTRODUCTORY: Immigration and status of beech, 358. Escarpment and plateau beech, 359. Beech and oak, 361. Beechwood soils, 363. The three types of beechwood, 366.

BEECHWOOD ON CALCAREOUS SOIL (FAGETUM CALCICOLUM): Distribution, 366. Slope and soil, 366. Structure and composition, 367. Tree layer, 367. Shrub layer, 368. Field layer, 368. Dominants, 369. Other abundant species, 369. Characteristic species, 370. Biological spectrum, 371. Bryophytes, 371. Larger fungi, 371. Cotswold beechwoods, 372.

SUCCESSION: Chalk scrub, 372. The two seres, 372. Habitats of juniper and hawthorn seres, 372. The scrub stage, 373. Shrubs of chalk scrub, 373. Trees of chalk scrub, 374. Structure of chalk scrub, 374. Grazing and rabbit attack, 375. Field layer, 375. Marginal species, 375.

YEW WOODS OF THE SOUTH DOWNS: Distribution, 376. Origin and development, 376. In juniper scrub, 376. In hawthorn scrub, 377. The role of ash, 377. Structure of mature yew wood, 377. Localisation of yew woods, 380. Degeneration from biotic attack, 380.

SERAL ASHWOOD, 380. Structure and composition, 382. Ash scrub, 382. Calcicolous coppice, 382.

DEVELOPMENT OF BEECHWOOD, 383. Beech invasion of juniper scrub, 383. Of hawthorn scrub, 383. Later stages, 384.

Scheme of the two seres, 384.

References, 385.

CHAPTER XIX

BEECHWOOD ON LOAM (FAGETUM RUBOSUM), *pp.* 386–407

Slope and soil, 386. Structure and composition, 387. Role of *Rubus*, 388. Species of the field layer, 388. Bryophytes, 389. Larger fungi, 389.

SUCCESSION, 390. The central sere, 390. The Chiltern commons, 390. Shrub invaders, 390. Tree invaders, 391. Development of woodland, 391. Ash-oak wood, 391. Planted oakwoods, 392. Mixed oak and beechwood, 392. Mature beechwood, 393. Succession on the South Downs plateau, 394. On the more calcareous loams, 396. Ash-oak wood, 396. On intermediate loams, 397. (Ash-)oak-hazel wood, 397. On the least calcareous loams, 397.

Invasion of beech, 398. Modes of invasion, 398. Pioneers and semi-pioneers, 398. Elimination of ash and oak, 400. Effect on shrub and field layers, 401. Causes of even age, 401. Planted and natural woods, 402. Later stages of development, 402. Bare stage, 402. *Oxalis* and *Rubus* stages, 402.

Regeneration of the beech consociation, 403. Colonisation of gaps, 403. Size of gaps, 404. Gaps and full-mast years, 405.

Scottish beechwoods on fertile loam, 406. Beechwood on sandy soil of high base status, 407.

References, 407.

Chapter XX

BEECHWOOD ON SANDS AND PODSOLS (FAGETUM ARENICOLUM OR ERICETOSUM), pp. 408–426
SUMMARY OF BRITISH BEECHWOODS

FAGETUM ERICETOSUM IN SOUTH-EASTERN ENGLAND. Soils of the Chiltern plateau, 408. Beechwood on podsolised silts, 409. Mycorrhiza, 410. Structure and composition, 410. Beechwood on sandy soils in south-eastern England, 411. Burnham Beeches: situation and soil, 411. Podsol profiles, 412. Composition and structure, 412. Succession, 414. Epping Forest and similar woods, 415.

PLANTED AND SUBSPONTANEOUS SCOTTISH BEECHWOODS, 416. Shelter belts, 417. *Deschampsia* and *Holcus* communities, 417. Subspontaneous beechwood, 418. Structure and composition, 419. Succession from conifer plantations, 419.

SUMMARY OF BRITISH BEECHWOODS, 421. The status of beech, 424. References, 426.

Chapter XXI

ASHWOOD ON LIMESTONE (FRAXINETUM CALCICOLUM), pp. 427–443

Distribution of climax ashwood, 427. Situation and soil, 428. Composition and structure, 428. Tree layer, 429. Shrub layer, 429. Field layer, 430. Ground layer, 431. Succession, 432. Two upland ashwoods, 433. Ling Ghyll Wood, 433. Colt Park Wood, 434. An upland ashwood on basic igneous rock, 436. Structure and composition, 437. Succession, 439. "Retrogressive succession" on limestone, 440. Ashwood on other calcareous soils, 441. Ash-oak wood (1) on deep calcareous soils, 442; (2) in wet climates, 442. Ash-birch woods, 442.

References, 443.

Chapter XXII

PINE AND BIRCH WOODS, pp. 444–459

History and present distribution, 444. Relation to oakwood, 445.

HIGHLAND PINEWOODS. Ballochbuie Forest, 446. Field layer, 446. Bryophytes, 447. Rothiemurchus Forest, 447. Glenmore Forest, 449. High level pine and birch, 449. Loch Maree, 450. Characteristic species of old native pinewoods, 451.

SCOTTISH BIRCHWOODS, 451. Caithness birchwoods, 451. Trees and shrubs, 452. Field layer, 452. Vegetation of wetter places, 453. Birch-juniper wood at South Kinrara, 453. Grassy field layer, 454. Variants of birchwood, 454. Altitudinal and latitudinal distribution, 454.

Seral birchwood, 455. "Birch invasion", 456. Subspontaneous pinewood, 457. References, 458.

CHAPTER XXIII

ALDERWOOD (ALNETUM GLUTINOSAE), pp. 460–471

Distribution, 460. Fen carr, 460. "Swamp carr", 463. Young carr, 463. Succession, 463. Hydrosere at Sweat Mere, 464. Valley fen woods, 467. Alder carr at Cothill, 468. Other examples of alderwood, 470. Birch and alder, 471
References, 471.

CHAPTER XXIV

SCRUB VEGETATION OR BUSHLAND (FRUTICETUM), pp. 472–484

Climatic scrub, 472. Seral, subseral and climax scrub, 472. The Burren district of Co. Clare, 473. Succession on the Burren, 473. Mixture of species, 473. Limestone heath, 473. Closed hazel scrub, 474. Progression to ashwood, 474.

Subseral scrub, 475. Thicket scrub, 475. Woodland scrub, 476. Flora, 476. "Marginal" species, 477. In hedgerow and coppice, 478. List of marginal species, 479.

Thicket scrub on Chalky Boulder Clay, 480. Woody plants, 480. Field and ground layers, 480. Rabbit effects, 481. Conditions of colonisation, 481. Development of the scrub, 482.
References, 484.

PART V

THE GRASSLANDS, pp. 485–576

CHAPTER XXV

NATURE AND STATUS OF THE BRITISH GRASSLANDS, pp. 487–498

Grassland as biotic plagioclimax, 487. The British Isles a hemicryptophyte region, 487. Turf or "sole" of grassland, 488. Effects of overgrazing and undergrazing, 489. "Rough grazings", 489. Types of rough grazing, 489. Siliceous grassland, 490. Invasion by heath and peat plants, 490. Bent-fescue community, 490. Nardetum and Molinietum, 490. Limestone grassland, 490. Arctic-alpine grassland, 491. Other grasslands, 491.

Common features of grassland habitats, 492. Climatic factors, 492. Edaphic factors, 492. Habit forms, 492. The chief grass communities, 493. Acidic grassland, 494. Basic grassland, 495. Neutral grassland, 495. Footpaths and road verges, 496.
References, 498.

CHAPTER XXVI

ACIDIC GRASSLANDS, pp. 499–524

I. AGROSTIS-FESTUCA (BENT-FESCUE) GRASSLAND, 499. "Siliceous grassland", 499. "Grass-heath", 499. Invasion by heath, 500. Soil, 500. The dominants, 500. Flora, 501. Pteridietum, 502. Ulicetum, 504.

GRASS HEATH, 506. Floristic list, 507. *The grass heaths of Breckland*. Climate and soil, 509. Vegetation, 509. Species of selected areas, 510. Species peculiar to Breckland, 511. Succession, 512. Soil, 512. The sere, 512.

II. NARDUS GRASSLAND, 514. Distribution, habitats and growth form, 515. Marginal Nardetum on the Moorfoot Hills, 516. Welsh Nardeta, 516. Nardetum on Cader Idris, 518.

III. MOLINIA GRASSLAND, 518. *Molinia* bog and *Molinia* meadow, 519. Habitat factors, 519. Vegetative structure, 520. *Molinia* peat, 520. Southern Scottish Molinieta, 520. Pennine Molinieta, 520. Cornish Molinieta, 521. Highland Molinieta, 522. Irish and other Molinieta, 522.
References, 523.

Chapter XXVII
BASIC GRASSLANDS, pp. 525–558

CHALK GRASSLAND: distribution and soil, 525. Water content, 527. Rooting depth, 528. Floristic composition, 528. Constant species, 530. Exclusive species, 533. The grasses, 534. The fine-leaved fescues, 534. *Bromus erectus*, 535. The Avenae, 536. *Brachypodium pinnatum*, 536. *Arrhenatherum elatius*, 536. *Briza media*, *Koeleria cristata* and *Trisetum flavescens*, 537. Other grasses, 537. Herbs, 538. Bryophytes, 540. Lichens, 541. Structure of the vegetation, 541.

SUCCESSION, 545. On bare chalk rock, 545. In quarries, 546. On spoil banks, 546. "Primitive" chalk grassland, 548. Stabilisation by grazing, 550. Formation of loam, 551. Chalk heath, 551.

Inferior Oolite grassland, 552.

GRASSLAND OF THE OLDER LIMESTONES AND THE BASIC IGNEOUS AND METAMORPHIC ROCKS, 552. Features of the older limestones, 553. Composite list of species, 554. Limestone heaths, 557. Limestone swamps, 557.
References, 557.

Chapter XXVIII
NEUTRAL GRASSLAND, pp. 559–576

"Permanent grass", 560. Treatment the important factor, 560. Commons, greens and verges, 561. The grasses of neutral grassland, 561. Non-gramineous species, 564. Leguminous plants, 565. "Miscellaneous" species, 565. Meadow and pasture, 566. Characteristic species, 567. Thames-side meadows and pastures, 568. Magdalen meadow, 570. General flora of neutral grassland, 571. The "kernel" of neutral grassland, 571. Juncetum effusi, 573. Grades of pasture, 574.
References, 576.

PART VI
THE HYDROSERES, pp. 577–720
FRESHWATER, MARSH, FEN AND BOG VEGETATION

Chapter XXIX
THE HYDROSERES, pp. 579–595
VEGETATION OF PONDS AND LAKES

The hydroseres, 579. Conditions of life in water, 579. Oxygen, 579. Silt, 579. Rivers, 580. Habitat classification, 580. The silting factor, 581. Life form, 581. Zonation of aquatic vegetation, 582. Stages of the hydrosere, 582. Summary of life form groups, 583. Forms of submerged leaves, 583.

PONDS ON THE SOFTER ROCKS. Foul waters, 584. Lowland ponds, 584. Trent valley ponds, 584. Bramhope ponds, 585. Hagley Pool, 586. White Moss Loch, 588. Submerged communities, 588. Reed swamp, 590. Norfolk Broads, 592.
References, 595.

CONTENTS

Chapter XXX
THE CUMBRIAN LAKES, pp. 596–621

ESTHWAITE WATER, 597. Deep water communities, 597. Shallow water communities, 600. Floating leaf communities, 603. Reedswamp communities, 603. Variations in habitat and succession, 606.

THE OTHER LAKES, 606. Dissolved substances, 607. Light, 607. Temperature, 608. Aeration, 608. Silt, 608. The two types of lake 609. Potash and fine silt, 609. Subaqueous peat, 609. Deep water communities, 609. Succession, 614. Shallow water communities, 615. (*a*) Mainly submerged communities, 615. (*b*) Floating leaf communities, 616. (*c*) Reedswamp communities, 616. The two main hydroseres, 617. Causal factors, 618.

The most primitive lakes and tarns, 618. Transition to bog, 621. References, 621.

Chapter XXXI
THE VEGETATION OF RIVERS, pp. 622–633

Current, 622. Silting, 622. Speed of current, 623. Floods, 623. Turbidity, 624. Solutes and reaction of the water, 624. Very rapid current, 624. Moderately swift current, 626. The Dove and the Wharfe, 626. The Upper Thames, 626. Moderate current, 627. Medium to slow current, 628. Current very slow or negligible, 631. Turbid rivers, 631. Riparian community, 631. Shading effect, 632. Summary, 632.

References, 633.

Chapter XXXII
MARSH AND FEN VEGETATION, pp. 634–647

Definitions, 634. Marsh, 634. Esthwaite marshes, 635. Cornish marshes, 637. Silting and peat formation, 638.

Fen, 639. Esthwaite fens, 639. Succession on the Esthwaite fens, 642. (1) Area of rapid sedimentation, 642. (2) Area of moderate sedimentation, 643. (3) Area of slow sedimentation, 646. Transition to bog, 646. The limits of fen, 646.

References, 647.

Chapter XXXIII
THE EAST ANGLIAN FENS. NORTH-EAST IRISH FENS, pp. 648–672
SUMMARY OF THE LATER HYDROSERES

THE EAST ANGLIAN FENS, 649. Surviving fragments, 649. Wicken Fen, 651. East Norfolk fens, 651. Soil of Wicken Fen, 653. Vegetation, 653. Cladietum, 653. Glycerietum, Phalaretum, 654. Typical species, 654. General list, 656. Unevenness of surface, 656. Oxyphilous communities, 656. Development of carr, 657. Carr at Wicken Fen, 657. Carr in East Norfolk, 660. "Swamp carr", 660. Anthropogenic fen communities, 661. (1) "Mixed sedge" (Cladio-Molinietum), 661. (2) "Litter" (Molinietum), 661. (3) Fen droves, 662. (4) "Mixed fen", 662.

NORTH-EAST IRISH FENS, 662. Zonation, 663. Societies of the Lower Fen, 664. Middle Fen, 664. Upper Fen, 666. Nature of the succession, 666. Relation to "moor", 667. Vegetation of "ramparts", 667. Fen scrub or carr, 668.

SUMMARY OF THE LATER HYDROSERES, 668. Marsh, 668. Fen, 669. Carr, 669. Oakwood, 669. Molinietum and transitory carr, 670. Raised bog, 670. Blanket bog, 670. Correlation with climate, 670.

References, 672.

Chapter XXXIV

THE MOSS OR BOG FORMATION, pp. 673–699

Terminology, 673. The word "moor", 673. "Moss" or "bog", 674. Valley bog and Raised bog, 675. Blanket bog, 676. Bog or moss communities, 676.

(1) SPHAGNETUM, 676. Structure of *Sphagnum*, 677. Variation of habitat and structure, 677. Acidity and mineral requirements, 677. Oxidation-reduction potential, 678. Habitats of various Sphagna, 678. Role of Sphagna in vegetation, 678. Structure of raised bog, 679. Sphagneta in Great Britain, 679. *Sphagnum* bogs of wet heath, 680. Raised bogs, 680. Tregaron bog, 682. Small bog at Loch Maree, 682. Raised mosses of Lonsdale, 686. Other raised mosses, 686.

IRISH RAISED BOGS. Distribution, climate and peat, 686. Structure and development, 687. Hollow-hummock cycle, 687. Vegetation of seral stages, 688. Structure of hummocks, 689. *Peat profiles of raised bogs*: Athlone bog, 690; Edenderry bog, 691. Profile tables, 692. Physiognomic dominants, 696. Vegetation of drainage channels, 696.

Sphagnetum of blanket bog, 697. Sphagneta of the northern Pennines, 697. Sphagnetum in Cornwall, 698. Flora of Sphagnetum, 698.

References, 699.

Chapter XXXV

THE MOSS OR BOG FORMATION (*continued*), pp. 700–720

(2) RHYNCHOSPORETUM ALBAE, 700. (3) SCHOENETUM NIGRICANTIS, 700. (4) ERIOPHORETUM, 701. (*a*) Eriophoretum vaginati, 701. Soil, 701. Floristic composition, 703. Other areas, 703. (*b*) Eriophoretum angustifolii, 704. Succession, 704. Retrogression of Eriophoretum, 705. (5) SCIRPETUM CAESPITOSI, 707. Scirpetum of Argyllshire, 707. Inverliver Forest, 707. Climate, 707. Vegetation, 708. Scirpetum in the northern Highlands, 708. Exmoor and Bodmin Moor, 710. Cleveland Moors, 711. Wicklow Mountains, 711. (6) MOLINIETUM CAERULEAE, 713. Callunetum vulgaris, 713.

IRISH BLANKET BOG, 714. Connemara and western Mayo, 714. Peat, 715. Flora and vegetation, 715. Blanket bog in north-western Mayo, 716. Hummock structure and flora, 716.

SUMMARY OF THE BOG OR MOSS FORMATION, 718. (1) Blanket moss or bog, 718. (2) Valley bog, 719. (3) Raised moss or bog, 719.

References, 720.

PART VII

HEATH AND MOOR, pp. 721–772

Chapter XXXVI

THE HEATH FORMATION, pp. 723–742

Conditions of establishment, 723. Distribution, 723.

LOWLAND HEATH (Callunetum arenicolum), 724. Southern heaths, 724. Soil—the podsol profile, 724. Hindhead Common, 725. Structure and Flora, 726. *Erica cinerea*, 726. *Ulex minor*, 727. *Pteridium*, 727. Esher Common, 728. Cavenham Heath, 729. Sherwood Forest, 729.

SUCCESSION—*The burn subsere*, 729. Fires at Hindhead, 730. The pioneers—algae and lichens, 730. Mosses, 731. Vascular plants, 731. Undershrubs, 732.

CONTENTS

Summary, 732. *Gravel or sand subsere*: at Hindhead, 733; at Crockham Hill, 733. *Felling subsere*: at Oxshott, 734.

Wet heath communities, 734. Molinietum, 735. Succession, 735. Chard Common, 735. Zonation, 737. Sphagnetum, 740. Summary of zonation, 741. Other wet heath species, 741.

References, 741.

Chapter XXXVII

THE HEATH FORMATION (*continued*), pp. 743–772

"Heath" and "heather moor", 743. Floristic differences, 743. Depth of peat, 747.

UPLAND HEATHS OR MOORS. Callunetum, 747. "Upland Heaths" of Somerset, 747. General flora, 748. The Cleveland heaths, 750. Burn subsere, 751. Scottish heaths, 751. Flora, 752. Burn subsere, 752. "Heather moors" of the Pennines, 753. Water and humus content of soil, 753. Chemical nature of the peat, 754. Floristic composition, 754. Firing of heather moor, 756. Burn subsere, 756. Heather moors of the Wicklow Mountains, 757. The Mourne Mountains, 758. BILBERRY MOOR (Vaccinietum myrtilli), 758. Highland Vaccinieta, 758. Pennine Vaccinieta, 761. "Vaccinium summits" and "edges", 761. Northern Pennines, 761. Mourne Vaccinieta, 761. Wicklow Vaccinieta, 762. Vaccinietum succeeding "moss", 762. Mixture of *Vaccinium* with *Calluna* and *Eriophorum*, 762.

DISTRIBUTION OF THE HEATH FORMATION IN THE BRITISH ISLES, 763. Upland heaths, 763. Lowland heaths, 763. East Anglian heaths, 763. South-eastern (Lower Cretaceous and Wealden) heaths, 763. Heaths of the London basin, 764. Heaths of the Hampshire basin, 764. South-western heaths, 764. Midland heaths, 764.

Status of the heath formation, 765.

SUMMARY OF PEAT COMMUNITIES, 766. Subaquatic peats, 766. Ionic content of waters, 766. Reedswamp, 766. Fen, 767. Succession in fen, 767. Carr, 767. Moss or bog, 768. Raised moss or bog, 768. Blanket bog, 769. Callunetum, 770. Vaccinietum, 771. Synopsis, 771.

References, 771.

PART VIII

MOUNTAIN VEGETATION, pp. 773–813

Chapter XXXVIII

THE UPLAND AND MOUNTAIN HABITATS. MONTANE AND ARCTIC-ALPINE VEGETATION, pp. 775–796

Upper limit of forest, 775. Montane vegetation, 775.

TALUS AND SCREE VEGETATION, 775. The scree succession, 776. Bryophytes, 776. Pioneers, 776. The parsley fern (*Cryptogramma crispa*), 776. Other ferns, 777. Bryophytes and flowering plants, 777. Formation of vertical strips, 777. The heath community, 778. Destruction of seral stages, 779. Stable screes, 779. Marginal invasion, 779. Climax vegetation on screes, 779.

ARCTIC-ALPINE VEGETATION, 780. The arctic-alpine zone, 780. Arctic-alpine species, 781. Life forms of arctic-alpines, 781. Chamaephytes, 781. Bryophytes and lichens, 782. Effect of rock and soil, 783. Richer flora of basic rocks, 783.

ARCTIC-ALPINE GRASSLAND, 783. Floristic composition, 784. Festucetum ovinae on Cader Idris: Dry facies, 785. Lower Festucetum: Moist facies, 786.

UPPER ARCTIC-ALPINE ZONE, 787. Exposed habitats, 787. Mountain top detritus, 787. Protected habitats, 787. Chomophytes, 787. Formation of mountain top detritus: the succession, 787. Moss-lichen associes, 788. *Rhacomitrium* heath, 788. Status of the community, 788. Habitat conditions, 789. Life forms, 789. Recruitment, 789. Flora of moss-lichen associes, 789. Adaptations to habitat, 793. Vascular flora, 793. The moss-lichen associes on Cader Idris, 793. Flora of *Rhacomitrium* heath, 794. *Rhacomitrium* heath on Cader Idris, 795.

References, 796.

Chapter XXXIX
ARCTIC-ALPINE VEGETATION (continued), pp. 797–813

Snow patch communities.—Effect of snow lie, 797. Anthelietum juratzkanae; Polytrichetum sexangularis, Selicetum herbaceae, 797.

VEGETATION OF PROTECTED HABITATS, 798. Lithophytes, chomophytes and chasmophytes, 799. Succession, 799. Screes, 799. Humidity, 800. *Lithophytes*, 800. Lichens, 801. Liverworts, 802. Mosses, 803. Vascular plants, 804. On open ledges, 804. *Chomophytes*, 805. Mosses, 805. Liverworts, 805. Lichens, 806. Vascular plants, 806. *Shade chomophytes*, 806. Vascular plants, 807. Mosses, 807. Liverworts, 808. Lichens, 808. *Basic rocks on Cader Idris*, 808. Lichens, 808. Liverworts, 808. Mosses, 809. Vascular plants, 809. Acid rocks, 810. *Hydrophilous chomophytes*, 810. Societies of bryophytes, 811. Mosses, 811. Liverworts, 812. Lichens on wet rocks, 812. Representative vascular hydrophytes, 813.

References, 813.

PART IX
MARITIME AND SUBMARITIME VEGETATION, pp. 815–903

Chapter XL
INTRODUCTORY. THE SALT MARSH FORMATION, pp. 817–843

Maritime vegetation, 817. The habitats, 817.

THE SALT MARSH FORMATION, 819. Relation to tides, 819. Salt content of ground water, 819. Zonation and succession, 820. Ecological factors, 820. Submersion and exposure, 820. Effects on different zones, 821.

SALT MARSH COMMUNITIES—variation in detail, 821. General zonation, 823: (1*) Algal communities and Zosteretum, 823. (1) Salicornietum herbaceae, 824. (1a) Spartinetum townsendii, 826. (2a) Glycerietum maritimae, 829. (2b) Asteretum tripolii, 830. (3) Limonietum vulgaris, 831. (4) Armerietum maritimae, 831. Green algae, 833. Soil fungi of Glycerietum and Armerietum, 833. "General salt marsh community", 834. (4a) Obionetum, 835. (4b) Suaedetum fruticosae, 838. Other Mediterranean species, 838. (5) Festucetum rubrae, 838. Mosses and algae, 838. (6) Juncetum maritimi, 839. Transition to land vegetation, 839. Salt-marsh pasture, 840. Drainage channels, 840. Formation of "pans", 841. Algal communities of the salt marsh, 841.

References, 842.

Chapter XLI
THE FORESHORE COMMUNITIES. COASTAL SAND DUNE VEGETATION, pp. 844–867

THE FORESHORE COMMUNITIES, 844. Tidal drift, 844. Characteristics, 844. Flora, 844.

THE SAND DUNE FORMATION—Conditions of existence, 845. Supply of sand, 845. Effect of growing plants, 845. Parallel dune ranges, 846.

MOBILE DUNE COMMUNITIES—Sea couch-grass consocies (Agropyretum juncei), 847. Foredunes, 847. Associated species, 847. Marram grass consocies (Ammophiletum arenariae), 848. Limitations and powers of *Ammophila*, 848. Features of Ammophiletum: dynamics and physiognomy, 849. Flora, 850. Variability of flora, 851. Role of mosses in fixation, 851. Nature of dune sand, 851. White and yellow dunes, 852.

FIXED DUNE COMMUNITIES, 852. Variability of vegetation, 852. Mosses and lichens, 853. Grey dunes, 853. Phases of succession, 855. Flora of typical grey dune, 855. Vascular plants of fixed dunes, 855. Degeneration of *Ammophila*, 856. "The Hood" at Blakeney Point, 857. Calcareous dunes, 857. Dog's Bay, 857. Rosapenna, 858. Dune grassland, 859. Castle Gregory, 859. Tramore, 859. Acidic dunes at Walney Island, 860. Dune heath: Ayreland of Bride, 860. Subseres, 861. South Haven peninsula, 861. Vegetation of "slacks", 861. Pools, 861. Dune marshes, 862. *Salix repens* dunes, 862. Effects of disturbance, 864. Ultimate vegetation, 864. Dune scrub, 864. Hippophaëtum, 864.

Dune soils, 865.
References, 867.

CHAPTER XLII

SHINGLE BEACHES AND THEIR VEGETATION, *pp*. 868–894

Nature of shingle, 868. Sand and shingle, 868. Types of shingle beach, 868. Morphology and development of the shingle spit, 870. Mobility of shingle, 873. "Camm" formation, 873. Structure of a shingle spit, 878. Shingle and sand, 879. Water content, 879. Humus, 879.

FLORA AND VEGETATION, 880. *Suaeda fruticosa*, 880. Distribution at Blakeney Point, 880. Dissemination and establishment, 881. Effects of shingling over, 882. Distribution on Chesil Beach, 885. Preference for shingle habitats, 885. Other flowering plants on Blakeney shingle, 885. Halophytes, 887. Psammophytes, 888. Habit of shingle plants, 888. Vegetation of lateral banks, 888. Vegetation of Chesil Beach, 890. Other characteristic shingle species, 890. Dungeness, 891. *Shingle lichens and mosses*, 891. Blakeney Point, 891. Lichens of Chesil Bank, 893. Mosses, 893.

References, 894.

CHAPTER XLIII

SUBMARITIME VEGETATION, *pp*. 895–902

Tolerance of sea salt, 895. Submaritime species, 895. Brackish water and marshes, 895. Spray-washed rocks and cliffs, 896. Maritime and submaritime grasslands, 898. Worm's Head, 898. Festucetum rubrae pruinosae, 899. Festucetum on sea stacks, 900. Effect of grazing, 900. *Plantago* sward, 900. Grassland of the Sussex cliff tops, 901.

References, 902.

INDEX, *pp*. 903–30.

FIGURES IN THE TEXT

Fig. 1.	The physical setting of the British Islands	page 2
2.	The continental shelf	5
3.	Orography	7
4.	Structural elements of the British Islands	9
5.	Physiographic regions of England and Wales	17
6.	Surface geology of the Weald	20
7.	Section through the Weald from north to south	22
8.	Section from the Midland Plain to the Weald	24
9.	Surface geology of East Anglia and part of the East Midlands	27
10.	Actual mean annual temperatures	31
11.	Mean annual rainfall	32
12.	Mean duration of sunshine	37
13.	Mean daily duration of sunshine in June	38
14.	Directions of winds	42
15.	Prevailing winds, January and February	44
16.	Prevailing winds, July and August	45
17.	January isotherms	48
18.	January isotherms, Western Europe	49
19.	July isotherms	50
20.	Mean minimal temperatures, January	51
21.	Mean maximal temperatures, July	52
22.	Annual ranges of mean temperatures	53
23.	Meteorological stations referred to in the text	56
24.	Mean number of rain days per annum	69
25.	Mean monthly temperatures at Berlin, Cambridge and Valencia Island	71
26.	Mean monthly maxima and minima for the same stations	72
27.	Brown earth profile	87
28.	Podsol profile from a Breckland sand	89
29.	Scottish podsols derived from Boulder Clay	91
30.	Rendzina profile	93
31.	Surface geology of East Anglia and part of the East Midlands	108
32.	Soil map of East Anglia and part of the East Midlands	109
33.	Section from the Midland Plain to the Weald	110
34.	Section through the Weald from north to south	114
35.	Grassland and heather on Moel Siabod, North Wales	131
36.	Effect of rabbits on Callunetum	137
37.	Zonation of vegetation round an isolated rabbit burrow	138
38.	Table of British post-glacial history	150
39.	Tree pollen in North Norfolk (East Anglia) during the pre-Boreal, Boreal and Atlantic periods	155
40.	The North Sea in pre-Boreal times	156
41.	Tree pollen in East Anglia during the Boreal, Atlantic and Sub-Boreal periods	157
42.	Tree pollen of Boreal, Atlantic and post-Atlantic periods from Tregaron Bog in the Teifi valley (Cardiganshire)	158
43.	Late Boreal tree pollen	160
44.	Early Atlantic tree pollen	161
45.	Profiles of peat from three Scottish localities	167

FIGURES IN THE TEXT

Fig.		page
46.	Layer of Sub-Boreal pine stumps	167
47.	Stratigraphy of peat mosses in Lonsdale (N. Lancs.)	168
48.	Scheme of a prisere	218
49.	The conception of subclimax	223
50.	Deflected succession—plagiosere and plagioclimax	226
51.	Oakwoods and related types	273
52.	Vegetative and flowering periods of field layer dominants in Hertfordshire oak-hornbeam woods	275
53.	Societies of field layer in Gamlingay Wood (Cambs.)	284
54.	Subseres and plagioseres leading to Quercetum roboris	298
55.	Oak-hornbeam coppice with standards	306
56.	Bramble-bracken and mercury societies in oak-hornbeam wood	308
57.	Subseres on Essex glacial gravels leading to Quercetum sessiliflorae	312
58.	Birkrigg and Keskadale oaks	321
59.	Profile of the Keskadale oakwood	323
60.	Killarney oak-holly wood, profile of "Type 1"	330
61.	Killarney oak-holly wood, profile of "Type 2"	334
62.	Killarney oak-holly wood, profile of "Type 2A"	335
63.	Killarney mixed wood, profile of "Type 3"	336
64.	Zonation of vegetation on the border of the upper lake, Killarney	339
65.	Distribution of spontaneous beechwood in southern England	359
66.	Beechwood in the Goodwood area (western South Downs)	360
67.	Beechwood of part of the Chiltern Hills (Bucks.)	362
68.	Beechwood of another part of the Chilterns (Oxon.)	363
69.	Calcicolous beechwood soils (rendzinas)	364
70.	Three plateau beechwood soils (Chilterns)	365
71.	Yew woods at Butser Hill (Hants.)	378
72.	Yew woods at Kingley Vale (Sussex)	379
73.	Yew wood on Downley Brow (Sussex)	380
74.	Beechwood in the Goodwood area	394
75.	Part of the Goodwood area on a larger scale	395
76.	Belt and profile transects across front of beech associes	399
77.	"Calliper" invasion of ash-oak associes by beech	400
78.	Stages of field layer in Fagetum rubosum	403
79.	Colonisation of gap in beechwood (reproduction circle)	404
80.	Profile of reproduction circle	405
81.	Chiltern plateau soils	409
82.	Subseres from heath to oak- and beechwood	414
83.	Successions from conifer plantations to beechwood and heath	420
84.	Height growth, field layers and soil characters of beechwood	422
85.	Summarised seres from grassland and heath to beechwood	423
86.	Rothiemurchus Forest	448
87.	Transect at Calthorpe Broad (Norfolk)	464
88.	Hawthorn thicket scrub on Chalky Boulder Clay	483
89.	Zonation on the Wicklow Mountains	505
90.	Development of Festuco-Agrostidetum in Breckland	513
91.	Zonation on the Moorfoot Hills	517
92.	Branch of *Molinia caerulea* in spring	519
93.	Rooting depths of chalk grassland plants	529
94.	*Linum catharticum*	530
95.	*Galium verum*	531
96.	*Poterium sanguisorba*	532
97.	Profile of herbage, pastured and rabbit grazed	542
98.	Profile of herbage, soil somewhat leached, moderately rabbit grazed	542
99.	Profile of herbage, severely rabbit grazed	542

FIGURES IN THE TEXT

Fig. 100.	Profile of severely grazed herbage, larger scale	page 543
101.	Chart quadrat, moderate rabbit grazing	544
102.	Chart quadrat, after exclusion of rabbits for 6 years	544
103.	Chart quadrat, after exclusion of rabbits for 12 years	545
104.	Vegetation of a small pond at Bramhope, near Leeds	587
105.	Vegetation of White Moss Loch	589
106.	Profile transect of White Moss Loch	591
107.	The fenlands of East Norfolk	593
108.	Shrinkage of Broads	594
109.	Southern half of the Lake District	596
110.	Esthwaite Water	598
111.	Wave-cut terrace on lee shore	599
112.	Strickland Ees and Fold Yeat Bay, Esthwaite Water	601
113.	Profile sections of Fold Yeat Bay	602
114.	Head of Esthwaite Water in 1914–15	604
115.	The same in 1929	605
116.	Decrease of light intensity with depth of water	607
117.	Monthly temperatures of lakes and ponds	608
118.	Relation of different aquatic communities to silting	610
119.	Thwaite Hill Bay, Ullswater	614
120.	Scheme of the main trends of aquatic succession in the Lake District	618
121.	Diagram of a primitive tarn	620
122.	Aquatic vegetation in the River Lark (Suffolk)	629
123.	Effect of shade on different species in the River Itchen (Hampshire)	632
124.	North Fen, Esthwaite in 1914–15	640
125.	The same in 1929	641
126.	The Fenland	648
127.	The fenlands of East Norfolk (Broads region)	650
128.	Part of the Bure valley, East Norfolk	651
129.	Hoveton and Salhouse Broads (River Bure)	652
130.	Part of the Yare valley	652
131.	Level of bush colonisation at Wicken Fen	655
132.	Bush colonisation in a corner of Wicken Fen, 1923–24	658
133.	Progress of bush colonisation in 1934	659
134.	Table of fen society dominants at Lough Neagh	665
135.	Climatic relations of hydroseres	671
136.	Quadrat of "regeneration complex" in Tregaron Bog	681
137.	Profiles of regeneration complex and Scirpetum in Tregaron Bog	683
138.	Diagram of small raised bog at Loch Maree	684
139.	Diagram of hollow-hummock cycle ("regeneration complex")	688
140.	Plateau of the Peak district (Derbyshire)	702
141.	Seral structure of hummock in the Scirpetum of the Wicklow Mountains	712
142.	Diagram of large hummock in the blanket bog of western Mayo	717
143.	Hydroseres in relation to fen, bog, Callunetum and forest	720
144.	Heath podsol profile	725
145.	Invasion of Callunetum by *Pteridium*	728
146.	Redevelopment of Molinietum after burning	736
147.	Wet tussock heath and channels	737
148.	Profile transect through a tussock	738
149.	Vegetation of the Mourne Mountains	759
150.	Vegetation of the summit of Slieve Donard (Mourne Mts.)	760
151.	Tabular scree with strips of vegetation	778
152.	Ranges of salt marsh communities at Scolt Head, Norfolk	822
153.	Zonation of angiospermic communities in the salt marsh at Ynyslas, Dovey estuary	824

xxxiv   FIGURES IN THE TEXT

Fig. 154. Zonation of algal communities in salt marsh at Canvey Island *page* 825
155. Zonation of algal communities in the Ynyslas salt marsh   826
156. *Salicornia herbacea* at Ynyslas   827
157. *Spartina townsendii*   828
158. *Glyceria maritima*   829
159. *Armeria maritima*   832
160. Root systems of salt marsh plants   836
161. Root systems of *Limonium vulgare* and *Triglochin maritimum*   837
162. Braunton Burrows, North Devon   846
163. Rates of leaching and decrease of pH values with age of sand dune soils   866
164. Increase in organic content with age of dune soils   866
165. Two shingle spits on the Solent   869
166. The Chesil Bank shingle bar   870
167. Diagram of apposition shingle beach   871
168. Blakeney Point and Scolt Head Island   872
169. Development of Blakeney Point, 1886–1911   874
170. Development of Blakeney Point, 1913–37   875
171. Diagram of a main shingle bank and hooks   876
172. Blakeney and Chesil shingle banks showing their relation to the marine terrace   876
173. Sketch of part of the "camm", Chesil Bank   877
174. Profiles of Blakeney and Chesil shingle banks   878
175. Zones of *Suaeda fruticosa*, Blakeney main bank   881
176. Origin of *Suaeda* zones in response to inward travel of beach   882
177. Branch of *Suaeda fruticosa*   883
178. Contoured strip of Blakeney main bank from crest to marsh   884
179. Contoured chart of ravine "can" in the camm of Chesil Bank   886

PLATES

Frontispiece. The country behind Beachy Head, Sussex, with the eastern South Downs

Plate 1.	Effect of prevailing winds	*facing page* 41
2.	Effects of rabbits	134
3.	Extreme devastation by rabbits	135
4.	Effect of rabbits on Callunetum	136
5.	Differential effects of rabbits	137
6.	Effects of excluding rabbits	138
7.	Rabbits and the sea aster	139
8.	Localised effect of rabbits	140
9.	Effects of excluding rabbits	141
10.	Biotic effects	145
11.	Peat sections in raised mosses	168
12.	After the fenland flood of January, 1915	187
13.	Typical pedunculate oakwoods (Quercetum roboris)	270
14.	Pedunculate oakwood	271
15.	Pedunculate oakwood	280
16.	Dry pedunculate oakwood on sandy glacial loam	281
17.	Oakwood in the extreme south and the extreme north of Great Britain	294
18.	Succession to pedunculate oakwood on clay pasture, Didcot, Oxon.	295
19.	Wistman's Wood, Dartmoor (Quercetum roboris)	298
20.	Detail of Wistman's Wood	299
21.	Wistman's Wood, field layer and epiphytes	
22.	Epiphytes in Wistman's Wood	*between pages* 300 *and* 301
23.	Edges of Wistman's Wood	
24.	Dartmoor woods, limiting conditions	
25.	Pennine sessile oakwoods (Quercetum sessiliflorae)	314
26.	Ravine sessile oakwoods	315
27.	Field layer of sessile oakwood	316
28.	Field layer of sessile oakwood	317
29.	Devonshire sessile oakwood	
30.	Devonshire sessile oakwood detail	*between pages*
31.	Quercetum sessiliflorae on the Quantock Hills	318 *and* 319
32.	Quercetum sessiliflorae (scrub) on the Quantock Hills	
33.	Birkrigg oaks, Cumberland (Quercetum sessiliflorae)	321
34.	Killarney woods (Quercetum sessiliflorae)	328
35.	Detail of Killarney woods, "Type 1"	329
36.	The Upper Woods, Killarney	332
37.	Bryophyte stratum in the Upper Woods, Killarney	333
38.	Local yew wood, Killarney	338
39.	*Arbutus unedo* at Killarney	339
40.	Sessile oakwoods at Lough Conn	340
41.	Irish Quercetum sessiliflorae	341
42.	Connemara island scrub	343
43.	Quercetum ericetosum and oak-birch heath	354
44.	Oak-birch heath	355
45.	Beech forest (Fagetum sylvaticae)	362
46.	Beech forest from the air	363
47.	Yew layer in Fagetum calcicolum	368

PLATES

Plate		facing page
48.	Fagetum calcicolum, summer and winter	369
49.	Fagetum calcicolum	370
50.	Fagetum calcicolum saniculosum	371
51.	Cotswold beechwoods	372
52.	Juniper scrub on chalk	373
53.	Chalk scrub	374
54.	Composition of chalk scrub	375
55.	Yew wood and yew-ash wood at Kingley Vale	376
56.	Kingley Vale yew woods	377
57.	Interior of yew wood, Kingley Vale	378
58.	Yew scrub	379
59.	Effect of wind on yew	380
60.	Yew in various regions	381
61.	Beech invasion of juniper scrub (Chiltern escarpment)	382
62.	Immature escarpment beechwood	383
63.	Mature Fagetum rubosum (Chiltern plateau)	
64.	Fagetum rubosum, field layer	between pages 388 and 389
65.	Mature Fagetum rubosum (South Downs)	
66.	Fagetum rubosum, detail of field layer	
67.	Fagetum rubosum, succession	396
68.	Ash-oak wood seral to Fagetum rubosum	397
69.	Fagetum rubosum, reproduction circles	404
70.	Fagetum rubosum, end of reproduction circle	405
71.	Fagetum ericetosum	412
72.	Fagetum ericetosum	413
73.	Ashwood on limestone (Fraxinetum calcicolum)	428
74.	Succession to Fraxinetum calcicolum	429
75.	Fraxinetum on basic igneous rock	436
76.	Fraxinetum calcicolum in south-east Devon	440
77.	Fraxinetum calcicolum	441
78.	Highland pine forest	446
79.	Rothiemurchus Forest	447
80.	Highland pine forest	448
81.	Native pines of contrasted habit	449
82.	Loch Maree pinewood	450
83.	Zonation of oak, pine and birchwood	451
84.	Highland birchwoods	452
85.	Highland birchwoods	453
86.	English seral birchwoods	456
87.	Subspontaneous pine in south-east England	457
88.	Alder carr in East Norfolk	460
89.	East Norfolk carrs	461
90.	Old carr at Wicken Fen	463
91.	Limestone pavement, limestone heath and hazel scrub	474
92.	Climax hazel scrub and heath on the Burren (Co. Clare)	475
93.	Hawthorn thicket scrub	482
94.	Pteridietum	504
95.	Sheep-grazed Ulicetum gallii	505
96.	Nardetum strictae	516
97.	Nardetum strictae, Moorfoot Hills	517
98.	*Molinia caerulea*	518
99.	Molinietum on the Southern Pennines	519
100.	Escarpment of the South Downs	526
101.	Chalk grassland	527
102.	Chalk grassland and rabbit effects	544

PLATES

Plate 103.	Chalk grassland and rabbit effects	*facing page* 545
104.	Limestone grassland and Nardetum	553
105.	Neutral grassland	570
106.	Submerged, emersed and floating leaves	582
107.	Floating leaf and reedswamp	583
108.	Reedswamp in the Norfolk Broads	595
109.	Zonation in the Thames	627
110.	Silted and non-silted river communities	630
111.	Glycerietum maximae in the Lark	631
112.	Fen vegetation	656
113.	Fen carr	657
114.	Fen carr	660
115.	Raised bog	686
116.	Irish raised bog—regeneration complex	687
117.	Irish raised bog—hummocks and pools	688
118.	Irish raised bog	689
119.	Eriophoretum vaginati on the Southern Pennine plateau	704
120.	Retrogressive Eriophoretum vaginati	705
121.	West Irish blanket bog	714
122.	Detail of Connemara blanket bog	715
123.	Hummocks in blanket bog	718
124.	Facies of Lowland Heath	726
125.	Development of Callunetum	727
126.	Development and destruction of Callunetum	728
127.	Species of south-western English heaths	764
128.	Scree vegetation	778
129.	Vegetation of mountain top detritus	788
130.	Arctic-alpine vegetation of Ben Lawers	804
131.	Arctic-alpines (Ben Lawers)	805
132.	Arctic-alpines (Ben Lawers)	806
133.	Blakeney Point from the air	
134.	"The Marams" at Blakeney Point, from the air	
135.	Maritime vegetation at Blakeney Point—sand dune, shingle and salt marsh	*between pages* 818 *and* 819
136.	Maritime vegetation at Blakeney Point—salt marsh and shingle spits	
137.	Bare mud and Salicornietum	824
138.	Salicornietum and *Spartina townsendii*	825
139.	Spartinetum townsendii	828
140.	Glycerietum maritimae	829
141.	Glycerietum and Armerietum	834
142.	General salt marsh and Obionetum	835
143.	Upper salt marsh	840
144.	Channels and pans in salt marsh	841
145.	Foreshore and foredunes	846
146.	Foreshore and foredunes	847
147.	Foredunes on Blakeney Far Point	848
148.	Active Ammophiletum	849
149.	Active Ammophiletum	850
150.	Fixed dune with moss carpet	851
151.	"Slacks" or dune valleys	862
152.	Dune scrub (Hippophaëtum)	863
153.	"Blow-outs"	864
154.	Various dune features	865
155.	Chesil beach from the air	870

xxxviii   PLATES

Plate 156. Chesil beach   *facing page* 871
      157. Blakeney beach   884
      158. Chesil beach   885
      159. Shingle vegetation at Blakeney Point   887
      160. Shingle beach vegetation   890
      161. Apposition shingle beaches   891
      162. Maritime and submaritime vegetation   898

PART I
THE BRITISH ISLANDS AS ENVIRONMENT OF VEGETATION

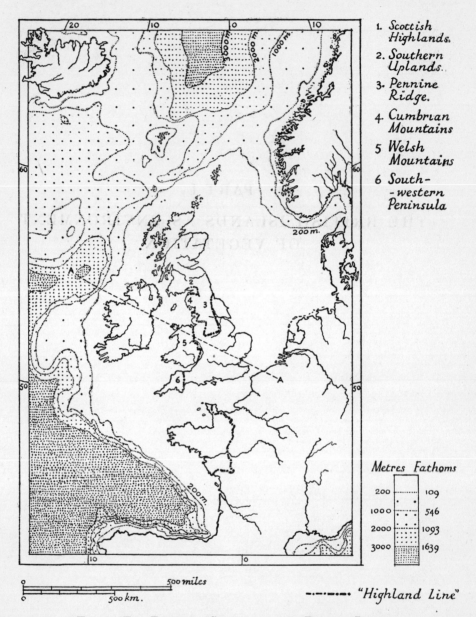

FIG. 1. THE PHYSICAL SETTING OF THE BRITISH ISLANDS

Map showing their situation on the Continental Shelf of North-western Europe. The line A–A marks the section shown in Fig. 2. The "Highland Line" separates the Palaeogenic region of the north and west of Great Britain from the Neogenic region of the midlands, south and east of England. The numbers 1–6 refer to the main Palaeogenic areas of Great Britain described on pp. 8 and 10. Ireland (pp. 10, 11) is almost entirely Palaeogenic.

Chapter I

PHYSICAL FEATURES AND GEOLOGICAL HISTORY

The continental shelf. The British Islands, including the Orkneys and Shetlands, stand on what is known as the "continental shelf" of north-western Europe (Figs. 1 and 2). The depth of the seas (English Channel and North Sea) separating the British archipelago from the continent nowhere reaches 200 m.,[1] and rarely exceeds 100 m. In the southern part of the North Sea, the Straits of Dover and the eastern part of the English Channel, indeed, there are no depths reaching 100 m., and much of the sea floor is less than 40 m. below the surface. In the northern part of the North Sea, and in the western Channel towards the Atlantic, as well as between Great Britain and Ireland, somewhat greater depths occur, exceeding 100 m., but very rarely reaching 200 m. Only in the "Norway Deep", a wide trough running southward from the Arctic Ocean and bending eastward close to the southern Norwegian coast into the Skager Rak, are there depths of over 400 m., and in the Skager Rak itself nearly reaching 800 m. Everywhere else the floor of the shallow seas surrounding Great Britain and Ireland slopes very gently northwards and westwards down to about 200 m. and then much more steeply to 2000 m., a depth reached at 150 km. from the north- and south-western Irish coasts. Beyond the 3000 m. contour the ocean floor slopes more gently to 4000 m., and ultimately—westward and southward—to the great Atlantic depths exceeding 5000 m. To the north of Scotland a ridge of sea floor, none of which is more than 800 m. and most of it less than 600 m. from the surface, stretches north-north-west to Iceland, which is surrounded by a wide belt of relatively shallow sea extending to the east coast of Greenland. The Faroes are situated on this ridge, and round them is a considerable belt of sea less than 200 m. deep (Fig. 1). This "Scoto-Icelandic Rise" of the sea floor is an ancient tectonic feature, probably dating at least from mid-Tertiary times, and indicating the former connexion of the British Isles with Iceland and Greenland.

Compared with the great ocean depths it is therefore a very shallow layer of water which covers the continental shelf and isolates the British Isles, thinning to a mere film, as it were (Fig. 2), between south-eastern England and the opposite coasts of north-eastern France, Belgium, Holland, Germany and Denmark, coasts which are composed of the same geological strata and were separated quite recently from the corresponding rocks of south-eastern England.

[1] Sea depths are always given in fathoms (a fathom is 6 ft. = about 1·829 m.) in British charts. In order to use round numbers a fathom is here taken as 2 m. For the present purpose the errors thus introduced are unimportant.

Endemic species. In every sense, then, the British Isles are "continental islands" very recently separated from the continent by comparatively slight subsidence of the land aided by marine erosion. This is reflected in the flora and fauna, the great bulk of which is composed of continental species. Forty years ago the current statement was that we had no endemic species of vascular plants, but with closer study of the indigenous flora and the increasing tendency to recognise as species smaller units than those of the older taxonomists, we may reckon as endemic between a dozen and twenty flowering plants (some of which are probably hybrids), apart from the polymorphic Rubi and the apomictic Hieracia, of which many British forms have not been recorded from the continent. Doubtless also there are a number of endemic geographical races—what are now called "subspecies" by modern zoologists—not as yet sufficiently known. Compared with islands like Corsica, or the Balearics, for example, the endemic element is almost negligible; and taken as a whole the British flora of vascular plants is essentially an impoverished fragment of the European, though it contains several distinct and interesting elements of more remote geographical affinities than those of the great mass of species common to the British Isles and the lowlands of western and central Europe.

Of mosses there are six or perhaps seven "good" species recognised by Mr H. N. Dixon (*in litt.*) as endemic and about as many more that are better ranked as "subspecies". Of liverworts Mr W. E. Nicholson writes that nine species have claims to be considered endemic, though some of them may not be entitled to specific rank. These are all extreme western (Atlantic) forms. The country is particularly rich in relict Tertiary liverworts formerly thought to be endemic but now known to occur in various other, sometimes widely scattered, parts of the world. Among fresh-water algae, according to Professor Fritsch, it is doubtful if any British forms are really endemic.

The same is true of the fauna. Among mammals the Irish stoat (*Mustela hibernica*) and the Irish hare (*Lepus hibernicus*) are good endemic species, and there are a few shrews, voles and field mice confined to islands off the British coasts. The Scottish Red Deer (*Cervus elaphus scoticus*) is an example of the fairly numerous endemic subspecies. There is only one endemic species of bird, the grouse (*Lagopus scoticus*), though there are a number of endemic subspecies. Of fishes there are several endemic species and subspecies, e.g. of char (*Salvelinus*), whose ancestors seem to have been cut off in isolated freshwater lakes, their descendants showing distinct characters which separate them from their nearest allies. There are also a good many species and varieties of insects recorded as confined to the British Isles.

Palaeogenic and Neogenic. But though Great Britain is now but a slightly detached island with its south-eastern portion almost continuous with the mainland of Europe, it is not so in origin, and its northern and western portions have an independent and far older geological history. The land of the British Isles consists in fact of two components, geologically widely distinct. The first, lying to the north and west, once formed part of an ancient land which stretched far to the north-west right up to the end of Mesozoic times, when most of what is now England lay below the waters of the great Cretaceous sea—a sea extending westwards and northwards to northern Ireland and western Scotland, southwards and eastwards to France, Denmark, and right across Central Europe to the Crimea. The bordering mountains of the old north-western land (sometimes known as

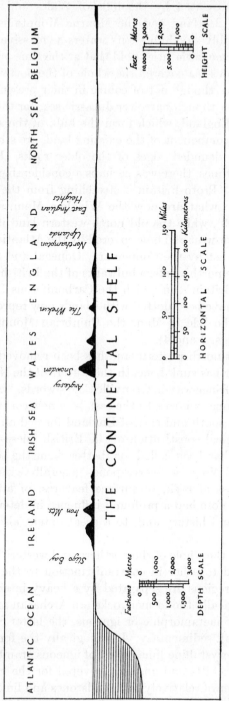

Fig. 2. The Continental Shelf

Section (A–A in Fig. 1) showing the extreme shallowness of the seas surrounding the British Islands compared with the Atlantic depths, and the much greater height of the Irish and Welsh Palaeogenic mountains as compared with the English Neogenic hills.

"Atlantis"), now represented by the hills of Wales and Cumbria, the Southern Uplands of Scotland and the Mourne Mountains of northern Ireland (Fig. 3), probably stood above its waters as massive promontories and islands, though some authorities hold that at the time of its maximum extension the Cretaceous sea covered the whole of the north-western land. These same mountains, though not of course in their present shape, must have fringed the earlier though narrower Jurassic sea, for the whole of the Mesozoic deposits of England, which form the bulk of the second, south-eastern, and younger component of the existing land, are strata laid down on the upturned and denuded edges of the older rocks. Right back in Lower Carboniferous times, there was perhaps a considerable island, called by Mackinder (1907) "Proto-Britain", stretching from the English midlands across Wales to what are now the Wicklow Mountains of south-eastern Ireland (Fig. 3), while the old north-western land itself, then extending far to the north-west and now covered by the Atlantic, was perhaps fringed to the south-east by the Connemara, Donegal and Scottish highlands (Fig. 3), now the north-western bulwarks of the British Isles. The sea surrounding "Proto-Britain", in which the Carboniferous Limestone was laid down, may then have been dotted with islands now represented by the Mourne Mountains, the Isle of Man, the Cumbrian Mountains and the Southern Uplands (Figs. 3 and 4).

Most of this ancient north-western land has been removed by westward "continental drift" or has sunk beneath the waters of the North Atlantic, but the south-eastern fringe was, in Grenville Cole's words, "caught into the new continent of Europe as it rose to the east"—a new continent of which the midlands and the south and east of England formed an integral part.

This twofold geological constitution of the British Isles—"Palaeozoic" and "Neozoic" as it has been called, or (better from the tectonic standpoint) *Palaeogenic* and *Neogenic*—corresponds generally with differences of structure and hardness of rock, of surface features, of altitude, and of climate, and has therefore had a profound influence in determining differences of population and history and, to a great extent also, of flora and vegetation.

For many purposes the whole of the northern and western rocks (Palaeogenic) may be classed together in contradistinction to the south-eastern (Neogenic). The two regions are separated by a heavy interrupted line in Figs. 1 and 4. Geologically the former rocks are Archaean and Palaeozoic, either sedimentary or metamorphic or igneous, the latter Secondary and Tertiary and nearly all sedimentary. Lithologically the former consist of hard grits, slates and crystalline limestones, of igneous granites and basalts and of metamorphic schists and gneisses, covered for the most part with shallow soils; the latter of relatively soft sandstones and limestones, marls, gravels, sands and clays, forming deeper soils. Physiographically the former are mainly hilly or mountainous, sometimes with abrupt relief: the latter

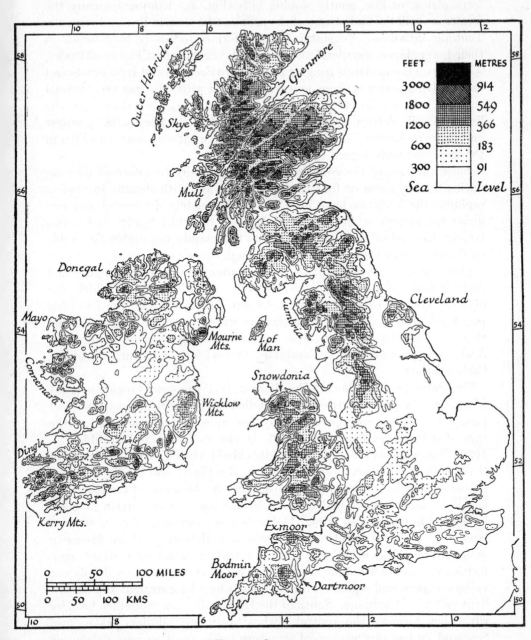

FIG. 3. OROGRAPHY

Most of the ground above 600 ft. and all of that above 1200 ft. (366 m.) (except in Cleveland) lies to the north and west of the "Highland line" (Fig. 4), i.e. is Palaeogenic.

form plains, or low, gently sloping hills (Fig. 3). Ethnographically the former are still the main home of the peoples who were in Britain before the Teutonic invasions. Vegetationally the former naturally bear forest on their lower slopes, moorland and mountain vegetation at higher altitudes, with bog on the lowlands fringing the western Scottish and Irish coasts and on high-lying plateaux: here, however, we meet with a distinct vegetational contrast between the calcareous soils produced by the limestones and the lime-free soils derived from the siliceous rocks. The deep-soiled younger and softer rocks once bore almost continuous forest, with marsh and fen in the low-lying waterlogged basins.

Since the face of the country has been gradually transformed by man during the last three or four thousand years, and with steadily increasing rapidity, the lowlands have been almost completely deforested and put under the plough, while the lower slopes of the older harder rocks, also largely denuded of their natural forests but mainly unsuitable for arable cultivation, have been chiefly used as rough sheep grazings.

It would serve no relevant purpose to enter here upon the lengthy and difficult task of describing the probable conditions under which the different systems of Palaeogenic rocks were laid down, and the following pages will deal mainly with the factors which have led to the existing structure of the country. The sequence of Palaeogenic systems is Archaean (pre-Cambrian), Cambrian, Ordovician, Silurian, Devonian, Carboniferous.

The Palaeogenic region. "The Palaeozoic [Palaeogenic] region consists of a series of separate districts of high relief, each a deeply dissected ancient peneplain with an epigenetic river system more or less readapted to the rock structure now exposed" (A. M. Davies in Evans and Stubblefield, 1930). There are six main districts of this kind included within the island of Great Britain (numbered 1–6 in Fig. 1; cf. also Figs. 3 and 4). The northern portion (Scotland) is almost entirely composed of the older and harder rocks and includes by far the greater portion of high-lying land in the British Isles. (1) The largest (northern) portion of Scotland—the "Scottish Highlands"—consisting mainly of very hard Archaean metamorphic rocks with large igneous intrusions (some of Tertiary age), forms an almost continuous mass of high ground intersected by deeply cut valleys or glens and separated into two parts by a long straight and narrow "rift valley"—Glenmore. South of the Highlands is another much broader rift valley forming the central lowland plain or "Midland valley" of Scotland, of complex geological structure made up of various Palaeozoic sedimentary and igneous rocks. South of this again is (2) another large area of high ground—the Southern Uplands—composed of Ordovician and Silurian rocks and occupying most of the area of southern Scotland. This is continuous with (3) the Cheviot Hills and the Pennine Hills of northern England, of Devonian and Carboniferous age. The Pennines, which are

FIG. 4. STRUCTURAL ELEMENTS OF THE BRITISH ISLANDS

The Palaeogenic region is dominated by (1) the Caledonian Folds running NNE–SSW or NE–SW in the Highlands, Southern Uplands, Cumbria, North Wales, northern and south-eastern Ireland; (2) the Armorican Folds of southern Ireland, South Wales and the south-western English peninsula, extending ("posthumously") eastward to the Weald in south-eastern England. Exposures of igneous rocks are confined to the Palaeogenic region.

The strike of the Neogenic rocks has a general direction SW–NE, bending N and NNW in northern England. The east and south-east of the Neogenic region are dominated by the outcrop of the Chalk (dotted).

entirely Carboniferous, form a broad ridge with a north and south axis dominating the structure of the north of England and terminated abruptly to the south by a fault cutting off the English midland plain, which is formed of the softer Secondary strata through which rise islands of hard Palaeozoic and Archaean rocks. To the west of the Pennine axis, and separated from it by another rift valley, lie (4) the lofty Cumbrian Mountains (Fig. 3) forming a detached group composed of Silurian rocks (Skiddaw slates) to the north and Ordovician to the south, with abundant igneous intrusions. To the west of the English lowland plain Wales (5) is almost entirely mountainous, though its average elevation does not nearly reach that of the Scottish Highlands (Fig. 3). The rocks (largely metamorphic) are mainly of Cambrian, Ordovician and Silurian ages. Finally (6) the south-western peninsula of England is also formed of Palaeozoic rocks of Devonian and Carboniferous age with great bosses of Tertiary granite. This region contains some elevated land (Exmoor, Dartmoor, etc.) but of considerably lower average altitude (Fig. 3).

Wales

South-western peninsula

Ireland   Ireland is separated from Scotland by the narrow and relatively deep North Channel, from northern England by the broader, and in its eastern part shallow, Irish Sea, and from Wales and the south-western English peninsula by the narrower and deeper St George's Channel. The North and St George's Channels, with the intervening western portion of the Irish Sea, occupy what is probably a rift valley, to the formation of which the separation of the two islands—Ireland and Great Britain—is primarily due.

The rocks of which Ireland is composed, like those of the mountains of Scotland, northern England, Wales and the south-western English peninsula, are almost entirely Archaean or Palaeozoic (sedimentary, metamorphic and igneous). The average elevation of the land is, however, slight, since the greater portion of the country is occupied by the low-lying central Irish plain of Carboniferous Limestone (Fig. 4). The higher mountains all lie near the coasts in more or less isolated groups. On the east coast the most important are the Mourne Mountains in Co. Down (Ordovician and igneous), and the Wicklow Mountains considerably farther south (Cambrian, Ordovician and igneous): on the west coast the Donegal highlands in the extreme north-west, the Mayo-Connemara complex, about the centre of the west coast, and the Kerry mountains in the extreme southwest. These include all ages of Palaeogenic rocks, from Archaean to Carboniferous, except the Cambrian. Besides the western mountains there are considerable mountain masses along the north and along the south coasts, and isolated ridges of older rocks rising through the central plain, especially in its southern portion. All these, except the southern mountains, which are mainly Devonian (Old Red Sandstone) and those of Antrim in the north-east (Tertiary basalt), are composed of very hard

Cambrian (only represented in Wicklow), Ordovician, Silurian, and Carboniferous metamorphic and igneous rocks.

The central Irish plain, entirely formed of Carboniferous Limestone, stretches continuously east and west for about 200 km., from the Irish Sea to the Atlantic: north and south its extreme length is more than 300 km. This plain of Carboniferous Limestone, worn down by continuous dissolution of the soluble material and by glaciation, rarely exceeds 150 m., and much of it is less than 80 m., above the sea. Most of its surface is covered with calcareous boulder clay, deposited from the Pleistocene ice-sheets, and considerable parts are thickly dotted with raised peat bogs.

The great tectonic folds. Nearly all the ancient rocks, the main outlines of whose distribution have been briefly indicated, have been heavily folded and afterwards very extensively worn down by subaerial denudation, so that the existing surface features generally bear little or no relation to the tectonics. Several distinct series of foldings have been determined, and the two most important, with the rocks they affected, are traceable (see Fig. 4) on the two sides of the narrow sea separating Great Britain and Ireland.

The Caledonian folds
The first series, known as the *Caledonian Folds*, which strike north-east and south-west, occurred at the close of Silurian times and were never obscured by later crumplings: they are responsible for the fundamental structure of the Highlands of Scotland and of north-western Ireland (Donegal). The deposits of the pre-Cambrian seas were "worked up and overthrust against the floor on which they had been laid down, but portions of that early Archaean floor also became involved. This floor, the oldest land mass in the Britannic region, forming a portion of what has been called the Huronian continent, is revealed in the worn-down gneissic band of the Outer Hebrides, and from Cape Wrath, the north-western extremity of Scotland, southward to the Sound of Sleat between Skye and the mainland" (Grenville Cole, 1914, p. 4). Surmounting the crystalline gneisses and schists are huge fort-like blocks of massive, still pre-Cambrian, Torridon Sandstone, an unaltered sedimentary rock, accumulated in the hollows of the underlying gneiss. Eastward lies the main mass of the Scottish Highlands, consisting of mica-schists, limestones and quartzites, known as the Dalradian series, originally sedimentary rocks whose age is now regarded as also pre-Cambrian and pre-Torridonian. These were probably metamorphosed before the Caledonian folding and then further rolled out and rearranged, including many igneous intrusions. To the south the Scottish Highlands are abruptly limited against the Midland Valley by the great "Highland boundary fault"—or rather series of faults—also the result of the Caledonian folding.

The highlands of northern and western Ireland obviously continue this same series of rocks to the south-west (Fig. 4). Donegal, with its folded mica-schists and dominating crests of quartzite, repeats the features of Argyll and of the Perthshire highlands, while Kintyre is continued in north-east Antrim. A conspicuous Caledonian axis with a core of granite runs south-westward for 100 km. to Castlebar in Mayo, while the quartzites and metamorphic rocks of the mountains of western Mayo and Connemara also recall the pre-Cambrian rocks that are folded and overfolded in the Scottish Highlands. South-westward of Connemara the Caledonian folds are lost beneath the Atlantic (Figs. 3 and 4).

The rocks of these highland regions of Scotland and Ireland have been remarkably added to by Tertiary lavas. The complex volcanic piles of the Inner Hebrides (Skye, Mull and smaller islands), some worn down to their crystalline cores, are matched by the great basaltic plateau of Antrim in north-east Ireland. These are evidence of the vast disturbances of the earth's crust in early Tertiary times which raised the mighty mass of the Alps in central Europe and drove volcanic vents through the oldest pre-Cambrian rocks of north-west Britain. It has been suggested that the fiords of this region, with their varied directions and remarkable rectangular bends, are the result of cracking during these "Alpine" disturbances, the lines of weakness so produced being afterwards excavated by water and later by ice. Immense denudation has since occurred, stripping the upper volcanic deposits from the surface of the Highlands and exposing the crystalline cores of the lavas solidified under huge pressure below. The Tertiary lavas, both of Antrim and Mull, have preserved underlying Mesozoic and even fossiliferous Eocene strata from destruction.

Just as the very ancient rocks of the Scottish Highlands are continued south-westward into northern and western Ireland, through the whole of which the Caledonian folds can be traced, so the Ordovician and Silurian rocks of the Southern Uplands of Scotland, consisting mainly of shales, slates, fine-grained sandstones and grits passing into quartzites, were also involved in the Caledonian folds, whose course can be traced through the Southern Uplands across the North Channel to the Mourne Mountains of Co. Down and on into the heart of Ireland (Fig. 4). The Silurian grits and quartzites are hard rocks, but it is mainly where they are reinforced by igneous dykes and cores that they form mountainous topography, as in the Lake District (Cumbrian Mountains), the Isle of Man and the Mourne Mountains. A huge mass of overlying rock has evidently been removed from these mountains also.

The great mountain mass of North Wales (Snowdonia) can be matched in the Wicklow Mountains across St George's Channel. The rocks of both are Cambrian and Ordovician, reinforced by igneous cores, and in both the main trend of the folds is again north-east and south-west (Fig. 4). In the low-lying island of Anglesey as in the low-lying Outer Hebrides, both worn down almost to sea-level, Archaean rocks are exposed; and it may have been across Anglesey that the main axis of a North Wales-Wicklow Caledonian fold ran south-westward through what is now St George's Channel.

The second important series of earth movements which built
The Armorican folds the mountains of the British Isles is that which caused the
Armorican Folds, occurring at the end of Coal Measure times but
with important "posthumous" revivals in post-Jurassic, post-Cretaceous
and post-Oligocene times, and conspicuously seen in South Wales, the
south-western English peninsula, and southern Ireland (Fig. 4), as well as
in Brittany ("Armorica") across the English Channel. The Armorican folds
run east and west, and in southern Ireland have produced a series of closely
set mountain ridges. These folds involved the Carboniferous as well as the
earlier Palaeozoic rocks, but the former have been largely worn away,
exposing the resistant conglomerates of the Old Red Sandstone (Devonian)
series which now form the mountain ridges of the Dingle promontory, the
Slieve Bloom, Galty and Knockmealdown mountains and the Reeks of
Kerry (cf. Figs. 3 and 4). In the synclinal valleys between the ridges
Carboniferous Limestone still remains, and where the hard Devonian
conglomerates have been worn through along the crests of the anticlines
the Silurian slates below are exposed, so that the abrupt scarps of heather-
covered or bare sandstone rise above basin-shaped depressions occupied
by arable land. The Galty and Slieve Bloom ridges also include cores of
the old "Caledonian" mass on which these Palaeozoic strata were laid
down.

In South Wales the same system of east and west folds can be seen about
Milford Haven. Farther east it was less intense and formed the broad
synclinal basin, holding the great South Wales coalfield. South of the
Bristol Channel the anticlinal ridge of the Mendip Hills in north Somerset,
running a little south of east, and involving Upper Devonian as well as
Carboniferous Limestone, is another Armorican feature, and farther south
the broad syncline which runs across Devonshire, including Devonian
slates and sandstones and Lower and Middle Carboniferous rocks, determines the structure of the south-western peninsula. In south Devon and
Cornwall the sedimentary rocks are interrupted by great masses of granite,
intruded at the close of the Armorican movements, and forming the uplands of Dartmoor and Bodmin Moor (Figs. 3, 4 and 5).

Thus these two great systems of Palaeozoic foldings—Caledonian and
Armorican—determine the primary tectonics of the northern and western
parts of the British Isles which formed the south-eastern fringe of the
ancient north Atlantic continent, a fringe that was joined by much later
deposits to the continent of Europe. But the existing structure of the
mountains was not determined by the primary folds. The
Existing sculpture enormous mountain masses that must have been formed by the
great Caledonian folding were worn down to a peneplain, which
was again upraised and then dissected. The rough equivalence in height
of the existing summits is evidence of this. Only in the Scottish Highlands,
and rarely there, do the summits rise above 4000 ft. (c. 1220 m.), but a

considerable number exceed 3000 ft. (915 m.) and most of the Highland plateaux lie between 2000 and 3000 ft. A very large number of summits, not only in Scotland, but in the Lake District, Wales and Ireland, exceed 2000 ft. (610 m.), but only a few exceed 3000 ft. Much the greater proportion of the Scottish Highlands as well as great areas of the southern Scottish and of the Welsh uplands exceed 1500 ft. (457 m.) in altitude. Further evidence is afforded by the synclinal structure of some of the highest mountains, the existing summits representing a wedge of harder more resistant rock remaining from the trough of the bend, while the stratigraphically lower, but softer, beds on each side have been worn away. Many summits also are formed of granite masses originally solidified far below the surface. All this implies immense denudation, the removal of thousands of metres of rock above the existing mountains, and the carving out of existing summits from the plateaux so produced.

To the southward of the Scottish Southern Uplands, and eastward from the Cumbrian Mountains—probably the worn-down remnants of the old Caledonian folds, upraised again centrally, perhaps by an "Alpine" movement, and showing a strikingly radial epigenetic river system—lies the Pennine axis which dominates the structure of northern England, and runs for more than 300 km. southward, to terminate abruptly by an east and west fault at the Midland plain (Fig. 5). The Pennine Hills are composed of massive Carboniferous rocks laid down after the Caledonian foldings but in which indications of Armorican folds are still apparent. Later a north and south axis of elevation raised the complex Pennine ridge, faulted on its western, but dipping gently below later strata on its eastern, boundary, with a basis of Carboniferous Limestone exposed in the northern half (Cumberland and north-west Yorkshire) and to the south (Derbyshire), but covered by the overlying Millstone Grit in the intervening stretch (south-west Yorkshire and north Derbyshire). On either flank lie the Coal Measures, forming the coalfields of Northumberland, Durham and south Yorkshire on the east and of Lancashire on the west.

Palaeogenic soils. The soils produced by the Palaeogenic rocks, though differing from one another widely in many lithological and pedological details, are for the most part strikingly alike in their effect on vegetation, as we shall see more fully in Chapter v. They are predominantly shallow, and either bear rather low oak or birch forest, with floristically poor subordinate heathy vegetation. Pastured they support rough grassland grading to bog in badly drained areas; and are generally unsuitable for arable cultivation. The Palaeogenic limestones are, however, an exception. Limestone soils and soils derived from basic igneous rocks bear a much better type of grassland. Some of the shales too are fertile, and there are other exceptions such as some of the Devonian rocks. The Herefordshire plain, for example (Fig. 5), though exclusively Palaeozoic, is very fertile, especially where its rocks are markedly calcareous. The Herefordshire plain indeed, though geologically

Palaeogenic, belongs physiographically and agriculturally rather to the Neogenic series.

The Neogenic region. We now come to the second great component of the structure of Britain, the "Neozoic" or "Neogenic", forming the rest of England, which is essentially lowland. This is composed of much softer sedimentary rocks—shales, clays, sandstones, limestones and marls—the harder of which form long escarpments facing north-west and separated by strike valleys or plains (Fig. 5). The successive seas in which these beds were deposited lay to the south-east of the old north-western continent but penetrated it deeply, leaving relics of their deposits in north-eastern Ireland and in the western islands of Scotland, after removal of great intervening tracts of Neogenic sediments between these outliers and the main existing outcrops. In these seas the south-eastern bulwarks of the ancient north-western continent, now represented by the Welsh and Lake District mountains, the Pennines and the Scottish Southern Uplands, must have stood out as mountainous islands or peninsulas.

The natural division between the Palaeogenic and Neogenic rocks is formed by the Coal Measures, deposited in great low-lying swamps, in which grew the extensive forests that formed the material fossilised as coal. Thereafter general elevation of the land took place, forming the great "Hercynian" continent which persisted through Permian and Triassic times and extended over most of western and central Europe including the British Isles. The climate was arid during the whole of this long epoch, the contemporary deposits being strikingly similar to those of modern desert regions, with widespread breccias formed from the angular material of great deltaic fans, wind-blown sands, and accumulations of salts precipitated from evaporating salt lakes. Immense subaerial erosion, largely wind erosion, wore down the Hercynian continent during this very long period of exposure.

Later, the continent was invaded by the Jurassic seas, in which the Jurassic limestone scarps (Fig. 5) and clays were formed, and again by the much more extensive Cretaceous sea. In this last the Chalk—the most extensive Neogenic limestone—was laid down, bringing the Mesozoic period to a close. The Chalk now dominates the south and east of England (Fig. 4), forming such features as the North and South Downs, the Hampshire uplands, the Berkshire Downs, the Chiltern Hills, the East Anglian Heights, and the Lincolnshire and Yorkshire Wolds (Fig. 5), just as the Carboniferous limestone, formed in the earlier extensive sea towards the close of the Palaeogenic period, dominates much of the north and west. The Tertiary period which followed saw far less extensive changes of land and sea, and Britain began to acquire its existing shape, though its final detachment from the continent took place much later. The Tertiary deposits are much less extensive and less important geographically than those of the earlier periods.

Permian and Trias. The earliest of the post-Carboniferous rocks are the Permian deposits, more or less contemporaneous with the extensive Armorican earth movements which raised the Hercynian continent, and showing markedly different facies in different English regions. In the west, midlands and south they are largely dominated by red sandstones and breccias, formed as the result of erosion by wind and water from the Carboniferous rocks on which they lie unconformably. The aggregate area of Permian rocks is not great, the most extensive development being east and west of the Pennines. To the west of the northern Pennines are the Permian sandstones of the Eden valley. To the east of the Pennines is a narrow belt of Magnesian Limestone (Fig. 4), formed in a westward extension of the "Zechstein Sea" which stretched over Germany and the southern Baltic region. The basal sands and breccias of this limestone rest unconformably on the Carboniferous rocks, from the detritus of which they were derived. Outliers of the Permian occur as far west as Tyrone in northern Ireland.

Desert conditions persisted through Triassic times, and the Bunter conglomerates and sandstones which form the lower division of the British Trias produce a poor coarse-grained soil, naturally bearing dry oakwood and heath, as in Sherwood forest in Nottinghamshire, but no conspicuous relief. The Keuper marls of the Upper Triassic, on the other hand, with which gypsum and salt beds are often associated, form much of the Midland plain of England (Fig. 5), a flat or gently undulating country of good agricultural land. The Trias extends from the lower Severn valley and the centre of England in Warwickshire north-westward through the "Midland Gate", between the Pennines and north Wales, to form the plains of Cheshire and east Lancashire (Fig. 5), with a detached area farther north, on the shores of the Solway Firth and in the Eden valley (Fig. 5), and a few unimportant outliers in north-eastern Ireland. To the east of the Pennines the Trias appears in Nottinghamshire and farther north in Yorkshire and Durham. Far to the south it appears again in east Devon, to the east of the Carboniferous rocks. The conditions of formation and general nature of the Trias (New Red Sandstone) are strikingly parallel to that of the Old Red Sandstone of the Devonian period, barren sandstones in each case contrasting with fertile marls.

The later Secondary rocks. The main outcrops of the two remaining systems of Secondary rocks—the Jurassic and the Cretaceous—run right across England from the coast of the Channel to that of the North Sea with a south-west—north-east strike, their northern extremities bent round in a northerly direction parallel with the Pennine axis (Figs. 4 and 5). Their general south-easterly dip is due to uplift of the edges of these rock systems (probably as a result of Tertiary "Alpine" earth movements coming from Central Europe) against the great resistant mass of the north-western Palaeogenic complex.

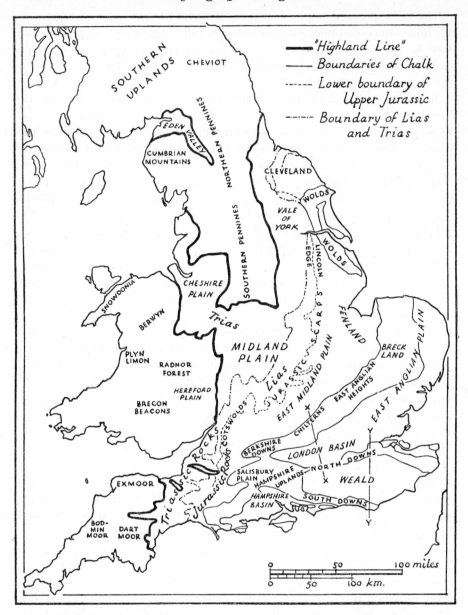

Fig. 5. Physiographic Regions of England and Wales

Compare Fig. 3. In the Palaeogenic region the chief hill and mountain complexes are marked. South-east of the "Highland Line" the features formed by the successive outcrops of the Mesozoic rocks are shown—the Trias and Lias (Cheshire and Midland Plains), the Cotswolds, the Jurassic scarps and the Cleveland Hills, the Chalk outcrops (North and South Downs, Hampshire Uplands, Salisbury Plain, Berkshire Downs, Chilterns, East Anglian Heights, and Wolds), the Weald, the Hampshire and London basins, and the East Anglian Plain. The lines X–X and Y–Y refer to the sections in Figs. 7 and 8.

The most prominent physiographic features of this Midland region are the limestone escarpments which form low hill ranges very rarely exceeding 1000 ft. (304 m.)—usually much less—and separated by plains of shale, clay, or marl. This region is sometimes called by geographers "the region of the limestone scarps" (Fig. 5).

Jurassic. The rocks of the Jurassic system which come next after the Trias and form a well-marked outcrop running across England from the Dorset to the Yorkshire coast, were laid down in a shallow land-locked sea transgressing across the Triassic land surface, and consist of alternating clays, shales and limestones, with occasional sandy beds. The lowest strata are the argillaceous limestones, shales, clays and calcareous sands of the Lias, which forms a broad belt of low-lying land across the Midlands with escarpments here and there formed by the more resistant beds. The Middle Jurassic limestones (Oolites) have produced bold and lofty escarpments in places, but are separated by clays and shales of much greater aggregate thickness. The most prominent of the limestones is the Inferior Oolite, which forms the escarpment of the Cotswold Hills in Gloucestershire, rising steeply to more than 1000 ft. (304 m.) and overlooking the Lias and Trias floor of the lower Severn valley (Fig. 5). Native beech forest occupies the calcareous escarpments and valley sides in parts of this region. North-eastward the Inferior Oolite becomes physiographically less prominent, except locally, till Lincolnshire is reached, where the prominent though not lofty "Lincoln edge" runs almost due north and south for many kilometres; and in north-east Yorkshire the sandstones, ironstones and calcareous grits of this formation rise to more than 1500 ft. (457 m.) in the Cleveland Hills (Fig. 5), bearing heath and moor. Above the Inferior Oolite comes the Great Oolite, and then various thin limestone bands (Forest Marble, Coral Rag, Purbeck Marble, etc.) collectively known as Corallian limestones, with much thicker intervening shales and clays of which the most extensive are the Oxford and Kimeridge Clays, forming country which is generally flat but locally in the south and east Midlands highly diversified on a small scale by narrow limestone outcrops.

Cretaceous. Next come the very extensive Cretaceous rocks, whose main escarpments run parallel with the Jurassic from the Dorset to the Yorkshire coast and whose outcrop occupies most of south-eastern England. The transition from the Upper Jurassic to the Lower Cretaceous is marked by a change from dominant limestones and marls to dominant sands and clays.

Wealden. The general outlines of land and sea were similar in the Jurassic and Cretaceous periods (though the Jurassic sea was much less extensive), but a great fresh-water lake was formed in the Wealden area of Kent, Surrey and Sussex, extending across what is now the English Channel to the Boulonnais. East and west "posthumous" Armorican axes of uplift, occurring after the Cretaceous rocks had been laid down, raised the various anticlinal folds which make up the Wealden dome and subordinate anticlines, and subsequent denudations stripped off the upper deposits, exposing the underlying Wealden beds in a descending series towards the central axis, where the uplift was greatest. In one place indeed (in east Sussex) the underlying Upper Jurassic rocks are exposed (Figs. 6 and 7). In these *synclinal* structure has been detected, and it is believed that the crumpling which resulted from the synclinal movement was the cause of the superposed Wealden anticlinal structures. The raising of the Wealden anticline resulted in the Cretaceous system sending, as it were, a great arm eastward across Hampshire, Surrey, Kent and Sussex, occupying nearly the whole of those south-eastern counties and separated from the main SW-NE Cretaceous outcrop of the south and east Midlands by the London basin. The Wealden rocks are not represented in the main Cretaceous outcrop of the east Midlands because the lake in which they were formed was restricted to the south-east. The Wealden sands form low hills, the Ashdown Sand of the Forest Ridge in the centre of the Weald, separated by Wadhurst Clay from the Tunbridge Wells Sand on each side of the central axis, but with highly irregular outcrops owing to complex faulting. These three formations are collectively known as the Hastings Beds. The whole is fringed by a belt of Weald Clay (Figs. 6 and 7).

Lower Greensand. Next above the Wealden beds come other Lower Cretaceous rocks which, unlike the Wealden Beds, are represented in the main Cretaceous outcrop. The Lower Greensand forms a bold escarpment on the northern and western edge of the Weald, rising to nearly 1000 ft. (304 m.) at Leith Hill, Hindhead and Blackdown in Surrey, overlooking the lower ground of the Weald (Fig. 7), but is much less prominent elsewhere. The Hythe beds of this formation, consisting of ferruginous sands with hard chert beds and sometimes (as in Kent) calcareous sandstone (Kentish Rag), cap the prominent Greensand hills to the north of the Weald, while the stratigraphically higher Folkestone Sands do not make so conspicuous a feature. Above the Lower Greensand comes the Gault, forming a narrow belt of heavy bluish clay of very constant lithological character throughout south-eastern England. To the north of the Weald this forms a valley between the heights of the Lower Greensand and the Chalk, and is succeeded by the so-called "Upper Greensand" which forms in many places a well-marked terrace below the Chalk escarpment, fringing the Gault round the Weald and along the main escarpment (Figs. 7 and 8). This is a calcareous glauconitic sandstone, and is very fertile. Though of no great thickness and often forming a very narrow outcrop the Upper Greensand is quite an important agricultural feature in parts of Kent, Surrey, Sussex, Berkshire, Wiltshire, Dorset and east Devon. Westwards it began to be formed lower down in the stratigraphical succession, replacing the Upper Gault and forming a micaceous sandy limestone making fertile agricultural land.

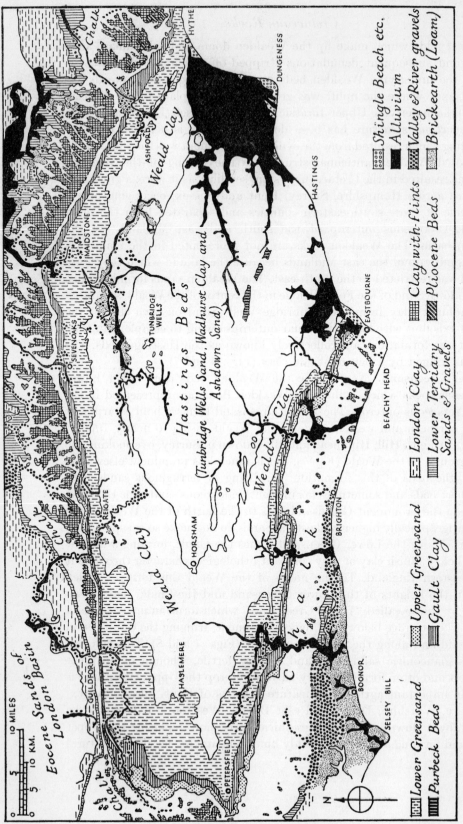

FIG. 6. SURFACE GEOLOGY OF THE WEALD

The narrow outcrops on the edge of the Weald bring about very rapid change of soil from place to place. The "Hastings Beds" occupying its centre are much faulted and exhibit a complex distribution of outcrops of sands and clays (not shown). Thus the vegetation of the whole area is very varied (cf. Fig. 7). Glacial drift is absent and the "solid" geology is therefore but little masked by surface deposits (contrast with Fig. 9). The most important of these are the Clay-with-flints on the northern Chalk, alluvium on Romney Marsh (north of Dungeness), and locally brick earth and various gravels.

The Chalk

The Upper Cretaceous sea. The transition to Upper Cretaceous was marked by the extension of the narrow gulfs and lakes of the Jurassic and Lower Cretaceous into a vast sea in which the Chalk was deposited. At "the close of the Cenomanian (Lower Chalk) the greater part of England, western Scotland and north-eastern Scotland were submerged. Cornwall, western Devonshire and part of the Welsh upland were probably promontories of a western land"—the old Palaeozoic north-western continent—"which included the south of Ireland. The Pennine uplands, together with much of Scotland may have been a peninsula connected with a westward extension of the Scandinavian massif. The submergence seems to have reached its maximum during the deposition of the lower part of the Upper Chalk. Almost all the British area was then probably under water" (Boswell in Evans and Stubblefield, 1930).

The Chalk. The Chalk formation dominates the structure of eastern and southern England (Figs. 4 and 5), except the south-western peninsula. The main outcrop runs in a broad belt from the shores of the Channel in Dorset, with outliers in east Devon, right across England in a north-easterly direction to the borders of Cambridgeshire and Suffolk, then swings northwards to the north coast of Norfolk, and reappears north of the Wash with a north-north-westerly trend forming the Wolds of north Lincolnshire and south-east Yorkshire, where it is cut off by the sea. North-westward of the main escarpment and its immediately neighbouring outliers the chalk has been entirely removed by denudation except for remote outliers in Antrim (north-eastern Ireland) where it was protected by the overlying sheet of Tertiary basalt, and in Morven (Argyllshire), Mull and Arran in western Scotland, similarly protected by Tertiary lavas. Pebbles of chalk and flints containing Upper Chalk fossils have even been found in the Pleistocene deposits of Aberdeenshire on the other side of Scotland. These occurrences indicate the very wide extension of the Upper Cretaceous sea over the British area.

The main Chalk escarpment runs roughly parallel with the Jurassic escarpments and makes a much more continuously prominent feature in the landscape, but the outcrop broadens out in the wide area of Salisbury Plain in Wiltshire, extending eastward through Hampshire and into the North Downs of Surrey and Kent and the South Downs of Sussex on the two sides of the Wealden dome (Figs. 4, 5 and 6). Between the central portion of the main outcrop, which forms the Chiltern Hills of Oxford and Buckinghamshire, and the North Downs lies the complex syncline of the London basin filled with Eocene beds, beneath which the Chalk is continuous (Fig. 8). South of the Hampshire Chalk is the similar Hampshire basin (Fig. 5), also occupied by Tertiary deposits, the underlying Chalk appearing again at the surface in the Isle of Wight where it is sharply folded in a narrow anticline. All these folds and several other considerable folds and faults in the Jurassic-Cretaceous region of southern England bear witness to the

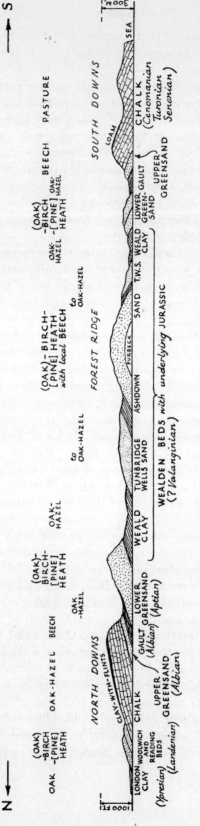

Fig. 7. Section through the Weald from North to South (Y–Y in Fig. 5)

Generalised and simplified, dip of strata exaggerated. The rapid succession of rocks of different lithological type—sands, clays and chalk—with corresponding dominant vegetation is shown.

dominance here of the east and west "posthumous" Armorican foldings which occurred towards the end of the Cretaceous and in Tertiary times.

The physiognomy of the Chalk country is very characteristic. There are no crags or cliffs except on the coast: the hills have softly rounded contours, though the slopes of the escarpments and valley sides are sometimes steep—up to nearly 40°. With the exception of a small portion of the Lower Greensand on the north-west of the Weald, the Chalk forms the highest hills in south-eastern England, but save in one or two places, as at Inkpen Beacon in Berkshire, these do not reach 1000 ft. (304 m.). The valley bottoms are often destitute of streams, primarily because of the porousness of the rock. Springs are frequent at the foot of the escarpment.

Clay-with-flints. Much of the Chalk outcrop which forms plateaux and gentle dip-slopes is masked by overlying deposits of varying thickness, chalky boulder clay in the more northern areas, and in the southern material representing insoluble residue accumulated through the dissolution of a great thickness of chalk, but often mixed with the fragmentary remains of overlying deposits more or less rearranged by the action of water or ice. This material is generally very poor in calcium carbonate and bears very different vegetation from the shallow highly calcareous soil directly derived from the Chalk. It includes the extensive and variable surface deposits known as "clay-with-flints" which covers wide areas of the Chalk outcrop in most of the south-eastern counties (Figs. 6–8).

Eocene. At the end of the Cretaceous period the Wealden anticline was raised, probably before the uppermost beds of the Chalk had been deposited, and the complementary London and Hampshire basins in which the Eocene beds were laid down were formed to the north and south-west. The Eocene rocks of these basins are sands and clays. The London Clay, formed under marine conditions, is the central and most extensive deposit: it is underlain by the sands of the "Lower London Tertiaries" (Reading, Woolwich, Oldhaven and Thanet beds) which fringe the basin both south and west of London and often form low hills, and overlain by more sands (Bagshot and Barton)—the two sets representing fluviatile and lacustrine phases before and after the central marine period (Fig. 8). The Bagshot Sand forms low hills, as at Harrow, Hampstead, Highgate and Epping Forest to the north of London, above the flat or undulating plain of London Clay. To the south-west, in west Surrey and east Berkshire, there is an extensive area of these sands in the Bagshot-Sandhurst district (north-west corner of Fig. 6, and Fig. 8).

Oligocene. Oligocene beds are almost restricted to a narrow belt in the Isle of Wight and south Hampshire, on the southern edge of the Hampshire basin syncline. During the succeeding Miocene period Britain was a land surface, and deposits of that date are absent. It was at this time that the "Alpine" movements spreading from Central Europe "buckled" the Wealden beds along the trend lines of the old Armorican folds. To the north

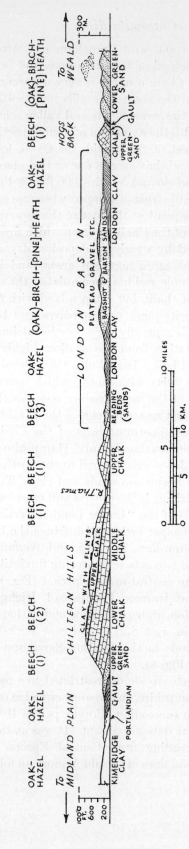

Fig. 8. Section from the South-Eastern Edge of the Midland Plain (in the wide sense) to the North-Western Corner of the Weald (somewhat simplified; X–X in Fig. 5)

Starting from the Upper Jurassic rocks (Kimeridge Clay and Portlandian) the section traverses the Chalk of the Chiltern Hills, the surface of which is largely covered by Clay-with-flints, crosses the Thames valley and then traverses the western part of the London basin occupied by London Clay with extensive overlying Eocene sands capped by Plateau Gravels. The Chalk of the North Downs is crossed at its narrowest outcrop (Hog's Back) which, with that of the Lower Greensand, here attains no great altitude. Compare the higher elevations farther west in Fig. 7. The natural vegetation of the various strata is indicated.

the Mesozoic rocks were tilted against the great western Palaeogenic massif, with a north-eastward strike in the Midlands, changing to a northerly direction farther north, parallel to the east side of the Pennine ridge.

Pliocene. The London basin is continued north-eastwards along the East Anglian coast, but here the Eocene strata are covered by Pliocene deposits which extend over the eastern part of the East Anglian plain bordering the North Sea. These are essentially littoral beds consisting of shelly sands with occasional clayey strata. They are known as "Crag" and are mainly exposed on the slopes of valleys where the overlying Pleistocene deposits have been removed. They are nearly horizontal and form no conspicuous features.

The Pleistocene ice age. By the later part of the Tertiary period the existing tectonic features of the British Islands were established, but the Pleistocene ice age and its sequel—the melting of the ice—brought about many surface changes and greatly affected the physiography of the country.

The increasing cold of the later Pliocene period culminated in the great ice age of the Pleistocene. An immense ice-cap accumulated in Scandinavia and later smaller ones in the Scottish Highlands, the Southern Uplands, the Cumbrian Mountains, the Southern Pennines, Wales and Ireland. From these centres the ice moved southwards, the Scandinavian ice covering the North Sea and invading the eastern shores of England, while most of the English lowlands were overwhelmed by ice from the north and west. There were however several areas remaining unglaciated among the mountains of northern Britain and the ice covering was not continuous throughout the Pleistocene period. Modern glaciologists have been able to distinguish four distinct phases of advance of the ice-sheets, though whether these correspond with the four glaciations which have long been recognised in Central Europe is not certain. The earlier glaciations mainly affected the east, the later the west of Britain. During the intervening "interglacial" periods the ice retreated, and the interglacial deposits provide evidence that a temperate flora lived in areas which before and afterwards were covered by ice. How far vegetation may have survived the glacial phases in unglaciated "refuges" is still a matter of dispute, owing to the paucity of direct evidence, but opinion is tending to the belief that such survival was considerable and important, though it is likely that the great bulk of existing vegetation immigrated from the continent after the last ice-sheets had disappeared from the lowlands.

Glacial deposits. The retreat of the ice-sheets left a mantle of glacial drift over a large part of the country. The most extensive variety of this drift is a dense tough fine-grained deposit known as "till" or boulder clay. Boulder clay is now regarded as a deposit formed from the mineral material locked up in glaciers and ice-sheets and freed when the ice melted. Its lithological character is determined by the rock from which its matrix was derived and over which it generally lies, but it often contains boulders derived from distant rocks over which the ice had travelled. Boulder clay is a *primary* glacial deposit, as contrasted with *secondary* deposits rearranged and sometimes redeposited by running water after melting out. The latter include moraines, eskers, and various bedded sands and laminated clays and loams. *Moraines* are the deposits of ice-worn rock fragments laid down at the sides and ends of glaciers. The dry grass-covered gravel ridges known as *eskers* represent the courses of subglacial streams which laid down bands of coarse material. They are particularly numerous and conspicuous in the central Irish plain, the softer boulder clay around them having been removed by erosion since the disappearance of the ice.

Boulder clay is absent in the Highlands, except on the margins and in the lower parts of the great straths (valleys), but there are innumerable well-developed moraines. In the midland valley of Scotland, on the other hand, moraines are rare but there is plenty of boulder clay and fluvio-glacial gravel. Moraines occur again in the Southern Uplands, in and around the mountains of Wales, and in the north of England as far south as York. In the mountains and valleys of Wales, however, as well as in the bordering lowlands there is much primary glacial drift of various kinds (gravels and boulder clays) as well as numerous moraines, and the primary deposits extend well to the south of Worcester in the western midlands. Other areas of boulder clay derived from the glaciers of the Cumbrian Mountains and of the Pennines occupy much of the northern and north midland lowlands. Others again derived from North Sea ice occur along the east coast. The highly calcareous "chalky boulder clay", derived from the Chalk and the Jurassic limestones, extends over several counties southwards into Essex and Hertfordshire where it gives place to the non-glacial "clay-with-flints" already described and just appearing in the south-west corner of Fig. 9. The Chalky Boulder Clay, in places overlying or interbedded with glacial sands and gravels, determines the soil character over very large tracts of East Anglia and the eastern Midlands. The enormous proportion of the surface covered by the Chalky Boulder Clay together with sandy and gravelly drift, and the extent to which they mask the underlying rocks in eastern England can be seen from Fig. 9, which also shows the larger areas of recent alluvium (river silt, peat, estuarine and coastal deposits). The pre-Pleistocene rocks (unshaded) are indicated by initials.

In the unglaciated regions south of the Thames it has been suggested that the freezing of the soil during the ice age and the subsequent melting and release of the frozen water in great volume led to the excavation of the deep and steep-sided but now dry "coombes" characteristic of the southern chalk.

Thus the Pleistocene ice not only modified many surface features by direct action, gouging out valleys, damming lakes and diverting the courses of rivers, planing down irregular rock surfaces and giving rise to the well-

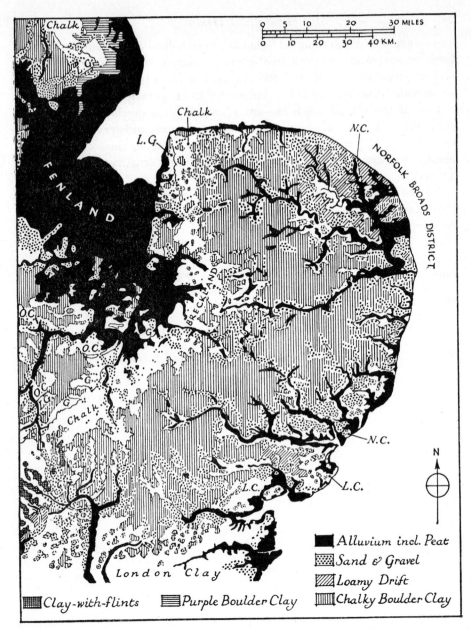

FIG. 9. SURFACE GEOLOGY OF EAST ANGLIA AND PART OF THE EAST MIDLANDS

This part of England is very largely covered with glacial drift, masking the solid geology. The chief kind of drift is the Chalky Boulder Clay, replaced, in the south-west corner of the map (Hertfordshire) by clay-with-flints. There is also a considerable amount of sandy and gravelly drift, especially over the Pliocene beds near the coast, and a certain amount of loamy drift. The peat and silt of the Fenland and of the Norfolk Broads region and the alluvium of the river valley (including coastal mud, sand and shingle) are also shown. The Tertiary and Cretaceous rocks appear only along the sides of the river valleys where they have been exposed by erosion, in southern Essex (London Clay), and along parts of the chalk outcrop near the escarpment. The Jurassic Oxford Clay, also partly covered by drift, appears in the west.

N.C. = Norwich Crag (Pliocene); L.C. = London Clay, G. = Gault, L.G. = Lower Greensand (Cretaceous); O.C. = Oxford Clay (Jurassic).

known smoothed and rounded "glacial topography" of much of the Palaeogenic region; it also covered great areas of country with a mantle of boulder clay or (more locally) of sand and gravel. The fresh basis for soil formation so provided has determined or influenced the vegetation according to the nature of the glacial deposits—gravel, sand, loam or clay—and to their content in calcium carbonate. Thus the glacial sands and gravels bear heath or corresponding woodland, the non-calcareous boulder clays oakwood, the chalky boulder clay ash-oakwood (see Chapters XIII, XVII, XX, XXI and XXXVI).

Recent alluvium, peat, estuarine mud, blown sand and shingle beaches that have been formed in post-glacial times are dealt with in later chapters.

REFERENCES

COLE, GRENVILLE, A. J. In *The Oxford Survey of the British Empire*, Vol. I. *The British Isles*, etc. Chapter I. Physical Features and Geology. Oxford, 1914.

EVANS, J. W. and STUBBLEFIELD, C. J. (edited by). *Handbook of the Geology of Great Britain*. London, 1930.

MACKINDER, H. J. *Britain and the British Seas*. Second edition. Oxford, 1907.

STAMP, L. DUDLEY and BEAVER, S. H. *The British Isles*. London, 1933.

WILLS, L. J. *The Physiographical Evolution of Britain*. London, 1929.

Chapter II

CLIMATE

Climate and the distribution of vegetation. Some elements of what is called climate, or the factors that act upon plants and animals primarily through the air, are certainly the paramount influences determining the distribution of the major plant communities of the world. In a small region like the British Islands, where the whole country is, as we shall see, very definitely under the influence of a particular climatic complex, though admittedly to different degrees in different parts, it has sometimes been thought that variations of climate within these islands have quite minor differentiating effects on the vegetation; and the first attempts (in *Types of British Vegetation*) at analysing the factors responsible for the distribution of British plant communities laid all the stress on differences of soil. The differentiating effects of climate are, however, far from negligible. As we shall see in Chapter IX, different climatic plant formations can undoubtedly be distinguished within the country, but very little has been done in the way of serious quantitative analysis of the effects on vegetation of the variations of British and Irish climates, though Salisbury (1932) and others have discussed the effects of climate on the distribution of particular species or groups of species. All that can be attempted here is to describe the climatic factors which chiefly affect vegetation and then to give an outline picture of the climate of these islands and of its main regional variations, indicating, quite generally, the relations of these to the plant formations.

Microclimates. In regard to the effects of narrowly localised climatic differentiation on small communities the influences of what is called "microclimate" are still not sufficiently realised.[1] The differences between the climate and vegetation of a protected ravine and an exposed ridge are strikingly obvious, but, as Kraus showed long ago, such local differentiation may extend to quite minute areas, such as the space protected by a rock. Our knowledge of actual microclimates is very slight, and the reason is not far to seek. Local differentiation of soil is obvious to the eye, and correlation with local differentiation of vegetation at once suggests itself to the investigator. But the elements of climate are invisible, and until they are actually measured the extent of local differentiation cannot be estimated and often remains entirely unsuspected. Almost the only microclimates which plant ecologists have studied are those of the different layers of vegetation in stratified communities such as forest.

[1] Except by the entomologists, who in recent years have published important researches on the microclimates inhabited by different insect populations.

Heat and moisture. The elements of climate which have a direct effect in differentiating vegetation are in the first place heat and moisture. The temperature of the air acts directly on plant function, and also directly determines the amount of water vapour the air can hold. Moisture acts in two ways: first the amount of precipitation is the ultimate source, and the most important direct source, of water supply to land vegetation; and secondly, the humidity of the air conditions the water loss by evaporation, both directly from the soil and indirectly (much more copiously) from the bodies of plants (transpiration). The humidity of the air depends not only on its absolute content of water vapour, but also on its temperature. The hotter the air the more water vapour it can hold, and therefore the "drier" it is with a given water-vapour content. Thus temperature and moisture act upon plants both separately and conjointly. Sufficient heat and sufficient water supply are primary requirements of all land vegetation. Different plants, it is true, require different amounts of each, but the luxuriance of the "average" plant (mesophyte) increases with the increase of both (within limits of course) in a balanced ratio. The relation of the temperature of the air to its content in water vapour on the other hand, controls the *loss* of water from the aerial shoots of land plants.

Temperature and rainfall. Fig. 10 shows very roughly and generally the distribution of actual annual temperatures over the British Isles. This differs widely from a map on which the temperatures are "reduced to sea-level", because the factor of altitude far outweighs the factor of latitude. It is true that the highest temperatures occur on the south and part of the west coast and in places to which the warm south-westerly winds have direct access, and there is also a patch of warmth immediately to the west of London, where the summers are hottest. But in the north and west of Great Britain the great areas of high ground are the leading factor in determining actual temperature, and are much colder than the coasts, which, like much the greater part of England and Ireland, have a mean annual temperature of 46–50° F. (7·8–10° C.). The average "lapse rate" as it is called, i.e. the decrease of temperature with increase of altitude, is generally taken as 6·5° C. per km., which is equivalent to 1° F. for 280 ft. of vertical rise. The mean annual temperature on the summit of Ben Nevis, the highest mountain (1342 m.), is 31·5° F. (−1° C.). No *detailed* map of actual annual temperatures would be possible on anything but a very large scale indeed, and indeed data are not available, because of the multitudinous and often great variations resulting from land relief, aspect, etc., within very short distances. Deviations from the mean in particular years at given stations are, however, slight.

With rainfall it is otherwise. Deviations from the mean in particular years are very considerable, and the higher mean rainfalls are determined partly by altitude, and partly by western position, as can be clearly seen in Fig. 11. Figs. 10 and 11 should be compared with Fig. 3 (p. 7), which shows the orography of the British Islands. The great influence of western position may be illustrated by the fact that certain areas in Inverness and Argyll in the Western Highlands at an elevation of 1000 ft. (304 m.) have a rainfall of 80 in. (c. 2000 mm.), while places at a similar elevation in eastern Scotland have less than 40 in. (c. 1000 mm.) (Bilham, 1938).

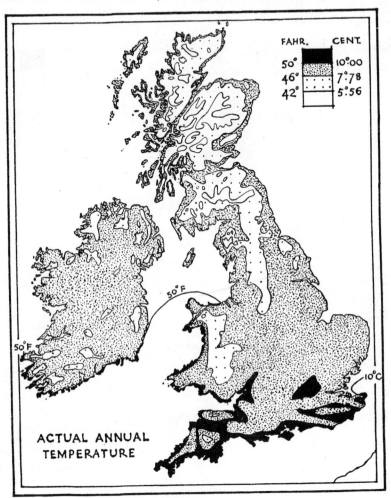

FIG. 10. ACTUAL MEAN ANNUAL TEMPERATURES

The influence of south coast sunshine (cf. Figs. 12, 13) and access of warm air from the southwest in determining higher, and of high altitude in determining lower temperatures is apparent (cf. Fig. 3, p. 7). From Bartholomew's *Physical Atlas*, Part III, *Meteorology*.

Ratio of precipitation to evaporation. Since, as we have seen, vegetation is necessarily dependent on precipitation and evaporation, determining water supply and water loss respectively, and different kinds of vegetation are adjusted to excess of one or the other, it is clear that the ratio between them is a possible means of expressing the total relation of various major plant communities to water. Many years ago it was shown that the vegetation of the eastern and central United States followed this ratio (P/E), as shown by lines of equal P/E plotted on a map and compared with a map of the distribution of vegetation, more closely than, for example, precipitation alone.

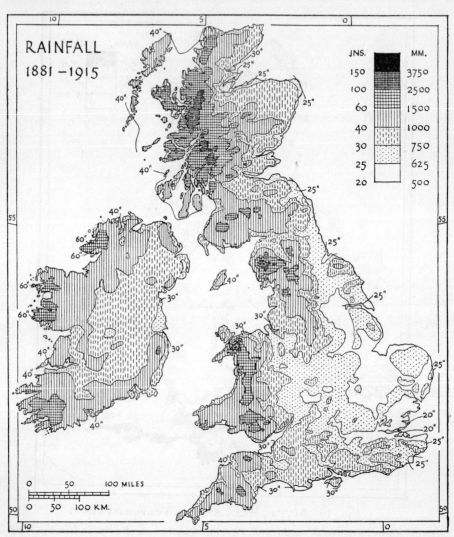

FIG. 11. MEAN ANNUAL RAINFALL

The effect of (1) high altitude (cf. Fig. 3) and (2) western position in determining higher precipitation is well shown. Adapted from the Royal Meteorological Society's *Rainfall Atlas of the British Isles*.

In this ratio E is taken as the quantity evaporated from a free surface of water, and thus the amount of water lost is directly compared with the amount received as precipitation (P) during a given period. The conditions under which water evaporates from a free surface are not, however, identical with evaporation from wet surfaces. Among other things the effect of wind is different. For this reason the amount of water evaporated from the surface of a Livingston atmometer (a freely exposed inverted unglazed porcelain cup whose inner surface is in contact with water from a reservoir below) gives a better measure of the "pull" of the air on the water contained in plant bodies.

Saturation deficit. In recent years this "pull", i.e. the "evaporating power of the air", has been increasingly measured by what is called *saturation deficit*. This is the amount by which the partial pressure of water vapour in a given sample of air, measured in millibars (or millimetres of mercury), falls short of the pressure at saturation point, and it is a direct measure of evaporating power. It has been shown that this quantity (S) has a fairly simple relation to E (evaporation from a free water surface). Another such measure is the depression of the wet-bulb thermometer reading below the reading of the dry bulb. *Relative humidity* is the vapour pressure of a sample of air, at a given temperature, expressed as a percentage of the vapour pressure at saturation point at the same temperature. This quantity can be calculated from the amount of depression of the wet-bulb reading, according to the temperature, and is the ordinary method of expressing air humidity used by meteorologists; but the saturation deficit is more satisfactory when we are considering the effect of air humidity on vegetation, because it is an absolute measure of evaporating power, whatever the temperature. It can be easily calculated from the difference between the wet- and dry-bulb readings when the temperature is known, and wet- and dry-bulb readings are taken at least once a day at most meteorological stations, while very few measure the actual amounts of water lost by evaporation.

The "Meyer ratio". The ratio of precipitation to saturation deficit (often called the "Meyer ratio" because Meyer (1926) first used it as a climatic index) is generally written N/S (German *Niederschlag-Sättigungsdefizit*), though there is no reason why it should not be written P/S in English. This ratio, as we have seen, shows a high degree of correlation with P/E, and recently an attempt has been made (Fraser, 1933) to map it for the British Isles. Fraser's map (1933, p. 104) shows the areas of high "N/S" corresponding roughly with the regions of high rainfall (over 1000 mm.), except in the south-western peninsula. In other words, where precipitation is high the Meyer ratio is also high, and it has not yet been clearly shown, in the British Isles at least, that it is more closely correlated with the distribution of vegetation. Where there is an annual rainfall of more than about 60 in. (1500 mm.) over considerable flat areas "blanket bog" (see Chapter XXXIV) seems to be the climatic plant formation, but it may well be true that where the air humidity is unusually and constantly high, this may determine the existence of blanket bog with an annual rainfall of no more than 50 in. (*c.* 1250 mm.) or even less, as for example in western Mayo on the west coast of Ireland. In the south-western English peninsula where the rainfall is relatively high, the higher temperatures of the south may increase the saturation deficit so as to exclude blanket bog except at the higher elevations such as Dartmoor, where precipitation is higher, temperature lower, and blanket bog actually developed. The other N/S "isopleths" shown by Fraser are not at all obviously correlated more closely with vegetation than the isohyets (lines of equal precipitation). Thus he marks a pear-shaped area of low "N/S" ratio in central England, but nothing is known of any corresponding vegetational difference. It must be remembered that Meyer's ratio is not an expression of a definite physical quantity, and it can only be of value so far as it can be *shown* to correspond to the distribution of definite entities such as climatically determined soils or vegetation.

It is even doubtful whether the available data justify the attempt to map P/E or P/S values in the British Isles. At six meteorological stations hourly observations of air humidity are taken: at most they are made at 9 a.m. only. There is, it is true, some evidence that the mean of the readings at this hour for a given month may fairly correspond with the mean of all the 24-hour readings for the month, but the evidence is not nearly extensive enough to justify secure generalisation, and unless the correspondence is universally true reliance on 9 a.m. observations alone may be very misleading. As Bilham (1938) remarks, "to derive really satisfactory information about humidity we need continuous records from dry- and wet-bulb thermographs".

When we consider the actual relation of air humidity to plants we encounter further uncertainties. Plants lose water by evaporation over the whole 24-hour period, though to different extents at different times of the day and night, not only because of the variations in air humidity but also because of the opening and closing of their stomata. They only lose water freely, however, when their aerial shoots are developed and exposed to the air, and it is quite conceivable that a climate in which the winter is particularly wet (or particularly dry) compared with the summer may show a mean *annual* humidity which will give a false impression of the total effect of the evaporating power of the air on vegetation during the actual growing season.

Diurnal variation of saturation deficit. Some idea of the diurnal variation of saturation deficit can be obtained from the following figures from Kew Observatory extracted from Bilham (1938, Table XL, p. 211), for the months of April, July and October, respectively at the beginning, middle and end of the growing season in a wide sense:

Table I. *Diurnal variation of saturation deficit in millibars at Kew*

Hours	0	2	4	6	8	10	12	14	16	18	20	22	Day
April	1·6	1·3	1·2	1·2	2·1	3·5	4·8	5·4	5·4	4·4	3·0	2·1	2·8
July	2·7	1·1	1·7	2·4	4·4	6·7	8·6	9·8	10·0	8·8	5·8	3·6	5·3
October	1·1	1·0	0·9	0·9	1·2	2·2	3·5	4·1	3·6	2·2	1·6	1·3	1·8

The January values are much lower, not rising above 1·8 at 14 hr. (2 p.m.). The drying power of the air in April is here clearly seen to be markedly greater than in October, but it is less than in September (Table II).

Seasonal variation. The monthly means of saturation deficit at the six stations (see map, Fig. 23, p. 56) at which hourly observations are made are given below (Bilham, 1938, Table XLII, p. 214). They are placed in descending order of the yearly means:

Table II. *Seasonal variation of saturation deficit*

	Jan.	Feb.	Mar.	Apr.	May	June	July	Aug.	Sept.	Oct.	Nov.	Dec.	Year
Kew	1·2	1·6	1·9	2·8	3·7	4·7	5·3	4·6	3·4	1·8	1·4	1·2	2·8
Falmouth	1·6	1·8	1·8	2·3	2·6	3·0	3·3	3·0	2·5	1·9	1·8	1·6	2·4
Glasgow	1·2	1·4	1·7	2·4	3·1	3·6	3·7	3·2	2·5	1·9	1·5	1·2	2·3
Aberdeen	1·5	1·6	1·8	2·1	2·4	3·0	3·3	3·2	2·7	1·9	1·7	1·4	2·2
Valentia	1·3	1·4	1·7	2·0	2·4	2·9	2·7	2·6	2·5	2·0	1·6	1·3	2·0
Eskdalemuir	0·9	0·9	1·3	1·8	2·2	2·9	2·6	2·3	1·9	1·4	1·3	0·9	1·7

Five of the six observatories are at quite low altitudes, Eskdalemuir alone being as much as 825 ft. (*c.* 250 m.) above the sea.

While the highest mean monthly value (July) at Kew is much higher than at any of the other stations, the winter values are lower than at Falmouth, so that the yearly mean is not very much higher. This is doubtless due to the greater amount of winter sunshine and the higher winter temperature at the south-western station (cf. Fig. 17). The midsummer values at Valentia and Eskdalemuir are practically the same, the extreme Atlantic situation of the former being compensated by the high altitude of the latter, which shows the lowest winter values and the lowest yearly mean of all six stations. None of the stations is situated in the extreme "blanket bog" climate, though the two last come near it.

The highest humidity records throughout the year are those for the summit of Ben Nevis, where the difference between the dry- and wet-bulb readings is only 1·9° F. at 14 hr. (2 p.m.) in June, equivalent to a saturation deficit of 1·5 mb., never more than 0·8° in August, equivalent to about 0·6 mb., and the yearly mean is about 0·7° F. (about 0·5 mb.).

Mist and fog. Mist or fog is formed when the air near ground-level is cooled below dew-point (the temperature at which water vapour condenses to the liquid form from air with a given vapour content) so that minute droplets of water condense and remain suspended. Fogs may be classified according to the particular conditions which lead to this result. *Sea fogs* are caused by a current of moist air passing over colder water so that it is cooled below its dew-point. Eddy motion involving overlying layers of air may then propagate the formation of fog upwards so that a stratum of considerable depth is formed. Such sea fogs are fairly frequent off our south-western coasts in summer and early autumn when the excess of air temperature over sea temperature is at a maximum. *Coastal fogs* are formed when a wet wind from the sea passes over land cooled below dew-point by radiation, and in this way a sea fog may penetrate inland. The effect is intensified if the coast slopes more or less steeply upwards, and this is the cause of the high frequency of fog at Beachy Head, a chalk cliff (150 m. high) on the Sussex coast, which is one of the foggiest places in England, with an average of eighty-eight fogs in the year, more than six per month in summer and autumn. The frequent fog on the much higher elevation of Dartmoor in South Devon (500–600 m.), which is not far removed from the Channel, is due to a similar cause—the wet air from the south-west impinging on the cooled upland. Seventy-nine morning fogs were observed at Princetown in 1935, and it is estimated that Dartmoor has an average of at least 100 days with fog in the year. The western coasts have much lower frequencies, the highest being at the south-western extremities of England and Wales, i.e. at Scilly (twenty-six) and at St Ann's Head, Pembrokeshire (twenty-nine). The Atlantic coasts generally are singularly free from fog, Valentia having only three, Malin Head six and Lerwick (Shetlands) fourteen in the year. The east coast stations show somewhat higher figures.

Inland fogs are caused by the cooling of the soil surface through radiation, usually on winter nights under anticyclonic conditions, i.e. when the air is still, but on a clear night in early autumn the intensity of radiation often produces fog in the early morning, soon dissipated as the sun's heat increases. The maximum frequency of inland fogs (over fifty per annum) is in England and corresponds with a broad belt of country stretching from north Yorkshire to the neighbourhood of London, largely coincident with the "continental area" (see p. 66). Cumbria, Wales, south-west England and East Anglia show much lower figures.

These figures compare quite favourably with those from the continental coasts of the narrow seas. Thus La Hève has 77 days of fog, Le Havre 52, Ostend 53, (Brussels 69), Flushing 73, Borkum 52, (Hamburg 86), Flensburg 54.

Thus the reputation of Britain, so widespread on the continent, as having a very foggy climate seems undeserved, at least compared with the other coasts and coastal lands of north-west Europe.

What has been said applies only to the number of days with more or less transient fog. Persistent fog is much rarer. Thus during the 10 years ending 1930 there were sixty-two persistent fogs at Kew, and forty or more at a number of other stations in the inland area of maximum fog. At Valentia there were none, only one at Holyhead, and three at Stornoway, Malin Head and Portland Bill (Bilham, 1938).

Sunshine. Duration of bright sunshine for the year is shown in Fig. 12, and the climate is certainly decidedly deficient as compared with most of Europe. With the exception of the south coast no part of the British Isles receives as much as 40 per cent of the possible sunshine in the year. The majority of inland stations range between 25 and 35 per cent. On the southern Pennines and over a large part of the Scottish Highlands the amount is less than 25 per cent, the lowest recorded percentage (17) being on the summit of Ben Nevis. Nearly the whole of France and much of

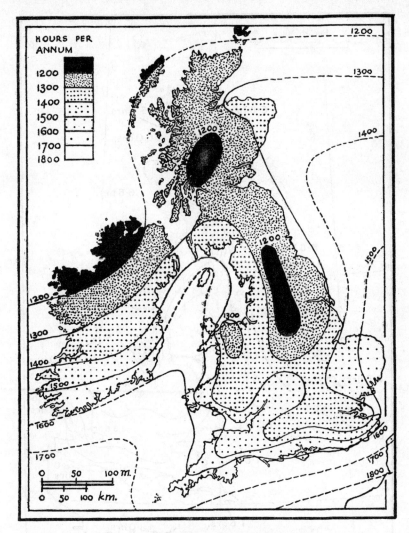

FIG. 12. MEAN DURATION OF SUNSHINE

The effect of high altitude and north-western position in reducing sunshine is clear.

This map (after H. N. Dickson) is based on ten years' records prior to 1891. Recent maps based on 30 years' data (1906–35) show higher values on the south coast and lower in the Scottish Highlands, though the general distribution is broadly similar.

Central Europe receive more than 45 per cent of possible sunshine. In June (Fig. 13) the British distribution follows similar lines.

It is difficult to estimate the precise effect of either fog or sunshine, as such, on vegetation. It is probable that the temperature factor with which they are correlated is the major cause of differences, such as increased luxuriance on the south coast and the disappearance of many species in the very sunless regions of the northern mountains. Any specific effect of the

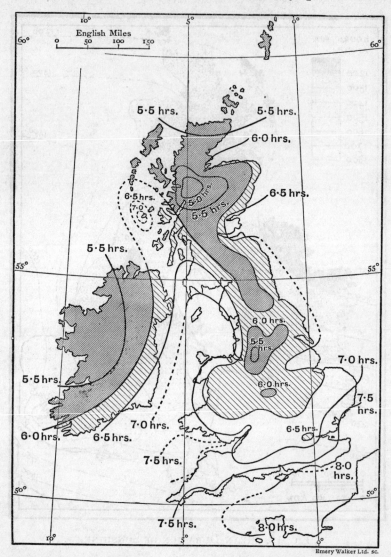

FIG. 13. MEAN DAILY DURATION OF SUNSHINE IN JUNE (1906–1935)

From Bilham's *Climate of the British Isles*, 1938, Fig. 62, by permission of the author and Messrs Macmillan and Co. Ltd.

cutting down of insolation, i.e. apart from the lowering of temperature, by sunlessness or fog, is difficult to demonstrate, but it is quite possible that both intensity and quality as well as duration of sunshine may affect the distribution of particular species. Fog of course corresponds with zero saturation deficit, though it is not always formed when the air is saturated with water vapour.

Frost. Temperatures below the freezing-point of water (32° F. = 0° C.) are obviously important for vegetation, since vegetative processes are brought to a standstill when water freezes. The "grass minimum", i.e. the lowest temperature shown by an unprotected thermometer in contact with the blades of grass on a lawn, is often several degrees below that indicated by a screened thermometer, and both these temperatures have to be taken into account when we are considering the effect of frost on vegetation. At the Radcliffe Observatory, Oxford, for the 50 years, 1881–1930, the mean minimum was below freezing-point on the grass in every month from December to March inclusive, whereas in the screen the mean minimum was above freezing-point throughout the winter. Frost was recorded on the grass at least once during the 50 years for every month in the year except July, whereas in the screen there were no frosts in June, July or August. An absolute minimum of $-2\cdot4°$ F. ($-19\cdot1°$ C.) was recorded on the grass in February, whereas the lowest temperature in the screen during the 50 years was $5\cdot9°$ F. ($-14\cdot5°$ C.) in January (Bilham, 1938, p. 233).

The frequency of ground frosts is considerable at many stations in the English midlands, and they continue to occur quite late in the spring. Among the highest frequencies in the midlands and south-east are 126 at Berkhamsted, 112 at Cambridge, 102 at Birmingham, 101 at Greenwich and Kew, and 91 at Oxford. The Oxford monthly maximum (15) actually occurs as late as March. At Belvoir Castle in Leicestershire the monthly occurrences (1896–1930) were as follows (Bilham, 1938, p. 236):

January	16·3	May	4·1	September	1·2
February	15·7	June	0·9	October	7·0
March	14·5	July	0·0	November	13·9
April	11·5	August	0·0	December	14·4

The late spring frosts are a well-known feature of this midland climate, and are due to the combined effects of cold winds, mainly from the east and north, which are frequent in these months (March to May), clear air, and persistence into the spring of the low winter soil temperatures. These spring frosts frequently do much damage to fruit crops at the end of April and during May. In May 1935 a very cold spell in southern England lasted from the 12th to the 19th of the month, doing enormous damage to vegetation. At Burnham Beeches in the Thames valley the whole of the young beech foliage was killed and most of the young oak foliage as well. A week later the woods showed a general brown tint from the mass of dead beech and oak foliage, against which the uninjured birch and hazel leaves stood out in bright green. On the high ground of the neighbouring Chiltern Hills the beech foliage was much less injured, evidently because the cold air had settled chiefly on the low-lying flat ground of the Thames valley.

The primary cause of the low grass temperatures, especially at night, is the loss of heat by radiation from the earth's surface when the sky is clear. A layer of cloud reflects heat back to the earth and itself radiates heat downwards. Wind also checks radiation from the earth, mixing the chilled air with warmer currents and eddies. Consequently the lowest temperatures on the grass are reached in calm anticyclonic conditions with a clear sky. As we shall see in the next chapter the lowest winter temperatures (apart from the northern mountain regions), as well as the highest temperatures in summer, occur in the English midlands, and especially the east midlands, because this area is least affected by the normal westerly wind current, and radiation is greatest.

The other meteorological condition which gives *general* low temperature in the midlands and south of England is a distribution of pressure that brings cold polar winds to this region, as when an anticyclone establishes itself to the north or north-west of these islands, where the barometric pressure is usually low. This was the cause of the severe May frost described above.

It is doubtful, however, whether frequency, severity or duration of frost determines the distribution of any of our major plant communities, though increased incidence of frost probably restricts the eastern range of western species which cannot tolerate temperatures that are continuously or extremely low.

Snow. Snow plays little part in the general climate of the British Islands, and its effect on vegetation is probably negligible except at high altitudes. Snowfall is not rare except in the extreme south-west, but its occurrence is highly uncertain and irregular. In the south-east many winters pass with little or no snow, though in others there may be fairly frequent and abundant falls and occasionally considerable duration of lie. The average numbers of snow days are given for a number of stations in the tables (IV–X) illustrating regional climates (pp. 57–67). The average duration of lie at low altitudes is certainly quite small, but it increases rapidly with altitude, though only at the higher levels, probably above 3000 ft. (c. 900 m.), is it at all likely seriously to restrict the active vegetative season of plant life. Unfortunately we have little or no data on this subject, but the duration of snow lie on the summit of Ben Nevis, described on p. 65, must restrict the possible vegetative season to about $2\frac{1}{2}$ months. There are (or were), in fact, two examples of semi-permanent snow beds in the highest mountains—one in the corrie of Allt-a-Mhuillin on the north-east side of Ben Nevis, which, after lasting for many years, disappeared in 1935, and one in the great corrie of Braeriach in the Cairngorms (eastern Highlands), which melted in September 1933, after persisting, it is believed, for more than 50 years (Bilham, 1938). In a gully of the Snowdon range at 3000 ft.

PLATE 1

Phot. 1. Moderately wind-cut oak on Inkpen Beacon, Berks. Inland exposure at relatively high altitude, about 1000 ft. (c. 300 m.), the highest ground in the neighbourhood. Fully exposed to prevailing south-west wind (from left). *N. T. Porter.*

Phot. 2. Extremely wind-cut oak near Sidestrand, Norfolk, close to the east coast. Fully exposed to north-east wind (from left). Close-set shoots clothing obliquely inclined trunk. Crown very poorly developed. *N. T. Porter.*

EFFECT OF PREVAILING WINDS

altitude snow remains till midsummer, and there are probably other similar examples. The vegetation of such areas would certainly repay careful investigation.

Wind. Where winds are violent and persistent the growth of tall woody vegetation is severely limited. If trees establish themselves at all they form no more than wind-cut scrub, which may not be able to set seed; and wind no doubt is one of the main determinants of the primitive upper limit of forest on mountain slopes. On well-drained siliceous soil heath replaces forest in situations exposed to constant violent winds.

Direction. The prevailing direction of wind is from the south, south-west (usually) or west at all stations in the British Islands (Fig. 14), though winds from the other quarters are by no means infrequent, and there is sometimes a seasonal maximum from east, north-east or north. The south-westerly maximum frequencies are often in the winter months and the north-easterly practically always in the spring and early summer (April to June). Cf. Pl. 1, phots. 1 and 2.

Velocity. Table III shows the mean velocity of the wind and the frequencies of winds of different velocities at Lerwick (Shetland Islands) and Richmond (Kew Observatory) in Surrey, an extremely exposed oceanic station and an inland one, both at low altitudes.

Table III. *Velocities in miles per hour*

			Percentage frequencies				
		Mean for month	>38	38–25	24–13	12–4	<4 m.p.h.
LERWICK	January	22·2 (max.)	8	31	40	17	4
(6 years)	July	11·4 (min.)	1	10	46	39	5
KEW	January	8·2	0	1	26	54	19
(12 years)	March	8·9 (max.)	—	—	—	—	—
	July	6·6	0	<0·1	10	62	28
	September	6·2 (min.)	—	—	—	—	—

It will be noted that the maximum mean monthly velocity at Lerwick is in January and the minimum in July, while at Kew the maximum, which is substantially less than the Lerwick minimum, is in March and the minimum in September: also that while well over half the winds at Kew have a velocity of between 4 and 12 m.p.h. and there are very few indeed of more than 24, at Lerwick in January nearly one-third exceed 24 m.p.h. At the Butt of Lewis, the northernmost point of the Outer Hebrides, 29 per cent of the winds for the whole year exceed 24 m.p.h., and in January the percentage of these strong winds is 48. The other exposed oceanic coasts show percentage frequencies of violent winds comparable with those at Lerwick. Spurn Head, the most exposed station on the English east coast, has a markedly lower frequency of gales.

Gusts of much greater velocity than that of the general wind commonly occur while strong winds are blowing. The highest recorded velocity was at Scilly, where a gust of 111 m.p.h. was noted on 6 December 1929, and several stations have recorded gusts of over 100 m.p.h. Gustiness is more marked at inland than at maritime stations, a fact which tends to bring the possible maximum violence of the wind on the coast and inland closer to equality.

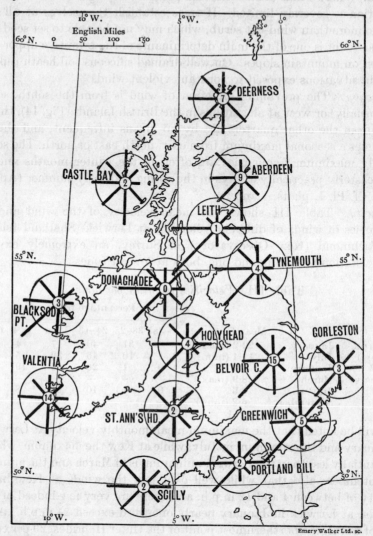

FIG. 14. DIRECTIONS OF WINDS

The average annual percentage frequencies of winds from each direction are shown. The difference between the radii of the large and small circles represents 12½ per cent. The figures in the small circles show the percentage frequencies of calms. From Bilham, 1938, Fig. 14, by permission of the author and Messrs Macmillan and Co. Ltd.

The north-west coast of Ireland is the stormiest region in the British Isles, with forty gales a year, of which about thirteen are classed as "generally severe"; at least half the stations in the region recorded wind velocities of 55 m.p.h. Gales are least frequent on the east coast of Britain and in south-east England. A larger proportion of these gales than of the whole range of winds blow from the westerly quarters, except in the north-east of Scotland where gales from the south-east are numerous. In all regions gales are overwhelmingly most numerous in winter, though in March there is a slight bump in the descending curve, and in September a slight hollow in the ascending curve on most of the English coasts, though not on the north-west coast of Ireland.

Wind at high altitudes. It is unfortunate that records of the frequency, intensity and direction of winds at the higher altitudes, say from 1000 to 4000 ft., are almost lacking. Apart from the coasts it is at such altitudes that wind is likely to have the most decisive effects on vegetation. The fact recorded on p. 65, that on the summit of Ben Nevis during 13 years there was an average of 261 gales a year of more than 50 m.p.h., suggests that at high altitudes gales are an even more formidable factor than on the stormiest coasts. The "arctic-alpine" climate of the higher mountains, which undoubtedly determines the corresponding arctic-alpine vegetation (Chapter XXXVIII), is made up of a complex of factors—low temperatures, usually high precipitation and air humidity, and great length of snow lie, as well as violent winds. In exposed situations at the higher levels the last-named must certainly be of prime importance.

"Arctic-alpine" climate

General characters of the climate. Having now briefly considered the various elements of climate and some features of their occurrence in the British Isles, we may pass to a general estimate of the climatic complex which they form in combination.

The weather of the British Isles is so extremely changeable that the country is sometimes said to *have* no climate. But such a statement is not of course founded on any tenable scientific conception of what is meant by climate, which does not imply long periods of stable uniform weather. Climate, as it affects soil, plants, animals and man, depends on the average conditions of certain crucial meteorological factors, with due consideration also of their range and of the occurrence and duration of extreme conditions. It is generally agreed that the two basic factors in climate are heat and moisture, and in regard to these the British climate is definitely *temperate* and definitely *damp*. Extreme temperatures are rare and never of long duration, and the moisture content of the air is on the average high.

Distribution of barometric pressures. The fundamental conditions which underlie and determine these characteristic features of British climate are of course the distribution of barometric pressure over north-western Europe and the neighbouring oceans. There are three great semi-permanent centres of high or low pressure in the region of the northern hemisphere lying to the north of the 30th parallel of north latitude and between longitudes 40° E. and 40° W. of the Greenwich meridian, i.e. the region of which the British Isles are the centre. These are (a) the "Icelandic low", an almost permanent region of low pressure in the neighbourhood of Iceland, which shifts a little northwards during the summer; (b) the belt of high pressure across North Africa, with a high centre near the Azores, which also moves

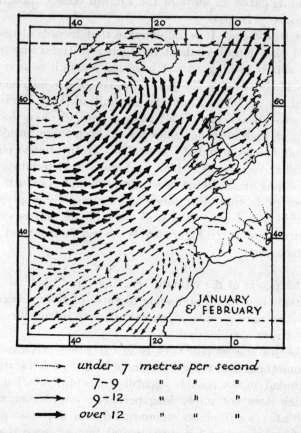

FIG. 15. PREVAILING WINDS—JANUARY AND FEBRUARY

The predominant winds are south-westerly. The thickness of the arrows indicates the average strength. Note the diversion of strong winds up the Norwegian coast, north of the continental anticyclone; also the circulation round the "Icelandic low". From Bartholomew's *Physical Atlas*, Part III, *Meteorology*.

northwards in the summer, the axis of the high-pressure belt then lying over the Mediterranean; (c) the great high-pressure area typically dominating Eurasia in winter, and replaced by a corresponding low-pressure area in summer.

The spells of "settled" fine weather which we sometimes experience in the British Isles—and anticyclonic conditions occur with considerable frequency at all times of the year—are generally due to comparatively slight extensions of these great high-pressure areas—in winter the westward extension of the continental anticyclone, in summer the north-eastward extension of the Azores "high". Local anticyclones of much smaller area occasionally form over or in the immediate neighbourhood of these islands, but our weather is more often determined—especially in winter—by the

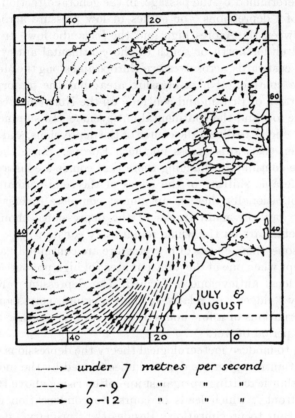

FIG. 16. PREVAILING WINDS—JULY AND AUGUST

The winds are still south-westerly and westerly. Note their lesser strength, and that they now cross Great Britain eastwards to the continent: also the persistence of the "Icelandic low" and the prominence of the "Azores high". From Bartholomew's *Physical Atlas*, Part III, *Meteorology*.

influence of the great continental or oceanic systems. Bilham writes (1938, pp. 161–2): "The most striking form of thermal abnormality to which our climate is liable is a spell of very cold weather in winter...in normal circumstances the prevalence of westerly winds, blowing over a warm sea, gives our islands a winter climate which is abnormally mild for their latitude. We live in fact on the edge of a continental area whose winter climate we do not ordinarily share, but which we are liable to share when the warming agency, the westerly wind stream, breaks down. Winter is quite definitely the season of greatest contrast between the British Isles and the climates of the land areas which lie to eastward." The general direction of the normal wind stream may be seen in Figs. 15 and 16, representing the winter and summer conditions respectively.

Passage of depressions. What may almost be called the "normal" weather is determined by the passage, in the general direction west to east, of a series of "depressions" or eddies of low pressure across the North Atlantic between the Azores high and the Icelandic low. In winter these are diverted either north or south of the continental anticyclone. Some travel south-eastwards, across western Europe and along the Mediterranean, south of the continental high-pressure area: but the majority cross the northern portion of Britain and continue up the Norwegian coast between the Icelandic low and the continental anticyclone. In summer there is no continental "high" to divert the course of the depressions, and their general course is then eastwards across northern Europe. Since the barometric pressures are continually changing there is of course a constant and sometimes considerable shifting of the edges of the semi-permanent pressure areas, and occasionally there is an almost complete reversal of pressure distribution; but the main tracks of the depressions, though infinitely various in detail, are on the whole remarkably constant.[1]

At all seasons, then, eddies of low pressure are continually crossing these islands, except when one of the adjacent areas of high pressure extends over them, or a local anticyclone is formed. The depressions are commonly separated by "ridges" of relatively high pressure, corresponding to the familiar alternations of a few hours or a day of fine weather with a somewhat longer period of wet.

According to modern meteorological theory the depressions or "lows" of the North Atlantic region are due to eddies formed by the moist warm air above the Atlantic drifting up against and often rising above the cold air of the "polar front", which leads of course to condensation of the water vapour and thus to precipitation. Besides the "primary" depressions or cyclones, secondary eddies are constantly formed on the edges of the primary lows, and these "secondaries" are responsible for much of the wet weather of the British Islands. Whether a low is formed on the actual polar front or as a subsidiary eddy it will always tend to give precipitation, for the warmer moist southerly and south-

<small>Cyclonic rainfall</small>

---

[1] It will be understood of course that the areas of high and low pressure appearing on seasonal or annual barometric maps represent the distribution of *average* pressures for the periods concerned. The pressure is constantly shifting, more or less rapidly and extensively, at any given station, and there are many occasions during the year or during a season when the actual distribution is quite different: for example the barometer may be high over Iceland and low in the region of the Azores. Nevertheless the daily barometric charts frequently reproduce very closely the average conditions shown in the seasonal or annual charts.

westerly winds in the right-hand (southern) rear sector of the cyclone are obstructed and lifted into the colder upper strata where they impinge upon the cold drift from the north-east on the other side of the low-pressure centre, thus giving precipitation mainly on the *left* side of the central track, the condensed water falling as rain, or if it is cooled below freezing-point, as snow. All precipitation so caused may be classed under the general heading of *cyclonic precipitation* in contrast with the *orographical precipitation* caused by the impingement of wet air on high land. The former **Orographical rainfall** may affect any part of the country according to the course of the depression, while the latter, depending on land relief, is far more prominent in the west, where it contributes much the greater part of the annual precipitation. Nevertheless cyclonic rainfall is subject to orographical influence and will give more copious precipitation over high ground than over low.

A third type of condensation is the cause of the *convectional* **Convectional rainfall** *rain* characteristic of thunderstorms. These occur mainly in the summer months, when the surface of the ground is heated by the sun and raises the temperature of the superjacent air: this expands and rises into colder strata, where it is cooled to condensation point. Convectional rain is commonly much heavier, i.e. the rate of fall is greater, than orographical or cyclonic rain, and it contributes an important fraction to the summer rainfall in the midlands and east of England, where the summer temperatures are highest and thunderstorms are commonest. Cyclonic precipitation, even the most severe, scarcely ever exceeds 25 mm. per hour, and is commonly very much less, while convectional rain often reaches 100 mm. per hour, and for short periods may fall at a rate of 150 or even 250 mm. per hour (Salter, 1921). But "thunderstorms giving heavy rains also occur not infrequently in association with shallow depressions" (Bilham, 1938).

The barometrical conditions described, with the lowest pressures lying to the north-west of our islands—whether in the Icelandic low itself or in a depression on the north-eastward track—determine the moisture-laden winds from the directions between south and north-west, and at most stations mainly south-west and west. These do not, it is true, blow continuously like a trade wind or monsoon, but they form a clear majority of all the winds in every part of the country. These winds come of course from the Atlantic and carry abundant water vapour taken up by evaporation from the surface of the sea, particularly in the summer when they blow from more southern latitudes (Figs. 15 and 16). This feature of the British climate has been even more strongly marked during the last few decades, owing to a slight shifting of the distribution of barometric pressures.

Effects of the westerly winds. The effect of the prevalence of westerly winds is to keep the air above the British Islands relatively damp, and (in the winter) relatively warm during the time that they blow.[1] A corresponding south-westerly current in the ocean itself, the "North Atlantic Drift", contributes to the warming of the shallow seas round Great Britain. This double effect of warm-air and warm-water currents determines the climate

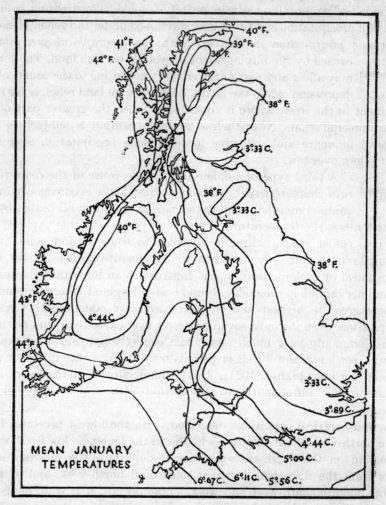

FIG. 17. JANUARY ISOTHERMS (reduced to sea level)

Effect of the western coastal climate on the winter temperatures: the isotherms run largely north and south. From Bartholomew's *Physical Atlas*, Part III, *Meteorology*.

of the whole west coast of Europe—the so-called Atlantic climate—right up to, and even round, the northern extremity of Scandinavia in lat. 72° N.,

[1] It is true that sometimes the south-west winds reaching these islands are cold. This happens when a large low-pressure system to the west is entered by a swirl of polar air from the north or north-east: travelling counter-clockwise round the centre of the depression this very cold air impinges on our shores from the south-west. But for much the greater periods the south-west winds are warm, coming from more southern latitudes.

and is strikingly illustrated in the course of the January isotherms (Figs. 17 and 18). Fig. 18, which shows the course of the January isotherms of 0, 3, 5, 6·67 and 10° C. in western Europe, demonstrates very strikingly the enormous effect of the south-westerly winds on the Atlantic coastal climates. The mean temperature of the Outer Hebrides in January is the same as that

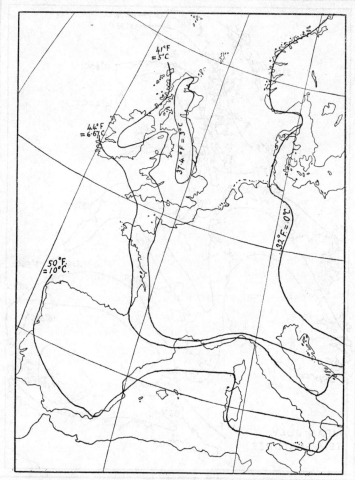

FIG. 18. JANUARY ISOTHERMS—WESTERN EUROPE

The mean winter temperature of south-western England, western Wales, southern and western Ireland, and the Hebrides is the same as that of the south coast of France. From *Types of British Vegetation*, Fig. 4.

of parts of the south of France, 15° of latitude farther south, while Cornwall and the south-west of Ireland enjoy the same winter temperature as the favoured Riviera coast.

The mild winters of our south-west and west account of course for the successful cultivation of Mediterranean and subtropical plants, not only in the gardens of Cornwall and south-west Ireland but in protected places in western Wales, on the coast of Galloway in south-western Scotland and even in the Hebrides.

In summer the effect of the south-west oceanic winds is to *cool* the British Isles, especially the west coasts, as can be well seen in Fig. 19, the July

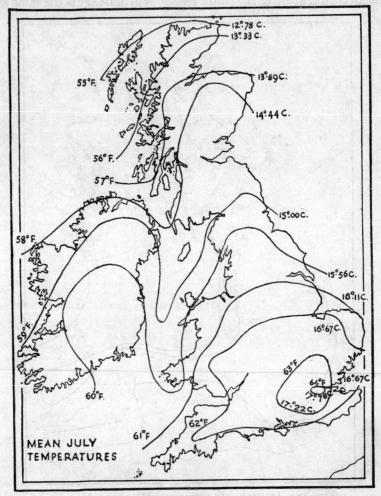

FIG. 19. JULY ISOTHERMS (reduced to sea level)

In contrast to the January isotherms (Fig. 17) those for July run mainly east and west, bending northwards over the land and southwards over the sea. For a given latitude the July temperature is therefore higher over the land than over the sea, and generally higher on the east than on the west coast. Note the area of high temperature in south-eastern England. From Bartholomew's *Physical Atlas*, Part III, *Meteorology*.

isotherms bending sharply southwards over the North Sea, the Irish Sea, and the Atlantic; and in passing from east to west they cut the coasts at successively lower latitudes. The influence of the cooling west winds of summer is felt right across the lowlands of northern Europe, though with progressively diminishing intensity, so that they carry a sub-Atlantic climate far over the continent.

Figs. 20 and 21, showing the distribution of mean minimal temperatures in January and mean maximal in July (corrected to sea level), again bring

FIG. 20. MEAN MINIMAL TEMPERATURES, JANUARY (reduced to sea level)
The isotherms are concentric with the land masses, the lowest temperatures in the eastern English midlands. From Bartholomew's *Physical Atlas*, Part III, *Meteorology*.

out very strikingly the moderating effects of the sea on extremes of temperature. The areas of lowest mean minima in winter and of highest mean maxima in summer are almost central, but lie rather nearer the east coast in England, and are separated by belts of country leading from valleys which open to the west, thus admitting the prevalent west winds with their moderating effects.

Fig. 22 shows the annual ranges of temperature between the means of the warmest and coldest months and again illustrates the effect of the sea winds in producing the mild oceanic and maritime climates. The annual range is low on the extreme western coasts and increases towards the east, the

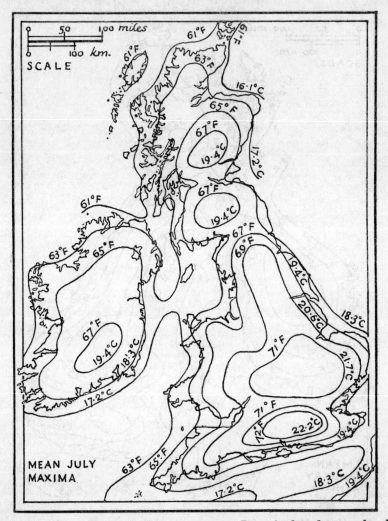

FIG. 21. MEAN MAXIMAL TEMPERATURES, JULY (reduced to sea level)

The isotherms are again concentric with the land masses, the highest temperatures in the south-east and the midlands. The overlap of the highest July maxima with the lowest January minima (Fig. 20) roughly indicates the area of most "continental" climate. From Bartholomew's *Physical Atlas*, Part III, *Meteorology*.

greatest ranges occurring in the eastern and south-eastern midlands of England (cf. pp. 57 and 66, where figures are given for particular stations). On most of the east coast of England the range is below the maximum, and on the east coast of Scotland, opposite the wider part of the North Sea, very considerably below.

Effect on rainfall. The impingement of the moisture-laden winds on high-lying land, especially near the west coasts, forces the air upwards and leads to condensation and the so-called "orographical" rainfall referred to on p. 47. This is the cause of the very high precipitation on the western mountains (cf. Fig. 11).

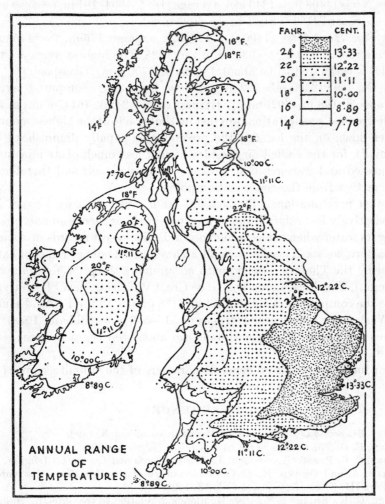

FIG. 22. ANNUAL RANGE BETWEEN MEAN TEMPERATURES OF THE WARMEST AND COLDEST MONTHS

The areas of greatest range are inland but eastwards of the centres of the land masses, and in places touching the east coast of England. From Bartholomew's *Physical Atlas*, Part III, *Meteorology*.

Highest precipitations. The three wettest districts in the British Isles are (1) the western and north-western Scottish Highlands, (2) the English Lake District, and (3) Snowdonia in North Wales. With the exception of the Cairngorm massif in the eastern Highlands, these three districts contain the highest British summits. The following figures are taken from Bilham (1938, p. 78):

(1) *Sgurr na Ciche* (3140 ft. = 957 m.), estimated to exceed 200 in. (= 5080 mm.). Loch Quoich (650 ft. = 198 m.) has an annual average of 165 in. (= 4178 mm.).

Ben Nevis (4406 ft. = 1343 m.), average (1885–1904) 161 in. (= 4084 mm.); extreme record (1898) 240·1 in. (= 6099 mm.).

(2) *Stye Head Pass* (1400 ft. = 426 m.); average 175 in. (= 4445 mm.); extreme record 250 in. (= 6350 mm.) in 1928. The highest average in this district is estimated to be about 185 in. or 4700 mm. (Glasspoole).

(3) *Glaslyn* (2500 ft. = 762 m.), below Snowdon: computed average (Glasspoole) 198 in. (5029 mm.). Highest record 242 in. (6147 mm.) in 1909.

The greatest precipitations are just on the lee side of the highest summits; farther down on the lee side the precipitation rapidly diminishes ("rain shadow"), for the chilled air, besides having lost much of its moisture, is compressed and warmed as it descends to lower levels and therefore departs farther from the condensation point.

Lowest precipitations. Flat coastal lands, even on a west coast, have comparatively low rainfall, the wet air passing over them and only precipitating its water when it meets high ground. The lowest records of all are on flat eastern coasts. There is a small area about 53 square miles in extent, bordering the Thames estuary, with an annual average of less than 20 in. (508 mm.). The driest spot seems to be Great Wakering, near Shoeburyness, where the computed average is 18·4 in. (468 mm.). In the Fenland south of the Wash there are several averages of 515–530 mm. (Bilham, 1938).

Thus the highest British rainfalls are about ten times as great as the lowest.

[For further remarks on general characters of the rainfall see pp. 67–8.]

REFERENCES

Atlas of Meteorology. Bartholomew's Physical Atlas, Vol. **3**, 1899.
BILHAM, E. G. *The Climate of the British Isles.* London, 1938.
BROOKS, C. E. P. and GLASSPOOLE, J. *British Floods and Droughts.* London, 1928.
BUCHAN, A. and OMOND, R. T. The meteorology of the Ben Nevis observatories. Part III. *Trans. Roy. Soc. Edinb.* **43**, 1905.
FRASER, G. K. Studies of Scottish Moorlands in relation to tree growth. *Bull. For. Comm.* No. 15, 1933.
KENDREW, W. G. Chapter v. "Climate," in Oxford Survey of the British Empire, vol. **1**. *The British Isles.* Oxford, 1914.
MACKINDER, H. J. *Britain and the British Seas.* Second Edition. Chapter XI. "The climates of Britain." Oxford, 1907.
SALISBURY, E. J. "The East Anglian Flora." *Trans. Norfolk and Norwich Natural Hist. Soc.* 1932.
SALTER, M. DE CARLE. *The Rainfall of the British Isles.* London, 1921.
STAMP, L. DUDLEY and BEAVER, S. H. *The British Isles.* Chapter IV. "British Weather and Climate." London, 1933.

Chapter III

REGIONAL CLIMATES

CLIMATE AND VEGETATION; CLIMATE AND AGRICULTURE

For our purposes the differences in the quantitative values of the leading elements of climate between different parts of the British Isles are best illustrated by meteorological data from a selection of stations, chosen, so far as they are available, to represent characteristic climates. Most of the stations referred to in the text are shown in Fig. 23. Monthly means for actual stations give a better picture of climate in relation to vegetation than averaged figures representing "districts". It would of course be impossible to reproduce, even if they were available, the objective pictures obtained by plotting day to day, or hourly, or continuous automatic records, and a middle course between the hopelessly voluminous "raw" data and the over-abstraction of multiple averaging, is here adopted.

Maritime climates. The climate of the British Isles is generally, and rightly, described as "maritime" or "oceanic", but there are considerable differences, as we have seen, in the intensity of the maritime influence. It will be clear from what has already been written, and from the figures, that the prevalent moisture-laden west winds have a much greater effect on the climate of the western parts of the country, particularly on that of elevated land, on which they first impinge, than on the eastern lower lying and flatter parts. Nevertheless, the climate of the southern and also of the eastern seaboard, particularly the northern part, is still markedly maritime as compared with inland areas.

By considering pairs of stations, one in each pair coastal, and the other inland but not very distant from the first, and as far as possible under comparable conditions, Bilham (1938) shows that where the mean temperature for the whole year is nearly the same at the two stations it is higher inland during the summer half-year and lower during the winter half-year. Similarly, the inland mean maxima are higher in the summer and lower (though only by a fraction of a degree) in the winter. In general the daily minima on the coast exceed the inland values in all months. Thus the all-round effect of the coastal position is a consistent reduction of temperature range, roughly equivalent to what is generally known as "mildness" of climate.

The contrast in this respect between coastal and inland climate is most clearly seen on the east coast of England. It is here that the effect of the "sea breeze" is most apparent. This is due to the passage of air from sea to land to replace the ascending air which has been warmed by the heating of

Fig. 23. Meteorological Stations referred to in the Text

the land surface, and on the east coast its direction is contrary to that of the prevailing winds; whereas on the west the two types of wind have the same direction and reinforce one another. The cooling effect of the sea breeze on hot summer afternoons is well known to visitors at east coast seaside resorts.

Extreme Atlantic climate. Five stations (Table IV), ranging through 10° of latitude, have been chosen to represent what may be called the extreme Atlantic climate of the western seaboard, including two isolated groups of islands where the oceanic influence is naturally at a maximum. The leading

Table IV. *Extreme Atlantic climate*

	Mean of hottest month	Mean of coldest month	Range*	Mean max. of hottest month	Mean min. of coldest month	Rainfall mm.	Rain days	Snow days	Sunshine (per cent of possible)
	\multicolumn{9}{c}{Shetland Islands (Sumburgh Head 59° 52′ N. and Lerwick 60° 10′ N.)}
	Aug.	Feb.		Aug.	March				
°F.	52·8	38·3	14·5	57·1	34·4	964	260	24·8	24
°C.	11·5	3·5	8·0	14·0	1·3				
	\multicolumn{9}{c}{Stornoway (Outer Hebrides) 58° 11′ N.}
	July	Feb.		July	Feb.				
°F.	54·6	39·2	15·4	60·2	34·2	1266	263	30·8	27
°C.	12·5	4·0	8·5	15·6	1·2				
	\multicolumn{9}{c}{Malin Head (Donegal) 55° 23′ N.}
	Aug.	Jan. and Feb.		Aug.	Feb.				
°F.	56·9	40·9	16·0	61·0	37·1	812	236	10·8	30
°C.	13·9	5·0	8·9	16·1	2·7				
	\multicolumn{9}{c}{Valencia Island (Kerry) 51° 56′ N.}
	Aug.	Feb.		Aug.	Feb.				
°F.	58·9	44·3	14·6	63·9	39·7	1414	252	5·2	31
°C.	14·4	6·8	8·1	17·7	4·2				
	\multicolumn{9}{c}{Scilly Islands (St Mary's) 49° 56′ N.}
	Aug.	Feb.		Aug.	Feb.				
°F.	60·8	45·3	15·5	65·0	41·7	809	207	3·1	38
°C.	16·0	7·5	8·6	18·3	5·3				

* Difference between means of warmest and coldest months.

features of the temperature records are at once apparent. The mean summer maxima are low and the mean winter minima very high for the latitude, while the difference between mean maximum and mean minimum for any month is relatively slight. Hence the annual range of temperature, as measured by the difference between the mean temperatures of the warmest summer and the coldest winter month, is very small, varying only from 8° C. in the Shetlands to 8·9° C. at Malin Head in Northern Ireland. The temperatures themselves are of course higher in the south, but not a very great deal. Thus the mean maximum for August (the warmest month) is only 65° F. (18·3° C.) in Scilly (49° 56′ N.) and 57·1° F. (13·9° C.) in the

Shetlands (60° N.), more than 600 miles (1120 km.) farther north. Extreme high records are 70° F. (21·1° C.) for the Shetlands and 82° F. (27·7° C.) for Scilly. The mean minimum is 41·7° F. (5·4° C.) for January in Scilly, and no lower than 34·5° F. (1·4° C.) in the Shetlands. There are decidedly lower extreme records in the northern islands, 16° F. ($-8·9°$ C.) in the Shetlands and 11° F. ($-11·6°$ C.) at Stornoway in the outer Hebrides, but such temperatures are rare, as can be seen from the mean minima. In Scilly the lowest extreme record (January) is 25° F. ($-3·9°$ C.), frost being quite negligible in the winter climate of the islands.

The comparatively slight change in temperature with latitude, between summer and winter, and, it may be added, between day and night, are the great characteristics of these extremely oceanic climates. Correspondingly the vegetation, treeless in fully exposed places because of the violent winds bearing salt spray, is characteristically evergreen heath and grassland on well-drained slopes, "blanket bog" where water cannot readily escape. The annual rainfall at these Atlantic stations is moderate to rather heavy, with a marked December maximum and a spring or early summer minimum; but on the whole well distributed throughout the year. The lowest (809 mm.), at Scilly, is due to the very slight elevation of the islands, which lie some distance to the west of the nearest land. The highest (1414 mm.), at Valencia Island, is doubtless influenced by the proximity of the mountainous south-western extremity of Co. Kerry. The number of rain days, i.e. days on which at least 0·01 in. (0·25 mm.) falls, is high throughout, varying from 207 at Scilly to 263 at Stornoway. On much of the Atlantic seaboard of Ireland and of north-west Scotland they exceed 225 (Fig. 24, p. 69). The average rain per rain day varies from 3·4 mm. at Malin Head to 5·6 mm. at Valencia Island. This small quantity of rain per rain day is due to the persistent drizzle which is so characteristic of the British climate with both orographical and often also with cyclonic precipitation. If the data were available the average quantity of precipitation per *hour* of rain would doubtless be even more striking. The long duration and extremely slow rate of fall is of course due to the constant high moisture content of the air combined with slow and slight changes of temperature. It is very noteworthy that the areas with the greatest number of rain days—more than 250 per annum (Fig. 24)—correspond with blanket bog areas—Kerry, Connemara, western Mayo and Donegal in Ireland, western Ross-shire, Sutherland and Lewis in Scotland, and that these are not necessarily the areas of highest rainfall (cf. Fig. 11). They are rather areas of low saturation deficit, the air being kept constantly moist by the *frequent* rain, not necessarily large in amount. Blanket bog, however, extends outside these areas to regions in which the number of rain days is less than 225.

In accordance with the lower temperatures the number of days on which snow falls is greatest at the northern stations—24·8 and 30·8 at Shetland and Stornoway respectively, only 10·8 at Malin Head, 5·2 at Valencia and

3·1 at Scilly, with a late maximum, often in March. The snow lies a very short time, the mornings on which it is seen on the ground averaging 12·6 in the Orkneys and 0·6 at Valencia Island (no data for the other stations). This means that at the southern stations snow hardly ever lies even for a few hours.

The extreme mildness of the Atlantic climate is conspicuously seen in the success with which subtropical plants such as palms and tree ferns, as well as a wide range of exotics from warm temperate regions can be grown in the open in protected stations on the south-west and west coasts—some of them even as far north as the west coast of Scotland. Another result of this climate is the descent of several high mountain species to sea-level, where they occur mixed with lowland forms.

Table V. *West coast of England and Wales*

	Mean of hottest month	Mean of coldest month	Range*	Mean max. of hottest month	Mean min. of coldest month	Rainfall mm.	Rain days	Snow days	Sunshine (per cent of possible)
	\multicolumn{9}{c}{Southport (Lancs) 53° 39′ N.}
°F.	July 59·4	Feb. 38·7	20·7	July 65·4	Jan. 34·6	813	189	9·8	35
°C.	15·2	3·7	11·5	18·5	1·4				
	\multicolumn{9}{c}{Holyhead (Anglesey) 53° 18′ N.}
°F.	Aug. 58·7	Feb. 41·7	17·0	Aug. 63·4	Feb. 38·0	888	201	7·0	35
°C.	14·8	5·3	9·4	17·4	3·3				
	\multicolumn{9}{c}{Aberdovey (Merioneth) 52° 7′ N.}
°F.	July and Aug. 60·9	Feb. 41·4	19·5	July 65·6	Jan. 38·1	1181	—	—	35
°C.	16·0	5·2	10·8	18·6	3·3	(Gogerddan)			
	\multicolumn{9}{c}{St Ann's Head (Pembroke) 51° 41′ N.}
°F.	Aug. 59·0	Feb. 42·3	16·7	Aug. 62·8	Feb. 38·9	897	201	5·9	36
°C.	15·0	5·7	9·2	17·1	3·8				

* Difference between means of warmest and coldest months.

The percentage of the possible bright sunshine, low everywhere in British as compared with continental climates, is particularly low in the north, 24 in the Shetlands, 27 at Stornoway, higher at Valencia (31) and markedly higher at Scilly (38), where it is but little lower than at the stations on the south coast of England that show the highest percentages (39–41) in the British Isles.

West coast of England and Wales. We may now take four stations on the west coast of England and Wales (Table V). Here the climate may be described as only "moderately oceanic", since there is a narrow sea between England and Ireland, which intervenes between this and the open ocean.

The summer temperatures are somewhat higher and the range between the mean temperatures of the warmest and coldest months, somewhat greater, varying between 9·2 and 11·5° C. as compared with 8 and 8·9° C. for the extreme Atlantic stations. But the climate is still very mild, the snow days at Southport being 9·8 (with morning lie 7·6 days), at Holyhead 7 and at St Ann's Head in Pembrokeshire only 5·9. This last is the most oceanic of these stations, with the lowest temperature range, the coolest summers and the warmest winters, because it lies quite open to the Atlantic towards the south-west, the direction of the prevailing winds. The rainfall is quite moderate, varying from 813 to 897 mm., except at Aberystwyth, where, with mountains close behind, it is 1181 mm. Bright sunshine is rather higher than at most of the extreme Atlantic stations, being 35 or 36 per cent of the possible.

South coast climate. The south coast of England possesses a similar climate, decreasing very gradually in mildness as we pass eastwards along the English Channel, away from the open Atlantic (Table VI). Falmouth has a somewhat cooler winter than Scilly, and an earlier and rather higher summer maximum, so that the annual range of temperature is greater, 9·5 against 8·6° C. The highest temperature record is almost the same and the lowest not very much lower, 83 and 20° F. (28·3 and −6·6° C.) for Falmouth, against 82 and 25° F. (27·8 and −3·9° C.) for Scilly. The snow days are only 5·2, and mornings of snow lie only 0·5 in the year. Rainfall is 1107 mm., the rain days 207, the percentage of possible bright sunshine 39. This is really an Atlantic climate, and the vegetation corresponds. The gardens are famous for the successful cultivation of subtropical plants. Oakwood grows in the sheltered valleys, heath on the shallow-soiled rocks of the hills and low plateaux.

The climates of Falmouth, Sidmouth, Bournemouth and Brighton, at intervals along the coast, form a series of decreasing oceanicity: steady increase in range of temperature between the warmest and the coldest months: Falmouth 9·5° C., Sidmouth 10·7° C., Bournemouth 12° C., Brighton 12·3° C.; decrease of rainfall—1107, 823, 785 and 705 mm.; and of rain days—207, 196, 172 and 163. The number of snow days also increases eastwards: Falmouth 5·2, Totland Bay (north-west coast of the Isle of Wight) 7·6, Eastbourne 9 (in 1935), Dungeness 11·6; and of days on which ground frost occurs: Falmouth 48·3, Totland Bay 68·6. The extreme records of temperature available over a long period are: Falmouth 20 and 83° F. (−6·6 and 28·3° C.), Totland Bay 14 and 86·6° F. (−10 and 30·3° C.), Eastbourne 15 and 89·5° F. (−9 and 31·9° C.), Dungeness 11 and 82° F. (−11·7 and 27·8° C.). The sunshine records are very uniform and the highest in England, 39–41 per cent of the possible maximum, with from 20 to 24 per cent of possible mid-winter sunshine.

Dungeness, the most easterly station, is an extensive promontory of shingle jutting out into the Channel where it narrows to the Straits of

South Coast Climate

Dover. Of the series of south coast stations it is farthest from the Atlantic influence, but its climate is not less oceanic than that of Brighton, which is situated in the concavity of a wide shallow bay. Its annual range of temperature is the same as that of Brighton, 12·3° C., but its July mean

Table VI. *South coast of England*

	Mean of warmest month	Mean of coldest month	Range*	Mean max. of warmest month	Mean min. of coldest month	Rainfall mm.	Rain days	Snow days	Ground frost days	Sunshine (per cent of possible)
				Falmouth (Cornwall) 50° 9′ N., 5° 4′ W.						
°F.	July 60·5	Jan. and Feb. 43·4	17·1	July 65·8	Feb. 39·4	1107	207	5·2	48·3	39
°C.	15·8	6·3	9·5	18·7	4·1					
				Sidmouth (Devon) 50° 41′ N., 3° 29′ W.						
°F.	July 60·3	Jan. 41·0	19·3	July 67·0	Jan. 36·2	823	196	—	—	—
°C.	15·7	5·0	10·7	19·4	2·3					
				Bournemouth (Hants) 50° 43′ N., 1° 53′ W.						
°F.	July 61·5	Jan. 39·8	21·7	July 69·7	Jan. 34·8	785	172	—	—	40
°C.	16·4	4·3	12·0	20·9	1·5					
				Totland Bay (Isle of Wight)						
°F.	Aug. 61·4	Feb. 40·7	20·7	July 67·4	Feb. 36·3	738	162·3	7·6	68·6	—
°C.	16·4	4·8	11·5	19·6	2·4					
				Brighton (Sussex) 50° 49′ N., 0° 8′ W.						
°F.	July and Aug. 62·2	Feb. 40·0	22·2	July 68·9	Jan. 35·8	705	163	—	—	39
°C.	16·8	4·4	12·3	20·5	2·1					
				Eastbourne (Sussex) 50° 46′ N., 0° 17′ E.						
°F.	Aug. 61·7	Feb. 41·0	20·7	Aug. 67·2	Feb. 36·6	786	167	9	—	40†
°C.	16·5	5·0	11·5	19·5	2·5					
				Dungeness (Kent) 50° 55′ N., 0° 58′ E.						
°F.	Aug. 61·4	Jan. 39·3	22·1	Aug. 67·3	Jan. 34·7	619	169	11·6	—	—
°C.	16·4	4·0	12·3	19·5	1·5					

* Difference between means of warmest and coldest months.
† The latest figures give an average of about 41 per cent for Eastbourne and Worthing. In 1911 47·7 per cent was recorded at Eastbourne.

maximum is a little lower (66·6 against 68·9° F.). The rainfall is lower (619 against 705 mm.) owing to the flat shingle promontory and low-lying marshland (Romney Marsh) behind.

East coast climate. The east coast of Great Britain is removed altogether from the direct influence of the Atlantic, but the climate is markedly maritime (Table VII). The range between summer and winter temperatures is greater on the southern stretch of coast opposite the narrower part of the North Sea and falls again as this opens out. For the northern half of the east coast of Great Britain the range is as low as on the south coast of Dorset and east Devon (Clacton 12·9, Yarmouth 12·6, Skegness 11·9,

Table VII. *East coast of England and Scotland*

	Temperature								
	Mean of hottest month	Mean of coldest month	Range*	Mean max. of hottest month	Mean min. of coldest month	Rainfall mm.	Rain days	Snow days	Sunshine (per cent of possible)

Clacton (Essex) 51° 47′ N., 1° 9′ E.

	Mean hot	Mean cold	Range	Max hot	Min cold	Rainfall	Rain d	Snow d	Sun %
° F.	July and Aug. 61·2	Jan. 37·9	23·3	July 67·7	Jan. 33·9	489 (Shoeburyness 479 mm.)†	149	—	39
° C.	16·2	3·2	12·9	19·8	1·0				

Great Yarmouth (Norfolk) 52° 37′ N., 1° 43′ E.

° F.	July 60·5	Jan. 37·8	22·7	Aug. 66·3	Jan. 33·7	622	183	16·7	—
° C.	15·8	3·2	12·6	19·0	0·9				

Skegness (Lincs) 53° 9′ N., 0° 21′ E.

° F.	Aug. 59·1	Jan. 37·6	21·5	Aug. 65·9	Jan. 32·7	571 (Spurn Head, 500 mm.)	186	—	37
° C.	15·0	3·1	11·9	18·7	0·4				

Tynemouth (Durham) 55° N., 1° 27′ W.

° F.	July 57·8	Jan. 38·6	19·2	July and Aug. 64·2	Jan. 33·6	621	179	22·6	—
° C.	14·3	3·6	10·6	17·8	0·8				

Leith (Midlothian) 55° 58′ N., 3° 10′ W.

° F.	July 58·6	Jan. 39·1	19·5	July 65·8	Feb. 33·8	592	182	17·8	—
° C.	14·7	4·0	10·8	18·7	1·0				

Aberdeen 57° 10′ N., 2° 6′ W.

° F.	July 56·7	Jan. 37·8	18·9	July 63·1	Feb. 33·1	748	214	34·0	31
° C.	13·7	3·2	10·5	17·2	0·6				

* Difference between means of warmest and coldest months.
† Now said to be too low, but the neighbouring Great Wakering rainfall is computed as 468 mm.

Tynemouth 10·6, Leith 10·8, Aberdeen 10·5, comparing with Sidmouth 10·7), though the climate is distinctly less "mild" since both summer and winter mean temperatures are more than 2° C. lower. The rainfall is markedly lower and includes, on flat Essex and Yorkshire coastlands, some of the lowest in the British Isles. Thus Clacton has 489 mm. and the neighbouring Shoeburyness 479 mm., for long taken as the lowest annual record in the country.[1] The lower winter temperatures bring a larger number of snow days—16·7 at Yarmouth, 22·6 at Tynemouth and 34 at Aberdeen.

Bright sunshine is high in the south (Clacton 39 per cent) and lower in the north (Aberdeen 31 per cent).

Northern inland climate. The Scottish inland climate is of course mainly dominated by the mountains. The climate of the Highland valleys whose floors are of low elevation is strikingly mild for so northern a latitude, especially if the south-west winds have access. Thus Fort Augustus near the centre of Glenmore, the great rift valley which runs south-south-west and north-north-east right across Scotland from sea to sea, has a January and February mean of 37·7° F. (3·2° C.) and a July mean of 56·7° F. (13·7° C.), giving a range of only 10·5° C., equal to that of the maritime climate of Aberdeen on the east coast and considerably less than that of many places on the east and south English coasts. The rainfall is only moderately heavy (1123 mm.) with a typical winter maximum and spring and summer minimum, and a large number of rain days (223)—almost an Atlantic climate. The mean annual duration of sunshine is low, 19 per cent of the possible, with very low records in the foggy winter (December 6 per cent, January 8 per cent), and a markedly higher percentage from April to June (26 and 27 per cent) than in July and August (21 per cent).

Glasgow lies in the "Midland valley" much farther south and not far from the west coast. The climate is not dissimilar from that of Fort Augustus, though not quite so wet, with a range of 10·8° C., a rainfall of 945 mm., 202 rain days and 24 per cent of sunshine, only 7 per cent in December, and a higher percentage in April and May than from July to September. Only one month (July) is without any ground frost, though in the winter it only occurs on about one day out of three.

Perth, in the eastern Highlands but at a low altitude and less accessible to maritime influence, has a range of 11·9° C., and a rainfall of 782 mm.

Inland stations in the north of England have higher temperatures and a higher range between winter and summer—Manchester 12·2° C. and York 12·6° C. Manchester, to the west of the great Pennine ridge, has a rainfall of 872 mm. with a well-marked secondary maximum in late summer, and 196 rain days. The winter is very foggy, with a low record of bright sunshine (6 per cent. in November, 3 per cent in December, 6 per cent in January and 24 per cent for the year), but this is partly due to smoke. York, in the drier plain to the east of the Pennines, has more sunshine in winter and 28 per cent for the year, a rainfall of only 618 mm. and 186 rain days. The extreme temperatures are greater, with single records of 92° F. (33·3° C.) in September, and −4·5° F. (−20·3° C.) in December. The climate is more like that of the midland plain (p. 66), though with less sunshine.

Scottish mountain climate. Owing to the maintenance for 20 years (1884–1903) of a well-equipped meteorological station on the summit of Ben

---

[1] This Shoeburyness average is now said to be too low, but a figure of 468 mm. is computed for the village of Great Wakering in the same neighbourhood (see Chapter II, p. 54).

Nevis in the western Highlands, the highest point in the British Isles, at which hourly observations of many different climatic factors were made, we have a very complete knowledge of the climate and weather of this mountain station.

The accompanying table (VIII) brings together some of the more important data and compares them with the records from Fort William at the

Table VIII

		Ben Nevis (1884–1903) 56° 48′ N., 5° W. Alt. 4409 ft. (1344 m.)	Fort William (1886–1915) 56° 49′ N., 5° 7′ W. Alt. 176 ft. (53·6 m.)
Temperatures:			
Mid-temp. of hottest month	°F. °C.	41·1 (July) 5·0	57·2 (July) 14·0
Mid-temp. of coldest month	°F. °C.	23·8 (Feb.) −4·6	38·8 (Feb.) 3·8
Range	°F. °C.	17·3 9·6	18·4 10·2
Mean max. of hottest month	°F. °C.	44·6 (July) 7·0	64·0 (July) 17·7
Highest max. of hottest month	°F. °C.	53·3 (Aug. 1899) 11·8	69·6 (Aug. 1899) 20·8
Mean min. of coldest month	°F. °C.	20·5 (Feb.) −6·5	43·2 (Jan.) 6·2
Lowest min of coldest month	°F. °C.	14·2 (Jan. 1895) −9·9	37·0 (Feb. 1895) 2·7
Extreme records:			
Maximum	°F. °C.	66·0 (28 June 1902) 18·8	86·0 (25 June 1888) 30·0
Minimum	°F. °C.	1·0 (6 Jan. 1894) −17·2	16·0 (28 Jan. 1910) −9·0
Rainfall (mm.):			
Mean annual		4084	2002
Highest		6099 (1898)	2756 (1877)
Lowest		2739 (1886)	486 (1920)
Rain days (mean)		263	240
Max. rain in 24 hr.		185 (3 Oct. 1890)	114 (11 Oct. 1916)
Snow days: Mean		169·6	—
Highest		206	—
Humidity of air (difference between dry and wet bulb) °F.:			
Highest hourly mean		1·9 June, 14 hr. (2 p.m.)	6·2° F., May and June, 15 and 16 hr. (3–4 p.m.)
Lowest hourly mean		0·3 (all hours in Jan.)	—
Mean for year 12–15 hr.		0·8	—
Sunshine (per cent of possible):		17·0	31·0
Most frequent direction of wind		N.	W.

head of Loch Linnhe, close to the foot of the mountain. The temperatures at Fort William are not very different from those of Fort Augustus farther up Glenmore, but the climate is more oceanic and much wetter.

The figures in Table VIII speak for themselves. The summer temperatures at the top of the mountain are very low, and it is remarkable that the range between the mid-temperatures of the warmest and coldest months is rather less than at sea-level. The figures of snowfall and snow lie are very striking. Snow falls in every month of the year, the mean number of days of snowfall for July and August, the months with fewest snow days, being 3·2 and 2·9 respectively. But the fallen snow melts again in those months and in September. It begins to accumulate in October, and the mean depth steadily increases till the end of April or the beginning of May, when it reaches 78 in. (198 cm.). Melting and sublimation then begin to exceed snowfall, and the depth steadily decreases till the accumulated snow disappears shortly after the middle of July. There are thus about $2\frac{1}{2}$ months during which the ground is free from snow, representing a possible vegetation season for plants. Unfortunately we know very little about the actual vegetation of such areas in Britain, but the importance of this long snow lie, both in protecting the plants during the long severe winter, and in shortening the vegetative season, is obvious enough (cf. Chapters XXXVIII and XXXIX, pp. 775, 781–2, 797).

The normal mean minima are above freezing-point from June to September inclusive, but the normal extreme minima show frosts for every month in the year.

The climate of Ben Nevis is very wet indeed, the rainfall being 161 in. (4084 mm.), more than double that of Fort William. The rainfall recorded in 1898 was 6099 mm. (240·1 in.). The other highest rainfalls recorded from the Highlands, Cumbria and North Wales are noted on p. 54.

An equally striking index of the wetness of the Ben Nevis climate is the constantly high humidity of the air. The mean differences between the readings of the dry- and wet-bulb thermometers for given hours of the day are extraordinarily low. For the whole year the figure between 19 hr. and 5 hr. (7 p.m. and 5 a.m.) is 0·6° F. (saturation deficit, 0·4–0·5 mb.), and from 12 to 15 hr. (12 noon to 3 p.m.) only 0·8° F. (saturation deficit 0·6 mb.). The highest difference for a particular hour in a particular month is 1·9° F. (saturation deficit, 1·5 mb.) for 14 hr. (2 p.m.) in June, no other mean reaching this figure, though the mean difference exceeds 1° F. (saturation deficit, 0·8 mb.) all day and night in June and during much of the day in May, July and September. In August (a wet month) the figure never exceeds 0·8° F. (saturation deficit, 0·6 mb.) from 10 to 16 hr. (10 a.m. to 4 p.m.). At Fort William, on the other hand (by no means a dry climate), the highest mean difference is 6·2° F. at 15 and 16 hr. (3 and 4 p.m.) in May and June, and more than 5° F. during July afternoons.

The mean percentage of cloud for the year is 84, departing but little from this value either in summer or winter. June (77 per cent) has the lowest monthly cloud value. The summit of Ben Nevis was clear of fog or mist twelve times in April, fourteen in May, fourteen in June, eleven in July, nine in August and nine in September and fewer times in the autumn and winter months.

During 13 years there was a total of 3405 gales with a velocity of more than 50 miles per hour, the fewest during July. The most frequent wind is north, while at sea-level it is west (Buchan and Omond, 1905).

From these figures there emerges a vivid picture of the wetness, fogginess and windyness of the climate. It is also clear that May and June are the driest and brightest months, but at this time the summit is still covered with snow.

English midland climate. In spite of the small size of the country, no place in England being more than about 70 miles (say 112 km.) from the nearest sea, there is some approach to continental conditions in the English midlands, and especially in the eastern midlands. This can be seen from Table IX.

Table IX. *English Midlands*

	Mean of hottest month	Mean of coldest month	Range*	Mean max. of hottest month	Mean min. of coldest month	Rainfall mm.	Rain days	Snow days	Ground frost days	Sunshine (per cent of possible)
	\multicolumn{10}{c}{Belvoir Castle (Leics.), 52° 53′ N., 0° 48′ W.}									
	July	Jan. and Feb.		July	Feb.					
°F.	60·3	38·7	21·6	69·0	33·1	638	205	19·8	99·5	34
°C.	15·7	3·8	12·0	20·5	0·6					
	\multicolumn{10}{c}{Oxford, 51° 56′ N., 1° 16′ W.}									
	July	Jan.		July	Jan.					
°F.	61·9	38·4	23·5	70·2	33·5	631	168	16·8	91·4	34
°C.	16·6	3·6	13·0	21·2	0·9					
	\multicolumn{10}{c}{Nottingham, 52° 56′ N., 1° 9′ W.}									
	July	Jan.		July	Jan.					
°F.	61·0	37·6	23·4	69·7	32·6	598	199	—	—	30
°C.	16·1	3·0	13·0	20·9	0·3					
	\multicolumn{10}{c}{Coventry (Warw.) 52° 25′ N., 1° 30′ W.}									
	July	Jan.		July	Jan.					
°F.	61·5	37·9	23·6	70·7	32·5	640	187	—	—	31
°C.	16·3	3·3	13·1	21·5	0·2					
	\multicolumn{10}{c}{Kew (Surrey), 51° 28′ N., 0° 19′ W.}									
	July	Jan.		July	Jan.					
°F.	62·7	38·9	23·8	71·0	34·6	606	167	13·4	100·8	33
°C.	17·0	3·9	13·2	21·6	1·3					
	\multicolumn{10}{c}{Cambridge, 52° 12′ N., 0° 8′ E.}									
	July	Jan.		July	Jan.					
°F.	61·9	37·6	24·3	71·7	31·9	556	163	—	111·9	35
°C.	16·6	3·0	13·5	22·1	0·0					

* Difference between means of warmest and coldest months.

The figures are seen to be strikingly close together though the stations are well scattered over the midland plain. Cambridge, to the east of the midlands, has the greatest difference (51° C.), about equal to that of York, between the absolute extremes recorded, and the greatest range between the mean temperatures of January and July (13·5° C.), which contrasts with 8° C. in the Shetlands and 8·1° C. at Valencia Island; also the greatest number of frost days, the lowest rainfall and number of rain days, and the highest percentage of sunshine of the inland stations, though the rainfall is higher than on the flat coast lands and the sunshine less than on the south coast. (For the closely related climate of Breckland see p. 70.)

Two stations in the western midlands close to the Welsh Marches lying in the rain shadow of the Welsh mountains, have fairly low rainfalls and rather lower temperature ranges (Shrewsbury 586 mm. and 11·9° C., and Hereford 668 mm. and 12·7° C.). The inland stations in the southern counties show no great differences.

In the south-west, owing to greater nearness to the Atlantic and sharper land relief, the rainfalls are higher but the temperature ranges about the same; Bath has a range of 12·3° C. and a rainfall of 757 mm., Cullompton in Devon a range of 12·2° C. and a still higher rainfall (897 mm.).

Irish inland stations. Dublin, on the east coast, has a maritime climate with a temperature range of only 10·3° C. between the July and January means, and extreme records of 30·5 and −10·5° C. The rainfall is moderate (695 mm.) with 198 rain days and 18·3 snow days, 35·5 days of ground frost and 30 per cent of possible sunshine. The figures for five inland Irish stations are given in Table X.

Table X

	*Range of temperature ° C.	Rainfall	Rain days	Snow days	Sunshine per cent
Armagh	10·4	807	215	9	29
Birr Castle	10·5	829	211	11	30
Kilkenny	10·8	842	198	—	—
Cahir	11·0	923	—	—	—
Killarney	9·3	1390	231	—	—

* Difference between means of warmest and coldest months.

The first four show the mild insular climate with moderate rainfall, gradually increasing towards the more hilly south. Killarney in the mountainous region of the south-west has a much higher rainfall and almost an Atlantic climate (cf. p. 327).

Seasonal distribution of rainfall. If we consider the British Islands as a whole the rainfall is remarkably well distributed through the year—there are no strongly marked wet and dry seasons. This is because of the persistence of the south-westerly wind drift and of the passage of depressions in the same general direction throughout the year. Nevertheless, the seasonal variation is quite well characterised both for the country at large and in the different climatic regions. Thus we may say that over the country as a whole the spring months are relatively dry and the autumn months relatively wet. In the west the winter is notably wet, while in the east there is a late summer increase of precipitation which may be quite marked, though in the south-east, especially, it does not equal the autumn maximum. These seasonal features are due to the great preponderance of orographical rainfall in the west, at its height in winter when the general south-westerly wind drift is most intense. A larger proportion of the summer rains are cyclonic and not affected by land relief; and there is a greater prevalence of convectional ("thunderstorm") summer rainfall in the east. The higher rainfalls of the west, particularly in winter, are not solely due to elevation of the land, for the sea-level west-coast stations show markedly higher records than those of the east coast, and the high percentage of winter rainfall in the west is not confined to high-level stations. This is supposed to be due to the friction exerted by the wind on the land surface, which checks the passage of the moving air, forming a cushion near the surface above which the wet winds tend to rise, cool, and thus precipitate their moisture.

Deviations from monthly and annual means. The rainfall of individual years may fluctuate widely from the mean values for a long series of years. A given year commonly shows marked monthly excesses at some stations and marked deficiencies at others. In any particular year some stations always show a deviation in one direction or another of as much as 40 per cent of the normal value. Departures of 100 per cent for individual months commonly occur at one station or another, and in extreme cases the deviation may be as great as 300, 400 or even 500 per cent. Nevertheless, such deviations from the average—both monthly and annual—are not so great or so numerous in the British Isles as in continental climates, and this is due to the predominance and constancy of the orographic rainfall. Correspondingly, the deviations from normal are more frequent in the east than in the west and generally less marked in Ireland as a whole than in central and eastern England (Salter, 1921).

In this sense then, in spite of their excessively changeable *weather*, the British Isles may fairly be said to have a more constant *climate* than many continental regions.

CLIMATE AND VEGETATION

There have been no very serious attempts, as already remarked, to correlate climatic factors quantitatively with the distribution of British natural and semi-natural vegetation, and it is difficult to say how far such attempts might be successful. It may be that the interactions and compensatory effects of the various factors are so complicated as to defy detailed analysis. But certain broad correspondences are obvious enough.

Climatic regions. We cannot construct climatic-vegetational "provinces" with anything like well-defined boundaries that are at all natural—there is too much gradual transition, and the specific effects of contrast between the western and the eastern and between the southern and the northern climates are too largely masked by the almost complete concomitant changes in the soil types. But if we eschew the attempt to assign quantitative values and refuse to tie ourselves down to boundary lines we may quite fairly recognise three or four regions in which both climate and vegetation are sufficiently distinct.

(1) **South-eastern region.** The south-eastern half of southern Britain (i.e. of England and Wales) *including* the Midland Plain, is on the average drier and warmer in summer than the rest, and in this district, especially

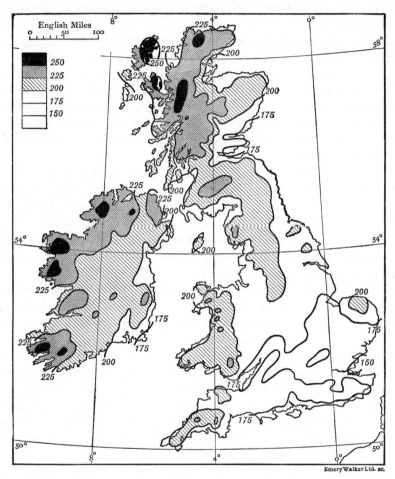

FIG. 24. MEAN RAIN DAYS PER ANNUM

The highest numbers of rain days are in areas near the west coasts of Scotland and Ireland which bear blanket bog vegetation, and are not always coincident with areas of highest rainfall (cf. Fig. 11, p. 32). From Bilham, 1938, Fig. 62, after Glasspoole, by kind permission.

towards the south, we find, on the whole, the finest development within the islands of the Central and West European deciduous summer forest—the best growth and the greatest number of species belonging to this plant formation. In the south-eastern counties themselves 11 per cent of the land is occupied by woodland as compared with 5·4 per cent for England as a whole (see Chapter IX, pp. 202, 203). This greater extent of woodland is not wholly, though it is partly, due to the large areas of sandy soil, which cannot be economically cultivated, in the south-east, and to the considerable extent of beechwood still remaining in the Chalk areas. There is a notable vigour of growth and regeneration in the woody vegetation of the south-east that is less marked in the north and west. In the south-east also (in the wide sense used above) the higher summer temperatures enable many exotic plants to flower and fruit, as for example in the Botanic Gardens at Kew and Cambridge, which cannot do so, or do so more rarely, in the cooler summers of the north and west; and for the same reason cereal crops ripen earlier in the south and east. Many continental species belonging to Central Europe occur in the south and east and not in the west.

On the basis of agriculture and pasture we might perhaps distinguish an eastern from a western subregion in the south-eastern region (taken in the wide sense used above), the former pre-eminent for wheat, the latter for pasture (cf. pp. 74–6).

Thus, though the British climate as a whole is rightly described as oceanic or maritime, there is a distinct, though not a very great, deviation in the direction of "continentality" in the eastern midlands of England and adjacent parts of East Anglia. Watt (1936, pp. 119–26) has some interesting remarks on the nature and limits of this tendency. It exists also, though to a less degree, in east-central Ireland, i.e. the district behind Dublin, and in the eastern Highlands of Scotland.

Breckland climate. In the east of England the Breckland of south-west Norfolk and north-west Suffolk (Figs. 5, 9 and Chapter XXVI, pp. 509–14), together with the adjacent Fenland, has the lowest rainfall of any inland area (average of seven stations in Breckland 591 mm., lowest mean 551 mm.), a high summer temperature and a relatively high percentage of sunshine, all factors making for dryness. The plant communities, as communities, are not specially distinctive, but they contain several species not found and others but rarely found elsewhere in Britain. Some of these are characteristic plants of the steppes of southern Russia, which here reach their farthest north-western limit (cf. Chapter XXVI, p. 512). The effect of the extreme climate on vegetation is reinforced by the coarse sandy soils.

Watt (1936) has pointed out that though Cambridge (the nearest station to Breckland for which the meteorological records are at all complete) has high summer mean and maximum temperatures and low winter means and minima (Figs. 25 and 26), as we have seen on p. 66, its late winter maxima are higher and its summer minima lower than those of the extremely oceanic Valencia Island (Fig. 26). The high late winter maxima together with the continuation of mean minima below freezing-point into the spring (up to and including May) means that snow never lies for long and that the soil, where unprotected by bracken litter or lichen cover, is exposed to frosts which may kill young bracken fronds, and even rhizomes to a depth of at least 10 cm., as late as the beginning of June. This is a small sample of the kind of analysis that is required on an extensive scale before we can begin to understand the actual relations of the climatic factors to vegetation.

(2) **Western region.** The greatest contrast to these south-eastern and eastern climatic conditions is seen in western Ireland and western Scotland. Here in the extreme Atlantic climate we have mild winters, cool summers and high air humidity—a climate which differs very little through more than 10° of latitude. The natural vegetation of flat and gently sloping impermeable soil is blanket bog, and only on rocky and broken well-drained

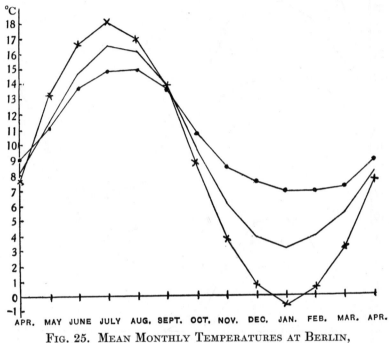

FIG. 25. MEAN MONTHLY TEMPERATURES AT BERLIN, CAMBRIDGE AND VALENCIA ISLAND

× — × — × Berlin ——— Cambridge —·—·— Valencia Island

This shows a typical continental temperature curve (Berlin), a typical oceanic curve (Valencia Island) and an intermediate curve (Cambridge). From Watt, 1936.

ground is heath and scrub developed, with oakwood (in the extreme north of Scotland birchwood) on valley slopes where it is well protected from wind. Blanket bog also occurs on the relatively impermeable soils of the plateaux of the higher hill masses away from the west coast. Here the climate is similarly wet, though distinctly colder.

Many species whose general distribution is in the western (Atlantic) parts of Europe occur in the west and not in the east of the British Isles, but a large proportion are found in the southern coastal counties of England whose climate as we have seen (pp. 60–1) is not far removed from the Atlantic type in its general mildness, though distinctly drier and sunnier and with a greater temperature range. Many such species are present in Ireland, but these are actually fewer than on the English south coast, probably because they have never reached Ireland on account of its relatively early isolation.

The south-west is notable for the successful cultivation in the open air

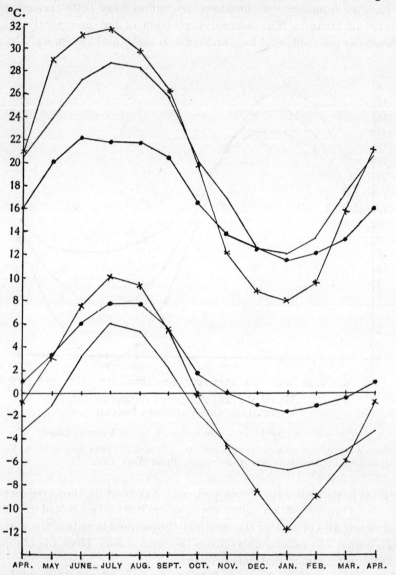

FIG. 26. MEAN MONTHLY MAXIMA AND MINIMA FOR THE SAME STATIONS
× — × — × Berlin ——— Cambridge —·—·— Valencia Island

The maxima for January, February and March are higher at Cambridge than at Valencia Island, and the Cambridge minima for April and May are below freezing point. From Watt, 1936.

of many subtropical evergreen plants, for which the gardens of South Devon, Cornwall and Scilly, and of Co. Cork and Co. Kerry in Ireland, are famous. Here the climate is very equable and the winters are the mildest in the country, frost being a negligible factor. The Cornish winter has been called "a languid spring". To a less extent these characteristics hold for the other coasts, particularly for the west coast of Wales, the south coast of England, and also for the western coast of Scotland, and least for the east and north. Even the east coast, however, as we have seen, has a definitely maritime climate, and many species of exotic trees and shrubs flourish better than inland.

(3) **Northern region.** The northern climate is marked by a lower range of temperatures at all times of year, and the late cool summer is the leading effective factor in limiting the vegetation. Most of the centre and north of Scotland is occupied by the mass of the Scottish Highlands, of considerable average elevation, and this of course reinforces the effect of the comparatively high latitude. The floors of many of the larger glens, however, are cut down to quite low altitudes, almost to sea level. Cloud and mist are very prevalent. The vegetation of this region may fairly be considered to belong largely to the northern coniferous forest formation rather than to that of the deciduous summer forest (see Chapter IX), but the two interdigitate extensively, oakwood occurring in the bottoms of the sheltered valleys, birch and pinewood higher up the slopes. There is plenty of seminatural oakwood in the valleys of the southern Highlands, but north of the pass of Drumochter oak is seen very much less, save in specially favourable localities, such as are numerous, for example, round the Moray, Cromarty and Dornoch Firths, along the bottom of Glenmore, and in the lower parts of the larger tributary glens. North of the 58th parallel of latitude oak plays no part in the landscape. The sparse natural woods of the hill sides are nearly all dominated by birch with very few associated trees and shrubs, especially in the far north. In the east and centre, particularly, there still exist forests of native pine—in some places ascending to 2000 ft. (*c.* 600 m.)—which was at one time much more abundant and widespread. Planted exotic conifers make extremely fine growth on deep soil in some of the sheltered valleys. A few remnants of native pinewood, at one time well grown and extensive, still exist in the north-western Highlands (cf. Chapter XXII). Heath and heather moor are particularly characteristic of the drier and more continental eastern Highlands, and the surviving native pinewoods of this region are typically associated with these Calluneta.

The bulk of the British "arctic-alpine" vegetation, mainly developed above 3000 ft. (*c.* 900 m.) with its characteristic extreme climate and correlated soil conditions, occurs in this Scottish Highland region (see Chapters XXXVIII–XXXIX). Exposure to strong winds, high humidity, intense erosion, and severe winters with long snow lie are the effective climatic factors. The western coastal strip of the Highlands, together with the western islands at much lower altitudes, lies in the region of extreme Atlantic climate, and here we may have a mixture of arctic-alpine and Atlantic vegetation.

Within each of the three climatic regions, south-eastern, western and northern, that we have distinguished there are of course many minor variations, and the transition from one region to another is for the most part quite gradual, except where it is reinforced by an abrupt change of soil or of relief.

CLIMATE AND AGRICULTURE

The same considerations hold in the relation of climate to agriculture, but here greatly modified not only by the varying soils but also by changing markets and other economic and political conditions. Owing to the great and sudden changes in the demands on the food supply during the war, and the rapidly altering economic conditions of the post-war period, agriculture is still far from anything like a position of equilibrium. It has therefore been thought best to base the following account on the pre-war conditions, and it has been reproduced from *Types of British Vegetation* (pp. 27–31) with little alteration.

Grassland. The mild climate of the British Isles with its comparatively cool summers and well-distributed rainfall is particularly well suited to grass (see Chapter xxv), and indeed the country is (or was) deservedly famous for the richness and beauty of its pastures. Direct results of this were the flourishing condition of the pastoral industries, the great wool trade of the Middle Ages, and the fine quality of British beef and mutton. The damper climate and milder winters of western England and of Ireland are the best suited to grass growing, and it is here as well as in the western midlands and, owing to favourable soil, in some areas further east (Leicestershire), as well as in parts of the north, that the bulk of the pasture is found and the cattle-raising industry particularly flourishes. In several of the western and midland counties there were in 1910 more than twenty head of cattle, and in Ireland twenty-three, to the hundred acres of land of all kinds. Sheep can stand more exposure than cattle, and the great sheep regions were the hill-side pastures, such as the Southern Uplands of Scotland, the hill sides of Wales and the Chalk Downs of southern England. In these areas there were from 50 to 100 sheep to every 100 acres of land of all kinds (1910).

In some of the midland and western midland counties such as Leicestershire, Warwickshire, Shropshire, Worcestershire and Gloucestershire, the area of permanent pastures exceeded (in 1910) half the total area of the county and was more than double the total area of arable land. In Somerset (south-west) the permanent pasture occupied about two-thirds of the area of the county and was more than four times the area of arable land.

Arable land. These figures contrast very strikingly with those for the eastern counties, e.g. Cambridgeshire, where the proportions were reversed, the arable land exceeding two-thirds of the whole area and being more than three times that of the permanent pasture.[1] In Norfolk, Suffolk, Essex, Huntingdonshire and Lincolnshire the area of arable land exceeded, and sometimes considerably exceeded, half the total area of the county, while it approached or exceeded twice that of the permanent grass.

Wheatland. This eastern district with the large proportion of arable land is also the region where most wheat is grown, the area under wheat in any given year varying from one-sixth to one-quarter of the total area under arable crops, since it was generally grown in the standard "four-course" rotation. Eight of these eastern counties, including Lincolnshire, provided about 42 per cent of the total area of wheat in Great Britain, while the total of their land areas is only 11 or 12 per cent of the whole island.

Considerably more than half of the rest of the wheat area, i.e. about 33 per cent of the whole, was found in the eastern midland, the western midland and the south-eastern groups of counties, whose total land areas make up between 19 and 20 per cent of the land area of the island. Here again the average area under wheat in a given year was about one-sixth or one-fifth of the total area under arable crops, but, as already pointed out, the total arable area was very much smaller proportionally than in the eastern counties. A great deal more wheat used to be grown in the midland and south-eastern counties, but with the great decline in prices which occurred between 1875 and 1895 this region suffered a more severe reduction in its wheat area, as in its total arable area, than the eastern counties.[2] Though not perhaps quite so favourable for wheat, the climate of much of this region does not differ very markedly from that of the eastern counties (see data on pp. 62, 66, 67). The midlands, however, contain a large proportion of heavy clay land, and this "strong soil", though quite good for wheat, and at one time considered as typical wheatland, requires a great deal of labour and does not pay to plough when the price of wheat is low. Consequently much of this land has been gradually laid down to permanent pasture. The south-eastern counties, so far as they remain rural, are now largely devoted to special crops such as fruit and hops.

Most of the 25 per cent of wheatland remaining to be accounted for occurs in the north and north-east of England, while the west and north of Scotland show areas which are practically negligible, the Orkneys and

---

[1] If the small patches of permanent pasture round the villages, which it is necessary to maintain for the grazing of farm animals and of cows for the local supply of milk, etc., were subtracted, the total would be very greatly reduced.

[2] The decrease in the area under wheat between 1875 and 1895, which was 42·75 per cent in the eastern and 52·56 per cent in the north-eastern counties, was nearly 60 per cent in the south-east and over 60 per cent in the midlands, though the wheat area was much smaller in the latter regions than in the former at the beginning of this period.

Shetlands, Caithness and Sutherland (extreme north), Argyll and Bute (extreme west), as well as some of the counties situated in the Southern Uplands, growing no wheat at all. According to W. G. Smith (1904, p. 627) "wheat ceases to be a regular crop of the farm at an altitude where the mean July temperature is below 56° F. (13·3° C.) and the rainfall exceeds 32–34 in. (about 800–850 mm.) per annum".

The total amount of wheat grown in Ireland is insignificant, but the province of Leinster, containing the driest part of the country and having the warmest summers, possessed the largest acreage.

It is thus seen very clearly that most of the British wheat is grown in the regions of low rainfall and comparatively warm summers, while the regions of high rainfall and the coolest summers grow no wheat at all.

Barley. Barley, of which the total British acreage (in 1910) was rather below that of wheat, is not so dependent on climatic conditions, but a large proportion is grown in the wheat-producing counties of England, though Wales and the east of Scotland had a good deal more barley than wheat, and no Scottish county showed absolutely none.

Oats. Oats, the third great cereal crop, mature well in far damper and cooler summers than either wheat or barley. In some years the acreage in Great Britain has been nearly double that of wheat, but the proportions have varied a good deal. The distribution of the crop is very different. While the eastern counties grow considerable quantities of oats, the acreage in 1910 was well below that of either barley or wheat. In the south-west, on the other hand, the acreage under oats approximately equalled that under wheat and barley together, while in the north-west it was nearly three times as great. In Wales and south-eastern Scotland it was nearly twice as great, in eastern and central Scotland more than three times as great in 1910, while in northern and western Scotland the oat crop, though absolutely smaller than in the east and centre, was and is the only important cereal crop. Its limits are only fixed at the higher altitudes in the north by the early autumn frosts, which kill the plant before the grain is mature.

In Ireland the crop of oats enormously exceeds all the other cereal crops put together. The Irish acreage under potatoes is also very large, greatly exceeding the acreage of all the cereals together, except oats.

The rye crop is quite insignificant compared with the other cereals, though a little was grown in nearly every county except those of the extreme north of Scotland.

Sugar beet. Sugar beet is entirely a post-war and a heavily subsidised crop. In 1936, 348,659 acres (c. 141,077 hectares) in England and Wales were under sugar beet, a very slight increase from 1930. About a quarter of the whole acreage was in Norfolk, and much more than half in that county taken together with parts of Suffolk, Cambridgeshire and Lincolnshire. The Fenland, with its relatively hot summers, accounts for a large proportion of this essentially "continental" crop. Though some is grown as far west as Pembrokeshire in south-western Wales, the acreages in most of the northern and western counties are insignificant.

REFERENCES

BILHAM, E. G. *The Climate of the British Isles.* London, 1938.
Book of Normals, The. Sections IV, IV*b*, V, Meteorological Office (Air Ministry), 1923–4.
BUCHAN, A. and OMOND, R. T. The meteorology of the Ben Nevis observatories. Part III. *Trans. Roy. Soc. Edinb.* **43**, 1905.
SALISBURY, E. J. The East Anglian flora. *Trans. Norfolk Norw. Nat. Soc.* 1932.
SALTER, M. DE CARLE. *The Rainfall of the British Isles.* London, 1921.
SMITH, W. G. Botanical survey of Scotland. Parts III and IV. *Scot. Geogr. Mag.* 1904.
WATT, A. S. Studies in the ecology of Breckland. I. Climate, soil and vegetation. *J. Ecol.* **24**, 1936

Chapter IV

SOIL

Nature of soil. Soil may be defined as the superficial "skin" covering the earth's crust and directly affected, indeed created, by surface agencies—primarily by the "weathering" agencies, heat, frost, wind and above all water, and secondly by the vegetation which grows upon it. For the student of vegetation it is the surface material in which plants grow—and this does not always correspond exactly with the soil of the pedologist—containing the underground parts of the higher plants as well as algae, fungi and bacteria. Besides the plants the soil of course also possesses a fauna, which may be very rich and varied in soils where the general life conditions are favourable, ranging from Protozoa to different kinds of worms and insects, and including also small mammals such as moles and mice which spend a great deal of their time beneath its surface. Surrounded by the mineral and the dead organic constituents of soil, and the air and water it contains, these various living organisms form a veritable microcosm, in which chemical, physical, and biological changes are constantly going on; and the different elements of this microcosm, especially in a mature and stable soil which has arrived at a state of approximate equilibrium, show the most complex system of mutual relationships.

During the past quarter of a century our knowledge of the soil as a separate subject of study has increased enormously, so that a new and flourishing branch of science—soil science or *pedology* as it is called—has come into existence. But we are still far from anything like a thorough knowledge of the structure, processes and general economy of the soil microcosm.

Stratification. A fundamental feature of all natural soils is their *stratification*, i.e. the existence of a number of more or less easily recognisable and definitely characterised superposed layers extending downwards from the surface to the *subsoil* or unaltered parent rock below. The formation of these layers is directly due, of course, to the surface agencies mentioned, which have a maximum effect at the surface, and to the differentiation set up by the processes they initiate. The term *profile* is applied by pedologists to the whole series of these layers, as they can be seen on the vertical sides of a pit dug through the soil down to the subsoil. The term *horizon* is used for the chief divisions of the profile, and *zone* for the subdivisions of the horizons.

Three main factors may be said to go to the making of a soil: first, disintegrated "rock", in the geological sense, which is usually the parent material; secondly, vegetation; and last but not least, climate, i.e. the sum of all the particular aerial weathering agencies which act on the soils of a particular geographical region year after year.

Climatic soil types (world groups). The recognition of the preponderant effect of climate in differentiating soil types has been one of the chief advances of modern pedology. It has been shown that in fact particular types of soil are characteristic of particular climates, and that under the continuous action of a given climate very different original rocks will tend to produce the same kind of soil, while in widely differing climates the same kind of rock will form very different soils. These climatic soil types (generally known to pedologists as *world groups*) are not of course uniform and constant throughout a climatic region. They vary with the parent material provided—some rocks can never produce the proper climatic soil type—with steepness of slope, with smoothness or irregularity of surface, and especially with drainage. But nevertheless the *tendency* of a given climate to produce a given type of soil is quite unmistakable.

A knowledge of the climatic types (or world groups) of soil is of importance to the student of vegetation because the great climatic vegetational types of the world are also determined by climate and hence correspond, though only in a very general way, with the soil types. And the kind of vegetation borne by a soil has a direct reciprocal effect on the soil itself.

Action of water and heat. The reason why climate is of such paramount importance in determining soil type is that the two leading factors in climate, water and heat, separately and conjointly, are also two of the leading factors in the processes that create soils. Apart from the activity of water, as rain or as streams and rivers, in eroding rock and soil, transporting the products and laying down "rainwash" or alluvium—thus preparing the material for the formation of new soils—rain water percolating through the surface layers of soil is always carrying down soluble substances, and also the finer, more easily moved insoluble particles, to lower levels. Thus the surface layers of permeable soils are *leached* and eluviated so that they become coarse-grained and poor in soluble salts, while the lower layers are correspondingly enriched. In climates with very low rainfall this process is at a minimum or in abeyance, so that the surface layers produced from the physical disintegration of the parent rock remain relatively unaltered.

Heat accelerates all chemical processes in the soil, so that these are carried out more quickly in hot climates. And heat also accelerates the evaporation of water, while hot air can hold much more water vapour than cool air. Thus with a given rainfall the water which falls evaporates much faster from the soil and from the vegetation in hot climates, so that there is correspondingly less to percolate through the soil. The ratio of precipitation to evaporation (P/E) is, as we saw in Chapter II, an important index of climate because it integrates the effects of water and heat, so far as the latter affects the evaporation of the water of precipitation. Where this ratio is high, particularly where it is above unity (or over 100 per cent as generally written), leaching and eluviation will be vigorous; where it is low, feeble or non-existent. Indeed, when evaporation exceeds precipitation an upward movement of water, bringing soluble salts with it, will occur through the capillary spaces of the surface layers (though only through a shallow horizontal zone) and these will be enriched. Alternating heavy precipitation and vigorous evaporation thus act on the soil in opposite ways and tend to maintain the stability of its salt content and distribution.

Mineral materials. Disintegrated, i.e. weathered, rock (in the common meaning of hard rock) and the softer sediments (also "rocks" in the geological sense)—clays, sands, marls, etc., the materials of which were originally derived by the weathering of igneous rocks, afterwards transported and deposited by water and at a later date again weathered ("secondary weathering")—form the primary raw material of soil and at first determine its nature almost entirely. Thus we speak of clay soils, silt soils, sandy soils, etc., and we can in fact classify many of the soils of a country like England, for example, by reference to their texture, directly traceable to the structure of the rocks from which they are derived (see p. 96). And the differences in texture have great importance for vegetation because they have primary effects on the water economy of soils and are also often correlated with their chemical properties. The analysis of the mineral part of a soil into "fractions" of different sizes, the different proportions of which determine its texture and its designation as a clay, silt, loam or sand, is called "mechanical analysis". It is carried out (after rejection of the stones and coarse gravel) by drying the "fine soil" at 100° C., burning away the organic matter at a red heat, pounding the residue in a mortar, and then washing it through a series of sieves of different standard mesh to separate the coarser particles, while the finer are separated by the different rates at which they sink in water. In the new international terminology the *coarse sand* fraction comprises particles between 2 and 0·2 mm. in diameter, the *fine sand* between 0·2 and 0·02 mm., the *silt* between 0·02 and 0·002 mm., while the *clay* consists of particles below 0·002 mm. (2μ). All natural mineral soils consist of mixtures of particles of different sizes, but according to the preponderance of one or other of these fractions a given soil is known as a coarse sand, a fine sand, a silt or a clay. A *loam* is a soil with a good mixture, favourable to plant life, of the different fractions (see pp. 97–8).

Soils derived from limestone rocks form a class apart, because the main material of the rock (calcium carbonate) is very soluble in rain water in which carbon dioxide is dissolved, and only in impure limestones is there present any substantial proportion of the insoluble or very slightly soluble mineral material of which most rocks largely consist and which forms the mineral residue or skeleton of an ordinary soil.

This mineral basis or residue consists of course of materials directly derived by weathering from the original igneous rocks, or more indirectly by the weathering of "sedimentary" rocks, themselves formed of the materials from the original igneous rocks, and transported by water or other agency to a new position. The main rock-forming minerals are the *felspars, hornblende, augite,* and *mica,* composed of complex aluminosilicates in which iron is very commonly present and into whose composition sodium, potassium, calcium, magnesium and other elements may enter; and *quartz,* which is pure silica (crystalline silicon oxide).

Chemical weathering. Chemical decomposition of these original minerals is largely hydrolytic, and occurs chiefly under acid or alkaline conditions when the rate of decomposition is accelerated by the higher concentration in the soil water of hydrogen or hydroxyl ions respectively.

The complex silicates are broken down, and calcium, magnesium, potassium and sodium are removed in solution. At the same time a portion of the silica is separated from the complex silicate. The residue remaining is still essentially an alumino-silicate, which now contains a much lower silica content than the original mineral, and is at least partially crystalline, although it may vary very considerably in chemical composition and in physical properties, some of it forming the very fine particles of colloidal clay. This, together with the colloidal humus (organic) material derived from the decomposition of vegetation, forms a complex of which the components are so intimately associated that it is extremely difficult to separate them without materially affecting their physical and chemical properties. *This complex is the reactive portion of the soil.* It is often called the "weathering complex", and has the property of adsorbing basic ions from solution with the result that its composition and properties depend on the concentration of the ions in the solution with which it is in contact.

Function of calcium. Normally in this country calcium is the dominant metallic ion and when present in quantity it imparts a physical and chemical stability to the whole weathering complex, aggregating the fine colloidal particles into *compound particles,* thus giving a granular or "crumb" structure to the very fine-grained soils which, without calcium, would be unfavourable to many forms of life.

Action of hydrogen ions. The most active agents in chemical change are the free hydrogen ions which are derived both from soil water with CO_2 in solution (weak carbonic acid, H_2CO_3) and from the organic acids of the humus, and which, when present in excess, render the soil acid, and promote hydrolysis. In "neutral" soil water in which the numbers of hydrogen and hydroxyl ions are about equal, the water contains approximately 10^{-7} g. per litre of free hydrogen ions; in slightly acid soils 10^{-6}, and so to the very acid soils which contain 10^{-4} to 10^{-3}, or in extreme samples even more. These negative indices are used as a measure of acidity (pH). Thus,

pH value $\qquad$ pH 7 marks a neutral, pH 3 a very acid soil, while a soil of pH 8 (in which there is an excess of hydroxyl ions) is distinctly alkaline. It is to be noted that a diminution of one integer in the pH value means a tenfold increase in free hydrogen ions. The pH value may be determined colorimetrically or electrometrically. Most British soils, except limestone soils saturated with calcium carbonate, those derived from certain basic igneous rocks, and saline soils, are more or less acid, even though the "rock" from which they are derived is alkaline. Percolating water containing excess of hydrogen ions is active in leaching out salts, especially calcium, from the upper layers and carrying the bases away. The results of this process are particularly marked in the north and west where the rainfall is heavy, and the rocks are mostly poor in bases to begin with. According to Pearsall the average pH value of northern English soils under natural vegetation is about 5.

Thus calcium and hydrogen are the two great antagonistic chemical agents at work in the soil.

Soil reaction may have marked effects on the vegetation. Most if not all species of plants have what is known as a "pH range", outside which they do not normally occur, and very often a narrower range within which they grow most freely and luxuriantly. For the great majority of species the range extends across the neutral point, usually for a greater distance on the acid than on the alkaline side. This corresponds with the fact that the majority of soils, except in very dry climates, are more or less acid. There are, however, some species, of which most of the bog mosses (Sphagna) are marked examples, which cannot tolerate any but acid soils—alkaline solutions poison them. These plants are commonly known as "oxyphilous" or "calcifuge" (properly "basifuge"), because calcium is by far the most abundant basic ion in most soils. Other species are "oxyphobe", i.e. unable to tolerate any considerable degree of soil acidity. There can be little doubt that excess of free hydrogen ions has a deleterious effect on their roots, but the ways in which different ions affect plants are still very obscure. Other factors, such as water content and temperature, besides the presence of pairs of "antagonistic ions" which balance one another, often profoundly modify the response of plants to soil reaction. Pearsall (1938) has recently pointed out that the "oxidation-reduction potential" of the soil is probably of first importance in its effect upon vegetation, particularly because it conditions the formation of nitrates, and this is by no means always directly correlated with pH value, though on the whole when it is low the pH value is also low.

A good supply of "exchangeable bases" (i.e. chemically active basic ions), particularly calcium, is one of the primary conditions of a "good" soil, i.e. a soil favourable to plants apart from the extreme oxyphilous species. But such soils may nevertheless have a relatively low pH value. Thus much of the clay-with-flints of the plateau of the Chiltern Hills (Chapter XVIII, p. 365, Fig. 70 and Chapter XIX, pp. 386–7) is rich in exchangeable calcium and supports luxuriant varied vegetation, though it is decidedly acid in reaction (pH 4–5). On the other hand the soils of the podsol group and allied types are characteristically highly acid, very poor in bases, and bear quite distinctive specialised vegetation.

On the whole the pH value (which is easily determined) is one of the most useful indices of the potentialities of a soil for vegetation, because the highly acid soils are unfavourable to most plants, supporting nothing but specialised types, and the same is true of highly alkaline soils. Between these two extremes many species have distinctive pH ranges. But soil reaction is by no means the only factor to be considered.

Organic materials. Besides the mineral basis the other constituent that goes to the making of a soil is the organic material mainly derived from plant debris—dead roots and rhizomes and the surface litter of dead leaves and shoots—much of which becomes incorporated with the mineral constituents and gradually disintegrates and decays. This process is brought about partly by the action of earthworms (which inhabit in abundance the soils most favourable to animal life, drag down dead leaves into their burrows and pass them through their bodies, evacuating "worm casts" of amorphous organic material), partly by other invertebrate soil animals and by soil fungi; but mainly, especially in the later stages of decay, by the successive attacks of different kinds of soil bacteria, which ultimately reduce the plant debris to simple substances—carbon dioxide, water, and soluble salts, nitrates, phosphates and sulphates of the various bases, thus reproducing the materials which the plants originally used as the sources of their nutriment.

Humus. To this organic substance in process of disintegration and decay the name *humus* is given. In the soils most favourable to plant and animal life the turn-over of humus is rapid. The materials from which it is formed are supplied in quantity (in deciduous woodland mainly from the litter produced by the annual leaf fall) and broken down rapidly. This quick turn-over is associated with moderate or high temperature, adequate but not excessive moisture, good aeration and an abundant supply of basic ions, though the soil reaction may be moderately acid. Humus of this type is known as *mild humus* or *mull*. In opposite soil types unfavourable to most plant and animal life and inhabited by specialised plants such as members of the heath family, the decay of humus is very slow, or is indefinitely arrested after it has reached a certain stage, while the original plant structure may be still largely preserved. Basic ions (electrolytes) are in very short supply and the soil reaction is extremely acid. This type of humus, sometimes called "raw humus", is known by the Danish name *mor* (now adopted as an international term). The distinction between extreme types of *mull* and *mor* has long been known, since they bear quite different kinds of vegetation. Recent research (Romell, 1932, 1935; Pearsall, 1938) has made it probable that they represent quite different physico-chemical and biological systems. Mull is inhabited by earthworms, which play an important part in the initial breakdown of the plant debris, while earthworms are absent from mor. Mull contains large numbers of

Mull (mild humus)

Mor (raw humus)

aerobic bacteria whose gelatinous capsules form a significant part of the humus colloids (see below), while mor contains predominantly fungal hyphae. Among the bacteria of mull are nitrifying (nitrate-producing) organisms, which are absent from mor. The soil faunas are also quite different. "Romell suggests that mull represents a delicate dynamic equilibrium in which the soil fauna plays an important part in destroying the soil fungi. Mor tends to be self-stable unless disturbed" (Pearsall, 1938, p. 199). It used to be thought that mor was always deficient in oxygen, but it has been shown that this is not so, except under waterlogged conditions. The outstanding chemical differences are the content in basic ions and the degree of acidity. There is evidence that pH 3·8 is a critical limit between the two types.

Peat. When humus is formed under waterlogged conditions it accumulates indefinitely and forms a pure organic soil (*peat*, or "turf") above the mineral substratum. *Fen peat* is formed where the ground water drains from, or is in contact with, a limestone or marl, and thus contains abundance of calcium which neutralises the organic acids and may render the reaction distinctly alkaline. Abundant irrigation with water rich in bases always leads to the formation of fen peat (see Chapter XXXII). *Acid peat* (*moor, moss* or *bog peat*) is formed where basic ions are in very short supply (see Chapter XXXIV). The common feature of the two types is the deficiency of oxygen.

Functions of humus. The humus of the soil, like the clay, forms a colloid complex, but it has more of the character of an emulsoid (i.e. of a colloid in which the particles—disperse phase—are liquid) than colloid clay: it has very high water-holding and adsorptive capacity and requires a higher concentration of electrolytes to coagulate it. Primarily it shows a high degree of acidity, owing to the organic acids formed during decomposition, unless the tissues from which it is derived have a high base content. The humus derived from conifer needles is much more acid than that formed by the leaves of most broad-leaved trees, and tends to form mor. Thus the pure humus of spruce has a pH value of 3·8, of pine 4·5, of beech 6·6, and of dog's mercury 7·4. This is one cause of the prevalent formation of mor from conifer litter. But mor is also formed from other kinds of litter when the organic acids are not neutralised by basic ions derived from the mineral soil, where this is very poor in bases, or where excessive leaching washes down the basic salts. Thus mor is characteristically formed in a wet cool climate where P/E is high. This kind of humus is essentially mobile, because the excess of hydrogen ions promotes hydrolysis and thus leads to humus and other substances being carried down to considerable depths, and also to the brown colour of streams where they drain from areas of acid peat. On the other hand, when calcium ions are dominant in the soil under fairly dry conditions the humus is relatively immobile to water, though in a state of constant chemical change.

British Soil Types

Mild humus or mull is thus an extremely important constituent of "good" soils. Together with the clay fraction it is the seat of the active chemical processes of the "weathering complex", the home of the soil flora and fauna, and the centre, so to speak, of the complex biochemical processes of the soil microcosm. Owing to its constitution and continual disintegration it is also the most important immediate source of food supply to the higher vegetation.

The humus content of different soils varies from less than 1 per cent, as in raw mineral soils as yet unoccupied by vegetation, to almost 100 per cent in pure peat. In good forest soils there is often about 15–20 per cent, in meadow soils much more, owing to the number of fine roots penetrating the soil very thoroughly and giving rise to humus when they die.

Formation of mature soil. On the soil complex formed by these two constituents (inorganic and organic) the climatic factors gradually work, moulding the soil into its ultimate form. A well-defined climatic soil type cannot, however, be produced unless there is a sufficient variety of original constituents to be acted upon and sorted out. Thus a very pure limestone ($CaCO_3$) or quartzite (SiO_2) cannot form a typical climatic soil because of the poverty of ingredients. Again, a mature climatic soil can only be formed on flat or gently sloping ground, and where the soil is not permanently waterlogged. On steep slopes, erosion and the "run-off" of rain water prevent the process of soil development proceeding beyond the initial stages, because the first products are constantly being removed, so that the soil remains permanently immature. In waterlogged soils the absence of percolation and of free oxygen also prevents the ordinary soil-forming forces from having free play and produces peculiar conditions and specialised vegetation. It is only the *mature soils*, on which the climatic factors have been able to work their full effect, that exhibit the characteristic features of the great world groups.

British soil types. The following are the most important British soil types:

Climatic	Depending on special water or subsoil relations
Brown Earth	Meadow soil
	Fen peat
Podsol	Raised bog- or moss-peat
Blanket bog- or moss-peat	
	Rendzina

The three main climatic soil types in the British Isles are known respectively as *Brown Earths* (or *Brown Forest-soils*), *Podsols*, and *Blanket bog peats*. The other well-marked soil types (second column) are dependent on other conditions such as local ground water or the nature of the parent rock.

Brown Earths. The Brown Earths are characteristic of the climatic region of which the climax vegetation is Deciduous Summer Forest, i.e.

the temperate suboceanic region of moderate, well-distributed rainfall, usually from 500 to 1000 mm. in the year (say 20–40 in.), fairly cold winters and summers of moderate warmth. This is the climate of the central portion of the eastern United States and of most of western and central Europe. The southern part of Britain (except the extreme south-west), the midlands, and eastern Ireland, as we have seen in Chapter III, have this kind of climate. The most typical Brown Earths are formed from clays and loams, but they may be formed from any rock which contains a sufficient variety of mineral elements, e.g. impure sandstones and limestones and certain metamorphic rocks.

The profile of a typical Brown Earth (Fig. 27) is free from calcium carbonate, any present in the original rock being leached out by percolating rain water, since in the "Brown Earth climate" the P/E ratio is sufficiently high. Other easily soluble salts, such as those of potassium and magnesium, are also often partially leached from the surface layers and tend to accumulate in the lower layers, together with finer clay particles. Nevertheless, the "base status" of the surface layers of a typical Brown Earth is relatively high, i.e. there is a good supply of basic ions including calcium, though not of calcium carbonate. The leached upper layers of a soil are called collectively the "eluvial" or "A" horizon, while the lower layers, in which the substances carried down accumulate, form the "illuvial" or "B" horizon. Owing to the partial washing out of bases and formation of humus in the upper layers the A horizon is normally more acid than the B horizon, though the whole profile is usually somewhat acid in most Brown Earths, even when the parent rock is alkaline in reaction. There is in fact usually a gradient of decreasing acidity and increasing salt content as we penetrate to lower levels. But sometimes the surface layers have a higher pH and a higher salt content, owing apparently to the carrying up of salts, either by the plants themselves or by capillary rise of water. The conditions which determine this inversion of the normal leaching gradient are not, however, fully understood. The humus is mild, and well incorporated with the upper layers of the mineral soil, which show "crumb structure", composed of aggregations of the finer particles cemented by a mixture of humus and clay colloids, forming the so-called "weathering complex" in which all the active chemical processes (as of "base exchange") take place (see p. 81). The A horizon is relatively well aerated and inhabited by a rich microflora and fauna, including earthworms. Below the B horizon comes the so-called *gley* horizon (G), not shown in Fig. 27, which is grey or grey-green in colour, streaked, speckled or blotched with red-brown, an effect due to the alternation of patches containing ferrous and ferric salts. The gley horizon is due to the presence of water below (held in the parent clay of a typical Brown Earth). In wet weather there is a downward movement of percolating rain water through the profile into the clay reservoir below, in dry weather an upward movement *from* the clay. The alternate wetting and drying, leading to decreased and increased aeration respectively, gives the alternation of reduced and oxidised iron salts. Below the gley comes C, the parent "rock" or subsoil, here most typically a clay or clay loam.

Thus the Brown Earth profile has two wet zones, one in A where water is held by the humus and one between G and C; and the alternate movements of water between these render the profile stable, maintain its base status

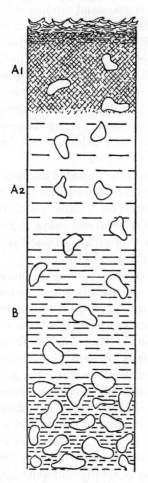

FIG. 27. BROWN EARTH PROFILE

This section is taken from the Chiltern plateau, where it bears beechwood, the parent material being Clay-with-flints. A, Eluvial horizon. (A1, surface layers stained with humus, A2, lighter). B, Illuvial layer not sharply marked but showing increasing fineness of texture downwards. After Watt, 1934.

and prevent podsolisation (see below), which is associated with a one-way (downward) movement of water.

The name "Brown Earth" is taken from the brown or reddish brown colour imparted to the soil by the hydrated ferric oxide derived from the partial decomposition of the original iron-containing complex aluminosilicates, but in the upper layers this may be more or less masked by the blackish brown of the humus.

Most of the good agricultural land of Great Britain is of the Brown Earth type, but ploughing of course destroys stratification and tends to make the soil homogeneous to the depth which the plough reaches. Manuring increases the base content, especially of the surface layers. The typical Brown Earth profile must be sought in natural or semi-natural woodland where the ground has long been undisturbed.

Meadow soils. Meadow soils occupy flat ground, with the fluctuating water table never very far below the surface. Typically this ground coincides with the alluvium of river valleys and is grassland traditionally cut for hay, but now often used as pasture. This riverside alluvium is subject to winter flooding, which brings fresh silt to its surface and maintains its fertility. The soils are in fact often rich in bases and very fertile. The flooding of the alluvial grasslands is now usually mitigated or prevented by dredging of the river beds, regulation of the water flow by sluices, and the cutting of drainage ditches, but the ground water does not lie at any great depth. With a persistently high water table marsh species increase in abundance, but the true meadow soils are neither acid nor deficient in oxygen, owing to the constant movement of the water.

The great characteristic of true meadow soils is the impeded drainage and the fluctuating water table. This leads to impeded leaching and the prominent development of the gley horizon. The abundant humus is not carried down but remains near the surface, while the mineral soil below is grey or bluish grey in colour. If this kind of soil is thoroughly drained so that leaching begins, it tends to pass over to the Brown Earth type.

Fen peat. When soil is constantly waterlogged oxygen is permanently deficient, and in the absence of silting humus consequently accumulates and does not disintegrate, forming a pure organic soil, or *peat*. If the water has drained calcareous or other basic rocks typical fen peat is formed, rich in bases, with a pH value between 7 and 8, as in the East Anglian fens, and bearing typical fen vegetation, though some fens are neutral or somewhat acid (see Chapters XXXII, XXXIII). When the surface rises above the groundwater level by continuous accumulation of plant debris, aeration increases and a greater variety of species can flourish. If the climate is sufficiently humid the peat, when it has risen above the influence of the ground water, is no longer neutralised by basic salts, develops acidity, and is colonised by an oxyphilous vegetation, which may form the starting point of a "raised bog" (see Chapters XXXIV, XXXV) producing a totally different kind of peat (pp. 92–3).

Podsols. The climatic soil type called *podsol* (Fig. 28), a Russian word meaning "ash", from the ashen colour and friable consistency which often mark the strongly leached soil of the A horizon, is characteristic of the cooler and wetter climate of the northern and western parts of the British Isles, as it is of north-west Europe generally. The P/E ratio is markedly higher than in the Brown Earth climate, and consequently the downward movement of water through the soil is on the average much stronger than the upward, and leaching is extreme, washing basic salts out of the surface layers. The prevailing lower temperatures check the disintegration of the plant debris, and since the acids produced are not neutralised, a layer of highly acid raw humus (mor), called the $A\,0$ horizon, tends to form on the surface of the soil. Below this the A horizon, often white or grey in colour and thus known as the "bleached layer", sometimes stained chocolate with the humus compounds carried down by the acidified rain water, is also very acid, and almost destitute of basic ions. The B horizon of a typical

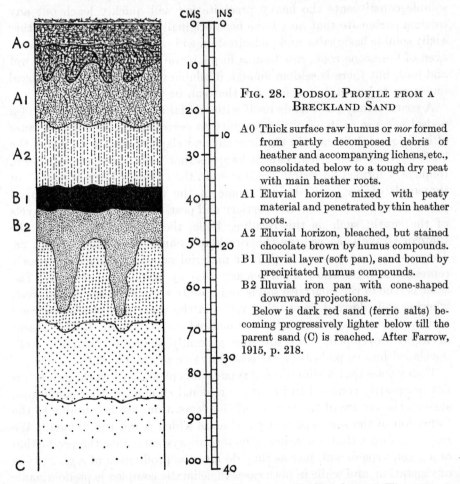

Fig. 28. Podsol Profile from a Breckland Sand

A0 Thick surface raw humus or *mor* formed from partly decomposed debris of heather and accompanying lichens, etc., consolidated below to a tough dry peat with main heather roots.

A1 Eluvial horizon mixed with peaty material and penetrated by thin heather roots.

A2 Eluvial horizon, bleached, but stained chocolate brown by humus compounds.

B1 Illuvial layer (soft pan), sand bound by precipitated humus compounds.

B2 Illuvial iron pan with cone-shaped downward projections.

Below is dark red sand (ferric salts) becoming progressively lighter below till the parent sand (C) is reached. After Farrow, 1915, p. 218.

podsol is sharply marked and often shows two distinct layers (Fig. 28), an upper in which the humus substances brought down from above are precipitated, and a lower in which the iron salts are thrown down. The former ($B1$) is generally dark chocolate brown or nearly black in colour, the latter ($B2$) reddish brown. Either or both may become very hard, forming the so-called *moorpan* or "hard pan", which is occasionally thick and resistant enough to stop the downward growth of tree roots, preventing their penetration to the richer subsoil, and thus confining them to the impoverished A horizon. Below the B horizon is often a well-marked gley zone, though upward movement of water from this is feeble or absent.

Podsols may be formed from any parent material except limestone and other hard basic rocks such as basalts and dolerites, which by constantly providing an alkaline buffer as the rock is dissolved by carbonated rain water prevents the acidification of the surface soil. From anything like loose or easily permeable soil containing a substantial proportion of insoluble constituents the heavy precipitation will quickly leach out any calcium carbonate that may have been originally present, as well as other easily soluble basic salts, and podsolisation will supervene. On flat surfaces, even of limestone rock, raw humus forms if the climate is sufficiently wet and cool, but there is seldom enough insoluble residue present in a typical limestone to produce a surface soil that can be podsolised.

A coarse sandy rock lends itself with special ease to the formation of a podsol because of its easy permeability to percolating water and because of its original deficiency in the finer silt and clay particles with which the basic ions are associated. Thus it happens that podsols are, or have been, frequently formed on coarse sands even in the "Brown Earth climate" of the south of England. Indeed, some of the finest and best-developed podsols are to be found on the Tertiary and post-Tertiary sands and gravels of the south, such as the Reading Beds, the Bagshot Sands and the Plateau Gravels, as well as the Breckland Sands illustrated in Fig. 28. These coarse sandy soils, with finer material and basic salts very poorly represented, often show a particularly sharply marked $B1$ horizon.[1] The same is true of the quartzite sands and granite soils of the north and west, whereas soils derived from other rocks in the real "podsol climate" are mainly "iron podsols" with $B2$ well marked and the humus B ($B1$) absent or feebly developed. They are also generally shallower than the well-developed humus podsols of the coarse quartz sands (Fig. 29).

Thus we see that a climatic soil type (soil typical of a "world group") is not necessarily confined to its proper regional climate if other conditions, such as the nature of the rock or of the vegetation, are favourable to the formation of the soil type under a climate which is not too different. We must not forget that a number of factors always enter into the production of a given type of soil, just as they do into the production of a given type of vegetation, and while in both cases the climatic complex is predominant, a change in a climatic factor may be compensated by parent rock or by vegetational characters and produce identical or closely comparable results. Exactly the same is true of vegetation itself. Climate is the general predominating influence determining the type of vegetation, but within a given climatic region special edaphic conditions may change the vegetation

---

[1] It has, however, been suggested that most, if not all, of these southern podsol profiles are really "fossil podsols", formed in post-Glacial times when the precipitation was heavier than it is at present (see Chapter VII).

type and approximate it to one which is characteristic of an adjacent type of climate.

The podsols correspond generally with the Northern Coniferous Forest region, and with oceanic heaths and "moors" which may result from deforestation or may represent the climax vegetation where the habitat is subject to violent winds. Podsols are not in general favourable to the growth of broad-leaved trees (apart altogether from the formation of

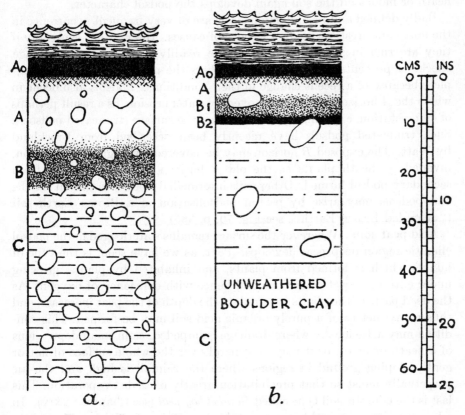

FIG. 29. SCOTTISH PODSOLS DERIVED FROM BOULDER CLAY

In *a* the illuvial layer (*B*) does not form a definite pan but is coffee-coloured and somewhat coherent: in *b* there is a similar coffee brown layer (*B*1) and a well marked humus-iron pan (*B*2) below. In both *a* and *b* there is a heavy surface layer of peaty *mor* (*A*0). After Watt.

moorpan) because of the impoverished surface soil, but spruce, pine and birch (which belongs ecologically to the northern coniferous type of forest) can and do form forests on this type of soil.[1] Podsols are also unfavourable to the majority of shrubs and herbaceous plants, many of which cannot tolerate extreme acidity. The heath and moor plants which are at home on this kind of soil and dominate the vegetation of our podsolised northern and

[1] Spruce however does not flourish on very shallow soil.

western uplands form a well-marked ecological group which may be called "oxyphils" or "acidicoles". For the same reasons podsols are not good agricultural soils, but it is possible by heavy dressings of lime and other appropriate manuring substantially to improve the base status of the soil so as to approximate to the Brown Earth type and make it possible for crops. When such land is left derelict or used for unregulated "extensive" grazing, in which the grasses are "undergrazed", it rapidly reverts to heath or moor and the soil again develops the podsol character.

Badly defined and incomplete podsols are of very frequent occurrence in the north and west, owing to a variety of causes. Heavy soils, especially if they are rich in salts, do not podsolise readily, and impeded drainage checks or prevents podsolisation. In others the process may not be complete because of a recent change in their conditions. Podsols also occur in which the A horizon has been removed by water erosion, as a result perhaps of deforestation, exposing the B horizon on the surface (truncated podsols). Such truncated podsols have recently been described from Breckland by Watt. The exposed B horizon may be covered by a grass vegetation, invaded by heath plants as the old B horizon is leached, and a fresh secondary podsol formed. Other soils intermediate between Brown Earths and podsols may arise by partial podsolisation of a Brown Earth soil ("degraded Brown Earth", see Fig. 70, p. 365).

Acid peat soils. Wherever the surface remains waterlogged in the podsol climatic region deep peat develops. Peat, as we have seen, is the type of humus which is formed from plants, and inhabited by plants, growing under extremely wet conditions and hence with deficiency of oxygen. As the dead parts of the plants accumulate the depth of the peat increases, and thus many metres of a purely organic acid soil may be formed. This condition may arise locally where drainage is impeded, as in small depressions of a heath or moor, or it may be general over the whole surface of flat or gently sloping ground in regions where the rainfall is high and the air perpetually moist so that precipitation greatly exceeds evaporation. This last is the climatic soil type called *Blanket bog peat* (see Chapter XXXIV). In a less wet climate raised bog may establish itself in a fen basin and produce a similar type of highly acid peat (*Raised bog peat*, see Chapter XXXIV). If there is regular drainage through the mineral substratum lying below the peat podsolisation will occur in it, A and B horizons being developed. The peat itself may then be regarded as an immensely exaggerated "$A\,0$" zone, and the whole soil complex as an extreme type of podsol. There are all transitions between this condition and an ordinary podsol, according to the relative permanence of the surface moisture and the consequent depth of the $A\,0$ zone of wet acid peat.

This type of peat soil is inhabited by a limited selection of species which can tolerate the combination of waterlogging, involving deficiency of oxygen, and extreme acidity. Of these the chief are various species of bog moss

(*Sphagnum*), certain species belonging to the sedge family (Cyperaceae), some belonging to the heath family (Ericaceae), and a few others (see Chapters XXXIV, XXXV).

Where the surface of the bog or "moss" is drying out owing to natural or artificial drainage, it may be colonised by some of the oxyphytes of less wet soils; thus the common ling or heather (*Calluna*), already present in bog, rapidly increases, and if the bog is thoroughly drained becomes

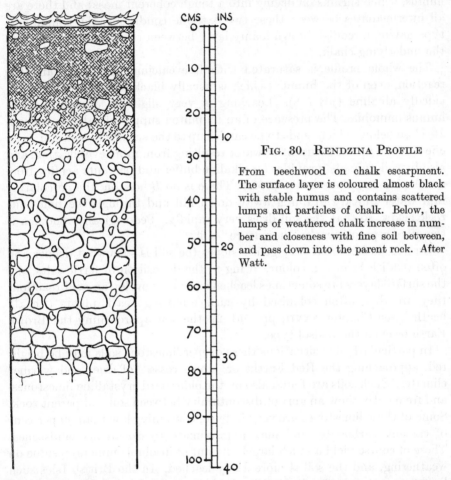

FIG. 30. RENDZINA PROFILE

From beechwood on chalk escarpment. The surface layer is coloured almost black with stable humus and contains scattered lumps and particles of chalk. Below, the lumps of weathered chalk increase in number and closeness with fine soil between, and pass down into the parent rock. After Watt.

dominant. Moss peat cut into "turfs" and dried has always been the staple fuel of the regions where it is the prevailing soil type. It is of little use for agriculture, unless it is thoroughly drained and heavily manured, because of its wetness, extreme acidity and lack of nutritive salts.

Rendzinas. A very characteristic type of soil not primarily determined by climate, but by the nature of which underlying rock, is the so-called *rendzina* formed from limestone (Fig. V). This is well seen in the shallow grey and black soils derived from the chalk, the Jurassic Limestones, the

Permian Magnesian Limestone, and other pure and relatively soft limestones. These soils are rarely more than 12–16 in. (30–40 cm.) thick and often much less: they occur most typically on fairly steep slopes. The very young chalk soils are extremely shallow and greyish white, the dense dark humus of the rooting layer passing almost directly into disintegrated chalk. On steep northern grass-covered chalk slopes, where the conditions are constantly moist, the soil is deeper and extremely rich in black colloidal humus, which shrinks on drying into a tough coherent mass; and there are all intermediates between these two extreme conditions, the commonest type having a reddish brown loamy layer between the surface humus and the underlying chalk.

The whole profile is saturated with free calcium carbonate, and the reaction, even of the humus (which is usually black or very dark), is decidedly alkaline (pH 7–8). Leaching is very slight or absent and the humus immobile. The presence of an unlimited supply of calcium carbonate close below, which tends to be carried up to the surface in solution as an effect of the upward current of water resulting from rapid evaporation, and also by soil animals, provides an alkaline buffer and prevents acidity from developing anywhere in the profile. There is no B horizon, the percolating rain water quickly draining away down hill and through the limestone rock below, so that the soil dries very rapidly. Pedologically rendzina is therefore a permanently "immature" soil.

On flat surfaces and very gentle slopes the soil is somewhat deeper and often reddish brown in colour, owing to the deposition of ferric salts, and the surface layers are sometimes leached and may develop an acid reaction: they are then often colonised by surface-rooting heath plants ("chalk heath", see Chapter XXVII, pp. 551–2), the soil approaching the Brown Earth or even the podsol type.

In particularly dry situations these deeper limestone soils may be bright red, approaching the Red Earths or "terra rossa" of drier and warmer climates. Such soils are found also on the older hard crystalline limestones, and frequently show an abrupt discontinuity between soil and parent rock. Some of these limestones are very impure, containing less than 50 per cent of calcium carbonate, and may approximate to calcareous sandstones. These of course yield a much larger amount of insoluble mineral residue on weathering, and the soil is more easily leached. In the British Isles such limestones are mostly situated in the podsol climate where leaching and raw humus accumulation is favoured; and a mixture of "calcicole" and "calcifuge" plants is very often found in the vegetation. The grassland of the basic igneous rocks, such as dolerite and basalt, is very nearly allied to limestone grassland, but a close comparison of the soils has not been made. The British limestone soils require much further study in relation to climate, nature of rock, slope, exposure and vegetation.

The shallow, highly calcareous soils bear characteristic vegetation: in the south beechwood, in the north and west ashwood. But they are very largely grassland and form an excellent and, when rainfall is sufficient, a bright green pasture including a characteristic set of grasses and herbs (Chapter XXVII, pp. 530–9), many of which, however, are "dry soil" plants not confined to calcareous soil. On south and west slopes at the end of a hot dry summer the southern chalk pastures may suffer severely from drought and show a much impoverished flora. In the wetter and cooler climate of the north and west the limestone pastures remain green throughout the year if the winter is mild.

Relations of climate, soil and vegetation. We have seen that climate is the predominant factor which determines the great world groups of soil, any one of which may be made, under the influence of a particular climatic type, out of material derived from very different kinds of rock. In a similar way climate determines the great vegetational types of the world, and as a result there is a rough correspondence between the climatic soil type and the climatic vegetational type. Thus, broadly speaking, the brown earths bear deciduous summer forest while podsols bear northern coniferous forest, heath or moor. Soil and vegetation are obviously well adapted to each other or they would not continue to co-exist. Each has a definite effect upon the other and contributes to its character, but certainly neither depends absolutely upon the other: both are determined, more or less independently, in the first place by climate, and the character of each may be modified by the nature of the underlying rock from which the soil is derived. Thus there is a network of interlocking relationships between the various components of the climate-soil-vegetation complex:—

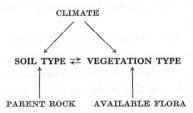

We now have to consider the effect of the underlying rock on this complex.

For various reasons mature, fully developed soil profiles are far from occurring over the whole surface of the country. Anything like typical examples, indeed, have to be very carefully sought. We have seen that the conditions of a steep slope do not admit of the maturation of the profile—the soil remains permanently immature. Hard rocks which weather slowly have only been exposed over most of the country to a climate of anything like the existing type during the few thousand years since the Pleistocene glaciation, and even during that time the climate has undergone significant changes (see Chapter VII), so that in many cases the soils have not had time to develop a mature profile. Soils with impeded drainage, and thus wholly or partially waterlogged, cannot develop the profile characteristic of the climatic type. Again, many profiles have been partly eroded and thus truncated, so that, for example, the *B* horizon, which was formed at some little depth, may appear on the surface; and these actually occupy the greatest area of the country. Others, again, have been disturbed by various agencies, of which by far the most important and widespread is cultivation. When a soil is ploughed the *A* horizon is partly or wholly destroyed as a stratified *A* horizon, though the materials remain; and when it is manured its base status is artificially raised. Thus a podsol may be converted into something resembling a Brown Earth. If the land goes out of cultivation the soil-forming climatic factors are again allowed free play and the proper climatic soil type tends to be re-established. We can therefore only look for natural soil types under natural, or at least semi-natural, vegetation which has long been undisturbed.

Among the natural soils themselves there are very great variations according to the nature of the parent rock. The effect of the underlying rock on the soil, and therefore on the vegetation, is at its maximum in the early stages of soil development, before this has proceeded far and has put, so to speak, a cushion of stratified "horizons" between the parent rock and the plants. But even on mature soils, which have developed the characteristic climatic profile, the influence of the rock, or rather of the characteristic mineral materials derived from the rock, on the vegetation and on the details of the profile itself may be very evident. Thus a sandstone and a clay may both develop a Brown Earth soil bearing deciduous summer forest, but the difference in parent material will be evident from the greater proportion of the sand fraction in the profile of the one and of the clay fraction in the other, and also in the presence of some different species of plants in the vegetation.

Soil texture. The effect of the parent rock on the soil which it produces depends partly on the chemical composition of the rock minerals, but very largely on the "grain" of the rock, giving rise on weathering to particles of characteristic size; and it is this which is called *soil texture*. Soil texture is of the greatest importance to plants because it affects water economy and the access of air, and through these the chemical processes of the soil. A classification based upon texture is therefore of primary importance in considering the relation of the soil to the vegetation it bears.

We have already seen (p. 80) that a soil can be separated by mechanical analysis into "fractions" after it has been dried at 100° C., the organic matter burned away at a red heat, and the mineral residue pounded in a mortar. This procedure destroys the *structure* of the soil, including the compound particles, which it reduces to their constituent simple particles; but it reveals the proportions of simple particles of different sizes which make up any particular soil, and according to the preponderance of one or other of these the soils have received their common names.

Soils characterised primarily by texture

Gravel. A soil with a preponderance of large particles above 2 mm. in diameter, usually mixed with sand below 2 mm. and some finer particles as well. Percolation of water and aeration are extremely free. A gravel soil has the characters of a coarse sand in extreme form. It is unfavourable to plant life because of its poverty in nutrients and its dryness, unless the ground water is high and carries nutrient salts as in some alluvial gravels.

Sand. Preponderance of particles with a diameter between 2 and 0·2 mm. (coarse sand) and between 0·2 and 0·02 mm. (fine sand): particles typically of quartz. Percolation and aeration free. Water-holding capacity and power of raising water slight, because the capillary spaces between the particles are too wide: hence a dry soil unless the ground water is high; and a warm, "early" soil because owing to its dryness it "warms up" quickly in spring. Light and easy to work. Typically poor in nutrients, partly because of deficiency in the finer particles with which the bases are associated, and because the free percolation of rain water leads to very thorough leaching. Easily podsolised and developing acidity owing to poverty in bases. In oceanic and suboceanic temperate climates sands often bear heath or heathy woodland, characteristically of birch and pine, with raw humus formation. Sandstones with sufficient of the finer fractions to maintain their base status develop the Brown Earth profile and support deciduous summer forest with mild humus, though coarse sand grains may be abundant in the soil and affect the nature of the subordinate vegetation. Deforested soils of this nature tend to podsolise and develop heath, but they make good agricultural soil for certain crops.

Coastal blown sands support specialised vegetation adapted to cope with their mobility. On stabilisation they resemble sandy soils derived from coarse sandstones and may develop heath.

Silt. Preponderance of particles between 0·02 and 0·002 mm. in diameter. Intermediate between sand and clay in size of particles. A favourable soil for vegetation because it has considerable water-holding capacity, while percolation, capillary rise of water, and aeration are all fairly free. The name is derived from the prevalent texture of alluvial soils laid down on the flood plains of rivers. These soils probably originally bore alder and oakwood but were mostly converted to meadowland; impeded drainage is common. They were traditionally mown for hay but make very good pastureland, and have been extensively used for market gardening.

Clay. Particles below 0·002 mm. in diameter numerous enough to give character to the soil, which is dense and plastic. Any soil with more than 30–40 per cent of such particles would be called clay. Particles typically of hydrated alumino-silicates. Has all the qualities opposite to those of sand. Percolation of water very slow; aeration defective; water-holding capacity very high. A wet heavy soil, difficult to work, and "late" because it warms up slowly owing to the high water content. Under continued drought the clay colloid shrinks, cracks and eventually "bakes" hard. Physically unfavourable to many plants, and root systems tend to be shallow owing to difficulty of aeration away from the surface; but chemically favourable because usually rich in bases associated with the complex silicates, the basic ions free or adsorbed by the colloid particles. Much improved by incorporation of mild humus, which "opens" and lightens the clay, flocculating the clay particles so as to form "crumb structure" and facilitating movement of water and aeration.

Supports typical damp oakwood: deforested and grazed clay soils bear shallow-rooting grasses. Much wheat formerly grown on the midland, southern and eastern clays, but cost of heavy working makes wheat growing unprofitable on clay land when the price of wheat is low. Mostly now in (poor) permanent pasture of relatively unproductive grasses (e.g. *Agrostis*).

Loam is a mixture of particles of many sizes belonging to the different "fractions". It is the soil most favourable to plant life because it combines the good qualities of the extreme types. Thus the clay (and humus) give consistency to the soil, retain water and provide a supply of plant "foods", the particles of medium size permit of capillary rise and thus tend to counteract excessive leaching, while the sand particles facilitate aeration. The constitution of loam has a wide range according to the relative preponderance of one fraction or another: thus we have "clay loams", "medium loams", and "sandy loams". With an adequate supply of bases, especially calcium, plenty of mild humus, and good drainage the medium loams are the ideal soils for all except the most specialised plants adapted to extreme edaphic conditions.

The loams naturally bear oakwood and sometimes beechwood, but they are of course the favourite soils for arable cultivation.

Soils whose character depends primarily on chemical composition

Calcareous soils. Calcium carbonate when present in excess, as in the chalk or other limestone rendzinas described on pp. 93–5, gives a character to the soil which generally outweighs the effect of texture, both pedologically and in its influence on vegetation. The complex alumino-silicates of the clay residue have an unusually high proportion of silica, the finer clay particles are flocculated by the lime, the humus is saturated with bases and immobile. The soil is therefore alkaline in reaction, is not tolerated by the extreme oxyphytes or "calcifuges", and is inhabited by a characteristic vegetation. Where the soil is directly derived from an underlying limestone it is typically light, dry and shallow. The vegetation includes a number of more or less xerophytic species (often including the dominants of the communities) which are not necessarily in any way limited to calcareous soils. But other species are also always present, often in great abundance, which are normally absent from, or very infrequent on non-calcareous soils of the same region. These are the so-called "calcicoles" which are evidently species definitely adapted to life in soils with a basic reaction, since they often occur also on soils derived from basic rocks other than the limestones, such as basic igneous rocks and unleached calcareous sandstones and marls whose textures vary widely.

Calcareous soils bear beechwood in the south and south-east and ashwood in the north and west, characteristic types of scrub, and on the limestones characteristic grassland. The shallower soils derived from chalk and the harder limestones are apt to be too shallow and dry for good arable cultivation, but the calcareous loams and marls are very favourable.

Saline soils. Salty soils containing much sodium chloride (common salt) and other similar salts are also alkaline in reaction. In this country most saline soils are maritime, laid down in tidal estuaries and liable to flooding by the sea at high tide, though there are local inland saline areas in the neighbourhood of salt springs. Saline soils vary in texture from fine clay to sand according to their position in relation to the tidal flow, and bear the very specialised *halophytic* or salt-marsh vegetation (see Chapter XL). The higher stabilised salt marshes are used for grazing sheep and sometimes cattle. When the tide is excluded, and the salt marsh drained, the excess of salt is leached out and good pasture (or arable) land produced.

The sand of sandy beaches, above high-tide mark but in immediate proximity to the sea, is also generally impregnated with salt blown up in spray, but when the sand is blown farther from the sea the salt is quickly leached out by percolating rain water. It is probable, however, that salt spray affects both soil and vegetation at some distance from the coast where prevalent winds constantly blow from the sea.

Organic soils. The deep and pure organic soils formed by plant remains whose decay is arrested under conditions of oxygen deficiency resulting from waterlogging, have already been described. They are all called *peats* and are traditionally used as fuel by the inhabitants of the regions where they abound, but the two main classes of peat—fen peat and bog peat—differ sharply in their characters and in the vegetation they bear. *Fen peat*, which is formed in lake and river basins not subject to constant silting, and where the drainage water is alkaline, is usually structureless and black, rich in ash and proteins, but with no cellulose left, and bears fen vegetation dominated by sedges, grasses and rushes, later by certain shrubs and eventually by alderwood; while *bog* or *moss peat*, which is highly acid in reaction and rich in celluloses and hemicelluloses, is often brown in colour and with plant structure more or less well preserved: this is dominated by bog moss (*Sphagnum*) or by species belonging to the sedge family. The accompanying species of fen and moss are also for the most part characteristic, and very few indeed are common to the two formations (*Molinia caerulea* is one of these few). Very little work has been done on the soil or vegetation of peats of intermediate reaction, i.e. neutral or slightly acid peats, but they are sometimes dominated by *Molinia* or *Myrica gale*.

Well-drained fen peat makes good agricultural land, especially for certain crops: moss peat requires heavy manuring as well as drainage, and the most extensive areas of it occur in wet regions, where adequate drainage and really successful agriculture are difficult or impossible.

Dry peat (German *Trockentorf*) is a compacted variety of *mor* (p. 83) formed on the surface of mineral soils poor in bases, in this country usually by heath plants and their associated mosses and lichens, or by the needle litter of planted coniferous forest, but also under unfavourable conditions of climate or soil by the leaf litter of deciduous woodland.

Any influence which checks disintegration of organic material, such as lack of oxygen, poverty in nutritive salts, lack of moisture, or low temperature, tends to the formation of mor, for all these are unfavourable to the life of the soil organisms by whose activity normal disintegration and decay of organic matter is brought about. Thus a layer of raw humus is formed on the floor of forest in cold climates or under very dry surface conditions, and also on the surface of heaths developed on poor sandy soil in a cool oceanic climate.

The nature of the litter which forms the material of humus is also, as we have seen, an important factor. The dead parts of different plants vary greatly in their resistance to the processes of decay. Thus oak leaves decay more quickly than those of beech, and the fallen needles of the evergreen conifers, as well as the foliage and twigs of the heaths, are much more resistant than either. The more resistant the litter the more it tends to accumulate and the greater the formation of raw humus.

Raw humus or mor is always acid in reaction, usually strongly acid, owing to the presence of organic acids liberated during the initial processes of its formation from litter, and not neutralised by bases.

REFERENCES

FARROW, E. P. On the ecology of the vegetation of Breckland. I. General description. *J. Ecol.* **3**, 211–28. 1915.

PEARSALL, W. H. The soil complex in relation to plant communities. II. Characteristic woodland soils. *J. Ecol.* **26**, 194–209. 1938.

ROBINSON, G. W. *Soils, their origin, constitution and classification*. Second edition. London, 1936.

ROBINSON, G. W. Soils of Great Britain. *Proc. Third International Congress of Soil Science* (Oxford, 1935). London, 1935.

ROMELL, L. G. Mull and duff as biotic equilibria. *Soil Sci.* **34**, 1932.

ROMELL, L. G. Ecological problems of the humus layer in the forest. *Bull. Cornell Agric. Exp. Sta.* No. 170. 1935.

Chapter V

DISTRIBUTION OF ROCKS AND THE SOILS THEY PRODUCE

Distribution of outcrops and soils. The textures of the soils immediately derived from the more consolidated rocks and shortly described in the latter part of the last chapter are directly related, as was explained, to the lithology of the parent rock; and those of the less consolidated deposits, mostly Pleistocene and "recent" organic, alluvial, and maritime soils, depend on the nature of the material and the local conditions of their formation. The rapid change of soil from place to place in the British Isles, and especially in England, is due to the great range of geological strata which are exposed on the surface. There can scarcely be a country in the world with a greater range of geological formations represented within so small an area.

The south and east and most of the midlands of England, as we saw in Chapter I, are occupied by Secondary (Mesozoic) and Tertiary strata, covered by Quaternary deposits, largely glacial drift over a very great part of the eastern and midland area (cf. Fig. 31). These we class together as *neogenic*. To the west and north-west of the Secondary outcrops are the older (with igneous) rocks which make up the hill and mountain regions of south-western England, Wales, northern England and Scotland. The Irish rocks are nearly all Palaeozoic, including not only the mountain masses of the north, west and south, but also the low-lying central Irish plain based on Carboniferous Limestone. All these we call *palaeogenic*.

The variety of rock and soil is so great that nothing remotely approaching a detailed description of their distribution could possibly be entered upon here, even if the necessary detailed knowledge were available. But it will be useful to give some summary notion of the situations of the more important and extensive outcrops and soils, with indications where possible of their natural or semi-natural vegetation and of the agricultural uses to which they are put. From many of them, however, the natural vegetation has disappeared altogether.

Relation to agriculture. Agriculture in the widest sense, and indeed land utilisation generally, depends so much on the varying needs and powers of the human populations that differences of soil and even of climate are far from being the sole determining factors. There are of course climates, and also soil types, which cannot possibly support certain crops; but the adaptability both of plants and of human skill is very great, and under the stimulus of extreme need or of specially good markets regions may be devoted, with a considerable measure of success, to particular crops, or to a particular kind of agriculture or grazing regime, which are far from being naturally well suited to them. Thus while the natural vegetation of a particular region is a close reflexion of the conditions of climate and soil, this is not necessarily true of the agriculture, though climate and soil are always important, and may be limiting, factors.

The truth of these statements is strikingly seen in the changes which have taken place, for example, in the early part of the nineteenth century, again towards its close, and yet again since the war of 1914–18. Land which was at one time devoted mainly to sheep walk was later used for cereal growing with sheep rearing combined in one system, and later again was laid down as permanent pasture, at first mainly for the raising and fattening of beef, but now more and more for the production of milk. All these changes were consequent on the fluctuations of markets, and each kind of agriculture in its turn was successful in accordance with the economic conditions of the time.

Order of treatment. We shall begin with the Quaternary deposits, which are the most widespread, "being scattered irregularly throughout the country and resting indifferently on any of the rocks from the oldest upwards", and pass down the geological series. This involves considering the Quaternary, Tertiary and Secondary rocks and soils of the south and east before the Palaeozoic of the north and west, the inverse order to that adopted in describing the structural history of the country in Chapter I, and corresponds with the treatment of brown earths before podsols in Chapter IV, and of the deciduous summer forest before the moorlands and mountain vegetation in later chapters. This treatment involves a certain amount of repetition of the material of Chapter I, but the point of view is different; and it will be noted that, whereas in Chapter I the greater proportion of space is devoted to the older (Palaeogenic) rocks, which are fundamental for the physical structure of the country, here it is the younger deposits, which produce the greatest variety of soils, bringing about the greatest differences of vegetation, that receive the fuller treatment.

NEOGENIC ROCKS
Quaternary deposits

Recent maritime soils. Coastal sand dunes occur wherever sand accumulates in shoals on a flat shore, and is carried up above tide marks so that it dries and comes under the action of the wind which drives it shorewards. When the sand is arrested or trapped owing to topographical features or other causes and is continuously deposited in sufficient quantity it is colonised by specialised plants which bind it and form dunes. Dune areas occur mainly adjacent to estuaries and in association with salt marsh. The vegetation is described in Chapter XLI. There are a certain number of dune areas on the east coast of England and Scotland, very few on the south English coast, and many more on the west coasts of England, Wales, and Ireland. Fixed dune is utilised traditionally (in Scotland) for golf courses ("links"), often as rabbit warren, and sometimes as rough grazing.

Shingle or pebble beaches thrown up by storms out of reach of normal high tides are also colonised by vegetation allied to that of sand dunes. The plants are rooted in the sand that is nearly always present between the stones or in the fine dark humus produced largely from organic tidal drift (see Chapter XLII). The best known extensive shingle areas are Orfordness in Suffolk, Dungeness in Kent, Pevensey Beach in Sussex, and the Chesil Beach in Dorset, but there are many other smaller examples, especially along the east coast, while a narrow shingle beach fringes much of the south English coast.

The silt deposited in tidal estuaries, passing to sand towards the mouth and to fine clayey mud in bays and towards the upper limits of the tidal zone, is gradually raised in level with the aid of the vegetation, but remains normally flooded by the highest spring tides (see Chapter XL). These deposits, on which salt marsh develops, are very widespread along the flat sides of estuaries all round the coast. The stabilised salt marshes themselves (*prés salés*) may be used as sheep or cattle pasture. The most extensive areas of this origin which have been reclaimed from the sea by dykes and draining and now carry cultivated grassland (or occasionally arable), are Romney Marsh in Kent, one of the most famous sheep-fattening pastures in the country: the "Marshland" south and west of the Wash in north Cambridgeshire and south-east Lincolnshire, round the silted up estuaries of the Witham, Welland, Nene and Ouse; and finally the very fertile natural and artificial "warplands" of the Humber estuary where the Rivers Ouse and Trent meet the incoming tide laden with chalk detritus from the east coast of Yorkshire and the flood is allowed to run over the land and deposit its rich silt. The average value of the very various crops grown on this enriched warpland may be as much as £16 to £18 an acre (Ruston in Maxton, 1936, p. 90).

River alluvium. On the flood plains of rivers, beyond the influence of the tides, the alluvium takes the form of fine clayey silt originally bearing reedswamp, fresh-water marsh vegetation, or wet alluvial forest, but now more or less drained and forming admirable pasture or water meadow. So far as the drainage is still impeded the silt develops typical meadow soils (p. 88). Well-drained river alluvium in the neighbourhood of large cities, which are often situated on the alluvial plains of the larger rivers, is generally used for market gardening or for fruit, and in Kent often for hops. In hilly country where the fall of the streams is much steeper and the currents are therefore more rapid and have more carrying power, much coarser material—sand and gravel—is transported, while mountain torrents carry and deposit coarse gravel and even boulders. The "Valley Gravels" and "Plateau Gravels" of the south of England are sometimes glacial drift but generally old river alluvium, the former belonging to the existing drainage system, the latter to rivers (partly Pliocene) which are now extinct.

Peat soils: (1) Fen peat. The general conditions of formation of peat soils have already been cursorily described (pp. 88, 92) and are again dealt with in relation to the vegetation in Chapters XXXII–XXXV. The largest existing area of fen peat is the East Anglian Fenland in central Cambridgeshire and adjacent parts of Lincolnshire, Huntingdonshire, Norfolk and Suffolk. This is almost all drained and under cultivation: a few fragments (now preserved) still exist, as at Wicken Fen (Cambs) and Woodwalton Fen (Hunts). A smaller area, much of which however still bears the distinctive natural vegetation, occurs in east Norfolk along the middle courses of the Bure, Yare and Waveney. Similar fenland exists round Lough Neagh in north-east Ireland, and smaller patches occur on the borders of various lakes and rivers and in "fen basins" formerly occupied by lakes in many parts of the country. Many areas of acid peat are underlain by fen peat, pointing to the former existence of fen conditions on the sites of existing bog.

Drained fen peat makes an excellent agricultural soil. Of the Cambridgeshire fenland four-fifths are arable and two-fifths of the arable land are under potatoes and sugar beet. Wheat is the chief cereal.

(2) Bog or moss peat. Acid peat soils cover large areas in the west of Scotland and the west of Ireland, as also on the plateaux of many hill masses in Scotland, Ireland, Wales and the west and north of England, especially at about 2000 ft. (c. 600 m.) altitude. This deep "blanket" peat, as we saw in Chapter IV, is a "climatic soil" which may be considered as an extreme development of the podsol type with the A 0 horizon very thick and permanently waterlogged. Local acid peat areas (raised bogs) are formed on the sites of old fens when the vegetation has grown up away from the influence of alkaline ground water. These are commonest in Ireland, but many exist also in Scotland; and at one time they were numerous in England and Wales, though most of the English bogs have been destroyed by peat cutting, draining and cultivation. Areas on the Somerset Levels, Foulshaw Moss and other "mosses" in north Lancashire, south Lancashire, Cheshire and Yorkshire, are remnants of extensive but local bogs built upon old fens, though few remain anything like intact. Evidence has recently accumulated that local bog or moss has developed in past times on the edges of the East Anglian fenland. The climatic limits within which raised moss, forming local acid peat, rather than forest, succeeds fen peat are not yet clear, but the former is characteristic of the less dry climates.

Blanket peat is very unsuitable for cultivation, partly because of the climate. In western Ireland and Scotland a few miserable crops of potatoes and oats are grown, with the common reed (*Phragmites*) as a prevalent weed. Something more can be done with raised bog peat by draining and heavy manuring. The traditional use of peat is of course as fuel, but its antiseptic and strongly absorbent properties have led to its increasing employment for medical and industrial purposes.

Glacial drift. The Pleistocene ice-sheets which covered the greater part of the country for tens of thousands of years, and finally retreated about ten or twelve thousand years ago, left a thick ground moraine lying on the exposed surfaces of the older rocks. This "glacial drift" has been partly removed by subsequent erosion, but much of it still remains. The greater portion is of fine texture and is generally known as "boulder clay", because it contains boulders carried by the ice-sheets from distant areas; but some of the drift consists of sands and gravels.

Irish drift

In Ireland boulder clay covers the greater part of the great low-lying central plain. Like the Carboniferous Limestone on which it is based it forms a highly calcareous soil, and it is to this, together with the moist equable climate, that the Irish grasslands, justly famed for cattle-raising, owe their excellence. Conspicuous among the glacial soils are the dry grass-covered gravel ridges known as *eskers*, and in various places also *drumlins* of boulder clay, with steep rounded flanks, furnish loamy and often stony soils. Glacial drift also occurs along the flanks of the mountain chains, as on the Leinster hills, where it has imported limestone material into a region otherwise poor in lime. Other glacial deposits of various constitution overlie the older rocks in far too many places to enumerate.

Scottish drift

In Scotland there are no extensive areas of limestone, and a great deal of the widespread glacial drift is likewise almost destitute of lime. The soils of the lowlands up to about 1000 ft. (c. 300 m.) have been almost entirely deposited by moving water or ice, while above this limit glacial drift frequently occurs in depressions or on terraces among the hills; but boulder clay proper is absent from the Highland region, except in the wide straths (valleys). Most of the glacial boulder clay of the north, deficient in lime as it is, tends in the moist cool climate to form an acid peaty soil which on the hills bears the various prevailing types of "moorland" vegetation—Callunetum (Chapter XXXVII) on the drier, Molinietum, Nardetum (Chapter XXVI), and on the deeper wetter peats Scirpetum caespitosi (Chapter XXXV).

The central Lowland plain or "Midland Valley" of Scotland, which stretches from the Firth of Clyde on the west coast to the Firths of Forth and Tay on the east, with a considerable extension north-eastward along Strathmore (the Great Valley), includes the greater part of the agricultural area of Scotland, as well as the largest cities, the coal mines and most of the industrial country. Its soils are almost all of glacial or alluvial origin, and though some of them near the east coast are very fertile, they consist to a great extent of tough, heavy glacial "till" or boulder clay, generally unproductive as farmland and largely left as wet grassland or peat bog. The most fertile land, where some of the best farming is carried on, is largely on loamy glacial drifts derived partly from the Old Red Sandstone, as in the Lothians round Edinburgh, especially at Dunbar (where the narrow strip of coastland is said to be the most expensive and highly farmed area in the world), in the Fife peninsula, in Strathmore, and on parts of the low lying eastern coastal plain farther to the north, where the boulder clay is derived from metamorphic rocks of high base status, and, though podsolised, makes good agricultural land.

South of the Midland Valley the Ordovician and Silurian hills of the Southern Uplands, which rarely rise more than 2000 ft. (c. 600 m.) above the sea, are likewise partly drift-covered, and the same is true of the adjoining Coal Measure sandstone hills of Northumberland and Durham. The low-lying Trias of south Lancashire and Cheshire is also largely overlain by glacial drift, partly sandy, and on this many lowland peat mosses were developed. The mixture of these light soils with peat is the basis of the famous South Lancashire potato-growing region. In Cheshire there is more clay and more grass, and here the dairy herds used to support the great Cheshire cheese industry, though this is now largely replaced by the direct sale of liquid milk (Orr in Maxton, 1936). The Trias, Lias and Jurassic soils of the Vale of York and of the English Midland plain are also largely overlain by glacial drift, much of it a tough non-calcareous boulder clay; but in many places the clay is replaced by loamy, sandy or gravelly drift. Some of it is excellent arable land, growing good crops, but most is now permanent pasture. This country was almost continuously covered with oak forest up to Roman and even Saxon times.

<small>English Midland drift</small>

Eastwards the non-calcareous drift passes into the Chalky Boulder Clay (Fig. 31), which is, on the contrary, highly calcareous, having been formed by the ice-sheets grinding over the Chalk and the Jurassic limestones. This deposit occupies wide areas in Lincolnshire, south Norfolk, Suffolk, Essex, Cambridgeshire, Huntingdonshire and Bedfordshire, and extends eastwards into Rutland, Northamptonshire and Leicestershire. The heavy calcareous clay forms wide sheets on the higher ground in parts of this area and sometimes extends into the valleys on a level with the river alluvium, but it has been removed by erosion from the steeper slopes and escarpments. The extent to which it covers the underlying rock can be appreciated from Fig. 31. The natural woodland is of the ash-oak type and the scanty remains have long been treated as (ash-) oak-hazel copse with oak standards and the ash coppiced. A comparison of Fig. 32 with Fig. 31 demonstrates that the Chalky Boulder Clay gives rise to "clay", "heavy" and "medium" agricultural soils. The heavy soil has been largely arable, and can grow good wheat; but owing to the cost of labour and the low price of wheat much of it has gradually been laid down to permanent pasture. Some is now almost derelict grassland, colonised by scrub and young trees. On the other hand some of the soils of medium texture, both calcareous drift and derived

<small>Chalky Boulder Clay</small>

directly from underlying rocks, are extremely favourable for both arable and pasture land. Of the latter the world-famous bullock-fattening Leicestershire pastures, based on Lias and the overlying Chalky Boulder Clay, which extend into Rutland and Northamptonshire, share with a very few others the distinction of being the best pastureland in Great Britain. The nature of the drift is generally closely related to that of the underlying or adjacent rocks, and this explains the great variation in texture and in lime content which it exhibits. In many places, especially towards the south-east, the Chalky Boulder Clay is replaced by non-calcareous glacial sands and gravels (Fig. 31).

Breckland. In south-west Norfolk and north-west Suffolk lies an area of sandy soil mainly derived from glacial drift, partly calcareous, partly leached and acid. The rainfall is low. Parts of this "Breckland" bear more or less calcareous grassland but most of it is very infertile (see pp. 509–14 for an account of the vegetation): much is heathland, and great areas have lately been planted with conifers by the Forestry Commission. The name is supposed to be derived from the occasional "brecks" or breaks made for crops. The principal farm crop is now sugar beet and it is one of the few regions where rye is still grown (Carslaw in Maxton, 1936).

Clay-with-flints. The glacial drift nowhere reaches the Thames valley, and is replaced on the chalk plateaux of the Chilterns, the North Downs of Surrey and Kent and parts of the Hampshire uplands by the "Clay-with-flints" (Fig. 31, SW corner and Figs. 33 and 34). This is non-calcareous and usually produces Brown Earth soils: in some places it is markedly acid though with a high base status, and varies a good deal in texture. It is supposed to represent the insoluble residuum of a much greater thickness of chalk from which the lime has been removed by solution. Clay-with-flints naturally bears oakwood, and sometimes beechwood, but has been much used as medium arable land.

Tertiary deposits

The south-eastern basins. The Tertiary beds of the British Isles are almost confined to the south-east of England, occupying basins in the underlying Chalk, which dips below their edges. There are two such basins —the Hampshire basin in the south, abutting on the English Channel, and the London basin to the north-east of it, round the lower course of the Thames (Fig. 5, p. 17): these are occupied mainly by Eocene strata.

East Anglia. Pliocene strata overlie the Eocene fringing the Chalk in East Anglia, and are confined to that region. They lie nearly horizontal, and though of no great thickness occupy a considerable area, but are much overlain with glacial drift of similar composition (Fig. 31). They are mainly sand and gravel (Pliocene "crag") with very thin bands of clay, and are largely covered with heath. The farming is highly diversified.

The "Cromer Forest Bed" at the top of the Pliocene contains the latest

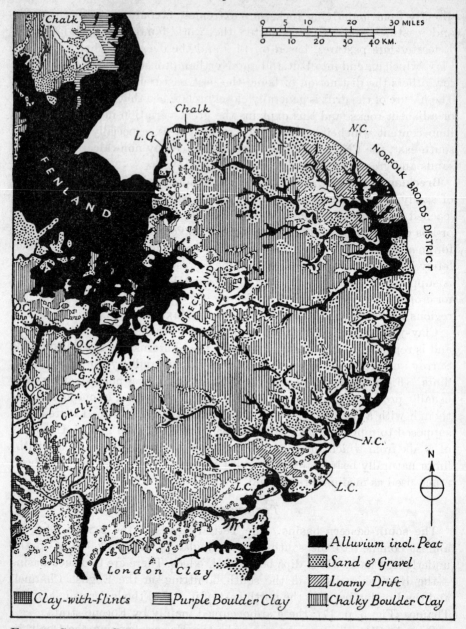

Fig. 31. Surface Geology of East Anglia and part of the East Midlands

Post-tertiary deposits cover by far the greater portion of the country. Of these the peat and silt of the Fenland and the Norfolk Broads region, the Chalky Boulder Clay, and glacial and alluvial sandy and gravelly deposits are the most important. The *coastal* sand, shingle and mud are included with peat, silt and river alluvium (black). *Underlying rocks*: N.C.=Norwich Crag (Pliocene), L.C.=London Clay (Eocene), G.=Gault, L.G.=Lower Greensand (Cretaceous), O.C.=Oxford Clay (Jurassic). Reprint of Fig. 9.

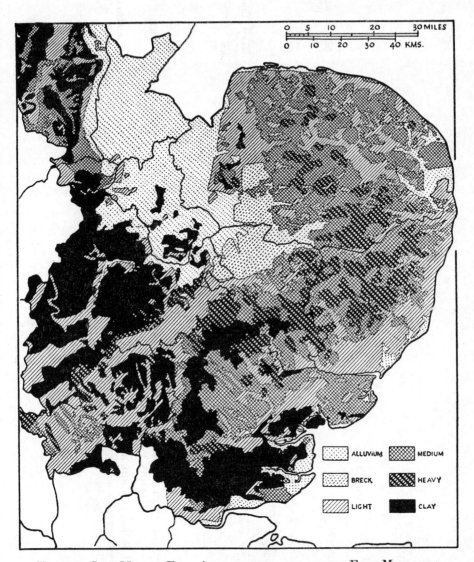

FIG. 32. SOIL MAP OF EAST ANGLIA AND PART OF THE EAST MIDLANDS

Comparison with Fig. 31 shows that the Chalky Boulder Clay gives medium and heavy loams and clay soils, the Oxford and London Clays clay soils: the Chalk, the Lower Greensand and the loamy, sandy and gravelly drifts of eastern Norfolk and Suffolk give light soils. Reproduced by kind permission from Report 19, Farm Economics Branch, School of Agriculture, Cambridge University, 1931.

Fig. 33. Section from the edge of the Midland Plain to the edge of the Weald, crossing the Thames Valley and the London Basin (somewhat simplified)

The main Chalk outcrop is traversed, bearing clay-with-flints and beech forest, and then the Eocene beds of the London basin including the London Clay and Bagshot Sand. To the south the Cretaceous beds are again encountered on the northern edge of the Weald. The natural vegetation of the various strata is indicated. Beech (1) = escarpment type, Beech (2) = plateau type, Beech (3) = sandy type of beechwood (see Chapter XVIII). Reprint of Fig. 8.

pre-Glacial British flora, and this shows that the climate just before the onset of ice must have closely resembled the existing climate.

Oligocene beds are confined to southern Hampshire and the Isle of Wight and are unimportant ecologically, while Miocene strata are altogether absent from the British Isles.

London and Hampshire basins. Of the Eocene beds of the London basin, the Bagshot Sand is an extremely coarse infertile sandy soil, forming an extensive tract to the south-west of London on the borders of Surrey, Hampshire and Berkshire (Bagshot-Woking district) and isolated patches to the north of London. It bears sandy oakwood in places, but mainly heath, largely covered with birch and subspontaneous pine. This tract of country is, however, becoming more and more suburbanised. The corresponding beds of the Hampshire basin show a greater complexity of alternating sands and clays. These Eocene sands, especially the Bagshot, often show typical podsols with a well-developed humus B horizon. The infertility of some Bagshot Sand areas in the Hampshire basin, e.g. near Wareham in Dorset, is apparently caused by definite soil toxicity whose nature is not understood. Planted pines are unable to form mycorrhiza and their growth is arrested at once. It is said that in certain places in and about the New Forest (south-west Hampshire) which consists of oak and beech, much of it planted, but is largely heathland colonised by birch and pine, neither birches nor pines can establish themselves spontaneously. On the other hand favourable Eocene soils near Southampton are famous for their early strawberries.

The London Clay (Figs. 31 and 33), underlying much of London and most of the immediately surrounding country, is the thickest and most extensive of the Eocene strata. It yields a typical stiff clay soil of Brown Earth type, and naturally bears damp oakwood. In the old oak forest region of south Essex three-quarters of it is under permanent grass, and this is a densely stocked dairying region. Below it, geologically, are the comparatively thin alternating bands of sand and clay ("Lower London Tertiaries") which give variety to the fringe of country between the Chalk and the London Clay plain, the sand as usual bearing birch, subspontaneous pine and heath. On the north coast of Kent and the south coast of Sussex, overlying the Chalk, these beds, partially covered with brick earth, form narrow maritime plains composed of fine well-tempered loams, which are extremely fertile. They used to grow wheat and fatten bullocks, but this gave way to dairying, and farm agriculture is now largely replaced by market gardening which supplies early produce to London and for the rest of the season to the numerous and crowded seaside resorts of the south coast.

Secondary rocks

The Secondary or Mesozoic rocks occupy more than half the area of England if we ignore the drift by which they are so largely covered.

Speaking generally they dip regularly to the south-east, one under another, so that their outcrops are arranged in a succession of bands, crossing England from north-east to south-west. The north-eastern ends of these bands bend northwards and north-north-west, parallel with the Pennine Palaeozoic ridge, and occupy the eastern part of northern England, running out successively to the sea on the north-eastern coasts of Yorkshire and Durham (see Figs. 4 and 5, pp. 9 and 17).

The Chalk. The youngest member of the series is the Chalk, a very pure and relatively soft limestone, whose outcrop determines the general topography and land relief of most of south-eastern and eastern England. The arrangement of the general Chalk outcrop has been sufficiently described in Chapter I (pp. 21, 23). The chalk soil proper is of the thin rendzina type described in Chapter IV (pp. 93–5), but it must not be supposed that this soil occurs over most of the outcrop. Much of the Chalk area is covered with the superficial deposits already mentioned, in the northern portion especially Chalky Boulder Clay (Fig. 31), and in the south loams of various origins and depths, giving Brown Earth soils of various kinds, much of which is the "Clay-with-flints" already described (Figs. 33–34). Where the Chalk itself forms the immediate subsoil the natural "climax" vegetation is beechwood in the south-east, but towards the south-west this passes into ashwood. The down grassland is very characteristic and occupies most of the area on shallow chalk soil (see Chapter XXVII).

The overlying loams and clays and the deeper chalk soils, naturally dominated by oak or sometimes in the south by beech, are used as arable land in which wheat is (or was) regularly brought into the crop rotation. The comparatively dry and sunny climate favours the crop, and the "wheat zone" (Chapter III, p. 75) which provided nearly half the wheatland of Great Britain, though its area does not exceed 10 per cent of the total area of the country, includes the northern and much of the central portion of the chalk outcrop (the Yorkshire and Lincolnshire wolds, west Norfolk and Suffolk, south Cambridgeshire and adjoining parts of Hertfordshire and Essex). The characteristic agricultural rotation was based on the so-called "Norfolk four-course system", in which wheat is followed by a root crop, this by barley, and the barley by a "seed crop", generally leguminous, such as clover or a mixture of clover and rye-grass. Very often there were only three crops, roots, cereal (mainly barley) and clover. This rotation depends essentially on the "folding" of sheep on the green crops ("roots" and legumes), so that the ground is trodden and manured by their droppings as well as by the nitrogen accumulated by the legumes. This is or was the characteristic "corn and sheep" agriculture of southern and eastern England. The typical chalk pasture of the southern parts of the outcrop (Berkshire, Wiltshire including Salisbury Plain, part of the Hampshire Downs, and the South Downs in Sussex) supported great flocks of sheep, and in the old "down farming" each farm included arable and sheep pasture, the sheep feeding sometimes in the folds and sometimes on the pastures. The Chalk of the North Downs and of the eastern counties has very much less open sheepwalk, often practically none at all, and is almost entirely arable, though in the eighteenth century the Yorkshire and Lincolnshire chalk wolds were also sheepwalk.

Gault and Greensand. Below the Chalk (in the south) comes a narrow terrace of (often highly calcareous) Upper Greensand (Albian) forming particularly good arable land, and then a belt of Gault, a stiff clay which is also somewhat calcareous and naturally bears damp oakwood, but is now mainly grass pasture (Fig. 34). Next below the Gault is the Lower Greensand (Aptian), varying a good deal in lithological character. Along parts of the northern edge of the Weald it forms a conspicuous sandstone escarpment (Fig. 34) as high as that of the Chalk: in other places, as on the south of the Weald, the sands are much softer and rise but little above the clays on either side. The sandy non-calcareous soil bears a climax vegetation of dry oakwood (Quercetum ericetosum, Chapter XVII), but this is frequently replaced by heath (heather, gorse and bracken) which is freely colonised by birch and subspontaneous pine, as on the Eocene sands. The massive sandstones of the Lower Greensand hills give acid soils which are very poor agriculturally, but where the sand contains finer particles and even thin beds of limestone, as in Kent ("Kentish Rag"), they are very fertile, and are intensively cultivated for fruit and hops, as in the lower Medway valley.

The Greensand of the "main" Cretaceous outcrop in the east midlands rarely makes a conspicuous feature. On the borders of Bedfordshire and Buckinghamshire there is however a well-marked escarpment bearing heath and largely planted with pine (Woburn and neighbourhood); and in Bedfordshire the low-lying Greensand with associated valley gravels is fertile and intensively cultivated for various "market-garden" crops.

Wealden beds. The Weald itself (Figs. 6 and 34) is occupied by Lower Cretaceous beds peculiar to this south-eastern area (see Chapter I, pp. 18, 19). Below the Lower Greensand comes a belt of Weald Clay, 10 miles wide at the western end of the Weald and 4 miles or more on the north and south, forming a plain overlooked on the north side by the bold escarpment of the Lower Greensand. The Weald Clay is mostly a heavy soil, although it may contain narrow bands of limestone and some loamy beds, and bears typical damp oakwood which once formed part of the great forest of Anderida. Most of the area is now permanent pasture used for dairying. Oak (*Q. robur*) is still the typical hedgerow tree of this soil, which was called by the early geologist, William Smith, "the Oak-tree Clay".

The centre of the Weald ("High Weald") is occupied by the much faulted "Hastings Beds", divisible into Tunbridge Wells Sand and Ashdown Sand, separated by Wadhurst Clay, and themselves interbedded with narrow clay bands. The sands form high ground, particularly the Ashdown Sand, which makes the so-called "Forest Ridge" in the centre of the Weald, at

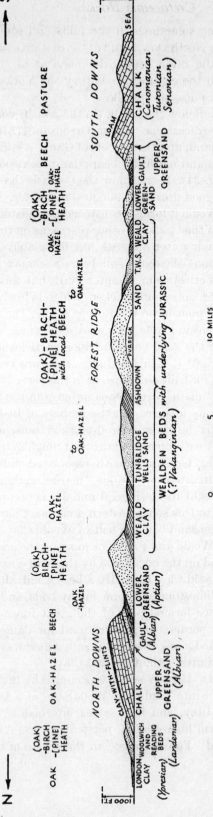

FIG. 34. SECTION THROUGH THE WEALD FROM NORTH TO SOUTH
Simplified and generalised, with the dip of the strata and the slopes exaggerated. The natural vegetation, depending mainly on the soil texture of the various strata, is indicated. Reprint of Fig. 7.

Crowborough Beacon rising to a height of 792 ft. (c. 241·3 m.) and still largely covered by oak- and birchwood and in places by heath and subspontaneous pine. The Tunbridge Wells Sand contains a good proportion of fine particles and tends to puddle like clay. It largely bears the damp type of oakwood, with extensive hornbeam coppice in east Sussex. The centre of the Weald, owing to faulting and the variety in texture of its soils, shows a good deal of relatively sharp relief on a small scale, especially in the eastern part not far from the Channel coast; and, since it still bears a great deal of semi-natural woodland, is one of the most picturesque regions of south-eastern England. Its woodlands are also remnants of the forest of Anderida, which originally extended from near the sea in east Sussex right across the Weald into Hampshire. Its soils are very ferruginous and ironworks were abundant there in the Middle Ages (see Chapter VIII, p. 182).

Jurassic rocks. Geologically below, i.e. to the north-west of the main outcrop of the Cretaceous, come the rocks of the Jurassic system, forming in the midlands a wide outcrop and on the whole a very regular belt of soils, from the middle and north Yorkshire coast of the North Sea to the Dorset coast of the Channel, running parallel to the Cretaceous, and occupying about half of the midland plain (see Fig. 5, p. 17). In the south midlands (e.g. north Oxfordshire) the higher Jurassic beds are very various, consisting mainly of clays with subordinate limestones and some few sands, presenting varied scenery on a small scale, and of varying agricultural value. The thickest member of the series is the Oxford Clay whose climax vegetation is damp oakwood of which a certain amount still remains in the usual modified form. The Oxford Clay is now largely under permanent pasture and forms much of the clay plain of the east midlands. The purest limestone is the Coral Rag (Corallian), which like other highly calcareous soils bears a good deal of ash, and the same may be said of the other thin bands of Upper Jurassic limestones—Forest Marble, Cornbrash, Purbeck Marble, etc.—collectively known, together with the Coral Rag, as Corallian limestones. The Kimeridge Clay, like the more extensive Oxford Clay, bears damp oakwood and is now mainly under permanent pasture. There is very little heath, corresponding with the paucity of sands, on the Upper Jurassic rocks.

Oolites. Next come the Oolites, which are partly composed of comparatively hard rock forming hill ranges, the most important of which, the Cotswold Hills in Gloucestershire and the Cleveland Hills in north-east Yorkshire, attain heights of over 1000 ft. (c. 300 m.) and 1500 ft. (c. 450 m.) respectively. The Oolite hills of the south, notably the Cotswolds, are relatively pure oolitic limestones, though with more insoluble residue than the Chalk. The long dipslopes of the Cotswolds are an agricultural region, famous at one time, like the Chalk, for its "rough grazing", a region of very old settlement and for long one of the main seats of the woollen industry, depending on the great flocks of sheep which were pastured there.

As the oolite outcrop is traced northwards the limestones become much less pure, and some are softer (though others furnish valuable building stone), so that they often fail to make a conspicuous feature in the landscape till the low but bold "Lincoln Edge" is reached. In north-east Yorkshire (Cleveland) the corresponding rocks are hard but mainly non-calcareous (Fig. 5, p. 17).

The limestones of the Cotswolds carry good semi-natural beechwoods on part of the escarpment and many of the valley sides, with much calcareous grassland, while the Cleveland valley sides show ash-oakwoods on the limestone, but mainly oakwoods (Quercetum sessiliflorae) on the siliceous rocks: their summit plateaux bear extensive heath and moor.

Lias. Next below comes the Lias, whose greatest extension is in the midlands, where it forms part of the central plain. It consists of clays, marls, shales and more or less pure limestone bands, and locally some sand and sandstone. This predominantly flat country has a few low escarpments formed by the narrow belts of limestone. The clays and marls give fertile agricultural country with rich pasturelands. Oakwood (ash-oak-hazelwood) and ashwood on the limestone exposures seem to be the natural vegetation, of which very little is left.

Triassic rocks. The Trias is stratigraphically the lowest formation of the Secondary rocks. The outcrop is parallel to that of the Cretaceous and Jurassic. It forms a broad expanse, broken by islands of older rocks, in the very centre of England (Fig. 5), forming about half of the Midland Plain, and extending north-westwards to the Cheshire and south Lancashire coasts between the southern end of the Pennines and the mountains of North Wales. The northern portion of the outcrop, on the eastern flank of the Pennines, strikes due north and occupies the whole coast of Co. Durham. There are also outlying areas in the Vale of Clwyd in North Wales and the Eden valley in Cumberland, both very fertile but mainly devoted to grass and dairying. The south-western portion of the outcrop is somewhat interrupted and extends from the Midland Plain to the south coast of Devon, where it forms the striking red cliffs of Sidmouth, Budleigh Salterton and Dawlish. In central Somerset and east Devon the Triassic marls provide a variety of very fertile soils. Some of it is devoted to market gardening, some to cider orchards, and much to dairying, with crops grown on the four-course rotation—excellent malting barley on the lighter and wheat on the heavier soils.

Speaking broadly the main central area of Trias forms low ground—to the north the Vale of York, to the north-west the Cheshire Plain, and in the centre the central plain of England in Staffordshire, south Derbyshire, Nottinghamshire, Leicestershire, Worcestershire, and Warwickshire. It has two main divisions, the Bunter, consisting of sandstones, and the Keuper, mainly of marls, the continental "Muschelkalk" being absent from this country. The Bunter Sandstone forms coarse poor soils bearing dry woodland of oak and birch as well as heath, as in Sherwood Forest in Nottinghamshire and Delamere Forest in Cheshire. The Keuper marls often form fine meadow- and pastureland. But these rocks are much overlaid with very various glacial drifts which form the surface soil. The tendency to replace arable by permanent grass, and bullock fattening by dairying, has been very marked throughout this area in recent times.

Thus the Midland Plain of England (if we reckon it as starting at the foot of the main Chalk escarpment) is formed mainly of wide bands of clays and marls (broken only by relatively unimportant bands of limestone) belonging to the Jurassic and the Triassic systems, though the Trias is interrupted by two or three considerable areas of Carboniferous (the midland coalfields), and here and there by small patches of Archaean rocks rising from beneath it. This great clay plain was for long covered with dense oak forest, and it was not till Saxon times that it was settled and the midland kingdom of Mercia founded.

Permian: Magnesian Limestone. The Magnesian Limestone of Permian age which lies at the top of the Palaeozoic series is sometimes included in the Mesozoic, and it is indeed a comparatively soft limestone more like those of the Secondary systems, though stratigraphically it is often united with the great underlying series of strata as the Permo-carboniferous system. It crops out in Yorkshire and other northern counties between the Trias and the great Carboniferous basins, and yields a good light dry arable soil of rendzina type. Its natural vegetation has the usual "calcicole" character, including abundance of ash.

The other Permian beds, of no great extent, consist of various red sandstones, marls, breccias and conglomerates, yielding very various soils.

Palaeogenic Rocks

General characters. Nearly all the Palaeozoic rocks share with the Archaean (pre-Cambrian) and metamorphic, as well as with most of the igneous rocks, the general character of hardness due largely to the immense pressure of superincumbent strata to which they have been subjected. This results in extremely slow weathering and consequent soil formation. Setting aside the very widespread glacial drifts, which have already been described, these old indurated rocks, as we saw in Chapter I, form the whole surface of Scotland, almost the whole of Ireland and Wales, the extreme south-west, and a large part of the north of England. Between them they form practically all the land in these islands over 1500 ft. (c. 450 m.) above the sea, as well as the lower lying land of all the north and west, and they are mostly situated in the regions of wet climate and cool summers (cf. Figs. 3 and 4, pp. 7 and 9 with Fig. 11, p. 32). It is therefore not easy to separate the effect of climate from that of the parent rock in producing their characteristic soils and vegetation. One outstanding effect of the heavy rainfall is, however, to denude the higher levels and steeper slopes of the scanty soil that is formed by the slow weathering of the hard rock. Thus the climatic and lithological factors tend to work in the same direction, combining to produce bare rock or very shallow soil.

Excepting the limestones and certain other basic rocks, nearly all these soils have much in common. They are shallow and mostly acidic, with a strong tendency to the formation of raw humus on the surface and to podsolisation. The rocks themselves include mudstones, gritstones and sandstones, slates, gneisses, schists, granites, syenites, quartzites, dolerites, basalts, etc. The softer rocks, which are usually eroded to form valleys, are generally shales, the Palaeozoic equivalent of clays.

The natural climax vegetation of all the non-calcareous rocks up to an altitude of 1000 ft. or so, except in the north of Scotland, is sessile oakwood (Quercetum sessiliflorae), with birchwood (Betuletum pubescentis) and pinewood (Pinetum silvestris) in the Scottish Highlands, forming a zone above and to the north of the oakwood. In the wettest places alder (*Alnus glutinosa*) is abundant or even dominant. On the limestones ashwood (Fraxinetum excelsioris) is quite characteristic, again passing into birchwood at the higher altitudes.

Woodland is, however, very sparsely present in the north and west. This is of course largely due to destruction of forest by felling and grazing, but much of the land surface, even at low altitudes, within the region of the very wet western climate, cannot support good forest, because of the constant formation of wet acid peat where drainage is impeded. The natural dominants are various sedges and grasses, largely *Scirpus caespitosus*, *Eriophorum* spp., and *Molinia caerulea*. Much the greater area of the drier soils of the hillsides which have been deforested is maintained as grassland by grazing. *Calluna* dominates well-drained slopes in the drier regions of the eastern Highlands; *Molinia* and *Nardus* the less well drained areas, particularly in the Southern Uplands, the Pennines and the Welsh hills; and *Agrostis* and *Festuca* the drier, grazed hillsides. At higher altitudes *Vaccinium myrtillus* replaces *Calluna*, and on the loftier mountains, most richly when they consist of certain basic rocks such as mica schists and chloritic schists in the central and eastern Highlands, the characteristic arctic-alpine vegetation appears. Descriptions of these various kinds of vegetation, and of their relations to habitat and soil, so far as these are known, are given in later chapters.

The thin acid soils produced by most of these rocks are not favourable for agriculture, and even in the lower lying regions below 1000 ft. this is restricted to the bottoms and lower slopes of the valleys, where rain wash and alluvium have accumulated, and is almost absent from the wild mountain regions of the Highlands of Scotland and of the west. The really fertile regions, of Scotland for example, as already mentioned, occur only on certain areas of glacial drift near the east coast. For the rest, and with a few striking exceptions, it is only where the rewards of specialised agriculture are great owing to the proximity of a good industrial market, that skilful intensive farming has attained success even on the most unpromising soils.

Carboniferous system: Mountain Limestone, Millstone Grit and Coal Measures. The Carboniferous rocks, the uppermost system of the Palaeozoic, occupy a larger area in the British Isles than any other of the great geological systems. The lower division is usually represented by the massive Mountain Limestone—which, together with the Chalk, forms by far the greater part of all our limestone soils. The upper division consists of the equally massive Millstone Grit and above this of the alternating sandstones, shales and coal seams of the Coal Measures. The contrast between the soils and vegetation of the limestone and sandstone, the former occupied by ashwood and the characteristic calcareous grassland giving excellent grazing for sheep, the latter by sessile oakwood or by birchwood, and by grazed siliceous grassland, heath or moorland, is strikingly seen in the Pennine region. The centre of the Pennine anticline is formed of Mountain Limestone, which comes to the surface in the south and also in the north, while the Millstone Grit lies on the flanks and covers the limestone in the centre of the ridge. To east and west of the southern part of the Pennine Hills are the Coal Measures of the Yorkshire and Lancashire basins, and farther north the Northumberland and Durham coalfield on the eastern side. These coalfields largely determined the positions of the northern industrial districts whose industries are based on coal. The agricultural industry of the lower parts of the valleys was stock raising, now largely replaced by dairying; but the higher lying and more exposed parts are devoted to sheep. The Mountain Limestone occurs again in North and also in South Wales, where it forms a rim round the great South Wales coalfield. Other outcrops of Mountain Limestone occur in the lower Wye valley, and again south of the Severn, of which the most conspicuous forms the Mendip ridge. Coal Measure sandstones and shales occur again round several coalfields south and south-west of the Pennines, the Staffordshire, north Worcestershire and Leicestershire coalfields interrupting the midland Triassic plain, the Wrexham coalfield on the borders of North Wales, and, farther south, the great Glamorganshire coalfield, and the smaller ones of the Forest of Dean and Bristol on either side of the Severn estuary. All these, so far as can be ascertained, bore sessile oakwood, damp and dry on the shales and sandstones respectively.

Culm measures. Carboniferous rocks occupy about half the surface of Devonshire, where they are collectively known as the "Culm measures" because they contain thin bands of impure anthracite or "culm". They consist of sandstones and shales with occasional limestones, but Mountain Limestone, Millstone Grit and Coal Measures proper are absent. The Culm measures give rise to a much diversified country with quite sharp relief on a minor scale and swiftly flowing streams in the small valleys, whose sides are often clothed with oakwood. The soil in the north-west of Devonshire is largely derived from shale, giving a cold wet clay loam: towards the east it is a reddish sandy loam. Both are deficient in lime and phosphates but respond readily to treatment. There is a good deal of cereal growing in places, but most of the corn is fed to stock.

Calciferous sandstone. To the north of the Pennines, in north-eastern Cumberland, Durham, and Northumberland, the Mountain Limestone is replaced by the "Bernician" or Calciferous Sandstone series, consisting of sandstones, grits and shales of very variable character, with some interbedded limestone zones. This country, like the adjacent Coal Measures of Durham, is occupied by various types of moorland. The Carboniferous rocks of the central valley of Scotland, between the Southern Uplands and the Highlands, are also described as calciferous sandstones.

Central Irish plain. The great Carboniferous Limestone plain of central Ireland is largely covered, as we have already seen (p. 105), with calcareous glacial drift formed by the scraping of ice over the limestone. The deeper soils so formed bear highly productive cattle pasture, and permit the entrance of oak into the ashwoods. The shallower limestone soils in the west bear the remains of old ashwoods and in places extensive hazel scrub, accompanied by the characteristic limestone plants (see pp. 473-5).

Limestone pavements. In the extreme west, on and near the windswept coast of Co. Clare, the limestone rises in largely bare terraced hills with fissured pavements on the flat terraces. Similar pavements also occur in the mountain limestone districts of north-west Yorkshire. The fissures are inhabited by a characteristic vegetation including many woodland species. In areas sufficiently protected from wind the terraces are also colonised by woody plants as soon as some soil is formed, very largely by nearly pure hazel scrub. In the north of England the ash comes into the succession, rooting in the fissures and shading the intervening pavement, which becomes covered with mosses, so that soil accumulates and a regular ashwood develops. Where humus accumulates from surface growth of lichens and mosses and leaching is extreme, as on flat surfaces, especially at high altitudes, the raw humus may become colonised by *Calluna* and other calcifuge plants, and thus a heath may develop with the limestone only a few inches below.

Devonian system. The Devonian system, which underlies the Carboniferous, occupies considerable areas north and south of the Carboniferous in north and south Devon and Cornwall, consisting mainly of grits and slates with numerous interbedded limestones. The diabases, tuffs (volcanic ash), and slaty volcanic rocks forming the red soils of the Ashprington series near Totnes in south Devon support some of the best arable and grassland in the district. The agriculture of this south Devon area is most diversified, including arable, with roots, cereals (mainly oats) and grass in the rotation, long leys, permanent pasture and cider orchards, with local industries for the production of flowers and fruit, as in the Tamar valley. The south Cornish soils of this formation are seriously deficient in calcium, except those derived from local basic igneous rocks which are very fertile. The serpentine of the Lizard peninsula with its excess of magnesium is extremely infertile and is mostly covered with heath, large areas being dominated by the "Cornish heath" (*Erica vagans*), a Lusitanian species, here occurring in its only British station.

Tertiary granite intrusions. The two easternmost of the great Tertiary granite bosses intruded among these rocks and forming a series from Dartmoor to Land's End and the Scilly Isles, bear the elevated moorlands of Dartmoor and Bodmin Moor, occupied by "moorland" vegetation and blanket bog on the highest plateaux. The next (north of St Austell) is the centre of the Cornish "china clay" industry, the kaolin being dug from the metamorphosed edges, while the westernmost are quite fertile and mainly under cultivation, producing heavy crops and carrying a large number of stock, mainly dairy cattle and pigs. The valley sides of all this region bear oakwoods.

The Scilly Islands, with their extremely mild Atlantic climate (pp. 57–9) and the famous gardens of Tresco, full of luxuriant subtropical plants, are almost entirely devoted to early flower growing—narcissus, "arum lilies" and anemones—for which shelter has always to be provided because of the violent westerly gales.

Old Red Sandstone. The Devonian rocks of north Devon and the adjoining region of west Somerset are Old Red Sandstone, most of which is a relatively sterile soil, and the highest land (the plateau of Exmoor and the narrow ridge of the Quantock Hills) bears heath and moor, woods of the sessile oak occupying the valleys. Exmoor was largely "reclaimed" during the nineteenth century, but is of little value for cultivation. North of the Bristol Channel the Devonian rocks are again represented by Old Red Sandstone, which covers a large area on the Welsh Border and extends into South Wales through Breconshire in a narrowing belt, reaching the Carmarthen coast to the north of the Carboniferous basin. Part of this area of Old Red Sandstone, for example in the Herefordshire plain, yields a strong loamy or marly soil which is very fertile, with a fair proportion of arable land under stock-feeding crops, but mainly good permanent pasture with many orchards and hop yards. The limestone bands called "cornstones", which occur in it, form, however, the richest lands. The Herefordshire plain, in fact, though geologically Palaeozoic, belongs physiographically and agriculturally to the neogenic English lowlands.

In south-central Scotland the Old Red Sandstone occurs on both sides of the great central valley, consisting of conglomerates, flags, sandstones, shales and marls, with some intercalated volcanic beds and also calcareous material. It is the glacial drift derived from the easternmost of these beds that gives the specially good farmland already mentioned (pp. 105–6). Old

Red Sandstone also occurs extensively in north-eastern Scotland, including the Orkney and Shetland Islands, again consisting of sandstones, flags, shales and some minor limestones.

In Ireland the Old Red Sandstone forms most of the mountains of the extreme south-west, including Macgillicuddy's Reeks in Co. Kerry, the highest mountains in Ireland (Carrantuohill, 3414 ft. = 1040 m.). From this region the outcrop extends eastward to Waterford on the south-east coast, following the lines of the old Armorican folds, and often forming considerable mountains with Carboniferous slates on their flanks. To the north, also, numerous ridges of Old Red Sandstone rise like islands above the plain, exposed by the anticlinal uplifts and the continuous wearing away of the Carboniferous Limestone from the ridges. These hill masses may reach considerable heights, as in the Slieve Bloom and Galty ranges (Galtymore, 3015 ft. = 920 m.). In the centre of these anticlines relatively soft Silurian shales are exposed and cultivated. Farther north the Old Red Sandstone exposures are fewer and smaller, but Co. Tyrone in Ulster has a considerable area. In contrast with the fertile limestone soils of the central plain, these sandstone hills are barren, supporting mainly heath and moor. In the low-lying valley of Killarney, with its mild and wet climate, the Old Red Sandstone supports oakwoods of *Quercus sessiliflora* accompanied by luxuriant holly and arbutus and a rich epiphytic vegetation of bryophytes (see Chapter XVI, pp. 327–40).

The older Palaeozoic rocks. For the present purpose the three lower systems of the Palaeozoic—Silurian, Ordovician and Cambrian—may be briefly treated together, since the soils they produce under the generally uniform cool damp climate, and the vegetation they bear, show a certain uniformity wherever the drainage is comparatively free, though there are a few notable exceptions. These rocks consist of mudstones, flags, slates, shales, sandstones, grits and quartzites with very occasional bands of limestone.

Silurian rocks form most of the Southern Uplands of Scotland with a broad continuation in north-east Ireland, and smaller areas in the south and west. They also form the mountains of the southern part of the Lake District and a considerable extent of eastern Wales, with smaller outcrops on the Welsh Marches including narrow ridges of limestone. Ordovician rocks form a band on the northern edge of the Silurian in the Southern Uplands and a similar narrower band in north-east Ireland. Part of this north-east Irish Silurian and Ordovician area is low-lying, hummocky and forms good arable soils. Ordovician rocks are, however, much more extensively developed in the Leinster chain in south-east Ireland, the core and the highest summits of which are formed of granite and bear wet moorland vegetation including blanket bog. The clays worn from the Ordovician foothills furnish good agricultural land.

The northern part of the Lake District mountains and the bulk of the central and western Welsh highlands are made of Ordovician rocks, largely metamorphosed into slates, and this system, together with the Silurian, is again slightly represented in Cornwall. The Welsh Ordovician highlands are largely rough grazings used for sheep walk. Both systems are associated with many and various igneous intrusions, both acidic and basic, and consequently show extensive metamorphism.

The Cambrian system, at the base of the Palaeozoic, covers a much smaller total area than any of the four systems above them. The largest exposures are in North Wales, the Southern Uplands, the north-west Highlands and the Leinster Chain of south-east Ireland, while there are a number of small outcrops in the English midlands, appearing from below the Mesozoic rocks.

Archaean rocks. Pre-Cambrian ("Archaean") rocks form almost the whole of the Scottish Highlands, to the north-west of the great "Highland boundary fault" which runs from the coast of Aberdeenshire to the Firth of Clyde. This huge mass of the most ancient rocks consists mainly of schists, gneisses and quartzites, with some slates, and is thus chiefly metamorphic, with various intrusive igneous rocks, but also includes some sandstones and limestones. It represents a vast period of sedimentary deposition, perhaps longer than the whole of the younger series previously described, and has since undergone vast deformations—folding, overthrusting and crumpling—as well as immense denudation—so that its structure is extremely complex. The soils are mostly impossible for agriculture except in the large valleys. Though lime-containing rocks are not wholly absent in the western region and on the high plateaux where the rainfall is over 60 in. (1500 mm.) and often a good deal higher, the mantle of deep, highly acid peat passes uniformly over everything alike and is uninfluenced by changes in the rock type. These "moors" or blanket bogs of the western areas are dominated largely by *Scirpus caespitosus* (see Chapter xxxv). In the extreme north the rainfall is a good deal lower (30–40 in. = c. 750–1000 mm.), and here, as in the eastern Highlands, heath (*Calluna*) is much more widely dominant.

The central and eastern Highlands, to the south and east of the rift valley of Glenmore, with its chain of lakes, through which runs the Caledonian Canal, likewise consist mainly of schistose metamorphic rocks, with continuous areas of high elevation. Here the climate is not so wet as in the west, and since the glacial deposits are limited to the valleys, the influence of the rock on the vegetation is often conspicuous. This is especially seen where the prevalent acidic schists, rich in quartz, are replaced by chloritic and mica schists, which introduce edaphic and physiographic features favourable to a much greater variety of species and to more luxuriant growth. They are more friable, thus forming soil more readily, the sharper micro-relief affords protection from wind, and the base status is notably higher. On the higher mountains possessing such features the greatest wealth of "arctic-alpine" vegetation is met with, including considerable tracts of natural grassland (see Chapter XXXVIII). On the mountains formed of acidic rocks *Calluna* is generally dominant up to about 2000 ft. (*c.* 600 m.), *Vaccinium myrtillus* at higher levels. Similar contrasts are seen in the mountains of North Wales.

In the western Scottish Highlands, then, we see the influence of extreme climate rather than the nature of the rock clearly determining soil type, while lithology comes into play in the more moderate climates of the centre and east.

In England and Wales Archaean rocks are comparatively rare, but they are exposed in a number of places in the extreme west, for example at the Lizard in Cornwall, in Pembrokeshire, and in Anglesey, while isolated inliers arise from beneath newer strata in the English midlands, as in Charnwood Forest in Leicestershire. Here the drier climate favours the development of heath or heathy grassland, with dry acidic oakwood remaining locally, vegetation not notably different from the heaths of the southern neogenic sands.

Retrospect. On looking back over this brief and inadequate summary of the soils produced by the various members of the geological series, from the recent deposits to the Palaeozoic and Archaean, it is clear that the lithology of the newer rocks—Tertiary and Secondary—has a much greater differentiating effect on the vegetation than that of the older, always excepting the limestones. The great majority of the Palaeozoic and Archaean soils, produced from intensely consolidated, very hard rocks which weather with difficulty, are shallow, comparatively "sterile" and unsuitable for agriculture proper, though there are a few notable exceptions; and, so far as they are used by man, are rough sheep grazings, "deer forest" or "grouse moor". The vegetation tends to be uniform over wide areas including rocks of very various lithology, the differentiating factors being rather water content of soil, slope, aspect and altitude, though limestones and other basic rocks, which are more readily soluble, always bring about a distinct change in the plant covering. The monotony is, however, greatly intensified by the cool damp climate of the north and west, which also contributes uniformity to both soil and vegetation, tending to the surface accumulation of raw humus, to podsolisation, and to the dominance of oxyphytes—in the extreme case to the development of deep acid peat and its characteristic vegetation.

Distribution of the main lithological types. While it is impossible to summarise the distribution of the soils mentioned in the preceding pages, it will be useful to indicate here some of the more important areas occupied by the various kinds of soil.

ORGANIC SOILS. *Fen peats* of lake basins and upper parts of old estuaries fed by waters rich in bases. The East Anglian Fenland, the Norfolk Broads region of east Norfolk, parts of the Somerset levels, the region round and to the south of Lough Neagh in north-eastern Ireland, and numerous smaller areas throughout the country, except in the regions of exclusively acidic rocks.

Acidic peats of the blanket bogs of west Scotland and Ireland and of elevated plateaux, as on the Pennines, Dartmoor, the Wicklow Mountains, and large parts of the Scottish Highlands; and likewise of the raised bogs built on ancient fen—very common in the central Irish plain and in parts of Scotland; formerly common, especially in the west of England and in Wales, but these are partly or wholly destroyed by drainage, removal of peat and cultivation (Tregaron Bog in Cardiganshire and some of the "mosses" in the north-west are surviving examples in a more or less natural condition, see pp. 682, 686).

SANDS AND GRAVELS. Blown sand deposits at various places on the coast, especially numerous on the west coasts. "Valley" and "plateau" gravels (in the south of England) overlying various rocks of Eocene, Pliocene or Quaternary age. Glacial sands and gravels of very various and widespread but local distribution (not in the extreme south of England). Pliocene "crag", etc., east Norfolk and Suffolk. Bagshot Sand, mainly south-west (and a little to the north) of London and in the Hampshire basin. Lower Eocene sands: fringing the London and Hampshire basins. Lower Greensand: fringing the Weald and in the south-east midlands. Hastings Sand: in the centre of the Weald. Sands derived from Bunter (Sherwood and Delamere Forests), and from various older sandstones and from granites.

SILTS, LOAMS, BRICK EARTH, ETC. Mixed and silt soils—alluvium of estuaries and river flood plains, glacial, inter-glacial, and Pleistocene extra-glacial brick earths, and varying loams derived from rocks of different ages.

CLAYS. Non-calcareous Boulder Clay or "till" of large areas in Scotland and the north of England. London Clay of the London and Hampshire basins. Weald Clay; Gault Clay (sometimes calcareous) of the Weald and main Cretaceous outcrop Kimeridge and Oxford Clays of the east and south midlands. Much of the so-called Keuper Marl: midlands and north. Clays and clayey loams of Devonian age in Hereford and Devon.

CALCAREOUS CLAYS AND MARLS. Chalky Boulder Clay: mainly derived from and largely over-lying chalk and Jurassic limestones in the east midlands and East Anglia. Calcareous boulder clay: central Irish plain. Liassic marls and Keuper Marl (calcareous): parts of the midlands and south-west England, and numerous minor marl formations.

SHALLOW SOILS OF THE OLDER NON-CALCAREOUS ROCKS (Archaean, Palaeozoic, metamorphic and igneous): very various in detail but with much in common, derived from slates, grits and the finer grained igneous rocks, the softer shales however giving soils akin to clays: most of the north and west of Great Britain, and coastal regions of Ireland.

LIMESTONE SOILS (largely rendzinas). Chalk: main outcrop from east Devon and Dorset right across England to Lincolnshire and east Yorkshire (Wolds); Wiltshire (Salisbury Plain), Hampshire, North and South Downs on the two sides of the Weald. Oolitic limestones: Cotswold Hills and northwards and southwards, interruptedly along the Lower Jurassic outcrop. Minor Upper Jurassic limestones to the south-east of the last. Magnesian Limestone: Nottinghamshire, Yorkshire, etc., east of the Pennines. Carboniferous (Mountain) Limestone: north and south Pennines, North and South Wales, Mendips, central Irish plain (much covered by Calcareous Boulder Clay), Co. Clare, Co. Sligo, etc. Devonian limestones: Devonshire. Silurian limestones: borders of South Wales.

REFERENCES

EVANS, J. W. and STUBBLEFIELD, C. T. *Handbook of the Geology of Great Britain.* London, 1929.

MAXTON, J. P. (edited by). *Regional Types of British Agriculture.* London, 1936.

WOODWARD, H. B. *The Geology of England and Wales.* Second edition. London, 1887.

Chapter VI

THE BIOTIC FACTOR

Having dealt with the chief "inorganic" factors in the British environment of plant life we must consider a no less important influence—the effect of animals on vegetation, now generally known as the *biotic factor*.

The ecosystem. The conception of vegetation as a series of systems in approximate equilibrium, or in progress towards equilibrium, with their whole environment will be explained in Chapter x (p. 228). It will there be pointed out that what we may call the *ecosystem* consists of both organic and inorganic components which may be conveniently grouped under the heads of climate, physiography and soil, animals and plants. In ecology our interest is centred in the organisms of the system—the *biome* as they have been called collectively—and the inorganic components are regarded as "factors of the habitat" conditioning the existence, structure and development of the particular biome.

Animal and plant ecology. But owing to the great differences between the life requirements of animals and plants it is for many purposes necessary to consider and study them independently. The methods and principles of animal ecology—a branch of science which has made great progress during the last twenty years—are in many respects necessarily different from those of plant ecology. The ideal subject of study is certainly the ecosystem as a whole, including all its components, animals and plants together with the inorganic factors, but there is as yet no single complex ecosystem of which we have anything approaching a satisfactory knowledge. Of one aspect of the ecosystem—the interrelationships of the organisms constituting the biome—our knowledge is indeed still in a very rudimentary state; and it can only be developed and perfected by independent work on the animals and plants as well as on the inorganic factors, and on the relations of all these components to one another. Synthesis of the results, involving a really scientific description of the ecosystems of the world, is a work of the future, scarcely likely to be completed before most of the "natural" ecosystems have been destroyed, broken up, or at least profoundly modified by human activity.

The "biotic factor". For these reasons it is still the soundest and most fruitful method in plant ecology to take the plant community as the primary object of study, and the associated animals as factors of their environment equally with the inorganic climatic and edaphic factors. In animal ecology animal communities must be studied with their vegetational background whenever the animal community depends on plants as the essential basis of its environment, as it very often, though by no means always, does. In plant ecology it is the totality of the direct and indirect effects of animals on the plant community which we speak of as the biotic factor—more exactly the complex of biotic factors.

The "virgin" ecosystem. In "virgin" ecosystems, in which man plays little or no part as a component, the animals and plants have developed together and have become fitted to one another and to the inorganic components till they are all in approximately stable equilibrium. Primitive man, just like any other animal, originally formed, and in some few parts of the world still forms, a component, nicely adjusted to the system as a whole, of the ecosystems of those regions which he inhabits. But with his increasing control over "nature" the human animal became a unique agent of destruction of the original ecosystems, as he cleared and burned natural vegetation and replaced it with his pastures, crops and buildings. Limited at first to the regions where civilisation originally developed, this destructive activity has spread during recent centuries, and at an increasing rate, all over the face of the globe except where human life has not yet succeeded in supporting itself. It seems likely that in less than another century none but the most inhospitable regions—some of the more extreme deserts, the high mountains and the arctic tundra—will have escaped. Even these may eventually come, partially if not completely, under the human yoke.

<small>Effect of human activity</small>

But the destruction has not everywhere been equally intense and far-reaching. Where the cleared ground has been occupied by buildings or by planted crops or is continually trampled by man and his domesticated animals, the natural vegetation of the original ecosystem has quite disappeared with no opportunity to return. But wherever it has a chance it tends to re-establish itself, just because it is the natural product of the particular climate and soil and of the sum of organisms present in the particular region. And where man has not destroyed but has *used* the natural vegetation for his own purposes, the particular use to which he has put it tends to the establishment of a new ecosystem, the result of the original factors of climate and soil together with the modifying factors which he has introduced.

Thus a natural forest which is exploited on the "selection system", only certain trees being removed and enough seed parents left to secure regeneration, is an example of a natural ecosystem but slightly modified by human activity; and the same is true of a not too heavily grazed natural grassland. A savanna or "savanna forest" which is repeatedly burned is more highly modified, being inhabited only by woody plants, grasses and herbs which can maintain themselves in spite of the fires, and the same is true of a heather moor preserved for grouse and regularly fired every few years.

The pastures of the British Isles (apart from the "long leys" ploughed and sown with grass and cut or grazed for a number of years before re-ploughing) are more highly modified but still semi-natural ecosystems.

Originally derived from forest land the pasture or meadow cannot be recolonised by woody plants because of the continual grazing; and we have here a plant community conditioned on the one hand by climate and soil and on the other by the constant factor of grazing.

Anthropogenic factors. Semi-natural communities, such as pastures whose essential nature is determined and maintained by human activity, are sometimes called "biotic climaxes" or "biotic subclimaxes", because they are held in a condition of equilibrium by a biotic factor, and so long as this factor is present are unable to reach the climatic climax. All the factors mentioned—felling, mowing, grazing and fire—are sometimes loosely termed "biotic factors", though they would be more properly called *anthropogenic factors*, and the climaxes they produce *anthropogenic climaxes*. Among the anthropogenic factors the grazing factor is biotic in the strict sense.

Importance of the biotic factor. The vegetation of a country like the lowlands of the British Isles, and up to a considerable height on the hills, is enormously influenced even where it has not been essentially determined by these factors, and the recognition of their ubiquitousness and importance has been one of the chief advances in the study of British vegetation during the twenty-eight years which have elapsed since *Types of British Vegetation* was written. The authors of that book, though they were far from ignoring the biotic factor, did not fully appreciate its almost universal prevalence and far-reaching effects; and it was largely the work of Farrow (1916, 1917) on the influence of rabbits on the vegetation of Breckland that called general attention to its significance. Watt's work (1919, 1923) on the influence of mice, birds and other small animals in the destruction of tree seeds and seedlings and the consequent prevention of oak and beech regeneration in our woods was also an important contribution in the same direction.

Creation of grassland by grazing. The history of the transformation of British vegetation by human activity (chiefly during the last 2000 years, and with increasing rapidity), of which some account is given in Chapter VIII, is mainly a history of the destruction of forest and its replacement either by arable crops or by grassland. The change from forest to grassland may properly be considered as due to the biotic factor of grazing. Forest in our climate can be converted into grassland by grazing alone, and though the great majority of our existing "permanent pastures" as opposed to "rough and hill grazings" have actually been sown to grass, there can be no question that the original process of conversion was carried out, sometimes intentionally, sometimes without deliberate purpose, by grazing cattle or sheep or by felling combined with grazing. The creation of pasture by sowing with grass seed usually followed the employment of land already under the plough for the growing of cereals or root crops. It is merely a method of getting more quickly a better grassland than that which would naturally arise by grazing arable land allowed to go fallow. In a word, grassland of some sort is the inevitable fate in our climate of most of the land which is regularly grazed.

Cattle and sheep. The chief agents in this transformation have been sheep and cattle, with horses to a much smaller extent, and goats as a comparatively insignificant subsidiary. Cattle do not bite so close as sheep, and they have a different effect on grassland, but there are no adequate studies of these differences. One instance, not really well established, may be mentioned. In the years following the war more cattle and fewer sheep were grazed on the South Down pastures. At the same time the heaths—*Calluna* and especially *Erica cinerea*—increased on the deeper and less purely calcareous soils of the Downs; and it was said that while these plants could not establish themselves under continuous sheep grazing they were able to do so when the pastures were grazed by cattle alone.

Variations in grazing intensity. The incidence of grazing varies very much in intensity. As a result of local or temporary agricultural conditions, or of inefficient farming, pastures are often very much undergrazed, and then woody plants establish themselves, though they do not increase to the extent of forming continuous scrub or incipient woodland unless the land is allowed to become quite derelict. But pastures suffer very much from undergrazing, not only through the invasion of woody plants but through the increase of coarse innutritious herbs and grasses which are neglected by the grazing animals, and the accumulation of a "mattress" of dead herbage; so that undergrazing initiates a vicious circle and the pasture eventually becomes valueless and derelict. The shrubs which are let in by temporary undergrazing will be eaten off when the land is again more adequately stocked unless they are protected by spines or thorns, and this is one reason why the commonest and sometimes the sole scattered bushes on pastures which have been somewhat neglected are armed species such as hawthorn (much the commonest), blackthorn, gorse, bramble and rose.

Sheep grazing and heather. On sandy and other acid soils the first woody invaders of neglected grassland are the heaths, and especially *Calluna*. On the slopes of the southern Pennines, as Adamson has described (1918), there is a regular reversible change between grassland and heath dependent on the presence or absence of sheep grazing. An elegant example of this process was observed in 1917 on the northern slopes of Moel Siabod above Capel Curig in North Wales (Fig. 35). Adequate stocking with sheep had been limited by the "mountain wall", a stone wall impassable to sheep and following the contour of the hillside at about 1600 ft. (say 500 m.). On the other side of this wall was a fairly dense Callunetum, while in the area grazed by sheep grasses were dominant and *Calluna* practically absent. The wall, however, had broken down here and there and the broken portions had recently been replaced by a wire fence which ran straight along the hillside, sometimes above and sometimes below the ruined wall, since it did not, like the latter, follow the sinuosities of the contour line (Fig. 35). The result was that in some places the sheep were stopped by the fence before they reached the wall, but in other places they could pass through the gaps in the wall and were only stopped by the fence some yards beyond. In the areas between the fence and the wall from which the sheep were excluded but which they had formerly grazed, the vegetation was still dominated by grasses, which had grown tall, but among which vigorous plants of *Calluna* had established themselves: in the areas between the wall and the fence, on the other hand, to which the sheep now had access but from which they were formerly excluded, the old Callunetum was severely eaten back and obviously moribund.

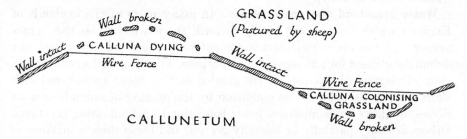

FIG. 35. GRASSLAND AND HEATHER ON MOEL SIABOD, NORTH WALES
Effect of fresh grazing and withdrawal of grazing (see text).

Effect of sheep grazing on bracken and mat-grass. Another effect of grazing by sheep, especially by ewes and lambs, is to facilitate the spread of species which they will not eat, such as bracken (*Pteridium*) and mat-grass (*Nardus*). The "rough grazings" of Scotland, which, including the deer forests, are said to occupy 62 per cent of the total land area (Stamp and Beaver, 1933), were almost exclusively pastured by sheep before the increasing reservation for deer began. There are said to be twelve million sheep in the country now, and there were certainly many more a century ago. "In the past [i.e. the eighteenth century and earlier] cattle as well as sheep prevented the natural regeneration of trees, and then heather began to feel the effects of the constantly increasing pressure of sheep grazing. Nearly two centuries ago sheep replaced cattle on the more remote hills and rough grazings....From that time dates the beginning of the bracken menace. Previously the bracken was kept in check, as it has many uses for man and beast. Since then it has virtually got out of control.... The destruction of heather and the overgrazing of the better areas has led to the extensive spread of mat-grass (*Nardus stricta*). In many parts of the Southern Uplands and in the southern parts of the central Highlands this grass is still spreading and presents a very difficult and serious problem.

It is a fact of great significance that these two regions show the densest sheep population in Scotland" (Fenton, 1937). Much the same thing is taking place on the hill grazings of Wales, both with *Nardus* (Roberts, 1935) and bracken (Davies, 1936).

In some plot experiments in Boghall Glen, Midlothian, at an altitude of about 1000 ft. (300 m.), the late W. G. Smith showed that if the Nardetum is fenced against grazing it tends to return to heath. It has since been established that if the Nardetum is fenced and regularly cut, or manures are applied and the plots grazed, an *Agrostis-Festuca* community (see Chapter XXVI, p. 499) tends to replace the Nardetum. The introduction of Galloway cattle will also reduce the *Nardus* and without diminishing the availability of the pasture for sheep, while the lambs improve in size and quality (Fenton, 1937).

Waste grassland and grass verges. In many parts of the lowlands of England which are not too highly farmed, there are, besides the "grass verges" of the roadsides, numerous patches of ground which are not definitely utilised for any agricultural purpose. These, as well as the verges, show all degrees of intensity of the grazing factor. Some are covered with short turf, being held in this condition by the grazing of animals such as horses, donkeys and sometimes goats belonging to neighbouring cottagers. Others are only partially or casually grazed, and these show a mixture of grasses, various colonies of perennial herbs, and often shrubs and young trees. In others again, where the grazing factor has for some time been absent or negligible, scrub or incipient woodland has established itself.

Biotic zonation on common land. On unenclosed common land in the English lowlands, particularly where it adjoins a village, there is often a more or less regular spatial diminution of the grazing factor, and a consequent change in the vegetation as one passes away from the village. Near the houses the common is grazed intensely enough to keep it in the state of grassland, and woody plants are absent, except perhaps for occasional trees surviving or planted for shelter. Here one may see the animals of the villagers, either tethered or free to wander. At some little distance isolated shrubs or patches of scrub occur, largely thorny or spinose species, but sometimes others, such as hazel or field maple; and these may shelter young trees, though they are more often of the type of "thicket scrub" (Salisbury, 1918; see Chapter XXIV), alternating with more or less grazed grassland. On the more distant edge of the common regular woodland may exist, the characteristic woodland ground vegetation usually degraded and mixed with weeds by the trampling of occasional beasts which have wandered farther than usual from the village. Finally, this "biotic zonation" may be completed by undegraded woodland, but this is generally preserved and fenced against the grazing animals.

The zonation described is usually complicated by temporal fluctuations in the intensity of grazing. The number of animals turned out on the common probably varies a good deal from time to time. When they are few the result is the same as in a neglected and understocked pasture: there is successful invasion by woody plants. When the grazing again increases the animals are unable to destroy well-established clumps of shrubs, particularly the armed species, and these persist with grazed grassland right up to their edges, so that the vegetation becomes sharply differentiated into alternating patches of scrub and grass. This is a common and conspicuous characteristic of such land. If grazing has been long enough in abeyance regular stages of succession through scrub ("woodland scrub", Salisbury, 1918) to woodland may be observed (Adamson, 1932). The actual distribution and state of the vegetation may be further modified by cutting of the shrubs, either for small wood or to increase the available pasturage, and not infrequently also by fire. It will thus be readily appreciated that the actual nature and distribution of the highly modified and fragmentary vegetation of such an area cannot be explained in detail without long-continued observation of the course of action of these diverse and fluctuating factors in each particular instance; but enough has been written to make clear the ubiquitousness of the grazing factor, which can never be neglected in any attempt to interpret the semi-natural vegetation of a country whose climate is not only supremely well adapted to the pastoral industry but in which miscellaneous grazing is universal through most of the year.

Rabbits

Apart from man's flocks and herds and his domesticated grazing animals, rabbit grazing is perhaps the most widespread and effective biotic factor in modifying our semi-natural vegetation.

History of the rabbit in Britain. The rabbit (*Oryctolagus cuniculus*), originally a native of south-western Europe where it was well known in classical times, was probably introduced into England by the Normans, perhaps in the twelfth century, and references to it became frequent in the thirteenth. At first rabbits were apparently scarce and fetched relatively high prices, but by the middle of the fifteenth century "they may well have become as abundant as they are now" (Barrett-Hamilton, 1910–21), judging from the numbers provided at certain feasts; and in the sixteenth Gesner describes their immense abundance ("copia ingens cuniculorum") in England. In the seventeenth century rabbits were highly valued as a source of profit for food and fur, and so they have remained until quite recent years. In Scotland they were little known till the nineteenth century, when their numbers rapidly increased so that they are now as abundant and ubiquitous as in England (Barrett-Hamilton) except at high altitudes. They occur, usually in great numbers, on almost every island and islet round the Scottish coast, apparently always introduced by man according to available records. In Ireland, as in England, to judge from early references, they were probably introduced by the Normans and about the same time. The rabbit is "now (1912) probably the most vigorous, prolific and abundant mammal in the islands, exclusive of the Brown Rat, the true mice and the common shrew" (Barrett-Hamilton, *op. cit.*, p. 190).

Present distribution. The difference between the human attitude towards the rabbit in the seventeenth century and in the twentieth is very striking. A distinguished botanist from one of the Dominions recently remarked after visiting part of the Breckland: "Well, in my country the authorities would long ago have served notices on these landowners requiring them to destroy the vermin on their estates without delay"; and indeed in many parts of the country rabbits are a real and serious pest to the crops and to young plantations, besides decreasing or destroying the value of rough and hill grazings. Simpson (1908) makes it clear that the right course, if it is desired to avoid the great destruction that now takes place, is to confine rabbits entirely to properly managed and securely fenced warrens, which can be very profitable, and to enforce by law the extermination of all outside the warrens. Though fairly described as "ubiquitous" and able to burrow in any dry soil, even tough clay or a surface coal seam (Simpson, 1908), their incidence is very unequal, partly because they prefer a light dry soil for their burrows and partly because it is only in certain districts that they are actually protected and preserved since they afford good "rough shooting" throughout the year. But on most parts of the chalk downs, on most areas of sandy heath, and on sand-dunes, they exist in countless thousands, and locally they really dominate the vegetation. Plates 2—9 illustrate the effects of rabbits.

Effect of rabbits on vegetation. Rabbits do not range very far from their burrows: they never go farther for their food than they need. Thus there is a well-marked concentric biotic zonation of vegetation round the collections of burrows that are locally scattered on the southern chalk. Immediately around the burrows, often to a distance of some yards, the soil is quite bare, the white chalk brought up by the constant scratchings of the rabbits often showing up conspicuously on a distant hillside. A little distance off the plants of the turf are eaten down to a height of about a centimetre over a considerable zone. Farther away the rabbit grazing is not so severe, and at a distance of 100 yards or so it may be scarcely noticeable, though there is evidence that it may be nibbled enough to prevent the growth of the taller grasses at considerably greater distances. In many such regions the collections of burrows are so close together that the severely grazed areas overlap and then the entire region is rabbit devastated, sometimes continuously or nearly continuously over many hundreds of acres. In such an area no woody plant can establish itself, and those present before the rabbit pressure became overwhelming are uniformly trimmed from the ground up to a height (about 50 cm.) which a rabbit can reach standing on its hind legs. This sharp undercutting of the existing clumps of scrub is absolutely characteristic of heavily rabbit-

PLATE 2

Phot. 3. Severely rabbit grazed slope of chalk down (middle distance). The grassland in the foreground is not so heavily grazed. The scrub is mostly hawthorn and its further spread is prevented by the rabbits. The white patches mark the positions of burrows. Near "The Bake", Wylye, Wiltshire. *R. J. L.*

Phot. 4. Detail of Phot. 3. Two rabbits are seen (centre and left), uneasy in the presence of the photographer: many more were previously feeding on the area. The stems of the hawthorn scrub are bare up to the height a rabbit can reach. The white patches of chalk stones mark the sites of burrows. *R. J. L.*

EFFECTS OF RABBITS

PLATE 3

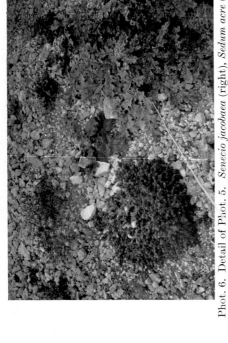

Phot. 6. Detail of Phot. 5. *Senecio jacobaea* (right), *Sedum acre* (left), *Arenaria serpyllifolia* (top).

Phot. 8. Head of Wascombe Bottom, Butser Hill, Hants, once occupied by yew wood (Watt). The bareness is maintained by rabbits. Arable cultivation on plateau above fenced against rabbits.

Phot. 5. Bared area, Windmill Hill, near Chalton, Hants. Mats of *Sedum acre*, with *Senecio jacobaea*, in foreground. *Atropa belladonna* behind. Relict beechwood in distance.

Phot. 7. Area adjoining Phot. 5, with relict beechwood and scrub. Rabbit-resistant plants in the foreground include *Sambucus nigra* (left), *Solanum dulcamara*, *Urtica dioica*, *Verbascum thapsus*, *Senecio jacobaea*, *Reseda luteola*.

infested areas, so that they can be recognised at a glance. The turf itself is eaten down so close to the soil that it can afford no pasturage for any other grazing animal (Pl. 2, phot. 4). The species of which it is composed are those which can survive this process, for the most part hemicryptophytes that continually respond to the severe nibbling by putting out fresh vegetative shoots on or in the actual soil surface. Under such conditions these plants show repeatedly branched but very short shoots, the whole rising not more than 1–2 cm. above the soil. In extreme cases the number of species per unit area is considerably reduced. Thus 2 sq. m. of severely rabbit-eaten chalk down turf on rather loamy soil, the herbage of which was less than 2 cm. high, contained only eight species, a much smaller number than in an average area of the same size and type (Tansley and Adamson, 1925):

Agrostis tenuis	r	Prunella vulgaris	f
Euphrasia nemorosa	r	Ranunculus bulbosus	o
Festuca ovina	a	Taraxacum erythrospermum	o
Lotus corniculatus	a	Thymus serpyllum	f

All these species were eaten down, including the two that were abundant, but all could maintain themselves nevertheless (see Fig. 99, p. 542).

Species not eaten by rabbits. There are certain species which rabbits will not eat, and these may be seen conspicuously erect and flowering in the middle of a rabbit-ridden area, often actually among the burrows. The species most generally abundant in sandy or other non-calcareous areas much frequented by rabbits is the bracken fern (*Pteridium aquilinum*), and this is quite untouched, so that it gains a great competitive advantage in such places. Many of the others are herbs of open soil and not proper constituents of grassland turf, so that they colonise by preference the areas where the turf is very thin or where bare soil has been exposed by much scratching. The following are common rabbit-resistant plants seen on the chalk or overlying loam in such situations:

Arenaria serpyllifolia	Helianthemum nummularium (vulgare)
Campanula glomerata	Myosotis arvensis
Cerastium vulgatum	Nepeta (Glechoma) hederacea
Cirsium arvense	Sedum acre
C. lanceolatum	Senecio jacobaea
C. palustre	Teucrium scorodonia
Centaurium umbellatum	Urtica dioica
Cynoglossum officinale	Verbascum thapsus

and in Breckland[1] Farrow (1917) cites:

Conium maculatum	Urtica dioica
Solanum nigrum	U. urens

Erica cinerea and *E. tetralix* are refused while *Calluna* is severely eaten back. Broom (*Sarothamnus scoparius*) and bramble (*Rubus fruticosus* agg.)

[1] *Myosotis collina*, *Alchemilla arvensis* and *Stellaria media* are also said to be avoided by rabbits in Breckland (Godwin).

are also said to be refused. The common elder (*Sambucus nigra*) is notably avoided by rabbits, and clumps of elder may very often be seen growing in the bare soil between rabbit burrows. The deadly nightshade (*Atropa belladonna*), the stems of which are sometimes gnawed without killing the plants, is locally though not at all generally abundant in similar situations on the chalk. Rabbits are immune to the poisonous alkaloids it contains.

If an area of rabbit-ridden chalk turf is adequately fenced against rabbits (and other grazing animals) the first result is the growth in height of the constituent species released from the constant nibbling. In a year or two the razed lawn-like turf is replaced by the aspect of a prairie (Pl. 9, phots. 18, 19). Thereafter the naturally taller growing grasses and herbs gradually acquire dominance over those of dwarfer habit which tend to be suppressed in the competition. In the neighbourhood of woodland shrubs and trees begin to colonise the ground, which passes through the stage of loose scrub to that of woodland (Tansley, 1922).

Effect on competition between species. The work of Farrow (1916, 1917) on the effect of rabbits on the vegetation of Breckland is the most important English study of this biotic factor. Farrow (1916) showed that rabbits had converted a tract of Callunetum on Cavenham Heath into grassland dominated by common bent and sheep's fescue, and he demonstrated the stages through which this change had occurred (Fig. 36 and Pl. 4, phots. 9, 10). He described the biotic zonation which occurs round the burrows, and showed how this is modified by the varying degrees of palatability of the different gregarious species present in the area. Thus fescue and bent, ling, sand sedge, bracken fern, form a series of decreasing palatability, the last-named being practically untouched by the rabbits, so that its vigorous growth and vegetative spread tend to oust the other species. Sand sedge (*Carex arenaria*) is eaten, but only when no other food is available, so that it is liable to spread round the burrows and to replace ling, which is suppressed by constant nibbling (Fig. 37). The bent-fescue community, though it can replace ling and maintain itself even when nibbled down close to the surface, cannot compete with the much taller sand sedge and bracken, which thus constantly advance over the rabbit-infested area. Watt (1936, 1937) has since shown that rabbit pressure does not determine Breckland vegetation quite so exclusively as Farrow supposed, and that other factors such as variations of soil, and exposure to frost, as well as the action of wind in initiating cycles of succession, play a leading part in the distribution of vegetation in other parts of Breckland; but the main phenomena seen on Cavenham Heath during the second decade of this century were quite convincingly demonstrated by Farrow to be due to the interaction of the overwhelming rabbit pressure on a small number of competing species with different degrees of palatability and different powers of competition. On the sandy soils of Breckland, as on the chalk, there are a certain number of scattered plants (ragwort, black night-

PLATE 4

Phot. 9. Degenerate Callunetum in process of conversion into Agrostido-Festucetum. Rounded heather bushes nibbled down by rabbits. Cavenham Heath, Breckland. *E. P. Farrow.*

Phot. 10. Moribund *Calluna* bush occupied by *Leucobryum glaucum* (centre) and *Cladonia coccifera* (left). Same locality. *E. P. Farrow.*

EFFECT OF RABBITS ON CALLUNETUM

PLATE 5

Phot. 11. Two areas of Caricetum arenariae (white) around rabbit burrows in Breckland Callunetum. The sand sedge is advancing into the heather (foreground) as rabbit pressure on the latter increases. *E. P. Farrow.*

Phot. 12. Incipient grass-heath (Festuceto-Agrostidetum) round a rabbit burrow in Caricetum arenariae. See Fig. 37 (p. 139) for explanation of this zonation. *E. P. Farrow.*

DIFFERENTIAL EFFECTS OF RABBITS

shade, hemlock, nettles) which are untouched by rabbits; and these stand up conspicuously from the razed turf or open soil.

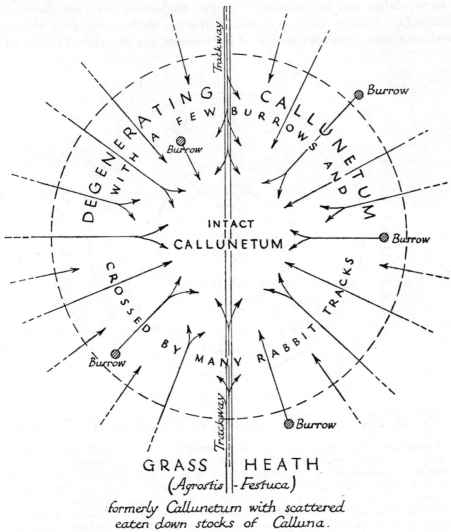

FIG. 36. EFFECT OF RABBITS ON CALLUNETUM

The diagram represents an isolated patch of heather surrounded by rabbit-eaten grassland ("grass heath"), which was formerly Callunetum. The remaining patch of Callunetum is penetrated by the rabbits on all sides, so that it is rapidly degenerating and giving way to grassland. Rabbit burrows are most frequent in the area outside the Callunetum, but a few have been established within it. The trackway through the centre is largely used by the rabbits. Modified from Farrow, 1916.

Farrow (1917) also showed that uniform severe rabbit grazing prevented the formation of inflorescences (Pl. 6, phot. 14), and thus cut down the supply of seed for regeneration.

Degradation of vegetation. The general effect of excessive rabbit pressure is to degrade vegetation to a lower form. Thus chalk grassland is impoverished and the number of species diminished (compare the very short list of species cited on p. 135). On steep north-facing chalk slopes, where mosses, especially species of *Hylocomium*, are abundant in the turf

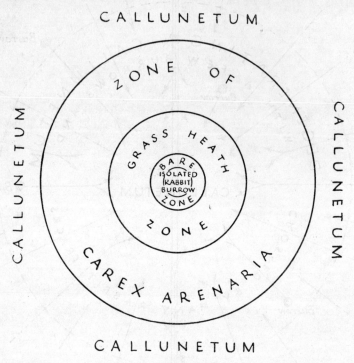

FIG. 37. ZONATION OF VEGETATION ROUND AN ISOLATED RABBIT BURROW

The zone immediately round the burrow is kept bare by the constant scratching of the rabbits. Next comes a grass heath zone which can maintain itself under maximum rabbit grazing; though the grasses are constantly eaten down close to the soil. Outside this is a zone of *Carex arenaria* (sand sedge) which has been able to establish itself in place of the original *Calluna* (heather) by which it is surrounded because the sand sedge is *relatively* unattractive to rabbits. The Caricetum cannot extend outwards because of the competition of the heather unless the rabbit pressure on the latter increases, nor inwards towards the burrow unless the pressure decreases. Cf. Pl. 5. After Farrow, 1917.

owing to the cool damp conditions, all the flowering plants may be eaten down, leaving a covering of luxuriant moss which the rabbits do not eat.[1] On the Breckland sands, as we have seen, Callunetum may be converted into grassland. The ultimate result may be actually to destroy the vegetation altogether. Thus a collection of rabbit burrows on a steep chalk slope may become the focus of a bare area covered with chalk stones (Pl. 3, phot. 5), the action of gravity and rainwash leading to a creep down the

[1] Pl. 102, phot. 248, p. 544.

PLATE 6

Phot. 13. On the right of the rabbit-proof fence the turf (bent-fescue) is eaten down close to the soil. On the left the herbage is luxuriant and includes the sand sedge. Cavenham Heath, Breckland. *E. P. Farrow.*

Phot. 14. Rabbit-proof cage with flowering *Calluna* and harebell (*Campanula rotundifolia*) inside. Outside there are no inflorescences. An eaten down *Calluna* bush on the right, in line with the front of the cage. Cavenham Heath, Breckland. *E. P. Farrow.*

Effects of Excluding Rabbits

PLATE 7

Phot. 15. *Aster tripolium* at Blakeney Point with its leaves severely eaten by rabbits. *Suaeda maritima*, also present, appears to be untouched. *W. Rowan.*

Phot. 16. Rabbit tracks across a *Salicornia-Pelvetia* marsh at Blakeney Point, Norfolk. These are permanent tracks made by rabbits coming from the neighbouring sand dunes to reach plants of *Aster tripolium* on the other side of the marsh. *W. Rowan.*

RABBITS AND THE SEA ASTER

slope of a mixture of the finer bare soil and chalk stones originally scratched up by the rabbits, the whole surface having become unstable (Tansley and Adamson, 1925). In Breckland the thinning or destruction of the vegetative cover may give the wind a purchase on the sand, so that it blows away and covers the adjacent vegetation (Farrow, 1917).[1] The same thing may be seen on fixed coastal dunes where rabbit scratching on a ridge exposed to the wind but covered with vegetation may, by destroying the plants which have fixed the sand, initiate wind erosion and start a "blow out".

Rabbits and maritime vegetation. The rabbits which inhabit coastal sand dunes in immense numbers exert very considerable effects not only on the dune vegetation itself but also on that of the neighbouring salt marshes. Rowan (1913) observed that on the salt marshes at Blakeney Point in Norfolk they ate *Salicornia herbacea* and *Glyceria maritima*, but especially *Aster tripolium*. The last named is in fact one of their chief maritime foods, and to reach it they make permanent tracks, visible at all times of the year, through a large *Salicornia-Pelvetia* marsh (Pl. 7, phot. 16). As a result the inflorescences of *Aster* are quite rare in the vicinity. On the dunes the very abundant *Senecio jacobaea* is apparently quite untouched (except for one solitary plant which Rowan found eaten and surrounded by rabbit dung), but *Cakile maritima*, *Silene maritima*, *Carex arenaria* are eaten, as well as the flowers only of *Calystegia soldanella*, which is eaten so assiduously that not a single flowering specimen can be found within the rabbit-infested area. They also dig up from the sand and devour the rhizomes of *Glaux maritima* in winter, even crossing water to reach a sandy island in a tide-filled "low" where the plant was growing, though it was invisible under the sand. In the shingle lows they eat *Limonium vulgare* and especially *L. binervosum*. But the most striking effect of the rabbit activity is seen on the *Suaeda fruticosa* which fringes the shingle banks on the edge of the marshes. On the approach of winter they "mow" whole areas of young plants almost to the ground, and prevent bushes which would otherwise grow 3 ft. high from reaching a stature of more than 6 in. They bite off so many branch tips that the drift lines of the higher spring tides are often lined with a continuous bank of *Suaeda* tips bitten off by the rabbits and carried away and deposited by the water. *Obione portulacoides* is treated in the same way, though whether the rabbits actually eat it or not is uncertain (Rowan, 1913; Oliver, 1913).

Other animals

Cattle, sheep and rabbits[2] are the only animals in this country whose effect on vegetation is decisive enough to determine and maintain specific semi-natural plant communities over wide areas in a state of approximate

---

[1] Watt, however (1937), says he has not been able to connect the origin of Breckland "blow outs" with rabbit attack on the vegetation.

[2] Perhaps the common vole should be added to the list (see p. 141).

equilibrium. But there are many others which are destructive to vegetation, and may endanger the existence of a "crop" such as a young plantation when nature's equilibrium has been upset. Only a few of the more striking of these can be mentioned here.

Effect of rodents on woodlands. Rabbits themselves do much harm to woodlands, eating or nibbling the tree seedlings, but the extent of the damage varies a great deal owing to their very unequal abundance. In some places rabbits are few or absent, while on sandy soils in the neighbourhood of unfenced rabbit warrens, and also on the chalk, the damage they do may be very severe indeed, and new plantations have to be protected by rabbit-proof fences.

Mice and voles. Mice and voles, however, are much more *evenly* distributed, and it is these small rodents, which spend much of their lives in tunnels just below the surface humus, emerging at intervals on to the floor of the wood, that probably do more damage to tree seedlings than any other animals. They gnaw through the roots of seedlings below the soil and the stems above, and are responsible for a very high seedling death-rate. They are indeed probably the chief agents in preventing the natural regeneration of our deciduous woods (Watt, 1919, 1923). For besides destroying the seedlings they also eat the nuts of oak, beech and hazel.

Watt (1919) found that of a total of 275 acorns which had fallen from the trees in October 1914, on ten selected plots in a Cambridgeshire woodland, all had disappeared by the following January. Those which were exposed vanished most rapidly, and those concealed under 2 in. of humus disappeared concurrently with the appearance of mouse tunnels in the humus. In an experiment with sowing acorns in May in a Somerset oakwood on controlled plots, protected (1) from rabbits alone, (2) from rabbits and mice, (3) from rabbits, mice and birds, Barrington Moore (1933) found that in the plots exposed to mice all acorns were destroyed within a month, the mice forestalling any rabbit attack. Where mice were excluded but birds had access more than half the acorns exposed to view on bared mineral soil were destroyed by birds (probably wood pigeons), but completely buried acorns were not dug up, nor were acorns disturbed when they were concealed by ground vegetation and surface litter.

Mice and voles then are the most important enemies of both seeds and seedlings of the heavy seeded trees, such as oak and beech. Three generally distributed species are concerned: the long-tailed field mouse (*Apodemus sylvaticus*) which lives in woods, is nocturnal in its habits, a seed eater, and probably responsible for the destruction of acorns and beech mast; the bank vole (*Clethrionomys glareolus*) which is omnivorous, its most frequent diet being probably seeds, fruits, bulbs, tubers, etc. (Barrett-Hamilton, 1910–21, pp. 414–15); and the common vole (*Microtus agrestis*) which is primarily a grass eater but also destroys seedlings and barks young trees. It is very likely the last that is mainly responsible for the destruction of

PLATE 8

Phot. 17. The closely eaten turf and the numerous (rabbit-resistant) plants of *Senecio jacobaea* in the middle distance are the result of the attacks of rabbits coming from the wood behind. The tall uneaten grass and other herbage in the foreground are not reached by the rabbits. *Ewen Cameron.*

LOCALISED EFFECT OF RABBITS

PLATE 9

Phot. 18. Corner of rabbit-proof enclosure at Downley Bottom, Ditcham Park, Feb. 1910, after one year's enclosure. Note difference in the turf inside and outside the fence. *C. J. P. Cave.*

Phot. 19. The same corner in Sept. 1911, taken along a line at right angles to Phot. 18. Note the tall grass (largely *Avena pratensis*) inside, and the isolated plants of *Cirsium palustre* (rabbit-resistant) outside the fence.

Effects of Excluding Rabbits

tree seedlings, but it is rarely that the animals can be observed and identified at work (Elton, personal communication).

Effects of voles on grassland. The effect of the common vole on grassland is marked. At the peaks of the fluctuating curves of vole population they certainly destroy a great deal of grass and kill tussocks of *Juncus* (Elton). *Microtus* also alters the composition of hill grassland. Summerhayes, as the result of experiments (1932–7) not yet published,[1] found in two separate localities, one in Wales and one in Scotland, that complete exclusion of voles from a number of plots mainly occupied by Molinietum led to greater luxuriance of the dominant and on the whole to diminution of the associated plants. This he attributes to increased competition of the dominant, which tends to form a taller and more uniform stand and to accumulate a greater mass of litter when freed from the attacks of the voles. Their activity consists in making innumerable tunnels just below the surface, gnawing through the roots and rootstocks of the grasses and thus keeping the community more open, allowing more room for a varied flora. Most remarkable is the effect on the mosses, which are not eaten at all by the voles. In the vegetation exposed to voles the mosses were numerous and important, but in the enclosures they were shaded out by the close-growing luxuriant *Molinia*, while the increase of litter made their growth more difficult, so that they had disappeared almost entirely at the end of the experiments. This phenomenon is quite parallel to the differential effects of rabbits on vegetation (pp. 136–9). The effects of small rodents (rabbits, mice, and voles) on woods are clearly different in kind from their effect on grassland. In the one case they prevent the appearance of the climatic or edaphic climax and determine a substituted community of whose environment they are a constant factor, and which they *maintain* in a position of equilibrium: in the other they damage the climax community and may prevent its regeneration, so that the wood can only be maintained by fresh planting.

Interference with "the balance of nature". It has often been pointed out (e.g. by Watt, 1919, p. 201) that the great increase of these "ground vermin" and their consequent destructive effect on the vegetation is a result of human interference with "the balance of nature" by the destruction in the interests of "game" (mainly pheasants) of carnivorous birds and animals which formerly kept the small rodents in check. This sequence of events is in fact a good example of the *indirect* effect of human activity in destroying a natural "ecosystem".

Enemies of rodents. The carnivorous enemies of mice and rabbits (as well as of pheasants) are said to be—among mammals—the polecat, stoat, weasel, fox, marten, badger, hedgehog and shrew: among birds the owls, buzzard, gulls, rook, black crow, hooded crow and kestrel (Watt, 1919), but the relative importance of the predaceous species has not been properly

[1] Very kindly placed at my disposal by Dr Summerhayes.

established. Some are obviously now so rare that their effect is insignificant. The stoat is said to be the most serious enemy of the rabbit, and foxes kill many of their young, but Simpson (1908) declares that poachers are more destructive than either! None of them now exercises much influence in keeping the rabbits in check.

Hares. The hare (*Lepus europaeus*) is very widespread throughout the country, locally quite abundant, and certainly very destructive, but since it is not gregarious and does not exist in such enormous numbers as the rabbit its *general* effect on vegetation is probably negligible. Both the common hare (*Lepus europaeus*) and the mountain hare (*L. timidus*) nip off tree seedlings.

Squirrels. The squirrels, both the wild red squirrel (*Sciurus vulgaris*) and the naturalised American grey squirrel (*S. carolinensis*) are also destructive animals, destroying the young shoots of trees such as pine and oak, and eating and storing the pine seeds and acorns. The red squirrel is also said to kill trees by barking them at a considerable height. But apart from damaging young plantations the squirrels cannot be said to exert widespread effects on vegetation.

Deer. Much the largest British mammals are the deer, of which two species are native and a third, which is commonly kept in parks, often escapes and temporarily establishes itself in the more extensive woods till it becomes a nuisance and is exterminated. These again are forest animals which form natural components of the original forest ecosystems, but under the unbalanced conditions of the modern countryside they are often very destructive to crops and plantations because they are driven to seek food, especially in winter, outside their proper habitats.

The two species of deer still native in the British Isles are the red deer and the roe. Both were formerly widely spread through the forests of the whole of the British Islands, but they are now much restricted in distribution.

Red deer. The red deer (*Cervus elaphus*) is much bigger than the roe deer and thus is by far the largest wild British mammal still extant, though the Scottish deer (subsp. *scoticus*) are smaller than those of central Europe. Owing to increasing reservation of "deer forests" during the last half century and strict preservation for stalking, there were said to be 150,000 head in the Scottish Highlands and islands in 1923. Outside this area there was a wild herd of about 400 which were hunted with hounds on Exmoor in North Devon and perhaps 150 on the Westmorland fells, besides a few about Muckross (Co. Kerry) in south-west Ireland (Cameron, 1923).

Though primarily a forest animal through most of its European range, the Highland red deer live almost entirely on the upland moors ("deer forests" which often contain no woodland), since few woods survived the use of the area as sheepwalk in the latter part of the eighteenth and increasingly throughout most of the nineteenth century. With the collapse of the Highland sheep industry towards the end of the century, more and more of the old upland sheep farms were reserved for deer, so that there are now between three and four million acres (say 13,000 sq. km.) devoted to deer alone. In the summer the deer feed on the young vigorously growing shoots of heather (*Calluna*) and on the grasses. The winter they spend in woods by preference, partly for shelter from the severe winter climate, and partly because there alone can they obtain food from the twigs and bark of the trees, also eating acorns and beech nuts. If they cannot get access to woods they will descend into the valleys and eat anything they can find on arable land and in gardens. It is said that when deer have once acquired a taste for farm crops they will risk their lives to gratify it. The practice of artificial feeding in winter is very common in the overstocked "forests" and is indeed often necessary to preserve the life of the herd. The result is that such native Highland woodland (mainly very small areas of birch with a little pine, see Chapter XXII) as remain within the deer forests suffer severely from the deer. Thus Crampton (1911) wrote of Caithness: "The most mischievous effect of overstocking with deer is seen in the gradual disappearance of the birchwoods, which everywhere in the Highlands is fast proceeding. Young birch never reaches maturity in places where the deer have access, and the older woods are gradually dying through age." Young plantations have always to be fenced against deer for the same reason. The best authorities on deer forests advocate the planting of hardwoods in the glens to the extent of one-fifth of the area of the "forest", thus replacing the old vanished woods and affording shelter and food for the deer in winter (Cameron, 1923). Under such conditions, and if the deer forest is not overstocked, a proper equilibrium such as must formerly have existed could no doubt be re-established.

Roe deer. The roe deer (*Capreolus capreolus*) is a much smaller animal than the red deer, and was at one time widely spread through Britain. It has always been wild in Scotland and is fairly often met with in the more remote regions of the northern Highlands. It was at one time exterminated through most of Scotland, but with the extensive planting of the eighteenth century it increased and spread considerably. In England there are said to be still a few in the Cumbrian mountains, but the roe was never indigenous in Ireland. Their food is varied, like that of the red deer, and they are said to eat fungi and various fruits, such as the berries of the rowan (*Sorbus aucuparia*), as well as leaves, twigs, grasses and other herbs. They do a good deal of damage to young conifer plantations, partly by rubbing their bodies against the young trees, and also by rooting up seedlings and sapling trees with their antlers.

Fallow deer. The fallow deer (*Cervus dama* or *Dama dama*), the species most commonly kept in English parks, is not now native to Britain, though its remains are found in interglacial deposits. Like the rabbit it was probably introduced by the Normans from southern Europe, and since then it has been a constant tame inhabitant of the country, often kept in parks from which it sometimes escapes and establishes itself in the wilder country such as parts of the Highlands and the more extensive woodlands in England. It is as destructive to crops and plantations as the native species.

Birds. Many different birds damage various crops and may become serious pests when the balance of nature is upset, while others attack the enemies of crops; but like the mammals they fit into natural ecosystems, and only cause continuous widespread destruction when their natural enemies are systematically destroyed. It is quite impossible to consider here the various effects of different species under various conditions.

"Bird cliffs." One effect of birds in so changing the soil as to determine a definite plant community which would not exist as such if the birds were absent, must however be mentioned. Certain sea cliffs or isolated "stacks" of rock on the coast are inhabited by vast numbers of sea birds, and if there is flat ground at the foot of the cliff not reached by the tide the accumulation of dung and urine on this produces a substratum so rich in organic nitrogen that very few species of plants can inhabit it, although those which do are exceedingly luxuriant. This peculiar vegetation at the foot of such "bird cliffs" has been described by Scandinavian observers. *Cochlearia officinalis* is a leading dominant recorded as occurring in such situations. These communities have scarcely been studied on our own coasts. Vevers (1936) however mentions a community consisting only of *Melandrium dioicum* (d), *Poa annua* (f) and *Urtica dioica* (o) as occurring on soil rich in guano at the foot of bird cliffs on Ailsa Craig, a rocky island in the Firth of Clyde.

Invertebrates. It is impossible to deal here with the effects of the multitude of invertebrate animals which influence vegetation in various ways, either by feeding on it, by cross-pollinating plants, by distributing their fruits and seeds, or by modifying the soil. Such aspects of the environment of plant communities have not yet been studied in relation to whole communities, though there is of course much information in existence about the depredators, pollinators, and seed distributors of various plant species.

Effects of fire

An important anthropogenic factor which is quite widespread, though not so frequent in this country as in drier climates, is fire. When vegetation is dry enough fire spreads very quickly, and under a strong wind the vegetation may be partially or entirely destroyed over a wide area if it is set alight intentionally or accidentally. In arid countries where there is a recurrent long dry season fires are a constant factor of the environment, and the vegetation of a perpetually fire-swept area becomes adjusted to this special condition, and is composed only of those plants which can survive the passage of the fires. This is the condition for instance in much of the savanna country of Africa and India.

PLATE 10

Phot. 20. A flock of sheep descending from the chalk pastures which they have created: note the parallel sheep tracks. To the left is a belt of chalk scrub separating the rough grazings of the Downs from the enclosed and cultivated land of the Weald. Mount Harry, near Lewes, Sussex. Photograph from *The Times*.

Phot. 21. An abundant crop of beech seedlings from the full-mast year 1922, cut over and their leaves stripped by mice or voles. *A. S. Watt.*

BIOTIC EFFECTS

In Britain the southern heaths are the areas most often swept by fires accidentally started. These commonly occur every few years and are largely determined by dry summers. The general floristic poverty of the dry southern British heath is probably largely due to fire, less frequent plants originally present having been destroyed and not returned. The species of the existing heath flora are either those which are commonly not killed by the fires and shoot again from the unburned parts, or those which are abundant in the near neighbourhood, and quickly recolonise the burned areas (cf. Chapter XXXVI, pp. 729, 735). The natural succession from heath to woodland (at first birch and pine) is prevented by the fires which destroy the tree seedlings. Thus the vegetation is maintained in the heath stage by the frequent fires and such heath represents a "fire climax". The fire-swept areas differ in different years but they overlap, with the result that a portion of heath which has been burned a second time after a short interval shows a younger stage of redevelopment of the vegetation than a portion which has escaped for a longer period (Chapter XXXVI, p. 730).

The northern upland heaths which are preserved as "grouse moors" are regularly fired every 10 or 15 years with the object of obtaining new vigorous growth of the heather (*Calluna*) on which the grouse feed. Various stages of succession in the "burn sub-sere" have been described (Chapter XXXVII). Late in the summer or autumn land dominated by tall grasses, and even fenland, often dries sufficiently to be swept by fires accidentally started.

Forest fires are not common in the British Isles, because the climate is damp and the trees, except pines, do not burn very readily. The heath fires in the south, however, have not infrequently destroyed areas of planted and subspontaneous pinewood.

Gorse (*Ulex europaeus* and *U. gallii*) is often deliberately burned on the upland bent-fescue hill grazings (Chapter XXVI) to obtain fresh growth of the grasses and thus improve the pasture; and other grassland areas, for example on the South Downs patches of *Brachypodium pinnatum*, an unpalatable grass, are occasionally burned by shepherds with the same intention.

REFERENCES

ADAMSON, R. S. Notes on the natural regeneration of woodland in Essex. *J. Ecol.* **20**. 1932.
ADAMSON, R. S. On the relationships of some associations of the Southern Pennines. *J. Ecol.* **6**, 97–109. 1918.
BARRETT-HAMILTON, G. E. H. *A History of British Mammals* (incomplete). London, 1910–21.
CAMERON, A. G. *The Wild Red Deer of Scotland*. 1923.
CRAMPTON, C. B. The vegetation of Caithness. British Vegetation Committee. 1911.
DAVIES, WILLIAM. "Grasslands of Wales" in Stapledon, *Survey of Agricultural Wastelands of Wales*. 1936.

Farrow, E. P. On the ecology of the vegetation of Breckland. II. Factors relating to the relative distributions of *Calluna*-heath and grass-heath. *J. Ecol.* **4**, 57–64. 1916.

Farrow, E. P. III. General effects of rabbits on the vegetation. *J. Ecol.* **5**, 1–18. 1917.

Fenton, E. Wyllie. The influence of sheep on the vegetation of hill grazings in Scotland. *J. Ecol.* **25**, 424–30. 1937.

Moore, Barrington. Oak woodlands on clay in south-western England and scarcity of natural regeneration of oak. *Forestry*, **7**, 1933.

Oliver, F. W. Some remarks on Blakeney Point, Norfolk. *J. Ecol.* **1**, 4–15. 1913.

Ritchie, J. *The Influence of Man on Animal Life in Scotland.* 1920.

Roberts, R. Alun. In *J. Ecol.* **23**, 251. 1935.

Rowan, W. Note on the food plants of rabbits on Blakeney Point, Norfolk. *J. Ecol.* **1**, 273–4. 1913.

Salisbury, E. J. The ecology of scrub in Hertfordshire. *Trans. Herts Nat. Hist. Soc.* **17**. 1918.

Simpson, J. *The Wild Rabbit in a new aspect.* Third edition. 1908.

Stamp, L. Dudley and Beaver, S. H. *The British Isles.* London, 1933.

Tansley, A. G. Early stages of the redevelopment of woody vegetation on chalk grassland. *J. Ecol.* **10**, 168–77. 1922.

Tansley, A. G. and Adamson, R. S. Studies of the vegetation of the English Chalk. III. The chalk grasslands of the Hampshire-Sussex border. *J. Ecol.* **13**, 177–223. 1925.

Vevers, H. G. The land vegetation of Ailsa Craig. *J. Ecol.* **24**, 424–45. 1936.

Watt, A. S. On the causes of failure of natural regeneration in British oakwoods. *J. Ecol.* **7**, 173–203. 1919.

Watt, A. S. On the ecology of British beechwoods with special reference to their regeneration. *J. Ecol.* **11**, 1–48. 1923.

Watt, A. S. Studies in the ecology of Breckland. I. Climate, soil and vegetation. *J. Ecol.* **24**, 117–38. 1936.

Watt, A. S. II. On the origin and development of blow-outs. *J. Ecol.* **25**, 91–112. 1937.

PART II
HISTORY AND EXISTING DISTRIBUTION OF VEGETATION

PART II

Chapter VII

PRE-HISTORY

Post-glacial history. During the last two decades very great advances have been made in our knowledge of the post-glacial history of European vegetation, i.e. of the process of revegetation of central and northern Europe after the retreat of the immense ice-sheets which spread over much of those regions during the "Great Ice Age" (Pleistocene glaciations) and of the major changes in the vegetation which have since occurred. These advances have depended largely upon the method of counting the numbers of pollen grains of different species of plants (particularly of trees) that have been preserved at successive levels in the thick deposits of peat which are so distinctive a feature of the fens and bogs of the lowlands as well as of the bogs and moors of many of the high-lying plateaux of central and north-western Europe. The pollen grains of many genera, when embedded in peat, retain their characteristic form for thousands of years; and when they are such as can be easily carried by wind for considerable distances we can use the percentage occurrences of the pollen of different species in a given layer of the peat ("pollen analysis", see Godwin, 1934a) as an index of the composition of the vegetation which existed in the surrounding country when that particular peat layer was being formed.

Pollen analysis

In this way it has been possible to reconstruct the post-glacial history of the central and north European forests in considerable detail. The results for different regions are on the whole strikingly concordant and have been successfully correlated not only with the recent geological history but also with the records of Palaeolithic, Mesolithic, Neolithic, Bronze Age and Iron Age Man.

Geo-chronology. Further, by means of an ingenious method based on the counting and correlation of the layers of silt deposited in certain Swedish lakes, the Swedish geologist De Geer was able to construct a chronology of the whole period that has elapsed since the retreat of the ice,[1] and to relate this to the distinct climatic periods inferred from the history of the forests and from the changes in the nature of the peat formed at different stages of development.

---

[1] The alternating coarse and fine layers of silt ("varves") occurring in these lacustrine deposits were shown by De Geer to represent summer and winter deposition respectively, so that each pair of layers would represent the deposits of a single year. On this basis he calculated the number of years which had elapsed since the retreat of the ice from various latitudes in Sweden, and thus founded his "geo-chronology" of the post-glacial period (De Geer, 1910). Though doubt has sometimes been thrown on his interpretation it is adopted here since it accords well with the probable periods of time occupied by the changing climates of the post-glacial period.

Date	CLIMATIC PERIODS according to BLYTT and SERNANDER		FOREST VEGETATION etc.	HUMAN INFLUENCE ON VEGETATION	ARCHAEOLOGICAL PERIODS
B.C. 9000	SUBARCTIC	Cold & dry	Dryas vegetation		End of PALAEOLITH.
8000	PREBOREAL	Fluctuations of Climate	Birch & Pine: some Hazel & Oak and a little Elm & Alder in eastern England		
7000	BOREAL	Warm & dry	Birch and then Pine dominant Hazel maximum Expansion of mixed Oak forest		MESOLITHIC
6000					
5000	ATLANTIC	? "Climatic optimum"	Recession of Pine and great increase of Alder		
4000		Warm & wet	Mixed Oak forest dominant (Sphagnum vigorous in wetter climates) Lime relatively abundant		
3000					
2000	SUB-BOREAL	Drier	Local increase of Pine & Yew Disappearance of Lime Entry of scattered Beech	? Extension of upland grassland under human influence: increase of agriculture	NEOLITHIC
1000			Bog growth checked		BRONZE AGE
0	SUBATLANTIC	Cool & wet	Formation of younger Sphagnum peat ? Spread of Beech	Increasing destruction of forest: great increase of agriculture & pasture spread of weeds of cultivation	IRON AGE HISTORICAL PERIOD
A.D. 1000	RECENT	? Warmer and drier			
2000					

FIG. 38. TABLE OF BRITISH POST-GLACIAL HISTORY

"Pollen analysis" has lately been extended to the British Islands, and the results show that the course of events in this country corresponds very well on the whole with what occurred on the continent, though it was different in some important respects. Thus *Picea* and *Abies* (spruce and silver fir) which characterise some of the later phases of post-glacial history on the continent were quite absent from this country, while *Fagus* and *Carpinus* (beech and hornbeam) were less important here. Much work still remains to be done before we can construct a satisfactory detailed post-glacial history of vegetation in different parts of the British Isles, but the main outlines of the story are perfectly clear.

A further word of caution is necessary here. The assumption of more or less *exact* correspondence between the archaeological periods and the climatic periods inferred from the prevalent forests and the nature of the peat—for example, the synchronism of the Bronze Age with the Sub-Boreal and of the Iron Age with the Sub-Atlantic—and the dating of their limits in De Geer's chronology to the nearest century or even half-century cannot be regarded as at all securely established. But the order of magnitude of the time lapses assigned to these periods may be taken as trustworthy.

Thus we are now in a position to present an outline history of British vegetation since the Ice Age, incomplete, it is true, in many respects, for much further work is required to fill gaps in our knowledge, but nevertheless furnishing a good "pre-historical" preface to an account of the later changes brought about by human agency, mainly during the historical period, and of the existing distribution of plant life.

Effect of the Pleistocene glaciations. In late Pliocene times, perhaps somewhere about a million years ago, when the rough outline of the land mass which is now the British Isles was already blocked out as a north-western projection from the continent of Europe, the climate of the region (which had cooled considerably since earlier Tertiary times) and the plants inhabiting it cannot have been very different from what they are to-day. Most of the species known from the late Pliocene deposits are still represented in the country, only a few, for example the common spruce (*Picea abies*), being certainly present then and not now. The Pleistocene glaciations, however, drove out or destroyed most of this flora. An immense ice-sheet formed in the Scandinavian highlands, and from this the ice extended southwards into central Europe, south-eastward into the plains of southern Russia, and south-westward across the North Sea, invading the eastern coasts of England. Subsidiary ice-fields were formed in the mountains of Scotland, Ireland, Wales and Cumbria, and from these last glaciers and ice-sheets descended the valleys and covered the Midland Plain of England, as well as overriding much of the Pennine ridge, on which smaller snowfields supplying local glaciers were also developed. In East Anglia the northern (Scottish) ice met the Scandinavian ice-sheet coming from the north-east (Wright, 1937, map, Fig. 35, p. 61). Remains of glacial deposits occur as far south as the valleys of the lower Severn and of the Thames, and it has often been said that the whole of the vegetation to the north of this line must have been completely obliterated. Recent work in the north of England (Woodhead, 1929; Raistrick, 1931) has, however, shown that considerable areas of high ground on the Pennines were free from ice during the period of maximum glaciation, and it is likely that plants could and did maintain themselves on such "nunatakker" (Blackburn, 1931).

Interglacial periods. It is important to realise that maps such as Wright's (1937, Fig. 35) do not show the extent of the ice existing *at any one* time, but the extreme *limits* of glaciation, i.e. *all* the ground covered by ice at any time during the Pleistocene glaciations. In the British Isles it has now been pretty well established (Boswell, 1932) that there were four successive main glaciations, the first two affecting chiefly the east, the last two the west of the country, and three interglacial periods. But glacial geologists will not commit themselves to the assertion that the four British glacial phases corresponded with the four well-recognised major glaciations (Gunz, Mindel, Riss, Würm) of central Europe. Simpson (1929–30) has proposed a theory of glacial and interglacial periods according to which the former should occur in pairs, each pair corresponding with a marked hump on the curve of solar radiation. The two members of a pair would be separated by a warm wet interglacial period corresponding with an actual maximum of radiation. Two successive pairs of glaciations would be separated, on the other hand, by a cool dry interglacial corresponding with a depression in the radiation curve. The scheme put forward by Simpson corresponds very completely with Penck and Brückner's scheme of the four great Pleistocene glaciations of central Europe. Unfortunately our knowledge of British interglacial beds and their fossils is far too scanty[1] to allow of any attempt at correlation with the types of climate required by Simpson's scheme. The best known interglacial floras (Reid and Chandler, 1923) are from the West Wittering, Selsey and Clacton deposits, assigned to the middle interglacial of Boswell's series, between the Chalky-Jurassic Boulder Clay and the Upper Chalky Drift (Acheulean period), and here the flora consisted, as in immediately pre-glacial times, mainly of species that are present to-day, accompanied by a very few that are no longer present.

Of the three interglacial periods thus recognised (Boswell, 1932) it is certain that at least one *followed* the period of maximum glaciation, and this fact is important in the history of the vegetation. Temperate vegetation establishing itself during an interglacial period may well have survived the

[1] In Denmark very complete vegetational cycles, presumably corresponding with climatic cycles within the interglacial periods, have been traced, but in the British Islands no continuous peat deposits extending through the interglacial phases are known, so that the evidence remains fragmentary.

last glacial phase, which was very partial. During the interglacial phases the ice receded and the glacial deposits left behind were subjected to enormous erosion and rearrangement by the great floods of water released in the summer by the melting of the ice after the climate had become warmer, so that the form and even the original nature of these deposits are now often barely recognisable.

The last glaciation. The vegetation of the last interglacial period was probably again displaced in Ireland, Scotland, Wales and parts of northern England by the last advance of the ice, perhaps corresponding with what is known on the continent as the "Würm II" (or "Bühl") glaciation of central Europe. In the British Isles, however, this final glaciation, as revealed by the distribution of the comparatively little altered newer drifts, did not extend so far to the south-east as the previous glaciations. In fact it seems to have left the whole of the midlands, the south, and most of the east of England free from ice, as well as considerable regions of the north. Besides these it is reasonable to suppose that there were now more ice-free areas among the mountains of Scotland, Wales and Ireland than during earlier glaciations.

The climate of the ice-free portion of England during the last glaciation was presumably fairly severe. It is perhaps reasonable to suppose that English vegetation then consisted *mainly* of dwarf shrub "tundra", alternating probably with moss and lichen tundra, such as exists to-day in the neighbourhood of the continental ice-sheets in Greenland, together with a number of "moorland" plants and very likely other cold temperate species. But it is quite likely that in the south a richer flora, including trees, may have survived from the last interglacial through the final phase of glaciation. And on the local ice-free areas of the northern and western mountains a number of "arctic-alpine" species, some of which still exist there as "relicts", were doubtless present.

Possible survival of species

Relict species. A. J. Wilmott (1929, 1935) has pointed out that there are a number of places (in Scotland and the north and west of England and in Wales) where there still exist collections of plants whose presence together in these places but not in the surrounding regions is difficult to interpret except by supposing that they are survivors in those particular (supposedly ice-free) spots from a warmer period antecedent to the last glaciation. We do not really know what the climate was like during this last glaciation, and it is quite possible that there were many species which survived it in relatively favoured areas. Furthermore, there are in southern Ireland and in the south-western counties of England a number of species occurring in south-western Europe but not always in western France (the so-called "Lusitanian" flora). It is easier to suppose that these species survived the last glaciation than to assume subsequent sporadic immigration across several hundred miles of sea from the Iberian peninsula. The difficulty of accepting spontaneous trans-maritime post-glacial immigration is even greater in the case of the animals common to the two regions. It has, however, been suggested that some of these "Lusitanian" species may have reached these islands along with pre-historic visitors from the south (cf. pp. 165–6).

The general evidence for the reality of the existence of such "relict" groups of species, far removed from the main area of their distribution, has been considerably strengthened during the last ten years, for example by Scandinavian work. It is certain of course that the existing dominants and general characters of the great types of British vegetation have been established since the last glaciation, but whether the dominant trees actually entered the country afresh across the continental land bridge in post-glacial times, or whether some of them at least maintained themselves in the south during the last glacial phase, and have therefore come down to us from the preceding interglacial period, is still uncertain.

Late glacial (Sub-Arctic) period. The general improvement of the climate which led to the final recession of the ice and its disappearance from the British Isles was taking place, according to De Geer's geochronology, from about 18,000 to about 7500 B.C. During by far the greater part of this period (known as the sub-Arctic) a cold climate still prevailed, as indicated by the typically arctic plants such as *Dryas octopetala* and dwarf birch and willow, which characterise the deposits. During the later part an increasing number of species were presumably immigrating from the continent into Britain (which was then a peninsula), just as they were also pressing northward into Scandinavia; and towards its close (in what is called pre-Boreal time) shrubs and trees began to come in (or to spread), as revealed by the presence of their pollen grains in contemporary peat deposits. The sub-Arctic species such as the little dwarf birch (*Betula nana*), which still occurs here and there in the Scottish mountains and is common in Scandinavia, may well have been present throughout the last glaciation, and so may species of the shrubby arctic willows, such as *Salix lapponum*, *S. reticulata*, and the very small *S. herbacea*, which have a similar distribution to-day.

Pre-Boreal period. The appearance in quantity of birch and pine, with some hazel and oak, and also traces of alder and elm (Fig. 39, bottom), indicates increasing warmth, and marks what is called the pre-Boreal period (possibly 8000–7500 B.C.). As the climate improved the existing British species of birch, *Betula pubescens* and *B. pendula* (= *B. alba* or *B. verrucosa*) probably advanced freely and fairly rapidly across the broad bridge of land between the continent and England, where are now the Straits of Dover and the southern part of the North Sea (Fig. 40), unless indeed they had survived the last glacial phase. The small winged fruits of the birches are specially well adapted for rapid dispersal by wind, as are the plumed seeds of the willows; and several of the shrubby willows as well as various birches are able to flourish in comparatively severe climatic conditions. The northern forms of birch, for example *B. tortuosa* (which however seems to have been a latecomer in northern Scandinavia), to-day extend farther towards the Pole than any

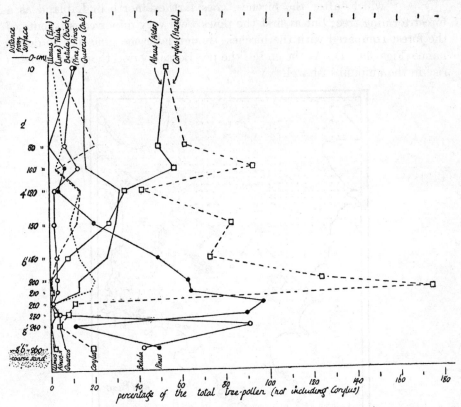

FIG. 39. TREE POLLEN IN NORTH NORFOLK (EAST ANGLIA) DURING THE PRE-BOREAL, BOREAL AND ATLANTIC PERIODS

In the pre-Boreal period, shown at the foot of the diagram, pine and birch pollen are in greatest quantity, though hazel, oak, alder and elm are also present. The Boreal period is marked by a great increase in pine (cf. Fig. 41), and thereafter of hazel, whose pollen at one time exceeded that of all the other trees together by 75 per cent (Boreal "hazel maximum"). At a depth of 150 cm. from the present surface the descending pine curve crosses the ascending alder curve (cf. the levels 350–90 cm. in Fig. 41); and this point is usually taken as the beginning of the Atlantic period, which is marked here, as in Fig. 41, by prominence of oak, alder, and hazel. Lime, which here enters before the end of the Boreal, is present in some quantity throughout the Atlantic. The increase of pine and birch towards the top of the diagram may indicate the beginning of the Sub-Boreal.

This peat-bed is on the coast and covered at high tide. The present surface is 1·5 ft. (45 cm.) below Ordnance datum. Judy Hard, Brancaster, Norfolk. From Godwin, 1934c.

other tree, forming the northernmost belt of woodland everywhere in western Lapland, and species of shrubby willows are very abundant in the same region. Even some forms of our own birches, usually considered as belonging to *B. pubescens*,[1] extend farther up the mountain sides and farther

[1] But see p. 444, footnote.

156   *Pre-history*

north in Scotland than any other tree except the rowan or mountain ash (*Sorbus aucuparia*), which is scarcely gregarious.

Pine   Next to the birches came the common pine (*Pinus sylvestris*), which, after the birches, goes farthest north in Lapland as a forest-forming tree; but at first the pines were very minor constituents of the forest compared with the birches, though later the pine rose to dominance (Figs. 39, 41). At the end of the pre-Boreal period there was a sharp rise in the amount of hazel.

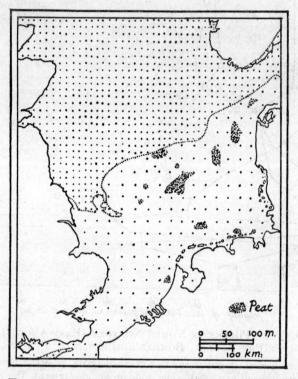

FIG. 40. THE NORTH SEA IN PRE-BOREAL TIMES

The area with widely spaced dots, south of the existing 50 m. contour of depth was then a land surface as shown by the dating of the peat dredged from the various places shown. After Jessen, 1935.

The actual physical features of the country were in main outline the same as they are to-day, though the physiognomy of the landscape as determined by the vegetation must have been very different. The relief of the land was more marked: the chalk downs of the south-east were somewhat higher and more extensive (since the escarpments had not been cut back so far), and were continuous with the French chalk across the Channel and the Straits of Dover. The low-lying country of the eastern counties extended right across the North Sea, so that England was joined not only to north-eastern France and the Netherlands of Holland and Belgium but also to Jutland (Fig. 40). Rivers and streams were larger and more numerous, lakes and meres and the marshy areas of the low-lying plains and basins much more extensive; and we may picture the margins of these colonised by willow thickets, while the birches and pines spread over the uplands, replacing the tundra and moorland vegetation of late glacial and sub-Arctic times.

Boreal period. The climate continued to grow warmer, till mean summer temperatures substantially higher than those obtaining to-day were

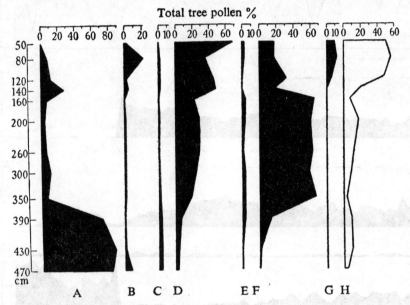

FIG. 41. TREE POLLEN IN EAST ANGLIA DURING THE BOREAL, ATLANTIC AND SUB-BOREAL PERIODS

A, Pine. B, Birch. C, Elm. D, Oak. E, Lime. F, Alder. G, Beech. H, Hazel. The Boreal period at the bottom of the diagram is marked by a great pine maximum, the other tree pollen present being insignificant in quantity. The onset of the Atlantic period is marked by the sharp recession of pine and concomitant increase of alder and oak. Lime is present in small quantity throughout this period. The Sub-Boreal period which follows saw a second, though much smaller, increase of pine accompanied by birch, and also of oak; while alder recedes, though still maintaining itself in considerable quantity. Beech now appears for the first time, and in fairly substantial amount. Hazel is present throughout, but the common Boreal hazel maximum does not appear. Wilton Bridge on the Little Ouse. From Godwin, 1935.

reached,[1] and at the same time drier. This, the so-called Boreal period, is believed to have lasted for some 2000 years till about 5500 B.C. It was marked by the spread of the hazel (*Corylus avellana*) which must have multiplied exceedingly, probably forming great masses of pioneer scrub on the margins of the pine forest as well as undergrowth of the forest itself.

[1] It has been suggested that on neighbouring parts of the continent the mean July temperature was 2° C. higher than it is now.

158   *Pre-history*

Hazel maximum   The numbers of hazel pollen grains preserved in certain horizons of the Boreal peat often greatly exceed those of all the other trees present taken together, and this striking "hazel maximum" (Fig. 39), of which there may be more than one, occurs also in many other parts of Europe. The birch and hazel penetrated to Ireland and

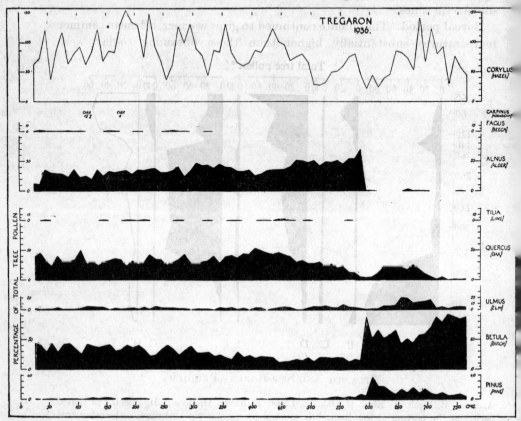

FIG. 42. TREE POLLEN OF BOREAL, ATLANTIC AND POST-ATLANTIC PERIODS FROM TREGARON BOG IN THE TEIFI VALLEY (CARDIGANSHIRE)

The historical sequence reads from right to left. Note the differences from the East Anglian diagrams. The Boreal period is here marked by dominance of birch, pine being present in much smaller amount, with elm, oak and abundant hazel. The Boreal-Atlantic transition is again marked, as in Figs. 39 and 41, by sharp increase of alder and recession of pine (and here of birch). Birch persists in some quantity throughout the Atlantic, gradually increasing towards its close, and accompanied in abundance by oak, alder and hazel. Lime pollen is also present in small amount (discontinuously) throughout this period, while beech enters later. Cf. Godwin and Mitchell, 1938, text-fig. 5, p. 442.

to the extreme north of Scotland, including the Shetland Islands. Here the birch was probably dominant, but in some places there was a marked hazel maximum at the same time (Erdtman, 1928).

Dominance of pine. The great feature of the early part of the Boreal period is the increase of pine (Figs. 39, 41), which was much the commonest tree in eastern England at this time, the number of birches decreasing, at least relatively. In competition under equally favourable conditions the pine will always beat the birch because it casts a much deeper shade, under which the birch cannot grow; and by the middle of the Boreal period the early birch forests must have been to a large extent replaced by pine. The same is true of central Ireland but not of the north (Erdtman, 1928). The pine penetrated into Scotland also, but there it seems scarcely ever to have exceeded the birch in abundance during the Boreal period, while for the most part its pollen is present in much smaller proportion (Fig. 43). In Wales also birch very much exceeded pine in the Boreal period (Fig. 42, right-hand end).

The "Lower Forestian". The remains of the Boreal birch and pine forests in the Scottish peat bogs (Fig. 45) were distinguished by James Geikie, and later by Lewis (1905–11) who first thoroughly investigated them, as the "Lower Forestian" or "Lower Forest" horizon (Lewis, 1911). Samuelsson (1910) also examined the Scottish peat and published a critical consideration of Lewis's findings, confirming them in this respect and correlating the "Lower Forestian" with the Boreal, and the "Upper Forestian" (see p. 167) with the Sub-Boreal periods of Blytt (1882) and Sernander (1908), though Lewis himself had not taken that view.

During the middle part of the Boreal period oak and elm, which, as we have seen, were already present in the pre-Boreal, increased considerably in amount and though pine and birch on the whole remained dominant, the ratio of their pollen to that of the deciduous trees decreased. But at the end of the period the dry climate had desiccated the surface of growing peat so as to allow of its invasion by pine and to create a second maximum of pine pollen (Godwin).

Invasion of xerophytes. At this time also (late Boreal), there was probably a great north-westward movement of plants adapted to a dry continental climate from south-eastern and central Europe. This invasion reached southern Sweden, where, on the Baltic coasts and islands, many such species still survive, and in smaller numbers it also reached southern and eastern England, where some of the same species are still to be found maintaining themselves on certain well-drained soils such as the chalk downs and the East Anglian sands (cf. pp. 511–12).

Towards the close of the Boreal period the climatic conditions of north-western Europe had improved so much that there must have been an abundance of (relatively dry) forest growth and in the succeeding (Atlantic) period, with increase of moisture and maintenance of a relatively high temperature, the climate reached a very favourable condition which is generally called the "climatic optimum". At this time many species extended farther northward (in Scandinavia, for example) than at present, and the same was probably true of Great Britain.

Atlantic period. Towards the end of the Boreal period substantial

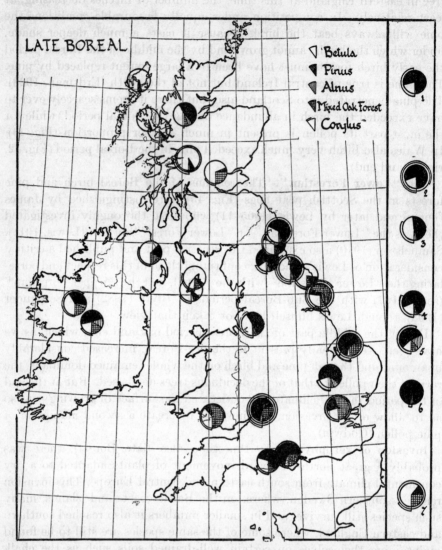

Fig. 43. Late Boreal Tree Pollen

The sectors of the circles show the percentages of the total pollen contributed by different trees in peat of different sites. The total amount of hazel (reckoned separately as a percentage of total tree pollen) is shown as a border to the circle. [The vertical series of circles on the right of the diagram represent analyses of comparable age from lowland stations on the North Sea border of the continent at corresponding latitudes.] It will be seen that birch increases and pine falls off as we pass to the north-west, while mixed oak forest pollen (oak, elm and lime) is generally more abundant in the south and east. Compare Fig. 44. From Godwin, 1934b.

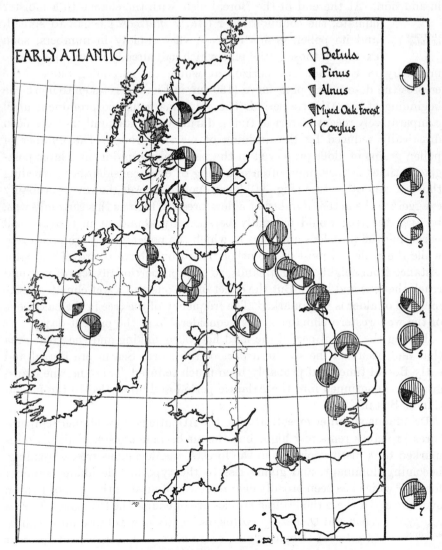

FIG. 44. EARLY ATLANTIC TREE POLLEN

Symbols as in Fig. 43. Alder is here the dominant pollen, while pine has almost disappeared except in Scotland, and birch has diminished except in Ireland. The decrease in proportion of mixed oak forest pollen shown is due to the overweighting of the alder fraction as a result of the preponderance of alder close to fens and bogs. Hazel has greatly decreased in proportion, while still maintaining itself in fair quantity. From Godwin, 1934b.

amounts of elm (probably *Ulmus glabra*, the wych elm, better known as "*Ulmus montana*"),[1] and of oak, were present, with lime (*Tilia cordata*) in addition. At the end of the Boreal also, with the change to a moister climate, the common alder (*Alnus glutinosa*) began to spread, and its pollen grains increase very rapidly in numbers, soon exceeding those of the pine which had already begun to decrease markedly in England. The horizon at which the ascending alder curve crosses the descending pine curve (150 cm. in Fig. 39) is taken to mark the beginning of the Atlantic period. The alder has remained prominent until comparatively recent historical times, during which its numbers have been drastically reduced by the draining of marshy ground. The number of pollen grains of alder preserved in the peat during middle and later postglacial times is very high, often equalling or even considerably exceeding those of any other tree (Fig. 44). These pollen records, however, probably exaggerate the actual number of alders present, high as this certainly was, because the tree would naturally be specially abundant in the wet soil bordering the peat fens and bogs in which the pollen grains are preserved, while the pollen of most of the other trees would have to travel a greater distance from neighbouring uplands, so that proportionally less of it would reach the fens and bogs. But the effect of the onset of a wet climate on the increase of alder is very marked. The frequency of the tree in the oakwoods of the wet western climates is very considerable at the present day.

Marginal note: Increase of alder

During the Boreal period the land had been sinking and the great area that had occupied the southern part of the North Sea in pre-Boreal and early Boreal times had probably been much reduced. This "marine transgression" continued into the Atlantic period and finally led to the insulation of Britain.

Deciduous summer forest. During the Atlantic period summer temperatures probably remained high, but the air became moister. This period is marked by a distinct change in the forests, the deciduous trees spreading, becoming dominant, and giving rise to the type of "deciduous summer forest"[2] which has been broadly characteristic of most of the area of Britain ever since. Of these the leading species are of course the two oaks, *Quercus robur* and *Q. sessiliflora* (or, as we are now told we must call it, *Q. petraea*). Elm and oak, as we have seen, were already present in pre-Boreal times, but increased markedly in numbers at the beginning of the Atlantic period. There is evidence in southern Sweden that the sessile oak spread only in a later (Sub-Atlantic) period, so that it may have

Marginal note: Oak, elm and lime

---

[1] The status of the other species of elm now growing in Britain, e.g. *Ulmus nitens*, *U. minor*, and *U. procera*, is still uncertain, since to-day they nowhere appear as undoubted elements of natural or semi-natural vegetation. It may be that one or more of them were native species in Boreal or later prehistoric times. *U. nitens* and *U. minor* may well have been constituents of natural woodland, particularly in the east and south-east of England.

[2] Known on the continent as Quercetum mixtum.

been only *Q. robur* which dominated the Atlantic oakwoods. The oaks became the dominant trees of the typical "Atlantic" forest, and with them were associated an elm (probably *Ulmus glabra*) and the small-leaved lime (*Tilia cordata*), which did not become an important forest component before the Atlantic period. Of these species the lime was very abundant at certain times and places during the Atlantic, and the elms were apparently more abundant in both England and Ireland than they are now. Of the history of the ash (*Fraxinus excelsior*), a far more important member of the deciduous forest to-day, we have too little knowledge. The hazel, though much diminished from its early Boreal superabundance, has maintained itself in quantity up to the present day, and has probably always formed an important part of the undergrowth of the oak forest as well as independent scrub in certain western regions, notably the limestone areas of Co. Clare in the west of Ireland (see pp. 473–4).

Increase of *Sphagnum* bogs. The marked increase of humidity in the Atlantic period encouraged the growth of *Sphagnum* bogs, which invaded and destroyed some of the woods that had spread over the drying fen peat in Boreal times. But oak forest was established as the dominant climatic type, and the vegetation thus began to take on the main outlines of the character it would have to-day if it were not for the activity of man. The higher temperatures probably enabled many species to range farther north than they do now.

Sub-Boreal period. The Atlantic period may have lasted for 3000 years, from 5500 to 2500 B.C. or thereabouts, then giving way to the drier Sub-Boreal. Up to about 2500 B.C. (Clark, 1936, p. 53) Britain was inhabited only by epi-Palaeolithic or Mesolithic peoples who were food gatherers, not cultivators or pastoralists, and can have had little effect on the natural vegetation. With the arrival of the Neolithic cultures, and increasingly after the coming of metals (soon after 2000 B.C. in Britain) men began to till the soil and to keep flocks and herds.

Neolithic culture

At this time Britain had definitively become an island, as a result of the continuance of the marine transgression which began in the Boreal; and the mixed oak forest, with varying proportions of elm and lime, which had long been established in the country, must have covered most of the lowlands except the fens and marshes. On the clays and loams at least there was a dense growth of trees, to deal effectively with which was probably a very laborious and lengthy task for the primitive axe, whether of stone or metal.[1] The forests were doubtless avoided too because they harboured dangerous animals, such as wolf and lynx, enemies alike of man and of his flocks. Grazing and primitive agriculture were thus probably mainly confined to the drier chalk and oolite and some other limestone

[1] It is said that now in New Guinea the felling with a stone axe of a tree large enough to make a canoe, together with the necessary shaping and hollowing, occupies a skilled craftsman for many months.

uplands, but were probably also practised on some of the sandy and other drier areas. A map of the Neolithic remains (excluding axes and arrowheads) in England and Wales (Bowen, after Fox and Chitty, in Darby, 1936, p. 13) shows the largest and densest concentration on the Yorkshire Wolds and Salisbury Plain (Chalk), and on the Cotswolds (Oolite), lesser ones at the southern end of the Lincolnshire Wolds and on the South Downs of Sussex (Chalk), on the Mendips and on the Southern Pennines (Carboniferous Limestone), as well as on the much older rocks of the coastal lands of low elevation (and therefore low rainfall) in Pembroke and Anglesey.

Distribution of Neolithic settlement

Forest *v*. Grassland on chalk and oolite. It is hard to believe that the chalk and oolite uplands were not covered with forest in the warm damp climate of Atlantic times, since this was essentially a forest climate in which deciduous forest attained its maximum extension in this country and in western Europe generally. Oak will not, it is true, form good trees on the shallow chalk and limestone soils, but ash, yew and beech will and do. We are still uncertain when beech first appeared in southern England, but it was certainly later than the other trees mentioned above and it first spread in eastern England in the drier Sub-Boreal period, and was followed by hornbeam (Godwin, 1935). If beech were present it would become, as it later became, the dominant tree of the chalk uplands, whether on the shallow chalk soils or on the loams with which they are often covered, as well as on the other lighter soils (see Chapters XVIII, XIX). If beech were absent yew and ash and the shrubs that are abundant on calcareous soils would be the natural dominants of the chalk. Are we to believe that Neolithic man cleared such vegetation to make room for his dwellings and camps, his corn crops and sheep? On the other hand, Neolithic long barrows dated about 2000 B.C., such as are frequent on the chalk, certainly do not seem very likely to have been constructed in the midst of forest, though Sir Cyril Fox (1933, p. 49) suggests that they might have been, and it must be remembered that they were funerary structures, which some prehistoric peoples certainly did make in forest land.

Perhaps the answer to the puzzle may be found in the following considerations. Assuming that the Neolithic people of south Britain were essentially an open country people, they were most likely to find suitable sites for colonisation on the soils which were not covered with woody vegetation, i.e. on the poorer sands, on recently formed terraces of alluvial river gravel (in both of which their implements are found), and also on hill tops and wind-swept crests such as the narrower ridges of the chalk downs, where the soil is very shallow and tree growth very scanty or absent altogether, even in a climate thoroughly favourable to trees. But at about this time, i.e. during the third millennium before Christ, it is believed that the climate became markedly drier, the Atlantic period passing into the Sub-Boreal; and this might well cause the easily drying soils of much

wider areas of chalk and oolite to become less suitable for tree growth.
Seedlings would have less chance of establishing themselves, those which did begin to grow would suffer or die from drought, and the forest of the upper slopes and on the more porous soils would begin to thin out. The presence of man on the ridges would work in the same direction, for his sheep and cattle would eat off tree and shrub seedlings and he himself could easily clear loose scrub and the weaker saplings. In this way, with increasing dryness of climate, the open areas on the shallower chalk and limestone soils would steadily spread at the expense of forest. It would be a cumulative process, for the less suitable for tree growth the terrain became the more the human inhabitants would increase and flourish, and the more they increased (and with them their flocks and herds) the faster the forest would be pushed back. Thus by the time the Sub-Boreal climate was well established much of the chalk and oolite plateaux and slopes would be bare of trees and have become pasture grassland with a little arable.

<small>Spread of grassland</small>

We may thus form a plausible picture of what was happening to the vegetation on the chalk and limestone uplands towards the end of the third millennium before Christ, though anything like direct evidence cannot be obtained. But we know that in the north and west and in the fenland, where thick peat had been formed throughout the period of Atlantic climate, conditions became dry enough in the Sub-Boreal to allow of a fairly extensive advance of pine forest over the peat (Godwin, Godwin and Clifford, 1935). And an equivalent increase of dryness in the south may well have led to conditions definitely unfavourable to tree growth on light shallow soils with very free drainage.[1]

The Sub-Boreal period, which in Britain may have begun during the Neolithic culture period, is marked by the invasions of the "Beaker" peoples, so-called from their use of goblet-shaped pots of specialised form and decoration. A map of the distribution of beakers shows a great preponderance in the southern half of England and northwards along the east coast as far as the Moray Firth, suggesting invasions from the south and east, just as a map of the earlier Megalithic monuments shows concentrations in Ireland and the west of Great Britain, though also in Lincolnshire on the east coast and in Caithness in the north of Scotland (Bowen, after Fox and Chitty, in Darby, 1936, pp. 6, 7). We cannot trace any effect upon the vegetation at large of the Megalithic invaders who came by sea from the west coast of Europe and the Mediterranean, though it has been suggested that they may have involuntarily

<small>"Beaker" settlements</small>

---

[1] Some archaeologists believe that such soils may have become too dry to support pasture grassland. We do not of course know just how dry the Sub-Boreal climate was, but we do know that on the damper soils deciduous forest existed throughout the period and it seems unlikely that there was aridity enough on the chalk generally to prevent the maintenance of closed grass vegetation.

brought some of the western Mediterranean or Iberian species of plants which are still found wild in Ireland and the south-west of England. The beaker finds are massed mainly on the southern chalk, in East Anglia (Pliocene sands, fenland islands and lighter glacial drift), the Yorkshire chalk wolds, the limestone of the southern Pennines, and the Scottish east coast.

Distribution of Bronze Age cultures. If we compare a map of the massed finds of the whole of the Bronze Age, from say 2000 to 500 B.C. (Fox, 1933, Map C), we see the same main concentrations, but also considerable extensions to the west, including many finds in Cornwall, Wales and Ireland. Much more of the surface of the British Isles was now apparently suitable for occupation, probably because of the drier climate, and it seems to have supported a more widely scattered and larger population. Among the regions which are blank or have yielded but few Bronze Age remains are some mountain areas, such as the Scottish Southern Uplands, the northern Pennines, and the mountains of central and southern Wales, as well as large parts of the Midland Plain, the Thames valley, and most of the Weald. The northern and western upland areas may well have been too inhospitable, while the central and southern lowlands referred to were covered with dense forest. The Cotswold oolite country, however, from some unknown cause (perhaps because the Neolithic culture persisted there), is also singularly destitute of Bronze Age relics, in strong contrast with its apparently intensive occupation in Neolithic times. Salisbury Plain, on the other hand, is densely covered in both periods.

All this evidence suggests that the incipient creation and control of the chalk and other dry grasslands which seems likely to have been begun by the Neolithic peoples was maintained and extended by the Bronze Age folk, who also spread to and occupied far wider areas.

Spread of pine and birch. For the rest, as has already been said, pine and birch again increased their range, apparently forming continuous pine forest on the Atlantic coast of Ireland and invading the margins of many bogs and fens, in contrast with the extension of oak and alder in Atlantic times. The native pine forests still existing, especially in the more continental parts of the Scottish Highlands (see Chapter XXII), are probably the relics of these Sub-Boreal forests which have been able to maintain themselves on the more porous soils and in less wet regional climates through Sub-Atlantic times. It is held by some that *Pinus sylvestris* has remained a native in the south of England even up to the present day, but the existing pinewoods of the south are generally considered as subspontaneous and derived from seventeenth- and eighteenth-century planting. The deciduous forest (Quercetum mixtum) was, however, widely maintained, and lime again occurs in some profiles of the early Sub-Boreal.

Sub-Atlantic period. During the last millennium before Christ—the beginning is put at about 700 to 550 B.C.—the climate suffered a marked

Sub-Atlantic Period

deterioration, becoming much wetter and decidedly cooler. Peat bogs extended over considerable parts of Ireland, Scotland and northern England, destroying great areas of Sub-Boreal pine and birch forest, whose

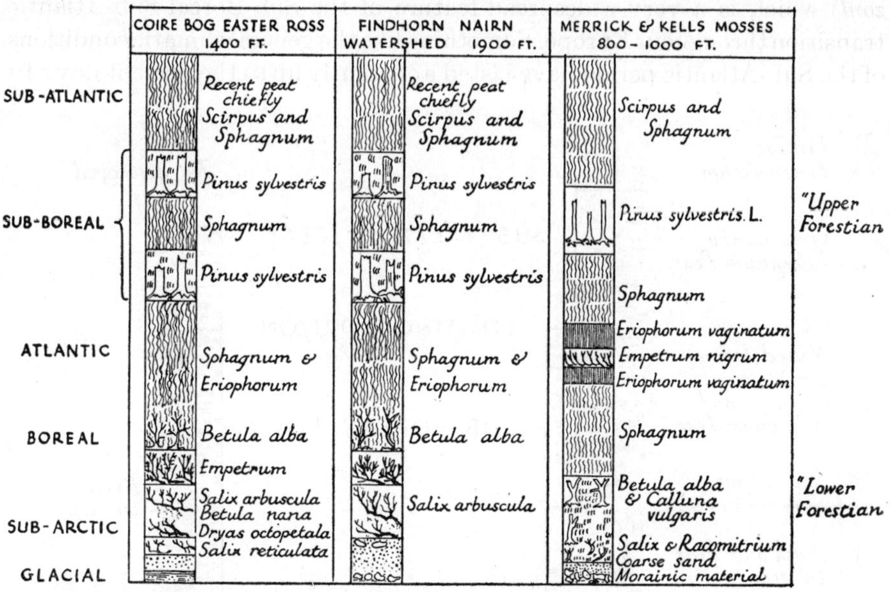

	COIRE BOG EASTER ROSS 1400 FT.	FINDHORN NAIRN WATERSHED 1900 FT.	MERRICK & KELLS MOSSES 800–1000 FT.	
SUB-ATLANTIC	Recent peat chiefly Scirpus and Sphagnum	Recent peat chiefly Scirpus and Sphagnum	Scirpus and Sphagnum	
SUB-BOREAL	Pinus sylvestris / Sphagnum / Pinus sylvestris	Pinus sylvestris / Sphagnum / Pinus sylvestris	Pinus sylvestris L.	"Upper Forestian"
ATLANTIC	Sphagnum & Eriophorum	Sphagnum & Eriophorum	Sphagnum / Eriophorum vaginatum / Empetrum nigrum / Eriophorum vaginatum	
BOREAL	Betula alba / Empetrum	Betula alba	Sphagnum	
SUB-ARCTIC	Salix arbuscula / Betula nana / Dryas octopetala / Salix reticulata	Salix arbuscula	Betula alba & Calluna vulgaris / Salix & Racomitrium	"Lower Forestian"
GLACIAL			Coarse sand / Morainic material	

FIG. 45. PROFILES OF PEAT FROM THREE SCOTTISH LOCALITIES

Showing the "Lower Forestian" (birch) separated from the "Upper Forestian" (pine) by moss peat. The horizons are interpreted (left-hand column) in accordance with Samuelsson's criticisms (1910) and modern views. After F. J. Lewis, 1905, 1906.

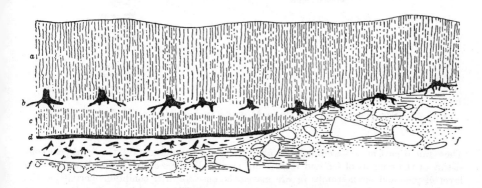

FIG. 46. LAYER OF SUB-BOREAL PINE STUMPS

On the moraine marginal to a peat moss and invading the (drying) peat of the moss. Remains of Boreal forest, etc. on the underlying moraine. From a sketch by F. J. Lewis, 1911.

remains may be found embedded in the base of the new peat. This is the "Upper Forestian" horizon described by Lewis (1907) from the Scottish peat mosses (Figs. 45 and 46). To this time also belongs the "young"

Sphagnum peat described by Rankin (1911) from the mosses of Lonsdale in north Lancashire (Fig. 47 and Pl. 11) and separated from the Sub-Boreal peat below by an unconformity, or "limiting horizon" (German *Grenzhorizont*) which is a very widespread feature of the Sub-Boreal-Sub-Atlantic transition throughout Europe. Something like the general climatic conditions of the Sub-Atlantic period have lasted apparently up to the present day. In

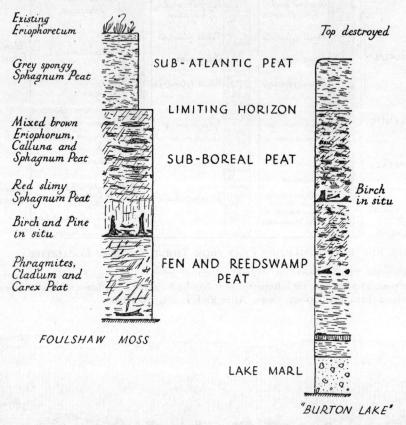

Fig. 47. Stratigraphy of Peat Mosses in Lonsdale (N. Lancashire)

These raised peat mosses were built up above fens (see Chapter XXXIV), the basal peat consisting of the remains of fen and reedswamp plants. In the section of Foulshaw Moss the layer of pine and birch stumps *in situ* marks the onset of drier conditions permitting the invasion of trees, which were later engulfed in bogmoss. The mixed brown Sub-Boreal peat consists of bogmoss, cottongrass and heather. The "limiting horizon" (*Grenzhorizont*), which is an actual "unconformity" in the peat, marks the close of the Sub-Boreal period. The uppermost layer of grey spongy *Sphagnum* peat was formed in the wetter Sub-Atlantic period and continued down to the nineteenth century when the moss became drier and was occupied by a surface vegetation of cottongrass, etc. In the generalised section through the peat of the old "Burton Lake" basin the shell marl deposited on the lake floor is overlaid by reedswamp and fen peat and this again by the same types of Sub-Boreal peat with birch stumps *in situ* at the base and (above the limiting horizon) Sub-Atlantic peat. After Rankin, 1911.

PLATE 11

Phot. 22. Sub-Boreal birch stumps underlain by Atlantic and overlain by Sub-Boreal peat. Above is seen the "limiting horizon" (*Grenzhorizont*) separating the Sub-Boreal from the darker spongy Sub-Atlantic peat, the top of which is destroyed. "Burton Lake" area, Silverdale, N. Lancs.

Phot. 23. "Corduroy road" probably marking the *Grenzhorizont*, against which the trowel is held by W. M. Rankin. The surface peat has dried as a result of draining and is now occupied by heather. Foulshaw Moss, N. Lancs. *Mrs Cowles.*

PEAT SECTIONS IN RAISED MOSSES

other words, the existing British climate resembles the Sub-Atlantic more closely than it resembles any other post-glacial climate, though it is believed that there were ameliorations about 300 and about 1000 A.D., and there may be a tendency towards increasing dryness of climate at the present time.

Iron Age settlements. Invaders who used iron for tools and weapons entered the country about 550 B.C. Early Iron Age settlement seems to have been mainly restricted (Fox, 1933, p. 16, and Bowen in Darby, 1936, p. 17) to what we have called in Chapter III (p. 69) the south-eastern climatic region, indicating invasion from both south and east. These peoples appear to have shown the same soil preferences as those of their Bronze Age and Neolithic predecessors, but it seems that before the end of the millennium the edges of the dense forests of the heavier lands were to some extent attacked, though most of the Weald and of the Midland Plain are still mainly bare of relics indicating human occupation.

REFERENCES

BLACKBURN, K. B. Possible glacial survivals in our flora. *Trans. North. Nat. Un.* 1931.
BLYTT, A. Die Theorie der wechselnden kontinentaren und insularen Klimate. *Bot. Jb.* 11. 1882.
BOSWELL, P. G. H. The ice-age and early man in Britain. *Brit. Ass. Rep.* 1932 (York).
BOSWELL, P. G. H. in Discussion on the origin and relationships of the British flora. *Proc. Roy. Soc.* B, **118**, 202–7. 1935.
BOWEN, E. G. Prehistoric South Britain (Chapter I in Darby, *Historical Geography of England before* A.D. 1800). Cambridge, 1936.
BROOKS, C. E. P. The climate of prehistoric Britain. *Antiquity*, **1**. 1927.
CLARK, J. G. D. *The Mesolithic Settlement of Northern Europe.* Cambridge, 1936.
CRAWFORD, O. G. S. and KEILLER, A. *Wessex from the Air.* 1928.
DE GEER, G. A geochronology of the last 12,000 years. *Int. Geol. Congr.* Stockholm. 1910.
ERDTMAN, G. Studies in the micropalaeontology of post-glacial deposits in northern Scotland and the Scotch Isles, with especial reference to the history of the woodlands. *J. Linn. Soc. (Bot.)*, **46**, 449–504. 1924.
ERDTMAN, G. Studies in the post-Arctic history of the forests of north-western Europe. I. Investigations in the British Isles. *Geol. Fören. Stockh. Förh.* 1928.
ERDTMAN, G. Some aspects of the post-glacial history of British forests. *J. Ecol.* **17**, 112–26. 1929.
FOX, CYRIL. *The Personality of Britain.* 1932.
GODWIN, H. Pollen analysis. I. Technique and interpretation. *New Phytol.* **33**, 278–305. 1934a.
GODWIN, H. Pollen analysis. II. General applications. *New Phytol.* **33**. 1934b.
GODWIN, H. Pollen analysis of peats. In Steers, *Scolt Head Island.* Norwich. 1934c.
GODWIN, H. in Discussion on the origin and relationships of the British flora. *Proc. Roy. Soc.* B, **118**, 210–15. 1935.
GODWIN, H., GODWIN, M. E. and CLIFFORD, M. H. Controlling factors in the formation of fen deposits, as shown by peat investigations at Wood Fen, near Ely. *J. Ecol.* **23**, 509–35. 1935.
GODWIN, H. and MITCHELL, G. F. Stratigraphy and development of two raised bogs near Tregaron, Cardiganshire. *New Phytol.* **37**. 1938.

JESSEN, K. The composition of the forests in northern Europe in Epipalaeolithic time. *K. danske vidensk. Selsk. Biol.Med.* **12**. 1935.

LEWIS, F. J. The plant remains in the Scottish peat mosses. *Trans. Roy. Soc. Edinb.* **41, 45, 46, 47**. 1905–7, 1911.

RAISTRICK, A. The glacial maximum and retreat. *Trans. North. Nat. Un.* 1931.

RANKIN, W. M. The lowland moors of Lonsdale. In *Types of British Vegetation*, pp. 247–59 (Figs. 16, 17). 1911.

REID, CLEMENT. *Submerged Forests.* Cambridge, 1913.

REID, E. M. R. and CHANDLER, M. E. J. The fossil flora of Clacton-on-Sea. *Quart. J. Geol. Soc.* **79**. 1923.

SAMUELSSON, G. Scottish peat mosses. *Bull. Geol. Inst. Uppsala*, **10**. 1910.

SERNANDER, R. On the evidences of post-glacial change of climate furnished by the peat mosses of Northern Europe. *Geol. Fören. Stockh. Förh.* **30**. 1908.

SIMPSON, G. C. Past climates. *Mem. Manch. Lit. Phil. Soc.* **74**. 1929–30.

SIMPSON, G. C. World climate during the Quaternary Period. *Quart. J. R. Met. Soc.* 1934.

WILMOTT, A. J. Concerning the history of the British flora. *Contributions à l'étude du peuplement des Îles Britanniques.* 1929.

WILMOTT, A. J. in Discussion on the origin and relationships of the British flora. *Proc. Roy. Soc.* B, **118**, 215–222. 1935.

WOODHEAD, T. W. History of the vegetation of the southern Pennines. *J. Ecol.* **17**, 1–34. 1929.

WRIGHT, W. B. *The Quaternary Ice Age.* Second edition. London, 1937.

Chapter VIII

THE HISTORICAL PERIOD

Belgic settlements. About 100 B.C., i.e. not very long before Caesar's first incursion (55 B.C.), southern England was invaded by Belgic tribes who settled in and dominated the south-east during the next century and a half till the final Roman occupation which began with the Claudian invasion of A.D. 43. These Belgae founded quite powerful small kingdoms in what is now Essex and Hertfordshire, with others in the south. They practised a more advanced agriculture than their predecessors, probably using wheeled ploughs, and founded commercial capitals on the lower ground in place of the hill forts of the earlier Iron Age peoples. They carried on an export trade to the continent, including corn and cattle, and Caesar found Kent densely populated (for those times) and was able to use the corn grown there for his troops.

The Roman occupation. The Romans continued and greatly extended the founding of towns in the lower country, enlarging and transforming several of the Belgic capitals and those of other tribes to the west and north, and constructing new cities. Their civil activity extended westwards to Exeter (Cornwall was reached by sea and only the coast occupied), Monmouthshire, and the Welsh Marches northward to Chester: on the east side of the Pennines northward to York and even beyond, nearly to Hadrian's Wall. The midlands, still mainly covered, we may confidently assume, with dense oak forest, were very little touched (Leicester being the only considerable Roman town), but two great intersecting roads were made through them, one from London to Wroxeter and the other from Cirencester to Lincoln, afterwards called by the Saxons "Watling Street" and the "Fosse Way" respectively, besides "Ermine Street" (the "Old North Road") from London to York, which skirts the eastern midlands and must also have passed for the most part through dense forest. The largest of the Roman towns was London, which became a great commercial centre with a large foreign trade in which corn is said to have been the most important export.

Romano-British agriculture. It may be presumed that this activity considerably increased the agricultural area in the south of England. On the one hand there were numerous Roman villas, believed to be occupied by landowners who farmed the surrounding land with semi-serf native labour. These villas were largely concentrated on the Hampshire uplands and the dipslopes of the North and South Downs;[1] the Cotswolds, and near Bath, Yeovil, and Ilchester in Somerset;[2] but there were many others outside

---

[1] Chalk.   [2] Oolite and other Jurassic limestones, i.e. well-drained soils.

these regions. Besides the villa farms there were native villages and isolated farmsteads, both surviving from pre-Roman times and thickly scattered over the southern chalk plateaux (Gilbert in Darby, 1936, p. 66). "In Romano-British times practically the whole of Salisbury Plain, Cranborne Chase, and the Dorset uplands were under the plough" (Crawford and Keiller, 1928).

Irregular corn plots dating from Neolithic and early Bronze Age times may still be detected associated with hut circles on some of the western moorlands. But by far the best preserved indications of ancient agriculture are the so-called "Celtic fields", the very wide extent of which was strikingly revealed, largely with the aid of air photography, by the pioneer work of O. G. S. Crawford (1928). The fields, generally strip-shaped or roughly square, are bounded by low banks called lynchets, thrown up by the plough. These are often perfectly preserved, and in other places show up plainly in photographs from the air. Holleyman (1935) has shown that on the Downs north of Brighton, between the valleys of the Adur and the Ouse, out of a total area of about 65 sq. miles, excluding the river valleys and the coastal plain, "at least 23 per cent (14·5 sq. miles) were under cultivation in Celtic times, although the actual area still extant is only 11·45 sq. miles or 18 per cent of the whole", the difference having been destroyed by later cultivation or building. Nearly all the lynchet areas examined by Holleyman are on the lower summits and on spurs facing south, very rarely on the steep main escarpment which faces north. It is believed that the Celtic fields were in cultivation throughout the Roman occupation.

Sheep grazing. Pastoralism may also be presumed to have increased under the Roman peace. There is evidence that a woollen industry existed, fulling mills of Roman age having been discovered in the Cotswold region and on the North Downs, and there was apparently a weaving factory at Winchester. It is believed that both raw wool and textiles were exported, though this trade naturally cannot have been on anything like the scale it assumed in the Middle Ages (Gilbert in Darby, 1936).

While the extension of arable land was probably largely at the expense of woodland, a good deal of which must also have been cleared for military purposes in the course of the first Roman invasions and before the Roman occupation was completed, the raising of sheep on the southern chalk and oolites depended on pre-existing grassland which had existed and been used as pasture from Bronze Age or even Neolithic times.

Forest. Large forest regions, however, still remained: for example, most of the Midland Plain (mainly clay or loam soils) with oak forest, and the Weald (sands and clays) with more varied woodland including birch and perhaps pine. The iron-smelting industry of the Weald, which began before the Roman invasion, was carried on during their occupation, since at least nine "bloomeries" have been found with Roman coins and pottery. Most of these are in the eastern Weald near Roman ports on the Channel coast, whence the iron was probably exported. The furnaces depended on fuel from the abundant wood of the Weald forest; and birch, oak, hazel and ash have been identified in the charcoal. Since there are no remains of buildings it has been suggested that the charcoal burning and smelting was a seasonal occupation only. It may be presumed that care was taken both then and later not to destroy the forest by clear felling, so that the supply of wood might be maintained. The same consideration applies to the Forest of Dean (Gloucestershire) and its neighbourhood, where charcoal from oak, birch, hazel, willow and elder has been identified; and coal from the local Coal Measures was also used for smelting the considerable quantity of iron produced in that region (Gilbert in Darby, 1936). The paucity of Roman or British settlements in the midlands, especially north-west of the harder rocks of the Jurassic outcrop, but also to the east, itself suggests the persistence of forest on the Triassic and Jurassic clay lands. The same may be said of the London Clay of south-west Essex and elsewhere in the neighbourhood of London, and of much of the great drift and Boulder Clay area stretching north and north-eastwards through East Anglia and part of the east midlands. It is of course impossible to be at all sure of the extent of woodland remaining in Roman times, but there was probably a good deal more than is shown in the Ordnance Survey Map of Roman Britain (second edition, 1931). Where the villas, villages and other settlements were thickest, there no doubt little woodland remained, and every farm would of course correspond with a cleared area. But the people must have had wood for fuel and for construction, and we may reasonably suppose that a certain amount of woodland was allowed to exist within easy reach of the settlements. And there are few if any of the lowland soils, except the wettest marshes and fens, incapable of bearing forest in the Sub-Atlantic climate. Outside the settled regions there remained much forest which was probably not exploited or even penetrated.

Upper limits of forest. The altitude reached by forest on the hillsides was certainly greater than the highest levels attained by the woods which still exist. In the Lake District the original level was probably about 2000 ft. (600 m.) (Pearsall, 1936), while the relict high level oakwoods which still exist do not occur above 1700 ft. (*c.* 500 m.). In the eastern Highlands open birch and pine wood in places still reaches and even slightly exceeds 2000 ft. In Sub-Boreal times it reached 3000 ft. (*c.* 900 m.), and in the Sub-Atlantic climate probably somewhat less. In the south, where altitudes of more than 1000 ft. (300 m.) are very rare, woodland may now occupy the highest plateaux, which are commonly loam-covered; but isolated and exposed ridges and summits, particularly with thin soil, which are likely to have been among the first parts of the chalk and oolite uplands to be occupied by Neolithic man, tend to be bare of trees, and were the favourite sites at a later date of the Iron Age hill forts. By this time most of the chalk uplands, as we have seen, were almost certainly bare of trees because of the extensive upland agriculture and sheep-grazing.

In the north again, for example in the Lake District, it was the higher land, in this case mostly above 2000 ft., that was mainly occupied by prehistoric man, as is shown by the remains he has left, which coincide closely with what are now the best high-level grazing grounds (Pearsall, 1934). It is the continuous grazing through the centuries that has led to the replacement by grassland of the original forest on the slopes, which with the valleys, we may presume, were all filled with trees. On the Pennines especially, and in parts of the Scottish Southern Uplands and of central and southern Wales, where there are extensive plateaux at about the 2000 ft. level, these are covered with deep peat, accumulated in Sub-Atlantic times, but containing in some places remains of birch (more rarely of pine) which had colonised them in the drier Sub-Boreal period. In the regions of sharper high-level relief, such as Snowdonia and the Lake District, these peat-covered plateaux are uncommon.

Effects of the Saxon settlement. With the coming of the Saxons in the fifth century A.D. changes began which have continued ever since and finally led to a complete alteration in the aspect of England as a whole. Broadly speaking the change has consisted in the clearance of forest from the loams and ultimately from the clays of the lowland plains and the lower valleys of the mountain regions, and their utilisation for agriculture. Most of the Celtic inhabitants were exterminated or driven westwards by the Saxon invaders, and their cultivated fields on the chalk plateaux left derelict, since the newcomers were valley cultivators. The plateaux were then invaded by forest, some of which has remained, more or less modified, to the present day. In this forest the lynchets of the old Celtic fields can often still be detected. Great Ridge Wood on the Wiltshire downs is one of the largest examples extant of such a secondary plateau wood, though it has now been extensively replanted, and the almost adjacent Grovely Wood, a Royal Forest in Norman times, is another.

Distribution of Saxon settlement. Wooldridge has recently shown (1933, 1936) that the Saxon invaders, like their Belgic predecessors, but on a much wider scale, tended at first to occupy the lighter soils, except the coarse and acid sands, in the south-east of England. These loam soils were both easier to till and more immediately fertile than the heavy clays, which were then, and to a considerable extent long remained, covered with thick forest. Many of these loam terrains were already arable under the Belgic settlers and their Roman successors. In a wide sense they include the "intermediate" soils of the East Anglian and Essex drift;[1] the river

---

[1] The Chalky Boulder Clay of the East Anglian and east midland region, which Wooldridge marks as "intermediate", is actually over much of its area a heavy impermeable soil which must have borne dense ash-oak forest originally, though the lime present does modify it in certain respects (cf. Figs. 31 and 32, pp. 108–9).

alluvium of the Thames valley; soils associated with the Chalk and the Jurassic limestones, both at the foot of the scarps and on the dipslopes; the Tertiary and post-Tertiary soils of the southern and eastern coastal strips; and a few others. The three great barriers to early Saxon settlement in the lowlands were the heavy clayland forests, the larger areas of marsh and fen and the most barren tracts of sand. Though the actual variation of soil type in the lowlands is considerably greater than Wooldridge (in Darby, 1936, pp. 89–100) suggests, his analysis of its relation to Saxon settlement is on the whole both sound and illuminating. Finally, of course, the wild and infertile country of the mountainous Palaeogenic rocks to the west—the so-called Highland zone—which was held by the Celtic tribes, resisted penetration through the following centuries.

The earlier Saxon settlements were practically confined to the eastern half of England with a western limit running north and south, in marked contrast to the early Iron Age and the Belgic settlements, which in the south extended farther west, but in the east not so far north, and were thus limited by a north-east to south-west boundary. The different trends of the western boundaries correspond with the different directions of the earlier and later series of invasions. Later the Saxon expansion extended over the Midland Plain, where the powerful kingdom of Mercia was eventually established, and westwards into Cheshire, the Hereford plain and Devonshire, so that Saxon power was established over the whole of lowland England and in places pushed across the less mountainous areas of the Palaeogenic (or "Highland") border. After the foundation of Mercia the original continuous forest of the Midland Plain must have been riddled with agricultural settlements, though considerable tracts of forest certainly remained until much later, indeed till near the end of the Middle Ages.[1] In the south-east, in spite of the continuous invasions and successive occupations of the region for so many centuries, certain areas still remained quite sparsely settled or almost unoccupied in Saxon times—the forest of south-west Essex, the Chiltern plateau, the sterile Bagshot area to the south-west of London, and nearly the whole of the Weald. The very sparse occupation of these regions can indeed still be seen in the distribution of Domesday "vills" at the end of the eleventh century; and even down to the present century they have been thinly inhabited till the recent great expansion of building, aided by the multiplication of motor cars, threatens to cover them with houses. The Chiltern and Essex areas have remained largely forest, but the Bagshot sands and overlying plateau gravels and the Forest ridge of the Central Weald (Ashdown Sand) have for long been largely

---

[1] Though the transference of agriculture from the chalk and limestone uplands to the clay plains must necessarily have involved a great deal of forest clearance, the Saxons were not altogether oblivious of the importance of preserving woodland. Richardson (1921) calls attention to the law and custom of the West Saxons who instituted fines for burning or cutting down trees.

heathland. Both of these, however, with the exception perhaps of a very few places, can and often do support tree growth (birch, pine, oak and beech), and there is a very rapid reproduction and growth of pine and birch when such a sandy area is left to itself. It is probable that the open heathland was initiated and maintained by human factors—here mainly fire.

The Danes and Norsemen. The invasions and settlements of the Danes and Norsemen in the later part of the ninth century, though it alienated for a time practically the whole of eastern and northern England from Saxon control and left a permanent mark on the population and the place names, is unlikely to have had any important effect on the vegetation. There was much slaughter and devastation, but the Scandinavians seem to have settled almost entirely on land previously occupied.

Effects of the Norman Conquest. It was otherwise with the Norman invasion of 1066, and the changes which Norman domination brought about were very marked. Considerable areas had been plundered and laid waste during the unsettled century before William the Conqueror arrived, but much of the Norman devastation in the years following the conquest was evidently deliberate and systematic, intended to overawe the population. Large proportions of many counties were thus laid waste, and the agricultural wealth of the country greatly decreased. Added to this were destructive raids by the Welsh on the border counties and by Irish pirates in the south-west. This destruction fell mainly on crops and farm animals and houses, and except for local damage by fire can hardly have injured the natural vegetation. Indeed cultivated land which became derelict would be colonised by wild vegetation until it was reclaimed for agriculture.

The Royal Forests. A more intentional return of land to the wild was made by the formation of the Royal Forests for hunting. "Forest law" was introduced into England from Normandy, and the areas included in the Royal Forests were placed outside (*foris*) the Common Law of the rest of the country. It has often been pointed out that a "forest" in this sense was not necessarily a "forest" in the sense that the ground was covered with trees, and that the terms "afforestation" and "disafforestation" do not mean planting with trees and clearing forest, but placing a tract of land under, or freeing it from, forest law. No one knows the actual number or extent of the Royal Forests of the eleventh and twelfth centuries:[1] some estimates place their total area as high as one-third of the country.

---

[1] Only five Royal Forests are actually mentioned in the Domesday Survey. This neglect was because they were not liable for geld (tax). These five are the New Forest in Hampshire, Windsor Forest in Berkshire, Grovely Wood in Wiltshire, Wimborne in Dorset, and some Oxfordshire woods, including Shotover, Stow Wood, Woodstock, Cornbury and Wychwood. Among other actual forests subject to forest law were Sherwood, Selwood, Dean, Andred, Arden and the Chilterns. Upland areas which cannot have been tree covered but were also under forest law were the Peak (Derbyshire), Dartmoor (Devon), Exmoor (Somerset) and the Yorkshire Wolds (Darby, 1936).

The area of Royal Forest seems to have reached its maximum in the middle of the twelfth century: afterwards there was a good deal of disafforestation and corresponding extension of cultivation.[1] Even from the beginning there were different degrees of "afforestation", and it is certain that cultivated areas existed within some of the "forests", since seven whole counties were placed under forest law, and it is inconceivable that they contained no arable land. Cultivation within the forest boundaries was however discouraged by fines levied in the Forest Courts, and these were a constant source of bickering between the king and his subjects (Darby, 1936, p. 175).

The net result must obviously have been a considerable increase in the area of woody vegetation at the expense of the cultivated area. Some of the old Royal Forests, or parts of them, such as the New Forest, the Forest of Dean, Windsor Forest, Epping Forest, are still largely woodland,[2] and several are still crown lands. It is interesting to note that in the thirteenth century large tracts of the Hampshire and Wiltshire chalk uplands which were certainly cultivated in Romano-British times, were Royal Forest, as for example Grovely Wood in Wiltshire (p. 176 n.). The later woodlands which made this sort of country worth afforesting had no doubt invaded it during Saxon times when the Celtic fields were abandoned and cultivation moved to the valleys. These plateau woods are very common on the southern chalk. Most of them are oak-hazel woods on fairly deep loam, others, as on the western part of the South Downs and on the Chiltern plateau, are dominated by beech. There is no evidence to show where this beech came from. The simplest theory is that it had entered England in prehistoric times (late Atlantic or sub-Boreal, see p. 164), and colonised the chalk escarpments of the south-east, extending to the Chilterns and the Cotswold oolite escarpment—probably also a good deal farther west and north; and that in Saxon times, when the upland cultivation was abandoned, beech spread from the escarpments over the plateaux wherever the soil suited it. But we must await evidence from pollen analysis in southern England for facts as to the history of this and other species in the region.

Disafforestation and extension of cultivation. The disafforestation, which probably began in the later part of the twelfth century and made greater progress in the thirteenth, proceeded throughout the Middle Ages *pari passu* with the extension of agriculture and the reclamation of lands that had been devastated and abandoned during the early Norman persecutions. Human colonisation of waste land and forest proceeded piecemeal, but its cumulative effect was very great and gradually changed the face of the country. Place-names ending in *-hurst* and *-ley* witness the human occupation of woodland: *dene* (*-den*) was the name given to a particular spot in the

[1] In 1228 twenty-three English counties are mentioned as still possessing Royal Forests (Darby, 1936, p. 178, footnote 2).
[2] Many of course have been replanted.

forest resorted to by a village to feed its swine, afterwards occupied and cleared and forming the starting point for a new hamlet. These three terminations of place-names are frequent in the Weald of Kent—for long, as we have seen, practically continuous forest. The Scandinavian termination -*thwaite*, meaning a forest clearing, is similarly common in the north.

A time came when the tillage reached its limit if it was not to restrict pasturage and the necessary supply of wood for fuel, fencing, building, implements and tools. A well-adjusted rural economy must maintain a proper balance between arable, pasture and woodland, corresponding with the three primary needs for vegetable and animal food and for wood. Early in the thirteenth century the enclosure of forest for tillage by the lord of the manor became a burning question, and the Statute of Merton (1235) established the right of the lord to enclose forest provided he left enough pasture for his free tenants (Darby, 1936, p. 189).

Pasturage in forest. It is clear that in regions which at the time of the Norman Conquest were still forest the increasing pasturage of the next centuries was in that forest, and this, apart from "assarting", i.e. clearance for tillage, must have led eventually to the disappearance of the woodland. The feeding of swine in oak- or beechwood may have actually assisted regeneration by stamping a proportion of the acorns and beechnuts into the soil and by the destruction of small vermin such as mice and voles (besides snails and slugs) which to-day make regeneration impossible because they destroy the nuts and seedling trees. But the pasturing of cattle through woods always leads to the disappearance of the natural ground vegetation, consolidation of the surface soil by trampling, the introduction of grass vegetation and the coming of conditions which make the establishment and growth of tree seedlings impossible, quite apart from their destruction by the actual browsing of the beasts. Thus while the influence of man immediately after the Conquest favoured the spread of natural vegetation by checking agriculture and by the "afforestation" of great areas, the human factor soon began to work in the opposite direction, to the detriment of forest, and this has continued right down to the present day.

The three kinds of land utilisation. The village community which was the type of rural English social organisation from 500 to 1500 A.D. depended on three main kinds of utilisation of land:

(1) *Ploughland* cultivated in strips as "common fields" with autumn sowings of wheat and rye and spring sowings of barley and oats in the following year, the crops being protected by fences. The cattle were grazed over the stubble, and thus provided a certain amount of manure. In the third year the land was fallowed and ploughed twice in the summer. This open field system was prevalent in central England from Durham to the Channel and from Cambridgeshire to the Welsh border (Darby, 1936, pp. 192–4).

(2) *Grassland*, which (apart from the upland chalk and limestone areas

already considered) must have been originally derived from forest by felling and pasturing or sometimes by pasturing alone. Grassland is of two kinds, *meadow* and *pasture*, the former typically on alluvial flats bordering on rivers, with high-ground water, the grass mown for hay, and the aftermath usually grazed; the latter on the drier soils and generally continuously grazed, though sometimes also "put up" for hay in certain years.

It will be noted that in addition to the grassland continuously used as pasture more or less important ancillary grazing areas were the stubble of the corn crops, the aftermath of the meadow hay-crops and to a considerable extent forestland.

(3) *Woodland*, the remaining portions of the original forest useful primarily as the source of wood for construction, for fencing, for tools and implements, for fuel, and for a variety of other purposes; secondly, for feeding swine and to some extent for pasturing cattle, though this last, if persisted in, led to its destruction. Besides these uses to the agricultural communities the Royal Forests were hunting grounds for the King, and there were various "chases" for the nobles and squires.

Regional variations in rural economy. The typical village communities were not, however, spread over the whole country, and the type of rural economy varied both with race and with physiography. Kent, inhabited by Jutes, was a country of hamlets rather than villages, and never had the open field system, which was also absent from East Anglia proper.

In the hill country of the north and west of England agriculture was at a minimum. Here the people depended almost entirely on stock raising, their diet consisting of flesh, milk, cheese and butter, with very little bread, fruit or vegetables, nearly all the land being pasture grassland. Instead of compact villages there were isolated homesteads, or very small groups of houses forming hamlets. In Wales, according to Giraldus Cambrensis, the people lived on the edges of the valleyside woods in huts made of intertwined branches; again their land was nearly all pasture and their diet almost wholly animal, though they gladly ate fruit and vegetables when they could get them.

In the Fenland and other extensive fen and marsh areas there was again a different economy. Here the people lived on the fen borders and islands and subsisted largely on fish and waterfowl of which there was great abundance. Presumably they used reeds for thatching and grasses and sedges for litter, as they still do, or did until very recently.

Except for the final complete or almost complete enclosure of the open fields and common "waste"[1] in the later eighteenth and early nineteenth centuries and the concomitant alienation of the common rights in arable land of the village communities, and the extreme depletion of woodlands, it

---

[1] There are a few areas of "waste"—mostly heath, scrub or grassland—in which common rights of pasturage and collection of dead wood still exist and one place (Laxton in Nottinghamshire) where an open field is still cultivated in common.

is noteworthy that the rural economy described lasted, in its essential features, for many centuries and even in some parts down to the beginning of the twentieth.

Corn growing and export. During the later thirteenth and the fourteenth centuries both corn growing and sheep raising increased very considerably because there was a great expansion of farming for profit in addition to the primitive subsistence farming. Some landlords leased their demesnes for a money rent, others converted them into sheep runs. The rise of the towns and the increase of population enlarged the local markets for cereals and there was often a surplus for export. "The period 1250–1350 was the most prosperous century of medieval agriculture and the one during which the largest quantities of corn were exported" (Pelham in Darby, 1936, p. 238). In 1377 the population of England is estimated at about two million, the vast majority rural. The densest population was in south Lincolnshire and Norfolk—about fifty people to the square mile. The area of lowest wheat prices and therefore, we may infer, of highest wheat production, stretched in a wide belt north-eastwards from the Severn estuary to the East Anglian coast, the north-eastern part of this belt roughly corresponding with the principal wheat area of modern times. Hull, Lynn, Yarmouth, London, Sandwich, Chichester, Southampton, and to a less extent Bristol, were important wheat ports. Increasing enclosure began in places to lead to the disintegration of common fields, as well as still further diminishing the waste and the common rights to wood and pasture.

Sheep raising and wool production. In the middle of the fourteenth century the sheep population of England alone is estimated at eight million (or four sheep to each head of the human population), and this was of course concomitant with the rise of the wool and later of the cloth trade. A great deal of wool from many parts of the country was exported.[1] The Cistercian abbeys of east Yorkshire and Lincolnshire exported so much that they must each have kept flocks of many thousand sheep. As the export of cloth increased that of raw wool declined. At the end of the century the weaving industry "burst the bonds of medieval town life and became established in rural areas, where it steadily maintained itself till the great expansion of Tudor times". The most important rural areas in which broadcloth was made were Somerset and Suffolk, but there were several others. Export was considerable and Bristol the greatest western port, but its exports were far exceeded by those of London. The sheep were kept not only on the hill grasslands, but also all over the lowlands, where they were often folded during the winter on the corn fallows. On the South Downs it is interesting to note that in 1341 the greatest wool production was on the

---

[1] 80 per cent of the wool export passed through Hull, Boston or London in 1333–6. Southampton was also an important wool port, being the channel of export not only for the products of the Downs, the Hampshire Uplands and Salisbury Plain, but also for those of the Cotswolds and the Welsh Marches.

eastern Downs, which are still the main Sussex sheep region and are bare of trees, in contrast to the western Downs which show only local concentrations of wool production. This suggests that then, as now, the latter were largely covered with woods. At the beginning of the fifteenth century most of the Sussex wool was said to be produced within 15 miles of Lewes (Pelham in Darby, 1936), the capital of east Sussex, and the centre of a bare Down area.

Timber and fuel. Wood, both for timber and fuel, and also oak bark for tanning were shipped from the Weald through the port of Winchelsea throughout the Middle Ages, both coastwise to London and other English ports and also to Flanders and the Netherlands (Pelham in Darby, 1936, p. 322). This shows that in the fourteenth and fifteenth centuries the Weald still had more than enough forest for its own use. On the other hand, on the north-east coast timber in various forms (including oak boards for wainscot, and "deles") was beginning to be imported from the Baltic in exchange for English cloth, suggesting a distinct dearth of woodland in the north of England (*ibid.* p. 326). There is evidence that timber was imported in quantity at the end of the thirteenth and in the fourteenth centuries, and that import began as early as the eleventh (Richardson, 1921, p. 163).

Origin of "coppice-with-standards". At some time during the Middle Ages the exploitation of English oakwoods as coppice with standards must have been established. We may picture this as resulting from free (but not clear) felling of the trees, giving an increase in vigour of the underwood, largely hazel, owing to the admission of more light. Thus an "open canopy" oakwood would be formed, consisting of spreading trees with a continuous shrub layer below which would respond well to regular coppicing at intervals of 10–15 years. The oaks ("standards") would be felled when they were mature, and would yield both good planks (though of no great length) and curved timber useful in shipbuilding. We see evidence of the standardisation of this form of exploitation in the series of English statutes, beginning in 1544, which enacted that twelve standards were to be left in each acre of wood felled, for there is about room for twelve mature well-grown oaks *in open canopy* on an acre of land.

Depletion of forest resources. Richardson (1921) remarks that "it would be hasty to conclude that there was any general shortage of timber in the fifteenth century", though he quotes a Frenchman writing about 1460 to the effect that wood was much dearer in England than in France, and that "by reason of the great extent of cultivation there are hardly any woods". This may have applied to the area immediately round London. But in the sixteenth century there certainly was an increasing shortage, though the midlands at this time were still fairly well wooded (Leland), the remains of the old oak forest being most abundant in the "clay vales"; and the deer which inhabited them were more than a nuisance to the farmers.

"In the first half of the sixteenth century wood fuel was supplied from

Sussex, Kent and even Essex to the cross-channel ports—the Weald, with its ample stocks of fuel was becoming the great (iron) foundry of England" (Richardson, 1922, p. 178). The iron founding included the casting and export of cannon to foreign countries. The trees of the Forest of Dean (in Gloucestershire) were said to be considerably thinned on account of the local iron industry, charcoal still having to be used for smelting, in spite of the presence of coal. Sherwood Forest in Nottinghamshire, in which there were an "infinite" number of deer, is also said by Camden to have been much thinner than it formerly was. The poor of London were increasingly using coal brought by sea from Newcastle, and great distress was caused when the trade was interrupted in 1643. "Sea coal" was not liked because of the smell, but by the middle of the seventeenth century wood was clearly the fuel of the wealthy alone (Richardson, 1922).

The first suggestion of planting. Another evidence that the lack of timber was beginning to be felt in the sixteenth century is found in Bodenham's statement (1571) that in the north of Spain, where most of the iron that supplied Europe was mined and smelted, there was no dearth of fuel for charcoal, because whenever an oak was felled six saplings, specially grown from acorns, were planted. This is the first suggestion in England we have been able to trace that it is necessary to plant trees in order to maintain forests that have been too heavily exploited.

General and regional developments. For the rest enclosure continued to increase, the advantages over common field cultivation being very apparent both for arable and pasture, and the cloth-weaving industry not only flourished but greatly expanded, with the result that the number of sheep increased and land that had formerly been arable was converted to sheep pasture. Corn was no longer exported, at least in any quantity, and in some places the people lacked bread when there was a bad harvest and were even driven to make flour from acorns. Silting of the estuaries on the south-east coast was particularly rapid during the sixteenth century, some of the "Cinque Ports" on that coast lost touch with the sea, and the reclaimed salt marshes supported great flocks of sheep and herds of cattle, as did the marshland behind King's Lynn on the Wash, which was still a considerable port. In the West Riding of Yorkshire, Halifax and its neighbourhood became (and still remain) a great cloth-weaving centre. Gloucester and Somerset, however, retained their pre-eminence as the most important cloth-producing counties and rivalled Kent and Surrey in density of population, though the industry was spread far and wide through the country. Cattle were also well represented though they were nothing like so numerous as sheep. The east Leicestershire cattle pastures were already famous, as they are to-day.

Dearth of timber in Scotland. In Scotland the shortage of wood was apparently felt a century earlier than in England, and it appears from contemporary charters that already in the twelfth century the southern part of Scotland was not well wooded. It is not easy to understand why this should have been so. There is no doubt that in Roman times, the valleys of the Southern Uplands, the Midland Valley and the Highland valleys were well wooded. This is evidenced from references to the Scottish forests in Tacitus' account of Agricola's campaign in the first century of our era and from the many contemporary references to the great "Caledonian Forest". And Gaelic place names indicating the presence of woodland are widespread (Watt, 1900). But at the beginning of the modern period the people are said to have disliked woods because they were punished for cutting them! The Royal Forests were held by the Scottish kings for venery and chase or else specially granted by them to subjects, and fines were inflicted for cutting. It was also said that the woods harboured dangerous beasts and that trees did not go well with agriculture. But all these circumstances were equally true of England. And yet as early as 1457, more than a century before there was any talk of planting in England, an act was passed requiring all the king's tenants to plant trees; and after 1468 most leases of land bound the tenants to plant and appointed men to see that they did (Murray, 1935). In 1503 the preamble of an Act of the Scottish Parliament stated that the woods of Scotland "are utterly destroyed". This was doubtless a considerable exaggeration, but it must imply a severe shortage; and in that century increasing fines and eventually the death penalty were to be inflicted for cutting wood. All this applies to the Southern Uplands and the midland valley, not to the Highlands, which certainly contained considerable virgin forests at that time.

One of the earliest actual records of Scottish planting dates from 1473, and in the following century there are records not only of fines imposed but of orders of the courts that tenants should plant three birches for every one damaged. Before the end of the sixteenth century, "in 1586, James, Lord Ogilvie, is found writing to Sir David Lindsay of Edzell—'concerning my planting—your thowsand young birkis sall be richt welcom'" (Watt, 1912). A sixteenth-century Parliament made definite planting laws with fines for non-observance. But much destruction of wood continued, and in 1679 the Englishman Kirke writes that the Scots have no woods and destroy those they have!

The seventeenth century. During the seventeenth century (Baker in Darby, 1936) the English forests and woodlands gradually diminished. "To the writers of the late sixteenth and of the seventeenth centuries the great enemies of English woods were iron and glass manufacture" (Richardson, 1922). In 1629 destruction of *timber* is attributed to iron works and tanneries, and this may well have been true, the high woods being converted into coppice. Before the end of the seventeenth century the iron-smelting industry was waning in the Weald, while it was increasing rapidly in other areas long before coke came into use for smelting. The industry *preserved* woodland of a sort, because in its absence the land would be used for pasture and tillage, whereas coppice was maintained in the west midland counties for use in the iron works (Yarranton, *England's Improvement*, 1677, quoted by Richardson 1922). During the Civil War there was much destruction of forest, and the Duke of Newcastle alone is said to have lost timber to the value of £45,000 in the north of England and Derbyshire.

The statutes intended to preserve timber by enacting that twelve standards (oaks) should be left to the acre of forest felled were not generally effective because various districts were excepted and licences were granted to vested interests: also enforcement was spasmodic and evasion common. Thus destruction of timber continued, and such woods as were preserved for charcoal and for tanning were converted into coppice. Already in the later sixteenth century Harrison in his *Description of England* had urged that every owner of 40 acres of "champaign soil" should be required to plant one acre of wood, but he despaired of getting any such project carried out.

More and more the timber-using industries, including the shipbuilders, were supplied from abroad, chiefly from northern Europe, though in the seventeenth century there was some importation from New England, while Irish timber was also exploited for various purposes. About 1680 Pepys mentions the dependence of the navy on imported timber. By the end of the century many prominent people had become reconciled to the loss of the forests, arguing that the necessary timber could always be imported from northern Europe (as indeed it was), though some, such as John Evelyn, deplored the loss of native timber. The seventeenth century may in fact be said to mark the definite exhaustion of the last forest reserves, turning England into a country almost entirely dependent on foreign lands for its timber. And besides the diminution of the woodland area, the woods suffered serious deterioration in quality during the seventeenth and eighteenth centuries, high forest giving way to coppice (oak coppice being used for charcoal and tanning bark), while a large proportion of the standing trees were deformed and decayed.

Early Scottish planting. We have already seen (p. 183) that the shortage of timber in Scotland (i.e. the Lowlands) was felt much earlier than in England, and that systematic efforts to enforce planting began long before any serious attention was given to the subject in southern Britain. It seems, however, that little effect was produced in either country, destruction of woodland far exceeding any planting that was done up to the end of the sixteenth century. At this time the sheep-raising industry in the south of Scotland was becoming over-developed and sheep farmers were getting into trouble for allowing their stock to graze in woods and plantations. About the turn of the century Sir Duncan Campbell of Breadalbane sowed acorns and pine seed and planted young pine and birch extensively— besides chestnut and walnut. His successor, Sir Colin, sent ripe pine cones to his relatives for extraction and sowing of the seed, and there is good evidence that they took great interest in making the "firr trees" grow. Thus began in Scotland the active interest in planting which developed so strongly at a later date. Towards the end of the century Sir Hugh Campbell of Cawdor planted up the "Old Wood", which by 1726 was "very thick, consisting of birch, alder and some young oak, together with 400 mature oaks fit for sale and use" (Murray, 1935; see also Watt, 1912).

The influence of Evelyn. Meanwhile a taste for planting, based more on curiosity and interest in the growth of different kinds of trees than in the provision of timber for the needs of the country, was being developed and spread by John Evelyn and his friends in the south of England. Evelyn described and planted all kinds of trees he could obtain and make grow, British and exotic. To his influence may perhaps be ascribed the beginning of the curious passion for planting exotic conifers that became so engrained in his countrymen, trees which for the most part—with the exception of pine on sandy or rocky soils—are so entirely out of harmony with the English landscape. But to Evelyn's interest in planting for its own sake must also be ascribed the impetus towards the beginnings of the large-scale planting which developed in the eighteenth and nineteenth centuries.

Land utilisation. During the seventeenth century rough quantitative measures of land utilisation are easier to come by, though still nothing like exact statistics. At the end of the century King's estimate for England and Wales gives (in approximate percentage figures):

Arable	23
Grassland	31
Woodland, Forest, Parks and Commons	16
Heaths, Moors, Mountains etc.	25

Though anything like strict comparison with current figures is quite impossible, we shall be safe in saying that between then and now the arable area has increased, but not very much, grassland has greatly increased, while much woodland has disappeared and "mountain and heathland" have substantially diminished. In other words, the trend towards grassland (pasture), already far advanced in the seventeenth century, has now gone farther still (cf. Chapters IX and XXV). Towards the end of the century many foreigners admired the richness and beauty of the English pastures. The decrease of mountain and heathland means enclosure, and often ploughing and sowing with grass seed, i.e. "laying down to grass", more or less intensive grazing, and sometimes manuring, that is to say the conversion of wild open land on the hill slopes into enclosed and more or less tended "permanent pasture".

Davenant at the end of the century estimated the total number of sheep as 12 million and cattle $4\frac{1}{2}$ million. Of rabbits he thought there were a million, but it may well be doubted whether the last estimate was more than a guess (see Chapter VI, pp. 133–4). Taking the means of the very diverse estimates of the human population it may have increased from $3\frac{1}{2}$ million at the beginning to 6 or $6\frac{1}{2}$ million at the end of the century.

Enclosure of common fields continued: in the midlands some of the enclosed area on heavy soils was at first devoted to sheep and cattle pasture. In East Anglia there was an increase of arable, and late in the century in the midlands also, following a substantial rise in the price of wheat and the resumption of export on a considerable scale.

The variety of crops grown increased considerably, and besides a number of special crops, included market gardening and fruit growing—cherries, apples and pears being largely localised in such counties as Kent, Gloucestershire, Worcestershire, Herefordshire and Somerset, which are still famous for these fruits.

The draining of the Fens. Perhaps the most notable event of the seventeenth century affecting the nature of a large area of England was the draining of the Fenland (Darby, 1913, Chapter XII). This name is generally applied to the flat area of approximately 1300 sq. miles lying at about meantide level in the counties of Cambridgeshire, Huntingdonshire, Lincolnshire and Norfolk and traversed by the Rivers Ouse, Welland, Nen and Witham and their tributaries, emptying into the wide bay of the Wash. The northern part nearest the sea with a soil of marine and estuarine silt is often distinguished as the marshland, while the southern part with a peat soil and interspersed with islands of hard land is the fenland proper. Attempts at drainage of parts of this area, and especially efforts to keep the sea out of the marshland by means of sea walls had been made with partial success from the time of the Romans. Towards the end of the sixteenth century Camden reported that the marshland bred abundance of cattle and at one place pastured 30,000 sheep. But it was still often inundated by the sea and most of the peat area further inland was still a waste of reeds and sedges, continually overflowed by water in the winter, and sometimes in the summer, and interspersed with permanent meres (e.g. Whittlesea Mere, Soham Mere), situated, like the similar Broads of the east Norfolk fen region, near the bordering upland. The fenmen depended on the abundant fish and waterfowl, used the reeds for thatching, and cut the sedge (presumably *Cladium*) and the "lid" (*Glyceria maxima* according to Miller and Skertchly) for litter (see Chapter XXXIII).

Schemes for draining the fens, always opposed by the fenmen who saw the threat to their livelihood, were much discussed and debated at the end of the sixteenth and the beginning of the seventeenth centuries. But it was not till 1631 that a company of thirteen "co-adventurers" associated with the Earl of Bedford began the work, making several straight cuts in the southern and western fens (the so-called "Bedford level"), the largest of which, the "old Bedford River" 70 ft. wide, ran in a straight line for 21 miles, from Earith to Salter's Lode, shortening the course of the Ouse by several miles. This method was advised by the Dutch engineer Vermuyden

PLATE 12

Phot. 24. Small tree of *Salix alba* which has produced adventitious roots at and below the flood level. *R. H. Compton.*

Phot. 25. Blanket of *Cladophora flavescens*. Here and there it is seen raised by the stubbles of the last cereal crop. *R. H. Compton.*

AFTER THE FENLAND FLOOD OF JANUARY 1915

to increase the fall of the water in a given distance. There was difference of opinion as to the success of this first venture, and the Civil War interrupted the work; but later it was resumed, and evidently with success, for presently we read of the most varied crops as well as of many sheep and cattle on the reclaimed land. Further similar cuts and drains followed, finally extending throughout the Fenland. The Lincolnshire fens, to the west of the Wash, were for the most part not drained until the eighteenth century.

The uneven "hassock" (the stools of the coarse fen plants) was pared off and burned, and this process was continued with the peat below, so that the surface of the soil was considerably lowered. This paring away of the surface of the peat, together with shrinkage from drying and wasting of the organic soil by bacterial action, has in fact considerably lowered the surface of the fen peat, so that whereas the general level of the peat fens was said in the sixteenth century to be 5–7 ft. above that of the silt marshland nearer the sea it is now about 10 ft. below. This lowering of the peat surface level, together with the continued silting up of the river outfalls into the Wash, seriously interfered with the effectiveness of the drainage. The levels of the peat drains and dykes were carried down by the shrinkage of the peat well below the water-level of the embanked rivers, and it soon became necessary to pump water from the ditches into the rivers. This was at first done by power derived from windmills, but these proved unsatisfactory in the long run. In 1820 the first steam engine was used for pumping, and steam power (now replaced by Diesel engines) was eventually adopted throughout the fenland.

In the latter part of the seventeenth century and throughout the eighteenth floods were continually occurring in wet years, causing widespread inundations, with consequent ruin and misery to the farming population which had now settled on the land. Partial flooding in one area or another still occurs every few years owing to cracking or subsidence of the banks after continuous heavy rains,[1] but the transformation of the fenland into a rich agricultural district which was substantially effected in the seventeenth century has not been seriously endangered since the eighteenth, and has added greatly to the agricultural wealth of eastern England. There are a few fragments of "unreclaimed" fen still remaining, all situated in bays of the Fenland near the edge of the uplands. The vegetation of these is described in Chapter XXXIII.

The eighteenth century. During the eighteenth century (East in Darby, 1936) the conversion of waste land to tillage proceeded further, more than 2 million acres (3125 sq. miles) being added to the agricultural area. In 1795 only a fifth of England and Wales was said to be "barren land", and this included the "rough" and "hill grazings" of which by far the greater part were in Wales and northern England (the so-called "Highland zone" of Palaeogenic rocks).

[1] Pl. 12, shows some of the temporary botanical results of one of these floods.

Enclosure. In 1700 more than half the agriculture was still carried on in open fields. By the end of the century the great bulk of these had been enclosed, the few remaining in the eastern midlands finally disappearing before the middle of the nineteenth century. Alongside of this went the enclosure of the common land used for pasture, and the consequent decrease of the area bearing semi-natural vegetation. The Cotswold uplands were enclosed between 1759 and 1789. In Wiltshire, owing to the rise in wheat prices, more of this crop was grown, and numbers of sheep were now kept mainly for folding on the cropped land, because of the value of their dung as manure and of their constant treading of the soil, rather than for the sake of the fleeces and meat. In other parts of the country and especially on the hills of Wales, the north of England and the Southern Uplands of Scotland, vast flocks of sheep were still pastured. In the latter part of the century the Scottish Highlands were also invaded by sheep farming *en masse*.

English agriculture was most advanced in the counties round London (an increasingly great market) and in East Anglia; and from the latter corn was shipped to Holland when prices were good. "Towards the end of the century Norfolk became the premier farming county", though a little time before much of its land had been waste or unproductive heathland. Here was established the famous four-course rotation which with various modifications has been the basis of arable farming in the English lowlands ever since. For the rest, both the nature and the quality of the farming was infinitely various in different parts of the country, following differences in climate, soil, markets, and the skill and traditions of the farmers. During the century the population increased from something over 6 to more than 9 millions, and of these nearly 78 per cent still lived in the country.

Plantations. The urge to plant trees initiated by Evelyn and his friends in the latter half of the seventeenth century came to full fruition in the eighteenth and nineteenth, not only by the planting of ornamental and "specimen" trees in parks and grounds, but by regular plantations for timber. Of the English planting, unfortunately, no general account is available. It was carried out by individual landowners, seldom on any very large scale; but in both England and Scotland it included the native hardwoods, oak, ash, elm, and a little beech, the European sycamore (*Acer pseudo-platanus*), introduced long before, as well as a number of other trees and shrubs; and, of European conifers, the larch (*Larix decidua*), first planted at Dawyck in 1725, spruce (*Picea abies*) and silver fir (*Abies alba*) (Watt, 1912). Larch was advocated as a substitute for oak, and was found useful for many purposes.

Scotland was on the whole in advance of England in planting activities, and Watt's paper (1912), and Murray's (1935) give good summary accounts of their progress.

Destruction of the Highland forests. The treelessness of Scotland in the later Middle Ages applied, as has been already said, only to the "Lowlands", i.e. including the Southern Uplands. The Highlands at that time possessed in their valleys and glens magnificent forests which were largely untouched and consisted of oak at the lower elevations and in the more sheltered and favourable places, birch and pine at higher altitudes and on the poorer soils. These were "discovered" by the exploiters about the beginning of the seventeenth century, and their wastage lasted for more than 200 years. Many were burned for charcoal, the iron ore being brought to them for smelting. The pine from some of the Argyll forests was sold to Irish companies and iron ore brought to use up the birch. What was left from this wholesale destruction was afterwards largely cleared (felled and burned) for the great expansion of sheep farming at the end of the eighteenth century continued through most of the nineteenth. The sheep farmers of the Southern Uplands, many of whom had already made fortunes there, offered very high rents for the Highland grazings—in some cases twenty times as much as the crofters had been paying—and the areas available for grazing were increased by clearing the forests of the lower slopes and glens (Cameron, 1923).

Eighteenth century planting in Scotland. Side by side with this destruction of native forest, the planting movement initiated at the end of the sixteenth century gathered a great impetus in the eighteenth, which was the Golden Age of tree planting in Scotland, in spite of difficulties caused by the forfeiture of the Highland estates of participants in the Stuart rebellions of 1715 and 1745. Some of these estates were planted by the Commissioners appointed to manage them. Among the large landowners who undertook large scale planting were Lord and Lady Hamilton, who in 1705 planted for profit 800 acres of pine on very poor sandy soil and with great success. This set an example which was widely followed. In the north of Scotland Sir Alexander Grant began operations in 1716 and continued to plant throughout his long life, reaching a total of 50 million trees. *Pinus sylvestris* was the favourite tree with most planters, and in one case it is said that one-half were Scots and the other half English, by which is presumably meant var. *scotica*, the native Scottish form, and some longer needled form obtained from England. In 1736 prizes were offered to farmers who should plant the largest number of oak, ash and elm before a certain date. In the middle of the century lords and lairds vied with one another in making plantations. During the later part the planting of European larch became common. By 1826 John, Duke of Atholl, had planted over 14 million larch, choosing as far as possible rocky ground and putting in 2000 to the acre. Wet ground was drained and planted with spruce. A little beech was planted in some districts.

This sort of activity continued well into the nineteenth century. Much intelligence was often shown in the planting methods employed, some of which have scarcely been improved upon. By the end of the eighteenth century, however, the movement had been carried too far and land had been planted which was much too valuable for trees (Murray, 1935).

Meanwhile other demands on the use of the land competed with tree planting. There was the great increase of sheep farming and corn growing during the French wars, the former, as already mentioned, being responsible for almost completing the destruction of the native forests of the Highlands. It was obvious that sheep paid much better than plantations, and the profits to be gained from the latter were heavily over-estimated by some too enthusiastic surveyors for the Board of Agriculture (Murray, 1935).

Nineteenth century planting in Scotland. At the beginning of the nineteenth century there was a great shortage of timber for shipbuilding. Larch, which had already been introduced in the previous century, was recommended as a quicker and more easily grown substitute for oak and was largely planted. But before the middle of the century many factors adverse to planting began to make themselves felt. Importation of timber became easier and cheaper, teak began to displace oak, and eventually iron largely replaced wood in shipbuilding. With the introduction of chemical tanning oak bark lost its value. And many other competing interests, such as agriculture and the new industries in which much capital was invested, became prominent. But with all these factors against it the Scottish record of planting was creditable, and a high degree of skill and knowledge was shown. Two western American conifers, the Douglas fir (*Pseudotsuga*) and the Sitka spruce (*Picea sitchensis*), were introduced and promised to be important timber trees, though even now their value is not fully tested (Murray, 1935).

None of these plantations of exotic species can be regarded as seminatural vegetation, and except under larch the ground is almost or quite bare, being covered with a thick more or less resistant litter of dead needles, which sometimes leads to deterioration of the soil. Only pine and native hardwood plantations in what may be regarded as "natural" habitats show a natural undergrowth. There are, however, still some fragments of natural wood in the Highlands—oak on the better soils of the larger valleys, pine, very local, and birch, widespread but increasingly exiguous and degenerate (see pp. 203, 343–9 and Chapter XXII).

The nineteenth and twentieth centuries in England. Great as was the alteration in the face of the country during the Middle Ages and the beginning of the modern period, the nineteenth century and the first part of the twentieth have seen much greater and much more rapid changes. Industrialisation of much of the midlands and north, which had made considerable progress during the eighteenth, now became overwhelming and completely altered the face of much of the country as well as the structure of society. The towns increased, often many fold, in size and population, which was more than trebled during the century, in contrast with the 50 per cent increase in the preceding one. The natural vegetation of considerable areas was blighted and agriculture rendered impossible by the smoke which spread over them like a pall. In the rest of the country, during the interval between the Battle of Waterloo and the Repeal of the Corn Laws, agriculture, especially wheat growing, was greatly stimulated by the high prices consequent on the increase of population and the prohibition of import; but this stimulus disappeared on the withdrawal of protection and the development of the great wheat-growing areas of the United States and Canada, and agriculture declined heavily during the later part of the century. The result was the laying down of more and more land to "permanent grass", and much of this pasture was increasingly neglected and has become of little value. Arable cultivation has been stimulated from time to time, notably in the present century during the Great War when many thousand acres were reploughed, not always with the best results, and by the recent "wheat quota" and guaranteed prices. There are now good prospects of greatly improving the pastoral industry as the result of recent research in manuring, recognition of the greater nutritive value of young grass, and the breeding of improved strains of pasture plants. It is impossible here to describe the details of all these changes.

Modern enclosure and building. Most of the unenclosed "waste" and commonland still remaining was enclosed and cultivated when farming was prosperous in the first half of the nineteenth century, and some extensive schemes of "reclamation" were undertaken in very unpromising districts, e.g. Exmoor, an elevated plateau in west Somerset, in the middle of the century. But on the whole the greatest recent destruction of natural and semi-natural vegetation has been due to building. All the great cities and many of the towns extended their boundaries. London has grown to a colossal size, extending its suburbs over nearly the whole of Middlesex and penetrating far into Essex, Kent and Surrey. Besides the extension of the suburbs of the city itself many new "building estates" have been opened up in rural districts since the war, and a large part of the sand country near London, especially to the south-west, as well as much of the chalk, especially near the coast, is being covered with houses.

The Forestry Commission. Lastly we have the post-war activities of the Forestry Commission which was entrusted by Government immediately after the Great War with the task of attempting to minimise the risk of shortage of essential timber supplies in time of war. The Commission is planting extensive areas of moorland, heathland and derelict woodland, as well as some grazing land and semi-natural woods of native trees which are not derelict though they have little or no commercial value, mainly with Scots pine, Douglas fir, Norway and Sitka spruce and Japanese and European larch, and thus completely altering the character of the landscape in many parts of the country. Strong protests have been made, especially where the Commission has invaded districts, notably the Lake District, whose beauty is much cherished. But few people have apparently realised how completely the Commission is destroying the beauty of the still wooded Highland glens by substituting close plantations of exotic conifers for the natural oak, birch and pine.

Efforts to preserve the countryside. To set against these potent and widespread causes of destruction of natural and semi-natural vegetation there are some forces working in the opposite direction, though on a much smaller scale. In the later nineteenth century feeling was aroused against the enclosure of the few remaining pieces of common (nearly always rough grazing) land, and by one means or another—largely by the efforts of the Commons and Footpaths Preservation Society—several such enclosures were prevented, thus preserving open spaces for the refreshment of the people which are also areas of semi-natural vegetation. At the beginning of this century the National Trust for the Preservation of Places of Historic Interest and Natural Beauty was established, and this body has acquired many areas bearing natural or semi-natural vegetation—mountain, heath, fen or woodland—on the score of their beauty and interest. These are either presented or bequeathed to the Trust by public-spirited landowners or bought by public subscription. Each is under the control of a voluntary local committee, which usually raises the funds necessary for upkeep. It is much to be desired that a systematic survey of the plant and animal life of those Trust properties which contain natural or semi-natural vegetation should be undertaken and published.

The recently formed Council for the Preservation of Rural England also does what it can to protect the amenities of the countryside, among which natural and semi-natural vegetation necessarily takes a foremost place. The formation of National Parks has been advocated, notably by Stapledon (1935) in a very comprehensive form. He desires that wide natural geographical areas should be demarcated and their exploitation regulated so as to maintain a well-balanced rural economy within their boundaries, preserving the best of the varied features of the life of the countryside and also providing the means of recreation for town dwellers. The movement for the creation of National Parks in some form seems likely to achieve a positive result in the immediate future.

Country planning. Some sort of national planning of a systematic "lay-out" of the whole country, in which the various interests are duly considered and adjusted—the rural as well as the urban, the spiritual and aesthetic, as well as the industrial and commercial—is now indeed very urgent if we are not to hand on to our descendants a progressively disorganised and wasted land of which no one can eventually be proud. The recent Town and Country Planning Acts have taken the first tentative steps in planning, but unfortunately merely permissive laws are quite ineffective against powerful vested interests, and the economy of the country must be considered as a whole. A comprehensive survey of the land in all its aspects is a necessary prerequisite of any such general planning on sound lines.

REFERENCES

BAKER, J. N. L. The Seventeenth Century (Chapter XI in Darby, 1936).
CAMERON, A. G. *The Wild Red Deer of Scotland.* 1923.
CRAWFORD, O. G. S. *Air Survey and Archaeology.* Second edition. 1928.
CRAWFORD, O. G. S. and KEILLER, A. *Wessex from the Air.* 1928.
DARBY, H. C. (edited by). *An Historical Geography of England before 1800.* 1936.
EAST, G. The eighteenth century (Chapter XIII in Darby, 1936).
FOX, CYRIL. *The Personality of Britain.* 1933.
GILBERT, E. W. The Roman occupation (Chapter II in Darby, 1936).
HOLLEYMAN, G. A. The Celtic field system in South Britain. *Antiquity,* **9**, 443. 1935.
Map of Roman Britain. Second edition, Ordnance Survey. 1931.
MURRAY, J. M. An outline of the history of forestry in Scotland up to the end of the nineteenth century. *Scot. For. J.* **49**, 1935.
PEARSALL, W. H. Woodland destruction in northern Britain. *Naturalist,* pp. 25–28. 1934.
PEARSALL, W. H. The Botany of the Lake District. *Scientific Survey of Blackpool and District.* 1936.
PELHAM, R. A. The fourteenth century (Chapter VII in Darby, 1936).
RICHARDSON, H. G. Some remarks on British forest history. *Trans. Roy. Scot. Arbor. Soc.* **35**. 1921. Part II. The sixteenth and seventeenth centuries. *Trans. Roy. Scot. Arbor. Soc.* **36**. 1922.
STAPLEDON, R. G. *The Land Now and To-morrow.* London, 1935.
TAYLOR, E. G. R. Camden and Leland's England (Chapter VIII in Darby, 1936).
WATT, H. BOYD. Scottish Forests and Woodlands in Early Historic times. *Ann. Anderson. Nat. Soc.* **2**, 89–107. Glasgow, 1900.
WATT, H. BOYD. Early tree planting in Scotland. *Trans. Roy. Scot. Arbor. Soc.* **26**, 12–31. Edinburgh, 1912.
WOOLDRIDGE, S. W. The Saxon settlements (Chapter III in Darby, 1936).
WOOLDRIDGE, S. W. and LINTON, D. L. The loam terrains of south-east England in their relation to its early history. *Antiquity,* **7**. 1933.

Chapter IX

DISTRIBUTION OF THE FORMS OF VEGETATION

The almost complete destruction of primitive forest and the radical though fluctuating changes in land utilisation described in the last chapter will have made it clear that over nearly the whole of England we have now only much modified remnants of the original covering of plants that had adjusted itself to the sub-Atlantic climate 2000 years ago; and the same is true of the lower lying parts of the "Highland zone" of the west and north of the British Isles up to a considerable height on the hills. Only in the higher mountains, in some waters, fens and bogs and on parts of the sea coast is there now any vegetation which can fairly be called virgin.

Nevertheless, much of the country is still occupied by communities of native plants, though no longer moulded by "nature" alone, and it is this kind of vegetation that we call "semi-natural". In the face of man's unintermittent but irregular activities Nature is always trying to restore some sort of equilibrium, and given sufficient time she succeeds in establishing new communities, into the organisation of which the human factor may enter to a greater or lesser degree.

Natural and semi-natural vegetation. By natural vegetation we mean of course that which is primarily due to "nature" rather than to man. To take extreme cases: a virgin forest is clearly natural, while a wheat or root crop (also examples of vegetation) is clearly not. But a great deal of existing vegetation is intermediate between two such extremes. If we leave out all genuinely virgin, i.e. untouched, communities on the one hand, and all plantations of exotic trees and field crops on the other, we find that according to the returns of the Ministry of Agriculture about 42·4 per cent of the total area of England (excluding water), and about 41·6 per cent of Wales, are under "permanent grass" which has been "laid down" for pasture or hay and is sometimes manured, but nevertheless consists of native species and behaves in many respects like a natural plant community. In 1925 only 7·7 per cent of Scotland, but on the other hand as much as 58·4 per cent of Ireland, was reported to be occupied by "permanent grass".

The next type of semi-natural vegetation in point of area consists of "rough grazings", largely hill and mountain side, which are essentially natural communities, though they have often replaced destroyed woodland and are held in their existing form by grazing. In 1936 about 11·5 per cent of England and 34·2 per cent of Wales: in 1925, 48·5 per cent of Scotland and 15·2 per cent of Northern Ireland (no figures being available for the Irish Free State) were recorded as "rough grazings". Both permanent grass and rough grazings may fairly be called "semi-natural", though the first is clearly less "natural" than the second, and is usually included under "cultivated land" in agricultural statistics.

Thus about 54 per cent of England, 56 per cent of Scotland, 75 per cent of Wales and probably about the same proportion of Ireland are occupied by semi-natural communities mainly dominated by grasses and partly by heath plants.

To this we must add that portion of "waste land", mainly mountain country at high altitude, which is so infertile or remote as to be inaccessible to grazing. So far as it is vegetated at all such land is occupied by completely "natural" plant communities. It is impossible to obtain figures for this fraction of the country. In the Scottish Highlands, however, it is not a negligible proportion of the whole, and would probably raise the Scottish percentage area under natural and semi-natural vegetation to a figure not very far short of the Welsh and Irish.

The woodlands, in so far as they consist of native trees, may also usually be regarded as semi-natural. In a great many the existing trees have been planted, but in structure and behaviour the wood resembles a natural wood more or less closely: in others there has been little or no planting, but the wood is modified by cutting undergrowth and felling trees. The proportion of land under woods and plantations is so small, however (see pp. 196–7), that its addition does not greatly affect the percentage area of the country bearing natural and semi-natural vegetation.

Thus we may conclude that between one half and two thirds of England, and something like three quarters of Wales, Scotland and Ireland are occupied by natural and semi-natural plant communities.

We must admit two main categories of semi-natural vegetation. The one began as entirely natural and has been modified by man's activities, as in an exploited natural forest which is allowed to regenerate by itself, or a pastured heath which may be fired periodically: the second consists of communities deliberately initiated by man for his own purposes, but consisting of native plants, and often behaving very much as they would if they had come into existence spontaneously. In the result the difference of origin may sometimes be hard or even impossible to detect. A planted oakwood, for example, is difficult to distinguish after the lapse of time from a fragment of natural oak forest if it is exploited in the same way;[1] nor does a sown pasture necessarily differ very much from grassland on the same soil which has replaced forest as a result of continuous pasturing.[2]

"**Permanent grass.**" The "permanent grassland" which now occupies

---

[1] See Woodruffe-Peacock (1918).

[2] Modified fragments of original forest and of old grassland commonly contain more of the less frequent but characteristic species of the forest or grassland flora, species which have never entered the planted woods or sown grassland: comparatively recent plantations and leys are floristically poor.

about 42 per cent of the total land area of England and Wales has for the most part been "laid down" by seeding; but the actual sown mixture of grasses, or of grasses and clovers, always tends to change in the direction of a mixed herbage in accordance with the nature of the soil and with the amount and nature of the pasturage. Apart from manuring, these two factors actually determine the composition of permanent grassland, which rarely contains any species that are not native, and whose actual origin (whether from seeding or from natural colonisation) has little effect on its ultimate flora. "Weeds" of various species derived from neighbouring vegetation may enter such grassland if the grass sod becomes thin and weak owing to neglect, but they can be greatly diminished in quantity or exterminated by an intelligent pasturing or manuring regime. The great majority of our permanent pastures are not in fact treated with artificial fertilisers, but the treading and manuring by stock is of the first importance. Thus, though permanent grassland generally belongs to the second of the two categories of semi-natural vegetation, it makes little difference in the long run whether it was sown or arose naturally: it is the subsequent treatment which is all-important.

The chief forms of natural and semi-natural vegetation are woodland, grassland, marsh and fenland, moss or bog, heath and moor, the so-called "arctic-alpine" vegetation of the higher mountains, and the maritime vegetation of salt marsh, sand dune and shingle beach. In point of area occupied the other forms are insignificant compared with grassland. They are of very various status, comprising many different kinds of plant community, which are dealt with in subsequent chapters. Here we shall briefly consider their general characters and distribution.

Forest and woodland. The British Islands have a smaller proportion of their area occupied by woodland than any other country in Europe. The area of woodland, including plantations, for the whole of Great Britain was in 1924 a little over 5 per cent (England 5·1, Scotland 5·6, Wales 5·0; Ireland had much less), the countries next in order being Holland with about 8·5 per cent and Denmark with about 9 per cent. These low percentages contrast with those of France (18 per cent), Germany (24 per cent), Sweden (55 per cent), and Finland (60 per cent).

The progressive destruction of the once widespread British forests throughout medieval and modern times has been described in Chapter VIII, and after the last forest reserves had disappeared in the seventeenth century there was no concerted effort to replenish the national supplies of timber until after the Great War, when the Forestry Commission was established with that object.

The main cause of this continuous destruction of forest was undoubtedly the suitability of the climate for grassland, i.e. for pastureland, the related multiplication of flocks and herds and the enormous development of the wool and textile industries. Not only in the west, where the milder and

Woodland

damper climate often permits continued growth of the grasses, and thus the pasturing of stock in the open almost throughout the winter, and is less suitable for cereal crops, but also throughout the midlands, the land was more and more devoted to grazing. The whole country eventually became preoccupied, first with the textile, later with other flourishing industries, and with commerce and shipping which facilitated the free importation of timber, and continued to neglect its native woodlands. By the nineteenth century the remnants of these were in a scandalously derelict condition, and were quite inadequately supplemented with new plantations.

Census of woodlands. According to a census made by the Forestry Commissioners in 1924 about 5·3 per cent of the land area of Great Britain was under woods and plantations. Of this total woodland area, amounting to 2,958,672 acres (about 1,183,470 hectares), 6·9 per cent were classified by the Commissioners as "uneconomic, including amenity woodlands and shelter belts" and were not further characterised; and 16·1 per cent were said to be "felled or devastated". Of the rest deciduous woodland occupied 44 per cent (high forest 14·9, coppice 17·9, scrub 11·2), conifers 22·7 per cent and "mixed conifers and hardwoods" 10·2 per cent of the total area.

Table XI is simplified and rearranged from the Forestry Commission's census:

Table XI

	Thousands of acres			
	England	Wales	Scotland	Great Britain
Deciduous:				
High Forest	338·5	44·0	60·9	443·4
Coppice	485·2	35·3	8·1	528·7
Scrub	87·4	34·9	208·4	330·7
Conifers	195·2	46·9	429·7	671·8
Mixed	220·4	22·1	59·2	301·7
Devastated or felled	194·7	62·2	221·2	478·1
Amenity and shelter belts	109·5	8·0	86·8	204·3
Totals	1630·9	253·4	1074·3	2958·7
Total land areas	32037·0	5098·8	19069·7	56205·7

It will be seen that nearly two-thirds of the "scrub" was in Scotland, and scrub formed a much larger proportion of the Welsh than of the English woodlands. In "scrub" is included the dwarf oak- and birchwoods of the siliceous hillsides in the "Highland" regions. The great bulk of the coppice was in England and much the larger proportion of deciduous high forest.

Status of deciduous woodland. The deciduous "high forest" and coppice, whether planted or not, may be regarded as semi-natural, and much of the "scrub" may be reckoned as strictly natural vegetation. It is impossible to say how the "mixed conifers and hardwoods" should be classified without further characterisation, but the coniferous woodlands (except some of the Scottish pinewoods) were of course all plantations, mainly of common ("Scots") pine, larch and spruce, with a small proportion of other exotic conifers. These last are now rapidly increasing owing to the activities of the

Commission, partly at the expense of the deciduous woods, but largely planted on grassland or heath. The coniferous plantations cannot be considered as semi-natural communities. The ground vegetation is in most cases radically altered from the pre-existing plant cover: it is usually greatly impoverished, and in the young plantations, as soon as canopy is established, almost absent.

But though there is very little natural forest left, the deciduous plantations, if they are left long enough, eventually become semi-natural. The ground vegetation of a planted oakwood is gradually assimilated to that of a natural wood, and though certain species indigenous to the original native forests may be absent from many plantations, the community as a whole is substantially a semi-natural forest community. Thus the planted deciduous woods, often on old forestland, though sometimes on old grassland or heathland or even arable, belong, like most of our "permanent grassland", to the second category of semi-natural plant communities described above. Other semi-natural woods are probably directly derived from original native woodland.

The traditional form of "cultivation" of oakwood in this country—hazel coppice with oak standards, in which most of our oakwoods are still maintained and which will be fully described in Chapter XII—involves certain modifications in the ground vegetation; but though species alien to the forest flora frequently enter this type of wood, yet the undergrowth, albeit modified by the periodic cutting of the coppice, retains the general character of forest vegetation.

We saw in Chapter VII that, as the climate improved after the final retreat of the ice, forest trees spread through the regions from which the ice had departed, becoming abundant roughly in the following order: birch and pine; elm, oak and alder, lime; and finally beech and hornbeam. The first five, together with ash, of the date of whose spread we are uncertain, and with other smaller trees of little importance as forest formers, spread through the whole, or almost the whole, of the two islands. In the extreme north of Scotland the oaks probably never became really dominant except on the lowest slopes of the glens, and the natural forests of most of this area, though now little more than relics, still consist mainly of birch and locally of pine. Ever since early Atlantic times, however, oaks[1] have been the dominant trees of most of the English, Welsh, Irish and southern Scottish forests up to a certain altitude, on all soils except the wettest marsh, fen and bog lands, the shallow-soiled limestones, and perhaps some of the poorer heaths.

---

[1] It is possible that during the periods of wettest climate alder may have become so abundant in the oak forests as to displace the oak from its position of dominance. It is impossible to be sure whether the enormous abundance of alder pollen in some of the peats, often greatly exceeding oak pollen in amount, is due to this cause or to the great extent of fen and marshland.

Deciduous summer forest. Thus at the present day we have the much modified and now very sparse native woodlands of the greater part of the British Isles still representing the north-western extremity of the European summer forest formation, with one or other of the two species of oak dominant over most of its area. On the shallow limestone soils, however, the oaks do not flourish, and here we have the deciduous forest dominated by beech in south-eastern and south-central England, and by ash on similar soils farther to the north and west, i.e. outside the region of the existing dominance of beech.

Northern coniferous forest. In the north-central and northern Highlands of Scotland, except along the lowest slopes of the larger glens, where there is often oakwood, the only existing natural forests are of birch and pine, the latter, especially dominant on the eastern side, as in the valleys of the Dee and the Spey, though there are remains of native pinewoods in many other places, some of them quite recently destroyed. These most northern regions of Britain belong in fact climatically to the coniferous forest formation of northern Europe, or more correctly to the transitional zone between this and the deciduous summer forest. The northern coniferous forest, as is well known, is represented in the extreme north of Scandinavia (Lapland) by pure birchwood, since the conifers of the main Scandinavian forest—the common spruce (*Picea abies*) and the common pine (*Pinus sylvestris*)—do not reach the polar tree limit in this western longitude. The northern coniferous forest of Eurasia is thus marked by a fringe of birch on its north-western edge, and this descends to lower latitudes on the Atlantic coasts of Europe, cutting across the north of Scotland.[1] Brockmann-Jerosch has shown that the polar limit of forest is depressed towards the equator in the neighbourhood of the great oceans; and its northern climatic belt, represented by birch forest, is similarly swung southwards towards the Atlantic.

Heath. The third great type of vegetation represented in the British Isles is heath (Chapters XXXVI and XXXVII). This is an oceanic and sub-oceanic plant formation dominated by dwarf ericaceous undershrubs, mainly the common ling or heather (*Calluna vulgaris*) but also by other species of the heath family. In the countries bordering the Atlantic and the North Sea this formation is prevalent on porous soils and on well-drained slopes and plateaux, from the west coast of Norway southward to the "Landes" of south-western France and to the Iberian peninsula, where it gives place to the taller growing but related *maquis* of the western Mediterranean coasts. The dominant heaths (and especially *Calluna*) depend on

[1] On the mountains farther south, also, e.g. in central and southern Scotland and in the north of England, there may be seen, here and there, the remains of a corresponding *altitudinal* zone of birch forest (see Chapter XXII, pp. 454–5). This similarly corresponds with the coniferous altitudinal zone which normally succeeds the deciduous zone in ascending the higher European mountains; but our mountain birchwoods—under our oceanic climate—have no coniferous trees.

moist air,[1] and they flourish on soils poor in mineral salts where they are free from the competition of more exacting, taller and more quickly growing plants such as trees and shrubs. In the exposed coastal regions *Calluna* is ubiquitous, but it attains its greatest luxuriance and becomes widely dominant on the well-drained slopes of subcoastal mountains in a less extremely oceanic climate. Here dominance of trees is prevented by frequent strong winds. On lowland plains of porous sandy soil, where the growth of trees is prevented by other factors such as grazing and burning, heath is again dominant. In such situations the moist air and cool summers lead to the luxuriant growth of fruticose lichens (*Cladonia* sens. lat.) with associated mosses, and thus to the formation of a thick layer of acid surface humus which disintegrates but slowly ("raw humus"). Heather seedlings establish themselves in this humus even on limestone plateaux where the surface soil develops acidity from the accumulation of humus and the leaching out of soluble salts from the mineral soil immediately below. The heath formation is developed largely within the climatic region of summer deciduous forest and also of northern coniferous forest. Above a certain altitude it is wind and possibly the low summer temperatures that inhibit forest. At lower levels it is partly the nature of the soil, but largely biotic factors (including especially fires) that decide between heath and forest.

The lowland inland heaths are almost confined to coarse sandy acid soils where the dominants are free from the competition of more exacting plants. These heaths represent a stage, sometimes more or less stabilised by moderate grazing, and often by recurrent fires, in the development of forest. In the absence of grazing such heaths are readily colonised by birch, subspontaneous pine and also by oak and beech (see Chapters XVII and XX), and existing heaths often occupy the sites of destroyed or degenerate woodland. On some very poor and dry sands, however, the soil conditions may not perhaps be good enough for the development of closed forest, and heath, often bearing scattered and poorly developed trees, may possibly in such circumstances maintain itself indefinitely apart from grazing and fire.

Moss or bog. The fourth plant formation[2] represented in the British Isles is the "moss" or "bog" formation. This is closely related to heath, from which it differs, so far as habitat is concerned, by its limitation to permanently wet soil. The poverty of the soil in mineral salts and its corresponding acidity moss shares with the heath formation, and if a moss dries out, from whatever cause, heath normally develops. Moss may be initiated in shallow pools or on the edges of lakes whose water is poor in

[1] *Calluna vulgaris*, in spite of its xeromorphic features, so far from being a "xerophyte" as is often stated, has been shown by Stocker and others to be a hygrophilous mesophyte, and this corresponds with its dominance in the oceanic climate of western Europe, and its diminution eastwards, till in Russia it becomes restricted to woodland where alone the air is sufficiently damp.

[2] See Chapter X, p. 229.

mineral salts, but most of it originates on constantly wet soil similarly poor in salts, or among constantly wet vegetation. Bog mosses (*Sphagnum*) are the normal initiators of moss vegetation, and their continual compact upward growth leads to the progressive death and partial decay of the lower and older shoots which are covered by the upper and younger ones. By the consolidation and partial disintegration of this dead material *peat* is formed and often reaches a depth of many feet. Wetness and acidity are the indispensable conditions for the initiation and maintenance of moss, and the wetness may be secured by permanent water, at or just below the surface of the soil, or by constantly damp air, rarely falling very far below saturation point, which keeps the soil from drying.

When moss vegetation begins in standing water or on wet soil or vegetation it cannot grow more than a certain height above the water-level nor can it spread laterally unless the air is more or less constantly damp; for though the bog moss has considerable powers of carrying water with it the upper surface of the moss eventually dries out and is killed by dry air. In a climate in which the air is almost continually damp, however, and soil drainage is defective, the bog moss, and other bog plants which accompany and follow it, can extend indefinitely, smothering and superseding other kinds of vegetation. Thus moss or bog depends on much the same climatic factors as heath, primarily damp air, but it also requires wet soil, while heath requires good drainage.

For these reasons the moss formation is found, as a climatic formation,[1] in the same regions as heath, along the northern parts of the Atlantic coasts of Europe, and also in the similar cool damp climates of the mountains. But it does not extend so far south as heath, since the increasing temperature decreases the average humidity of the air. In the British Isles it covers very extensive tracts in western Scotland and Ireland, right down to sea-level ("blanket bog", see Chapter XXXIV), on mountain plateaux and gentle slopes which have a high rainfall and are badly drained. If conditions become drier, as by draining of the soil or decreasing wetness of the climate, moss passes into heath, the heath plants colonising the moss peat; and many transitions between the two actually occur.

Eastwards, where the climate is less damp, moss or bog arises in pools or in badly drained hollows of heathland, or where acid water, poor in mineral salts, seeps out on a slope. Such limited local bogs are often found on heaths; and their vegetation, of *Sphagnum* and associated plants, may equally well be regarded as a wet heath community or as incipient moss whose development is strictly localised.

Another type is the "raised bog", usually arising in fen basins through colonisation of the fen vegetation by *Sphagnum* and plants of similar requirements, and is distinguished by the marked convexity of its surface. This is the typical "Hochmoor" of the Germans and is very well repre-

[1] See Chapter X, p. 229.

sented in the central Irish plain, in parts of Scotland, and formerly also in England. Raised bog is formed in less wet climates than those necessary for the development of blanket bog.

Arctic-alpine formation. The fifth plant formation (Chapter XXXVIII) is altitudinal, occurring mainly towards the summits of the higher British and Irish mountains, for the most part more than 3000 ft. (*c.* 900 m.) above seal-level, though the actual limits of altitude are very various. The vegetation of these high levels is quite distinctive, and depends on the peculiar climatic features and also on the characteristic physiography and soil of the mountains, which exclude the great majority, though not all, of the lowland species, and to which the arctic-alpine species are adjusted. The heath and moss formations usually extend up to a considerable altitude, 2000 or 3000 ft., but at these levels the arctic-alpine vegetation typically becomes increasingly prominent. The name "arctic-alpine" refers to the combination of arctic and alpine species, or of species common to arctic and alpine habitats, in mountain vegetation. Outside Britain some occur in the extreme north of Europe at sea-level, a few only in the Alps and other high European mountain ranges, while most are both arctic and alpine.

This vegetation is only at all fully represented by numerous species on those mountains which consist of rock containing minerals comparatively rich in basic, especially calcium, salts. The rocks which are poor in salts are also poor in flora and bear a uniform and monotonous vegetation, impoverished heath or moor types, with only a few of the most widespread arctic-alpine species.

Many of the arctic-alpine species descend to much lower altitudes along the courses of streams, and in the extreme Atlantic regions of western Scotland and western Ireland they may appear at sea-level.

Distribution of the climatic formations. Thus we have four formations (or five if we include heath) mainly determined by climatic factors represented in the British Isles: deciduous summer forest, northern coniferous forest, heath, moss or bog, and the arctic-alpine formation. We may now indicate their general distribution in the country.

The constituents of the deciduous forest originally entered the country from the south-east, and it is this region which has a climate most nearly resembling that of the adjacent parts of continental Europe and which even now bears the greatest area of semi-natural deciduous woodland. Kent, Surrey, Sussex, Hampshire and Berkshire had in 1905 no less than 11·5 per cent of their combined area covered by woodland (representing more than a quarter of the whole area of English woodland) as compared with 5·3 per cent in the area of England as a whole.[1] With the adjoining counties of Essex, Hertfordshire and Buckinghamshire this region includes much the greater portion of the semi-natural English beech and hornbeam.

---

[1] These figures include plantations and subspontaneous pine, but the bulk of the woodland is deciduous. The Forestry Commission's 1924 census gives the following

It has a richer woodland flora than the regions farther north and on the whole substantially freer spontaneous seeding and establishment of young trees. The midlands are poorly wooded now, though at one time they were certainly covered with oak forest, and as we pass from the midlands towards the west and north with its cooler and damper climate we come to country of much greater average altitude, where the woodland is confined to sheltered valleys. The rocks are mostly hard so that they weather slowly and with certain exceptions poor in minerals which give rise to nutritive salts. The combination of these rock characters with the greater elevation and cool moist climate results in the formation of a shallow and poor mineral soil overlain by humus which disintegrates slowly, remains constantly damp, and develops a markedly acid reaction.

In these conditions deciduous woodland does indeed exist, but save in specially favourable situations and on exceptional soils, as in the bottoms of river valleys, it is seldom well grown and regenerates with difficulty. On the Welsh, Pennine and Cumbrian hillsides the sessile oakwoods consist of poor and twisted trees: in the more exposed situations they are not much more than oak scrub; and the accompanying flora is few in species, and predominantly heathy, while the associated trees and shrubs are very few. On the northern Scottish hills the very scattered birchwoods belonging to the northern coniferous forest are accompanied by rowans (*Sorbus aucuparia*), and there are very few other native trees. In Caithness aspen (*Populus tremula*), alder (*Alnus glutinosa*) and bird cherry (*Prunus padus*) practically exhaust the list. Correspondingly there are few species of shrub: hazel (*Corylus avellana*), blackthorn (*Prunus spinosa*), the sallows (*Salix atrocinerea* and *S. aurita*) and the juniper (*Juniperus communis*).

The more extensive woods of native Scots pine have very few associated trees or shrubs, but an undergrowth of bilberry (*Vaccinium myrtillus*), cowberry (*V. vitis idaea*), and heather (*Calluna*) where enough light penetrates the canopy, with characteristic associated mosses (*Hylocomium*) and a small number of mostly rare herbs which do not occur in the south (see Chapter XXII).

On the other hand, the heath and especially the moss formation (to-

woodland percentages for these counties: Sussex 14, Surrey 12·3, Hampshire 11·6, Kent 11·1, Berkshire 8·7.

The percentages of total wooded areas occupied by the various forms of deciduous woodland were as follows:

	Kent	Sussex	Hants	Surrey	Berks
Deciduous high forest	4·5	10·2	13·0	11·4	20·2
Coppice-with-standards	66·5	56·9	41·4	36·2	27·8
Coppice	8·6	5·9	2·2	4·7	3·9
Scrub	4·7	3·6	1·8	6·0	4·9
Total deciduous	84·3	76·6	58·4	58·3	56·8

The very high proportion of deciduous woodland, particularly of coppice-with-standards will be noted, and also that this declines sharply as we pass westward from Kent.

gether with allied grasslands) cover many hundreds of square miles in the north and west, the former on well-drained, the latter on badly drained soils. In the grassland also, which occupies much of the country, the prevalent climatic conditions are very evident from the occurrence in abundance even over limestone rocks of species which affect acid soils.

Finally, the upper stretches of the higher mountains of the north and north-west are the seat of the arctic-alpine communities which we here group together as the arctic-alpine formation. This formation, however, as has already been said, only occupies comparatively limited areas, even on the higher mountains, mainly on rocks rich in nutritive mineral salts.

Grassland. The grassland which covers so much of the area of the British Islands is very various in origin. We cannot speak of a climatic grassland formation in Britain, as we can of the climatic forest and moss formations. Most of the existing grassland, though not quite all, is certainly due to man's activity, primarily to his clearing of forest and to the grazing of his flocks and herds. But the conversion of forest to grassland could not have taken place without the co-operation of the British climate, which is pre-eminently favourable to its development.

Climatic requirements. It is well known to plant geographers that grassland is specially favoured by rainfall well distributed through the growing season of the grasses. Thus the climates of the great continental grassland areas of South Russia, the Prairie region of the United States and Canada, and the Grass Veldt of South Africa, are characterised by frequent rainfall during spring and early summer when the grasses are making their vegetative growth, though the total for the year is low. In the British Isles there are on the average frequent comparatively light precipitations and high average relative humidity throughout the year. The almost continuous surface moisture thus maintained secures the best conditions for the growth of our ordinary meadow grasses, because the surface layers of the soil are kept constantly though only slightly damp. The British climate is, however, notable for very frequent changes of weather and there are wide differences between different years. The effect of three weeks' or a month's drought in May is at once seen in the poverty of the grass crop, while a severe drought in July or August, such as occurred in 1911, 1921 and 1933, results in the complete browning of the pastures in late summer. A wet late summer, on the other hand, will maintain the vivid green of the pastures till late autumn; and an "open" winter, especially in the very mild climate of the south-west, will keep many grasses vegetating throughout the year, while a "hard" winter kills the subaerial shoots. But in an "average" season most of the grass keeps green throughout the summer, turns dull green but not brown in late autumn, and produces vigorous new growth in late April and May. A number of species are, however, remarkably "winter green" in our climate and these afford valuable winter pasture for sheep and cattle.

Relation to grazing. The characteristic "life form" of our meadow grasses fits them peculiarly well for continuous vegetation when they are continuously grazed. They belong to the class of "hemicryptophytes" which produce their new shoots from buds situated at or very near the surface of the soil. The eating off of the shoots stimulates the outgrowth of the buds, leading to the production of close turf by the constant fresh branching at the soil surface. The compactness of the turf is increased by the constant treading of the grazing animals, whose dung meanwhile manures the soil. When the grasses are mown for hay instead of being grazed this natural manuring is of course absent and has to be replaced artificially, or the crop will rapidly deteriorate, but some compensation may be supplied by grazing the aftermath of the hay crop. In alluvial water meadows traditionally cut for hay the winter flooding from the rivers brings nutritive silt.

Origin of British grassland. Grassland of one sort or another now covers more than half of the entire land area of England and three-quarters or more of Wales and Ireland. Much of this was gradually converted from forest land, though recently a great deal has been laid down from arable. When sheep and cattle are grazed through woods they eat off and trample down the tree seedlings and woodland ground vegetation, at the same time carrying the seeds of plants from the open country into the woodland. The woodland is thus prevented from regenerating, and when the existing trees die they are not replaced. The ground thus becomes occupied more and more completely by the invading plants so that a continuous turf is gradually substituted for the original woodland vegetation, and eventually the forest is replaced by pasture. This was probably the fate of much of our original British forest, when the flocks and herds of the country began to increase greatly in numbers and were pastured in the forests. The disappearance of forest was of course much accelerated by the actual clearing of forest for military purposes, for the sake of the timber, for fuel, and for conversion to arable land. Under such conditions natural woodland could only be maintained by special protection, either for the irreducible minimum requirement of timber and small wood, or for hunting, as in the Royal Forests and "chases", or later for fox coverts and pheasant preserves.

Arable and grassland. The relative proportions of arable and grassland have fluctuated very much from time to time according to the demand for, and consequently the prices of, cereals, meat and milk. The area under the plough was at its maximum just before the repeal of the Corn Laws in 1846. With the free importation of foreign wheat the price rapidly fell and more and more of the former arable land was allowed to "tumble down" or was "sown down" to grass; and grass crops, or mixed grass and clover, in the form of temporary "leys" as they are called, for hay or for grazing, were also sown and allowed to remain for a few years or for a single year.

These leys, however, are properly included in arable, not in permanent grassland.

Permanent grassland. "Permanent grassland" occupied in 1930 about 41 per cent and in 1936 42·4 per cent of England and Wales. It is nearly all enclosed in fields of various sizes and most has been laid down from seed mixtures, though some also is doubtless grassland originally converted from forest in the way described. In either case it tends to come to a state of equilibrium depending on climate, soil and agricultural treatment. Some of the species sown usually die out and others enter, so that the permanent grassland on a given soil tends to approximate to a single type, whatever its origin, provided it is uniformly treated. The common character of this enclosed permanent grassland is that it is trodden and manured by stock which spend part of their time and get part of their food elsewhere, and though of course they also permanently remove foodstuffs in building up their own bodies, they contribute manure to the herbage which they graze. The grass fields may or may not be manured with dung or "fertilisers" in addition.

Chalk and limestone grassland. Besides this enclosed man-made grassland, there are certain well-marked types of more natural grassland below the altitudinal forest limit and also used as pasture. First, there is the grassland of the shallow calcareous soil derived from the chalk of the south-east of England and the similar soils of the older limestones in the north and west. This very well-characterised plant community has a remarkably constant floristic composition. Its origin is something of a puzzle. It exists within the best forest region of southern England, and most of it is now certainly a biotic climax due to grazing. The chalk grassland of the "Downs", e.g. the South Downs in Sussex, the Berkshire Downs, and Salisbury Plain in Wiltshire, as well as other areas in adjoining counties, has been used as "sheepwalk" from time immemorial. But most of its extent at least is readily colonised by shrubs and trees if grazing is withdrawn, and some of it is actually occupied by beechwood, though some few areas may be too dry or too much exposed to wind for the establishment of woody vegetation. The late Neolithic (c. 2000 B.C.) and Bronze Age barrows (c. 1000 B.C.) which occur on this grassland in many places seem to show, as we saw in Chapter VII, that it was open country at least in Sub-Boreal times; and it is difficult to believe that Neolithic man would or could have cleared great areas of well-grown forest when the climate of southern Britain in his time was at first, i.e. in late Atlantic times, presumably at least as favourable to the development of forest as it is at the present day. In "Celtic" times (i.e. in the period immediately preceding the Christian era) a good deal of this land was ploughed, the barrows for the most part being left untouched. A possible explanation of the course of events has been suggested in Chapter VII (pp. 164–5). The grassland of the oolitic limestones, as for instance on the Cotswold Hills in Gloucestershire, is closely similar to that of the chalk, and like the chalk grasslands is now a biotic climax due to grazing.

Of the older limestones the Carboniferous (Mountain) Limestone is by far the most important and extensive, being developed over wide extents of country, including considerable parts of the Pennines, areas in North and South Wales, the Mendip Hills in Somerset, and extensively also in Ireland. Much of this limestone is covered with pasture ("natural pasture" as the older surveyors called it) which, like that of the chalk, is very good sheep-grazing land. It belongs to quite the same type of vegetation and contains many of the same species. Much of it was probably at one time covered with forest. On the other hand, some of it, unlike the chalk and oolite grassland, lies above the present altitudinal limit of forest, and even though this limit was probably substantially higher in pre-Sub-Atlantic times there were almost certainly slopes and plateaux too windswept for effective colonisation by trees on the shallow limestone soil. On these it may be supposed that the limestone grassland established itself, to spread later over much wider areas with the deterioration of the climate and the increase of grazing.

We may thus reasonably consider the British limestone grassland as in origin a natural plant community, though greatly extended, and also distinctly modified, by the factors which have been indicated.

Siliceous grassland. Besides the limestone grassland many "siliceous" (i.e. non-calcareous) hillsides, which are the great majority, bear more or less good sheep pasture—notably the Southern Uplands of Scotland, and many of the Welsh hills, and these indeed are great sheep-rearing regions. Practically all the British mountains are largely or wholly bare of woodland, even on their lower slopes which are well below the upper climatic limit of forest; and they are largely covered by pasture grassland. The better drained and less acid of the siliceous hillsides are dominated by bent and sheep's fescue (species of *Agrostis* and *Festuca ovina*), the more acid and peaty by *Nardus*, *Molinia* or *Deschampsia flexuosa*, which are of little value for pasture. The *Agrostis-Festuca* grassland, like the limestone grassland, occupies land once very largely if not wholly covered with forest. The soil is acid though not extremely so, and the community may easily pass into *Calluna* heath and back again into grassland as the incidence of grazing decreases or again increases. The *Agrostis-Festuca* community (which used to be called "grass heath") may similarly replace *Calluna* heath on sandy soils in the lowlands of southern England owing to the influence of grazing. The wetter types of grassland, on the other hand ("grass moor"), develop a distinct though usually thin peat soil, and occupy less well-drained habitats and usually higher altitudes than the bent-fescue community.

Thus all these types of grassland within the forest region are mainly biotic climaxes[1] determined by grazing: in the British climate, wherever

[1] See pp. 224–5.

grazing is important, grassland tends to supersede other types of vegetation.

Alluvial grassland. Alluvial grasslands developed on flood plains of rivers are natural in some parts of the world. Floods, and especially floating ice carried down by floods, may prevent the development of trees and shrubs, and grassland consequently becomes the natural climax in such situations; but it is doubtful whether any of the British alluvial grassland has this status, though it is characteristically enough developed on many of our flat alluvial tracts. It is probable, however, that marshy woodland (the German "Auenwald") would appear in such situations if the vegetation were allowed to develop naturally, and that mowing or pasturing has created and maintained this grassland.

Maritime and submaritime grassland. Minor types of grassland are maritime and submaritime. Of the latter the grassland of cliff tops and slopes facing the sea are characteristic.[1] They have not been thoroughly studied, but it would seem that though they often owe their existing form, which is commonly a close turf, to grazing, they are natural in origin, arising where other plant communities cannot flourish because of a combination of strong winds and sea spray. Maritime grassland proper is again a biotic climax,[2] the culmination in pastureland of the succession from salt marsh (see Chapter XL). Dominated by *Glyceria maritima* or by some form of *Festuca rubra* freely mixed with maritime (halophytic[3]) herbs, this community forms a most nutritious pasture for either sheep or cattle, on which stock can actually be fattened—the famous *prés salés* of the French. Fixed sand dune may also be covered by grassy turf, and this community again is certainly a biotic climax grazed by sheep or closely nibbled by rabbits, the dunes often forming extensive rabbit warrens.

Fen and marsh. On the flat shores of lakes and the upper parts of estuaries out of the reach of maritime influence, where the water is nearly neutral or alkaline without being salt, and in wet valley bottoms and depressions where the conditions are similar, the fen formation develops (Chapters XXXII and XXXIII). It normally succeeds the aquatic formations of fresh water which culminate in reedswamp. Fen is developed only where the water table is close to the ground-level, generally where it stands a little below the soil surface in summer. It is dominated in the earlier stages by sedges, grasses and rushes and later colonised by characteristic shrubs and trees (fen "carr"). The fen plants form black peat, usually structureless, which differs widely from moor or bog peat in its general characters and reaction. It appears that fen may develop into climax forest, but this tendency is scarcely more than indicated in the limited and scattered, often exploited, fens which are all that still exist in this country.

---

[1] See Chapter XLIII, pp. 899–903.   [2] See pp. 224–5.

[3] Greek ἅλος, salt. The halophyte is a characteristic life form generally more or less succulent, and able to tolerate large quantities of sodium chloride and similar salts in the soil.

The fen vegetation is so distinctive and the habitat so sharply defined that it is natural to consider fen[1] as an edaphically determined formation. But the dominant sedges and grasses have been commonly cut as a crop for litter and thatching, and the development of woody plants has thus been prevented, so that most existing fen is anthropogenic.

Most of the more extensive fenlands of the British Isles have been drained and cultivated, but fragments of so-called "primitive" fen, i.e. areas which have been maintained as fen by sedge cutting or are partially colonised by bushes and trees, exist here and there, as in East Anglia (especially east Norfolk) and round the shores of Lough Neagh in north-eastern Ireland.

Where waterlogged soil is mainly mineral instead of consisting wholly of peat the corresponding physiographic unit with its vegetation is usually called marsh. Marsh occurs where silting spreads a mineral soil over a flat waterlogged surface, and owing to the localisation of such effects marsh is much more restricted in area than fen, and is usually found in narrow strips bordering rivers and liable to flooding and silting.

Maritime formations. The two maritime formations—salt marsh and sand dune (Chapters XL and XLI)—correspond to two habitats which are often developed in proximity but have very distinct characters and a totally different flora and vegetation. Both are developed on flat parts of the coast and in estuaries where deposition rather than erosion is taking place, that is to say where the land is gaining on the sea. The former (Chapter XL) is the vegetation of the maritime silt laid down by the tides in protected estuaries and always covered by the highest spring tides: the latter (Chapter XLI) comprises the vegetation of sand thrown up by the sea above the limit of high tides, and blown inland after it has been dried by exposure to the air.

Both sand dune and salt marsh show a well-marked zonation of communities corresponding with progressive succession;[2] but while the characteristic sand dune vegetation may eventually be replaced by communities representing progress towards the climatic climax,[3] there is not sufficient evidence, in this country at least, that the salt-marsh succession ever passes beyond the limits of the halophytic vegetation characteristic of the formation, unless it is definitely withdrawn from the influence of the tides. The later stages of salt marsh are, however, commonly stabilised by pasturing (see p. 840). Stabilised sand dunes are commonly used as golf links or rabbit warren.

Shingle beach (Chapter XLII) is a habitat closely allied to sand dune, and though the vegetation shows some differences it must be reckoned as belonging to the same formation, since the vegetation is commonly based on sand between the stones. Where shingle is thrown on to salt marsh, however, a characteristic vegetation is often developed which really belongs to the salt-marsh formation, since it is rooted in soil reached by sea water.

[1] Together with marsh (see pp. 634, 639).   [2] See pp. 216–18.   [3] See pp. 217–18.

Sea-cliff vegetation. The vegetation of sea cliffs has not been adequately studied. It is of mixed character, including halophytic communities where it is under the influence of the salt spray, and also open "migratory" communities such as grow on bare unstable soil.

REFERENCES

Agricultural Statistics, 1936, p. 71, Part I. Ministry of Agriculture. 1937.
Forestry Commission: *Report on Census of Woodlands* (1924). 1928.
STAMP, L. DUDLEY and BEAVER, S. H. *The British Isles*. London, 1933.
WOODRUFFE-PEACOCK, E. A. A fox-covert study. *J. Ecol.* 6, 110–25. 1918.

PART III

THE NATURE AND CLASSIFICATION OF VEGETATION

Chapter X

THE NATURE OF VEGETATION

Plant communities. Plants are gregarious beings, because they are mostly fixed in the soil and propagate themselves largely in social masses, either from broadcast seeds (or spores), or vegetatively by means of rhizomes, runners, tubers, bulbs or corms; sometimes by new shoots ("suckers") arising from the roots. In this way they produce *vegetation*, as plant growth in the mass is conveniently called, which is actually differentiated into distinguishable units or *plant communities*. This last term, which has come into general use for any recognisable unit of vegetation, expresses the unity of life impressed upon an aggregate of plants by living together under common conditions, i.e. in a common habitat. It also expresses the fact that plants living together exhibit interdependence as well as dependence on a common habitat and thus resemble in many respects the members of a human community.

Life form. Plant communities may differ from one another not only in the species but in the types of plant body (*life form*) of which they are composed, and different life forms are an expression of adjustment to different kinds of habitat. Characteristic life forms are, for example, the large evergreen tree of the equatorial rain forest, the deciduous tree of the "summer forest" of medium temperate regions such as western and central Europe, the coniferous needle-leaved tree of colder regions, the small tree or shrub with small hard leathery leaves (*sclerophyll*, from the Greek σκλήρος, hard) of the Mediterranean and similar climatic regions; and among herbaceous vegetation the reedswamp type with tall slender stems or leaves which grows in water by the sides of ponds, lakes or slow rivers, the meadow-grass type with soft leaves and perennating buds in the surface layer of soil, the rosette herbs with a basal rosette of leaves on an abbreviated stem, from which arises a short or tall inflorescence; and so on. Other life forms are represented by various types of plant body among the lower plants—the mosses, algae, fungi, lichens, etc.—each group containing several different ones.

Very simple plant communities may consist of a single life form to which the few species present all conform, and the simplest of all consist of a single species of plant; but the great majority contain several associated life forms, and the most complex have many life forms represented among the multitude of species present. The particular set of life forms present in a well-defined plant community and their arrangement within its bounds constitute the primary "morphological" features of the community and give it a more or less definite *structure*.

Open and closed communities. When the individual plants forming a community (either the shoots or the root systems) are in lateral contact the community is said to be *closed*: when there is free space between the plants which fresh individuals may colonise, we have an *open* community. In the mature vegetation of all the more favourable habitats of the earth's surface closed communities are the rule. An apparently "open" community, i.e. one in which the shoots of the individual plants are separated by open spaces, may in reality be "closed" because the root systems may occupy the whole of the upper layers of the soil and there is no space for the root systems of additional individuals.

Dominance. As a rule certain species are *dominant* in the community, nearly always the tallest species, which, in a closed community, by the fact of their stature, control many of the conditions affecting the other members; and the life form of the dominant species is directly related to and characteristic of the general conditions of life in which the particular community exists. Where there is more than one dominant species these are usually of the same life form, as in a forest when dominance is shared by several species of tree, or in a grassland by several species of grass.

Layering. All but the simplest communities are *layered*, i.e. there are in the community as a whole different layers, strata, storeys or tiers of plants, as they have been variously called,[1] the shoots of which rise to different heights. Thus in an English deciduous wood there are commonly four such layers or strata of vegetation—the tree layer, the shrub layer, the herb or "field" layer and the moss or "ground" layer. In each layer there are commonly one or more dominant species of characteristic life form. One or more layers may be absent in any given example of a type of community as compared with the full development. This happens particularly when an upper layer cuts off so much light that a lower one cannot exist, or when the roots of the dominants exhaust the water of the upper layers of soil. The layers are not always sharply marked off from one another. Part of the active leaf-bearing shoots of a given layer, e.g. of the herb layer in a forest, may be developed on the level of the moss layer below, or may rise into the shrub layer above. Again, the seedlings of trees growing up within a forest at first form part of the herb layer, and then of the shrub layer, before their leafy shoots rise to the height of the tree canopy. The subaerial layering of a plant community has its counterpart in a layering of root systems in the soil. Broadly, the plants belonging to the highest subaerial layers have the deepest root systems because of the greater size of their bodies. Thus tree roots on the whole penetrate more deeply than shrub roots, and these than herb roots, but the correspondence is not always exact.[2] Each layer of a well-integrated many-layered community occupies,

---

[1] *Stratum* is the best word for international use, though "layer" is used in the text as the most familiar English term.

[2] The layers of a many-layered plant community are very important units, since the individuals composing it share a common "partial habitat" and differ in species

and its existence is determined by, a "partial habitat" depending on the spatial relations of the particular layer.

Ecotones. Plant communities vary very much in the clearness and sharpness with which they are bounded and constituted. Between two adjacent communities there is very commonly a "tension belt" or *ecotone* (Greek οἶκος, house (habitat), and τόνος, strain, tension) which shows a blending of the constituents of the two communities. A true ecotone occupies a habitat intermediate between the habitats of the communities on either side.

Much vegetation too, particularly in countries subject to varied human activity, is difficult or impossible to separate into distinct, well-characterised communities—it often presents all grades of mixture of the elements of several communities in which dominance and layering is confused, obscure or totally absent. This is commonly a sign of a recent change of conditions, or of constant interference, i.e. of the immaturity of the vegetation, which has not, so to speak, had time to "sort itself out" into definite communities. As a general principle, the longer vegetation is let alone and left to develop naturally, the more it tends to form well-defined communities, and the more these develop relatively constant and well-defined "structures" in relatively stable equilibrium with their conditions of life.

Habitat. In the study of vegetation the term *habitat* is applied to the whole complex of environmental factors which differentiate units of vegetation, and is not used in the limited sense of a particular soil or situation. It is obvious that plant communities are in the first place related to habitat. This is equally true of the largest units and of the smallest. On the one hand it is true of the immensely complex congeries of species and life forms

---

and in life form from those of the other layers. Each layer may in fact be regarded as a separate subordinate community within the complex many-layered community of which it forms part. In some cases the lower layers of such a complex community are but little influenced by the dominants of the uppermost layer, such as the trees of a forest, because the crowns of the trees are too sparse to cast any considerable shade. In such circumstances the subordinate layers may be developed in much the same form whether the trees are present or not, as in many open heathy woodlands. Such a forest community is clearly very incompletely integrated, the layer communities of which it is composed being virtually independent entities. For these reasons it has been proposed (Lipmaa, 1935) to consider the "one-layered association" as the fundamental unit of vegetation. Such a practice, however, would neglect the typical dependence of the lower layers of a complex community upon the presence of the dominants of the upper layers, and would thus ignore the essential feature of integration which is the justification of the conception of a community. We have to recognise of course that some communities are much less integrated than others, though the *tendency* to integration is universal in all stabilised vegetation. It remains true that the layer or stratum has always a status of its own and requires independent study because it is composed of species and life forms different from those of other layers, whether or not it is dependent for its existence on the dominants of higher strata.

represented in the tropical (equatorial) rain forest, or of the much less complex temperate deciduous ("summer") forest which we have in this country, or again of the temperate continental grasslands—for example, the prairies and "plains" of North America. These are three of the great "climatic" plant communities, each of which exists only within a given range of climatic conditions that are the essential features of its "habitat". On the other hand, it is equally true of small, locally restricted communities, such as the vegetation of a small pool, or of a dry rock exposure, whose existence is limited by very special conditions of water supply, soil or exposure. Fundamentally, the vegetation of the world is a mosaic of plant communities whose distribution is determined by a corresponding mosaic of habitats, and within this mosaic the smaller pieces make up distinctive "patterns" or combinations, and these again larger combinations which represent the primary divisions of the whole and are distributed, more or less zonally, according to the distribution of the climates which control them.

Historical factors. But besides the *habitat factors*, the complexes of environmental factors affecting vegetation, which are spatial, there are *historical factors*, which are temporal, and play a considerable part in determining the nature and distribution of vegetation. The ultimate elements of vegetation are of course the collections of individuals of available species, the sum total of the actually existing products of plant evolution at any given moment. These are necessarily the materials from which plant communities are constructed, and it is the history of these which is the basic historical factor determining vegetation. The existing species have been produced and moulded under the influence of the varied and changing environments of the past, and it would only be by a study of these past environments in conjunction with a study of past floras and vegetations that we could arrive at a complete understanding of the constituents of existing vegetation. This is a study which has now begun to make some progress, but it is a difficult and obscure branch of investigation, because the data are hard to come by, and often difficult to decipher and interpret. Many of them have been destroyed beyond recall, so that we can never hope to attain more than a fragmentary knowledge.

Secondly, there are the mass migrations of species in the past: in the temperate regions of the northern hemisphere, especially, the mass migrations of existing species during very recent geological time, brought about by the great changes of climate related to the last (Pleistocene) Ice Age. These are of a different order from the changes due to evolution. So far as they affect British vegetation they have been touched upon in Chapter VII.

Succession. There is a third historical factor which may be said to be of a different order again and which we must consider here in more detail. On any given area, subject to given conditions of climate and soil, natural vegetation is different according to its actual age in that area. When a bare space is presented to colonisation by plants these settle upon it and establish themselves in a more or less definite sequence. All the plants which actually settle, at whatever time, must of course be able to tolerate the prevailing climatic conditions, the fundamental soil characters depending on the nature of the underlying rock, the nature and extent of the water supply, and other conditions such as the local land relief obtaining in the area. And only those species can settle which can arrive, i.e. those whose seeds or spores are transported to the area. But of these some will come much more quickly than others because their seeds or spores travel more easily, or are available in greater numbers, and of these only some will be able to establish themselves—those which are adapted to life in the bare habitat. These will be the first colonists or *pioneers* of the succession that will ensue. The habitat will then be progressively modified, partly by inorganic forces independent of the organisms living upon it, partly by the activities of the plants themselves. Thus, if the primary habitat is bare rock the surface of this will disintegrate, slowly or rapidly, partly by subaerial weathering, partly by the action of the plant colonists. If it is composed of loose soil the surface layers will be modified by the same two agencies, the contribution of the plants being partly the shelter they afford to new-comers which could not tolerate completely exposed conditions, and partly the gradual accumulation in the soil, and gradual breakdown, of the organic substance of their dead bodies or parts, which form the soil *humus*. These changes will enable other kinds of plants, which could not have succeeded on the bare area, to establish themselves. As the available space becomes filled, a new factor, *competition*, comes into play, resulting in the failure to establish themselves or the later death of some individuals; and this is effective largely through the overshadowing of lower growing plants by taller ones, sometimes also by competition of the roots for soil constituents, especially for soil water.

In these ways the first plant communities to appear are gradually suppressed and superseded by others which develop later, so that a series of communities is initiated, usually of progressively larger and taller plants. This *succession* of communities culminates in a *climax* community dominated usually by the largest and particularly by the tallest plants which can arrive on the area and can flourish under the particular conditions which it presents, thus bringing the vegetation into equilibrium with the whole of its environment. The community immediately preceding the climax, particularly where this is forest, is often characteristic and is known as the *preclimax*. Thus birchwood is preclimax to oakwood on light soils in Britain, and ashwood or ash-oakwood to beechwood. The preclimax type often occupies *as climax* an adjoining (less favourable) climatic region in which the more exacting climax dominants cannot establish themselves. In exceptionally favoured areas within a climatic region the climax itself may be replaced by a more exacting community called the *post-climax*, which may become the climax throughout an adjacent (more favourable) climatic region.

Any concrete example of succession, i.e. any definite series of communities leading to a climax, is called a *sere*, and the complete succession from bare habitat to climatic climax is a primary sere or *prisere* (Fig. 48).

Climax vegetation. The great climax communities of the world, such as the various types of forest mentioned on p. 213, clearly take a leading place among the multitude of plant communities covering the earth's surface. They represent the kinds of vegetation we visualise when we think of

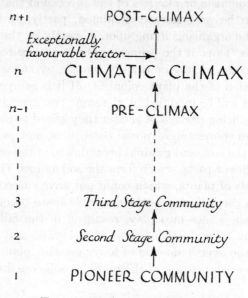

Fig. 48. Scheme of a Prisere

great geographical divisions, wet tropics, Mediterranean, desert or prairie, since they represent both the permanent forms of vegetation and also the forms with the completest and most complex adaptation to the conditions of life obtaining in the various climatic regions of the world. In all regions where the climate is most favourable to plant life the climax communities are dominated by trees, as in the equatorial "rain forests", the "monsoon forests", the "laurel forests" of warm-temperate humid climates, the temperate deciduous forests, the northern coniferous forests (the last two represented in Britain), and other less important types. In less favourable regions, where there is deficiency of either of the two great factors—water and heat—on which plant life depends and which are most unequally distributed over the surface of the earth, trees are replaced as the dominant climax types by less exacting life forms, such as dwarf shrubs, or by grasses or sedges, or even by mosses or lichens, according to the particular climate.

The priseres. But the natural vegetation of the world is of course far from consisting exclusively of climax types. Many agencies are constantly at work to destroy them, wholly or partially, and thus to produce the conditions for new successions. Completely new bare soils are formed by rivers in flood laying down tracts of shingle, gravel or silt; by the high tides in estuaries and on protected coasts spreading layers of salt mud far above the low-tide marks though below the level reached by the highest spring tides; by the wind blowing inland the sea sand washed up above the low-tide mark; by the retreat of glaciers and ice-sheets exposing ground moraine; by the accumulation of talus fallen from mountain cliffs and rocks; by the eruption of volcanoes spreading lava, scoriae or volcanic ash over surrounding country. In the long run the most important of all the ways in which new habitats are formed is the emergence of new land above the surface of the sea as the result of earth movements, and in certain parts of the world by the building up of coral reefs and coral islands. All these, except in the most unfavourable regions, are colonised by pioneer communities of various kinds, followed by regularly succeeding populations; and thus considerable areas are always occupied by different stages of various priseres.

Xeroseres. All the natural processes referred to result in the formation of new subaerial surfaces which become colonised by plants, some quickly, others, as for example certain types of lava and all rock exposures in dry climates, very slowly indeed. On wet silt, in the crevices between talus blocks, and on some few other bare soils where there is loose earth and adequate moisture, trees may be the first or almost the first colonists, and in drier loose soils the pioneers are often annual flowering plants. But on rock faces, the pioneers are confined to *lithophytes* (Greek λίθος, stone), plants which can exist on relatively dry, hard and exposed surfaces—at first lichens and terrestrial algae. These are succeeded by mosses, perennial herbs, and eventually shrubs and trees, as a soil is gradually formed and deepened by the addition of organic substance, derived from the decay of the plants, to the disintegrated mineral surface. On dry *soil*, as opposed to rock, the pioneers (*xerophytes*, Greek ξέρος, dry) are more various in life form. Eventually, in all climates favourable to tree growth this *xerosere*—on rock surfaces a *lithosere*—culminates in climax forest. The process is essentially a progress from the pioneer vegetation which can exist on bare rock or on dry exposed mineral soil (lithophytes or xerophytes) to higher types of life form, of larger size, greater complexity and more exacting requirements. Conditioning, and also conditioned by, this progress in the type of vegetation represented by the series of plant communities, there is a corresponding development of the habitat, represented primarily by the deepening of the soil and its enrichment by organic material which gives it a greater water-holding capacity. Thus each successive population of plants prepares the way for its more exacting successor, and the habitat gradually changes and improves from an extreme dry type of surface, fit only for the non-exacting pioneers, to a "medial" moist type in which so-called *mesophilous* (Greek μέσος, middle) plants (*mesophytes*) can flourish.

Hydroseres. Another type of sere, beginning on submerged soil, is met with on the margins of lakes and slow rivers. These are colonised first by water plants (*hydrophytes*), and then, when the water is shallow enough, by reeds and bulrushes and other plants of similar life form. Accumulation of the products of decay of aquatic and reedswamp plants eventually raises the submerged soil surface above the water and enables marsh and fen plants to get a footing: by their partial decay and the formation of peat from their remains the soil surface is raised still higher. Shrubs and trees which can tolerate a saturated soil settle down between the fen plants and gradually close up to form a fen or marsh scrub or woodland (now generally known in this country as *carr*), the soil level gradually rising all the time by the deposition upon it of dead leaves and twigs. The process of raising the soil is often substantially accelerated by flood silting along the lower courses of rivers and along lake shores adjoining the mouths of streams which bring down silt, adding mineral constituents to the organic soil, or even replacing it by an almost purely mineral soil. Eventually the surface layers of the soil become dry enough for the growth of mesophytic trees which cannot tolerate waterlogged soil, and this *hydrosere* (Greek ὕδωρ, water), as it is called, beginning in water, culminates, like the xerosere, in the establishment of climax forest. Here the development of the soil is from a state of complete saturation with water to a moist condition in which air can penetrate between the constituent soil particles, the reverse direction from that of the xerosere, but resulting in a similar climax. Both types of sere are thus *mesotropic* (Greek μέσος, middle, and τρόπος, turn), resulting in a soil of medial condition in regard to its water content, the most favourable condition for the growth of the highest life forms, the mesophilous trees.

Within the main categories of xerosere and hydrosere, different varieties of seres depending on peculiarities of the initial bare substratum are recognised: e.g. the *halosere* on soil impregnated with excessive salt, the *psammosere* on loose sand, the *lithosere* on rock faces, etc.

Subseres. But the surface of the earth as we see it to-day is not only the scene of the development of vegetation on new soils: it is enormously and increasingly influenced by the activity of man. Man not only clears forests, drains marshes, plants crops and builds cities, he pastures his flocks and herds over wide areas of natural vegetation and thus not only modifies natural grassland but forest also, sometimes converting it into grassland because the tree seedlings are eaten off and the soil trampled by cattle so that the trees cannot reproduce themselves. In numerous other ways he may modify natural vegetation without actually destroying it altogether. Various aspects of these effects of human activity, as they are seen in the British Isles, are dealt with generally in Chapters VI, VIII and IX.

Whatever happens to it, by natural processes or by human agency, vegetation always strives, so to speak, to resume its progress towards the climax, because under the given conditions the climax represents the position of relatively greatest equilibrium between the available vegetation and its environment. Every partial destruction of vegetation, as well as every presentation of a new bare soil, initiates a new sere whose normal goal is the climax type corresponding with the climatic region. In contrast with the full primary sere, the prisere, from bare soil to climax, any shorter sere beginning from partially destroyed climax vegetation or from some earlier seral phase in which progress has been arrested, by human activity or otherwise, is called a secondary sere or *subsere*.

In virgin regions, or where the inhabitants do little or nothing but gather vegetable food and hunt or fish, the country is largely covered with climax vegetation, unless, indeed, the region is constantly subject to the action of violent natural agencies such as floods or eruptions, when all stages of primary and secondary seres will be found, occupying areas which have been more or less recently denuded of vegetation, completely or partially, and alternating with the climax type where the ground has been undisturbed for a considerable time. In countries which have been long settled by agricultural human communities, on the other hand, the land becomes a patchwork of the most various plant communities. Cultivated crops and plantations alternate with regular pastures and patches of climax vegetation more or less modified (e.g. tended and exploited forest), together with all stages of various subseres initiated in different ways, by the most various human activities, such as intermittent felling of trees, grazing, burning or quarrying.

Conditions in England. In a country like England, for example, except on the higher hills of the north and west, on the saturated fens, marshes and boglands, on the sea coast, and in a few other areas, the natural climax vegetation has been deciduous summer forest for several millennia, but it is many centuries since any considerable portion has been in a really natural or "virgin" condition. Most of the forest has long been cleared altogether and replaced by permanent pasture or arable crops, but a little has been left as modified, more or less tended, and more or less replanted woodland, and much has been converted, by felling, cutting and sporadic or intermittent pasturing, into "waste land" or "rough grazings"—grass or heath—with a certain amount of scrub, and sometimes trees, which represent attempts of the climax forest to reassert itself. This partial redevelopment of forest, or, in formal language, the progress towards the climax of the subseres initiated by man, is hindered not only by fresh direct human interference, but also by two other causes: first, the introduction of conditions—for example casual pasturing, nibbling by rodents, or the hardening of the soil surface by trampling and exposure—which prevent the regeneration of trees; and secondly, the increasing paucity of trees which can act as seed parents owing to the general clearance of forest in the country at large.

Subclimax vegetation. Finally, there is another fact which must be taken into account before we can fully appreciate the principles underlying the actual distribution and development of vegetation. It is only on relatively favourable soils and in the presence of relatively favourable conditions of life that we get the full development of climatic climax vegetation, in England deciduous summer forest. Over most of the area of England, there is little doubt, oak forest would again extend, perhaps in a couple of centuries, if the country were completely abandoned by man and his flocks and herds, while beech would become dominant over most of the chalk of the south-east and perhaps on many of the sands and loams as well. But there are many areas on which we should not expect to find either oak or beech forest if we revisited the country after two centuries of desertion. Where the soil had remained constantly wet, owing to the persistence of marshy tracts accompanying rivers and lakes (a kind of ground which would of course be greatly extended when all river regulation and drainage work was suspended) there would still be only marsh and fen communities, whether of herbaceous plants or of trees like the alder and the willows which can establish themselves in constantly wet soil. Such communities may be, as already explained, stages in the hydrosere leading to the climatic climax (in this case oak forest), but the drainage relations in low-lying regions are such that many areas would probably remain indefinitely as marshland bearing marsh, fen or carr vegetation because the hydrographic system of the region would maintain standing or very slowly moving water above or very close to the soil level in spite of constant building up by the accumulation of vegetable debris. Such vegetation, held in a phase below that of the climatic climax by some more or less permanent factor (in this case permanently wet soil), was called by Clements (1916) a *subclimax*, because it represents a stage inferior to the climatic climax in the progressive succession of the hydrosere (Fig. 49).

Under soil conditions which are at the other extreme, on steep rocks or on the tops of narrow ridges which maintain their essential form and character for a very long time, the regional (climatic) climax vegetation cannot establish itself, because the habitat is too dry or too much exposed to wind, or both. Here one will find rock plants, with patches of xerophilous grasses and other herbs where a shallow soil has collected between the rocks or covers the top of an exposed ridge, maintaining themselves indefinitely and forming another type of subclimax—a stage which might form part of a xerosere. The arresting factors in both these types of subclimax are *edaphic factors* (Greek ἔδαφος, the ground), i.e. dependent on *soil*.

Where the configuration of the ground, as distinct from the nature of the soil, is unfavourable, climatic climax may again fail to develop. This is the case on very steep slopes where the characteristic elements of climax forest cannot maintain themselves because the soil is not stable; and on slopes of unfavourable aspect, either too damp and cold or too hot and dry, or exposed to violent winds. In the last cases the actual arresting cause is the local modification of the general regional climate depending on a *physiographic factor*. But it is clear that this type of subclimax cannot be sharply separated from those previously mentioned, for the nature of the land relief affects not only the local climate but also the soil.

Pasture. Another cause of the appearance of subclimax vegetation is the change brought about by an external "artificial" factor, for example the

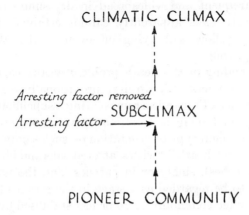

FIG. 49. THE CONCEPTION OF SUBCLIMAX

arrest of the normal prisere by the intervention of man, usually through the agency of his grazing animals (Chapter VI). All pasture in regions where the climatic climax is forest is of the nature of subclimax, and in the British Isles pasture is by far the most extensive and important type. This is true in a general sense, as we have already seen in Chapter IX, even of the artificial "permanent" grasslands "laid down" by man, as well as of the extensive "rough grazings"—the chalk downs, limestone pastures and non-calcareous (siliceous) pastures. The meadows mown for hay, and the fens cut for litter and thatching are also subclimax vegetation due to human activity. Much vegetation not regularly used as pasture is also subject to the effects of more or less casual grazing and also to constant nibbling by small rodents, especially rabbits and mice (see Chapter VI).

All such vegetation is determined by the influence of *biotic factors*, which when directly due to human intervention, may be called *anthropogenic*.

Heathland. Another good example of subclimax is heathland, as it is developed on sandy soil in many parts of southern England. This well-known plant community, dominated by the common ling (*Calluna*), and sometimes also by the purple bell heather (*Erica cinerea*), is nearly always found on sandy soils with acid soil water, relatively poor in bases. On comparatively well-drained slopes of our higher hills, as we saw in the last chapter, especially on the eastern side of Ireland, Wales and Scotland, largely used as "grouse moors", where wind prevents the growth of trees, it is probably a climatic climax community; but in southern England it usually forms a stage in the sere which begins on bare sandy soil and culminates in forest. This sere is, however, very commonly arrested in the stage of heathland, either by pasturing or burning, and sometimes perhaps by deficiency of water or other edaphic factors. Heaths are very often pastured, heath fires are frequent and widespread in dry summers, but there are some heaths where oaks apparently cannot establish themselves, the seedlings turning yellow and dying off even in the absence of these factors (cf. also p. 356).

A full understanding of the heath problem awaits closer investigation and especially experiment, but we may provisionally regard most of the heathlands of southern England as representing a subclimax determined by biotic (i.e. animal or human factors), and perhaps sometimes by edaphic factors. A very common type of vegetation on such sandy soils is what has been called "oak-birch heath",[1] with scattered oaks and birches (sometimes also pines and beeches), singly or in patches, and the heath community between. This is to be regarded as a stage in the sere leading from heathland to forest, a stage commonly more or less stabilised by sporadic felling and occasional burning. Here we have a partially stabilised mixture of two or more communities—successive communities of a sere—belonging to heath and forest.

Climaxes determined by different factors. The subclimaxes briefly described in the preceding pages are seen to be due to various kinds of ecological factors which we may broadly designate as *edaphic, physiographic* and *biotic* (including *anthropogenic*). The first two kinds of factor are clearly integral and permanent constituents of the factor-complexes determining the particular types of vegetation. They actually differentiate the vegetation from the climatic climax, and stabilise it in a particular form in equilibrium with the particular factor-complex. The forms of vegetation so determined are thus quite legitimately themselves considered as climaxes, which we may call *edaphic* and *physiographic* climaxes when the corresponding factors are clearly the differentiating determinants. Where the action of man is regular and long continued, as in regular grazing, mowing, burning or coppicing the vegetation, then biotic factors also may be properly considered as determining corresponding climaxes, since they introduce a

[1] *Types of British Vegetation*, pp. 100–3, and see Chapter XVII, pp. 354–7.

permanent influence which stabilises the vegetation, keeping it in a condition of equilibrium with all the forces to which it is subject, including this constant human action. Such vegetation may thus properly be called a *biotic* or *anthropogenic* climax.

The word subclimax may be retained as a general term, when we are considering the prisere as a whole, for the various climaxes determined by edaphic, physiographic and biotic factors are normally dominated by life forms inferior to (i.e. simpler than) and therefore "below" the climatic climax. Fig. 49 illustrates the conception of a subclimax in which the arresting factor may be removed, thus permitting progress to the climatic climax.

Deflected seres (plagioseres). But though they are "below" the climax of the full climatic prisere the various "subclimaxes" rarely if ever correspond exactly, though they do correspond in a general way, with particular stages of the full prisere. Thus the constant intervention of man by grazing, cutting or burning does not stop succession altogether but rather *deflects* it into a new course, which proceeds, subject to the biotic factor, till it reaches a new position of equilibrium which is different from any stage of the prisere (Godwin, 1929a). For example, the characteristic sheep pasture of the chalk downs, dominated by the sheep's fescue (*Festuca ovina*) and the red fescue (*F. rubra*)—a stable community so long as pasturing continues—is not, as a community, by any means identical with any stage in the chalk prisere which ultimately leads to beech forest. Though most of the species of grasses and herbs which enter into the prisere at the herbaceous stage of development may be present in chalk pasture, yet the dominants, the whole structure of the community, and the proportions of the species composing it are different (see Chapter XXVII). The typical and very characteristic chalk pasture community would never have arisen had it not been for pasturing. Again, on East Anglian fenland, when the "saw sedge" (*Cladium mariscus*) is cut at frequent intervals (2 years) an apparently stable community is produced dominated by the purple moorgrass or "flying bent" (*Molinia caerulea*) and this is different from any stage in the normal fen succession (Godwin, 1929b).[1] These "deflected successions" or *plagioseres* (Greek πλάγιος, slanting, sideways) as they may be called, are characteristic results of man's activity, and their end products, varying with the precise form of exploitation, are characteristic biotic climaxes or *plagioclimaxes*. They usually correspond, *in a general way*, with some stage of the prisere, but have special characters of their own. When they are released from the human factor a subsere is initiated which rejoins, as it were, the prisere, progressing to the normal climax after having traversed the deflected route. The accompanying diagram (Fig. 50) is intended to make these considerations clear.

Plagio-climaxes

[1] Chapter XXXIII.

It will be evident from what has been written that in our efforts to perceive the uniformities in the almost infinitely varied processes of the development of vegetation we have to introduce a number of conceptions, some of which are still unfamiliar, while they have not all reached completely satisfactory formulation. Nevertheless, it should be clear that general laws of development do exist, and that this must be so, when we consider that the methods of growth, reproduction and dispersal of plants, the conditions of their competition, and the types of combinations of factors which actually affect them in their different habitats are essentially the same everywhere, in spite of the immense diversity of their forms and of the detailed conditions of their existence.

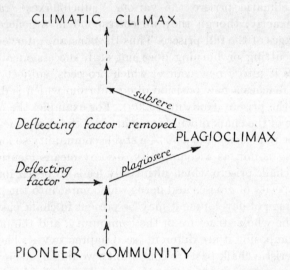

Fig. 50. Deflected Succession—Plagiosere and Plagioclimax

Nothing can be more certain than that vegetation, like all life, is dynamic or kinetic, and not static, and that it can only be understood by considering each plant community we observe in nature as something which has had an origin and will have a fate. Climax communities are relatively permanent because they are in approximate equilibrium with all the conditions in which they exist. When these conditions change, they too will give place to something else. And even climatic climax communities, though seemingly permanent so long as the existing climatic complex persists, may contain within them the seeds of their own decay. They may, for example, gradually bring about changes in the soil on which they are growing, changes that will ultimately prevent their regeneration. But though we may suspect their occurrence, of the details of such events we have as yet very little knowledge.

The nature of a plant community

Interrelations of the members. A plant community may be defined as any collection of plants growing together (of one species or of several or many species) which has, as a whole, a certain unity. In the simplest case this unity is created in the first place by the fact that the plants in question exist together, that is in a common habitat, which has, so to speak, "selected" the species which can flourish in or tolerate its conditions, and that the vegetation produced shows a certain uniformity. But the mere fact that a number of organisms habitually live in juxtaposition in itself creates interrelations between them; and in more complex communities, where the life forms of different members are widely divergent, the members may develop considerable and intimate functional interrelations. Some of the members of the more complex communities could not exist in such communities in the absence of certain other members. For example, many of the species rooted in the humus soil of a forest could not exist without the trees which shelter them, nor without the vegetable mould provided by the decay of their leaves; nor could parasites exist without the presence of their "hosts", nor epiphytes without the plants on whose surface they find a footing. The roots of the trees themselves may enter into a very intimate relationship (*symbiosis*) with certain fungi (*mycorrhiza*), a relationship which has been clearly shown in certain cases to benefit both the tree and the fungus. Other flowering plants, such as the heaths and the orchids, cannot exist without entering into a similar but even more intimate symbiosis with other kinds of fungi. Thus relations of dominance and dependence, or of mutual interdependence, often exist between different members of a plant community in addition to the dependence of all upon the general conditions of the habitat.

The role of competition. Competition also is constant between certain of the members of a closed plant community, both between individuals of the same species and between those of different species. But in climax communities competition between its members merely serves to keep the composition of the population and the structure of the community substantially constant, the various classes of members being in approximate equilibrium with one another, and the community as a whole with its habitat. Thus competition between the members for light or for water maintains the community in a stable condition, preventing any one species from breaking away from its existing status in the community—as it may do if other members of the community are removed—and fresh species do not normally enter a climax community.

In seral communities the case is otherwise, for as long as fresh plants are entering any community (so soon as the available space becomes nearly filled) active competition must take place between the new entrants and the plants already on the ground. This ordinarily leads to suppression of earlier populations by later ones, and thus to the normal succession of seral communities. A seral community is not in equilibrium with its environment when we include in the conception of environment the new invaders which are coming to dispute the territory. As each seral community vegetates and is succeeded by another the habitat is continuously modified—in the xerosere by increasing, and in the hydrosere by decreasing, soil moisture. At the same time the conditions to which the subaerial parts of the plants are exposed are also modified as soon as the community becomes nearly closed, and increasingly as taller take the place of less tall dominants. The humidity of the air between the shoots is increased, light is intercepted, extremes of temperature are diminished, and shelter from wind is established. These changes will enable shade plants to establish themselves in the community as well as affecting the lower shoots of the dominants and all members of the community which do not reach the full height of the canopy.

The ecosystem. It will be clear from what has been written that the key concept which must govern all our efforts to formulate the phenomena of vegetation in a rational system is the idea of progress towards equilibrium, which is never, perhaps, completely attained, but to which approximation is made whenever the factors at work are constant and stable for a long enough period of time. This leads us to include the units of vegetation in the general conception of physical systems of which the universe is composed—systems which mark positions of relative if only temporary stability in the general flux. From time to time the systems are partially or wholly destroyed, but so long as conditions remain the same and the original components are present these will always tend to re-establish a system of the same type as before. A unit of vegetation considered as such a system includes not only the plants of which it is composed, but the animals habitually associated with them, and also all the physical and chemical components of the immediate environment or habitat which together form a recognisable self-contained entity. Such a system may be called an *ecosystem* (Tansley, 1935), because it is determined by the particular portion, which we may call an *ecotope* (Greek τόπος, a place), of the physical world that forms a *home* (οἶκος) for the organisms which inhabit it. A prisere is the gradual development of such a system, the climax represents the position of relative equilibrium which it ultimately attains, and a subsere is the redevelopment of the same type of system after partial destruction.

The naming of plant communities

The hierarchy of communities. It is obvious that plant communities are not all of equal rank, since they vary in size and complexity from the great climatic vegetational types such as equatorial rain forest or prairie grassland, which cover many thousands of square miles of the earth's surface, down to small collections of plants growing in some special habitat of very restricted area. Beginning with the great climatic climax communities we must in fact recognise a series or hierarchy of communities of decreasing size and status, of which the smaller are commonly included in the larger.

There is still, unfortunately, a certain amount of discrepancy between the applications by different schools of phytogeographers of the terms currently used to designate these various categories of community. The terms employed here are applied in a way which corresponds fairly closely with the commonest modern usage among English-speaking writers, nor does it differ widely from the usage of the more conservative schools of continental Europe.

Formation and Formation-type. The representative vegetation of each of the great climatic climax types of the vegetation of the world shows more or less wide differences according to the geographical region in which it is developed. Thus the Indo-Malayan equatorial rain-forest climaxes differ in important respects from the African, and these again from the American, though they all show a fundamental correspondence. The deciduous summer forest of the eastern United States differs from, though it is obviously closely related to, that of western and central Europe, containing many of the same genera. Each of these climax forests, belonging to a specific geographical (and climatic) region, we shall recognise as a *climatic plant formation*. The *type* of formation, characterised by the same dominant life forms, common to all the formations developed in similar climates throughout the world we designate a *climatic formation-type*. A climatic formation-type may equally well be called a *climatic climax* type of vegetation. Thus we have the formation-type of equatorial rain-forest, and the formation-type of summer deciduous forest, but of the former we distinguish between the Indo-Malayan, African and American formations, and of the latter between the European, American and East Asian formations. Similarly with the relatively stable communities determined by edaphic or biotic factors, we can speak of the reedswamp or of the meadow formation-type, each marked by the dominance of a characteristic life form, but divisible into geographical formations.

Climax, formation and stability. When we are considering succession we think in terms of climaxes, of change from one to another, and of temporary or permanent arrest or deflection of this change. When we are studying a particular community we think in terms of dominant and associated species, their life forms, behaviour, and inter-relations. It is this latter study that calls for the conception of "formation". A plant formation is a unit of vegetation *formed* by habitat and *expressed* by distinctive life forms. In so far as it has relative stability the adult formation (mature ecosystem) is a climax, though the *degrees* of stability are very different, as for instance between climatic or sharply marked permanent edaphic climaxes, which are very stable, and communities such as reedswamp, or seral heath, which very readily give way to others as the result of quite normal autogenic[1] changes of habitat. Thus a prisere passes through several formations, or positions of relative stability, marked by different complexes of habitat factors and by distinctive life forms. On the other hand quite transitory communities leading from one phase of stability to another are naturally regarded as developmental phases of the next formation and are marked by mixtures of life forms with incomplete dominance.

Association. A formation is composed of *associations*, each with different dominants and at least some different subordinate species. The difference of formation-types is a difference of dominant *life forms*; of associations a difference of dominant *species*—a "floristic" as distinct from a vegetational difference. If the dominant species throughout the formation are everywhere the same the formation consists of but a single association. The existence of different associations within a formation is related to the different ecological requirements of the dominants—for dominants of the same general life form rarely have exactly the same ecological requirements—or to an historical cause.

Consociation. Where a single species dominates a portion of an association the community so formed is known as a *consociation*. Thus we have consociations of oak and consociations of beech in the oak-beech association of the European deciduous forest. Where the two (or more) dominants mingle we have the association not segregated into consociations. The technical name of the consociation is formed by the stem of the generic name of the dominant with the suffix *-etum*, followed by the "trivial" or "specific" name in the genitive: thus the European beech consociation is called *Fagetum silvaticae*, which is really a shortened form of *Fagetum Fagi silvaticae*.

Society. Within an association or a consociation there are subordinate communities dominated by species which are not general dominants of the main community, and these are called *societies*. Thus we may have ash and alder societies on wet ground in an English oakwood, and these commonly contain species which do not occur or occur less commonly in the main woods. In grassland communities societies of herbs, often with conspicuous flowers, are characteristic; for example societies of the horseshoe vetch (*Hippocrepis comosa*) in the southern English chalk grasslands.

The subordinate layers or strata of vegetation in complex associations are generally largely composed of what may be called *layer societies*. For example, the shrub and field layers of an English oakwood are rarely uniform throughout; and societies of hazel, dogwood, hawthorn, sallow, etc., may often be distinguished in the shrub layer, and of mercury, primrose, bluebell, wood anemone, etc., in the field layer. *Aspect societies* are those whose dominants vegetate actively during a part only of the growing season, as for example the wood anemone society in early spring, and the

---

[1] *Autogenic*, a term applied to a sere or seral changes which are effected by the activity of the vegetation itself, as opposed to *allogenic*, produced by factors external to the vegetation (Tansley, 1928).

bluebell society later. "Prevernal" aspect societies, vegetating and flowering in early spring, before the leaf canopy appears, are characteristic of deciduous summer forest, and middle or late summer societies are characteristic of many grasslands. Very often, however, herbaceous societies have more than one dominant, and each of these may mark a seasonal aspect, or on the other hand the periods of their active vegetation may to some extent overlap. The root systems of two or more species respectively dominating a society at different seasons, or all at the same season, frequently occupy different layers of the soil, so that there may be little or no direct competition between the different dominants. A good example is the *Holcus mollis-Scilla non-scripta-Pteridium aquilinum* society of English oakwoods on light soil, the three species vegetating at different seasons and their root systems often occupying different layers of the soil (see p. 281).

Owing to the particular ecological requirements or tolerations of certain society dominants portions of the association habitat which differ slightly or markedly from the average conditions are often occupied by corresponding societies, and subordinate species with similar requirements may therefore occur exclusively or more abundantly in such societies. Again, the vegetation of the society dominant may itself bring about special conditions within the society area and thus lead to the exclusion, or alternatively to the special abundance, of particular subordinate species.

The name *clan* is sometimes applied to small aggregations of subordinate species, brought about by locally active social vegetative growth or gregarious establishment of seedlings.

Many of the terms used in this chapter are those of Dr F. E. Clements (1916, 1928) to whom is due the foundation of the system adopted in the present work. But Clements uses formation, association and consociation solely for communities of the climatic climax, which is the only climax he recognises. In this book, however, they will be used also for the well-marked edaphic and biotic climaxes (or subclimaxes): thus submerged fresh-water aquatic vegetation, reedswamp, heath and grassland will be considered as *formations*, and their communities, dominated by characteristic species, as *associations*. These "subclimax" or "plagioclimax" formations and associations are sharply characterised by life form, habitat and floristic composition, they are stable so long as the habitat factors remain unchanged, and the only way in which they differ in principle from the climatic climax vegetation is that in addition to the climatic factors there are edaphic or biotic factors, not depending or only partly and indirectly depending on climate, permanently at work to stabilise the community. Finally, not least among the advantages of this usage is the fact that it is much more closely in accord with the older and more conservative European application of the terms formation and association. It applies to a larger number of communities than those represented in the climatic climaxes alone, but to a very much smaller number than the multitude of those which have been called "associations" by a considerable number of modern "plant sociologists" in Sweden and elsewhere.[1]

Seral communities. To seral communities Clements applies a set of terms corresponding with those belonging to the climax, but with the termination *-es*. Thus *associes* is used for a seral community corresponding with the association, *consocies* with the consociation, *socies* with the society. A *colony* is a number of individuals which have just invaded a seral area. These terms will be used in the same sense in the present work, except that they will not be applied to stabilised edaphic or biotic "subclimaxes", which are here considered climaxes, as already explained, and have therefore a claim on the terms ending in *-ion*.

The following table gives the hierarchy of plant communities recognised in this book:

Climax units	Seral units
Formation-type	
Formation	
(climatic, edaphic or biotic)	
Association	Associes
Consociation	Consocies
Society	Socies
Clan	Colony

[1] During the last 30 years the term "association" has been used in at least four different senses. After its widespread adoption in the first decade of the present century it was at first mainly used for relatively large units of specific floristic composition very nearly in the sense in which it is employed in the present work. This may be called the first meaning. Very soon, however, it began to be used, quite vaguely, for almost any unit of vegetation in the same way that "formation" had been used twenty years earlier. This is the second meaning, for which there is no justification: "community" is the recognised term in English for any unit of vegetation of undefined status. In 1916 Clements restricted the term "association" to the large floristic units that can be recognised within his "formations" (climatic climaxes only). This is the third meaning. About 1920 Osvald and Du Rietz in Sweden began to use "association" for very small units, recognition of which is indispensable for detailed study but for which the term "associations" is unsuitable. This is the fourth meaning, which necessarily led to the creation of an enormous number of "associations" (for example, 164 in the dwarf shrub heath of south-eastern Norway, Du Rietz), which may only be valid within a very limited region. It was based on an illusory belief that these "micro-associations" were in some sense fundamental, and is very inconvenient in practice and quite out of accord with the original use. The highly specialised usage of the Uppsala school was unfortunately followed by several continental workers. By applying the typical phyto-sociological term "association" to these very minor phenomena of vegetation attention is concentrated on them to the neglect of larger units, to which awkward compound designations have to be applied. It is not of course intended to suggest that "micro-associations" are unimportant phenomena unworthy of study, but only that the diversion of the term "association" from its original use to these minor phenomena is in every way undesirable. In 1930 however Du Rietz proposed to substitute the term *sociation* for the units which he had hitherto been calling "associations", and at the International Botanical Congress at Amsterdam in 1935 this proposal was adopted. The Congress also recommended the use of *association* in the first meaning mentioned above.

REFERENCES

CLEMENTS, F. E. *Plant Succession.* Carnegie Institute of Washington, 1916.

CLEMENTS, F. E. *Plant Succession and Plant Indicators.* Carnegie Institute of Washington, 1928. (The first part of the work published in 1928 is a reprint of most of *Plant Succession.*)

DU RIETZ, G. E. Classification and nomenclature of vegetation. *Svensk bot. Tidskr.* **24**. 1930.

GODWIN, H. The subclimax and deflected succession. *J. Ecol.* **17**, 144–7. 1929 a.

GODWIN, H. The "sedge" and "litter" of Wicken Fen. *J. Ecol.* **17**, 148–60. 1929 b.

LIPMAA, T. La méthode des associations unistrates et le système écologique des associations. *Proc. Int. Bot. Congr.* Leiden, 1935.

TANSLEY, A. G. The classification of vegetation and the concept of development. *J. Ecol.* **8**, 118–48. 1920.

TANSLEY, A. G. Succession: the concept and its values. *Proc. Internat. Congress of Plant Sciences, Ithaca,* 1926. 1928.

TANSLEY, A. G. The use and misuse of vegetational terms and concepts. *Ecology,* **16**. 1935.

Chapter XI

DIFFERENT METHODS OF CLASSIFYING VEGETATION
RAUNKIAER, MOSS, CRAMPTON

In the last chapter the method of classifying vegetation adopted in this book has been explained. It is essentially based on a recognition of the fact that vegetation is kinetic or dynamic and not static, and that if we are to consider it intelligently we must draw a primary distinction between climax vegetation, which is in relative equilibrium with all the conditions to which it is subject, and seral vegetation, which is on its way to change into something else. It must also be recognised that a given community may change its status from seral to climax or vice versa when there is an effective change in its environmental conditions.

It is desirable here to consider briefly certain other methods of classifying vegetation which have been suggested and used from time to time, because they bring out the relative importance of various points of view without which we cannot obtain a clear and comprehensive perspective of the phenomena. The following remarks in no way attempt an exhaustive review of the types of classification that have been proposed: only three are considered because they throw light on the principles involved in the scheme here adopted. We leave on one side all classifications which treat plant communities as static entities and also those which are based on the distributions of individual species (other than dominants) or of collections of species—"geographical elements"—which occur in different parts of the world. These last concern the distribution of *flora* rather than of *vegetation*.

Importance of life form. But *life form* is of primary significance in classifying vegetation, as we saw in the last chapter. The life forms of the dominants of communities are characteristic features of the first rank. Those of the great climatic formations are directly related to the climates that determine the different formation-types, which are characterised by their dominant life forms. And the life forms both of dominant and subordinate species of all communities of whatever size or status are features of significance.

Raunkiaer's life form classes. The most widely used and one of the most significant life form systems is that of Raunkiaer, which applies primarily to vascular plants and is based on the relation of the perennating buds to the soil level. According to this relation species are grouped into *life form classes*. The precise limits Raunkiaer used are necessarily arbitrary, but convenient enough. All plants which bear their perennating buds more than 2 m. above the soil are known as *phanerophytes*. Between 2 and 8 m. (shrubs and small trees) they are called *microphanerophytes*: above 8 and above 25 m. respectively (medium and tall trees) they are known as *meso-*

and *megaphanerophytes*. Plants with their perennating buds between 25 cm. and 2 m. above the soil surface (dwarf shrubs) are called *nanophanerophytes*: those with buds between soil level and 25 cm. *chamaephytes*: those whose buds are at or in the actual soil surface *hemicryptophytes*; and those with perennating buds buried in the soil (underground rhizomes, tubers, corms and bulbs) *geophytes*. The phanerophytes dominate the climax vegetation in all the favourable climates, including of course the deciduous summer forest of this country. *Helophytes* (marsh plants), *hydrophytes* (water plants), epiphytes, parasites and succulent plants are reckoned as separate life form classes, but the helophytes and hydrophytes are sometimes grouped with the geophytes as *cryptophytes* because their perennating buds are hidden in the soil or below the water.

Phytoclimates. Raunkiaer (1907, 1934) used his system to characterise different "phyto-climates" by the proportions of the *total flora* of a given region belonging to the different life form classes, and usually with great success. When the percentage of one or more of the life form classes present in a given region significantly exceeds that of the same life form class in the flora of the world at large the phytoclimate of that region is said to be "phanerophytic", "chamaephytic", "hemicryptophytic", and so on.[1]

Phytoclimate and climax dominants. This often gives quite a different result from that indicated by the life form of the climax dominants. For example, the natural climax dominants of most of Britain are phanerophytes, but a clear majority of all the flowering plants inhabiting the country are hemicryptophytes, and this is about double the percentage in the world as a whole, while the percentage of phanerophytes in the British Isles is much less than that in the world flora. The "phytoclimate" is therefore not phanerophytic but hemicryptophytic. The dominants of tropical rain forest are of course phanerophytes, and the phytoclimate is also phanerophytic, because a much larger percentage of the total flora of those regions than of the world at large consists of shrubs and trees with buds more than 2 m. above the soil. Here the two methods give the same result.

The hemicryptophytic character of the British climate is strikingly demonstrated by the readiness with which hemicryptophytes come to dominate the vegetation directly the originally dominant phanerophytes are removed and prevented from returning. As we have already seen in Chapters II and IX and shall see again in Chapter XXV, the climate of the British Isles is from this point of view a "grassland climate", and the greater portion of its area is now actually grassland. As soon as the trees are removed and prevented from returning by the grazing of animals, the hemicryptophytic grasses and associated herbs establish themselves, though if grazing is removed the shrubs and trees can and do come back and re-establish their dominance because they can overshadow and kill the

[1] The list of percentages of different life forms in any flora is called its *biological spectrum*.

pasture plants. And since trees and shrubs can grow and flourish in the British climate the mere fact that they grow taller than the grasses gives them undisputed dominance when left alone. Most of the area of the British Isles has thus primarily a forest climate: the greater part lies within, though on the north-western border of, the natural climatic area of central and west European deciduous summer forest; but it is not a country inhabited by a large number of native species of woody plants. This is partly because a number of species which were here in late Pliocene times, under a climate probably not very different from the present, were never able to return after the Pleistocene glaciations. But it is also because it is not a climate that gives an *overwhelming* advantage to forest as does that of the wet equatorial regions. This is probably because of the low summer temperatures and the prevalence of late frosts.

Thus we see the different results we may get when we consider the dominant life form alone and when we take into account the life forms of all the species of a flora. Neither method can be neglected, but the former must be used in primary vegetational classification because dominance determines the essential character of the vegetational type.

Moss's scheme. In his great work *Pflanzengeographie auf physiologischer Grundlage*, published in 1898 (translated as *Plant Geography on a Physiological Basis*), A. F. W. Schimper pointed out that the larger units of vegetation (the great plant formations) were determined by climatic factors (climatic formations) and that the vegetation so determined was sorted out into smaller units by the influence of different soils (edaphic formations). In trying to classify British vegetation on natural lines the late C. E. Moss (1910) based his scheme on the view that the British climate was so uniform as to justify the primary classification of vegetation *within the country* according to the soil types characteristic of different kinds of geological (or rather lithological) formation: and these primary edaphic divisions he called "plant formations". Moss's "formations" therefore corresponded with Schimper's "edaphic formations", but Moss stressed the developmental or successional aspect. Within each formation he recognised the "progressive succession" of different plant communities ("subordinate plant associations" he called them) leading up to a "chief association", which was permanent and stable; and "retrogressive succession" from the "chief association" leading down to subordinate associations. Thus on chalk down the "chief association" was beechwood, while chalk scrub and chalk grassland were "subordinate associations" resulting from retrogression of beechwood: on siliceous hillsides there was a "chief association" of oakwood dominated by *Quercus sessiliflora*, and corresponding retrogressive "subordinate associations" of scrub and grassland. The main cause of retrogression Moss recognised of course in the human factor, felling and pasturing, but he believed that other factors also might

Edaphic formations

"Progressive" and "retrogressive" succession

be at work. Because of the almost complete human control of British land vegetation most of it exhibits mainly "retrogressive succession", but here and there examples of progressive succession may be seen.

On salt marsh and sand dune there were well-marked "subordinate (but here progressive) associations" leading up to equally well-marked "chief associations". In each case the distinctness of the "habitat", i.e. the soil type, occupied by the whole set of communities was emphasised as corresponding to and determining the distinctness of the whole plant formation, within each of which a successional series of associations was developed. Moss's ideas were made the foundation of the classification adopted in *Types of British Vegetation* (1911).

At the time they were put forward these conceptions undoubtedly represented a real constructive advance on previous methods of classification, and they certainly furnished a perfectly good, workable, objective classification of British vegetation. They were nevertheless justly subjected to considerable criticism. In the first place climatic differences within the British Islands do actually determine, as we have seen in Chapter IX, distinct climatic formations in different regions. Again the "habitats" of the "formations" were interpreted too narrowly as represented by the lithological soil types characteristic of particular kinds of geological formations (sandstones, shales, limestones, etc.). While these soil types do on the whole bear, in the British Isles, distinctive series of communities, this is not necessarily so in other regions. Thus beech forest, which is largely (though not wholly) confined to chalk and limestone in England, is not at all so confined on the continent, and in order to arrive at a natural classification of plant communities it is always necessary to consider them over the whole geographical range of their occurrence. Furthermore, we cannot ignore the life forms of the dominants in defining and characterising a plant formation, which is a *vegetational* unit. We cannot, for example, separate chalk and limestone grassland from other types of grassland showing most of the same characteristics, while associating it *in the same plant formation* with a particular type of scrub and forest. Again, as Clements (1916) pointed out, the so-called "retrogression" of forest to scrub and grassland is, in most cases at least, dependent on the artificial destruction of the dominant trees, and is not a process comparable with "progressive succession".[1]

Nevertheless, the bond between the grassland, scrub and forest of chalk and other limestones, and again between these types of community as developed on the older siliceous rocks, is perfectly real and close; and, as Moss clearly saw, is of the same general nature as that existing between the communities successively developed on sand dunes, on saltmarsh, or on fen peat. In both cases, in fact, we have successional series of communities,

[1] There are however instances of true "retrogressive succession" not dependent on human intervention.

or, as we now call them, particular types of *sere*. Each sere is to a greater or less extent controlled throughout by the characteristic edaphic features of the habitat, which retains certain general distinctive characters throughout though it undergoes modification during the progress of the sere. But the actual relation of the various communities or stages of the sere is radically different in the two cases. It is not a question of the difference between "progression" and "retrogression". In the "progressive successions" we have the development of a mature formation through the stages of the prisere. In the seres which have forest as their "chief association", the destruction of this, whether by grazing alone, or by direct clearing followed by grazing, results in the establishment of a pasture community—a biotically determined formation—dominated by grasses and stabilised by grazing. When the grazing factor is partly or entirely removed the community is colonised by woody plants and a subsere is initiated, succession progressing again to give scrub and eventually forest. The pastured grassland is not exactly equivalent to any stage in the normal prisere; it corresponds in a general way with the stage antecedent to colonisation by woody plants, but it is itself a biotic climax which should be regarded as a distinct formation. While we may admit the possibility of forest "spontaneously retrogressing" to grassland or heath, we have no certain evidence that this has taken place independently of climatic change.

The classification adopted in *Types of British Vegetation* is in effect a classification, not into plant formations as the term was originally understood or as it is used in the present work, but into series of communities corresponding with different types of edaphic habitat, which Moss called formations, and which actually occur in the British Islands. This seral classification is of essential value and must be represented in any natural scheme of the plant communities of a country, but it should not be put in the place of the natural plant formations, characterised by dominants of specific life form.

Crampton's stable and migratory types. The last basis of classification that we need to consider here is that put forward by the late C. B. Crampton (1912). Crampton, who was a geologist by profession, attracted to the study of vegetation by the close connexion between it and the geodynamic features of the Scottish regions he surveyed, proposed a primary division of vegetation into *stable* and *migratory* types. Stable vegetation is developed on mature and stable land surfaces, and is in stable and permanent, or quasi-permanent, equilibrium with the climatic and edaphic factors of the environment. In other words, it remains constant and does not normally change its characters. Migratory vegetation, on the other hand, occupies habitats subject to perpetual change as the result of the constant influence of active geodynamic factors, as on the sea coast, in the valleys of active rivers, and on mountains. In such situations the smaller habitats are continually being destroyed and renewed, so that the plant populations are constantly "on the move" in detail, though the vegetation belonging to the whole series of habitats maintains its general nature. A sand-dune complex, for example, is accumulated and colonised by plants in one place, and in another eroded by wind so that the vegetation is destroyed: a layer of salt mud is deposited and colonised by salt-marsh plants, but a little later partly or wholly carried away and redeposited elsewhere by a change of direction in the tidal scour: a sea cliff is continuously, though slowly, undercut by the waves, so that the plants which can maintain themselves for a time on the cliff face are ultimately dislodged and destroyed by a land slip, to be replaced by fresh individuals on the newly cut face. River alluvium is often alternately laid down and removed by changing flood currents, with consequent destruction, followed by colonisation in a fresh place, of pioneer alluvial vegetation. Again, talus accumulates below mountain rock faces and is partially colonised by plants, but much of the vegetation is destroyed by fresh falls except in places where the rain of rock ceases, so that a progressive succession can be initiated and continued.

The vegetation of the coastal, streambelt and mountainous regions subject to the operation of active geodynamic agents is thus characteristic in constitution and behaviour, and it was the whole of these communities which Crampton proposed to separate as "migratory" vegetation, because the communities are constantly forced to "migrate" from place to place according to the exigencies of the changing habitats. The distinction is undoubtedly natural, corresponding alike with the nature and the habitats of the vegetation, but as the foundation of a general scheme of classification it is one-sided, since it is based primarily on the common violence of the geodynamic factors in primitive habitats. Expressed in terms of the developmental concept of vegetation migratory communities are the beginnings of seres which may culminate in stable vegetation. In all active geodynamic regions or zones these beginnings are continually being destroyed and recommencing in new places; but they can and do progress towards a climax if they escape the destructive influences which constantly menace them. From the developmental point of view they are all early stages of seres which can progress to later stages only with progressive stabilisation of the habitat; and the "migratory" habitats are in fact often gradually stabilised as they pass out of the sphere of influence of the active geodynamic factors. The common characters of "migratory" vegetation are thus derived from its common subjection to dynamic physiographic factors, and the prevalence of the action of these in certain belts of country gives the vegetation of these belts a certain permanent character *as a whole*, though the individual communities are constantly shifting and changing in detail.

Thus we may conclude that the various ideas underlying the attempts at constructive classification discussed in this chapter find their proper places in the developmental system which we owe to Clements and which is adopted with certain modifications in this book.

We now pass on to consider systematically the various kinds of vegetation existing in these islands. Owing to the paucity of xeroseral stages the mature climax and other woodlands are first dealt with in the following chapters (Part IV), and what is known of the seral and subseral stages added. The pasture grasslands, which so largely replace them, are considered next, in Part V, while the remaining types of vegetation, whose seral phases are usually available, are dealt with developmentally in Parts VI–IX.

REFERENCES

CLEMENTS, F. E. *Plant Succession*. Carnegie Institution of Washington, 1916.
CRAMPTON, C. B. The geological relations of stable and migratory plant formations. *Scot. Bot. Rev.* 1912.
MOSS, C. E. Fundamental units of vegetation. *New Phytol.* **9**, 18–53. 1910.
RAUNKIAER, C. *The Life forms of Plants*. Oxford, 1934 (especially Chapter II, originally published in Danish, 1907).
SCHIMPER, A. F. W. *Pflanzengeographie auf physiologischer Grundlage*, Jena, 1898, translated as *Plant Geography on a physiological basis*, Oxford, 1903.
TANSLEY, A. G. (ed.). *Types of British Vegetation*. Cambridge, 1911.
TANSLEY, A. G. The classification of vegetation and the concept of development. *J. Ecol.* **8**, 118–48. 1920.

PART IV
THE WOODLANDS

Chapter XII

NATURE AND STATUS OF THE BRITISH WOODLANDS

DOMINANT AND OTHER TREES. THE MORE IMPORTANT SHRUBS

Before describing the British natural and semi-natural woodlands in detail, it will be well to consider briefly their general nature and status and the roles of the chief trees and shrubs which compose them.

Natural, semi-natural and artificial woods. Of the great forests which at one time covered a large part of the British Islands only fragments remain, and most of these have been very much modified, directly or indirectly, by human activity. It was pointed out in Chapter IX that the vegetation known as "semi-natural" may have one of two origins: either it may be natural vegetation more or less modified by man, as when he exploits a native forest without completely clearing it, or pastures a natural heath or grassland; or it may be a vegetation initiated by man, but of a type which might naturally exist in the given environment, and eventually coming to show little or no trace of its artificial origin. Examples of this second kind of origin are an oak plantation made on arable or grassland where at one time grew natural oakwood, or permanent pasture composed of native species of grasses "laid down" on arable land and eventually approximating to grassland that would be produced on similar soils by pasturage alone. But we must clearly exclude from the category of semi-natural vegetation plant communities composed of or dominated by alien species planted and maintained by human agency, such as a wheat crop or a Douglas fir plantation, for these introduce dominant species and accompanying conditions altogether alien to the natural vegetation of the country. To an increasing degree the modern woodlands of Britain are coming to be composed of planted conifers of foreign origin. Scientific study of their ecology is naturally of the utmost importance to successful forestry, but it lies outside the scope of this book, and is indeed still in its earliest infancy.

Confining our attention then to natural and semi-natural communities in the two senses defined we find that a number of our woodlands are dominated by native trees, oak, beech, ash, birch, pine and alder, and that most of these woods are semi-natural communities, while a very few may fairly be regarded as completely natural, either quite virgin or so little modified by human interference that they are essentially the same things as they would have been if the country had not been occupied by man. Of the much larger number of semi-natural woodlands it is unfortunate that we can rarely distinguish with any certainty between those which have been planted and those which are the direct descendants of natural woodlands that have been continuously exploited for many centuries. We are thus usually obliged to consider the two categories together, but this is of less importance than might appear at first sight because of the tendency to convergence between them. An oak plantation on a soil and in a climate which would naturally support oakwood becomes in time substantially similar to, if not indistinguishable from, a natural oakwood exploited in the same way (cf. Chapter IX, p. 198). What is more important is to realise clearly and to keep constantly in mind the artificial conditions introduced and maintained by human occupation of the country and affecting most of our semi-natural woods in different degrees.

Effects of enclosure by agricultural land. First of all the woods which still exist in the predominantly agricultural regions are sharply limited and enclosed by arable fields and pastures. This enclosure and proximity have two effects: the natural spread of the woods is prevented and they are laid open to invasion by "weeds" wherever open well-lighted soil occurs within the wood. Partial destruction of the canopy by felling, or of the shrub layer by over-coppicing, would quickly be repaired by regeneration under natural conditions, but under the existing conditions described below such regeneration is slow or fails to occur. If the floor of the wood is exposed, especially on a clay or clay loam, the woodland humus is destroyed and the surface tends to "bake" and crack, rendering it unfit for many woodland plants and for the germination or establishment of shrub and tree seedlings, while the establishment of various light-demanding species is encouraged—species which may cover the soil with a thick tangle of vegetation altogether alien to the woodland flora, and incidentally may prevent the seeds of the trees from reaching the soil at all.

Lack of seed parents. A further and very important effect of the occupation of the vastly greater proportion of the country by grassland and arable instead of by trees is the great diminution in the number of seed parents, and hence of the seed available for colonising any open grassland or heath which is not so heavily pastured as to destroy all seedlings of woody plants.

Destruction of carnivorous animals. Another effect of the first importance is the interference with the "balance of nature" brought about by the wholesale destruction of carnivorous birds and mammals by the gamekeepers. Of the woods which remained after the widespread destruction of the original forests the larger were maintained for hunting (the "Royal Forests" and "Chases"), and the smaller for the supply of timber and underwood to meet the needs of the countryside. But with the great rise of population and the increasing predominance of the towns the native woodlands, even when supplemented by plantations, became quite inadequate to the demands for timber, which was increasingly imported. With the extermination of the larger wild mammals—wolves, deer, pig— hunting was reduced to specialised and artificialised fox-hunting, which, together with the even more artificial pursuit of pheasant shooting, became almost the only blood sports available to the wealthier country residents.

As the result of these changes the deciduous woods have lost their connexion with the natural economic life of the countryside and are now maintained mainly as fox coverts and pheasant preserves, though the coppice still has a certain use as small wood for fencing and similar purposes.

The activities of the gamekeepers in charge of pheasants are largely directed to combating the natural enemies of the eggs and young birds, and this involves systematic destruction of the carnivorous birds and mammals such as hawks, owls and crows, stoats, weasels, badgers, hedgehogs and shrews. The fox alone is almost exempt, because it is preserved for hunting, though its raids on the pheasants and on the farmers' poultry sometimes lead to "illicit" shootings. The severe reduction in numbers of all these carnivorous animals has led to an enormous increase of the small herbivorous mammals which form a large part of their food, notably voles, mice and rabbits, and to a less extent squirrels. The rabbits are even preserved for "rough shooting" on many estates, and unfenced or imperfectly fenced "rabbit warrens" are maintained on suitable soils. These rodents destroy an enormous number of the seeds and seedlings of trees and thus largely sterilise the power of the wood to regenerate (see Chapter VI, pp. 140–2).

Thus the capacities of the native woody vegetation to occupy and maintain itself in the areas which naturally belong to it are seriously impaired and often nullified by limitation of the ground open to colonisation, sometimes by deterioration of the soil, and especially by diminution in the number of seed parents and by destruction of the seeds and seedlings.

Effects of methods of exploitation. Other more special effects on the composition of woods are due to the manner in which they are exploited. By far the commonest culture-form in the south, especially the south-east, and parts of the midlands of England is coppice with standards. The great majority of these woods are hazel coppice with oak standards. Their nature and origin together with their vegetation, which differs distinctly though not fundamentally from that of "high forest", are described in Chapters VIII and XIII (pp. 181, 270–1). When high forest is selectively felled and left to regenerate naturally the effect of the removal of certain trees depends of course upon what trees are removed. If only mature trees are taken out and conditions for regeneration are good, the general composition of the wood will not be altered. But if in a mixed wood—for example, of oak and beech—the mature trees of one species alone or mainly of one species, are felled, it is clear that apart from the immediate change, the proportion of the species chosen for felling will also be reduced because of the diminution in the number of seed parents. It is more than probable that the composition and even the dominants of some of our woods have been changed by this sort of exploitation (cf. pp. 249–50, 424). Such selective felling may be done in order to obtain the timber which is in demand at the time, or on the other hand with the object of cutting out as far as possible a species whose timber is not wanted at all, so as to give more room for the growth of the useful trees. In the latter case the cutting will include young growth as well as mature trees of the unwanted species, and may result in its partial or complete extermination from the wood.

The dominant trees of the British woodlands

The two oaks. The two native British species of oak, *Quercus robur* L. (*Q. pedunculata* Ehrh.) and *Q. petraea* (Matt.) Liebl. (*Q. sessiliflora* Salisb.), are the most abundant and characteristic of forest-forming British trees and dominate the most extensive woods. It is true that a great many of the existing trees, probably the great majority of pedunculate oaks, have been planted, but we cannot doubt that they occupy for the most part the place of native oak forest, which, as we know from definite records and from many descriptions and allusions, covered very great areas within the historical period. Oak forest, as we can deduce from the occurrence of its pollen in post-glacial peat, has been the predominant British forest community on most soils at the lower altitudes since it established itself through the greater part of the British Isles perhaps 7000 or 8000 years ago in late Boreal and early Atlantic times.

The pedunculate oak. Of the two species *Q. robur*, at least in the south and midlands of England, affects the deeper, and of these the damper and heavier, soils, though it grows fairly well also on relatively poor and acid sands, provided they are not too dry. It is the dominant oak on moist clays and loams, and can colonise and even establish dominance in woods where the summer ground-water level is not far below the surface. While it is most characteristic of the heavier soils with a relatively high content of mineral salts and often a considerable proportion of calcium carbonate in the subsoil, it can tolerate a low salt content and a comparatively high degree of acidity. For example, it is the dominant tree and the only oak in Wistman's Wood on Dartmoor—undoubtedly a natural wood—on a soil derived from granite boulders and with a pH value of about 5. In correspondence with its hereditary deep-rooting habit, the pedunculate oak fails to form good trees on shallow limestone soils, and it is absent from very acid peats.

Growing in close canopy the tree makes a long straight bole and narrow crown, and on good deep soils in protected stituations may reach a height of over 100 ft. (30 m.), extreme heights of 140 ft. being recorded. A common height of mature well-grown oaks in close canopy is 70–80 ft. (21–24 m.). The "zigzag" branching, due to the persistent growth of single lateral buds making a comparatively wide angle with the parent axis while the terminal bud of the axis dies, is well known. It is best seen in oaks grown in the open whose lateral branches are well developed and whose spread equals or exceeds their height which is rarely greater than 50 or 60 ft.

Climatically it is a tolerant tree, reaching 63° N. lat. in Norway; and though it does not attain so high a latitude farther east, the pedunculate oak can clearly withstand considerable "continentality" of climate, for it extends right across central Russia to the Ural Mountains. South-eastwards it reaches the Caucasus and Asia Minor, and southwards parts of North Africa, but like all our native deciduous trees it avoids the Mediterranean climatic region. In the lowlands of the British Isles it nowhere reaches its climatic limits, but it does not normally extend much above 1000 ft. (300 m.). It is in fact a lowland tree in Europe generally, characterising on the continent a zone below that dominated by the beech.

The sessile or durmast oak. *Quercus sessiliflora* (or *Q. petraea* as we are now told we must call it) is more sensitive climatically, since eastwards it does not reach Russia, and northwards it only extends to 60° N. lat. in Norway; but this species also can comfortably tolerate the whole range of the British and Irish lowland climates. It is also more limited by soil than *Q. robur*, usually avoiding the heavier clays and loams, but often becoming abundant or dominant on the lighter, drier and more acid soils of the south of England. It is said to inhabit the acid London Clay in the dry climate of north-eastern Kent (Wilson, 1911). While the sessile oak thus occurs freely on many sands and sandstones in the south and the midlands of England, and so far bears out its continental reputation of preferring drier soils, it is also dominant on the shallow siliceous soils of the western and north-western hills to the exclusion of *Q. robur* (except where the latter is planted). These soils are, it is true, usually well-drained, but the climate is often excessively wet,[1] and most of the western Querceta sessiliflorae (or petraeae) are probably essentially native woods. The height growth is about the same but the zigzag branching not so well marked as in *Q. robur*. The woods of *Q. sessiliflora* on the western and northern siliceous hillsides are low in stature, the trees often only 30 or 40 ft. high.

The two species frequently grow together in the south-east, especially on deep sands. They are closely allied and hybridize freely, producing a great variety of intermediate forms. They are however perfectly "good" species, and where either forms pure woods (as *Q. robur* does on the clays and loams of the midlands and south of England, and *Q. sessiliflora* on the western siliceous hillsides), there is no difficulty in distinguishing the species by their leaf characters. The greater part of their geographical range is the same and their average requirements must be closely similar. The natural distribution of the two oaks has been considerably confused because of the extensive planting, and on the whole it is *Q. robur* which has been most planted. Elwes and Henry say that all the specimens of this species they received from Ireland were from planted trees.

The oaks, as is well known, are on the whole light-demanding trees, and

[1] Recent (unpublished) experiments have shown that first year seedlings of the sessile oak tolerate wet soil better than those of the pedunculate oak.

except when young and growing in close stand do not cast a very deep shade. Consequently both shrub and field layers are relatively rich, and, owing also to the wide variety of soils on which oakwoods grow, the total number of species of the oakwood communities considerably exceeds that of any other type of British woodland. The oakwood flora is the typical British woodland flora.

The common culture form of oakwood in the midlands and south is coppice (mainly hazel coppice, but often containing a variety of other shrubs) with standard oaks growing in open canopy or even more isolated; and the periodic increase of illumination due to the cutting of the coppice favours the growth of an abundance of "marginal" woodland plants, as well as the periodic increase of the typical gregarious "prevernal" species, such as mercury, wood anemone, primrose, bluebell, etc., which give beauty and character to our spring woodlands.

The beech. The beech (*Fagus sylvatica* L.) is more restricted geographically than either of the oaks, though its total range is not very much less than that of *Quercus sessiliflora*. It requires considerable atmospheric moisture and therefore avoids all strongly continental climates with hot dry summers, being absent from the Polish plains, and poorly developed in the "continental wedge" of north Germany. Like the two oaks it is absent from the Mediterranean region, except in the mountains where the climate is cooler and moister. Northwards it is well developed in southern Sweden, but only just gets into Norway. It is absent from most of south-western France but reappears on the northern Spanish mountains. On the continent generally, and especially towards the south, it is mainly a tree of the lower slopes of the mountains, sometimes mixed with the silver fir (*Abies alba*), dominating an altitudinal zone above that of the oaks and below the spruce forests. It is present in some abundance on the north-west German plains.

The graceful habit of the beech is well known. The branches are markedly sympodial, and the contrast between the foliage shoots developed in shade, which often show very complete and beautiful "leaf-mosaics" with flat plates of foliage set at right angles to the maximum incident light, and the sun shoots with their obliquely ascending stems and smaller and thicker leaves is very striking. From a distance beechwood can always be recognised by these slender, switch-like shoots rising above the massive canopy.

Present distribution of beech. In Britain the beech is now apparently confined as a native tree to the south and mainly to the south-east of England. It is not native in Scotland, or in the north of England, or in the greater part of the midlands. It is very doubtfully indigenous in south-west England (though it apparently just extends into south Wales), and entirely absent from Ireland. But this restriction does not represent its climatic limits, for not only does the tree flourish perfectly well when planted in favourable situations in all these regions, it regenerates freely from planted trees in Aberdeenshire, and even sets ripe seed at low altitudes as far north as Caithness in the extreme north of Scotland. The causes of its present limitation are not accurately known. Lindquist suggests destruction of the flowers by late frosts in Scandinavia, and this may also be a factor in parts of north Britain. And it has been plausibly suggested, though there is no satisfactory proof, that the spread of the beech into the more extremely oceanic regions is hampered by the coolness of the summers leading to depression of the power of ripening mast. The absence of the beech from Ireland may well be due to its having had insufficient time to get there since its arrival in Britain, at least before human activity interfered with the unrestricted spread of trees, for it was certainly late among post-glacial invaders. But we now know for certain that beech wood was once more widely spread in England and Wales than at present, though its former limits of distribution have not yet been determined. Its restriction to the south-east may be due to deterioration of climate combined with human interference, and some of the scattered trees to the north and west of the existing native beechwoods may be relicts of a much wider extension.

Soil preferences. The soil preferences of the beech are often misunderstood in this country. It is sometimes said that beech is confined to chalk and other limestone soils, and it is a fact that by far the greater part of our native beech occurs in the chalk regions of south-eastern England. But much of this, including the finest growth, particularly on the plateaux of the Chiltern Hills and of the South Downs, is rooted in overlying loams which are often poor in or destitute of calcium carbonate and may be markedly acid in reaction. Beech will actually flourish on a considerable variety of soils, and it flourishes best, as most trees do, on a deep, well-drained, fertile loam containing a good supply of mild humus.

Beech avoids heavy and especially waterlogged soils, in this respect resembling the sessile oak, from which, however, it differs in its power of making good growth on shallow limestone soils. On this last type of soil it easily beats all its native British competitors, on the one hand by its power of spreading its feeding roots in very shallow soil, and on the other because it casts a very deep shade. Beech also grows and becomes dominant locally on sands and sandstones, both in the south of England, and sub-spontaneously in Aberdeenshire and elsewhere. On podsolised sands it is also locally dominant, but here it has difficulty in regenerating because of the surface crust of raw humus formed under its shade on these soils, partly from the beech leaf litter and partly from mosses, so that it is doubtful if the tree can maintain itself under these conditions beyond the first generation.

Relations of beech and oak. The relations of beech with oak are not entirely clear. Under conditions equally favourable for both species beech can obviously overcome oak in direct competition, because its height growth is equal or even superior (100–120 ft., extreme 140 ft.), and because it casts a much deeper shade. On the loam-covered plateaux of the South

Downs beech actually suppresses ash-oak wood and forms the climax stage of the succession. But in some places oak (*Q. robur*) maintains itself in the canopy side by side with beech, at least for a long time, and many other similar soils are occupied by pure oakwood (in the form of oak-hazel coppice), sometimes containing isolated beeches. It is probable that human action has largely contributed to this result. Some authorities in Germany believe that man has favoured oak at the expense of beech, particularly in the north-west of that country, and that this largely accounts for the decrease of the latter tree, which was more abundant in that region in former times, as shown for instance by the records of pollen in the later post-glacial peat. During the Middle Ages, and later, it is clear that hazel coppice with oak standards was particularly favoured in southern England, and it has since been often planted. Much of it is found on the "clay-with-flints" and other loamy soils overlying the chalk, some of it on soils which can (and do elsewhere) bear dominant beech. But the evidence tends to the conclusion that on the heavier soils oak has a real advantage over beech, and we may thus envisage a series of forest types, passing from pure climax beechwood, through types developed on soils where the two species compete on almost equal terms and in which therefore both would be represented in the climax, to others in which the oak is naturally dominant. For these reasons it seems correct to speak of the "oak-beech association" in Clements's sense, of which pure oakwood and pure beechwood are consociations.

The shrub layer of beechwood is at the best very poor and sometimes non-existent, but on calcareous soils the scanty shrub flora contains a larger proportion than oakwood of the "calcicolous" shrubs, usually suffering severely from the deep shade, and rarely or never flowering. The field layer also is often very poor and indeed practically absent from the densest (particularly the young) beechwood—but on calcareous soils and where more light reaches the ground it is rich in species, of which a considerable number are "calcicoles" and some saprophytes. A few species are recorded only from such beechwoods.

Mosses are mainly confined to the bases of the tree trunks and to bare soil immediately round the roots of the trees, the heavy and rather persistent beech-leaf litter excluding them from the general floor of the wood.

Sharply contrasting with the calcicolous beechwoods are those on acid (often podsolised) sands, which have scarcely a single species in common with the former type.

The two birches. The two native species of birch, *Betula pendula* (*B. alba* L., *B. verrucosa* Ehrh.) and *B. pubescens* Ehrh. are very abundant in the British Isles on many though not on all soils. *B. pendula*, the so-called silver birch, is commonly the more abundant in the south and does not reach the north of Scotland or the north of Ireland, while the polymorphic *B. pubescens*, which is also locally abundant in the south, is much the commoner tree in the north and west. The birches of the scanty birchwoods which are the only forest type represented in the extreme north of Scotland have generally been regarded as forms of *B. pubescens*, but Dr E. F. Warburg informs me that he considers they belong to a distinct species. The distribution of the British birches is in accord with their continental occurrence: *B. pendula* (*alba*) is said not to extend beyond the 65th parallel in Scandinavia and the 60th in north-east Russia, while *B. pubescens* goes farther north. Extreme northern forms (*B. tortuosa*, etc.), analogous with our north Scottish forms, reach Iceland, the North Cape at 71°, the Kola Peninsula at 69°, and 67° in north-east Russia. These northern birches may form the polar forest limit, and in northern Scandinavia, especially, the broad belt of birch forest north of the last conifers is a marked vegetational feature. Southwards in Europe the birches, like the beech and the two oaks, avoid the Mediterranean climatic region, but *B. pendula* goes considerably farther south than *B. pubescens*. Both extend through Siberia to the Pacific, and the two together have a general distribution very similar to that of *Pinus sylvestris*. On the continental mountains *B. pendula* may be dominant up to 500 m., and *Betula pubescens* up to 1000 m. in the Hercynian region of central Germany; both birches ascending much higher in the Alps.

In Britain *B. pendula* grows on a wide variety of soils, but is rare on chalk: it is often, however, frequent on the less pure limestones and forms an upper altitudinal zone above the ashwoods of the Carboniferous Limestone in the north of England (northern Pennines). *B. pubescens* is associated in the same way with *Quercus sessiliflora* on the siliceous hills of the west and north. The birches are pioneers (along with pine) in the passage from heath to forest. Both species are extremely common on the sandy soil of heathland; in some places one species is prevalent, in some the other, and often the two are mixed and accompanied by numerous hybrids. In this respect they recall the behaviour of the two oaks on the same soils, and like these can be easily distinguished by leaf characters, which, however, naturally become much confused in the hybrids. *Betula pubescens* is a pioneer, and sometimes forms woodland, on drying raised bogs. It also occurs and is sometimes dominant in the fen woods ("carrs") of East Anglia, where the peat is distinctly alkaline in reaction. Besides all these situations where birches are or may be dominant, both species, but especially *B. pendula*, are very common associates of the oak in oakwoods on a great variety of soils, though not when the oaks are in close canopy. They are among the commonest native English trees, especially on sandy soils, owing to their free seeding, easy dispersal, rapid growth and unexacting requirements. But they cannot withstand competition from species casting a deep shade.

Birch forest forms an altitudinal zone above, and a latitudinal zone to the north of, oak forest in the British Isles. It may also form a distinct phase in the succession to oak forest on light or sandy siliceous soils (e.g. on heaths). Thus it is "pre-climax" to oak in Clements's sense, in much the same way that ash is pre-climax to beech.

The ash. The ash (*Fraxinus excelsior* L.) is another of our commonest native trees, occurring throughout the British Isles except the extreme north of Scotland. In Europe it has nearly as wide a distribution as that of the pedunculate oak, extending eastward beyond the middle of European Russia, and with much the same limits as the oak towards the Arctic and the Mediterranean.

As a dominant forest-forming tree it forms pure ashwoods, which appear to have no exact parallel on the continent, on limestones of various age, as for example on hillsides of the Carboniferous Limestone in Somerset, Derbyshire, the northern Pennines, Wales and Ireland; and on the narrow outcrops of Upper Jurassic limestones in the south and east midlands. It is also an important associate in oakwood, tending to become abundant or dominant wherever the soil is highly calcareous. For example, on the Chalky Boulder Clay of eastern England the oakwoods contain abundant ash, which is, however, practically always coppiced along with the hazel and other shrubs: in the sessile oakwoods of the impure Silurian limestones on the borders of Herefordshire ash is very abundant and often better grown than the oaks; and on the calcareous Upper Greensand of the south and south-west it is equally abundant with or more abundant than oak. Many of these woods are in fact properly called ash-oak woods, the ash quite holding its own with the oak, and some are practically pure ashwoods. Its light demand is but little higher than that of oak.

On soils which are not specially calcareous ash increases and often forms distinct societies wherever the soil of an oakwood is specially damp, as in local depressions or along stream sides. It is also often abundant in wet alluvial woods, as on the continent (Auenwälder of the Germans), and in developing carr on the alkaline peat of the East Anglian fens (Bruchwälder). In the oakwoods of the north-western hills, as in the Lake District with its very wet climate, ash is often exceedingly abundant. In such habitats it seems to resemble alder in its requirements.

Finally, in certain parts of the chalk region of southern England ashwood forms a definite successional stage leading to beechwood. Here ashscrub or pure ashwood occurs on the shallow chalk soils (rendzina), and ash-oak wood following pure ashwood is developed on the loams overlying the chalk. In both cases beechwood forms the climax of the succession. On the South Downs generally ash is extraordinarily vigorous and ubiquitous. Comparing this habitat with the climax ashwoods of the Carboniferous Limestone outside the beech region we see that ash is preclimax to beech on calcareous soils just as birch is to oak on the siliceous. On parts of the Chalk formation, however, while never entirely absent, it is by no means abundant, and this seems to be correlated with slight differences of climate, the ash showing its greatest vigour of growth and reproduction in the more humid regions nearer the sea. This climatic moisture may compensate for the soil moisture which the tree affects in non-calcareous habitats.

Ash demands a good supply of mineral salts and avoids all markedly acid soils; it is notably absent both from the wet acidic humus of the north-western moory soils and from the dry acidic sands of the south-east.

Owing to the light shade cast by the dominant the ashwoods of the Carboniferous Limestone have shrub and field layers varied and rich in species, which are largely calcicolous, many the same as those of the southern calcicolous beechwoods.

The common pine. The common pine (*Pinus sylvestris* L.), generally known in this country as the "Scots pine" or "Scotch fir", is one of the hardiest and least exacting of trees. It has a very wide range in Eurasia, from Scotland to the Lena valley in Siberia, and from Lapland to the Mediterranean mountains, and includes several different varieties or geographical races. It is unable to compete with such trees as spruce, oak or beech on the better soils, and is supposed to find difficulty in establishing itself in the more extremely oceanic climates. It is not now native in Ireland, though it was abundant in post-glacial periods such as the Boreal and Sub-Boreal when the climate is believed to have been drier; and it has disappeared as a native tree from several of the coastal regions of western Europe during the last few centuries. In Great Britain it is now confined, as a native tree, to the Scottish Highlands.[1] The pine forms extensive pure woods in certain forests of the central and eastern Highlands, for example Ballochbuie Forest at the foot of Lochnagar in Aberdeenshire, the Black Wood of Rannoch in Perthshire, Rothiemurchus Forest and neighbouring forests and Locheil Old Forest in Inverness-shire (now destroyed). Remains of old pine forest also occur in several places in the northern Highlands. These forests contain the apparently endemic variety *scotica* (Elwes and Henry), of *P. sylvestris*, with short grey needles and small cones. Unlike the common pines, planted and subspontaneous, of southern England, whose mature crowns are usually flat-topped, the Scottish variety (or at least some forms of it) maintains the pyramidal growth for a long time and eventually forms an ample rounded crown (Pl. 81, phot. 188). Other trees seem to assume a spreading habit from the first. The pyramidal variety is like other northern and also alpine forms of the common pine which are characterised by the pyramidal shape: the flat-topped forms of the south are liable to be broken down by the weight of snow which accumulates during the long and severe winters of the sub-arctic and sub-alpine climates. The Scots Pine is known to have occupied much more extensive areas in Scotland within recent historical times, but was largely cleared before the modern planting of conifers began. It is noteworthy that its present greatest abundance is in the most continental regions of the Scottish climate, namely, the central and eastern Highlands.

[1] This is the generally accepted view, but it is becoming increasingly probable that a good deal of the southern pine is relict from former native pinewoods.

P. sylvestris was present in quantity and was almost certainly dominant, at least on the lighter and drier English soils, in Boreal times, and it certainly lingered on when pine forest had been largely displaced by deciduous forest in the next (Atlantic) period. During the Sub-Boreal it seems to have made a fresh advance in several places, for its pollen, and also stumps and cones, are abundant in the Sub-Boreal peat of such widely separated regions as the East Anglian fens, the west of Ireland and the Scottish mountains. Thereafter it again decreased, but may well have persisted at least into early historical times. Old unrotted pine stumps uncovered some years ago in a dried-out bog at Sutton Park near Birmingham were in very recent peat attributed to the Roman period. Between Durham and Gateshead, on the eastern flank of the northern Pennines, John Leland (about 1540) records "Firres" or "Fyrres" (Taylor in Darby, 1936, p. 347). These may have been the last remains of native Scots pines in the north of England. It has been suggested that the *Pinus sylvestris* of the Wealden and Tertiary sands of south-east England, where it flourishes and springs freely from self-sown seed (certainly often derived from planted parents), may be descended partly from this southern native stock; and it is possible that, as with the beech (p. 249), some of the scattered pines in the west of England and in Wales may be relics of original native pinewoods. It is noteworthy that both in Belgium and Denmark, in neither of which is the pine now indigenous, the native tree is known to have disappeared only in quite recent historical times.

Habitats. Though it can colonise sporadically very various soils, including even fen peat and chalk, the pine is much more at home on the southern English sands than anywhere else in Britain except in the old Scottish pine forests. In both of these habitats it is associated with the heath formation and can form close forest. This is in accord with its usual habitats on the continent, where it is specially characteristic of the sandy, heathy soils of north-west and north Germany, and also of the drier and shallower soils of the mountain slopes where the more exacting species such as spruce, larch and silver fir cannot make good growth. It also commonly forms a stage in the succession to climax conifer forest, as for example in the spruce-pine-birch forest of central Sweden, where it follows birch and is followed by spruce (*Picea abies*), though the succession is a good deal telescoped, the stages being not at all sharply separated. Through much of its range the common pine may be considered preclimax to spruce.

The climax forests of *Pinus sylvestris* on the north German plain as well as the south-eastern English subspontaneous, and some of the native Scottish pinewoods, are undoubtedly mainly attached to sandy and gravelly soils, poor in nutritive salts and acid in reaction, and here it is associated with a heathy field layer often dominated by *Vaccinium myrtillus* or, where the soil is extremely poor, only with *Cladonia* and a few scattered xerophytic plants. But the common pine is not excluded by calcareous soils since it is characteristic of the poor marls of the French Champagne country, and isolated individuals often sow themselves on the English chalk by seeding from planted parents. On the better, more loamy soils, pine begins to give way to other trees, and in Britain pinewood, whether native or subspontaneous, never now establishes itself abundantly except on heath.

The requirements of the common pine are very close to those of the birches, the species being constantly associated. The ecological requirements of the oaks are also not far removed from those of the pines, as can be seen from the fact that pinewood may pass into oakwood and that oak and pine are often mixed, both in Europe and in North America. On sandy soils in south-eastern England beech may also enter into the mixture (compare the account of "oak-birch heath" on pp. 354–7).

The common alder. The common alder, *Alnus glutinosa* Gaertner, extends through most of Europe, except the Arctic, the Mediterranean region (though it is said to grow in Sardinia) and the Russian steppes. It also occurs in Asia Minor, north Persia and even western Siberia. To judge from the records of its pollen in the peat alder was very much more abundant through most of the post-glacial period, especially the Atlantic and the early Sub-Atlantic, than it is now. This is not difficult to understand when we realise that the extent of marshy ground must have been many times greater before systematic draining of the land was undertaken. In periods of wet climate it may also have spread widely through the general woodland, as it still does in Wales and western Scotland. The alder still grows in wet places in woods dominated generally by other trees, and by streams and lakes where humus and nutritive salts are available, widely distributed through almost the whole of the British Isles, and it still forms small alderwoods where such habitats occur. In the wet climates of the west, as in Wales and the western Highlands, alder may be extremely abundant throughout the oakwoods. Some of the most notable alderwoods still existing in England are the "carrs" of east Norfolk in the region of the Norfolk Broads (corresponding with the "Bruchwälder" of Germany) in which it grows on wet peat or humous soil rich in mineral salts. It is absent from the acid peat of raised bog, though it often forms a narrow belt bordering the stream or ditch (lagg) of open water which is commonly found on the edges of such areas, and whose soil is richer in nutrient salts.

In the Norfolk carrs the dominant alder is often accompanied by ash, often also by the pubescent birch and sometimes by subspontaneous pine, while isolated pedunculate oaks may also occur. Of the shrubs of the alder carrs the most abundant is the grey sallow (*Salix atrocinerea*), while *Frangula alnus* and *Viburnum opulus* are also characteristic. In the field layer the common stinging nettle (*Urtica dioica*) is the most abundant plant (as also in the German "Bruchwälder"), together with the common flag (*Iris pseudacorus*), meadowsweet (*Filipendula ulmaria*), neither usually flowering, *Carex paniculata*—whose great tussocks are rather constantly present in alderwoods, however small—*Dryopteris thelypteris*, and a variety of other marsh or fen plants.

Where the ground-level is rising by the accumulation of humus above the water table alderwood is invaded by oaks and probably eventually passes over into oakwood with the transition to drier conditions.

Other trees

Yew. None of the remaining trees of our rather scanty arboreal flora dominates any considerable tracts of British woodland. But there are a few which achieve local dominance in some situations. Of these one of the most striking is the yew (*Taxus baccata* L.), whose broadly pyramidal habit and flat horizontal plates of dark foliage are well known: it was once more widely spread in Europe than now. Schoenichen (1933) estimates that there are still about 25,000 yews in Germany. Including young growth there are certainly far more than that in the much smaller area of the British Isles. This tree is very local throughout its European range, avoiding the regions with markedly continental climate (for it is decidedly hygrophilous), and in central and southern Europe following certain mountain ranges, from which it extends eastward to the Caucasus, north Syria and north Persia. In many, though not all, of these regions it is confined to limestone. In northern Europe yew is fairly abundant on the south Baltic coast, and extends westwards through southern Scandinavia to the British Isles where it occurs locally in southern Scotland, England, Wales and Ireland. In some limited areas it is abundant, mainly though not exclusively on limestone. By far the most important of these areas is the Chalk of south-eastern England, and especially the western Sussex downs, where it is very abundant indeed. The yew frequently colonises chalk scrub, and on the South Downs of west Sussex and Hampshire, to a less extent also on the North Downs of Surrey, it forms pure local woods, excluding other trees owing to the dense shade of the foliage, and persisting for a long time because of the long life of the individual trees. This may exceed 800 years, though it is probable that the age of yews has often been overestimated. Little or no vegetation can exist under the shade of close yew wood. An account of these yew woods is given in Chapter XVIII, pp. 376–80. Very many of the yew trees in Britain, even in the chalk area, are however isolated individuals, generally either in the chalk or limestone scrub or in ashwood or beechwood, because of the common preference for limestone soils. Under the shade of beech the yew can survive and grow but takes on a shade habit and does not flower. Occasional yews occur in oakwood, and the tree can apparently sow and establish itself easily in the British climate in suitable situations.

Hornbeam. The hornbeam (*Carpinus betulus* L.) has a more limited distribution, both in the British Isles and on the continent, than any of the trees hitherto considered, though it goes farther east and especially southeast (extending to the Caucasus, Asia Minor and north Persia) than the beech or the sessile oak. Its distribution most nearly resembles that of the beech. Northwestwards it only extends to the extreme south of Sweden and the south-east of England; and westwards it does not reach the Atlantic coast. It is much more of a lowland tree than the beech and is mainly important in north-west, north-east and central Germany, where it sometimes forms pure or nearly pure hornbeam woods. Though hornbeam can grow on a considerable variety of soils it is chiefly associated in Britain with one or other of the oaks on moist silty sands or sandy clays, and occurs in the greatest abundance in Kent, east Sussex, Essex, Hertfordshire and Middlesex, where it is nearly always coppiced. Coppiced hornbeam sometimes dominates the shrub layer, either as a co-dominant with hazel or to the exclusion of the latter. Only isolated trees are allowed to grow up; but it is probable that hornbeam would in no case dominate the oaks with which it is associated since its height growth is not so great, though it casts a deeper shade. The wood of hornbeam is difficult to work and very hard. It was formerly employed for many purposes for which metal is now used. Hornbeam was, and still is, much used for arbours and hedges, since it stands constant trimming extremely well. As a firewood hornbeam is in the first rank (Miller Christy, 1924). Like the beech, hornbeam was a late post-glacial arrival in England.

Holly. The holly (*Ilex aquifolium* L.) is one of the very few British evergreen trees, and by far the most abundant. It is distributed throughout the British Isles except the extreme north and is often abundant in the woods, notably so in the west owing to the dampness and mildness of the climate. Holly occasionally forms an almost continuous subordinate tree layer, and sometimes a shrub layer when its growth is limited by deep shade or insufficient soil moisture. In the Killarney oakwoods it is characteristically subdominant to the oaks (Chapter XVI, p. 328), locally in oakwoods elsewhere and in the Sussex beechwoods, but it may occur in almost any wood as far as the north of Scotland, except on very wet soil. Occasionally it forms pure local woods whose origin and status are not known.

The maples. The sycamore (*Acer pseudo-platanus* L.) is not a native, but it has been very freely planted in Britain during the last two centuries or so, though it is said to have been introduced much earlier: locally it springs very abundantly from self-sown seed, and owing to its free growth and deep shade it may become locally dominant in various kinds of woodland. In some of the South Down beechwoods on deep loams sycamore may even become co-dominant with beech, recalling the beech-maple forests of eastern North America. The field maple (*Acer campestre*) is not a gregarious tree and does not reach the north of Scotland. It is specially abundant on calcareous, and avoids acid soils. Though nearly always coppiced it can, when let alone, form a good tree 50 or 60 feet high.

Wych elm. Wych elm (*Ulmus glabra* Huds. = *U. montana* Stokes), a magnificent widely branching and lofty tree when well grown, is specially abundant and sometimes forms societies in the limestone ashwoods and ash-oakwoods. For the rest it is scattered as an occasional tree through the oakwoods, occurring more frequently in the north and west than in the south.

Other elms. The other species of elm met with in Britain scarcely form part of any natural or semi-natural woods, and their origin and status is obscure. The smooth-leaved elm (*U. nitens* Moench) is a good species commonest in eastern England though not now entering into the composition of surviving semi-natural woods; and the small-leaved elm (*U. minor* Miller) locally forms pure scrubby communities in parts of the east. The English elm (*U. procera* Salisbury) and the Cornish elm (*U. stricta* Lindley) do not enter into semi-natural vegetation at all.

Rosaceous trees. Rowan or "Mountain Ash" (*Sorbus aucuparia* L.) is widely distributed throughout the British Isles, but in the south is no more than an occasional constituent of oak or other woods, especially on the lighter soils. In the north and west it is more abundant, and in Scotland it is a very frequent constituent of birchwood and is sometimes very abundant, even rising to local dominance (Rowan-birchwood, Anderson, 1932). It is extremely resistant to wind, and individuals ascend to higher altitudes (2600 ft. = c. 800 m.) than are reached by any other British tree.

Two other species of *Sorbus*, the whitebeam, *S. aria* (L.) Crantz, and the wild service, *S. torminalis* (L.) Crantz, occur mainly in the south, the former being abundant in the south-eastern chalk scrub and sometimes persisting in the canopy of beechwood, the latter rather local in oakwoods and ash-oakwoods: neither is dominant in any type of woodland or scrub. Other species of *Sorbus* closely allied to *S. aria*, of very limited distribution and some very rare, occur on the hills of the north and west, and in Ireland.

The crab apple (*Pyrus malus* L.) is occasional and sometimes frequent in oakwoods, especially in the south of England, while the pear (*P. communis* L.) is rare and probably always derived from planted trees whose seeds have been dropped by birds.

The wild cherry or gean (*Prunus avium* L.) is also occasional, frequent or locally abundant in oakwoods and also in the woods on limestone soils; in parts of the Chiltern beechwoods it is especially conspicuous and may actually enter the canopy. The bird cherry (*P. padus* L.) is quite frequent in the woods at higher altitudes in some parts of the north and west, but does not occur wild in the south. The true cherry (*P. cerasus* L.) is apparently indigenous in the south, and occasionally met with in woods. The plum (*P. domestica* L.), like the pear, is probably always bird sown from planted trees.

The limes. Of the two species of lime *Tilia cordata* Miller occurs with some frequency and *T. platyphyllos* Scopoli more rarely in the limestone woods of the west. Neither is now wild in the south-east nor in Scotland. Pollen of *T. cordata* is so abundant in the peat formed in the Atlantic period that we must suppose the trees to have formed an important constituent of the oakwoods at that time.

Willows. Of the willows most native species are shrubs, but the crack and white willows (*Salix fragilis* L. and *S. alba* L.) are good-sized trees commonly planted along the sides of embanked rivers and canals, where their roots are useful in holding the soil together. There is however every reason to believe that both species are native in wet and marshy soils (*S. fragilis* more widely spread and commoner than *S. alba*), though much the greater number of existing individuals are planted trees.

Poplars. Aspen (*Populus tremula* L.) is a widely distributed tree, but avoids the more extreme soil types. It forms local societies in various woods throughout the country. This tree is markedly gregarious as well as quick-growing, and suckers freely: it may take the place of birch as the first tree to occupy the ground after felling. The closely allied American aspen (*P. tremuloides* Michaux) has a similar role, and is also dominant in a considerable belt of woodland on the edge of the prairie region. *P. nigra* L., the black poplar, is widely but very thinly scattered in Britain and fairly often planted, though nothing like so commonly as the hybrid "Black Italian" poplar (*P. serotina* Hartig). *P. nigra* is however almost certainly native, at least in the valley fen woods of East Anglia. *P. canescens*, the grey poplar, is also thinly scattered in damp oakwoods in the south and in the same valley fen woods, probably native but very often planted.

The more important shrubs

Of the fairly large number of native shrubs, only those that are most important in British vegetation will be mentioned here. Shrubs play two distinct roles in vegetation, forming the shrub layer of the less deeply shaded woods and the dominant layer of "scrub" (Chapter XXIV). Several species may be found in either situation.

Probably the two most abundant and characteristic British shrubs are the hazel and the hawthorn (whitethorn or "may").

Hazel. The hazel (*Corylus avellana* L.) has a wide distribution in Europe, reaching to 63° N. lat. in Norway, well into central Russia, and south-eastwards to the Caucasus. In the British Isles it extends to the extreme north of Scotland, but avoids the more acid and, except on limestone, also the drier soils. It is best known as the dominant constituent of the shrub layer of the lowland oakwoods (generally Quercetum roboris, but also Q. sessiliflorae) where it is habitually coppiced at intervals of 10–12 years in the south-east, in the ordinary culture form of coppice with oak standards. Much of this hazel coppice has been planted, so that the shrub layer is almost pure hazel, but the hazel is abundant in nearly all such oakwoods and in all probability was naturally dominant in the shrub layer of the original lowland oak forest. In the Quercetum sessiliflorae of the northern and western siliceous hillsides hazel is not so abundant, though it is by no means rare, and this is no doubt related to the infrequency or absence of coppice with standards in those parts of the country. Hazel is locally very frequent in roadside and streamside scrub in the north and west. On calcareous soils it flourishes very well: thus it is abundant in the ashwoods on the Mountain Limestone of Derbyshire, where it often forms pure copses, and it is a frequent constituent of chalk scrub in the south, where it is sometimes a pioneer in colonising grassland.

In the Boreal period hazel spread in this country before the beginning of the main period of dominance of the deciduous trees, and its pollen occurs in enormous abundance at certain horizons of the Boreal peats, often exceeding in quantity that of all the trees put together. It is indeed plausibly held that at this time hazel must have formed extensive pioneer scrub in advance of the formation of woodland proper. Pure hazel scrub is not uncommon to-day in places where the conditions are unfavourable for trees, and on the windswept limestone of Co. Clare in the west of Ireland such scrub is of great extent (see p. 473).

The hazel usually grows as a shrub, i.e. it produces multiple stems which grow up side by side. This habit is much accentuated by coppicing, to which the shrub is very well adapted, the stools living for a very long time under repeated cutting of the coppice shoots. Occasionally however only a single stem is produced, which if uncut and growing in close canopy may reach a height of 20 or 25 ft., with considerable branches only towards the top, thus forming a small tree. Neglected hazel coppice may reach a similar height.

Hawthorn. By far the most abundant and widespread species of hawthorn in Britain is the form with a small fruit and usually a single carpel (*Crataegus monogyna* Jacquin.). The *C. oxyacantha* of Linnaeus is believed by some botanists to refer to the large-fruited shrub with two or three carpels, usually known as *C. oxyacanthoides* Thuiller, and nearly confined in England to the south-east. *C. oxyacanthoides* is quite local, though often abundant where it occurs, as on the London Clay near London and parts of the Chalky Boulder Clay in Cambridgeshire. This species is of comparatively restricted sub-Atlantic distribution in Europe, whereas *C. monogyna* extends well into Scandinavia, central Russia, Siberia, the Caucasus, Armenia, Syria, the Himalayas and North Africa. *C. monogyna* is very rare in the north of Scotland, and there is much less of it even in southern Scotland than in England.

Hawthorn is far more abundant in the open than under the shade of trees. It is in fact very much the commonest and most widespread of the pioneer shrubs, colonising a great variety of soils though avoiding acid peats. The ubiquity and abundance of hawthorn in the south of England and the midlands is doubtless partly, and perhaps very largely, due to its almost universal use to form "quickset" hedges. Hawthorn often forms pure or nearly pure scrub alike on clays and loams, on chalk and the older limestone hillsides, and not infrequently on sands. When the bushes are closely set their interlacing rigid spinose branches render the hawthorn scrub practically impenetrable. The spine-bearing habit, of course, is an important help to a shrub in colonising grassland which is lightly pastured, and though hawthorn (whose spines are formed only by the tips of some of the branches) is less resistant to the attacks of grazing or nibbling animals than such shrubs as juniper or gorse, whose leaves are themselves spinose, nevertheless it is not so severely eaten back as shrubs that are free from spines altogether. This may be seen in moderately pastured grassland where isolated hawthorns may survive while no spineless shrubs are present. Apart from this degree of protection, however, no doubt hawthorn is favoured by its abundant fruiting and the free distribution of its fruits ("haws") by birds.

In unpastured grassland and on really favourable soil hawthorn makes dense scrub with clean tall stems 20–25 ft. high, and small crowns in lateral contact, forming a miniature wood which shades the soil so deeply that little can grow beneath it (Plate 93, phot. 226). Isolated in the open a thick bush of much less height is produced. Under the shade of a tree canopy the shrubs are more loosely branched, and bear much sparser foliage.

Blackthorn. Blackthorn or sloe (*Prunus spinosa* L.) is another abundant shrub in the British Isles, only absent in the extreme north of Scotland, though not nearly so important a constituent of woody vegetation as hawthorn. It also flourishes on a wide variety of soils, and often forms local patches of pure scrub, mainly vegetatively owing to its power of suckering from the roots. In this way it spreads out on to the grassland from the edges of woods directly the grazing pressure is lifted, and forms thickets by this process of vegetative "edge colonisation". Blackthorn does not fruit nearly as freely as hawthorn, so that its reproduction other than vegetative is much more limited. Nor does it grow so tall or survive shading so well. Blackthorn is a common constituent of mixed spinose "thicket scrub" in company with such species as hawthorn, gorse, bramble and rose.

Gorse, furze or whin. There are three species of *Ulex* native to the British Isles, *Ulex europaeus* L., *U. gallii* Planchon and *U. minor* Roth. In common with the other species of the genus they are essentially western European plants, not penetrating far eastwards into the continent. *U. europaeus*, the common gorse, furze or whin, a fairly free-growing shrub generally 4–6 ft. high but sometimes taller, is by far the commonest and most widespread. It occurs throughout the British Isles except the extreme north of Scotland, very largely on disturbed soil, as by waysides and on the edges of commons and heaths. At one time it was much planted for cattle fodder, the spinose shoots being cut up in specially designed mills; and is now often used as fox covert. The common gorse travels most readily along paths and tracks, it is believed because ants, attracted by the "caruncle", a succulent orange outgrowth, drag the seeds along the ground, following their own tracks, which often coincide with rabbit runs or with the paths made by human beings. From the sides of such tracks the gorse may spread over heath or grassland on clays and loams, but especially on sands, wherever it can find room, but it scarcely invades dense Callunetum. Though not incapable of growing on calcareous soils it generally avoids them, and on the chalk downs the distinction between the pioneer scrub of juniper on the shallow chalk soils and of gorse on the non-calcareous loams on plateaux, or on rainwash in valley bottoms is often very conspicuous. Hawthorn occurs on both.

While *Ulex europaeus* is essentially a lowland plant, though here and there it ascends to considerable altitudes, *U. gallii*, which is far more abundant in the west, less so in the north-west, and does not penetrate far into Scotland, typically occupies a higher zone on the siliceous hillsides (Fig. 89). It is of lower growth and has shorter spines which are usually yellowish or brownish green, as contrasted with the bluer green of *U. europaeus*. *U. gallii* flowers from July to September or into October, the pods dehiscing in the following spring, while the first flower buds of *U. europaeus* often begin to open in October and a few flowers are produced, if the weather is warm enough, throughout the winter till April, when flower buds open *en masse*, covering the bushes with blossom in a favourable season. By June the common gorse is in fruit, the pods dehiscing in July. In the west of England the two species often grow together, but no hybrids have been recorded, probably because their flowering seasons scarcely overlap.

The third species, *U. minor* Roth., is characteristic of the heaths and commons of the southern and south-central counties, where it is frequently abundant, though it is also recorded from East Anglia, north-west Wales and south-west Scotland. It occurs in two forms, a small prostrate form and an erect form which may grow 2 or 3 ft. high. The spines are shorter and weaker and the flowers much smaller than those of the other two species.

The seedlings and young shoots of the species of gorse are soft and very nutritious, but the mature stems of *U. europaeus* and *U. gallii* are sharp, stiff spines, and the leaves, though less rigid, are also spinose. The result is that once a bush is established it is well protected from destruction by grazing animals, which may however keep it trimmed down into a compact hemispherical cushion by continually eating off the soft young shoots. On hillsides heavily pastured by sheep the bushes of *U. gallii* in heathy grassland are all eaten down into these rounded cushions (Pl. 95). If the cushion becomes too wide or too high for the sheep to reach the centre a tuft of freely growing uneaten branches arises from this, and the bush is able to increase in height.

Juniper. Another pioneer shrub with somewhat spinose leaves is the common juniper (*Juniperus communis* L.) which has a wide distribution throughout the British Isles, though with many gaps. In the south it occurs mainly on heaths and on chalk downs but is rather frequent in the pine and birchwoods of the Scottish Highlands. On the chalk downs it plays much the same part as gorse on the siliceous soils, and a full account of its behaviour will be found in Chapter XVIII (pp. 372–5). Juniper occurs in two growth forms, one erect and strict in habit, the other low and spreading. These often grow side by side and intermediate forms between them also occur. The prostrate juniper with shorter, broader, imbricating, appressed leaves (*J. communis* var. *montana* = *J. nana* Willd. or *J. sibirica* Burge) occurs in exposed situations in the mountains of the north and west, and descends to sea level in Skye and Kerry. Juniper is extremely intolerant of weak light, so that it is rapidly killed when heavily shaded by trees.

Roses and brambles. Important among the thorny shrubs and undershrubs, largely because of their ubiquitous and abundant occurrence, are the roses and brambles, which are armed with more or less thickly set, often recurved "prickles", arising from the surface tissues of the stems and petioles, instead of with the stem or leaf spines of the species hitherto noticed. Several species of rose form bushes of some height, but they are not gregarious and do not dominate scrub.

The brambles on the other hand produce widely arching shoots, the tips of which enter the ground and take root, forming new centres of growth, so that the plants may form small compact patches of impenetrable thorny thicket. The brambles also often scramble over other shrubs with the help of their recurved prickles and the shoots may then reach a considerable height. Both genera, but especially the brambles, may enter into the composition of the mixed thorny "thicket scrub" already mentioned, along with gorse, blackthorn and other species. Besides playing these roles in the open the brambles dominate the field layer, reaching a height of 2 or 3 ft., in certain types of woodland, for example in beechwoods on loam and in some oakwoods (see pp. 281, 307, 388, 403).

The brambles belong to the fruticose section (*Eubatus*) of the genus *Rubus*. Some of the more widespread and abundant species in the south of England are *R. ulmifolius* (*rusticanus*), *R. leucostachys*, *R. dumetorum*, *R. corylifolius*, and *R. suberectus*. The dewberry (*R. caesius* L.) is a dwarf trailing species growing especially in dry grassland both on the chalk and on loamy soils and also in fens.

Of the wild roses the common dog rose (*Rosa canina* L.) is the commonest and most widespread, and is a strongly growing robust species. The field rose (*R. arvensis* Hudson) with white flowers, also very common, is of lower growing habit: the sweetbrier (*R. rubiginosa* L.) and the allied *R. micrantha* Smith are low-growing, weak-stemmed plants which often colonise chalk grassland. *Rosa spinosissima* L. sometimes forms dwarf scrub on calcareous sand dunes and in some other situations.

All the thorny shrubs described owe part of their ability to establish themselves to their possession of spines or prickles. With the exception of the species of gorse they all have fleshy fruits and are bird-sown.

Sallows and osiers. Of shrubs which are not thorny the willows (*Salix*) are among the most prominent in British vegetation, particularly the "sallows" belonging to the section *Vetrix*. *Salix caprea* L., the goat willow or great sallow, is the largest species, frequent in a wide variety of woods, from marshy places to comparatively dry soils. The grey sallow (*S. atrocinerea* Broterus) is on the whole commoner. Particularly abundant in wet situations, it is by no means confined to them, though it seems to require more water than *S. caprea*. It is one of the leading dominants of developing fen carr. *S. aurita* L. is a low-growing sallow with rugose leaves and large stipules occurring particularly in woods on rather acid soil, and almost the only sallow in the extreme north of Scotland.

Various varieties of the creeping willow (*S. repens* L.), a dwarf species spreading by underground stems, and thus often gregarious over considerable areas, are found alike on wet heaths, on fens (var. *fusca*) and in the "slacks" (wet hollows) of sand dunes, where it can grow up through and bind the sand deposited by the wind.

The commonest osier, *S. viminalis* L., is rather doubtfully wild, but is commonly planted in wet alluvial ground to form osier beds for basket making. Other native species, for example the purple osier (*S. purpurea* L.), not uncommon along riversides and in some fens, and also hybrids between the common osier and the sallows, are planted for the same purpose. The remaining species of shrubby willow occur in similar situations, along riversides and in marshy ground, but are of little importance in British vegetation. *Salix lapponum*, *S. nigricans*, *S. myrsinites* and other rarer species sometimes form scrub in the Scottish mountains.

Other shrubs. Apart from the heath family, some British species of which are undershrubs dominating communities of the heath formation which will be considered in Chapters XXXVI and XXXVII, none of the other British shrubs is more than very locally dominant, but some are of sufficient importance to notice here.

The elder (*Sambucus nigra* L.) is a very common shrub along roadsides and on the edges of woods, but especially abundant in waste places, as about farmyards, vacant building lots and on shingle beaches, sand dunes and such situations, where it suffers little or no competition from other shrubs. It does remarkably well on shallow chalk soil, even where the chalk is quite bare as a result of rain wash on a steep slope or in the middle of rabbit burrows; and in such situations small patches of pure elder scrub may occur, and are not touched by rabbits.

"Calcicolous" shrubs. There is a set of shrubs, which while not confined to calcareous soils, are especially abundant on chalk in the south of England, as in chalk scrub, in hedges on chalk and limestone soil, and on the edges of calcicolous beechwood and ashwood; so that their occurrence together in quantity is a fairly certain indicator of highly calcareous soil. These are the dogwood (*Cornus sanguinea* L.), privet (*Ligustrum vulgare* L.), wayfaring tree (*Viburnum lantana* L.), spindle tree (*Euonymus europaeus* L.) and the common buckthorn (*Rhamnus catharticus* L.), usually in that order of frequency. They are commonly associated with the dominants of chalk scrub, hawthorn or juniper, and very often also with field maple, hazel and elder, as well as with the characteristic trees of chalk scrub, whitebeam and yew. Dogwood may form locally pure scrub over small areas, and so may privet. These shrubs are southern in their British distribution, becoming scarce or disappearing altogether in the north. Thus dogwood scarcely reaches Scotland, privet is probably introduced in Scotland, spindle tree does not go north of the Firth of Forth and is rare even in southern Scotland, buckthorn occurs mainly in the southern half of England.

The alder buckthorn (*Frangula alnus* Miller), formerly included in the genus *Rhamnus* as *R. frangula* L., has quite a different habitat, being characteristic of woods and heaths on damp peaty or acid soils. It is of great interest that the two buckthorns grow together on the alkaline peat of the East Anglian fens, where they may dominate young fen carr (see p. 657), the damp peat meeting the requirements of the one, the alkaline soil of the other. *Viburnum opulus* L., the guelder rose, which is frequent in damp woods and copses, is also characteristic of fen carr.

Box (*Buxus sempervirens* L.) is an evergreen shrub which may attain the stature of a small tree, occurring here and there, apparently wild, on the chalk and oolite of the south of England. Doubt has often been thrown on its nativity: it is absent from the north of France except where it has been planted; and since its wood is valuable it is one of the plants that may have been introduced to Britain by the Romans. A boxwood formerly existed on the chalk at Boxley in Kent, of which the first published record is by Ray in 1695, and another still exists on the oolite at Boxwell in the southern Cotswolds, which we know to have been present since the thirteenth century. This Gloucestershire wood is regularly coppiced, different plots being taken in rotation. It has been much extended and is certainly at least partly planted: whether it was originally a native wood is uncertain. There is also abundant well-grown box scrub at one place on the chalk of the Chiltern escarpment near Little Kimble in Buckinghamshire. At Boxhill near Dorking in Surrey, on the chalk of the North Downs, box is abundant on the edge of yew and beechwood. The late Count Solms-Laubach, who was a connoisseur of such things, was convinced that the Boxhill wood is a fragment of old native forest, and that is certainly the impression it gives. For the rest box occurs very locally in and on the edges of chalk beechwoods, and a good deal, if not all, of this sporadic box has certainly been planted, sometimes as cover for pheasants.

The sea-buckthorn (*Hippophäe rhamnoides* L.) is a spiny shrub whose habitat on the continent is mainly shingle banks on the edges of mountain torrents. In Britain it is confined to the sea coast, where it is locally dominant on fixed sand dunes, and occasional on cliffs. It is also often planted near the sea.

Woody climbers. Of the four British species of woody climbers ivy (*Hedera helix* L.) and honeysuckle (*Lonicera periclymenum* L.) are by far the most widespread and abundant. The ivy, climbing on vertical surfaces by adhesive roots, occurs on trees in every kind of wood, except those on the more acid soils, provided it can obtain a good supply of water from soil which is not waterlogged. It also climbs the faces of rocks, sometimes attaining huge dimensions. In beechwoods, particularly, it is locally dominant in the field layer, creeping below the surface of the soil and sending up short aerial shoots which never flower.

Honeysuckle is equally abundant and is a true liane, its looped woody stems hanging from the branches of the trees and shrubs, and sometimes attaining considerable thickness. It is perhaps more abundant on sandy than on calcareous soils. Like ivy the honeysuckle sometimes occurs in the field layer of a wood, creeping below the surface and sending up short vegetative shoots.

Bittersweet or woody nightshade (*Solanum dulcamara* L.) though not so abundant as the last two, is also widespread and particularly in wet soil. It is specially characteristic of wet woods, such as fen carr and alderwood, of riversides, and of hedgerows bordered by wet ditches.

Traveller's joy (*Clematis vitalba* L.) is specially characteristic of the southern chalk and oolite, where it often covers the hedges or the scrub with its festoons of white flowers and plumed fruits. It is abundant only in the south, rare in the north of England, and absent from Scotland. Like honeysuckle its growth is that of a liane, and its stems are sometimes as thick as a man's arm.

REFERENCES

ANDERSON, MARK L. *The Natural Woodlands of Britain and Ireland.* Department of Forestry, Oxford, 1932.

CHRISTY, MILLER. The hornbeam (*Carpinus betulus* L.) in Britain. *J. Ecol.* 12, 39–94. 1924.

ELWES, H. J. and HENRY, A. *Trees of Great Britain and Ireland.* Edinburgh, 1906–13.

HEGI, W. *Illustrierte Flora von Deutschland.* Munich, 1906–31.

SCHOENICHEN, W. *Deutsche Waldbäume und Waldtypen.* Jena, 1933.

TAYLOR, E. G. R. "Leland's England." (Chapter IX in Darby, H., *An Historical Geography of England before* A.D. 1800. Cambridge, 1936.)

WILSON, M. Plant Distribution in the woods of north-east Kent. Part I. *Ann. Bot.*, 25. 1911.

Chapter XIII

OAKWOOD. INTRODUCTORY

PEDUNCULATE OAKWOOD (QUERCETUM ROBORIS)

Oak forest in the British Isles. Deciduous oak forest is clearly the climatic climax community of the greater part of England, of many of the valley sides of Wales, Ireland, southern and central Scotland, and of the deeper soils towards the bottoms of the Highland glens except in the extreme north. Geographically it is part of the deciduous forest formation of western and central Europe: historically it is the direct descendant of the so-called mixed oak forest (Quercetum mixtum) which spread and established itself in the British Isles during Boreal and early Atlantic times (Chapter VII, pp. 161–3).

Nothing but the smallest fragments remain, scattered on remote hillsides, of the original natural oak forest which once covered most of the country; but the occurrence of oak as the predominant tree in many of the subfossil forests below tide level round the southern English coasts—forests which were presumably submerged by the main post-glacial subsidence that gave rise to the North Sea as we now know it, and to a lesser extent by later submergences: the prevalence of oak pollen in most of the English peat which was formed from the Atlantic period down to very recent times; and the documentary evidence that from Roman times to the later Middle Ages great tracts of country were covered with oak forest, furnish sufficiently convincing grounds for the general conclusion. Furthermore, the great majority of the semi-natural woodlands of England and Wales are still dominated by oak: oak grows well, ripens acorns, and springs naturally from seed under suitable conditions, the widespread failure to regenerate being demonstrably due either to local conditions unfavourable to the seedlings, such as hardening of soil surface, competition of the plants of the undergrowth, or most generally to the attacks of ground vermin (mice, voles or rabbits).

Semi-natural oakwood. Most of the existing oakwoods dominated by the pedunculate oak on the deeper soils of the midlands and south of England are in the form of "coppice-with-standards" (oak standards and hazel coppice), a culture form easily derivable from natural oak forest in whose undergrowth hazel is dominant (see pp. 270–1). Much of the existing coppice-with-standards has been planted, and this usually differs from natural oakwood by the relative scarcity of trees other than oaks and of shrubs other than hazel. But in spite of such deviations this type of wood does not differ *essentially* in constitution and structure from a natural oakwood, and may be fairly regarded as representing the natural climax forest, though in a modified form. Many such woods still exist on old forest land, originally no doubt colonised by the invading trees and their associated shrubs and herbs, so that these sites have been continuously occupied by the trees and their associated vegetation from the time of the original colonisation: a few have been planted on land which had been previously cleared—perhaps centuries before—and used for pasture or for arable crops. These last will become colonised after plantation with a miscellaneous collection of species, among which will be woodland shrubs and herbs. The non-woodland plants will be more or less transient, but the true woodland species will remain, so far as they find the right soil and shade conditions; and the plantation will thus after a certain time approximate to the condition of natural woodland (Woodruffe-Peacock, 1918), though certain species, characteristic of old woodlands, may never enter them. Whichever origin a particular wood has had, whether it is planted on an old forest site or on grassland or fallow, provided it has been planted with the climax dominant, the resulting plant communities will not differ essentially from those of a natural spontaneously arising wood which has been exploited in the same way, except that the trees are more even-aged; and both, irrespective of their origins, are therefore properly regarded as "semi-natural" woods.

The two British oaks. The British oakwoods are dominated, like those of central and western Europe, by the pedunculate and sessile oaks, *Quercus robur* L. (*Q. pedunculata* Ehrh.) and *Q. sessiliflora* Salisb. (*Q. petraea* (Matt.) Liebe.), of which some account has been given in the last chapter (pp. 246–8). On the whole the former is the dominant tree on the softer rocks of the south, east and midlands of England and on many of the soft alluvial soils of the broader river valleys of the remaining parts of the British Islands. The sessile or durmast oak, on the other hand, is the dominant tree of the harder, largely Palaeozoic, metamorphic or igneous rocks forming the hills of the west and north. In the fragments of native oakwood still remaining on Dartmoor (Wistman's Wood, Black Tor Copse, Piles Wood, see pp. 298–9), however, the species is *Q. robur*. On many of the soft coarse-grained sands of the south and east, such as the Tertiary and Quaternary sands or the Lower Greensand, *Q. sessiliflora* may be either dominant alone or co-dominant with *Q. robur*, and the former has even been recorded as dominant on acid London Clay (Wilson, 1911), though *Q. robur* is certainly the usual oak on this geological formation.

Wherever the two species grow together hybrids between them occur freely, and these hybrids show a great variety of combinations of the characters of the two parent species, especially often combining the small sharply folded crescentic leaf auricles of *Q. robur* with the branched hairs on the lower surface of the leaf of *Q. sessiliflora*. It was this circumstance that led many of the older British botanists to recognise a single species only of British oak, since, as they said, all transitions occur between the extreme types. Such transitions, however, are not found where either parent exists alone in pure formation over a considerable area. The characters of the two species are quite well defined and easy to recognise, though there is considerable variation within each specific type.

The very extensive planting of oak has to some extent confused the natural distribution. Thus the pedunculate oak, which seems on the whole to have been the more commonly planted, now bulks largely on some of the Scottish hillsides and in some of the Cornish and Devonshire valleys, though *Q. sessiliflora* is almost certainly the native tree in many of those localities.[1] In the Forest of Dean (Gloucestershire), which has been extensively replanted, it is difficult to be sure which of the species was originally dominant. The commonest oak there now is *Q. robur*, but *Q. sessiliflora* is frequent, and is said to have been the original oak in this forest, as indeed is most probable by comparison with neighbouring areas and with rocks of the same formation in the north of England. Moss (1907) was of opinion that *Q. sessiliflora* was the natural dominant on shallow soils over hard rocks, and this view corresponds quite well on the whole with its distribution in Britain, though it is often dominant also on deep sands and gravels. Salisbury (1916, 1918) held that the sessile oak flourishes better than the pedunculate on more acid soils, and that *Q. robur* succeeds better on nearly neutral or slightly alkaline soil; but the limits are certainly not very sharp.[2] The general requirements as well as the habit of the two trees, which often flourish side by side, are so similar that they cannot be correlated at all closely with accompanying vegetation. Though the apparently more oxyphil tendency of the sessile oak leads to the very frequent occurrence of more "heathy" ground societies and of species having a lower pH range in the Quercetum sessiliflorae, there are, nevertheless, many examples of such societies associated with *Q. robur*, while less oxyphilous species occur in certain types of Quercetum sessiliflorae (Chapter xv).

PEDUNCULATE OAKWOOD (QUERCETUM ROBORIS)

General distribution. Semi-natural woods dominated by *Quercus robur* L. are typical and abundant on almost all deep soils in the south, the east and the midlands of England. They extend also into the west and north, though possibly not Ireland (p. 247), where, however, they are confined to the alluvial and certain other soils derived from the relatively soft rocks of the valleys and plains. Such habitats are very much scarcer in the mountainous regions—absent indeed from large tracts of country—so that woods of this type become very rare or completely lacking in some of the highland areas.

[1] See also Chapter XVI, p. 349.
[2] So far as the first-year seedlings are concerned recent experiments have shown that both species do somewhat better on slightly acid soils and that there is no significant difference between them in this respect.

Former extent of oak forest. There can be little doubt that the soils on which pedunculate oakwood now occurs were once (from Atlantic times onward) almost continuously covered with it. It seems that it was the Saxon invaders who first seriously began to clear and cultivate this type of country in the valleys and plains, the "Celtic" cultivation having been almost exclusively upland, though often on similar soils such as the "clay-with-flints", and the Romans having had little influence on the type of agriculture in Britain (see Chapter VIII). As time went on more and more of the midland plains and of the southern plains and valleys, and eventually much of the Weald, were cleared and brought under the plough or used as pasture, though extensive woods were maintained for the pannage of swine and for timber supply, besides the great Royal Forests and Chases which were reserved for hunting. Thus at the time of the Domesday survey (1086) there was pannage for 30,729 swine in Hertfordshire and for 92,991 in Essex, which implies very large areas of forest. Practically the whole of this must have been oak (or oak-hornbeam) forest, and we may reasonably suppose that it occupied a great proportion of the areas of the two counties, though they had been settled regions for many centuries. To-day only 6·4 per cent of the area of Hertfordshire and only 3·1 per cent of Essex, which is more than twice the size of Hertfordshire, is occupied by woodland.

Origin of coppice-with-standards. The traditional English method of growing the oak, as standard trees in open canopy with coppice below, has been continued from the Middle Ages to the present day, though it has long been quite uneconomic. Such woods are now mainly used for preserving pheasants and as fox coverts, and very often the rents received from the "shooting tenants" far exceed the annual value of the timber and underwood, which indeed are often almost completely neglected. Woods of the pedunculate oak in close canopy are rare, and no example has been studied ecologically with any thoroughness.

The original advantage of the oak grown with lateral space to spread its crown is said to have been the curved form of the timber and the "knee-pieces" obtained from the main spreading, and smaller sharply bent branches, used largely for shipbuilding. The coppice, mainly of hazel, was useful for a great variety of local purposes, such as hurdles, fencing, firewood, hop poles and bean and pea sticks, and oak bark was valuable for tanning leather. During the Middle Ages and up to the seventeenth century great quantities of coppice wood, and a good deal of timber too, were used in making charcoal for iron smelting (see Chapter VIII, pp. 182–3). It is easy to see how from the earliest utilisation of the native oakwoods this form of standardised exploitation would gradually arise. Felling of a number of trees would greatly stimulate the growth of the shrubby underwood in which hazel was almost certainly prominent. The young oak saplings which were left would have more room to develop their crowns and

PLATE 13

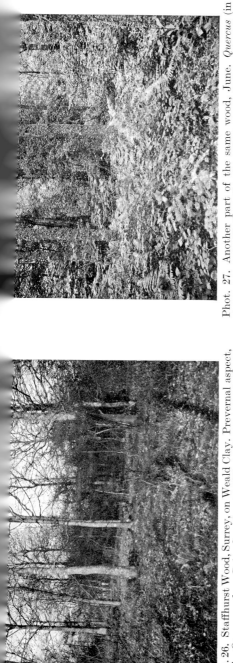

Phot. 26. Staffhurst Wood, Surrey, on Weald Clay. Prevernal aspect, April. *Quercus robur*, *Primula vulgaris*. Discontinuous shrub layer.

Phot. 27. Another part of the same wood, June. *Quercus* (in close canopy), *Hedera*, *Corylus*, *Crataegus*, *Rubus*, *Pteridium*, *Digitalis*, etc. S. *Mangham*.

Phot. 28. Coppice (*Corylus*) with standards (*Q. robur*) on rainwash in a valley. Ditcham Park, Hants. One oak in young leaf, April. Beech dominant on chalk slope to left.

Phot. 29. First-year coppice (*Corylus*) with standards (*Q. robur*), Chevening Park, Kent, on Clay with flints. Thistle and bracken in fully lighted foreground. S. *Mangham*.

TYPICAL PEDUNCULATE OAKWOODS (QUERCETUM ROBORIS)

PLATE 14

Phot. 30. Staffhurst Wood, Surrey, on Weald Clay. Oaks in close canopy about 35 years old. The wood is said to be part of the old Wealden oak forest, and it regenerates naturally. *Corylus* and *Crataegus* very abundant. *Prunus spinosa* in flower, April.

Phot. 31. Young yew in Staffhurst Wood. *Corylus*, *Crataegus* and *Primula vulgaris*.

PEDUNCULATE OAKWOOD

would spread out laterally, acquiring the typical "open canopy" development casting much less shade, so that the shrub layer could develop more vigorously and become practically continuous. This shrub layer would itself be cut for small wood, and many of the shrubs, including hazel, would readily sprout from the stools, acquiring a useful maturity in from 10 to 15 years, at the end of which period it would be cut again. The trees would mature in from 80 to 150 years according to the kind of soil, and would then yield on felling the useful type of curved timber, with "knee pieces" where the branches made sharp bends, and straight timber also from the trunk and uppermost branches. Experience of these results would gradually lead to standardisation of the double rotation, for coppice and standards respectively.

After the demand for such timber had begun to exceed the supply, as happened at latest during the course of the sixteenth century, oak and hazel were planted in imitation of the "natural" coppice-with-standards which had arisen by the response of the original forest to this particular form of treatment, though it is probable that the demand was always met to a far larger extent by importation than by planting. Such planting continued till the nineteenth century, often on the sites of old woods though sometimes on other land, but it never kept pace with the ever-increasing demand. Natural regeneration, which had been effective in the earlier forest, entirely failed to satisfy the exacting demand for timber, and failed also because of over-exploitation, leading to exposure of the soil and drying and hardening of the surface, as well as through the increase in the destruction of seeds and seedlings by ground vermin such as mice and voles, whose natural enemies were being increasingly hunted, trapped or shot by man. At the same time the proportion of land under trees had fallen far below that obtaining in continental countries. As a result the needs of the country eventually came to be supplied almost exclusively from imported timber.

"Damp" and "Dry" oakwood. In *Types of British Vegetation* a primary distinction was drawn between "Damp" and "Dry" oakwoods belonging to the Quercetum roboris, and each was related to corresponding shrub and grassland communities. "Neutral grassland" corresponded with "Damp oakwood", and "Grass heath" with "Dry oakwood". The "Damp oakwoods" were said to be developed on clays, loams and fine-grained sandstones: the Dry mainly on coarser sands and sandstones. The Damp oakwoods were characterised by mild humus with a neutral or only slightly acid reaction: the Dry by humus with a higher degree of acidity. The Damp oakwoods showed a fairly rich shrub layer typically dominated by hazel, were usually in the form of coppice-with-standards, and had a rich and abundant field layer, especially in coppiced woods: the Dry oakwoods were more often in close canopy, were much poorer in shrubs, hazel being typically absent except where planted, the field layer was much poorer in species, with a few specially characteristic, while the great majority of the Damp oakwood species were absent.

There can be no doubt that this separation is good as far as it goes, and that it has been found useful by students of the subject. Nevertheless, the separation is not taken as primary in this book, because it has seemed better to treat the pedunculate oakwoods as a whole and to exhibit the facies characterised by the differences of the shrub and field layers, and the correlated soil differences, as forming parts of a continuous series. Doubtless it might be contended that the same argument would apply to the whole of the oakwoods, and that it would be better to treat the Querceta roboris and the Querceta sessiliflorae together, since they also show transitional types, and the same ground vegetation may exist in both. But the difference of dominant species is *usually* associated with differences of soil and accompanying vegetation as well as with geographical distribution, and there is here on the whole a better case for primary separation.

There is, however, the further complication that owing to the widespread planting of oak it is impossible to be certain that some of the dry pedunculate oakwoods were not originally dominated by the sessile oak; and the two species are apparently naturally associated on many deep sandy soils. Though *Q. sessiliflora* is more generally attached to acid soil, the requirements of the two species are not very different, so that their soil ranges overlap and *Q. robur* will not only flourish when planted but will also regenerate naturally on acid soils, on which it is in places (e.g. on Dartmoor, see p. 298) undoubtedly the natural dominant.

The difficulties described are of the kind which we constantly meet in attempting to classify biological phenomena—they are so complex that, while we can recognise well-marked "types" representing the *prevalent* combinations of different factors, we cannot draw sharp boundaries.

Ash-oak wood. Another distinct type recognised in *Types of British Vegetation* is the "Ash-oak wood" on the deeper soils containing a considerable percentage of calcium carbonate, such as the calcareous boulder clays of eastern England, and certain marls and calcareous sands. These woods are marked by the abundance of ash (which is, however, generally coppiced along with the hazel), by the prevalence of field maple (*Acer campestre*) and of the so-called "calcicolous" shrubs, such as dogwood (*Cornus sanguinea*), wayfaring tree (*Viburnum lantana*), spindle (*Euonymus europaeus*), and privet (*Ligustrum vulgare*), by the presence of a number of herbaceous species generally characteristic of calcareous soils, and by the corresponding absence of calcifuge species. This type of wood was first described by Moss (1907) as "oak-hazel wood", but since much of the typical oak-hazel wood on other soils does not conform to it the type was renamed "ash-oak wood", or "(ash-)oak-hazel wood" for the common coppiced form in which the ash is coppiced with the hazel. This type of wood is here included in the Quercetum roboris as a separate facies (p. 288).

Fig. 51 shows the relations of the principal facies of both pedunculate and sessile oakwoods to one another and to allied types of woodland.

Soils. The typical soils on which the Quercetum roboris is developed are the clays, silts, loams and marls which are derived from Secondary, Tertiary and post-Tertiary rocks and occupy so much of the south, east

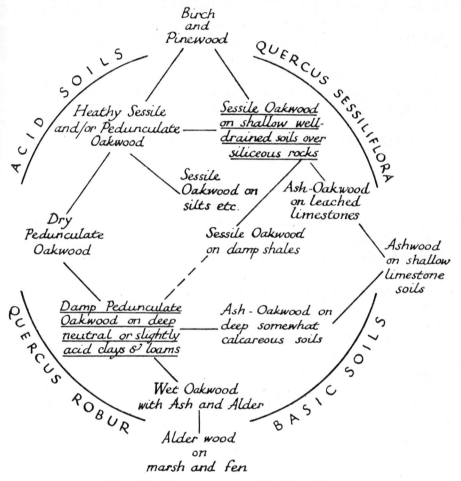

FIG. 51. OAKWOODS AND RELATED TYPES

The lines connecting the different types indicate affinities and often transitions between the types.

and midlands of England. The more mature soils belong to the "Brown Earth" type (see Chapter IV, pp. 85–8). Many of the rocks from which they are derived are poor in or almost destitute of lime, while others contain a considerable amount: in mechanical constitution the lighter loams and silts grade gradually into the impure and fine-grained sands and thence into the coarse sands. All these changes are reflected in the vegetation, but it is not until we come to the highly calcareous soils on the one side, particularly those based on limestone rock, and the typical coarse sands on the other, that we find any marked change in the dominant trees. Some of the more extensive pedunculate oakwood soils are those derived from the Keuper Marls of the Triassic, the Lias and the Oxford and Kimeridge Clays of the Jurassic, the Weald Clay and the Gault of the Cretaceous, the London Clay of the Eocene, the Boulder Clays, the "Clay-with-flints" (overlying the Chalk) and many tracts of alluvium belonging to the post-Tertiary deposits. These clays and loams occasionally give an acid reaction but those with a considerable proportion of calcium carbonate are quite alkaline. The leaching of the surface layers ("A horizon") by the percolation of rain water, however, together with the formation of humus, normally gives an acid reaction to the upper levels of the typical oakwood soil.

On continuously or almost continuously wet soil, by streamsides, etc., the oakwood grades into marshy woodland, of which the typical edaphic climax seems to be the alderwood (Alnetum glutinosae) (see Chapter XXIII).

Water content. The natural water contents of the oakwood soils range from about 20 to about 60 per cent of the dry weight of the soil. A record of 15·24 has been obtained from the clay-with-flints at the end of a long drought when *Mercurialis* was strongly wilted (Crump in Adamson, 1922, p. 181). Most of the values range from 25 to 45 per cent, and particular dominants of the field layer can often be correlated with the soil water content.

Humus content. The humus content of the upper layers of soil ranges from about 5 to about 20 per cent of the dry weight. Most of the records lie between 8 and 16 per cent. The pH values of the surface layer range from about 5 to 6, higher values generally occurring as lower soil levels are reached. The litter of fallen leaves is typically very sparse in summer because of their rapid decay. Below this there is sometimes a layer of humus with a few mineral particles, but especially on clay soils this is usually absent, or represented by a mere film. The proportion of humus contained by the mineral soil rapidly decreases downwards, though streaks and pockets of humus derived from the partial decay of roots and rhizomes may be found down to considerable depths.

Light. In regard to the amount of light available for the undergrowth a sharp distinction must be drawn between the *light phase* before the leaves of the deciduous trees and shrubs unfold in early May and the *shade phase* after the foliage is expanded (Salisbury, 1916). It is during the light phase, of course, that the characteristic "prevernal" species vegetate and flower (Fig. 52).

No readings of light intensity are available from oakwood in close canopy. In open canopy it is mainly the coppiced shrub layer which is effective in reducing the light reaching the field layer.

Table XII gives the light values recorded by Adamson (1912, p. 353, and 1922, p. 183) from oak-hazel coppice in Cambridgeshire and Hampshire

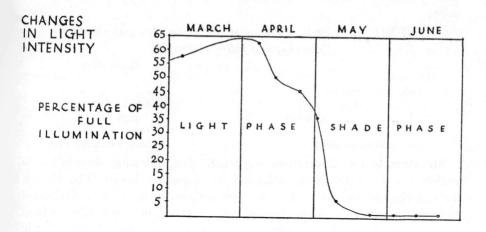

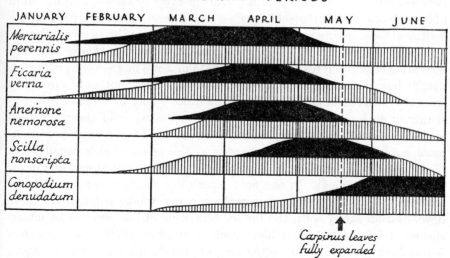

FIG. 52. VEGETATIVE AND FLOWERING PERIODS OF FIELD LAYER DOMINANTS IN HERTFORDSHIRE OAK-HORNBEAM WOODS

The period of active vegetation of each dominant is shown by vertical lines, the periods of flowering in black. Above is the curve of light intensity as percentage of full illumination. The flowering of the first three societies is "prevernal", of *Scilla* vernal, of *Conopodium* aestival. The vegetative shoots of *Ficaria*, *Anemone* and *Scilla* are short-lived, disappearing during June, while those of *Mercurialis* and *Conopodium* last through the summer. After Salisbury, 1916.

and by Salisbury (1916, p. 93) from oak-hornbeam coppice in Hertfordshire:

Table XII. *Average light intensities under coppice with standards.*
Percentages of full illumination

	Light phase	Shade phase
Adamson:		
Oak-hazel (Hants)	68	8·4
,, (close)	37	3·2 (min. 1·4)
,, ,, (Cambs)		min. 0·5
Salisbury:		
Oak-hornbeam (old)	41–49	About 1·0 (min. 0·16)

Structure of the Quercetum: layering. The following description is mainly based on the Damp oakwood developed on loams (Pls. 13–15). This is the commonest and most typical Quercetum roboris. Oakwood typically consists of four "layers" or "strata" of vegetation—whose foliage systems reach different heights above the ground level: (1) the *dominant trees* forming the roof or canopy of the wood, the top of which may vary from 25 to 100 ft. (c. 7·5–30 m.) in height, but usually falls between 40 and 70 ft. (c. 12–21 m.); (2) the *layer of shrubs*, varying from about 6 to 15 ft.; (3) the *field layer*—a useful term commonly used on the continent—consisting mainly of herbs, but sometimes also of dwarf shrubs, from 6 in. to several feet in height, and finally (4) the *ground layer* typically consisting of bryophytes and rarely more than 2 or 3 in. (5–8 cm.) in height, usually less. Below the upper layer of dominant trees another of subsidiary trees (1a) not reaching the height of the canopy may be represented, consisting of such species as crab apple (*Pyrus malus*), wild cherry or gean (*Prunus avium*), holly (*Ilex aquifolium*), rowan (*Sorbus aucuparia*), or more rarely field maple (*Acer campestre*)—this species being usually coppiced with the shrubs—but the subsidiary trees are often absent and never more than occasional. In the field layer (3) a distinct upper and lower stratum (3a and 3b) are recognisable in some woods, particularly those developed on sandy soils, the former consisting of tall herbs or of undershrubs, 3–5 ft. (0·9–1·5 m.) high, such as bracken (*Pteridium*), rose-bay willow-herb (*Epilobium angustifolium*), or bramble (*Rubus fruticosus* agg.), while the height of the lower stratum is measured in inches or centimetres.

Seedling trees soon after germination belong of course to the field layer; and as they grow up into saplings they pass into the shrub layer before reaching the height of the tree canopy, thus having to compete in turn with the other plants of the two middle layers.

Without the dominant trees or at least dominant shrubs we do not of course have a wood at all, since tall woody plants are the distinguishing feature of this type of community and the conditions of shade and moisture brought about by their presence are necessary for the existence of the characteristic shade plants of the field and moss layers. Any or all of the other layers may, however, be absent.

If the trees are absent the shrubs become the dominants of the community, which is then generally called *scrub* if it is left to grow without interference (see Chapter XXIV) or *coppice* if it is periodically cut. Many of our coppices are actually destitute of standard trees. The degree of development of the upper layers has a very marked effect on the presence and development of the lower, mainly owing to interference with the light supply, but partly as a result of root competition. Thus a dense tree canopy (1), casting deep shade, greatly reduces or even eliminates the shrub layer (2), and sometimes the field layer (3) also. A dense shrub layer acts in the same way on the field layer, as can be seen in "thicket scrub" (p. 475) or old close coppice. A thick close upper field layer (3a) may eliminate 3b and 4, and similarly a continuous carpet of 3b may greatly reduce or suppress 4.

Lateral spacing. The lateral spacing of the trees is of great importance. In a naturally developed climax wood the dominant trees compete with one another for light and often for soil water also. The weaker are killed or "suppressed", often lingering with poorly developed crowns beneath the dense canopy formed by their successful competitors. In a young oakwood, while the phase of active competition of the densely crowded saplings and young trees endures, the ground is typically almost or quite bare, though a certain number of bryophytes may appear. As the saplings grow up the majority are shaded out and the number existing on a given area progressively decreases, till in a mature wood from 60 to 100 only remain to the acre (150–250 to the hectare). At the same time the canopy "opens out", i.e. the crowns of the remaining trees cast a less dense shade and the wood becomes progressively occupied by the plants which will form the lower layers of the mature wood. Artificial thinning of the wood, by decreasing the density of the canopy, of course hastens and increases this process of colonisation. A mature oakwood in close canopy, with about sixty trees to the acre, has a sparse shrub layer, but the field and ground layers are often well developed, depending partly on the nature of the soil.

Structure of coppice-with-standards. The artificial coppice-with-standards type of oakwood, on the other hand, to which the great majority of our Querceta roboris belong, has a very much smaller number of standard trees, twelve mature trees to the acre (thirty to the hectare) being a full complement. With this number each tree has room to spread its branches to the full extent without interfering with its neighbours, the crowns of adjacent trees just touching so that a continuous light shade is cast by the canopy. The shade is much lighter than that of a mature oak wood in close canopy, and a practically continuous shrub layer is developed. This is coppiced at more or less regular intervals of from 10 to 15 years, and the development of the field layer is closely related to the coppicing. Towards the close of the period of rotation, when the shrubs of well-grown coppice are tall and close, the field layer is very sparse or sometimes apparently almost absent; and the herbs of which it is composed are mostly unable to flower owing to weak light. In the first

Periodicity of field layer

spring after coppicing the underground parts of those perennial herbs which have survived the period of deep shading begin to shoot more vigorously and dormant seeds germinate, while the opening of the coppice allows a much freer entry of wind-borne seeds and spores. Other plants may spring from seeds brought in on the boots and clothes of men engaged in the work of coppicing. In the second or third year a carpet of vegetation has usually arisen and the plants of the field layer are commonly at their most vigorous and often produce the conspicuous sheets of vernal and prevernal flowers for which our coppice woods are famous. Thereafter the vigour of the field layer gradually diminishes in close coppice, as the ground becomes more and more deeply shaded, till towards the end of the period of rotation the soil again appears almost bare, except for a few species which can tolerate the deepest shade. In coppices which are not closed, i.e. where the individual coppice stools are widely separated, or where the shrubs are not well grown, the shading is much less and the field layer may flourish throughout the period of rotation.

Composition of field layer in coppice. Salisbury has pointed out that the flora of the field layer of coppice is composed of three elements: (1) the shade plants of the wood, (2) the so-called "marginal" plants which flourish in half shade and (3) wayside and arable weeds.

The plants of (3) are most favoured in the early years of the period of rotation by the full illumination, those of (2) in the intermediate years and of (1) towards the close.

The dominant tree layer. *Quercus robur* is typically almost exclusively dominant in the tree layer of our pedunculate oakwoods. The original oakwoods (the "Quercetum mixtum" of Atlantic times) certainly contained a considerable mixture of other species, and the purity of the modern oakwood—modern in the sense of nineteenth century—is doubtless mainly due to planting. The following is a generalised list of the trees associated with the oak in these woods[1]:

Quercus robur	d	Populus tremula*	o, l
Fraxinus excelsior*	f	Carpinus betulus*‡	la
Acer campestre*	f–a	Trees of the second layer	
Betula pendula (alba)*	o–f†		
B. pubescens*	r–o†	Ilex aquifolium	o–a
Prunus avium	o–f	Pyrus malus	o–f
Alnus glutinosa*	o	Sorbus torminalis*	l
Ulmus glabra (montana)	o	S. aucuparia	r–o
Populus canescens*	r, l	Taxus baccata	o, l

* Often coppiced. † Especially on the lighter soils.

‡ This tree is confined to certain areas in the south-east of England, where it is nearly always coppiced: in those areas it replaces hazel partly or completely as the main constituent of the coppice.

[1] The letters following the names of species in this and all subsequent lists have the following meanings: d, dominant; a, abundant; f, frequent; o, occasional; r, rare; l, local; v, very.

Most of the trees in the above list are often cut with the coppice, and many of them spring freely from the cut stools, e.g. the oak itself, which was at one time extensively coppiced and the bark of the coppice shoots used for tanning; ash, which in many woods is regularly coppiced; birch, alder, aspen, wild service and hornbeam.

In natural oakwoods oak suppresses birch and aspen, and these trees are thus to be regarded as seral trees not properly belonging to the climax consociation. But in thinned or coppiced woods they frequently reach the general height of the canopy.

Shrub layer. *Corylus avellana* is by far the most abundant shrub (Pl. 14, phot. 29). It is often nearly as pure in the shrub layer as *Quercus robur* in the tree layer, and probably for the same reason, that it has been planted. There are, however, a number of other shrubs frequently present:

Corylus avellana	d	Viburnum opulus	f
Crataegus monogyna	f	Euonymus europaeus	o
Prunus spinosa	f	Salix atrocinerea	f
Cornus sanguinea	f	S. aurita	o, l
Sambucus nigra	o	S. caprea	f
Ligustrum vulgare	l		

All of these, except the two spinose shrubs (the hawthorn and the blackthorn), spring freely from the cut stools.

Woody climbers. Two of the four woody climbers native to Britain occur typically in the oakwood. The ivy (*Hedera helix*) is one of the most abundant and is seldom absent from woods where the moisture and pH value are adequate. It requires a good water supply and avoids strongly acid soils. Climbing by means of adhesive roots it is only able to ascend wide surfaces such as are provided by rocks or the trunks of standard trees. In default of these it trails on the ground, rooting at intervals, and may occasionally become dominant in the field layer, as it often does in beechwoods.

The common honeysuckle (*Lonicera periclymenum*), perhaps the most widespread of all, occurring on a very wide range of soils, is a twiner which climbs the shrubs and also trails on the ground, rooting at the nodes, when it can get no support.

Field layer societies. Extensive societies of gregarious herbs are the most striking feature of the coppiced oakwoods. These nearly all flower profusely in early spring (*prevernal aspect*, March–April) before the bursting of the winter buds of the deciduous trees and shrubs, or in mid-spring (*vernal aspect*, April–May) during the bursting of the buds and the expansion of the leaves. Many of them have conspicuous flowers, and it is in recently cut coppice that they reach their most luxuriant development, often producing continuous sheets of brilliant colour. In some species (*Ficaria verna*, *Scilla non-scripta*, *Anemone nemorosa*) active vegetation is confined to the spring, the aerial shoots dying off in May or June; in others,

(*Mercurialis perennis, Primula vulgaris, Sanicula europaea*) the leaves persist throughout the summer (cf. Fig. 52).

The dominants of these prevernal and vernal societies are by no means always strictly segregated in well-defined separate communities. Many of them commonly occur mixed with one another and with other species in varying proportions over a wide range of habitat conditions, and large areas of woodland soil are covered with a mixture of species in which none is predominant. Nevertheless, most of the species are *characteristic* of particular conditions, though the factors involved show so many different combinations that the limiting conditions for the different species under differing circumstances have never been worked out with any satisfactory precision.

The following are the principal society dominants with some of their more usual associates.

Societies of lighter and drier soils

1. **Pteridietum.** The common bracken fern (*Pteridium aquilinum*) is very commonly present in oakwood (both pedunculate and sessile) on the lighter non-calcareous soils. Deep shade restricts its vegetative powers, so that its fronds are separated by considerable intervals, but under moderate shade it may be completely dominant, as in Sherwood Forest (Hopkinson, 1927).[1] Here under dry soil conditions the bracken and oakleaf litter decays so slowly that the ground is often covered with litter and raw humus to a depth of several inches, and few other species can find entrance. Access of light greatly stimulates the luxuriance of *Pteridium*, and in two years from clearance of a tree or shrub canopy the bracken may increase so much as to form dense masses of foliage excluding all other growth from about the middle of June, when its fronds are fully expanded. During the earlier months, however, provided the litter is not too heavy, other species can flourish in its company.

In oakwoods whose canopy casts at all a deep shade the bracken fronds appear only at intervals, and are relatively feeble in development, but in many open canopy woods casting moderate shade they are more numerous, and the petiole bends over diaphototropically so that the lamina of the frond forms a gently curved plate whose closely set, thin and delicate pinnules shade a large proportion of the soil. In bright light where the trees are widely separated, or on quite open ground, the laminae are much thicker and more leathery in texture, and are directed obliquely upwards:[2] on a favourable soil and in the absence of effective competition the fronds are then so closely set as to shade the soil very deeply, and completely exclude other growth after they are fully developed. The soil under such a *Pteridium* society is covered with a thick litter of slowly decaying fragments of the dead fronds, passing down into a peaty humus. Woodland regeneration appears impossible under such conditions.

[1] This however is "mixed oakwood" (Chapter XVII).   [2] Pl. 16, phot. 35.

PLATE 15

Phot. 32. Oakwood grown in close canopy but recently thinned. Sparse shrub layer of hazel and sallow. *Anemone nemorosa* dominant in field layer. Prevernal aspect. Gravetye Manor, Sussex. Photograph from *The Times*.

Phot. 33. Remains of oakwood, Ashdown Forest. The two oaks in front in open canopy, coppice growing up behind in close canopy, *Pteridium*, etc. Photograph from *The Times*.

PEDUNCULATE OAKWOOD

PLATE 16

Phot. 34. *Quercus robur, Betula pubescens, Pteridium, Holcus mollis, Anthoxanthum odoratum*, etc., July. Gamlingay Wood, Cambs. *S. Mangham.*

Phot. 35. *Quercus robur, Pteridium, Holcus mollis*, May. Gamlingay Wood, Cambs. *R. S. Adamson.*

Dry Pedunculate Oakwood on Sandy Glacial Loam
(Cf. Fig. 53)

The bracken seems to be originally a woodland plant of moderate shade, and it often marks the sites of woods which have been destroyed; but when it is freed from control in the open it becomes a pestilent weed, spreading widely and preventing almost all other growth (see Chapter XXVI, p. 502).

Pteridium-Holcus-Scilla. Woodhead (1906) was the first to describe the "complementary society" formed by *Pteridium* with *Holcus mollis* and *Scilla non-scripta* in the Quercetum sessiliflorae (petraeae) of West Yorkshire. The three species are not only seasonally complementary in their subaerial development, *Holcus* succeeding *Scilla*, and *Pteridium* succeeding *Holcus*, but their underground parts, as described by Woodhead, avoid competition by inhabiting different layers of the soil. In deep soils, however, the roots of *Pteridium* often descend to much lower levels than Woodhead found in his shallow soils derived from coal-measure grit and shale, and then its roots enter into competition with those of *Scilla*: in such conditions the two species may not flourish together. Throughout the country, however, *Holcus mollis* is very commonly associated with *Pteridium*, and alternatively with *Scilla*,[1] in woodlands on light, rather acid, soils dominated by either species of oak. *Viola riviniana*, *Teucrium scorodonia*, *Potentilla erecta*, *P. sterilis* and *Anemone nemorosa* are frequent members of this community.

Rubus-Pteridium. Another widespread society on similar soils is that which the bracken forms with the bramble. The bramble is another dominant and aggressive plant also belonging to the upper field layer, but the "fiddleheads" of bracken easily grow vertically upwards through the tangle of arching bramble shoots, where these are not too dense, so that the two plants, of very different life form, may share dominance from midsummer to autumn. The trailing form of the honeysuckle (*Lonicera periclymenum*) is often abundant in the *Rubus-Pteridium* society, which owing to the closeness with which it occupies the ground is very poor in species.

2. **Anemone nemorosa.** The wood anemone (*Anemone nemorosa*) (Pl. 15, phot. 32), flowering in March and April, the foliage disappearing by the end of May or a little later (Fig. 52), is one of the most widespread of the oakwood pre-vernal field layer dominants. It ranges from sands to heavy clays, though it avoids the driest and the wettest soils, and is more characteristic of the lighter and medium loams than of either the coarse sands or the heaviest clays. In the oak-hornbeam woods of Hertfordshire on clay-with-flints Salisbury (1916) records this society, with *Scilla non-scripta* and *Conopodium majus* as the most abundant associates, on a somewhat moister soil than that on which *Pteridium aquilinum* is dominant. In Tubney Wood (oak-hazel) near Oxford, on a sandy soil of leached "calcareous grit" but a relatively high ground water-level, the dominant *Anemone* is accompanied by *Scilla* (abundant), by *Primula vulgaris*, *Anthoxanthum odoratum*, *Fragaria vesca* and *Mercurialis perennis* (frequent)

[1] Pls. 16, 27 (phot. 57), 28, 31 (phot. 64), and Fig. 53.

and a large number of other species. In Hell Coppice (Shabbington Woods, near Oxford) on the heavy Oxford Clay the dominant *Anemone* is accompanied by *Festuca rubra* and *Anthoxanthum odoratum* (abundant) by *Luzula pilosa, Viola riviniana, Carex sylvatica* and *Ajuga reptans* (frequent) and a large number of other less frequent species. In both woods the ground layer of mosses is well developed.

3. **Scilla non-scripta.** The bluebell or wild hyacinth (*Scilla non-scripta*), apart from its frequent association with *Anemone* and with *Holcus mollis*, is by itself a widespread dominant, mainly on light soils. Unlike most of the woodland dominants it has no rhizome, each plant arising from a separate bulb, but it often covers the soil in continuous sheets and its masses of deep blue flowers are one of the most glorious features of English oakwoods on light soil in April and May. The bluebell frequently picks out isolated patches of sand in a woodland (as for instance the fragments of "Reading Sands" met with here and there in the general covering of clay-with-flints on the Chiltern plateau), forming practically pure societies exactly coinciding with the sandy soil, while it is scarce or absent on the surrounding heavier soil.

Societies of heavier soils

4. **Mercurialis perennis.** Of the societies which affect heavier and more basic soils that dominated by dog's mercury (*Mercurialis perennis*) is one of the most clearly marked. Though most characteristic of the chalk beechwoods (see Chapter XVIII) it is often found in the oakwoods, especially on the heavier loams. It avoids acid soils except where there is an abundant supply of water (Salisbury, 1916, p. 104) or of nutritive salts (De Silva, 1934). Under optimum conditions the rhizomes of the dog's mercury send up closely packed shoots often a foot (30 cm.) or more high and almost excluding other plants. The mercury is one of the earliest woodland species to start growth after the winter rest and in a favourable season it begins to flower in March or even February in the south English woods. Where the mercury shoots are not too closely set *Arum maculatum, Adoxa moschatellina, Ranunculus auricomus, Ficaria verna* are among the species most characteristically associated. The last-named may also form societies under similar soil conditions.

5. **Sanicula europaea.** The wood sanicle (*Sanicula europaea*) forms societies on medium loams under similar conditions to those of the mercury societies. This species is also specially characteristic as a dominant of chalk beechwood societies, but is often abundant in oakwoods on loams.

6. **Primula vulgaris.** The primrose (*Primula vulgaris*) is one of the commonest and best known of our woodland species. It occurs on a wide range of soils extending from light or medium loams to heavy clays, but avoids the coarser and more acid sands and very calcareous soils. The primrose is commonly a frequent member of societies such as those of sanicle and anemone, but sometimes is so abundant as to assume domi-

Societies of the Field Layer

nance. In two sharply defined areas of calcareous boulder clay in Cambridgeshire and adjoining counties the oxlip (*Primula elatior*) completely replaces the primrose. On the edge of these areas is a well-defined zone of hybrids between the two species and it is thought that the primrose is gradually encroaching on the oxlip area—"hybridising it out of existence" (Miller Christy, 1922).

More local societies. Other abundant or locally abundant species of the oakwood flora may rise to dominance here and there, forming local societies, e.g. the enchanter's nightshade (*Circaea lutetiana*), the ground ivy (*Nepeta hederacea*) or the yellow archangel (*Galeobdolon luteum*). Adamson (1912) describes a society dominated by wild strawberry (*Fragaria vesca*) in an oakwood on calcareous boulder clay belonging to facies (4) (Fig. 53), with *Geum urbanum*, *Primula*, *Circaea* and *Nepeta* abundantly associated.

Societies of very damp or wet soils

7. **Ranunculus repens.** On soil with a really high water content and subject to winter inundation, Salisbury (1916, p. 103) records a society dominated by the creeping buttercup (*Ranunculus repens*). This species is not uncommon in woods on rather heavy soils.

8. **Filipendula ulmaria.** The meadow-sweet (*Filipendula ulmaria*) is described by Adamson (1912) as dominant in (ash-)oak-hazel wood on calcareous boulder clay in Cambridgeshire, accompanied by *Angelica sylvestris*, *Cirsium palustre*, and *Rumex viridis*; and in the wettest patches, where the soil remains nearly saturated throughout the year, by various sedges (*Carex* spp.) and rushes (*Juncus* spp.), the ragged robin (*Lychnis flos-cuculi*), and the purple loosestrife (*Lythrum salicaria*). Where there was sufficient light *Deschampsia caespitosa*, which is a characteristic large tufted grass of wet or "puddled" areas in open woods and grasslands on loamy soils and clays, was co-dominant with the meadow-sweet (Fig. 53).

9. **Urtica dioica.** On very damp soil rich in decaying leaf litter the common stinging nettle (*Urtica dioica*) is often dominant. This plant is well known to frequent localities with a soil rich in combined nitrogen, such as cow-byres and other places about farmyards, and Olsen (1921) has shown that it requires good supplies of nitrates and is hence an index of high nitrifying power in the soil. For this nitrification ample water supply and not too high acidity (pH 5–7) are required. Where the conditions favour it the nettle forms practically pure societies, excluding other plants by its vigorous growth.

Societies of fully illuminated areas

10. **Melandrium, Galeobdolon, and Euphorbia amygdaloides.** When coppice has been recently cut a society dominated by the red campion (*Melandrium dioicum (rubrum)*, *Lychnis diurna*) sometimes develops, and this is often associated with the yellow archangel (*Galeobdolon luteum*), the wood spurge (*Euphorbia amygdaloides*), and other species which benefit

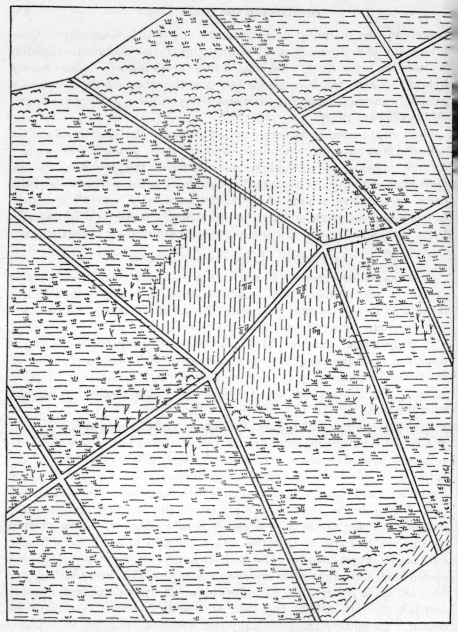

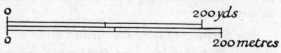

— Filipendula ulmaria society
⩊ Filipendula-ulmaria-Deschampsia-caespitosa society
~ Fragaria vesca society
/ Mercurialis perennis society
||| Pteridium aquilinum society
||| Holcus mollis society
⋇ Carex spp.
Υ Juncus spp.
π Polytrichum gracile

markedly from the increased light. These plants, which form nothing but vegetative shoots in the deeper shade, flower freely and produce a brilliant show of colour in such open coppice. The red campion, though extremely abundant in many parts, is entirely and rather inexplicably absent from certain counties of southern England.

11. *Epilobium angustifolium.* The rose-bay willow-herb (*Epilobium angustifolium*), a plant of very wide distribution in the temperate climates of the northern hemisphere, with its bright rose-coloured flowers, is one of the best known of the various species called "fireweeds" in North America, and very often colonises recently cleared areas in woods, or burnt places on heaths, in great quantity. This plant propagates itself very freely by underground runners (suboles) and also by means of its plumed seeds, which are produced in great numbers and are carried long distances by the wind. Through the great flights of its seeds this willow-herb may suddenly appear in places from which it was previously quite absent. Thus in the early years of the century it appeared and flowered in profusion on some extensive vacant building areas in central London. When at all deeply shaded the willow-herb does not flower, but its vegetative shoots persist for a long time in moderately open woods. The species is most characteristic of light acid sandy soils, but is not absent from the heavier, nor from calcareous, soils.

12. *Digitalis purpurea.* The biennial foxglove (*Digitalis purpurea*), one of our handsomest and most conspicuous flowers, resembles the rose-bay in its habit of suddenly appearing in great numbers in open spaces which have been recently cleared or burned, where it also produces glorious shows of colour. It can, however, flower freely in deeper shade than is tolerated by the *Epilobium*. The foxglove is generally reputed to be a marked calcifuge, and is most abundant on light soils.

Both these species, which are tall plants belonging to the upper field layer, are commonly accompanied in their favourite habitats by the characteristic plants of light acid soils: *Pteridium aquilinum, Holcus mollis, Galium saxatile, Hypericum pulchrum*; also by species which are generally abundant on these soils, such as *Potentilla erecta, Veronica officinalis, Teucrium scorodonia, Viola riviniana, Conopodium majus*, etc.

Species of the field layer

The following lists of herbaceous species, bryophytes, and larger fungi include those which occur commonly or characteristically in oakwood,

---

FIG. 53. SOCIETIES OF FIELD LAYER IN GAMLINGAY WOOD (CAMBRIDGESHIRE)

The general soil of this (ash-)oak-hazel wood is chalky boulder clay, and the field layer is dominated mainly by *Filipendula ulmaria* (damp to wet oakwood type). The pear-shaped area consists of sandy loam, and the field layer there is marked by the dominance of *Pteridium aquilinum* and *Holcus mollis* (dry oakwood type). In this hazel is sparse and there is much birch, and no ash, which is abundant (coppiced) through the rest of the wood. Cf. Pl. 16. From Adamson, 1912.

though they do not of course include every species which may be present. The species of lighter and drier soils are on the whole those which require more light for their vegetative processes and especially for flowering. On the whole also, the plants of the lighter and drier oakwood soils are those adjusted to a relatively low pH value, i.e. to the relatively high acidity which is commonly met with in the sands. The factors mentioned do not, however, run by any means exactly *pari passu*. Thus we may have permanently moist sands and highly calcareous sands, also highly acid clays and clays with a high percentage of calcium carbonate, while very clayey soils which bake hard and crack in summer then present a physiologically dry and unfavourable habitat. Different species show every degree of variation in the combinations of their requirements and the ranges of their tolerance; and again the absence of a factor usually required may be compensated by the presence of another, so that the same species may flourish in two habitats which appear very different, without necessarily having a wide range of tolerance.

(1) *Dry oakwood facies* with distinctly acid soil (pH 5–6), usually sandy; shrub layer poor or absent, and when present relatively seldom coppiced. Closely related to the heathy type and to the commonest type of Quercetum sessiliflorae. Many of the species are intolerant of deep shade (Pl. 16).

Dominants

Upper field layer (aestival species):
Pteridium aquilinum,
Rubus fruticosus (often semi-evergreen),
Epilobium angustifolium (open places)

Lower field layer:
Scilla non-scripta (vernal)
Agrostis tenuis (aestival)
Holcus mollis (aestival)

Associated species

Anemone nemorosa	la	Melampyrum pratense	la
Conopodium majus	f	Oxalis acetosella	f
Convallaria majalis	r and l	Polygonatum multiflorum	r and l
Digitalis purpurea	la	Potentilla erecta	f–a
Galium saxatile	f	P. sterilis	f
Hieracium boreale	l	Sedum telephium	r–o
H. vulgatum	l	Solidago virgaurea	o–f
Hypericum pulchrum	f	Teucrium scorodonia	a
Lathyrus montanus	o	Veronica officinalis	o–f
Lonicera periclymenum	f–a	Viola riviniana	f–a
Luzula pilosa	o–f		

Ground layer (bryophytes):

Catharinea undulata	o	Eurhynchium praelongum	f
Dicranum scoparium	f	Polytrichum formosum	f
Dicranella heteromalla	f	P. gracile	o

The following are among the more frequent species of larger fungi:

Amanita mappa	f–a	Marasmius erythropus	o–f
A. phalloides	f	Psatherella gracilis	o–f
Clavaria cinerea	o	Tricholoma chrysites	la
Hebeloma crustulineforme	f–a	T. sulfureum	f
Inocybe geophila	a		

Species of Damp Oakwood Facies

Tricholoma sulfureum is apparently exclusive to oakwoods, while *Cortinarius acutus* (rare) and *Marasmius undatus* (very rare) are apparently exclusive to dry oakwoods (Harley).

(2) *Medium to damp oakwood facies.* The central type of Quercetum roboris. Shrub layer well developed: hazel commonly the dominant species, usually coppiced and often planted. Soil typically a loam with a wide range of texture: reaction typically somewhat acid to neutral, say pH 6–7 (Pls. 13–15).

Dominants

Lower field layer (Prevernal species):
- Anemone nemorosa
- Ficaria verna
- Mercurialis perennis
- Primula vulgaris
- Sanicula europaea (vernal-aestival)

Vernal species:
- Galeobdolon luteum ⎱ sometimes dominant
- Melandrium dioicum ⎰ in recently
- Nepeta hederacea   cut coppice
- Circaea lutetiana (aestival)

Associated species

Abundant or frequent:
- Ajuga reptans
- Anthoxanthum odoratum
- Arctium minus
- Arenaria trinervia
- Arum maculatum
- Brachypodium sylvaticum
- Bromus ramosus (asper)
- Carex sylvatica
- Cirsium palustre
- Epilobium montanum
- Euphorbia amygdaloides
- Fragaria vesca
- Geranium robertianum
- Geum urbanum
- Hypericum perforatum
- Lathyrus montanus
- Listera ovata
- Luzula pilosa
- Melampyrum pratense
- Myosotis arvensis
- Oxalis acetosella
- Poa nemoralis
- Stachys officinalis (betonica)
- S. sylvatica
- Stellaria holostea
- Veronica chamaedrys
- V. montana
- V. officinalis
- Vicia sepium
- Viola riviniana

Frequent to occasional or local:
- Adoxa moschatellina
- Allium ursinum
- Angelica sylvestris
- Aquilegia vulgaris
- Asperula odorata
- Athyrium filix-femina
- Cardamine flexuosa
- C. pratensis (Kent)
- Carex helodes (laevigata)
- C. remota
- Daphne laureola
- Dipsacus sylvestris
- Dryopteris dilatata
- D. filix-mas
- D. spinulosa
- Epipactis latifolia
- Helleborus viridis var. occidentalis
- Luzula forsteri
- Lysimachia nemorum
- L. nummularia
- Milium effusum
- Myosotis sylvatica
- Narcissus pseudonarcissus
- Orchis maculata
- O. mascula
- Ranunculus auricomus
- Rumex viridis (condylodes)
- Scrophularia nodosa
- Stellaria nemorum
- Teucrium scorodonia
- Valeriana officinalis
- Vicia sylvatica

Rare and local:
- Danaa (Physospermum) cornubiensis (south-western)
- Gagea lutea
- Lathraea squamaria
- Neottia nidus-avis
- Phyteuma spicatum (south-eastern)
- Polygonatum multiflorum

Commonest dominants of the ground layer (bryophytes):

Catharinea undulata   Thuidium tamariscinum

Associated species, sometimes dominant locally

Brachythecium purum	Hylocomium triquetrum
B. rutabulum	Hypnum cupressiforme
Eurhynchium praelongum	Lophocolea bidentata
E. striatum	Mnium hornum
Fissidens taxifolius	M. undulatum
Hylocomium splendens	Plagiochila asplenioides
H. squarrosum	

The following are among the more characteristic species of larger fungi.

Boletus scaber	o	Lactarius fuliginosus	o
B. subtomentosus	f	L. quietus	f
Cantharellus cibarius	f	L. subdulcis	f
Claudopus variabilis*	la	L. vellereus	f–la
Collybia fusipes*	o	Russula adusta	o–f
Cortinarius elatior	o–f		

* Attacking dead oakwood.

Cortinarius hinnulius (occasional), *Lactarius vellereus*, and *Tricholoma sejunctum* (occasional) are apparently exclusive to oakwood (Harley).

The lichen *Peltigera canina* is not infrequent in grassy woods.

(3) *Damp to wet facies* (insufficiently investigated): soil a heavy clay or with very high ground water.

Local dominants

Deschampsia caespitosa	Ranunculus repens
Filipendula ulmaria	Urtica dioica

Associated species

Ajuga reptans	Juncus bufonius
Carex sylvatica	J. conglomeratus
C. paniculata	J. effusus
C. pendula	J. inflexus (glaucus)
Circaea lutetiana	Lysimachia nummularia
Cirsium palustre	Scrophularia aquatica
Galium aparine	Stachys sylvatica

The rushes usually occur in the better lighted places. Several common species of the damp and even of the dry oakwood, e.g. *Anemone nemorosa*, *Mercurialis*, *Nepeta hederacea*, *Oxalis acetosella*, *Poa nemoralis*, and *Viola riviniana*, may extend into the wet woods.

No bryophytes or fungi have been recognised as characteristic of wet oakwoods. *Catharinea undulata* and *Thuidium tamariscinum*, on the whole the two most abundant mosses of damp oakwoods, occur also in the wet.

(4) *Calcareous oakwood facies*—"*ash-oak*" wood. When coppiced = (ash-)oak-hazel wood: on marls, calcareous clays, calcareous sands or thin loams over chalk. Soil alkaline but with the surface layers often somewhat acid. Ash very abundant, usually coppiced. Shrub layer well developed and containing several species other than hazel. Field layer marked by abund-

Calcareous Facies

ance of several calcicolous species in addition to those of the damp oakwood.

The shrub layer may contain the following species:

Corylus avellana	va–d	Ligustrum vulgare	la
Fraxinus excelsior*	va	Euonymus europaeus	f
Acer campestre*	f–a	Sambucus nigra	f
Cornus sanguinea	f–a	Viburnum lantana	f

* These trees are commonly coppiced along with the shrubs.

The field layer consists of most of the same societies as occur in the damp oakwoods, especially *Mercurialis perennis* with *Fragaria vesca, Brachypodium sylvaticum* and *Circaea lutetiana*: on the damper soils societies of *Allium ursinum, Filipendula ulmaria* (Fig. 53) and *Carex pendula* are met with. In two well-defined areas in the eastern counties the oxlip (*Primula elatior*) is dominant (replacing *P. vulgaris*) in the ash-oakwood on chalky boulder clay (Miller Christy, 1922). The oxlip is confined to these areas in Britain.

The associated species are also largely the same as those of the damp oakwood; but some of the following calcicolous (or more strictly basiphilous) species are likely to be present:

Campanula trachelium	Lithospermum purpureo-caeruleum
Clinopodium vulgare	(south-western)
Colchicum autumnale (mainly western)	Paris quadrifolia
	Vicia sylvatica (r–o)
Epipactis latifolia	Viola hirta
Hypericum hirsutum	V. sylvestris
Iris foetidissima (mainly south-western)	

Of bryophytes *Eurhynchium striatum, Thuidium tamariscinum, Porotrichum alopecurum, Mnium undulatum* and *Hylocomium triquetrum* are characteristic.

Watt (1924, p. 175) records the following bryophytes as frequent or abundant in the ash-oak associes which precedes beechwood in seres 2 and 3, i.e. on the more or less calcareous loams, of the South Downs plateau (see Chapter XIX):

Brachythecium purum	Hylocomium loreum
B. rutabulum	H. triquetrum
Catharinea undulata	Mnium undulatum
Eurhynchium praelongum	Porotrichum alopecurum
E. striatum	Thuidium tamariscinum
Fissidens taxifolius	

REFERENCES

ADAMSON, R. S. An ecological study of a Cambridgeshire woodland. *J. Linn. Soc.* (*Bot.*), **40**, 339–87. 1912.

ADAMSON, R. S. The woodlands of Ditcham Park, Hampshire. *J. Ecol.* **9**, 114–219. 1922.

CHRISTY, MILLER. *Primula elatior* Jacquin: its distribution in Britain. *J. Ecol.* **10**, 200–10. 1922.

DE SILVA, B. L. T. The distribution of "calcicole" and "calcifuge" species in relation to the content of the soil in calcium carbonate and exchangeable calcium, and to soil reaction. *J. Ecol.* **22**, 532–53. 1934.

HOPKINSON, J. W. Studies on the ecology of Nottinghamshire. I. The ecology of the Bunter sandstone. *J. Ecol.* **15**, 130–71. 1927.

MOSS, C. E. Geographical distribution of vegetation in Somerset: Bath and Bridgwater District. *Roy. Geogr. Soc.* 1907.

MUKERJI, S. K. Contributions to the autecology of *Mercurialis perennis* L. *J. Ecol.* **24**, 38–81, 317–39. 1936.

OLSEN, C. The ecology of *Urtica dioica*. *J. Ecol.* **9**, 1–18. 1921.

SALISBURY, E. J. The Oak-hornbeam woods of Hertfordshire. Parts I and II. *J. Ecol.* **4**, 83–120. 1916.

SALISBURY, E. J. The ecology of scrub in Hertfordshire: a study in colonisation. *Trans. Herts. Nat. Hist. Soc.* **17**. 1918.

TANSLEY, A. G. *Types of British Vegetation*, pp. 76–83. Cambridge, 1911.

WATT, A. S. On the ecology of British beechwoods with special reference to their regeneration. Part II. The development and structure of beech communities on the Sussex Downs. *J. Ecol.* **12**, 145–204. 1924; **13**, 27–73. 1925.

WILSON, M. Plant distribution in the woods of north-east Kent. Part I. *Ann. Bot., Lond.*, **25**, 857–902. 1911.

WOODHEAD, T. W. The ecology of woodland plants in the neighbourhood of Huddersfield. *J. Linn. Soc. (Bot.)*, **37**, 333–406. 1906.

WOODRUFFE-PEACOCK, E. A. A fox-covert study. *J. Ecol.* **6**, 110–25. 1918.

Chapter XIV

PEDUNCULATE OAKWOOD (*continued*)

REGENERATION. SUCCESSION
QUERCETUM ROBORIS UNDER EXTREME CONDITIONS

General failure to regenerate. Regeneration within the wood, i.e. the successful germination of acorns, establishment of seedlings and growth of young trees to replace those which are removed or die by accident, disease or old age, is now not only very far from general but usually quite lacking in British oakwoods. The floor of a wood may be thickly littered with viable acorns in late October, and sometimes even covered with first-year seedlings in the following summer, but older seedlings, saplings and young trees are quite rare in our semi-natural woods. Various suggestions as to the causes of this failure to regenerate, which must obviously be a new phenomenon, since the original oakwoods maintained themselves for some thousands of years, had often been made, but the subject was first systematically investigated in this country by A. S. Watt in 1914–15 and 1919 (Watt, 1919). He distinguished three periods of danger to successful regeneration of the oaks: (1) the stage of the ripe acorn before germination, (2) the period of germination, (3) that of the establishment of the first-year seedling and its growth into a sapling.

Causes of failure. (1) *Dangers to the fallen acorn before germination*. These are two, the danger of being eaten by an animal and the danger of drying up and thus losing viability. Acorns which are lying freely exposed on the surface of the soil commonly disappear with startling rapidity, often within a few days. Even when they are concealed in the herb cover or in the leaf litter or humus they may also disappear quite soon, though not so rapidly or universally.

Of fourteen different mammals and birds known to eat acorns in this country Watt considered that pheasants, wood pigeons, jays, rabbits, mice and voles are the principal offenders. In woods used for preserving pheasants these birds may be the main agents of destruction: on light soils where rabbits abound they may be largely responsible. But it seems that mice and voles (see Chapter VI, p. 140) are the most widespread destroyers of acorns, since they are more evenly distributed on all kinds of soil. Mice and voles not only run over the surface but burrow below the humus and in the surface layer of soil, and can thus find acorns which escape rabbits and birds.

(2) *Conditions necessary for germination*. Ripe acorns which escape destruction by animals have, so to speak, two chances of germination. They may germinate immediately before they have lost 20 per cent of

their weight by evaporation of water: alternatively if they lose from 20 to 27 per cent, but not more, they may reabsorb the necessary water from a film of liquid water with which they may come into contact in the soil. If the loss exceeds 27 per cent the power of germination is lost altogether. For this reason, unless there is continuous wet weather, acorns lying exposed on the ground will fail to germinate because they will lose more than the critical proportion of water. This is an important cause of the general failure of regeneration in dry oakwood where the surface of the soil is level and devoid of a continuous layer of loose litter and humus. Seedling oaks are often found locally where fallen leaves have collected and formed humus in depressions of the soil. Acorns falling on a close mat of grass or moss will also be desiccated and destroyed. In damp oakwood, on the other hand, where there is a continuous layer of litter and humus so that the fallen acorns are soon covered and are frequently in contact with films of water, they are not only more likely to escape destruction by birds or rodents (unless mice are very abundant) but the conditions for germination are good, and regeneration is more likely to be generally successful. A tall growth of close herbage prevents the acorns reaching the soil at once and thus leads to their desiccation.

(3) *Dangers to the seedling*. These are mainly two: destruction or crippling by animal nibbling, and insufficient light. Mice and rabbits are again the most important enemies. The mice and voles often merely bite off the top of the seedling and leave it lying on the ground: they may also bite through the stem below the surface of the humus, apparently to remove the obstruction in their burrow. Rabbits on the other hand eat off the entire shoot, to judge from the disappearance of seedlings on the sandy soils where rabbits abound. Watt found that in exposed areas on sandy soil only 1 per cent of seedlings remained after nearly 4 years; in partially protected areas 6·4 per cent; and in areas fully protected by close meshed wire netting 27 per cent. On unprotected clay areas 8·5 per cent had survived.

A mildew (*Oidium quercinum*) very commonly attacks oak seedlings (first the leaves and then the stem) and may be observed on a variety of soils and under all conditions of illumination. The seedlings may, however, withstand the effects of the fungus and grow normally if they are exposed to full light. It seems that the fungus has more effect in destroying the seedlings on sand than on clay soil.

The intensity of light required under different conditions for the vigorous growth of oak seedlings has never been exactly determined; but it is certain that the oak is a "light demander" through all stages of its life history, and that seedlings flourish best in the open or in the lightest of shade. For this reason regeneration is impossible under the shade of parent trees in close canopy. A thick growth of bracken will seriously enfeeble or kill seedlings by cutting down the light, and a mass of dead fronds will drag down seedlings and prevent successful growth. The frequent dominance of bracken in the field layer of dry oakwoods is thus inimical to regeneration in this type of community.

On a review of the facts described by Watt, it is apparent that the attacks of animals, and particularly of the widespread mice and voles, are one of the main causes of general failure of oakwood to regenerate; and there can be no doubt that the constant war carried on by gamekeepers against the carnivorous birds and mammals which prey upon the small rodents has contributed in an important degree to preventing regeneration in the existing English oakwoods. (See also Barrington Moore's observations described in Chapter VI, pp. 140–1.)

SUCCESSION

In climax communities occurring on land the whole of which has long been systematically utilised there is little or no opportunity for studying primary succession, and this is notably true of our pedunculate oakwoods, which exist for the most part as sharply defined enclosures in the midst of arable or pasture land. Fragments of secondary successions (subseres) may, however, occasionally be seen on derelict ploughland or pasture and on common land bearing heath or grass heath, as well as on the wide grass "verges" which border many of our country roads.

On abandoned arable land—"Broadbalk Wilderness". The history of the land known as "Broadbalk Wilderness" at the experimental station at Rothamsted in Hertfordshire is an example of such a subsere (Brenchley and Adam, 1915). The soil is a heavy loam containing about 3 per cent of calcium carbonate, and before 1882 the area carried a wheat crop every year. In that year the wheat was not harvested and was allowed to seed. Four years later very few wheat plants remained, the cultivated crop having been speedily ousted by invading plants. Of the details of this invasion unfortunately no records were kept, but in 1914 one-half of the area, which had been left quite untouched,[1] was reported to be occupied by "a dense thicket of trees and shrubs". This thicket "is really an oak hazel wood with which are associated various species of *Rubus*...". "The thickness of the wooded area checks the development of the undergrowth, but *Hedera helix* occurs throughout, and in the spring *Arum maculatum*, *Sanicula europaea*, and *Viola canina* [? = *V. odorata*[2]] were found in the interior. Towards the outskirts *Geum urbanum*, *Brachypodium sylvaticum*, *Mercurialis perennis*, *Heracleum sphondylium*, *Arrhenatherum elatius*, *Stachys sylvatica* and *Urtica dioica* were noted."

In May 1938 the oakwood in Broadbalk Wilderness was again listed by

---

[1] On the other half the woody plants were grubbed up as soon as the tendency to woodland development became clear, and this treatment has since been periodically repeated. Brenchley's paper (1915) deals mainly with the fluctuations of herbaceous plants on the portions from which the woody plants have been regularly removed.

[2] *Viola odorata* was the only violet present in 1938.

Dr A. R. Clapham and Mr H. Baker. The number of woody species had greatly increased since 1914, when only oak, hazel, bramble and ivy were recorded. Of trees there were now four or five species.

Acer campestre	f	Fraxinus excelsior	one
A. pseudoplatanus	ld	Quercus robur	ld

The sycamores were the largest trees, though some were dying. In the shrub layer the following were present:

Acer campestre	f	Ligustrum vulgare	lf
Cornus sanguinea	o	Prunus spinosa	o
Corylus avellana	f–a	Salix caprea	r
Crataegus monogyna	o	Sambucus nigra	r
Ilex aquifolium (marginal)	o		
Rubus fruticosus (agg.)	la	Rosa sp.	o
Hedera helix*	sd	Solanum dulcamara	o

* Dominant in the field layer through most of the wood.

All the herbs recorded in 1914 were still present but none in general abundance, and the only additions were *Galium aparine* (o) and *Conopodium majus* (r). The species of greatest frequency was *Mercurialis perennis*, which was locally abundant at the south end where ivy was absent. The youth of the wood and correspondingly dense growth of the young trees and shrubs, together with the general dominance of ivy in the field layer, no doubt account for the poor development of the field layer. *Heracleum sphondylium* and *Anthriscus sylvestris* were present on the edge of the wood.

"**Geescroft Wilderness.**" Another piece of land at Rothamsted, "Geescroft Wilderness", on which cultivation was abandoned about the same time, is of heavy loam with only 0·005 per cent of calcium carbonate, acid in reaction, and nearly always waterlogged. This area has never developed a continuous cover of woody plants, but is dominated mainly by *Deschampsia caespitosa* accompanied by a large number of herbs and grasses (seventy-seven species in all), among which marsh plants are prominent, and by a number of common woodland and grassland bryophytes. In 1913 it was "studded with a few small trees and shrubs": *Ulex europaeus, Corylus avellana, Quercus robur, Acer campestre, Sambucus nigra, "Ulmus sativa", Fraxinus excelsior, Crataegus* and species of *Rubus, Rosa* and *Prunus*. (Brenchley and Adam, 1915.)

These few facts are interesting as establishing (1) the rapidity with which young oakwood develops on a "good" soil left to itself (Broadbalk) and (2) the much slower succession on the acid and waterlogged soil (Geescroft) which here clearly passes through the stage of wet grassland dominated by *Deschampsia caespitosa*.

Succession from arable and heath land. Salisbury (1918) has described and given lists of the species occurring in four examples of "woodland scrub" (see p. 475) in Hertfordshire. These, as he says, are certainly to be regarded

PLATE 17

Phot. 36. Edge of High Wood near Burwash, E. Sussex. Oak standards in open canopy. Dense coppice of hornbeam (right and centre) and hazel (left). *R. J. L.*

Phot. 37. Oakwood on the River Cassely, Sutherlandshire. Shrubs absent, *Agrostis tenuis* dominant in the field layer. Note the tree trunks covered with lichens. *A. S. Watt.*

OAKWOOD IN THE EXTREME SOUTH AND THE EXTREME NORTH OF GREAT BRITAIN

PLATE 18

Phot. 38. Loose scrub of hawthorn and dog rose colonising neutral pasture, oakwood behind. Gault Clay near Didcot, Oxon. *R. J. L.*

Phot. 39. Scrub of gorse and hawthorn colonised by young oaks (next stage of succession). Same locality. *R. J. L.*

SUCCESSION TO PEDUNCULATE OAKWOOD ON CLAY PASTURE
DIDCOT, OXON.

as examples of stages in the succession from arable land (two examples) and heath (two examples) to oakwood—Quercetum roboris with hazel or hornbeam or both as the dominant shrubs. In one case (Bricket Wood scrub with 143 species), owing to periodic cutting and fires, the scrub is more or less stabilised, but in the others it is "clearly transitional to the type of woodland characteristic of the soils on which these scrubs are encountered". The number of species of vascular plants found in the scrub areas is large, 125, 96, 143 and 99 respectively. The herbaceous species represent mixtures of arable weeds and grassland and heath species which have persisted, along with "marginal" and "shade" species of the woodlands. The two latter categories, as Salisbury shows, correspond very well with those of the climax woodlands. Particularly striking is the close correspondence with the flora of the coppiced woodlands, in which the so-called marginal species, i.e. those which do not permanently tolerate deep shade, are numerically an important feature. These scrub floras are large because each represents a mixture of three different communities, the original flora of arable weeds, or of pasture or heath, the "semi-shade" (marginal) vegetation of wood edges and hedgerows, and the shade flora of the climax forest. The lowest number of species (96) was recorded from an area which was arable land so recently as 20–25 years before it was examined.

Abandoned pasture. No detailed studies have been made of the subsere initiated by the abandonment of pasture land on the heavier soils, but Adamson (1912, see ref., p. 289) says that on the chalky boulder clay of western Cambridgeshire "pastures which have been allowed to run wild for some years are becoming colonised by shrubs which form a more or less dense scrub; this scrub is mostly composed of hawthorns, springing from the surrounding hedges, but in older stages there appear also hazel, oak, maple and ash". (See Ross's observations described in Chapter xxiv, pp. 480–4.) Examples of this subsere may be observed, indeed, in all parts of the south and midlands of England, and it is clear from the most superficial observation that *Crataegus* is nearly everywhere the pioneer shrub. That *Quercus robur* can also directly colonise such grassland may be seen by the frequent abundance of established oak seedlings on the edge of an abandoned pasture bordered by an oakwood.

Plate 18 shows phases in the colonisation of an old pasture field by woody vegetation on Gault Clay near Didcot, Oxon. One end of this field is regularly pastured by a few cattle and kept clear of shrubs, but the beasts rarely use more than half the field, so that a loose scrub of *Rosa* and *Crataegus* in the centre is succeeded at the other end by denser scrub of *Crataegus* and *Ulex europaeus* freely colonised by oaks from a neighbouring small wood.

Grass verges. On the broad grassland verges of country roads bordering oakwoods colonisation by woody plants may often be observed and various stages of the succession traced. Thus, on the edge of Hell Coppice on the Oxford Clay near Oxford the verge consists of grassland with *Festuca rubra* and *Carex diversicolor* (*flacca*) dominant[1] and in several places scrub has developed, the pioneer being *Prunus spinosa* which suckers readily from shrubs on the edge of the wood. This blackthorn scrub is accompanied by

Crataegus monogyna	Rosa spp.
Corylus avellana	Pyrus malus
Cornus sanguinea	Fraxinus excelsior (saplings)
Salix atrocinerea	Quercus robur (saplings)
Ligustrum vulgare	

thus establishing the basis of a damp oakwood vegetation.

Developing oakwood on clay-with-flints. A brief account is given by Adamson (1922, pp. 201–3) of oakwood (Quercetum roboris) developing in "grassland" on the clay-with-flints on the plateau of the South Downs (Hampshire). The soil was a light brown fine-grained loam with a high proportion of fine sand and coarse silt, total carbonates 2 per cent and pH about 6. Flints were numerous from a depth of 3–4 in. downwards. The feeding roots of the young oaks began at a depth of 3 in. and reached a maximum depth of 6 in. The colonisation had occurred from some large adjacent oaks, and consisted of young trees of all ages and sizes, ranging from seedlings to saplings 25 or 30 ft. in height, crowded in the centre and more scattered on the outskirts. Associated with the dominant oaks were *Prunus avium* and *Ilex aquifolium* (frequent), *Betula pendula* (*alba*), *Fraxinus* and *Crataegus* (occasional). Seedlings of all these species except the ash were noted. In the central part of the young wood, where the oaks were crowded, the ground was mostly bare, but the following undershrubs and herbaceous plants were present:

Rubus leucostachys	r	Fragaria vesca	r
R. macrophyllus	o	Mercurialis perennis	r
Ajuga reptans	o	Nepeta hederacea	r
Arctium minus	r	Primula vulgaris	o
Cirsium palustre	r	Veronica chamaedrys	f
Digitalis purpurea	o	V. officinalis	o
Epilobium montanum	o		

all of which occurred also in the outer fringe; and the bryophytes

Brachythecium rutabulum	f	Hypnum cupressiforme	f–a
B. salebrosum	—	Mnium hornum	l
*B. velutinum	r	*Porotrichum alopecurum	o
Catharinea undulata	r	*Lophocolea bidentata	o
*Fissidens bryoides	l		

of which those marked with an asterisk were confined to the centre. In the outer fringe where the oak canopy was not continuous there was a very much larger number of species, including *Pteridium*, which was locally dominant, sixty-six Phanerogams and fifteen bryophytes. The phanerogams included very few grasses, none of which was abundant, and a large number of marginal woodland species. The "grassland" thus colonised was really a

---

[1] Probably due to the addition of lime to the soil of the verge.

mixed community of herbs with a few species of grass, and no one species generally dominant.

It is impossible to be sure of the history of this so-called "grassland"—very likely it arose on the site of older woodland which had been cleared, but a comparison of its composition with the lists of plants (Brenchley and Adam, 1915, pp. 198–9) colonising the portion of Broadbalk Wilderness from which the woody plants have been systematically removed as they come back, suggests that the natural development of oakwood on bare loams is not preceded by a community in which grasses are overwhelmingly dominant (unless the area has been pastured), but by a mixed community of herbs in which grasses are represented. (In Broadbalk, however, some species of grass has always been dominant.) This community is very soon colonised by woody plants if seed is available in the neighbourhood. The protection of the herb cover is probably necessary for the successful establishment of the shrub and tree seedlings, but no direct evidence is available. Shrubs and trees colonise the area together, and the climax community is not preceded by a shrub community as such, except in so far as the more rapid establishment and development of the shrubs leads to a transitory phase dominated by bushes, shortly succumbing to the trees, the latter invading and developing more slowly, though they become tall and dense enough to kill out most of the shrubs comparatively soon.

Deflected succession. The portion of Broadbalk Wilderness from which the woody plants were systematically removed affords an excellent example of Godwin's "deflected succession" (see Chapter x, p. 225), the deflecting factor being the prevention of the development of trees and shrubs, not, as is usually the case, by pasturing, but by grubbing up.

Lists of the flora are available from 1886 (four years after the wheat crop was abandoned), 1895, 1903 and 1913. Forty species of vascular plants were present in 1886, forty-nine in 1895, fifty-seven in 1903, and sixty-five in 1913. Of the last list eighteen occurred in that year as weeds in the wheat crop on the adjoining field, and nine more had been present as weeds in the crop in 1867 but had disappeared from the cultivated area in 1913. In 1886 and also in 1895 dominance was shared by *Agrostis tenuis* and *Medicago lupulina*, both persistent weeds in the wheat crop. In 1903 these species had lost dominance, "being replaced by *Dactylis glomerata*, which formed 35 per cent of the herbage in that year, *Lathyrus pratensis* coming a good second with 18 per cent. In 1913 both of these dominants had retired in their turn, their place being taken by *Arrhenatherum elatius* and *Centaurea nigra*" (? *C. nemoralis*) (Brenchley and Adam, 1915).[1] The last named had steadily increased since 1886, while *Arrhenatherum* was first recorded in

---

[1] It is noteworthy that the same association of the false oat-grass and the black knapweed is characteristic of the grassy road verges on calcareous soil in Cambridgeshire and other counties. On these verges also the woody plants are kept down, here by annual cutting and mowing (see p. 561).

1903. Of the other species some have remained comparatively constant throughout the 27 years, others have disappeared altogether, and others again are newcomers at various times. It is doubtful if this deflected sere had in 1913 reached a position of equilibrium. The only woody plant which was not removed was the ivy (*Hedera helix*), which in 1913 occupied a considerable portion of the area, forming dense carpets on the ground (the largest about 25 ft. across) practically excluding other species; on the edges of these patches the ivy competed for dominance with the herbs. Ivy was present more or less all over the area.

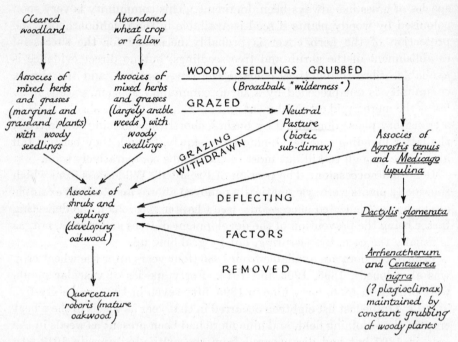

Fig. 54. Scheme of some Subseres and Plagioseres leading to Quercetum roboris

Fig. 54 is an attempt to represent diagrammatically the various types of succession that have been described.

QUERCETUM ROBORIS UNDER EXTREME CONDITIONS

Wistman's Wood. Some account must be given of a very small wood (Wistman's Wood) which has probably as good a claim to be considered a "virgin" wood as any in the British Isles. It is situated in South Devon on a Dartmoor valleyside in a region of high rainfall (estimated at 67–70 in.) and in a very exposed position, between 1200 and 1400 ft. alt., on shallow and markedly acid soil: under conditions in fact which we commonly associate with the sessile oak. Nevertheless it is a pure *Q. robur* wood,

PLATE 19

Phot. 40. Part of the central division of Wistman's Wood seen across the valley of the West Dart, showing restriction of the wood to the belt of "clatter". Above the clatter is the gentler slope of the moor with Littaford Tor on the skyline. *R. J. L.*

Phot. 41. Southern end of Wistman's Wood showing bracken dominant on the clatter outside the wood. The "lammas shoots" arising from the crowns of the trees are well seen. *R. J. L.*

WISTMAN'S WOOD, DARTMOOR

PLATE 20

Phot. 42. Young oaks on the upper edge of the wood, with bracken and bilberry. Slope of the moor behind. *R. J. L.*

Phot. 43. General view inside the wood, oak trunks and branches covered with epiphytes. *Luzula maxima* dominant in the field layer: *Dryopteris dilatata* and *D. spinulosa*. *R. J. L.*

DETAIL OF WISTMAN'S WOOD, DARTMOOR

certainly emphasising the fact that this species can successfully maintain itself in such a habitat. It is probably a remnant of extensive valleyside woods most of which were destroyed by tin miners, and owes its preservation to the trees being rooted between granite blocks.[1]

Two other woods on Dartmoor, Piles Wood and Black Tor Beare or Copse, the upper edge of which is said to reach an altitude of 1530 ft., occupy similar situations and have a similar vegetation. The trees, however, are less deformed and attain a greater height though a lesser girth than those of Wistman's Wood: correspondingly they are more closely placed. This is supposed to be due to the fewer spaces available for root systems between the large granite blocks at Wistman's. The oldest trees in Wistman's Wood have been estimated on fairly good grounds to have an age of about 500 years. Those in the other two woods, where cutting once took place, are probably not so old.

Hansford Worth (in Christy and Worth, 1922) suggests, with much plausibility, that these woods survive not only because they are of little value as timber and would be very laborious to remove, but also because cattle found little to eat among the great boulders which the trees colonised; and finally, because the trees could not have withstood the violent winds to which they are exposed unless their roots, and often also their trunks and the lower parts of their branches, had been wedged between rocks. From this last point it would seem to follow that the habitat of these woods represents the extreme limits of altitude and exposure at which oaks can maintain themselves in the Dartmoor region.

Situation and nature. Wistman's Wood (Pls. 19–24), comprising three small fragments of oakwood of a total extent of 4 acres (Christy in Christy and Worth, 1922) lies on the west-facing slope (left bank) of the valley of the West Dart two miles north of Two Bridges, and is the most remarkable of the three small oakwoods which occur in similar situations on the great elevated granitic mass of Dartmoor.

The three fragments of which Wistman's Wood consists lie between altitudes of about 1200 and 1380 ft. (365–420 m.).[2] They are all based on the conglomeration of angular masses of granite, locally known as "clatter", which occupies the strip of hillside between the "grass moor" above and the valley bottom below. It is evident from the small outlying fragments of scrub that the wood once formed a continuous belt along the side of the valley coincident with the strip of clatter, which is now dominated for the most part by *Pteridium* and *Vaccinium myrtillus*, with *Ulex gallii* occasional to abundant (Pl. 19).

[1] It is interesting to compare Wistman's Wood with Birkrigg and Keskadale oaks (*Q. sessiliflora*) described by Leach from about the same altitude (1000–1500 ft.) on Cumberland hillsides, and evidently also living in extreme conditions (see Chapter xv, pp. 320–4).

[2] These figures were determined by aneroid readings checked with the contour lines marked on the 1 in. Ordnance Map and giving good agreement.

The floor of the wood is extremely irregular, being formed of the uneven surfaces of the great masses of granite with deep crevasses between, the whole masked by a thick covering of bryophytes and vascular plants, so that most of the wood can only be penetrated by cautious crawling over the granite masses. Here and there are small areas of comparatively level ground.

Soil. The soil is very various in depth, corresponding with the inequalities of the rock foundation. Apart from the crevasses, in which the oak roots doubtless descend to considerable depths, it is very shallow and consists in some places of pure humus above the rock, while in others there is an admixture of mineral soil. There is generally a surface layer of litter about an inch deep. The reaction is always markedly acid. In one instance where the soil was only 2 in. (5 cm.) deep over the rock, though it bore *Deschampsia flexuosa, Luzula pilosa, Teucrium scorodonia, Pteridium, Dryopteris dilatata* and *Rubus*, the pH value was 4·8: in another, where it was apparently pure humus, more than 8 in. (2 dm.) deep and bore *Holcus mollis, Luzula pilosa, Potentilla erecta, Vaccinium myrtillus, Rubus* and *Pteridium*, the pH value of the top 3 in. was 4·3: in a third sample the top 3 in. showed pH 5·1.

Habit of trees. The patches of wood are composed almost entirely of dwarf trees of *Quercus robur*. Most of the oaks are much contorted, with thick procumbent trunks and main branches lying on or between the rocks, and rising only 8–15 ft. (2·4–4·5 m.) in the vertical height of their tallest aerial shoots. A few individual trees are 20–25 ft. (6–7·6 m.) high with thick erect trunks and wide-spreading crowns. One isolated tree is only 10 ft. (3 m.) high but its crown has a spread of 25 ft. (7·6 m.). A few relatively protected trees within the upper edge of the south wood are symmetrically developed—one pyramidal, about 18 ft. (5·5 m.) high and 16 in. (40·6 cm.) in girth near the base, another about 25 ft. high and 19 in. (48·2 cm.) round at the base. Hansford Worth (Christy and Worth, 1922) measured 26 trees occupying an area of 6500 sq. ft. with an average distance between the trees of 13·5 ft. (4·1 m.). These varied from 9 to 20 ft. in height (average 14 ft. 7 in.) and from 19 to 102 in. in girth (average 49 in.). The tallest oak seen in any part of the wood was 26·5 ft. (*c.* 8 m.) high, and the greatest girth 106 in. (*c.* 270 cm.). There is no distinct shrub layer (Pl. 20, phot. 43).

Flora. The following is a list of the vascular plants:

Trees and shrubs

Quercus robur	d	Ilex aquifolium	r
Sorbus aucuparia	o–f	Salix atrocinerea	o

Woody climbers

Hedera helix	f	Lonicera periclymenum	f

Phot. 44. Detail of field layer: *Luzula maxima*, *Oxalis acetosella*, *Corydalis claviculata*, *Hedera helix*, *Dryopteris dilatata*. R. J. L.

Phot. 45. An old oak with epiphytes: thick mantle of bryophytes, *Polypodium vulgare* (right and top), *Vaccinium myrtillus* and *Sorbus aucuparia* (left). R. J. L.

WISTMAN'S WOOD: FIELD LAYER AND EPIPHYTES

PLATE 22

Phot. 46. Another oak with epiphytes: *Vaccinium myrtillus, Dryopteris dilatata, Polypodium vulgare.* R. J. L.

Phot. 47. *Sorbus aucuparia* 10 ft. (3 m.) high, epiphytic on an oak, the base of whose trunk is seen in the bottom left-hand corner. R. J. L.

EPIPHYTES IN WISTMAN'S WOOD

PLATE 23

Phot. 48. Northern, fully exposed fragment of wood (alt. 1350 ft.). Lammas shoots are seen arising from the tree crowns. Clatter with bracken dominant in the foreground. *R. J. L.*

Phot. 49. Wind-cut oak with horizontal main branch directed eastwards at top of the northern fragment of wood (to the right of Phot. 48). Alt. 1380 ft.—the highest point of the wood. *R. J. L.*

EDGES OF WISTMAN'S WOOD

PLATE 24

Phot. 50. Patch of clatter (lichen- and moss-covered) above and north-east of the northern fragment of Wistman's Wood. Young shrubby oaks and one thin rowan bent away from the wind. Limiting conditions for colonisation, alt. 1410 ft. *R. J. L.*

Phot. 51. Protected fragment of wood in the valley of the West Okement (north-western edge of Dartmoor). The large tree is *Quercus sessiliflora. Betula pubescens, Sorbus aucuparia, Corylus, Salix atrocinerea,* alt. 1000 ft.

DARTMOOR WOODS, LIMITING CONDITIONS

Field layer

Vaccinium myrtillus	d (mostly on rocks)	Holcus mollis	ld
		Sedum anglicum	a, ld on rocks
Luzula maxima	a, ld in deeper shade	Teucrium scorodonia	o
Rubus fruticosus (agg.)	f	Galium saxatile	l, largely on rocks
Dryopteris dilatata	f		
D. spinulosa	o–f	Oxalis acetosella	o
D. filix-mas	lf	Corydalis claviculata	o
Pteridium aquilinum	l (in gaps)	Luzula pilosa	o
Potentilla erecta	o	Deschampsia flexuosa	o

There have also been recorded from the wood by Harris (1921):

Anthoxanthum odoratum	Geranium robertianum	o
Athyrium filix-femina	Scilla non-scripta	o
Blechnum spicant	Stellaria holostea	f
Digitalis purpurea		

Ground layer. The boulders and trunks and branches of the trees are alike covered with a thick carpet of bryophytes. No list is available; presumably it would be long. The following are recorded: *Antitrichia curtipendula* in large masses, hanging down to a length of a foot or more and producing fruit in abundance (H. N. Dixon), *Eurhynchium myurum, Dicranum scoparium, Hylocomium loreum, Ulota crispa*; and of lichens the following (Christy and Worth, 1922, p. 303) on the authority of Lorrain Smith and Paulson:

Cetraria glauca	Karmelia fuliginosa var. laetevirens
Cladonia cervicornis	Lecanora intumescens
C. coccifera	Opegrapha herpetica
C. fimbriata	Parmelia caperata
C. furcata	P. omphaloides
C. rangiformis	P. perlata
C. squarrosa	P. physodes
C. sylvatica	P. saxatilis
Evernia prunastri	Ramalina siliquosa
Graphis elegans	Usnea florida var. hirta

Vascular epiphytes. The prevalence of procumbent trunks and horizontal branches covered with a thick mantle of bryophytes leads to the accumulation of humus on their upper surfaces and unusually favourable conditions for the growth of epiphytic vascular plants (Pls. 21, 22). Seven species were recorded, a considerably greater number than has been met with in any other British wood and a very much larger proportion of the total vascular flora:

Polypodium vulgare	the most abundant	Oxalis acetosella	o
		Sedum anglicum	r
Dryopteris dilatata	o	Sorbus aucuparia	o
Luzula maxima	a	Vaccinium myrtillus	f

It is obvious that the habitat provided by the layer of humus accumulating on the upper surfaces of the horizontal trunks and branches is not very different from the similar layer on the rocks of the wood floor.

"**Lammas shoots.**" A peculiarity of the oaks in and around Wistman's Wood is the frequent production, especially by the younger oaks, of long stout sappy annual shoots, often 1 ft. (3 dm.) and up to 2 ft. (6 dm.) in length,[1] comparable in appearance and character with, but much larger and more vigorous than the "lammas shoots" often formed about midsummer by oaks growing in more normal habitats. One of the longest and stoutest of these was 10 mm. in diameter at the base and 5 mm. half-way up: it bore 14 large leaves. The explanation is probably as follows. When the oak buds first unfold in May the weather is often still so severe that many are killed and little growth can be made. By midsummer the increased warmth leads to the accentuation of the formation of "lammas shoots", into which goes most of the growth energy of the tree. Most of these shoots are killed (by frost or wind) in the first year, but a few persist and lead to rapid height growth of the young saplings. As noted by Miller Christy and Hansford Worth the younger oaks are perfectly healthy plants, though stunted by the severe conditions of their existence. They are, however, very extensively attacked by gall flies, several kinds of gall occurring on most of them.

Regeneration. It is clear that the wood actually does regenerate, for there are numerous young saplings just outside it, especially along its upper (most protected) edge,[2] but also here and there in the clatter at some little distance from the wood; and acorns are quite freely produced on the older trees. Not all of these can ripen, for those seen in the most exposed (northern) patch of wood were still quite small—not half grown—on 15 September 1932; but clearly some do. There are no seedlings on the lower side of the wood towards the stream where there is more grass and more animals feed. Dung of the ponies, sheep and cattle pastured on the moor was seen right up to the edge of the wood, but no evidence was found that they eat off the oak seedlings. The highest altitude (430 m.) and the most exposed position reached by oaks from the wood is a patch of clatter 150 m. to the north-east of the northern patch of wood. In this are three good shrubby oaks 4 ft. (1·2 m.) high, one with an erect trunk 4 in. (1 dm.) thick at the base, and one rowan 7 ft. (2·1 m.) high, but thin and evidently suffering (Pl. 24, phot. 50).

REFERENCES

ADAMSON, R. S. The woodlands of Ditcham Park, Hampshire. *J. Ecol.* **9**. 1922.

BRENCHLEY, WINIFRED E. and ADAM, HELEN. Recolonisation of cultivated land allowed to revert to natural conditions. *J. Ecol.* **3**, 193–210. 1915.

CHRISTY, MILLER and WORTH, R. HANSFORD. The ancient dwarfed woods of Dartmoor. *Trans. Devon. Ass. Sci. Lit. and Art.* **54**, 291–342. 1922.

HARRIS, G. T. Ecological notes on Wistman's Wood and Black Tor Copse, Dartmoor. *Trans. Devon. Ass. Sci. Lit. and Art.* **53**, 232–45. 1921.

SALISBURY, E. J. The ecology of scrub in Hertfordshire. *Trans. Herts. Nat. Hist. Soc.* 1918.

WATT, A. S. On the causes of failure of natural regeneration in British oakwoods *J. Ecol.* **7**, 173–203. 1919.

[1] Pl. 19, phot. 41 and Pl. 23, phot. 48.   [2] Pl. 20, phot. 42.

Chapter XV

SESSILE OR DURMAST[1] OAKWOOD
(QUERCETUM PETRAEAE OR SESSILIFLORAE)

ENGLISH WOODS

Soil conditions. Oakwoods dominated by the sessile oak (*Quercus petraea* (Matt.) Liebl. or *Q. sessiliflora* Salisb.) are characteristic of the older siliceous rocks forming the hills of the west and north of England, of Wales and Ireland, including sedimentary and metamorphic strata derived from all the Palaeozoic formations, such as grits, sandstones, mudstones, slates, schists and shales, with associated igneous rocks. Woods of *Quercus sessiliflora*, to name some of the chief examples, are recorded from metamorphic rocks in Co. Galway: from Ordovician rocks in the Lake District, in north and central Wales and in Co. Wicklow in Ireland: from the Silurian and Malvernian (Archaean) of Herefordshire: from Old Red Sandstone in Co. Kerry, in Hereford and in Somerset: from the Millstone Grit and Coal Measures of the Carboniferous on the Pennines; and from various Palaeozoic strata in Devonshire and Cornwall. They also occur on leached soil derived from the Palaeozoic limestones, e.g. the Carboniferous of North Wales, and the Silurian limestones of Herefordshire.

As Moss (1913) pointed out, these "siliceous soils are usually much shallower than sandy soils [derived mainly from the Secondary, Tertiary and post-Tertiary sands and sandstones]: they contain less silica: they are frequently more finely grained in texture". These differences, particularly shallowness of soil, Moss held to be causally correlated with the dominance of *Q. sessiliflora*, instead of its congener, *Q. robur*. Salisbury (1916) took the view that dominance of *Q. sessiliflora* is typically correlated with relatively high soil acidity, and that its prevalence in the west is due to the greater intensity of leaching and acid humus formation in the wetter western climate, as well as to the general presence in the west of siliceous rocks poor in bases.

In default of careful experiments on the behaviour of the two species of oak in relation to varying depths and acidities of soil, it is impossible to make dogmatic statements.[2] But it is to be noted that *Q. sessiliflora* is also

---

[1] This name appears to have originated with the woodmen in the New Forest (Nichols, 1791) and to have been applied to *Q. sessiliflora* from a mistaken idea that the species is characterised by dark coloured acorns ("dun mast", corrupted to "dur mast"). It has obtained a fairly wide currency (*New English Dictionary*, 3, 725, Oxford, 1897).

[2] Culture experiments with first-year seedlings of the two species show that both make better height growth on the more acid soils, and that there is no significant difference between them in this respect.

dominant or abundant on soils of the midlands and south-east of England which have relatively high acidity as a common character and are derived from Secondary, Tertiary or post-Tertiary rocks, e.g. the Bunter Sandstone of Sherwood Forest in Nottinghamshire (Hopkinson, 1927), the Lower Greensand of Kent and Surrey, the Reading Beds (Eocene) of Burnham Beeches in south Buckinghamshire, and the same formation in Hertfordshire, the London Clay of north Kent (Wilson, 1911), the Bagshot Sand of Hampstead (Tansley, 1913), and the glacial sands round Danbury in Essex (Adamson, 1932). Hopkinson (1927) suggests that, in Sherwood Forest, where both species occur, the increasing abundance of *Q. sessiliflora* is correlated with increasing acidity of the soil. But on some of these acid soils *Q. robur* also occurs abundantly and may be dominant, and it cannot always be stated with certainty that either species to the exclusion of the other is the natural dominant on such a soil. *Q. robur*, as we have seen, is certainly the natural dominant of some very acid soils.

It has been explained in Chapter XIII that substantially the same field layer societies may exist in woods dominated by the pedunculate and by the sessile oak, but that those of the Quercetum sessiliflorae conform much more often to the "heathy" types of the more acid "Dry oakwoods", while the characteristic societies of the Damp ("oak-hazel wood") type are less commonly seen in the sessile oakwoods. Another distinguishing feature is the greater variety and luxuriance of the shrub layer in Quercetum roboris, and to this is related the much greater frequency of coppice-with-standards in those parts of the country where the pedunculate oak mainly flourishes. This form of exploitation is, however, by no means unknown in Quercetum sessiliflorae, since sessile oak standards often co-exist with good hazel coppice. The sessile oak itself is, or was, coppiced wholesale (originally for fuel and tanning bark) in many of the typical sessile oak-woods on acid soils, where hazel is not so abundant. On the other hand the great majority of sessile oakwoods in this country are in "high forest" with close or semi-open canopy and relatively few shrubs (Pls. 25–28)[1] though hazel is often the most frequent of these (Pl. 29, phot. 61; Pl. 30, phot. 63).

Hertfordshire sessile oakwoods. The woods dominated by the sessile oak which occur in the south-east of England, i.e. in exactly the same climate as the much commoner woods of pedunculate oaks described in Chapter XIII, certainly resemble the latter much more closely than do the sessile oakwoods of the older siliceous rocks of the north and west. The best account of these south-eastern Querceta sessiliflorae is to be found in Salisbury's study of the oak-hornbeam woods of Hertfordshire (1918 a).

---

[1] These differences were the origin of Moss's early distinction (1907) between "oakwoods" and "oak-hazel woods" in Somerset, the former being Quercetum sessiliflorae in high forest on sandstones with shallow soil, the latter Quercetum roboris with hazel coppice on shales and marls.

Soils. The woods in Hertfordshire in which the sessile oak is associated with hornbeam occur on the gravels and sands of the Woolwich and Reading beds (Lower London Tertiaries), on glacial gravels, and also on glacial Boulder Clay and London Clay. On these two last, however, the upper layers of the soil are not typical clays, but are distinctly loamy or even sandy, in some cases perhaps because of the eluviation of the finer particles which are carried down to lower levels, leaving the coarser constituents nearer the surface. There is evidence also that the uppermost stratum of the London Clay is sandy in original constitution, as it certainly is on and near Hampstead Heath, on the northern outskirts of London, where this uppermost (Claygate) bed, bearing sessile oak-hornbeam wood, forms a passage bed to the overlying Bagshot Sand (Tansley, 1913).

In the Hertfordshire oak-hornbeam woods the soils on which the sessile oak is dominant have a substantially higher range of the gravel-sand fractions, though they are always wetter in winter than those dominated by the pedunculate oak. The humus content is lower, but the acidity of the surface humous layer is probably higher.

Light intensities. In the woods where the dominant sessile oak is associated with a well developed shrub layer of hornbeam coppice the light intensity during the light phase (early spring) is considerably lower than in similar woods of pedunculate oak, owing to the closer tree canopy. In the shade phase (summer) on the other hand the light intensity is higher than in the pedunculate oakwoods. The result is that the prevernal and vernal societies of field layer species so characteristic of the typical oak-hazel coppice are absent or poorly developed in these sessile oak-hornbeam woods. The summer flora of the oak-hornbeam coppice is extremely poor, but in the close canopy woods where the hornbeam coppice layer is very sparse or absent, it is rich in species tolerant of shade, to the exclusion of the gregarious vernal species whose abundant development depends especially on the periodic stimulus given by the cutting of the coppice.

Composition and structure. *Tree layer.* The tree layer is dominated by *Quercus sessiliflora*. *Q. robur* and hybrids may be present, but in that case planting may be suspected. In addition to dominating the shrub layer *Carpinus betulus* is frequent as a tree. Of other trees *Betula pubescens* is the most abundant, *B. pendula* (*alba*) occurring in much less quantity. *Sorbus aucuparia* is only occasional but very characteristic, while *S. torminalis*, *Pyrus malus*, *Alnus glutinosa* and *Ilex aquifolium*, though very rare, are also characteristic, i.e. they do not occur in the pedunculate oakwoods of the region.

Ash, wych elm and gean, on the other hand, characteristic of the pedunculate oakwoods, are here rare, or locally common only in hollows where the soil is damper, of finer texture, and with a higher content of basic salts.

Shrub layer.

The shrub layer is dominated by the coppiced hornbeam (Fig. 55) which never attains the density it has in the *Q. robur-Carpinus* woods. The two hawthorns (*Crataegus monogyna* and the large-fruited *C. oxyacanthoides*) are frequent, together with abundant brambles, roses, and sallows. *Frangula alnus* and *Viburnum opulus*, like the wild service

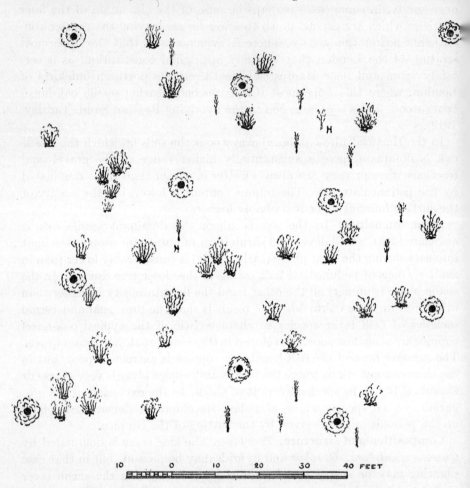

Fig. 55. Oak-hornbeam Coppice with Standards

Chart showing spacing of trees and shrubs. There are 10 oak standards in the space of about 10,000 sq. ft., or about 43 to the acre. C = hornbeam coppice, H = hornbeam saplings. From Salisbury, 1918 *a*.

tree and the holly, though very rare, are characteristic of these sessile oakwoods. The former indeed is characteristic of woods on this type of soil throughout the country. Hazel, dogwood, elder, field maple and blackthorn are chiefly found in hollows where the soil surface approaches the chalk.

Field layer. This is strikingly scanty in individuals though it includes a comparatively large number of species. The most generally distributed of these are:

Anemone nemorosa	Primula vulgaris
Epilobium montanum	Pteridium aquilinum
Holcus mollis	Rubus fruticosus (agg.)
Luzula pilosa	Teucrium scorodonia
Oxalis acetosella	Viola sylvestris

The continuous sheets of prevernal and vernal gregarious species characteristic of ordinary coppiced woods with (pedunculate) oak standards are conspicuously absent, but a number of societies of the field layer can be recognised, and of these by far the most extensive are those dominated by the brambles and the bracken fern.

Rubus society. The areas not bearing hornbeam coppice are often occupied by a dense growth of brambles, including *Rubus köhleri, rhamnifolius* and *leucostachys*. This *Rubus* society can flourish in a wide range of soil conditions and can endure a diminution of the light intensity in the shade phase down to about 10 per cent of full illumination. The bramble is usually accompanied by the honeysuckle and the bracken fern (compare Chapter XIII, p. 281), and frequently by *Holcus mollis, Anemone nemorosa* and *Oxalis acetosella*. But the society is very poor in species, mainly owing to the smothering growth of the dominant. In the Hertfordshire woods Salisbury thinks that the establishment of the *Rubus* society results from the nibbling of the young coppice shoots by rabbits, which ultimately kills the stools. He has found the decayed remains of the stools of the coppice shrubs among the bramble. If the shrubs are able to shoot vigorously the bramble seedlings, which appear in large numbers the first year after coppicing, are killed out as the shrubs grow in height and close up.

Pteridium society. The bracken fern is sometimes co-dominant with the brambles, and on the lighter soils it is dominant alone, forming extensive societies. Like the bramble society it is poor in accompanying species. The closely growing fronds reduce the light intensity very severely, and even vernal species have difficulty in establishing themselves among the bracken owing to the persistence of the dead fronds of the previous season. *Pteridium* itself does not, according to Salisbury, form a closed society where the light intensity falls below about 11 per cent of full illumination, though individual plants are met with down to 4·1 per cent. *Pteridium*, like *Rubus*, forms societies only where shrubs are absent: it is well known to be avoided by rabbits, and this immunity, together with its steady spread and the deep shade which it casts, gives the plant a great advantage in competition with other species. The slow decay of the fronds leads to the accumulation of a thick layer of raw humus of peat-like character, and this may reach a depth of 5 or 6 in. The characteristic associates of bracken are much the same as those of the bramble society, with the addition of the bluebell (*Scilla non-scripta*).

Smaller societies. Other societies of the field layer, though of much less extent, are those dominated by *Scilla non-scripta*, *Nepeta hederacea*, *Mercurialis perennis* and *Ficaria verna*. The bluebell society Salisbury regards as a local variant of the bracken society, developed where the diminished light of the shade phase does not allow of the complete dominance of the bracken. *Holcus mollis* and *Galeobdolon luteum* are the only conspicuous accompanying species. The *Nepeta* society occurs on soils of slightly finer texture and higher humus content than those of the bluebell society. The associates of the ground ivy are dog's mercury and creeping bugle (*Ajuga reptans*). The mercury society is restricted to damper situations or to places where the underlying chalk is close to the surface, occurring typically on the slopes of depressions and valleys where the soil is

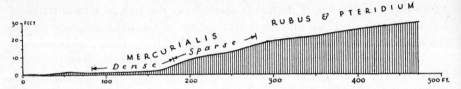

FIG. 56. BRAMBLE-BRACKEN AND MERCURY SOCIETIES IN OAK-HORNBEAM WOOD

Bramble and bracken occupy the higher ground of a slope, giving way to sparse mercury lower down and dense mercury on the flat ground of the valley, correlated with a higher base status of the soil. After Salisbury, 1918*a*.

usually of finer texture (Fig. 56). It is correlated with low soil acidity due either to a dilute soil solution or a relatively high content of mineral salts. Several common woodland plants are associated with the mercury, notably sanicle and wild strawberry, where the soil above the chalk is shallow. At the bottoms of the valleys or depressions, where the soil is always damp in spring and the shade in summer considerable, the mercury society passes gradually into a society dominated by the lesser celandine (*Ficaria verna*). *Arum maculatum* is a frequent associate and *Ranunculus auricomus* is confined to this society: a variant in which *Adoxa moschatellina* is frequent or even co-dominant was met with in two woods.

"Path society." Besides those already described Salisbury distinguishes two distinct "societies" (though they are hardly societies in the accepted sense) which are exposed to much higher illuminations than are those of the field layer of the closed wood. Paths through a wood always show a distinct vegetation, both because of the greater light intensity and also because of the higher soil water content corresponding with higher percentages of silt and clay than are present in the general sessile oakwood soils, and more comparable with those of the pedunculate oakwoods. The soil is moist all the year round and in winter the paths and rides are often flooded. The slight depression of the paths below the level of the surrounding wood floor, together with the compression of the soil by traffic and the consequent decrease of permeability and eluviation are together responsible for this effect. The flora is rich, numbering more than eighty species, of which the two most characteristic are the land form of water starwort (*Callitriche stagnalis*) and the water-pepper (*Polygonum hydropiper*).

The most important associates are other more or less damp-loving plants:

Ajuga reptans	f–a	Mercurialis perennis	f
Cardamine flexuosa (London Clay)	–	Mentha arvensis	o–a
		Nepeta hederacea	f–a
Circaea lutetiana	f–la	Oxalis acetosella	f–la
Fragaria vesca	f–a	Peplis portula	lf–a
Galium palustre	f–a	Poa annua	f–la
Geum urbanum	a	Prunella vulgaris	a
Gnaphalium uliginosum	f	Ranunculus repens	o–a
Juncus bufonius	lf	Stellaria uliginosa	f
Lysimachia nemorum	a		

Many of the more abundant species of the pedunculate oakwood are almost confined in the sessile oakwoods to the path "society", and this is clearly correlated with the higher water content and lower acidity of the soil.

"Marginal society." The most characteristic species of the wood margins, according to Salisbury, are *Succisa pratensis*, *Stachys officinalis* (*betonica*), *Teucrium scorodonia* and *Potentilla erecta* among the commonest, and, of the less frequent, *Carex pallescens*, *C. pilulifera*, *Hieracium boreale* (agg.) and *Luzula maxima*. Of species which are here rare but "significant" Salisbury mentions *Carex strigosa*, *Deschampsia flexuosa*, *Gnaphalium sylvaticum*, *Hypericum androsaemum*, *Lathyrus montanus*, *Serratula tinctoria*, *Sieglingia decumbens* and *Solidago virgaurea*.

Bryophytes. *Ground layer.* The bryophyte and especially the liverwort flora is richer both in species and individuals than in the pedunculate oakwoods of the county, doubtless because of the generally damper soil, the deeper shade of the shade phase and the smaller amount of coppicing. Fifty species of moss and thirty-six liverworts were recorded in all. But most of the commoner mosses of the sessile oakwood are also frequent in the woods of the pedunculate oak, though *Polytrichum formosum*, *Dicranum scoparium* and *Fissidens bryoides* are more generally abundant in the former. The damper and more shaded parts of the woods are frequently occupied by a moss carpet composed of most of the commoner species. The following appear to be specially characteristic of the sessile oakwoods: *Dicranella heteromalla*, *Leucobryum glaucum*, *Mnium punctatum*, *Polytrichum juniperinum*, and in one wood *Tetraphis pellucida* and *Plagiothecium undulatum*. By the sides of the woodland streamlets a society composed of *Dicranella heteromalla* and the liverwort *Calypogeia fissa* is characteristic, with *Plagiochila asplenioides* and *Scapania nemorosa* often associated. The mosses *Eurhynchium praelongum* and sometimes *Porotrichum alopecurum* are particularly abundant where the chalk is close to the surface.

The following are the most frequent liverworts:

Alicularia scalaris	Lophocolea bidentata
Aplozia crenulata	L. heterophylla
Calypogeia fissa	Pellia fabbroniana
Cephalozia bicuspidata	Plagiochila asplenioides var. humilis
Diplophyllum albicans	Scapania nemorosa

Metzgeria furcata and *Radula complanata* were not seen on the bark of the sessile oaks, though the former is quite common on tree trunks in the *Quercus robur* woods and the latter is not infrequent. *Madotheca platyphylla* is confined to the tree trunks where the soil is of low acidity. The liverwort flora numbers considerably more species than in the sessile oakwoods of Derbyshire, though it is far inferior to that of the west of England.

Algae. The terricolous algal flora resembles that of English heaths, and is strikingly similar to that described by Petersen for soils with an acid reaction. The commonest and most characteristic alga is *Zygogonium ericetorum* which occurs in the "path society", often with the liverwort *Alicularia scalaris*. *Mesotaenium violascens*, the *Dactylococcus* stage of *Scenedesmus obliquus*, and *Gloeocystis vesiculosa* are sparsely distributed among mosses on the tree trunks or among the leaves of *Bryum* on the ground. *Hormidium flaccidum*, the only common terricolous alga of the *Quercus robur-Carpinus* woods, and characteristically frequent on neutral or alkaline soils, also occurs, but less abundantly.

The lichen flora is much richer both in species and individuals than that of the *Quercus robur-Carpinus* woods, and the same may be said of the fungi, of which over 450 species have been observed in all. The following are distinctive: *Boletus parasiticus, Craterellus cornucopioides, Nyctalis asterophora, N. parasitica, Scleroderma vulgare, Stropharia aeruginosa* and *Thelephora laciniata*.

Effect of coppicing. The effect of coppicing is much less marked in the Hertfordshire sessile oakwoods than in those of the pedunculate oak. This is largely due to the much denser tree canopy of the former, and hence the vegetation of areas in which the trees have been felled is really more comparable with the coppiced areas of the latter. But nowhere in the durmast oakwoods are seen the sheets of flowering *Anemone nemorosa* or *Ficaria verna* characteristic of the pedunculate oak coppices in the first year after the shrubs are cut; and such species as *Conopodium majus* and *Stellaria holostea* are much less abundant.

One of the most striking features after coppicing has already been described—the great development of bramble seedlings. These are checked and finally smothered as the coppice shoots develop from the stools, and only establish quasi-permanent societies when the shrubs are absent or moribund. The other frequent or abundant constituents of the field layer in recently coppiced or felled areas are:

Agrostis stolonifera
Ajuga reptans
Anthoxanthum odoratum
Arenaria trinervia
Centaurium umbellatum
Fragaria vesca
Galeobdolon luteum

Holcus lanatus
H. mollis
Hypericum humifusum
Lysimachia nemorum
Veronica montana
V. officinalis

SUCCESSION

Salisbury (1918 a, pp. 41–2, and 1918 b) investigated three areas of neglected land on which scrub was developing in the neighbourhood of *Quercus sessiliflora* woodland. These were acid but heavy soils _{In Hertfordshire} (London Clay and Boulder Clay) and bore *Calluna* and several other heath plants. The scrub was varied and rather rich in species, none of which was generally dominant. Gorse, hawthorn, blackthorn, hazel, shrubby willows, brambles and roses were all represented on one or more of these areas. Of the colonising trees *Betula pendula* (*alba*) was the commonest, and was succeeded by hornbeam and by the two oaks and their hybrids. All three areas were very wet, even in summer, and to this Salisbury attributes the greater prevalence of young trees of *Quercus robur* than of *Q. sessiliflora*, and the large proportion of hazel accompanying the hornbeam.

Adamson (1932) has recorded active development of woody vegetation, apparently leading to *Q. sessiliflora-Carpinus* woodland, on glacial sands, loams and clays associated with Boulder Clay in southern
In Essex   Essex, round the village of Danbury, between Chelmsford and Maldon.

The three subseres. Three subseres may be recognised: (1) on ground that had been cleared of woodland and carried a mixed herb community; (2) on ground dominated by *Agrostis tenuis* (with *Festuca rubra*) not forming a closed turf, but long decumbent shoots; and (3) where the succession is checked by grazing.

(1) In the first subsere the shrubs and trees directly seed the herbaceous community. *Rubus caesius* and other brambles, *Rosa* spp., *Crataegus* and *Salix atrocinerea*, with *Ulex europaeus* and *Sarothamnus scoparius* on the less clayey soils, are the shrubs; on gravel *Betula pendula* (*alba*), but with *B. pubescens* usually more abundant, and *Quercus sessiliflora*, with *Q. robur* occasionally, are the trees. Birch and oak soon attain dominance and form closed thickets 10–15 ft. in height. Within these thickets *Rubus* and *Rosa*, with *Pteridium*, may form very dense undergrowth, while *Crataegus* occurs fairly commonly, and also *Salix atrocinerea*. *Ulex* and *Sarothamnus* become drawn up and are gradually suppressed. *Corylus* and *Carpinus* also appear but are not plentiful in these young tree thickets.

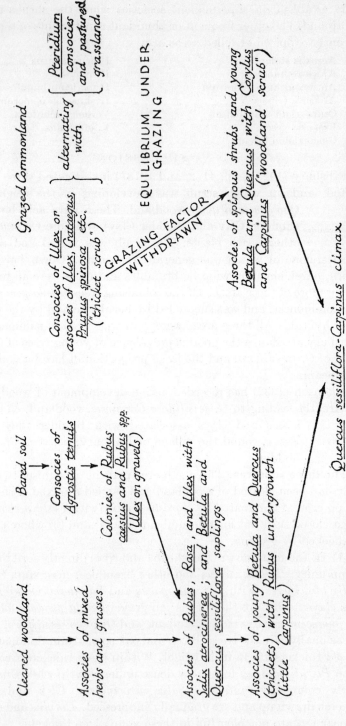

Fig. 57. Subseres on Essex Glacial Gravels leading to Quercetum sessiliflorae

(2) The *Agrostis-Festuca* grassland is invaded by *Rubus*, especially *R. caesius*, and to a less extent by *Rosa* spp. The trailing shoots of the Rubi spread through and over the grasses and establish local dominance. These *Rubus* "islands" form a nidus for the establishment of various woody plants, of which *Salix atrocinerea*, *Betula* spp. and *Quercus sessiliflora* are the commonest. *Crataegus* is always present and, rarely, *Corylus* and *Carpinus*. *Fraxinus* and *Salix caprea* are confined to permanently wet places. On gravelly and other light soils *Ulex* and *Sarothamnus* largely replace the Rubi as pioneers. The *Rubus* "islands" with their associated bushes and young trees gradually increase in size and density till they unite and the trees assume dominance over the bushes.

On the site of a hutment camp, removed in 1919, where almost all vegetation had been eliminated, the ground in 1930 was covered with a dense growth of *Agrostis tenuis* with patches of *Rubus* occupying more than a third of the area, and scattered plants of

Betula pendula (alba)		Rosa arvensis
B. pubescens		Rosa spp.
Carpinus betulus	r	Salix atrocinerea
Crataegus sp.		Sarothamnus scoparius l
Quercus sessiliflora		Ulex europaeus

(3) On the commons, where continual grazing by cattle and goats and occasional burning occur, the vegetation is mainly dominated by alternating *Pteridium* and *Ulex*. *Rubus*, *Rosa*, *Prunus spinosa*, and *Crataegus* are associated with the gorse scrub, and very occasionally small trees of *Betula* and *Quercus*, but there is no obvious sign of advance in the succession, and the gorse scrub appears stable under existing conditions. This is the typical "thicket scrub" of Salisbury (1918*b*). On the remoter parts of these grazed commons, however, there are thickets of hawthorn and blackthorn, to some extent bound together by brambles and briars, and only fringed by gorse; in these are young oaks. In the larger patches are also *Betula*, *Salix atrocinerea*, *Corylus* and *Carpinus*. Here we have Salisbury's "woodland scrub", clearly a stage of development to the climax woodland.

Fig. 57 represents diagrammatically the successions observed round Danbury by Adamson.

It is noteworthy that neither *Corylus* nor *Carpinus* is ever a pioneer in these successions. Neither can establish itself until the conditions of woodland shelter are secured, though there are abundant seed parents available.

THE PENNINE OAKWOODS

The best general description of the Quercetum sessiliflorae of the English older siliceous rocks is given by Moss (1913) for the region of the Southern Pennines (see Fig. 5, p. 17), and is fairly typical of the sessile oakwoods of the north and west. The Pennine oakwoods are developed on the hillsides formed by the non-calcareous rocks of the Carboniferous series—Yoredale shales, Millstone Grit (hard compact sandstone) and Coal Measures (shales and sandstones). They extend to an altitude of about 300 m., not reaching 350 m., and are almost exclusively dominated by *Quercus sessiliflora*.

Structure and composition. The height and girth of the oaks is only moderate and is markedly poor towards the upper limit of altitude (Pl. 25, phot. 53). The trees very seldom grow in actual close canopy (cf. Pl. 25, photos. 52, 53) and enough light normally reaches the forest floor to permit of a closed carpet of vegetation. The shrub layer is seldom luxuriantly developed and the underwood is not regularly coppiced. The following data are adapted and rearranged from Moss (1913). He recognises four facies of these woods distinguished mainly by the vegetation of the field layer, though also partly by the associated trees. These are (1) upland heathy woods the field layer of which is dominated largely by *Vaccinium myrtillus*, (2) those of drier, less humous and less acid soils, with a field layer dominated by *Holcus mollis*, *Scilla non-scripta* and *Pteridium*, (3) those with damper mild humus and no general dominants in the field layer, and (4) those of stream sides and flushes.

Tree layer

Quercus sessiliflora	d	P. avium (lower levels only)	o
Betula pubescens	a		
Ilex aquifolium	a	*Wet places*	
Ulmus glabra (montana)	f–a	Fraxinus excelsior	r–la
Sorbus aucuparia (upland and heathy woods)	f–a	Alnus glutinosa	o–la
		Salix fragilis	l, o
Prunus padus	r–a		

Shrub layer

Corylus avellana	o, la	R. arvensis	f
Salix atrocinerea	o–a	Pyrus malus (as shrub)	r
S. caprea	o–a	Rubus fruticosus (agg.[1])	f–a
S. aurita	la	R. idaeus	r–la
Crataegus monogyna	o–la	Acer campestre (as shrub)	l
Prunus spinosa	r–o, la	Sambucus nigra	o–la
Rosa canina (agg.)	f–la	Viburnum opulus	o

Field layer

The societies of the field layer fall into much the same categories as those described for the Quercetum roboris (see pp. 280–8), but the more oxyphilous are much more prevalent, and many of the upland woods have a typically heathy field layer.[2] On the other hand the "central types" of medium and damp oakwood with nearly neutral humus characteristic of Quercetum roboris (p. 287) are very much less common, and are confined to the damp shales of some of the valleys.

(1) **Heathy woods.** Birch and rowan are almost the only associates of the oak in heathy oakwoods; shrubs are rather rare or almost absent. The

[1] *R. lindleianus*, *R. selmeri* and *R. dasyphyllus* are said to be the most abundant species. [2] Cf. Chapter XVII (Quercetum ericetosum).

PLATE 25

Phot. 52. Oakwood on a Pennine valley side. Note the paucity of shrubs. *Pteridium aquilinum* and *Holcus mollis* dominant in the field layer. *W. B. Crump.*

Phot. 53. Upland oakwood near the altitudinal limit. The trees are small and poorly grown and there are no shrubs. *Vaccinium myrtillus* and *Deschampsia flexuosa* dominant in the field layer. *W. B. Crump.*

PENNINE SESSILE OAKWOODS (QUERCETUM SESSILIFLORAE)

Phot. 54. Oakwood in a Pennine clough. *W. B. Crump.*

Fig. 55. Oakwood at the Devil's Bridge, Cardiganshire. *Quercus sessiliflora* dominant. *Ulmus glabra* on the left, *Fraxinus* on the right.

RAVINE SESSILE OAKWOODS

bilberry, w(h)imberry or whortleberry (*Vaccinium myrtillus*), called in Scotland the blaeberry, is the commonest dominant of the heathy oakwoods, and the following plants are associated:

Frequent or abundant species

Deschampsia flexuosa	a–d	Carex pilulifera	o
Galium saxatile	a	Polygala serpyllifolia	o
Melampyrum pratense	la	Lathyrus montanus	o
Calluna vulgaris	la	Dryopteris dilatata	o–la
Potentilla erecta	f	Blechnum spicant	o–f
Teucrium scorodonia	f	Solidago virgaurea	o–f
Luzula multiflora	f	Pteridium aquilinum	r–a
L. pilosa	o	Digitalis purpurea	r–a

Rare and local species

Molinia caerulea	l	Hieracium vulgatum	r–o
Holcus mollis	l	Pyrola minor	r
Vaccinium vitis-idaea	r, la	P. media	vr

(2) **Societies of drier soils.** In woods developed on dry, moderately acid soils birch, rowan and holly are the associated trees: the shrubs (*Salix* and *Corylus*) are fairly abundant, as also *Rubus* and *Rosa*, *Lonicera* and *Hedera*.

The wood soft-grass (*Holcus mollis*) is the most general and characteristic dominant (Pl. 25, phot. 52). It is often associated with *Pteridium* or *Scilla*, either of which may be dominant or co-dominant with *Holcus*, the first in the aestival, the second in the vernal phase (cf. p. 281).

The following species are associated:

Galium saxatile	a	Anthoxanthum odoratum	o
Potentilla erecta	f	Rumex acetosella	o
Teucrium scorodonia	f	Digitalis purpurea	o
Viola riviniana	f	Galeobdolon luteum	o
Polygala serpyllifolia	f	Luzula pilosa	o
Arenaria trinervia	f	Dryopteris dilatata	o
Conopodium majus	o	D. filix-mas	l
Campanula rotundifolia	o		

(3) **Damp soil with mild humus.** The damper woods with mild humus, mostly situated on rain wash or alluvium, or on shale in valley bottoms, come closest to the central types of Quercetum roboris. Most of the trees and shrubs of the sessile oakwood are present except *Betula*, *Ilex* and the ericaceous undershrubs.

The field layer is composed in the main of many of the species enumerated in Chapter XIII, pp. 287–8, but the formation of well-marked societies appears to be less common, probably owing to the absence of coppicing, a practice which provides conditions favourable to the dominance of single species over considerable areas (cf. the Hertfordshire woods, p. 305). Moss in fact (1911, 1913) does not recognise definite field layer societies in this habitat. The rarity of such species as the primrose and the wood sedge and the much greater abundance of ferns provide a contrast with southern woods on similar soils.

Abundant

Ajuga reptans
Athyrium filix-femina
Dryopteris dilatata
D. filix-mas
Galeobdolon luteum
Luzula pilosa

Luzula sylvatica
Mercurialis perennis
Oxalis acetosella
Scilla non-scripta
Stellaria holostea
Vicia sepium

Locally abundant

Allium ursinum
Arum maculatum
Bromus sterilis
Circaea lutetiana
Dryopteris montana

Equisetum sylvaticum
Ficaria verna
Milium effusum
Sanicula europaea
Veronica montana

Frequent

Anemone nemorosa
Cardamine flexuosa
Epilobium montanum
Fragaria vesca

Heracleum sphondylium
Melandrium dioicum (rubrum)
Viola riviniana

Occasional

Bromus ramosus (asper)
Carex helodes (laevigata)
Deschampsia caespitosa

Festuca gigantea
Holcus lanatus
Melica uniflora

Local

Asperula odorata
Carex remota
Geum urbanum
Hieracium boreale

Hieracium vulgatum
Lactuca muralis
Myosotis sylvatica
Orchis mascula

Rare

Campanula latifolia
Carex sylvatica
C. strigosa
Dryopteris linnaeana
D. phegopteris
Epipactis latifolia
Festuca sylvatica

Lathraea squamaria
Listera ovata
Platanthera chlorantha
Poa nemoralis
Polystichum aculeatum
Primula vulgaris

Very rare

Aquilegia vulgaris
Gagea lutea
Neottia nidus-avis

Paris quadrifolia
Vicia sylvatica

(4) **Stream sides and flushes.** The alder (*Alnus glutinosa*) and ash (*Fraxinus excelsior*) become more abundant and the oak is rare. *Salix fragilis* is occasional at lower altitudes and *Prunus padus* locally abundant. This type is transitional to alderwood, but is included here for the sake of completing the series.

PLATE 27

Phot. 56. General view of oakwood (midsummer). Shrubs absent. *Pteridium*, etc. Forest of Dean. *C. G. P. Laidlaw.*

Phot. 57. Oaks in close canopy. *Pteridium aquilinum, Holcus mollis. C. G. P. Laidlaw.*

FIELD LAYER OF SESSILE OAKWOOD

PLATE 28

Phot. 58. *Scilla non-scripta* in flower. Young *Pteridium* fronds. Late spring.
C. G. P. Laidlaw.

Phot. 59. *Scilla* in fruit, *Pteridium*. Midsummer. C. G. P. Laidlaw.

FIELD LAYER OF SESSILE OAKWOOD

Here again Moss does not describe societies of the field layer, for which he lists the following species:

Abundant

Athyrium filix-femina
Caltha palustris
Chrysosplenium oppositifolium
Cirsium palustre
Equisetum sylvaticum
Filipendula ulmaria
Geum rivale
Iris pseudacorus

Juncus articulatus (lamprocarpus)
J. effusus
Lysimachia nemorum
Myosotis scorpioides (palustris)
Petasites hybridus (ovatus)
Valeriana officinalis var. sambucifolia

Local

Angelica sylvestris
Cardamine amara
Chrysosplenium alternifolium
Crepis paludosa

Dryopteris phegopteris
Epilobium palustre
Stellaria nemorum

Rare

Carex pendula
Cirsium heterophyllum
Dryopteris linnaeana
Geranium sylvaticum

× Geum intermedium
Trollius europaeus
Valeriana dioica

Abundance of ash under high rainfall. In the southern Pennine oakwoods ash is not recorded by Moss as a member of the general community, but only as locally abundant in wet places. This increase of ash on wet soil is in fact a perfectly general phenomenon in all our oakwoods of both pedunculate and sessile types (Fig. 51, p. 273). But in regions of really high rainfall ash frequently becomes an abundant general constituent of oakwood, with its abundance in no way restricted to wet soil. This is well seen in some of the Cumbrian Querceta sessiliflorae. The following list is taken from Naddle Low Forest on the southern shores of Haweswater, on Skiddaw slate, a metamorphic rock of Silurian age. The wood is rather open, and floristically rich for a sessile oakwood:

Quercus sessiliflora d.

Betula pubescens	f–a	Prunus padus	o
Fraxinus excelsior (with many seedlings)	a–f	Pyrus malus	r
		Sorbus aucuparia (seedlings f)	o
Populus tremula	lf		
Corylus avellana	a	Rubus idaeus	f
Crataegus monogyna	o	Salix aurita	r
Prunus spinosa	r	S. caprea	r
Rosa canina	o	Viburnum opulus	r
R. villosa	r	Lonicera periclymenum	f
Agrostis stolonifera	o	Anthoxanthum odoratum	f–a
Ajuga reptans	o	Arrhenatherum elatius	o
Alchemilla alpestris	r	Athyrium filix-femina	o
Anemone nemorosa	o–lf	Carex pallescens	o
Angelica sylvestris	o	C. sylvatica	r

Conopodium majus	f	Lotus uliginosus	o
Corydalis claviculata	r	Luzula pilosa	o
Crepis paludosa	r	Lysimachia nemorum	f–o
Dactylis glomerata	r	Melampyrum montanum	r
Deschampsia flexuosa	lf	Mercurialis perennis	l, r
Digitalis purpurea	o	Oxalis acetosella	a
Dryopteris dilatata	o	Poa pratensis	r
D. filix-mas	f	P. trivialis	o–f
D. linnaeana	o	Potentilla erecta	la
D. montana	f	P. sterilis	f
D. phegopteris	o	Primula vulgaris	o
Epilobium montanum	o–lf	Pteridium aquilinum	ld
Filipendula ulmaria	lf	Rumex acetosa	o
Fragaria vesca	r	Scilla non-scripta	a
Galeopsis tetrahit	r	Scrophularia nodosa	r
Galium aparine	o–r	Solidago virgaurea	la
G. saxatile	l	Stachys betonica	f
Geum urbanum	o	S. sylvatica	o
Geranium robertianum	o–f	Stellaria graminea	r
G. sylvaticum	f	S. holostea	f
Holcus lanatus	r	Succisa pratensis	o
H. mollis	a	Vaccinium myrtillus	r
Hypericum pulchrum	r	Valeriana officinalis	o
Hypochaeris radicata	r	Veronica chamaedrys	f
Lapsana communis	r	V. officinalis	r
Lathyrus montanus	o	Viola riviniana	f

Bryophytes

Brachythecium purum	f	Hypnum schreberi	o
Camptothecium sericeum	o	Leucobryum glaucum	r
Dicranum sp.	o	Mnium hornum	a
Diplophyllum albicans	a	Mn. undulatum	f
Frullania tamarisci (on ash)	r	Orthotrichum affine	o
Hylocomium loreum	a	Plagiothecium sylvaticum	o
H. splendens	a	P. undulatum	f
H. squarrosum	f	Polytrichum formosum	o–f
H. triquetrum	f	Scapania nemorosa	o
Hypnum cupressiforme	a	Thuidium tamariscinum	a
H. molluscum	r		

It is clear from the rather long list of species that the soil of this wood is not generally wet, and since it is not at all calcareous either, the abundance of ash must be attributed to the high rainfall, which is here about 80 in. (2000 mm.).

Malvern oakwoods. The sharply defined ridge of the Malvern Hills only attains a height of 300 to 400 m., but is so narrow (in some places less than a quarter of a mile wide, that its contours are abrupt and mountainous. It runs due north and south and separates the Midland Plain with its clays and marls and woods of the pedunculate oak, on the east, from the predominant Quercetum sessiliflorae of the western hills on the Welsh Marches and in Wales itself. The Malverns themselves, composed of Archaean meta-

PLATE 29

Phot. 60. Halstock Wood on side of the East Okement valley, northern edge of Dartmoor. Near side of valley deforested and pastured. *Pteridium*. R. J. L.

Phot. 61. Interior of Halstock Wood. *Quercus sessiliflora, Corylus avellana, Hedera helix, Rubus* sp., etc. R. J. L.

DEVONSHIRE SESSILE OAKWOOD

PLATE 30

Phot. 62. Edge of Halstock Wood with the East Okement. Left to right, *Salix atrocinerea, Sorbus aucuparia, Quercus sessiliflora* (and *Quercus sessiliflora* behind). *R. J. L.*

Phot. 63. Edge of Halstock Wood. *Ulex europaeus, Corylus, Fraxinus, Salix atrocinerea, Betula pendula, Quercus sessiliflora* behind. *R. J. L.*

DEVONSHIRE SESSILE OAKWOOD: DETAIL

PLATE 31

Phot. 64. Oakwood 15–16 m. high in protected valley at 210 m. alt. *Holcus mollis* and *Pteridium* locally dominant, *Scilla non-scripta*, *Vaccinium myrtillus*, *Polytrichum formosum*. R. J. L.

Phot. 65. Oakwood 6–8 m. high, probably old coppice, at 250 m. alt., less protected: *Vaccinium myrtillus* dominant, *Pteridium*, *Blechnum*, *Sphagnum*. R. J. L.

QUERCETUM SESSILIFLORAE ON THE QUANTOCK HILLS

Phot. 66. Close scrub of sessile oak about 3 m. high at 280 m. alt. *Calluna* and *Pteridium* in front. R. J. L.

Phot. 67. Fully exposed scrub of sessile oak about 2 m. high near the summit of a ridge at 300 m. alt. *Pteridium aquilinum*. Behind in the valley the oakwoods and scrub shown in Phots. 64–66. R. J. L.

QUERCETUM SESSILIFLORAE (SCRUB) ON THE QUANTOCK HILLS

morphic rocks (largely syenite), are mostly covered with grassland and scrub of *Ulex gallii*, but the southern portion is well wooded in places, and the steeply tilted Silurian (May Hill) sandstone and limestones (Woolhope, Wenlock, and Aymestry), which flank the chain to the west, are also largely covered with wood to their summits (maximum elevation about 250 m.). These Malvern woods are largely dominated by *Quercus sessiliflora* (Salisbury and Tansley, 1921), but show a considerably wider range of composition and structure than those of the Pennines.

In the first place many of them have been regularly exploited as coppice-with-standards, like the pedunculate oakwoods of the Worcestershire plain to the east of the hills; and the hazel, perhaps sometimes planted, is abundant, or at least common, in all of them. Secondly, two of the wooded Silurian ridges are limestones and the varied Archaean rocks of the Malvern Hills themselves are locally rich in basic minerals. The resulting soil composition is naturally reflected in the vegetation, which contains numerous "calcicolous" species, especially on the Wenlock (the purest) Limestone, but also on the Aymestry Limestone (less pure), and the Malvernian of the Hills. The May Hill Sandstone, on the other hand, bears much more typical Quercetum sessiliflorae, though even here *Betula pubescens*, much the most abundant associate of the sessile oak in the Pennine and most of the northern woods, is practically absent. *Betula pendula* (*alba*) on the other hand, is frequent in all the woods, and the wych elm and the yew also occur in all, but are especially abundant on the Wenlock Limestone. Ash again, present in all the woods, is most abundant on the Wenlock, where the oak is not well grown on the shallow soil of the steepest escarpments. The woodland of these escarpments approaches nearest, both in dominance of ash and in the rest of the vegetation, to the typical calcicolous ashwoods described in Chapter XXI (cf. Fig. 51). On the gentler dipslopes the soil is deeper and the surface leached: here the oak standards are better, and though the calcicolous element is still marked, a number of more typical oakwood plants occur in abundance. This last type of woodland approximates in fact to the (ash-)oak-hazel wood described in Chapter XIII, p. 272.

Succession. There is little opportunity for the study of woodland succession in the Malvern area, but even on the Archaean rock of the Hills themselves it is noteworthy that ash takes the place of birch as the tree which accompanies the oak. It is probable indeed that it would be the forerunner of oak, both on the Archaean and on the Silurian limestones. On the former there is, in places, pure ashwood on soil practically identical with that which bears oakwood, and on the latter we may fairly conjecture that original ashwood, as on other limestones, tends to give place to oakwood as the soil deepens and leaches.

Other sessile oakwoods in the West of England. Plates 27 to 32 illustrate sessile oakwoods in the Forest of Dean (Monmouthshire), on the northern edge of Dartmoor and on the Quantock Hills in Somerset, with some of their typical undergrowth. Plates 31 and 32 show three facies of sessile oakwood at different altitudes in the Quantock Hills—oak scrub with *Calluna* and *Pteridium* at the highest (Pl. 32), *Vaccinium myrtillus* where there is continuous shade, *Holcus mollis* on the deeper soils where there is more shelter and the oaks are better grown (Pl. 31).

Sessile oakwoods on Mountain Limestone. Although typical woodland of the Carboniferous or Mountain Limestone (by far the most widespread and important limestone of the west and north) is pure ashwood (Chapter XXI) and oak is characteristically absent from calcicolous ashwood; yet *Quercus sessiliflora* is known to occur on soils directly derived from this limestone, e.g. in north Lancashire, in North Wales, at Symond's Yat in the Wye Valley, and in Co. Kerry in south-west Ireland. Sometimes the sessile oak is merely an admixture with the ash, but in other places it is definitely dominant, forming real Quercetum sessiliflorae, with ash no more than occasional, though often with calcicolous elements associated in greater or less degree.

So far as is known, the presence, abundance or dominance of the sessile oak on soils derived from limestones in this country always depends on leaching of the surface soil, as on the Silurian limestones previously described (p. 319). It is noteworthy that most of the Quercetum sessiliflorae on such soils occurs in regions of high rainfall. In the Killarney woods, for example (Chapter XVI), sessile oakwood is developed alike over Old Red Sandstone and Carboniferous Limestone and many associated species are common to the two soils.

Sessile oakwoods at the highest altitudes. Leach (1925) has described two woods, the Birkrigg and Keskadale oaks, occupying 8 and 19 acres respectively, and situated in similar positions (separated by about 1200 m.) on the southern slopes of two adjacent mountain ridges in Cumberland (Fig. 58 and Pl. 33), at 1150–1400 ft. (350–426 m.) and 1000–1500 ft. (305–457 m.) respectively. The soil is "shallow, stony, argillaceous", formed by the weathering of the shaly rock belonging to the "Skiddaw series" of the Lower Silurian. The pH value of the soil is approximately 5. The average slope of the ground bearing the Birkrigg oaks is about 30°, while the Keskadale oaks are situated on a steeper slope of about 40°.

Both woods are almost entirely composed of dwarf trees of *Q. sessiliflora* which grow from 6 to 12 ft. apart, are about 12 ft. (3·6 m.) high (or a little more along the upper margin of the Keskadale wood just below the crags, Fig. 59), and with a girth of 36 in. (90 cm.). The Birkrigg wood is bounded along its upper margin by a narrow belt of open oak scrub.

The oaks in both woods have multiple stems springing from a common stool, suggesting that they have at one time been coppiced. The Keskadale oaks indeed are said to have been coppiced for tanning bark in the latter

PLATE 33

Phot. 68. General view from the south. In the foreground the scree is still partially active and prevents the development of continuous wood. *W. Leach.*

Phot. 69. Interior of Birkrigg Oaks showing mu stems. Bilberry dominant, bracken abundant i field layer. *W. Leach.*

Phot. 70. Oaks sprouting from the base. *Calluna* dominant outside the wood. *W. Leach.*

Phot. 71. Oak whose top has died sprouting fr the base. Bilberry dominant in the field lay *W. Leach.*

BIRKRIGG OAKS, CUMBERLAND (QUERCETUM SESSILIFLORAE)

part of the eighteenth century, but there is no tradition of coppicing of the Birkrigg oaks. Leach shows that the production of multiple trunks results from destruction of the old shoots either by burning or disease, and fresh suckering from the stools. Fires not infrequently occur in the more open parts of the wood where the heather and bracken growing between the oaks provide combustible material, while many of the oak shoots are very unhealthy and others quite dead. In such trees, though to a less extent than after burning, there is a marked tendency for young shoots to arise from the bases of the trees. The shoots of all the oaks are considerably attacked by fungi, the most conspicuous being *Dichaena quercina* in its oidial stage. It

FIG. 58. SKETCH SHOWING THE POSITIONS OF THE BIRKRIGG OAKS (to the left) AND THE KESKADALE OAKS (to the right). See text

From Leach, 1925.

is uncertain whether the fungi are the actual cause of the death of the shoots, or whether the limiting conditions of altitude (exposure) is a cause predisposing to fungal attack. The prevalence of fungi and particularly of galls on the oaks of Wistman's Wood (p. 302) suggests that this is so, and that the altitude of all these woods is really limiting to oaks in the climate of western England.

Sorbus aucuparia is occasional in both woods, but there are no other associated trees.

The field layers of the two woods are practically identical, forming an open community; the fine soil from the open intervals is often washed away from the surface. *Vaccinium myrtillus*, however, which is commonly dominant in the shade of the Birkrigg oaks, is only locally frequent in the Keskadale wood. *Pteridium*, or occasionally *Calluna*, tends to dominance where the lighting is good (Pl. 33, phots. 69–71).

	B.	K.		B.	K.
Agrostis canina	o–f	f–a	Potentilla erecta	o	o–f
Anthoxanthum odoratum	f	f–a	Pteridium aquilinum	f	f–a
Blechnum spicant	o	o	Quercus sessiliflora (seed-	r	
Calluna vulgaris	o	o–la	lings)		
Campanula rotundifolia	o	o	Rubus sp.		o
Deschampsia flexuosa	f	f–a	Sorbus aucuparia (seedlings)	o	o
Dryopteris filix-mas	o		Teucrium scorodonia	o	o
Festuca ovina	o	o	Vaccinium myrtillus	f–d	lf
Galium saxatile	o–f	o–f	Veronica humifusa	o	o
Hypericum pulchrum	r	o	Viola riviniana	o	o
Oxalis acetosella	o–f	o–f			

Bryophytes are abundant, numerous in species, and in the more shaded parts of the wood locally dominant:

	B.	K.		B.	K.
Andreaea petrophila	+	.	Plagiothecium sylvaticum	+	+
Antitrichia curtipendula	+	+	Polytrichum aloides	.	+
Brachythecium purum	+	.	P. formosum	+	+
B. rutabulum	.	+	P. gracile	+	.
Bryum capillare	+	+	P. juniperinum	+	.
Camptothecium sericeum	+	.	P. piliferum	.	+
Campylopus fragilis	.	+	Rhacomitrium heterostichum	+	+
C. schwarzii	+	.	Rh. lanuginosum	+	+
Dicranum fucescens	+	.	Rh. sudeticum	+	.
D. majus	+	+	Thuidium tamariscinum	+	.
D. scoparium	+	+	Ulota crispa	+	+
Eurhynchium praelongum	+	+			
Hylocomium loreum	+	+	Diplophyllum albicans	+	+
H. splendens	+	+	Lophocolea bidentata	+	.
Hypnum cupressiforme	+	+	Lophozia alpestris	+	.
var. ericetorum	.	+	L. barbata	+	+
var. filiforme	+	+	L. hatcheri	+	.
Hypnum schreberi	+	+	L. floerkii	.	+
Mnium hornum	+	.	Marsupella emarginata	+	.
Plagiothecium denticulatum	+	+	Scapania curta	.	+

The most abundant species are *Dicranum scoparium*, *Hypnum schreberi* and *H. cupressiforme*. Of the last named the variety *filiforme* completely covers the lower parts of most of the tree trunks, while epiphytic cushions of *Ulota crispa* are abundant everywhere on the upper branches, together with various lichens. Outside the wood the ground is dominated by *Calluna* or *Pteridium* with *Vaccinium myrtillus*, which is everywhere abundant, sometimes assuming local dominance.

The effect of the strong winds which sweep up the hillside is shown in the evenness of the upper surface of the canopy. The Birkrigg oaks grow on a very uniform slope and are of approximately uniform height, but those of the Keskadale wood increase in height very considerably as the base of the crags is approached, to fall again in the much stunted trees which grow upon the crag itself. The deflection of the wind upward as the crags are approached is well shown in the accompanying diagram (Fig. 59).

Succession. The subsere initiated by burning shows, in the first year, seedlings of the following species:

Calluna vulgaris	f	Anthoxanthum odoratum	o–f
Erica cinerea	o–f	Deschampsia flexuosa	o
Galium saxatile	f	Potentilla erecta	o

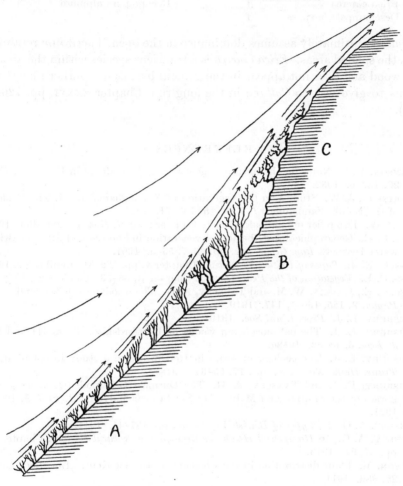

FIG. 59. PROFILE OF THE KESKADALE OAKWOOD

Showing taller growth where there is relative protection from wind. From Leach, 1925.

and the following springing from stocks not killed by the fire:

Quercus sessiliflora	a	Erica cinerea	o
Vaccinium myrtillus	a	Deschampsia flexuosa	o
V. vitis-idaea	f	Sorbus aucuparia	o
Pteridium aquilinum	o–f	Blechnum spicant	o

Thus the bilberry tends at once to resume its dominance in the field layer after a fire, as noted also on the Cleveland moors and elsewhere, and after five or six years the following composition of the field layer is reached:

Vaccinium myrtillus	d	Anthoxanthum odoratum	f
V. vitis-idaea	f–a	Galium saxatile	o–f
Calluna vulgaris (young)	f	Potentilla erecta	o
Erica cinerea	f	Lycopodium alpinum	o
Deschampsia flexuosa	f		

Calluna ultimately assumes dominance in the open, *Vaccinium* retaining it in the shaded areas. *Erica cinerea* is a transient species within the area of the wood and does not appear in the general list: as on southern heaths it seems to give way to *Calluna* in the long run (Chapter XXXVI, pp. 726–7, 734).

REFERENCES

ADAMSON, R. S. Notes on the natural regeneration of woodland in Essex. *J. Ecol.* 20, 152–6. 1932.

HOPKINSON, J. W. Studies on the vegetation of Nottinghamshire. I. The Ecology of the Bunter Sandstone. *J. Ecol.* 15, 130–71. 1927.

LEACH, W. Two relict upland oakwoods in Cumberland. *J. Ecol.* 13, 289–300. 1925.

MOSS, C. E. Geographical distribution of vegetation in Somerset. Bath and Bridgwater District. *Roy. Geogr. Soc.* pp. 36–7, 53–6. 1907.

MOSS, C. E. In *Types of British Vegetation*, Chapter V, pp. 122–31. Cambridge, 1911.

MOSS, C. E. *Vegetation of the Peak District*, Chapter II, 38–58. Cambridge, 1913.

MOSS, C. E., RANKIN, W. M. and TANSLEY, A. G. The woodlands of England. *New Phytol.* 9, 125, 130–1, 147. 1910.

SALISBURY, E. J. *Proc. Linn. Soc.* 1916.

SALISBURY, E. J. The oak-hornbeam woods of Hertfordshire. Parts III and IV. *J. Ecol.* 6, 14–52. 1918a.

SALISBURY, E. J. The ecology of scrub in Hertfordshire: a study in colonisation. *Trans. Herts. Nat. Hist. Soc.* 17, 53–64. 1918b.

SALISBURY, E. J. and TANSLEY, A. G. The Durmast Oakwoods (Querceta sessiliflorae) of the Silurian and Malvernian Strata near Malvern. *J. Ecol.* 9, 19–38. 1921.

TANSLEY, A. G. In *Types of British Vegetation*, pp. 137–41. 1911.

TANSLEY, A. G. In *Hampstead Heath, its Geology and Natural History*, Chapter IV, pp. 87, 92. 1913.

WILSON, M. Plant distribution in the woods of north-east Kent. *Ann. Bot., Lond.*, 25, 889. 1911.

Chapter XVI

SESSILE OAKWOODS (*continued*)

WELSH, IRISH AND SCOTTISH WOODS

Welsh oakwoods. The Welsh oakwoods have not been closely studied, but like practically all others on the siliceous hillsides of the west those which have been examined are sessile oakwoods. The few trees of *Quercus robur* that are sometimes seen are always either near the bottom of a river valley or in situations where they may obviously have been planted.

Most of the sessile oaks have slender trunks and are poorly grown, and some have multiple stems, evidently springing from stools coppiced long ago or from the bases of trees the tops of which have been destroyed by fire or otherwise. Ash and alder are very frequent associates, the latter not only on waterlogged soil but also often scattered freely through the oakwood. Birch and rowan are also pretty constant constituents, and wych elm is often seen. Of shrubs hazel is the commonest and is sometimes abundant. The following is a composite list of the commoner vascular plants, put together from superficial observation of several valleyside woods in north and north-central Wales during the late summer, so that the transient vernal species are lacking:

Quercus sessiliflora d.

Acer pseudoplatanus	r–a, subspontaneous	Fraxinus excelsior	o–f
		Sorbus aucuparia	f
Betula pubescens	f	Alnus glutinosa	la
Fagus sylvatica	o, subspontaneous from planted trees[1]	Ulmus glabra	o
Corylus avellana	f–a, sometimes forming a shrub layer	Rubus fruticosus (agg.)	o–f
		R. idaeus	o
		Salix aurita	o–f
Crataegus monogyna	r–o	Viburnum opulus	r
Prunus spinosa	r–o	Lonicera periclymenum	o–f
Agrostis tenuis	o, ld	Epilobium montanum	o
Anthoxanthum odoratum	o	Festuca rubra	l
Athyrium filix-femina	f–a	Galium saxatile	o–f
Blechnum spicant	o–a	Geranium robertianum	o, l
Dactylis glomerata	o	Holcus mollis	o, ld
Deschampsia flexuosa	f, ld	Hypericum pulchrum	f
Digitalis purpurea	o–f	Lathyrus montanus	o
Dryopteris filix-mas	o–f	Melandrium dioicum (rubrum)	l
D. linnaeana	o		
D. spinulosa	o	Melampyrum pratense	f

[1] It is possible that some of the scattered beeches seen in north and central Wales are really native, i.e. that they are isolated descendants of native beech which was formerly more widespread. See Chapter XVIII, p. 358.

Oxalis acetosella	f–a, ld	Sieglingia decumbens	o
Polypodium vulgare	f	Solidago virgaurea	f
Potentilla erecta	o–f	Stachys officinalis	o
Prunella vulgaris	o	Succisa pratensis	f
Pteridium aquilinum	o–a	Teucrium scorodonia	f
Rumex acetosella	o	Vaccinium myrtillus	l
Sedum telephium	r	Veronica montana	o
S. anglicum	o	Viola riviniana	f

The occasional local dominance of grasses probably indicates grazing within most of the woods.

Among the commonest mosses are *Dicranum majus*, *Polytrichum formosum* and *Hylocomium loreum*.

In one Welsh wood the pH value on a slope with *Dicranum*, *Oxalis*, and *Deschampsia flexuosa* was 5·7: in the bottom with many ferns and seedling ashes it was 6·3.

At the Devil's Bridge. The sessile oakwoods clothing the sides of the steep gorge of the Rheidol near the Devil's Bridge show similar facies. The trees are thin, poorly grown *Quercus sessiliflora* with a few seedlings. On the steep slope of a tributary stream both *Fraxinus* and *Ulmus glabra* are frequent and the oaks much better grown (Pl. 26, phot. 55). *Betula pubescens* is locally abundant, *Sorbus aucuparia* occasional and *Corylus* the commonest shrub.

An almost flat, slightly grazed area showed *Deschampsia flexuosa* dominant and the following short list of accompanying species:

Anthoxanthum odoratum	o	Lysimachia nemorum	o
Blechnum spicant	o	Melampyrum pratense	o–f
Carex limosa (wet place)	o	Oxalis acetosella	f–la
C. remota (wet place)	o	Potentilla erecta	f–a
Festuca rubra	f–a	Rubus fruticosus (agg.)	o
Galium saxatile	o–f	Scilla non-scripta	l
Lonicera periclymenum	o	Solidago virgaurea	o

The following mosses were the most conspicuous:

Brachythecium purum	f	Hypnum schreberi	f
Dicranum majus	a	Plagiothecium undulatum	o
D. scoparium	f	Polytrichum formosum	f
Hylocomium loreum	a		

On a rather deeply shaded stream bank (not grazed) were:

Athyrium filix-femina	Geranium robertianum
Chrysosplenium oppositifolium	Hypericum pulchrum
Circaea lutetiana	Rubus idaeus
Digitalis purpurea	Teucrium scorodonia
Dryopteris filix-mas	Valeriana officinalis

and the following bryophytes:

Dicranum majus	Plagiochila asplenioides
Eurhynchium myosuroides	Polytrichum formosum
Hylocomium loreum	Thuidium tamariscinum

On the very steep west-facing slope of the gorge (very slightly grazed) there were practically no grasses except *Molinia*, which was luxuriant and increased in abundance down the slope. *Vaccinium myrtillus* appeared on the rocky outcrops, with a little *Calluna*. *Blechnum* was frequent to abundant, *Melampyrum* more abundant, while most of the other species were very sparsely present, except *Galium saxatile* and *Potentilla erecta*.

At the bottom of the gorge on the alluvial bank of the Rheidol, still shaded by the oakwood, several more exacting species appeared:

Asperula odorata	Holcus mollis
Dactylis glomerata	Oxalis acetosella
Epilobium montanum	Prunella vulgaris
Fragaria vesca	Sanicula europaea
Geum rivale	Valeriana officinalis

It will be seen from the foregoing lists that except on stream sides where there is accumulated rainwash or alluvium the field layer of most of the Welsh woods consists of the less exacting species occurring in facies (1) and (2) of the Pennine oakwoods (pp. 314–15).

THE KILLARNEY WOODS (Plates 34–39)

In Ireland, the Atlantic woods of Killarney in Co. Kerry (south-west Ireland) have recently received close attention. Though Killarney is not actually on the coast the climate is practically oceanic, with very mild winters, the mean January temperature being as much as 5·9° C. (mean minimum 2·8° C.), the July mean 15·2° C., giving a range of only 9·3° C. between the means of the warmest and the coldest months. The rainfall is heavy (1390 mm. or 54·7 in.[1]), and the mean atmospheric humidity high.

The woods on Carboniferous Limestone have already been mentioned (p. 320), but most of them are based on Lower Devonian rocks of the Old Red Sandstone series. The following account is mainly taken from data (as yet unpublished) most kindly and generously supplied by Dr J. S. Turner, Dr A. S. Watt and Dr P. W. Richards, who in August and September 1936 spent a month in the woods with a party from the Cambridge Botany School.[2]

It is generally agreed that these woods are quite natural, and that so far as is known there has been no planting (except of exotics which are locally obvious, as for example on Ross Island and in the Muckross demesne) nor sowing of acorns. Felling has taken place from time to time, but the woods have been badly neglected for a long while, so that the natural behaviour of the vegetation has had free scope.

The oak-holly wood. *Quercus sessiliflora* is the dominant tree, with a

---

[1] 87 inches (2209 mm.) at the Upper Lake where Derrycunihy Wood, described below, is situated (Praeger, 1934).

[2] The other members of the party were A. Burges, Verona M. Conway, G. C. Evans, Mrs Richards, H. A. Turner, E. F. Warburg.

subordinate layer of *Ilex aquifolium*, giving the characteristic glittering effect of light reflected from the shining leaves familiar to those who know the beech-holly woods of the humid climate of Vizzavona in Corsica or of Sainte Baume in Provence. The margins of the Killarney woods are now the main Irish locality of *Arbutus unedo*, the "strawberry tree", a western Mediterranean tree which occurs also in the Iberian peninsula and in isolated stations in western France, but in the British Isles is only found in south-west Ireland, the nearest continental station being in the Côtes-du-Nord in Brittany. Formerly the Irish distribution of the tree was much more extensive, but its widespread use as firewood has severely restricted its occurrence. *Arbutus* is however still abundant on the edges of the Killarney woods (Fig. 64 and Pl. 39, phots. 82–4), and even locally dominant on rocky ridges, sometimes reaching a height of 30 ft. (c. 9 m.) and a basal girth of 6 or 8 ft. (say 1·8–2·4 m.).[1] This is a much greater size than *Arbutus* commonly attains in the Mediterranean region, where it is usually a shrub. The explanation is probably that in Ireland its growth is not checked by the dry Mediterranean summer. In sheltered ravines in Corsica *Arbutus* attains much finer growth than in the mâquis though the trees are not so large as in Ireland.

Ilex and *Arbutus* are both laurel-leaved trees, a type characteristic of the woody vegetation of mild oceanic climates; and the "laurel character" of the Killarney woods is enhanced not only by the abundance of ivy, but, as Rübel points out, by the subspontaneous luxuriance of the Caucasian and Balkan *Rhododendron ponticum*, which not only establishes itself freely from self-sown seed, but even occurs as an epiphyte on branches of *Prunus lusitanica* (Rübel, 1912). The last named is another Atlantic laurel (Portugal and the Atlantic Islands), which is not native to Ireland, but grows luxuriantly when planted. *Rhododendron* does in fact invade untouched woods at Killarney, though slowly (Turner), while it propagates itself vegetatively as well as by seed on the disturbed margins, as near Queen's Cottage by Upper Lough Leane (known as the "Upper Lake"), in the Muckross demesne, and throughout the woods on Ross Island (limestone), in company with *Prunus lusitanica, P. lauro-cerasus, Laurus nobilis* and *Viburnum tinus*.

All these "laurels" are of course widespread in the British Isles, holly and ivy as abundant natives, *Arbutus* and the others very commonly planted, flourishing in parks and gardens, and bearing witness to the oceanic character of the British climate as a whole. But they are all, together with other laurels, most abundant and luxuriant in the more extreme oceanic climates of the west. Here, in the Killarney woods, while the deciduous oak is the actual dominant, the holly is definitely dominant in the second layer, and evergreens as a whole bulk largely in the physi-

---

[1] Scully records an extreme girth of 14 ft. (more than 4 m.), and Watt saw an *Arbutus* 40 ft. (12·2 m.) high at Glengariff.

PLATE 34

Phot. 72. Derrycunihy Wood ("Type 1") on Old Red Sandstone. Molinietum on alluvium in foreground. *R. J. L.*

Phot. 73. Wood on Carboniferous Limestone, Ross Island. *Sorbus anglica* in foreground. *R. J. L.*

KILLARNEY WOODS (QUERCETUM SESSILIFLORAE)

PLATE 35

Phot. 75. The same: closer view of holly stratum. Ivy and bryophytes covering the oak trunk. *Mrs Cowles.*

Phot. 74. *Quercus sessiliflora* dominant in the canopy, *Ilex aquifolium* in the second layer, *Pteridium*, etc. on the edge of the wood. *Mrs Cowles.*

ognomy of the vegetation. The relatively hygrophytic evergreen yew is present in varying abundance.

Soils and physiography. The work of Dr Turner's party was primarily carried out in Derrycunihy woods near the Upper Lake.[1] The soils here are formed from the debris of rocks of Lower Devonian age consisting of alternating beds of hard gritty sandstone and soft slaty shale. These form a series of steps or terraces, often about 20 ft. wide and separated by vertical cliffs, 20–60 ft. high, on the hillside, and very well seen on the slopes of Cromaglen Mountain. Streams which have cut down into this formation eat away the shales, undercutting the sandstone and bringing it down in blocks which lie on and become embedded in the alluvial fan of loamy soil at the bottom. As the stream cuts back and the valley is widened, the slope becomes gentler. In the earlier stages of this soil development ash and wych elm and such plants as wild strawberry occur, suggesting a fairly high base status.[2]

When the slope has become gentler it bears the oakwood described below as Type 1. The soil is now loamy, with 12 per cent of clay and 54 per cent of coarse and fine sand. Embedded in this are fragments of sandstone and shale, and on the surface sandstone boulders. The litter is from half an inch to an inch thick, mainly formed of holly and oak leaves, and the litter-covered portions of the surface not occupied by boulders are mostly free from moss. The A horizon is quite shallow, only half an inch to an inch in depth, dark in colour, and with 10 per cent of humus. The pH value is 3·8, whereas that of the B horizon is 4·6 and of the C horizon (subsoil) 5·7. Podsolisation progresses with lapse of time and in Type 2 the A horizon varies from 2 to 9 in. in depth, is almost white, and contains only about 1 per cent of humus: in Type 3 it is still deeper. The sesquioxide percentages are markedly higher in the B horizons of Types 2 and 3 than in those of Type 1. The slope is much gentler and more uniform, the terraces are largely obliterated, and the sandstone fragments have weathered down and now contribute from 70 to 80 per cent of sand to the soil (A. S. Watt).

Structure of the oakwoods—Type 1. Of the four types of sessile oakwood distinguished at Killarney, Type 1 (Pl. 35 and Fig. 60) on the least sandy soil, is the finest, with the dominant oaks up to 70 ft. (21 m.) high, in open canopy, i.e. the crowns in lateral contact but not interfering with one another's development, and casting a deep shade, so that even on a clear sunny day the woods are dark and humid, the sun flecks few and small. The ground is covered with boulders partly buried in the soil and averaging 2 or 3 ft. (6–9 dm.) in diameter, but some much larger forming miniature cliffs 10 ft. (3 m.) high. The boulders are almost entirely covered with a

---

[1] Plate 34, phot. 72. The particular wood studied is called Derry-na-heirka.
[2] The almost complete absence of *Fraxinus*, *Ulmus*, *Alnus* and *Corylus*, not only on the more mature (podsolised) soils at Killarney, but even on the comparatively young soil bearing Type 1 oakwood, is noteworthy.

thick continuous carpet of bryophytes with a few lichens, which extends up the tree trunks (Pl. 36, phot. 77) and covers all but the topmost branches of the oaks.

The dominant oaks are mature or over-mature, from 150 to 200 years old, and varying in girth at breast height from 2·5 to 9 ft. (0·75–2·7 m.)

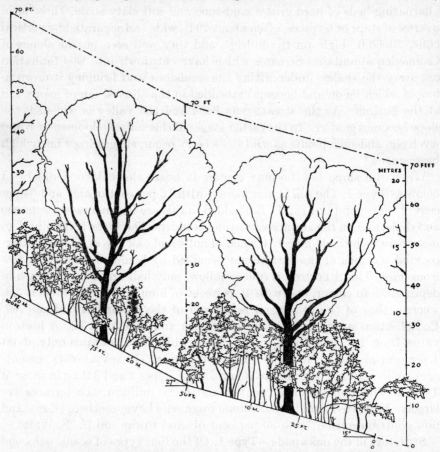

FIG. 60. KILLARNEY OAK-HOLLY WOOD. PROFILE OF "TYPE 1"

Trees of *Quercus sessiliflora* in open canopy, 20 m. high, dominate the wood. Below these is a practically pure continuous layer of holly (*Ilex aquifolium*). S = *Sorbus aucuparia* (rowan). From J. S. Turner (unpublished). Drawn to true scale, uniform in Figs. 60–64.

with the mode at 6·5 ft. (2 m.), widely and rather uniformly spaced (about thirty-five trees to the acre), and of "semi-pioneer" form. The main branches are rather spindly and bear no heavy masses of foliage. Regeneration within the wood is totally lacking, even in gaps, nothing more than seedlings, one or two years old, occurring in clearings or near paths.

Below the oak canopy is a continuous layer of holly (*Ilex aquifolium*), about 875 to the acre, 20–30 ft. (6–9 m.) high, large and well grown, pyramidal or flat-topped, often suckering from the base (Pl. 35, phot. 75). The flat-topped form may be due to lopping branches from the young trees to encourage suckering and thus feed the red and roe deer which come down into the woods in winter. Small seedlings of holly are frequent, with some older plants. Other trees are sparse or rare, but yew (*Taxus baccata*) is characteristic, scattered thinly over the area. The trees are large (about 50 ft. or 15 m. high) and old, with very poor foliage and thickly covered with epiphytes. Seedlings are much more frequent than those of the oak, but no saplings were seen. *Sorbus aucuparia* and *Betula pubescens* are both rare, with occasional seedlings, usually growing in small gaps in the canopy associated with dead or felled oaks.

Lonicera periclymenum and *Hedera helix* are occasional, climbing on the oaks and hollies.

The field layer is very poor, both in species and individuals, consisting only of a few plants rooted round the bases of the moss-covered boulders. The soil between the boulders is largely bare with a light leaf litter. None of the herbs appear to flower, but nearly all possess good means of vegetative propagation. The growth of *Vaccinium*, *Luzula* and *Pteridium* is very poor. Seedlings of *Vaccinium*, *Hedera* and *Lonicera* are occasional, the seeds presumably coming from outside.

Agrostis tenuis	r	Polypodium vulgare	f as epiphyte
Blechnum spicant	o–f		
Dryopteris aemula	f	Pteridium aquilinum	o
Luzula maxima	f	Vaccinium myrtillus	f
Oxalis acetosella	f		

Bryophytic communities on boulders. The bryophytic covering of the boulders which nearly cover the ground is exceedingly rich and luxuriant (Pls. 36, 37).[1] Two communities can be distinguished:[2] (1) An open community consisting chiefly of small liverworts, of which *Diplophyllum albicans* is the only constant species. These are well provided with rhizoids which enable them to colonise even steep rock faces. (2) A closed community of tall mosses, chiefly *Thuidium tamariscinum*, *Polytrichum formosum* and species of *Hylocomium*, often mixed with the fern *Hymenophyllum tunbridgense*. The species of this community occur as young recently established plants in the open community, in which they can obtain an anchorage; and they eventually shade out the pioneer liverworts. The heavy mats of the closed community are easily detached from the boulders, especially when wet; and after heavy rains these mats may sometimes be found lying loose on the ground. In this way fresh rock surfaces are exposed for colonisation by the open community.

On fallen logs. A parallel succession occurs on fallen logs. The epiphytic

[1] Plates 36 and 37 are from photographs taken in the upper woods at a somewhat higher altitude (see below), but the bryophytic species are mostly the same as in Type 1.

[2] The following account of the bryophytes is based on data very kindly supplied by Dr P. W. Richards. Cf. Richards (1938).

species (*Eurhynchium myosuroides*, *Metzgeria furcata*) may continue to grow for a time, but they do not survive indefinitely the different conditions of light and moisture. When decay of the wood of the fallen trunk or branch has progressed to a certain point two characteristic liverworts which never occur on boulders, *Cephalozia catenulata* and *Nowellia curvifolia*, appear; and eventually the mosses of the closed boulder community.

Epiphytic bryophytes. On a large oak three communities may be found: (1) a basal community consisting of species which can establish themselves directly on the bark (e.g. *Eurhynchium myosuroides*) and others which grow up from the neighbouring boulders round the base of the tree (e.g. *Thuidium tamariscinum*). The tree-base community extends upwards for a height of 1 to 2 m. from the ground. (2) A trunk community consisting chiefly of cushions of the *Eurhynchium*. (3) An open community of *Frullania* spp. and *Ulota crispa* on the branches above.

The *Eurhynchium* community grows up from the trunk on to the branches, invading and suppressing the *Frullania-Ulota* community, except on the higher, more exposed branches where the latter is the climax "associule".[1] The *Eurhynchium* community cannot develop beyond a certain point, because when the cushions reach a certain size they become unstable and drop off. The rate of succession seems to depend on the inclination of the surface of the branch. On large horizontal branches the *Eurhynchium* community reaches its highest development, forming a dense mat of *Eurhynchium*, *Plagiochila spinulosa*, *Hymenophyllum unilaterale*, etc. This may be colonised also by species of the closed boulder community: thus *Thuidium* was found several times on trees, and *Polytrichum formosum* occurred on a branch 34 ft. (10·3 m.) above the ground. There is therefore a tendency towards the same climax on rocks and on living and dead tree trunks, but it is not reached on trees because of the precarious hold of the larger mosses on the substratum.

Below is a list (Richards) of the mosses and liverworts in a sample plot (100 × 25 ft.) in Type 1 of the Derrycunihy woods. The filmy ferns are included as integral parts of the bryophytic communities.

	Boulder communities		Log community	Epiphytic communities
	Open	Closed		
Filmy ferns:				
Hymenophyllum tunbridgense	.	la–lf	r–lf	+
H. unilaterale	.	.	.	a
Total 2				
Mosses:				
Campylopus flexuosus	r	r	.	.
Dicranum majus	.	o	.	.
D. scoparium	.	r	o	a
Eurhynchium myosuroides	lf	lf	a–ld	a
E. myurum	.	.	.	+

[1] The terms *associule*, *consociule* have been suggested by Clements to designate associes or consocies of minor life forms which show a succession (*serule*) that is quasi-independent of that of the major dominant life forms (*sere*).

PLATE 36

Phot. 76. The Upper Woods. Limit of altitude about 600 ft. (180 m.). *R. J. L.*

Phot. 77. View in the Upper Woods. A thick carpet of bryophytes covers the boulders and tree trunks. A thick stem of *Hedera helix* climbing the tree on the left. Field layer represented only by isolated plants. *Digitalis purpurea, Juncus* sp. *R. J. L.*

THE UPPER WOODS, KILLARNEY

Phot. 78. Boulder in the Upper Woods thickly covered with *Hylocomium brevirostre*. R. J. L.

Phot. 79. Another boulder covered with bryophytes and filmy ferns: *Eurhynchium myosuroides* (left centre and left), *Hymenophyllum peltatum* (left centre and bottom), *Bazzania trilobata* (bottom right), *Thuidium tamariscinum* (right), *Dicranum majus* (right), *Hylocomium loreum* (top, right). R. J. L.

The Killarney Woods

	Boulder communities		Log community	Epiphytic communities
	Open	Closed		
Mosses: *continued*				
Eurhynchium praelongum	o–f	o–f	o	+
Heterocladium heteropterum var. fallax	a	.	.	.
Hylocomium brevirostre	lf	co–d	a	a
H. loreum	lf	co–d	lva	+
H. splendens	.	r	.	.
Hypnum cupressiforme	.	.	.	+
var. ericetorum	.	r	lva	+
var. filiforme	.	.	.	+
var. resupinatum	.	.	.	+
Mnium hornum	.	r	.	.
Mn. punctatum	r	.	.	.
Plagiothecium elegans	f	r	.	.
P. silvaticum	vr	.	.	.
P. undulatum	o	lf	o	.
Polytrichum formosum	lf	f	.	+
Porotrichum alopecurum	.	lf	.	.
Pterygophyllum lucens	o	r	.	.
Sematophyllum micans	f–a	.	.	.
Sphagnum plumulosum	.	la	.	.
Thuidium tamariscinum	f–o	co–d	va	a
Ulota crispa	.	.	vr	a
Total mosses 26	14	17	9	13
Liverworts:				
Calypogeia fissa	r	.	.	.
C. trichomanis	o	r	.	.
Cephalozia bicuspidata	o?	vr?	.	.
C. catenulata	.	.	vr	.
C. media	vr	.	.	.
Cololejeunea microscopica	.	vr	.	.
C. minutissima	.	.	.	+
Diplophyllum albicans	f–a	vr	.	.
Drepanolejeunea hamatifolia	.	.	.	+
Frullania germana	.	.	.	a
F. tamarisci	r	vr	la	a
Harpalejeunea ovata	.	.	.	+
Harpanthus scutatus	f	.	.	.
Lejeunea flava	.	vr	.	.
L. patens	vr	vr	.	.
L. planiuscula?	o	.	vr	+?
Lepidozia pinnata	.	.	vr	.
L. reptans	vr	vr	.	.
Lophocolea bidentata?	.	vr?	vr	.
L. fragrans	r	.	.	.
Metzgeria furcata	vr	.	r	+
Microlejeunea ulicina	.	.	.	+
Nowellia curvifolia	.	.	r	.
Plagiochila punctata	.	.	.	a
P. spinulosa	.	r	o–f	a
Saccogyna viticulosa	o	vr	.	.
Scapania gracilis	vr	.	.	.
S. nemorosa	o	.	.	.
S. umbrosa	r	.	.	.
Telaranea nematodes*	.	r	.	.
Total liverworts 30	16	12	8	10
Total bryophytes 56	30	29	17	23

+ = present but frequency not determined.
* A species and genus previously only known from tropical South America.

The communities distinguished by Richards are thus seen to be quite distinctive. Only five mosses and one liverwort are common to all four, while thirteen mosses and nineteen liverworts occur in only one out of the four. The two boulder communities are the richest floristically, liverworts preponderating in the open and mosses in the closed. Twelve mosses and 16 liverworts occur on the boulders and not on bark or wood, while four mosses and nine liverworts with *Hymenophyllum unilaterale* show the opposite distribution. All the liverworts in the closed boulder community

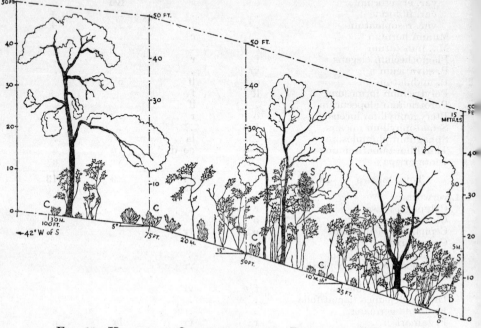

FIG. 61. KILLARNEY OAK-HOLLY WOOD. PROFILE OF "TYPE 2"

(On more podsolised soil with a greater proportion of sand.) The dominant oaks are more numerous and slender and only reach about 13 m. in height, with interrupted canopy. Holly layer also discontinuous. S = *Sorbus aucuparia*, B = *Betula pubescens*: both more frequent than in "Type 1". C = *Calluna*. From J. S. Turner (unpublished). Drawn to true scale.

are recorded as rare or very rare in the sample plot. It is evident that though many are present they find little scope for development within this "climax associule" dominated by large mosses.

Type 2. Type 2 (Figs. 61, 62), on more podsolised soil with a greater proportion of sand, forms most of the woodland at Derrycunihy and indeed in the Killarney district generally. The oaks are very various both in age and form, and do not exceed 45 ft. (say 13 m.) in height: though much more closely placed (85 to the acre) they cast less shade. The holly layer (940 to the acre) is not quite continuous and only reaches a height of 4 or 5 m. Both birch and rowan are more frequent. More light accordingly reaches the field layer, in which *Calluna* is dominant in the more open places, averaging

5 ft. (1·5 m.) in height. *Vaccinium myrtillus* is much better grown than in Type 1, reaching 2 or 3 ft. (60–90 cm.) in the more shaded spots. *Pteridium* is frequent, the fronds reaching 4 ft. in height. *Molinia caerulea* is local, while *Dryopteris dilatata*, *D. filix-mas* and *Galium saxatile*, none of which occurs in Type 1, are rather rare. *Agrostis tenuis* appears to be absent.

The greater openness of this type is probably partly due to age and partly to removal of trees. Regeneration from seedlings occurs, though it is not very frequent.

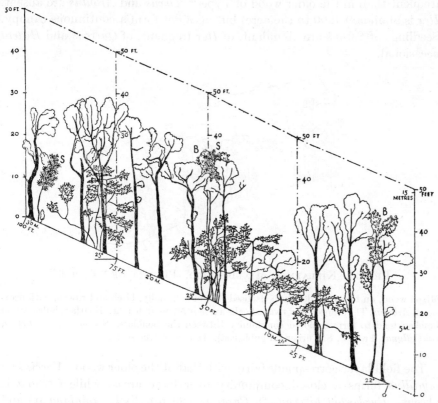

FIG. 62. KILLARNEY OAK-HOLLY WOOD. PROFILE OF "TYPE 2A"

(Probably a young stage of "Type 2".) Oaks numerous, slender. Holly layer not continuous. S = *Sorbus aucuparia*, B = *Betula pubescens*; both more abundant than in "Type 2". From J. S. Turner (unpublished). Drawn to true scale.

Of the bryophytes, shade-loving species such as the Lejeuneae are rarer, and light-demanding species, such as *Hylocomium splendens*, commoner. The open boulder community is less extensive, apparently because the moss carpet is more readily colonised by *Vaccinium* and *Calluna*. The epiphytes tend to grow at lower levels. Thus *Hymenophyllum unilaterale*, which in Type 1 occurs only on the branches of the oaks above about 35 ft., in Type 2 commonly grows to within a foot or two of the ground.

What appears to be a young stage of Type 2, on quite similar soil, but more uniform, and very likely the result of general colonisation of a cleared or burned area, has numerous slender, straight-stemmed oaks (200 to the acre) with few large trees (Fig. 62). The maximum height recorded was 45 ft. (13·7 m.), the mean 35 ft. (10·6 m.). The girths ranged from 1·5 to 7 ft. (0·45–2·1 m.), the mode being only 2·5 ft. (0·76 m.). The canopy is lighter but more uniform than in Type 2 as described above.

Sorbus and *Betula*, of the same height as the oaks, are rather more frequent than in the older wood of Type 2. *Taxus* and *Arbutus* are absent. *Ilex* is abundant (200 to the acre) but does not form a continuous canopy. Seedlings of *Sorbus* are abundant, of *Ilex* frequent, of *Quercus* and *Hedera* occasional.

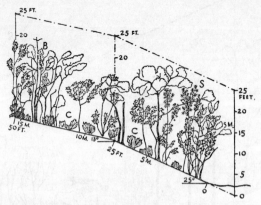

FIG. 63. KILLARNEY MIXED WOOD. PROFILE OF "TYPE 3"

Mixed wood, on the most deeply podsolised soil, of oak, holly, birch and rowan, with occasional *Arbutus*. The stunted and twisted oaks average 5·2 m. in height. Heather, bilberry and bracken tend to form a closed community between the boulders. S=*Sorbus*, B=*Betula*, C=*Calluna*. From J. S. Turner (unpublished). Drawn to true scale.

The field layer corresponds fairly with that of the older wood. *Vaccinium myrtillus* forms a closed community over large areas, while *Calluna* is absent. *Euphorbia hiberna* (f), *Carex remota* (o), *Viola riviniana* (r) and *Asplenium adiantum-nigrum* (r) occur here and also in clearings of Types 1 and 2, apparently in response to better illumination.

Type 3. Type 3 (Fig. 63) is a low mixed wood in which the oaks, though much more numerous per unit area (280 to the acre), are not dominant. They average 17 ft. (5·2 m.) in height. The wood is really a mixture of *Quercus*, *Ilex*, *Betula* and *Sorbus*, with occasional *Arbutus*. The oaks range from 7 to 34 in. (about 18–90 cm.) in girth with a mode of 18 in. (45 cm.), and are stunted and twisted in growth. The hollies (700 to the acre) average 11·5 ft. (3·5 m.) in height and form smaller crowns. The birches and rowans average 50 and 70 to the acre respectively.

Calluna (3–4 ft., 0·9–1·2 m.), *Vaccinium myrtillus* (1·5–2 ft., 0·45–0·6 m.) and *Pteridium* (2–3 ft., 0·6–0·9 m.) tend to form a closed community between the boulders. Besides the species of Type 2, among which *Luzula maxima* is abundant but does not flower, and *Blechnum spicant* is also abundant, with fertile fronds, the following also occur:

Deschampsia flexuosa	o	Potentilla erecta	o
Erica cinerea	r	Rubus fruticosus (agg.)	r
Hypericum androsaemum	vr	Saxifraga spathularis	o
H. pulchrum	r	Solidago virgaurea	r
Melampyrum pratense	o	Viola palustris	vr

The bryophytic flora shows a further change brought about by increased illumination and exposure, Types 1, 2 and 3 forming a series in this respect. The epiphytes descend to a still lower level, *Hymenophyllum unilaterale* and *Polypodium* occurring here on boulders.

The soil shows the deepest podsolisation, but the mechanical fractions, humus content and pH value of the A horizon scarcely differ from those of Type 2. The surface is covered with a layer of typical raw humus 2–4 in. (5–10 cm.) thick, derived from *Calluna* and *Vaccinium* with some oak and holly litter.

Seedlings of *Calluna*, *Vaccinium*, *Potentilla* and *Ilex* are occasional, of *Sorbus* abundant. One seedling each of *Quercus*, *Taxus* and *Crataegus monogyna* was found. Thus it appears that regeneration of the community is slow but probably adequate.

Lichen flora. Lichens are not nearly so prevalent as bryophytes on the tree trunks or on rocks in the woods, and they are least prominent in Type 1. Though they are numerous in species and present everywhere, many occur in the rudimentary or "leprous" condition, which makes identification difficult. The following species (probably not a complete list) from the Derrycunihy woods were determined by Mrs Porter for Dr Turner's party. No attempt was made to separate communities, but three general habitats are distinguished.

On tree trunks

Arthonia ilicina
Biatorina premnea
Buellia disciformis
Cladonia fimbriata
C. pyxidata
Diplochistes bryophilus
Graphis elegans
Lecidea parasema
Leptogium sinuatum
Lobaria laetevirens
L. pulmonaria
Normandina pulchella
Opegrapha atra
    var. parallela
O. vulgata
Pannaria rubiginosa and var.
    conoplea
Parmeliella atlantica
P. corallinoides
Parmeliella microphylla
P. plumbea
Peltigera canina
Pertusaria faginea
P. globulifera
P. pertusa
P. wulfenii
Phaeographis inusta
Porina carpinea
Pyrenula nitida
Sphaerophorus globosus
Thelotrema lepadinum
Stenocybe septata
Sticta dufourii
S. fuliginosa
S. limbata
S. sylvatica
S. thouarsii

On upper branches

Evernia prunastri
Lecanora subfusca var. chlorona,
   f. geographica
Leptogium rubiginosum
Lobaria scrobiculata
Parmelia caperata
Parmelia perlata
P. subaurifera
P. sulcata
Physcia stellaris
Ramalina farinacea
Usnea florida var. hirta

On rocks

Baeomyces rufus
Bilimbia aromatica
Buellia myriocarpa
Cladonia cervicornis
C. digitata
C. macilenta
C. rangiformis
C. sylvatica
C. uncialis
Dermatocarpon aquaticum
Lecanora parella
L. calcarea
Lecidea contigua and var. calcarea
L. rivulosa
Lobaria laetevirens
Opegrapha saxicola
Parmelia saxatilis
P. omphalodes
Peltigera canina
Pertusaria ceuthocarpa and
   f. variolosa
Pertusaria concreta
P. sulphurea
Placodium citrinum
Rhizocarpon obscuratum
R. petraeum
Stereocaulon coralloides
Verrucaria aquatilis
V. laevata
V. rhodosticta

The upper woods. The photographs on Plates 36 and 37 were taken in woods at a somewhat higher altitude (300 ft.) than Derrycunihy (about 100 ft.), nearer the altitudinal tree limit (about 500–600 ft.). These upper woods have not been closely studied, but the dominant oaks are contorted and of comparatively low stature. The development of the bryophytic ground layer is extraordinarily rich and luxuriant.

Local yew wood. Plate 38 illustrates an apparently natural local yew wood on limestone pavement in the Muckross demesne close to the lower lake, with a ground layer of *Porotrichum alopecurum*.

Status of *Arbutus*. The "strawberry tree", *Arbutus unedo* (Pl. 39, phots. 82–84), for which the Killarney woods are famous, is essentially a pioneer, establishing itself in the crevices of rocky ledges and between boulders, marginal to oakwood, in acid humus (pH 3·8). It persists in the *Calluna-Ulex gallii* scrub which follows, and in the *Ilex* scrub which comes next (Fig. 64), but cannot maintain itself under the canopy of *Quercus sessiliflora*. Its prominence in some of the holly scrub may be accounted for by the fact that it is scarcely grazed, while the holly is, and that it largely avoids burning because it inhabits projecting rocky ledges. When *Arbutus* attains any considerable height it tends to fall over and become decumbent and old trees are sometimes found projecting laterally from the margin of oakwood. Exceptionally it may maintain itself erect in an oak canopy: thus at Glengariff an *Arbutus* 40 ft. (12·2 m.) high was seen in the middle of an oakwood, but it was rooted in the summit of a rocky ridge. It is never met with in Type 1, and only marginally in Types 2 and 3 (A. S. Watt).

PLATE 38

Phot. 80. Yew wood (regenerating) in the Muckross demesne. Ash saplings. Limestone pavement covered with *Porotrichum alopecurum*. A. Burges.

Phot. 81. Detail of Phot. 80. Cushion of *Porotrichum alopecurum*. A. Burges.

LOCAL YEW WOOD, KILLARNEY

Phot. 82. *Arbutus* on the side of a rock ridge, *Betula pubescens* on the right. Oakwood to the left, behind. A. Burges.

Phot. 83. Zonation of Molinietum (water's edge), bare rock, Callunetum, and Ilicetum with *Arbutus*. A. Burges.

Phot. 84. Rocky island with *Calluna* and *Ulex gallii*, crowned by *Arbutus* and *Ilex*, and rising from *Molinia* marsh. A. Burges.

Arbutus unedo at Killarney

The Killarney Woods

This habitat and behaviour correspond very well with the Mediterranean habitats. The well-drained soil of the rock crevices and the full exposure to light on the one hand and the absence of severe frosts in winter on the other satisfy the requirements of this tree or shrub of the *maquis*, while the cool damp summers enable it to make more luxuriant growth than is possible during the Mediterranean summer drought.

The establishment of *Arbutus* seems to take place only in the rock crevices. The tree is probably unable to colonise the *Calluna-Ulex* heath, though it persists through that into the next stage (Watt).

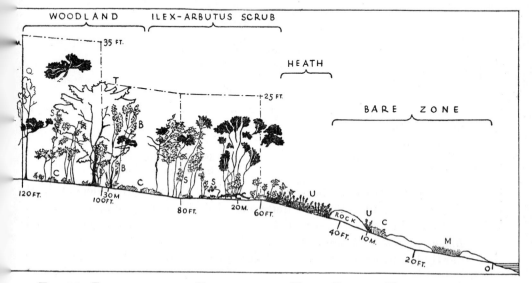

FIG. 64. ZONATION ON THE BORDER OF THE UPPER LAKE AT KILLARNEY

Partly representing succession to woodland. On the right is a zone of bare rock on the edge of the water with *Molinia* established in the fissures. Higher up *Arbutus* comes in (the first *Arbutus* is not shown in the transect) as a woody pioneer, and is succeeded by *Calluna* and *Ulex gallii*, forming a heath zone, through which *Arbutus* persists. *Ilex* follows and the holly-arbutus scrub zone is formed, reaching a height of 5–6 m. In this other trees (*Taxus, Sorbus* and *Betula*) establish themselves, and finally *Quercus sessiliflora*, of which a young tree is seen on the extreme left. M = *Molinia*. C = *Calluna*. U = *Ulex gallii*. S = *Sorbus*. B = *Betula*. T = *Taxus*. Q = *Quercus*. *Arbutus* is in black, and may reach a height of over 9 m. From J. S. Turner (unpublished). Drawn to true scale. Cf. Pl. 39.

Status of *Rhododendron ponticum*. This laurel-leaved shrub flourishes exceedingly well in the Killarney climate, and has been much planted in different places. From these it has invaded the oakwoods even of Type 1, increasing vegetatively by suckering from the tips of its procumbent branches. It seems to have displaced holly in some areas of woodland, casting a deeper shade, so that the field layer is practically absent beneath it and the moss layer much reduced, leaving the soil almost bare (Turner). Seedlings and young plants of *Rhododendron* are occasionally found at a considerable distance from the nearest adult shrub, but most of the woods away from the plantations of the shrub are still free from it. It seems likely, however, that the invasion is progressive. If this replacement of *Ilex* by *Rhododendron* is really progressive it is an interesting example of an apparently rare phenomenon—the successful invasion of an *undisturbed* community by an exotic species.

Succession. The general successional sequence appears to be as follows:

Quercetum sessiliflorae ilicosum
↑
Ilex-Arbutus scrub
↑
Calluna-Ulex gallii
with <i>Arbutus</i>
↑
Calluna-Molinia-Arbutus
colonising the crevices
↑
Bare rock

Vascular epiphytes. Rübel (1912) noted the following species as epiphytes in the Killarney woods:

Geranium robertianum
Hymenophyllum spp.
Ilex aquifolium
Polypodium vulgare
Rhododendron ponticum
Saxifraga spathularis

Glengariff woods. The woods on the low slopes above Glengariff, at the head of Bantry Bay, are quite like those of Killarney in general aspect. Holly again forms the second layer, *Arbutus* is present on margins and *Rhododendron ponticum* spreads freely where it gets the opportunity. They mostly belong to the second and third types recognised at Killarney.

Woods at Pontoon. Of other Irish woods dominated by *Quercus sessiliflora* those near Pontoon at the southern end of Lough Conn in Co. Mayo were superficially examined with the following results:

Quercus sessiliflora dominant, 45–50 ft. (13·7–15·2 m.), diameters 8–10 in. (20–25 cm.), rather contorted, in rather open canopy. Seedlings f to a: no saplings seen.
Betula pubescens, with seedlings and saplings, f to a.
Sorbus aucuparia f. Seedlings a to va, and saplings.
Ilex aquifolium o, seedlings lf.
Fraxinus excelsior saplings l.
Lonicera periclymenum o.
Hedera helix climbing trees and on rocks.
Corylus avellana l.

Field layer

Agrostis tenuis	a	Festuca gigantea	r
Anthoxanthum odoratum	la	Galium saxatile	f
Blechnum spicant	ld on rocks	Hypericum androsaemum	r
Calluna vulgaris	o–a in openings	H. pulchrum	f
		Melampyrum ericetorum	r
Carex sylvatica	o	M. pratense var. hians	f
Dryopteris aemula	f	Oxalis acetosella	l round rocks and stumps
Erica cinerea	o–f		

PLATE 40

Phot. 85. Oakwoods on the shores of Lough Conn, near Pontoon, eastern Mayo. *R. J. L.*

Phot. 86. Interior of a Lough Conn oakwood in close canopy on shallow soil. Trees badly grown, many with ivy. Shrubs absent. Bilberry (with much bracken) dominant in the field layer. *R. J. L.*

SESSILE OAKWOODS AT LOUGH CONN

PLATE 41

Phot. 87. Interior of another oakwood (Massbrook demesne) on flat alluvium close to Lough Conn. The trees have multiple stems (old coppice) and cast deep shade. *Oxalis acetosella* dominant in field layer, with very luxuriant *Dryopteris filix-mas*. (Full list on pp. 341–2.) R. J. L.

Phot. 88. Fragment of oakwood near Maas, Co. Donegal. (See list on p. 342.) R. J. L.

IRISH QUERCETUM SESSILIFLORAE

Polypodium vulgare	on rocks	Sieglingia decumbens	a
Potentilla erecta	o–f	Succisa pratensis	f
Primula vulgaris	r	Veronica chamaedrys	o
Pteridium aquilinum	f–va	V. officinalis	o
Rubus fruticosus (agg.)	va	V. riviniana	f

Bryophytes

Dicranum majus	(tree trunks and stumps)	H. triquetrum	ld
D. scoparium	f	Hypnum cupressiforme	a on stumps and stones
Eurhynchium myosuroides	(tree trunks)	H. schreberi	f
		Plagiothecium undulatum	o
Hylocomium brevirostre	a on stumps and stones	Polytrichum formosum	f
		Scapania sp.	o
H. loreum	o–f	Thuidium tamariscinum	o–f
H. splendens	o		

In local wet places there were moss societies of *Polytrichum commune* and *Sphagnum cymbifolium* with *Carex muricata*, *Lysimachia nummularia* and *Viola palustris*. This wood, with its very mixed flora in which the less exacting grasses were abundant, and which was probably sometimes grazed, showed evidence of relatively good soil in places (ash, hazel, primrose, etc.).

In another wood (Pl. 40, phot. 86) the soil was shallower and poorer. The oaks (in close canopy) were badly grown, crooked, and not more than 35 feet (10·6 m.) high. There was a complete absence of shrubs and the field layer, with very few species, was dominated by *Vaccinium myrtillus* with much *Pteridium*.

In the Massbrook demesne on flat ground very little above the level of the lake there is a wood of *Quercus sessiliflora*, the trees 50–60 ft. (15–18 m.) high, rather crooked and badly grown, often with multiple stems, some having evidently been coppiced a long time ago. *Betula pubescens* was locally dominant (having probably colonised cleared gaps many years before), and *Crataegus monogyna* occasional, but there were no other trees or shrubs.

The shade was deep and the field layer poor in species but very finely developed, *Oxalis acetosella* covering the ground in sheets, with *Hedera helix* frequent and locally very abundant, while *Lonicera periclymenum* was occasional to locally frequent. Two species of *Dryopteris* were extremely luxuriant, *D. dilatata* with fronds 45 in. (1·14 m.), and *D. filix-mas* 55·5 in. (1·4 m.) long (Pl. 41, phot. 87).

The following were all the species seen in this deeply shaded field layer:

Circaea lutetiana	l	Oxalis acetosella	d
Dryopteris dilatata	a–va	Poa nemoralis	o
D. filix-mas	f	Primula vulgaris	r–o
D. aemula	o–f	Polypodium vulgare	(epiphytic on trees)
Hedera helix	f–lva		
Hypericum pulchrum	o	Rubus fruticosus (agg.)	o–f
Lonicera periclymenum	o–lf	Veronica chamaedrys	r
Luzula maxima	vl	V. officinalis	r
Lysimachia nemorum	o		

Of mosses *Hylocomium triquetrum* was locally dominant on the ground in one place, with *Eurhynchium praelongum* occasional to locally abundant. On the trees *Thuidium tamariscinum* often covered the bases, passing upwards into a community composed mainly of *Eurhynchium myosuroides* to 4 ft. (1·2 m.) above the ground, while *Hypnum cupressiforme* var. *filiforme* ascended to 10 or 12 ft.

Donegal woodland and scrub. There is not much natural woodland left in Donegal. The scrub is composed of the same species as before, thick hazel scrub, usually with a good deal of ash and alder and sometimes oak, being not uncommon. One such example (Pl. 41, phot. 88), without ash or alder, close to the sea on a steep rocky slope near Maas, was almost a birch-oak wood and included the following species:

Betula pubescens	a–ld	Quercus sessiliflora	a
Corylus avellana	f–a	Salix atrocinerea	o
Fraxinus excelsior	one seedling	S. caprea	o
Ilex aquifolium	o	Sorbus aucuparia	r–o
Hedera helix	o	Lonicera periclymenum	a

Field layer

Ajuga reptans	f	Oxalis acetosella	f–la
Athyrium filix-femina	f–a	Potentilla erecta	l
Blechnum spicant	lf	Prunella vulgaris	f
Brachypodium sylvaticum	o	Pteridium aquilinum	o
Carex sylvatica	r	Sanicula europaea	f–la
Dryopteris aemula	o	Scilla non-scripta	la
Geranium robertianum	la	Solidago virgaurea	o
Hedera helix	a	Veronica chamaedrys	o
Lathyrus montanus	o	V. officinalis	o
Lonicera periclymenum	f	Viola riviniana	f
Mnium undulatum	f	Thuidium tamariscinum	a

In north-western Donegal in the neighbourhood of Dunfanaghy, Creeslough, and Carrigart, and along the shores of Mulroy Bay, ash, alder, hazel and oak, in that order, are the most abundant constituents of the natural woodland and scrub along the sides of sheltered creeks and valleys. Sallows are very common, mainly *Salix atrocinerea*, but also *S. aurita*, and there is a good deal of birch (*Betula pubescens*) about the woods. Plantations of conifers and beech are frequent, and sycamore is very common, planted and subspontaneous. There is also a little holly and hawthorn. Rowan is infrequent. No *Ulex gallii* was seen in Donegal and very little in Mayo—a great contrast to the south: the same contrast between north and south is seen in the west of Great Britain. *Ulex europaeus* on the other hand is very common by roadsides and spreads up the hills, sometimes forming a regular scrub. This may have been originally planted for cattle feeding, as it is said to have been in North Wales.

PLATE 42

Phot. 90. Closer view of one of the islands. The scrub here is mostly *Salix aurita*. Also *Pteridium*, *Calluna*, *Erica cinerea* and *tetralix*, *Molinia caerulea*, etc. Sparse reed swamp of *Cladium mariscus*. R. J. L.

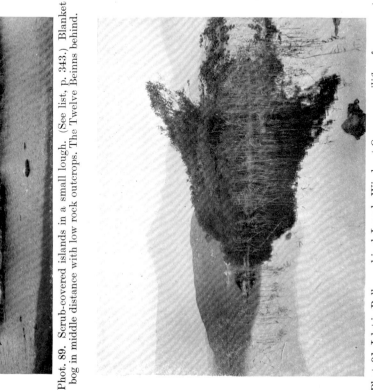

Phot. 89. Scrub-covered islands in a small lough. (See list, p. 343.) Blanket bog in middle distance with low rock outcrops. The Twelve Beinns behind.

Phot. 91. Islet in Bally-na-hinch Lough. Wind-cut *Quercus sessiliflora* forms most of the scrub, including the horizontal and obliquely ascending branches on the right. *Ilex* and *Salix aurita* in centre and to left. *Viburnum opulus* on water's edge. Sparse reed swamp of *Phragmites*.

An example of the scrub by Mulroy Bay between Carrigart and Milford included the following woody plants:

Alnus glutinosa	a	Lonicera periclymenum	f
Betula pubescens	la	Quercus sessiliflora	a–d
Corylus avellana	a	Salix atrocinerea	a
Crataegus monogyna	o	S. aurita	a
Hedera helix	o	Sorbus aucuparia	o

In front of the wood was *Ulex europaeus, Calluna, Erica cinerea, Pteridium, Dryopteris montana, D. dilatata, Athyrium filix-femina,* and *Blechnum spicant.*

Thus the sessile oak extends to the extreme north-west of Ireland, though it is now rarely dominant in semi-natural woods.

Connemara and Mayo. There is very little woodland in the western parts of Galway (Connemara) and Mayo, though fragments of scrub occur here and there on sheltered valley sides. Sheltered scrub on a steep south facing side of the gorge of the Owenmore River in western Mayo had the following composition:

Betula pubescens	o	Lonicera periclymenum	o
Corylus avellana	va–d	Rubus fruticosus	f–a
Crataegus monogyna	o–f	Salix atrocinerea	o
Fraxinus excelsior	r	S. aurita	a, ld
Ilex aquifolium	r	Sorbus aucuparia	r

The sessile oakwoods present in the few well-wooded valleys are often rich in ash and alder, no doubt an expression, as in Wales, of the wetness of the climate.

Island scrub. The rocky islets in the many small lakes of Connemara usually bear wind-cut scrub (Pl. 42, phots. 89–91), while the rocky shores of the lake, presenting an apparently identical habitat, are almost or quite destitute of shrubs. This is no doubt because the islands are protected from grazing. Rowan is generally present in the island scrub, sometimes birch or yew and rarely oak (Pl. 42, phot. 91). A composite list from these islets includes the following trees, shrubs and woody climbers:

Betula pubescens	o	Rubus fruticosus (agg.)	o
Hedera helix	f	Salix atrocinerea	o
Ilex aquifolium	f	S. aurita	f
Juniperus communis	f	Sorbus aucuparia	f
Lonicera periclymenum	o	Taxus baccata	o
Quercus sessiliflora	r	Viburnum opulus	r

The windward side is generally occupied by *Calluna, Erica cinerea* and *Ulex gallii,* with *Pteridium,* often browned by exposure to the wind.

HIGHLAND OAKWOODS

Oakwood is certainly the natural vegetation of the larger valleys and glens of the Scottish Highlands,[1] except probably in the extreme north, and

---

[1] The same is almost certainly true of the valleys of the Southern Uplands, but no information is available about any semi-natural oakwoods that may still exist in this very poorly wooded region.

though many of these woods have been replaced by planted conifers, some still remain. In the southern and central Highlands, however, they have scarcely been examined ecologically. Their vegetation corresponds generally with that characteristic of the sessile oakwoods of the older siliceous rocks of England, Wales and Ireland, but the dominant oaks seldom conform to the type of *Quercus sessiliflora* which is the common form in Great Britain and in Ireland, though many must be assigned to that species. Some on the other hand are typical *Q. robur*; and many are probably hybrids between the two species (see p. 349).

Several fragments of oakwood in the more northern Highlands have recently been examined superficially, and since many of these have been and are being destroyed by the Forestry Commission a brief account may be given here.

(1) On better soils. Along the steeply sloping shores of the lochs of Glenmore—the "Great Glen" which runs north-east and south-west across the Highlands from sea to sea (see Fig. 3, p. 7)—oak forest was probably once continuous and some fragments of this still remain, both in Glenmore itself and on the lower slopes above the streams of the tributary glens which debouch into it. The lower lying of these oakwoods contain much ash in places, with alder, hazel and a little wych elm, as well as birch and rowan. Though the soils have not been analysed these oakwoods possess a fairly rich flora indicative of good soil conditions, approximating in fact to the sessile oakwood of the better soils of the Pennines (pp. 315–16). The following list refers to two such woods, one on the slope above Loch Ness near the mouth of Glen Moriston (Portclair Forest) most of which has been replanted with conifers by the Forestry Commission, though a little has escaped; the other on the north side of Loch Leven, a sea loch which connects with Loch Linnhe at the south-western end of Glenmore.

	Portclair (lower slopes)	Loch Leven
Quercus robur (and seedlings)	d	.
Q. sessiliflora (and seedlings)	.	d
Alnus glutinosa (and seedlings)	.	+
Betula pendula	+	+
B. pubescens	+	+
Corylus avellana	+	f
Fraxinus excelsior (and seedlings)	+	+
Ilex aquifolium (and seedlings)	.	+
Sorbus aucuparia (and seedlings)	+	+
Hedera helix	.	+
Lonicera periclymenum	+	+
Agrostis canina	.	+
A. tenuis	+	.
Anemone nemorosa	+	.
Angelica sylvestris	+	+
Anthoxanthum odoratum	.	+
Athyrium filix-femina	.	+

	Portclair (lower slopes)	Loch Leven
Blechnum spicant	+	+
Calluna vulgaris	+	+
Centaurea obscura	.	+
Cirsium palustre	.	+
Dactylis glomerata	+	.
Deschampsia caespitosa	+	+
D. flexuosa	+	.
Digitalis purpurea	+	.
Dryopteris filix-mas	+	+
D. montana	.	+
Erica cinerea	.	+
Festuca ovina	+	+
Filipendula ulmaria	.	+
Fragaria vesca	+	.
Galium saxatile	+	+
Geranium robertianum	+	.
Holcus lanatus	+	.
H. mollis	.	+
Hypericum pulchrum	+	+
Lathyrus montanus	+	.
Luzula multiflora	.	+
Lysimachia nemorum	.	+
Melampyrum pratense	.	+
Mercurialis perennis	+	.
Molinia caerulea	.	+
Myrica gale	.	+
Oxalis acetosella	+	+
Polygala serpyllifolia	.	+
Potentilla erecta	+	+
Primula vulgaris	.	+
Prunella vulgaris	+	.
Pteridium aquilinum	+	.
Rosa sp.	+	+
Rubus fruticosus (agg.)	+	+
Sanicula europaea	.	+
Sieglingia decumbens	+	+
Solidago virgaurea	+	+
Stachys sylvatica	+	.
Succisa pratensis	+	+
Teucrium scorodonia	+	+
Vaccinium myrtillus	.	+
Valeriana officinalis	+	+
Veronica chamaedrys	+	+
V. officinalis	+	+
Viola sp. (probably riviniana)	+	+
Brachythecium purum	.	+
Catharinea undulata	.	+
Dicranum majus	+	.
Diplophyllum albicans	.	+
Eurhynchium myosuroides	+	.
Hylocomium splendens	+	.
H. triquetrum	+	.
Hypnum cupressiforme	.	+
Pellia sp.	.	+
Plagiochila asplenioides	+	.
Polytrichum formosum	.	+
Thuidium tamariscinum	+	+
	48	57

Total number of species   75
Species common to the two woods   27

About one-third of these species, though not by any means always the same in the two woods, are suggestive of much better soil conditions than those indicated in the woods described below.

(2) On intermediate soils. In the lower part of Glen Garry (opening into Glenmore about the middle of its length) there are similar oakwoods, giving place farther up the valley to birchwood (500 ft. = 152 m.), and then to pine, before blanket bog is reached at an elevation of only 600 ft. (182 m.)—the rainfall at Invergarry at the mouth of the glen is about 80 in. (c. 2000 mm.). Lists were made of two examples of these oakwoods (one in process of destruction, the other to be destroyed by the Forestry Commission). One (Bolinn Hill near Loch Garry) was dominated by *Quercus robur*. The field layer of the lower slopes was dominated by *Pteridium*, of the upper by *Agrostis tenuis*, suggesting grazing. The other (Glen Luie (also doomed)) was dominated by *Quercus sessiliflora* (not quite typical) very well grown and 70–80 ft. (21–24 m.) high, with some birch of both species and rowan seedlings. A third wood on the north side of Glen Loy (farther south-west again) is of closely similar character. The following species were recorded:

	Bolinn Hill	Glen Luie	Glen Loy
Agrostis canina	.	o	a–va
A. tenuis	ld	.	a–va
Anemone nemorosa	+	o	.
Anthoxanthum odoratum	+	o	a–va
Blechnum spicant	.	lf	f–la
Calluna vulgaris	.	f–la	r
Cerastium vulgatum	+	.	.
Deschampsia caespitosa	.	.	r
D. flexuosa	.	f	a–va
Digitalis purpurea	+	.	.
Dryopteris montana	.	.	o
Erica cinerea	.	f	.
Festuca ovina	+	ld	f
Galium saxatile	+	o	f
Holcus mollis	ld	.	l
Hypericum pulchrum	.	o	r
Lonicera periclymenum	.	f	.
Luzula multiflora	.	o	r
Melampyrum pratense	+	.	.
Molinia caerulea	l	f	ld
Oxalis acetosella	+	l	l
Potentilla erecta	+	a	f
Pteridium aquilinum	ld	ld (sparse)	l
Pyrola sp.	.	l	.
Rumex acetosa	+	.	.
Scilla non-scripta	+	.	.
Solidago virgaurea	.	o	.
Succisa pratensis	+	r	r
Teucrium scorodonia	.	o	.
Trientalis europaea	+	f	.
Vaccinium myrtillus	.	va	l
V. vitis idaea	.	r	.
Veronica chamaedrys	+	.	.
V. officinalis	+	o	lf
Viola riviniana	+	f	.

Highland Oakwoods

	Bolinn Hill	Glen Luie	Glen Loy
Brachythecium purum	f–a	.	.
Catharinea undulata	.	r	.
Dicranum majus	+	o	o
D. scoparium	.	stumps	.
Eurhynchium myosuroides	+	stumps	l
Frullania tamarisci	.	stumps	.
Hylocomium loreum	f	r	la
H. splendens	la	va	ld
H. squarrosum	r	.	.
H. triquetrum	+	o	.
Leucobryum glaucum	.	.	o
Polytrichum commune	.	.	o
P. formosum	.	o	o
Plagiothecium undulatum	.	.	l
Thuidium tamariscinum	a	o	ld
	28	35	29

Total number of species	50
Common to the three woods	14
Common to two of the three woods	14

These three woods have poorer floras and are nearer the upland heathy type, though they still contain a wider selection of species than do those of the poorest soil types on which the pine and most of the birchwoods are developed (Chapter XXI).

Of seventeen species of herbs, ferns and mosses noted in a hasty inspection of another oakwood of the same type on the north side of Loch Garry nine occur in all three of these lists, five in two and three in one only.

(3) **On poor soils.** The oakwood on Craigendarroch ("oakhill") in the Dee valley close to Ballater, with a steep slope, a south and south-west aspect and a well-drained soil (altitude 800–1000 ft. = 240–300 m.) is composed of *Quercus robur*, not casting deep shade, only about 30 ft. (9 m.) high and 7–10 in. (18–25 cm.) in diameter. In the upper part of the wood there are some trees of *Q. sessiliflora* and hybrids. There are numerous seedlings as much as 2 or 3 ft. (0·6–0·9 m.) high. The other trees present were:

Betula pendula	la	Populus tremula	la
B. pubescens	l	Sorbus aucuparia	o

all four species with seedlings; and there were a few old isolated pines.

In the field layer *Calluna vulgaris* was dominant and the following species were present:

Anemone nemorosa	l	Melampyrum pratense	o
Anthoxanthum odoratum (near paths)	a	Oxalis acetosella	r
		Pteridium aquilinum	o–ld
Ajuga reptans	la	Teucrium scorodonia	a
Campanula rotundifolia	o	Trientalis europaea	lf
Deschampsia flexuosa	a	Potentilla erecta	o
Erica cinerea	la	Vaccinium myrtillus	l–ld
Festuca ovina	f	V. vitis idaea (under aspens)	l–la
Galium saxatile	l		
Luzula pilosa	o	Veronica officinalis	f–a
L. sylvatica	o	Viola (? riviniana)	a

A well-marked moss layer consisted mainly of three dominant species:

Hylocomium splendens	ld	Thuidium tamariscinum	o
H. triquetrum	ld		
Hypnum schreberi	ld	Total 24 species.	

On the bases of the birch trunks were *Hypnum cupressiforme* and its var. *filiforme*. On tree bark generally were *Evernia prunastri* and *Parmelia physodes*.

This is a heathy oakwood with only two or three species indicating a rather better soil than that of the poorest heath types, on which pine and birchwood characteristically occur, but the dominance of *Calluna*, the local dominance of the Vaccinia and of *Hypnum schreberi* together with the Hylocomia, bring the wood very close to the typical Highland pinewood (Chapter XXI). And Craigendarroch shows more than half the species in common with the South Kinrara birchwood of which a list is given on pp. 453–4.

The oakwoods of the lower slopes in Glenmore and its tributary glens described in the preceding pages pass up into birch and the remains of old pinewoods and the passage is marked by the disappearance of the more exacting species and their replacement by the increasing dominance of heath plants.

Disappearance of oak in the far north. Farther north there are still oakwoods or indications of the existence of former oakwoods in the most protected places at low altitudes. Thus on the east coast, on the sheltered slopes above the shores of Cromarty and Dornoch Firths, there are both sessile and pedunculate oaks associated with a vegetation which is mainly heathy and including such species as sweet vernal grass, wood sorrel, tormentil, cow-wheat and *Luzula pilosa*. Inland in Strathoykell there is pedunculate oak in several places by the roadside near the river, sometimes accompanied by hazel. Farther north again, in the lower part of Helmsdale (Strath Ullie), there are isolated oaks (both species and hybrids), some quite well grown, both on the roadside and in copses of *Betula*, *Sorbus*, *Salix aurita* and *Corylus*. These are associated mainly with heath plants, but also with more exacting species like primrose and germander speedwell. These scattered oaks are quite well grown up to Kildonan (altitude 200–300 ft. = 60–90 m.), but above this point and in Strath Halladay and Strath Naver, opening on the north coast, no oaks were seen, nor between Alltnaharra and Lairg. Over the Caithness border on the east coast there are oaks in the valleys of Langwell Water and Berriedale, but these may possibly have been planted. Crampton mentions no oak in Caithness, and there is probably no natural oakwood in the north of either Caithness or Sutherland.

Oaks and hybridism. On the west coast of Sutherland and Ross by some of the sea lochs there is still some oak, probably the remains of a fringe of oakwood on the lowest slopes close to the shore: for example by Loch Broom near Ullapool, with ash, alder, birch, rowan and aspen; and again on the south-west side of Loch Maree and on the north side of Loch Carron, with ash, hazel and birch. Practically all of this oak is *Quercus robur*, and there is no reason to doubt its nativity.

With regard to the species of the Highland oaks, a great mixture of forms occurs. Out of many scores of trees examined all through the areas described in the preceding pages scarcely any were seen whose leaves conformed to the typical *Q. sessiliflora* familiar on the sands of south-eastern England and on most of the western hills of Great Britain and Ireland. This has obovate leaves with regular shallow lobing, a flat, dark green and glossy upper surface, branched hairs on the lower surface, long yellow petioles and midribs, and the edges of the blade diverging gradually, without any trace of basal auricles. It is certain however that the aggregate *Q. sessiliflora* comprises forms departing more or less widely from this type, particularly in having less regular lobing, shorter petioles and the margin folded on each side near the base of the blade. Most of the Highland *Q. sessiliflora* show departures from type of this nature. But apart from these there is certainly a great deal of *Q. robur* in the Highlands, far more than in the *Q. sessiliflora* regions of western Britain and Ireland; and there are also many trees that may well be hybrids—some with the basal auricles of *robur* combined with the branched hairs of *sessiliflora* (the clearest form of hybrid met with in the south), others with the basal marginal folds grading to the *robur* type of auricle, others again without auricles, but with glabrous lower leaf surfaces.

The interpretation of this puzzling mixture of forms must be largely speculative. The most probable view is that besides many variations from typical *sessiliflora* there is genuine native *robur* along the river sides and on some of the deeper soils towards the bottoms of the valleys, and that where this meets the *sessiliflora* of the hillsides extensive hybridisation has occurred. But there was certainly also much planting of oak in the eighteenth and nineteenth centuries, and since either species may have been planted almost anywhere—both probably being present in all the larger Highland valleys—the ultimate result would be greatly increased hybridism and the existing complex mixture of forms.

REFERENCES

PRAEGER, R. LL. *The Botanist in Ireland.* Dublin, 1934.
RICHARDS, P. W. The bryophyte communities of a Killarney oakwood. *Annales Bryol.* **11**, 108–30. 1938.
RÜBEL, E. A. The Killarney Woods. *New Phytol.* **11**, 54–7. 1912.
    Most of the information in this chapter is unpublished.

Chapter XVII

MIXED OAKWOOD[1] ON SANDY SOILS

QUERCETUM ROBORIS ET SESSILIFLORAE OR QUERCETUM ERICETOSUM

"OAK-BIRCH HEATH"

Mixture of the two oaks. In his study of the Hertfordshire oak-hornbeam woods, as we have seen in Chapter xv, Salisbury (1916, 1918) found marked differences between the woods in which the common oak on the one hand, and the sessile oak on the other, was associated with the hornbeam. These differences included the average edaphic factors in the two types of wood, the prevalent associated shrubs, and especially the societies and associated species of the herbaceous flora. Though comparatively few species were absolutely confined to one or the other type of wood, the frequencies of a considerable number were found to be markedly different in the two communities, so that their "facies" are very distinct. Intermediate between the two types Salisbury found a few woods in which the two oaks were present together, and here the edaphic factors and the associated vegetation were also intermediate. He suggests, however, that some "intermediate" woods in which *Quercus robur* is definitely dominant really belong to the *Q. sessiliflora* type and were originally dominated by that species, of which a few self-sown trees still remain, the wood as a whole having been replanted with *Q. robur*. This suggestion is indeed borne out by the edaphic characters and the general vegetation of the woods in question. On the other hand *Q. robur* "may be present in considerable amount" in woods now dominated by *Q. sessiliflora*, but here the pedunculate oak is restricted especially to "hollows and lower slopes of the valleys where the water content is higher and the soil consists of finer particles, and is richer in mineral salts" (Salisbury, 1918, p. 27). These moister and less acid parts of the *Q. sessiliflora* woods belong in fact to the *Q. robur* type. In such places ash may accompany the pedunculate oak, and also wild cherry and wych elm, trees characteristic of the *Q. robur-Carpinus* woods (Chapter xv, p. 305). It is generally true that while the sessile oak is very rare in and usually quite absent from typical pedunculate oakwoods, *Q. robur* is often met with in woods dominated by *Q. sessiliflora* on the deeper soils of the south of England, though it is absent from the typical sessile oakwoods on the shallow soils of the siliceous rocks of the north and west except

[1] This type of oakwood is not to be confused with the "Quercetum mixtum" of Atlantic times (Chapter vii) in which the oaks are dominant, accompanied by several other deciduous trees, and which is the stock from which all the types of modern British oakwood are historically derived.

on wood edges and in other situations where it is likely to be due to planting.

Quercetum ericetosum. Apart from these local developments and the mixtures introduced by planting we have to recognise the existence of woodland dominated by either of the oaks, or by both together, on many of the deep sandy soils of the midlands and south-east; and the widespread occurrence of this type cannot be explained by assuming that the pedunculate oak has been planted. It is evident that while *Q. sessiliflora* does not naturally occur (at least in the south-east of England) on heavy and wet soils or on soils with a high proportion of basic salts—though it is sometimes dominant on the acid London Clay of north-east Kent (Wilson, 1911)—*Q. robur* does occur naturally, regenerates freely and may be dominant on coarse sands with an acid soil solution, as indeed it does and is on the acid peaty soils of Dartmoor (Chapter XIV, p. 298). It is not here a question of habitat modified in the direction of that of typical *Q. robur* woods, as in Salisbury's Hertfordshire woods just mentioned; but rather of a habitat in which both species of oak may grow and flourish side by side under apparently identical conditions, regenerating and hybridising freely. This may occur on the Triassic Bunter Sandstone, on the Wealden sands, on the Lower Greensand, on the Reading Beds of the Lower Eocene, on the Bagshot sands and related Upper Eocene beds, and on Quaternary sands and gravels. Either species of oak may be dominant, or the two may be co-dominant, and we cannot, at least with our present knowledge, assign any cause other than the chances of dispersal to the distribution of the two on this type of soil. Hybrids between the two oaks are nearly always present.

This "Quercetum roboris et sessiliflorae", or "Quercetum ericetosum", as we may call it, is closely related to Salisbury's Hertfordshire sessile oakwoods (Chapter XV, pp. 304 ff.), to the "dry" facies of the pedunculate oakwood (Chapter XIII, p. 286), and also to the heathy facies of Quercetum sessiliflorae (Chapter XV, pp. 314–15) with all of which it has many species in common. It is often genetically related to the heath formation (see Chapter XXXVI), and its soil is typically a sand or gravel which is often podsolised, even in the south-east. In succession it is based on heath or grass heath as a starting point, and generally maintains "heathy species", i.e. species (whether ericaceous or not) adjusted to a distinctly acid well-drained soil with surface raw humus, throughout the sere up to the mature woodland. This mature woodland is usually, as has been said, dominated by one or both of the two oaks, but locally it progresses to beechwood (see Chapter XX, Fig. 82, p. 414). Birch is always prominent in the succession, and in the south-east also subspontaneous pine. Both these trees may form pure communities on this type of soil; and either or both commonly maintain themselves here and there among the oaks of the mature woodland where the canopy is open, birch commonly filling the gaps made by clearings.

Soils. The soil of the mixed oakwood is typically a deep sand with a low clay fraction, but sometimes with a large proportion of silt, a soil very different in composition from those of the woods with which we have hitherto been dealing. The following data, selected from those given by Hopkinson (1927, pp. 138–9) and from Hall and Russell (1911) illustrate the composition of the soils on which these mixed oakwoods grow, though only those from Mereworth Woods on the Hythe Beds of the Lower Greensand in west Kent are actually taken from a wood.

	Bunter Sandstone (Notts) (average of 7 agricultural soils)	Folkestone Beds (Chilworth, Surrey)	Bagshot Sand (Brookwood, Surrey)	Hythe Beds (Mereworth Woods, Kent)		Ashdown Sand (Forest ridge, Sussex)
Coarse sand	64·3	65·9	16·6	14·3		0·3
Fine sand	22·6	23·7	64·2	27·8		35·2
Silt	6·1	4·4	11·0	35·9 {27·0 coarse / 8·9 fine	51·0	{35·2 coarse / 15·8 fine
Clay	0·27	0·9	1·0	8·2		5·4
Loss on ignition	3·3	2·6	4·4	5·1		5·9
Carbonates (as calcium carbonate)	0·14	nil	nil	0·09		nil
Lime as CaO	not given	0·05	0·08	0·16		0·21
Total potash (K_2O)	0·044	0·025	0·06	0·17		0·03
"Available potash"	0·002	0·010	0·006	0·008		0·017
Total phosphates (P_2O_5)	0·116	0·081	0·017	0·056		0·019
"Available phosphates"	0·025	0·004	0·003	0·007		0·003
Nitrogen	0·117	0·03	0·086	0·121		0·13

It will be seen that the Bunter and the Folkestone Beds are by far the coarsest sands, corresponding very closely in mechanical analysis, and on the whole the poorest soils, though the Bunter is much richer in phosphates than the Cretaceous and Eocene sands. This however may be due to phosphate manuring, since the data are taken from agricultural soils while the other four are natural soils. Though soil reaction data are not given it is certain that all five of these soils must be very acid. The Hythe Beds and the Ashdown Sand have a considerable clay fraction and are rich in silt, but the bulk of this is coarse silt.

Sherwood Forest oakwood. The type of mixed oakwood dealt with in this chapter has been very little studied, Hopkinson's account (1927) of the old oakwoods of Sherwood Forest on the Bunter Sandstone in Nottinghamshire being the best available description.

Only remnants of the extensive oakwoods of the Royal Forest of Sherwood now remain, and a great deal of planting has been carried out at various times since the early part of the eighteenth century. The fragments of old forest still extant are very simple in structure, consisting of two layers only, the tree canopy and the field layer (Pl. 43, phot. 92). The tree layer consists of the two species of oak (with hybrids) in varying proportions, together with *Betula pubescens* where the light is sufficient. The only other spontaneous trees are *Sorbus aucuparia* (r to la) and *Pyrus malus* (r), except in the river valleys and adjoining the Permian marl where alder and ash are met with. There is no shrub layer, shrubs being practically absent from most of the woods. Hawthorn is however recorded as frequent, blackthorn as rare, elder as rare to locally abundant, guelder rose from one wood only, goat willow (*Salix caprea*) only on wet soil. A few others are present only on the margins, and suppressed trees occur here and there in shrubby form. Of climbers honeysuckle is recorded as rare to frequent, decumbent on the ground and not flowering where it cannot climb, and ivy (usually absent from this type of wood) as rare to occasional.

The field layer consists of practically pure bracken, with a very few other species barely existing in the thick litter of bracken debris and largely etiolated. From 7·3 to 53 per cent of the diffuse light reaches the bracken layer, but only 0·4–10 per cent reaches the soil below, the mean value being less than 5 per cent. Here and there, where local conditions interfere with the overwhelming dominance of the bracken, some other species are better developed. The following is the general list:

Pteridium aquilinum	d		
Agrostis tenuis	o–f	Potentilla erecta	o
Deschampsia caespitosa	r	Rumex acetosella	o
D. flexuosa	f–lsd	Scilla non-scripta	r
Digitalis purpurea	o–f	Teucrium scorodonia (in open places)	lf
Dryopteris spinulosa	o, l		
Galium saxatile	o–f	Vaccinium myrtillus	r–o
Holcus mollis	f–a	Viola riviniana	o

Of these it is only the two "heath grasses", *Deschampsia flexuosa* and *Holcus mollis*, which play any considerable part in the vegetation. The former is unable to establish itself in the thick litter formed by the undecayed dead bracken fronds, but where the canopy is broken up and there is less litter it can enter and become dominant, its tussocky habit rendering the soil difficult for *Pteridium*. Thus pure Deschampsietum becomes the community of open spaces. *Holcus mollis* is the only really abundant species, other than the bracken, in the oakwood itself, for it can maintain its rhizomes in the dry humus and push its aerial shoots through the thick litter.

Regeneration and succession. Seedlings and saplings of the oaks occur along the sides and margins of the woods, but invasion of the grass heath and *Calluna* heath which occupies much of the Bunter area does not seem to take place, partly no doubt owing to rabbits, to whose activities Hopkinson makes frequent reference. The scrub communities which exist on the edges of the woods, especially along the wide verges of the roads through the forest, seem to play no part in succession to forest (see Chapter XXIV).

Birch (*Betula pubescens*), on the other hand, is ubiquitous in colonising not only felled oakwood or felled plantations on this area but also grass heath or heath, though it is often eaten back by rabbits. This is in accordance with the behaviour of the tree (and of *B. pendula*) on similar soils throughout the country, and is largely due, no doubt, to mass colonisation by the abundant and easily distributed fruits (cf. Chapter XXII, pp. 452, 456). There seems to be no evidence in the Sherwood Forest region that oak is anywhere actually succeeding birch. On drift gravels which have a layer of quartzite pebbles closely packed and 12 to 18 in. below the surface, and thus present an obstacle to the establishment of the deeper rooting trees, there is pure birchwood in close canopy (Pl. 86, phot. 208) and this may well be the climax type on such a soil.

"**Oak-birch heath.**" Oakwoods on this type of soil have not been closely studied in the south of England, though many such exist. The soils in question are actually largely covered with a mixed vegetation in which heath proper (Callunetum, Ericetum or the two mixed), grass heath dominated by *Agrostis tenuis* and *Festuca ovina* or by *Deschampsia flexuosa*, Pteridietum, gorse scrub or mixed thorny scrub, birchwood, pinewood, oakwood and (locally) beechwood, are all represented. Very often the vegetation is extremely fragmented. In *Types of British Vegetation* (pp. 89–92, 99–103) this "oak-birch heath", as it was called in that book, was considered as a transitional phase from dry oakwood (of the kind dealt with in this chapter) to heath, and the change was supposed (following Graebner in *Die Heide Norddeutschlands*) to be determined primarily by degeneration (podsolisation) of the soil as a result of the washing out of soluble salts from the surface layers, often accompanied by the accumulation of humous substances and iron salts at a lower level (pan). This was supposed to enfeeble the growth of the trees and to permit of the entrance of heath plants into the wood, at first those which could endure shade, and eventually, as the trees died and were not replaced by young growth, the light-demanding *Calluna*, thus leading to the conversion of woodland into heath. It is not here intended to deny that such a process may take place, but it has become clear that many of our heaths can be colonised by trees and replaced by woods when grazing and rabbit attack are suspended, and English lowland heath must in general be considered as a seral community preceding woodland, and when it is stable as a biotic or fire climax on light and especially sandy soils. Certainly some heaths are very old, alluded to in early Norman records as "bruaria", but many and perhaps all the inland English heaths probably occupy the place of earlier woodlands which had been destroyed, so that, strictly speaking, their change back to woodland where it occurs is subseral. Heath as a climax association is oceanic or suboceanic, establishing itself in coastal (and mountain) regions where trees cannot become dominant because of the climatic conditions, preeminently wind, but heath easily colonises sandy inland areas where trees are prevented from establishing dominance by other factors, as felling, grazing or burning: and such heaths may be recolonised by forest if such factors are removed. There may be lowland, non-coastal but suboceanic regions in northern Europe where the climate and soil do not permit of the

PLATE 43

Phot. 92. Part of Sherwood Forest, Notts, on Bunter Sandstone. Very old relict oaks (no regeneration). *Betula pubescens*. Shrubs absent. *Pteridium, Deschampsia flexuosa. J. W. Hopkinson* (1927).

Phot. 93. Half open oakwood with *Pteridium* (also *Vaccinium myrtillus* and *Polytrichum*). Hythe Beds, near Rake, Sussex.

Phot. 94. Oak-birch heath, Ashdown Forest, Sussex, on Ashdown Sand. *Quercus robur, Fagus sylvatica, Pinus sylvestris, Betula pubescens, Pteridium, Calluna. S. Mangham.*

Phot. 95. Young woodland at Toy's Hill, Kent, on Hythe Beds (Lower Greensand). *Betula pubescens, Fagus sylvatica, Sorbus aria, Pteridium, Calluna. S. Mangham.*

QUERCETUM ERICETOSUM AND OAK-BIRCH HEATH

PLATE 44

Phot. 97. Developing Ericetum cinereae. *Betula pubescens, Pinus sylvestris, Quercus robur.* Crockham Hill Common, Kent, on Hythe Beds.

Phot. 99. *Quercus sessiliflora* (right), *Q. robur* (left).

Phot. 96. *Calluna, Erica cinerea, Pteridium, Betula pubescens, Pinus sylvestris, Quercus robur.* Crockham Hill Common.

Phot. 98. Pteridietum in foreground. *Betula pubescens, Quercus robur* (centre), *Pinus sylvestris* (left), *Fagus sylvatica* (right). St Leonard's Forest, Sussex.

establishment of continuous forest, and where heath, with or without scattered trees, is the real climax, but it is very doubtful if such exist in England.

Cyclical alternation of forest and heath. The possibility of a cyclical alternation of forest and heath, independently of anthropogenic factors, must not be neglected. When certain types of forest, for example beechwood, establish themselves on a level permeable soil the firm continuous layer of raw humus that is formed on the surface by accumulated and compacted beech-leaf litter with the accompanying mosses does not provide a suitable nidus for the germination and growth of the seedling trees. When the old trees die there are no successors to replace them, the wood decays, and on the open ground exposed to sun and wind the compacted humus may break up and disintegrate, the area being then recolonised by heath. If this heath is again recolonised by trees beechwood may develop afresh and again in course of time decay. This is a speculative but seems a perfectly possible picture of what may happen in the light of the phenomena that can be observed in such areas. The closer study of the processes involved in the development of particular types of vegetation render increasingly probable the occurrence of such cyclical changes in which the conditions brought about by the development of a particular plant community may eventually lead to its destruction. Thus we should have to recognise a series of temporary climaxes rather than a permanent climax which could maintain itself indefinitely.

Nature of "oak-birch" heath. The "oak-birch heath" (which might indeed be called the "beech-oak-pine-birch heath", for all these trees occur in it in south-eastern England (Pl. 43, phot. 94; Pl. 44, phot. 98) and any one may be locally dominant) is really, as has been said, a congeries of many different communities, which, taken together, as they commonly exist together, determine the characteristic scenery of the commons and "wastelands" of many of the southern sands; but they ought also to be considered individually. They are mostly subseral communities potentially capable of developing into sandy oakwood (Quercetum roboris et sessiliflorae) and locally—perhaps temporarily—into beechwood, but often arrested or changed by the varying incidence of biotic factors, mainly felling, burning and rabbit attack and by the local supply of seed parents. They cannot however be understood unless their underlying potentiality is appreciated, and the constant efforts, so to speak, of the higher forms of vegetation to supplant the lower, are taken into account. It is true that these efforts are thwarted again and again, and may remain ineffectual for long periods, so that many of the lower communities acquire a considerable degree of permanence. Under the existing unnatural conditions described in Chapter XII, particularly the enormous diminution in the number of available seed parents as a result of general deforestation of the country and the perpetual biotic interference with the course of succession, no consistent progress to the climatic climax can take place. But it remains probable that if the inhibiting factors were removed these areas, as indeed most of the country, would as a whole eventually return to oak forest, though the possibility of the occurrence of soil and vegetational changes which would permanently or temporarily inhibit such a return cannot be left entirely out of account. The progress of podsolisation on certain soils to such an extent as to render oaks incapable of successful establishment is one of these possibilities, and the formation under closed beech canopy of a surface layer of compact raw humus rendering successful regeneration of beech impossible is another.

Meanwhile the vegetation of these sandy areas presents an extraordinarily varied picture,[1] but a picture whose variety of detail, constantly repeated, results in a general unity of impression. Larger or smaller areas may be occupied by true heath, with *Calluna* or with *Erica cinerea* dominant, or with the two mixed.[2] Abutting on these there may be great stretches of dominant bracken, which is constantly striving to gain entrance into the Callunetum, but whose enfeebled rhizomes where they do penetrate can only produce a low-growing frond every few feet, unless the heath is severely eaten back by rabbits so that the Pteridietum can gain mastery of the soil and obliterate the Callunetum altogether. Alternating with these are areas of grass heath, dominated by bent and fescue or by silver hair-grass, often if not always determined by traffic, by grazing, or by the partial shade of relatively sparse bracken or birch: or areas of dense "thicket scrub", either dominated by *Ulex europaeus* in patches or lines, often obviously radiating from disturbed ground in the neighbourhood of villages or roadsides, and extending along paths or tracks, invading undisturbed heathland slowly and with difficulty if at all; or mixtures of thorny shrubs— gorse, bramble, hawthorn, blackthorn—forming a thick impenetrable tangle (thicket scrub) under which neither herbs nor tree seedlings can establish themselves.

Imposed upon this variety of low-growing communities come the trees, first and in greatest numbers the birches, which may invade the open Callunetum or the grass heath in countless numbers, with individuals of all ages thinly or thickly scattered, while the heath plants maintain themselves between the growing trees for a long time, the *Calluna* being drawn up into "leggy" bushes, while the bracken, which can endure considerable shade, often becomes dominant in the field layer. Only where the young birches are very thickly set is the shade deep enough to obliterate all undergrowth, and a pure birchwood results (Pl. 86, phot. 207). Locally, in the neighbourhood of pine plantations or previously established subspontaneous pinewoods, this tree invades the heaths in the same numbers as the birches (Pl. 87, phot. 210). In one area of Callunetum on a Kentish

[1] Pl. 43, phots. 94, 95, and Pl. 44 illustrate various aspects of oak-birch heath in south-eastern England.   [2] Pl. 44, phots. 96, 97.

common close to such a subspontaneous pinewood careful search of the heather revealed an average of one pine seedling to every square metre growing among the thick heather. Many of these failed to grow clear, but that some succeeded was evident from the numerous young pines of every age scattered about the heath and in places closing up to form pinewood. In the old mature subspontaneous pinewood the ground was mostly bare of plants, covered with thick pine-needle litter, but here and there were bilberry, silver hair grass, occasional "leggy" *Calluna*, and dwarfed "shrubs" of oak and beech. During the war the parent wood and most of the young pines were felled, and 12 years later the area was occupied by dense young birchwood, presumably because available pine seed parents had been so largely destroyed.

Next come the oaks of both species, of which occasional seedlings may be found about the heath. Oaks of various ages, from saplings to nearly mature trees, occur in the heath, but usually in much smaller number than the birches or pines. Nevertheless in places they may close up sufficiently to form an open oakwood with birches or pines or both accompanying the dominant trees, and several species of the heath flora maintaining themselves according to their capacity to endure shade.

Of associated trees, mountain ash (*Sorbus aucuparia*) and whitebeam (*S. aria*) are occasional but rather characteristic. Holly (*Ilex aquifolium*) is often frequent. Occasionally yew (*Taxus baccata*) is met with. Of shrubs juniper sometimes occurs in the open heath or in open parts of the woods, *Frangula alnus* in damper places and a few others here and there.

Finally beech may come into the picture and in places form pure beech-wood (Chapter xx), killing out the other trees by the shade which it casts. The ground beneath it is often almost bare of plants, except for cushions of the characteristic *Leucobryum glaucum*; and bilberry sometimes forms a sparse field layer which does not flower. Often the colonies of beech remain quite small or the species is only represented by isolated trees in the open oakwood.

Such are the characters of "oak-birch heath", named from the very prevalent and characteristic phase in which birch and oak are colonising the heath, but are constantly checked by the varying and intermittent incidence of felling, burning or grazing.

REFERENCES

HALL, A. D. and RUSSELL, E. J. *The Agriculture and Soils of Kent, Surrey and Sussex.* London, 1911.

HOPKINSON, J. W. Studies on the vegetation of Nottinghamshire. I. The ecology of the Bunter Sandstone. *J. Ecol.* **15**, 130–71. 1927.

SALISBURY, E. J. The oak-hornbeam woods of Hertfordshire. Parts I—IV. *J. Ecol.* **6, 8**. 1916, 1918.

TANSLEY, A. G. In *Types of British Vegetation*, pp. 89–92, 99–103. 1911.

WILSON, M. Plant distribution in the woods of north-east Kent. Part I. *Ann. Bot., Lond.*, **25**, 857–902. 1911.

Chapter XVIII

BEECHWOOD

INTRODUCTORY. BEECHWOOD ON CALCAREOUS SOIL (FAGETUM CALCICOLUM) CHALK SCRUB AND YEW WOOD. SERAL ASHWOOD

Oak forest, either of the pedunculate or of the sessile oak, or occasionally of the two species together, is, as we have seen in the preceding chapters, the natural type of woodland at lower altitudes and on most soils over the greater part of England and Wales, southern and central Scotland and a considerable part of Ireland. On the shallow-soiled limestones, however, oak cannot flourish and ashwood is the characteristic type in the greater part of the country; but in the south-east, on the Chalk, and also on part of the Inferior Oolite of the Cotswold Hills in the south-west midlands (Gloucestershire), beechwood is the characteristic forest type. Outliers occur even farther west, on the massive "Mountain Limestone" of Carboniferous age on the Welsh marches and in South Wales itself (Fig. 65).

Immigration and status of beech. As we saw in Chapter VII the beech probably came into England in Atlantic or Sub-Boreal times: there is now good evidence that it existed in Cambridgeshire (Godwin, 1935) in the early Bronze Age (Sub-Boreal), but the records of beech pollen grains from the peat of other areas are either not well enough dated or are too scanty to be decisive. Nevertheless, such evidence as exists suggests that native beech existed a good deal farther west than it does now, perhaps as far as Cornwall and throughout Wales; and some of the scattered beech met with in these regions may well be descendants of native trees. In other words there has probably been a recession of beech from the west during the last few millennia. Whatever may have been its earlier distribution the beech now behaves only in the south-east as if it were a native tree, and its distribution in this part of the country rounds off very naturally its present distribution on the continent. In the north and west, in Scotland, Cornwall and Ireland, the beech can ripen seed, and in places propagates itself spontaneously, but outside the south-eastern area with the western extension referred to, it scarcely behaves like a native tree. So far as can be ascertained the *woods* of beech in the north and west (except perhaps in South Wales) are planted or subspontaneous: even where the tree can propagate itself naturally the seedlings mostly come from planted parents. Many of the actually existing beechwoods of the south-east are certainly planted also; but natural regeneration is general and abundant except where ascertainable causes prevent it, and a native nucleus of beechwood has probably persisted in the south throughout historical times. On the plateau of the Chiltern Hills in Buckinghamshire and south Oxfordshire, the most extensive existing area of English beech forest, this has been preserved largely on account of the chair-making industry, still carried on in the woods (Pl. 45, phot. 102).

Escarpment and plateau beech. On the Chalk of the south-east the most characteristic of the natural beechwoods occur on the relatively steep escarpments and valley sides where the soil is shallow (Pl. 45, phots. 100–1;

FIG. 65. DISTRIBUTION OF SPONTANEOUS BEECHWOOD IN SOUTHERN ENGLAND

F = Beech on chalk and limestone soils and on loams, f = beech on sand, ? = occurrences which are doubtfully native. The dotted area is the Chalk outcrop. The three main beech areas, (1) the chalk fringing the Weald, (2) the Chiltern Hills, and (3) the Cotswold Hills, are conspicuous, with western outliers in the Wye valley and South Wales. In passing from the Chilterns along the main chalk outcrop south-westward into Wiltshire and north-eastward into Bedfordshire and Cambridgeshire the beechwoods have more and more the aspect of plantations, though it is impossible to be sure that some of them may not be descendants of native woods; and it is practically certain that native beech once extended much farther north and west than it does now. The spontaneous occurrences are noted on conservative lines.

Pl. 49), varying according to Watt's records from 8 inches to a maximum of 20 in. (20–50 cm.), and with a depth of about 12–16 in. (30–40 cm.) over the chalk rock. In this habitat beechwood is more "constant" than in any other. These are the characteristic "beech hangers", and possess the most specialised ground vegetation, determined by the deep shade and the shallow calcareous well-drained soil with its persistent litter and characteristic humus. But in some places, notably on the Chilterns, and on the western stretch of the South Downs in west Sussex and Hampshire, beech-

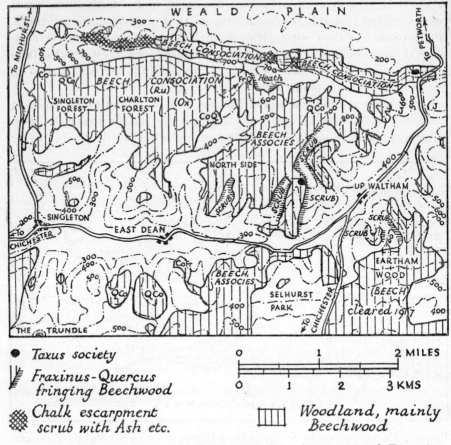

FIG. 66. BEECHWOOD IN THE GOODWOOD AREA (western South Downs)

Woodland area (nearly all beechwood) vertically ruled. Q=*Quercus*, Fr=*Fraxinus*, Co=*Corylus*, J=*Juniperus* (scrub), F→FrQ marks an area where beech is invading and suppressing ash-oak wood. (*Ru*)=field layer of beech consociation dominated by *Rubus fruticosus* (final stage). (*Ox*)=field layer dominated by *Oxalis acetosella*.

At the top of the map immediately south of the Weald Plain is the escarpment of the Chalk clothed with calcicolous beechwood and chalk scrub. This is separated from the plateau beechwoods by a broad ancient trackway with Bronze Age tumuli. Singleton and Charlton Forests (partly at least planted) represent the most mature types of plateau beechwood (consociation). Northside is a younger and certainly natural wood in which oak and ash are still represented. In the right centre of the map are numerous seral stages in which chalk scrub is succeeded by ash and ash-oak wood and these by beechwood.

To the south of the valley in which lie the villages of Up Waltham, East Dean and Singleton are other examples of seral stages and of the beech associes, as well as some oak-hazel wood. At the "Trundle" in the extreme south-west corner of the map, are remains of a Neolithic and also of an early Iron Age settlement on the same site, the latter probably the capital of the Regni, the people who cultivated these plateaux now occupied by beechwood. [Most of the woods shown have since been clear felled.]

woods extend over the chalk plateaux, which are covered by a considerable depth of soil. This plateau soil may be a more or less calcareous loam, derived directly from the Chalk (Sussex), or on the other hand it may be a non-calcareous and sometimes a markedly acid loam (clay-with-flints), often several feet deep. On the Chiltern plateau, besides the clay-with-flints there are outliers of Tertiary sands and rearranged glacial deposits of coarse texture also bearing beechwood.

Beech and oak. Unlike the shallow escarpment rendzinas, these plateau soils, which are primarily brown earths, will bear excellent oak; and on the North Downs of Surrey and Kent, on the Hampshire uplands, and to a less extent on the South Downs and Chilterns, typical oak-hazel wood occurs. On the western South Downs and on the Chilterns the plateau soils are however mainly covered with beechwood. Fig. 66 shows the distribution of several types of beechwood in a region of the West Sussex downs, and Figs. 67 and 68 on different parts of the Chilterns. While some of this beechwood has been planted, much is certainly a natural growth which has followed and suppressed ash-oak wood or oakwood in succession. On the South Down plateaux much of this land was at one time arable, as can be seen from the remains of lynchets within the woodland, and the various stages of recolonisation by woody plants—chalk scrub, ash-oak wood and finally beechwood—have been clearly traced by Watt (1924, 1925). What actually determines the dominance of beech or of oak on these plateau soils is still not entirely clear, but it seems probable that it is largely a question of the type of exploitation, partly also that on the heavier soils beech is at a relative disadvantage compared with oak. It is not uncommon, especially on the Chiltern plateau, to find oak maintaining itself mixed with the beech, though in other places oak is clearly suppressed, and the climax forest of the plateau (as well as of the escarpment) is then composed of practically pure beech.

The factors determining the distribution of natural oak and beech forest in Britain may therefore be summarised as follows. Beech is now confined as a native tree to the south and mainly to the south-east, and it is only in this region that it directly competes with oak, though it may show considerable powers of reproduction much farther north. The causes of its existing restriction are not certain. It is intolerant of waterlogged soils and apparently at a relative disadvantage on heavy clays. Oak, at least as a competitive tree, on the other hand, is excluded from very shallow soils such as the chalk rendzinas, so that there beech holds undisputed dominance. On loams and deep sands oak and beech flourish alike and beech has the advantage because it grows at least equally tall and casts a deeper shade, though on easily podsolised soils beech is at an ultimate disadvantage because it tends to create surface conditions unfavourable to germination and establishment of seedlings. The comparative scarcity of beech wood and the prevalence of oakwood on these last soils (loams and deep sands) may also be attributed partly to the slowness of beech in the production of viable nuts and in migration, partly to the deliberate culture of oak.

The ground vegetation of the plateau beechwoods does not differ essentially from that of the oakwoods on the same soil, though it is apt to

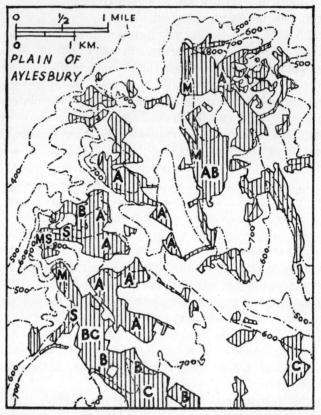

Fig. 67. Beechwood of Part of the Chiltern Hills (Buckinghamshire)

Woodland areas (nearly all beechwood) vertically ruled. To the north west is the main chalk escarpment (here very sinuous), overlooking the plain of Aylesbury. The escarpment is partly occupied by chalk grassland and scrub and partly by calcicolous beechwood. M = Mercury, S = Sanicle wood. A marks the plateau woods with high base status and the finest beech growth, C the plateau woods with podsolised soil and poorest growth, B = intermediate woods. On the whole the A woods are nearest the escarpment, the C woods farthest away. The white areas are clearings for arable and pasture. Contours in feet.

be more uniform and poorer. It is doubtful indeed if there are any species of flowering plants which clearly differentiate these two types of forest. The ground vegetation is in fact primarily related to the type of soil rather than to the dominant trees, but it is also affected of course by the shade cast by the trees, which is much deeper under mature beech than under mature oak, and also (perhaps the most important factor) by the competition of the

PLATE 45

Phot. 100. Escarpment beechwood (Fagetum calcicolum). Stoner Hill, near Petersfield, Hants.

Phot. 101. Another view from the same locality showing characteristic habit of beech.

Phot. 102. Chair "bodgers" in the beechwoods (Fagetum rubosum) of the Chiltern plateau. Note the field layer dominated by bramble. Photograph from *The Times*.

BEECH FOREST (FAGETUM SYLVATICAE)

PLATE 46

Phot. 103. Air photograph of the area surrounding Wormsley Hall (Oxon.) on the Chiltern Hills,* showing the distribution of "escarpment" and "plateau" beechwoods. The "plateau" beechwoods (largest crowns and finest growth) are semi-open in places owing to the taking out of merchantable timber. The escarpment woods show smaller crowns and denser stocking, and in places (well seen on the south facing slope of the southern tributary valley) appear grey in the photograph, because the foliage is yellowing from drought on the shallow soil. Below (to the right) of this strip (at $E^\times$ in overlay) the canopy is open owing to windfall and fungal attack. Here there is abundant ash regeneration. Aircraft Operating Co., Ltd. (from *C. R. Robbins*, 1931).

* Rectangular area in Fig. 68.

BEECH FOREST FROM THE AIR

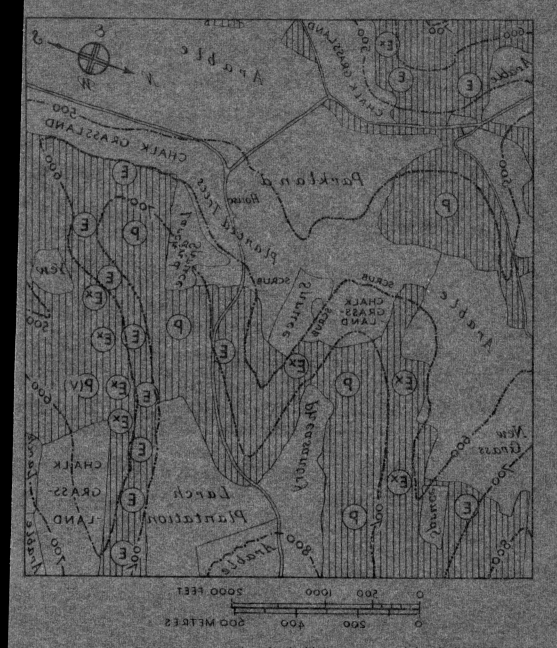

The vertically ruled areas are occupied by beechwood.
A dry valley runs southward across the top of the map from left to right, the bottom of the valley is occupied by arable cultivation. Three tributary valleys (shown by the V-shaped contour lines), separated by spurs, run eastward from the bottom of the map into the main valley.

The summits of the spurs and also the bottom of the southern tributary valley are occupied by 'plateau' beechwood—*P* and *P*(*L*)—the steeper slopes, at the base of the Upper Chalk, by 'escarpment' beechwood (*E*). At *Ex* the outcrop of a bed of marl causes water seepage and the beeches suffer from windfall and fungal attack. Here ash regenerates abundantly. Semi-natural chalk grassland ('rough grazing') replaces escarpment beechwood here and there.

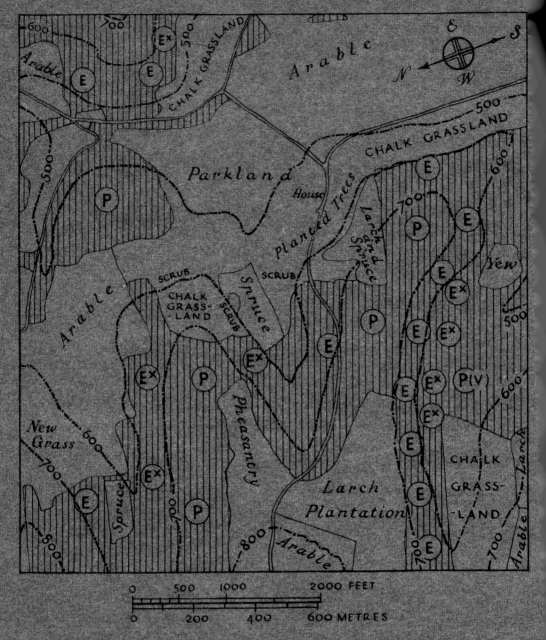

The vertically ruled areas are occupied by beechwood.
A dry valley runs southward across the top of the map from left to right: the bottom of the valley is occupied by arable cultivation. Three tributary valleys (shown by the V-shaped contour lines), separated by spurs, run eastward from the bottom of the map into the main valley.
The summits of the spurs and also the bottom of the southern tributary valley are occupied by "plateau" beechwood—P and $P(V)$—the steeper slopes, at the base of the Upper Chalk, by "escarpment" beechwood (E). At E^x the outcrop of a bed of marl causes water seepage and the beeches suffer from windfall and fungal attack. Here ash regenerates abundantly. Semi-natural chalk grassland ("rough grazing") replaces escarpment beechwood here and there.

beech roots in the surface layers of soil, which is more severe than that of the feeding roots of the oak, since these run at deeper levels.

Beechwood soils. The profiles of typical beechwood soils are shown in Figs. 69 and 70. Fig. 69 represents the rendzina soils of the steeper chalk

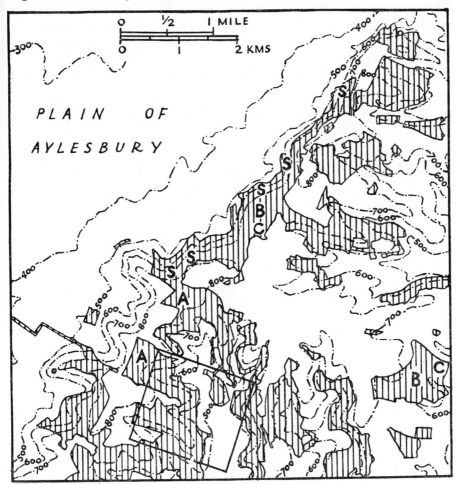

FIG. 68. BEECHWOOD OF ANOTHER PART OF THE CHILTERNS (OXFORDSHIRE)

Topography, ruling and symbols as in Fig. 67. The part of the escarpment which is not woodland is occupied by chalk pasture and chalk scrub, into which beech is advancing in several places. The rectangular area ruled off corresponds with the air photograph of Pl. 46. Contours in feet. The belt between the 400 and 500 ft. contours roughly corresponds with the Upper Greensand terrace, which is good agricultural land.

slopes, (a) of the sanicle, (b) of the mercury woods (see p. 369). Fig. 70 shows a series of beechwood soils from the Chiltern plateau, (a) a brown earth with good mull humus, (b) and (c) with increasing tendency to formation of mor and podsolisation. The process is greatly aided by the nature of beech leaf litter, since this does not decay so rapidly as oak leaf litter, especially on the more acid soils where the ratio of silica to calcium is higher in the beech leaves, which are then more resistant to decay. When they are wet the dead beech leaves stick together so as to cover the soil with a laminated stratum of raw humus which may reach a considerable thickness. The absence of the disintegrating action of earthworms, which are rare or absent in these acid soils, is another influence tending in the same direction. Below the raw humus layer (*mor*) the mineral soil shows a whitish "bleached" layer (*A*, eluvial layer, see p. 86), and then comes a dark layer (*B*, illuvial layer) in which humous substances, together with iron salts and

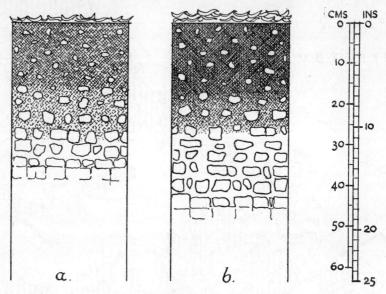

FIG. 69. CALCICOLOUS BEECHWOOD SOILS (RENDZINAS)

(*a*) From sanicle, (*b*) from mercury wood. Note the shallowness of the soil and the absence of a *B* horizon. After Watt.

alumina, carried down from the surface layers, have accumulated (Fig. 70*c*). This is the "podsol profile" described in Chapter IV, rarely more than slightly developed on these plateau soils, but always associated, where it appears at all, with plants characteristic of acid soils. *Calluna, Polytrichum* and other plants of these proclivities occur in and about the plateau beechwoods with this type of humus and soil profile. (For further details see the explanation of Fig. 70.)

On many of the sands of south-eastern England the beech occurs and behaves like an indigenous tree, springing up freely from seed, suppressing its competitors and locally forming pure beechwood. On some of the better sandy soils, especially on slopes, the flora is not very dissimilar to that of the loams (Chapter XIX), but on the whole the sands are poor (some of them very poor) in mineral salts, the soil water is markedly acid, and the

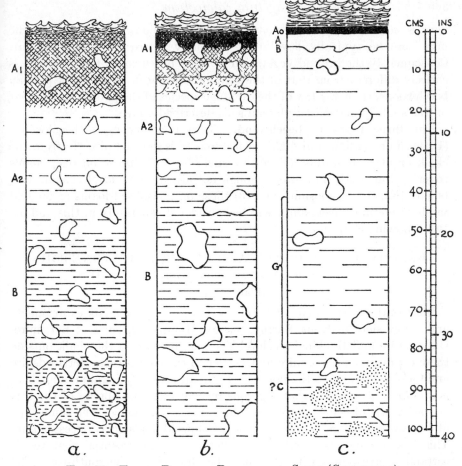

FIG. 70. THREE PLATEAU BEECHWOOD SOILS (CHILTERNS)

(*a*) is a typical brown earth with high base status and supporting very well grown beech (Pl. 66, phot. 144). The leaf litter is loose and produces good friable mull. The top 7 in. of mineral soil (*A* 1) is coloured by incorporated humus. This is followed by about 13 in. of pale soil (*A* 2) with scattered flints. The *B* horizon is not well marked but is probably represented by the next 12 in. of tenacious red-brown loam. The lowest layers have much sand and many flints. The *p*H increases from 4·3 at the surface to 6·0 at 30–33 in.

In (*b*) the leaf litter shows a tendency to become laminated below and the humus more compacted and deeper in colour than in (*a*). Below the humus the colour quickly disappears. At about 13 in. the soil becomes stickier in consistency (? *B* horizon). The *p*H is 4·6 at the surface, 4·2 at 9–12 in. and 4·6 again below. This type does not support such good beech growth, nor certain species of the field layer which occur in (*a*).

(*c*) shows definite though very shallow podsolisation. The leaf litter becomes strongly laminated and converted into a layer of peaty black humus (*A* 0) which also colours the top quarter inch of mineral soil. Below this comes about 1¼ in. of grey, almost white soil ("bleached layer", *A*) limited below by a very thin undulating illuvial horizon (*B*). Thus with the change from the brown earth to the podsol type the illuvial layer completely alters its position and character. In this soil there is a well marked gley horizon (*G*) from 16 in. to about 31 in., mottled grey, yellow and red. The lower soil contains sandy pockets (? *C*). The *p*H is 3·85 in the raw humus, with a maximum of 4·65 at 37–40 in. This soil supports only the poorest type of beech growth, and characteristic oxyphytes in the field layer. After Watt.

flora is correspondingly "calcifuge" or "oxyphilous". A podsol profile may here be typically developed, especially where the ground is flat. Consequently the ground vegetation of these beechwoods (Chapter xx) is entirely different from that of the chalk and of most of the chalk plateau beechwoods, though a few of the calcifuge species of the podsol sands occur on the slightly podsolised plateau soils described above.

The three types of beechwood. These three types of beechwood: (1) those on shallow and highly calcareous soils derived directly from the chalk or other limestone (Fagetum calcicolum), (2) those on deep loams covering the chalk, and often, though not always, devoid of calcium carbonate (Fagetum rubosum), and (3) those on podsolised loams or on acid sands (Fagetum ericetosum), will be considered separately in this and the two following chapters.

BEECHWOOD ON CHALK AND OTHER LIMESTONE SOILS
(FAGETUM SYLVATICAE CALCICOLUM)
(Plates 47–51)

Distribution. This type of beechwood occurs on the chalk escarpments and valley sides of Kent, Surrey, Sussex, Hampshire, parts of Berkshire, Oxfordshire and Buckinghamshire. The beechwoods on the chalk outside these counties, e.g. Wiltshire and Dorset to the south-west, Hertfordshire and Cambridgeshire to the north-east, are all probably planted, and there is a notable absence of that vigorous regeneration and spread of the tree characteristic of the chalk of the counties first mentioned. The same type of beechwood recurs however on the Cotswold Hills in Gloucestershire. This has not been studied in any detail, but in essential features it certainly corresponds with the chalk beechwoods of the south-east. The beechwood on the Mountain Limestone of Symonds Yat and neighbouring parts of the Wye valley near the Welsh border is of somewhat similar type, though the harder, more massive rock and the damper western climate introduce certain differences which have not been investigated. There is also some beechwood which may well be native on the Mountain Limestone and also on Old Red Sandstone to the north of Cardiff;[1] and a few beech shrubs, which are certainly not planted, in the clefts of the limestone cliffs of Craig-y-Castell near Crickhowell and Craig Cilau in Breconshire (marked by the two westernmost letters F in Fig. 65). This is the westernmost beech in Britain which has any claim to be considered native.

The chalk escarpment and valleyside beechwoods of the part of Hampshire next to the Sussex border have been described by Adamson (1922) and those of part of the Chiltern escarpment in Buckinghamshire by Watt (1934). The two areas show an essential similarity in the climax woodland.

Slope and soil. The slopes on which these beechwoods (or beech

[1] My attention was called to these by Mr H. A. Hyde of the Welsh National Museum at Cardiff.

"hangers" as they are called in Sussex and Hampshire) occur, vary from about 18° to 35°. On gentler slopes the soil is deeper and the character of the wood changes, very often including oak. On steeper slopes the soil tends to be washed away and the ground vegetation is almost absent. The soil (Fig. 69), a "rendzina" in the terminology of modern pedologists (see pp. 93–5), is generally covered during winter and spring with a continuous layer of beech leaf litter except where the wind sweeps it away, and varies from 8 in. (20 cm.) on the steeper slopes to a maximum of about 20 in. (50 cm.) on the gentler. An average depth is 12–16 in. (30–40 cm.). The top stratum, 2–4 in. (5–10 cm.) deep, consists of very dark, almost black humus (mull) derived directly from the decay of the leaf litter. Even this often contains tiny visible chalk particles, and normally effervesces vigorously with hydrochloric acid. Below, there is from 6–8 in. (15–20 cm.) of brown or pale coloured mineral soil in which the size and frequency of chalk fragments rapidly increase downwards till the fissured and disintegrated surface of the chalk itself is reached. The reaction, even of the surface humus, is markedly alkaline owing to the large amount of calcium carbonate it contains. The pH values of this shallow soil vary from 7·2 to as much as 8·4 (though it is usually less than 8), and the surface layers have a high humus content. The losses on solution and on ignition are therefore very high. The minute insoluble residue consists mainly of particles of silt size (0·05–0·002 mm.) but there is a considerable clay fraction. The structure and composition of this characteristic soil depend upon the slope, which maintains it in an "immature" condition. The surface humus and underlying soil are densely filled with slender ramifying beech rootlets.

Structure and composition. Four layers or strata of vegetation may be distinguished, as in most English woodland types, and sometimes a fifth is present, but all of these are never well developed in the same wood. In the escarpment beechwoods the tree layer is sometimes present alone, as in young woods or on very shallow soils. There is occasionally a local subordinate tree layer, of yew or of holly, but the shrub layer is absent or very sparse, except in openings. The field layer is absent in very young woods, but increases as the canopy opens, and the moss layer is exceedingly patchy, usually indeed confined to the bases of the tree trunks and the upper surfaces of the main spreading roots diverging from the bases of the trunks.

(1) *The tree layer* consists almost entirely of the dominant beech, whose roots spread horizontally in the shallow soil, which they exploit very thoroughly: anchoring roots also enter fissures in the chalk rock below.

The dominant beeches of the escarpment woods never attain the same height and girth as those of the deeper soils on the plateau. Thus on the Ditcham Park estate on the South Downs of the Hampshire-Sussex border (Adamson, 1922, pp. 122, 143), the average girth at 3 ft. (0·9 m.) from the ground of 60 mature trees taken in groups of twelve from each of five escarpment woods was 1·61 m., and the maximum girth observed 2·6 m., while on the plateau the average girth of twenty-four mature trees was 2·26 m. and the maximum girth observed 3·8 m. This limitation of growth doubtless depends on the shallowness of the escarpment soil restricting the water supply, the feeding roots of the trees being necessarily confined to the thin layer of humus and mineral soil above the surface of the chalk rock. The roots spread out horizontally in this layer and exploit the whole of the soil very thoroughly, thus reducing the water available for subordinate vegetation. The surface "run off" also of course diminishes the supply entering the soil, especially on the steeper slopes during heavy rain. Mycorrhiza is abundant in the humus layer, the beech roots showing a short spring period of active growth and infection by the mycorrhizal fungus. As the shallow soil dries up in early summer this activity is arrested (Harley, 1937).

Occasional ashes (*Fraxinus excelsior*), relics of an earlier stage of the succession, reach the height of the canopy, but these are always thin, poorly developed trees, "drawn up" to the light by the deep shade of the surrounding beech. The whitebeam (*Sorbus aria*), also a relic of an earlier stage, maintains itself more successfully in competition with the dominant. On the Chilterns the wild cherry or gean (*Prunus avium*) is present in considerable numbers, reaching the height of the canopy, and in late April or early May the masses of white cherry blossom are a conspicuous and beautiful feature of this landscape. Oaks are typically altogether absent.

A subordinate tree layer, existing entirely below the beech canopy, is sometimes formed by the yew (*Taxus baccata*) (Pl. 47), or by the holly (*Ilex aquifolium*), locally richly developed and forming a continuous stratum. Under the conditions of deep shade neither of these species flowers, but the evergreen habit of both doubtless enables them to take advantage of the "light phase" before the beech leaves unfold.

(2) *The shrub layer* is always poorly developed, and very often absent altogether. The commonest shrubs are elder (*Sambucus nigra*) and field maple (*Acer campestre*). The spindle tree (*Euonymus europaeus*) and the hazel (*Corylus avellana*) may also occur. The dogwood (*Cornus sanguinea*) and the wayfaring tree (*Viburnum lantana*)—characteristic plants of chalk scrub—are usually confined to openings in the woods, the very few which exist under the deep shade of the beeches showing feeble growth. The evergreen box (*Buxus sempervirens*) is even more local than yew and holly and can survive deep shade. It is certainly often planted, but may possibly be native, especially at Box Hill on the escarpment of the North Downs. A fourth evergreen, the ivy (*Hedera helix*), is often present, both climbing the trees and creeping on the soil, its vegetative shoots enduring very deep shade. Traveller's Joy (*Clematis vitalba*), the most characteristic woody climber of calcareous soil in the south of England, is a light demander, and is rarely found in the interior of beechwoods.

(3) *The herb* or *field layer* varies very much in development. In some

PLATE 47

Phot. 104. Yew in beechwood on chalk slope near White Lodge, Arundel Park, Sussex. Pioneer beech to right, close canopy to left. Field layer absent. *R. J. L.*

Phot. 105. Dense layer society of yew in calcicolous beechwood. Duncton Down, Sussex. *R. J. L.*

YEW LAYER IN FAGETUM CALCICOLUM

PLATE 48

Phot. 106. Slope of Oakham Bottom, Ditcham Park, Hants, on Middle Chalk. Deeply shaded foreground dominated by *Hedera helix* (creeping) with *Sanicula* and *Viola sylvestris*. Tall *Epilobium angustifolium*, *Cirsium palustre* and *Euphorbia amygdaloides* in lighted opening. *R. S. Adamson.*

Phot. 107. Slope of Oakham Bottom in winter, with beech leaf litter. Suckers from the surface beech roots, in centre. *C. J. P. Cave.*

FAGETUM CALCICOLUM, SUMMER AND WINTER

beechwoods it is almost or entirely absent. This may be due to weakness of the light reaching the floor of the wood, as is notably the case in a young self-sown wood where the close-set saplings cast an extremely deep shade, but other factors are often more important. Thus on the shallowest soils both herb and shrub layers are frequently absent, and this is largely due to deficiency of soil moisture increased by the competition of the beech roots, which exploit very thoroughly the whole of the thin layer of soil covering the chalk. Again, where the leaf litter and underlying humus are very thick, vegetation is entirely absent except for occasional colourless saprophytes.

That light is not the only factor determining absence of ground vegetation is shown by the observations of Adamson (1922, p. 128), who found bare areas with a light value during the summer of as much as 4·72 per cent, while other areas with sanicle and ivy gave average values of 2·18 and 1·65, and minimum values of 1·4 and 1·0 respectively. The lower summer limit of light intensity for ground vegetation seems to be in the neighbourhood of 1·0 per cent of full illumination. It must be remembered of course that the woodland plants which vegetate wholly or partly in the spring, before the leaf canopy of the trees is fully developed—and these are most of the more abundant species—receive much more light during this season. Thus in escarpment beechwoods in April Adamson obtained light values ranging from 17·5 to 63 per cent, with an average of 38·9.

Dominants In mature beechwoods on not too shallow soils, where the canopy has more or less "opened out", the field layer is much more fully developed and sometimes covers the ground with an almost continuous carpet of vegetation (Pls. 48–50). By far the most abundant plants are dog's mercury (*Mercurialis perennis*), wood sanicle (*Sanicula europaea*), and frequently ivy (*Hedera helix*), the last often forming an uninterrupted sheet in the deepest shade, growing as a chamaephyte, and not necessarily climbing the trees.[1] Watt (1934) found mercury characteristically dominant on the deeper soils of the gentler slopes, and sanicle dominating the shallower soils and the steeper slopes, but the two are sometimes mixed, though the tall growth of closely set mercury shoots where the plant is growing under optimum conditions tends to exclude other species.[2] Mercury and sanicle, both hemicryptophytes, vegetate largely during the "light phase", i.e. before the leaf canopy is developed, but both retain their leaves during the summer, while the ivy of course is an evergreen.

Other abundant species Of other abundant species the sweet woodruff (*Asperula odorata*) is very characteristic, and so is the early-flowering wood violet, *Viola sylvestris*.[3] The other wood violet, *Viola*

[1] Pl. 48, phot. 106.

[2] M = mercury woods, S = sanicle woods, in Figs. 67, 68; and see Pl. 49, phot. 110, and Pl. 50.

[3] The small-leaved violet in Pl. 49, phot. 111; it also occurs in other calcareous (e.g. ash) woodland.

riviniana, also occurs, but is not characteristic, having a very wide distribution on light soils generally, in oakwoods, and on grass heaths. *Viola hirta* (Pl. 49, phot. 111), more abundant on chalk grassland, occurs also in many chalk beechwoods. Other common and locally dominant species are the wild strawberry (*Fragaria vesca*),[1] the yellow deadnettle or archangel (*Galeobdolon luteum*), wood anemone (*Anemone nemorosa*), and the cuckoo-pint (*Arum maculatum*). These are all forms which leaf early, and do the whole or a large part of their active vegetation and flowering from March to May, before the canopy of beech foliage is fully formed. Enchanter's nightshade (*Circaea lutetiana*), another commonly abundant species, flowers in summer, though it develops its leaf rosettes in April and May, and the wood spurge (*Euphorbia amygdaloides*) shoots in April and flowers in May. All these plants are more or less gregarious, forming pure or mixed societies in the field layer. They propagate largely from shoots growing in the surface layer of soil (hemicryptophytes), or from more deeply buried rhizomes (geophytes): in *Galeobdolon* and *Euphorbia* the perennating buds are above the soil surface (chamaephytes).

Besides the conspicuous early spring flora composed of the abundant gregarious species described there are others not so common which also flower in spring or early summer and are rather characteristic of beechwood and other woods on limestone soils, e.g. the columbine (*Aquilegia vulgaris*), the green hellebore (*Helleborus viridis* var. *occidentalis*) and Solomon's Seal (*Polygonatum multiflorum*). *Daphne laureola*, the spurge laurel, a calcicolous evergreen dwarf shrub (nanophanerophyte) flowering in early spring, is occasional in the escarpment woods and frequent to abundant in the Cotswold beechwoods (Pl. 51, phot. 115).

Finally there are a number of late spring or summer plants, some of which are confined or almost confined in this country to the calcicolous beechwoods, while others occur in most semi-natural deciduous woods on a wide range of soils. Of the latter more widely distributed and abundant species not at all peculiar to beechwoods, the germander speedwell (*Veronica chamaedrys*), the creeping bugle (*Ajuga reptans*), the lesser burdock (*Arctium minus*), the wood sedge (*Carex sylvatica*), herb Robert (*Geranium robertianum*) and the grasses *Poa nemoralis*, *Brachypodium sylvaticum* and *Bromus ramosus* (*asper*) may be mentioned.

Characteristic species More characteristic of calcareous woods are *Lactuca muralis*, and the elegant melic grass (*Melica uniflora*). Most characteristic of all, however, and either confined or nearly confined to escarpment beechwoods are certain orchids: the common helleborine (*Epipactis latifolia*), occasionally found in other woods, the white helleborine (*Cephalanthera latifolia* or *grandiflora*), a remarkably constant species in the escarpment beechwoods, and the rare narrow-leaved helleborine (*C. longifolia* or *ensifolia*). The very rare red helleborine (*C. rubra*) found only in one place on the Cotswold

[1] Pl. 49, phot. 111.

PLATE 49

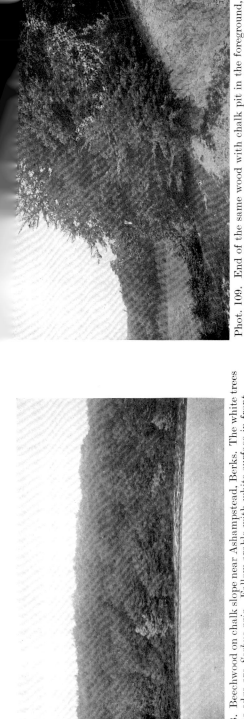

Phot. 108. Beechwood on chalk slope near Ashampstead, Berks. The white trees on the edge are *Sorbus aria*. Fallow arable with white surface in front.

Phot. 109. End of the same wood with chalk pit in the foreground, showing extreme shallowness of the soil.

Phot. 110. "The Miscombe", Ditcham Park, Hants, on escarpment of the Lower Chalk. *Mercurialis perennis* dominant. *Acer campestre* (as shrub), *Hedera helix*.

Phot. 111. Detail of field layer on shallow chalk soil near Watlington, Oxon. *Mercurialis perennis* (sparser and lower growing than in a typical "mercury wood"), *Fragaria vesca*, *Viola hirta*, *V. sylvestris*.

FAGETUM CALCICOLUM

PLATE 50

Phot. 112. *Sanicula europaea* dominant in field layer, but many other species present. Crowell Hill Wood, Oxon. Escarpment of Lower Chalk, Chiltern Hills. R. J. L.

Phot. 113. Top of escarpment showing abrupt cessation of sanicle dominance at the boundary of the plateau Clay-with-flints. Pulpit Hill, Bucks. Escarpment of Lower Chalk, Chiltern Hills. R. J. L.

FAGETUM CALCICOLUM SANICULOSUM

oolites, occurs rather as a marginal or limestone scrub plant. The colourless saprophytic bird's nest orchid (*Neottia nidus-avis*),[1] with its fellow saprophyte the yellow bird's nest (*Monotropa hypopitys*) which is allied to the heaths, are also very characteristic. These all flourish in the deep beech humus and in deep shade, the colourless saprophytes, of course, being independent of light altogether, so far at least as photosynthesis is concerned.

Biological spectrum. The biological spectrum (see Chapter XI, p. 235 n.) of the most widely spread and characteristic species of the field layer of the escarpment beechwoods is remarkably constant. The percentages of species occurrence are approximately: chamaephytes 8, hemicryptophytes 65, geophytes 22, therophytes 5. The dominance of the tall, close-growing mercury involves the greater frequency of the hemicryptophytes of this type, with leafy flowering shoots (hemicryptophyta scaposa), in the mercury woods; while the rosette-forming hemicryptophytes (hemicryptophyta rosulata) with foliage leaves confined to a basal rosette, which are unable to compete with mercury, are more frequent in the sanicle woods (Watt, 1934).

Bryophytes. (4) *The ground* or *moss layer* is never continuous and is often restricted or almost restricted to the lower parts of the tree trunks, the upper surfaces of the main roots which diverge horizontally from the bases of the trunks, and the ground immediately around. Over much of the floor of the wood, where it is covered with the leaf litter or occupied by a thick carpet of a field layer dominant such as mercury, mosses are altogether absent.

The two most abundant and widespread species are *Hypnum cupressiforme* on the bases of the beech trunks, and *H. molluscum* on the soil. Other frequent or abundant and fairly constant species are *Neckera complanata, Brachythecium rutabulum, Anomodon viticulosus, Eurhynchium striatum, Mnium undulatum*, and *Camptothecium sericeum*. Of liverworts the commonest species are *Frullania dilatata, Metzgeria furcata, Madotheca platyphylla* and *Lophocolea bidentata*. The bryophytes are much more in evidence on the floor of the sanicle woods than in the mercury woods, where they are generally sparse or almost absent because they cannot establish themselves between the densely packed shoots of the field layer.

Larger fungi. The following are among the more characteristic species:

Collybia radicata	a	*Mycena pelianthina	r–o, l
Hygrophilus eburneus	o–f	*M. pura	f–a

* Apparently dependent on free calcium carbonate in the soil, for they occur also on chalk grassland and in plateau woods where the chalk is very close to the surface.

Cortinarius calochroous (o–f) is apparently exclusive to escarpment woods. *Marasmius peronatus* (f–a), on the other hand, occurs also in the

[1] Pl. 51, phot. 116.

other extreme type, the plateau beechwoods with podsolised soil, the common character of the habitat being perhaps the deep accumulations of beech leaf litter found in both (Harley).

Cotswold beechwoods. The semi-natural beechwoods developed on the escarpment and plateau of the Inferior Oolite of the Cotswold Hills are essentially calcicolous and resemble most closely the escarpment woods on the Chalk, but with some features of the chalk plateau woods that possess more calcareous soil. Pl. 51 illustrates some of the vegetation.

SUCCESSION—CHALK SCRUB (Pls. 52–54)

The succession from chalk grassland through scrub to beechwood that occurs when pasturing and rabbit grazing (which hold the grassland communities stable) are withdrawn or greatly diminished, has been studied mainly on the escarpment of the Chilterns, but there is every reason to believe from observations on the South Downs and the Cotswolds that the phenomena everywhere follow the same general lines, except that on the Chilterns ashwood is not a normal phase of the succession.

The two seres. On the Chiltern escarpment Watt (1934) has distinguished two closely allied seres which he separates by means of a number of habitat and vegetational characters, and which correspond with the mature sanicle and mercury beechwoods respectively. These he names the *juniper* and the *hawthorn seres*, from the characteristic dominants of the scrub stages. The outstanding seral distinction is that beech directly colonises the juniper scrub, whereas the hawthorn scrub may be succeeded by seral ashwood before the beech comes in. The ash consocies is, however, never well developed on the Chilterns (probably, according to Watt, because the climate is too continental and the habitat too exposed and because under these conditions rabbit-grazing has more serious results), though indications of it are not wanting in sheltered situations. On the South Downs and elsewhere ashwood probably represents a normal seral stage between scrub and beechwood.

Habitats of juniper and hawthorn seres. The crucial distinction between the habitats is the firmer and shallower soil of the juniper sere, with an average depth of 30 cm., a much higher percentage of calcium carbonate, a sparser leaf litter in the mature beechwood, a lower humus content of the surface soil, a lower water-holding capacity, rendering the soil more liable to summer drought; and probably, as a result, a lower degree of nitrification. The juniper sere is typically developed on more exposed and steeper slopes, and characteristically on the harder strata of the Upper Chalk, though none of these physiographic characters is decisive by itself.

On the other hand the hawthorn sere is typically developed on gentler, more sheltered slopes of the softer rock of the Middle Chalk. The average depth of the more yielding soil is greater (50 cm.), the lime content lower (though still high), and in the mature beechwood the leaf litter is thicker,

PLATE 51

Phot. 114. *Fagus sylvatica* dominant but sparse (recently thinned). Seedlings and saplings of *Crataegus*, *Ilex*, etc. forming a low shrub or upper field layer. Buckholt Wood, Inferior Oolite. *C. G. P. Laidlaw*.

Phot. 115. *Daphne laureola* in Buckholt Wood. *Uehlinger*.

Phot. 117. *Pyrola minor, Hieracium serratifrons*, etc. in deep humus. Pope's Wood, Cranham, Glos, on Inferior Oolite. *C. G. P. Laidlaw*.

Phot. 116. *Neottia nidus avis, Hieracium serratifrons, Sorbus aria.* Buckholt Wood. *C. G. P. Laidlaw.*

COTSWOLD BEECHWOODS

PLATE 52

Phot. 119. Buckthorn, dogwood, holly and rose protected by juniper. Chalton Down, Hants.

Phot. 121. Dense juniper scrub [...]

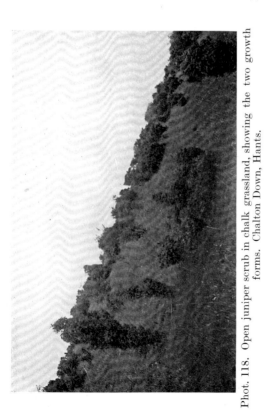

Phot. 118. Open juniper scrub in chalk grassland, showing the two growth forms. Chalton Down, Hants.

Phot. 120. Yew, buckthorn, holly and rose protected by juniper. Near Figsbury Ring, Wilts. *R. J. T.*

and the humus content greater. The water content is greater and more eustatic, the nitrifying power probably higher, and the conditions altogether more favourable for luxuriant growth.

The scrub stage. Juniper scrub,[1] with its rigid, dark, evergreen bushes, looking almost black from a distance, dotting the grassland, or aggregated into more or less dense masses, is a familiar sight in most of the chalk region of southern England. The bushes have two distinct growth forms which occur side by side, the one erect, strict, and averaging 6–8 ft. (1·8–2·4 m.) high, the other of lower and widespreading habit. According to Watt, however, intermediate forms predominate, averaging about 4 or 5 ft. in height. The juniper frequently occurs in the hawthorn scrub and is there commonly taller in growth (as much as 12 ft. or 3·6 m. in sheltered places with deeper soils), but its leaves are paler in colour and it tends to suffer from the competition for light of the hawthorn and other shrubs. In exceptional cases the erect form of juniper has been recorded as reaching 20 ft. (6 m.) in height.

The hawthorn is by far the most widespread dominant in English scrub, not only in the succession to beechwood but also to oakwood, since it flourishes on a very wide range of soil types. On the relatively shallow soils of the chalk escarpments and valley slopes it favours, as we have seen, the rather deeper, less compact and more humous soils of the gentler slopes, and it is actually somewhat less frequent in the juniper sere than juniper is in the hawthorn sere.

Shrubs of the chalk scrub. Of the accompanying species of shrub there is a relatively large number with a high degree of constancy: chalk (and other calcicolous) scrub indeed is by far the richest in species of British scrub communities. On the Chiltern escarpment the highest degree of constancy is shown by the dogwood (*Cornus sanguinea*),[2] privet (*Ligustrum vulgare*),[3] blackthorn (*Prunus spinosa*),[4] buckthorn (*Rhamnus catharticus*)[5] and wayfaring tree or mealy guelder rose (*Viburnum lantana*).[6] All of these, except perhaps blackthorn, are very characteristic species, i.e. they are far more constant and abundant in chalk scrub than in any other British plant community. Other, somewhat less constant, but still quite characteristic species are the spindle tree (*Euonymus europaeus*) and elder (*Sambucus nigra*). The field maple (*Acer campestre*) and hazel (*Corylus avellana*) are frequently present and flourish especially on the deeper soil towards the bottoms of the slopes. *Clematis vitalba* is characteristic and often luxuriant.

Species of *Rosa* are also characteristic of chalk scrub, including the dwarf *R. rubiginosa* (sweetbrier) and *R. micrantha*, besides various forms of the common dog rose (*R. canina*).[7] The small species are often pioneers in colonising chalk grassland. The fruticose Rubi (brambles) are on the whole

---

[1] Pl. 52. [2] Pl. 52, phot. 119. [3] Pl. 53, phot. 122. [4] Pl. 54, phot. 124.
[5] Pl. 52, phots. 119, 120; Pl. 54, phots. 124–5. [6] Pl. 54, phot. 125.
[7] Pl. 52, phots. 119, 120; Pl. 53, phot. 122.

less characteristic and abundant, though they frequently occur. The dewberry (*Rubus caesius*) is also not uncommon.

Trees of chalk scrub. Of trees the whitebeam (*Sorbus aria*)[1] is one of the commonest and most constant members of chalk scrub, where its heavy foliage with brilliant white undersides makes a beautiful feature. This tree frequently maintains itself in the canopy of the mature beechwood (p. 384). The gean (*Prunus avium*), less abundant, but characteristic especially of the Chiltern area, behaves in somewhat the same way. Holly (*Ilex aquifolium*) is a moderately constant, and rowan or mountain ash (*Sorbus aucuparia*) an occasional, constituent of the scrub, especially in the juniper sere.

The yew (*Taxus baccata*) occupies a special position. It colonises chalk scrub very freely, the dark foliage standing out conspicuously from the bright green background of the shrubs.[2] Owing to the very deep shade which it casts, excluding competition from other trees and shrubs, and the persistence of individual trees to a great age, the yew sometimes forms pure societies, especially on the South Downs, as at Kingley Vale near Chichester and in the coombes on Butser Hill (see Figs. 71, 72, pp. 378–9).

Seedlings of the pedunculate oak (*Quercus robur*) are widespread in escarpment grassland and scrub, but they are not at all common. Occasionally small oak trees occur in the scrub, but they never reach a height of more than 10 or 12 ft. (3 or 4 m.) at the outside, being quite unable to form good trees on these shallow soils. On the Chiltern escarpment, which is on the edge of its natural area of distribution in Britain, the hornbeam (*Carpinus betulus*) is also occasionally met with in the chalk scrub, but it has no significant role in the community. Subspontaneous pine (*Pinus sylvestris*) derived from planted seed parents is also sporadic, occurring occasionally in the scrub.

Ash and beech, on the other hand, have a high constancy in the scrub, and both are of first importance as dominants of the next stages in the succession. On the South Downs ash is very abundant in the scrub, but in the Chiltern area it has a low average frequency, though it is seldom entirely absent. It has a large number of enemies and parasites, both insects and fungi. Beech is equally widespread, and in the Chiltern area much more abundant than ash. The roles of these two trees in the succession are dealt with on pp. 380–3.

Structure of chalk scrub. The height of the scrub varies from the dwarf juniper scrub growing under the most unfavourable conditions of steep exposed slopes with shallow soil, and not more than 18 in. or 2 ft. (0·4–0·6 m.) high, to the luxuriant hawthorn scrub 3 or 4 m. high, or even taller, in sheltered places on deeper soil.

The actual density of the adult scrub varies from scattered bushes in the

[1] Pl. 52, phot. 121.
[2] Pl. 52, phots. 120, 121.

PLATE 53

Phot. 122. Juniper protecting privet (in fruit) and rose (right) in severely rabbit-grazed chalk grassland. Pulpit Hill, Bucks. *R. J. L.*

Phot. 123. Patches of scrub near Wylye, Wilts. Juniper scrub protecting young ash in left foreground. On right gorse poorly developed. On farther slope chalk scrub with juniper (opposite white arable field)—detail shown on Pl. 54: in bottom gorse-hawthorn scrub on loam (opposite darker arable field). Rabbit-grazed grassland. *R. J. L.*

CHALK SCRUB

PLATE 54

Phot. 124. Left to right: yew, juniper, blackthorn, buckthorn (tall, with bramble below), juniper. Near Wylye, Wilts. *R. J. L.*

Phot. 125. Left to right at back: juniper (tall), *Viburnum lantana*, hawthorn, juniper, ash, buckthorn. Low juniper in front. Near Wylye, Wilts. *R. J. L.*

COMPOSITION OF CHALK SCRUB

chalk grassland to dense impenetrable thickets. The density is apparently correlated partly with exposure to wind and partly with severity of grazing, for when grazing or rabbit pressure is lifted the intervals between the old scattered bushes are colonised by hawthorn or juniper and other shrubs springing up among the ungrazed grass. The prevalent uniformity of chalk scrub and the general correspondence of the constancy and frequency of the other shrubs in the hawthorn and juniper seres is no doubt partly due, as Watt (1934) suggests, to the incidence of grazing and rabbit attack, since the chalk grassland in which the scrub develops is commonly used for grazing (sheep walk), and is locally heavily infested with rabbits. While all the shrubs are attacked (except apparently the elder), not all suffer equally, and the more resistant tend to increase in abundance and prominence relatively to the less resistant. Thus a potential hawthorn scrub may be replaced by a community of the more resistant juniper (with or without elder) and if rabbit pressure is severe enough the juniper seedlings themselves are destroyed, only the older scattered bushes lingering on. This is the explanation of the moribund appearance of much of the scattered juniper scrub of the chalk downs. Under very severe rabbit pressure the vegetation eventually retrogresses to closely nibbled grassland, often with clumps of elder, which readily colonises disturbed soil round rabbit burrows (Pl. 56, phot. 128; Pl. 58, phot. 134).

<small>Grazing and rabbit attack</small>

In chalk grassland which is not so heavily grazed the clumps of juniper bushes form an effective protection for the less resistant shrubs, and vigorous examples of dogwood, spindle-tree, buckthorn or wayfaring tree, or saplings of ash or beech, may be found surrounded by a protective girdle of juniper (Pls. 52–54).

Field layer. The field layer of chalk scrub is naturally very variable in development. Under the deep shade of young dense scrub the soil is quite bare except for occasional mosses or isolated and starved examples of chalk grassland plants. In loose scrub enclaves of chalk grassland are enclosed between the clumps of bushes and some of the grassland species grow more luxuriantly in these owing to the partial protection they obtain. Mosses, notably the Hylocomia, which occur abundantly in the cooler and damper grassland, as on northern slopes, may become a conspicuous feature of such enclaves, even assuming dominance in heavily rabbit-attacked areas because the rabbits do not eat them.

Plants which affect partial shade though they can grow in the open, such as *Brachypodium sylvaticum* and *Viola hirta*, may invade these enclaves and other grassland on the margin of scrub. Among the marginal plants commonly found in and about scrub are *Nepeta hederacea*, *Hypericum hirsutum*, *Inula conyza*, *Clinopodium vulgare*, *Origanum vulgare*, and sometimes *Helleborus foetidus* or *Atropa belladonna*. All these marginal species can grow in the open, though not in the most exposed situations, and in woods, though not in the deeper shade, and so

<small>Marginal species</small>

far as scrub can be said to have a field layer of its own it is represented by these marginal or "wood-edge" species. Under the shade of older scrub where woodland conditions are approached a few true woodland species often occur. Of these the commonest is mercury, which in such situations may form considerable pure societies.

YEW WOODS OF THE SOUTH DOWNS
(Figs. 71–73 and Pls. 55–7)

Distribution. The yew, as we have seen, is a frequent invader of chalk scrub; but it seldom forms pure woods except on the western South Downs in west Sussex and the adjoining part of Hampshire. These are situated on the sides and round the heads of the dry chalk valleys or "coombes" (Figs. 71, 72 and Pls. 55–6, 60) or on sheltered chalk slopes, particularly on slopes running south-west and north-east, whose south-western ends are occupied by beechwood in the shelter of which yew can develop freely in the scrub (Fig. 73). These yew woods are either surrounded by grassland and isolated from the forest areas of this richly wooded district, or they may occur on the edges of progressive forest. Watt (1926) has made a detailed study of the origin and fate of some characteristic yew woods and of their relationship to scrub and forest. His conclusions are based on the careful examination of ten separate examples.

Origin and development. Watt finds that yew does not normally colonise open grassland, partly owing to exposure of the seedlings to the destructive attack of rabbits which infest most of this region, and partly perhaps because close turf is unfavourable for their successful establishment. Yew woods originate in chalk scrub, and mainly in juniper scrub, which is more efficient than hawthorn scrub in protecting the young yews. Yew invasion may be general in closed juniper scrub and a relatively quick transition to yew scrub then follows. In open juniper scrub invasion is more sporadic (Pl. 58, phots. 134–5), and results in the formation of clumps consisting of yew families girt by scrub. A yew family consists of a central older yew, with a richly branched spreading crown (pioneer form), and surrounded by its offspring of younger yews whose cleaner stems and lopsided crowns are due to the shading effect of their parent. In the scrub round these clumps still smaller and younger yews establish themselves. This grouping of young straighter stemmed yews round the pioneers can usually be made out even in the pure yew wood which eventually results after the destruction by shading out of the initially protecting scrub, whose dead remains are found under the dense yew canopy. During the earlier stages of development the growth of the chalk scrub is actually favoured by the protection which the taller yews afford, the shrubs growing more luxuriantly and to a greater height than when they are fully exposed to wind.

PLATE 55

Phot. 126. General view of the head of Kingley Vale, Sussex, from east-south-east. Beyond the clumps of trees and shrubs in the bottom of the valley the oldest yew wood is seen, with a few relict whitebeams showing white among the dark yews and cut off to the right by the bare gully marked × in Fig. 72. To the left the yew wood passes into yew-ash wood. *R. J. L.*

Phot. 127. Approximate continuation to the left of Phot. 126. The southern end of the yew-ash wood is seen on the right, and is adjoined to the left (south) by a dark strip of pure yew wood, and then by advanced yew scrub (with relict whitebeams) occupying most of the width of the picture. Cf. Fig. 72. *R. J. L.*

YEW WOOD AND YEW-ASH WOOD AT KINGLEY VALE

PLATE 56

Phot. 128. The oldest yew wood at the head of the valley with relict whitebeams. Rabbit-grazed turf with elder and mullein in right foreground.

Phot. 129. The old yew wood from above showing its sharp limitation to the valley slope. Loose hawthorn scrub on the plateau beyond. Loose chalk and flints in foreground, severely rabbit grazed.

Phot. 130. Fine old oak and yew on loam in the valley bottom, with fringe of elder and clematis. Moderately pastured grass in front.

KINGLEY VALE YEW WOODS

Yew and juniper are the most wind-resistant elements of the scrub: and in the most exposed situations and on shallower soil isolated examples of these two species are found alone, both severely wind-trimmed (Pl. 59, phot. 137), nearly every yew being accompanied by a living or dead juniper. Most of the yews are small, rarely more than 18 in. (45 cm.) high, and closely nibbled, but often quite old. One which was only 7 in. (18 cm.) high showed fifty-five annual rings! Under these extreme conditions of exposure to wind and rabbit attack no other chalk shrubs can establish themselves, nor can even juniper and yew form anything like a closed scrub.

In hawthorn scrub
Hawthorn scrub develops on the more protected slopes and the deeper soils of the same valleys, and is much richer in species than juniper scrub. It is colonised by yew in essentially the same way, but much less generally, since the hawthorn forms a less efficient protection than the juniper. The yew families are usually fewer in number and enlarge more slowly, though the individual trees may be larger, so that the whole process of development of the yew wood occupies a longer time. As in the juniper sere the accompanying shrubs increase in variety and luxuriance owing to the increased protection afforded by the yews as they develop, though the shrubs are again eventually killed out when the yew wood becomes closed.

The role of ash. When ash (the normal tree colonist of chalk scrub on the South Downs) has entered the scrub the development of yew wood is checked for a time, the ash growing along with the yew and forming an intermediate stage of yew-ash wood (Fig. 72 and Pl. 55, phot. 126), in which *Mercurialis perennis* may form a field layer. The yew outlasts the ash since it has a very much longer life, but in places the ultimate development of pure yew wood is very slow indeed, relicts of the original scrub still surviving between yews estimated as 500 years old.[1]

Both the nursing scrub and the yew wood which follows it establish themselves first in the most protected situations, i.e. the places least exposed to strong winds, and progress both along, and up and down, the slopes, faster along the slopes and faster up than down. The rate and direction of spread can be deduced with some certainty from the distribution of yews of different ages and habits of growth.

Structure of mature yew wood. The yews of the mature wood are uneven-aged in consequence of the mode of development described. The largest and oldest in the slowly developing yew-ash wood may attain an extreme height of about 60 ft. (18 m.) and a basal girth of 18 ft. (5–6 m.)

---

[1] The largest yews on the chalk slopes at Kingley Vale were calculated by Watt, on the basis of ring counts of felled trees and comparative measurements, as being about 700 years old. On the deeper soil in the bottom of the valley one of the largest yews was estimated as 800 years. But if the rate of growth falls off considerably in later life their age may be considerably greater, perhaps as much as 1000 years. It is probable however that the age of old yews is often considerably overestimated.

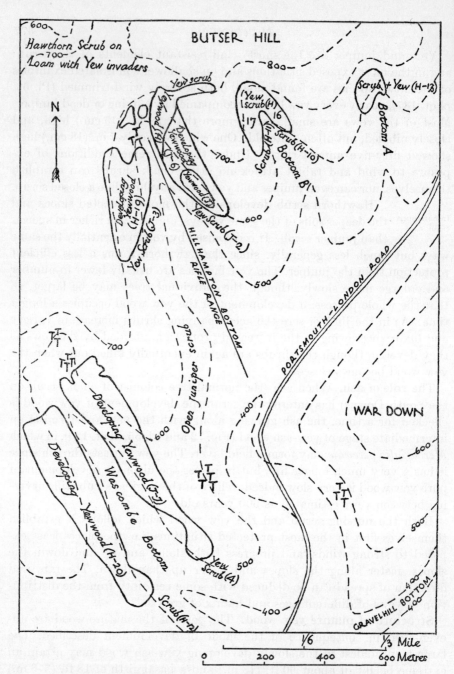

FIG. 71. DEVELOPMENT OF YEW WOODS IN FOUR COOMBES ON THE SOUTHERN SLOPES OF BUTSER HILL (HANTS.)

The two smallest coombes or "bottoms" (A and B) contain chalk scrub in which yew is present and yew scrub. Hillhampton Bottom has developing yew woods on its two slopes, but the head of the valley is partly bare, with relict yews, and is devastated by rabbits (cf. Pl. 60, phot. 139). The head of Wascombe Bottom is completely bare except for relict yews and a little degenerate scrub (Pl. 3, phot. 8, p. 135): the sides bear extensive developing yew woods. (H) means that the developing yew wood belongs to the hawthorn sere, (J) to the juniper sere. The numbers refer to lists in Watt's paper (1926). T = outlying clumps of yew presumably occupying the sites of former scrub. Contours in feet. From Watt, 1926.

PLATE 57

Phot. 133. An old yew measuring 6·15 m. in girth. *Uehlinger.*

Phot. 131. Interior of old yew wood at head of valley. Oldest yews estimated as at least 500 years. Ground bare. *Uehlinger.*

Phot. 132. Dead stems of *Clematis vitalba* below the yews. *Uehlinger.*

INTERIOR OF YEW WOOD AT KINGLEY VALE

PLATE 58

Phot. 134. Open and closed yew scrub at Kingley Vale, south-east of the gully × (Fig. 72). Rabbit burrows, elder and nettles in foreground at foot of gully, which extends uphill to the left out of the picture.

Phot. 135. Detail of yew scrub from top left corner of Phot. 134. Young yew and moribund juniper (centre, light coloured above dark yew: to the left juniper above yew). The yew scrub has been preceded and its growth protected by juniper scrub.

YEW SCRUB

or more, with deeply fluted trunks like clustered columns, owing to the aggregate stems of which pioneer trees are often formed. In the pure yew wood suppression of all other plants is sometimes complete,[1] but occasional

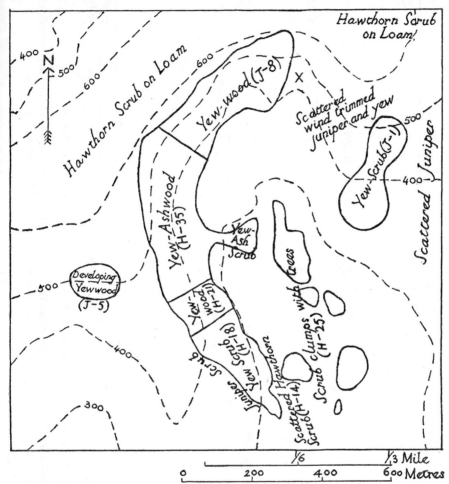

FIG. 72. YEW WOODS AT KINGLEY VALE (SUSSEX)

The oldest yew wood at the head of the valley is cut off sharply to the east by a bare windswept gully (X) with isolated wind-cut juniper and yew (Pl. 59, phots. 136). South of the oldest yew wood is yew-ash wood, yew wood, and yew scrub on the protected western slope of the valley. On the north-eastern slope of the valley, where the wind is not so severe as in the northern corner, is pure yew scrub in succession to juniper scrub (Pl. 58, phots. 134–5). (H) = hawthorn, (J) = juniper sere. Contours in feet. From Watt, 1926.

holly and rare hawthorn and blackthorn, drawn up by the competition for light, may find a place in the canopy, above which rare ash and occasional whitebeam raise their crowns.[2] All these species are strictly speaking relicts of earlier stages of the succession. There is no shrub or field layer and mosses are rare. Gaps formed by the death of ash or old yews may be difficult to

[1] Pl. 57, phot. 131.   [2] Pl. 55, phot. 126 and Pl. 56.

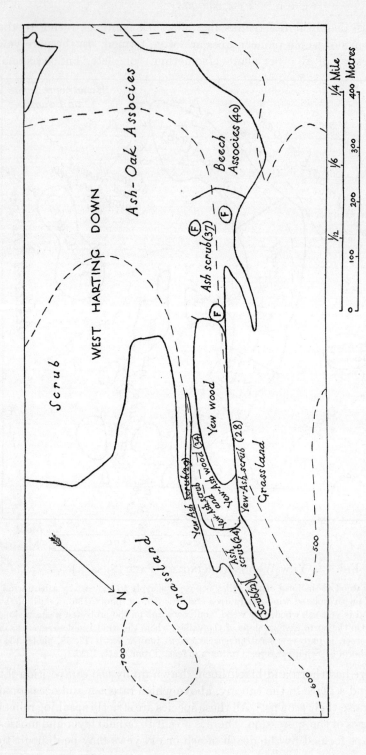

Fig. 73. Development of Yew Wood on Downley Brow (Sussex)

Here the yew wood has developed along the slope in the direction SW → NE. It is colonising ash scrub (left) and is protected from the prevailing south-westerly winds by the advancing beech associes (right). From Watt, 1926.

PLATE 59

Phot. 136. The wind-swept and rabbit-ridden gully marked × in Fig. 72. The straight line of small wind-cut yews ascending the head of the gully are in the direct path of the most violent wind. The yew scrub to the right is little affected. *R. J. L.*

Phot. 137. Old wind-cut yew in the upper part of the gully (Prof. H. C. Cowles, 1911). These yews are but little altered after 25 years. *Mrs Cowles.*

EFFECT OF WIND ON YEW

PLATE 60

Phot. 139. Head of Hillhampton Bottom, Butser Hill (left-hand combe in Phot. 137). Yew scrub with ashes and yew wood. See Fig. 71.

Phot. 141. Yew scrub on steep side of Burrington Coombe, Mendip Hills, Somerset (Carboniferous Limestone). Hawthorn scrub on opposite face. *M. Tomlinson.*

Phot. 138. Three combes on the southern slopes of Butser Hill, Hants (chalk), showing limitation of yew woods and scrub to the valleys (taken from just out of map in Fig. 71 to the south). Hawthorn scrub in foreground.

Phot. 140. Abundant yew in ash-oakwood on dipslope of Wenlock Limestone near West Malvern, border of Worcester and Herefordshire.

fill, largely owing to rabbit pressure, and either remain bare or become occupied by a few rabbit-resistant plants, notably elder.

Localisation of yew woods. For the most part, as has been said, the yew woods are confined either to the heads and sides of the little dry chalk valleys or coombes (Figs. 71, 72 and Pl. 60, phot. 138), or to adequately sheltered chalk slopes (Fig. 73). On exposed ridges and slopes and on much of the open plateau, neither scrub nor yews can establish themselves. But where the plateau is covered by a layer of loam scattered hawthorn scrub is often present, and this also may form a stage in succession to yew wood (Chapter XIX, p. 396).

By careful comparison of the relationships of the different stages of development that have been described above, Watt shows that the situation of the yew woods depends on the distribution of the nursing scrub, which itself starts in the most sheltered places, i.e. at the heads of the valleys, and progresses along the sides, followed by the yew colonisation. Consequently the oldest yew woods (and the oldest yews) are found at the heads of the coombes, and both are progressively younger as they are traced along the sides towards the mouth.

Degeneration from biotic attack Correspondingly it is from the heads of the valleys that degeneration of the yew woods begins (Fig. 71 and Pl. 60, phot. 139). Owing to the attacks of ground vermin (rabbits and mice) the trees cannot recolonise the gaps formed by the death of the oldest members, nor can scrub fit to protect young yew establish itself. That this is largely due to rabbit pressure is shown by the occupation of the gaps almost solely by rabbit-resistant plants, among which *Sambucus* is often prominent. The ripe yew seeds are also often badly damaged, apparently by birds and mice. Under these conditions the gaps increase by the progressive death of old yews and eventually the entire head of the coombe may become bare, with a few isolated yews representing the remains of the former wood (Pl. 3, phot. 8, p. 135). In some cases, however, clumps of scrub partially recolonise such an area, and in these young yews may again establish themselves. We must suppose that in the absence of rabbit pressure scrub and yew would again colonise the gaps and there would be adequate regeneration of the yew wood.

SERAL ASHWOOD

As already noted (p. 374) there is a striking difference between the abundance of ash on the South Downs and adjacent parts of Hampshire and its sparseness on the Chiltern escarpment, though its constancy is high in both areas. On the South Downs the ash consequently plays an important part in the development of the woody vegetation, typically dominating the first woodland stage of the succession, though it is normally soon overtaken by beech or yew, while on the Chilterns its role as a wood former is insignificant. No detailed studies of the colonisation of chalk slopes by ash have been made on the South Downs, but some examples of the ash consocies in both areas have been examined by Watt.

Structure and composition. In a sheltered valley at Wainhill near Princes Risborough, ash sapling invaders of hawthorn scrub are present in quantity. At Aston Hill (on the Wycombe-Oxford road) an ash consocies occurs, showing various stages in development from scrub to woodland. The ashes are young (saplings and poles), the offspring of old trees which have been felled and whose stumps are still recognisable. The tallest trees are about 60 years old and 60 feet high, with a very slow rate of increase in diameter during the last 20 years of growth. Beech invaders are present. The commonest shrub is privet, but blackthorn, elder, wayfaring tree and bramble have high local frequencies. Near Ashridge, in the northern Chilterns, a mature consocies is dominated by ash 66 ft. (20 m.) in height with hawthorn, hazel, elder, bramble and spindle tree forming the shrub layer. In the field layer, besides relicts from chalk grassland, the more abundant species of the immature and mature consocies are the following:

	Immature consocies	Mature consocies
Fragaria vesca	f–a	—
Viola hirta agg.	f	—
V. sylvatica (probably V. sylvestris)	f	—
Geum urbanum	o–f	o–f
Mercurialis perennis	a–ld	d
Urtica dioica	ld	o–la
Circaea lutetiana	f	la
Asperula odorata	f–a	—

The first three species (hemicryptophytes with basal leaf-rosettes) have given way to the dominance of mercury with its tall leafy shoots. The most frequent mosses are *Brachythecium rutabulum*, *Eurhynchium praelongum*, *E. striatum*, *Fissidens taxifolius*, and in the immature wood *Hylocomium triquetrum*. Thus the seral ash consocies, so far as can be told from the available data, shows many of the species characteristic of the mature beechwood belonging to the mercury sere.

Ash scrub. In many places on the South Downs "ash scrub" occurs, i.e. chalk scrub in which ash is dominant or becoming dominant and which would develop into ashwood were it not for the intervention of yew or beech. This mostly belongs to the hawthorn sere, and *Crataegus* is the most abundant woody plant other than ash. Another fairly constant shrub is *Cornus sanguinea*, with the woody climbers *Clematis* and *Hedera*. *Euonymus, europaeus, Viburnum lantana* and *Corylus* also occur but are less constant.

Calcicolous coppice. Coppice composed largely of calcicolous shrubs, together with abundant coppiced ash, is often met with on the escarpments of the North and South Downs, though apparently absent from the Chilterns. It is almost certainly derived from "ash scrub", i.e. chalk scrub in which ash is becoming dominant, and which is no longer dominated by

PLATE 61

Phot. 142. Dense juniper scrub on Chinnor Hill, Oxon., invaded by beech from the plateau above. The succession is checked but not stopped by rabbit pressure. *R. J. L.*

Phot. 143. Loose juniper scrub on Pulpit Hill, Bucks, similarly invaded. Note slight effect of wind on the young beeches. *R. J. L.*

BEECH INVASION OF JUNIPER SCRUB (CHILTERN ESCARPMENT)

PLATE 62

Phot. 144. Young beechwood on Chinnor Hill, Bucks. The preceding juniper killed out (dead shrubs visible). Live holly (left) and yew (right) persist. Ground bare. *R. J. L.*

Phot. 145. Immature escarpment beechwood above Graffham, Sussex. Pioneer beech in centre, semi-pioneer to right. Younger beeches on extreme right and left. *H. Godwin.*

IMMATURE ESCARPMENT BEECHWOODS

the spiny pioneers, juniper and hawthorn. Hazel is often an important constituent, but is accompanied much more abundantly by other species of shrub, such as dogwood, spindle, wayfaring tree, privet, buckthorn, field maple, elder, etc., than in the hazel coppice associated with oakwood (Adamson, 1922).

DEVELOPMENT OF BEECHWOOD
(Pls. 61, 62)

Beech is the only permanently successful wood-forming invader of chalk scrub both on the Chilterns and on the South Downs, and this is probably also true of the North Downs of Surrey and Kent. It follows ash in the succession, but also colonises scrub on the side of existing forest exposed to the strongest winds (nearly always from the west or south-west), a thing which ash cannot do, and directly colonises juniper scrub (Pl. 61).

Beech invasion of juniper scrub. On the Chiltern escarpment the beech often directly colonises juniper scrub on exposed slopes, rapidly killing the pioneer, which is very intolerant of shade, and immediately forming closed beechwood. From hawthorn scrub to closed beechwood the transition is more gradual. On the Chiltern escarpment the seed parents are mostly trees of the extensive plateau woods, so that progress is from the top of the escarpment downwards. Much of this invasion is relatively recent, young beechwood, often with dead juniper scrub in the deep shade (Pl. 62, phot. 144), being in full occupation of ground which is marked on fairly modern maps as open grassland and recorded in the estate "terriers" as sheepwalk.

Of hawthorn scrub In the hawthorn sere there is much greater variation in the mode and rapidity of colonisation, so that isolated beech pioneers, beech families, and groups of various sizes and degrees of maturity are met with in the scrub. "The state of the subsidiary vegetation varies similarly: beech with dead scrub below, clumps of beech with a bare floor, a mixture of grassland and woodland species, a mixture of woodland species which have not yet entered into full competitive relations, and finally local patches of vegetation characteristic of mature beechwood. The influence of the invading beeches extends to the enclaves of grassland or open scrub, and under lateral shade true woodland species (e.g. *Mercurialis perennis*), and wood-edge species (e.g. *Nepeta hederacea*) replace the heliophilous species of grassland. In the juniper sere the transition from scrub and grassland is usually abrupt: in the hawthorn sere gradual, with the developmental stages characterised by a mixed flora" (Watt, 1934, p. 252).

In the developing beechwood of the hawthorn sere the proportions of the shrubs present in the chalk scrub gradually change. Juniper is killed out and hawthorn decreases in abundance. The smaller pioneer roses, such as the sweetbrier and *Rosa micrantha*, are eliminated: the brambles, hazel and elder maintain their status; and the larger roses—such as *R. arvensis*, *R. canina*—and the ivy increase in frequency. The raspberry (*Rubus idaeus*), not recorded from the scrub, is sometimes present.

Development of Beechwood

Later stages. In the field layer woodland species make their appearance mixed with relicts from earlier stages, but the woodland flora is not yet consolidated and there is no very clear distinction between the juniper and the hawthorn seres, corresponding with the adult "sanicle" and "mercury" beechwoods, except in regard to a few species. Thus the sanicle itself has a much higher constancy in the developing "sanicle" wood, and the mercury is both more constant and more abundant in the developing "mercury" wood. *Bromus ramosus* (*asper*) is both more constant and more frequent in the immature sanicle wood, *Circaea lutetiana* slightly so in the immature mercury wood, and *Cephalanthera latifolia* (*grandiflora*) is more constant in the young sanicle wood. But marked differences do not extend to any great range of species.

Though there is much bare ground in the young sanicle woods there are fewer species of bryophytes than in the mercury woods, contrary to the state of things in the mature woods, but in both types of developing wood the occurrence of bryophytes is quite sparse.

As the woods come to maturity the dominant beech is still accompanied in the canopy by a few ashes. The whitebeam also remains in the canopy of the sanicle wood, but the taller beeches of the mercury wood usually suppress the tree, which cannot here rise above the shrub layer, and in this it is only occasional. The shrub layer is very sparse, and but few species show high constancy. Hawthorn and bramble are the most widespread but are not at all abundant. *Cornus*, *Hedera*, *Rosa arvensis*, and *Viburnum lantana* are more constant in the juniper, and *Sambucus* in the hawthorn sere. All these species have fleshy fruits and are probably bird sown; and all, except hawthorn and elder, are able to reproduce vegetatively.

Seres of woody vegetation on chalk soils (based on Watt's data)

"JUNIPER-SANICLE" SERE	"HAWTHORN-MERCURY" SERE
Beech consociation	**Beech consociation**
Canopy 60–80 ft. (18–24 m.)	Canopy 70–90 ft. (21–27 m.)
Trunks often somewhat rough and crooked	Trunks smooth and straight
Field layer with *Sanicula* generally dominant: hemicryptophyta rosulata and geophyta rhizomata characteristic	Field layer with *Mercurialis* generally dominant: hemicryptophyta scaposa characteristic

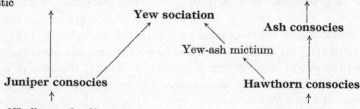

Juniper consocies	Hawthorn consocies
↑	↑
[Chalk grassland]	[Chalk grassland]
Soil up to 12 in. (30 cm.) deep, firm, very highly calcareous. Slope relatively exposed	Soil up to 20 in. (50 cm.) deep, less firm, less highly calcareous. Slope relatively protected

REFERENCES

ADAMSON, R. S. The woodlands of Ditcham Park, Hampshire. *J. Ecol.* **9**, 114–219. 1922.

GODWIN, H. Vegetation phases reconstructed from the pollen analysis of peat, in discussion on the origin and relationship of the British flora. *Proc. Roy. Soc.* B, **118**, 214. 1935.

HARLEY, J. L. Ecological observations on the mycorrhiza of beech. *J. Ecol.* **25**, 421. 1937.

ROBBINS, C. R. An economic aspect of regional survey. *J. Ecol.* **19**. 1931.

WATT, A. S. On the ecology of the British beechwoods with special reference to their regeneration. Part I. The causes of failure of natural regeneration of the beech (*Fagus silvatica* L.). *J. Ecol.* **11**, 1–48. 1923. Part II. The development and structure of beech communities on the South Downs. *J. Ecol.* **12**, 145–204. 1924; **13**, 27–73. 1925.

WATT, A. S. Yew communities of the South Downs. *J. Ecol.* **14**, 282–316. 1926.

WATT, A. S. The vegetation of the Chiltern Hills, with special reference to the beechwoods and their seral relationships. Part I. *J. Ecol.* **22**, 230–70. 1934. Part II. *J. Ecol.* **22**, 445–507. 1934.

Chapter XIX

BEECHWOOD ON LOAM
(CHALK PLATEAU TYPE)

FAGETUM RUBOSUM[1]

(Plates 63–70)

In contrast with the escarpment beechwoods, which occur on the chalk escarpments and valley sides of most of the south-east of England, those developed on the chalk plateaux are mainly confined to two regions, the western part of the South Downs in Sussex and south-east Hampshire, and the Chiltern Hills in Oxfordshire and Buckinghamshire to the north of the Thames valley. Between the two, on the Hampshire chalk uplands and in Berkshire, there is considerable abundance of beech but no extensive plateau beechwoods, and the region has not been investigated ecologically. The chalk plateau beechwoods of the South Downs and Chilterns have been thoroughly studied by Watt (1924, 1925, 1934), and those of the South Downs to some extent also by Adamson (1922). Though some of them are certainly planted, in others the stages of natural development from grassland and scrub have been carefully and thoroughly investigated and there can be no doubt that beech is the natural climax dominant.

Slope and soil. The ground of the plateau beechwood is either flat or slopes gently away from the edge of the escarpment, following, more or less, the dipslope of the upper chalk. Where the outcrop of the chalk is intersected by valleys the steeper slopes are covered by the shallow chalk soil (rendzina) and bear woods of the escarpment type, but the gentler slopes bear deeper soils, and show transitions to the plateau type. The plateau soil also often extends over the edge of the escarpment and over the rims of the valleys where the slope is not too steep, so that plateau conditions may be carried down, as it were, on to the upper part of the slope.

The soils of the chalk plateaux have already been described on pp. 363–6 (Figs. 69, 70). Apart from outliers of sandy Tertiary beds, and local post-Tertiary deposits of sand and gravel which occur here and there on the plateaux, the soils which bear beechwood may all be classed as "loams", some heavy and sticky in texture, approximating to clay soils, others much lighter and more friable. The loams of the western South Downs plateau are for the most part relatively shallow and derived directly from the chalk: the top 6 in. of some of the shallower (average 12·5 in. = 31·6 cm.) are distinctly alkaline (pH 7·7), of the deeper (18 in. or 45·7 cm. to 26 in. or

[1] Beechwood with fruticose Rubi (brambles) ultimately dominant or characteristically present in the field layer. The employment of an adjective ending in *-osus* to characterise a community with special abundance or subdominance of a genus or group is a useful practice of some continental authors.

66 cm.) on the average distinctly but by no means extremely acid, and sometimes actually alkaline (up to pH 7·2 and 7·1). The Chiltern clay-with-flints, on the other hand, is a much deeper soil and always acid, varying in different types of wood from an average of 5·2 to 4·1. Watt (1934) showed in fact that there was a complete series of plateau soils from the most alkaline on the South Downs to the most acid on the Chilterns, that the woods borne by this series of soils show a corresponding range, and that those on the deepest and most acid of the South Down soils correspond closely in many respects with those on the least acid soils of the Chiltern plateau. For details the reader is referred to Watt's papers (1924, 1934): here a general account only will be given.

On passing from the central types of this series (pH 5·1, 5·2), which are brown earths (Fig. 70a, p. 365), to the more alkaline soils, the water content decreases and the humus progressively increases; and these characters become most marked on the escarpment soils which are rendzinas (Fig. 69, p. 364) described in the last chapter, except that in the shallow chalk soils of the "sanicle beechwood" the humus content is again lower.[1] Passing in the other direction, from the central to the more acid types of plateau soil, the humus again increases, but here in the extreme type (*podsols*) we have raw peaty humus which is known as *Trockentorf* by the Germans, and for which the Danish word *mor* is now to be adopted as an international term. *Mor* is a compacted humus very acid in reaction and concentrated on the surface with accompanying podsolisation below (see pp. 83–4). The beechwoods on this extreme soil type are described in the next chapter. The "central" plateau soil types (brown earths) are on the whole developed near the edge of the escarpment, the more acid farther away (A—C, Figs. 67, 68, pp. 362–3).

The best brown earth soils of the Chiltern plateau show a reddish brown close-textured illuvial horizon at a depth of about 50 cm. The surface layer is loose, well mixed with humus, and much frequented by earthworms and wood mice. In the somewhat more acid soils the surface litter shows a tendency to mat below, the beginning of the trend towards the formation of *mor* (Fig. 70b).

Mycorrhiza is normally abundant in the humus of the deeper plateau soils. Infection of the young beech roots takes place side by side with growth, which continues on from the spring period to which it is restricted on the shallow escarpment soils (p. 368), and is only interrupted by periods of extreme drought. There is considerable variety in the form of mycorrhizal growth, including the "diffuse", "loose weft" and "pyramidal" types (Harley, 1937). The two former are ectotrophic and admit of free continued growth of the root system. They are characteristic of the best plateau soils bearing the beechwoods described in this chapter.

Structure and composition. The finest height growth of the beech occurs

[1] See Fig. 84, p. 422.

in the central type (Pl. 63, phot. 146 and Fig. 84), mature trees reaching 107 ft. (32·6 m.), and the average girth of the dominant beech in the best Chiltern woods 5 ft. 4½ in. (1·61 m.). Of other trees oak (*Quercus robur*) varies from occasional to locally abundant and reaches a height equal to that of the beech. Ash is present and very well grown on the South Downs; on the Chilterns ash is confined to the "central", i.e. the least acid soils, where it reaches the height of the canopy. The naturalised sycamore (*Acer pseudoplatanus*) is locally dominant. Wild cherry or gean (*Prunus avium*) is occasional to locally frequent in the Chiltern woods, and its masses of white bloom form a beautiful feature in spring before the beeches are in full leaf. Hornbeam occurs occasionally on the Chilterns. The whitebeam (*Sorbus aria*) is rare to local in the plateau woods.

A second layer of trees, entirely shaded by the canopy, is generally absent, but on the South Downs yew and holly are sometimes present, and the latter is locally abundant. On the Chilterns holly is not more than occasional to frequent, though its constancy is high, and yew is almost absent from the plateau. The greater frequency of these evergreens in Sussex is almost certainly due to the more oceanic climate. Shrubs are very sparse indeed or absent altogether, as in all beechwoods: less than half a dozen species are recorded.

Role of Rubus. The undershrub *Rubus fruticosus* (agg.)[1] is, however, one of the most characteristic and important structural elements of the plateau beechwoods. Bramble is here, as in the chapters on oakwoods, reckoned as part of the field layer, or as forming an upper field layer. Both in stature and mode of growth it differs widely from a typical shrub and it comes into direct competition with other species of the field layer. In the central types of mature plateau beechwood bramble is dominant in this layer, often forming an almost continuous cover (Pl. 65, phot. 148), and having important effects on the subordinate layers besides seriously hindering the regeneration of the beech (see Watt, 1934, pp. 483–4, 488, 496–8).

Species of the field layer. The herbaceous species of the field layer are numerous and varied, but few have more than a high local frequency. Where *Rubus* is excluded or enfeebled by the deepest shade from the tree canopy, *Oxalis acetosella* is one of the most characteristic dominants (Pl. 64, phot. 147), and in other places *Scilla non-scripta* or *Mercurialis perennis*. It is noteworthy that plants which are usually reckoned as at least relatively calcicole, such as *Asperula odorata, Fragaria vesca, Galeobdolon luteum* and *Sanicula europaea,* flourish on the markedly acid soils of the central types of plateau beechwood. This is probably because these soils possess a good supply of exchangeable calcium and other anions (see De Silva, 1934), though they contain no free calcium carbonate, and the apparent preference

[1] Common species in the Sussex plateau woods are said to be *R. vestitus* and *R. rudis* (W. Watson).

PLATE 63

Phot. 146. Hailey Wood on Clay-with-flints, plateau of the Chiltern Hills (Oxon.). Well grown beech in close canopy, exceeding 100 ft. (30 m.). Shrubs absent. Upper field layer of bramble (2–3 ft.), *Deschampsia caespitosa*, *Epilobium angustifolium*. *Asperula odorata* and many other species are present in the lower field layer. Regeneration on left in a gap. *R. J. L.*

MATURE FAGETUM RUBOSUM (CHILTERN PLATEAU)

PLATE 64

Phot. 147. Edge of plateau wood near Pulpit Hill, Bucks, on Clay-with-flints. *Rubus fruticosus* (agg.) and *Oxalis acetosella* dominant. *R. J. L.*

FAGETUM RUBOSUM, FIELD LAYER

PLATE 65

Phot. 148. Mature beechwood on loam (sere 2), age about 140 years, height 100 ft. (30 m.). The rule is 3 ft. (91 cm.) long. *Rubus fruticosus* (agg.) dominant in the field layer, 18 in. (45 cm.) high. *Hedera* and bryophytes on the tree trunks. Charlton Forest, Sussex. *H. Godwin.*

Phot. 149. Mature beechwood on loam (sere 3—more carbonates than in sere 2), age about 130 years, height 90 ft. (27 m.). *Ilex aquifolium* forms a sparse shrub or low tree layer. *Rubus* sparser and less luxuriant (12 in. or 30 cm.) than in Phot. 148. Charlton Forest, Sussex. *H. Godwin.*

MATURE FAGETUM RUBOSUM (SOUTH DOWNS)

PLATE 66

Phot. 150. *Monotropa hypopitys* with *Oxalis acetosella* and oak and beech litter in Kingston Wood, Oxon. (Chiltern plateau). R. J. L.

Phot. 151. *Ithyphallus impudicus* with *Rubus* and *Oxalis* and beech litter. Kingston Wood, Oxon. (Chiltern plateau). R. J. L.

FAGETUM RUBOSUM, DETAIL OF FIELD LAYER

of the species mentioned for calcareous soils is determined by the exchangeable bases and not by the free calcium carbonate. It is of course well known that so-called "calcicoles" do not depend on the presence of calcium carbonate as such, but either on alkaline reaction or on abundance of exchangeable bases.

The tall wood-grasses *Bromus ramosus*, *Milium effusum*, *Festuca gigantea* and *Elymus europaeus* are more often met with in the beechwoods of the Chiltern plateau than on the South Downs, and *Deschampsia caespitosa* is notably more abundant, sometimes even locally dominant on close-grained soils.

Along with the relatively "calcicole" species mentioned above are others more characteristic of acid soils, such as the bracken fern (*Pteridium*) which is locally dominant on some of the more sandy soils, the foxglove (*Digitalis purpurea*), and in more open places *Agrostis tenuis* and *Holcus mollis*. These two grasses increase in frequency on the more acid soils, where they are joined by the strongly oxyphilous silver hair-grass (*Deschampsia flexuosa*).

Bryophytes. The bryophytes of the ground layer show similar relations. There is a very large diminution in number from the seventy-two species of the escarpment wood to thirty species in the central types on the plateau, taking the Chilterns and South Downs together. This is largely owing to the dominance of *Rubus* and the thicker and more continuous beech-leaf litter on the plateau. On the Chilterns *Eurhynchium striatum*, *Fissidens taxifolius*, *Mnium rostratum* and *Mn. undulatum* are confined to the "central" soils. In the more acid types there is a further heavy diminution in number of species, till in the extreme (podsolised) type the original list is very greatly reduced, but the number of species is somewhat augmented by oxyphilous mosses and liverworts (see Chapter xx). The bryophytes of the beech bark are confined to the basal portions of the trunks, in contrast to their much higher ascent on the trees in moister climates.

Larger fungi. The following are characteristic:

Collybia butyracea	o–f	*Mycena galopus	f–a
C. platyphylla	f	*M. sanguinolenta	f–a
Laccaria laccata	f–a	Russula atro-purpurea	o–f
var. amethystina	f–a	R. cyanoxantha	o–f

* Dependent on the *Rubus* layer probably because of the moist air between the bramble shoots.

Cantharellus cibarius, *Lactarius quietus* and *L. subdulcis*, frequent species of damp oakwood, also occur, though not so commonly, in the central types of plateau beechwood. *Russula cyanoxantha* and *R. nigricans*, on the other hand, are more frequent in these beechwoods, though they also occur in the damp oakwoods (Harley).

SUCCESSION

While the series of phases in the development of beechwood on the chalk escarpments dealt with in the last chapter have been most closely studied on the Chilterns (Watt, 1934), certain of the seres leading to the mature plateau beechwood have been worked out in detail only on the South Downs (Watt, 1924). The Chiltern plateau beechwoods are more sharply limited by arable cultivation and more closely exploited than those of the South Downs, so that few areas exist in which their seral development can be followed throughout. But on the occasional "commons" used for grazing enough can be seen to determine the main characters of the seres.

The central sere. So far as the data allow the following account deals primarily with the sere leading to what we have called the central type of plateau beechwood, that developed on a deep somewhat acid soil rich in exchangeable bases and supporting the finest tree growth. It is unfortunate that the South Down data for the earlier stages of this sere are very scanty. The grassland from which it starts would approximate to the "neutral" type (see Chapter XXVIII), but with a distinct tendency to soil acidity. The heaths (*Calluna* and *Erica cinerea*) and gorse (*Ulex europaeus*) would be prominent as pioneer woody colonists, while "calcicolous" shrubs would not be strongly represented. In the later stages of the sere ash-oak wood would be followed by oakwood and this by beechwood. On the more calcareous loams the grassland increasingly resembles chalk grassland (Chapter XXVII) in composition, the heaths are absent, and calcicolous shrubs play an increasing part in the sere, while in the later stages ash enters relatively earlier and becomes more prominent than oak.

The Chiltern commons. On the grazed commons of the Chiltern plateau the earlier stages of the "central" sere can be studied. These commons support a mixed community of grasses and herbs, which is generally an "acidic" grassland (Chapter XXVI), though not very extreme, and with a strong element of "exacting" species, i.e. species dependent on a good supply of basic ions. This grassland is invaded by heather (*Calluna*) and bracken (*Pteridium*), the latter covering large parts of most commons, the former occupying smaller areas. The growth of both bracken and heather is tallest and most vigorous on the soils corresponding to the central types of woodland, and these soils are also richer in the more "exacting" grasses and herbs.

Shrub invaders. Twenty-six species of shrubs and climbers invade this grassland, as compared with forty-eight on the escarpment, and of these only ten are at all common. In order of frequency they are:

Ulex europaeus	Sambucus nigra
Rubus fruticosus	Rosa canina
Crataegus monogyna	R. arvensis
Prunus spinosa	Solanum dulcamara
Lonicera periclymenum	Corylus avellana

Of these the gorse is often locally abundant or locally dominant and may cover considerable areas. Its abundance is doubtless partly determined by its former extensive use as cattle fodder. The high local frequency of bramble is associated with overhead canopy, and Watt found no evidence that it could assume or maintain dominance in the open. Hawthorn and blackthorn, though of high constancy and frequency and sometimes forming small dense thickets, do not invade so freely as on the South Downs plateau when grazing pressure is lifted. On the more fertile soils field maple, dogwood, buckthorn and wayfaring tree occur, though they are rare and local.

Tree invaders. Most of the trees occurring on the commons appear to have been planted, the conditions for regeneration being unfavourable. Only on one common has there been a general colonisation by beech and oak from parent trees of pioneer form, now included in the young wood. The oak, which is found on all the commons, is easily the most successful: ash and beech come far behind in point of numbers, though they are found on all or nearly all the commons. Sycamore is not so widespread as any of these but is locally successful. Other scarcer invaders are birch, hornbeam, holly, aspen, gean, whitebeam, rowan and yew. The grass facies is the most difficult for trees to invade directly, and only oak seedlings were seen: the closed turf mat may perhaps prevent the taproot getting a firm hold before spring droughts. Bracken communities are not much easier to colonise, but the Callunetum is successfully invaded by oak, birch, aspen, pine, hawthorn, blackthorn and elder. Of these oak is the most frequent and in some places birch.

But scrub is by far the easiest community for tree invasion: at least five out of six developing woodlands established on the commons have passed through a scrub stage, dead remains of hawthorn and blackthorn being found beneath the older trees, and young trees occur in the marginal zone of scrub. Gorse scrub is rarely colonised by trees.

Development of woodland. The commonest mode of development is round a pioneer oak or beech (which may or may not have been planted). Younger ash, oak and beech establish themselves round the pioneer. Below them there may be dead scrub, and, further out, young ash colonising scrub. All three species reproduce themselves, ash the most freely and beech the least. Sycamore also freely invades the scrub, and gean (*Prunus avium*) occurs in most of the developing woods but has a low average frequency. In this seral ash-oak wood there is no continuous shrub layer formed, though a number of species are present including *Corylus, Cornus, Ligustrum, Euonymus*, of varying but usually low constancy and frequency, with *Acer campestre, Rosa arvensis, Salix caprea* and *Viburnum opulus*.

Rubus fruticosus (agg.) and *Lonicera periclymenum* in the upper field layer show a nearer approach to continuity,[1] while the following are

---
[1] Cf. this society in Dry Oakwood (Chapter XIII, p. 281).

selected species of the very mixed herbaceous field layer, arranged by Watt under three heads:

Exacting species

Asperula odorata
Circaea lutetiana
Nepeta hederacea
Mercurialis perennis

Sanicula europaea

Catharinea undulata
Mnium undulatum

Tolerant species

Deschampsia caespitosa
Oxalis acetosella
Pteridium aquilinum

Eurhynchium praelongum

Oxyphilous species

Anthoxanthum odoratum
Conopodium majus
Digitalis purpurea
Galium saxatile
Holcus mollis
Potentilla erecta

Potentilla sterilis
Scilla non-scripta
Stellaria uliginosa

Dicranella heteromalla

The exacting species predominate in number of individuals. In the woods on the most fertile soil (those which show the greatest height growth of beech) there is a higher frequency of *Mercurialis* and *Scilla*, and the former shows great luxuriance, reaching 20 in. (50 cm.) in height.

Planted oakwoods. The oak wood which one would expect to succeed the developing ash-oak wood described is only represented in the area by woods which are almost certainly planted, some on arable land. They may, however, be taken as equivalent to the next stage in development. They conform neither to the typical Damp Oakwood nor to the typical Dry Oakwood described in Chapter XIII, being less rich and complex in structure than the former and not so simple as the latter.

Quercus robur is dominant, sycamore abundant here and there, gean and hornbeam occasional, and birch occurs in all those examined. No continuous shrub layer is formed. The most prominent woody plants are *Corylus, Crataegus, Viburnum opulus, Ilex, Rosa arvensis, Hedera* and *Ligustrum*. Again we have the prevalence of *Rubus* and *Lonicera*, the former sometimes forming a continuous canopy which affects the herbaceous field layer. There are, however, practically the same number of species as in the developing woodland, and again the prevalence of exacting species, with tolerant species often of quite high frequency, and a greater number of oxyphilous species.

Mixed oak and beechwood. Unlike the oakwoods, the existing examples of which are probably all planted, the immature beechwoods (called by Watt the "beech associes") appear to be natural, though often modified by human interference. They consist, on the central types of soil, of a mixture of oak and beech, which often results rather from a mixture of pioneer trees and their descendants than from invasion of mature oakwood by beech. *Corylus, Crataegus, Ilex, Rosa arvensis, Sambucus, Sorbus aucuparia, Taxus,* and *Viburnum opulus* are all present, but they do not form a continuous shrub layer.

Where there is enough light *Rubus fruticosus*, again associated with *Lonicera*, forms a continuous layer. The herbaceous species of the field layer shows a reduction in number compared with the oakwoods, and the same is true of the bryophytes. Exacting species are still in evidence, the following being confined to the central type of soil:

Anemone nemorosa	Fragaria vesca
Ajuga reptans	Galeobdolon luteum
Angelica sylvestris	Galium aparine
Brachypodium sylvaticum	Geranium robertianum
Bromus ramosus	Nepeta hederacea
Circaea lutetiana	Primula vulgaris
Festuca gigantea	Sanicula europaea

while *Asperula odorata* and *Geum urbanum* belong essentially to this type, and among bryophytes *Eurhynchium striatum* and *Fissidens taxifolius* are confined to it. Oxyphilous species also occur.

The frequency of species varies according to the density of the canopy, but in most woods it is low. "The effect of beech appears to be first to kill out a number of species by its heavy shade, and then on opening out of the tree canopy to allow them to come in again, but with a frequency determined by the combined shade of the tree and *Rubus* layers" (Watt, 1934).

One of these woods, Hillocks Wood on the Hampden estate on the Chilterns, was formerly known as "Hillocks Scrubs". The name "scrubs" is applied both on the Chilterns and on the South Downs to communities of shrubs, but also to coppiced woodland. Thus "High Scrubs" occurs twice as the place name of an existing forest of oak and beech which has almost certainly grown up from real scrub or coppice, and "the Scrubs" on the South Downs is now a high forest of ash and oak, containing evidence of the change from "scrub" to forest.

Mature beechwood. On the whole, and in spite of modifications brought about by felling, the Chiltern plateau beechwoods are mature, with mature structure and a full complement of species. The older "age classes" predominate, suggesting inadequate regeneration. Their general structure has already been sufficiently described (pp. 387–9) and the more significant species mentioned. The extreme paucity or absence of shrubs and the presence of a continuous layer of *Rubus* are the most important features. The list of species in the field layer of the central type has much in common, as might be expected, with that of the oakwoods on the better brown earths. The deep leaf litter, providing abundant humus, supports many fungi, as well as the colourless saprophyte *Monotropa hypopitys*, on all types of plateau soil, but more frequently on the most acid (Pl. 66, phots. 150–1). *Monotropa* is not so common, however, as in the sanicle woods of the escarpment, and *Neottia*, characteristic of the latter, appears to be absent from the plateau. *Holcus mollis* and *Scilla non-scripta*, which occur chiefly on the margins where the shade is not so deep, are regarded by Watt as relicts of the oakwood stage of the succession.

Succession on the South Downs plateau. The dynamic aspect of the plateau seres culminating in beechwood on the South Downs has been

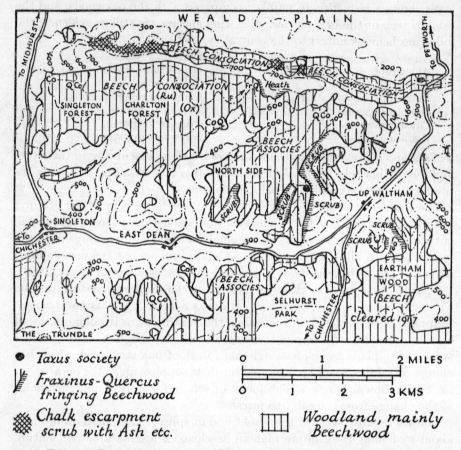

FIG. 74. BEECHWOOD IN THE GOODWOOD AREA (WESTERN SOUTH DOWNS) WITH SERAL STAGES

On a basis of more or less calcareous grassland (unmarked) scrub has colonised many of the spurs and broad ridges. This is fringed with ash-oak wood which invades the scrub. Beechwood ("beech associes") has already covered most of the area and invades and suppresses the ash-oak wood. Reprint of Fig. 66.

admirably analysed by Watt (1924), working in the area shown in Fig. 74. In this map the main stages of succession are distinguishable: (1) grassland, (2) scrub, (3) ash-oak wood, fringing (4) "beech associes" (immature beechwood), (5) beech consociation (mature beechwood). In Fig. 75 part of the area is shown on a larger scale, and it is seen that the scrub (*Crataegus* and *Prunus spinosa*) colonises the broad ridges of Oxen Down and Heath Hill.

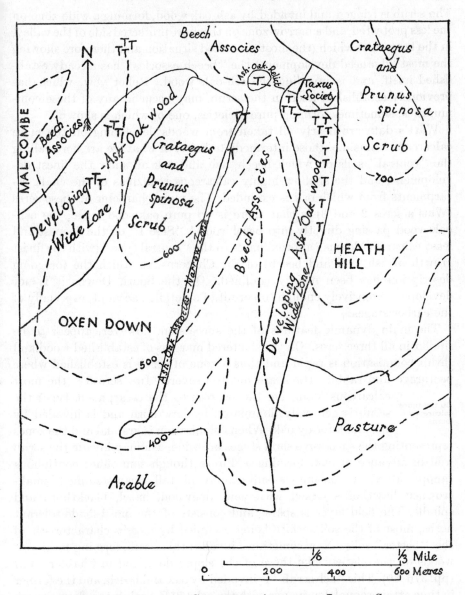

Fig. 75. Part of the Goodwood Area on a Larger Scale

(Right centre of Fig. 74.) Oxen Down and Heath Hill, to west and east, are parallel loam-covered chalk spurs, mainly grassland but colonised by scrub, and fringed by developing ash-oak wood on the sides of the central valley (Deepcombe) as well as to north-west of Oxen Down. Immature beechwood ("beech associes") occupies the plateau to the north-west and also fills the bottom of Deepcombe. The beechwood invades the ash-oak wood, while this invades the scrub. T = *Taxus baccata*.

The scrub is fringed and invaded by ash-oak wood, forming a wide zone on the less protected, and a narrow zone on the more protected side of the valley, in the bottom of which (the most protected situation and therefore showing the most advanced development) the "beech associes" has already established itself, and is invading the ash-oak wood. Relict yew, which has previously established itself in the scrub, may occur in any of these communities, sometimes forming pure societies, one of which is shown.

Watt's data are nearly all taken from woods developing on the more calcareous loams, i.e. those intermediate between what we are here calling the "central" soils on which beech, ash and oak all show the finest development, and the shallow highly calcareous rendzinas of the chalk escarpments from which oak is excluded. The main character of these seres (Watt's seres 2 and 3) is that there is no pure oakwood stage, the ash-oak wood passing directly into beechwood.[1] "Sere 1" on the deepest and least calcareous loams,[2] approximates to the "central type" (with the finest growth of ash, oak and beech) of the Chilterns, to which the foregoing description has been mainly confined. On the South Downs it is not developed extensively enough to permit of a detailed account, especially of the earlier stages.

The main dynamic features of the succession, however, appear to be similar in all three seres. On the sheltered margins of established woodland shrub colonisation is rapid and thus a zone of scrub is established where pasturage and rabbit attack are not too severe. In "sere 3" (the most calcareous loam, i.e. the nearest to the escarpment type) the scrub is generally dominated by hawthorn and is invaded by ash and later by oak. When colonisation is gradual and the zones representing the successive seral stages are wide, ash may invade the scrub well in advance of oak, forming a diffuse though sometimes continuous canopy above a continuous sub-canopy of tall clean-stemmed, small-crowned hawthorns mixed with yew, dogwood, hazel, blackthorn and spindle. The field layer is sparse and consists of the most shade-tolerant herbs, most of the soil surface being occupied by mosses characteristic of the "better" soils. Yew sometimes invades this scrub, forming local yew woods on the plateau like those of the slopes described in Chapter XVIII (pp. 376–80). This "ash scrub" is invaded by oak and beech, and the former in time attains equal dominance with the ash (50 ft.), while the slower beech invaders, which may reach a height of over 70 ft., establish "families" round the pioneer beeches.

Ash-oak wood. Thus an ash-oak wood, with beech locally dominant, comes into existence. With assumption of dominance by the trees the shrubs become sparser and alter their proportions. Their form, too, changes, the hawthorn for example becoming scarcer, more diffuse in habit and its foliage thinner. The herbs of the field layer show an increase in number of

---

[1] Second column in Fig. 85, p. 423.   [2] Third column in Fig. 85.

PLATE 67

Phot. 153. Blackthorn scrub invaded by ash, oak and beech (centre and right). Beechwood on right, heather and bracken in front. "Sere 2." Near Newhouse Farm, East Dean, Sussex.

Phot. 155. Carpet of beech seedlings from 1922 mast year. Lamb Lea, near East Dean, Sussex. *A. S. Watt.*

Phot. 152. Mature beechwood fringed by colonising ash. Prehistoric trackway in front, with ant hills, and Bronze Age barrow on left. Northern edge of Charlton Forest, Sussex.

Phot. 154. Carpet of beech seedlings from 1922 mast year. Ditcham Park, Hants. *A. S. Watt.*

FAGETUM RUBOSUM—SUCCESSION

PLATE 68

Phot. 156. Ash-oak associes invaded by beech. Lamb Lea, near East Dean, Sussex. Pioneer oaks in centre and on right, younger ash and oak behind. Ash saplings in foreground, leaning towards light. Beech invading from the left at the back. *H. Godwin.*

ASH-OAKWOOD SERAL TO FAGETUM RUBOSUM

species and individuals, such characteristic woodland plants as sanicle, dog's mercury and enchanter's nightshade making their appearance, and eventually the field layer becomes uniform and characteristic of typical woodland. Mosses are well represented both in number of species and in frequency, the commonest being:

Brachythecium purum	Hylocomium loreum
B. rutabulum	H. triquetrum
Catharinea undulata	Mnium undulatum
Eurhynchium praelongum	Thuidium tamariscinum
E. striatum	

On intermediate loams

In the less calcareous "sere 2" the grassland contains a larger admixture of non-calcicolous species, but the most conspicuous change is the invasion of the heaths, *Calluna vulgaris* and *Erica cinerea*, which may become locally dominant where grazing and rabbit attack are not too heavy.[1] In the scrub stage hawthorn is still the dominant species, with juniper and the characteristically calcicolous liane *Clematis vitalba* decidedly less frequent. The most striking change is the addition of non-calcicolous shrubs, raspberry, sallow, and especially the common gorse (*Ulex europaeus*). Among trees *Betula pendula* (*alba*) occasionally appears.

With the formation of the ash-oak wood hazel becomes the dominant species of the shrub layer, while the trees, ash and oak, attain a larger size when mature. In the field layer mercury is very luxuriant, sometimes reaching 60 cm. in height, and *Circaea* occasionally reaches 1 m. The effect of this luxuriance is to suppress the lower growing plants, including the mosses.

(*Ash-*)*oak-hazel wood*. The deeper soil increases the competitive ability of the oak, so that it tends to become definitely predominant. In woods of this type which were regularly exploited the greater value of oak timber, at least in past times, caused ash to be coppiced and invading beech cut out, so that the wood became an oak-hazel wood of standard form but with abundance of ash in the coppice layer. Such woods commonly show, by the frequency of calcicolous shrubs (*Viburnum*, *Euonymus*, *Cornus*, etc.) and herbs (*Mercurialis*, *Sanicula*, *Arum*, *Viola sylvestris*, *Hypericum hirsutum*, etc.), their difference in origin and status from the oak-hazel woods derived from damp oakwoods on clays and loams with little or no free calcium carbonate (cf. Chapter XIII, p. 272).

On the least calcareous loams

"Sere 1", in which the loam is deepest and least calcareous, and all three of the leading trees, ash, oak and beech, show the tallest height growth, approximates closely, as has been said, to what we have called the "central type" already described for the Chilterns, but there are no data for the earlier stages on the South Downs. The soil is somewhat acid, the grassland would be heathy, passing to pure heath, the

[1] Pl. 67, phot. 153 shows this sere, with *Calluna*, scrub of *Prunus spinosa*, and invasion of ash, oak and beech.

scrub with *Ulex europaeus* often dominant. "Calcicolous" shrubs would be much less in evidence, and ash would play a less important part in the sere.

Invasion of beech. Isolated beeches may invade scrub, and even develop beech "families" (i.e. the original pioneer colonists with their offspring) in this stage of the sere; and if the further progress of scrub is held up beech becomes dominant up to its outer edge and thus comes to abut directly on grassland. Colonisation by beech, or by any tree, cannot occur in very dense scrub which casts too deep a shade, nor under a dense growth of ash saplings. Where the conditions are favourable, as between the shrubs of loose scrub, so that the seedlings obtain a suitable soil *nidus*, protection, and also sufficient light, the degree of colonisation depends very largely on the proximity of seed parents in adequate number. The beech is certainly a slow-moving tree: indeed, one set of estimations by Watt puts its spread at only 100 ft. (30 m.) a century, though it must have migrated a good deal faster than this at times. But its slowness is doubtless one reason why beechwood is preceded by ashwood or by ash-oak wood in regions where beech dominates the climax community. For this reason also ashwood may become established as a pseudo-climax in small areas somewhat remote from mature beechwood, just as it forms the normal climax on limestone soils outside the existing beech region.

Modes of invasion The normal definitive invasion *en masse*, ultimately leading to climax beechwood, occurs in the phase of ash-oak wood (Pl. 68, phot. 156). The conditions in this community are optimal for colonisation by the beech—a good soil *nidus*, good protection and sufficient light. Invasion may take place along a continuous front, as shown in the belt transect illustrated in Fig. 76; but it very often occurs along diverging lines from single pioneer beeches. Between the lines of invasion a wedge of ash-oak wood is enclosed, and the additional shelter provided by the beeches at first actually increases the luxuriance of the ash-oak wood, and especially of the oaks, which become more numerous in the near neighbourhood of the invaders. The process is well illustrated in Fig. 77. Eventually the *enclave* of ash-oak wood is invaded by the offspring of the maturing beeches, and gradually elimina

The early stage climax beechwood which results from this invasion is characterised by the very unequal age and form of the beeches which compose it. All stages of development occur, from seedlings to the old pioneers.

Pioneers and semi-pioneers The pioneers[1] are trees of comparatively low stature, with short boles and richly branched widely spreading crowns, bearing witness to the open conditions in which they were developed. The true pioneer form is characteristic of trees grown in grassland or scrub. In the relatively well-lighted conditions of ash-oak wood, with some degree of lateral shading, the beeches have a "semi-pioneer" form, intermediate between the pioneer and the tall, small-crowned tree of close canopy

[1] F_1 in Fig. 76.

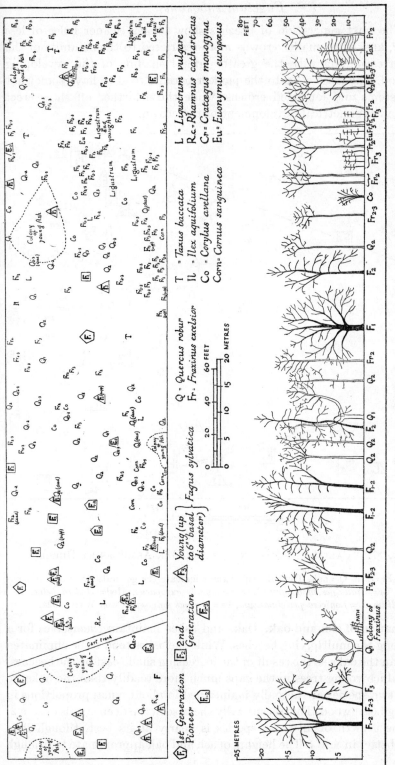

Fig. 76. Belt and Profile Transects across an Advancing Front of "Beech Associes" (left) Invading the Ash Oak Associes (right)

Note the increased number of oaks and decreased number of ashes as the beech invades: also the uniform height of ash and oak which is much surpassed by the (non-pioneer) beeches. From Watt, 1924.

forest. The next generation of beeches show longer cleaner boles of less diameter, and with smaller crowns more restricted to the summits of the trees, corresponding with the greater lateral shading of the developing beechwood. The trees close to the pioneer parents tend to have somewhat oblique stems and lop-sided crowns, while those farther off show erect trunks and symmetrical development of the crown.

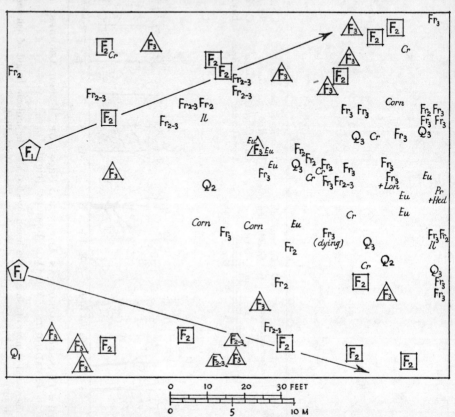

FIG. 77. "CALLIPER INVASION" OF ASH-OAK ASSOCIES BY BEECH

F = beech, Fr = ash, Q = oak. The subscript figures indicate the generations. Corn = *Cornus sanguinea*, Cr = *Crataegus monogyna*, Eu = *Euonymus europaeus*, Hed = *Hedera helix*, Il = *Ilex aquifolium*, L = *Lonicera periclymenum*, Pr = *Prunus spinosa*. From Watt, 1924.

Elimination of ash and oak. Oaks and ashes maintain themselves for a time among the multiplying beeches. While the ashes tend to be eliminated faster than the oaks as the result of the increasing shade cast by the beeches, they produce young trees in the gaps much more readily than the oaks, so that the two species are actually maintained in about equal proportions in the young beechwood. This is especially the case in the deeper soils of sere 2, where the growth of all three species is nearly at its best, though it is actually better in sere 1. The heights of ash and oak approximate to, though they do not quite attain, that of the beech, and both are substantially taller than when they are growing in the ash-oak associes.

In sere 3 (the most calcareous plateau type) the dominance of beech is more quickly attained. On these shallower soils, where the chalk is closer to the surface, as on the pure chalk soil of the escarpments, the competitive power of the beech, as against the two other trees, is greater and more decisive, though its ultimate absolute height growth is less. As the beechwood increases in age the beeches not only become more numerous in proportion to the remaining oaks and ashes, but, as the pioneers die, the trees show less diversity in age, greater uniformity in height and diameter, and ultimately a greater absolute stature.

Effect on shrub and field layers. The shrub layer of the young beechwood is less luxuriant and less numerous in individuals than that of the ash-oak associes, though most of the "calcicolous" species prominent in the earlier stages of the succession are still present. The field layer shows a similar diminution, but several species, such as mercury, sanicle, primrose, bluebell and bracken, are locally abundant, and wood violet, wild strawberry and germander speedwell frequent. This is of course while the structure of the wood is uneven and oak and ash are still present. Under *pure* beech and while the wood is still young there is an almost complete absence of herbaceous plants.

Thus in the progress towards adult beechwood the subsidiary layers are gradually eliminated, and the purer and more uniform the developing beechwood becomes the more closely it approximates to a single layered community, often consisting (apart from mycorrhiza and other soil organisms) of the tree layer alone. It is only later, when the canopy opens out, that new subsidiary layers are established.

Causes of even age. The initial cause of the generally even-aged character of naturally developed beechwood in this country is to be found in the occurrence of "full mast" years only at considerable intervals. The beech seedlings surviving from the partial crops of mast produced in other years are insignificant in number owing to wholesale destruction of the seeds and seedlings by various enemies. It is only when very large amounts of seed are produced that enough seedlings survive the attacks to secure adequate regeneration of existing woods and colonisation of new areas.[1] Thus the stands of even-aged beech can all be traced back to particular full-mast years, and the number of trees originating in other years is very small. This would not be so if the seedlings derived from "partial mast" years survived in greater numbers. Their current destruction is largely due to the disturbance of the "balance of nature" caused by the war waged by gamekeepers against the carnivorous birds and animals, which under completely "natural" conditions would keep down the

---

[1] The great crop of seedlings produced by the "full mast" year 1922 is shown in two separate localities on Pl. 67, photos. 154, 155.

numbers of ground vermin such as mice, voles and rabbits that destroy the seeds and seedlings. Hence the natural or "virgin" beech forests still existing in parts of Europe consist of trees showing extreme variation in size and age, from young saplings to aged veterans.

Planted and natural woods. The even-aged character and almost uniform spacing of mature English beechwoods is largely responsible for the prevalent notion that they have all been planted, but their origin from single full-mast years together with the natural processes of competition and suppression of the weaker trees leads to this homogeneous character and to the survival of an approximately constant number of trees of uniform stature per unit area. The same result is of course obtained by planting on the same type of soil, and many of the existing beechwoods on the Chilterns, and some on the South Downs are in fact known to be planted woods. Of the great majority no records exist (or have been discovered), but many beechwoods, especially on the South Downs, are certainly natural, since the phases of natural development described above can be clearly and unmistakably traced.

Later stages of development. The later stages of development, after the ash-oak associes has been definitely superseded by the invading beech, are marked by the increasing elimination of the remaining trees of the former community, of its numerous shrubs, and of almost the whole of the ground vegetation. Thus the earliest phase of the climax beechwood (consociation) is a wood in which the soil is practically bare, with a very few enfeebled plants surviving here and there from the earlier communities.

Bare stage   This result is primarily due to the very deep shade cast by the closed canopy of the beeches, but it is intensified by the thorough exploitation of the surface layers of soil by the superficial root systems of the trees, which severely reduces the water supply and perhaps also the supply of nutrients available for other plants. This *bare* stage (Fig. 78) is characteristic of all young, densely packed, communities of woody plants, but in beechwood it is maintained for a great part of the life of the community owing to the factors of deficient light and water supply just mentioned.

As the trees become older and fewer in number by the suppression of the weaker individuals, the canopy "opens out" and more light reaches the ground. The floor of the wood is now colonised by mosses, herbs and a few shrubs, and the characteristic vegetation of mature beechwood is developed. This associated flora thus arises *de novo* and is not genetically related to that of the preceding ash-oak wood, though it is of course composed of many of the same species.

Oxalis and *Rubus* stages   On the central type of plateau soil on the South Downs (sere 1) the bare stage, according to Watt, gives place to the dominance of *Oxalis acetosella* when the beech consociation is between 50 and 60 years old, and this "*Oxalis* phase" lasts about 10 years: in sere 2 it does not begin before the 80th year and lasts about 20 years (Fig. 78). In both seres *Oxalis* is succeeded by bramble, which becomes dominant and apparently lasts during the rest of the life of the wood. On the most calcareous plateau soil (sere 3) *Oxalis* and *Rubus* come in together about the 120th year and neither is dominant (Fig. 78).

On the Chiltern plateau, as we have seen, the bramble is also dominant in the field layer of the mature beechwood. The time of its entrance cannot be exactly determined, but it is already present and may be locally abundant in the woodland developing on the commons and sometimes forms a continuous layer in the planted oakwoods which represent the preclimax stage of the sere. Presumably it would disappear in the young beechwood and reappear later in the mature stages. In the "A" woods, which come nearest to the South Downs plateau woods of sere 1, it normally forms a

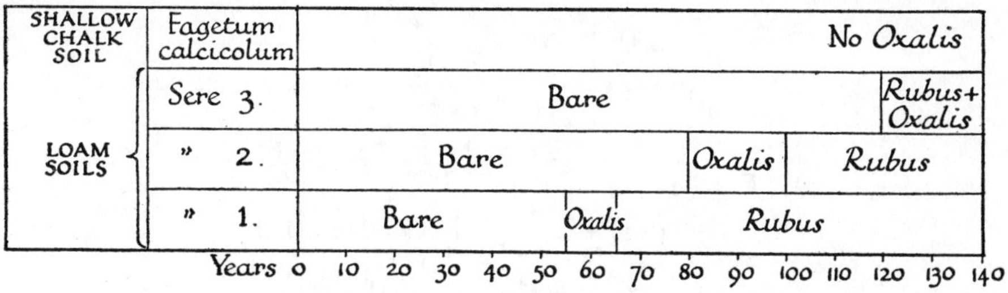

SHALLOW CHALK SOIL	Fagetum calcicolum		No *Oxalis*			
LOAM SOILS	Sere 3.	Bare				Rubus + Oxalis
	" 2.	Bare		*Oxalis*	*Rubus*	
	" 1.	Bare	*Oxalis*	*Rubus*		

Years 0 10 20 30 40 50 60 70 80 90 100 110 120 130 140

FIG. 78. STAGES OF FIELD LAYER IN FAGETUM RUBOSUM

The initial bare stage gives way to dominance of *Oxalis acetosella* and this to dominance of *Rubus fruticosus* (agg.). With increasing calcareousness of the soil these stages occur at increasingly later periods in the life of the wood, and in the highly calcareous soil of Fagetum calcicolum neither species ever becomes dominant. After Watt, 1924.

continuous "upper field layer", and here it makes the most luxuriant growth, 18–24 in. (45–60 cm.) high under continuous canopy, and 4 ft. (1·2 m.) in clearings. *Oxalis* is also a most characteristic plant of the Chiltern plateau, locally dominant where the shade is too deep or the competition of the tree roots too great for *Rubus* (see Pls. 64, 66).

Regeneration of the beech consociation. We may distinguish *regeneration*, in which the mature quasi-organism regenerates lost parts, from the process of its original *development* in succession from other communities such as scrub or a preclimax tree community. The process of regeneration may also be regarded as the occurrence of a short subsere which is consequent on the destruction or disappearance of a small portion of the canopy, often the crown of a single old tree which has been felled or has died of old age or as the result of fungal attack, and whose removal has thus formed a small gap in the consociation.

Colonisation of gaps. Watt (1924) studied in great detail the colonisation of such gaps left by the disappearance of a single beech tree, and found that the exact sequence of events depends mainly on three factors: (1) the age of the wood, (2) the size of the gap, and (3) the time when the gap is formed in relation to the occurrence of a full-mast year. In the earlier stages of development of the mature wood, when the beech is still accompanied by many ashes and oaks, the shrub and herb flora immediately available to colonise a gap is considerably greater than later, when the beech is practically pure and the lower layers very sparse or absent. In these earlier stages a wide variety of herbs and shrubs occur in the gaps. For the most part these do not seriously interfere with beech regeneration. In the *Rubus* stage of the central soil types, however, where the bramble enters about the time when the beech canopy comes to the seed-bearing age, it may fill the gap with such a luxuriant growth that regeneration is pre-

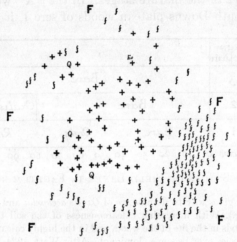

FIG. 79. COLONISATION OF GAP IN BEECHWOOD (REPRODUCTION CIRCLE)

Beech seedlings establish themselves only at the periphery of the gap, since the nuts mostly fall vertically downwards from the parent trees. The centre of the gap becomes filled with young ash which is later shaded out by the growing beech.

Eu = *Euonymus europaeus*, F = parent beech, f = young beech. + = young ash, Q = young oak. From Watt, 1924.

vented unless the bramble is cut over. Under natural conditions, however, some ash and oak would probably find their way up through the bramble cover, presently suppressing it, at least partially; and under the shelter of ash and oak beech can again grow up. Thus ash and oak would remain permanent constituents of the beechwood.

Size of gaps. The size of the gap has an important effect. The beech mast produced by the trees in high forest drops for the most part vertically to the ground, so that in a wide gap the periphery only is seeded, the centre being often occupied, especially in the younger stages of development, by a core of ash seedlings, among the dense growth of which an occasional beech seedling has little chance of growing into a tree (Figs. 79, 80). Eventually, however, the taller growth and deeper shade of the peripheral beech

PLATE 69

Phot. 157. Advanced stage of reproduction circle in sere 3. Young ash and beech leaning inwards. Fallen stems on the ground. Northside, Goodwood estate, Sussex. *H. Godwin.*

Phot. 158. Later stage of reproduction circle (sere 3). Suppressed ash and beech (thin stems). Dead fallen stems on the ground. Surviving ash (white stem) in centre. Northside, Goodwood estate, Sussex. *H. Godwin.*

FAGETUM RUBOSUM—REPRODUCTION CIRCLES

PLATE 70

Phot. 159. End of succession in reproduction circle with falling and dead ash stems in young beech consociation, Northside, Goodwood estate, Sussex. The field layer is absent in this and the two preceding photographs owing to the dense stocking and deep shade of the young beechwood. *H. Godwin.*

FAGETUM RUBOSUM—END OF REPRODUCTION CYCLE

saplings, combined with gradual extension of the canopy of the parent trees across the gap, will eliminate the ashes. In a narrow gap, on the other hand, the whole floor may easily be seeded by beech mast, and the seedlings and saplings will then meet with practically no competition from any other species of tree.

Plates 69 and 70 illustrate the later stages of competition between beech and ash in a reproduction circle. Phot. 157 shows ash and beech in the pole stage leaning inwards, with some fallen dead individuals; Phot. 158

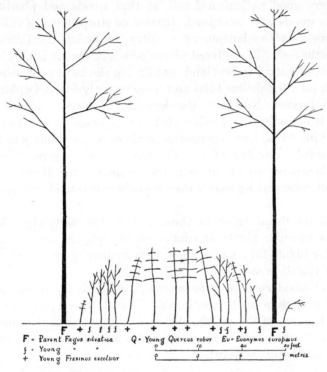

FIG. 80. PROFILE OF REPRODUCTION CIRCLE

Note the bending inwards of the young ash away from the shade of the parent beech.
From Watt, 1924.

suppressed ash and beech still standing, and Phot. 159 the final elimination of ash in a young beech consociation. The soil is bare throughout these phases.

Gaps and full-mast years. When the formation of a gap coincides with a full-mast year regeneration takes place much more readily and certainly, because the gaps will be much more fully seeded and because there will not be time for a serious competitor, such as *Rubus*, to fill the gap with luxuriant growth before the beech seedlings can establish themselves. This is a very important point for the forestry of beechwoods managed on the "selection" system. Mature trees should be taken out in the winter following a full-mast season, so that plenty of fertile mast is present in the soil, ready to colonise immediately the gaps formed by the fellings.

Scottish beechwoods on fertile loam. Planted beechwoods on deep fertile loam derived from Old Red Sandstone in east and north-east Scotland (Watt, 1931, pp. 154–6) show considerable resemblance to the central types of chalk plateau beechwood described in this chapter, though *Rubus* is not dominant in the field layer, and is not indeed recorded. This loam is very good agricultural soil, so that woods and plantations are scarce. Six woods were examined, situated on steep slopes of valleys cut in the soft rock, at the bottom of a valley, or facing the Cromarty and Beauly Firths—all "in sheltered places or where the air is humid".

Beech is not native to Scotland, but during the last two centuries it has been much used for shelter belts and more recently for underplanting coniferous plantations. Northern Aberdeenshire is clearly within its climatic range, for it not only grows well on this soil (average 92 ft. = 28 m.) with a smooth white bark, but regenerates with some freedom, young growth being recorded from five of the six woods. Certainly no other tree in the neighbourhood could successfully compete with beech, and these planted but regenerating woods may therefore be considered as a pseudoclimax.

There is no shrub layer in these woods, but sixty-eight species of herbaceous vascular plants and fourteen bryophytes were recorded. Of the vascular plants fifty had a fairly high frequency or occurred in three or more of the six woods.

The great wood rush (*Luzula maxima*), which is far commoner in the north and west than in the south-east of Great Britain, was abundant to dominant in most of the woods. *Sanicula europaea* was dominant in one wood and *Stellaria nemorum* in another. Among the other most frequent or well distributed species were:

Agrostis tenuis	Mercurialis perennis
Ajuga reptans	Oxalis acetosella
Anemone nemorosa	Primula vulgaris
Galium aparine	Ranunculus repens
Geranium robertianum	Veronica chamaedrys
Geum urbanum	V. montana
Holcus mollis	V. officinalis
Melandrium dioicum	Viola sylvatica

Asperula odorata, *Lysimachia nemorum*, *Stellaria holostea* and *Vicia sepium* were present in one only.

Of plants usually indicating more acid soils were *Digitalis purpurea*, *Galium saxatile* and *Holcus mollis*. Of ferns *Dryopteris filix-mas* occurred in all, *D. dilatata* in four, and *Blechnum spicant* in three: bracken and common polypody in one each.

Of bryophytes *Mnium hornum* was commonly dominant. The other most widely distributed species were:

Catharinea undulata
Eurhynchium praelongum
Hypnum cupressiforme
Mnium undulatum
Polytrichum formosum

A relatively high percentage of chamaephytes (16) Watt considers as reflecting the mildness of the maritime climate, while the predominance of hemicryptophyta scaposa (37) shows affinity with the chalk plateau woods of the "central" type and with the mercury woods of the chalk escarpments.

Other planted and subspontaneous Scottish beechwoods on sandy soils or on glacial "till" derived from granitic rocks are described in the next chapter.

Beechwood on sandy soil of high base status. Most sandy soils are poor in nutritive salts to begin with, and, moreover, readily podsolised when the surface is horizontal; but some, especially on slopes and broken ground, are evidently richer in electrolytes judging by the vegetation they bear. In the south-east of England beechwood may occur on these sands of higher base status, for example, in the Weald and on the Lower London Tertiaries at the base of the Eocene. None of these woods is of any great extent and none has been properly investigated, but the field layer is sometimes dominated by bramble, so that the wood is Fagetum rubosum, and may contain a considerable variety of the more exacting herbs. Thus their vegetation shows affinity, more or less close, with that of the chalk plateau woods described in this chapter.

REFERENCES

ADAMSON, R. S. The woodlands of Ditcham Park, Hampshire. *J. Ecol.* **9**, 114–219. 1921–22.

DE SILVA, B. L. T. The distribution of "calcicole" and "calcifuge" species in relation to the content of the soil in calcium carbonate and exchangeable calcium, and to soil reaction. *J. Ecol.* **22**, 532–53. 1934.

HARLEY, J. L. Ecological observations on the mycorrhiza of beech. *J. Ecol.* **25**, 422. 1937.

WATT, A. S. On the ecology of British beechwoods with special reference to their regeneration. Part I. The causes of failure of natural regeneration of the beech (*Fagus silvatica* L.). *J. Ecol.* **11**, 1–48. 1923. Part II. The development and structure of beech communities on the South Downs. *J. Ecol.* **12**, 145–204, 1924; and **13**, 27–73, 1925.

WATT, A. S. Preliminary observations on Scottish beechwoods. *J. Ecol.* **19**, 137–57, 321–59. 1931.

WATT, A. S. The vegetation of the Chiltern Hills with special reference to the beechwoods and their seral relationships. *J. Ecol.* **22**, 230–70, 445–507. 1934.

WATT, A. S. and TANSLEY, A. G. "British Beechwoods" in Rübel, *Die Buchenwälder Europas*. 1932.

Chapter XX

BEECHWOOD ON SANDS AND PODSOLS
(FAGETUM ARENICOLUM OR ERICETOSUM)
SUMMARY OF BRITISH BEECHWOODS

FAGETUM ERICETOSUM IN SOUTH-EASTERN ENGLAND

Soils of the Chiltern plateau. The non-calcareous soils which cover so much of the plateau of the Chiltern Hills are, as we have seen, very various in nature. Most of them come under the general designation of "Clay-with-flints", which is supposed to represent the insoluble residue resulting from the gradual solution of a great thickness of the chalk rock during much of post-Cretaceous time. The very small proportion of insoluble constituents existing in the original chalk consists mainly of finer particles, so that when the calcium carbonate is dissolved away the residue is a clayey loam or a silty soil containing aggregations of flints (massive concretions of silica) which are abundant in several horizons of the chalk. The clay loams are typical "brown earth" soils (Chapter IV, pp. 85-8), rich in electrolytes, the "central types" of chalk plateau soil bearing the finest beechwoods, which were described in the last chapter. But a good deal of the non-calcareous soil of the Chiltern plateau is not clay-loam or silt, but much more sandy in constitution, and cannot possibly have originated simply from the solution of chalk. Some of this sandy soil represents remnants of the sandy beds at the base of the Eocene (such as the Reading Beds) which have been left on the dipslope of the chalk after the general denudation; and some is probably the remains of later sands and gravels. Much of this material has been subject to a good deal of rearrangement by the action of running water during the Pleistocene and post-Pleistocene period. The result is a series of plateau soils of textures varying from heavy clay loams through silts and sandy loams to nearly pure sands. The first occur mainly near the chalk escarpment, while the silts and the more sandy soils are found on the dipslope nearer to the outcrop of the Lower Eocene beds on the edge of the London basin (cf. Fig. 81).

The more sandy soils, which are generally poorer in bases to begin with, are also more easily leached, so that they may become strongly podsolised, the surface layers very poor in bases and increasingly acid in reaction. These sands and gravels may bear beechwood, but the accompanying vegetation differs radically from that of the good loams of the typical clay-with-flints.

Of the plateau loams Watt (1934) distinguishes three types on the Chilterns, as we saw in Chapter XVIII (Fig. 81 reproduces their profiles): the "A" type described in the last chapter as the "central" plateau type, i.e.

the best "brown earth" soils usually lying near the escarpment and
bearing the best beechwoods; the "C" type, which are silty and
podsolised; and the "B" type which are intermediate.[1] Here we
shall be concerned only with Watt's "C" type soils which tend to
bear heathy beechwood similar to that of the sands, gravels and other

Beechwood on podsolised silts

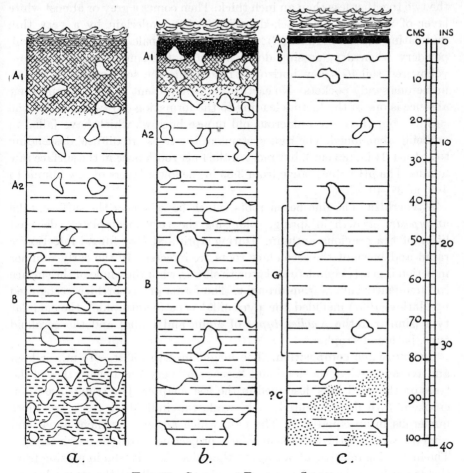

FIG. 81. CHILTERN PLATEAU SOILS

The three profiles illustrate the change from the Brown Earth type (*a*) to the Podsol type (*c*) with corresponding change in distribution and kind of humus (mull to mor). The *c* type bears the woods described in the present chapter. For more detailed description see Fig. 70 (p. 365) of which this is a reprint. After Watt, 1934.

podsolised soils. The subsoil here is less fertile to begin with, lacking the considerable clay content with which are associated the exchangeable bases (electrolytes) characteristic of "A" soils.

The thick layer of loose beech-leaf litter in a particular wood of this

[1] Cf. Figs. 67, 68, pp. 362–3.

type examined by Watt (Fig. 81c) is underlain by a stratum of laminated litter from half an inch to an inch thick formed by many layers of dead beech leaves, moist and compacted at first, but becoming very dry in summer drought, disintegrating below into a black peaty humus ($A\,0$) which passes down into black humus-coloured mineral soil (horizon $A\,1$),[1] the two together less than an inch thick. Then comes a grey or almost white layer of mineral soil (bleached horizon, $A\,2$), underlain by a very thin undulating illuvial horizon (B); and below this a pale yellow compact soil, powdery when dry, passing down into a mottled more tenacious grey-yellow-red soil (G = gley horizon). Below this the soil (probably C) has numerous sandy pockets. No earthworms are present in the soil, and their absence is one of the factors leading to the formation of raw humus on the surface; but there are numerous old mouse burrows which have fallen in on being abandoned. The root systems of the trees are diffuse, descending to at least 34 in. (86 cm.), but most of the finer rootlets lie in the surface raw humus. The pH values range from 3·85 in the lower layers of raw humus to 4·65 at 37–40 in.

Mycorrhiza. The fine beech roots, which lie very near the surface, take an *upward* growth in spring, colonising the very incompletely decayed litter of the previous autumn. During April and May fungal infection is rapid and most of the secondary roots are infected. In the early summer infection is nearly complete and growth slower, while the whole of the surface soil frequently suffers from drought. The forms of mycorrhiza recognised by Harley (1937) included the pyramidal type (ectotrophic), the nodular type (similar to the *Knollen-typus* of Melin and Laing) and the coralloid type, both endotrophic.

Structure and composition. The height growth (60 ft. = 18 m., or less) and average girth of the dominant beeches fall very much below that of the trees on the good soils described in the last chapter. The trunks are often crooked, with dark-coloured bark, while the individual trees are more numerous in an equal area. The oaks which accompany them equal and sometimes exceed the beeches in height, projecting above the beech canopy. The increasing relative advantage to the oaks which is seen in passing from the somewhat calcareous to the non-calcareous soils of the South Downs plateau is thus most marked on these podsols. Other accompanying trees and shrubs are few, only ten species of woody plants in all, compared with twenty-five in the richest woods described in the last chapter. Ash, whitebeam and hornbeam are absent, and gean finds no place in the canopy. *Salix aurita* (not found in the other woods) is recorded from a gap. The holly (*Ilex aquifolium*) is still constant and has a slightly higher frequency. It is conspicuous as a low semi-procumbent bush.

In the field layer the more exacting species are largely absent, while

---

[1] $A\,1$ is not distinguished in Fig. 81c—it is included in $A\,0$, and $A\,2$ is marked simply as A.

oxyphilous species are characteristic, e.g. *Calluna vulgaris*, *Deschampsia flexuosa*, *Hypericum pulchrum*, *Carex pilulifera*. The caespitose hemicryptophytes show the greatest preponderance among the life forms. Among the bryophytes are *Hypnum schreberi*, *Leucobryum glaucum*, *Cephalozia bicuspidata*, *Diplophyllum albicans*, *Frullania tamarisci*, *Lepidozia reptans*, while *Dicranum scoparium* increases in constancy and frequency, *Mnium hornum* and *Polytrichum formosum* in frequency.

Beechwood on sandy soils in south-eastern England. The natural occurrence of the beech and beechwood on sandy soils in south-eastern England generally has already been mentioned in Chapter XVIII (pp. 364–6). These sands are usually poor in electrolytes to begin with, and owing to their coarse texture become very thoroughly leached by percolating rain water, especially on horizontal surfaces. Thus they easily form podsols much deeper and more completely developed than those of the more silty soils of the Chiltern plateau just described. In Aberdeenshire planted beechwoods or shelter belts which can regenerate occur on various more or less sandy soils, many of which are podsolised, and these also have a strongly marked heathy field layer (Watt, 1931). None of the sandy beechwoods of the south-east has been studied with the care and thoroughness which Watt has devoted to those of the "rendzina" chalk soils, of the "brown earth" loams and of the Aberdeenshire podsols, but enough is known to furnish the data for a brief general account.

The southern sandy beechwoods occur mainly on the Hastings Sands, on the Lower Greensand, on the sands at the base of the Eocene known as the Lower London Tertiaries which are developed on the fringes of the London and Hampshire basins, on the Bagshot and related sands of the Upper Eocene lying above the London Clay in the same basins, and on "Plateau Gravels" overlying these beds. They are mostly small in extent, one of the largest being the woodland to the north of Slough on the northern side of the Thames valley known as Burnham Beeches (now belonging to the Corporation of the City of London and dedicated as a public resort) with adjacent woods, situated on Reading Beds (Lower Eocene) and superimposed "plateau gravel" overlying the chalk. Of the origin of Burnham Beeches we know very little, but though the wood has been interfered with to a considerable extent the main phases of its development are perfectly clear.

Burnham Beeches

Situation and soil. Most of the beechwood of this area lies on a low plateau of the Reading Beds composed of coarse gravelly sand overlain in places by "plateau gravel" (marked in the newer geological drift maps as "glacial gravel"), and intersected by shallow valleys whose streams have cut down to underlying loamy soil (occupied by oak-hazel wood of the damp oakwood type), or even to the chalk. By no means the whole of the plateau woodland is dominated by beeches, much of it being occupied by mixed oak and birch (really a stage in the succession), with open areas largely covered with bracken.

The soil itself is a coarse sand or gravel which shows in places a typical podsol profile, with characteristic compact surface humus (*mor*), underlain by a bleached horizon and sometimes a deep chocolate brown illuvial layer below (p. 89). The *p*H values are low. Thus close to an old pollarded beech the soil surface was level and compact, with *Dicranum* and *Leucobryum*, and sparse fronds of *Pteridium* not more than 6 in. to a foot high. A section showed on the surface 3–4 in. (7·6–10 cm.) of litter and raw humus composed of the two mosses, bracken debris and beech leaves. Below this came 8 in. (20 cm.) of pinkish grey bleached sand and gravel (horizon *A*) with the lowermost 2 in. purplish: then 2 in. of purple-stained sand with the bottom inch more compacted (horizon *B*), underlain by yellow and orange sand and gravel subsoil (*C*). In another place the surface layer of *Leucobryum* together with the raw humus (into which the lower layers of its cushions decompose and which was filled with fine beech roots) was 11 in. (27 cm.) thick with a bleached layer (*A*) below, but no very clear *B* horizon was reached.

[Podsol profiles]

A more sharply developed podsol (old quarry near "Egypt") was seen under 4 in. of bracken litter, which covered 8 in. of bleached sand (*p*H 3·7) stained with humus in its upper part (*A* 1), underlain by a hard, dark, chocolate-brown, well-compacted pan $\frac{3}{4}$ in. (1·9 cm.) thick (*B* 1) and this by an inch of deep orange sand (? *B* 2), *p*H 3·6, followed by the lighter orange gravel subsoil (*C*) with a *p*H value of 4·1.

These examples were from glacial gravel. On the Reading Beds typical podsolisation was not seen, the soil was much finer grained and had a *p*H value of 4·1 at a depth of 6 in. (15 cm.). Here the beech made better height growth (70 ft. = 21 m.), and regenerated, bramble was dominant in the field layer, and a more varied flora was present (*Ilex, Lonicera periclymenum, Teucrium scorodonia, Melampyrum pratense, Potentilla erecta, Holcus mollis, Deschampsia flexuosa*) though still markedly oxyphilous.

Composition and structure. In many places the beechwood is practically pure, in others mixed with oak, while birch springs up in quantity in the larger gaps. Of the few associated trees and shrubs *Sorbus aucuparia* is characteristic, and also *Sorbus aria*, though the latter is much more sparsely distributed than on the chalk. Holly (*Ilex aquifolium*) is occasional to frequent. The alder buckthorn (*Frangula alnus*) is also characteristic though not abundant, and *Salix atrocinerea* is occasional: juniper occurs here and there. Honeysuckle is the characteristic climber, though never very abundant, while ivy is absent. Of the trees and shrubs occurring in most English woodlands, ash, hazel and field maple are characteristic absentees from these woods, appearing directly we descend into one of the shallow valleys with heavier and better soil. In the pure beechwood accessory woody plants are practically absent, owing to the deep shade cast

PLATE 71

Phot. 160. Beechwood on gravel overlying Reading Beds, adjoining Burnham Beeches, Bucks. Crooked stems of no great height. Field layer absent. Soil surface covered with beech leaf litter and with large cushions of *Leucobryum glaucum*. *H. Godwin.*

FAGETUM ERICETOSUM

PLATE 72

Phot. 161. Old pollard beeches at Burnham Beeches, Bucks, on gravel overlying Reading Beds. Shrub and field layers absent. Ground layer of mosses interrupted by leaf litter. Younger beechwood in background. Photograph from *The Times*.

Phot. 162. Mature beechwood (subspontaneous) 104 years old and not exceeding 20 m. in height, with smooth white bark. Based on drained heath (glacial till), Aberdeenshire. Shrub layer of *Sorbus aucuparia*. Field layer of *Vaccinium myrtillus*. A. S. Watt. (See pp. 418–21.)

FAGETUM ERICETOSUM

by the beeches together with the unfavourable soil conditions. The beeches themselves are of very various age, including very old pioneers and pollarded trees (Pl. 72, phot. 161) and saplings. They are never of very lofty growth, rarely exceeding 60 ft. (18 m.), and are frequently bent or twisted (Pl. 71).

The field layer is absent from considerable areas and is always poor in species and individuals. The bilberry (*Vaccinium myrtillus*) is often dominant in this type of beechwood, though not at Burnham Beeches. Of other gregarious species *Pteridium aquilinum*, often dominant in open areas, occurs also under beech where the shade is not too deep: *Holcus mollis* and *Deschampsia flexuosa* are locally dominant, *Melampyrum pratense*, though local, is characteristically abundant in some places, *Teucrium scorodonia* occasional to frequent, while *Luzula pilosa*, *Carex pilulifera*, *Viola riviniana* and *Veronica officinalis* are occasional.

The ground layer, which is rarely continuous, consists of characteristic bryophytes: *Polytrichum formosum* (the commonest woodland species of *Polytrichum*), *Dicranella heteromalla*, *Dicranum scoparium*, *Mnium hornum*, and some few others: *Diplophyllum albicans*, a very widely distributed and abundant leafy liverwort of acid soils; and notably the moss *Leucobryum glaucum*, which forms large silvery green cushions (Pl. 71, phot. 160). Under the old beeches the surface of the soil is characteristically bare of vascular plants and covered by a thick compact layer of raw humus formed of the matted and partially decomposed beech leaf litter or of the dead lower layers of the moss cushions (see p. 412). This compact raw humus forms a peculiarly unfavourable surface for the germination of seeds and is also largely responsible for the poverty and sparseness of the flora over the flat ground on which it prevails. Indeed, it is mainly where the soil surface is somewhat broken that the phanerogamic vegetation, sometimes with beech seedlings, occurs.

The following species of larger fungi are characteristic of beechwoods on these acid and podsolised soils:

*Amanita mappa	f–a	*Paxillus involutus	—
*Amanitopsis fulva	f	Russula fellea	o
Boletus chrysenteron	f	R. lepida	f–a
B. edulis	a	R. ochroleuca	f–a
Cortinarius elatior	f	*Scleroderma aurantium	—

* Especially on the very acid sandy soils.

Some of these occur in all types of acid beechwood, but the Russulae increase particularly on the very acid sandy soils, as in Burnham Beeches, while the Boleti on the other hand are more frequent on the acid silty soils of the chalk plateau.

On slopes where water oozes out of the overlying sands and gravels trees often fail to develop, and in the light gaps so produced bogs are formed which show the typical flora of wet heath (see p. 734). *Sphagnum* is often dominant, and is accompanied by its associated plants, such as *Erica tetralix*, with *Molinia caerulea* on soil which is rather less saturated.

On the flat ground the smooth surface *mor* prevents the establishment of beech seedlings even in gaps, and it is clear that beech does not regenerate here. On the other hand the slopes of the shallow valleys show quite different conditions. Rain-wash keeps the surface loose and open, the

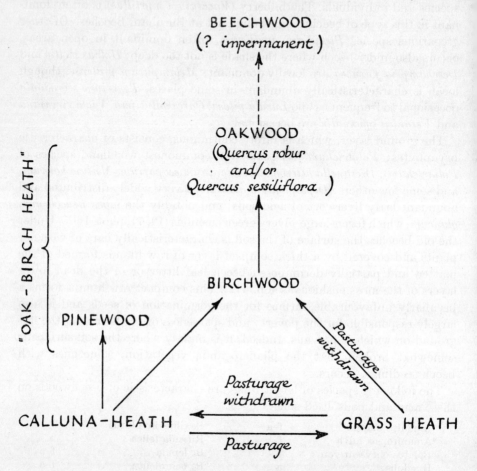

FIG. 82. SUBSERES FROM HEATH TO OAK- AND BEECHWOOD

compact continuous layer of raw humus does not form, the associated herbaceous flora is richer and more abundant and largely composed of different species, while beech regeneration occurs freely and even abundantly. It is noteworthy that in some such areas bramble is dominant in the field layer, recalling the beechwoods of the Chiltern plateau (cf. p. 407).

Succession. To the south of Burnham Beeches lies Burnham Common on the same soil as the woodland of the Beeches. This is a typical open heathy south English common, and the line between it and the woodland is more or less sharp and artificialised by wood cutting and planting; but the fundamental zonation representing the succession from heath to beechwood is still clearly traceable.

The Common itself is dominated mainly by *Calluna vulgaris* with *Erica cinerea* and *E. tetralix*, partly by heath grasses (*Festuca ovina*, *Agrostis tenuis* and *Deschampsia flexuosa*, in damper places *Molinia caerulea*), with dwarf gorse (*Ulex minor*) and occasional petty whin (*Genista anglica*). The process of colonisation is the same as on "oak-birch heath" on similar soils in the south-east of England (see Chapter XVII). The only difference is that here beech in quantity ultimately comes into the succession, probably because of the proximity of the Chiltern beech forests.

The first tree colonisers of the heath are the birches (*Betula pendula* (*alba*), *B. pubescens* and hybrids) which sow themselves in the heath and close up to form a young birch wood in which the heaths are shaded out. Subspontaneous pine also sows itself freely on the heath, though it does not here form pure pinewood as it does on many of the south-eastern sands. A fringe of birch and pine representing this stage of the succession occurs on the southern edge of the woodland. Behind this the oaks (*Quercus sessiliflora*, *Q. robur* and hybrids) come in and then beech, which, as already described, here forms at least a temporary climax, though, as we have seen, on the level ground of the low plateau it fails to regenerate, and the woodland of this flat area will probably in time revert to heath (cf. p. 355).

Fig. 82 represents the succession from grass heath or *Calluna* heath to oakwood or beechwood as seen at Burnham and on various other south-eastern English sandy soils. Cf. Chapter XVII, pp. 354–7.

Epping forest and similar woods. Paulson (1926) describes the beechwood of Epping Forest in south-western Essex on glacial gravel overlying Bagshot Sand, with allusions to other woods in Essex and Middlesex on the same type of soil. Here we have the same contrast that we have seen at Burnham Beeches between pure beechwood (or mixed beech, oak and birch, representing seral stages) of the higher ground and the oak (here also hornbeam) of the loams and clays below.

The surface is covered with the litter of dead beech leaves (2–4 in. thick), except where the fallen leaves are swept away by the wind, and in such places the ground layer is continuous, consisting of:

Campylopus puriformis	Hypnum cupressiforme
Dicranum majus	Leucobryum glaucum
D. scoparium	Mnium hornum

Below the litter or the moss carpet the horizontal surface roots of the trees form a closely ramifying interlaced layer with abundant mycorrhiza closely adhering to the half-decayed leaves or mosses, which are full of fungal hyphae. During the long summer drought of 1921 this layer of surface roots was killed, and did not regenerate till the autumn of 1922. The connexion of sporophores with the mycorrhiza has been demonstrated, according to Paulson, in the following fungi:

Amanita rubescens	Russula chloroides
Amanitopsis vaginata	R. emetica
Boletus chrysenteron	R. lepida
Hydnum repandum	R. nigricans
Lactarius blennius	Scleroderma aurantium (vulgare)

The following larger fungi were noted in addition:

Amanita mappa	Pholiota squarrosa
A. phalloides	Russula adusta
Collybia maculata	R. citrina
C. platyphylla	R. fragilis
Lactarius quietus	R. furcata (var. of R. cyanoxantha)
Pleurotus ostreatus	R. ochroleuca

The mycelium of *Collybia platyphylla* "forms white cord-like strands spreading out radially to a distance of 3–5 ft.", and destroying dead wood or beech mast with which it comes in contact. The parasitic fungi *Armillaria mucida*, *Fomes applanatus* and *Polyporus adustus* are associated with the death of a number of trees.

Lichens are not conspicuous because of the weak light. They generally occur on the main roots spreading from the base of the tree or upon moss clumps. The following are the commonest:

Cladonia fimbriata	Lecanora conyza
C. furcata	Parmelia physodes
C. squarrosa	P. saxatilis
C. sylvatica	

Below the layer containing the surface roots the moist beech leaves form a peaty mass highly acid in reaction (Epping Forest, pH 4·5; Kenwood, Hampstead, pH 3·6–3·7).

Of flowering plants the colourless saprophytes *Neottia nidus-avis* and *Monotropa hypopitys* are very rare but recur at long intervals in some of these woods. Paulson makes no mention of other phanerogamic plants in the beechwoods of Epping Forest, but in Pinner Wood, Middlesex, a considerable woodland flora is found, and here regular regeneration is said to take place. In Epping Forest beech seedlings occur locally in considerable quantity and establish themselves on the outskirts after a full-mast year such as 1922; and in the more open woods saplings and young trees of various ages are found.

PLANTED AND SUBSPONTANEOUS SCOTTISH BEECHWOODS

Several examples of the successful growth and regeneration of beech on acid and podsolised soils have been studied by Watt (1931) in Aberdeenshire. The woods are all composed of planted trees or derived from planted parents, but they may be fairly compared with those on the southern podsols, which they resemble in the associated heathy vegetation.

Shelter belts The shelter belts of beech which have been widely planted throughout the north-east of Scotland for the protection of arable cultivation and of conifer plantations, as well as round the borders of estates, consist mainly of mature trees now about a century old. They are for the most part planted on glacial drift derived from various geological formations, but most of these soils may be described as sands or sandy loams. The soil reaction is decidedly acid (pH 3·6–4·75), and under laminated raw humus derived from beech leaf litter distinct podsolisation occurs. The shelter belts are mostly quite narrow, 50–100 ft. (15–30 m.) wide and the vegetation is therefore exposed to wind. They consist solely of the canopy and a low-growing field layer dominated by grasses—in the more exposed places *Deschampsia flexuosa*, and in the better protected *Holcus mollis*. Where the wind has free play the beech leaf litter is blown away, and the ground is occupied by a low-growing mat of *Deschampsia*, the growth form of which is determined partly by shade and partly by the root competition of the beech trees, the grass growing more luxuriantly in plots surrounded by trenches cut through the surface-feeding beech roots. *Holcus* is dominant in the more sheltered situations, and also where litter is retained, as on the lee side of the wood, or where it is experimentally held down by wire netting. Again, where the surface beech roots are cut the better soil conditions so induced enable *Holcus* to compete successfully with the less exacting silver hair-grass.

Deschampsia and Holcus communities

The height growth of the beech is greater and seedlings are more frequent in the *Holcus* type of shelter belt, but regeneration within the belt is rare in both and almost restricted to gaps. Though there is no shrub layer, isolated shrubs of eight species occur in both types, with ivy, privet and elder in the *Holcus* type alone.

From the *Holcus* type eighty-two species have been recorded, including woody plants and bryophytes; from the *Deschampsia* type sixty-two. Fifty-four of these are common to both types, and the remainder are of low constancy. The characteristic differences depend on the different frequencies of the species occurring in the two types. Thus *Vaccinium myrtillus*, *Trientalis europaea*, *Calluna vulgaris* and *Potentilla erecta* are more frequent in the *Deschampsia* type; *Viola riviniana*, *Veronica chamaedrys*, *Deschampsia caespitosa*, and *Dactylis glomerata* in the *Holcus* type. *Oxalis acetosella* and *Anemone nemorosa* grow much more vigorously in the *Holcus* type. Of the twenty-one bryophytes *Catharinea undulata*, *Eurhynchium praelongum* and *E. myurum* occur in the *Holcus* but not in the *Deschampsia* type, while *Dicranum scoparium*, *Hylocomium splendens* and *Hypnum schreberi* are more constant or more frequent in the *Deschampsia* type.

The life-form spectra show a high percentage of chamaephytes as compared with the "normal" spectrum. According to Watt this is an expression of the relatively mild winter conditions near the seaboard. As between the two types there is a higher percentage of hemicryptophyta scaposa in the *Holcus* type, compensated by chamaephytes in the *Deschampsia* type. This difference is doubtless correlated with the more mesophilous conditions of the former. In fact the characteristic species of the *Holcus* type find the conditions in the *Deschampsia* community limiting or less than optimal, while in the *Holcus* type competition of these species tends to exclude the less exacting plants of the *Deschampsia* type.

Along the wood margins and under a more open canopy *Deschampsia flexuosa* is succeeded in dominance by *Vaccinium myrtillus*, and here trees may establish themselves freely. In the absence of grazing, rowan (*Sorbus aucuparia*) forms thickets in which occasional birch and beech occur, and in places the young beech may become locally dominant. Here we have the initiation of a vegetation essentially identical with that described below for the subspontaneous woods on glacial till.

Subspontaneous beechwood. In this region there are also extensive subspontaneous beechwoods arising from the invasion by beech of conifer plantations or of the rowan-birch woods derived from the selective felling of such plantations. The glacial tills on which these occur are composed largely of particles of the original minerals ground by the ice from the underlying granite and metamorphic rocks. Some of these soils contain an abundance of basic elements including calcium, though no free calcium carbonate. When cultivated this soil is very fertile, the main obstacle being the enormous number of large boulders it contains. The soil proper is about 2 ft. (60 cm.) deep above the unweathered glacial till, and is covered by a surface layer of raw humus, which however is fairly friable and disintegrates fairly fast. This humus layer lies on a loamy soil, stained above with humus but yellowish to reddish brown below, and resting immediately on the compact unweathered till. In other places the soil is much shallower, of lower base status, and podsolised.

The natural vegetation is probably a boulder-strewn "pine heath" and much of it remains in this condition, though the existing pines are all ultimately derived from introduced trees. Some of the land has been cleared of boulders, drained, and ploughed, and some has been planted with pine, larch, spruce and various other exotic conifers.

The beechwood is sometimes derived from the direct invasion of the exotic conifer plantations by the offspring of planted beech, since the conifers do not regenerate and the beech does. If the conifers are selectively felled the undergrowth of the conifer wood is left and forms a rowan-birch associes in which the birches reach 35–40 ft. in height and the rowans 20 ft. This is then invaded by young beech.

In the field layer *Deschampsia flexuosa* and *Vaccinium myrtillus* are generally dominant, but with an abundance of *Anemone nemorosa* as well as frequent *Oxalis acetosella*. With the increase of beech the heath plants are suppressed and *Anemone* and *Oxalis* rise to dominance, while *Mnium hornum* appears in the moss layer. In the adult wood *Oxalis* and *Anemone* become less abundant, and *Vaccinium myrtillus* (accompanied by some *Anemone*) again becomes dominant in the field layer.

Though this type of sere must be reckoned as belonging to the heath group, since it is developed on soil which in the "natural" state bears "pine heath" and *Vaccinium* is ultimately dominant in the field layer of the mature wood, the soil is obviously better than in the podsol type.

In the shallower podsolised soils the pH values are notably low but show a very distinct gradient corresponding with the successive horizons, ranging from an average of 3·9 in the raw humus of the surface to 4·9 in the glacial till at a depth of about 15 in. (38 cm.). Sometimes a hard iron pan ($B2$ horizon) formed above the till stops the downward growth of the tree roots, but even where pan is absent the compact till itself presents almost as effective a barrier, so that the total exploitable thickness of soil does not exceed 15 in., a depth insufficient for good growth of the deeper rooting trees.

Structure and composition Adult beechwood on this soil has less than sixty species in all, including nineteen bryophytes and twenty-five herbs and grasses. It does not exceed 20 m. (66 ft.) in height, and is relatively open, only thirteen adult beeches occurring in the unit area 100 ft. square (930 sq. m.), but the trees are well grown.[1] Accompanying native trees are whitebeam, holly and birch. The shrub layer is composed solely of rowan (*Sorbus aucuparia*). The field layer is almost entirely of bilberry (*Vaccinium myrtillus*), the accompanying herbs and grasses being little in evidence. Among these *Deschampsia flexuosa*, *Holcus mollis*, *Luzula pilosa*, *Oxalis acetosella*, *Galium saxatile* and *Potentilla erecta* are the commonest, with the broad buckler fern (*Dryopteris dilatata*). The moss layer is largely suppressed by the deep beech humus: it is represented most frequently by *Hylocomium triquetrum*, with *Hypnum cupressiforme*, *Mnium hornum*, *Brachythecium purum* and Polytricha: also by the ubiquitous woodland liverwort *Lophocolea bidentata*, and by *Calypogeia trichomanis*.

Though young beech occurs in these woods, Watt does not believe that beechwood can maintain itself indefinitely on this soil in the maritime climate. The bulky beech leaf litter continually accumulates under the given conditions, forming a thick layer of laminated humus which does not become incorporated in the mineral soil below. By preventing aeration this would probably ultimately prevent the normal growth of the trees and check regeneration, so that the trees would finally decay without leaving young growth to succeed them, and the area would revert to heath.

Succession from conifer plantations As mentioned above these Aberdeenshire subspontaneous beechwoods are developed by the invasion of beech from neighbouring plantations into planted conifer wood. These conifer woods consist mainly of pine and larch, with some spruce, silver fir and

[1] Pl. 72, phot. 162.

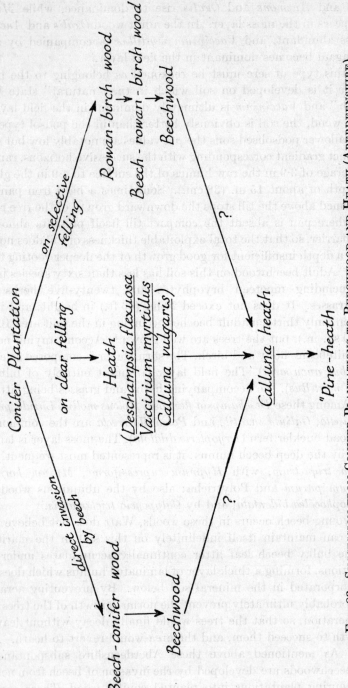

FIG. 83. SUCCESSION FROM CONIFER PLANTATIONS TO BEECHWOOD AND HEATH (ABERDEENSHIRE) Modified from Watt, 1931. For description see text.

other exotic conifers. They have a discontinuous shrub layer, chiefly of rowan (*Sorbus aucuparia*), but also of birch, a field layer whose dominants are *Deschampsia flexuosa* and *Vaccinium myrtillus*, with *Calluna* in the better lighted spots, and including altogether between fifty and sixty species: the ground layer numbers twenty-nine bryophytes. Most of the species of the field and ground layers, as would be expected, are characteristic of acid soils.

These conifer plantations do not regenerate naturally, for though seedlings of several species have been seen they do not reach maturity, except in the case of larch, whose regeneration is widespread but not abundant. Left to themselves the trees would ultimately die without leaving successors and the area would revert to heath. This would be colonised here and there by pine, forming the "pine heath" which, Watt holds, is the natural vegetation of this type of soil over much of eastern Scotland. Wherever adult planted beech adjoins such a conifer plantation the latter is invaded by the beech (though slowly and over a narrow zone only, in correspondence with the slow migration of the tree), and gradually converts it into a mixed wood in which beech alone becomes dominant. The flora of this mixed wood shows a very heavy reduction in the number of species of the field and ground layers (from fifty and twenty-nine in the conifer wood to twelve and fifteen in the mixed wood) as a result of the much deeper shade, the root competition of the beech, and the thick layer of litter and raw humus which accumulates.

If the conifer wood is selectively felled the rowan and birch of the shrub layer grow up to form a rowan-birch wood with fifty-one species in the field layer and twenty-six bryophytes in the ground layer; and this also is invaded by adjoining beech, again giving a mixed wood with a reduction to twenty-two species in the field layer and twenty-three in the bryophyte layer.

No mature beechwood derived either from the mixed beech-conifer wood or from the mixed beech-rowan-birch wood has been observed, but "consociation nuclei", i.e. centres of formation of mature beechwood, have been found; and there is little doubt that these would develop into mature beechwood having the characters already described (pp. 416–18), though the description is actually based on planted beechwoods.

Fig. 83 (slightly modified from Watt) represents the succession from conifer plantation to beechwood and heath on these weathered and sometimes podsolised Aberdeenshire tills of varying base status.

SUMMARY OF BRITISH BEECHWOODS

It will be useful here to present some summarised data relating to British beechwoods.

Fig. 84 shows the mean height growths of beech on different types of soil in southern England and illustrates the marked superiority of the brown-

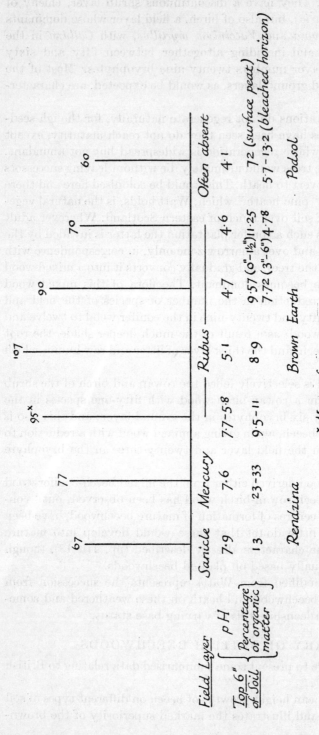

FIG. 84. HEIGHT GROWTH OF BEECH WITH CORRESPONDING FIELD LAYER AND SOIL CHARACTERS

The pH values show a continuously descending series from the rendzina soils of the escarpment through the brown earths to the silty loams, sands and gravels. The humus content is lowest in the "central" brown earths, which are the deepest soils and bear the finest beech. It rises markedly in the rendzinas (left). To the right it reaches a very high figure in the surface layer of mor. With increasing podsolisation the difference in organic content between this and the immediately underlying (bleached) layer rises enormously. Data condensed from Watt, 1934. Height growths in feet.

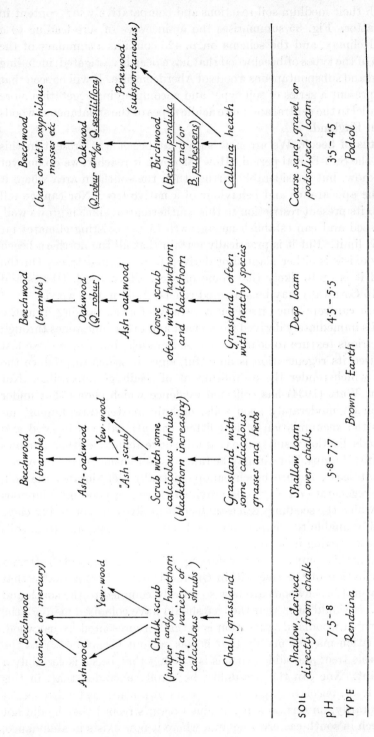

Fig. 85. Summarised Seres leading from Grassland or Heath in Southern England to different kinds of Beechwood

Data partly from Watt.

earths with their medium soil reactions and comparatively low content in organic matter. Fig. 85 summarises the main types of sere leading to a beechwood climax, and the scheme on p. 425 contains a summary of the vegetation of the types of beechwood that have been investigated, including the planted and subspontaneous woods of Aberdeenshire. It will be seen that these last present a series of soil types and accompanying vegetation more or less parallel to the central and more acidic types of the spontaneous woods of southern England.

The status of beech. We are now sure that beech was present in this country in the Sub-Boreal period. How far north it reached we do not yet certainly know, but considerably farther than the southern area where it now has the appearance and behaviour of a native tree. Nor can we tell what led to its present restriction to this southern area, since it grows well, sets good seed and can establish new growth in the existing climates far outside this limit. But it is practically certain that all the northern beech which now exists is either planted or derived from planted trees. On the other hand it is quite likely that some of the western beech (Wales and Wiltshire to Cornwall) may be scattered relicts of old native woods.

The beech can reproduce itself on a variety of soils, ranging from the shallow soils immediately derived from chalk and other limestones through loams of various texture to coarse sands and gravels, but where these last are podsolised its regeneration is doubtful, since it seems to produce the conditions which render the establishment of seedlings impossible. Not only so but Watt (1934) has collected evidence which shows that under beech, even on moderately good soils, organic matter (raw humus) increases by the accumulation of leaf litter, the pH value is lowered and incipient podsolisation results; also that mice have abandoned their burrows under this raw humus. Thus it may be taken as established that, on certain substrata at least, a pure beech canopy actually deteriorates the soil. Beech can regenerate on clay soils provided there is a good mild humous surface in which the seedlings can establish themselves, but on heavy clays it is probably unable to compete on equal terms with oak, and from soils liable to waterlogging it is absent.

Considering this comparatively wide range of suitable soils and the former wider distribution of the tree within Great Britain it is remarkable that existing natural beechwoods are now so strictly confined to the south and south-east. It would seem that the tree should have colonised many of the loams on which woodland vegetation is actually represented by oakwood. The explanation may be mainly that it was cut out because oak was the more valuable tree; and added to this is the fact that beech is certainly a slow migrant. Nor can the possibility be wholly excluded that, in this climate at least, beech may put an end to its own continued existence, at least temporarily, on certain soils. Julius Caesar's record that he did not see the beech in south-eastern England, where it now exists in abundance,

TYPES OF BRITISH BEECHWOODS

	Southern English (native)			Scottish (Aberdeenshire) (planted and subspontaneous)	
	Soil	Vegetation		Soil	Vegetation
(1)	Shallow chalk and oolite soils (rendzinas) markedly alkaline	Varied "exacting" herbaceous flora. Hemicryptophyta scaposa prominent on the deeper, H. rosulata on the shallower soils (Fagetum calcicolum)	Rendzina		
(2)	Shallow soils of harder limestones	Varied: not fully known, probably similar to (1) and (3)			
(3)	Loams, slightly alkaline to markedly acid, rich in electrolytes (brown earths)	Varied "exacting" herbaceous flora. Hemicryptophyta scaposa prominent. Rubus layer characteristic. (Fagetum rubosum)	Brown Earth	(1) Fertile loams derived from Old Red Sandstone (brown earth)	Varied "exacting" herbaceous flora. Hemicryptophyta scaposa prominent; also Chamaephyta
(4)	"Good" sandy soils (not podsolised)	Little known, probably similar to, but poorer than (3). Rubus layer frequent		(2) Glacial till rich in electrolytes	Anemone nemorosa, Oxalis acetosella: Vaccinium myrtillus ultimately dominant (Rubus sometimes abundant under oak)
(5)	Podsolised silts and sands with thick surface mor: extremely acid	Oxyphytes characteristic: Deschampsia flexuosa, Vaccinium myrtillus, Leucobryum, etc. (Fagetum ericetosum)	Podsol	(3) Podsolised glacial till, etc. more or less sandy	Subspontaneous beechwood Vaccinium myrtillus dominant. Exposed planted shelter belts Deschampsia flexuosa and Holcus mollis dominant [Vaccinium myrtillus]

remains unexplained; though it has been suggested (Woodruffe Peacock) that the tree was at first present in quantity only farther north in eastern England and that it spread to the south-east at a later date.

REFERENCES

HARLEY, J. L. Ecological observations on the mycorrhiza of beech. *J. Ecol.* 25. 1937.
PAULSON, ROBERT. The beechwood: its canopy and carpet. *Trans. S.E. Union Sci. Soc.* pp. 24–37. 1926.
WATT, A. S. Preliminary observations on Scottish beechwoods. *J. Ecol.* 19, 137–57, 321–59. 1931.
WATT, A. S. The vegetation of the Chiltern Hills with special reference to the beechwoods and their seral relationships. *J. Ecol.* 22, 230–70, 445–507. 1934.

Chapter XXI

ASHWOOD ON LIMESTONE
(FRAXINETUM CALCICOLUM)

Distribution of climax ashwood. We have already seen in Chapter XVIII that ashwood may be a seral stage (consocies) leading to the development of beechwood on the shallow calcareous soils of the Chalk and Oolite escarpments and valley sides in the south of England. Outside the southern and especially the south-eastern region, in which beech behaves as the ultimate dominant, ashwood is the climax forest type (consociation) on the shallow soils of the limestones on which oak cannot flourish. With beech absent the "preclimax" ash is left in possession of the ground.

On the Chalk itself, south-west of the region in which beech is the dominant tree, ashwoods occur, as in Wiltshire, Dorset and in the Isle of Wight; and also on other limestones, such as the Jurassic and Permian (Magnesian) limestones westwards and northwards throughout the country. By far the most important and extensive limestone of the north and west of Britain and in Ireland is the massive Lower Carboniferous or Mountain Limestone which forms conspicuous hill ranges, and sometimes outcrops in great horizontal sheets at quite low altitudes, as in the central plain of Ireland. In Somerset it forms the Mendip Hills and some smaller areas farther north extending into Gloucestershire: in South Wales it appears as a rim round the South Wales coalfield in Glamorgan and Brecon, and crops out also on the Welsh border in the Wye valley: in North Wales it occurs in Flint and Denbighshire and also in Anglesey. But the most important mass of Carboniferous Limestone in Great Britain is that which forms the core of the southern Pennines in Derbyshire, and of the northern Pennines in west Yorkshire and north Lancashire. Farther to the north the Lower Carboniferous rocks are represented largely by the Calciferous Sandstone series, with interbedded limestone bands of subordinate importance. In Ireland the Carboniferous Limestone forms the basis of the great central Irish plain. The limestone of the Irish plain is largely overlaid by boulder clay (glacial drift) and by clay formed from the recent direct weathering of the rock. These clays often contain much lime and support the excellent cattle pastures for which Ireland is famous, but in the west the limestone itself forms extensive plateaux and terraced hills, as in Sligo and east Galway and most extensively in Clare.

Fragments of ashwood, together with calcicolous scrub, occur in most or all of these Mountain Limestone areas, and on various less extensive limestone outcrops belonging to other geological horizons. The vegetation of these woodlands was briefly described by Smith and Rankin (1903) (central Pennines of west Yorkshire), and by Lewis (1904) (northern Pennines), but was more thoroughly investigated by Moss on the Mountain Limestone of the Mendips and of that to the north extending into Gloucestershire, also on the oolites around Bath (1907), and on the Mountain Limestone of Derbyshire (southern Pennines) (1913). It was these observers who first definitely recognised the existence of pure natural ashwood in this country. The best development of ashwood occurs in the Somerset and Derbyshire areas studied by Moss. Since his work was done, more than a quarter of a century ago, this type of woodland has been neglected by ecologists, so that it is not possible to do much more than summarise Moss's data and those of his conclusions which still seem justified in the light of modern knowledge. Quite recently however a few fresh observations have been made on upland ashwoods, one on basic igneous rocks (pp. 433–40).

Situation and soil. Most of the ashwoods of the Mendip region and of the Derbyshire dales (Pl. 73), as well as those on other limestones and basic igneous rocks (Pls. 75–7), occur on more or less steep hillsides; and in such situations the soil is typically shallow, so that oak is excluded, just as it is from the soils of the chalk escarpments and steeper valley sides. The ground surface of the Mountain Limestone woods is typically much more uneven than that of the Chalk, and large boulders are frequent, because of the much greater hardness of the rock. The ashwoods on Mountain Limestone are often developed on the scree or talus which has accumulated by falls from the limestone cliffs. The actual soil has never been properly investigated, but presumably it is highly calcareous to the surface: Moss (1907) says that the Somerset limestone soil is "a red marl sometimes very shallow, stony and dry, and in such situations the ash often occurs to the exclusion of every other arboreal species". Taken together with what we know of the ashwoods on chalk, there can be no doubt (as was first recognised by Moss) that on such shallow, highly calcareous soils the ash is the natural climax dominant where it is not exposed to the competition of the beech. On some of the flat Mountain Limestone pavements of west Yorkshire and north Lancashire ashwood is developed with the trees rooted in the fissures (Pl. 74, phots. 167–8). Some account of such a wood is given on pp. 434–6. The deep and frequent fissures in the rock, which contain a rich and largely "woodland" vegetation even when the surface of the pavement is quite bare and there are no trees, play an important part in the development of this pavement ashwood.

On the deeper soils which accumulate towards the bases of the limestone slopes and at the bottoms of the valleys the ashwood is less pure, and in most districts the pedunculate oak is occasional to frequent (ash-oak wood, see pp. 272, 396, 442), though not apparently in Derbyshire. These woods on deeper soils are usually more or less exploited and alien trees are often planted in them.

Composition and structure. The typical purity of ashwood has already been alluded to. The shade is much slighter than that cast by the oak, so

Phot. 163. Ashwood on the steep rocky slopes of Dovedale, Derbyshire (Carboniferous Limestone), showing limitation of woodland to the valley. The plateau, shown on the left, is occupied by enclosed fields (pasture). Planted conifers are seen in the ashwood on the right. *W. J. B. Blake, Longton, Staffs.*

Phot. 164. Bottom of ashwood on the steep rocky side of a Derbyshire dale (Carboniferous Limestone). In the foreground is a pool with water crowfoot (*Batrachium* sp.) surrounded by marsh plants. *W. B. Crump.*

ASHWOOD ON LIMESTONE (FRAXINETUM CALCICOLUM)

PLATE 74

Phot. 165. Ash-hawthorn scrub in foreground, ashwood in the valley behind. Ebbor Gorge, Mendip Hills, Somerset (Carboniferous Limestone). *W. B. Crump.*

Fig. 166. Ashwood covering an isolated hill of Carboniferous Limestone. Austwick, W. Yorks. *W. M. Rankin.*

Phot. 167. Ash colonising fissures in eroded limestone pavement. Chapel-le-Dale, W. Yorks. *W. M. Rankin.*

Phot. 168. More advanced stage of development of ashwood on limestone pavement.

that sufficient light penetrates the canopy to permit the development of an abundant subsidiary vegetation. This may take the form of a rich shrub layer or of a rich field layer, but when the former is densely developed the latter is reduced, as in all woods. Typically, however, both are well developed.

Tree layer. The most characteristic and frequent tree accompanying the ash is the wych elm (*Ulmus glabra*); and in the south (Somerset) the whitebeam (*Sorbus aria*) and the yew (*Taxus baccata*), neither of which is more than occasional or rare in Derbyshire. The field maple (*Acer campestre*) and the aspen (*Populus tremula*) are rather characteristic, the latter no doubt a relict from an earlier stage of the succession. The bird cherry (*Prunus padus*) is occasional to abundant in the ashwoods of Derbyshire: it does not occur in Somerset. The birches are quite absent except at high altitudes (see p. 442).

Shrub layer. Ashwoods show the great variety of shrubs characteristic of limestone soils with which we have already become familiar in the composition of chalk scrub, but some of these are confined to the south of England and others are rare and local so far north as Derbyshire. The commonest shrubs are hawthorn and hazel, which are abundant almost everywhere. There are a good many species of fruticose *Rubus* listed by Moss for the Derbyshire ashwoods, but they are rather fewer and less abundant than in the oakwoods of the district. *R. caesius* is abundant locally, *R. idaeus* rare. Of roses various forms of *Rosa canina* occur, *R. arvensis* is abundant. *R. micrantha* and *R. spinosissima* are local; and they do not occur at all in the oakwoods. *Ribes alpinum* is rare, while the red and black currants (*R. rubrum* and *R. nigrum*) and the gooseberry (*R. grossularia*) are possibly bird-sown from gardens.

Of the more characteristically "calcicolous" shrubs dogwood (*Cornus sanguinea*), spindle (*Euonymus europaeus*), buckthorn (*Rhamnus catharticus*), privet (*Ligustrum vulgare*) and elder (*Sambucus nigra*) are more or less frequent: wayfaring tree (*Viburnum lantana*) is absent from the Derbyshire ashwoods. Of the woody climbers, ivy (*Hedera helix*) is abundant in both, honeysuckle (*Lonicera periclymenum*) abundant in Somerset but only occasional in the Derbyshire ashwoods; traveller's joy (*Clematis vitalba*), occasional in Somerset (not nearly so common as on the south-eastern chalk) is absent from Derbyshire. The goat and grey sallows (*Salix caprea* and *S. atrocinerea*) both occur.

Though the shrubs of exploited ashwoods are of course constantly cut, the woods are very rarely kept on the uniform coppice-with-standards system characteristic of the exploited oakwoods of the south and midlands. The ashwoods on the steeper slopes are often left very much alone, mature trees being cut only occasionally, and in these woods the density of trees and shrubs varies greatly from place to place according to soil and situation, the shrub layer naturally being thickest in gaps.

Field layer. According to Moss the herbaceous flora of the ashwoods is the richest of any woodland community in Britain because there are more species occurring in ashwood but absent from oakwood than vice versa, and also because many species that do occur in oakwoods are more abundant in ashwoods; and this may perhaps be true, both because of the generally favourable effect in a damp climate of a considerable amount of calcium carbonate in the soil and also because of the comparatively light shade cast by the dominant tree.

Moss recognised three groups of communities in the Derbyshire ashwoods, corresponding with the varying water content of the soil:

(1) *The marshes* by the sides of streams, in the bottoms of some of the streamless valleys, and round springs. For these he lists the following plants, without commenting on their respective roles in the community.

    Caltha palustris   Orchis maculata
    Cirsium heterophyllum   Petasites hybridus (ovatus)
    Epilobium hirsutum   Phalaris arundinacea
    Filipendula ulmaria   Phragmites communis
    Geum rivale   Scirpus compressus
× G. intermedium   Sparganium erectum (ramosum)
    Mentha aquatica   Trollius europaeus
    Myosotis scorpioides   Valeriana officinalis

(2) *Communities of damp soil.* On damp soil societies of bear's garlic or ramsons (*Allium ursinum*) and of lesser celandine (*Ficaria verna*) are characteristic, and the accompanying flora is rich, including the following species:

    Agropyron caninum   Fragaria vesca
    Anemone nemorosa   Galeobdolon luteum
    Aquilegia vulgaris   Geum rivale
    Arum maculatum   × G. intermedium
    Asperula odorata   Melandrium dioicum
    Bromus ramosus   Myosotis sylvatica
    Campanula latifolia   Orchis maculata
    Carex sylvatica   Oxalis acetosella
    Cirsium heterophyllum   Platanthera chlorantha
    C. palustre   Polemonium caeruleum
    Deschampsia caespitosa   Trollius europaeus
    Dryopteris filix-mas   Valeriana officinalis
    Elymus europaeus   V. dioica

(3) *Communities of drier soils.* The most widespread society of drier soil is dominated by dog's mercury (*Mercurialis perennis*). With this is often associated the moschatel (*Adoxa moschatellina*) which finds shelter for its delicate prevernal shoots among those of the more vigorous mercury, while its roots exploit the surface layer of soil, those of mercury going deeper. In the driest parts of the ashwood, societies of ground ivy (*Nepeta hederacea*) are the commonest; and on old screes incompletely colonised by the trees and shrubs societies of upright herbs, such as *Hypericum hirsutum*, *Urtica dioica*, *Teucrium scorodonia*, forming close herbaceous thickets in

summer. Locally, societies of lily of the valley (*Convallaria majalis*) and stone bramble (*Rubus saxatilis*) occur, and in them *Melica nutans* and *Epipactis atropurpurea* are sometimes found.

Among species mentioned by Moss as occurring in the Derbyshire ashwoods, but not in the oakwoods of the same region, the following may be cited:

Asplenium adiantum-nigrum	Hypericum montanum
A. ruta-muraria	Helleborus viridis var. occidentalis
A. trichomanes	Pimpinella major
Cystopteris fragilis	Polemonium caeruleum
Dryopteris robertiana	Polypodium vulgare
Geranium lucidum	Phyllitis scolopendrium
G. sanguineum	Viola hirta
Hypericum hirsutum	V. sylvestris

The following species of the field layer recorded by Moss for the ashwoods of Somerset are said to be absent from the oakwoods:

Carex digitata	Melampyrum pratense
Convallaria majalis	Polygonatum multiflorum
Geranium sanguineum	P. officinale
Helleborus viridis var. occidentalis	Rubia peregrina
Hypericum montanum	Rubus saxatilis

This list and the preceding ones do not of course by any means exhaust the herbaceous flora of the ashwoods. Full lists will be found for Somerset in Moss (1907, pp. 54–6) and for Derbyshire in Moss (1913, pp. 79–86).

Ground layer. The following bryophytes are recorded by Moss for the Derbyshire ashwoods, a total of thirty-five species as compared with forty-six for the oakwoods:

Aneura pinguis	r	Lejeunea serpyllifolia	la
Bryum capillare	o	Lophocolea bidentata	r
Catharinea undulata	a	L. heterophylla	o
*Dicranella heteromalla	a	Lunularia vulgaris	r
Dicranum scoparium	r	Metzgeria furcata	r
Eurhynchium confertum	r	Mnium punctatum	r
E. striatum	o	M. stellare	o
Fegatella conica	o	M. undulatum	o
Fissidens adiantoides	o	Nardia scalaris	r
F. bryoides	r	Pellia calycina	o
F. taxifolius	o	Plagiochila asplenioides	o
Fossombronia pusilla	r	Plagiothecium denticulatum	o
Frullania dilatata	la	Porella platyphylla	o
F. tamarisci	la	Porotrichum alopecurum	o
Hylocomium rugosum	r	Radula complanata	o
H. splendens	o	Rhacomitrium lanuginosum	o
*Hypnum schreberi	o		
Lejeunea calcarea	l	Thuidium tamariscinum	o
L. rosettiana	o		

* These species are characteristic of very acid soils and in ashwood on limestone soil presumably occurred in exceptional habitats.

Succession. The succession leading up to ashwood has never been systematically and thoroughly studied, but from the data furnished by Smith and Rankin (1903) and by Moss (1907) and from a scheme given by Moss (1907, p. 50) we can reconstruct it in outline.

The grassland of the Carboniferous Limestone—the so-called "natural pasture" of the earlier writers (see Chapter xxvii, pp. 552–7)—covers most of the hill slopes which are in places occupied by ashwoods. The direct colonisation of this by woody plants does not appear to be described, though one would suppose that it must take place when grazing is withdrawn, as it certainly does in chalk grassland.

On flat summits and gentle slopes, as the soil over the limestone rock deepens and leaches, the grassland is invaded by heath plants, just as on the chalk plateaux; and the resulting mixture of heath plants and calcicolous plants is called a "limestone heath" (Chapter xxvii, p. 557). In Somerset, according to Moss, the limestone heath is invaded by shrubs, especially by hawthorn and hazel, which form the foundation of the "limestone copse" or "ash copse", into which ash and associated trees, such as wych elm, whitebeam and yew soon enter. By increasing dominance of the ash the "limestone copse" develops, in suitable situations, into ashwood.

The "scarwoods" of Smith and Rankin (1903, pp. 20–1) are developed mainly on the screes below the vertical cliffs of Mountain Limestone which form a feature of the Craven district of west Yorkshire. These scarwoods have a foundation of hazel with various other species of calcicolous shrubs —hawthorn, buckthorn, dogwood, elder, privet, wayfaring tree: in many of them ash is present, sometimes a few trees, in other cases so many as to form an ashwood with a shrub layer dominated by hazel. The occasional ashes which rise above the shrub layer of many of the scarwoods may represent invasion, the progressive development of woodland from scrub, or, alternatively, the trees remaining after removal of the majority. Since the ash is practically the only timber tree of these limestone dales it has been severely exploited in the past.

It is clear then that both in Somerset and Yorkshire the "limestone copse", or, as we may call it, "ash-hazel scrub" is a natural stage in succession to ashwood. Whether hazel or hawthorn is here actually the pioneer shrub colonist is not quite clear, but probably hazel.

The following twelve herbaceous species appear from the records to be common to the "limestone copse" or "ash copse" of Somerset and the "scarwood" of the Craven district:

Centaurea scabiosa	Mercurialis perennis
Filipendula hexapetala	Ophrys apifera
Geranium robertianum	Origanum vulgare
Helianthemum nummularium	Polystichum aculeatum
Hypericum montanum	Teucrium scorodonia
Lactuca muralis	Viola hirta

The succession may be represented as in the diagram below:

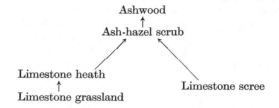

Two upland ashwoods. Dr J. L. Harley has examined two ashwoods in west Yorkshire, one (Ling Ghyll) in a rocky limestone ravine belonging to the Yoredale series, the other (Colt Park) developed on a horizontal pavement of Mountain Limestone, both lying at an altitude of about 1100 ft. (335 m.). Dr Harley's account is unpublished and I am greatly indebted to him for putting his data at my disposal.

The district is mainly devoted to sheep pasture, and the survival of these woods is doubtless due to the danger of sheep falling into the fissures at Colt Park and to the inaccessibility of the steep rocky slopes at Ling Ghyll.

Ling Ghyll wood. In the ravine wood (Ling Ghyll) the terrain varies from bare rock to a soil covering (pH 6·6–6·9) about 1 ft. deep. The canopy is very light, the trees not more than 40–50 ft. (12–15 m.) high with a close shrub layer dominated by hazel and a very rich flora (120 species in all).

Tree layer

Fraxinus excelsior	d	Sorbus aucuparia	o
Ulmus glabra	a	[Acer pseudo-platanus	r]
Prunus padus	a	[Larix decidua	r]
Betula pendula	lf	Alnus glutinosa	r
B. pubescens	lf		

Shrub layer

Corylus avellana	d	Salix caprea	o
Crataegus monogyna	a	S. atrocinerea	r
Prunus spinosa	f–la	S. phylicifolia	r
Rosa villosa	o		

The field layer contained the following:

General field layer

Ajuga reptans	f	Circaea lutetiana	r–o
Alchemilla vulgaris	f	Conopodium majus	f
Allium ursinum	o	Cynosurus cristatus	o–f
Anthyllis vulneraria	r	Dactylis glomerata	a–lsd
Arrhenatherum elatius	f	Epilobium montanum	o–f
Asperula odorata	f–la	Fragaria vesca	a
Avena pubescens	o	Galium aparine	l
Briza media	o	G. cruciata	f–la
Campanula rotundifolia	o	G. verum	o
Centaurea nigra	f	Geranium sanguineum	r

Geum urbanum	f	Pimpinella saxifraga	f–a
Heracleum sphondylium	f–a	Polygonum bistorta	r
Holcus lanatus	f–a	Primula vulgaris	f
Juncus acutiflorus	la	Prunella vulgaris	f
Lamium album	r–o	Rumex crispus	r
L. purpureum	f–a	Scilla non-scripta	o–f
Lathyrus montanus	o	Scrophularia nodosa	r
L. pratensis	a	Senecio jacobaea	o
Listera ovata	r	Taraxacum erythrospermum	r
Luzula campestris	o		
L. pilosa	o	T. officinale	r–o
Lychnis flos cuculi	lf	Teucrium scorodonia	a
Lysimachia nemorum	o	Urtica dioica	la
Melandrium dioicum	f	Valeriana sambucifolia	r–o
Mentha aquatica	o	Veronica chamaedrys	a
Mercurialis perennis	a–ld	V. officinalis	f
Myosotis sylvatica	o	Vicia sativa	o
M. versicolor	r–o	V. sepium	o
Origanum vulgare	f	Viola sylvatica (agg.)	a
Oxalis acetosella	a		

Marshy places

Ajuga reptans	a	Myosotis scorpioides	o
Anacamptis pyramidalis	r	Orchis mascula	o
Caltha palustris	a–ld	Pinguicula vulgaris	o
Carex panicea	l	Platanthera chlorantha	o
Equisetum palustre	l	Primula farinosa	o
Filipendula ulmaria*	a	Radicula nasturtium	la
Geum rivale	o	Veronica beccabunga	l
Juncus conglomeratus	ld		

Rocky places

Arabis hirsuta	f	Geranium robertianum	o
Asplenium adiantum-nigrum	r–o	Lactuca muralis	f
A. ruta-muraria	f	Phyllitis scolopendrium	f
A. trichomanes	a	Polypodium vulgare	o–f
A. viride†	f	Polystichum aculeatum	o
Cystopteris fragilis	r–o	P. lonchitis	o
Geranium lucidum	a–ld		

* Also occurs throughout the wood.
† Especially on the upper cliff.

The result of pasturing on the more even and gentle slopes accessible to sheep is to degrade the woodland to scrub consisting of *Crataegus* and *Prunus spinosa* with *Corylus* and *Sorbus aucuparia* quite frequent. In a later stage the hawthorn and blackthorn alone remain, often much dwarfed, with rowan in rocky places. The grassland produced is dominated by *Festuca ovina*, with *Potentilla erecta* and *Galium saxatile*, and on low-lying ground where the water table is close to the surface there is meadow with *Holcus lanatus* dominant and including *Lotus uliginosus*, *Prunella vulgaris*, etc.

Colt Park wood. Stages in the development of the woodland on limestone pavement can be traced in the neighbourhood of this wood. The

Ashwood on Limestone Pavement

"scales", as the bare horizontal surfaces of limestone pavement separated by deep fissures are called, are very slowly colonised, but the fissures, which vary from a few centimetres to nearly a metre in width, contain quite a rich vegetation largely composed of woodland plants. Harley found the following species in the clefts between the "scales" near the Colt Park ashwood:

Allium ursinum	o	Lactuca muralis	o
Arum maculatum	o	Melandrium dioicum	o
Asperula odorata	la	Mercurialis perennis	ld
Asplenium ruta-muraria	a	Myosotis collina	r
A. trichomanes	a	Oxalis acetosella	a
A. viride	o	Phyllitis scolopendrium	a
Cystopteris fragilis	r	Polypodium vulgare	r
Dryopteris filix-mas	lf	Senecio jacobaea	r
Epilobium montanum	a	Taraxacum officinale	r–o
Euphrasia vulgaris	r–o	Urtica dioica	ld
Galium cruciata	o	Veronica chamaedrys	a
Geranium lucidum	o	V. officinalis	o–f
G. robertianum	a	Viola sylvestris	o
Heracleum sphondylium	o–f		

[The following have also been recorded from the clefts of limestone pavements in other localities:

Actaea spicata	l	Listera ovata	o
Anemone nemorosa	f	Polystichum lobatum	r
Dryopteris rigida	la	Thalictrum minus	f]

Of woody plants in the fissures and on calcareous debris between the pavements Harley found:

Corylus avellana	l	Prunus padus	a
Crataegus monogyna	a	P. spinosa	ld
Fraxinus excelsior	l	Rosa villosa	r–o
Lonicera periclymenum	l	Rubus idaeus	r

[Woody species found elsewhere in the fissures of limestone pavement are *Hedera helix* and *Ribes petraeum*. Pl. 74, phots. 167–8 illustrate the development of ashwood on limestone pavement.]

Where *Fraxinus* has become dominant, forming the ashwood, the woody species are the same, with the addition of *Ulmus glabra* (frequent), *Hedera helix* and *Betula pendula* (both rare).

The field layer, which is thick and tall, consists of the herbs and ferns of the fissures, with the addition of:

Alchemilla vulgaris	o	Galium erectum	r
Anacamptis pyramidalis	vr	Geum rivale	o
Centaurea nigra	r	Geranium sp.	r
Chrysanthemum leucanthemum	r	Primula vulgaris	o
		Sesleria caerulea	r
Circaea lutetiana	r	Scilla non-scripta	o–f
Dipsacus sylvestris	r	Stellaria holostea	o
Fragaria vesca	o–f	Viola lutea var. amoena	vr

The soil of the wood is black, with a constant pH value of 7. This covers the surface of the pavement and fills up most of the crevices. Where these still remained open they were mostly too deeply shaded to support much vegetation unless they were exceptionally wide.

Where sheep have access to the "scales" a turf tends to be formed, and becomes part of the adjacent pasture. The following species occur:

Festuca ovina	d		
Achillea millefolium	o	Lotus corniculatus	la
Bellis perennis	o–f	Polygala vulgaris	o–f
Cerastium vulgatum	o–f, la	Potentilla erecta	f–la
Conopodium majus	o–f	Ranunculus acris	r
Dactylis glomerata	f–ld	R. bulbosus	r
Holcus lanatus	f–ld	Taraxacum officinale	r
Lathyrus pratensis	o		

In another place (Scar Close) about a mile from Colt Park, and lying below a moss or bog (Scar Close Moss), acid water drains down on to the pavement and acid peat is developed on the surface of the "scales" as the result of colonisation, perhaps of standing pools in wet weather, by *Sphagnum* and *Eriophorum*. On this peat are growing *Calluna, Erica cinerea, Vaccinium myrtillus, Pteridium*, and *Potentilla erecta*. This area is also colonised by trees and shrubs in the following order of abundance:

Fraxinus excelsior
Acer pseudo-platanus*
Crataegus monogyna
Corylus avellana
Prunus padus

Prunus spinosa
Sambucus nigra
Lonicera periclymenum
Rubus idaeus
Sorbus aucuparia

* From the shelter belts round neighbouring farms.

An upland ashwood on basic igneous rock. Price Evans has recently described an upland ashwood on the "Upper Basic" group of Ordovician volcanic ashes and calciferous lavas, with dolerite intrusions, at Benglog, a farm about six miles north-east of Dolgelley in Merionethshire. His observations are not yet published and he has been kind enough to allow me to use them. This wood is a little more than 13 acres (5·35 hectares) in extent and lies on the southern slopes of Craig-y-Benglog at an altitude of 900–1250 ft. (270–380 m.) with a southerly aspect (Pl. 75). It is developed partly on scree and partly on the rocky slope and appears to be of recent origin, probably having seeded from ashes in the oakwoods (now largely cleared) on the lower ground. It is said that within living memory (? 60–70 years) there were only a few small ashes on this hillside.

The woodland falls into two more or less distinct parts according to the nature of the ground, viz. a western part of moderately steep slope consisting of large block scree with occasional bedrock exposed, and an eastern part of steep slope whose upper portion consists largely of precipitous bedrock and the lower of scree in all stages of formation (Pl. 75, phots. 171–2). In

PLATE 75

Phot. 169. Benglog farm with ashwood on the southern slope of Craig-y-Benglog (mist-covered) above. Near Dolgelly. *E. Price Evans.*

Phot. 170. Western end of Benglog ashwood. *E. Price Evans.*

Phot. 171. Eastern end of Benglog ashwood on almost precipitous rock above and steep scree below: bare scree to the right. Colonising hazel in foreground. *E. Price Evans.*

FRAXINETUM ON BASIC IGNEOUS ROCK

between is a much smaller area of rather open land which passes into hill pasture towards the top. The soil is everywhere rather scanty, occurring in "pockets" between the blocks of the scree, on ledges or in depressions in the bedrock, or as a thin layer on mature scree. Towards the bottom of the slope there is in most parts an increase in depth of soil.

Structure and composition. The average height of the dominant trees is about 50 ft. (15 m.) and the girth about 3 ft. (c. 0·9 m.). The distance between the trees averages some 15 ft. (4·5 m.) so that the wood is rather open, including some quite open spaces, and contains numerous shrubs.

Trees

Fraxinus excelsior d

Betula pubescens var. glabrata	r–o	Quercus sessiliflora × robur	o
		Tilia cordata	vr
Quercus sessiliflora	o	Ulmus glabra	r–o

Shrubs

Corylus avellana	d	Prunus spinosa	o–f
Crataegus monogyna	o–f	Rosa canina	r–o
Euonymus europaeus	vr	Sorbus aucuparia	o
Hedera helix (climbing over rocks)	o	Lonicera periclymenum	r–o

The ground vegetation falls into three more or less distinct communities: (1) on very thin soil over rock or incompletely overgrown scree (pH 4·9), giving a dry habitat, (2) on comparatively thin soil of medium dampness, and (3) in wet or marshy places.

(1) *On rocks and screes incompletely overgrown*

R	Arabis hirsuta	l	Linum catharticum	o
	Asplenium trichomanes	o	Lotus corniculatus	o–f
	Cardamine hirsuta	o	R Melica nutans	r
	C. impatiens	o–f	Origanum vulgare	la
R	Cotyledon umbilicus-veneris	o	RT Polypodium vulgare	o
	Deschampsia flexuosa	o	Potentilla erecta	o–f
	Epilobium montanum	o	Rumex acetosella	o
	Euphrasia sp.	l	Saxifraga hypnoides	l
	Festuca ovina	o	Sedum anglicum	o–f
	Galium aparine	o	S. forsterianum	l
R	G. saxatile	f	S. purpureum	l
	Geranium lucidum	l	Teucrium scorodonia	la
	G. robertianum	o–f	Thymus serpyllum	la
	Hieracium pilosella	o	Veronica officinalis	l
	H. sylvaticum	o		
	Helianthemum nummularium f. tomentosum	r–o	R Neckera crispa	o
			Rhacomitrium lanuginosum	ld
	Jasione montana	l	Trichostomum tortuosum	o
R	Lathyrus montanus	o	Peltigera sp.	o

R = on rocks. T = on trees.

Fraxinetum calcicolum

(2) *Soil of medium dampness* (pH *up to* 5·6)

Agrostis tenuis	la	Hypericum montanum	o
Ajuga reptans	o	Hypochaeris radicata	l
Alchemilla vulgaris	l	Lapsana communis	o–f
Anthoxanthum odoratum	la	Melampyrum pratense	l
Arrhenatherum elatius	f	Melandrium dioicum	o
Asperula odorata	la	Mercurialis perennis	la–ld
Athyrium filix-femina	o	Nepeta hederacea	la
Bellis perennis	o	Oxalis acetosella	f
Blechnum spicant	o	Pimpinella saxifraga	o
Brachypodium sylvaticum	o–f	Poa trivialis	f
Campanula rotundifolia	l	Primula vulgaris	o–f
Cardamine impatiens	o–f	Prunella vulgaris	f
Carex binervis	o	Pteridium aquilinum	la
Caucalis anthriscus	o	Rumex acetosa	o
Centaurea nigra	o	Rubus fruticosus	o
Cerastium vulgatum	o	R. idaeus	o
Circaea lutetiana	la	Sanicula europaea	la
Conopodium majus	o–f	Scilla non-scripta	l
Cynosurus cristatus	o–f	Solidago virgaurea	o
Dactylis glomerata	f	Stachys arvensis	o
Digitalis purpurea	r–o	Stellaria holostea	o–f
Dryopteris filix-mas	o–f	S. media	o
D. montana	f	Succisa pratensis	o
Epilobium montanum	o–f	Urtica dioica	la
Fragaria vesca	o–f	Veronica chamaedrys	o
Geum urbanum	o–f	Vicia sepium	o
Holcus lanatus	o	Viola riviniana	f

In these long lists it will be observed that besides several "exacting" and many indifferent species there are a few attached to acid soils, doubtless to be explained by the local accumulation of humus in this wet climate. The *p*H values show marked though not extreme acidity, from 4·7 to 5·6.

(3) *Wet or marshy places*

Cardamine pratensis	l	Juncus conglomeratus	l
Cirsium palustre	o	Lotus uliginosus	o
Chrysosplenium oppositifolium	l	Lysimachia nemorum	o–f
Crepis paludosa	l	Mentha aquatica	o
Deschampsia caespitosa	o	Myosotis scorpioides	l
Ficaria verna	l	Ranunculus flammula	l
Fraxinus excelsior (seedlings)	o	R. repens	o–f
Galium palustre	l	Trollius europaeus	l
Juncus articulatus	l	Valeriana dioica	f

Calluna and *Vaccinium myrtillus* are absent from the general ground flora though they are present with *Erica cinerea* in small quantities on the rocks in the upper part of the wood. The varied nature of the ground flora is attributable to the operation of several important factors, viz. the presence of calciferous rocks, together with pockets of acid soil developed in the humid climate, the open nature of the woodland and the great diversity of the screes. In the upper part of the wood the soil has not collected in sufficient quantity to support healthy full-grown trees. Stag-headed, dead, and uprooted trees are rather frequent. This, together with the absence of seedling ashes of any size, seems to point to degeneration and possible reversion to hazel scrub. The situation of the woodland at the extreme upper limit of trees, where the winter conditions are often very severe (as in 1936–7 when the weight of snow caused many of the branches to be torn off the trunks) and the presence of goats (about twenty in number) which frequent the wood, devour the seedlings and strip the bark off the branches, are additional destructive factors. The hazel, which is the pioneer shrub, is abundant in most parts, but from the lower portion much has been cut to provide wood for fencing. At the bottom of the wood, where the slope is gentle and the soil deeper than in the upper parts, oaks become rather frequent.

Succession. By comparison of areas in which scree is overgrown to different extents the following stages in succession may be inferred:

(1) Bare scree (often in unstable equilibrium).

(2) Moss community in which *Rhacomitrium lanuginosum* is the pioneer and dominant species.

(3) Invasion of flowering plants: the following are pioneers on the moss or on humus formed by decayed moss and leaves between the stones.

Cardamine hirsuta	Geranium robertianum
C. impatiens	Saxifraga hypnoides
Epilobium montanum	Sedum anglicum
Galium aparine	S. forsterianum
Geranium lucidum	S. purpureum

eventually followed by such species as *Nepeta hederacea, Origanum vulgare, Rubus idaeus, Teucrium scorodonia, Urtica dioica*.

(4) Invasion of hazel and ash (seedlings are occasionally found on fine dry scree), and formation of the general field layer.

In *open* places on the scree the moss community (2), in which *Rhacomitrium* is completely dominant, is invaded by such plants as *Euphrasia brevipila* (?), *Galium sylvestre, Hieracium pilosella, Jasione montana, Linum catharticum, Lotus corniculatus, Potentilla erecta, Thymus serpyllum* and *Veronica officinalis*, a community eventually passing into a Festucetum ovinae, the climax on well drained and pastured calciferous soils at this elevation.

Bryophytes. On medium fine scree in the eastern portion of the wood the following occurred:

Antitrichia curtipendula	Hypnum cupressiforme
Dicranum scoparium	Rhacomitrium lanuginosum
Hylocomium splendens	
H. triquetrum	Frullania tamarisci

Mosses are abundant in the ground layer of the wood, forming a carpet on the boulders of the block scree and on the bedrock, in correspondence with the high rainfall. The most abundant and conspicuous are

Dicranum scoparium
Eurhynchium myosuroides
Hylocomium loreum
H. splendens
H. triquetrum

Polytrichum juniperinum
Thuidium tamariscinum
*Frullania tamarisci
*Madotheca platyphylla
Plagiochila spinulosa

and more locally

*Hypnum molluscum
Neckera crispa
Polytrichum formosum

*Trichostomum tortuosum
Metzgeria furcata

* Calcicolous or mainly calcicolous species.

The lower parts of the trees are covered with a selection of the same bryophytes.

Lichens such as *Cladonia* spp., *Evernia prunastri* and *Usnea florida* (towards the upper parts) are also abundant on the bark of the trees.

"Retrogressive succession" on limestone. In *Types of British Vegetation* (1911) and in Moss's *Vegetation of the Peak District* (1913) the existing scrub and grassland communities on limestone soils, as well as the corresponding communities of siliceous soils, are treated as if they were almost invariably the products of degeneration or "retrogression" of the corresponding woodlands. The failure of the woods to regenerate from one cause or another—excessive felling leading to exposure of the soil and diminution of seed parents, or the occurrence of so thick an undergrowth that the seeds of the trees fail to reach the ground or the seedlings are smothered before they can establish themselves—does no doubt lead to the disappearance of the woods as such, and to their replacement by scrub formed by the increase of the most vigorous members of the shrub layer. A continuance of the operation of similar causes, especially of grazing, may result in the disappearance of almost all woody plants from the area, which will ultimately be occupied by grassland. There can be no question that these are the general causes of that sweeping change in the vegetation which has given the vast majority of the lower hills of Wales, northern England and southern Scotland their grass-clad aspect.

Moss of course did not fail to recognise the existence of "progressive" succession, in which communities of woody plants succeed herbaceous communities. On the contrary he describes several examples in his "Geographical Distribution of Vegetation in Somerset" (1907); and indeed he was a pioneer in introducing the conception into this country, and in insisting on its general validity. But in his later work he was so much impressed with the widespread occurrence of "retrogressive succession" that he sometimes failed to visualise vegetation in terms of normal succession as a primary phenomenon, whereas it is essential to take this point of

PLATE 76

Phot. 172. Ashwood in a south-east Devon valley on calcareous Upper Greensand. Watercombe Wood, Branscombe. April. R. J. L.

Phot. 173. Interior of Seller's Wood, Branscombe. *Fraxinus* dominant, with *Hedera*. *Mercurialis perennis* and *Scilla non-scripta* dominant in field layer. *Phyllitis scolopendrium* and *Arum maculatum* in foreground. On calcareous Upper Greensand. April. R. J. L.

FRAXINETUM CALCICOLUM IN SOUTH-EAST DEVON

PLATE 77

Phot. 174. Knowle Coppice, Branscombe, on calcareous Upper Greensand. *Fraxinus* dominant, with *Hedera*. *Mercurialis perennis* mostly dominant in the field layer. April. *R. J. L.*

Phot. 175. Seral ashwood on escarpment of Inferior Oolite, Cotswold Hills. *Fraxinus* dominant, *Corylus*, *Euonymus europaeus*, Witcombe Wood, Glos. June. *C. G. P. Laidlaw.*

FRAXINETUM CALCICOLUM

view if we are to obtain a comprehensive understanding of the dynamic relations of any given set of related communities. It is noteworthy that there is no description of normal succession in Moss's book on the Peak District, except in the Introduction in purely general terms and a single case quoted from another author.

There can be no question of the widespread occurrence of this so-called "retrogressive succession" in this country; but, as Clements pointed out in his *Plant Succession* (1916), it cannot be put on a par with normal succession, but must be interpreted in terms of the latter. The general aspects of this topic have been sufficiently dealt with in Chapter x, but here it may be pointed out that the hawthorn scrub and limestone grassland described and figured in *Vegetation of the Peak District* are all regarded by Moss as examples of "retrogression". Many of them doubtless illustrate the "retrogression" of vegetation brought about by grazing, but careful and detailed investigation of the processes—both "progressive" and destructive—at work in each individual case is necessary before the particular phenomena involved can be assigned to their proper place in a general scheme, which must itself be based on normal progressive succession because that is the fundamental phenomenon underlying all vegetational processes. Watt's work on the woody vegetation of the Chalk and its overlying loams, described in Chapters xviii and xix, and Godwin's on that of the East Anglian fens described in Chapter xxxiii, are examples of the way in which such investigation enables us to understand the entire complex of processes involved.

Ashwood on other calcareous soils

Besides those on Mountain Limestone, calcicolous ashwoods occur on various limestones and other highly calcareous soils throughout the country, though they are never extensive. Plates 76 and 77 illustrate two of these. The Upper Greensand of south-east Devon is exposed on the steep sides of the little valleys which intersect the plateau near the Channel coast. Where it immediately underlies the thin layer of Chalk—here at its extreme south-western limit—the Upper Greensand is highly calcareous, and (together with the overlying chalk) bears nearly pure scrubby calcicolous ashwood. Away from the Chalk the Greensand is non-calcareous and highly acid, bearing oakwood of the heathy type. The ashwoods on the calcareous Greensand, shown in Phots. 172–4, contain a good woodland flora with a number of calcicolous species. *Mercurialis perennis* is the general dominant of the field layer, with *Allium ursinum* in the damper places, *Phyllitis scolopendrium* and a number of other forms. Where the soil is locally deeper, oaks (*Q. robur*) occur in these woods, which then have a general resemblance to the ash-oak woods on the Silurian limestones at Malvern described in Chapter xv (p. 319).

Phot. 175 illustrates a seral ashwood on the Inferior Oolite escarpment of the Cotswold Hills in Gloucestershire. Here the climax beech covers most of the escarpment, and ash, associated with the regular calcicolous shrubs, is preclimax to the beech.

Ash-oak wood

(1) *On deep calcareous soils*. It has already been mentioned (p. 428) that on the deeper soils derived from limestone, such as are met with on the lower gentle slopes and in the bottoms of the valleys, the common or pedunculate oak (*Quercus robur*), which cannot establish itself on the shallow soils of the steeper and more exposed slopes, is a frequent constituent of ashwood. On sufficiently deep soil, where oak can compete with ash on anything like equal terms, the oak may conceivably overshadow and tend to kill out the ash, or the two may share dominance of the community. This type of wood, in which a mixture of the plants characteristic of ashwood and of oakwood occur, is closely comparable with the "calcareous facies" of the damp oakwood occurring on the Chalky Boulder Clay and described in Chapter XIII, pp. 272, 288, and to some extent with the seral ash-oak wood leading to beechwood on the shallower and often somewhat calcareous loams of the chalk plateau of the South Downs (Chapter XIX, p. 396). When the beech does not come into the sere it is doubtful if a community in which oak became completely dominant would follow ash-oak wood in the succession, in spite of the greater competitive power of the oak. No cases have been examined which throw light on the point. The great majority of mature woods of this type, in the midlands and south, as has been indicated in previous chapters, are in fact (ash-) oak-hazel woods, the ash being coppiced along with the hazel and other shrubs and the oaks left as standard trees; but in some ash and oak share dominance in various proportions.

(2) *In wet climates*. The other type of woodland in which ash may be co-dominant or almost co-dominant with oak is associated not with calcareous soil but with the wet climates of high-lying regions in the extreme west and north-west, where *Quercus sessiliflora* is the regional dominant oak. An example of this type of wood has been mentioned and a list given in Chapter XV (p. 317). Specifically calcicolous plants are absent and this kind of ash-oak wood does not of course belong with the types described in this chapter.

Ash-birch woods. On some of the north-western limestones *Betula pubescens* enters into the composition of the ashwood at high elevations up to 1250 ft. (380 m.) with a high rainfall (50 in. or 1250 mm.). These ash-birch woods occur for example on the precipitous slopes of the narrow limestone valleys and on the face of the Pennine escarpment looking westwards towards the Lake District mountains (Lewis, 1904). The climate is extreme, the woods being exposed to violent winds and strong radiation as well as to high precipitation. Besides the ash, wych elm and birch, the sessile oak also occurs in these woods. Of shrubs hazel is extremely abundant, hawthorn, holly and sallow are frequent; but the other "calcicolous" shrubs and trees are for the most part absent. The herbaceous vegetation is of the calcicolous ashwood type, but impoverished.

At the highest levels of these woods ash disappears and *Betula pubescens* is practically pure, with *Sorbus aucuparia* associated. Here we have convergence with the highest level woods on siliceous rocks, described in Chapter XXII. The high altitude climate becomes the decisive factor and determines practically the same type of woody vegetation—birchwood—on calcareous and siliceous rocks alike.

REFERENCES

CLEMENTS, F. E. *Plant Succession*. Carnegie Institution of Washington, 1916.

LEWIS, F. J. Geographical distribution of vegetation in the basins of the rivers Eden, Tees, Wear and Tyne. *Geogr. J.* 1904.

Moss, C. E. Geographical distribution of vegetation in Somerset: Bath and Bridgewater district. *Roy. Geogr. Soc.* 1907.

Moss, C. E. *Vegetation of the Peak District*. Cambridge, 1913.

SMITH, W. G. and RANKIN, W. M. Geographical distribution of vegetation in Yorkshire. Part II. Harrogate and Skipton district. *Geogr. J.* 22. 1903.

Chapter XXII

PINE AND BIRCH WOODS

History and present distribution. Birch and pine woods are best considered together because the requirements of the dominants are closely similar, and they are generally found together on the same soils. The two species of birch, *Betula pendula* (*B. alba* or *B. verrucosa*) and *B. pubescens*,[1] and the common European pine, *Pinus sylvestris* (including the apparently endemic variety *scotica* E. and H.), known in Great Britain as the "Scotch fir" or "Scots pine", are hardy unexacting trees, some of the birches (*B. pubescens* and especially the northern forms mentioned in the footnote) even more tolerant than the pine. As climatic climax types indeed, birchwood and pinewood are properly considered as belonging to one climatic formation, the Eurasian formation of Northern Coniferous Forest, the formation-type developed in Europe, as in eastern North America, to the north of the zone of Deciduous Summer Forest. In Scandinavia the Northern Coniferous Forest is mainly dominated by spruce, *Picea abies* (L.) Karst. (=*P. excelsa* Link.). Northwards, in Swedish and Norwegian Lapland, the spruce gradually dies out and is replaced by pine and birch (*Betula tortuosa*), and eventually, towards the north, by pure birch, which here forms the polar forest limit. In Britain the spruce never returned after the Ice Age, and the chief Boreal tree invaders of the uplands were birch and pine. With improving climate these trees were replaced by the more exacting mixed deciduous forest, mainly dominated by oak, as the climax forest of the south penetrating into the northern valleys, while pine and birch remained as chief dominants of the uplands in the Scottish Highlands. Birch gives way to pine in direct competition on soils favourable to both, but maintains itself better (though making very poor growth) under generally adverse conditions of soil and climate.

There is no *certainly* native pine now left in England (cf. p. 254), but Leland in the sixteenth century (Taylor in Darby, 1936, p. 347) mentions having seen "Fyrres" or "Firres" near Gateshead in Co. Durham. It is probable that this refers to pine.

The Scottish pine forests were once very much more extensive than they are now, but judging from Erdtman's pollen diagrams (1924) there must generally have been more birch than pine wood in early post-glacial times,

---

[1] The birches of the north of England and Scotland have hitherto been considered as belonging to *B. pubescens* Ehrh., though recognised as including forms differing somewhat widely from the pubescent birches of the south. Dr E. F. Warburg, however, informs me of his conviction that some of the northern forms must be put into a distinct species.

the percentage of pine pollen in the peat of the Boreal being nearly always exceeded, often very considerably, by that of the birch. The Sub-Boreal forests were, however, typically composed of pine (cf. Figs. 45–6, p. 167). To-day there are a few tracts of native pine forest left in the central and eastern Highlands, such as the Black Wood of Rannoch in Perthshire, Ballochbuie Forest at the foot of Lochnagar in Aberdeenshire, and parts of Rothiemurchus Forest in Strathspey (Inverness-shire). These are characteristic examples of old native pine forest. Smaller patches, mostly mixed with birch, still exist in the north; while birch forest, though very much has disappeared, is still not infrequent as strips of woodland along the sides of the upland valleys, extending northward of the pinewoods almost to the northern coasts of Caithness and Sutherland. The groups of pine still left here and there in the northern Highlands, i.e. north-west of the Great Glen (Glenmore) which runs across Scotland from sea to sea, rarely form anything like closed pinewood. Very often the pines are seen towards the upper limits of the hillside woods, with pure birch below, not because that is the natural zonation but because the more accessible pine at the lower levels has been cut out. This is seen in most of the glens running eastwards into Glenmore, such as Glen Loy, Glen Maillie, Glen Garry, Glen Moriston and Glen Urquhart, and farther to the north in Glen Affric and Glen Strathfarrar, but these remains of old native pinewoods are being rapidly destroyed by the Forestry Commission. Again to the north-west there is still some native pine remaining, e.g. on the south-east side of Glen Carron, remnants of the old Forest of Achnashellach, most of which is being converted by the Commission into plantations of exotic conifers. Farther north in Strathoykell, and at Rhidorroch near Ullapool on the west coast, the same feature may be observed. Regeneration is very poor indeed or quite lacking, apparently owing to the almost universal sheep and rabbit grazing. The best existing natural pinewood left in the north-west Highlands is undoubtedly that on the south-western shore of Loch Maree, some of which is still regenerating well, and here one can still study the natural behaviour of native pine.

Relation to oakwood. In the glens running into Glenmore, the remains of old oakwoods described in Chapter xv (pp. 344–7) can still be seen below the pine- and birchwoods. Where the soil is fertile these are often associated with ash and have quite a rich flora. On wet ground near the river there is much alder which sometimes ascends into the hillside woods, especially in the wetter climates. In the valleys farther north the oak is represented only by a few trees in copses of hazel, alder and rowan, but apparently native oak is met with right up into Sutherlandshire,[1] though scarcely to the north coast. In ascending the hills this valley oakwood evidently once passed over everywhere to pine- and birchwood, which was thus zoned above the oakwood both latitudinally and altitudinally.

[1] Pl. 17, phot. 37.

Highland pinewoods

Ballochbuie Forest. Ballochbuie Forest, belonging to H.M. the King,[1] (Pl. 78, phots. 176, 178–9), is situated in a sloping basin on the southern side of the Dee valley between Balmoral and Braemar, with an elevation of 1000 ft. rising to about 1750 ft. (300–530 m.) at the north-western foot of the mountains Lochnagar (1150 m.) and Cairn Taggart. The lowest slopes near the river have been largely replanted, but the greater part of the forest, about 4 square miles (say 1000 ha.) in extent, consists of native pine mostly in open canopy interspersed with areas of open heath. The soil is morainic and gravelly but has not been closely examined.

Near the falls of Garbh Allt at 1200 ft. (c. 365 m.) there is continuous forest of pines in open canopy, about 60 ft. (18 m.) high[2] and commonly 18 in. (45 cm.) in diameter at breast height, though some are much bigger, up to 32 or 34 in. (80–85 cm.) in diameter.

This is not an unusual size for the larger mature trees of native Scots pine. In various other parts of the Highlands a number of individual big pines measuring 32–38 in. (80–95 cm.) in diameter were seen: these were commonly 50–60 ft. (15–18 m.) high and of widely spreading habit (grown in the open). Occasionally still greater dimensions are attained. The trunk of the famous "big pine" in Glen Garry measures 42 in. (107 cm.) in one direction and 48 in. (122 cm.) in the other. An epiphytic birch 11 ft. (3·3 m.) high grows on its trunk. And there is a record of a pine at Leckmelm near Ullapool which exceeded 110 cm. in diameter. The age of these very big mature pines (and of many smaller ones) is commonly 150–180 years.

Of associated trees only birch (*Betula pubescens*)[3] and rowan (*Sorbus aucuparia*) were occasional, aspen (*Populus tremula*) local, with alder (*Alnus glutinosa*) by the burn side.

Field layer. The field layer of the more open pinewood is largely dominated by *Calluna*, of luxuriant, bushy and irregular growth, 2–3 ft. high. Associated with this is abundant *Vaccinium myrtillus* and *V. vitis idaea*, concentrated especially round the bases of trees and covering the numerous hummocks formed by rocks and tree stumps. The rather sparse crowns of the trees allow a good deal of light to reach the field layer: where the trees are close and the shade is deeper *Calluna* is sparse and weak or absent altogether, its place being taken by the two Vaccinia. Young pine is conspicuously absent from the forest, and the only seedlings seen were in open soil on the edge of a roadside quarry. There can be no doubt that the dense tall *Calluna* and *Vaccinium* prevent general regeneration of the pine, but the ultimate cause is probably grazing, and is generally attributed to red deer from the "deer forest" on the hills above.

[1] I have to thank the courtesy of Sir Archibald Hardinge, the King's Private Secretary, in giving me permission to visit Ballochbuie, and Captain Ross, the King's Factor at Balmoral, for very kindly showing me the Forest.

[2] In the protection of a steep-sided glen they reach a height of about 80 ft. (24 m.).

[3] See footnote on p. 444.

PLATE 78

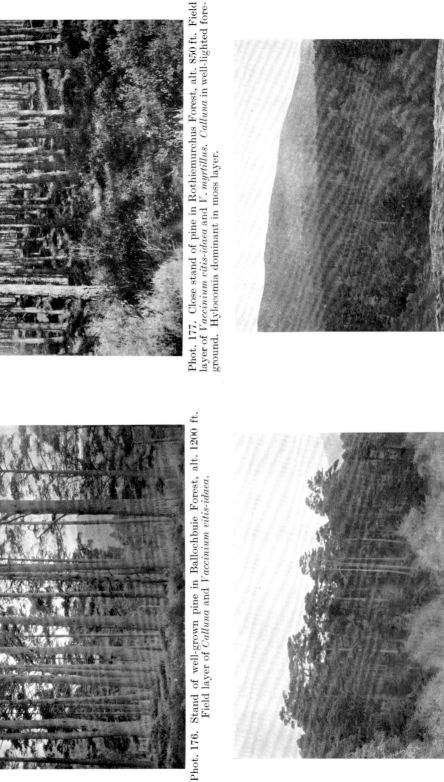

Phot. 176. Stand of well-grown pine in Ballochbuie Forest, alt. 1200 ft. Field layer of *Calluna* and *Vaccinium vitis-idaea*.

Phot. 177. Close stand of pine in Rothiemurchus Forest, alt. 850 ft. Field layer of *Vaccinium vitis-idaea* and *V. myrtillus*. *Calluna* in well-lighted foreground. Hylocomia dominant in moss layer.

Phot. 178. Well-grown pines in semi-open canopy about 70 ft. high. Birches in valley in front. Ballochbuie Forest.

Phot. 179. Upper part of Ballochbuie Forest, alt. 1250 ft. Forest limit (1500 ft.) on steep slope behind.

HIGHLAND PINE FOREST

PLATE 79

Phot. 181. Pine heath with luxuriant *Calluna*, *Juniperus* (both spreading and conical) up to 10 ft.—3 m., and fairly free pine regeneration.

Phot. 183. Larig Ghru. Upper limit of pines at about 1500 ft. Birches especially in the valley. *Calluna* and *Juniperus*. *A. S. Watt.*

Phot. 180. General view of pine forest with juniper and heather in foreground. Pines scattered on upper slopes to the right behind.

Phot. 182. North-west slope of Creag Fhiaclach at 1650 ft. with scattered pines, some dead (south-west corner of Fig. 86). *Vaccinium myrtillus* and *V. vitis-idaea*, *Calluna*, *Empetrum*. Lochan Eilein below.

The herbaceous flora is very sparse, three grasses being the most frequent species: *Agrostis canina*, *Deschampsia flexuosa* and *Holcus mollis*. These are local and occur especially along the sides of paths. Of other species the following were noted, but none, except perhaps *Blechnum*, was more than rare or occasional:

Blechnum spicant	Melampyrum pratense
*Carex distans	Molinia caerulea
Deschampsia caespitosa	*Nardus stricta
Equisetum sylvaticum	Oxalis acetosella
*Euphrasia sp.	Dryopteris linnaeana
Galium saxatile	Polystichum aculeatum
Hieracium sp.	Potentilla erecta
Juncus squarrosus	*Rumex acetosella
Listera cordata	Solidago virgaurea
*Luzula multiflora var. congesta	Veronica officinalis
L. pilosa	*Viola sp. (? riviniana)
L. sylvatica	

* Along pathsides.

Bryophytes. The only dominant plants other than the pines and heaths were mosses, of which the chief were:

Campylopus flexuosus	o	Mnium hornum	o
Dicranum majus	o–f	Plagiothecium undulatum	f
D. scoparium	f	Polytrichum commune	o
Eurhynchium myosuroides		Sphagnum cymbifolium	a–va
(bases of trees)	a	S. girgensohnii	va
Hylocomium loreum	a	S. obesum (especially in	
H. splendens	a	wet spots)	r
H. squarrosum	r	S. plumulosum	va
H. triquetrum	a	S. squarrosum (in wetter	
Hypnum crista-castrensis	f	spots)	o
H. schreberi	f		

The open heath, where only isolated trees are left, was dominated by *Calluna* and *Vaccinium vitis idaea* with *Digitalis*, *Empetrum* and *Erica tetralix*, in addition to several of the species recorded for the pinewood.

The leading characteristics of Ballochbuie Forest, namely, the dominance of Ericaceae, Sphagna and Hylocomia, and the absence of regeneration, are repeated in most of the existing Highland native pinewoods, as will be seen in the sequel.

Rothiemurchus Forest. Rothiemurchus Forest (Fig. 86) occupies a great basin of glacial sands and gravels partly enclosed by the Cairngorm Mountains, with gently undulating surface and an average elevation of 1000 ft. (c. 300 m.) above sea-level. Morainic terraces occupy the adjoining lower slopes of the mountains, and the native pine extends up to these and even above them (though not in close stands), reaching in places an elevation of more than 2000 ft. (over 600 m.), about the highest altitude of any sort of existing woodland in the British Isles. Some of the area, besides the Forest of Glenmore which adjoins it to the north, and which before the war consisted entirely of native pine, is now being replanted by the Forestry Commission, but there is still much native pine (Pl. 78, phot. 177 and Pl. 79).

Much of the "forest" is really "pine heath", dominated by *Calluna vulgaris* with *Vaccinium myrtillus* and *V. vitis idaea* abundant and *Arctostaphylos uva-ursi* and *Empetrum nigrum* towards the upper limits. In the more sheltered lower parts vigorous *Juniperus communis* is locally abund-

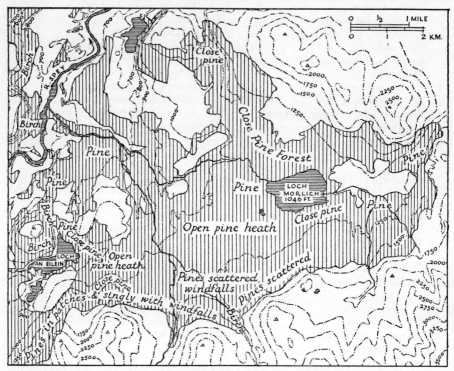

FIG. 86. MAP OF ROTHIEMURCHUS FOREST

with the adjoining Glenmore Forest (round Loch Morlich) showing the altitudinal limits of forest as marked by the Ordnance Survey in 1866–70. Nearly the whole (vertical ruling) is pine forest or "pine heath": the white patches within the forest are clearings. The northern stretch marked "Close pine forest" was clear felled during the war and is now being replanted by the Forestry Commission. The Cairngorm Mountains lie to the east and southeast. Contours in feet.

ant, sometimes low and spreading, sometimes conical in habit, the conical bushes reaching a height of 10 or 12 ft. (3–3·5 m.) (Pl. 79, phot. 181). The pines are thickly scattered and locally they grow in close canopy. Regeneration is quite free in some places, but over much of the area it is prevented by the thickness and height of the heather. The distribution and effect of rabbits has not been studied, but they are apparently very local in occurrence. Deer, especially roe deer, are said to be very destructive, rooting up seedlings and young saplings with their antlers.

The soil is often podsolised, showing a heavy surface layer of raw humus

PLATE 80

Phot. 185. Relict pine at Rosehall, Strath Oykell, Sutherlandshire. Birchwood across the river, blanket bog beyond.

Phot. 187. Close young pinewood in Glen Cassely, one of the stations farthest north. *A. S. Watt.*

Phot. 184. Close, finely-grown pine on the southern shore of Loch Morlich. Cairngorm Mountains behind. *A. S. Watt.*

Phot. 186. Old relict pine at Rosehall, in sheep pasture with Pteridietum.

HIGHLAND PINE FOREST

PLATE 81

Phot. 189. Old dome-shaped pine (25 ft.) and younger columnar pine (35 ft.) side by side. Ballochbuie Forest, Aberdeenshire.

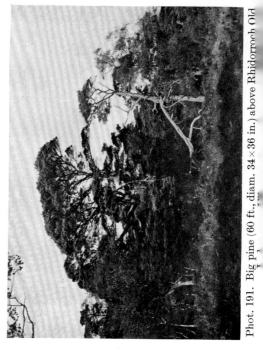

Phot. 191. Big pine (60 ft., diam. 34 × 36 in.) above Rhidorroch Old

Phot. 188. Old lofty pine, with rounded crown, in centre, young pyramidal trees to left. *Calluna* and *Molinia*. Black Wood of Rannoch, Perthshire.

Phot. 190. Old flat-topped pine and pyramidal younger tree side by side. Glen Loy, Inverness-shire.

(mor) consolidating to black peat several inches thick, underlain by more or less bleached grey or yellow sand stained with humus, and with a hard yellow-brown illuvial layer (pan) in places at a depth of about 12 in. (30 cm.).

A well-grown pinewood in close canopy (Pl. 78, phot. 177) east of Lochan Eilein at an elevation of 850 ft. (258 m.) had the trees 60–80 ft. (18–24 m.) high and 7–12 in. (18–30 cm.) in diameter, an old stump 12 in. thick showing 75–80 annual rings. This wood may or may not have been planted, but in any case it well represents the characters of a middle-aged pinewood in the region. The field layer was very uniform, with *Vaccinium vitis idaea* generally dominant, *V. myrtillus* abundant and *Calluna* locally frequent to locally abundant and flowering in the lighter openings. *Deschampsia flexuosa* was abundant but not in flower. The moss layer was luxuriant in the intervals of the field layer, consisting of *Hylocomium splendens* and *H. triquetrum* locally dominant, *H. loreum* rare to occasional, *Hypnum schreberi* frequent, with *Dicranum scoparium* and *Cladonia* spp. confined to old tree stumps. The two Vaccinia, *Calluna*, and the two Hylocomia probably formed more than 99 per cent of the vegetation below the trees. Under younger, closer set pine the dwarf shrubs were absent, the moss layer forming the whole of the ground vegetation. In some lighter patches *Pteridium* had invaded; and there was occasional juniper by the sides of paths.

Glenmore Forest. On the adjoining Glenmore estate, south of Loch Morlich at an elevation of 1050 ft. (320 m.), there is a little good native pine left (Pl. 80, phot. 184), including very fine timber trees, 70–90 ft. (21–27 m.) high, with thin bark, small groups of which are in close canopy. The diameters of eleven of these were measured as they came and six of them were 18 in. (46 cm.) or more in diameter, the largest being 25 in. (63 cm.) at breast height. When the trees are less than about 20 ft. (6 m.) apart the canopy is close and the lower branches are shed, giving a clean trunk for a height of 30 or 40 ft. (9–12 m.).

The field layer was as follows:

Vaccinium myrtillus	ld	Erica tetralix	o–lf
V. vitis idaea	ld on old stumps	Blechnum spicant	o–la
Calluna vulgaris	f	Luzula pilosa	r
Deschampsia flexuosa (not flowering)	a	Molinia caerulea	o

Calluna was dominant in the open.

The moss layer was composed of:

Hylocomium splendens	a	Hypnum sp.	o
H. triquetrum	a	Dicranum scoparium	on stumps
Hypnum schreberi	o		

High level pine and birch. At a higher elevation in Rothiemurchus Forest, on the steep rocky north-west slope of Craig Fhiaclach, where the pines reach an altitude of more than 2000 ft. (*c.* 600 m.), lists were made from small patches of close birch and close pinewood at 1600 ft. (487 m.). There were many dead pines, but some were quite well grown and fairly symmetrical. (Pl. 79, phot. 182 was taken from just above this point, at 1650 ft. = 503 m.)

Closed birch		Closed pine	
Betula pubescens[1] (not more than 12 ft. high)	d	Pinus sylvestris (30–40 ft. high)	d
Vaccinium vitis idaea	d	Vaccinium vitis idaea	d
V. myrtillus	f	V. myrtillus	f–a
Calluna vulgaris	f	Calluna vulgaris	la–ld
Polypodium vulgare	f on rocks	Empetrum nigrum	f–a
Luzula pilosa	o	Goodyera repens	r
Hylocomium splendens	va	Hylocomium splendens	a
H. loreum	o	H. loreum	f
Dicranum scoparium	f–la	Hypnum schreberi	f
		Dicranum scoparium	f
		Sphagnum spp.	f

Usnea and *Parmelia* sp. were abundant on the birch trunks.

Though the shade of the pines was continuous in the small patch listed it was considerably lighter than that of the dwarf birch thicket, because the habit of the pines was spreading and the foliage sparse, and this accounts for the abundance of *Calluna* and *Empetrum*. Otherwise there is little difference between the two communities, except for the presence of *Goodyera* under the pine. *Rubus idaeus*, small *Sorbus aucuparia*, and *Deschampsia flexuosa* were generally dispersed on this slope. Pl. 79, phot. 183 shows open stunted pine towards its upper limit in the pass of Larig Ghru.

Loch Maree. The pine forest on the south-western shore of Loch Maree towards the south-eastern end of the loch, and extending up a steep valley and adjoining slopes to the 1000 ft. (300 m.) contour, is one of the few native Highland pinewoods that is still freely regenerating. It is on very uneven morainic ground, and its composition as seen in a particular area was as follows:

Pinus sylvestris	d	Sorbus aucuparia (seedlings)	o
Betula pubescens[1]	o		
Calluna vulgaris	f–ld	Sphagnum spp.	va, ld
Vaccinium myrtillus	va–ld	Hylocomium splendens	va
V. vitis idaea (especially on and round stumps)	o	Dicranum majus	f
		Hypnum schreberi	f
Molinia caerulea (in wetter places)	f	Hylocomium loreum	o
		Hypnum crista castrensis	o
Pteridium aquilinum	o–la	Polytrichum commune	o
Blechnum spicant	lf	Rhacomitrium lanuginosum	o
Deschampsia flexuosa	o–lf		
Galium saxatile	r	Dicranum scoparium	o (on stumps)
Potentilla erecta	r		

[1] See footnote on p. 444.

PLATE 82

Phot. 194. Tall pine about 70 ft. (one with climbing ivy), and tall birch. Near the shore of Loch Maree.

Phot. 192. Free pine regeneration. Loch Maree, Ross-shire.

Phot. 193. Free pine regeneration. Loch Maree.

LOCH MAREE PINEWOODS

PLATE 83

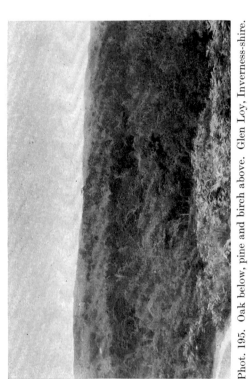

Phot. 195. Oak below, pine and birch above. Glen Loy, Inverness-shire.

Phot. 197. Birch below, relict pine above, owing to removal of lower pine. Rhidorroch Old Lodge, Glen Achall, Sutherlandshire.

Phot. 196. Birch below, pine above, probably due to removal of lower pine. Old pine by the river. Glen Strathfarrar.

Pine seedlings, some of them 4–5 ft. (1·5 m.) high, were occasional in open places. In another more open part of the wood regeneration was very free, with young pines of all ages (Pl. 82, phots. 192–3). Holly (*Ilex aquifolium*) is occasional to frequent in these woods, and ivy (*Hedera helix*) occurs. The occurrence of these two evergreens is no doubt due to the western climate.

A comparison of the lists given in the preceding pages shows a striking uniformity in general composition, the Loch Maree list, closely similar to those from Ballochbuie and Rothiemurchus, being very typical. The vegetation below the remaining pines in the relicts of old native pinewood seen in other places in the northern Highlands has been too much altered by grazing to give any trustworthy information about its original constitution.

Characteristic species of old native pinewoods. The following species are given by White (1898) as specially characteristic of, while some (*) are confined to, the old pine forests:

Corallorhiza trifida*	Pyrola media
Goodyera repens*	P. minor
Linnaea borealis	P. rotundifolia
Listera cordata	P. secunda
Moneses uniflora*	Trientalis europaea

With the exception of *Trientalis* and *Pyrola minor* these species are on the whole rare.

Scottish birchwoods

The widespread Highland birchwoods (Pls. 84–5) are composed of various forms of birch hitherto assigned to *Betula pubescens*, with very few associated trees and shrubs. Of these by far the commonest is the rowan, *Sorbus aucuparia* L., which occurs in most birchwoods. The birchwoods occupy the sides of valleys, but are now scantily developed. There is evidence that they were once much more extensive and that they have greatly decreased through the widespread sheep grazing (Pl. 87, phots. 198–9) and in quite recent times by the red deer browsing on the seedlings and saplings. The "balance of nature" is upset by the over-stocking of the so-called "deer forests", which are really open stretches of moorland and mountain side from which the deer descend into the woods for food and shelter, especially in winter (cf. Chapter vi, p. 143).

Caithness birchwoods. Crampton (1911, pp. 95–104) gives a good description of the birchwoods of Caithness in the extreme north-east of Scotland, which are developed on the most various soils along the sides of many of the valleys. Several of these woods consist of very old contorted trees, rarely as much as 20 ft. high, and very often attacked by *Exoascus turgidus*. Such trees are partly rotten, largely covered with lichens and often riddled by boring insects. They are easily blown down by the violent north-west winds, and whole woods succumb in this way, adequate regeneration being prevented by the ravages of the deer.

Trees and shrubs. The following is a list of trees and shrubs compiled from Crampton's account. The frequency letters are not in the original, but are inserted from the general indications given:

Betula pubescens[1]	d	Salix caprea	f
Sorbus aucuparia	a	S. aurita	o–f
Corylus avellana (forming pure woods)	o, ld	Prunus padus	o
Alnus glutinosa (forming local societies in the wettest places)	o, ld	Populus tremula (forming pure local societies)	la, ld
		Hedera helix	o
		Lonicera periclymenum	o

Of these species it is noteworthy that birch, rowan, bird cherry, and aspen are the deciduous trees which reach the highest altitudes in the coniferous forest zone of the Alps.

Crampton notes that while individual rowans occur as much as two miles into the moorland and far from any possible parent tree, seedling birch rarely occurs more than 100 yards from the parent trees, and many are usually found together. This is related to the fact that rowan seeds are dropped by birds, while birch fruits are carried in quantity by wind. Hence the dispersal of rowan is sporadic and long distance, while birchwood spreads as a whole, gregariously, and short distances at a time.

Field layer. The vegetation of the field layer is varied: in some woods it is dominated by grasses, in others by blaeberry (*Vaccinium myrtillus*), in others again by bracken. *Calluna* cannot exist under the shade of the birches when they are close set, but it is said to be an efficient nurse for the seedlings. The following lists are compiled from Crampton's account of the Caithness birchwoods:

Grasses

Agrostis spp.	f	Deschampsia flexuosa	a
Anthoxanthum odoratum	a	Festuca rubra	o
Brachypodium sylvaticum	o	Holcus mollis	a
Dactylis glomerata	o	Poa trivialis	o

Herbs

Ajuga pyramidalis	r	Polygala vulgaris	o
Conopodium majus	r	Potentilla erecta	a
Fragaria vesca	o	Primula vulgaris	o
Galium saxatile	a	Prunella vulgaris	o
G. verum	o	Scilla non-scripta	o
Lapsana communis	o	Solidago virgaurea	o–f
Listera cordata	f	Stellaria holostea	o–f
Luzula pilosa	f	Teucrium scorodonia	o–f
L. maxima	o	Trientalis europaea	f
Lysimachia nemorum	o–f	Veronica chamaedrys	o
Melandrium dioicum	o	V. officinalis	o
Oxalis acetosella	f	Viola riviniana	a

[1] See footnote on p. 444.

PLATE 84

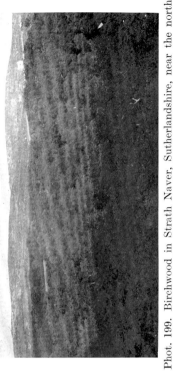

Phot. 198. Birchwood near Kinloch Rannoch, Perthshire: 900–1250 ft. (c. 270–380 m.) alt. Myricetum in bottom right-hand corner.

Phot. 199. Birchwood in Strath Naver, Sutherlandshire, near the north coast. 58° 30′ N.L. and less than 100 ft. alt. *Betula* sp., *Sorbus aucuparia. Corylus* and *Salix aurita* in foreground.

Phot. 200. Betuletum pendulae with juniper shrub layer. *Trientalis* and *Hylocomium* in field layer. South Kinrara, Speyside, Inverness-shire.

Phot. 201. Birchwood on the shore of Loch Shin, Sutherlandshire. *Pteridium, Oxalis, Agrostis tenuis*, etc. *A. S. Watt.*

HIGHLAND BIRCHWOODS

PLATE 85

Phot. 203. Heavily grazed birchwood (sheep and rabbits) degenerating. *Agrostis tenuis, Potentilla erecta, Hylocomium splendens.* On the right *Erica tetralix, Myrica gale,* and some birch seedlings. Above Loch Shin, Sutherlandshire. *A. S. Watt.*

Phot. 204. Late stage of birchwood degeneration due to sheep grazing. *Agrostis tenuis* abundant, 550 ft. alt. Near Loch Shin, Sutherlandshire.

Phot. 202. Birch-juniper wood at South Kinrara, Speyside, Inverness-shire. *Betula pendula, Juniperus communis.*

Ferns

Asplenium adiantum-nigrum	o	Dryopteris montana	o
Athyrium filix-femina	o	Polypodium vulgare	o
Blechnum spicant	f–a	Pteridium aquilinum	ld

Mosses

These surround the bases of the old trees and form mounds on the floor of the wood. The following are among the more conspicuous species:

Hylocomium splendens   Hypnum schreberi
H. loreum   Plagiothecium undulatum

Vegetation of wetter places. In wetter places in the woods the following species occur:

Ajuga reptans   Filipendula ulmaria
Caltha palustris   Juncus bulbosus (supinus)
Carex panicea   J. effusus
C. pulicaris   Molinia caerulea
C. stellulata   Ranunculus flammula
Chrysosplenium oppositifolium   R. repens
Cirsium palustre   Rumex acetosa
Equisetum arvense   Valeriana officinalis

Mosses

Aulacomnium palustre   Sphagnum spp.
Climacium dendroides   Thuidium tamariscinum
Polytrichum commune

It will be noted that nearly all the species may occur in oakwood of various types—in fact the flora represents a small selection from a wide range of species covering extremes of woodland development, from dry to wet and from extremely poor to quite fertile soil. Very few species are characteristically northern in distribution. One of these is *Ajuga pyramidalis* ("continental northern" according to Matthews, 1937), which is also a subalpine plant in Switzerland. *Trientalis europaea* and *Listera cordata* are "northern montane" (Matthews, 1937). Thus these north Scottish woodland communities are for the most part, though not entirely, negatively rather than positively characterised.

Birch-juniper wood at South Kinrara. Almost the same may be said of a birchwood much farther south, at South Kinrara (800 ft.), on sandy soil in Speyside, i.e. in the eastern Highlands close to the edge of Rothiemurchus Forest. This wood (Pl. 84, phot. 200; Pl. 85, phot. 202) showed the following composition:

Tree layer

Betula pendula, dominant in open canopy
B. pubescens, rare
Populus tremula, in actively regenerating societies
Sorbus aucuparia, rare
Prunus padus, very rare

Shrub layer

Juniperus communis, sub-dominant to the birch, forming a pure almost continuous shrub layer

Field layer (short turf)

Anemone nemorosa	f	Oxalis acetosella	l
Calluna vulgaris	f–a	Potentilla erecta	lf
Deschampsia flexuosa		Succisa pratensis	lf
(not flowering)	va	Trientalis europaea	la
Galium saxatile	f	Vaccinium myrtillus	o
Lathyrus montanus	l	V. vitis idaea	va
Luzula pilosa	a	Veronica chamaedrys	f
Melampyrum pratense	l	Viola sp. (? riviniana)	f

The field layer is kept in a short turf by the abundant rabbits. This birchwood possessed only one species of grass (*Deschampsia flexuosa*) but this was very abundant.

Moss layer

Hylocomium splendens }	co–d	Hylocomium loreum	r
H. triquetrum }		Dicranum scoparium	o
Cladonia spp. (C. sylvatica, coccifera, pyxidata, etc.)			l

Where the birches grow thickly the juniper and most of the field layer is shaded out and the moss layer, with a few scattered herbs, alone exists below the tree canopy.

Grassy field layer. The obvious difference between the lower layers of most of the Highland birchwoods and the pinewoods is the tendency in the former to dominance of grasses instead of Ericaceae. This seems to be correlated both with soil type and prevalence of grazing. The pinewoods are largely developed on coarse glacial sands and gravels and offer little to the grazing animal, while the birchwoods are mostly on finer grained soils on which grasses play a larger part, developing a turf under grazing. But the difference is by no means sharply defined and most of the species are common to the two types, with the addition in the birchwoods of a selection of common oakwood species.

Variants of birchwood. The "alder-birch", "birch-rowan" and "birch-aspen" woodlands described by Anderson (1932) are probably to be regarded as slight variants of the subalpine birchwoods, developed in northern England and the Highlands, the first on wet soil and in relatively wet climates, the two latter in relatively dry soil and climate.

Altitudinal and latitudinal distribution. Though rowan reaches higher altitudes (2600 ft. = 790 m.) as individual trees with roots anchored in rocky clefts, and is one of the most wind-resistant of all our native trees, birch is not only the most northerly wood-forming tree in Britain, but together with the Scots pine forms the extreme altitudinal woodland limit at about 2000 ft. (c. 600 m.). The continuous clearances of forest, especially during the later Middle Ages and down to the nineteenth century (cf.

Chapter VIII), the general pasturing of sheep on the hills, together with the more recent ravages of the red deer in the Highlands of Scotland, have greatly depressed the upper limit of forest throughout northern Britain, besides rendering the great majority of north English, Welsh, and Scottish hills completely bare of trees even on their lower slopes. But an underlying cause, antecedent to human agency, of the general depression and restriction of these upland woods is probably increased wetness and coolness of the summer climate during the Sub-Atlantic period, extending from something less than 1000 B.C. into the historical period and perhaps down to the present day. Such a change of climate not only favours the wet moorland plants, notably *Sphagnum*, *Scirpus caespitosus* and *Eriophorum*, which are enemies of forest, but is probably inimical in other ways to the growth of trees at relatively high altitudes. Brockmann-Jerosch has shown (1919) that the highest forest limits occur in the most continental regions of the Alps, and the highest woodlands in Britain are in the most "continental" parts of the Highlands of Scotland, which also contain the largest high mountain mass (the Cairngorms). It is in this region too that most of the larger native pine forests still persist. Pine, as Robert Smith (1900) observed, is also attached especially to the coarser and more permeable soils which bear *Calluna* and *Vaccinium*, birch on the whole to finer grained soils with grass vegetation predominating, at least under grazing.

Thus from the point of view of climatic distribution the central and north Highland birch and pinewoods represent relics of a south-westward extension of the European Northern Coniferous Forest formation, whose upper and northern limits, as well as the northern limits of Summer Deciduous Forest, dominated mainly by oak, are depressed both altitudinally and latitudinally in the extremely oceanic western climate.

As we have seen in Chapter XVI the oakwoods which still exist (though now largely replaced by planted conifers) in the valleys of the Scottish Highlands often have a fringe of birchwood (or birch and pinewood) along their upper edges, and this is also true of many of the oak- and ashwoods of northern England. Though their development is now very incomplete these upper fringing birchwoods clearly represent the altitudinal zone of "subalpine" woods, touching 2000 ft. (600 m.) in Perthshire (R. Smith) and Inverness-shire, corresponding with the latitudinal zone of birch in the extreme north. The peat pollen records as well as the considerable quantities of birch stumps and trunks buried in the recent peat of the upland mosses, both in Scotland and on the Pennines, show that these upland birch woods were once much more extensive than now. Pine is also found in this peat, but in much less quantity, especially on the Pennines.

Seral birchwood. Besides dominating the climax birchwoods described above the birches play an important part in woodland succession in several parts of southern Britain. In the valleys of the southern Scottish

Highlands the birches, and especially *Betula pubescens*, are abundant in all kinds of situations and are constant constituents of the more open oakwoods. A depleted oakwood often "fills up" with birch, and areas of moorland or grassland protected from excessive grazing by rabbits, sheep or deer may quickly be colonised with birch seedlings derived from parent trees in the neighbourhood. In the oakwood region a birchwood so produced would naturally be followed by oak colonisation, resulting in the establishment of climax oakwood, though under existing conditions in the Scottish valleys this seldom happens. But in the south of England it is much more frequent. Here also the birches, and especially *B. pendula* (*B. pubescens* in Sherwood Forest in Nottinghamshire), are constant associates of the oaks, especially, though not exclusively, on the more sandy soils. They colonise the open heathland characteristic of these soils very freely, and thickets of birch and older birchwood may often be seen developing in such areas (Pl. 86). Oaks colonise the same type of soil, but much more slowly, because they form less seed and the seed is less mobile. Consequently, while in places oaks may be seen directly colonising heathland, it is commoner to find young thickets of birch with scattered seedling and sapling oaks. If such vegetation is left to itself an oakwood will eventually result. But human interference nearly always occurs sooner or later, and we have to be content with observing fragments only of the natural succession.

Birch invasion. Various stages of this sere are represented in the very characteristic mixed communities collectively called "oak-birch heath" in *Types of British Vegetation* and described in Chapter XVII, pp. 354–357. It is extensively developed on the southern sands, especially on the Lower Greensand fringing the Weald in Surrey and Kent, on the Ashdown Sand in the centre of the Weald of Sussex (both of Cretaceous age), on the Reading Beds and Bagshot Sand of Eocene age, and on glacial and postglacial sands and gravels. Here we have areas of pure *Calluna*-heath or of grass heath (especially where the ground is grazed or trampled), others dominated by *Pteridium* or by *Ulex*, with scattered birches and oaks or little patches of developing or mature birch and oak woodland. Subspontaneous pine and sometimes beech may also invade such areas, and may develop considerable stretches of pure pinewood and smaller patches of pure beechwood.

The maintenance of "oak-birch heath" with its multiform facies is doubtless mainly due to perpetual human interference varying in nature and intensity from place to place and from time to time, and confusing the natural successional processes determined by the presence of different gregarious woody species.

Besides the important role which it plays in colonising heathland, birch (*Betula pubescens*) often sows itself in fenland and marshland, where it may even become dominant, since it can tolerate a wide range of water content and also of soluble mineral material in the soil. Hence it often accompanies

PLATE 86

Phot. 205. Young birch (*B. pendula* and *B. pubescens*) colonising heath land (*Erica cinerea*, etc.). Oak in background. St Leonard's Forest, Sussex, on Ashdown Sand.

Phot. 206. Birchwood (*B. pendula*) at Hesworth Common, Sussex, on Lower Greensand. *Erica cinerea*, *Calluna* and *Pteridium*.

Phot. 207. Young birchwood (*B. pendula* and *B. pubescens*) at Press Ridge Warren, Sussex, on Ashdown Sand. Pteridietum on well-lighted edge. *S. Mangham.*

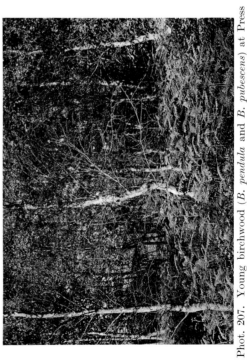

Phot. 208. Birchwood (*B. pubescens*) on glacial gravel. Bilhaugh, Ollerton, Sherwood Forest, Notts. *J. W. Hopkinson.*

ENGLISH SERAL BIRCHWOODS

PLATE 87

Phot. 209. Invasion of Callunetum by *Pinus sylvestris*. Wych Cross, Ashdown Forest, on Ashdown Sand. *S. Mangham.*

Phot. 210. Invasion of heath by pine from pinewood in the background. Hook Common on Bagshot Sand.

Phot. 212. *Molinia* society in subspontaneous pinewood at Esher Common, Surrey, on Bagshot Sand. *V. S. Summerhayes* (1924).

Phot. 211. Heath, with sparse low *Pteridium*, invaded by pine. Birch in right centre. Crockham Hill Common, Kent, on Hythe Beds.

Sub-spontaneous Pine in S.E. England

alder (*Alnus glutinosa*) in marsh and fen woods, though in natural succession it would be suppressed by the latter just as on drier soils by oak and pine. Birch, together with pine, also colonises drying bog (Chapter XXXIV). None of these successions has however been closely studied.

Subspontaneous pinewood. Among the gregarious species of oak-birch heath the common pine (*Pinus sylvestris*) is widespread and locally abundant on the south-eastern heaths, having been extensively planted in the south during the eighteenth and nineteenth centuries. From the plantations it has spread to the heaths, which it colonises freely in company with the birches, forming subspontaneous pinewoods which are often dense and extensive (Pl. 87). Many of these woods were cleared for trench and pit props during the war of 1914–18 when importation was difficult.

The subspontaneous pinewoods of the southern heaths cast a fairly deep shade, which Summerhayes (1924), in a particular example, determined as 17 per cent of the full open illumination; and the ground is thickly covered with a litter of dead pine needles, so that there is very little subordinate vegetation. Small beech and oak shrubs with flat plates of foliage and very thin leaves sometimes occur, and locally also similar shrubs of sweet chestnut (*Castanea sativa* Miller), since this tree is often planted on sandy soils and in hot summers the nuts ripen and germinate well.

On Esher Common — Summerhayes, Cole and Williams (1924) have described such a subspontaneous pinewood, the trees of which were 35–55 years old, on the Bagshot Sand of Esher Common in Surrey. Most of the soil surface was covered with pine needle litter which reached a thickness of 2 or 3 in. round the bases of the trees. This was underlain by raw humus up to an inch in thickness, with 1 or 2 in. of structureless peat below. Where the litter was thick there was little or no other vegetation. *Molinia caerulea*, which was dominant in the wetter parts of the heath on which the pinewood had developed, maintains itself well as low tufted herbage in the shade of the pines, though it seldom flowers (Pl. 87, phot. 212). *Erica tetralix* and *E. cinerea* survive longer than *Calluna*, which is suppressed and disappears. Seedlings of *Fagus* and *Quercus* and also of subspontaneous *Acer pseudo-platanus* occur but do not develop far. *Castanea sativa*, however, as already mentioned, may form shrubs from 4 to 10 ft. high.

In other similar pinewoods, e.g. on the Hythe Beds of the Lower Greensand in Surrey and Kent, *Vaccinium myrtillus* and *Deschampsia flexuosa* are the commonest species of the field layer and oak and beech shrubs are fairly frequent under the pine canopy.

Where little or no litter is present the following bryophytes are found in the raw humus:

Campylopus flexuosus	f	Leucobryum glaucum	la
Dicranella heteromalla		Webera nutans	
Dicranum scoparium		Calypogeia trichomanis	
Hypnum schreberi		Lophocolea cuspidata	o

and the lichens:

- Cladonia fimbriata var. radiata
- C. pyxidata
- C. squamosa
- Cladonia sylvatica
- C. uncialis
- Peltigera canina

The most abundant vegetation, however, consists of Hymenomycetous fungi, whose fruit bodies stud the ground everywhere in the autumn. The following species have been observed (Summerhayes, Cole and Williams, 1924):

Amanita rubescens	f	Lactarius rufus	
Boletus bovinus		L. serifluus	f
B. edulis		Paxillus involutus	a
B. subtomentosus	f	Polystictus abietinus (on stumps)	
B. variegatus	f		
B. versipellis		Russula emetica	f
Cantharellus cibarius		R. fragilis	
Clitocybe aurantiacus	f	R. ochroleuca	
C. brumalis		Tricholoma portentosum	
Hygrophorus hypothejus		T. terreum	

No one has observed the supersession of these subspontaneous pinewoods by another woodland type, and it seems probable that they would maintain themselves side by side with oak or beech or whatever kind of woodland may form the local climax on these soils. Subspontaneous pines often establish themselves in the open oak or oak-birch heathy woods, but they do not seem capable of suppressing oaks or any trees other than birch.

It has been suggested that some of the southern English pines may be descendants of those which formed forests in prehistoric times, and which certainly maintained themselves for a long while after that; but unless fairly complete pollen records become available for the whole of later post-glacial time in the south of England it will be impossible to decide this question one way or the other. The unrotted pine stumps found in a dried out bog near Birmingham and mentioned on p. 254, certainly suggest that native pine persisted in the south till comparatively recent times, and if so it is not unreasonable to suppose that the species may have perpetuated itself on the southern sands down to the present day.

REFERENCES

ANDERSON, M. L. *The Natural Woodlands of Britain and Ireland.* Department of Forestry, Oxford. 1932.

BROCKMANN-JEROSCH, H. Baumgrenze und Klimacharakter. *Ber. Schweiz. bot. Ges.* **26.** 1919.

CRAMPTON, C. B. *The Vegetation of Caithness considered in Relation to the Geology.* Committee for the Survey and Study of British Vegetation. 1911.

ERDTMAN, G. Studies in the micropalaeontology of post-glacial deposits in Northern Scotland and the Scottish Isles, with special reference to the history of the woodlands. *J. Linn. Soc. (Bot.),* **46.** 1924.

MATTHEWS, J. R. Geographical relationships of the British Flora. *J. Ecol.* **25,** 1–90. 1937.

References

SMITH, R. Botanical Survey of Scotland. Part II. North Perthshire District. *Scot. Geogr. Mag.* 1900.

SUMMERHAYES, V. S., COLE, L. W. and WILLIAMS, P. H. Studies on the ecology of English heaths. I. *J. Ecol.* 12, 287–306. 1924.

SUMMERHAYES, V. S. and WILLIAMS, P. H. Studies on the ecology of English heaths. II. *J. Ecol.* 14, 203–43. 1926.

TANSLEY, A. G. *Types of British Vegetation*, pp. 116-20, 141–3. Cambridge, 1911.

TAYLOR, E. G. R. in DARBY, H. *An Historical Geography of England before* A.D. 1800. Cambridge, 1936.

WHITE, F. B. *The Flora of Perthshire*. Edinburgh, 1898.

Chapter XXIII

ALDERWOOD (ALNETUM GLUTINOSAE)

Distribution. There is the strongest possible evidence from the records of the pollen preserved in post-glacial peat that the common alder, *Alnus glutinosa* Gaertner (= *A. rotundifolia* Miller), was far more abundant in the British Isles from the Atlantic period, when it rapidly spread through the country, until quite recent centuries, than it is to-day; and this is not in the least surprising when we remember that the waterlogged soil of undrained marshes and fens must have been very much more extensive before systematic drainage and reclamation of such land was undertaken. Indeed, we may picture the flatter lowlands of post-glacial Britain as studded with lakes and meres bordered by wide stretches of swamp, marsh and fen, which were probably largely occupied by alderwood; and during the periods of wetter climate the tree may well have spread through the general deciduous forest.

The common alder is still distributed throughout these islands, reaching the extreme north of Scotland and the extreme west of Ireland, but the very extensive alderwoods which must once have existed have mostly disappeared. The increase in frequency of the alder in the wetter climates of western Wales, Scotland and Ireland is very noticeable. In the hillside oakwoods of these regions it is a frequent and sometimes an abundant tree not limited to saturated or specially wet soil, whereas in the midlands, south and east of England it is practically confined to fens and marshes, marshy hollows and the margins of streams and ponds.

Alnus glutinosa has a wide distribution in Europe, and extends into western Asia and North Africa. Its habitat is the same everywhere, wet humous soils with a sufficiency of mineral salts: it avoids extremely acid peat bogs, though a belt of alder often occurs bordering the ditch or stream of open water ("lagg") on the edge of raised bog where the water is not so poor in electrolytes; but the actual range of pH values the tree can tolerate has not been worked out, nor are there any systematic ecological studies of alderwood available.

Fen carr. One of the best developments of alderwood still extant in England is seen in the region of the "Broads" in east Norfolk, where the tree is a main dominant of many of the characteristic fen woods or "carrs"; and similar woods occur in the valley fens of Breckland. But almost wherever the soil is waterlogged and not too acid, in wet hollows in oak- or ashwoods, on marshy land, by the sides of rivers, streams, lakes or ponds, alders are to be found, though rarely covering any considerable area because the suitable habitats are now very limited in extent.

The photographs reproduced in Pl. 88 are external views of alder carr

PLATE 88

Phot. 213. Salhouse Broad, River Bure, and Hoveton Great Broad, seen from the upland. The broader strip of peat between the river and Hoveton Broad bears carr mainly dominated by alder (cf. Fig. 129). F. F. Blackman.

Phot. 214. Cockshoot Broad with alder swamp carr behind fringed by Carex paniculata. Salix atrocinerea on the right. M. Pallis (1911).

Phot. 215. Carex paniculata, Cladium mariscus and Salix atrocinerea on the right. Alder swamp carr behind. Rockland Broad. J. Massart.

ALDER CARR IN EAST NORFOLK

PLATE 89

Phot. 216. Heron's Carr, Barton Broad. *Alnus glutinosa, Humulus lupulus.* J. Massart.

Phot. 217. Heron's Carr. *Alnus* and *Eupatorium cannabinum.* J. Massart.

Phot. 218. Swamp carr, Rockland Broad on the Yare. *Salix atrocinerea, Filipendula ulmaria.* J. Massart.

Phot. 219. Fen carr on the Bure, E. Norfolk. *Rhamnus catharticus, Calystegia sepium.* J. Massart.

East Norfolk Carrs

and those in Pl. 89 illustrate the tangle of vegetation that characterises these largely untouched fen woods.

Miss Pallis (in *Types of British Vegetation*, 1911) gives the following short general list of typical species in "ultimate carr", i.e. the "edaphic climax", woodland of the fens of east Norfolk:

Trees

Alnus glutinosa	d	Fraxinus excelsior	f
Betula pubescens	f	Quercus robur	o

Shrubs

Salix atrocinerea	a	Ribes rubrum	f
Frangula alnus	f	R. grossularia	o–f
Rhamnus catharticus	f	Viburnum opulus	f
Ribes nigrum	f	Ligustrum vulgare	o

Herbs and ferns

Carex paniculata*	a	Filipendula ulmaria*	f
Iris pseudacorus*	a	Carex acutiformis	o
Dryopteris thelypteris	a	Osmunda regalis	r
Urtica dioica	a		
Caltha palustris	f	Mnium hornum	a

* Not flowering.

The abundance of the stinging nettle is characteristic of these wet woods and of similar woods on the continent. It is specially luxuriant in Norfolk below herons' nests, where it obtains heavy supplies of combined nitrogen.

The following fuller list of vascular plants was made in a carr at Wheatfen Broad, north of Surlingham Broad near the River Yare (Ellis, 1933–4). This carr is certainly a natural and may well be practically a virgin wood. It is not dominated by alder, but consists of a mixture of trees and shrubs into which oak and sycamore have entered. Water stands in hollows among the trees and is covered with filamentous green algae (undetermined). The pH value of the water in the surrounding pools and dykes is between 7·5 and 8.

Trees and shrubs

Acer pseudo-platanus	r	Quercus robur	r
Alnus glutinosa		Rhamnus catharticus	a
Crataegus monogyna		Ribes nigrum	
Euonymus europaeus		R. rubrum	
Frangula alnus		Rosa canina	
Fraxinus excelsior	a	Rubus fruticosus (agg.)	
Ligustrum vulgare		Salix atrocinerea	a
Populus tremula	r	S. caprea	
Prunus spinosa		Viburnum opulus	a

Climbers

Calystegia sepium	Humulus lupulus
Hedera helix	Solanum dulcamara

Abundant or very abundant

Carex paniculata
C. remota
Dryopteris thelypteris
Galium aparine
Peucedanum palustre
Urtica dioica

Frequent

Angelica silvestris
Calamagrostis canescens
Caltha palustris
Cardamine flexuosa
C. pratensis
Carex acutiformis
C. riparia
Cirsium palustre
Epilobium hirsutum
Eupatorium cannabinum
Galium palustre
Glyceria maxima
Iris pseudacorus
Lycopus europaeus
Lysimachia vulgaris
Lythrum salicaria
Mentha aquatica
Myosotis scorpioides
Phragmites communis
Poa trivialis
Polygonum hydropiper
Ranunculus repens
Rumex conglomeratus
R. hydrolapathum
Sparganium ramosum
Valeriana officinalis

Occasional, local or rare

Bidens cernua
Cicuta virosa
Dryopteris filix-mas
Epilobium palustre
E. parviflorum
Filipendula ulmaria
Hypericum quadrangulum
Lysimachia nummularia
Phalaris arundinacea
Polypodium vulgare
Ranunculus sceleratus
Scrophularia aquatica
Sium erectum
Sonchus asper
Stachys palustris
Stellaria uliginosa
Veronica anagallis-aquatica
V. beccabunga

The air of the carr is constantly very moist, and there is a corresponding abundance of bryophytes and lichens on the masses of fallen branches and twigs and on the tree trunks and living branches, sometimes even covering the young growing shoot (Ellis, 1933–4). The following species were noted:

Mosses

Amblystegium filicinum
A. serpens
A. varium
Aula comnium androgynum
Barbula convoluta
Brachythecium purum
B. rutabulum
B. velutinum
Bryum argenteum
B. capillare
Catharinea undulata
Dicranella heteromalla
Eurhynchium praelongum
Fissidens bryoides
Funaria hygrometrica
Hypnum cupressiforme and var. filiforme
Hypnum cuspidatum
H. riparium
Leskea polycarpa
Mnium affine
Mn. hornum
Mn. punctatum
Mn. undulatum
Orthotrichum affine
O. tenellum
Plagiothecium denticulatum
Polytrichum aloides
Tortula muralis
Weisia viridula
Zygodon viridissimus

PLATE 90

Phot. 220. *Rhamnus catharticus* dominant. Field layer: *Dryopteris thelypteris, Symphytum officinale, Iris pseudacorus, Urtica dioica, Cladium mariscus* (relict). [Alder is absent from Wicken Fen, but the field layer is substantially similar to that of alder carr, cf. Pl. 89 and Pl. 114, phot. 281.] *H. Godwin.*

Liverworts

Frullania dilatata
Lophocolea bidentata
L. heterophylla
Marchantia polymorpha
Pellia epiphylla
Radula complanata

Lichens

Evernia prunastri
Parmelia caperata
P. fuliginosa
P. physodes
Parmelia sulcata
Ramalina farinacea
Usnea florida var. hirta

"Swamp carr." Miss Pallis (1911) distinguished "swamp carr", developed on floating peat, from "fen carr" on the solid peat of fen, but Godwin and Turner (1933, p. 243, footnote) are inclined to believe that the former, in which islands of floating peat bearing alders and sallows are often separated by free water, is due to the depression of the first-formed floating peat by the weight of the trees and shrubs, in the absence of consolidation of the soil by birch and oak. The ecology and genetic relationships of alderwood are still much in need of thorough investigation.

Young carr. The scanty remains of undrained fen still existing in the fenland of Cambridgeshire, of which Wicken Sedge Fen in the valley of the Cam is the best known, bear "carr" wherever the fen has been left uncut so that woody plants can colonise it. There is no alder remaining at Wicken, which is surrounded by drained and partly cultivated land, and alder has not appeared in the carr, presumably because there are no parent trees near enough to supply seed. The dominants of the carr at Wicken are *Salix atrocinerea*, *Rhamnus catharticus*, *Frangula alnus* and *Viburnum opulus*—all abundant or frequent shrubs of the Norfolk alder carr—and forming the bulk of the *young* carr in the region of the Broads. Thus the oldest carr now existing at Wicken is evidently an advanced "scrub", or shrub stage of the fen sere which in Norfolk leads to alderwood (Pl. 90; cf. pp. 657–9 and Pl. 114).

Succession. Though many of the alder carrs show little or no signs of invasion by drier types of vegetation there can be no doubt that they may give way to oakwood, and Godwin and Turner (1933) give an account of such a succession from reedswamp through carr to oakwood. Fig. 87 shows a transect along the line of this sere on the edge of Calthorpe Broad, Norfolk. Alder and *Salix atrocinerea* are pioneer woody plants colonising the "mowing marsh", which succeeds reedswamp. The sallow dominates the first-formed carr, but is the most transient species. The alder, which then becomes dominant, is joined by *Betula pubescens* and later by *Quercus robur*, which shades out the lower growing trees and shrubs and converts the carr into oakwood. "The earlier stages of carr are obviously floating on semi-liquid peat, and the intermediate stages of carr development are generally associated with increasing stability of the ground."

In "intermediate" carr the black peat surface is completely bare except for a few bryophytes (*Brachythecium rutabulum*, *Mnium punctatum*, and

Alderwood

Chiloscyphus pallescens) with occasional tufts of *Carex riparia*, *Calamagrostis lanceolata* and *Iris pseudacorus*. At a later stage the field layer becomes denser and has the following composition:

Agrostis stolonifera	a	Lysimachia vulgaris	r
Calystegia sepium	o	Osmunda regalis	
Carex riparia	a	Rosa sp.	o
C. stricta	o	Rubus caesius	o
Comarum palustre	o	R. fruticosus (agg.)	o
Iris pseudacorus	f	Solanum dulcamara	o
Lycopus europaeus	o		

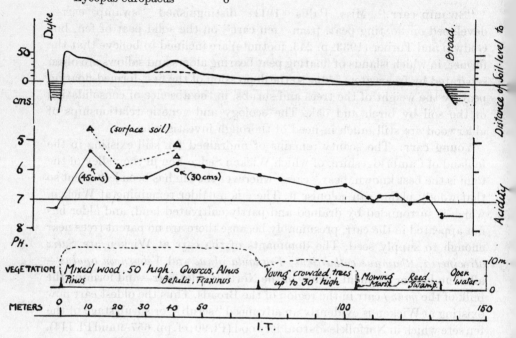

FIG. 87. TRANSECT AT CALTHORPE BROAD (NORFOLK)

Showing succession from reedswamp to oakwood with water levels and *p*H values. From Godwin and Turner, 1933, p. 247. Note the increasing acidity with progressive raising of the peat level.

The trees of the carr reach a maximum height of 5–6 m. (Godwin and Turner, 1933). Though most is shaded out some tall alder survives in the oakwood.

Hydrosere at Sweat Mere. The most complete hydrosere leading from reedswamp to alderwood and beyond that has come to my notice I owe to Dr A. R. Clapham, who observed it at Sweat Mere near Ellesmere in Shropshire in March 1938. The floristic composition of the successive phases is given in the table on p. 466, and it will be seen that the invasion, increase, and ultimate decrease in frequency of most species follows a very regular course. The maximum number of species occurs in the Alnetum, to which several are confined. With the rise of birch to co-dominance the mud becomes fairly firm and completely covered with a field layer of grasses and herbs, while most of the distinctively marsh species have disappeared. The two last stages (columns to the right of the thick line), in which the birch is dominant alone and is apparently succeeded by oak, are affected by partial drainage, and are marked by the local occurrence of bracken in the field layer. The acidity of the surface peat is very high and no varied oakwood flora has developed. *Dryopteris dilatata* and *Rubus fruticosus* (agg.) are remarkably persistent from the Caricetum to the oakwood.

The stages of the succession are as follows:

(1) TYPHETUM LATIFOLIAE forming a floating mat.

(2) CARICETUM PANICULATAE forming tussocks 3–4 ft. high, with their centres 2–10 ft. apart and standing water between: *Typha* mat beginning to break down.

(3) SALICETO-ALNETUM: incompletely closed carr, with *Typha* mat replaced by black mud: *Carex* tussocks still luxuriant.

(4) ALNETUM: closed carr with mud freely colonised by a variety of species.

(5) BETULETO-ALNETUM: *Betula* co-dominant with *Alnus*. Field layer closed, consisting mainly of woodland herbs (with grasses), and only a few marsh species persistent.

(6) BETULETUM (partially drained): colonisation by fresh tree species: *Pteridium* locally dominant.

(7) QUERCETUM (drained): flora poor, *Pteridium* locally dominant.

The species which are dominant or co-dominant in successive phases are printed in heavy type. The names of the less important species are indented.

Mosses and lichens were only noted in the Alnetum:

Brachythecium		Mnium punctatum	f (mud)
rutabulum	f (mud)	Lichens	
Hypnum		Evernia prunastri	f ⎫
cuspidatum	o (mud)	Parmelia sp.	f ⎬ on tree trunks
H. cupressiforme	f–a (trunks)	Usnea sp.	o ⎭
Mnium hornum	f (tree bases)		

and in the Betuletum:

Polytrichum formosum	o	Mnium hornum (stumps)	f
Thuidium tamariscinum	o		

The pH values showed a striking change from 7·3 in water between the *Carex* tussocks in stage (4) to 4·3 in the top two inches of the peat beneath an alder in stage (5), and 3·7 in the surface peat of stage (6).

The importance of the Alnetum (stage 4) as an autonomous community, marking the culmination of the earlier, more aquatic phases of development, is very clear from the great increase in the number of species that accompany

Alderwood

	1 Typhetum latifoliae	2 Caricetum paniculatae	3 Saliceto-Alnetum	4 Alnetum glutinosae	5 Betuleto-Alnetum	6 Betuletum pubescentis	7 Quercetum roboris
Typha latifolia	d	f	r–o				
Carex paniculata	o–lf	d	a	o–f			
Salix atrocinerea	o	f	a–**co–d**	o	o		
Epilobium hirsutum	o	lf	la	la‡	o		
Carex pseudocyperus	o	o	r				
Phragmites communis	r						
Alnus glutinosa		o	**co–d–ld** d	**co–d**	o		
Dryopteris dilatata		o–f*†	f	f–a*†	f	ld	o–ld
Solanum dulcamara		o–f*†	f	f			
Rubus fruticosus (agg.)		f*	f	f		f	f
Galium palustre		o*		f*‡			
Lythrum salicaria		r					
Betula pubescens			o	o–f	**co–d**	d	o
Lonicera periclymenum			o	f		f–a	f
Filipendula ulmaria			r	f*‡	f		
Carex acutiformis			o	o*			
Dryopteris spinulosa			r	r*†			
Caltha palustris				a‡	a		
Urtica dioica				la*‡	a		
Mentha aquatica				la‡	a		
Iris pseudacorus				la‡	l		
Cardamine pratensis				o‡			
Callitriche stagnalis				o‡			
Valeriana sambucifolia				o*			
Cirsium palustre				r†	f		
Geranium robertianum				o*			
Angelica sylvestris				r*			
Lychnis flos-cuculi				r†			
Viburnum opulus				r			
Digitalis purpurea				r†		la	o–f
Agrostis stolonifera					f		
Holcus lanatus					f		
Poa annua					f		
Ranunculus repens					f		
Juncus effusus					lf		
Melandrium dioicum (rubrum)					l		
Quercus robur					r	o	d
Populus canescens						o	
Ilex aquifolium						o	
Sorbus aucuparia						r	
Pinus sylvestris						o§	
Rubus idaeus						lf	lf
Oxalis acetosella						f–la	f–a
Pteridium aquilinum						ld	ld
Rumex acetosa							o
No. of species in each stage:	6	10	14	26	18	14	10

* On tussocks of *Carex paniculata*.   † On tree bases.   ‡ On bare mud.
§ The few scattered pines in the birchwood, according to Dr Clapham, "did not look as if they had been planted".

its establishment, while at the same time the Alnetum retains most of the species of the preceding phases. With the rise of birch to co-dominance (stage 5) and dominance (stage 6) the peat becomes consolidated, more than half the species of the Alnetum disappear, and a number of new ones come in, mainly trees and shrubs, but also bracken and woodsorrel. The short stages 2 and 3 are properly regarded as developmental phases of the Alnetum, while from stage 5 onwards there is a sharp change to a different type of community characterised by consolidated acid peat and an almost entirely different set of species. The part played by birch in the succession is noteworthy. Entering before the attainment of dominance by the alder, it ultimately supersedes the latter, apparently playing the leading part in consolidating the peat, as Godwin and Turner suggested, and introducing a definitely "dry land" vegetation. At Sweat Mere the establishment of Quercetum has not yet, however, brought about the development of a varied oakwood flora. The number of species present decreases progressively from stage 4 to stage 7.

Valley fenwoods. Farrow (1915, pp. 225–6) describes the flora of the valley fenwoods on the wet peaty ground of the bottoms of the shallow valleys of Breckland, for example that of the River Lark (Suffolk), as consisting of the following species:

Trees

Alnus glutinosa
Betula pendula (alba)
B. pubescens
Fraxinus excelsior
Populus canescens

Populus nigra
[P. serotina and P. italica, planted]
Salix alba[1]
S. fragilis[1] and hybrids
Ulmus glabra (montana)

Shrubs

Corylus avellana
Crataegus oxyacanthoides
Ligustrum vulgare
Prunus spinosa

Salix purpurea
S. triandra
S. viminalis
Solanum dulcamara

Herbs and grasses

Agrostis stolonifera (alba)
Ajuga reptans
Calamagrostis canescens
Carex disticha
C. hirta
C. paniculata
C. riparia

Comarum palustre
Deschampsia caespitosa
Epilobium palustre
Filipendula ulmaria
Galium aparine
G. uliginosum
Geum rivale

[1] The white and the crack willows (*Salix alba* and *S. fragilis*) grow in much the same habitats as the alder, but are absent from most of our natural alderwoods, and doubt has been cast on the nativity of *S. alba*. Since *Populus serotina* and *P. italica* are present in these woods there has evidently been some planting, though the basis is almost certainly natural. The two willows are usually seen as planted trees (mostly pollarded) along the courses of embanked rivers, where their roots are useful in binding the soil.

Alderwood

Gymnadenia conopsea
Juncus acutiflorus
J. conglomeratus
J. subnodulosus
Listera ovata
Lotus uliginosus
Lychnis flos-cuculi
Lycopus europaeus
Lysimachia vulgaris
Lythrum salicaria
Malva sylvestris
Mentha aquatica

Menyanthes trifoliata
Myosotis scorpioides (palustris)
Nepeta hederacea
Phragmites communis
Polygonum amphibium
Potentilla anserina
Scutellaria galericulata
Succisa pratensis
Veronica beccabunga
V. officinalis
Urtica dioica

Ferns, etc.

Dryopteris thelypteris
D. filix-mas
D. dilatata

Polystichum aculeatum
Equisetum spp.

The dominants of the Wicken carr and the young carr of the Norfolk Broads are here absent, other species of *Salix* are present and the shrubs and herbs include a number of species belonging to the damp oakwood, as well as some alien invaders.

Alder carr at Cothill. The following list[1] was taken from an alder carr by the Sandford Brook in the Cothill basin (north of the Abingdon-Marcham road in Berkshire). The short list of herbaceous species consists almost entirely of marsh or fen plants.

Woody plants

Alnus glutinosa	d	Betula pubescens	
Ligustrum vulgare		Populus tremula	
Betula pendula		Salix atrocinerea	

Field layer

Angelica sylvestris	o	Iris pseudacorus	la
Caltha palustris	o–f	Listera ovata	o
Carex acutiformis	d	Lotus uliginosus	o–r
C. paniculata	o	Lychnis flos-cuculi	o
Epilobium hirsutum	a	Orchis praetermissa	o
Equisetum palustre	f	Phalaris arundinacea	o
Eupatorium cannabinum	f	Stachys palustris	o
Filipendula ulmaria	o	Valeriana dioica	o–r
Hydrocotyle vulgaris	o	Vicia cracca	o
Hypericum quadrangulum	o		

Bryophyta

Amblystegium filicinum
Eurhynchium swartzii

Mnium affine
Mn. undulatum

[1] This and the two succeeding lists were kindly put at my disposal by Dr A. R. Clapham, who (with Mr H. Baker) is making an extensive study of the Cothill fen basin in north Berkshire, a few miles south of Oxford. This is based on Calcareous Grit, an Upper Jurassic formation of Argovian age, and contains a great depth of alkaline peat on which various fen communities are developed, including fenwood or carr of alder, ash and birch.

In a coppiced carr in the Ruskin Reserve (Cothill) on very wet, soft, deep peat, with drainage ditches and some pools of standing water, there was a much richer woody flora with a well marked calcicolous element.

Shrub layer

Alnus glutinosa	d	Rhamnus catharticus	r
Betula pendula	r	Rosa canina (agg.)	r
B. pubescens	o	Rubus fruticosus (agg.)	o
Crataegus monogyna	r	R. idaeus	o
and seedlings	o	Salix atrocinerea	f
Fraxinus excelsior		Sambucus nigra (seedlings)	o
saplings and seedlings	f	Sorbus aucuparia	r
Ligustrum vulgare	f–la	Viburnum opulus	o
Quercus robur	r		

Climbers

Lonicera periclymenum	o	Tamus communis	o
Solanum dulcamara	o		

Field layer

Agrostis stolonifera		Galium uliginosum
Asperula odorata		Hydrocotyle vulgaris
Brachypodium sylvaticum		Juncus subnodulosus
Carex panicea		Mentha aquatica
C. paniculata		Mercurialis perennis
Cirsium anglicum		Oenanthe lachenalii
C. palustre		Potamogeton coloratus (in ditches
Daphne laureola	la	and pools)
Dryopteris spinulosa		Schoenus nigricans
Equisetum palustre		Succisa pratensis
Eupatorium cannabinum		Valeriana dioica
Filipendula ulmaria		Viola riviniana

Bryophyta

Brachythecium rutabulum	Mnium hornum
Eurhynchium swartzii	Mn. punctatum
Fissidens adiantoides	Mn. undulatum
Hypnum cuspidatum	Sphagnum plumulosum
Mnium affine	(on old alder stumps)

A much drier carr in the same basin (between Morland's Meadow and Bessel's Meadow) on deep calcareous peat showed a very slight representation of marsh or fen plants, so that the flora is practically indistinguishable from that of a calcicolous ashwood:

Shrub layer

Alnus glutinosa dominant, coppiced (? planted)

Acer campestre	o	Fraxinus excelsior and	
A. pseudoplatanus		seedlings	o
(coppiced)	lf	Ligustrum vulgare	
Corylus avellana	o–f	(forming thickets)	a
Crataegus monogyna and		Populus canescens	
seedlings	r–o	seedlings	r
Euonymus europaeus	o–f	Prunus ?domestica	r

Rhamnus catharticus	r	Ulmus sp.	o
Rosa arvensis	o	Viburnum lantana	r
Rubus fruticosus (agg.)	lf	V. opulus	r–o
Sambucus nigra	o		

Climbers

Bryonia dioica	o	Humulus lupulus	f
Hedera helix	r	Tamus communis	a

Field layer

Angelica sylvestris	r	Equisetum palustre	o
Arctium minus	r–o	Melandrium dioicum	r
Brachypodium sylvaticum	ld	Mercurialis perennis	f–d
		Urtica dioica	r
Circaea lutetiana	lf	Viola riviniana	o
Deschampsia caespitosa	lf		

Bryophyta (not noted)

It is plausible to regard this carr as a later stage of succession than the one on soft wet peat, possibly leading to ashwood, but the frequent planting of alder confuses the natural sequences.

Other carrs in the Cothill basin show dominance or various mixtures of alder, ash and birch. Besides the species included in the three preceding lists, *Ribes rubrum*, *Rubus caesius*, *Salix caprea* and *Calystegia sepium* occur here and there, as in the East Anglian carrs. While these show clear indications of the alkalinity of the fen peat, their flora is not so markedly calcicolous as that of the carrs of the Cothill basin.

Other examples of alderwood. The alderwoods described above are all on fen peat of markedly alkaline reaction. Anderson (1932) also describes the alder as co-dominant with ash in woodland on wet calcareous soils in western Scotland and western Ireland. Such woods are evidently transitional between ashwood (calcareous) and alderwood (wet). The accompanying vegetation would decide whether they are truly intermediate or more naturally assigned to one or the other.

There are no lists or descriptions available from alderwoods on marshy soils of neutral or slightly acid reaction, but Rankin (in *Types of British Vegetation*, p. 251) briefly describes the type of alderwood marginal to the acid lowland peat bogs of north Lancashire (whose soil water, derived from the drainage from surrounding rocks, is certainly less acid than that of the bogs) as containing the following species associated with the dominant alder: birch, rowan, oak, yew, juniper, alder buckthorn (*Frangula*), sallows; with common reed, bog bean, marsh cinquefoil, *Carex paniculata* and other sedges, *Dryopteris dilatata*, *D. spinulosa*, and *Blechnum spicant*. *Gentiana pneumonanthe*, *Hottonia palustris* and *Osmunda regalis* formerly occurred in these woods, but are said to be extinct.

These narrow marginal alderwoods are part of the "lagg" vegetation associated with the margin of the typical raised moss or bog (see Chapter

XXXIV, p. 675). Rankin (1911, p. 261) also mentions the (usually) interrupted relict lines of "alder thicket" following the course of the streams, now obstructed by the growth of the *Sphagnum* bog, in the shallow valleys of the New Forest, and originally based on the stream alluvium. These contain many of the elements of typical alderwood, accompanied by wet peat species such as *Eriophorum angustifolium* and the very rare *E. gracile*.

Birch and alder. It will be noticed that in the lists from the fens, from Sweat Mere and from Cothill, birch and especially *Betula pubescens* is frequently associated with alder. *B. pubescens* has a very wide range of tolerance in respect of soil water, and also of pH value. It may even become locally dominant in fen carr where the pH is over 7, and in many marshy woods birch and alder are freely mixed. In such woods there may be Sphagna belonging to the *subsecundum* and *recurvum* groups which are indicators of relatively low acidity. The association of birch and alder in the wet places of upland birch woods is mentioned in Chapter XXI (p. 454).

Fragments of alderwood on wet soils are scattered far and wide through the lowlands of the country, and they contain various selections of such wet soil species as can endure the shade of the trees. The great stools of *Carex paniculata* are one of their most characteristic features.

REFERENCES

ANDERSON, L. M. *The Natural Woodlands of Britain and Ireland*. Department of Forestry, Oxford, 1932.

ELLIS, E. A. Wheatfen Broad, Surlingham. *Trans. Norfolk Norw. Nat. Soc.* **13**. 1933–4.

FARROW, E. P. On the ecology of the vegetation of Breckland. I. *J. Ecol.* **3**, 211–28. 1915.

GODWIN, H. and TURNER, J. S. Soil acidity in relation to vegetational succession at Calthorpe Broad, Norfolk. *J. Ecol.* **21**, 243 and Fig. 5. 1933.

PALLIS, MARIETTA. The river valleys of East Norfolk: their aquatic and fen formations. Chapter X in *Types of British Vegetation*. Cambridge, 1911.

RANKIN, W. M. The Lowland Moors ("Mosses") of Lonsdale (North Lancashire) and their Development from Fens. In *Types of British Vegetation*, p. 251. 1911.

Chapter XXIV
SCRUB VEGETATION OR BUSHLAND (FRUTICETUM)

Communities dominated by shrubs or bushes, generally known collectively as "scrub" in modern English accounts of vegetation, are intermediate, in more senses than one, between forest and grassland. In the first place the dominants are, broadly, of stature intermediate between trees and herbs. Very many shrubs have, or may have, multiple stems, a tendency much enhanced by coppicing; but others if left uncut form a single trunk, and there is in fact only a more or less arbitrary distinction between shrubs and small trees. Shrubs of less than 2 m. (say 7 ft.) but more than 25 cm. (10 in.) in height[1] fall into Raunkiaer's class of *nanophanerophytes*, but many of our shrubs grow taller than this and have to be included in the next class, the *microphanerophytes* (2–8 m., or 7–26 ft.), which also includes small trees. Woody plants less than 25 cm. in height are classed by Raunkiaer as *chamaephytes*, and are not included as dominants of scrub. In this book heath is not counted as scrub, though its dominants commonly exceed 25 cm. and are therefore nanophanerophytes.

Climatic scrub Woody plants of the stature of shrubs often form communities in situations climatically intermediate between the natural habitats of forest on the one hand and of grass and herb communities on the other, for example in the ecotone or "tension belt" between forest and prairie, or on mountains in the zone between subalpine forest and alpine grassland. Shrubs are also very commonly pioneers of the **Seral, subseral and climax scrub** woody vegetation when forest is pushing out into grassland or when forest vegetation returns to land which has previously borne it but from which it has been cleared. In these last cases the shrubs form *seral* communities, superseding grassland or heath and being in turn superseded by forest, whereas scrub belts on the upper altitudinal limits of forest or on the inner edges of forest in continental regions, where trees can no longer grow, are of the nature of climax communities which are relatively static.

In the British Isles most of the scrub is certainly subseral, and not only subseral but also often subclimax or plagioclimax, i.e. it represents vegetation which has come to occupy open ground at one time covered by forest, but where trees have not yet re-established themselves, and cannot do so under existing conditions. There is, however, a certain amount of windswept scrub in exposed situations where trees could not in any case succeed, for example on exposed coasts and in certain upland situations; and in some places local edaphic conditions may help to prevent the growth of trees

[1] Strictly shrubs whose uppermost perennating buds are situated between these distances above the ground level, but unless the upper shoots habitually die back in the winter the statements are equivalent.

though they are just good enough for shrubs to maintain themselves. None of this "climax scrub" has been thoroughly investigated, but a short description may be given of one area of extensive hazel scrub on the Carboniferous Limestone of western Ireland which is the best example of the kind that has been examined.

The Burren district of Co. Clare. Under the extreme climatic conditions of the west of Ireland the Carboniferous Limestone, which is extensively developed in Co. Clare to the south of Galway Bay, presents a very interesting series of habitats. In a dry climate the rock would only support a xerophilous vegetation, but in this oceanic climate, with its heavy rainfall, high relative humidity and strong winds, the pioneer xerophytes are quickly succeeded by other plant communities.

In this striking district, known as "The Burren", the nearly horizontal beds of limestone rise in terraced hills to heights of several hundred feet (Pl. 91, phot. 222). Limestone pavement is extensively developed, *Succession on the Burren* and is often bare or sparsely colonised by lithophilous lichens and mosses. The lines of seral advance are mainly two. Where the flat or nearly flat surfaces of limestone carry a thin layer of black humus formed by pioneer mosses and lichens, this becomes covered by a turf dominated by such (largely calcicolous) grasses as *Sesleria caerulea, Koeleria cristata* and *Festuca ovina*, accompanied by *Dryas* *Mixture of species* *octopetala* (ld), *Lotus corniculatus, Asperula cynanchica, Plantago lanceolata, Thymus serpyllum, Linum catharticum* and *Carex diversicolor*. But scattered in this calcicolous vegetation are such species as *Potentilla erecta, Hypericum pulchrum, Empetrum nigrum* and *Calluna vulgaris* (ld). Both *Dryas* and *Calluna* may form locally pure patches of turf on as little as 2 in. (5 cm.) of this humus above the limestone rock, the rootlets in contact with the rock (Black Head, 190 m. alt.). Grazing tends to stabilise this curiously mixed vegetation (Pl. 91, phot. 221).

Limestone heath. On the hills south of Ballyvaghan similar vegetation occurs. Extensive tracts of the flat terraces become dominated by *Calluna* which forms "limestone heaths" (see p. 557) showing the same mixture of calcifuges and calcicoles—on the one hand *Pteridium, Molinia* (va), *Potentilla erecta*, and *Hypericum pulchrum* associated with the *Calluna*, on the other several markedly calcicolous as well as "indifferent" plants.[1]

Besides the herbs and grasses a number of woody plants are present in this limestone heath, especially:

Corylus avellana
Crataegus monogyna
Euonymus europaeus
Hedera helix
Prunus spinosa
Rosa spinosissima

These grow largely in the fissures of the limestone pavement, and in a much fissured area tend to convert the heath community (developed on the thin

---

[1] For a short list of typical species from such a limestone heath in Yorkshire see p. 557.

humus above the flat limestone surfaces) into a scrub. Over considerable tracts which are relatively protected from wind, scrub is actually developed, and is dominated by the hazel. Exposure to wind and grazing keeps the hazel bushes sparse and low, 2–3 ft. = 60–90 cm., but in sheltered places a fairly tall and thick *Corylus* scrub occurs (Pl. 92, phot. 223).

Closed hazel scrub. The example photographed and listed grew in the shelter of a higher terrace bounded by a vertical cliff about 5 ft. high. On the lower damper ground the dominant hazel bushes (8–10 ft. = 2·4–3 m.) grew close but not very densely and considerable light reached the ground. The following species were present:

Corylus avellana (10 ft. = 3 m.)   d

Crataegus monogyna		Rosa arvensis	
Hedera helix		Rubus fruticosus (agg.)	
Lonicera periclymenum		Sorbus aucuparia (12–14 ft.)	
Prunus spinosa		Viburnum opulus	f, la
Brachypodium sylvaticum	a, ld	Lysimachia nemorum	r
Bromus ramosus (asper)	o–f	Poa nemoralis	o
Carex sylvatica	o	Potentilla sterilis	
Deschampsia caespitosa	la	Primula vulgaris	o
Fragaria vesca	la	Sanicula europaea	l
Geum urbanum	f	Veronica chamaedrys	o
Hypericum pulchrum	o	Vicia sepium	o
Lathyrus montanus	o–f	Viola riviniana	o
Hylocomium brevirostre		Hylocomium triquetrum	
H. squarrosum		Thuidium tamariscinum	

Here we have a perfectly good woodland flora, though not numerous in species.[1] At the eastern end this scrub was more exposed to wind and the hazel bushes were evidently suffering, the upper shoots drying and withering. Further east still, where the wind had even freer access, the hazels were killed to the base, and *Calluna* was successfully competing, so that a heath was formed (Pl. 92, phot. 224).

Progression to ashwood. Some miles to the south, in the interior of Clare, at lower altitudes and in well protected situations there are signs of progression to ashwood, though there do not appear to be any extensive, well developed ashwoods in the region. One example of the increasing dominance of ash was seen between Ennis and Corrofin. Here in a small valley with very uneven terrain there was a mixed scrubby wood partly

---

[1] About 8 miles to the east, at the base of Slieve Carran, an even finer hazel scrub 15–20 ft. (4·5–6 m.) high, was seen in 1908 (Brit. Veg. Committee, *New. Phyt.* **7**, 1908, p. 259). Besides several of the species in the list given above the woody plants included *Betula, Euonymus europaeus, Fraxinus, Ilex aquifolium, Rosa canina* and *Ulmus glabra*; and the herbaceous *Angelica sylvestris, Circaea lutetiana, Epipactis* sp., *Oxalis acetosella, Succisa pratensis, Veronica montana,* as well as *Ophioglossum vulgatum* and *Polystichum angulare*.

PLATE 91

Phot. 221. Limestone pavement at Black Head, Co. Clare. In foreground mat of *Sesleria caerulea*, *Dryas octopetala*, *Calluna vulgaris*. R. J. L.

Phot. 222. Terraced limestone pavement on the Burren, south of Ballyvaghan, Co. Clare, with heath developing in exposed positions. Thick hazel scrub in more sheltered positions. R. J. L.

LIMESTONE PAVEMENT, LIMESTONE HEATH AND HAZEL SCRUB

PLATE 92

Phot. 223. Closer view of hazel scrub protected by higher terrace on the left. *Corylus* dominant (10 ft.), *Sorbus aucuparia* (14 ft.), *Prunus spinosa*. (See list on p. 474.)

Phot. 224. Exposed end of hazel scrub belt with taller shoots killed back by wind. *Calluna* competing. R. J. L.

CLIMAX HAZEL SCRUB AND HEATH IN CO. CLARE

open and showing evidence of grazing. *Fraxinus* was becoming dominant, but was mixed with *Betula pubescens* (f–a), *Crataegus monogyna* (a), *Ulmus glabra (montana)*, *Corylus, Cornus sanguinea, Euonymus europaeus* (o) and *Prunus spinosa* (r). *Fagus sylvatica* (not native) evidently self-sown from a neighbouring small beech plantation, had developed a number of small closely nibbled bushes. A great deal of similar scrub was seen in this neighbourhood and in some of it *Corylus* was dominant.

Subseral scrub. Of English subseral scrub some is apparently static, while other scrub communities are obviously progressing to woodland. The static scrub was called by Moss (1913) "retrogressive", and supposed to be derived from woodland by disappearance of the trees and to be a precursor of further degeneration to grassland, so that it would represent an intermediate community in a "retrogressive succession" from forest to grassland; and this point of view was adopted in *Types of British Vegetation*. It is, however, more logical to consider apparently static scrub of this kind as a deflected and temporarily (or permanently) stabilised stage of the subsere which would normally have led back to woodland after the clearance or death of the original trees. If the trees are destroyed and grassland is established by grazing, the trees being prevented from returning by continued grazing or by a serious deficiency in seed parents, or both, shrubs, and especially spiny shrubs, may increase and close up to form a dense scrub which has every appearance of permanence. This may actually be self-maintaining, either by vegetative increase of species like the brambles and blackthorn, or by the filling of gaps caused by the death of individual bushes through the growth of seedlings derived from neighbouring individuals of species like gorse or hawthorn.

It is however probable that most of this apparently static scrub does not maintain itself indefinitely, but eventually gives way, by death of the individual shrubs, either to woodland, if grazing is lifted, or to grassland under heavy grazing. But if the grazing is light or intermittent shrubs and especially spinous shrubs may continue to invade the area, maintaining an uneven scrub vegetation with the bushes scattered or in clumps, mixed with grassland, and this is a very common and widespread condition.

Thicket scrub. The apparently static state just described has been distinguished by Salisbury (1918) as "thicket scrub", and contrasted with "woodland scrub" which is "dynamic" and obviously in process of giving rise to woodland. "Woodland scrub" was distinguished by Moss as "progressive", since it clearly forms part of an active sere, the more or less scattered shrubs being interspersed with tree saplings derived from seeds which have sown themselves among the shrubs, gaining protection and favourable conditions for germination and establishment from the shelter of the bushes. Thicket scrub, on the other hand, commonly forms dense thickets devoid of trees, and formed mainly or entirely of thorny species.

Of these the commonest are common gorse (*Ulex europaeus*), blackthorn (*Prunus spinosa*), hawthorn (*Crataegus monogyna*), brambles (*Rubus* spp.) and roses (*Rosa* spp.). The roses and especially the brambles often bind together the other shrubs into an impenetrable tangle. This kind of scrub is typically developed on commonland where there is usually more or less pasturing, but not intensive grazing, and spine-bearing shrubs are able to establish themselves. Along with the thorny species others which do not bear spines, such as hazel or dogwood, are fairly often associated, protected by their spiny neighbours, but they do not attain dominance unless they are sheltered from grazing.

Thicket scrub is often so dense as to exclude practically all herbaceous vegetation (except an occasional etiolated individual) from the soil beneath it, and this is especially true of the evergreen gorse scrub. Thicket scrub is thus commonly a one-layered community. Herbaceous ("marginal") plants, other than those of the pastured grassland or heath communities in which the thicket scrub is usually set, may grow up round the margins of the scrub because grazing (except in areas heavily infested by rabbits) rarely extends right up to the edges of the clumps of bushes. A particular example of thicket scrub (stabilised by rabbit grazing) is described on pp. 480–4.

Woodland scrub. "Woodland scrub" is of course a stage in some subsere (or prisere) leading to a particular type of woodland, and examples have been described in Chapters xiv, xv, xviii and xix. Other instances are the young Norfolk "carr", seral to alderwood (see Chapter xxiii), and the ash-hawthorn scrub, seral to ashwood, shown in Pl. 74, phot. 165.

Untouched examples of woodland scrub are rather rare in this country, because even when such an area is not grazed the young trees are nearly always cut and the shrubs frequently coppiced, especially in the midlands and the south-east. This treatment tends to induce the formation of a scrub approaching to thicket scrub by stopping the progress to woodland and by the promotion of low dense growth of the shrubs. Bricket Scrub, in Hertfordshire, described by Salisbury (1918), is an instance. Relatively heavy grazing and rabbit infestation will convert woodland scrub into thicket scrub. Active woodland scrub is typically loose, and presents a variety of habitats with varying degrees of shade, which can be colonised by herbs of varying requirements.

Flora. The total number of species occurring in areas occupied by woodland scrub, especially when it is partially stabilised by coppicing and grazing, is generally very large, because several different plant communities are represented. This point is well brought out in Salisbury's account (1918) of four Hertfordshire scrubs. In two of these (Kingsland and Markyate) the areas were arable land two or three decades before the observations were made. When such land is left derelict it is first colonised by herbs and grasses, followed by shrubs and trees (cf. Chapter XIV, pp. 294–5). Among the herbs and grasses we may distinguish the following categories: (1) the common *arable weeds* which are already present in the soil: naturally they are the most numerous and conspicuous element in the first vegetation that follows the abandonment of agriculture. In natural undisturbed succession to woodland the arable weeds would be completely eliminated by the establishment of woodland cover, but if the ground is kept partly open by grazing, burning, or removal of young trees, many of them persist for a long time, and some indefinitely. If the area is grazed (2) *pasture grasses* and many *accompanying herbs of pasture*, which have followed the arable weeds, are encouraged and will be largely represented in the flora of such an area. A third element is composed of the numerous (3) *marginal woodland plants* which are adapted to semi-shade conditions and which find their opportunities as soon as shrubs establish themselves. Finally there are (4) the true *woodland shade plants* which come in as soon as good continuous shade and corresponding surface soil conditions are established by the shrubs or young trees. Thus a patch of woodland scrub colonising an area at one time arable and still including pieces of open ground which are lightly grazed will contain five distinct vegetational elements, reckoning the trees and shrubs themselves as one, and a correspondingly large flora.

There is of course a considerable overlap between the species of three of these elements, viz. the grassland species, the "marginal" plants and the woodland shade plants. Some species, e.g. the wild strawberry (*Fragaria vesca*), can maintain themselves in all three habitats in certain circumstances, and whether they will do so probably depends largely on the actual competition they meet with in the different habitats in a particular case. The three habitats corresponding to deep shade, half shade and full light are sufficiently distinct and support distinct plant communities; but it is easy to see that many species which flourish best for example in half shade, may nevertheless be able to vegetate (though they may not flower) in deeper shade; and with some protection, but without overwhelming competition from the species of full light, they may also flourish in the open. More numerous are the species common to the "full shade" and the "marginal", and to the "marginal" and "open grassland" habitats respectively, but absent from the third in each case.

"Marginal" species. It is really only the "marginal" plants which properly form part of the woodland scrub community as such. In the subsere indicated the arable weeds, and open grassland or heath species are clearly relicts from earlier communities, and the full shade plants are advance guards of the true woodland flora which is not yet fully established.

Marginal woodland species are a very important element of the British flora. They include a number of very widely distributed and common plants because (in spite of the comparative rarity of really good examples of woodland scrub) the half shade habitats are of very frequent occurrence. For these habitats include not only scrub, but the margins of artificially bounded woods and also, in the south and midlands of England, the ubiquitous hedgerows, where the same conditions of half shade are realised. "We can regard the sharply defined edge of the wood and copse or the abruptly delimited hedgerow as presenting artificially restricted zones of potential scrub" (Salisbury, 1918, p. 54), and alongside the hedgerows we have the partial protection and partial shade which furnish the characteristic habitat of the marginal species.

in hedgerow

Another important habitat of "marginal" species in the south and midlands of England is the coppiced wood, as Salisbury also points out. A large number of species are in fact common to scrub, hedgerow and coppice. In coppice, as we have seen (p. 278) the periodic cutting of the shrub layer impresses a corresponding periodicity on the life of the field layer vegetation, but for most of the period between two cuttings the conditions are those of partial shade, and thus well suited to the marginal species.

and coppice

It is impossible to draw up a list of "marginal" species which is both comprehensive and accurate, just because of the fact, already mentioned, that so many of them have a wide range of tolerance in respect of light and moisture conditions. Nothing short of a systematic quantitative study of the range of each species in regard to light under different conditions of climate, soil and competition would make such a list possible, and even then more or less arbitrary limits would have to be drawn between the different habitats.

The following list is therefore necessarily a rough one, and very far from exhaustive. Doubtless several hundred species actually grow in the marginal habitat and may be found among scrub. Differences of opinion, largely determined by their individual experiences, may also be expected among those who are familiar with such floras. It should be pointed out here that most of the published lists of woodland species of the field layer, including those given in earlier chapters of this book, contain many marginal as well as "full shade" species.

In the list below species whose main habitat is the forest floor are as far as possible excluded, though they may often occur also in the "marginal" situations, as indeed may many species of grassland or heath. Even so the list is necessarily very heterogeneous, comprising plants of very different status and very different soil and moisture requirements, so that some of them would rarely or never be found actually growing together; but it includes at least a representative selection of marginal plants, most of which may be commonly found on hedgebanks, along wood edges, in coppice and among scrub.

General list of "marginal" woodland species, growing on wood margins, in "woodland scrub" or loose "thicket scrub", on hedgebanks and along the sides of hedgerows

Herbs

Actaea spicata
Agrimonia eupatoria
Anthriscus sylvestris
Arum maculatum
Astragalus glycyphyllos
Brachypodium sylvaticum
Campanula latifolia
C. patula
C. trachelium
Caucalis anthriscus
Centaurium umbellatum
Chaerophyllum temulum
Clinopodium vulgare
Digitalis purpurea
Dipsacus sylvestris
Epilobium montanum
E. obscurum
E. parviflorum
Equisetum arvense
Fragaria vesca
Galeobdolon luteum
Galium cruciata
G. mollugo
Geranium columbinum
G. dissectum
Geum urbanum
Gnaphalium sylvaticum
Heracleum sphondylium
Hypericum hirsutum
H. perforatum
H. pulchrum
Knautia arvensis
Lapsana communis
Listera ovata
Linaria vulgaris
Malva moschata
Melandrium album
M. dioicum
Myosotis arvensis
Nepeta hederacea
Potentilla procumbens
P. sterilis
Sedum telephium
Scrophularia nodosa
Stachys sylvatica
Stellaria graminea
S. holostea
S. nemorum
Valeriana officinalis
Veronica chamaedrys

Woody climbers

Clematis vitalba (calcicole)
Lonicera periclymenum
Solanum dulcamara

Herbaceous climbers

Bryonia dioica
Calystegia sepium
Tamus communis

Scramblers

Galium aparine
Lathyrus pratensis
Vicia cracca
V. sepium

Calcicoles

Atropa belladonna
Helleborus foetidus
Inula conyza
Ophrys muscifera

Rare species

Cephalanthera rubra
Cypripedium calceolus
Melampyrum cristatum
Danaa (Physospermum) cornubiensis
Sibthorpia europaea

Thicket scrub on Chalky Boulder Clay. Ross (1936) has recently studied some areas of thicket scrub in south-west Cambridgeshire on Chalky Boulder Clay overlying Gault, containing 36 per cent of clay and easily waterlogged. The content in calcium carbonate is variable, increasing with depth and sometimes reaching 4–8 per cent at 10 in. The scrub has established itself widely on derelict arable or pasture land in this district, of which a considerable extent has been abandoned from time to time because it is too expensive to cultivate when prices are low.

Woody plants. The scrub is dominated by hawthorn (*Crataegus monogyna*), which, as we have seen, is a very common pioneer shrub on a wide variety of soils, and definitely dominant on the deeper chalk soils and on the heavier loams and clays. It is accompanied throughout the area by the much more local large-fruited species *C. oxyacanthoides* (sometimes known as *C. oxyacantha*) which is present in considerably smaller amount. Besides the hawthorns the common dog rose (*Rosa canina*) is a prominent member of the community and *R. micrantha* is also widespread; *R. arvensis* is less common. The ubiquitous brambles (*Rubus fruticosus* agg.) are nearly constant but not so frequent as the roses. The much lower growing dewberry (*R. caesius*) is locally abundant. Another locally abundant woody species in this scrub is *Ulmus minor*, the small-leaved elm (a common species in East Anglia) which sometimes forms dense thickets by suckering from trees in neighbouring hedgerows. Ross found altogether 18 species of woody plants in the eight areas he investigated, including a few calcicolous shrubs, but only those mentioned above were prominent. Seedlings and saplings of ash and oak, the natural woodland dominants in this region (see Chapter XIII, p. 272), are no more than rare or occasional.

The hawthorn scrub may develop gradually on abandoned pasture, resulting in a scattered uneven-aged community, or rapidly on derelict arable land, giving a denser even-aged scrub. Rabbits are common, and in some of the areas very abundant in spite of the heavy clay soil, checking the development of the scrub rather severely, barking the *Crataegus* and *Ulmus* during the winter and sometimes killing them, as well as eating down the herbage.

Field and ground layers. The herbaceous vegetation occurring among the bushes is numerous in species, of which Ross records ninety-seven, with one Pteridophyte (*Ophioglossum vulgatum*), seventeen mosses, and one liverwort (*Lophocolea*). The climbers are few and rare, including only *Bryonia dioica*, *Tamus communis*, and *Lonicera periclymenum*. The general list of herbaceous species shows a great mixture, as might be expected, comprising weeds of arable land and wayside ruderals, many grasses and grassland plants, including a strong calcicolous element, and a considerable number of "marginal" species. The only true woodland plants are *Carex sylvatica* and *Mercurialis perennis*, both of which are quite rare. *Urtica dioica*, which, like *Mercurialis*, occurs only in the oldest scrub where the conditions are those of woodland, must also be counted here as a true woodland species. There are also a few species, such as *Fragaria vesca* and *Geum urbanum*, which are both marginal and woodland in status.

Rabbit effects. The herbage, where it is not eaten down by rabbits, commonly grows to the height of a foot (30 cm.) and sometimes 2 or 2½ ft. (60–75 cm.), especially round the bushes. In many areas however the effects of rabbit grazing are very evident and the height of the herbage is reduced to 10 cm., but they are certainly not so severe as on chalk grassland or sandy heath. Only in one place was the herbage eaten to about 2 in. (5 cm.).

Of the mosses, rather unexpectedly numerous in species, the most abundant are common grassland forms such as *Brachythecium purum* and *B. rutabulum*, *Hylocomium triquetrum* and *H. squarrosum*, with others that are more commonly found in woods such as *Fissidens taxifolius*, *Mnium undulatum* and *Thuidium tamariscinum*. The heavy damp soil together with the shelter afforded by the bushes evidently favours the bryophytes.

Conditions of colonisation. The overwhelming dominance of hawthorn in this boulder clay scrub is partly due to the enormous number of seed parents in the neighbourhood. Rabbits play an important part in its distribution, eating the fruits and voiding the stones. The drainage furrows remaining from the time when the area was ploughed are especially colonised, perhaps because they are damper, perhaps because the more luxuriant herbage affords greater protection to the hawthorn seedlings.

The accompanying woody plants, *Rosa canina* and *R. micrantha*, *Rubus caesius*, *Ulmus minor* and to a less extent *Prunus spinosa* and *Rubus fruticosus*, are much more prominent where there is general dense colonisation of an area than where invasion is gradual and sporadic. In the latter conditions *Crataegus* colonises the area almost alone.

The other woody species (*Viburnum lantana*, *Cornus sanguinea*, *Ligustrum vulgare*, *Rhamnus catharticus*, with ash and oak) only begin to establish themselves when the scrub has reached a height of about 5 ft., and their coming is directly related to the presence of seed parents in the immediate neighbourhood, i.e. in the hedges. Ash saplings have not been found more than 75 yards from the nearest possible parent trees, and they can establish themselves only when completely protected by hawthorn bushes, but here they could form the nuclei of young woodland. The great majority of seedlings not so protected do not survive the attacks of rabbits. Acorns and oak seedlings are similarly destroyed.

The only tree which Ross actually found suppressing the *Crataegus* is *Ulmus minor*, whose suckers have in some places reached a height of 30 or 40 ft. The small woods of *Ulmus minor* which are not rare in the district are lacking in many abundant and characteristic species of the field layer which occur in the semi-natural ash-oak wood, and it is very likely that they may have taken origin through past subseral colonisation of open ground by suckers of this elm. The status of *Ulmus minor* in the British flora is obscure.

Development of the scrub. In the younger areas, about 10 years old, the hawthorn bushes were from 1 to 3 ft. (30–90 cm.) high, their growth being probably checked by rabbits. The roses (*Rosa canina* and *R. micrantha*) were taller, up to 5 ft. (1·5 m.). In an area of scrub more than 20 years old the hawthorn bushes varied from 3 to 10 ft. (0·9–3 m.) in height, and the roses were much less abundant. Of the thirteen species of woody plants in this area only five (the three species of rose, with bramble and blackthorn—all spiny plants) occurred outside the protection of the hawthorns, except for one young oak. The herbage was about a foot (30 cm.) high. In another area of about the same age (Pl. 93, phot. 225) the hawthorn bushes were rather more widely spaced and grew mainly along the furrows of the old arable "lands" (i.e. the ridge and furrow formation) and the unarmed woody plants (*Cornus, Viburnum lantana, Ligustrum, Fraxinus, Quercus* and *Ulmus*) were again confined to the interior of the *Crataegus* clumps. *Rubus caesius* was abundant beneath their shade. Here the herbage was eaten down by rabbits to about 4 in. (10 cm.).

As the *Crataegus* scrub matures *Rosa* and *Rubus* gradually disappear. In the oldest examples dead remains of both are found, and a few long straggling living shoots growing up into the crowns of the hawthorn bushes, but these plants are obviously barely able to live. As the plagiosere progresses the arable weeds and open grassland species decrease and finally disappear, while marginal species, such as *Brachypodium sylvaticum*, increase to a maximum and then decline. Two woodland species, *Mercurialis perennis* and *Urtica dioica*, were present only in the oldest area examined. The mosses increase in abundance and variety up to a certain point, but in the oldest scrub (Pl. 93, phot. 226) they were very sparse, though the light reaching the ground was no less than in a younger area where they were abundant. Ross suggests that the accumulated litter of dead twigs may have a toxic effect.

Fig. 88*a* shows the distribution of the stems and crowns of the hawthorn bushes in an area more than 35 (probably more than 40) years old. The canopy is nearly closed, privet and rose occurring in the few gaps. In the oldest area, marked as scrub on a map of 1886, and therefore more than 50 years, probably at least 60 years old, the hawthorn bushes were about 18 ft. (5·4 m.) high and formed a completely closed canopy: a view beneath this is shown in Pl. 93, phot. 226). There were no other woody plants in this closed scrub except a little blackthorn and elder and one tree of *Pyrus malus*. The distribution of the *Crataegus* bushes is shown in Fig. 88*b*. The spacing of these was about the same, but more had multiple stems. Dead remains of *Rosa* were found under the canopy. Only 5 herbaceous species, (*Galium aparine, Mercurialis perennis, Primula veris, Urtica dioica, Viola hirta*) were recorded where the hawthorn canopy was fully developed, and

PLATE 93

Phot. 225. Thicket scrub about 20 years old on abandoned arable land on Chalky Boulder Clay, south-west Cambs. *Crataegus monogyna* and *C. oxyacanthoides*, *Rosa canina*. Hawthorns mainly in the furrows. *R. Ross.*

Phot. 226. View beneath closed canopy of hawthorn thicket scrub, more than 50 years old and about 18 ft. (5·4 m.) high. Deep shade, soil almost bare. Same locality. *R. Ross.*

HAWTHORN THICKET SCRUB

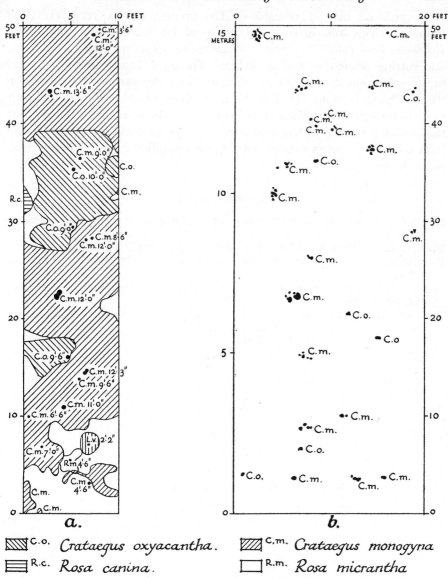

FIG. 88. BELT TRANSECTS OF HAWTHORN "THICKET" SCRUB ON CHALKY BOULDER CLAY IN S.W. CAMBRIDGESHIRE

a, From an area more than 35 years old with nearly closed canopy. *Crataegus monogyna* and *C. oxyacanthoides* (*oxyacantha*) are both present, the former in greater quantity. The only other shrubs present are privet and two species of rose. The heights of the bushes are given in feet and inches. The areas of closed canopy are shown by the diagonal lines.

b, From an area more than 50 years old with completely closed canopy, 18 ft. high. Only the two hawthorns are present: multiple stems are common. Pl. 93, phot. 226 is a view taken beneath the canopy. From Ross, 1936 unpublished.

these were all rare or occasional. The ground was covered with dead branches, twigs and litter and there were few mosses (the commonest *Brachythecium rutabulum* and *Amblystegium serpens*)—mostly on fallen and rotting branches—and no lichens. There was much dead wood, and sappy epicormic shoots had arisen from the lower parts of the hawthorn stems. The indications are that this scrub is senescent, and if this is so it looks as though the duration of its life is well under a century. It is evident that with the paucity of tree seed parents and so long as the rabbit pressure lasts these scrub areas cannot develop into woodland, and they are good examples of a plagioclimax.

REFERENCES

British Vegetation Committee's Excursion to the west of Ireland. *New Phytol.* **7**. 1908.

Moss, C. E. *The Vegetation of the Peak District*, pp. 93–102. Cambridge, 1913.

Ross, R. The ecology of hawthorn scrub in south-west Cambridgeshire (unpublished). 1936.

Salisbury, E. J. The ecology of scrub in Hertfordshire. *Trans. Herts. Nat. Hist. Soc.* **17**, 53–64. 1918.

Tansley, A. G. (ed.). *Types of British Vegetation*, pp. 7, 78–80, 83–4, 94, 130–1, 153–4, 171–3, 236–40. Cambridge, 1911.

Part V

THE GRASSLANDS

Chapter XXV

NATURE AND STATUS OF THE BRITISH GRASSLANDS

Grassland as biotic plagioclimax. While the majority of our deciduous woodlands are properly regarded as more or less modified climax communities, as seral woods clearly leading to climax, or as planted but roughly equivalent representatives of one of these, the vast bulk of our "permanent" and semi-natural grasslands, which occupy between them a far greater area of the country than any other type of vegetation (Chapter IX), are subclimax or (better) biotic plagioclimax vegetation, i.e. vegetation stabilised by pasturing (Chapter X). In other words, if pasturing were withdrawn their areas would be invaded and occupied, as they were originally occupied, by shrubs and trees.

Raunkiaer has pointed out, as already explained in Chapter XI, that the vegetation of the world can be classified in two distinct ways which give very different results. We may either divide it into regions according to the life forms of the dominant species of the climatic climax communities, such as tropical rainforest, temperate continental grassland, summer deciduous forest, northern coniferous forest, and so on; or we may calculate the percentages belonging to the different life forms represented in the total flora of an area, and divide our regions according to the life form or life forms which show a marked excess over their average percentage in the flora of the world. If we use this second method we find that the Mediterranean region, for example, in which the climax communities are dominated by evergreen sclerophyllous shrubs or small trees (*nanophanerophytes* or *microphanerophytes* in Raunkiaer's terminology) becomes a region of *therophytes* (annual plants), because the species of these are preponderant in number in the whole flora; and the temperate region of Central and Western Europe (including most of the British Isles) in which the climax communi-

The British Isles a hemicryptophyte region ties are those of the deciduous summer forest, dominated by *mesophanerophytes*, is a *hemicryptophyte* region because about 50 per cent of all the native species are hemicryptophytes—plants whose perennating buds are situated in or immediately in contact with the surface layer of the soil—against 27 per cent in the flora of the world as a whole.

Now in the grasslands of Britain the large majority of the species of flowering plants are hemicryptophytes. Thus in the chalk grassland of the South Downs 72 per cent of the whole number of species of flowering plants are hemicryptophytes, and 88 per cent of the thirty-three commonest species, making up the great bulk of the vegetation, belong to this life form (Tansley and Adamson, 1926, p. 31). Of our common meadow grasses the vast majority are hemicryptophytes. Thus the British grasslands are essentially hemicryptophytic communities, and we may conclude that the British climate is specially favourable to grassland because grassland is the most characteristic hemicryptophyte vegetation. If the dominant phanerophytic life forms, the trees and shrubs, are destroyed and prevented from returning by regular pasturing (or mowing) they are naturally replaced by hemicryptophytes. While the climate permits the establishment of forest on most soils it is also eminently suitable for grasses. Most of the grasses, and notably the meadow-grasses, require constant moisture during their main growing season (late spring and early summer) if they are to make good vegetative growth, and the predominantly moist climate and well distributed rainfall are excellently suited to their needs. This is most conspicuously seen in the more markedly Atlantic climate in the west of Great Britain and in Ireland (the "Emerald Isle") where the grasslands are particularly luxuriant. In years when rain falls at frequent intervals during the whole summer, as it often does, the meadow grasses remain green and continue to grow during the whole season. After a hot dry summer when the grasslands turn brown after the flowering period, the occurrence of heavy autumn rains leads to a resumption of growth in all but the coldest climates, and this is important for the grazing of stock. Deficiency of rain, low temperatures, and cold, drying winds during the primary growing season (April and May) are particularly inimical to grass vegetation and result in a poor hay crop and deficient pasture.

Turf or "sole" of grassland. The turf-forming grasses are specially well adapted to withstand and flourish under constant pasturing. They naturally form numerous lateral shoots ("tillers") from the base of the primary shoot axis and when the upward growth of these is prevented by the eating or mowing off of the tops, fresh buds are formed at their base, either in or just above the soil surface, and grow out laterally into new vegetative branches, till the whole surface of the soil is covered by a continuous felt-work of leafy shoots—the typical turf or "sole" of good pasture or a well-tended lawn. Flowering and the consequent expenditure of reserves on seed production is very largely prevented and the whole energy of the plants is devoted to vegetative growth. Fresh mineral food must of course be supplied to make up for what is removed by the stock or the mowing machine. In a pasture this is partly furnished by the dung of the grazing animals, often supplemented by extra supplies spread by hand in the winter, but since this is deficient in calcium and phosphorus, lime and phosphates should also be added. A mown lawn has to be constantly manured to replace the nitrogen lost, for example with sulphate of ammonia, and on many soils also with lime, phosphate and potash, to replace the losses involved in mowing. Otherwise the grasses languish, lose their competitive power, and are largely replaced by "weeds"—often deep-rooted herbaceous plants which draw their water and mineral supplies from lower levels of the soil, e.g. dandelion (*Taraxacum*), cat's ear (*Hypochaeris radicata*) and restharrow (*Ononis*), or by mosses which make very small demands.

Effects of overgrazing and undergrazing. In an "overgrazed" pasture the grasses are sometimes grazed down so close to the soil that many of their lateral buds are destroyed, and so much of the plant is eaten away as to impair its powers of vegetative regeneration. This result is seen in its extreme form in grassland heavily infested with rabbits. To keep a pasture or lawn "in good heart" an equilibrium must be established between the supply of mineral food and of water (in the form of frequent rain or from an artificial source such as a lawn "sprinkler") and the constant removal of aerial shoots. In an "undergrazed" pasture tall "rank" growth and flowering are not adequately prevented, so that fresh young shoots, which are by far the most nutritious for stock, are not produced, and no good turf is formed. Also the dead remains of the shoots of many grasses may form a mat or "mattress" which obstructs fresh growth. Then the grassland is invaded either by heath plants or by the seedlings of shrubs and trees, or by both together, and tends to "revert", as the agriculturist would say, to heath or woodland, or, as the ecologist would say, the normal succession to heath or woodland, interrupted by regular pasturing, reasserts itself. The earlier stages of this process can be seen in the pastures left derelict as the result of agricultural depression (cf. Chapter XXIV, p. 480).

"**Rough grazings.**" Besides the large area of "permanent" pasture, which is fenced so that grazing can be regulated, and is often manured in addition, there is a great area of "rough grazings", mostly on open commons in the lowlands and on hillsides and mountain sides, over which sheep and sometimes cattle or ponies are allowed to range in summer. The vegetation of these is largely grass, but partly heath. The grassland is for the most part only lightly pastured and is open to invasion by heath plants and in the more protected places at lower altitudes by scrub and even by trees. The palaeogenic and partly mountainous country of Wales, Scotland and the north of England which is less suitable for agricultural crops naturally bears a much larger proportion of this type of vegetation than the neogenic lowlands which form the greater part of England.

The whole of this hillside vegetation, so far as it is dominated by grasses, was grouped under the general name of "hill pasture" by Robert Smith, the pioneer of ecological survey in Great Britain, in his papers on the Edinburgh District and on North Perthshire (1900). He distinguished different types on the different soils of various hill ranges. In the subsequent surveys of the southern Pennines (west Yorkshire) by W. G. Smith, Moss, and Rankin (1903) "heath pasture", with grasses dominant but heath plants associated, was distinguished from "natural pasture" mostly on limestones, without heath plants and with calcicolous associates. F. J. Lewis (1904), in his work on the northern

Types of rough grazing

Pennines, used the term "grass heath" instead of "heath pasture". In *Types of British Vegetation* (1911) the comprehensive term "siliceous grassland" was used for this kind of vegetation, because it is developed on soil derived from rocks containing a large proportion of silica and deficient in basic minerals; while "limestone grassland" was used as roughly equivalent to the "natural pasture" of the earlier surveyors, because it is mainly developed on limestones, though also on other basic rocks. In the present work the vegetation is described as communities named from the dominant plants in each case, and the communities are grouped according to soil reaction, which on the whole corresponds very well with their natural affinities. Thus the soil of the "grass heath" or "siliceous grassland" is always acid in reaction, that of the "natural pasture" or "limestone grassland" and related types is primarily alkaline because the soil is derived from basic rocks. Owing however to strong leaching under the heavy rainfall of many of the districts in which limestone grassland occurs, and the related accumulation of acid humus, the surface layers even of limestone soil may show an acid reaction and characteristic heath plants are often present.

[margin: Siliceous grassland]

Invasion by heath and peat plants. Most of these grassland communities are based on an acid soil and many are developed in immediate proximity to peat moors, so that they are subject to invasion by heath and moor plants, and all transitions are found between the heath and moor communities proper and the grassland communities. The drier siliceous grasslands are invaded by heath plants (ling, bilberry, etc.) and easily pass into upland heath, while the moister may pass into communities dominated by various peat-forming plants which require a damp or wet soil. Grazing is the main and often the sole factor which maintains the dominance of the pasture grasses. This is notably the case with the drier and better drained grassland on the siliceous rocks, dominated mainly by the common bent (*Agrostis tenuis*) and the sheep's fescue (*Festuca ovina*), often with the red fescue (*F. rubra*), which form the basis of most of these hill pastures. This "grass heath" was perhaps all at one time covered with forest and is subject to invasion by heath and sometimes by scrub and trees. The bent-fescue community is also often invaded by the mat-grass (*Nardus stricta*) which has a rather wide range of edaphic habitats and is often dominant on the drier peats, giving rise to a community which has been called "grass moor". The purple moor-grass (*Molinia caerulea*) with an even wider range of habitats but a higher water requirement, is dominant on the peats which are wetter (the water completely stagnant) of many hill slopes of the north and west. While the *Nardus* grasslands are of little value for pasture, those dominated by *Molinia* are quite useful in the early summer (Chapter XXVI).

[margin: Bent-fescue community]
[margin: Nardetum and Molinietum]

Limestone grassland. Contrasting with the "siliceous grassland" there are in the north and west extensive areas of limestone and other basic rocks (igneous or metamorphic) which bear a distinctive grassland community, generally dominated by *Festuca ovina* or *F. rubra* or both and often referred to by the earlier surveyors of vegetation as "natural pasture", because they afford particularly good grazing for sheep and are as a whole less accessible to colonisation by woody plants. Much of this area too was probably formerly occupied by woodland or scrub, and owes its existing condition to grazing. With it must be classed the chalk and oolite grasslands of southern England (Chapter XXVII).

Arctic-alpine grassland. Closely related edaphically to the limestone grassland is the arctic-alpine grassland developed on basic rocks on the slopes of some of the higher mountains, mostly on southern aspects above 2000 ft. (610 m.). Like the limestone grassland this arctic-alpine community is usually dominated by the sheep's fescue (*Festuca ovina*), but usually the viviparous form, sometimes accompanied by an abundance of the alpine lady's mantle (*Alchemilla alpina*), and a mixture of arctic-alpine species with others from lower altitudes. The true arctic-alpine grassland is certainly a climax vegetation, not primarily dependent on grazing, and belongs of course to the arctic-alpine formation developed above the upper altitudinal limit of trees and shrubs (Chapter XXXVIII); but some of these grasslands, even at high levels, are grazed in summer by sheep.

Other grasslands. Less extensive types of grassland occur in a great variety of other situations. Some occupy man-made habitats like the grass "verges" of country roads, stabilised by cutting and sometimes by pasturing (see p. 561). Some are entirely natural seral types, such as the maritime *Glyceria maritima* and *Festuca rubra* communities of salt marsh (Chapter XL), which may be stabilised by pasturing, and the marram grass community of sand dunes (Chapter XLI): others are maritime rock and cliff communities constantly subjected to salt spray: others again submaritime cliff top (Chapter XLIII) and sand dune communities partly stabilised by grazing. Existing alluvial grasslands on the flood plains of rivers are probably always largely determined by grazing or cutting for hay; but other quite natural seral types exist on the borders of lakes and slow rivers and represent stages of the hydrosere (succession from fresh water), though these also may be stabilised by mowing or pasturing (Chapters XXXII, XXXIII).

Finally, streaks and patches of grassland may occur in heath and moor communities as the result of flushing by water relatively rich in mineral salts, along the lines of paths, or in places where sheep or cattle congregate as the result of manuring and trampling. At the sides of roads, too, the blowing of dust, especially when it contains lime, mineralises a wide verge. This encourages the growth of pasture grasses, and thus more intensive grazing, which leads to the development and maintenance of a good grass sward on each side of the road in place of the original moorland or heath vegetation (Stapledon, 1935). And there are almost endless other situations where the local conditions lead to development of communities dominated by grasses.

Common features of grassland habitats. The dominance of grasses in such a wide variety of situations is the most convincing evidence of the suitability of the British climate to this form of vegetation. The question now arises: can we find any common condition of these grass habitats, apart from the grazing factor, which will differentiate them as a whole from the habitats of the other forms of vegetation which are dominant in the same general climate? It must of course be remembered that though the great majority of our native grasses are hemicryptophytes there are a great many herbs also belonging to this life form, and that the hemicryptophytic grasses themselves vary a great deal in their soil and water requirements: further, that there are a number of grasses which are rhizome geophytes (such for example as *Phragmites*, *Ammophila* and species of *Agropyron*) whose habit of growth is very different from that of the hemicryptophytic meadow grasses. But most of these rhizomic grasses do not form grassland in the ordinary sense, and it remains true that turf and meadow grasses in the wide sense, such as the species of *Poa*, *Festuca*, *Agrostis*, *Avena*, *Dactylis*, *Alopecurus*, *Lolium*, *Cynosurus*, *Anthoxanthum*, etc. do have habit characters and habitat requirements sufficiently similar to be correlated with corresponding habitats.

Climatic factors

The general nature of the grassland habitats has been best expressed by Smith and Crampton (1914). In regard to climate they point out that in Britain the "cold temperate moist summers and open winters favour grassland from sea level to the highest elevations"; but that the moist cold temperate climate also favours leaching and consequent exhaustion of soluble mineral salts in the surface soils of quickly drained places, leading to soil acidity and accumulation of raw humus or peat which are the great physical enemies of grassland, apart from the "grass moors" dominated by *Nardus*, *Molinia* and *Deschampsia flexuosa*. Hence grassland establishes itself on finely divided thin residual soil (correlated with the shallow and fibrous roots of most grasses) developed from a soluble or smooth-weathering rock basis, such as we get on chalk downs and other limestone rocks, and on dolerite and other hard basic rocks; or alternatively where there is periodical flushing of sloping surfaces with waters containing alkaline salts in solution. Grassland developed in the former type of habitat is considered by Smith and Crampton as "stable", in the latter "migratory". These last, they say, "are mainly found on alluvial rain-washed or spring-flushed surfaces along the rivers and coastal belts and on the flanks of mountains". Such grasslands "depend on periodic flushing, flooding and renewal of surface fertility...". In the absence of this dynamic factor constantly renewing fertility of the soil, leaching impoverishes the surface layers of the well-drained soils and heath tends to come in, while on the

Edaphic factors

badly drained soils marsh and fen, or where soluble salts are deficient bog or "moss" are developed.

As we pass west and north in the British Isles the climatic factors hostile to forest are greatly intensified and lead to the increase of leaching and accumulation of raw humus, so that the heath, bog or "moss" vegetation becomes the most serious competitor of grassland: in the midlands and the east and south-east, with higher summer temperatures and lower rainfall, it is forest which naturally supervenes when grassland is left ungrazed. Superposed on these climatic and edaphic factors is the biotic factor of grazing, and this, as explained earlier in the chapter, everywhere favours grassland against the other plant formations, and may override the climatic and edaphic factors. Even the most strongly leached soils, in Wales, for example, remain grassland as long as they are grazed and not waterlogged.

Habit forms. Smith and Crampton divide grassland into five main types according to the habit of the dominant grasses. Of these the *turf-forming* type, characteristically the grazing type, is by far the most important, and next to this the related *meadow* type, consisting mainly of taller growing grasses, traditionally cut for hay, and not forming so close a turf as the first type, partly because of their freer natural habit and largely because their upward growing shoots are not constantly removed. The general vegetative economy of these two types has already been described. The *tussock* type is characteristically formed of coarse hard grasses whose dead basal parts decay with difficulty, accumulating soil and humus between the masses of shoot bases, so that a "tussock" is formed and gradually increases in size. The dominant peat grass *Nardus* belongs to this type, which is also represented by various species of sedge. The *stooled* meadow type (e.g. *Deschampsia caespitosa, Molinia* and various sedges) is an exaggeration of the tussock type. Here the grasses and sedges accumulate the silt deposited by frequent gentle flooding of the flat belts of land on the borders of lakes and slow rivers which are the typical habitats of the stooled meadow type. The tussock and stooled types, owing to the slow decay and constant accumulation of the dead shoots are much less suitable for grazing. Smith and Crampton's fifth type, the *lair* type, is quite specialised and local, originating from the congregation of cattle or sheep and the consequent trampling and manuring of the ground with highly nutritious organic substances.

The chief grass communities. The enormous variety in detail of the habitats and the composition of the British natural and semi-natural grasslands will be sufficiently clear from what has been already written, and in attempting to classify the various communities dominated by grasses it is obviously impossible to take account of every local variation. Still there are a number of important communities that are sufficiently plain, and may be summarised as follows. The first three may be considered together as "acidic" grasslands and are dealt with in more detail in Chapter XXVI.

Acidic grassland. (Chapter XXVI).

(1) *Agrostis-Festuca* **grassland.** This is dominated by species of bent (mainly *Agrostis tenuis*, but also *A. alba, A. canina* and hybrids) with sheep's and red fescue (*Festuca ovina* and *rubra*) and is developed on siliceous and sandy soils, well drained and distinctly (but not extremely) acid in reaction. Of the "natural" grasslands used for grazing ("rough grazings") it is by far the most widespread, occurring on many of the sandy heaths and commons of the lowlands and very extensively on the sides of hills and mountains composed of "siliceous" rocks, approximately up to the limit of former woodland, which it has replaced. The accompanying dicotyledonous plants are rather few, except for some woodland relicts (see Chapter XXVI). Among the few leguminous plants bird's foot trefoil (*Lotus corniculatus*) is the most widespread and is valuable as a pasture plant.

Subordinate types of grassland on more acid sandy and siliceous soil, with more tendency to peat formation, often replacing heath as the result of grazing, trampling or burning, are sometimes dominated by *Holcus mollis* (often a woodland relict), and especially by *Deschampsia flexuosa*.

(2) *Nardus* **grassland.** This is dominated by the mat grass or "white bent" (*Nardus stricta*) and is developed over siliceous soils of medium dampness where the conditions favour the accumulation of acid peaty humus, and on disintegrated peat washed down from the eroding edges of elevated peat moors. The sod is commonly 8–12 in. (20–30 cm.) thick, resting on an impermeable subsoil such as shale or heavy boulder clay from which it is easily separable. It is of much less value for grazing than the bent-fescue type, for though sheep nibble round the *Nardus* plants they fail to keep down the tough growth of the rootstocks and basal leaf sheaths, which accumulate a mat of dead material forming tussocks or stools. *Nardus* grassland is floristically very much poorer than the bent-fescue type, which it often invades when grazing is inadequate.

(3) *Molinia* **grassland.** Dominated by the purple moor-grass or "flying bent" (*Molinia caerulea*) this occurs on peaty soil which is kept much more continually wet than that of the Nardetum. There must, however, always be a certain slow movement and consequent aeration of the water to allow *Molinia* to become dominant. The flora, like that of the Nardetum, is usually relatively poor, but is marked by more species characteristic of boggy places. *Molinia* grassland has a certain grazing value for sheep in early summer, though it is much better if cattle are employed to keep it succulent, but *Molinia* itself is of no value for winter keep, though sheep may find green leafage of other species growing among and protected by the *Molinia* plants.

Grassland dominated by *Molinia* is also formed on fens under certain circumstances (Chapters XXXII, XXXIII) and in other places where it can obtain a constant moving water supply, such as in sandy heath where flowing underground water is available close to the surface.

Basic grassland. (4) **Grassland on chalk, limestone and other basic rocks** (Chapter XXVII). This is a very characteristic type, at the opposite extreme from the acid types, and corresponds with the "stable" grassland of Smith and Crampton (1914). It is no doubt "stable" in the sense that it does not change as a result of allogenic factors, but it is open to invasion by woody plants (chalk and limestone scrub, ash and beech) except at the higher elevations and in the most exposed places. The soil is shallow, finely divided, basic in reaction and very well drained, so that the vegetation is liable to suffer from drought. The most widespread and characteristic grass is *Festuca ovina*, often overwhelmingly dominant. On the Chalk *F. rubra* is almost as abundant, either mixed with *F. ovina* or itself dominant. And in parts of the south of England other grasses (such as *Bromus erectus* and *Brachypodium pinnatum*) may assume dominance. Associated with the sheep's fescue or other dominant, are a number of other grasses and herbs, some of which are commonly known as "calcicole" because they occur especially or almost exclusively on these basic soils.

Three main kinds of rock on which this grassland occurs may be distinguished: (1) the chalk and oolitic limestones of the south and east of England, (2) the harder Palaeogenic limestones of the north and west, of which by far the most important and widespread is the Carboniferous or "Mountain" Limestone, and (3) the hills and "bosses" of basic igneous or metamorphic rock specially characteristic of parts of Scotland, though by no means confined to that country. All three have characteristic rounded contours owing to smooth weathering, though this feature is most marked in the case of the chalk, which is the softest and most easily soluble in percolating rainwater carrying dissolved carbon dioxide, and which never develops crags and screes like the older and harder limestones. The flora and vegetation of the first and second are on the whole closely similar, though a number of species occur on the chalk which are rare or absent on the northern and western limestones. The vegetation of the third type, the basic igneous rocks, has not been closely studied, but it certainly has much in common with that of the harder limestones.

Calcareous grassland affords one of the best *natural* (i.e. unmanured) sheep pastures in the country. It has the reputation of being "dry, sweet and healthy", dry because of the porous soil, "sweet and healthy" because the dryness and the basic "buffer" prevents acidity developing and because the varied herbs and grasses of the community provide in the aggregate a ration that is rich in a variety of mineral nutrients.

(5) **Neutral grassland** (Chapter XXVIII). This collective term has been applied to grassland developed on soils which generally do not depart very widely from the neutral point, such as are derived from many lowland clays and loams (Tansley, 1911, p. 84). It is necessarily a comprehensive and

rather vague category, and the natural or semi-natural communities of which it is composed have not been subjected to close ecological investigation, though many agricultural studies of the effects of different regimes of grazing and manuring have been published. The majority of the grasslands on such soils are "permanent pasture" as opposed to rough grazing, i.e. it is enclosed and more or less intensively grazed, more rarely mown for hay. Most of it is derived from seed mixtures sown on ploughed land; but the grazing and manuring regime a has such decisive selective effect on the composition of the sod, some species dying out and others coming in, that the flora may be completely changed within an amazingly short time according to the treatment.

To this category belongs the great bulk of the grassland known to agriculturists as "permanent grass" and "long ley". It is mainly on village greens, and on some commons, on the edges of woods, and in alluvial meadows which are not too wet, that the more natural communities occur. Such land is practically always grazed (or mown)—otherwise it develops scrub and forest—but the grazing is rarely systematic. The soils on which these communities develop are often very fertile, and among the dominant grasses are the best meadow and pasture grasses, principally perennial rye-grass (*Lolium perenne*), cocksfoot (*Dactylis glomerata*), the rough stalked meadow grass (*Poa trivialis*), crested dogstail (*Cynosurus cristatus*), meadow fescue (*Festuca pratensis*), meadow foxtail (*Alopecurus pratensis*) as well as ubiquitous but less desirable grasses like common bent (*Agrostis tenuis*), sweet vernal grass (*Anthoxanthum odoratum*), and Yorkshire fog (*Holcus lanatus*). Among the grasses are many dicotyledonous herbs, which become more and more restricted in number when the land is enclosed and subjected to a proper grazing regime. The dominance of perennial rye-grass (*Lolium perenne*) and "wild" white clover (*Trifolium repens*) is the mark of the best pastures of this type. The less valuable have *Agrostis tenuis* in increasing quantity.

Footpaths and road verges. Everyone must have noticed the contrast between the vegetation of footpaths and that of the land they traverse, and graziers have found that horses and sheep often prefer to feed upon the footpath rather than upon other parts of the field. The most striking case is seen on footpaths crossing heath, where the edges of the path are inhabited by grass vegetation contrasting sharply with the dwarf shrubs of the heath itself; but footpaths crossing grass fields are marked by the darker green colour of their herbage, which is normally maintained throughout the winter, while the rest of the herbage may be more or less withered. The trampled grass verges bordering on highways show similar vegetation.

Bates (1935) and Davies (1938) have recently investigated the nature and causes of this phenomenon. Bates studied the vegetation of eight footpaths in Lancashire, Derbyshire and Norfolk and on very different soils. They included one traversing a hillside sheepwalk with a very acid soil (pH 4) and a high rainfall, and another through a larch plantation on the same soil: a third traversed a mixed wood, on a somewhat acid medium loam (pH 5): a fourth a neighbouring paddock where the soil was neutral owing to heavy liming. The fifth and sixth examples were on alluvium—one on silt land of excellent grazing value and the other on black riverside neutral soil. The remaining two traversed heath, one a chalk heath with neutral rather moist soil, the other a dry sandy heath (pH 5).

Bates found that the characteristic species were everywhere the same except on the sandy heath, where footpath and "surrounds" alike were dominated by *Agrostis* and *Festuca ovina*. The centre of a path, where much used, is of course bare, all plants being destroyed by the trampling. Adjoining this Bates describes an edging of *Poa pratensis*, and next a zone of *Lolium perenne*, followed by one of *Trifolium repens* (which is however absent in shaded situations such as woodland paths). The lateral boundaries are not sharply defined, the path vegetation merging gradually into the surrounding grassland or heath, for intermittent trampling occurs at the sides though it is much more concentrated in the centre. Of accompanying species *Cynosurus cristatus*, *Dactylis glomerata* and *Plantago* spp. were fairly constant.

Bates considered that this zonation was entirely due to the selective influence of trampling, *Poa pratensis* being the most resistant species, followed by *Lolium* and then *Trifolium repens*. This view he tested by planting ten species side by side in strips and treading a pathway across all the strips. In one experiment the path was also puddled when wet, and then all the species were killed except *Poa pratensis*, *Lolium perenne*, *Dactylis glomerata* and *Trifolium repens*. The growth of the three grasses was restricted, but they recovered and made fresh growth. The clover was also adversely affected but not destroyed. On the other hand *Agrostis tenuis*, *Alopecurus pratensis* and *Anthoxanthum* were completely destroyed by trampling; so were *Agrostis stolonifera*, *Agropyron repens* and *Festuca elatior* when the soil was puddled as well as trampled.

Davies (1938) insists on the point that trampling not only involves vertical pressure from above, but that there is a horizontal twisting action as the ball of the foot leaves the ground, and a similar action by the hooves of farm stock, which tends to break the surface, move, and therefore aerate the soil and 'earth up' the bases of the plants. He therefore holds that treading, unless it is too severe, benefits the majority of grassland plants. When the frequency of treading exceeds a certain point its selective action becomes effective, and beyond a certain degree of intensity all vegetation is destroyed. The two species *Lolium perenne* and *Trifolium repens* are well known to be the most valuable as pasture plants, and occur on the most fertile soils. Their occurrence or dominance on footpaths and grass verges would therefore indicate high soil fertility of these strips of grassland, and this may be partly due to the 'cultivating' action of the foot. The presence of ryegrass and white clover would in its turn attract stock, and the fertility would again be raised by the dropping of their dung along the path or verge.

Davies finds that *Poa annua*, not *P. pratensis*, is the prevalent grass where disturbance is great enough to kill other species. It "shows an amazing capacity to flower, set seed and re-establish itself, and though short-lived is not a typical therophyte, for when 'earthed up' it forms runners which live for more than a year and produce new tillers and roots at the nodes". *Poa annua* should therefore, Davies thinks, be classed as a hemi-cryptophyte along with the great majority of our meadow grasses.

Davies found a much greater variety of grasses on roadside verges than Bates found at the sides of footpaths, differing according to situation, soil and the natural or semi-natural vegetation of the surrounding land. He also criticises Bates' interpretation of the resistance of various species to differential intensity of treading in terms of leaf structure and life form.

In a later paper Bates (1938) replies that sheep tracks and grass verges cannot be directly compared with human footpaths because the effect of human treading is quite different from that of the hooves of sheep and cattle; and that it is difficult to see how the flat-soled human foot can have a 'cultivating' action on the soil. Its rotatory action rather grinds plants into the soil as well as bruising them. *Poa pratensis* (whose determination as the most resistant footpath grass has been confirmed by other authorities) is the characteristic grass of much trodden stable turf, while *Poa annua* is characteristic of unstable, shifting areas.

REFERENCES

BATES, G. H. The vegetation of footpaths, sidewalks, cart-tracks and gateways. *J. Ecol.* **23**, 470–87. 1935.

BATES, G. H. Life forms of pasture plants in relation to treading. *J. Ecol.* **26**, 452–4. 1938.

DAVIES, WILLIAM. Vegetation of grass verges and other excessively trodden habitats. *J. Ecol.* **26**, 38–49. 1938.

LEWIS, F. J. Geographical distribution of vegetation of the basins of the rivers Eden, Tees, Wear and Tyne. Parts I and II. *Geogr. J.* **23**. 1904.

Moss, C. E. In *Types of British Vegetation*, pp. 131–6 (siliceous grassland); pp. 154–60 (limestone grassland). 1911.

SMITH, ROBERT. Botanical survey of Scotland. Part I. Edinburgh district. Part II. North Perthshire district. *Scot. Geogr. Mag.* 1900.

SMITH, W. G. and CRAMPTON, C. B. Grassland in Britain. *J. Agric. Sci.* **6**, 1–17. 1914.

SMITH, W. G. and Moss, C. E. Geographical distribution of vegetation in Yorkshire. Part I. Leeds and Halifax district. *Geogr. J.* **22**. 1903.

SMITH, W. G. and RANKIN, W. M. Geographical distribution of vegetation in Yorkshire. Part II. Harrogate and Skipton district. *Geogr. J.* **22**. 1903.

STAPLEDON, R. G. "Permanent Grass", in *Farm Crops*, edited by W. G. R. Paterson. **3**, 74–136. 1925.

STAPLEDON, R. G. *The Land Now and Tomorrow*. London, 1935.

STAPLEDON, R. G. and HANLEY, J. A. *Grassland: its management and improvement*. Oxford, 1927.

TANSLEY, A. G. *Types of British Vegetation*, pp. 84–7 (neutral grassland), pp. 94–7 (grass heath), pp. 173–8 (chalk grassland). Cambridge, 1911.

TANSLEY, A. G. and ADAMSON, R. S. A preliminary survey of the chalk grasslands of the Sussex Downs. *J. Ecol.* **14**, 1–32. 1926.

Chapter XXVI

THE ACIDIC GRASSLANDS

I. *AGROSTIS-FESTUCA* (BENT-FESCUE) GRASSLAND

Grassland dominated by the bents, mainly *Agrostis tenuis* (= *A. vulgaris*), but also *A. stolonifera* (*alba*) or *A. canina*, together with sheep's and red fescue (*Festuca ovina* and *F. rubra*), all shallow-rooting grasses, probably covers a greater area in the British Islands than any other of the "natural" types of grassland. It is the typical community of the "grass-heath" or "siliceous grassland" of the earlier writers.[1] Its main development is on grazed and well-drained hillsides and the lower slopes of mountains composed of "siliceous" rocks—grits, sandstones, mudstones, slates, and the less basic kinds of igneous and metamorphic rocks; and of one or other of these the great majority of the hills and mountains of the north and west are formed. Davies (1936) estimates that in Wales *Agrostis-Festuca* grassland occupies 5·5 per cent of the total "agricultural area", i.e. including rough grazings. In some siliceous hill regions the percentage area is certainly much higher.

"Siliceous grassland"

But substantially the same "grazing community" is also developed on lowland sandy "heaths" and commons which are grazed or trampled, though here many of the associated species are different; and this last is the community to which the term "grass heath" is most appropriately applied. Commonly dominated, like the siliceous hill pasture, by bent and fescue, and with a similarly acid soil, grass heath contains a wide variety of more xerophilous species, in correspondence with the drier (sandy) soil and with the drier climate of the east and south-east of England.

"Grass heath"

Bent-fescue grassland is not determined by altitude since it may occur in the appropriate habitat at any elevation below the arctic-alpine zone. Characteristically it forms a lower belt on the hills and mountains up to about 1000 ft. (300 m.) or a little higher, above the enclosed pastures, and below the great areas of peat which commonly cap the plateaux and rounded summits. The bent-fescue grassland contrasts by its relatively brighter green (though it is not so vivid as limestone grassland) with the more neutral tints of the peat grasses and the darker colour of the ericaceous undershrubs.

---

[1] The earlier surveyors recognised only the *Nardus* and *Molinia* communities in "hill pasture", "grass heath" or "siliceous grassland", but it has become increasingly apparent that the most widespread semi-natural "grazing community" on these soils is dominated by bent and the fine-leaved fescues. Nardetum and Molinietum maintain themselves independently of grazing, and though dominated by a grass, Molinietum is closely allied, as we shall see, to the bog or moss formation.

Most if not all of this siliceous grassland is potential woodland, and was formerly covered with forest or scrub belonging to the Quercetum sessiliflorae or related communities, which still exist here and there side by side with the grassland under identical conditions of exposure and subsoil. Felling combined with grazing, and very often probably grazing alone, are the agencies which have converted the forest area into grassland. Woodland plants such as the bluebell (*Scilla non-scripta*), wood anemone (*Anemone nemorosa*), wood sorrel (*Oxalis acetosella*) and cow wheat (*Melampyrum pratense*) often occur in the grassland, especially in the shade of bracken, and doubtless represent relicts of former forest. "Grass heath" and bracken-covered slopes on the southern Pennines which are now quite treeless commonly bear Anglo-Saxon or Scandinavian place-names indicating the existence of former forest. And there is documentary evidence that the long Millstone Grit spurs and rock-terraces were covered with forest in Norman times (Woodhead, 1929, p. 26).

Invasion by heath. The heath formation often invades this grassland, the common ling (*Calluna vulgaris*) frequently colonising it when sheep grazing is diminished or excluded. This can be well seen along the line of the stone wall ("mountain wall") which commonly separates the grazing area from the moorland above. On the lower side is pure grassland, on the upper heather moor: if sheep are admitted to a limited area of the moor they quickly convert it into grassland; if they are excluded from the grassland this is rapidly colonised by *Calluna* (see Chapter VI, p. 130, and Fig. 35). Farrow (1916), working on the lowland sandy heaths of Breckland in Norfolk and Suffolk, showed that ling heath is converted into *Agrostis-Festuca* grassland ("grass-heath") by intensive rabbit attack, and rapidly recolonises the ground when rabbits are excluded. The grassland is thus clearly a "biotic climax".

Soil. The soil is typically thin over the subjacent rock, well-drained and moderately acid, but shows little tendency to accumulate raw humus and form peat, except locally where drainage is impeded. The few pH values recorded vary somewhat widely: pH 5·9 (Leach, Longmynd, an unusually high value), 5·0 (Malvern Hills), 4·9 (Clee Hills), and 4·2 in two different Welsh localities, widely separated but both with very high rainfall where leaching must be extreme and *Nardus* readily invades. At Cahn Hill near Aberystwyth the pH value varies between 4·3 and 4·7. Thus pH 4–5 is probably the usual range.

The dominants. The turf is mainly composed of *Agrostis tenuis* (which may be replaced locally by *A. canina*) and of *Festuca ovina*, which occur in various proportions. Thus Stapledon (1914) states that on the sheep-walks of mid-Wales *Agrostis* may constitute from 6 to 45 per cent, *Festuca ovina* from 15 to 56 per cent of the herbage. The sheep's fescue is a more xerophilous grass than the bent and thus tends to become dominant in drier situations. On the lowland sandy heaths of Breckland, Farrow (1917) has shown experimentally that luxuriance of *Agrostis* and its dominance over *Festuca ovina* is directly induced by increased water supply. The bent is of moderate value as sheep feed, and the fescue, though it produces no great bulk of herbage, is nutritious and sheep thrive well on it, though they depend largely on accompanying plants.

Flora. A long list of species may be compiled from the bent-fescue upland pastures, since such factors of the habitat as altitude, exposure, angle of slope, depth and water content of the soil, physical and chemical nature of the underlying rock, as well as the intensity of grazing, vary very much; and no good purpose would be served by enumerating all the species that have been found in this kind of grassland.

Of grasses accompanying the dominants *Sieglingia decumbens* is often present in quantity and may form from 5 to 20 per cent of the herbage in some of the bent-fescue pastures of mid-Wales. Another grass often occurring in this community is the almost ubiquitous sweet vernal grass (*Anthoxanthum odoratum*). *Deschampsia flexuosa* occurs only locally where peaty humus accumulates. The field woodrush, *Luzula campestris*, is locally abundant. Tufts of *Nardus* frequently occur in the bent-fescue grassland, and the invasion of this species from adjacent Nardetum occurs when grazing is diminished, especially if there is a long series of wet years.

Of dicotyledonous herbs the heath bedstraw (*Galium saxatile*) is generally abundant and the tormentil (*Potentilla erecta*) almost as common: other species met with are *Campanula rotundifolia*, *Lathyrus montanus*, *Linum catharticum*, *Lotus corniculatus*, *Polygala vulgaris*, *Teucrium scorodonia*, *Viola riviniana*. *Euphrasia officinalis* (agg.), *Hieracium pilosella*, *Leontodon autumnalis* and *Veronica officinalis* which have very often been recorded from this community, are, according to Davies (*in litt.*) more characteristic in Wales of enclosed fields which have been arable and have reverted to grassland. The yellow mountain pansy (*Viola lutea*) is locally abundant in some regions, forming a beautiful floral feature at midsummer.

There is often a great deal of moss among the herbage in winter and early spring, the commonest species being:

Dicranum scoparium
Hypnum cupressiforme
Hypnum schreberi
Hylocomium squarrosum

On the drier and less peaty soils overlying the granite of Bodmin Moor in Cornwall *Festuca-Agrostis* grassland covers a very large area (Magor, unpublished). The flora is poor, the following species alone being recorded from two localities at 800 and 950 ft. (243 and 290 m.) altitude respectively.

Co-dominant
Festuca ovina and Agrostis tenuis

Abundant or very abundant
Agrostis setacea
Deschampsia flexuosa
Nardus stricta
Potentilla erecta

Less abundant

Carex binervis	r–f	Sedum anglicum	vr
Galium saxatile	o–f	Sieglingia decumbens	lva
Molinia caerulea	o		

Species of *Hylocomium*, *Hypnum* and *Polytrichum* are frequent. These areas are colonised by *Ulex gallii* (f–a) and by *Calluna vulgaris* (r–a), *Vaccinium myrtillus* (o–f) and *Erica tetralix* (r).

On the Staddon Grit and locally also on the slates of the same district *Festuca-Agrostis* grassland occurs on the drier soils and is always associated with locally dominant *Ulex europaeus*, *U. gallii* or *Pteridium* (Magor).

Pteridietum (Pl. 94). The bracken fern (*Pteridium aquilinum*) is a very aggressive species which has spread and is still spreading over much of the *Agrostis-Festuca* grassland and has seriously diminished its grazing value. Dense Pteridietum is said to occupy as much as 3·4 per cent of the total "agricultural area" (i.e. arable and pasture including "rough grazings") of Wales (Davies, 1936). In Scotland it is recognised as a serious menace to the sheep farmers. When the fronds are really luxuriant and closely set they shade the ground so deeply as to inhibit the vegetation of grasses and herbs. In extreme cases the soil is covered with a thick layer of bracken litter and becomes bare of other vegetation. But where the *Pteridium* is not so dense the grasses and herbs of the grassland survive below the over-arching fronds of bracken. Since the *Pteridium* canopy is never fully developed till nearly the middle of June, and at higher altitudes and latitudes not till midsummer, the grassland plants are able to assimilate during the late spring and the beginning of summer. *Pteridium* requires a certain depth of soil for vigorous development, though its horizontally running rhizomes can grow at very various depths, from a few inches to 2 ft. or more. On the shallower soils it can scarcely compete successfully with other species, and it cannot penetrate waterlogged soils, stopping abruptly on the edge of marshy ground.[1] Its distribution is also limited by its demand for shelter. The bracken ascends to 1250 ft. on the Longmynd in Shropshire,[2] to 1500 or 1700 ft. on the southern Pennines, and to 2000 ft. in Scotland, in accordance with the greater height and mass of the mountains, but it is dominant only in valleys and on slopes sheltered from the more violent winds, often ceasing quite abruptly as the ridge is reached.

Pteridium is largely, if not primarily, a woodland plant, but is adaptable to a considerable range of light intensities, changing the habit of its fronds accordingly (see p. 280); and it is perhaps significant that the altitude it reaches corresponds more or less with the limit to which we may suppose forest at one time extended. It is also the fact that the woodland plants which often occur in siliceous grassland are very frequently (though not always) associated with the bracken. A notable instance of this is mentioned by Pethybridge and Praeger (1905), who describe, from the Wicklow

---

[1] Pl. 94, phot. 227.   [2] Pl. 94, phot. 228.

Mountains south of Dublin, glorious stretches of colour in April and May produced by sheets of bluebell, violet, germander speedwell, earthnut, lesser celandine and primrose, on areas which later in the season are covered by the fronds of the bracken. Just as these prevernal and vernal species vegetate and flower in the woods before the leafy canopy of the trees has unfolded, so they do here before the bracken fronds expand; and we may perhaps regard such vegetation as representing the lower layers of the forest community after the trees and shrubs have disappeared, the upper field layer, here composed of *Pteridium*, now dominating the community. But there is no doubt that in suitable soil the bracken spreads far and wide over siliceous grassland, where it attains a dominance much more overwhelming than in most woodlands. The normal method of spread is by the growth and branching of the rhizomes, but in certain years the germination of the spores and the growth of prothalli and young plants start fresh centres of invasion (Braid, 1937).

Not only does *Pteridium* thus invade *Agrostis-Festuca* grassland, but also Calluneta, Vaccinieta, Nardeta and even the drying peat of the edges of retrogressive Eriophoreta at relatively high altitudes, provided the situation is sheltered (Adamson, 1918, p. 104). This invasion is facilitated by rather deep sandy soil or drying peat, and is probably greatly aided by the burning of the moorland, since the bracken rhizomes, owing to their depth below the surface, are rarely injured by the fire, and they may easily occupy the ground before the moor plants have regenerated.[1]

Bracken is often cut for litter in the autumn, since it forms excellent bedding for cattle; but cutting after the fronds are dry and empty of reserves does nothing to check the growth of the rhizomes. Repeated cutting in successive years, if carried out after the fronds are mature but before the reserves have been transferred to the rhizomes, will of course diminish the luxuriance of the plant, but it takes a very long time to exterminate bracken in this way. Spraying with sulphuric acid, and more recently with sodium chlorate, has been employed to keep down bracken and is often recommended, but this procedure is both expensive and troublesome and is not free from danger. A better and quite effective method, though it takes longer to carry out, is to cut repeatedly during the season, as soon as new fronds appear. But this is seldom done owing to the cost of the labour required. Where bracken becomes overwhelmingly dominant and thus completely destroys the pasture, and where effective checks are too expensive, it is probably better to abandon the pasture and afforest the land (Stapledon).[2]

If we consider the bent-fescue community as replacing oak-birch wood of which *Pteridium* is an important constituent, then the Pteridietum is

---

[1] It is doubtful if bracken can invade vigorous undisturbed Callunetum (see p. 727)

[2] Quite recently considerable success in destroying bracken has been obtained by the use of various types of machine which crush the young fronds.

primarily a society which survives from the woodland association, protected from grazing because it is unpalatable, and finding the opportunity for fresh aggressiveness and dominance when it is exposed to the full light of the open.

Ulicetum. Gorse scrub composed of one or both of the two species, *Ulex europaeus* and *U. gallii*, is often a conspicuous feature in bent-fescue grassland, sometimes covering considerable areas. In the west the latter is the typical species. Locally the low gorse scrub is dense and forms a continuous canopy, but more usually the Ulicetum gallii consists of separate dwarf bushes more or less closely spaced, not more than 2 or 3 ft. high, and nibbled by sheep into a rounded form, set in a matrix of grassland (Pl. 95). According to Davies (1936) dense Ulicetum occupies about 0·2 per cent of the agricultural area of Wales.

Ulex europaeus, the common gorse, furze or whin, is mainly a plant of the lowlands, though isolated bushes may ascend to 2000 ft. in southern Scotland. Its commonest habitats are disturbed grassland on sandy or loamy soils belonging to the bent-fescue or neutral grassland types and mainly occurring on commons and waysides and in neglected pastures. It is also found on the edges of *Calluna* heath, often occupying in dense masses the disturbed ground between a village and a neighbouring heath, spreading along the lines of tracks and paths across the heath and sometimes invading the heath itself. The spread of the gorse along paths and tracks is probably due largely to transport of the seeds by ants, which bite and tear at the fleshy orange-coloured "caruncle", dragging the seeds along the ground. The common gorse cannot be regarded as a proper member of the heath association, and it is doubtful if it can invade *Calluna* heath unless the ground has been disturbed.

The common gorse was at one time much used as fodder. The soft young shoots are very palatable and nutritious, and formerly the hard mature shoots were ground up in special mills and used as cattle food in some parts of the country. In north and west Wales *Ulex europaeus* is said not to be native, but introduced (from Ireland) in the eighteenth century for cattle fodder, the plants being cut every alternate year (Alun Roberts). When this use was abandoned the gorse grew up, flowered and seeded, so that it has now spread widely along the roadsides and over the lower hill pastures, where it may be seen side by side with the native *U. gallii*. It is still occasionally planted to form hedges and rather widely in the midlands and south for fox coverts.

Ulex europaeus is a taller and freer growing shrub than *U. gallii* and commonly reaches a height of several feet, so that unless it is closely grazed or kept cut while still young its shoots are lifted out of the reach of grazing animals. Its main flowering season is April and May, though a few flowers open in mild weather from October to March, and quite a conspicuous show of blossom sometimes appears near the south coast in the

PLATE 94

Phot. 227. Pteridietum stopping abruptly at the foot of a well-drained slope. *Holcus lanatus* occupies the wet ground on the right. Wanister Hill, Co. Durham. *H. Jeffreys* (1916).

Phot. 228. A small valley on the Longmynd (Shropshire) with Pteridietum extending up the slope on the left but not on the right, which is more exposed to wind and occupied by Vaccinietum myrtilli. Callunetum covers the summit of the ridge above. *W. Leach* (1931).

PTERIDIETUM

PLATE 95

Phot. 229. Ulicetum gallii on the Wicklow Mountains south of Dublin. Cf. Fig. 89. *R. Welch.*

Phot. 230. Ulicetum gallii in *Agrostis-Festuca* grassland, Malvern Hills, Worcestershire.

SHEEP-GRAZED ULICETUM GALLII

middle of a mild winter. The pods ripen and explode so that the seeds are shot out during July.

When the two species grow together, as on the roadsides of Devon, Cornwall and Wales, they may be easily distinguished not only by the fact that they are never in flower at the same time, but also because the freer

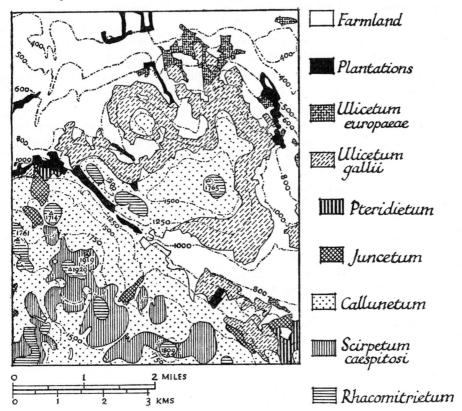

FIG. 89. ZONATION OF VEGETATION ON THE WICKLOW MOUNTAINS SOUTH OF DUBLIN

Ulicetum gallii forms a broad well-marked zone, mainly between 600 and 1300 ft. (Pl. 95, phot. 229). Below (to the north) are patches of Ulicetum europaeae between 400 and 700 ft. Pteridietum occurs locally in protected situations. Above is Callunetum (1250–1800 ft.—Chapter xxxvii), and the flat summit areas are occupied by Scirpetum caespitosi (above 1750 ft.—Chapter xxxv) often capped by Rhacomitrietum. Contours in feet. After Pethybridge and Praeger, 1905.

growing, taller bushes of *U. europaeus* are of a bluer green than those of the lower, more compact *U. gallii*, which are yellowish green.

On the bent-fescue grasslands of the northern and western hillsides *Ulex europaeus* usually occurs only in the lowest zone, especially along roadsides, tracks, and close to cultivated land, in the west often giving way above to *U. gallii* (Fig. 89 and Pl. 95). This species, as has been said, is of lower and more compact growth and on much-grazed land forms characteristic rounded cushions from 1 to 2 ft. high. This growth form is mainly due to constant nibbling of the young shoots by sheep or rabbits, but in the most exposed places also to the action of wind in drying off shoots protruding from the surface of the cushion. In Wales, Cornwall, and to a less extent in other western English counties, *Ulex gallii* often forms a well marked zone on the grassy hillsides, alternating with areas dominated by bracken, and in some years covering the hills in late summer with sheets of golden bloom. In Ireland, where its distribution is similar, it sometimes ascends to 2000 ft. (610 m.). On the hills of the north of England it is much less general, and in the midland and eastern counties, while not wholly absent, it is quite rare. The shrub flowers from July to September, and the pods ripen in the following spring.

On sandy or loamy commons scrub of *Ulex europaeus* may protect the seedlings of colonising trees, much as juniper does on the chalk escarpments, and thus form the first stage of the seral development of woody vegetation; but more usually it forms, either alone or in company with other spiny shrubs, a dense and apparently permanent scrub community, casting so deep a shade that practically no vegetation can grow beneath it (cf. Chapter XXIV). How long such scrub communities actually survive has never been determined, but it is probable that they are relatively short-lived and thus more or less "migratory", the bushes dying in some places and seedlings starting in fresh areas of disturbed soil.

Gorse scrub is very often burned to clear the ground for the pasture grasses, but it has considerable power of springing again from the basal unburned parts of the stem, at or just below the soil surface, after destruction of the aerial shoots.

GRASS HEATH

"Grass heath", as it is developed on the sandy commons of the English lowlands, is a very characteristic though very variable community. It is called "grass heath" both because it occurs on the same soils as "true" (i.e. *Calluna*) heath, by which it is invaded and which it may replace, and also because it is inhabited by the "heath grasses" that also occur in Callunetum wherever there is room: among these *Agrostis* spp., *Festuca ovina*, *F. rubra*, *Deschampsia flexuosa*, and sometimes *Holcus mollis* are the commonest. The relation of grass heath to lowland *Calluna* heath is essentially the same as the relation of the bent-fescue hill pasture to upland heath or heather moor. Grazing will convert the heath into grassland, cessation of grazing leads to recolonisation by heath and the formation of a surface layer of raw humus or dry peat. Where the dominance of *Calluna* is incomplete the grasses of the grass heath and many of the associated species occur in the intervals. For this reason any extended floristic list of Callunetum will include many grass heath species. An essential difference between the two communities is that grass heath does not form the surface layer of raw humus (dry peat or mor) characteristic of Callunetum.

On the drier sandy soils in a relatively dry climate grass heath may contain a great number of "arenicolous" species, some of which are confined, in Britain at least, to this habitat. East Anglian heaths, particularly those of Breckland, are among the driest communities in England, and include many species which are not found in Callunetum, and several which do not occur outside East Anglia. Grass heath communities are also often developed on grazed or rabbit-infested fixed sand dunes.

The soil of grass heath may initially contain considerable amounts of calcium carbonate (e.g. on fixed sand dunes where the sand contains shell fragments and on the calcareous drift of Breckland), but its highly permeable nature leads to rapid leaching, so that the surface layers become markedly acid in reaction, unless buffered by abundant calcium carbonate in the subjacent soil as in many areas of Breckland. And the numerous sands which are more purely siliceous show an acid reaction from the beginning. In the driest and in disturbed places the community may be open, with bare patches of sand between the plants. This gives room for ephemerals and other annuals and also for the invasion of weeds from neighbouring sandy arable land.

Floristic list. For these reasons any comprehensive floristic list will be a long one. The following is a composite list made up from various sources, and it must not be supposed that the species included would all be found together on a particular heath. The species confined or nearly confined to Breckland are given in a separate list (p. 512). Most of them are not in fact grass heath species.

General dominants:
    Agrostis tenuis   Festuca ovina (with F. rubra)

Dominant on the more acid heaths where thin peat is formed:
    Deschampsia flexuosa   Cladonia spp.

Locally dominant:
    Ulex europaeus (invading)   Sarothamnus scoparius (invading)
    Pteridium aquilinum (invading)   Festuca capillata
    Holcus mollis   Holcus lanatus

General list:
    Achillea millefolium   Campanula rotundifolia
    Agrostis canina   Carduus nutans
    A. setacea (south-west England)   Carex arenaria (ld in Breckland)
    A. stolonifera   C. binervis
    Aira caryophyllea   C. caryophyllea
    A. praecox   C. divulsa
    Alchemilla arvensis   C. pilulifera
    Allium vineale   Carlina vulgaris
    Anthoxanthum odoratum   Centaurium umbellatum
    Arenaria serpyllifolia   Cerastium arvense
    Calluna vulgaris (invading)   C. vulgatum

C. semidecandrum
Cirsium arvense
C. lanceolatum
Conopodium majus
Corydalis claviculata
Daucus carota
Dianthus armeria
D. deltoides
Erica cinerea (occ. invading)
Erodium cicutarium
Erophila verna (ephemeral)
Euphrasia officinalis
Filago minima
Galium saxatile
G. verum
Genista anglica
G. pilosa
Geranium dissectum
G. molle
G. pusillum
Gnaphalium sylvaticum
Hieracium boreale
H. pilosella
H. umbellatum
Hypericum humifusum
H. perforatum
H. pulchrum
Hypochaeris glabra
H. radicata
Jasione montana
Koeleria cristata
Kohlrauschia prolifera (Dianthus prolifer)
Leontodon autumnalis
L. nudicaulis
L. hispidus
Linum catharticum
Lotus corniculatus
Luzula campestris
L. multiflora
Malva neglecta (rotundifolia)
Medicago arabica
M. hispida var. denticulata
M. lupulina
Minuartia (Arenaria) tenuifolia
Moenchia erecta
Myosotis arvensis
M. collina
M. versicolor
Ononis repens
Ornithopus perpusillus
Plantago coronopus
P. lanceolata
Polygala vulgaris
Potentilla argentea
P. erecta
Radiola linoides (damp ground)
Ranunculus bulbosus
Rumex acetosella
Sagina apetala
S. ciliata
S. procumbens
S. subulata
Saxifraga tridactylites
Scilla autumnalis
Scleranthus annuus
Sedum acre
S. anglicum
Senecio jacobaea
S. sylvaticus
Spergularia rubra
Taraxacum erythrospermum
Teesdalia nudicaulis
Teucrium scorodonia
Thymus serpyllum
Trifolium arvense
T. dubium
T. filiforme
T. procumbens
T. subterraneum
Veronica arvensis
V. officinalis
Vicia angustifolia
V. lathyroides
Viola canina
V. riviniana

In heavily rabbit-grazed areas (e.g. parts of Breckland) a number of tall rabbit-resistant plants standing up far above the level of the razed herbage are characteristic, for example:

Conium maculatum
Senecio jacobaea
Solanum nigrum
Urtica dioica
U. urens

together with smaller plants such as species of *Myosotis*, *Arenaria serpyllifolia*, etc.

Societies of *Pteridium* and *Ulex* frequently invade the heath, the bracken often covering wide areas with dense growth.

The following are among the common bryophytes on acidic grass heaths:

Campylopus flexuosus
Ceratodon purpureus
Dicranum scoparium
Hypnum schreberi
Leucobryum glaucum
Polytrichum piliferum
P. juniperinum
Tortula muralis

and of lichens the commonest are

Cetraria aculeata
Cladonia spp.

The grass heaths of Breckland

Climate and soil. Breckland is a well-defined physiognomic region in south-west Norfolk and north-west Suffolk (see Figs. 5, 31). It is of low elevation, varying from about 50 to about 200 ft. above sea-level (15–60 m.) with an average annual rainfall (seven stations) of 591 mm. or a little less than 24 in. The lowest record in the area is from Kilverstone Hall and is 551 mm. or about 22 in. The soil is derived from chalky glacial drift and interglacial sands, with local gravels and loess, and is almost everywhere sandy. There is local dune formation on a small scale, with blowouts, and considerable areas are covered with wind-blown sand. Despite its comparatively uniform texture the soil varies a great deal chemically. Where the chalky drift comes close to the surface the calcium carbonate and pH value are high, while in other areas there is much less carbonate but a relatively high base status, and others again are destitute of calcium carbonate, poor in bases generally and show very low pH values. These differences are reflected in the vegetation, which is floristically rich in the two former and very poor in the last-named type.

Vegetation. The vegetation is largely grassland with societies of *Pteridium* and *Carex arenaria* and considerable stretches of Callunetum. Farrow showed that the very heavy rabbit pressure to which some areas are subjected suffices to convert Callunetum into grassland (see pp. 136–8), but this is by no means the only factor involved. The calcareous soils are not invaded by *Calluna* and bear quite a different flora from the highly leached and podsolised sands. Watt (unpublished)[1] has in fact distinguished seven different types of grassland, one of which (A) is scarcely distinguishable from chalk grassland, another (B) has a wide range of pH values, a high base status and a rich flora (eighty species of flowering plants), while others again (D–G) have a low base status and a poor flora, the poorest (G) with only ten species of phanerogamic plants and lichens dominant.

Types D–G may fairly be classed as "grass heath" since it is only these which *Calluna* invades. Consideration of the status of types A–C must await the publication of Watt's work.

[1] Dr Watt has most kindly put his data, as yet unpublished, at my disposal.

Acidic Grasslands

Composite soil samples 0–6 in. (15 cm.)

	D	E	F	G
Loss on ignition	2·7	3·34	5·73	4·91
Loss on ignition or top 1½ in.* (3·75 cm.)	4·19	5·30	7·93	7·52
Exchangeable calcium†	0·76	0·13	0·07	0·0
pH value	4·3	3·8	3·6	3·5

* Soil black because of fine dispersion of humus.

† Figures almost negligible: of the better types of grassland A has 51·2, B 34·4, C 6·3 per cent of exchangeable calcium.

Species of selected areas. The numbers in brackets refer to the "constancy" of the species: 5 = occurrence in more than four-fifths, 4 = more than three-fifths, 3 = more than two-fifths, 2 = more than one-fifth, 1 = less than one-fifth of the areas listed.

It is to be noted that *Carex arenaria* and *Calluna vulgaris* can and do invade all these four types, but the areas chosen for examination were those in which *Agrostis* and *Festuca* were dominant.

	D		E		F		G	
Spermaphytes								
Agrostis stolonifera / A. tenuis	f–a	(5)	o–a	(5)	o–a	(5)	o–la	(5)
Aira praecox	o–lf	(5)	o–la	(5)	lf–la	(5)	lf	(5)
Alchemilla arvensis	l	(5)	l	(3)	l	(2)		
Anthoxanthum odoratum	o	(1)						
Arenaria serpyllifolia	l	(3)	l	(1)				
Calluna vulgaris	r–o	(4)	r	(2)	vr	(1)		
Campanula rotundifolia	lf–la	(5)	r–lf	(2)	vr	(1)		
Carex caryophyllea	r–la	(3)						
Cerastium arvense	l	(1)						
C. semidecandrum	l	(4)	l	(3)	l	(2)		
C. vulgatum	l	(2)						
Cirsium arvense	o	(3)						
C. lanceolatum	r–o	(2)						
Erodium cicutarium	l	(1)						
Erophila verna	l	(2)						
Festuca ovina	a	(5)	f	(5)	o–f	(5)	o–f–la	(5)
Galium saxatile	o–a	(5)	o–la	(5)	a	(5)	o–lf–la	(5)
G. verum	o–f–la	(5)	r–o	(3)	vr	(1)	vr	(1)
Hieracium pilosella	r	(2)	vr	(1)				
Hypochaeris glabra	o	(5)	l	(2)				
Koeleria cristata	o	(3)	vr	(1)				
Luzula campestris	f–a	(5)	o–f–la	(5)	a	(5)	o–la	(5)
Myosotis collina	l	(5)	l	(3)	l	(2)		
Ornithopus perpusillus	r	(2)						
Poa pratensis	r–o	(1)						
Rumex acetosella	f–a	(5)	o–lf	(5)	f	(5)	o	(5)
Sagina ciliata	l	(4)	l	(2)				
S. procumbens	l	(2)						
Sedum acre	l	(3)						
Senecio jacobaea	o	(5)	r	(2)				
Stellaria media			l	(1)				
Taraxacum erythrospermum	o	(4)						
T. officinale	r	(4)						
Teesdalia nudicaulis	o–f–la	(5)	o–lf–la	(5)	l	(5)	l	(5)
Urtica dioica	l	(5)	l	(3)	l	(2)		
Veronica arvensis	l	(5)	l	(2)	l	(2)		
Vicia lathyroides	r	(1)						
Viola canina	o–f	(4)						
Total species of spermaphytes	**38**		**23**		**16**		**9**	

	D		E		F		G	
Bryophytes								
Brachythecium albicans	l	(5)	l	(2)	l	(2)		
Bryum sp.	l	(4)	l	(3)	l	(1)	l	(1)
B. capillare					l	(1)		
Cephaloziella sp.	o	(4)	l	(5)	l	(3)	l	(4)
Ceratodon purpureus	la	(5)	lf	(5)	la	(5)	lf	(5)
Climacium dendroides	l	(1)						
Dicranum scoparium	f–a	(5)	o–f	(5)	f	(5)	f–la	(5)
Hypnum cupressiforme	o	(5)	l	(3)	l	(3)	l	(2)
H. schreberi	lf	(3)						
Lophocolea bidentata	vr	(1)						
Lophozia barbata	o–f	(5)	l	(3)	l	(1)		
L. excisa	o	(5)	l	(2)	l	(1)		
Polytrichum juniperinum	o–lf	(5)	o–ld	(5)	la	(5)	lf	(4)
P. piliferum	lf	(4)	lf–la	(5)	lf	(2)	lf	(4)
Ptilidium ciliare	f–a	(5)	o–f	(5)	f	(5)	o–f	(5)
Tortula ruraliformis	l	(1)						
Species of bryophytes	15		11		12		8	
Lichens								
Biatora granulosa	vr	(1)					r	(2)
B. uliginosa			l	(5)	l	(5)	o	(5)
Cetraria aculeata	o	(5)	f	(5)	o–f	(5)	o–f	(5)
Cladonia alcicornis	r–o	(4)	o–l	(4)	r–o	(4)	r–o	(1)
Cl. bacillaris			l	(2)	l	(2)	o	(3)
Cl. coccifera	r	(2)	l	(3)	o, l	(2)	o, l	(3)
Cl. fimbriata	o	(5)	o, l	(4)	o	(3)	o	(5)
Cl. floerkiana			r	(1)				
Cl. furcata	f	(5)	f	(5)	f–a	(5)	f	(5)
Cl. pityrea	vr	(1)	r	(1)	r	(1)	vr	(1)
Cl. pyxidata	o	(5)	l	(5)	o	(4)	o	(5)
Cl. rangiformis	l	(4)	l	(2)	l	(3)		
Cl. sylvatica	f–a–d	(5)	d	(5)	d	(5)	d	(5)
Cl. uncialis	o	(4)	o	(5)	o–lf	(5)	o	(5)
Peltigera polydactyla	o	(5)						
Species of lichens	12		13		12		12	
Total of all species	65		47		40		29	

In these grass heaths, highly acid in reaction, very poor in bases and increasingly rich in humus, it will be seen that the phanerogamic and to a less degree the bryophytic flora decreases in passing from the less poor to the poorest type, while the number of species of lichens remains practically constant. Of the phanerogams most species are absent from G, but of those which remain nearly all are quite constant (5), though they may be only occasional and local in occurrence. These are of course the most unexacting and acid-tolerant species. The lichens on the other hand, mainly Cladoniae, are as numerous in species in G as in D, and *Cladonia sylvatica* is definitely dominant. All these types might indeed fitly be called Cladonietum sylvaticae with scattered flowering plants here and there in the *Cladonia*-mat, while the more calcareous grasslands A–C (not described here) are closed communities of grasses and herbs with a more or less interrupted under-storey of mosses and lichens.

Species peculiar to Breckland. Breckland is notable for the occurrence of a number of species which are entirely or nearly confined to this region within the British Isles. They are all species of definitely "continental" distribution, either with headquarters in the steppes of south-eastern Europe or with the centre of distribution not so definitely localised (Matthews, 1937).

The following is a list of British species either confined to Breckland or largely centred in and characteristic of that region:

‡*Artemisia campestris
*Carex ericetorum¹
Herniaria glabra²
*Holosteum umbellatum³
†*Medicago falcata
†M. minima
*M. varia (sylvestris)
*Muscari racemosum
Ornithogalum umbellatum

‡Phleum phleoides
*Scleranthus perennis
Silene conica⁴
‡*S. otites
Tillaea muscosa¹
*Veronica praecox
‡*V. spicata
V. triphyllos⁴
†*V. verna

* Confined or practically confined in Britain to Breckland.
† "More or less essentially steppe plants" (Matthews, 1937).
‡ "Definitely have their headquarters in the Sarmatian-Pontic steppes" (Matthews).
 ¹ "Continental northern" (Matthews).
 ² "Continental southern" (Matthews).
 ³ "Continental" (Matthews), perhaps introduced but possibly native in Breckland.
 ⁴ "Continental" (Matthews).

It is noteworthy that none of these species appear in Watt's list on p. 510, and that in fact all of them, except *Tillaea*, occur only in the better types of Breckland grass communities, whose soils contain calcium carbonate or have a high base status. They are not, therefore, grass-heath species and have been mentioned here only as a matter of interest. Consideration of the status of these grassland communities on the better soils must await the publication of Dr Watt's results.

Succession. Very recently also Watt (1938) has studied the development of the *Agrostis-Festuca* community in Breckland on very infertile sandy soil largely derived by leaching out of the lime from Chalky Drift containing much sand, and partly on sand which has been blown over the surface.

Soil. The development of this "Festuco-Agrostidetum" takes place on sand bared by "blow-outs" following degeneration of preceding vegetation. In the particular area studied this soil, which overlies Chalky Boulder Clay at a depth of 1·25 m., contains 93 per cent or more of "coarse" and "fine" sand, the silt and clay fractions being negligible. It is devoid of calcium carbonate and the surface layer has a pH value of 4·2.

The sere. In main outline the succession is simple (Fig. 90), but in detail very complex, each stage "slipping back" from the position it has won and progress starting again on a lower plane. The first effective colonist is *Polytrichum piliferum* (Fig. 90, "early stages"), which brings about the initial stabilisation of the loose blown sand. This is followed by *Cetraria* *aculeata* and later by *Cladonia sylvatica*. At the same time plants of *Festuca ovina* appear, scattered through the moss and lichen communities, and of *Agrostis* (*A. stolonifera* and *A. tenuis*) in rare patches. The moss accumulates sand, growing up through the freshly deposited layers, and the two lichens settle down on this substratum. Under the *Cetraria-Cladonia* mat the *Polytrichum* dies, and the mat, having lost its anchorage, disintegrates in whole or part, exposing the soil to partial erosion. Further

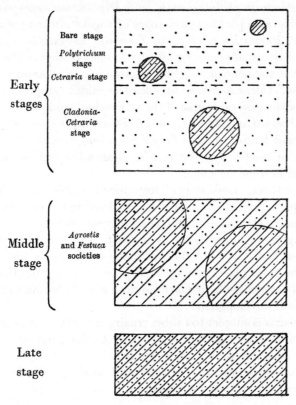

FIG. 90. DEVELOPMENT OF FESTUCO-AGROSTIDETUM IN BRECKLAND

Only the early progressive and "peak" stages are shown. The dots represent individuals of *Festuca ovina*, the diagonal lines *Agrostis* spp. (*Polytrichum* and the lichens are not represented by symbols.) From Watt, 1938.

development is conditioned by the presence of the higher plants, whose effect is to expedite the progress of the succession. In a narrow ring round each *Festuca* tuft, and throughout the patches of *Agrostis*, the dominance of *Cladonia* is established. The islands of *Festuca* and *Agrostis* are set in a background of *Cladonia-Cetraria*.

The next (Fig. 90, "middle stage") shows patches of abundant *Agrostis* with a little *Festuca* set in a background of much *Festuca* with a little *Agrostis*, the whole having an almost continuous carpet of *Cladonia*.

The third or adult stage (Fig. 90, "late stage") shows an intimate mixture of abundant *Agrostis* with numerous but small *Festuca* plants, set in a carpet of *Cladonia*. The humus content of the soil, very low in the early stages, now reaches 2·9 per cent.

The other lichens present during the succession are Cladonias (*C. furcata*, which rises to 13 per cent of cover in the *Cladonia-Cetraria* stage, *C. uncialis, crispata, coccifera, verticillata, pityrea, bacillaris, fimbriata, floerkeana*), *Alectoria chalybeiformis*, and *Diploschistes scruposus*, but they play no considerable part in the changes of vegetation; and of higher plants *Aira praecox*, *Rumex acetosella*, and *Teesdalia nudicaulis*, a few scattered individuals of which appear sporadically in one or other of the early stages. It appears that the poorest Breckland soils exclude all annuals except *Aira* and *Teesdalia*. *Carex arenaria* invades the middle stage in quantity but does not alter the general course of the succession, being subject to the same influence as *Festuca* and *Agrostis*; and *Luzula campestris* appears only in the adult stage, which also contains *Polytrichum juniperinum*, *Ptilidium ciliare*, *Lophozia excisa* and *Cladonia alcicornis*. These additions are perhaps due to the increased humus content.

None of the three "peak stages" described is stable. *Agrostis* may die in whole or in part, the lichen cover disrupts, leaving the soil open to erosion, and retrogressive stages, similar to the progressive, can be traced, followed by fresh progression towards the next peak. A series of dry years favours the retrogressions but is not their primary cause, since progressions and retrogressions take place side by side, and Watt thinks that the cyclic behaviour of each stage is primarily due to a still obscure cause related to cycles of lichen development.

The whole area is subject to rabbit grazing and the grassland climax is no doubt determined by this factor. Neither *Calluna* nor any other woody plant enters into the communities studied.

For extensive quantitative details and analyses of the data Watt's original paper (1938) must be consulted.

II. NARDUS GRASSLAND (NARDETUM STRICTAE)
(Pls. 96–7).

The very characteristic and widespread community dominated by the mat-grass or "white bent" (*Nardus stricta* L.) is intermediate in habitat and character between the bent-fescue grassland and the drier peat vegetation dealt with in Chapters XXXIV and XXXV. In *Types of British Vegetation* it was in fact described both as a typical form of "siliceous grassland" in the southern Pennines (Moss) and also, along with Molinietum, as a facies of "grass moor" in Scotland and elsewhere (W. G. Smith). While it is properly described here as a characteristic type of acid grassland, it has more intimate genetic relationships, as Adamson (1918) has shown, with acid peat vegetation than with bent-fescue grassland.

Nardetum

Distribution, habitats and growth form. *Nardus stricta*, which has a wide geographical distribution, extending from the mountains of southern Europe and the plains of western Asia to western Greenland, occurs in a great variety of habitats, whose common character seems to be a soil of acid and damp but not permanently wet raw humus or peat, with stagnant or nearly stagnant soil water. In this country it ranges from near sea level to quite high altitudes, attaining more than 3000 ft. in Scotland, though it is much more commonly dominant on the hills than in the lowlands, and decreases in abundance above 2000 ft. On many of our northern and western mountains it forms a zone between the bent-fescue grassland and the peat vegetation of the plateaux (Fig. 91). The whitish areas of Nardetum on the shoulders and plateaux of the hills between 1000 and 2000 ft. are conspicuous features seen from a distance. *Nardus* is very often associated with the silver hairgrass (*Deschampsia flexuosa*).

"The vegetative organs form very compact tufts (or tussocks) made up of a system of horizontal rhizomes almost on the surface of the soil. The leafy and flowering shoots are firmly enclosed in thick tough basal sheaths which persist long after the foliage is dead, and these form the double or triple series of comb-like teeth so characteristic of the rhizomes. The frequent branching of the rhizomes and the closeness of the shoots leads to considerable congestion, and as new organs are formed they become piled on the remains of the old ones; thus tussocks are formed which we have seen knee-high. Plant remains and wind-borne detritus are added, and hence a colony of *Nardus* generally forms a thick mat or sod which extends peripherally and ousts all but a few rivals such as *Juncus squarrosus*, *Vaccinium myrtillus* and in wet places *Molinia caerulea*. The humus mat furnishes, however, a substratum for shallow-rooted, humus-frequenting species, such as *Deschampsia flexuosa*, *Agrostis* spp., *Anthoxanthum*, *Luzula campestris* (vars.), *Galium saxatile*, *Potentilla erecta*, etc. The roots of *Nardus* are thick, and cord-like, and the finer lateral roots are mycorrhiza with an endotrophic fungus" (W. G. Smith, 1918, p. 6).

Adamson (1918, p. 103) says that in the southern Pennines the *Nardus-Deschampsia* grassland is perhaps the most extensive vegetation type, occurring at altitudes of 400–1700 ft. (120–520 m.) or more, and that its composition varies considerably with variation in the habitat. "On relatively gentle slopes or level ground—where peat can accumulate and where surface drainage is poor—*Nardus* becomes truly dominant almost to the exclusion of other species." In such situations 6 in. (15 cm.) of peat may be present. "On rather steeper slopes or where the peat is thin and drier, dominance is shared by *Deschampsia flexuosa*,[1] which fills the inter-

---

[1] The silver hair-grass (*D. flexuosa*), according to Moss, is so abundant in the Nardetum of the southern Pennines that in early summer, when its tall purple scapes rise above the general level of the herbage, they give tone and colour to the whole community.

stices between the tufts of *Nardus* and is not only frequently numerically far more abundant but owing to its freely branching superficial rhizomes has as much or more controlling effect on the other species of the association.... On steep slopes where only a thin peat is present...liable to periods of great drought, and in very exposed situations, *Nardus* is often quite absent and *Deschampsia* is present alone.... The distribution of the two plants is well brought out in their topographic relations on many of the lower grass-covered hills: the grassland dominated by *Nardus*, the escarpment face by *Deschampsia* alone, while below the 'edge', on the more moderate slope, co-dominance of the two plants occurs" (Adamson, 1918).

W. G. Smith (*Types of British Vegetation*, 1911, pp. 282–4) writes that the soil of the "grass moor in Scotland consists of a sod of about 6–9 in. thick, made up of shoot bases and rhizomes matted with mosses and decaying herbage, and resting on an impervious subsoil (glacial till or boulder clay) from which it is easily separable", but which is penetrated by the cord-like roots of the dominant. One of the chief habitats of Nardetum is the zone of redistributed peat derived by erosion from and deposited round the edge of a plateau peat moor (Pl. 96, phot. 232), and it may also occupy the lower summits which have been denuded of peat (Pl. 97, phot. 234): *Nardus* rarely colonised untouched peat.

Marginal Nardetum on the Moorfoot Hills. The best detailed account of the community is given by W. G. Smith (1918) from this redistributed marginal peat on the Moorfoot Hills in south-eastern Scotland (Fig. 91). The most abundant and constant plants associated with the dominant in this Nardetum were *Deschampsia flexuosa* and *Agrostis* spp. (*A. tenuis* the commonest, but also *A. canina* and *A. stolonifera*). The other less constant grasses were *Anthoxanthum odoratum*, *Festuca ovina* and *F. rubra*. *Galium saxatile* and *Potentilla erecta* had a high constancy, as in the bent-fescue grassland; and *Vaccinium myrtillus* an even higher one. *Calluna vulgaris*, *Juncus squarrosus* and *Luzula campestris* had a moderate constancy, while *Molinia caerulea* was much less constant. All of these varied from occasional to abundant in different samples. Less abundant species were *Carex goodenovii*, *C. binervis*, *Eriophorum vaginatum* and *Blechnum spicant*. Of mosses *Polytrichum commune* was not at all constant but was abundant where it occurred. *Aulacomnium palustre*, *Sphagnum*, *Dicranum* and *Campylopus* spp. were less abundant. These data were taken from nineteen samples in which *Nardus* was definitely dominant.

Welsh Nardeta. In Wales Nardetum is widely distributed on the hill slopes, mainly between 1000 and 2000 ft. (300–600 m.). Davies (1936) estimates that Nardetum together with Molinietum occupies 16·8 per cent of the Welsh "agricultural area". It invades insufficiently grazed bent-fescue grassland wherever the conditions favour the formation of acid peaty humus without being too wet. It is of little use as general sheep pasture, though the old wethers can graze it, and the disappearance in

PLATE 96

Phot. 231. Nardetum with *Juncus effusus* (Pennines).

Phot. 232. Upper margin of Nardetum invading the eroded edge of a peat-covered plateau. Moorfoot Hills, Southern Uplands, south of Edinburgh, 600 m. *Donald Macpherson.* From W. G. Smith (1918).

NARDETUM STRICTAE

PLATE 97

Phot. 233. Zonation of Nardetum with peat-covered plateau above and Callunetum on steeper slope below. Moorfoot Hills, 450–530 m. alt. Cf. Fig. 91. *Donald Macpherson.* From W. G. Smith (1918).

Phot. 234. Nardetum occupying the summit of a spur in the eastern Moorfoots. Flushed grassland with sparse *Pteridium* on the slope below, Callunetum on a rocky patch towards the left. In the foreground Nardetum on the near side of the valley. *Donald Macpherson.* From W. G. Smith (1918).

NARDETUM STRICTAE, MOORFOOT HILLS

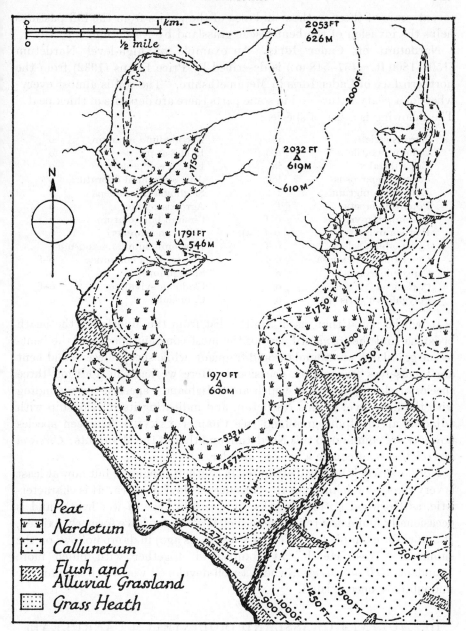

FIG. 91. ZONATION OF VEGETATION ON THE SLOPES OF THE MOORFOOT
HILLS (SCOTTISH SOUTHERN UPLANDS)

The peat-covered plateaux are eroded at their edges, and the redistributed peat below is colonised by Nardetum (about 1300–1800 ft.), on the whole following the contours pretty closely. Below this is Callunetum alternating with "grass heath" (*Festuca-Agrostis*), broken by flush and alluvial grassland in the valleys (cf. Pl. 97). After W. G. Smith, 1918.

recent years of these wethers and also of the ponies from the hill pastures helps the invasion of the bent-fescue grassland by Nardetum.

Nardetum on Cader Idris. An example of high-level Nardetum (1500–1800 ft. = 457–548 m.) is described by Price Evans (1932) from the northern face of Cader Idris in Merionethshire. "The soil is almost everywhere of a peaty nature and in some parts there are deposits of thick peat." The following is a list of species:

Nardus stricta	d	Lycopodium alpinum	o
Galium saxatile	a	L. clavatum	o
Erica tetralix	la	Luzula campestris	o
Juncus squarrosus	la	Anthoxanthum odoratum	o
Empetrum nigrum	f	Sieglingia decumbens	o
Potentilla erecta	f	Agrostis tenuis	o
Vaccinium myrtillus	o–f	Cirsium lanceolatum	r
Festuca ovina	o–f (la)	Hypnum schreberi	f
Lycopodium selago	o–f	Rhacomitrium lanuginosum	f
Deschampsia flexuosa	o–f	Polytrichum commune	o–f
Carex binervis	o	Sphagnum sp.	o
C. panicea	o	Cladonia sylvatica	o–f
Polygala vulgaris	o	C. uncialis	o

It will be noted that, as in Smith's list from the Moorfoots, the heath bedstraw and the tormentil, two of the most constant species of the bent-fescue grassland, are abundant and frequent, while sheep's fescue and bent are themselves frequent or occasional. Here we have in addition three species of clubmoss, two of which are "Highland" species, corresponding with the high level of this Nardetum and indicating its relationship with adjacent arctic-alpine grassland (see Chapter xxxviii). Thirteen species however are common to this Nardetum and that of the Moorfoots. *Cirsium lanceolatum* is of course a chance invader.

Nardus stricta is widely distributed in southern England, but now at least is very rarely dominant over any considerable lowland area. It is characteristic, as Horwood (1913) remarks, of a certain type of wet heath, and is occasionally found as a relict where such heaths may have formerly existed. It occurs for example (or did occur a few years ago) in damp hollows on the suburban "West Heath" at Hampstead, together with much more abundant *Molinia caerulea*, where a well-developed wet-heath community undoubtedly once existed.

III. *MOLINIA* GRASSLAND (MOLINIETUM CAERULEAE)

(Pls. 98–9)

Grassland dominated by the purple moor-grass or "flying bent" (*Molinia caerulea*) has several characters in common with the Nardetum, which it often adjoins. Like the *Nardus* community Molinietum is closely related to the main peat vegetation of the northern and western hills, to which it is often marginal, and both *Nardus* and *Molinia* themselves form

PLATE 98

Phot. 235. Tussock habit of *Molinia caerulea*.
T. A. Jefferies (1915).

Phot. 236. *Molinia* tussock with *Deschampsia flexuosa*. T. A. Jefferies (1915).

Phot. 237. Root system over 40 cm. long.
T. A. Jefferies (1915).

Phot. 238. *Molinia* seedlings colonising sun-cracks in bare peat. T. A. Jefferies (1915).

MOLINIA CAERULEA

PLATE 99

Phot. 239. *Molinia* sward in winter. *T. A. Jefferies* (1915).

Phot. 240. *Molinia* flush in Callunetum. *T. A. Jefferies* (1915).

Phot. 241. *Molinia* flushes in Eriophoretum. *T. A. Jefferies* (1915).

Phot. 242. *Molinia* invading Callunetum. *T. A. Jefferies* (1915).

MOLINIETUM ON THE SOUTHERN PENNINES

peat to a certain extent. Molinietum is indeed often regarded (as by Fraser, 1933) as part of the "moorland". *Molinia* freely colonises the drier parts of the "blanket bog" of western Ireland and Scotland (Chapter XXXIV), and its associates are predominantly wet peat plants, as can be seen from the lists given below. It invades and eventually replaces degenerate oak-birch woods in the wet climate of the Pennines, and under some conditions may also invade Callunetum. Like *Nardus* again, *Molinia* occurs on lowland heaths and moors, and much more abundantly; but unlike the mat-grass it also occurs and may become dominant not only on acid fens but on those which are alkaline in reaction (Chapters XXXII, XXXIII).

Molinia bog and Molinia meadow. Stapledon (1914) and Jefferies (1915, p. 94) distinguish between "*Molinia* bog" and "*Molinia* meadow", the former dominated by large *Molinia* tussocks, the latter more lawn-like, smooth, and uniform, and of some value as pasture. A well-known variety of *Molinia caerulea* is "forma *depauperata*" with smaller leaves and one-flowered spikelets, and this may perhaps always be the dominant of "*Molinia* meadow".

Habitat factors. In his excellent paper on the Molinietum of the Pennines near Huddersfield (Pls. 98–9), T. A. Jefferies (1915) showed that while *Molinia* is adapted to a wetter soil than *Nardus* (as indeed was already well recognised) the decisive factor of its habitat is a supply of fresh water, relatively, though not of course extremely, rich in soluble salts, and this is entirely in accordance with the occurrence of the grass in other regions. Stagnancy and increasing acidity of the water enable the cotton grass (*Eriophorum vaginatum*), the main dominant of the wet moors or mosses of the Pennine plateaux, to supplant *Molinia*, while decreasing water supply brings in the Nardetum, and better drainage, with increase of mineral salts and of pH value, will favour the development of grassland of the bent-fescue type.

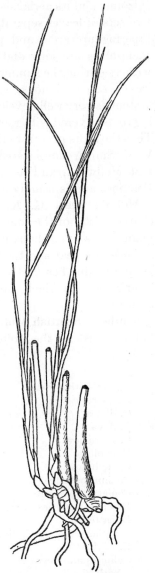

FIG. 92. BRANCH OF *MOLINIA CAERULEA* IN SPRING

The two leaf-bearing shoots are growing at the expense of the food in three leaf bases. The right-hand leaf base has lost the whole of its food store. From Jefferies, 1915.

Vegetative structure. *Molinia* has a condensed branching sympodial rhizome, and its aerial shoots are very characteristic (Fig. 92). Each bears two sets of leaves separated by a basal internode, swollen at the base and tapering upwards, and packed with reserve food material. The leaves themselves are long and thick, and taper to a fine point which usually withers during the summer—rather definitely a "mesomorphic" type. The leaves are deciduous, separating by a definite abciss layer. The root system[1] is very well developed, consisting of "cord roots" and finer branching roots covered with root hairs to their bases and reaching a depth of 2 ft. Thus the roots penetrate the soil below the surface peat, which, according to W. G. Smith, shows a strikingly constant depth of about 9 in. in the Moorfoot Molinieta, and according to Jefferies rarely exceeds 12 in. in the Pennines and is usually a good deal thinner.

Molinia peat. G. K. Fraser (1933) considers Molinietum specially characteristic of the southern Scottish moorlands and points out that the plant forms amorphous peat, in which, unlike the fibrous *Sphagnum* and *Eriophorum* peat and the "pseudo-fibrous" *Scirpus* peat, the decay of the plant remains has destroyed all trace of structure. This amorphous peat "varies from a black, mud-like peaty mass to a brown peat of spongy texture".

Southern Scottish Molinieta. W. G. Smith (1911, pp. 284–5) gives the following list for the Molinietum of Scottish "grass moor".

		Molinia caerulea	d		
Agrostis stolonifera	a		Hydrocotyle vulgaris	f
Carex canescens	f		Juncus articulatus	ld
C. dioica	f		J. effusus	ld
C. flava	f		Myrica gale	la
C. goodenowii	a		Narthecium ossifragum	f
C. panicea	a		Oxycoccus quadripetalus	f
C. stellulata	a		Parnassia palustris	f
Cirsium palustre	f		Pedicularis palustris	f
Comarum palustre	f		Ranunculus flammula	f
Deschampsia caespitosa	ld		Scirpus caespitosus	la
Erica tetralix	la		Triglochin palustre	f
Eriophorum angustifolium	la			
E. vaginatum	la		Aulacomnium palustre	f
Galium palustre var. witheringii	f		Sphagnum spp.	a

Pennine Molinieta. The Molinietum studied by Jefferies (1915) on the south Pennine moors, situated at about 1200 ft. (365 m.) above sea-level on the edge of the cotton-grass plateau, showed remarkably few species associated with the dominant, which occupies the ground very completely. Stretches occur in which no other species can be found except a few algae

[1] Pl. 98, phot. 237.

and liverworts.[1] The most constant associate is *Deschampsia flexuosa*. *Eriophorum vaginatum* (and more rarely *E. angustifolium*) occurs in more swampy, *Calluna* in drier, conditions, while *Vaccinium myrtillus* is frequent, especially on ridges, and occasionally colonises the *Molinia* tussocks themselves. Beyond these five species, according to Jefferies, none is more than occasional where *Molinia* is well developed. In addition to these Moss (1913) gives twenty-eight other species, nearly all wet heath or moorland forms and about half occurring in Smith's Moorfoot list. Among these are the sedges *Carex goodenowii*, *C. panicea*, *C. stellulata* and *C. flava*. And Stapledon (1914) mentions *Scirpus caespitosus* and *Erica tetralix* as typical constituents of the corresponding Molinietum in central Wales, where this community is particularly extensive, occupying, together with Nardetum, 16·8 per cent of the total "agricultural area" (Davies, 1936).

Cornish Molinieta. Magor (unpublished) describes Molinietum as "the typical wet facies of the grass moor" which is the main vegetation of the granite of Bodmin Moor in Cornwall. "It is developed extensively over the peat in all but the wettest parts of the bogs, and is rich in species, though actually made up of scattered societies and not very homogeneous." The following species were recorded from the "Pillar Marsh" near Roughton:

Molinia caerulea   d
Sphagnum spp.   ld

Very abundant

Juncus conglomeratus   Holcus lanatus
Festuca ovina

Abundant and locally subdominant

Erica tetralix   Eriophorum angustifolium
Juncus squarrosus

Abundant

Calluna vulgaris   Potentilla erecta
Narthecium ossifragum   Luzula multiflora var. congesta
Juncus bulbosus forma uliginosus   Aulacomnium palustre
Viola palustris

Frequent

Agrostis tenuis   Lotus uliginosus
Drosera rotundifolia   Menyanthes trifoliata
Eriophorum vaginatum   Polygala serpyllifolia
Galium palustre   Ranunculus flammula
Hydrocotyle vulgaris   R. hederaceus
Hypericum elodes   Rhynchospora alba
Juncus bufonius   Succisa pratensis

[1] Watson (1932) suggests that the floristic poverty of this Molinietum, as of other communities of the region, is largely caused by the smoke arising from neighbouring industrial centres, but Davies remarks that the Molinieta of central and western Wales, where smoke pollution cannot be a factor, are equally poor in species. Recent experimental work at Manchester finds that smoke pollution has relatively little adverse effect on vigorous pasture.

Occasional

Carex binervis
C. goodenowii
C. stellulata
Deschampsia flexuosa

Nardus stricta
Pedicularis palustris
Sieglingia decumbens
Vaccinium myrtillus

Rare

Eleocharis multicaulis
Orchis ericetorum
Pinguicula lusitanica

Ranunculus lutarius
Prunella vulgaris

In the same district, on the very shallow black gravelly soil overlying the Staddon Grit (Devonian) a Molinietum, floristically much poorer and obviously a good deal drier, is developed. This contains the following species:

Very abundant

Ulex gallii
Erica tetralix

Agrostis setacea

Abundant

Agrostis canina
Deschampsia flexuosa

Festuca ovina

Frequent

Erica cinerea
Calluna vulgaris

Carex diversicolor

Occasional

Polygala serpyllifolia

Orchis ericetorum

Rare

Eriophorum angustifolium

Highland Molinieta. Crampton (1911) reported Molinieta from the channels of wide flushes arising from springs beneath the peat of the Caithness heather moors, the flow from which had caused the peat surface to collapse. *Molinia* is dominant chiefly where a sudden decrease of slope causes partial arrest of the water flow, leading to silting, *Nardus* on the other hand where the slope is steeper and the soil consequently drier. Molinieta are said to be more extensive in Sutherland and the west of Scotland generally than in Caithness. Fraser (1933) describes *Molinia* flushes in the "moorland" (blanket bog) vegetation and also in the grassland of Argyllshire. These are intermediate between the somewhat richer *Juncus* flushes and the poorer *Eriophorum* flushes.

Irish and other Molinieta. Molinietum occupies extensive areas in and about the blanket bogs of the west of Ireland, for example in Connemara. Here it freely colonises the drier parts of the bog, and contains more

Calluna and *Erica tetralix*, but still retains some of the same associates as the wetter Rhynchosporetum (see p. 713).

The Molinietum which forms a well-marked belt on the edges of the raised bogs in north Lancashire and in the New Forest, between the wet bog or "moss" dominated by *Sphagnum* or *Eriophorum* on one side, and a reedswamp or alder wood lining a stream or marginal "ditch" on the other, is clearly a transitional community occupying a habitat of intermediate edaphic character, less acid than the "moss" and drier than either moss or reedswamp (Rankin in *Types*, pp. 251, 262–3).

Pearsall (1918) has described Molinieta which form stages in the hydrosere leading to raised bog in the vicinity of Esthwaite Water in the Lake District. These, together with those of the East Anglian fens, will be dealt with in Chapters XXXII and XXXIII, where the successions from fresh water are described.

REFERENCES

ADAMSON, R. S. On the relationships of some associations of the Southern Pennines. *J. Ecol.* **6**, 97–109. 1918.

BRAID, K. W. Herbicides, with special reference to sodium chlorate and its effect on bracken. *Scot. Farmer and Farming World*, **41**, No. 2131, p. 1484. 1933.

BRAID, K. W. The Bracken eradication problem. *Agri. Prog.* **14**, 38. 1937.

CRAMPTON, C. B. *The vegetation of Caithness considered in relation to the geology.* Com. for Survey and Study of British Vegetation, 1911.

DAVIES, WILLIAM. "The Grasslands of Wales" in Stapledon, *A Survey of the Agricultural and Waste lands of Wales.* 1936.

EVANS, E. PRICE. Cader Idris: a study of certain plant communities in south-west Merionethshire. *J. Ecol.* **20**, 1–52. 1932.

FARROW, E. P. On the ecology of the vegetation of Breckland. II. *J. Ecol.* **4**, 57–64. 1916.

FARROW, E. P. On the ecology of the vegetation of Breckland. IV. *J. Ecol.* **5**. 1917.

FRASER, G. K. Studies of Scottish moorlands in relation to tree growth. *Bull. For. Comm., Lond.*, No. 15. 1933.

HORWOOD, A. R. Vestigial Floras. *J. Ecol.* **1**, 100–2. 1913.

JEFFERIES, T. A. Ecology of the purple heath-grass (*Molinia caerulea*). *J. Ecol.* **3**, 93–109. 1915.

JEFFREYS, H. On the vegetation of four Durham Coal-Measure fells, I. *J. Ecol.* **4**, 174–95. 1916.

LEACH, W. The vegetation of the Longmynd. *J. Ecol.* **19**, 34–45. 1931.

MAGOR, E. W. Geographical distribution of vegetation in Cornwall: Camelford and Wadebridge district (unpublished).

MOSS, C. E. *Vegetation of the Peak District.* Cambridge, 1913.

PEARSALL, W. H. The aquatic and marsh vegetation of Esthwaite Water. *J. Ecol.* **6**, 53–74. 1918.

PETHYBRIDGE, G. H. and PRAEGER, R. LL. The vegetation of the district lying south of Dublin. *Proc. Roy. Irish Acad.* **25**, 124–80. 1905.

SMITH, W. G. "Grass moor association." In *Types of British Vegetation*, pp. 282–6. 1911.

SMITH, W. G. The distribution of *Nardus stricta* in relation to peat. *J. Ecol.* **6**, 1–13. 1918.

SMITH, W. G. Notes on the effect of cutting bracken (*Pteris aquilina* L.). *Trans. Proc. Bot. Soc. Edinb.* **30**. 1928.

STAPLEDON, R. G. *The Sheep Walks of Mid-Wales* (privately printed). 1914.

WATSON, W. The bryophytes and lichens of moorland. *J. Ecol.* **20**, 284–313. 1932.

WATT, A. S. Studies in the ecology of Breckland. III. Development of the Festuco-Agrostidetum. *J. Ecol.* **26**, 1–37. 1938.

WEISS, F. E. The dispersal of the seeds of gorse and broom by ants. *New Phytol.* **8**, 81–9. 1909.

WOODHEAD, T. W. History of the Vegetation of the Southern Pennines. *J. Ecol.* **17**, 1–34. 1929.

Chapter XXVII

BASIC GRASSLANDS

At the opposite extreme from the grasslands considered in the preceding chapter are those developed on basic soil, i.e. soil in which the pH value of the solution exceeds 7. The maintenance of such a condition requires a constant supply of alkaline salts, usually calcium carbonate, and this is typically maintained by the progressive solution of the underlying limestone rock through the agency of percolating rainwater. It is because of this alkaline "buffer" that these grasslands are relatively "stable", apart from invasion by woody plants; but grazing is the main stabilising factor.

CHALK GRASSLAND

The most sharply defined and typical of the basic grasslands is the characteristic chalk grassland developed on shallow soil directly derived from the Chalk (Upper Cretaceous) of the south and east of England. The Chalk is a very pure limestone, containing from 90 to 99 per cent of calcium carbonate and thus leaving a very small residue on solution. This residue forms a thin soil (rendzina, see pp. 93–5), varying in thickness from a mere film to a depth of about 30 cm. above the chalk rock, on slopes and narrow ridges. Twenty to thirty-five cm. (8–14 in.) is a very common depth to the weathered surface of the chalk. Increase in thickness is checked, in spite of the progressive solution of the underlying rock, by the constant slow creep of the soil towards the bottom of the slope. It must always be borne in mind, however, that much of the extensive chalk outcrop, where the ground is flat or nearly so on the wide plateaux and gentle dipslopes, is covered with a variable thickness of loam owing to accumulation of insoluble residue. This may take the form of a thick layer of "clay-with-flints" (formed from the residue of many times the depth of chalk together with debris from vanished Tertiary beds and sometimes Quaternary deposits) which overlies a great portion of the outcrop; but much of the area not mapped as clay-with-flints is actually covered with non-calcareous loam. Both clay-with-flints and loam yield a soil of totally different character from the chalk rendzina. It is mainly on the relatively steep slopes of the escarpments and valley sides and on the summits of the narrower ridges that a typical chalk soil and therefore typical chalk grassland is encountered.

Distribution and Soil

Typical chalk grassland, which forms by far the most extensive "rough grazing" of south-eastern England, occupies much of the chalk escarpment surrounding three sides of the Weald, and also a good deal of the main chalk escarpment which runs, with many irregularities and interruptions, north-eastward from the coast of Dorset to that of the North Sea. The steeper sided valleys and minor escarpments which are numerous in the broad outcrop of the Chalk in central southern England also bear this type of grassland. Wherever, in fact, the angle is too steep and the soil too thin for arable cultivation, and where the chalk slopes are not wooded, we find this characteristic plant community maintained by grazing.

The eastern and central Sussex Downs ("South Downs"), bounded on the north by the escarpment falling steeply to the Wealden Plain are one of the best developed and most typical chalk grassland areas. Plate 100, for which I am indebted to *The Times*, shows a long stretch of the central South Downs in Sussex. The western South Downs, extending into Hampshire, together with the escarpment facing east on the western edge of the Weald, are largely covered with beechwood, and the same is true of the "North Downs" of Surrey and Kent, though the grassland areas are not inconsiderable. In Dorset and especially in Wiltshire (Salisbury Plain, Marlborough Downs, etc.), Hampshire and Berkshire, there are great areas of chalk pasture. In south Oxfordshire and Buckinghamshire (Chiltern Hills) beech forest is more widespread, but much of the escarpment is grassland (Pl. 101, phot. 244). North-east of the Chilterns, in Bedfordshire, Hertfordshire and south Cambridgeshire the chalk outcrop diminishes greatly in height, the slopes are gentler and there is much less open downland than in the southern counties, though it is not entirely wanting (Dunstable Downs, Royston Heath). Practically absent in east Suffolk and Norfolk, chalk grassland reappears on the Lincolnshire and south-east Yorkshire Wolds, where the chalk outcrop is cut off by the sea.

The characteristic chalk rendzina soil is very constant in general character, whether it is under grass, scrub or trees, though it is of course affected by the kind of vegetation it bears. The typical "mature" chalk grassland soil (though the shallow chalk soil can never become *pedologically* mature since it is constantly wasted and rejuvenated) is richly humous in the surface layer and passes below into several inches of brown loam (often decidedly clayey just below the surface humus) containing particles of chalk, minute near the surface and becoming larger at lower levels till the weathered chalk surface is reached at a depth of 8 to 12 or 14 in. (20–35 cm.).

A "young" or "primitive" chalk grassland recently colonised by vegetation, subject to relatively rapid "creep" or checked in its development by dry conditions, as on steep southern exposures, is whitish grey and powdery when dry, almost to the surface, while the solid rock may occur at a depth of 3 to 4 in. or even less.

Another type of soil is formed below chalk grassland on steep cool northern slopes, and is well seen in several places along the north-facing escarpment of the South Downs. Here the soil is very dark, or even black, and very rich in humus, so that it shrinks strongly on drying to a hard dark cake. The vegetation, particularly when not heavily pastured, is

PLATE 100

Phot. 243. Panoramic view of the escarpment of the central South Downs (Sussex), entirely covered with chalk grassland (except for a little scrub), from the Devil's Dyke on the extreme left to Chanctonbury Ring (clump of planted trees on the summit of the Downs) in the distance on the right—between 8 and 9 miles (c. 13·5 km.). Below the Ring the escarpment is wooded. The western South Downs—20 miles (32 km.) from the observer—just appear on the extreme right in the far distance. A narrow belt of scrub has been left between the arable cultivation of the Weald and the Downs grassland. Photograph from *The Times*.

ESCARPMENT OF THE SOUTH DOWNS

PLATE 101

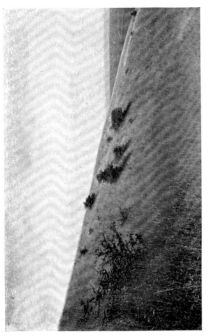

Phot. 245. Unpastured grassland on the north-facing escarpment at Beddingham Hill south-east of Lewes. Tall *Bromus erectus* and *Avena pratensis* dominant. *Pimpinella saxifraga* in flower. Scattered *Crataegus*.

Phot. 246. Continuous grass cover not maintained on steep (34°–39°) southern slope of Ramsdean Down, Butser Hill, Hants.

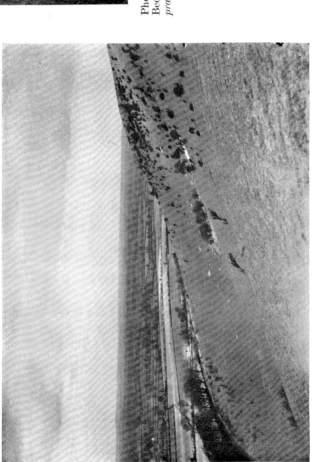

Phot. 244. Chiltern escarpment (Lower Chalk) with Upper Greensand terrace (light coloured arable fields) and the Plain of Aylesbury beyond. Scattered juniper scrub, wind-cut nearest the observer, and a group of elders associated with a rabbit burrow. Grassland in right-hand bottom corner not excessively rabbit grazed though not far from the nearest burrow. Photograph from *The Times*.

dominated by the taller grasses such as *Avena pratensis*, *A. pubescens*, *Bromus erectus*, *Arrhenatherum elatius*, and occasionally even *Deschampsia caespitosa* and *Festuca elatior*.

The average pH value of the top 2 or 3 in. of soil is about 7·5 in all three types. The CaO content is high, with an average of about 20 per cent; it decreases from the "young" grey to the dark humus soils, though it is still high in many of the last, and the average "loss on ignition" (humus content) correspondingly increases. Potash and magnesia, too, are in good supply throughout. Under these conditions, then, because of the alkaline buffer, leaching is ineffective in reducing the content in basic salts and substantially increasing the acidity, even of the top two inches, of most chalk soils, though in a few the pH value falls below the neutral point.

Table XIII. *Chalk grassland soils from Sussex Downs*

	Grey (6)	Brown (6)	Dark (7)
Mean CaO	29·1	18·1	15·0
Mean loss on ignition	24·6	25·9	31·3
Mean pH value	7·4	7·4	7·5

Table XIV. *Partial analyses of six typical chalk grassland soils*

	Grey and brown soils				Dark soils	
	A	B	C	D	E	F
No. of area (Tansley and Adamson, 1926)	3	22	12	5	21	19
Depth of sample	0–2″	0–3″	0–3″	0–5″	0–1″	0–3″
Loss on ignition	25·9	23·0	17·3	32·0	32·6	42·6
CaO	27·4	32·6	33·7	11·1	15·5	3·3
MgO	0·32	0·51	0·31	0·35	0·62	0·77
K_2O	0·58	0·28	0·26	0·29	0·33	0·35
pH	7·4	7·9	7·8	6·9	7·8	7·2

Water content. The water content of a shallow chalk soil, half way down a slope where the angle of inclination was about 23°, bearing typical chalk grassland vegetation, and its variations throughout the year, have been thoroughly investigated by Anderson (1927). "The surface soil was a brown friable loam; small lumps of chalk up to 12 mm. diameter occurred at a depth of 2·5 cm. and increased in size and frequency until the loam gave way to broken chalk at a depth of 20 cm. Solid chalk occurred at 25 cm." "The pH values ranged from 7·4 at the surface to 7·6 for the underlying chalk.... The upper 7 cm. of soil were abundantly occupied by roots, which were less frequent in the zone below, while the lowest zone considered was tapped only by roots of *Galium* (*verum*) and *Poterium* (*sanguisorba*)." This description represents quite a typical structure of southern English chalk grassland soil.

Determinations of the water content were taken at four levels, (1) 0–7·5 cm., (2) 15–22·5 cm., (3) 30–37·5 cm., (4) 68·5–76 cm., every few days from January 1924 to August 1925, covering periods of spring and summer drought as well as wet weather.

The average value of the water content (loss at 100° C.) of the surface 3 in. was 36·8 per cent, a value comparable with the averages of other well vegetated English soils of various textures. A maximum value of 61·29 per cent was recorded on 12 May 1924, after a period of heavy rain, and a minimum of 8·58 per cent on 6 July 1925, after a period of drought during which evaporation was considerable. Rising to 17·8 per cent on 7 July after a little rain it fell again to 8·8 and 8·9 per cent on the 13th and 19th as a result of further absence of rain. With good rainfall at the end of the month the percentage water content of the surface soil rose to something like normal (29·5 per cent) on 8 August. These extreme summer minimum values (comparing with 18·5 per cent in an oakwood) are probably an important factor in determining the composition of chalk grassland. The water content at greater depths shows much less extreme variation:

Table XV. *Percentage water contents of chalk grassland soil*
January 1924—August 1925

Depth in cm.	0–7·5	15–22·5	30–37·5	68·5–76
Mean	36·78	26·63	25·84	27·20
Maximum	61·29	35·59	34·05	34·93
Minimum	8·58	11·50	10·76	15·34
Range	52·71	24·09	23·29	19·59

Thus the water content at greater depths is comparatively eustatic, and there is some evidence that it is replenished from below by capillary rise of water. It is therefore clear that while the surface-rooting vegetation (Fig. 93, top line), which includes the dominant grasses and several species of characteristic chalk grassland herbs (Fig. 94), must be distinctly limited by summer droughts, the species whose roots penetrate deeply (Fig. 93, bottom line), like *Galium verum* and *Poterium sanguisorba* (Figs. 95, 96), can rely on a steady supply.

<small>Rooting depth</small>

Drought is in fact the great danger to the surface-rooting vegetation on these very well-drained shallow soils overlying the extremely permeable chalk rock, especially on narrow ridges and southern and south-western exposures which are open to the full heat of the summer sun and also to prevailing winds. In a dry summer the leaves of the shallow-rooting grasses become parched and withered, though the deeper-rooting grasses and dicotyledons may remain fresh and green, for their roots penetrate to the clayey layer referred to and some to the fissures of the chalk below.

Floristic composition. The chalk grassland flora is only moderately rich in species, the average number (including mosses) occurring on 20 areas taken at random being 46. The smallest list (26) was taken from a southern slope with very shallow soil, the largest (65) from a northern slope.

The occurrence of about 20 of the commonest species of chalk grassland, which form the vast bulk of the herbage of grazed areas, is on the whole

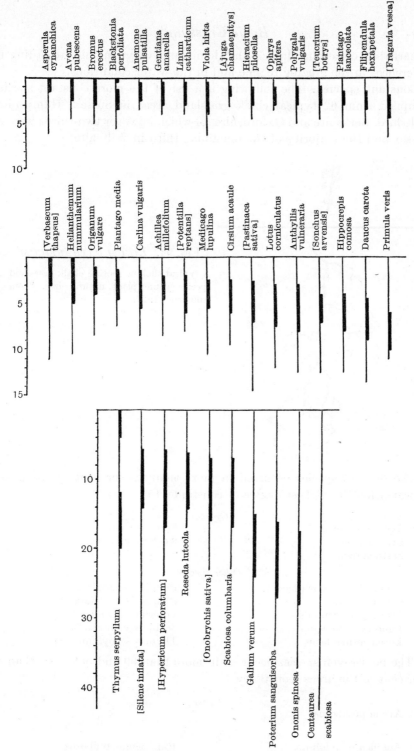

FIG. 93. ROOTING DEPTHS OF CHALK GRASSLAND PLANTS
(SHALLOW, MEDIUM AND DEEP)

The thin line indicates the maximum depth of penetration observed, the thick line the range of maximum development of feeding roots. The names of plants not belonging to the chalk grassland community proper are enclosed in square brackets. From Anderson, 1927.

remarkably constant, though the dominants vary, mainly according to differences in the grazing regime and in the water conditions.

Constant species. The following is a list of the more constant species compiled from 62 typical chalk grassland areas in Sussex, Hampshire, Wiltshire, Berkshire and Oxfordshire (of which however two-thirds were in Sussex and the majority of the remaining third in Wiltshire).

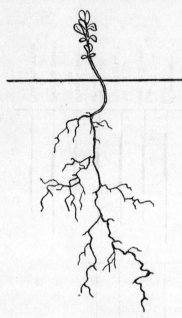

Fig. 94. *Linum catharticum*

A typical shallow rooting chalk grassland species (young plant, natural size). From Anderson, 1927.

The first 15 species occurred in more than 80 per cent of the areas, representing the highest degree of constancy (5):

Grasses

Avena pratensis   Festuca rubra
Briza media   Koeleria cristata
Festuca ovina

Other species

Carex diversicolor (flacca)   Pimpinella saxifraga
Cirsium acaule   Plantago lanceolata
Leontodon hispidus   Poterium sanguisorba
Linum catharticum   Scabiosa columbaria
Lotus corniculatus   Thymus serpyllum

The next seven species occurred in more than 60 and not more than 80 per cent of the areas (constancy 4).

Grass

Avena pubescens

Other species

Achillea millefolium   Ranunculus bulbosus
Asperula cynanchica   Trifolium pratense
Galium verum   Brachythecium purum

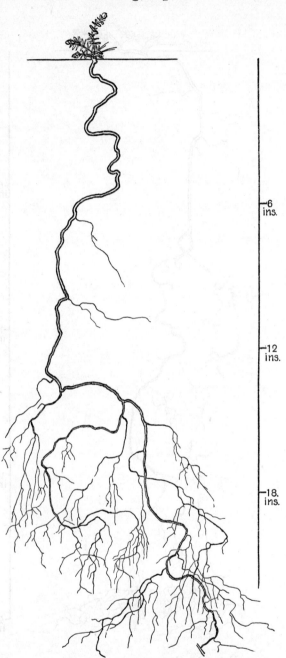

Fig. 95. *Galium verum*

A typical deep rooting chalk grassland plant (¼ natural size). From Anderson, 1927.

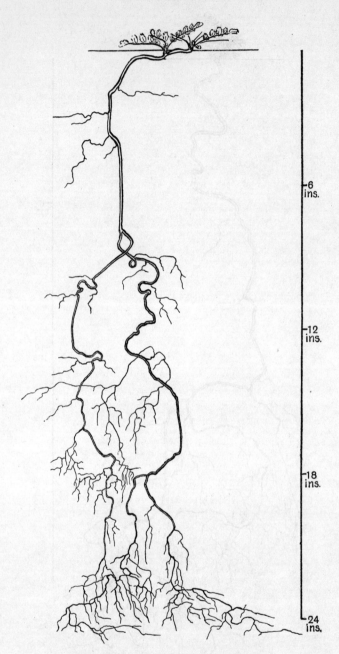

FIG. 96. *POTERIUM SANGUISORBA*

A typical deep rooting chalk grassland plant (¼ natural size). From Anderson, 1927.

A larger number (twenty-three species) occurred in more than 40 and not more than 60 per cent of the areas (constancy 3).

Grasses

Anthoxanthum odoratum
Bromus erectus
Dactylis glomerata
Trisetum flavescens

Other species

Bellis perennis
Campanula rotundifolia
Carex caryophyllea
Carlina vulgaris
Centaurea nemoralis
Cerastium vulgatum
Chrysanthemum leucanthemum
Euphrasia nemorosa
Filipendula hexapetala
Hieracium pilosella

Medicago lupulina
Phyteuma orbiculare
Plantago media
Polygala vulgaris
Primula veris
Prunella vulgaris

Camptothecium lutescens
Hylocomium squarrosum
H. triquetrum

Exclusive species. Of these forty-five species which fall into the three highest constancy classes, some are also widely distributed on other soils (mostly other types of dry or medium grassland) while others are either confined or nearly confined to chalk grassland. This last category comprises the so-called *exclusive species*, and 5 degrees of exclusiveness may be recognised. The highest degree (5) includes species never (or very rarely) found in communities other than chalk grassland, the next (4) those which are far more common in chalk grassland, but also occasionally occur in other communities, the next (3) those which are common in chalk grassland but often occur in other communities also. The two lowest degrees—(2) and (1)—include the species which are in no way specially characteristic of chalk grassland, and those which are accidental aliens in that community. Species of the higher degrees of exclusiveness may occur in chalk grassland with any degree of constancy, for a species confined or nearly confined to this community may be found in a high percentage of areas (e.g. *Poterium sanguisorba*, or *Scabiosa columbaria*), or on the other hand it may be a comparatively rare species (e.g. *Senecio integrifolius*, *Thesium humifusum* or *Aceras anthropophorum*).

The following is a list of species belonging to the two highest degrees of exclusiveness which occurred in, but were not found in any large percentage of, the areas used as data (constancy less than 3):

Anacamptis pyramidalis
Aceras anthropophorum
Anthyllis vulneraria
Brachypodium pinnatum
Campanula glomerata
Helianthemum nummularium
  (vulgare)

Hippocrepis comosa
Ophrys apifera
Senecio integrifolius
Thesium humifusum

In addition the following species, which were not found in any of the areas investigated, are known to be confined or nearly confined to chalk grassland in other areas. Most of them are rare or local:

Anemone pulsatilla	l	Orchis ustulata	r
Herminium monorchis	l	Picris hieracioides	o
Hypochaeris maculata	r	Polygala austriaca	vr
Linum anglicum (perenne)	vl	P. calcarea	la
Ophrys fuciflora (arachnites)	r	Seseli libanotis	r
		Viola calcarea	l
O. sphegodes (aranifera)	r		

The grasses:

Characteristic species:
(1) Of greatest constancy:
Festuca ovina, very abundant, very often dominant.
F. rubra, very abundant, very often co-dominant, but rarely completely dominant.
Avena pratensis, abundant, often dominant.
Briza media, moderately abundant.
Koeleria cristata, moderately abundant.

(2) Of less constancy:
Avena pubescens, abundant, occasionally dominant.
Bromus erectus, abundant, often dominant.
Trisetum flavescens, moderately abundant.

(3) Of low constancy:
Arrhenatherum elatius, sometimes dominant, but perhaps only in disturbed soil, though it may persist for many years after disturbance has ceased.
Brachypodium pinnatum, locally dominant.

Non-characteristic species:
(1) Of medium constancy:
Agrostis stolonifera, Anthoxanthum odoratum, Dactylis glomerata, all moderately abundant.

(2) Of low constancy:
Cynosurus cristatus, occasional to frequent.
Deschampsia caespitosa, on damp soils only.

(3) Of very low constancy:
Agrostis tenuis, only on slightly acid soil.
Brachypodium sylvaticum, near wood edges, very local.
Festuca pratensis, locally abundant on damp slopes.
Lolium perenne, occasional.
Sieglingia decumbens, only on slightly acid soil.

The fine-leaved fescues. The most widespread, abundant and important member of the chalk grassland vegetation is almost certainly *Festuca ovina*, but *F. rubra* follows it as a close second. The latter species is only known in typical semi-natural chalk grassland as the subspecies *genuina* Hack. It has been stated more than once that *F. rubra* subspecies *fallax* plays an important part in chalk grassland, but this is a mistake which presumably must have originated on account of the very close association of *F. ovina* and *F. rubra* subspecies *genuina* in the majority of chalk grasslands, creating an intimate mixture of the two which is distinctly tufted and at the same time contains unmistakable *F. rubra* wherever one chooses to look. A hasty examination of such a turf might give the impression that it was composed of a caespitose form of *F. rubra*, i.e. of the subspecies *fallax*. But careful and widespread search of the chalk grasslands of Surrey, Sussex, Hampshire, Berkshire and Oxfordshire has so far not revealed any specimens of *F. rubra* that are not more or less stoloniferous.

The ecological relationship of the two fescues to one another on the downs has not yet been fully studied, but it seems evident that the areas where *F. ovina* preponderates most markedly over *F. rubra* occupy the driest habitats on south-facing slopes, and also in a good many cases surfaces where the parent rock has been exposed and soil formation and the grassland succession are not complete. The more active part played by *F. ovina* as a colonist may well be correlated with its apparently greater powers of drought resistance.[1]

While in the driest situations *F. rubra* may be very inconspicuous or even absent, in the moistest *F. ovina* is present in fair quantity. The two species seem to withstand grazing equally well. Differences in their relative abundance are not correlated with the grazing factor except where grazing has recently become slight or has ceased. Here *F. rubra* (which is the larger plant when not grazed) on the whole has the advantage in the long grass that develops especially on a north or east aspect, becoming codominant or dominant; but *F. ovina* remains abundant. This predominance of *F. rubra* is not so strikingly seen where grazing has been for long in abeyance, for here the place of both the fescues tends to be occupied by taller growing species, of which the two most important are *Bromus erectus* Hudson and *Avena pratensis* L.[2] (sometimes replaced by *A. pubescens* Hudson).

Bromus erectus

Bromus erectus is intolerant of grazing but it can grow in very dry situations and may colonise bare chalk surfaces alongside of the sheep's fescue. In southern England it is mainly a plant of chalk and oolite grassland but of rather local occurrence, and this may be due to the grazing factor. Where it does occur *Bromus erectus* is very often dominant over considerable areas.[3] It is noteworthy that this

[1] Most of the information on this page from Hope Simpson.

[2] Pl. 101, phot. 238.

[3] Tüxen (1927) considers that the English chalk grassland belongs to the "Mesobrometum", a community in which it appears that *Bromus* is not necessarily present. Whatever we may think of such a nomenclature there is no doubt that the very constant floristic composition of English chalk grassland corresponds pretty closely with that described by Tüxen for the "Mesobrometum" of the Muschelkalk (Miocene) of Hanover.

grass is quite local on the western Sussex Downs where the rainfall is relatively high and the effect of rabbits very severe, but on the drier eastern downs where the grazing (sheep rather than rabbits) is less drastic, it is often dominant over considerable areas, as also in many localities in Berkshire and Wiltshire.

The Avenae *Avena pratensis*, which has a much wider distribution in Britain than *Bromus erectus*, since it extends to the north of Scotland and is not confined to calcareous soils, is nevertheless a very characteristic chalk grassland species. It is nearly as widespread on the chalk soils as *Festuca ovina*, but its average abundance is lower, though still considerable. Here and there it is perhaps co-dominant, and the same is true of *A. pubescens* though this species has a lower constancy than its congener. The factors which determine the occurrence and abundance of these two species of *Avena* are still obscure.

Another grass often closely associated with *Bromus erectus*, and locally dominant in chalk grassland, is the tor grass, *Brachypodium pinnatum*. On *Brachypodium pinnatum* many downs it appears exclusively in nearly pure, actively spreading, circular patches. Here it has all the appearance of a recent and aggressive invader of the community. It is believed to have increased substantially in abundance and distribution during the last half century, but is still confined to the south and east of England. Its rather harsh foliage is avoided by stock and by rabbits, so that the luxuriant patches of *Brachypodium*, bright yellowish-green in colour, often stand out conspicuously from the closely grazed surrounding grassland. The evidence suggests that this grass tends to become dominant where rabbit pressure is not excessive and sheep grazing is at a low ebb. Its recent spread is very likely connected with the reduction of the sheep flocks on the downs, for example in east Sussex. *B. sylvaticum*, a woodland and wood edge grass, sometimes occurs in chalk grassland, spreading on rather open soil from the borders of woods, or occupying sites where woodland has been felled, but it is not a normal constituent of the chalk grassland community.

Arrhenatherum elatius *Arrhenatherum elatius*, the false oat-grass, is another species which sometimes plays a conspicuous part in chalk grassland, though it is far from having a high constancy in any available set of chalk grassland lists, and is nearly always recorded as "local". It is said to be quite intolerant of grazing and trampling and is indeed apparently absent from all heavily pastured slopes. Nevertheless it attains dominance here and there on chalk grassland in Oxfordshire and Wiltshire, but probably only after disturbance of the soil. A more characteristic habitat of this grass is the highly calcareous roadside "verges" in chalk districts, where it is frequently dominant for long stretches, often in company with knapweed (*Centaurea nemoralis*="*C. nigra*" pro parte) and wild parsnip (*Pastinaca sativa*). This vegetation is generally cut over in the late summer or early autumn, but is not grazed except quite casually.

Briza media, Koeleria cristata and Trisetum flavescens The quaking grass (*Briza media*) is almost as constant in chalk grassland as *Festuca ovina*, but its average abundance is much less. It is by no means confined to chalk soils, occurring in many grasslands with poor soil, both dry and wet.

Koeleria cristata is another chalk grassland species of high constancy and relatively high average abundance. Though characteristic of chalk grassland it occurs on other soils of good base status, and ranges to the north of Scotland.

Trisetum flavescens, the golden oat-grass, is less constant but has a higher average abundance than *Koeleria*. In the south it is characteristic of chalk grassland, though occurring in many other good pastures, and it does not range so far north as the two preceding species.

The ten species of grass described are all that are really characteristic of chalk grassland as contrasted with neighbouring grassland communities—*Festuca ovina* and *F. rubra* because of their very high constancy and very great average abundance. Though extremely widespread and even co-dominant in the drier siliceous pastures and on sandy soils (Chapter XXVI), *F. ovina* is nowhere so markedly predominant as on limestone and chalk. Of the others *Bromus erectus* is nearly confined to chalk and oolite, while the Avenae, *Koeleria*, *Trisetum* and *Brachypodium pinnatum* are far commoner on chalk than elsewhere.

Other grasses. The other grasses occurring in chalk grassland are not in any sense "characteristic", because they are equally or more abundant in neutral grassland communities. Of these species the cocksfoot, *Dactylis glomerata*, has a relatively high constancy in chalk grassland (occurring in 72 per cent of the lists from the South Downs) and also a fairly high abundance. This is a very widespread and abundant plant of neutral grassland but tolerates relatively high alkalinity, and though a fairly tall grass it also stands grazing well.

The sweet vernal-grass, *Anthoxanthum odoratum*, a ubiquitous species of very wide edaphic tolerance, has about the same constancy and abundance in chalk grassland as *Dactylis*.

The crested dogstail, *Cynosurus cristatus*, is much less constant (32 per cent) and has a moderate average abundance. Timothy (*Phleum pratense*) has about the same constancy as *Cynosurus*, but is never more than an occasional constituent. *Poa pratensis* has about the same constancy as the last two species: on one ungrazed north-facing escarpment slope in Sussex it was very abundant in company with *Avena pubescens* and *Dactylis*: for the rest it is only local and occasional.

Lolium perenne (perennial rye-grass), a dominant of the best neutral grasslands, is decidedly rare on chalk soils. On the South Downs it is found occasionally on the deeper brown loams derived from the chalk.

Holcus lanatus, the "Yorkshire Fog", is an aggressive grass which spreads from centres of insemination in pastured grasslands by means of prostrate or semi-prostrate rosette-shoots. It occurs in about half the lists from the South Downs, varying from occasional to locally abundant. It becomes specially prominent when rabbit pressure is withdrawn from a rabbit grazed area (Pls. 103, phots. 252, 254). This species again is in no way specially attached to chalk grassland, having a very wide edaphic tolerance.

Deschampsia caespitosa is a species forming large rosettes or tussocks with long harsh leaves and tall spreading panicles. It is essentially a grass of wet places and especially of puddled surfaces. Together with *Festuca elatior*, which has also a high water requirement, its occurrence on chalk is limited to cool moist unpastured slopes of northern aspect, but in such situations it tends towards co-dominance with other grasses, notably *Festuca rubra*, in a thick herbage which eliminates most of the lower growing chalk grassland species.

Two grasses which are characteristic of acid soils, *Agrostis tenuis* and *Sieglingia decumbens*, are occasionally met with in chalk grassland, especially where surface leaching has taken place: in such cases the calcium content is much lower than usual. *Agrostis stolonifera*, on the other hand, is much commoner on the Sussex Downs, occurring with considerable frequency in about half the areas studied. In the more western chalk grassland areas however (Wiltshire, Berkshire, etc.) it has not been noticed. This is a species which occurs both in wet and in highly calcareous soils, as is true of several others, e.g. *Cirsium palustre*,[1] *Carex diversicolor*[2] (*flacca*), etc. Intermediates (presumably hybrids) between *Agrostis tenuis* and *A. stolonifera* are common and the autecology of the *A. tenuis*—*A. stolonifera* complex is by no means understood.

Herbs. Chalk grassland is well known for the relatively large number of highly or moderately exclusive species, and the attractiveness of many of their flowers.

Characteristic species, i.e. of high constancy or of high exclusiveness:

(i) Of greatest constancy (5), roughly in order of abundance:

Carex diversicolor (*flacca*), very abundant, sometimes locally dominant.

Poterium sanguisorba, moderately exclusive, very abundant, locally dominant.

Plantago lanceolata and *Cirsium acaule*,[3] abundant.

Thymus serpyllum, Linum catharticum, Lotus corniculatus, Leontodon hispidus, Scabiosa columbaria (exclusive), *Pimpinella saxifraga*, all moderately abundant.

On account of their high constancy and considerable or great abundance

[1] Pl. 102, phots. 248, 250.
[2] Pl. 102, phots. 247–9.   [3] Pl. 102, phots. 247, 249; Pl. 103, phot. 251.

Herbs of Chalk Grassland

all the above species may be regarded as characteristic of chalk grassland, though several are abundant in quite different communities.

(ii) Of somewhat less but high constancy (4):

Asperula cynanchica, almost exclusive, moderately abundant.
Hieracium pilosella, abundant on relatively open soil.
Phyteuma orbiculare, exclusive, locally abundant on the southern chalk.

(iii) Of medium constancy (3):

Carlina vulgaris, Medicago lupulina, Plantago media (the last abundant on relatively open soil), *Polygala vulgaris*. Two species which might be called "characteristic" are *Gentiana amarella*, which just falls below constancy 3 (reaching it on the South Downs), and is moderately though not highly exclusive, and *Viola hirta*, almost exclusive to calcareous grassland and calcareous woodland, and also only just failing to reach constancy 3 in the 62 areas on which the figures are based.

(iv) Of low constancy but very high or relatively high exclusiveness:

Aceras anthropophorum, exclusive, local.
Anacamptis pyramidalis, almost completely exclusive, locally sometimes abundant.
Anthyllis vulneraria, almost completely exclusive, locally abundant.
Blackstonia perfoliata, moderately exclusive.
Campanula glomerata, almost completely exclusive.
Centaurea scabiosa, moderately exclusive.
Daucus carota, moderately exclusive, locally abundant, especially near the sea.
Filipendula hexapetala, almost completely exclusive, locally abundant.
Gymnadenia conopsea, moderately exclusive.
Helianthemum nummularium (*vulgare*), almost completely exclusive.
Hippocrepis comosa, almost completely exclusive, locally abundant.
Ophrys apifera, almost completely exclusive, sometimes locally abundant.
Orchis maculata Sm. et auct. (= *O. fuchsii* Druce), locally abundant and fairly characteristic.
Origanum vulgare, moderately exclusive, but mainly near wood margins, in seral stages, or where rabbits are numerous.
Senecio integrifolius, exclusive, locally frequent.
Thesium humifusum, exclusive, local.

Non-characteristic species:

(i) Of high, but not the highest constancy (4) and low exclusiveness:

Achillea millefolium, Galium verum, Ranunculus bulbosus, Trifolium pratense.

(ii) Of medium constancy (3) and low exclusiveness:

Bellis perennis, Campanula rotundifolia, Centaurea nemoralis, Cerastium vulgatum, Euphrasia nemorosa, Galium erectum, Primula veris, Prunella vulgaris, Rumex acetosa, Succisa pratensis, Trifolium repens, Veronica chamaedrys.

The non-characteristic herbs of low constancy are very numerous and scarcely worth citing:[1] some are occasional grassland species, some alien invaders from woodland and arable.

Bryophytes.[2] The following are the most important bryophytes that have been found in chalk grassland. None is generally "characteristic" of the community except *Brachythecium purum*, because of its high constancy, though it is in no way exclusive, and *Camptothecium lutescens*, because it is a calcicole and moderately constant.

The most constant species, in approximate order of importance are:

Brachythecium purum
Hylocomium triquetrum
H. squarrosum
H. splendens

Dicranum scoparium
Camptothecium lutescens
Hypnum cuspidatum
Fissidens taxifolius

After these come a considerable number of species, falling roughly into two classes:

(*a*) Species generally absent, but playing an important part in one or two areas:

Neckera crispa
Rhacomitrium lanuginosum

Frullania tamarisci

and a set of acrocarpous mosses which play an important part in some of the driest places where the turf is short or open, viz. species of *Weisia, Trichostomum* and *Barbula*.

(*b*) Species which are probably present in many areas, but which never play a conspicuous part in the community and are rarely recorded:

Brachythecium rutabulum
Eurhynchium praelongum
Fissidens adiantoides
Hypnum chrysophyllum
H. molluscum

Mnium affine
M. undulatum
Thuidium abietinum
T. tamariscinum
Scapania spp.

Hypnum cupressiforme and its variety *elatum* are intermediate in status between (*a*) and (*b*).

Among the mosses in the foregoing lists, the only one of really high constancy is *Brachythecium purum*, a very common grassland species which is found as an understorey in the great majority of the chalk grasslands. Next come the Hylocomia, woodland or grassland mosses, which appear

---

[1] Complete lists from the South Downs will be found in Tansley and Adamson (1925 and 1926).

[2] Most of this information from Hope Simpson.

especially in relatively damp chalk grassland, particularly on northern exposures protected against the direct rays of the sun, and are often very luxuriant: *Dicranum scoparium*, a very common moss, more generally found on acid soils: its rather frequent occurrence in chalk grassland is presumably to be attributed to its loose attachment to the herbage; and *Camptothecium lutescens*, one of the few "calcicole" mosses really common in this community.

Of the rarer mosses, *Hypnum molluscum*, *H. chrysophyllum* and *Neckera crispa* are calcicole, the last named however usually inhabiting bare limestone rock. *Rhacomitrium lanuginosum* is on the whole a mountain and moorland moss, and has been recorded from two localities only, the relatively lofty War Down (802 ft. = 243 m.) and Butser Hill (889 ft. = 270 m.) in Hampshire, where the rainfall is high. The liverwort *Frullania tamarisci* occasionally plays a conspicuous part on northern exposures which are unusually damp and have a relatively numerous moss flora. Mosses are at a minimum or absent altogether on dry southern slopes.

Other bryophytes besides these have been recorded in chalk grassland but they are not worth mention here. Some of them, and of those mentioned above, are much commoner in seral stages leading to chalk grassland (e.g. *Thuidium abietinum*).

Lichens. Lichens are infrequent, locally conspicuous only where the turf is short, and not always there. *Cladonia rangiformis* (about the commonest) and *C. sylvatica* are the most important species. Other species of *Cladonia*, as well as *Peltigera* and *Collema*, are found more rarely. *Collema pulposum* is fairly common in places disturbed by rabbits.

Structure of the vegetation. The structure of the grassland naturally depends upon the grazing. If it is left ungrazed and is not situated on a dry southern or western exposure it is colonised by the taller growing grasses such as the Avenae, which are the most widespread on soils of various degrees of moisture, sometimes by *Arrhenatherum*, very occasionally by the meadow grasses *Festuca pratensis*, *F. elatior* and *Lolium perenne*, locally over wide areas by *Bromus erectus* (which is not dependent on shelter or moisture), frequently by *Dactylis glomerata*, and fairly often, on damp northern exposures only, by the wet-loving species *Deschampsia caespitosa*. In such situations the tall herbage tends to suppress the dwarfer species of grazed grassland and even the mosses, which become weak and etiolated. *Centaurea nemoralis* and *Rumex acetosa* hold their own well among the tall grasses, but *Poterium sanguisorba* is the only species, other than grasses, which may become locally dominant in such situations. A lower storey is formed by the partially suppressed plants of the lower growing species and by mosses.

On well grazed grassland *Festuca ovina* (often intimately mixed with *F. rubra*) is most frequently dominant, or there is a mixture of numerous species, including the markedly "calcicolous" grasses and herbs of re-

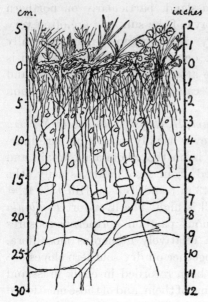

FIG. 97. PROFILE OF HERBAGE AND ROOT SYSTEMS ON PASTURED AND MODERATELY RABBIT GRAZED AREA

Herbage about 5 cm. deep. 0–7 cm. soil firmly held by roots of *Festuca ovina, *Avena pratensis and pubescens, Hieracium pilosella, Linum catharticum, Leontodon hispidus, Ranunculus bulbosus, *Plantago lanceolata, and most of the root systems of Carex diversicolor. 7–15 cm., soil lightly held by roots of *Asperula cynanchica, Thymus serpyllum, Trifolium pratense and some Carex diversicolor. 15–25 cm. loose soil with roots of *Poterium sanguisorba, *Cirsium acaule and some Lotus corniculatus. Occasional roots penetrate below 30 cm. * Recognisable in the herbage. From Tansley and Adamson, 1925.

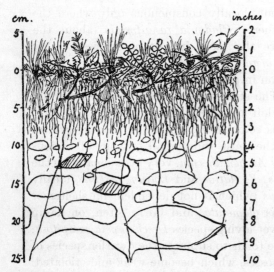

FIG. 98. PROFILE OF HERBAGE AND ROOT SYSTEMS ON SOMEWHAT LEACHED SOIL, MODERATELY RABBIT GRAZED

Herbage about 5 cm. deep. 0–9 cm. soil closely held by roots of Festuca ovina, Agrostis tenuis, Carex diversicolor, Euphrasia nemorosa, etc. 9–25 cm., loose soil mixed with chalk lumps and a few flints. Roots of Poterium sanguisorba and Cirsium acaule penetrate beyond 25 cm. From Tansley and Adamson, 1925.

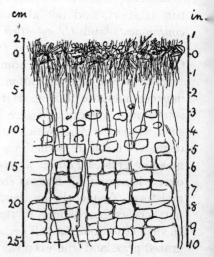

FIG. 99. PROFILE OF SEVERELY GRAZED HERBAGE EATEN DOWN TO 2 CM.

Festuca ovina, Lotus corniculatus, Thymus serpyllum, Plantago lanceolata, Prunella vulgaris, etc. Note the great reduction in the size of leaves and that the dense root layer extends here only to 4 cm. From Tansley and Adamson, 1925.

latively dwarf habit, but an absence of the non-calcicolous meadow grasses. If the herbage is 4 in. or so deep (moderate sheep grazing) there is very commonly an understorey of mosses in which *Brachythecium purum* is usually dominant or very abundant. If the herbage is reduced to an inch or less by intensive rabbit nibbling the species are not much reduced in numbers but the mosses are usually absent, presumably owing to increased dryness, except on steep northern slopes or in the shadow of woodland or tall scrub. In these last situations mosses are, on the contrary, luxuriant, especially the Hylocomia, and in rabbit-

<small>Effect of rabbits</small>

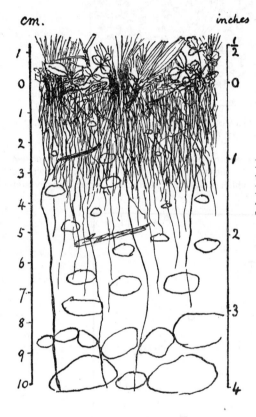

Fig. 100. A similar severely Rabbit Eaten area

Drawn to a larger scale. Herbage less than 2 cm. high: same species as in Fig. 99. Dense root layer barely extending to 4 cm. Soil shallower (brown loam). From Tansley and Adamson, 1925.

infested areas actually dominant, since they are not touched by the rabbits, and may form a continuous sward, through which here and there protrude the shoots of a few more or less rabbit-resistant flowering plants (Pl. 102, phot. 248).

Figs 97 and 98 give an idea of the structure of chalk grassland under conditions of moderate rabbit grazing, Figs. 99 and 100 under extremely heavy rabbit attack. Figs. 101–103 show the effect of the exclusion of rabbits in reducing the number of individual plants, after 6 and after 12 years. These three charts were made in the same enclosure but are not of the

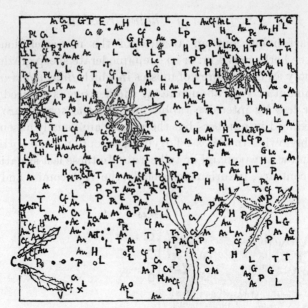

FIG. 101. CHART QUADRAT OF GRASSLAND UNDER MODERATELY HEAVY RABBIT PRESSURE (25 cm. square)

Herbage 4–5 cm. high. 441 shoots of 21 species are shown. The leaf rosettes of *Cirsium acaule* (C) and *C. palustre* (Cp) are sketched in. All the interspaces between the plants shown are filled with *Festuca ovina* (and ?*F. rubra*) which is definitely dominant. From Tansley and Adamson, 1925.

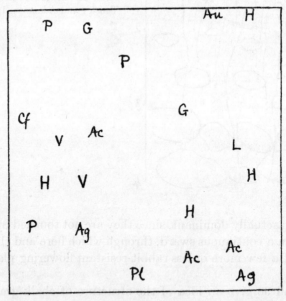

FIG. 102. CHART QUADRAT OF THE SAME SIZE AFTER EXCLUSION OF PASTURING AND RABBITS FOR 6 YEARS

Herbage 18–20 cm. high. 20 shoots of 10 species are shown. *Festuca ovina* (and ?*F. rubra*) still dominant filling all the space between the shoots of the other plants on the surface of the vegetation, though there are concealed bare spaces of soil. Note the enormous effect of competition in diminishing the number of shoots (and species) when all grazing is excluded. From Tansley and Adamson, 1925.

PLATE 102

Phot. 248. Grassland on north slope shaded by wood and exposed to rabbits. *Hylocomium squarrosum* and *H. splendens* dominant (untouched by rabbits). The sparse flowering plants include *Carex diversicolor* (abundant) and *Cirsium palustre* (both rabbit-resistant).

Phot. 250. Another quadrat 50 cm. square. Same species, with *Cirsium palustre* and *Viola hirta*.

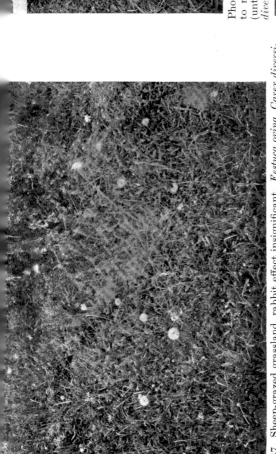

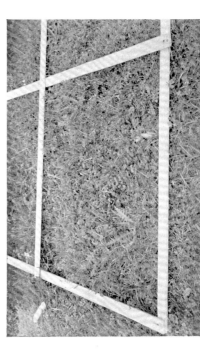

Phot. 247. Sheep-grazed grassland, rabbit effect insignificant. *Festuca ovina*, *Carex diversicolor*, *Galium verum* (centre), *Leontodon hispidus* (fruit), *Cirsium acaule*, *Poterium sanguisorba*, *Scabiosa columbaria* (fl.), *Spiranthes spiralis*. Base of juniper bush at the back.

Phot. 249. Quadrat 50 cm. square fully exposed to rabbits. *Festuca ovina*, *Poterium sanguisorba*, *Cirsium acaule*, *Carex diversicolor*.

CHALK GRASSLAND AND RABBIT EFFECTS

PLATE 103

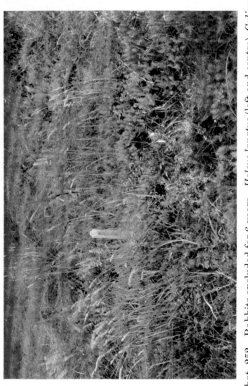

Phot. 252. Rabbits excluded for 6 years. *Holcus lanatus* (left and centre), *Galium verum* (right) dominant. *Poterium sanguisorba*, *Agrostis tenuis*, etc. Herbage at least 8 in. deep. *Festuca ovina* subordinate.

Phot. 254. Metre quadrat protected from rabbits for 12 years. Herbage 8–10 in. deep. *Poterium sanguisorba*, *Avena pratensis*, *Holcus lanatus*, *Helianthemum*, etc. Species and individuals much more numerous than in phot. 253. Cf. Figs.

Phot. 251. Uneven ground with rabbit hole and short turf of *Festuca ovina*. *Helianthemum nummularium* (rabbit-resistant) dominant and flowering; also *Cirsium acaule*, *Poterium sanguisorba*, *Galium verum*, etc. Herbage barely 3 in. deep.

Phot. 253. Rabbits excluded for 6 years. *Avena pratensis* (front) mainly dominant, with *Galium verum*, *Calluna vulgaris* (fl., centre) and *Potentilla erecta* (fl., right centre) have appeared. Herbage about 6 in. deep.

same quadrat. Pl. 103, phots. 251-4, illustrate the appearance of the herbage under these different conditions.

On southern and south-western exposures open to the full strength of the summer sun the vegetation is apt to be stunted whether it is grazed or not, the number of species is reduced, and mosses are absent or negligible.

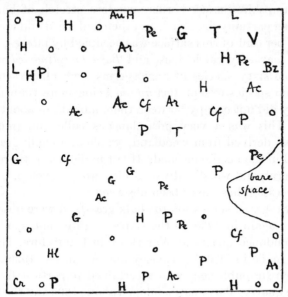

FIG. 103. CHART QUADRAT OF THE SAME SIZE AFTER EXCLUSION OF PASTURING AND RABBITS FOR 12 YEARS

Herbage 20-25 cm. high, 51 shoots of 14 species are shown. The herbage is less dense and there are many bare spaces on the soil though none on the surface of the vegetation. *Festuca ovina* is still very abundant but no longer dominant. Competition has proceeded further and has led to the proportional increase of *Galium verum*, *Poterium sanguisorba* and *Holcus lanatus*, which tend to local dominance (cf. Pl. 103, phot. 252). The increase in the number of species and individuals during the second six years shows that equilibrium is not yet reached. From Tansley and Adamson, 1925.

SUCCESSION

Succession from bare chalk to chalk grassland may be studied on the surfaces of chalk rock on the sides of road cuttings and chalk pits, and on the heaps of "spoil" consisting of larger or smaller fragments which are often accumulated when chalk is quarried.

On bare chalk rock The actual surface of the rock may be colonised in quantity by the strictly calcicolous moss *Seligeria calcarea*, whose embedded protonemata give a grey colour to the exposed rock surface. Accompanying the *Seligeria* are sometimes filaments of an alga, *Chroolepus* sp. But it is only where fine particles of chalk accumulate in the joints and bedding planes that higher plants can get a hold for their roots, and in such situations a number of species of spermaphytes occur. On these easily disintegrated chalk surfaces *Festuca ovina, Echium vulgare, Senecio jacobaea* and *Tussilago farfara* have been found as pioneer colonists (with *Bromus erectus* in Wiltshire) as well as a number of other species, mostly belonging to chalk grassland. No less than sixty species of flowering plants, besides eight species of moss, were seen on the horizontal floor of a quarry on the Hampshire-Sussex border, where a stretch of bare chalk had been partially dissolved, and then had dried and cracked. About 10 per cent of this surface was occupied by mosses, of which four species were markedly calcicolous, and *Barbula cylindrica* was the most abundant. The sixty species of angiosperms were all present as isolated individuals, no species except *Poa annua* being more than frequent, and they probably did not occupy between them more than about 1 per cent of the surface. This was a very miscellaneous collection (including three woody species) derived from woodland, arable and chalk grassland, all of which existed in the neighbourhood. If left undisturbed there can be little doubt that scrub and woodland would ultimately develop, but the course of such a succession has never been observed.

In quarries

Stages of succession to chalk grassland were observed in 1920 and again in 1936 on the slopes of three fine spoil banks in an abandoned quarry at War Down in Hampshire. The first observations were made in 1920 by Tansley and Adamson (1925), the later by Hope Simpson (unpublished) who ascertained important facts about the history of the quarry. This was abandoned about 1895 and the two youngest spoil banks probably date from that time. An older bank was estimated by a quarryman to have been made about 1880. If these dates are correct the newer banks were about 25 and 40 years old, and the older bank 40 and 55 years old when the observations were made. No rabbit grazing was noticed in 1920, but during the later years at least, i.e. some time after 1920, the banks were grazed by rabbits, which had made a warren in the quarry. In 1932 War Down was acquired and planted by the Forestry Commission and rabbits were excluded. Thus the plant communities developed for at least part of their history under the influence of rabbit grazing, from which they were freed during the last 4 years.

On spoil banks

The coarse quarry talus, consisting of larger blocks and fragments mixed with small to minute chalk particles, was found in 1920 to be colonised largely by *Echium vulgare* and *Senecio jacobaea*, while on the fine spoil banks *Festuca ovina* and *Tussilago farfara* were the leading pioneers. Some of the colonists of the latter were "weeds" of open soil and there were one or two chalk scrub plants, but the great majority of the species belonged to chalk grassland, with which the quarries were surrounded. Among the most abundant of these were *Festuca rubra, Lotus corniculatus, Agrostis stolonifera, Carex diversicolor, Thymus serpyllum*.

Hope Simpson found that the number of species present on the two younger slopes had approximately doubled between 1920 and 1936, and very nearly doubled on the older slope.[1] The change was for the most part a "filling up" of the community by new species and the process was evidently slow. The percentage areas that were bare in 1920 had considerably decreased in 1936, but even on the oldest bank the vegetation cover was not quite continuous after a presumed lapse of 55 years. The flora of the three banks was by no means uniform and each clearly showed fluctuations in its plant covering not related to the general successional process. Thus *Echium vulgare, Sonchus oleraceus* and *Tussilago farfara*, all plants of open soil, still persisted after the lapse of 16 years, and two of them had even appeared where they had not been present in 1920; while *Arctium* sp., *Cirsium arvense, Crepis capillaris, Epilobium angustifolium, Heracleum sphondylium*, and a few other species which are not chalk grassland plants, had freshly arrived. And a few species, some of which often occur in this community, such as *Anthoxanthum odoratum, Cirsium palustre, Ranunculus bulbosus*, had disappeared from the banks.

But apart from these fluctuations, a definite progress towards a more complete development of the community may undoubtedly be recognised on the three slopes taken together, just as the vegetation of the oldest bank represented in 1920 a more complete development of the community than that of the younger banks. Taking all three slopes together, *Agrostis stolonifera, Avena pratensis, Festuca ovina* and *F. rubra, Hieracium pilosella*, had not fluctuated much, while the following had increased on balance:

Arrhenatherum elatius
Asperula cynanchica
Carex diversicolor
Gentiana amarella
Leontodon hispidus
Linum catharticum
Lotus corniculatus
Plantago lanceolata
Thymus serpyllum
Brachythecium purum

nearly all species of high or considerable constancy in chalk grassland.

The following, of lower average constancy but including some highly constant and characteristic species, had freshly appeared:

Avena pubescens
Briza media
Galium erectum
Holcus lanatus
Koeleria cristata
Leontodon autumnalis
Medicago lupulina
Orchis maculata
Pimpinella saxifraga
Plantago media
Polygala vulgaris
Poterium sanguisorba
Prunella vulgaris
Trifolium pratense
T. repens
Trisetum flavescens

Hylocomium triquetrum

Highly characteristic species which are still absent are *Scabiosa columbaria* and *Hylocomium squarrosum*.

Woody plants were rare on the banks in 1920, *Clematis vitalba* appearing on one and *Rosa canina* on another. Both of these were still present in

[1] It is possible that the proportional increase of species was not quite so great as this because Hope Simpson's search was conducted in a more thorough and leisurely manner than the earlier one.

1936, though on different banks, while *Crataegus monogyna*, *Rubus fruticosus*, *Solanum dulcamara* and *Sorbus aria* had freshly appeared, though all were still rare. Their appearance represents a progress towards the development of chalk scrub which is checked or inhibited by rabbit grazing. Where sufficient seed parents of shrubs and trees are present in the neighbourhood there is no doubt that scrub and eventually forest would develop in the absence of grazing, before the grassland became mature. In other words, chalk grassland, as we know it, is a subclimax or plagioclimax determined by grazing.

The data that have been obtained from the War Down spoil heaps do not enable us to trace the details of development of this plagioclimax, but they do indicate the pioneers and the gradual "filling up" of the community in the presence of rabbits. The exclusion of rabbits during the four years from 1932 to 1936 may be responsible or partly responsible for the appearance of the new woody plants: also of *Arrhenatherum*, and perhaps of *Deschampsia*, *Heracleum* and *Orchis*.

"Primitive" chalk grassland. Table XVI on p. 549 gives the floristic composition and soil analyses of three closely adjacent small areas of chalk grassland, each of about 2 sq. m., on very shallow soil occupying irregular ground near the quarries at Buriton Limeworks in Hampshire (Tansley and Adamson, 1925). The chalk was probably bared at some stage in the working of the quarries and has been completely colonised by grassland species forming a "primitive" chalk grassland community which shows a simpler structure and composition than mature grassland. It appeared at the date of the visits to be practically unaffected by grazing, though it is of course possible, and indeed likely, that grazing had taken place at some time previously:

"A comparison of the three soils shows a steady increase from (*a*) to (*c*) in water content and in humus and a decrease in total carbonates and in calcium. The nitrate determinations, made some four months after the samples were collected, are probably rather an index of the abundance of nitrifying or nitrogen fixing organisms in (*c*) than a trustworthy measure of the amount of nitrates in the soil as it exists in the field.

"The characteristic feature of the vegetation of (*a*) is its extreme dwarfing, due no doubt primarily to drought. The shallow rooting *Festuca ovina* is markedly dominant, forming more than half of the whole herbage. The other grasses are few (four species) and not abundant. The rest of the vegetation is made up of a small selection of the common herbs of dry grassland, of which *Hieracium pilosella*, *Leontodon hispidus*, *Lotus corniculatus*, *Thymus serpyllum* and *Asperula cynanchica* were most conspicuous.... The only two species found on this area and not on (*b*) or (*c*) are *Taraxacum erythrospermum* and *Phleum pratense* var. *nodosum*, both plants of markedly dry soils. Of the seven mosses four are distinctly xerophilous forms, and three are marked by Watson as calcicolous.

"Primitive" Chalk Grassland

Table XVI. *Vegetation and soils of three small areas of "primitive" chalk grassland in close proximity*

(a) Turf scarcely continuous. Herbage ½–1 in. (1·25–2·5 cm.) in height. Slope to S.S.W. 18–20°. Soil 2½–3 in. (6·25–7·5 cm.) in depth to chalk *in situ*, grey, very dry and powdery, included in a mass of fine rootlets.
(b) Turf more continuous. Herbage 4 in. (10 cm.) in height. Very slight slope to south. Soil 4 in. (10 cm.) in depth to chalk *in situ*, grey-brown.
(c) Close herbage 5 in. (12·5 cm.) in height. Bottom between two hummocks. Soil 4 in. (10 cm.) to chalk *in situ*, brown.

Soil samples	Water loss of fresh soil on air drying	Water loss of dry soil at 100° C.	Loss on ignition	Total carbonates	Nitrates as NaNO₃	P₂O₅	K₂O	MgO	CaO	Insoluble residue	"Lime requirement"
(a)	17·0	3·0	6·2	71·1	0·0048	0·160	0·327	0·201	54·6	14·6	Nil
(b)	27·0	4·0	19·7	65·9	0·0037	0·164	0·640	0·183	32·1	5·6	Nil
(c)	33·8	4·6	28·6	62·8	0·0500	0·170	0·301	0·062	36·25	22·0	Nil

	(a)	(b)	(c)		(a)	(b)	(c)
Agrostis stolonifera	—	f	a	Plantago lanceolata	r	o	f
Asperula cynanchica	f	a	f	Primula veris	—	o	r
Avena pratensis	r	—	o	Ranunculus bulbosus	o	o	o
Bellis perennis	—	o	o	Taraxacum erythrospermum	—	f	f
Briza media	o	o	o	Thymus serpyllum	f	a	f–a
Campanula rotundifolia	—	f	o	Trifolium pratense	e	f	a
Centaurea nemoralis	r	f	f	T. repens	—	a	a
Cynosurus cristatus	—	—	f–la	Trisetum flavescens	a	o	—
Dactylis glomerata	—	—	—	Barbula cylindrica*	—	—	—
Euphrasia nemorosa	—	d	va	B. unguiculata	a	o	f
Festuca ovina	—	f	a	Brachythecium purum	o	—	o
Galium verum	f.la	a	f	B. rutabulum	f	a	—
Hieracium pilosella	—	—	—	Bryum capillare	f	f	o
Holcus lanatus	—	f	f	Camptothecium lutescens*	f	—	f
Leontodon hispidus	f	f	f	Fissidens taxifolius	—	—	—
Linum catharticum	o	—	o	Hypnum cuspidatum	o	o	o
Lolium perenne	—	o	o	H. molluscum*	—	f	—
Lotus corniculatus	f	f	a	Mnium undulatum	o	o	o
Medicago lupulina	f	—	—	Thuidium abietinum	a	—	o
Origanum vulgare	—	—	—	T. tamariscinum	—	—	—
Phleum pratense var. nodosum					25	25	31

* Mosses marked by Watson (1918) as calcicolous.

"Comparing (b) and (c) there is an increase in the number of species of grass from 5 to 6 and 8 respectively, and in (c) a great increase in the bulk of the grass herbage. This is to be correlated with the great increase in water content and the much greater nitrifying power of the soil of (c). *Agrostis stolonifera, Cynosurus cristatus, Dactylis, Lolium* and *Holcus lanatus* appear, *Trisetum* increases in abundance, while *Festuca ovina* progressively decreases, though it still remains abundant. Of the herbs *Trifolium repens* appears, *T. pratense* and *Plantago lanceolata* increase in abundance, while some species, such as *Asperula cynanchica, Campanula rotundifolia, Hieracium pilosella, Lotus corniculatus,* and *Thymus*, show their maximum frequency in (b), perhaps owing to the competition of the taller grasses in (c). *Euphrasia nemorosa, Origanum vulgare* and *Primula veris* appear for the first time in (c).

"Of the mosses, *Barbula cylindrica, B. unguiculata* and *Bryum capillare* decrease and disappear with complete closure and increasing depth of the turf; *Brachythecium purum*, one of the most ubiquitous of chalk grassland mosses though not a "calcicole" species, appears in (b) and increases in (c); while *B. rutabulum, Mnium undulatum* and *Thuidium tamariscinum* first appear in the damper conditions of (c)." (Tansley and Adamson, 1925, pp. 184–5.)

Stabilisation by grazing. Chalk grassland, like most of the other grasslands of this country, is maintained as grassland by pasturage or mowing for hay, the latter being very rare on the chalk. The downs have been used as sheep walks for many centuries and probably from Neolithic times, and horned stock are now frequently pastured upon them. The feeding of sheep in folds and of cattle on cake or in "tended" pastures between their visits to the downs may have some effect on the natural grassland community by the transference of foodstuffs through the dung to the down grassland, but such possible effects have not been investigated.

The extreme abundance of rabbits in some chalk grassland areas has an overwhelming effect on the vegetation (see pp. 542–5 and Chapter VI, p. 134), but this, though very widespread, is always local, the razing of the herbage to within half an inch of the soil surface never extending very far from a collection of burrows.

The grazing factor, then, is the primary factor which maintains and stabilises grassland, but the successional data given on previous pages seem to show that a vegetation dominated on the whole by grasses, though largely consisting of dicotyledonous herbs, comes into existence on bare chalk independently of grazing. Under natural conditions, i.e. in the absence of more or less intensive grazing, the grassland would pass more or less rapidly (according to the degree of proximity of parent shrubs and trees) into scrub and woodland, for woody plants may begin to come in as soon as the herbs and grasses, but no examples of this complete sequence have been observed. The succession to woodland on stabilised chalk grassland where grazing is insufficient to exclude the colonisation of woody plants has been described in Chapter XVIII.

Formation of loam. On the flat or gently sloping ground of the chalk plateaux and dipslopes non-calcareous soil, even where it is not formed of definite Quaternary deposits, may reach a considerable depth as a result of the gradual accumulation of residual material from the dissolution of the chalk. Loams of this origin occur in all stages of development and degrees of thickness. They show the characteristic "brown earth" rather than the "rendzina" profile, are distinctly leached in the upper layers and bear vegetation different from that of the shallow chalk soil. When this loam reaches a considerable thickness it is known as "clay-with-flints", from the numerous flints originally embedded in the chalk which form part of the chalk residue. A good deal of it however probably contains the remnants of Tertiary and Quaternary deposits. These loams cover very great areas of the chalk outcrop and bear beechwood or oakwood where they are not cultivated.

Chalk heath. The earlier stages of loam formation are constantly met with on flat or slightly inclined parts of the chalk plateau, close to an escarpment or chalk valley. The leaching of the upper layers gives rise to a slightly acid soil, which is inhabited by a mixture of typical chalk grassland plants, with other species indicative of acid conditions. Of the latter the two grasses *Agrostis tenuis* and *Sieglingia decumbens* are those which first appear in such situations, and they are followed by species like *Potentilla erecta* and *Calluna vulgaris*. Of the chalk grassland species some, like *Festuca ovina*, which are surface rooting, can tolerate the acidification of the surface layers, while others, such as *Poterium sanguisorba* have deep roots which descend to the calcareous soil below. Thus is developed the beginnings of a vegetation which has been called *chalk heath*, because it is marked by a mixture of calcicolous plants rooting in calcareous soil with indifferent and calcifuge heath plants rooting in the acid surface soil. Large areas of plateau and dipslope are dominated by *Calluna* or *Erica cinerea*, accompanied by such characteristic southern heath plants as *Ulex minor*.

An analysis of the vegetation of five chalk heaths shows that the main list of species is the same as that of the normal chalk grassland, but that the occurrence and abundance of such characteristic species as the Avenae, *Koeleria*, *Scabiosa columbaria*, *Pimpinella saxifraga* and *Phyteuma orbiculare* are considerably diminished. At the same time *Aira praecox*, *Calluna*, occasionally *Erica cinerea* and *Galium saxatile*, make their appearance, while the occurrence and abundance of *Agrostis tenuis* greatly increase, and to a less degree, though still considerably, the frequency of *Carex caryophyllea*, *Trifolium repens*, *Potentilla erecta*, *Veronica officinalis*, *Viola riviniana* and *Dicranum scoparium*. The CaO content of the upper layers of soil of these five chalk heaths varied from 0·53 to 2·0 per cent, and the pH values from 6·0 to 6·9, increasing with depth.

It seems possible that "chalk heaths" of the kind described develop into typical heaths by further leaching and the accumulation of acid humus at the surface, but nothing of the kind was obvious in the district investigated (South Downs). More typical heaths do occur (rarely) in the district, but they are situated on shallow sandy Quaternary deposits overlying the chalk. Such heaths occur much more extensively on the plateaux of other chalk districts.

Under natural conditions, i.e. in the absence of regular grazing, the incipient heaths, like the chalk grassland itself, would undoubtedly be invaded by shrubs and trees and progress to woodland.

INFERIOR OOLITE GRASSLAND

The Inferior Oolite of the Cotswold Hills in Gloucestershire is a less pure and harder limestone than the chalk, but bears much the same vegetation. Thus the Cotswold beechwood community does not differ widely from that of the chalk escarpment, and the natural grassland of the Cotswolds is almost identical with that of the chalk downs. Though they have not been intensively studied there can be little doubt that both beechwood and grassland communities are very close to those of the chalk. The general lists of grassland species and their relative frequencies are very much the same as on the chalk, though on the flat tops of the hills, even near the escarpment, there is a greater depth of soil, probably correlated with the lesser purity of the oolitic limestone giving a greater proportion of residual material; and the surface layers of soil are poorer in free calcium carbonate. With this goes a greater frequency of species like *Agrostis tenuis* and *Sieglingia decumbens*. Certain other species, unknown or rare on the south-eastern chalk, also figure more or less conspicuously on the oolite, for example *Solidago virga-aurea* and *Spiranthes spiralis*. *Bromus erectus* is very widespread and very frequently dominant in the grassland of the Cotswolds, and it is often closely associated with *Brachypodium pinnatum*; and this is also true of some of the Wiltshire and Berkshire chalk downs. The natural grassland of the oolite has not, however, been closely studied, and it is not possible to give a fuller description.

GRASSLAND OF THE OLDER LIMESTONES AND THE BASIC IGNEOUS AND METAMORPHIC ROCKS

The grassland of the older (Palaeozoic) limestones has much in common with that of the chalk, but also marked differences. The thin soil has the same fundamental relation to the rock below, whose continual solution acts as a basic buffer to the soil water: the turf is typically dominated by *Festuca ovina* (probably associated with *F. rubra*), with several of the same accompanying species, both grasses and dicotyledonous herbs, as in chalk grassland; and like it the grassland of the older limestones is traditional sheep pasture.

PLATE 104

Phot. 255. Limestone grassland on the steep western slope of Cross Fell (northern Pennines), with *Festuca ovina*, *Sesleria caerulea* and many calcicolous species, but also *Agrostis tenuis*. To the right, above, is a limestone "scar" and small scree: to the left, below, is Callunetum on a gently inclined grit slope. *Mrs Cowles.*

Phot. 256. Nardetum on slightly inclined flagstone platform above the limestone scar shown in Phot. 255. *Mrs Cowles, I.P.E.* 1911.

LIMESTONE GRASSLAND AND NARDETUM

Features of the older limestones. On the other hand the much greater hardness of the older limestones leads to the formation of crags—"scars" as they are called in the north—which often support a flora of their own: a large number of chalk species are absent from the older limestones, while one or two "calcicoles" (e.g. *Sesleria caerulea*) occur that are not found on the chalk; and finally the cooler and damper climate in which the older limestones are exposed leads to widespread formation of acid humus, so that a number of "calcifuge" species are usually present, rooted in this humus, and often mixed or alternating with the calcicoles.

The grasslands of the older limestones were briefly described by several of the pioneer students of British vegetation in the early years of the present century. Thus Robert Smith (1900b) gives a short list of the chief species occurring on a limestone exposure at Balnabodach, Loch Tummel, in Perthshire: W. G. Smith and C. E. Moss (1903) describe the "limestone hill pasture" (Mountain or Carboniferous Limestone) in the Harrogate and Skipton district of Yorkshire as "consisting chiefly of *Festuca ovina*", and give a list of species: F. J. Lewis (1904) gives a similar list of the "bright green closely cropped pasture" where the limestone is "free from peat" in the northern Pennines:[1] C. E. Moss (1907) describes though he does not give a list of the limestone pastures (also on Mountain Limestone) of the Mendip Hills in Somerset; and later (1911, 1913) deals with the grassland of the same geological formation in the southern Pennines of Derbyshire.

Very similar grassland occurs on the igneous and metamorphic rocks "rich in mineral salts, especially calcium and magnesium" (basalt, andesitic lavas and schists) described by Robert Smith (1900a) from the Edinburgh district, and by W. G. Smith (1904-5) from Fife.

More than one of the earlier surveyors point out that the "calcicole" plants growing on these geological formations typically occur where the soil is thin above the rock, and are thus associated with the "scars" of the limestone or with exposed knolls of basalt. In hollows between the knolls, where the surface is kept damp, humus accumulates and calcifuges such as *Calluna* and *Vaccinium myrtillus* are present, belonging to quite a different vegetation. And in the regions of higher rainfall the surface soil of the general limestone grassland itself becomes leached, accumulates acid humus, and contains a number of species which are not characteristic of limestone soils. Thus the turf may contain abundance of *Agrostis tenuis* (which, as we have seen, only occurs very locally in chalk grassland) so that the limestone grassland may even come to approximate to the bent-fescue pastures of the siliceous rocks. As we shall see in a later chapter this leaching and accumulation of acid humus may go so far on limestone plateaux as to lead to the development of typical *Calluna* heath immediately above the limestone rock, while "limestone heaths", with mixtures of calcicolous and calcifuge species, quite comparable with the chalk heaths

[1] Pl. 104, phot. 255.

previously described, are common in many limestone areas. Nevertheless where the drainage is free, as on the slopes of the rounded Mountain Limestone hills, a uniform pasture, not containing calcifuge species and approximating more or less closely to chalk pasture, though always with fewer species, is developed over wide areas.

The following composite list (Table XVII) is constructed from those given for the older limestones (mainly Mountain Limestone) and other basic rocks by the earlier workers mentioned above. The absence of species from the shorter lists does not necessarily mean that they do not occur in the corresponding areas, since the thoroughness of the listing varies a great deal. But at least a fair idea can be obtained of the dominant, characteristic, and most abundant species:

Table XVII

(1) Moss (1913), Derbyshire, Mountain Limestone.
(2) Smith and Rankin (1903), West Yorkshire, Mountain Limestone.
(3) Lewis (1904), Northern Pennines, Mountain Limestone.
(4) R. Smith (1900a), Edinburgh district (basaltic rocks).
(5) W. G. Smith (1904–5), Sidlaws, volcanic rocks.
(6) W. G. Smith (1904–5), Central Fife, basalt.
(7) R. Smith (1900b), N. Perthshire, limestone at Loch Tummel.

	1	2	3	4	5	6	7
Grasses							
Agrostis spp.	a	.	+	+	+	+	.
Anthoxanthum odoratum	o, la	+	.	+	+	+	.
Arrhenatherum elatius	la	.	.	.	.	.	.
Avena pratensis	o	+	.	.	.	.	.
A. pubescens	l	+	.	.	.	.	.
Brachypodium pinnatum	vr	.	.	.	.	.	.
B. sylvaticum	a	.	.	.	.	.	.
Briza media	o–a	+	.	.	.	.	.
Bromus erectus	?	.	.	.	.	.	.
Cynosurus cristatus	la	+	.	.	+	.	+
Deschampsia flexuosa	.	.	.	.	.	+	.
Festuca elatior	r	.	.	.	.	.	.
F. ovina	a–d	d	+	+	+	+	+
F. rubra	r	.	.	.	.	.	.
Holcus lanatus	.	.	.	.	.	.	+
Koeleria cristata	o	+	+	.	.	.	+
Poa pratensis	l	.	+	+	+	.	+
P. trivialis	.	.	.	.	+	.	.
Sesleria caerulea	.	+	+	.	.	.	.
Sieglingia decumbens	.	.	.	+	.	.	.
Trisetum flavescens	l	.	.	.	.	.	.
Herbs							
Achillea millefolium	a	.	.	.	+	.	.
Alchemilla vulgaris	o	.	+	.	.	.	+
Allium vineale	vr	.	.	+	.	.	.
Anacamptis (Orchis) pyramidalis	vr	.	.	.	.	.	.
Antennaria dioica	la	+	.	.	.	.	.
Anthyllis vulneraria	o	+	.	+	.	+	.
Asperula cynanchica	la	.	.	.	.	.	.
Astragalus danicus	.	.	.	+	.	.	.
Bellis perennis	o	.	.	.	+	.	.

Older Limestones and Basic Igneous Rocks

Campanula glomerata	o	.	.	.	.	.	.
C. rotundifolia	o	.	+	+	+	+	.
Carduus nutans	o	+	.	.	.	.	.
Carex caryophyllea	la	.	.	.	.	.	.
C. diversicolor (flacca)	o	.	.	.	.	.	.
C. ornithopoda	r	.	.	.	.	.	.
C. pilulifera	l	.	.	.	.	.	.
Carlina vulgaris	o	+	+	.	.	.	.
Centaurea nemoralis	a	.	+	.	.	.	.
C. scabiosa	o	.	.	.	.	.	.
Centaurium umbellatum	r	.	.	.	.	.	.
Chrysanthemum leucanthemum	a	.	.	.	.	.	+
Cirsium eriophorum	l	+	.	.	.	.	.
C. palustre	l	.	.	.	.	.	.
Clinopodium vulgare	r	.	.	.	+	.	.
Cochlearia alpina	l	.	.	.	.	.	.
Coeloglossum viride	r	+	.	.	.	.	.
Conopodium majus	o	.	.	.	+	.	.
Crepis capillaris	a	.	.	.	.	.	.
Erophila boerhaavii	la	.	.	.	.	.	.
E. verna	la	.	.	.	.	.	.
Euphrasia officinalis	a	.	+	.	.	.	+
Filipendula hexapetala	r	+	.	.	.	.	.
Galium cruciata	la	.	.	.	.	.	.
G. pumilum (sylvestre)	la	+	?	.	.	.	.
G. saxatile	r	.	+	+	+	+	+
G. verum	a	+	.	+	+	+	+
Gentiana amarella (agg.)	o–a	+	.	.	.	.	+
Geranium lucidum	la	.	+	.	.	.	.
G. sanguineum	r	.	.	+	.	+	.
G. sylvaticum	.	+	.	.	.	.	.
Gymnadenia conopsea	r	+	.	.	.	.	.
Helianthemum nummularium	a	+	+	+	+	+	+
Hieracium murorum	la	.	+	.	.	.	.
H. pilosella	a	+	+	.	+	.	+
Hippocrepis comosa	r	+	.	.	.	.	.
Hutchinsia petraea	l	.	.	.	.	.	.
Hypericum hirsutum	lf	.	.	.	.	.	.
Lathyrus montanus	r	.	.	+	+	.	.
L. pratensis	o	.	.	.	.	.	.
Leontodon autumnalis	la	.	.	.	.	.	.
L. hispidus	a	.	.	.	.	.	.
L. nudicaulis	a	.	.	.	.	.	.
Leucorchis (Habenaria) albida	.	+	.	.	.	.	.
Linum catharticum	a	+	+	.	+	.	.
Lotus corniculatus	a	+	+	.	+	.	+
Luzula campestris	a	+	.	+	.	.	.
Minuartia (Arenaria) verna	l	+	.	.	.	.	.
Ononis repens	.	.	.	+	.	.	.
Ophrys apifera	r	.	.	.	.	.	.
Orchis maculata	l	.	.	.	.	.	.
O. mascula	la	.	+	.	.	.	.
O. ustulata	r	.	.	.	.	.	.
Origanum vulgare	o–a	.	.	+	.	.	.
Oxalis acetosella	.	.	.	.	.	+	.
Picris hieracioides	o	.	.	.	.	.	.
Pimpinella saxifraga	o	.	.	+	+	+	.
Plantago lanceolata	a	.	.	.	.	.	+
P. media	la	+	.	.	.	.	.
Platanthera bifolia	vr	+	.	.	.	.	.
P. chlorantha	.	+	.	.	.	.	.
Polygala vulgaris	a	+	+	.	+	.	+
Potentilla erecta	.	.	.	+	+	+	.
P. verna	l	.	l	+	.	.	.

Poterium officinale	l	+	.	.	.	.	.
P. sanguisorba	a	+	.	.	.	.	.
Primula veris	a	.	.	.	.	.	.
Prunella vulgaris	o	.	.	.	.	.	.
Ranunculus bulbosus	a	+	.	.	.	.	.
Rumex acetosa	o	.	+	.	.	+	.
R. acetosella	r	+	+	.	+	.	.
Saxifraga granulata	la	+	.	.	.	.	.
S. hypnoides	l	+	+	.	.	.	.
Scabiosa columbaria	o	+	.	.	.	.	.
Sedum acre	a	.	+	.	.	.	.
Spiranthes spiralis	vr	.	.	.	.	.	.
Stachys (Betonica) officinalis	o	+	.	.	.	.	.
Succisa pratensis	r	.	.	.	.	+	.
Teucrium scorodonia	l	.	+	.	+	+	.
Thalictrum minus	l	.	.	.	.	.	.
Thymus serpyllum	a	a	+	.	+	+	+
Trifolium dubium	r	.	.	.	.	.	+
T. pratense	o	.	.	.	.	+	.
T. repens	r–o	.	.	+	.	.	+
Veronica chamaedrys	o	.	.	.	+	+	.
V. officinalis	.	.	+	.	.	+	.
Vicia angustifolia	vr	.	.	.	.	.	.
V. hirsuta	.	.	.	+	.	.	.
V. lathyroides	.	.	+	.	.	.	.
Viola hirta	o	.	.	+	.	.	.
V. lutea	la	la	+	+	+	+	.
V. riviniana	o	.	+	+	+	+	.
Botrychium lunaria	r–o	+	.	.	.	.	.
Ophioglossum vulgatum	la	+	.	.	.	.	.
Pteridium aquilinum	r	+	.	+	.	.	+
Totals	110	45	31	27	26	22	20

If these lists are compared with those of chalk grassland it is obvious at once that many of the southern species are absent, while others become increasingly sporadic or disappear altogether as we travel farther north. The much greater length of the Derbyshire list (110 species) is due partly no doubt to the more thorough investigation of the grassland there than in the more northern areas, but mainly to the failure of many species which occur, though often rarely, in the southern Pennines, to reach the northern Pennines or Scotland. Several plants, however, which are absent from the southern Scottish basalts come in again on the limestone in Perthshire.

Certain species, very characteristic of the southern chalk, are regularly found both on the limestones and on the basic igneous rocks in many of the northern areas, and of these *Helianthemum nummularium* (*vulgare*) is the most constant. Others, such as *Thymus* and *Pimpinella saxifraga*, occur in several, though not in all. *Astragalus danicus*, local on the chalk of the eastern counties of England, appears again on the basaltic rocks of the Edinburgh district.

Among positive characters of the northern limestone and other basic grasslands we may note the appearance of certain species which do not occur in the south: for example *Viola lutea* is abundant in several of the northern localities, and is found also in the bent-fescue grassland on siliceous rocks, but not in the south. *Sesleria caerulea* does not occur in the south, nor even in Derbyshire, and is very local in Scotland, but is a characteristic grass of the northern Pennine limestones. Besides these we have the frequent occurrence on limestone, already referred to, of species characteristic of acid soils, e.g. *Agrostis tenuis*, sometimes abundant, *Potentilla erecta*, and *Galium saxatile*, the last quite absent from chalk grassland.

Limestone heaths. These are formed in just the same way as the chalk heaths described on pp. 551–2. Smith and Rankin (1903) record the following species as occurring together in a limestone heath in west Yorkshire:

Calcicolous or tolerant	Calcifuge
Sesleria caerulea	Calluna vulgaris
Festuca ovina	Vaccinium myrtillus
Viola lutea	Nardus stricta
Thymus serpyllum	Polytrichum sp.

and Moss (1907, 1913) records many more from similar mixtures of heath and calcicolous vegetation on the Mountain Limestone of the Mendips in Somerset and of the Pennines in Derbyshire. (Cf. also p. 473.)

Limestone swamps. Unlike the Chalk, the Mountain Limestone often supports swamps, owing to high rainfall and locally obstructed drainage, and these contain a characteristic vegetation. Moss (1913) gives the following list for the limestone swamps of the southern Pennines (Derbyshire):

Caltha palustris	a		Filipendula ulmaria	a
Carex acuta	a		Geum intermedium	la
C. disticha	l		G. rivale	la
C. diversicolor (flacca)	o		Juncus compressus	vr
C. pendula	l		J. inflexus (glaucus)	la
C. strigosa	r		Mentha spp.	a
C. sylvatica	l		Orchis maculata	o–a
Chrysosplenium alternifolium	la		Parnassia palustris	la
			Pedicularis palustris	r
C. oppositifolium	la		Petasites hybridus	la
Cirsium heterophyllum	a		Polemonium caeruleum	la
C. palustre	a		Scirpus compressus	r
Epilobium hirsutum	a		Thalictrum flavum	r
Epipactis palustris	vr		Trollius europaeus	la
Eupatorium cannabinum	la		Valeriana dioica	la
Festuca arundinacea	vr		V. officinalis	la
F. elatior	r			

REFERENCES

Anderson, Violet L. Studies of the vegetation of the English chalk. V. The water economy of the chalk flora. *J. Ecol.* 15, 72–129. 1927.

Hope Simpson, J. F. The War Down spoil heaps (unpublished).

Lewis, F. J. Geographical distribution of the vegetation of the basins of the Rivers Eden, Tees, Wear and Tyne. Part I. *Geogr. J.* 1904.

Moss, C. E. Geographical distribution of vegetation in Somerset: Bath and Bridgewater District. *Roy. Geogr. Soc.* 1907.

Moss, C. E. "Limestone grassland association" in *Types of British Vegetation*, pp. 154–60. 1911.

Moss, C. E. *The Vegetation of the Peak District.* Cambridge, 1913.
Smith, Robert. Botanical survey of Scotland: I. Edinburgh District. *Scot. Geogr. Mag.* 1900a. II. North Perthshire District. *Scot. Geogr. Mag.* 1900b.
Smith, W. G. Botanical survey of Scotland: III. Forfar and Fife. *Scot. Geogr. Mag.* 1904–5.
Smith, W. G. and Moss, C. E. Geographical distribution of vegetation in Yorkshire. Part I. Leeds and Halifax district. *Geogr. J.* 22. 1903.
Smith, W. G. and Rankin, W. M. Geographical distribution of vegetation in Yorkshire. Part II. Harrogate and Skipton district. *Geogr. J.* 1903.
Tansley, A. G. and Adamson, R. S. Studies of the vegetation of the English Chalk. III. The chalk grasslands of the Hampshire-Sussex border. *J. Ecol.* **13**, 177–223. 1925. IV. A preliminary survey of the chalk grasslands of the Sussex Downs. *J. Ecol.* **14**, 1–32. 1926.
Tüxen, R. Bericht über die pflanzensoziologische Excursion der floristisch-soziologischen Arbeitsgemeinschaft nach dem Plesswalde bei Göttingen. *Mitt. d. floristisch-soziologischen Arbeitsgemeins. in Niedersachsen,* pp. 25–51. 1927.
Watson, W. The bryophytes and lichens of calcareous soils. *J. Ecol.* **6**, 189–98. 1918.

Chapter XXVIII

NEUTRAL GRASSLAND

The term "neutral grassland" was first used in *Types of British Vegetation* (p. 84) to include semi-natural grasslands whose soil is not markedly alkaline nor very acid, mostly developed on the clays and loams which occupy so much flat lowland country in the midlands and south of England as well as on many of the tracts of alluvium in the valleys of the north and west—broadly speaking the soils which also bear the lowland oakwood dominated by *Quercus robur* (Chapter XIII). Most of these soils are actually on the acid side of neutrality—perhaps the majority between pH 6 and 7—owing to slight leaching and humus accumulation, though many of the mother clays from which they are derived are distinctly alkaline. Their general characters, however, and the vegetation they bear, differ completely from those of the acidic sands and of the upland siliceous rocks which bear the bent-fescue community (Chapter XXVI) and Quercetum sessiliflorae (Chapter XV). Typically they are much richer in available nitrogen and the mineral elements necessary for nutrition, though bad treatment may result in serious impoverishment in any or all of these. And they are inhabited by many species of grass and a wide variety of other herbaceous plants not found on the extremely acid soils, nor on the thin basic soils dealt with in the last chapter.

On the other hand some alluvial meadow soils which bear "neutral grassland" vegetation, for example the Thames-side meadows described on pp. 568–70, are actually as alkaline as typical chalk or limestone soils. The pH value is by no means the only determining factor, and the term neutral was originally intended to apply (in fact is better applied) to characterisation of this grassland neither by markedly "calcifuge" nor markedly "calcicole" species, rather than to neutrality of the soil solution.

"Neutral grassland" is thus a very comprehensive term, including a great variety of vegetation, on one side passing into that of the more acid soils dominated by bent and sheep's fescue, on the other into the grassland of markedly calcareous soils with the characteristic calcicolous species. The alluvial soils often have a high water table and may be periodically flooded ("meadow soils", p. 88), passing towards marsh when they tend to be waterlogged, but some are well drained and fairly dry throughout the year. The soils also show a wide range of texture and of natural fertility.

The chief difficulty in the way of a satisfactory treatment of neutral grassland is the rarity of really "natural" examples, a rarity due to the fact that by far the greater part of this type of grassland is enclosed and subject to more or less intensive grazing (or to mowing for hay), often also to surface cultivation and manuring. Most of the "permanent grass" of our agricultural returns falls into the category of neutral grassland, and "permanent grass" represented, in 1919, 53 per cent of the farmlands of England and Wales, or nearly 40 per cent (in 1936 about 42 per cent) of the total area of the country. A great deal of this land was at one time arable, as may be seen from the "lands" (ridge and furrow) in many grass fields. This land has been sown down or allowed to "tumble down" to grass. There is, however, a certain amount of neutral grassland, as on the edges of old woods and on commons and village greens, which has never been sown or ploughed and which is doubtless derived from original forest land by felling followed by grazing, or by degeneration and ultimate disappearance of the woods through grazing alone, in the same way that bent-fescue grassland has been derived from the original forest on the hillsides of the uplands.

Treatment the important factor. It is important to recognise that the existing state of any particular piece of enclosed and tended neutral grassland is very little affected by its origin, whether it is derived from forest by felling and grazing or whether it has been ploughed and sown down, or has "tumbled down" to grass; nor when it has been sown does the composition of the original seed mixture greatly affect its present condition. Even the *original* nature of the soil, i.e. of the underlying rock, may make little difference. According to the climate, situation and soil and (most important of all) the treatment of the field, i.e. the kind, amount and duration of grazing and/or manuring, some of the sown species will die out, others will increase, and invaders from outside may enter and establish themselves till a condition of approximate equilibrium is attained between the plant population and all the incident factors. Of these factors the regime to which the field is subject is of decisive importance. The classical "park grassland" experiments at Rothamsted demonstrated the profound effect of differential manuring on the grassland flora, and more recently Stapledon and his colleagues have frequently shown that different kinds of grazing alone, in the absence of any manuring, will completely alter the character of the vegetation and the species present. The kinds and degrees of cultivation (or neglect) are almost infinitely various. Thus a permanent grass field may be intensively grazed, or it may be undergrazed, by sheep, cattle or horses, or by mixed stock, and at different times of year. It may be manured with dung, or with various artificial manures in any proportion, it may be harrowed or rolled. And a pasture field may be "put up" for hay at various intervals. Every combination of these factors will produce a somewhat different, often a widely or almost totally different result, and all these results will naturally also be conditioned by soil, climate and situation.

Theoretically it would be possible to ascertain the assemblage of grasses and other herbaceous plants that would naturally colonise a given soil type in the absence of any agricultural treatment except moderate grazing, and then the various modifications which would follow different kinds of grazing and manuring. In fact the occasional fragments of untreated and apparently unmodified grassland on clays and loams are too limited to enable us to draw any satisfactory general conclusions even if they had been closely studied ecologically, which they have not. We have to content ourselves at present with such information as can be gained from general observation, from recorded lists of species, and from data derived from studies whose primary object was agricultural.

Commons, greens and verges. Probably the least modified of the neutral grasslands are those on the edges of woods and on commons and village greens on medium or heavy soil. These last are, it is true, practically always grazed, but they are not subject to intensive exploitation; and in the remoter parts of the larger commons the grassland, which often passes into open scrub or woodland, is probably as near as we can get to "natural" neutral grassland.

The grass "verges" at the sides of the country roads are also instructive in this connexion. They are generally cut over in the early autumn when the hedges are trimmed, sometimes earlier in the summer, and locally they may be grazed by donkeys, horses or goats; but they are not manured, though their vegetation is frequently modified by the piling of road metal, or the throwing up of road scrapings or the material dug out from ditches, and, before the introduction of the modern motor road surface, by the road dust. The grassland of such verges would normally be quickly succeeded by scrub and woodland if parent trees are available in the neighbourhood, and here and there colonisation by shrubs and trees and the process of succession may actually be observed (see pp. 295–6). In general the succession is held up by the annual cutting of the verge and the grazing of the commons. Such habitats would undoubtedly repay much closer study than they have received. The effect of treading on verges and footpaths has been considered in Chapter xxv, pp. 496–8.

The grasses of neutral grassland. Of the grasses which dominate neutral grassland all the species are indigenous, though they are mostly better known as more or less "cultivated" plants in tended pasture or meadow, or sown in temporary "leys", than in "natural" situations. The perennial rye grass (*Lolium perenne*) is undoubtedly the most valuable agriculturally, both in pasture and for hay. It grows best on the richest soils, where it is typically co-dominant with "wild" white clover (*Trifolium repens*). The vigour of these two species on really good soils tends to exclude other species and thus to reduce the total flora. Perennial rye-grass is very much commoner in the best-tended pastures, and sown with white clover in temporary leys, than in "wild" situations. It is of tufted habit and when luxuriant its flowering stem may be as much as 2 ft. (60 cm.) in height.

Cocksfoot (*Dactylis glomerata*) is another valuable grass both for hay and in pasture, and is very widely distributed in a great variety of soils and situations, as much in "natural" situations, such as wood edges, commons and roadside verges, as in tended grassland. It is a strongly tufted grass of rather coarse habit, and its flowering stalks may reach a height of 4 ft.

Meadow foxtail (*Alopecurus pratensis*) grows in large, rather loose tufts and is of coarse tall growth, the vegetative shoots, which develop early in the season, as much as 2 ft. high, and the flowering stems with their terminal cylindrical spikes often exceed 3 ft. It flourishes best and is often dominant in meadow land regularly cut for hay, and is less prominent in pasture.

Timothy (*Phleum pratense*) is also a tall tufted grass, reaching a height of 4 ft., and flourishing best in hayfields on good, rather heavy soils, usually disappearing under pasture. It develops much later in the season than meadow foxtail.

Crested dogstail (*Cynosurus cristatus*) is a shorter grass and unlike the two preceding species flourishes better in pastures than in permanent hayfields. It occurs on a considerable range of soils, both dry and damp. Crested dogstail is a valuable grass on second-class pastures, where it is often dominant or co-dominant, taking the place of the perennial rye-grass of the best pastures. The hard glumes of the characteristic mature inflorescences are refused by stock, so that unless it is heavily grazed early in the season the dry inflorescences remain in the pasture.

The two meadow-grasses in the narrow sense (*Poa pratensis* and *P. trivialis*) are rarely dominant, but often grow freely mixed with other grasses, the former on rather drier, the latter on moister or even wet soils, but the two may be found together in alluvial meadows. *Poa pratensis* is the only British meadow grass (if we exclude *Holcus mollis*) with an underground creeping rhizome, while *P. trivialis* has creeping stolons. The former, though widely distributed, is not very productive and but seldom really abundant in pastures in this country, though it is highly esteemed in America as "Kentucky blue grass", and has invaded most of the prairies, where it is not native. *P. trivialis* is more valuable on suitable soils.

Several species of fescue occur in neutral grassland. *Festuca pratensis* (meadow fescue) is a widely distributed and valuable grass in the lowlands. *F. elatior*, as its name implies, is a much taller grass, coarser and more robust, though well taken by stock. It is said to be more hardy and to do better on poorer and drier soil than *F. pratensis*, and also does well on the heaviest clays.

The fine-leaved fescues, often dominant both in acidic and basic grasslands are not characteristic of the neutral type. Thus the red fescue (*Festuca rubra*) has very numerous varieties, some of which occur in salt marsh (and spray-washed maritime grassland), and one on sand dunes, and this species is also abundantly mixed with sheep's fescue on chalk downs. But some varieties occur in neutral grassland, both meadow and pasture, and especially the former. *F. rubra* is indeed occasionally locally dominant in pasture, forming a dense well-grazed turf of shortly creeping shoots: other forms grow in isolated tufts. The red fescues have much narrower, finer leaves than any of the species previously described, the leaves of some being bristle shaped like those of sheep's fescue.

The sheep's fescue (*F. ovina*), with bristle-shaped leaves, also has numerous varieties, some of which are now separated as species. The commonest form (*F. ovina* sensu stricto) is one of our most widely distributed and abundant grasses, being frequently dominant, as we have seen in the last two chapters, both in acid grasslands of the "bent-fescue" type, and in chalk and other limestone grasslands. It is essentially a plant of dry well-drained soils, and though it can flourish under very high rainfall it cannot stand water-logged soil nor the competition of more luxuriant grasses, and under such conditions becomes subordinate or suppressed altogether. Hence it is not at all characteristic of, and is usually absent from, neutral grassland, but occurs in some of the poorest and driest pastures where it is comparatively free from this kind of competition. It actually forms more than 10 per cent of the herbage in the *unmanured* park grassland at Rothamsted (p. 573). *Festuca ovina* differs from *F. rubra* in being strictly tufted, with no creeping shoots escaping from the base of the tuft ("extra-vaginal" shoots) and typically bristle-shaped leaves.

The common bent (*Agrostis tenuis*) is another very abundant and widely spread grass, probably the most numerous species of all in semi-natural grassland vegetation: on the poorest pastures, where it is very generally dominant over great areas, it sometimes forms 90 per cent of the herbage on a great range of soils, from sandy loams to heavy clays. It also occurs freely in medium pastures which are somewhat acid, mixed with the better grasses. In neutral pasture it is of limited grazing value, though, as we have seen, it is one of the staples, together with sheep's fescue, of upland sheep pasture, especially in regions of higher rainfall. It has rather short, somewhat scabrid leaves and matures late in the season: in the poor neglected pastures of which the common bent is so characteristic, its shoots become withered and quite useless for pasture in late summer.

The white bent, *Agrostis stolonifera* (*alba*), is commoner on moister and better soils, the typical form being of distinct agricultural value in wet meadows and on heavy soils.

The sweet vernal grass (*Anthoxanthum odoratum*) is another very widely distributed grass on a great variety of soils, though it is not a dominant species. It develops very early in the season, flowering in late April or early May before any of the other grasses. It is tufted, but of very variable size and habit. Sweet vernal grass is well grazed and quite useful in the upland bent-fescue sheep-walks.

The Avenae are not characteristic of neutral grassland, though both *Avena pubescens* and *A. pratensis* sometimes occur. The closely allied golden oat grass (*Trisetum flavescens*) is also much more characteristic of chalk soils, but occurs, sometimes abundantly, in good grassland on the lighter loams, where it does better under haying conditions than in pasture. It is a tufted, slightly creeping species with flowering shoots 1–2 ft. high. The false oat-grass (*Arrhenatherum elatius*) is a much taller, strong-growing, rather coarse grass, up to 4 ft. in height, which grows on a variety of soils, and is often dominant on the grass verges of roads, especially on calcareous soils. It persists well under cutting, and produces abundant hay though of poor quality, but cannot endure grazing.

Two very common grasses which, on the better grasslands, are weeds from the standpoint of the pastoralist, are the Yorkshire fog (*Holcus lanatus*) and the coarse, tufted, rosette-forming *Deschampsia caespitosa*. The former is a ubiquitous, very aggressive grass on good soils, and often forms a large proportion of the herbage of very good pastures, increasing under manuring. It is covered with soft hairs and is generally refused by stock, except the young shoots. *Deschampsia caespitosa* forms large tufted tussocks or stools with long narrow harsh leaves finely serrated along the margins. It affects waterlogged soil, especially if puddled so that water is held up in the surface layers.

Briza media, the common quaking grass, with its very loose panicle of ovate spikelets, is much more characteristic of chalk than of neutral grassland, but it occurs on various poor soils in both dry and wet situations. *Briza* is worthless agriculturally and rapidly disappears under manuring.

Non-gramineous species. The non-gramineous species of the herbage are generally classified by agriculturists into "leguminous" and "miscellaneous", because the former are extremely valuable pasture plants—some as valuable as the best grasses—while many though not all of the latter are worthless as food for stock. The Leguminosae contain on the average more protein, much more calcium, and less carbohydrate than the grasses, and are of quite comparable nutritive value, but in addition they are of great importance agriculturally because they fix the free nitrogen of the air and leave it in the soil in an available form. They are practically always represented, though not specially numerous in species and individuals, on most neutral grassland, but they are much encouraged by phosphatic and potash manures such as "basic slag" and "kainit", as the Rothamsted park grassland manuring experiments show.

Taking neutral grassland as a whole the "miscellaneous species" of herbs are very numerous. The actual number of species on any given area is more or less inversely proportional to the "fertility" of the grassland—the better the soil and the better it is grazed the fewer the species. The primary cause of this relation is that the best and the best managed soils encourage the growth of certain species—notably *Lolium perenne* and *Trifolium repens*—to such an extent that the miscellaneous species are crowded out and are unable to re-enter the community. But on the very poorest and most unfavourable soils the number of species is again reduced, because only a few are able to endure the conditions.

Leguminous plants. Of leguminous species the clovers (*Trifolium*) are some of the most widely distributed as well as the most important agriculturally. The three species most characteristic of neutral grassland are *T. pratense*, *T. repens* and *T. dubium*.

The red clover (*T. pratense*) is a free-growing tufted biennial or perennial. Though it is a slow coloniser it occurs naturally on many different types of soil, and is more abundant in old meadowland than in old pasture, though it is of use in both and a valuable accumulator of nitrogen. The cultivated varieties are often sown as pure crops.

The white clover (*T. repens*) is also found on a wide variety of the better soils, and propagates very freely by means of runners, especially in pastures. It is one of the most valuable agricultural plants, and is characteristically associated with *Lolium perenne* in the best pastures. The native strains ("wild white clover") are much more desirable than the so-called "Dutch clovers".

T. dubium (yellow suckling clover) is a somewhat tufted annual, with semi-erect shoots and small heads of minute yellow flowers, widely distributed in various kinds of neutral grassland. It is of some use in pasture though not nearly so valuable as the red and white clovers, being a smaller plant.

Medicago lupulina (the black medick or trefoil) is also an annual, with weak trailing shoots and yellow flowers, and has a strong superficial resemblance to *Trifolium dubium*. It is also very widely distributed, but is absent from the more acid soils in regions of high rainfall. Being a small plant it is not of great importance in mixed herbage, but may have considerable value when sown.

One of the most widely distributed of leguminous species is bird's foot trefoil (*Lotus corniculatus*), which occurs on a much greater variety of soils than the clovers and medick, not only in neutral grassland (though it is absent from the best soils) but in chalk and limestone grassland, where it is constant, in grass heath, and in many of the bent-fescue upland sheep pastures, where it is often the only legume present. It has a long tap-root and is thus able to obtain water from deeper soil layers when the surface becomes dry. Doubtless it has considerable value in pastures where it is naturally abundant, but *Lotus* is not sown agriculturally. Like white clover it responds freely to phosphatic manures.

The meadow vetchling (*Lathyrus pratensis*) is another leguminous plant common in neutral grassland, particularly hayfields, and on roadsides, and the same may be said of several species of vetch (*Vicia*).

"Miscellaneous" species. With the horde of "miscellaneous" species inhabiting neutral grassland it is not necessary to deal individually. Three or four are however worth mention.

The milfoil or yarrow (*Achillea millefolium*) is an abundant plant in many kinds of grassland, particularly on the lighter and drier soils. It is deep-rooting and extremely drought-resistant, the leaves remaining fresh and green at the end of hot dry summers when most of the herbage is dried up. When growing gregariously it forms a dense "sole", excluding other plants. It is frequently well cropped by stock, and is thus of considerable value.

The ribwort plantain (*Plantago lanceolata*) is another very abundant plant on almost all kinds of soil except acid peat, and like the milfoil, is deep-rooting and drought-resistant. It certainly has a value on many pastures, though it may increase so much at the expense of better pastoral plants as to become a troublesome weed. It is useless for hay because its leaves are not easily caught by the machine, and even when harvested are difficult to dry.

The "burnet saxifrage" (*Poterium sanguisorba*) is characteristic of calcareous soils, but is also found in some kinds of neutral grassland, both meadow and pasture. Its shoots are not much eaten by stock, except sometimes when quite young. Its congener *P. officinale*, is more characteristic of damp alluvial meadows.

Carex diversicolor (*flacca*) is the most widely distributed of the grassland sedges. It is a chalk grassland constant and also occurs, often abundantly, on low-lying and waterlogged soils.

Many of the "miscellaneous" species of neutral grassland are both palatable and rich in protein and minerals, especially calcium, so that they may be heavily grazed in winter. On the other hand they contain on the whole less dry matter than the grasses and do not shoot again so readily when grazed down.

Meadow and pasture. The broad classification of grasslands according to their use, i.e. into *meadows* which are regularly mown for hay, and *pastures* which are grazed, is ecologically of great importance, because the alternative uses lead to very considerable differences in the vegetation. Meadows are specially characteristic of alluvial soil in which the water table is seldom far below the surface, and which may be periodically flooded by the rise in the waters of neighbouring streams. Sometimes they are *irrigated*, i.e. artificially flooded by means of irrigation channels. Regularly inundated meadows are known as *water meadows*. Natural or artificial irrigation by river water fertilises the soil by addition of fresh mineral salts, and the alternate motion of the water from the streams to the meadows in time of flood and back to the river when the water level sinks keeps the soil well aerated. If there is any considerable fall in the course of the river bed the ground water below the meadow soil is also constantly moving. Added to the high initial fertility which is a common character of alluvial soils this combination of factors favours the dominance of moisture-loving grasses of tall luxuriant habit and the consequent relative suppression or absence of lower growing species. The flora of water meadows is remarkably constant because of the constancy of conditions dominated by the inundation factor and the regular mowing.

Grassland cut for hay is not of course by any means confined to alluvial meadows, any more than pasture is confined to the uplands. And much grassland is frequently used for the two purposes in different years, or in the same year by grazing the aftermath of the hay crop.

The three following lists contain the species which, according to Stapledon (1925), are on the whole characteristic of (1) old meadows, (2) old pastures, and (3) are equally common in both.

(1) *Plants specially characteristic of old meadows*

Grasses

Alopecurus pratensis	Festuca rubra
Bromus hordeaceus (mollis)	Holcus lanatus
Dactylis glomerata	Trisetum flavescens

Herbs

Allium vineale	Orchis maculata
Anthriscus spp.	O. morio
Cardamine pratensis	Platanthera bifolia
Centaurea nemoralis (nigra)	Polygonum bistorta
Chrysanthemum leucanthemum	Poterium officinalis
Colchicum autumnale	Rhinanthus spp.
Geranium molle	Taraxacum officinale
Heracleum sphondylium	Trifolium dubium
Lathyrus pratensis	T. pratense
Leontodon hispidus	Vicia cracca
Lychnis flos-cuculi	V. hirsuta
Malva sylvestris	

(2) *Plants specially characteristic of old pastures*

Agrostis tenuis	Crepis capillaris
Cynosurus cristatus	Filipendula ulmaria
Festuca ovina	Genista tinctoria
	Juncus spp.
Carex spp.	Linum catharticum
Cirsium arvense	Ononis spp.
C. lanceolatum	Poterium sanguisorba
C. palustre	Senecio jacobaea
Conopodium majus	

(3) *Plants equally common in old meadows and old pastures*

Agrostis stolonifera	Daucus carota
Lolium perenne	Galium verum
	Geranium dissectum
Achillea millefolium	Hypochaeris radicata
Ajuga reptans	Leontodon autumnalis
Alchemilla vulgaris	Luzula campestris (agg.)
Bartsia odontites	Plantago lanceolata
Bellis perennis	P. major
Cerastium vulgatum	Primula veris
Convolvulus arvensis	Prunella vulgaris

Pulicaria dysenterica
Ranunculus acris
R. repens
Rumex acetosa

Rumex acetosella
Veronica chamaedrys
V. serpyllifolia

Thames-side meadows and pastures With these lists it is interesting to compare the data obtained by H. Baker (1937) from some alluvial grassland bordering the Thames just above Oxford. These meadows are of great interest because it is known that one of them (Port Meadow, about 400 acres) has been continuously pastured by horses and cattle (not sheep) at least since the time of the Domesday survey (1085), while the others (Yarnton, Oxey and Pixey Meads, together about 270 acres) have been regularly mown for hay at least for centuries. The two grasslands may, therefore, be considered to have come into equilibrium with these contrasted biotic factors.

The soil of both is river alluvium of the same general type. It is subject to annual flooding, and before the flow of the river was regulated by locks and weirs it was probably submerged for six months or so in winter. Since the control the floods have been smaller but more frequent. The pH values of the Port Meadow soil vary from 6·6 and 6·8 at 3 in. depth in the middle area (B), which lies about 2 ft. above the summer water level in the adjacent river and is subject to the greatest leaching, to 7·9 and 8·0 at 6 in. in the lowest part (A) which slopes gradually from about 1 ft. above the summer water level to the river. All the "Meads" (cut for hay) show very constant values, between 7·3 in the surface soil in summer and 8·0 to 8·3 at 6 in. in the autumn and winter. The pH value of Thames water is about 8·0.

The list of ninety-five species occurring in these grasslands includes 26 found only in Port Meadow (grazed) and thirty-nine only in the Meads (mown), while thirty are common to both—striking evidence of the differential effect of the two biotic factors. In Port Meadow there is a further differentiation between the lowest lying region (A) which slopes from river level to about 1 ft. above, containing marsh species, and the highest area (C) whose surface is about 3 ft. above water-level. Nineteen of the twenty-six species occurring in Port Meadow only are confined to one of these two areas (in A 10, in C 9):

Port Meadow (grazing only)

A (wettest)	C (driest)
Alopecurus geniculatus	Koeleria cristata
Apium nodiflorum	Achillea millefolium
Hippuris vulgaris	Carex caryophyllea
Nasturtium officinale	Cirsium acaule
Ranunculus drouetii	C. lanceolatum
R. flammula	Leontodon nudicaulis
Sium erectum	Plantago media
Veronica anagallis-aquatica	Ranunculus bulbosus
V. beccabunga	Trifolium dubium
V. scutellata	

Nearly all the species limited to (A) are marsh plants which exist in the short turf in miniature forms, prostrate or about 1 in. high, closely cropped but flowering.

The remaining seven species occur in B alone (the intermediate area) or extend through two or all three of the areas, AB, AC or ABC:

Poa annua (A, B, C)
Bellis perennis (A, B, C)
Crepis capillaris (A, B)

Luzula campestris (B)
Potentilla anserina (A, C)
P. reptans (B)
Prunella vulgaris (A, C)

Hay Meads (mowing)

Drier	Wetter
Alopecurus pratensis	Agropyron repens
Arrhenatherum elatius	Festuca elatior
Bromus hordeaceus	F. loliacea
Briza media	Hordeum pratense
Carex hirta	Carex acutiformis
Chrysanthemum leucanthemum	Lychnis flos-cuculi
Linum catharticum	Lysimachia nummularia
Ophioglossum vulgatum	Oenanthe silaifolia
Rumex acetosa	Pedicularis palustris
Succisa pratensis	Rumex crispus
	Senecio aquaticus
	Tragopogon pratensis
	Triglochin palustre
	Valeriana dioica

Throughout

Bromus commutatus	Leontodon hispidus
Caltha palustris	Poterium officinale
Carex disticha	Rhinanthus crista-galli
C. goodenowii	Silaum silaus
C. panicea	Stellaria graminea
Equisetum palustre	Thalictrum flavum
Filipendula ulmaria	Vicia cracca
Lathyrus pratensis	

Most of the grasses confined to these hayfields are among the tallest and most abundant species of the mead vegetation. The great majority of the dicotyledons are tall scapose hemicryptophytes intolerant of heavy grazing.

The thirty following species occur both in Port Meadow and the hay meads:

*Agrostis stolonifera
Anthoxanthum odoratum
Cynosurus cristatus
Dactylis glomerata
*Deschampsia caespitosa
Festuca pratensis
F. rubra

Glyceria fluitans
*Holcus lanatus
Lolium perenne
Phleum pratense
Poa pratensis
P. trivialis

* The species marked with an asterisk occur throughout the entire alluvial area examined.

Cardamine pratensis	Lotus corniculatus
Carex diversicolor (flacca)	Myosotis scorpioides (palustris)
Centaurea nemoralis	Oenanthe fistulosa
Cerastium vulgatum	Plantago lanceolata
Galium palustre	Ranunculus acris
Hypochaeris radicata	R. repens
Juncus articulatus	Trifolium pratense
Leontodon autumnalis	T. repens

Magdalen Meadow. As an example of a low-lying alluvial meadow with a medium loamy soil and mixed treatment the flora of Magdalen Meadow, Oxford, may be given. This is not strictly a water meadow since it is not regularly irrigated, but the ditch which surrounds it connects with the River Cherwell, and the meadow is flooded during the winter or early spring of some years, rarely as late as May. The summer water table in the bordering ditches may fall to 3 ft. or more below the surface of the meadow. Thus there is considerable movement of water in the soil due to the rise and fall of the water table. The meadow is usually pastured in summer and autumn (deer or cattle), but is sometimes put up for hay. As a pasture it would probably be reckoned second or third class.

Flora of Magdalen Meadow (not exhaustive)

Grasses

Agrostis stolonifera	o	Holcus lanatus	la
Alopecurus pratensis	a, ld	Hordeum nodosum	
Anthoxanthum odoratum	la	(pratense)	la
Dactylis glomerata	ld	Lolium perenne	l
Festuca pratensis	lf	Poa pratensis	lf
F. rubra	la	P. trivialis	lf

Herbs

Ajuga reptans	o	Ranunculus acris	a
Bellis perennis	l	R. bulbosus (only on highest	
Cardamine pratensis	f	ground)	
Cerastium vulgatum	l	R. repens	l
Ficaria verna	o	Rumex acetosa	f
Fritillaria meleagris	la	Taraxacum officinale	o
Galium palustre	o	Trifolium pratense	l
Lychnis flos-cuculi	o–f	T. repens	o–f
Lysimachia nummularia	f–a	Veronica serpyllifolia	o
Nepeta hederacea	o		
Potentilla reptans	f	Ophioglossum vulgatum	o
Prunella vulgaris	o		

Alopecurus and *Dactylis* are co-dominant in some areas, forming herbage 2 ft. high in June. In other places the species of *Festuca* and *Poa* with *Hordeum nodosum* dominate herbage which is shorter and less coarse, but luxuriant. The fritillaries (Pl. 105, phot. 257) for which the meadow is famous are confined to these types of herbage and are absent from slightly higher lying, more closely grazed, areas of mixed grasses in which *Bellis* and *Cerastium* occur. Grazing between March, when their shoots appear above

PLATE 105

Phot. 257. *Fritillaria meleagris* ("snake's head") in neutral alluvial grassland. Magdalen Meadow, Oxford. *N. F. G. Cruttwell.*

Phot. 258. *Narcissus pseudo-narcissus* in Lias clay pasture, borders of Dorset and Devon. Photograph from *The Times*.

NEUTRAL GRASSLAND

the ground, and mid-June, when the seeds are distributed, suppresses the fritillaries.

General flora of neutral grassland. The following list (Table XVIII) contains 126 species in all. Of these eighty-two are recorded by Fream (1888) from the water meadows at North Charford, bordering the Christchurch Avon in south Hampshire (first column). These meadows are developed on a clayey alluvial loam above the Upper Chalk. Almost the same number (85) are recorded by Lawes, Gilbert and Masters (1882) from the very old park grassland at Rothamsted in Hertfordshire on a heavy loam (clay-with-flints) resting on the Chalk (second and third columns). An area of this park grassland was divided into a number of plots in 1856, and different plots were consistently treated with different manures for a long series of years, two of them being left without any treatment. Every year the hay is cut and sorted into the constituent species for each individual plot separately.

The first column (W) gives the Hampshire water-meadow species recorded by Fream, the second (P) the species from all the plots, manured and unmanured, of the park grassland at Rothamsted, the third (P^1) from the unmanured plots only. Of the letters in the third column a indicates that the species formed more than 10 per cent, b between 1 and 10 per cent and c less than 1 per cent of the whole bulk of hay from the two unmanured plots. It will be noticed that twenty-four species have appeared in various manured plots which are not present in the unmanured, though the latter contain more species than any of the *individual* manured plots. This is in accordance with the general principles that the poorer the land (within limits) the more numerous is the flora, and that one-sided manuring favours particular species.

Of the exclusively water meadow species a few (marked *) are definitely reedswamp or waterside species and scarcely belong to the meadow community as such; while in the park land some of the recorded species (marked §) are woodland or wood edge plants which have established themselves in the grassland.

Forty-one of the whole list of species occur on the water meadows only (thirty-three if we exclude waterside plants not belonging to the meadow community), forty-four on the Rothamsted park land only (thirty-nine if we exclude woodland and wood-edge plants), while forty-one are common to both lists.

These combined lists contain a representative selection of species, including a wet and a relatively dry habitat, and the 41 grasses and herbs which are common to the two lists include the most constant species of neutral grassland, which have a wide range of habitat in respect of soil moisture.

The kernel of neutral grassland. The following sixteen species—eight grasses and eight herbs—which occur in the Charford alluvial meadows, in

Table XVIII

Species	W	P	P¹	Species	W	P	P¹
Agrostis stolonifera	+	.	.	Lathyrus pratensis	+	+	b
A. tenuis	+	+	a	Leontodon autumnalis	+	+	c
Alopecurus geniculatus	+	.	.	L. hispidus	+	+	c
A. pratensis	+	+	b	Linum catharticum	+	.	.
Anthoxanthum odoratum	+	+	b	Lotus corniculatus	+	+	b
Arrhenatherum elatius	+	+	c	L. uliginosus	+	+	.
Avena pubescens	.	+	b	Luzula campestris	.	+	b
Briza media	.	+	b	Lychnis flos-cuculi	+	+	.
Bromus hordeaceus (mollis)	+	+	c	Lysimachia nummularia	+	.	.
B. racemosus	+	.	.	*Lythrum salicaria	+	.	.
Cynosurus cristatus	+	+	c	Medicago lupulina	.	+	.
Dactylis glomerata	.	+	b	Myosotis scorpioides (palustris)	+	.	.
Deschampsia caespitosa	+	+	c	Ononis repens	.	+	.
Festuca elatior	+	.	.	Ophioglossum vulgatum	.	+	c
F. loliacea	+	+	c	Orchis morio	.	+	c
F. ovina	.	+	a	Ornithogalum umbellatum	.	+	.
F. pratensis	+	+	b	Pimpinella saxifraga	.	+	b
F. rubra	+	.	.	Plantago lanceolata	+	+	b
*Glyceria maxima (aquatica)	+	.	.	P. major	+	.	.
G. fluitans	+	.	.	P. media	.	+	.
Holcus lanatus	+	+	b	Polygonum persicaria	+	.	.
Lolium perenne	+	+	b	Potentilla anserina	+	.	.
*Phalaris arundinacea	+	.	.	P. reptans	+	+	c
Phleum pratense	.	+	c	§P. sterilis	.	+	c
Poa annua	+	.	.	Poterium sanguisorba	.	+	c
P. pratensis	+	+	c	Primula veris	.	+	c
P. trivialis	+	+	b	Prunella vulgaris	+	+	c
Trisetum flavescens	+	+	b	Ranunculus acris	+	+	c
				§R. auricomus	.	+	.
Achillea millefolium	+	+	b	R. bulbosus	+	+	b
Agrimonia eupatoria	.	+	c	R. repens	+	+	b
Ajuga reptans	.	+	c	Rumex acetosa	+	+	b
Alchemilla vulgaris (agg.)	.	+	c	R. crispus	+	+	.
Anthriscus sylvestris	.	+	.	R. longifolius (aquaticus)	+	.	.
Bellis perennis	.	+	c	R. obtusifolius	.	+	.
Caltha palustris	+	.	.	§Scilla non-scripta	.	+	c
Cardamine pratensis	+	+	.	Scrophularia aquatica	+	.	.
*Carex paludosa	+	.	.	Scutellaria galericulata	+	.	.
C. caryophyllea	.	+	c	Senecio aquaticus	+	.	.
Centaurea nemoralis (nigra)	.	+	b	S. erucifolius	.	+	.
Cerastium vulgatum	+	+	b-c	Sonchus oleraceus	.	+	.
Chrysanthemum leucanthemum	+	+	c	Stellaria graminea	.	+	c
Cirsium arvense	.	+	.	§S. holostea	.	+	.
C. palustre	+	.	.	Symphytum officinale	+	.	.
Conopodium majus	.	+	b	Taraxacum officinale	+	+	c
Daucus carota	.	+	.	Thalictrum flavum	+	.	.
*Eleocharis palustris	+	.	.	Thymus serpyllum	.	+	c
*Epilobium hirsutum	+	.	.	Tragopogon pratensis	.	+	c
E. parviflorum	+	.	.	Trifolium dubium	.	+	.
E. tetragonum	+	.	.	T. pratense	+	+	b
Eupatorium cannabinum	+	.	.	T. procumbens	.	+	.
Ficaria verna	.	+	.	T. repens	+	+	c
Filipendula ulmaria	+	+	.	Valeriana dioica	+	.	.
Fritillaria meleagris	.	+	.	V. officinalis	+	.	.
Galium aparine	.	+	c	*Veronica anagallis-aquatica	+	.	.
G. palustre	+	.	.	*V. beccabunga	+	.	.
G. verum	.	+	c	V. chamaedrys	+	+	c
Geum rivale	+	.	.	V. officinalis	.	+	.
Heracleum sphondylium	.	+	c	V. serpyllifolia	.	+	c
Hieracium pilosella	.	+	c	Vicia cracca	+	+	.
Hypericum perforatum	.	+	.	§V. sepium	.	+	.
Hypochaeris radicata	.	+	.	Veronica scutellata	+	.	.
Juncus acutiflorus	+	.	.				
J. inflexus (glaucus)	+	.	.		82	85	61
Knautia arvensis	.	+	c				

* Reedswamp or waterside plants not forming part of the meadow community.
§ Woodland or wood-edge plants which have established themselves in the park grassland.

The total number of species is 126, or 113 if we exclude species not properly belonging to the grassland communities.

the Rothamsted park grassland, and in both the grazed and the mown alluvial meadows at Oxford may be taken as the kernel of the neutral grassland community.

Grasses

Anthoxanthum odoratum
Cynosurus cristatus
Deschampsia caespitosa
Festuca pratensis

Holcus lanatus
Lolium perenne
Poa pratensis
P. trivialis

Herbs

Cardamine pratensis
Cerastium vulgatum
Leontodon autumnalis
Lotus corniculatus

Plantago lanceolata
Ranunculus acris
Trifolium pratense
T. repens

It is a striking fact that none of these species, but only *Agrostis tenuis* and *Festuca ovina*, the two dominants of the acid hillside pastures and of the grass heaths, form more than 10 per cent of the *unmanured* plots at Rothamsted, emphasising their tolerance of poor soil and their inability to compete successfully with the more luxuriant grasses and herbs of manured or grazed neutral grassland.

Besides the plants in the preceding lists the following may also be met with fairly often in neutral grassland:

Achillea ptarmica
Bartsia odontites
damp Carex diversicolor (flacca)
Centaurium umbellatum
Cerastium glomeratum
Crepis capillaris
Euphrasia officinalis (agg.)
Galium cruciata
G. erectum
G. mollugo
Geranium pratense
Hypericum acutum (tetrapterum)
Lathyrus nissolia
Malva moschata
Narcissus pseudo-narcissus (Pl. 105, phot. 258)
Nepeta hederacea
damp Orchis maculata

Polygala vulgaris
Potentilla erecta
damp Poterium officinale
damp Pulicaria dysenterica
wet Ranunculus flammula
Rhinanthus crista-galli (agg.)
Saxifraga granulata
Senecio jacobaea
Serratula tinctoria
Silaum silaus
dry Spiranthes spiralis
Succisa pratensis
Trifolium medium
Vicia angustifolia
V. hirsuta
V. lathyroides
V. tetrasperma

Juncetum effusi. On flat heavy grassland used as pasture, and in which the soil is often waterlogged, species of *Juncus* (*J. effusus*, the commonest dominant, *J. conglomeratus*, often with *J. inflexus* (*glaucus*), *J. acutiflorus* or the annual *J. bufonius*) almost always occur and are locally dominant, and where the surface is wettest any of the following species may be found in the turf between the rushes:

Agrostis stolonifera
Alopecurus geniculatus
Deschampsia caespitosa

Glyceria fluitans
Poa trivialis

Bidens cernua
B. tripartita
Caltha palustris
Cardamine pratensis
Carex flava
C. goodenowii
C. leporina
C. panicea
C. remota
Callitriche stagnalis
Eleocharis palustris
Epilobium palustre
E. parviflorum
Galium uliginosum
Gnaphalium uliginosum
Hydrocotyle vulgaris

Lotus uliginosus
Mentha aquatica (agg.)
Pedicularis sylvatica
Polygonum hydropiper
P. persicaria
Ranunculus flammula
R. hederaceus
R. sceleratus
Stellaria uliginosa
Triglochin palustre
Veronica beccabunga
V. scutellata

Equisetum fluviatile (limosum)
E. palustre

The water meadows and this Juncetum have several species in common and form transitions from neutral grassland to marsh, but while the former are well aerated the water of the latter is usually stagnant, leading to decided differences in the floristic list. Acid soil conditions sometimes develop in the Juncetum, and occasionally bog plants, such as bog moss and cotton grass, may occur in the wetter places, forming a transition to bog (Chapter XXXIV).

Grades of pasture. Very much the greater area of enclosed English grassland is used as pasture, and the conditions of pasture are very different from those of meadow, resulting in decided differences in the flora, as we have already seen. The dung of the stock is an important manurial factor, especially if cake is fed to the animals or they have access to other sources of food extraneous to the pasture itself. In the best practice the dung is collected and spread evenly over the turf. On the other hand the pasture loses the increase in live weight of the animals fed upon it, and extra manure of some kind is required according to the soil and the kind and amount of grazing contemplated. The proper regulation of the grazing is of the utmost importance. The great desideratum is to secure a constant supply of fresh young shoots of the more nutritious grasses, and to avoid the production of harsh coarse growth and of flowering stems. A continuous turf or "sole" with a minimum of bare spaces (which are generally colonised by worthless "weeds") and an avoidance of the formation of "mat" (dead plant debris) should be aimed at.

From the agricultural point of view Stapledon (1925) classifies the "tended pastures" (i.e. excluding "rough grazings") in a descending series of grades of fertility: (1) fatting pastures, (2) dairy pastures, (3) general purposes and sheep pastures, and (4) "outrun" pastures. The fourth class are practically "untended" pastures, mostly on old arable land that has tumbled down to grass and cannot support stock unless the animals have repeated access to cultivated land or tended grassland.

(1) *Fatting pastures*, so called because they are capable of fattening beasts during the summer months without the addition of feeding stuffs. The best known are the famous pastures of Leicestershire and Northamptonshire (boulder clay and lias), but there are others on Romney Marsh (marine silt), in Blackmoor Vale, Dorset (alluvium), and elsewhere. The soil is always a loam or silt.

The great floristic feature of all the best pastures is the prominence of perennial rye-grass (*Lolium perenne*) and white clover (*Trifolium repens*), and the paucity of "weeds"—in other words, the small number of miscellaneous species. The total flora usually comprises less than twenty-five species of flowering plants (of which the majority are grasses), and sometimes no more than twenty. Of useful grasses (other than *Lolium*), *Cynosurus cristatus*, *Dactylis glomerata*, *Hordeum nodosum* (*pratense*) and *Trisetum flavescens* are usually well represented, while of the less valuable species *Agrostis* spp. and *Holcus lanatus* are sometimes present in considerable quantity. In the Leicestershire pastures rye grass and white clover constitute from half to nearly three quarters of the herbage, in the Blackmoor Vale from a quarter to a half, and on Romney Marsh from a third to nearly nine-tenths. Of the non-gramineous plants milfoil or yarrow (*Achillea millefolium*) is generally present.

(2) *Dairy pastures* occur on a greater variety of soil, and where this is calcareous the number of species present may be more than 30. The chief grasses are the same as in (1), but the bents may exceed 50 per cent of the herbage and *Cynosurus* may reach 20 per cent or more. In addition to white clover, which is often very abundant, red clover (*Trifolium pratense*), *Lotus corniculatus* and *Lathyrus pratensis* are more prominent.

(3) *General purposes and sheep pastures* cover a greater area than (1) and (2) together, occurring on all soil types and managed in every conceivable manner. They form the most important class for the country generally, and would especially repay more skilled management.

Lolium is very seldom dominant, at most co-dominant with *Cynosurus* and *Agrostis*, which are usually much more prominent. *Holcus*, *Poa trivialis*, *Festuca rubra*, *F. ovina* and *Dactylis* make up a good deal of the herbage. White clover is not so consistently dominant as in (1) and (2), though in some fields it is very abundant. Other Leguminosae are more plentiful and varied, and "weeds" contribute as much as 15–30 per cent of the herbage. The total number of species often exceeds 40, and may reach 50.

(4) The untended pastures, generally completely neglected, are very frequently dominated, especially on dry poor soil, by *Agrostis tenuis*, a shallow-rooting grass whose shoots tend to dry up in the summer and give very little feed. Its dominance may be so complete as to reduce the total number of species to less than 20. Thus the most restricted flora is found in the best and the worst pastures.

REFERENCES

BAKER, H. Alluvial meadows: a comparative study of grazed and mown meadows. *J. Ecol.* 25, 408–20. 1937.
FREAM, W. On the flora of water-meadows, with notes on the species. *J. Linn. Soc. (Bot.)* 24, 454–64. 1888.
JENKIN, T. J. "Pasture Plants". In *Farm Crops*, 3, 1–42. 1925.
LAWES, J. B., GILBERT, J. H. and MASTERS, M. T. Results of Experiments on the mixed herbage of permanent meadow. Part II. Botanical results. *Philos. Trans. Roy. Soc.* 1882.
STAPLEDON, R. G. "Permanent grass." In *Farm Crops*, 3, 74–136. 1925.
TANSLEY, A. G. "Neutral grassland." In *Types of British Vegetation*, pp. 84–7. 1911.

Part VI

THE HYDROSERES

Freshwater, Marsh, Fen and Bog Vegetation

Chapter XXIX

THE HYDROSERES.
VEGETATION OF PONDS AND LAKES

Hitherto we have been dealing with land vegetation: first the climax forests, or rather such remains of them as still exist, nearly all in a more or less modified condition, representing the natural climatic formations of the lowlands and lower hill slopes of the greater part of these islands; secondly the semi-natural grassland formation into which most of the forest land has been converted by the direct or indirect activities of man. Of the primary xeroseres leading to forest we have been able to recognise fragments here and there, though it is impossible to find complete examples owing to constant human interference with the natural sequences of vegetation. Various subseres initiated by felling, grazing or burning have also been described.

When we turn to the hydroseres, the other great primary type of successional sequence, we are able to present a more complete picture. The aquatic and subaquatic habitats are less easily accessible and utilisable than the plains and hill slopes of the dry land; and though man has not been backward in draining lakes and marshes and in canalising rivers he has left enough relatively undisturbed aquatic habitats in which the vegetation is essentially natural to make possible a fairly full reconstruction of the processes involved in primary succession.

Conditions of life in water. Life in water is subject to conditions radically different from those of terrestrial subaerial life. Aquatic organisms, unlike land plants, are in no danger of suffering from lack of one of the prime necessities of existence. At the same time green plants are exposed to the risk of deficiency of the necessary gases, oxygen and carbon dioxide, and also of mineral nutrients. The amounts of the two gases actually available to submerged plants growing in natural waters varies enormously according to a number of different conditions, and no general statement can be made except that oxygen is much more likely to be deficient than carbon dioxide. Rooted aquatic plants have been shown to be more dependent on nutrient ions absorbed from the soil than on absorption of these from the water surrounding their shoots. Rooted aquatic vegetation therefore varies very much in amount and composition according to the nature of the substratum below the water, and this usually depends on the presence or absence, and the composition, of *silt*. Floating plants are of course entirely dependent upon the water itself for all their supplies, and since different waters contain very different amounts of dissolved substances, floating vegetation may be scanty or altogether absent from these causes.

Oxygen

Silt

Decreasing light intensity limits the depth to which green plants can descend on a lake bottom, and some can exist with less light than others; but apart from light the two chief factors affecting aquatic vegetation are the presence or absence of sufficient dissolved oxygen and the deficiency or adequacy of the mineral nutrients derived from silt. Where abundant green vegetation has been able to establish itself photosynthesis amply covers the supplies of oxygen necessary for the respiration of both plants and animals, but in poorly vegetated lakes and rivers the oxygen supply depends largely on solution from the air by current and wave action. Silt is also one of the most important factors, except in ponds situated in rich mineral soil where abundant nutrients are always present and the floor is covered by a rich organic mud composed of rainwash mixed with humus derived from decayed plant remains. Unless the water is definitely fouled from any cause a luxuriant aquatic vegetation is maintained. A pool of similar size in a district of hard rock or sterile sand may support a very scanty vegetation of specialised type, or be almost barren of life. In rivers the rate of flow and the nature of the rocks through which the course of the stream lies determine the existence and nature of the silt, and the rate of flow is also one factor in the supply of dissolved oxygen.

Rivers. Rivers too, just because of the current, are subject to less variation than still waters in such factors as temperature and dissolved gases which are critical for the existence and luxuriance of vegetation; but when floods bring down a lot of silt, sweeping away existing vegetation, and increasing turbidity, the conditions may nevertheless change completely within a short time. The vegetation of rivers with a very slow current flowing through alluvial soils resembles that of canals and of ponds and small lakes situated in similar soils.

Habitat classification. The provisional classification of fresh-water plant communities adopted in *Types of British Vegetation* was mainly based on the above-mentioned factors, and ran as follows:

Communities of
(a) Foul waters.
(b) Waters rich in mineral salts:
   (i) Nearly stagnant waters.
   (ii) Slowly flowing waters.
(c) Waters poor in mineral salts.
(d) Quickly flowing streams:
   (i) Non-calcareous streams.
   (ii) Calcareous streams.

This classification according to habitat is on the whole natural so far as it goes, and has been followed by some subsequent writers, but it is not quite logically coherent, and is especially defective because it neglects to take proper account of the important "silting factor".

The silting factor. Pearsall (1918) pointed out that rooted water plants had been shown to depend mainly on solutes derived from the substratum rather than on salts dissolved in the water surrounding them, and that water poor in mineral salts may bring down vast quantities of sediment, especially when the streams are in flood. Deposited in a lake in the form of silt this is important as furnishing a favourable habitat for many water plants which do not grow on very coarse or on highly organic substrata. On the other hand most of our waters rich in mineral salts occur in geologically stable areas largely composed of relatively soft Secondary, Tertiary or Quaternary beds overlaid by more or less finely divided soil. Such waters, when in flood, and often normally, will contain a large proportion of material in suspension and in consequence lakes, pools, "broads" and the river beds themselves, if not constantly dredged, tend to become silted up. Thus they do not show the "primitive" features characteristic of "young" lakes, with rocky, stony or gravelly shores and bottoms. These "primitive" lakes occur in mountainous regions of hard rocks, often with steep marginal slopes undergoing relatively rapid surface erosion so far as the hardness of the rocks permits. The substrata are therefore rocky or consist of unstable stones or gravel. Such waters are as a rule themselves poor in dissolved mineral salts, but this is not the decisive factor or even necessarily an effective factor in these primitive aquatic plant habitats. Poverty in the finer inorganic silts and instability of the substratum are likely to be more important.

The characteristic aquatic plants of a primitive lake are therefore likely to be (1) those which can colonise rough and often unstable substrata, (2) those which can do without a copious supply of basic ions, and (3) those which flourish on preponderantly organic soils, since the lack of bases prevents the decomposition of humus, which tends to accumulate, just as it does on land in similar conditions.

Small mountain tarns remain in an extremely primitive condition and support a scanty flora of such specialised plants. The larger and deeper lakes of mountain districts may also remain largely primitive, but where there are important affluents bringing in silt which is laid down near their mouths, drifted by currents, and deposited on different parts of the lake floor, the vegetation becomes more various and includes many plants also found in the silted waters situated on the softer rocks of the lowlands. Nevertheless the lists from typical examples of the two kinds of water show very striking differences.

Life form. In any body of water with a well developed aquatic vegetation there is a great variety of life form among the species present. First of all there are a considerable number of species which are more or less amphibious—either in the sense that they live sometimes on land and sometimes in water, or that they can form leaves of different shape and structure, some adapted to an aerial and others to an aquatic life. The two types of leaf, submerged and aerial, may be borne on one and the same individual plant (as in water-lilies and some of the water crowfoots), or some individuals may lead an entirely submerged existence while others display their leaves in the air (as in other water crowfoots). Of amphibious species some are normally terrestrial but can exist in the water, some are aquatic but can exist on land, and others again inhabit a zone on the edge of a pond or lake which extends on both sides of the water's edge, and are apparently indifferent to the elements surrounding their shoots. The adjustments of form and structure in response to the different medium vary greatly in degree in different species. In some plants the change is comparatively slight, while others have air and water leaves so different in appearance and construction that they would not be suspected of belonging to the same species (Pl. 106, phots. 259-62). Most submerged water plants raise their flowers above the surface (phots. 259-60), but the reproductive processes of a few take place entirely under water.

Zonation of aquatic vegetation. Among water plants proper we can distinguish the three broad categories of "submerged", "floating leaf", and "reedswamp", and this division, correlated on the whole with depth of water and therefore showing a distinct zonation, can be recognised in most lakes, large ponds and slowly flowing rivers.[1] The completely submerged species occupy the deeper waters and descend as far as the decreasing light permits, the species with floating leaves occur in the shallower water near the margin, while the "reedswamp" species fringe the land, though they sometimes extend into water more than a metre in depth and also landwards into the wet soil where the water table is not far below the surface. The "floating leaf" and "reedswamp" plants frequently intermingle, e.g. waterlily and bulrush (Pl. 109, phot. 270), the former finding room (and protection) between the erect shoots of the latter where these are not too closely set ("open reedswamp").

There is a fourth category of species with largely submerged but partly emersed shoots, whose aerial leaves do not float, but rise freely into the air. These are intermediate between the species with floating leaves and the reedswamp plants, whose leaves are entirely or almost entirely aerial.

Besides these life forms there are the plants which are not rooted in the substratum but float freely in the water, and the vegetative shoots of these may be completely submerged or their leaves may float on the surface. The completely submerged forms include very many of the algae and a few flowering plants (*Ceratophyllum, Utricularia*,[2] *Lemna trisulca*), those with floating leaves including the other duckweeds (*L. minor, L. gibba*, etc., *Wolffia arrhiza*), the frogbit (*Hydrocharis morsus-ranae*), and the "water fern" *Azolla*.

Stages of the hydrosere. These different life forms characterise successive stages of the hydrosere. Submerged plants are the pioneers, and as the

[1] Pl. 107, phot. 266; Pl. 108.   [2] Pl. 106, phot. 259.

PLATE 106

Phot. 259. *Utricularia vulgaris* in aquarium. Free floating. Submerged leafy shoot with emersed inflorescence. *R. H. Yapp.*

Phot. 260. *Sagittaria sagittifolia* in aquarium. Submerged leaves band-shaped, emersed sagittate. *R. H. Yapp.*

Phot. 261. *Sagittaria sagittifolia*, submerged form (*vallisneriifolia*) with band-shaped leaves. River Cam. *J. Massart.*

Phot. 262. *Nuphar luteum, Potamogeton natans* (floating leaves). Emersed inflorescence of *Utricularia*, and shoots of *Scirpus lacustris*. Sutton Broad, Norfolk. *J. Massart.*

SUBMERGED, EMERSED AND FLOATING LEAVES

PLATE 107

Phot. 264. *Nuphar luteum.* R. H. Yapp.

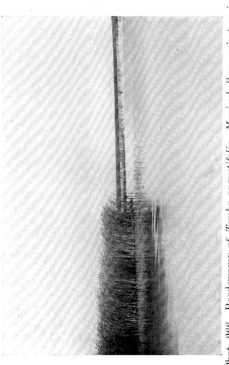

Phot. 266. Reedswamp of *Typha angustifolia, Myriophyllum spicatum* in front. Heigham Sound, Norfolk. F. F. Blackman.

Phot. 263. *Nymphaea alba.* R. H. Yapp.

Phot. 265. *Nymphoides (Limnanthemum) peltatum* (floating leaf), *Oenanthe fluviatilis, Alisma plantago-aquatica, Sparganium erectum* (emersed). River Thames, Medley, Oxford. A. H. Church.

soil level is gradually raised towards the water surface by the accumulation of organic debris resulting from the death of individual plants, or by inorganic silting, or by both together, they are succeeded by the plants with floating leaves, and these, in their turn, by the reedswamp dominants,[1] till the soil level reaches the water level and aquatic vegetation gives place to fen or marsh. But the particular communities entering into the sere depend upon a variety of conditions, such as the nature of the original substratum, the amount and nature of silting, exposure to current or wave action, and so on. As the water shallows it becomes quieter and warmer, increased protection is afforded and the conditions are more uniform, so that the floating leaf and reed swamp dominants are less various than those of the submerged communities.

Summary of life form groups. These life form groups may be summarised as follows:

(1) Wholly submerged plants:
   (a) Rooted or fixed.
   (b) Free floating (still waters).

(2) Plants with floating leaves (relatively still waters):
   (a) Rooted.
   (b) Free floating (still waters).

(3) Plants with partly emersed shoots.

(4) Plants (mostly tall) with functional leaves mainly or entirely emersed (reedswamp type).

Forms of submerged leaves. The submerged leaves of water plants are either entire and very thin, or "dissected", i.e. divided into filiform branches. All the living cells of the leaf are thus within a very short distance of the supply of gases dissolved in the surrounding water. Entire submerged leaves are frequently strap-shaped,[2] especially when growing in a river, and both the strap-shaped and the dissected type offer the least possible resistance to the water, the leaves streaming out with the flow of the current. This can be easily seen by looking over the parapet of a bridge at the submerged plants growing on the bed of a clear stream.

The aquatic plant communities of the British Isles have never been summarised, and it is still scarcely possible to write a general account of them. Several good descriptions of the vegetation of lakes and rivers have

[1] The plants with largely emersed shoots, such as the water plantain (*Alisma plantago-aquatica*, Pl. 107, phot. 265) and the arrowhead (*Sagittaria*, Pl. 106, phot. 260), mostly occur in silted waters just beyond or replacing the reedswamp, i.e. in fairly shallow water. Free floating vegetation, so far as vascular plants are concerned, is confined to waters rich in mineral salts, and a rich and varied plankton depends on the same conditions.

[2] Pl. 106, phots. 260, 261.

however appeared since the publication of *Types of British Vegetation*, and a selection from these will serve to illustrate the modes of occurrence of many of the communities and the general course of the hydroseres.

Examples illustrating the different habitats will be taken in the following order:—Ponds on the softer rocks; White Moss Loch; the Norfolk Broads; the Cumbrian Lakes; the most primitive lakes and tarns.

PONDS ON THE SOFTER ROCKS

Foul waters. Small stagnant ponds containing an excess of decaying organic matter, for example such as are constantly fouled by sewage or by the excrement of cattle, and also densely shaded pools overhung by thick trees or shrubs so that the bottom is thickly covered with decaying leaves and illumination is very low, may be entirely destitute of green plants. They are inhabited by anaerobic bacteria, saprophytic flagellata, and to some extent by blue-green algae, but the free oxygen necessary for respiration is present only in minimal quantities, and the substances produced by the anaerobic forms probably also inhibit the growth of algae or higher plants. The vegetation of such waters has not been studied ecologically in this country.

Lowland ponds. Ponds situated on strata relatively rich in mineral salts, especially if they receive rainwash from cultivated land, often become rapidly choked with vegetation unless they are constantly cleaned out. Flowering plants make up the bulk of the vegetation, and there are a few species of algae: bryophytes are generally absent, but two species of simple thalloid liverworts may occur—*Riccia fluitans* and *Ricciocarpus natans*. Both have a floating and also a terrestrial form, the former without rhizoids. The floating form of *Ricciocarpus natans* has long pendant scales on the ventral surface, and these give the floating thallus stability as well as increasing its photosynthetic and absorptive surface. The terrestrial forms bear rhizoids which penetrate the mud at the water's edge on which the liverwort grows. If the water level rises and the thallus is submerged, the plants continue to vegetate. Certain species of *Hypnum* sometimes occur in the same situations.

Trent valley ponds. A set of seven ponds described by Godwin (1923) in the alluvial gravels of the Trent valley near Trent Junction in south-east Derbyshire, including an oxbow cut off from the river Trent many years ago, possessed between them the following 32 species belonging to the aquatic communities proper.

Submerged and partially emersed vegetation

Alisma plantago-aquatica	Glyceria fluitans
Apium inundatum	Hottonia palustris
Butomus umbellatus	Lemna minor
Callitriche verna	L. trisulca
Elodea canadensis	Myosotis scorpioides (palustris)

Myriophyllum spicatum
Nasturtium officinale
Nuphar luteum
Oenanthe fistulosa
Oe. phellandrium

Polygonum amphibium
Potamogeton acutifolius
P. natans
Ranunculus circinatus

Reedswamp

Calamagrostis epigejos
Carex vulpina
Eleocharis palustris
Epilobium hirsutum
Equisetum fluviatile (limosum)
Glyceria maxima (aquatica)
Rumex hydrolapathum

Scirpus lacustris
Sium latifolium
Sparganium erectum
Thalictrum flavum
Typha angustifolia
T. latifolia

Besides the water plants proper, a number of marsh plants and other species of damp soil occurred round the edges of the ponds.

With the exception of the oxbow all these ponds were artificial, having been dug for ballast or other purposes at various dates during the nineteenth century. They afford therefore a fair sample, though by no means an exhaustive list, of the aquatic plants naturally colonising such habitats. Godwin points out that the most recently dug ponds were the poorest in species, and that most of the species occurred in one or more of the ponds, but not in all. Only one species indeed, *Alisma plantago-aquatica*, was found in all seven. Since the conditions were apparently fairly uniform in all the ponds this illustrates the largely chance distribution of the species constituting the actual flora of a small pond.

In some of Godwin's ponds zonation was very well shown. Thus in "Fletcher's Pond" there were three well-marked zones passing inshore from the open water: the first dominated by *Potamogeton natans*, the second by the yellow water-lily (*Nuphar luteum*), both rooted plants with floating leaves; the third by the common tall reedswamp dominant *Scirpus lacustris*, behind which, in shallow water, came a zone of the much lower growing *Eleocharis palustris*. In another pond with a considerable strip of very shallow water at its edge the reedswamp was represented by the horsetail *Equisetum fluviatile* (*limosum*), with *Eleocharis palustris* as an understorey and extending a little beyond the outer edge of the horsetail; *Scirpus lacustris* was here absent. These lower growing reed-swamp elements occur only in shallow water or in marsh, where the water level (at least in summer) is frequently below the soil surface.

Bramhope ponds. With these Trent valley ponds one may compare a set of eight ponds described by Norman Walker (1905), situated above the Bramhope railway tunnel about half-a-mile south of Bramhope, near Leeds in Yorkshire. These were small ponds lying close together, and owed their origin to surface drainage into excavations made in 1839 when the tunnel was built. Their floors sloped gently from the edge of the water to a maximum depth of 4 ft., and none had either inlet or outlet.

The total number of species of flowering plants was much smaller than in the Trent ponds, corresponding with the smaller area involved, and were largely different species. *Potamogeton natans* (which occurred in all the Bramhope ponds and in all but one of the Trent ponds), *Eleocharis palustris* and *Sparganium erectum* were however dominants of aquatic and reedswamp communities in both sets.

Aquatic communities

Juncus bulbosus (clayey ponds)
Myriophyllum spicatum (few and dwarf)
Potamogeton natans (dominant in all ponds)
P. alpinus (clear water above shallow mud only)
Ranunculus sp. (sect. Batrachium)
Glyceria fluitans (almost alone in small clayey ponds)

Reedswamp

Eleocharis palustris	d	Oenanthe fistulosa	d
Juncus effusus	d	Sparganium erectum	d
Carex goodenowii	ld	Juncus articulatus	o

In the Juncetum effusi were also:

Galium palustre   Peplis portula
Glyceria fluitans   Ranunculus flammula
Myosotis caespitosa   Senecio aquaticus

Juncus effusus is not a typical reedswamp plant, but occurs (mainly on waterlogged soil rather than in water) in a great variety of situations, very abundantly in low-lying clay lands where the water table is close to the surface. In the case of the Bramhope ponds it throve best on the banks of the pond outside the water line, but extended into the water to a depth of 6 in. In the pond shown in Fig. 104 *Oenanthe fistulosa* began among the *Juncus* at a depth of 3 in., the slender stolons burrowing through the soft mud and stopping at the stiff clay soil. At a depth of about 6 in. the *Juncus* and *Oenanthe* gave way to *Eleocharis palustris* and *Sparganium erectum*. The last-named extended to a depth of more than 18 in., and gave way in its turn to *Potamogeton natans*, which occupied the centre of the pond.

It is interesting to note that this set of small ponds possessed together about the same number of species as the much larger Fletcher's Pond in the Trent valley which was made about the same time (1839 and 1836).

Hagley Pool. With the two sets of north midland ponds just described we may compare the vegetation of Hagley Pool near Oxford, an old detached backwater belonging to the network of waterways of the Thames system where it passes through the alluvial plain based on Oxford Clay.

Here development is on a larger scale and the vegetation much older. Correspondingly the reedswamp, which is dominated by *Sparganium erectum* (Pl. 107, phot. 265; Pl. 108), is considerably more varied.

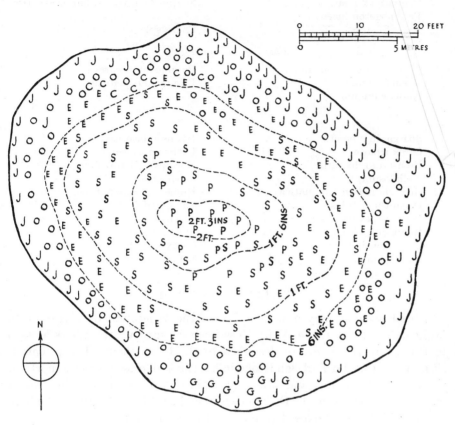

J = *Juncus effusus*
G = *Glyceria fluitans*
O = *Œnanthe fistulosa*
P = *Potamogeton natans*
C = *Carex goodenowii*
E = *Eleocharis palustris*
S = *Sparganium erectum*

Fig. 104. A small pond at Bramhope, near Leeds

Juncus effusus is the main dominant of the marginal reedswamp also extending on to the bank. *Sparganium erectum*, with *Eleocharis palustris*, is dominant in deeper water, and *Potamogeton natans* forms a well marked community in the centre of the pond. After N. Walker, 1905.

Submerged community

Myriophyllum spicatum	Ranunculus circinatus
Oenanthe fluviatilis	Scirpus lacustris (submerged form)
Potamogeton lucens	
P. perfoliatus	Chara sp.

Floating leaf community

Hydrocharis morsus-ranae	Potamogeton natans
Lemna minor	Ranunculus peltatus
Nuphar luteum	

Reedswamp

Sparganium erectum	d	Iris pseudacorus	f
Carex riparia	a	Lythrum salicaria	f
Equisetum fluviatile (limosum)	a	Rorippa (Nasturtium) palustre	f
Glyceria maxima (aquatica)		Rumex hydrolapathum	f
Oenanthe fistulosa	a	Sium latifolium	f
Menyanthes trifoliata	la	Stachys palustris	f
Ranunculus lingua	la	Stellaria glauca	f
Carex acutiformis	f	Carex vesicaria	o
Galium uliginosum	f	Sium erectum	o

WHITE MOSS LOCH

A brief account may now be given of the vegetation of a very much larger sheet of water, the White Moss Loch (or simply "the White Moss") in Perthshire (Figs. 105–6) described by J. R. Matthews (1914). This is situated on the Old Red Sandstone in a natural depression to the north of the Ochil Hills in Fife and 175 ft. (*c*. 53 m.) above sea-level. The area of water was recorded by the Ordnance Survey in 1901 as 16 acres (*c*. 6·5 ha.), but in 1913 it did not exceed 10 acres (*c*. 4 ha.), so that the process of centripetal encroachment of the swamp and marsh vegetation was quite rapid. The greatest diameter was about 540 yards (*c*. 0·5 km.). The loch has no natural inflow or outflow stream, but ditches conduct into and out of it a considerable quantity of water during the winter months. There was nowhere more than a metre's depth of water in summer, though in winter the loch might be nearly twice as deep. In most winters the duration of ice does not exceed a fortnight, though in 1910 (a cold late winter all over the country) the loch was ice-bound for six weeks. Mud consisting of fine sand mixed with much organic material covered the bottom to a depth of at least two metres.

Two submerged communities were distinguished and two main consocies in the reedswamp.

Submerged communities. Elodetum occupied the centre of the loch where the water was more than 3 ft. deep and contained the following species:

Elodea canadensis	d	Potamogeton filiformis	f
Nitella translucens	f	P. perfoliatus	o

The White Moss Loch

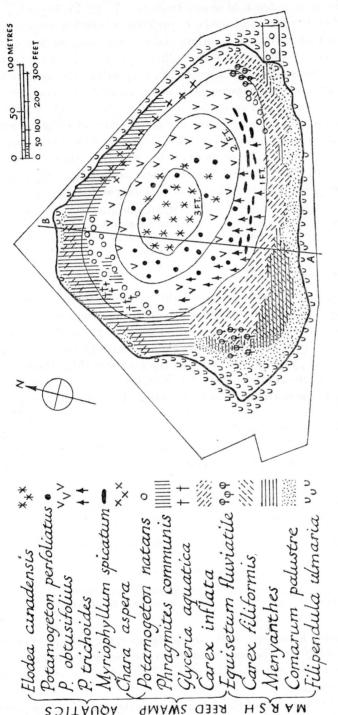

FIG. 105. WHITE MOSS LOCH (PERTHSHIRE) IN 1913

Willows and alders were planted about 1885 up to the sinuous line which was then close to the edge of the water. Zonation of marsh, reedswamp and aquatic vegetation is well shown. The reedswamp is mainly dominated by *Phragmites communis* (north) and *Carex inflata* (south). *Potamogeton natans* grows in about the same depth of water (30–60 cm.) as in the Bramhope Pond (Fig. 104). The deeper water in the centre of the loch is dominated by submerged species of *Potamogeton* and by *Elodea*. AB = transect line (Fig. 106). After Matthews, 1914.

The next zone, from a depth of about 18 in. to 3 ft. (c. 45–90 cm.), was largely occupied by the two pondweeds, *P. perfoliatus* in rather deeper, and *P. obtusifolius* in rather shallower water; and outside this, bordering the reedswamp in about 30–40 cm. of water, three local societies were represented. Towards the south side *Myriophyllum spicatum* was dominant, with an understorey of *Potamogeton perfoliatus*, *P. trichoides* and *Nitella opaca*: along the east side and extending into the reedswamp the floor was carpeted with *Chara aspera* var. *inermis*; and on the north *Potamogeton natans* was dominant, with species of *Nitella* as undergrowth. Since species of pondweed were the predominant plants in the zone of water from 1–3 ft. deep, the community (Matthews's "Shallow water association") may fairly be called a *Potamogeton* associes. The list of species was as follows:

Potamogeton natans	ld	M. alterniflorum	f
P. perfoliatus	a	Chara aspera var.	
P. trichoides	a	subinermis	ld
P. obtusifolius	a	Nitella opaca	f
P. gramineus	f	N. flexilis	f
Myriophyllum spicatum	ld		

Reedswamp. In 1913 reedswamp dominated by *Phragmites* on the northern, and by *Carex inflata* on the southern side surrounded the loch, except for a space of about 80 yards at the eastern extremity which was probably kept clear by wave action. Since then however (1936) *Phragmites* has filled up this gap, joining the *Carex* on the southern side. There were and are no reeds on the southern side, but a little of the sedge associated with *Phragmites* on the northern. The contour line of 1 ft. (30 cm.) depth of water ran near the outer margin of the reedswamp, sometimes a little inside it, sometimes a little outside. The reeds on the northern side have also advanced towards the centre since 1913, so that the two reed "islands" on the north-west are no longer distinguishable.

Phragmitetum

Phragmites communis	d	Equisetum fluviatile	
Glyceria maxima (aquatica)	ld	(limosum)	la
Carex inflata	a	Carex flava	o
Littorella uniflora	a	Echinodorus ranunculoides	o

Caricetum inflatae

Carex inflata	d	Sparganium erectum	ld
Equisetum fluviatile		Potamogeton natans	l
(limosum)	lsd		

The reed persisted for a time, as it commonly does, after the level of the soil had been raised above the water, but both the Phragmitetum and the Caricetum were succeeded on the outer margin by the lower growing plants of the circumjacent marsh.

In all there were 21 species of charads and vascular plants in the aquatic communities and reedswamp of White Moss Loch, a substantially smaller

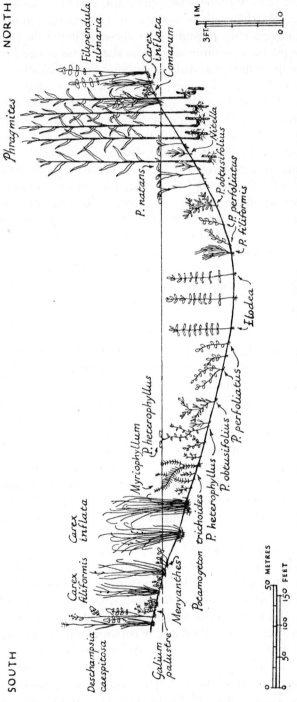

Fig. 106. Profile Transect across White Moss Loch from South to North AB in Fig. 105. After Matthews, 1914.

number than in the Trent ponds. Conspicuous absentees were the reedmaces, the water-lilies and the water plantain, clearly indicating lack of inorganic silting. Matthews gives no data on the composition of the water or of the soil, but one may presume that these, situated on the Old Red Sandstone, were poorer in soluble basic salts than those of the Trent valley or of the Bramhope ponds, and this, together with the absence of silting, probably accounts for the difference (for example, the presence of a species like *Littorella uniflora*) and for the greater poverty of the flora.

Besides the advance of *Phragmites* referred to above the principal change between 1913 and 1936 is the colonisation (mainly of the *Carex inflata* swamp) along the south side of the loch, and also to the east and west, by young willows and alders derived from trees planted about 1885 between the fence and the black sinuous line in Fig. 105, which marked the inner limit of the plantation. Some of the young self-sown trees have now established themselves almost up to the level of the 1 ft. depth contour line as it was in 1913 (Matthews *in litt.*). Advance of the vegetation has evidently been much less rapid than at the beginning of the century.

We may now pass to larger areas of freshwater, where the ecological factors affecting aquatic vegetation are more various and operate on a bigger scale.

THE NORFOLK BROADS

These are shallow lakes situated in the upper parts of the old estuarine area of East Norfolk now entirely filled with alluvium. Most of the existing broads are surrounded by fen peat, and lie in the north of the region round the courses of the Bure and its tributaries the Ant and the Thurne (Fig. 107). There are a few remains of broads in the Yare valley (Fig. 108). The lower courses of the rivers run through mineral alluvium, the region repeating the geographical and geological features of the "Fenland", south of the Wash, on a smaller scale. Unlike the latter it is, however, only partially drained, and some of it still bears natural vegetation. There is a good preliminary description by Pallis of the aquatic and fen vegetation in *Types of British Vegetation* (1911), but the region requires fresh study on modern lines, and the aquatic vegetation can only be cursorily dealt with here.

The broads are not expansions of the rivers themselves but lie to one side of the stream beds, enclosed by fen peat which has formed around them. The process of "Verlandung" is often very rapid, so that several broads have greatly shrunk or even been obliterated within comparatively recent years (Fig. 108). The water of the broads and rivers has a pH value of more than 7, since, like that of the Fenland, it comes mainly from the chalk.

Pallis (1911) pointed out considerable differences between the aquatic vegetation of the broads of the various river valleys, differences which are much less marked between the corresponding reedswamps and less still between the fens. Thus in some of the broads of the Bure valley which are filled up close to the surface with organic mud, and have little or no free communication with the rivers, the aquatic vegetation proper is extremely poor or altogether absent: in those connected with the Ant, Thurne and Yare, on the other hand, where the circulation of water is much freer,

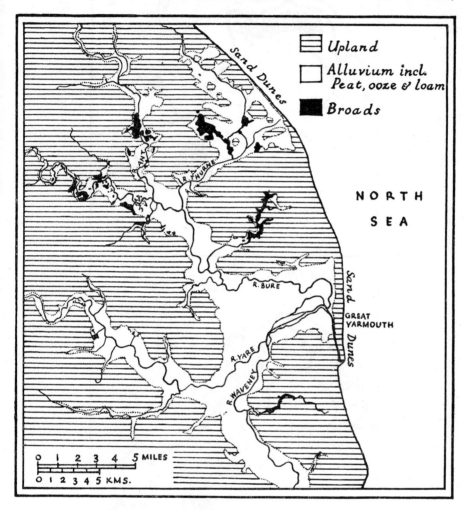

FIG. 107. THE FENLANDS OF EAST NORFOLK

The broads, shown in black, are confined to the upper parts of the old estuarine area where peat is formed. From Pallis, 1911.

aquatic vegetation is well developed, with *Potamogeton pectinatus* forma *interruptus* commonly dominant in the Thurne and Yare broads, and *Stratiotes aloides* in those of the Ant. Of abundant associated plants *Myriophyllum spicatum* and *Ranunculus circinatus* occur in the broads of the Ant and Thurne, *M. verticillatum* and *Ceratophyllum demersum* in those of the Ant and Yare. In the Thurne broads a series of charads and algae

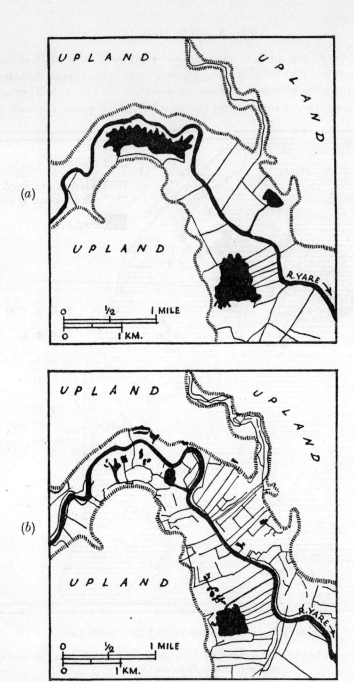

FIG. 108. SHRINKAGE OF BROADS

Part of the Yare valley, Norfolk. The broads (black) are entirely surrounded by peat. The straight lines are dykes cut in the peat. (*a*) As shown in the Ordnance Survey of 1816–21. (*b*) The same area as surveyed between 1879 and 1886. The peat-forming plants have encroached on the Broads and greatly reduced the areas of open water. Note the increased number of dykes.

Phot. 267. Floating clump of *Glyceria maxima*: reedswamp and carr behind. Rockland Broad. J. Massart.

Phot. 268. Island of *Phragmites communis*: edge of reedswamp on left. Rockland Broad. J. Massart.

Phot. 269. Open reedswamp of *Scirpus lacustris* with under-storey of *Nymphaea alba*. Sutton Broad. J. Massart.

Phot. 270. Reedswamp of *Typha angustifolia, Sium erectum, Cicuta virosa, Epilobium hirsutum*, etc. J. Massart.

REEDSWAMP IN THE NORFOLK BROADS

may be dominant, while *Hippuris, Stratiotes*, and the very rare *Naias marina* occur. The duckweeds and water-lilies, the frog-bit and various pondweeds occur mainly in the broads of the Yare and Ant, in the latter associated with open reedswamp (Pl. 108, phot. 269).

The open reedswamps fringing the broads of all four rivers are dominated by *Typha angustifolia*[1] and *Phragmites communis*, but in the closed reedswamps of the Yare these species are replaced by *Glyceria maxima (aquatica)*[2] and *Phalaris arundinacea*. Some of these differences are probably correlated with the stronger current in the Yare which brings down more inorganic silt, favouring such plants as *Nymphaea*, and in the fens *Glyceria maxima* and *Phalaris arundinacea* (cf. Chapter XXXII, p. 643 and Chapter XXXIII, p. 653). In the Ant broads there is also a very wide open reedswamp dominated by *Scirpus lacustris*,[3] with which are associated the water-lilies, duckweeds and several pondweeds, including *Potamogeton natans*, recalling the Trent pond vegetation. In some of the Bure and Yare broads, too, there is a reedswamp composed of the great tussock-forming sedges, *Carex paniculata* and *C. acutiformis* (Pl. 88, phots. 214, 215).

REFERENCES

ARBER, AGNES. *Water plants: a study of aquatic Angiosperms.* Cambridge, 1920.
GODWIN, H. Dispersal of pond floras. *J. Ecol.* **11**, 160–4. 1923.
MATTHEWS, J. R. The White Moss Loch: a study in biotic succession. *New Phytol.* **13**, 134–48. 1914.
PALLIS, MARIETTA. "The River Valleys of East Norfolk." Chapter X. In *Types of British Vegetation*, pp. 214–45. 1911.
PEARSALL, W. H. On the classification of aquatic plant communities. *J. Ecol.* **6**, 75–83. 1918.
Types of British Vegetation. Chapter VII, Aquatic Vegetation, pp. 187–96. 1911.
WALKER, NORMAN. Pond vegetation. *The Naturalist.* No. 585, pp. 305–11. October, 1905.
WATSON, W. The bryophytes and lichens of fresh water. *J. Ecol.* **7**, 71–83. 1919.

[1] Pl. 107, phot. 266; Pl. 108, phot. 270.   [2] Pl. 108, phot. 267.
[3] Pl. 108, phot. 269.

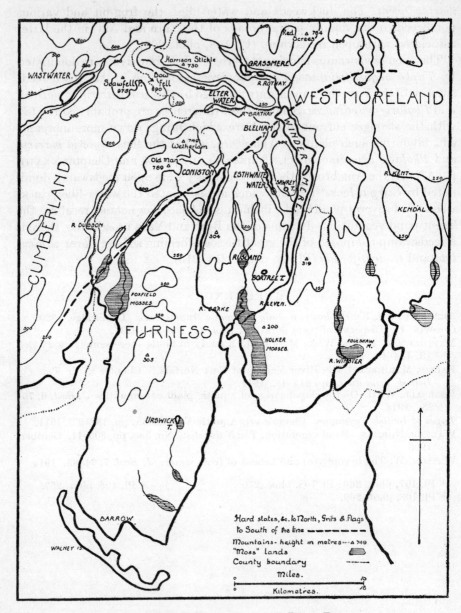

Fig. 109. Southern half of the Lake District

Showing the positions of Esthwaite Water and other lakes: also of the mosses (raised bogs) on the lower courses of the rivers (see p. 686). From Pearsall, 1917.

Chapter XXX

THE CUMBRIAN LAKES

Pearsall's very careful and thorough surveys of Esthwaite Water (1917–18) and of other lakes (1920) in Cumberland and Westmorland, i.e. in the region commonly known as the English Lake District (Fig. 109), provide a wide range of data relating to the conditions of life of aquatic, marsh, fen and bog vegetation in that region and by far the fullest account of the hydrosere that is available for any British waters.

In this chapter the aquatic communities of Esthwaite Water will first be briefly described with their characteristic habitats: then a general account of the aquatic vegetation of the lakes will be given, preceded by a notice of factors such as light and temperature and followed by a description of the observed successions.

Esthwaite Water. Esthwaite Water (Fig. 110) has only a quarter of the total residue of solid material found in the waters of the Norfolk broads (0·07 against 0·28 in parts per thousand), and this is distinctly higher than that of other lakes of the region (0·04–0·06). The "hardness" of the Esthwaite water is less than one-seventh of that of the Broads water (3·25 against 23·68). The organic residue of Esthwaite is relatively high and the water brown from peaty substances in solution. This is due to affluents coming from areas of acid peat and is correlated with relatively high acidity of the water.

In regard to the physiography of the aquatic habitats there is a marked difference between the eastern (lee) and the western (windward) shore, the former being far more subject to wave action, which cuts a sloping terrace, mainly below the water line (Fig. 111, AB), and continues it outwards into the lake (BC) by the accumulation of material washed back by the undertow of the waves. This gently graded terrace ends in a much steeper slope (CE). The terrace consists only of the coarser detritus (BCE), the finer sediments being carried down its steep slope and deposited farther out on the floor of the lake (EH). Besides this direct wave action a current is set up along the shore by the oblique incidence of the waves, and this scours the shallows clean and deposits silt in deeper bays. The affluents also bring quantities of silt into the lake and lay it down round their mouths. The kinds and degrees of silting are thus very various and they are of first importance in determining the vegetation.

Most of the leading aquatic communities of Esthwaite Water are well illustrated at Strickland Ees and the adjoining Fold Yeat Bay (Fig. 112).

Deep-water communities. (1) Of the entirely submerged species the charad *Nitella flexilis* (Figs. 112, 114) penetrates to the greatest depths, typically ranging from 1·8–3·6 m., with a light intensity of 0·05–0·102

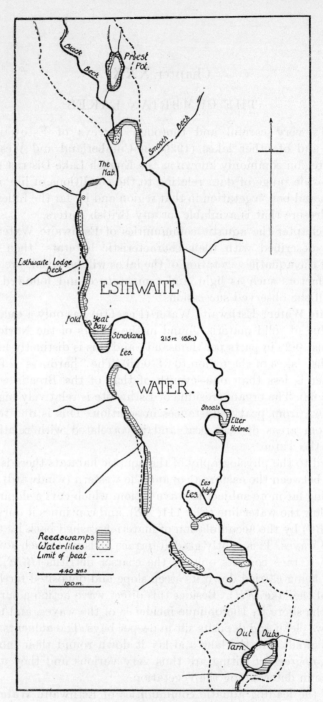

FIG. 110. ESTHWAITE WATER

Showing the positions of reedswamps and water lilies and the limits of peat formation: also Strickland Ees (Fig. 112) and North Fen (Figs. 114, 115). From Pearsall, 1917.

(dysphotic), on blue-grey clayey inorganic mud with an average organic content of 16·9 per cent.

(2) A similar consocies is formed by the moss *Fontinalis antipyretica* (Figs. 112, 114), typically pure, but with *Elodea canadensis* and *Potamogeton obtusifolius* occasionally associated, and higher average organic content (18·9 per cent) in its substratum.

(3) *Sparganium minimum* forms another consocies (Figs. 112, 113, 114) on loose, grey or yellow-grey mud which is still more organic (22·4 per cent) outside the *Nymphaea alba* community ((7), p. 603) and at a depth of 2·4–3 m. (occasionally at 1·2–1·8 m.), frequently with the same two species that accompany *Fontinalis*.

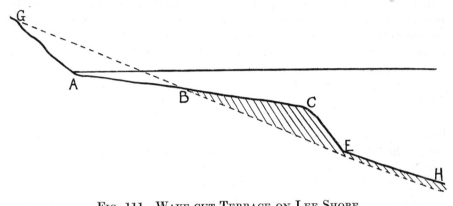

FIG. 111. WAVE-CUT TERRACE ON LEE SHORE

For description see text, p. 597. From Pearsall, 1917.

(4) A characteristic collection of species, which Pearsall calls the "linear-leaved associes" (Fig. 112), develops best at depths of 1·5–2·6 m. (light intensity 0·06–0·03—dysphotic) on blue-green clayey mud, like that of the *Nitella flexilis* community, but always with less than 15 per cent of organic matter (av. 13·7 per cent). This community shows two well-marked consocies dominated respectively by *Potamogeton pusillus* and *Naias flexilis* (average percentage of organic matter 7·3). The complete list of species is as follows:

Potamogeton pusillus	la	Callitriche autumnalis	la
P. pusillus subsp. lacustris	ld	Elodea canadensis	f
P. panormitanus	ld	Hydrilla verticillata var.	
P. obtusifolius	l	pomeranica	f
P. crispus var. serratus	o	Myriophyllum alterni-	
P. perfoliatus	r	florum	l
P. crispus	vr	Nitella flexilis	l
Naias flexilis	ld		

Pearsall remarks on the notable uniformity in life form of all the characteristic and more abundant species. "All have the pellucid linear leaves and delicate stems of the *Potamogeton pusillus* type. The elongate subspecies *lacustris* has this character still better marked: the variety *serratus* of *P. crispus* is bright green and translucent: *Hydrilla verticillata* is represented by one of its slenderest varieties; and the peculiar elongate form of *Elodea canadensis* also closely approximates to the type form of the community. The uniform light conditions under which the associes develops have probably a direct influence on its growth form, but the constant character of the substratum is also a marked feature of the habitat. The fragile growth form and the need for inorganic silt both indicate that shelter is of great importance in the development of this associes. Exposure to water movements would injure the plants and prevent the deposition of silt."

Naias flexilis is the only member which fruits abundantly and is dependent on seed for dispersal. The other characteristic species rely mainly on vegetative means of reproduction. The multitude of seeds produced by *Naias* probably enables it to colonise newly deposited inorganic silts before its associates.

(5) Of quite different life form—the rosette type—is the consocies of *Isoetes lacustris* (Figs. 118, 119), developed on a "primitive" substratum of rounded stones at depths of 1·5–2·7 m. The light intensity here varies from 0·03 to 0·06 (dysphotic). This consocies occurs only at the extreme southern end of the lake, where no fluvial silt is left to be deposited, since the main current in the lake is from north to south.

Along the foot of the steep slope bounding the wave-formed terrace described above, where fine silt is deposited, a consocies of *Myriophyllum alterniflorum* forms a narrow zone chiefly along the exposed eastern shore at a depth of 1·25–1·5 m. (Fig. 112). This is the most favourable habitat on the exposed shore, for below are bare stones devoid of silt, and shorewards is the wave-swept gravel of the terrace. The well-developed root system of *Myriophyllum* anchors the plant firmly and the finely cut leaves create the minimum resistance to wave-wash. The water milfoil is usually alone, but *Potamogeton heterophyllus* var. *longipedunculatus* may accompany it on the exposed eastern bank, developing a thick carpet of short vegetative shoots close to the substratum; and on the sheltered western shore *Myriophyllum* is accompanied by *P. alpinus*.

Shallow-water communities. (6) Another well-marked community of the rosette type of life form—a community found in relatively shallow water round the shores of most lakes among the older rocks—is the *Littorella-Lobelia* associes (Figs. 112, 114, 115), consisting only of the two species *Littorella uniflora* and *Lobelia dortmanna*. This usually occurs at depths of 0·3–1·2 m. (euphotic), but occasionally to 2 m., on gravelly substrata, sometimes overlaid by 3 or 4 cm. of black peaty mud, but never by inorganic silt. At such depths there is vigorous wave action which keeps the substratum unstable and prevents the deposition of silt. The mobile gravel is successfully colonised by families of *Littorella* (single plants of

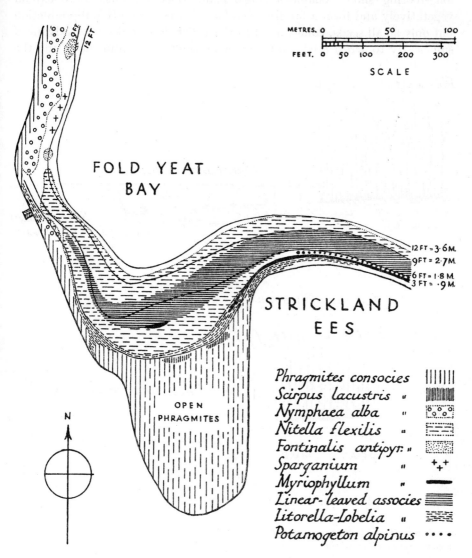

FIG. 112. STRICKLAND EES AND FOLD YEAT BAY, ESTHWAITE WATER

The greater part of the area of submerged vegetation is occupied by the *Nitella* consocies, which extends in places to a depth of 3·6 m. and the "linear-leaved associes", with *Sparganium minimum* and *Fontinalis* to the north. Next above come *Myriophyllum alterniflorum* and the *Littorella-Lobelia* associes, with *Nymphaea alba* on the terrace to the north (cf. Fig. 113). Behind is the Phragmitetum, occupying a large area in the shallow southward extension of Fold Yeat Bay, with patches of Scirpetum lacustris on its outer limit. From Pearsall, 1917.

which are uprooted by ducks and washed on to exposed shores), and the long-rooting shoots enable a single plant, once established, to extend vegetatively and form a family. *Lobelia*, with no vegetative reproduction but only small seeds, comes in when the *Littorella* colonies have advanced sufficiently to arrest seeds and silt among their leaves and roots, and the

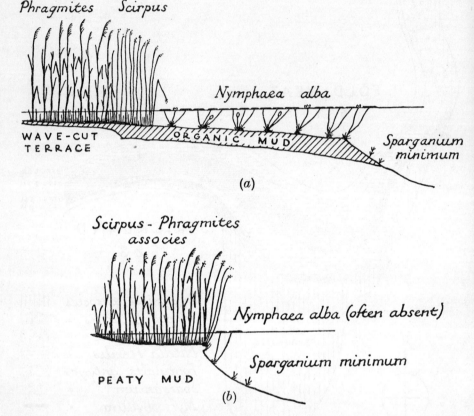

FIG. 113. PROFILE SECTIONS OF FOLD YEAT BAY

(a) *Phragmites* and *Scirpus* on the upper part of the wave cut terrace with *Nymphaea alba* below. *Sparganium minimum* on the steep slope to the floor of the lake.

(b) Terrace occupied entirely by reedswamp, which has almost entirely replaced *Nymphaea*. From Pearsall, 1917.

final sward of *Littorella-Lobelia* completely stabilises the gravel, in the interstices of which silt and plant remains are increasingly caught.

The flowers of *Littorella* are borne in scapes shorter than the leaves and are only produced when the plant is left above water: the inflorescences of *Lobelia*, borne on long scapes, are loose, few-flowered racemes of pale lilac flowers rising above the surface of the water.

Floating leaf communities. (7) *Nymphaeetum albae* (Figs. 112–115). The white water-lily consocies are the only floating leaf communities in this lake, and of these the one dominated by the common white water-lily is much the most widely distributed. It is characteristic of the western shore, where it can obtain the shelter that all floating leaf communities require, and may extend to an extreme depth of 2·5 m. The organic content of the substratum averages 23·5 per cent. Water-lilies are said to need abundant nitrates, but while the soils here are rich in nitrogen it appears to be all in the form of ammonia, no nitrates being determinable.

This community contains the following species:

Nymphaea alba	d	Ranunculus peltatus	
Nuphar luteum	lf	(not flowering)	r
(near the mouths of becks)		R. truncatus (not	
Lobelia dortmanna	l	flowering)	r

(8) *Nymphaeetum occidentalis*. This occurs in shallower water to a depth of 1·2 m. on black peaty mud, whose organic content is over 30 per cent, rarely more than 5 cm. thick, and lying on stones. The habitat is in the region of wave action so that no silt can accumulate, bases are deficient, and nitrates presumably in scanty supply. This is in accord with the distribution of the small white water-lily in upland tarns whose waters are acid.

Reedswamp communities. (9) *Phragmites-Scirpus* associes (Figs. 112–15). This is the characteristic reedswamp of Esthwaite, *Scirpus lacustris* being dominant towards the open water (to a depth of 1·3 m.) and *Phragmites communis* towards the land (to a depth of 0·92 m.). Reedswamp only occurs along stable sheltered stretches, not on loose gravel or on the primitive rounded stones of the lake shores. But if the gravel is stabilised, as by *Littorella*, *Phragmites* usually spreads and its substrata are generally organic, containing from 30 to 60 per cent of humus. Scirpetum often occurs alone in exposed situations, and its smooth elastic shoots probably resist wind and wave action better than those of *Phragmites*. The seeds of *Scirpus* can germinate under water.

(10) *Typhetum latifoliae* is confined to one situation at the mouth of a beck (see Figs. 114, 115), where abundant silt brought down by the stream favours rapid decay of organic matter (as shown by the quantities of marsh gas evolved) and thus probably the free production of nitrates.

(11) *Caricetum* occurs at the southern end of the lake in rather exposed places on peaty mud where silt is scanty. The organic content of the substratum probably always exceeds 60 per cent. The following species occur:

Carex lasiocarpa	ld	Equisetum fluviatile	
C. inflata	ld	(limosum)	f
C. vesicaria	a	Lobelia dortmanna	o
C. inflata × vesicaria	a	Nymphaea occidentalis	o

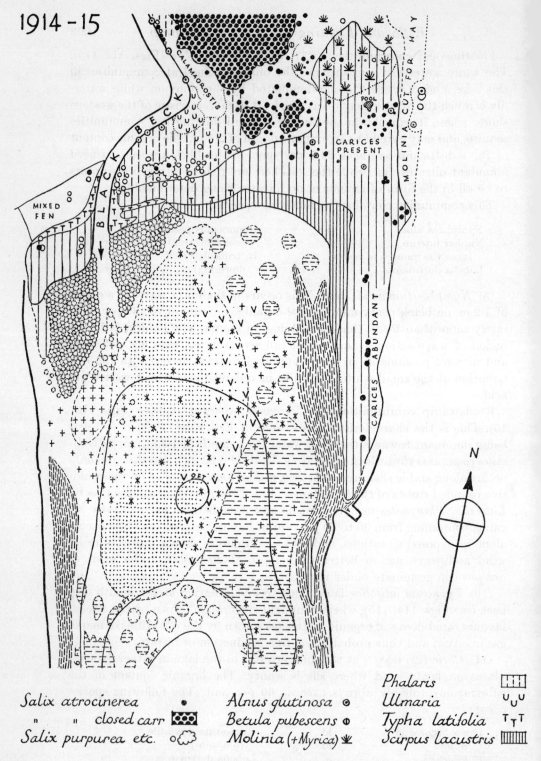

FIGS. 114, 115. COMMUNITIES SURROUNDING THE MOUTH OF BLACK

The aquatic communities include *Nitella flexilis*, *Elodea* and *Sparganium minimum* in the round the heavily silted mouth of the beck. Next come the reedswamps of Scirpetum and (The fen communities in the upper parts of the maps are dealt with in Chapter XXXII.)

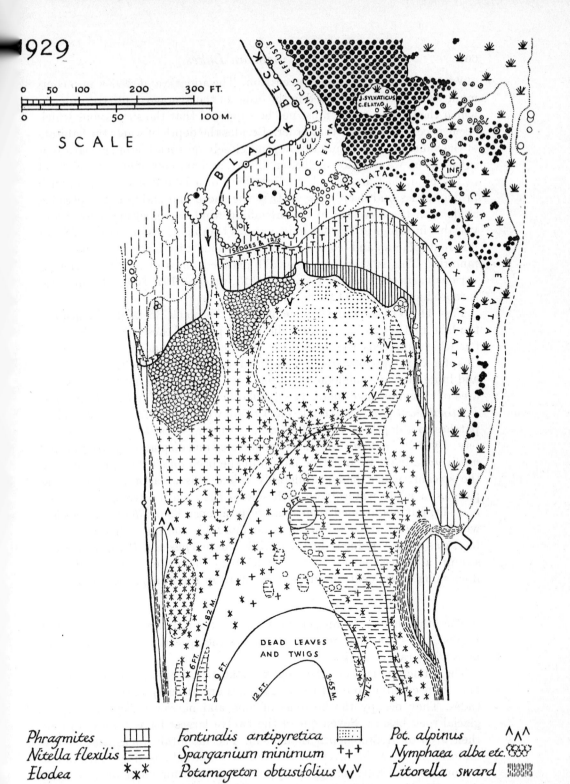

Fig. — Beck at the head of Esthwaite Water: 1914–15 and 1929

deepest water, followed by *Fontinalis* and then consocies of *Nymphaea alba* richly developed Phragmitetum with Typhetum extending outwards from the mouth of the beck. See p. 606. From unpublished field maps kindly lent by Dr Pearsall.

Variations in habitat and succession. The aquatic and reedswamp communities have been described in the order of the average depths of water in which they grow: but it must not be supposed that the successions which occur necessarily follow that order. Besides the depth of water the habitats of the submerged communities depend closely on such factors as nature of substratum, wave action or shelter, silting, and amount of organic material in the mud. The diversified winding shore of a lake like Esthwaite, with promontories and deep bays, so that the shore line makes every possible angle with the direction of the prevailing wind, while silting is local and variable in rate, leads to the most various combinations of these factors in different places, and correspondingly various distribution of the communities. This is shown in Pearsall's detailed maps of the vegetation of various parts of the shore (Pearsall, 1917, pp. 194, 197, 200). One of these is reproduced in Fig. 112.

Figs. 114, 115 show the debouchment of the Black Beck into the head of the lake in 1914–15 and 1929. (The upper parts of these figures show the fen which succeeds the reedswamp in succession and are considered in Chapter XXXII, pp. 639–46.) The beck brings down a lot of silt, both organic and inorganic, and immediately around its mouth is the most richly silted area of the whole lake shore. The dysphotic consocies of *Nitella flexilis* and *Fontinalis antipyretica* extended in 1915 down to the 9 ft. contour and in patches to a depth of 12 ft. (3·65 m.), occupying most of the floor of the bay. By 1929 *Fontinalis* was largely restricted and displaced by *Sparganium minimum* advancing from the mouth of the beck on silt brought down by the stream, which had considerably raised this area of the floor. The patches of white waterlily on each side of the mouth advanced over this new silt, but maintained their relative positions. The *Littorella* consocies, which in 1915 formed broad zones on inorganic soil on both sides of the bay, was greatly reduced in 1929, while the *Scirpus* and *Phragmites* reedswamp advanced some 50 to 100 ft. into the water. *Typha latifolia*, flourishing on rich silt, progressed eastward in the *Phragmites* zone.

The other lakes. In his general paper on the aquatic vegetation of the English lakes (1920) Pearsall supplements the special study of Esthwaite by a wider range of data.

The numerous lakes of the English Lake District lie in a small area of mountain country: but all the larger ones are at comparatively low altitudes. They occupy the bottoms of long and narrow valleys which are glacial rock basins. Seven out of the twelve largest have more than half their area of greater depth than 20 m. The mountain rocks are of very uniform physical character—a series of very hard slates, flags and grits of Ordovician and Silurian age: they are all poor in lime and weather very slowly. The lake waters are consequently singularly pure, containing only

0·03–0·07 g. per litre of dissolved material. Only two of them (Bassenthwaite and Esthwaite) are coloured with peaty material. For the twelve lakes the average content of the waters in the most important substances in parts per million is as follows (Pearsall, 1920, 1930).

Dissolved substances

Residue { Mineral   45·3
          Organic   8·3
Na$_2$O }
K$_2$O } 10·6
CaO   5·7
MgO   2·8

Fe$_2$O$_3$ }
Al$_2$O$_3$ } 0·05
CO$_2$ 5·2
SO$_3$ 11·2
Cl$_2$ 7·6
SiO$_2$ 2·4 to 0·1
NO$_3$ 0·2 to 0·02

The poverty in nitrates is very marked: phosphates as P vary from 0·02 to 0·001, the higher values occurring in early spring.

Light. In regard to the light factor, vegetation does not descend into the water below the point at which the light intensity (as determined by the rate of decomposition of potassium iodide) is reduced to 2 per cent of full illumination on a bright day, and sometimes not so far. This point

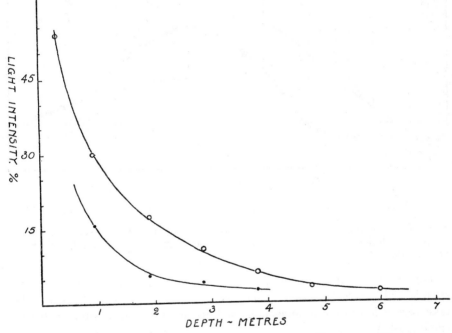

FIG. 116. DECREASE OF LIGHT INTENSITY WITH DEPTH OF WATER

The upper curve is based on data from Ullswater, Windermere and Derwentwater, the lower from Esthwaite. From Pearsall, 1920.

varies in different lakes (Fig. 116) ranging from a depth of 10 m. in the very clear Wastwater, where the vegetation descends as low as 7·7 m., to 4·1 m. in the peat-coloured waters of Esthwaite, where the vegetation reaches to 4 m. only (Fig. 116). The failure of light is therefore one factor limiting the downward penetration of the water plants and determining the depths at which certain species can exist; but variation of light intensity does not determine the general distribution of the vegetation.

Temperature. The temperatures of the lakes rise to maxima of about 17° C. in July and August, the means for these months being about 15·5° C. This compares with means of 20 to 22° C. in small ponds of the same district (Fig. 117). In April, May and June the ponds show means of 16, 20 and 21·5° C., while the lakes are very much colder with 4·9, 7·7 and 12·3° C.

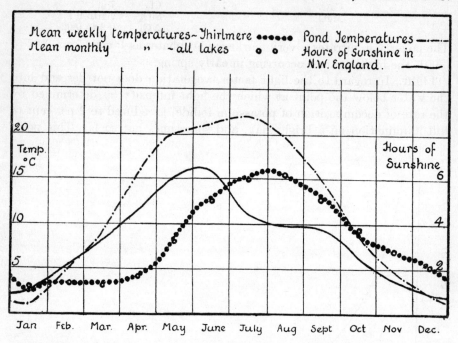

FIG. 117. TEMPERATURES OF LAKES AND PONDS THROUGHOUT THE YEAR

Note the earlier spring rise and much greater summer heights of the pond temperatures compared with those of the lakes, determining the earlier growth of pond vegetation. From Pearsall, 1920.

This marked difference in temperature in spring and early summer corresponds with earlier growth of vegetation in the ponds than in the lakes. The lake vegetation probably does not reach its maximum development before the end of August.

Aeration. The aeration of the lake waters is on the whole good. The oxygen content rises from July to August and September, but is highest in winter. The CO_2 content seems to be highest in August and September when the vegetation is most fully developed.

Silt. In these rocky lakes rooted vegetation is necessarily confined to places where the substratum is soft, i.e. in most cases where silt is present. Pearsall shows that different species react differently to silting, some succeeding best where it is rapid, others where it is slow or intermittent (Fig. 118). He concludes that "the silting factor underlies the distribution of all the types of aquatic vegetation found in this lake area, and that vegetation is only luxuriant in those places where silt is being deposited".

The decisive effect of silting on the vegetation, not only rooted but free floating, has been pointed out by Pearsall in another paper (1922), in which he has assembled an extensive series of data both from Great Britain and from other parts of the world, showing that abundance of free floating vegetation, both of flowering plants such as *Stratiotes*, *Hydrocharis*, *Ceratophyllum*, and Lemnaceae, and of plankton (Protococcaceae, Diatoms and Myxophyceae), is found only in waters rich in nitrates, silica, and calcium and magnesium salts. The waters poor in these salts, **The two types of lake** and in which the ratio $\frac{K+Na}{Ca+Mg}$ is high, are characterised on the other hand by a Desmid plankton, and by the absence of free floating vascular plants. Lakes of the latter type occur on hard rocks not covered by soil so that silting is absent, and correspond, on the whole, to the so-called "Highland" type. Where silt is abundant all lakes *tend* to have waters of the so-called "calcareous" type, whatever the underlying rock.

By comparing soils on the lake floor where silt is abundant with similar samples at the same depth where silt is scanty (e.g. of boulder clay associated with very little silt) it was shown that the silted samples contain more potash and phosphates and less organic matter than those with little silt, and further that the potash is predominantly associated with the finer particles of the silt. In the absence of inorganic silting, with its **Potash and fine silt** fresh supplies of bases, the organic matter derived from the decay of the plants which have colonised the substratum tends to accumulate, and to produce an acid peaty soil which is unfavourable to the **Subaqueous peats** more luxuriant types of vegetation, just as are the similar terrestrial moor and heath soils.

An unmistakable correlation was shown between texture and potash content on the one hand and the type of vegetation present on the other. Thus the substrata of communities of *Potamogeton pusillus* have a fine silt fraction of 25·2 per cent and a mean potash content of 0·044, those of *Juncus fluitans* 7·5 and 0·025, and of *Isoetes lacustris* 11·8 and 0·021. Since the sediments become finer in passing to deeper water and the differences of light intensity are certainly not decisive, Pearsall holds that the zonation of vegetation is primarily the result of these differences in soil conditions, and that the variations in zonation between one lake and another and between different parts of the same lake are due to the effects of different degrees of wave action and of silting.

The following is a summary of the aquatic plant communities of the lakes of the English Lake District abstracted from Pearsall's account (1921).

Deep water communities. These are usually developed in light in-

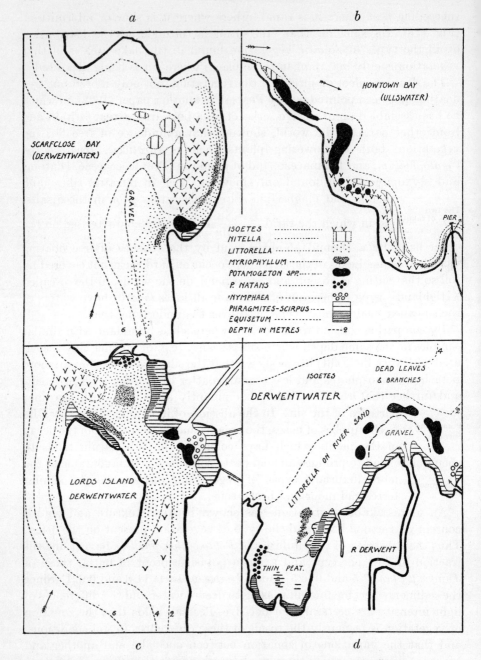

FIG. 118. RELATION OF DIFFERENT COMMUNITIES TO SILTING

On parts of the shore lines of different lakes the arrows show the silt-bearing currents (except in the south-west corners of *a* and *c*, where the barbed arrow represents wind and wave incidence across the lake). Submerged species of *Potamogeton* and *Nymphaea*, and *Scirpus-Phragmites* reedswamp occur near sources of silt; *Isoetes*, *Myriophyllum*, *Potamogeton natans*, and *Equisetum fluviatile*, away from them. In *c* wave action keeps the south-west shore of Lord's Island bare, and *Isoetes* develops on each side of the unsilted inlet behind the island. From Pearsall, 1920.

tensities of less than 15 per cent. The depth at which they occur depends on the degree of erosion and the abundance of fluvial silts. Thus in Wastwater, a rocky lake in which silts are scarce, they do not normally occur above 4 m., while in Esthwaite, where fluvial silt is abundant, they reach to within 1 m. of the surface.

(1) *Isoetes lacustris* consocies. This occurs normally on stones masked by a thin layer of silt, or on boulder clay: also on sand or occasionally thin peat. In most cases the soils are unsilted sterile glacial deposits with low potash content. With increase of silting associates begin to appear, the community becomes closed, and *Isoetes*, unable to alter its rooting level, dies. The following is the composition of this community:

Isoetes lacustris	d	Utricularia major	l
Nitella opaca	la	U. ochroleuca	l
Myriophyllum spicatum	la	Ranunculus spp.	r
Potamogeton perfoliatus	l	Sparganium minimum	r
Nitella flexilis	f	Fontinalis antipyretica	l
Chara fragilis	l		

Besides the above submerged plants, the following (normally terrestrial) mosses occur:

Climacium dendroides	r	Hypnum cuspidatum	r
Eurhynchium praelongum	r	Mnium cuspidatum	vr
E. rusciforme	r		

It is still uncertain whether they form integral parts of this deep water community or are surviving fragments of plants washed from the shore.

(2) *Juncus fluitans* consocies occurs in the more rocky lakes in well-defined areas near the mouths of streams where there is abundance of coarse silt or sand, relatively poor in potash but richer than the *Isoetes* substratum, and with only a moderate organic content. The following is the list of observed species:

Juncus bulbosus f. fluitans	d	Potamogeton pusillus	
Callitriche intermedia	la	subsp. lacustris	l
Myriophyllum spicatum	f	P. polygonifolius var.	
M. alterniflorum	r	pseudofluitans	r
Isoetes lacustris	f	Utricularia intermedia	l
Nitella opaca	f	Fontinalis antipyretica	l
N. flexilis	l	Chara fragilis	o

Juncus fluitans is locally subdominant in a very different habitat and with very different associates, such as *Lobelia dortmanna* and *Potamogeton natans*. This is in shallow water on a peaty soil with an average of 72 per cent of organic matter and a very low potash content. It seems probable that these unsilted peaty soils share an effective edaphic factor (scarcity of bases) with the coarser silts. A corresponding similarity is observed in terrestrial vegetation between dry peats and coarse sandy or gravelly soils, alike inhabited by heath plants.

(3) The *Nitella* associes is the only community typical of all the lakes, and it covers the most extensive areas on the deep water silts, developing in silted habitats where there is an accumulation of easily penetrable material ranging from fine sand to the finest muds, with an organic content which is never extreme. It is very variable, being in contact in different areas with most of the other deep water communities, both preceding and following (surrounding as it were) the *Potamogeton* and linear-leaved associes (see Figs. 115, 120). The following species are included:

Nitella opaca	d	Potamogeton praelongus	l
N. flexilis	ld	P. pusillus	r
Chara fragilis	o	subsp. lacustris	o
subsp. delicatula	la	P. perfoliatus	r
Myriophyllum spicatum	f	P. crispus	r
M. alterniflorum	l	Callitriche intermedia	r
Elodea canadensis	l	Eurhynchium rusciforme	r
Isoetes lacustris	l		

(4) The *Potamogeton pusillus* community, in which the subspecies *lacustris* is dominant, replaces the Nitelletum where fine sediments rich in potash are frequent and silting is rapid, as in Coniston, Windermere, Ullswater and Esthwaite. These fine sediments without much organic material are only deposited in deep water, so that the light intensity is always low (2–10 per cent), and this community is the only one which lives entirely in so weak a light. The community can range from fine rich silts to poor sands: its edaphic demands overlap those of *P. perfoliatus* and *P. praelongus*. The following species occur:

Potamogeton pusillus		Callitriche intermedia	o
subsp. lacustris	d	Myriophyllum spicatum	f
P. pusillus	o	M. alterniflorum	r
P. praelongus	l	M. verticillatum	r
P. perfoliatus	l	Isoetes lacustris	l
P. obtusifolius	r	Nitella opaca	f
P. crispus var. serratus	vr	N. flexilis	l
Elodea canadensis	l		

The next two communities (5) and (6) are found in Esthwaite only.

(5) The *Naias flexilis* consocies grows on very fine semiliquid mud, rich in potash and poor in organic matter. Here *Naias* plays the colonising role of *Nitella*, preceding *Potamogeton pusillus*.

Naias flexilis	d	Hydrilla verticillata var.	
Potamogeton pusillus	o	gracilis	o
P. panormitanus	o	Nitella flexilis	o

(6) The linear-leaved associes represents a development of the *P. pusillus* consocies resulting from the luxuriant silting conditions found in this lake, other linear-leaved species becoming associated. For a fuller description see pp. 599–600.

(7) The *Potamogeton obtusifolius* consocies occurs on fine silt, rich in

potash and with rather high organic content, especially in Esthwaite and Windermere, where it has replaced the *P. pusillus* consocies within a few years.

Potamogeton obtusifolius	d	Sparganium minimum	f
P. alpinus	o	Elodea canadensis	f
P. pusillus subsp. lacustris	r	Callitriche autumnalis	l
P. panormitanus	r	Myriophyllum spp.	o

(8) *P. perfoliatus* consocies:

P. perfoliatus (including vars. obtusifolius, lanceolatus and macrophyllus)	d	P. alpinus	r
		Myriophyllum spicatum	o
		Elodea canadensis	l
		Isoetes lacustris	l
P. praelongus	o	Nitella opaca	o
P. pusillus (agg.)	l	Chara fragilis	r
P. angustifolius	l		

(9) *P. praelongus* consocies:

P. praelongus	d	P. lucens	r
P. pusillus subsp. lacustris	o	Nitella opaca	f
P. perfoliatus	l	Myriophyllum spicatum	o
P. pusillus	r	Elodea canadensis	l
P. alpinus	r		

The *Potamogeton perfoliatus* and *P. praelongus* consocies are characteristic of well silted places in the lakes rich in silt (Coniston, Windermere, Ullswater and Esthwaite). *P. praelongus* grows on finer and more abundant silts in deeper water and therefore in lower light intensity than *P. perfoliatus*, within the habitat range of *P. pusillus*, and following that species or *Nitella*. *P. perfoliatus* on the other hand appears more often where silt is accumulating on the *Isoetes* habitat, though it may also succeed *Nitella*, *Juncus fluitans* or *P. pusillus* itself, but on coarser silts with a higher organic content. It appears that very rapid subaqueous silting favours the growth of the linear-leaved pondweeds, consolidation and stabilisation the establishment of the larger, broad-leaved species.

(10) The *P. alpinus* consocies succeeds *P. praelongus* or *P. pusillus* where silts are rich and very abundant, apparently only on silt which has been long stabilised (as in Windermere, Esthwaite and Derwentwater) and has accumulated a good deal of organic matter (mean 33·4 per cent), which takes a long time when silting is free. The following are characteristic species:

Potamogeton alpinus	d	P. obtusifolius	l
Elodea canadensis	f	Sparganium minimum	o
Myriophyllum spicatum	f	Nuphar luteum	o
M. verticillatum	l		

(11) The *Sparganium minimum* consocies grows in Derwentwater, Coniston, Esthwaite and Windermere in deep water, also on organic soils, and typically consists of scattered plants. Apparently no other submerged

plants are able to live in the habitat, though such places are nearly always being colonised by water-lilies.

(12) The *Fontinalis antipyretica* consocies shows no very definite relation to soil conditions or light intensity, but a decided preference for places near the mouths of streams. The soil is normally organic, but here and there the community may be sparsely developed on bare rock. The associates include:

<ul style="list-style:none">
<li>Potamogeton obtusifolius</li>
<li>Elodea canadensis</li>
<li>Utricularia major</li>
<li>U. ochroleuca</li>
</ul>

<ul style="list-style:none">
<li>Sparganium minimum</li>
<li>Juncus fluitans</li>
<li>Eurhynchium rusciforme</li>
<li>E. praelongum</li>
</ul>

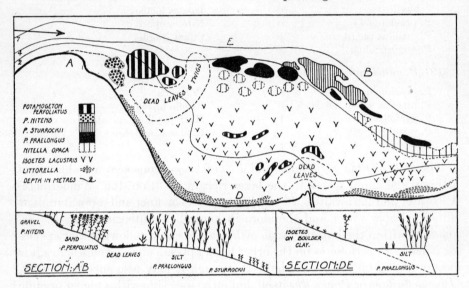

Fig. 119. Thwaite Hill Bay, Ullswater

Different species of *Potamogeton* occupy different parts of the silt terrace: *P. nitens* and *P. perfoliatus* on gravel and sand, *P. praelongus* on the finer silt at a greater depth. *Isoetes* occurs on unsilted boulder clay between the silt terrace and the shore line. *Littorella* lines the shore in shallow water ("*Potamogeton sturrockii*" is a synonym of *P. pusillus* subspecies *lacustris*). From Pearsall, 1920.

Succession. Among the deep-water communities three main lines of succession have been actually observed, or inferred on good grounds:

(1) On coarse soils in rocky lakes, poor in potash and ultimately organic:

Isoetes →*Juncus fluitans* →*Potamogeton natans*.

(2) On soils of intermediate character and most widespread in all the lakes:

Isoetes →*Nitella* →*Potamogeton perfoliatus* → $\begin{Bmatrix} \textit{Elodea-Sparganium minimum} \\ \textit{Potamogeton alpinus} \end{Bmatrix}$ →*P. natans* and/or water-lilies.

(3) On the finest, richest and least organic silts, chiefly in Esthwaite Water: *Naias* (or *Nitella*) →Linear-leaved species →*Nitella* →*Sparganium* →Water-lilies.

The proportion of potash decreases in (3) and the later stages of (2) and organic matter increases.

Changes due to silting factors appear to be usually reversible, while those due to increase of organic matter in the substratum are normally irreversible. The organic soils become more fibrous (less completely decayed) and yellow or brown rather than grey or black. In the later stages this organic matter accumulates more rapidly than the silt, and the parallel with raw humus formation on land is obvious. On the poorer and coarser subaqueous soils, the change (e.g. *Juncus fluitans* → *Potamogeton natans*) may in fact be completely identified with the formation of acid peat.

Shallow-water communities. In the Cumbrian lakes these normally occur in light of more than 15 per cent of full intensity, though they sometimes tolerate less. In the larger lakes they grow on soils that are, or have been, subject to wave erosion; or else on soils derived from those of deep water by the accumulation of silt and humus. The shallow water communities are not necessarily developed in succession to the deep water—they also colonise eroded shores *de novo*. The greater number of the shallow-water vascular plants in all freshwaters are partly emergent—either the inflorescence alone, as in *Myriophyllum*, or the upper surfaces of the floating leaves as well as the flowers, as in the water-lilies; or again, as in some of the ubiquitous plants of richer waters, such as *Alisma plantago-aquatica* and *Sagittaria*, the greater part of the photosynthetic shoots is developed above the surface of the water.

The shallow-water soils of the Cumbrian lakes are characterised by coarseness of sediment, poverty in potash, and a higher organic content than the soils of deep water, except where there is vigorous wave action or near the mouths of streams. There is a striking scarcity of pondweeds and water-lilies except in sheltered and silted areas.

(a) **Mainly submerged communities.** The first community is: (1) The *Littorella-Lobelia* associes. *Littorella uniflora* is the pioneer, spreading over and stabilising loose sand, gravel or morainic stones by means of its vegetative propagation. Here, on soil pulverised by wave action, it forms a continuous sward, collects sand and probably seeds, ultimately giving way to other communities. As the soil becomes more organic *Lobelia dortmanna* appears and tends to replace the *Littorella*.

Other dominants appear at a later stage, and the following communities may be distinguished, with progressive cessation of silting and increasing organic content.

(2) *Potamogeton* associes (average organic content 22·2 per cent):

P. gramineus (with var. longipedunculata)
P. nitens var. subgramineus
P. angustifolius
P. perfoliatus
P. alpinus

Any one of these species may be dominant, and the last two persist when the organic content of the soil is over 40 per cent.

(3) *Myriophyllum* consocies (average organic content 42·7 per cent):

M. spicatum	d	M. verticillatum
M. alterniflorum (Esthwaite)		Ranunculus peltatus var. truncatus

(4) *Juncus fluitans* socies (average organic content 86·6 per cent) only occurring where the soil is more or less peaty. This is no more than a socies of the *Littorella-Lobelia* community, since the *Juncus* is not dominant over any considerable area. Associated are:

Sparganium minimum	f	Scirpus fluitans	o
Utricularia major forma		Nitella opaca	o
gigantea	la	Isoetes lacustris	r
Apium inundatum	l		

Myriophyllum (f–o) and *Ranunculus* sp. (o) are probably relict.

These communities merge into one another and are often indistinct. At one end of the scale is the *Littorella-Potamogeton* type, with inorganic soil relatively rich in potash, at the other the *Lobelia-Juncus fluitans* type, with organic soil poor in potash.

(b) Floating leaf communities. These are all limited to the most sheltered places, being excluded by wave action, though *Nymphaea occidentalis* and *Nuphar intermedium* are able to endure somewhat more vigorous wave action than the common white and yellow water-lilies. For the rest the following four communities show correlation with increasing organic content of the soil, parallel with that shown by the submerged communities.

(1) *Nymphaea alba* consocies, average organic content 25·7 per cent.

2) *Nuphar luteum* consocies, average organic content 38·6 per cent.

(3) *Nymphaea occidentalis* consocies, average organic content 48·5 per cent.

(4) *Potamogeton natans* consocies, average organic content 74·2 per cent.

Plants with floating leaves are usually very local in the Cumbrian lakes, particularly the large ones, because continual wave action along the exposed shores prevents the deposition of silt. Only Esthwaite, small and rich in silt, has any large areas of water-lilies: in Derwentwater, also small but with rather scanty silting, they are generally replaced by the consocies of *Potamogeton natans* except near the mouths of streams, where silt and water-lilies occur.

(c) Reedswamp communities. The reedswamp dominants again show the same general relation to the organic content of the substratum and to silting, but less definitely, for the vigorous growth of their rhizomes enables them to persist in spite of changing conditions. The following are the main communities:

(1) *Typha latifolia* consocies, average organic content 34·0 per cent.

(2) *Scirpus-Phragmites* associes, average organic content 40·7 per cent.

(3) *Equisetum fluviatile* (*limosum*) consocies, average organic content 60·9 per cent.

(4) *Carex* associes, average organic content 89·4 per cent.

Typha latifolia (the great reedmace or bulrush) occurs in these lakes only at the mouth of Black Beck in Esthwaite (see Figs. 114, 115) a place of abundant fine silting and rapid change of soil level.

Scirpus lacustris (the "true" bulrush) and the common reed (*Phragmites vulgaris*), the commonest European reedswamp dominants, form the characteristic reedswamp (see Figs. 112, 115, etc.), but they are nevertheless quite local in the Cumbrian lakes, because of the resistant nature and also of the instability of most of the shores. *Scirpus* extends into the deeper water and also prefers the more organic substrata, *Phragmites* forms the bulk of the reedswamp inshore and persists in the fens after the soil has been raised above the water level. The associes occurs in open bays where wave action and silting are both slight; also on sheltered western shores, in the mouths of streams, and in extensive shallows.

Equisetum fluviatile and *Carex* spp. form reedswamp in closed bays and on very sheltered shores, where both wave erosion and affluents are lacking. This reedswamp is therefore characteristic of the more rocky lakes and rarer in the more silted lakes such as Esthwaite. *Carex inflata* and *C. vesicaria* are the most frequent dominants of the Caricetum, with hybrids between them, and less commonly *C. lasiocarpa*, which is locally abundant.

The reedswamps are typically very pure communities in which the aerial shoots of the dominant plants are so closely set that there is little room for subordinate vegetation. But the fully developed community is usually fringed with open reedswamp where the dominants are beginning to establish themselves. The following species, among others, occur in this ecotone or mictium:

Sparganium natans	la	Scirpus fluitans	o
Apium inundatum	la	Callitriche intermedia	o
Utricularia major (neglecta)	la	C. stagnalis	r

These plants occur in their own habitat conditions in such a mictium. Open reedswamp of *Scirpus lacustris* with *Nymphaea alba* is a common type of mictium in other waters, e.g. in the Norfolk Broads (Pl. 109, phot. 270).

The two main hydroseres. From Pearsall's extensive series of data it is apparent that there are two main *types* of sere in the shallower waters of the Cumbrian lake shores.

Where silting is rapid and the substrata inorganic we have the *silted type*:

(A) *Littorella* →*Potamogeton* →*Nymphaea* →*Phragmites*.

Where silt is scarce and the soil is organic the *peaty type* occurs:

(B) $Lobelia \rightarrow Myriophyllum \rightarrow Potamogeton\ natans \rightarrow \begin{cases} Equisetum \\ Carex. \end{cases}$

The Cumbrian Lakes

The actual successions are however very numerous, and the variations are brought about largely by *changes* in silting. Thus if silting is at first heavy and then stops altogether the general line of succession will shift from (A) to (B), and beginning with *Littorella* may end with *Carex*. On the other hand, if an increasingly organic soil becomes heavily silted the succession will change from (B) to (A) and end with *Phragmites*.

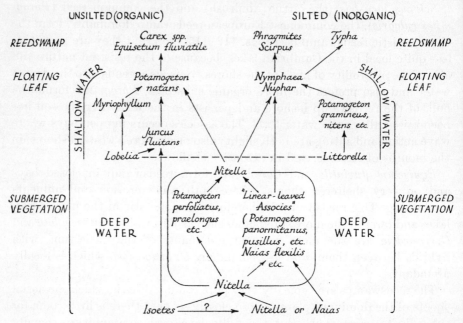

FIG. 120. SCHEME OF THE MAIN TRENDS OF SUCCESSION

In addition to somewhat excessive simplification, this diagram has one conspicuously unnatural feature in the separation on opposite sides of *Lobelia* and *Littorella*, which in fact commonly occur together in one associes, though the former increases as the soil becomes more organic, while the latter is the pioneer on inorganic silts. Such anomalies are inevitable when we attempt to exhibit the changes of three variants—succession in time, organic content of substratum, and depth of water—in a two-dimensional diagram.

The leading causal factors in the hydroseres are the increase of organic content in the substratum, and the raising of the soil level till it eventually reaches the surface of the water, so that the aquatic communities give way to fen. The succession brought about by these factors is checked or stopped by wave action, and altered by silting. Areas which are freely silted from the beginning bear a different set of communities from those which are not. Where silting is intermittent a "silted community" gives place, on cessation of silting, to a community characteristic of organic soil, while a fresh incidence of silting produces the reverse effect. Such changes are in fact reversible. But the changes in life form by which floating leaf and reedswamp dominants replace the submerged, occurring in the later stages of the sere, are irreversible, and so is the eventual predominance of accumulation of organic material.

Primitive Lakes and Tarns

The general direction of change is towards a more organic and acid soil with a decreased content of bases and especially potash, and is thus parallel to the formation of peaty and acidic moorland soils on land, a progression which as we shall see in Chapter XXXII is completed in the terrestrial "raised moss" of the Lake District lowlands.

By combining the observed seres in deep and shallow water we can construct a scheme such as that shown in Fig. 120, ignoring all but the main trends of succession.

THE MOST PRIMITIVE LAKES AND TARNS[1]

Pearsall has shown that the different lakes he studied form a series illustrating the evolution of glacial lake basins from a primitive type, with rocky floor, stony or peaty margins and little or no inorganic silt, to the richly silted types of which Esthwaite Water is an example.

The most primitive types are now to be found in the smaller mountain tarns where this development cannot take place, and of these we may distinguish (1) those with very little silt, what there is being inorganic and derived mainly from erosion of the shores; and (2) those in which the silt is peaty and derived from terrestrial sources, i.e. from the peat formed by adjoining land vegetation, while the accumulation of plant debris builds up a highly organic substratum.

(1) In this type silt is scarce, but what there is is inorganic. Any fluvial silt brought in by streams is coarse and sandy.

The vegetation is everywhere sparse. The species of deep water are:

Isoetes lacustris	la	Myriophyllum spicatum	l
Nitella opaca	la		

Where some inorganic silt is deposited we have:

Juncus fluitans (local, near streams)
Callitriche intermedia
Potamogeton polygonifolius var. pseudofluitans (entirely submerged)
P. pusillus subsp. lacustris (r)

In shallow water *Littorella uniflora* (lf) is the principal species. *Lobelia dortmanna* is very local.

(2) In this type the substratum becomes organic and inorganic silt is quite absent. When the lakes are very small (tarns) so that their shores are not eroded by wave action there is a good succession from the aquatic to the adjoining bog vegetation with gradual increase of acid peaty humus.

(a) Beginning in deeper water we have the *Juncus–Myriophyllum* open associes:

Juncus fluitans	l–la	Chara fragilis	l
Myriophyllum spicatum	la	Nitella opaca	l
Utricularia minor	f		

[1] Information kindly supplied by Dr W. H. Pearsall.

(b) The next community consists of *Sparganium minimum* (o–f), followed by a floating leaf community of

(c) *Nymphaea occidentalis* (a–d) and *Nuphar pumilum* (f–a). This is in turn followed by an open reed belt of

(d) *Equisetum fluviatile* (*limosum*), and this by

(e) Reedswamp of *Carex inflata* (d) with which are associated:

Eriophorum angustifolium	f	Potamogeton polygonifolius	l
Juncus fluitans	f	Utricularia minor	o
Menyanthes trifoliata	l		

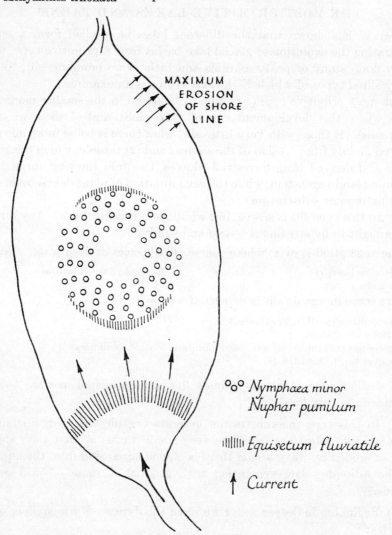

FIG. 121. DIAGRAM OF A "PRIMITIVE" TARN (LAKELET)

The shores are stony and barren with strong wave erosion on the leeward side. There is a development of characteristic reedswamp and small water-lily communities where slight silting occurs, i.e. beyond the mouth of the inlet stream and in the centre of the tarn just below the level of wave action. After a sketch supplied by Dr Pearsall.

In shallow water *Littorella* is rare or absent, *Subularia aquatica* local, while *Lobelia dortmanna* becomes locally frequent as the organic mud accumulates. This community often passes into open reedswamp of *Phragmites*, which is normally succeeded by open *Equisetum fluviatile*, often with *Scirpus lacustris* and *Potamogeton natans*. The closed reedswamp seems always to be formed by *Equisetum fluviatile* and *Carex inflata*. *Nymphaea occidentalis*, *Nuphar pumilum* and *Sparganium minimum* may occur in any of these reedswamps.

In tarns and lakelets of this type which are relatively advanced in development but subject to wave erosion of the shore, the water-lily and reed belt communities, (c) and (d), are often formed in a zone at some distance from the shore just below the zone of wave erosion, and also just beyond the mouth of a stream entering the tarn (Fig. 121). In both these areas a little silt is deposited and plant colonisation begins, while the shore zone is kept free from vegetation.

Transition to bog. As the peaty soil reaches the water level, bog plants such as *Sphagnum* spp., *Erica tetralix* and *Carex panicea* appear; and these are joined later by several other species of bog plants till the bog community is established. Among the bog species relicts of the reedswamp community (e) often linger. This is the most "organic" type of succession that occurs. Rather surprisingly the lowest pH value recorded is 5.

REFERENCES

PEARSALL, W. H. The aquatic and marsh vegetation of Esthwaite Water. *J. Ecol.* **5**, 180–202; **6**, 53–74. 1917–1918 a.

PEARSALL, W. H. On the classification of aquatic plant communities. *J. Ecol.* **6**, 75–83. 1918 b.

PEARSALL, W. H. The aquatic vegetation of the English Lakes. *J. Ecol.* **8**, 163–201. 1920 (1921).

PEARSALL, W. H. A suggestion as to factors influencing the distribution of free-floating vegetation. *J. Ecol.* **9**, 241–53. 1922.

PEARSALL, W. H. Phytoplankton in the English Lakes. I. The proportions in the waters of some dissolved substances of biological importance. *J. Ecol.* **18**, 306–20. 1930.

Chapter XXXI

THE VEGETATION OF RIVERS

The obvious difference between the conditions of life in a river and in a pond or lake is of course the existence in the former of a continuous current. Correspondingly the vegetation of a very sluggish stream is quite similar to that of a pond, both containing much the same zoned communities composed of the same species. But where the river current is considerable new factors come into play.

Current. In the first place the current tends to equalise conditions, such as temperature and content of dissolved salts and gases, in different parts of the river's course; but this action is far from being complete, because of the changing gradient and the various obstacles such as sharp bends which alter the rate of flow of the current, and also because of the varying nature of the strata through which the river flows. Secondly, the river current, like the waves and currents on a lake shore, erodes the banks, deposits silt here and removes it there—and not only silt, but also the vegetation itself, so that the river bed may be quite denuded of plants in some places and freshly colonised in others, and the vegetation of rivers continually subject to heavy floods remains sparse. Thirdly, the normal progressive process of "Verlandung", as the Germans call it, the "growing up" and ultimate obliteration of a pond or lake where there is no direct drainage through it, does not occur in a river. A river may become choked with vegetation so as to hold up the water and cause flooding: its course may be diverted, with the resulting formation of "oxbows", which if they are completely cut off from the stream, behave like ponds and ultimately fill up with vegetation; and extensive swamps and marshes, through which the river current winds its sluggish way, may be formed by the flooding of flat alluvial plains. But so long as water is supplied from its sources the river must continue to exist, and whenever any considerable body of water comes downstream the impact tends to clear the bed of vegetation which may have threatened to choke it. In rivers of medium or slow current the conditions in midstream are, however, very different from those near the banks, because the current is stronger and the depth of water usually greater. There is in fact a zonation of conditions, and consequently of vegetation, parallel to the direction of flow, i.e. to the banks, and therefore analogous to the zonation on the margin of a lake. Many such rivers are normally fringed with reed-swamp whose dominants have succeeded in colonising the shallower quieter water near the bank, though the current prevents its encroachment on the deeper water and thus holds up the progress of the succession.

Silting. In the last chapter we saw that silting had an all-important effect on the vegetation of lakes, and the same is true of rivers. Silt directly supplies most of the mineral nutrients for the rooted plants, as the salts dissolved in the whole body of water supply the free-floating plants and the algae, so that a richly silted river bed is the most prolific in vegetation. Silting is due to the slowing down of moving water which is carrying solid material in suspension, and thus depends on the velocity of the current, which is correlated with the nature of the river bed and with the amount and texture of the silt carried or deposited.

Speed of current

The general correspondence is shown in Table XVIII (Minnikin, 1920, Butcher 1933):

Table XVIII

Velocity of current per second	Nature of bed	Habitat
More than 4 ft. (1·21 m.)	Rock	Torrential
,, ,, 3 ft. (0·91 m.)	Heavy shingle	,,
,, ,, 2 ft. (0·60 m.)	Light shingle	Non-silted
,, ,, 1 ft. (0·30 m.)	Gravel	Partly silted
,, ,, 8 in. (0·20 m.)	Sand	Partly silted
,, ,, 5 in. (0·12 m.)	Silt	Silted
Less than 5 in. (12 cm.)	Mud	Pond-like

In a general way this change in the nature of a river bed with current velocity corresponds with the gradual decrease in average gradient of the bed of a river which rises in a mountainous region and flows through foothills and finally across plains towards the sea. But local topographical and geological features introduce many variations, and a river may increase and decrease the average rate of its current many times during its course. Bare rock and boulders are characteristic of the beds of mountain torrents, and as the gradient decreases and the rate of flow lessens the bed changes to gravel and sand, while the slowly moving waters of an alluvial plain have silty or muddy bottoms. But the currents of rivers are not constant even at any one spot and their beds are therefore unstable. Heavy precipitation in the mountains will add enormously in a few hours to the volume of water coming down and will correspondingly increase the rate of flow, shifting downstream the smaller particles that have been deposited. When the flood subsides new particles brought down from upstream or flowing in from the banks will sink to the bottom as the water clears. Thus the condition of the bed at any spot is always changing in different directions, in broad contrast with the bottom of a pond or lake where change is on the whole slow and in the same direction, though lake bottoms also show fluctuating changes near the mouths of affluent streams and as the result of varying wave action and shore currents. Any obstacle to moving water carrying suspended material causes it to drop part of its load, and vegetation itself, once established, is an important agency in trapping silt. Silt does not consist solely of inorganic particles derived from eroded rocks, but may be partly or wholly organic, formed from debris of water plants themselves, from eroded peat or humus, and from the soil

Floods

of manured agricultural land—not to mention sewage. A strong organic constituent normally increases the fertility of silt.

Turbidity. According to the amount of fine silt they carry rivers vary very much in clearness. Streams draining areas of limestone and of hard siliceous rock are in general clear, while those passing through agricultural country are more turbid. Flood water resulting from heavy rainfall will always bring soil into a stream and increase its turbidity. When a river is habitually very turbid it may be destitute of all submerged vegetation, owing to the diminution of light penetrating the waters, and perhaps partly to the unstable bed of silt and mud, as Butcher (1933) found in the river Tern in Shropshire.

Solutes and reaction of the water. The amount and the chemical nature of the solutes are of first importance to the plankton, to the algae, and also to the bryophytic vegetation of mountain torrents, which depend entirely on this source for their mineral nutrients; and the pH value of the water may also have a direct effect on the plants. On the whole, however, these factors, apart from the amount and nature of the silt with which they may be correlated, do not cause very marked differences in the rooted vegetation of vascular plants.

Thus Watson (1919) recognised that certain submerged mosses such as *Amblystegium filicinum*, *Fissidens crassipes*, *Hypnum commutatum*, *H. falcatum* var. *virescens*, *Orthotrichum rivulare*, *Philonotis calcarea*, and the thalloid liverwort *Pellia fabbroniana*, especially var. *lorea*, as well as the blue-green algae *Rivularia haematites* and *Scytonema* spp., are characteristic of calcareous streams. Butcher (1933) divides the rivers whose vegetation he investigated according to the proportion of lime dissolved in the water, but the rooted phanerogamic vegetation of a highly calcareous river like the Dove or the Itchen does not differ greatly from one which is moderately calcareous like the Tees, nor even from acid non-calcareous streams like those of the New Forest, providing that the silting factor is comparable. There are, however, a few species which are more characteristic of one or the other type. Thus *Myriophyllum spicatum*, *Potamogeton polygonifolius* and *P. alpinus* avoid the more calcareous waters, and *P. perfoliatus* and *P. crispus* the more acid.

The following data are mainly taken from Butcher's account (1933) of the vegetation of the different types of stream bed enumerated above and from Watson's paper on the Bryophytes (1919), supplemented by information about the Thames near Oxford kindly contributed by Mr H. Baker.

(1) **Very rapid current** (bed rocky, mountain torrent type). Vegetation consisting mainly of bryophytes, usually a mixture of species: leafy liverworts are often abundant and locally dominant. Flowering plants sparse or absent owing to lack of silt and of sufficiently secure rooting places.

Watson (1919) gives the following as the most characteristic liverworts, often occurring in pure patches and sometimes carpeting the rocky beds of mountain torrents:

Alicularia compressa	d or ld
Aneura sinuata	f
Aplozia riparia forma potamophila	lf
Chiloscyphus polyanthus var. rivularis	o
var. fragilis	f
Eucalyx obovatus var. rivularis	r
Marsupella aquatica	lf
Scapania dentata	a, ld
S. uliginosa and obliqua (alpine and subalpine)	lf

Among mosses the following are the most characteristic forms and are often locally dominant:

*Amblystegium filicinum	f	Fontinalis squamosa	f–a
A. fluviatile	o	Hyocomium flagellare	a
Brachythecium rivulare	a	*Hypnum commutatum	a
Bryum pseudotriquetrum	o, la	H. falcatum	o
Eurhynchium rusciforme	a, sd	H. riparium	f
Fontinalis antipyretica		Philonotis fontana	a
(and especially var. gracilis)	lf		

* Usually in calcareous streams.

Lichens are few, though they may be locally abundant. The commonest are crustaceous forms such as *Dermatocarpon aquaticum* and *Verrucaria submersa*. There are also various algae, of which *Sacheria mamillosa*, and the Myxophycean *Rivularia haematites* (calcareous), *Oscillatoria irrigua*, *Nostoc verrucosum* and *N. sphaericum* are frequent, *Cladophora glomerata* is abundant, and *Vaucheria sessilis* occasional (Watson, 1919).

Of flowering plants the following occur most commonly, but they are often absent altogether and never dominant:

Apium nodiflorum	f	Potamogeton pusillus	la
Callitriche spp. (commoner in slow rivers)	o–la	Ranunculus fluitans	f
		R. lenormandi	f
Glyceria fluitans	la	R. peltatus var. pseudo-fluitans	o, f
Montia fontana var. rivularis	la	Stellaria uliginosa	o–la
Potamogeton crispus	la		

The upper course of the River Tees is a good example of this type. Its current is swift and its bed of solid rock, with some stones and sand, in which none of the higher plants can retain a foothold for long when the stream is in flood. The vegetation is therefore confined to mosses and the larger algae, which grow in firmly anchored cushions. The most characteristic species are:

Fontinalis antipyretica	a	Grimmia fontinaloides	r
Eurhynchium rusciforme	a		
Amblystegium fluviatile	o	Lemanea fluviatilis	a
Cinclidotus fontinaloides	o	L. mamillosa	o

(2) **Moderately swift current.** Bed composed of stones or boulders. Vegetation consisting partly of bryophytes but also of flowering plants where a little silt is trapped between the stones and the floods are not strong enough to tear out the plants.

The Dove and the Wharfe. Butcher found the following six species of flowering plants in the Dove (Derbyshire) and the first three, besides the mosses and algae, also in the Wharfe (Yorkshire), both calcareous streams draining the Mountain Limestone:

Ranunculus fluitans	d	Apium nodiflorum	r
Potamogeton densus	o–f	Myosotis scorpioides	
Sium erectum	o–a	(palustris)	r
Veronica anagallis-aquatica	o		

and of non-vascular plants:

Fontinalis antipyretica	f	Batrachospermum spp.	o
Eurhynchium rusciforme	f	Lemanea spp.	f
Cinclidotus fontinaloides	o		

The upper Thames. Even a river like the Thames, whose course is mainly "lowland", has quite a swift current in some of its upper reaches. The Thames has a course of 160 miles (c. 257 km.), and between its source in the oolites of the Cotswold Hills to the estuary below London it traverses a considerable variety of geological strata. The bed of the Thames has of course been largely artificialised by embanking, dredging, and the construction of locks, weirs, and "lashers".[1] In this way the river is regulated, made more easily navigable, and the flooding of adjacent meadowland is minimised. The vegetation is, of course, correspondingly affected, but most of the natural habitats are still provided, the swiftness of the current and the silting of the bed varying from place to place in relation to weirs, bridges, etc., as well as to the fall of the general river course. In its upper reaches, where it flows through the Oolites, the current is generally swift and the bed stony, and here the submerged vegetation includes five out of the six species of flowering plants found by Butcher in the Dove:

*Apium nodiflorum	*Ranunculus fluitans
*Myosotis scorpioides	*Veronica anagallis-aquatica
Nasturtium officinale	
Oenanthe fluviatilis	
*Potamogeton densus	*Fontinalis antipyretica
P. pectinatus forma interruptus	Cladophora sp.
Ranunculus circinatus	Enteromorpha intestinalis

* Also in the "moderately swift" reaches of the Dove.

The plants grow in longitudinal streaks of varying width with their shoots trailing downstream and sometimes reaching a length of 6 ft. (c. 2 m.).

[1] A "lasher" is an artificial waterfall taking that part of the stream which is not conducted through the accompanying lock.

PLATE 109

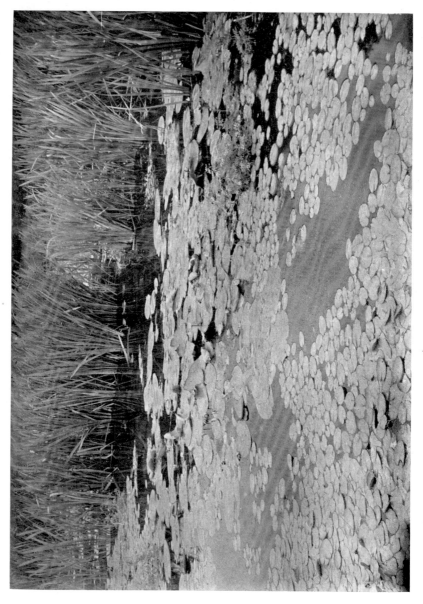

Phot. 271. *Nymphoides* (*Limnanthemum*) *peltatum*, *Nuphar luteum* (floating leaf), *Oenanthe fluviatilis* (emersed), *Sparganium erectum* (reedswamp). Medley, near Oxford. *A. H. Church.*

ZONATION IN THE THAMES

(3) **Moderate current.** Bed gravelly. The following species were observed by Butcher in various rivers in the north and north midlands:

Ranunculus fluitans[1]	f–d	Callitriche stagnalis	r–f
Sparganium simplex	f–d	Elodea canadensis	r–f
Potamogeton crispus	r–f	Myriophyllum spicatum	r–o
P. densus	r–f	Mimulus guttatus	o
P. pectinatus	o	Polygonum amphibium	f
P. pectinatus forma interruptus	lsd	Sagittaria sagittifolia	f
		Scirpus lacustris	f
P. perfoliatus	r–d	(the last three in the Tern)	
P. pusillus	r		

In the Thames above the lock at Medley, just above Oxford, with a medium current and a gravelly bottom covered with thin silt in midstream, the following occurred:

(i) *Submerged community*:

Sagittaria sagittifolia, dominant, but producing submerged strap-shaped leaves only (f. *vallisneriifolia*).

Potamogeton lucens, subdominant.

P. perfoliatus, abundant.

Oenanthe fluviatilis,[2] abundant (a typical south English species).

Nuphar luteum with submerged leaves only, arising by extension of rhizomes from the zone of floating leaf plants (iii) nearer the bank where the yellow water-lily was dominant.

Hippuris vulgaris, submerged plumose shoots coming from rhizomes in (iv).

Next, across the mouth of a silted bay, was

(ii) a zone of *Scirpus lacustris*. This bulrush grew in a narrow band parallel with the axis of the river, its rhizomes embedded in silt and gravel below 4–6 ft. of water, and aerial shoots rising 7 ft. above the surface. Inside the line of bulrush was a thick bed of silt with only 2 ft. of water above it and inhabited by

(iii) a zone of *floating leaf species*:

Nuphar luteum, dominant.

Nymphoides (*Limnanthemum*) *peltatum*, sub-dominant, competing with the *Nuphar* by means of its long rhizomes (Pl. 109, phot. 271).

Potamogeton natans, *Polygonum amphibium*, and *Callitriche stagnalis* all frequent.

(iv) *Reedswamp*, in a few inches of water:

(a) Zone of *Hippuris vulgaris* as the pioneer dominant, forming broad expanses of aerial shoots up to 18 in. (45 cm.) high, and also with its rhizomes spreading out beyond (ii) into deep water, where they bore sub-

---

[1] Pl. 110, phot. 272.   [2] Pl. 109, phot. 271.

merged plumose shoots; and with *Sagittaria sagittifolia* frequent in isolated patches among the *Hippuris*. The mare's-tail is not a very common plant in the Thames, and this position as pioneer of the reedswamp is commonly occupied by *Sagittaria*.

(*b*) The next zone is not here dominated by the typical tall reedswamp species, but by a number of smaller plants which ordinarily accompany the reed dominants where they can find room:

Mentha aquatica	a	Myosotis scorpioides	
Nasturtium officinale	a	(palustris)	f
Veronica beccabunga	a	Rorippa islandica	
Apium nodiflorum	f	(Nasturtium palustre)	o
Veronica anagallis-aquatica	f		

The taller species colonised this zone in patches, starting from transported rhizomes swept downstream in winter floods and rapidly spreading:

Sparganium erectum	f	Alisma plantago-aquatica	o
Iris pseudacorus	f	Glyceria maxima (aquatica)	o
Rumex hydrolapathum	o		

(4) **Medium to slow current.** Bed sandy and silted.

The examples described by Butcher are mostly from lowland streams rising from low hills, springs or marshy ground—the common type of river in the midlands and the east and south of England. The phanerogamic vegetation is very much richer owing to the better nutritive conditions and the lessened danger of detachment of rooted plants by the current, though the vegetation of any given stretch of river bed is more or less transitory (Fig. 122).

In the River Lark in East Anglia the fastest stretches contain several of the same species as in (2) and (3) above, with *Ranunculus fluitans* again dominant, but with the addition of *Oenanthe fluviatilis*, a species confined to the south. Where the current is slow Butcher found the composition of the vegetation to be as follows (Pl. 110, phot. 273):

Sparganium simplex	d	Potamogeton perfoliatus	f
Sagittaria sagittifolia	sd	P. praelongus	r
Elodea canadensis	f	P. pusillus	r
Potamogeton crispus	f	Callitriche stagnalis	o

In the medium and slow-flowing stretches of the Itchen (Hampshire) there were the following:

Elodea canadensis	f–d	Ranunculus fluitans	r
Hippuris vulgaris	a–sd	R. pseudo-fluitans	r
Sparganium simplex	a–sd	Sium erectum	r
Callitriche stagnalis	f–sd	Apium nodiflorum	r
Oenanthe fluviatilis	f–a	Lemna trisulca	r
Potamogeton densus	f	*Reedswamp*	
P. pusillus	f	Scirpus lacustris	a
P. crispus	r	Typha latifolia	r

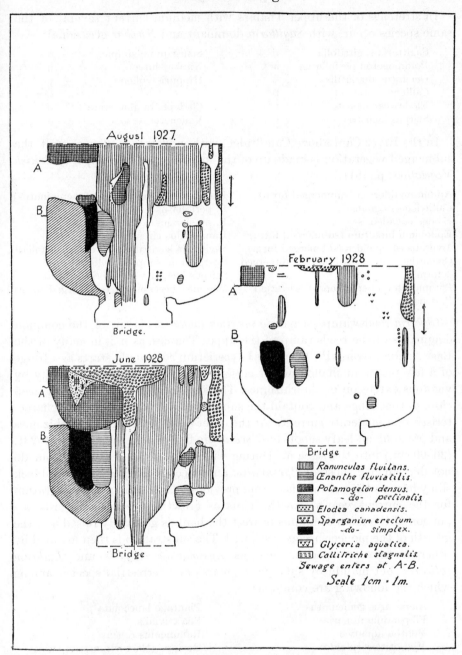

FIG. 122. AQUATIC VEGETATION IN THE RIVER LARK (SUFFOLK)

Showing the changes in the areas occupied by different species from season to season. From Butcher, 1933.

In stretches of the upper Thames with medium current several of the same species occur, with *Sagittaria* dominant and *Nuphar* occasional:

Sagittaria sagittifolia	d	Sparganium simplex	o
Potamogeton perfoliatus	a	Nuphar luteum	o
Oenanthe fluviatilis	a	Hippuris vulgaris	r
Callitriche stagnalis	f		
Elodea canadensis	f	Cladophora glomerata	o
Scirpus lacustris	f	Vaucheria sessilis	o

In the River Cam above Cambridge, with moderate to slow current, the submerged vegetation is made up of the following species (*Types of British Vegetation*, p. 191):

Apium nodiflorum (submerged form)
Callitriche stagnalis
Elodea canadensis
Epilobium hirsutum (submerged form)
Myosotis scorpioides (submerged form)
Oenanthe fluviatilis (and submerged form)
Potamogeton crispus (and f. serratus)
P. densus
Potamogeton lucens (and f. acuminatus)
P. perfoliatus
P. praelongus
Ranunculus circinatus
Sagittaria sagittifolia (f. vallisneriifolia)[1]
Sparganium simplex (f. longissima)
Scirpus lacustris (submerged form)
Veronica beccabunga (submerged form)

Glyceria reedswamp. *Glyceria maxima* (*aquatica*) is one of the common dominants of the reedswamp of the upper Thames, as it is in many of the East Anglian rivers (Pl. 111). The Glycerietum commonly starts as a fringe of a few plants in shallow water close to the bank, and grows rapidly by vigorous extension of the rhizomes. The stream may be several feet deep close to this fringe and contain the submerged aquatic community characteristic of a moderate current. At the outer edge of the fringe the rhizomes and procumbent leafy shoots of *Glyceria* are free-floating and extend 1–2 ft. (30–60 cm.) into the stream. During winter the aerial shoots die, but do not dry out like those of *Phragmites*, and become compacted into a black slimy humus, the rotting mass from previous years forming the substratum for the current year's growth. Thus as growth proceeds outwards the portion of the mass of humus nearest the bank is gradually raised and the growth of the pioneer becomes stunted. The substratum is then invaded by other reedswamp species, such as *Lysimachia vulgaris* and *Lythrum salicaria*, but mainly by subaquatic, marsh and terrestrial species, among which the following are common:

Alopecurus geniculatus
Filipendula ulmaria
Mentha aquatica
Myosotis scorpioides
Nasturtium officinale
Plantago lanceolata
Poa trivialis
Ranunculus repens
Rorippa (Nasturtium) amphibia
R. islandica (Nasturtium palustre)

Thus the growth of Glycerietum maximae depends much more on the accumulation of black humus produced by itself than on the trapping of silt, and at least the outer edge floats on watery organic ooze. The closely

[1] Pl. 106, phot. 261, p. 582.

PLATE 110

Phot. 272. Non-silted community in the River Tees at Neasham, Yorkshire. *Ranunculus fluitans* dominant. R. W. Butcher (1933).

Phot. 273. Silted community in the River Lark, Suffolk. *Sagittaria sagittifolia* (f. *vallisneriifolia*) and *Sparganium simplex* (submerged form) dominant. R. W. Butcher (1933).

SILTED AND NON-SILTED RIVER COMMUNITIES

PLATE 111

Phot. 274. Reedswamp of *Glyceria maxima*, summer. River Lark, Suffolk. *R. W. Butcher* (1933).

Phot. 275. The same view, winter condition. *R. W. Butcher* (1933).

GLYCERIETUM MAXIMAE IN THE LARK

matted rhizomes and decaying debris of *Glyceria* appear to be more resistant to the scouring of floods than the plants of *Sparganium erectum*, which, in the Thames, not uncommonly appears as isolated patches in a Glycerietum and may compete successfully with *Glyceria*.

(5) **Current very slow or negligible.** River bed of fine mud.

In the Lark, under nearly stagnant and muddy conditions, Butcher (1933) observed the following community:

Potamogeton lucens	d	Nuphar luteum	f
P. pectinatus	sd	Zannichellia palustris	o
Sparganium simplex	a	Callitriche stagnalis	o
Potamogeton natans	f	Scirpus lacustris	r
P. crispus	f	Sparganium erectum	r

With the appearance of the yellow water-lily and the floating-leaved pondweed this community is obviously beginning to approximate to that of a pond.

In very slow-moving or almost stationary water in the Thames, over fine silt and mud, we have the same dominants with different associated species:

Potamogeton lucens	d	Potamogeton compressus	o
P. pectinatus	sd	P. pectinatus forma	
Elodea canadensis	a	interruptus	o
*Sagittaria sagittifolia	a	Hippuris vulgaris (sub-	
*Oenanthe fluviatilis	a	merged)	o
Potamogeton pusillus	o	Ceratophyllum demersum	r

* Flowering.

Reedswamp, too, is well developed, and the following species occur in canalised portions with no current and a heavy deposit of fine mud:

Sparganium erectum[1]	d	Typha latifolia	f
Glyceria maxima[2]	d	Scirpus lacustris	f
Phalaris arundinacea	f	Alisma plantago-aquatica	f

Turbid rivers. When the water is turbid, cutting down the light intensity, and the substratum is unstable under flooding, the submerged vegetation may be almost absent, as in the River Tern in Shropshire. In such rivers there is however often a littoral reedswamp, which Butcher found in the Tern to consist of the following species:

Sparganium erectum	d	Myosotis scorpioides	f
Scirpus lacustris	sd	Typha latifolia	o
Sagittaria sagittifolia	f	Alisma plantago-aquatica	o
Butomus umbellatus	f	Rorippa amphibia	o

Riparian community. Parts of the river bank of the Thames (and of many other rivers) particularly in the concavities of bends subject to the scour of the stream in flood, commonly rise vertically 2–3 ft. (*c.* 60–90 cm.) above the average water level. This exposed wall, generally of clay and overlying alluvium, affords a station for a characteristic open community,

[1] Pl. 109. [2] Pl. 111.

which is transitory, since it is cut away at intervals by the river. The following species occur:

- Achillea ptarmica
- Barbarea vulgaris
- Brassica rapa
- Cochlearia armoracia
- Hypericum acutum (tetrapterum)
- Juncus articulatus (lamprocarpus)
- Lycopus europaeus
- Lysimachia nummularia
- Lythrum salicaria
- Polygonum amphibium*
- Potentilla reptans
- Scutellaria galericulata

* Terrestrial state with rhizomes extending into the water as the floating aquatic form.

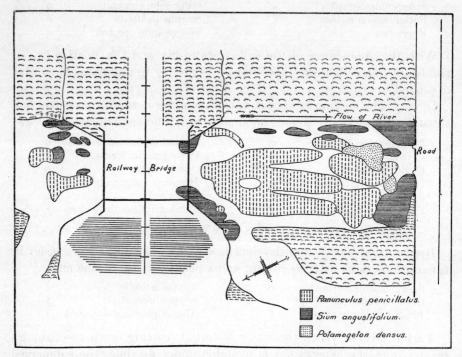

FIG. 123. EFFECT OF SHADE ON DIFFERENT SPECIES IN THE RIVER ITCHEN (HAMPSHIRE)

Sium erectum (angustifolium) grows farthest under the bridge, while *Ranunculus pseudo-fluitans (penicillatus)* diminishes in quantity as it approaches the bridge. From Butcher, 1927.

Shading effect. Differentiation of habitat by degrees of light intensity does not seem to be so conspicuous among aquatic as among terrestrial plants, but Fig. 123 illustrates the tolerance of shade by *Sium erectum* in contrast with other species.

Summary. Butcher summarises the river communities he has studied and of which examples have been given, as *torrential*, *non-silted*, *silted* and "*littoral*", the last-named being so-called because it approximates to the vegetation near the sides of ponds. The first is dominated by bryophytes, the second by *Ranunculus fluitans* (Pl. 110, phot. 272), *R. pseudo-fluitans* or *Myriophyllum spicatum*, while the dominants of the last two are much more varied (Pls. 109, 111).

As Butcher (1933) points out, the "ideal" river, which is richest in plant and animal life, is one which is silted, but not excessively, particularly not by silt which is preponderantly organic, since the layer of water immediately above thick organic silt becomes de-oxygenated, so that the fauna is neither varied nor plentiful.

REFERENCES

BAKER, H. Unpublished memoranda on aquatic vegetation of the Thames near Oxford.

BUTCHER, R. W. A preliminary account of the vegetation of the River Itchen. *J. Ecol.* **15**, 55–65. 1927.

BUTCHER, R. W., PENTELOW, F. T. K. and WOODLEY, J. W. A biological investigation of the River Lark. *Fisheries Investigation Series*, No. 3. Ministry of Agriculture and Fisheries, 1931.

BUTCHER, R. W. Studies on the ecology of rivers. I. On the distribution of macrophytic vegetation in the rivers of Britain. *J. Ecol.* **21**, 58–91. 1933.

MINNIKIN, R. C. *Practical River and Canal Engineering.* London, 1920.

WATSON, W. The bryophytes and lichens of fresh water. *J. Ecol.* **7**, 71–83. 1919.

Chapter XXXII

MARSH AND FEN VEGETATION

MARSH. THE ESTHWAITE "FENS"

Definitions. Words like *marsh, fen, bog* and *swamp* are rather loosely employed in ordinary language; and when words in common use are defined in a technical sense with the necessary limitations and exclusions, there is always some danger of misunderstanding. Nevertheless in the ordinary usage of these four terms there is a certain rough uniformity which corresponds well enough with the nature of the vegetation and soil, and invites an attempt to stabilise their meanings in an ecological sense.

In this book (as in *Types of British Vegetation*) the term *marsh* is applied to the "soil-vegetation type" in which the soil is waterlogged, the summer water level being close to, or conforming with, but not normally much above, the ground level, and in which the soil has an inorganic (mineral) basis; *fen* to a corresponding type (whose vegetation is closely similar) in which the soil is organic (peat) but is somewhat or decidedly alkaline, nearly neutral, or somewhat, but not extremely, acid. *Bog,* on the other hand (bearing a radically different vegetation), forms peat which is extremely acid. *Swamp* is used for the type in which the normal summer water level is above the soil surface: it is usually dominated by reeds (*Phragmites*) or by other tall grasses, sedges or rushes, often accompanied by dicotyledonous species of similar habit: the commonest kind of swamp is therefore called *reedswamp*.

In marsh, fen and swamp the water is telluric in origin, in bog it may or may not be.

Swamp or reed-swamp has already been described (Chapters XXIX–XXXI) as the last term of aquatic vegetation. It also forms a transition to the land vegetation of marsh or fen. When the soil level rises to or above the water level, marsh plants begin to colonise the swamp, and the reed-plants begin to give way to them, though certain species, notably the common reed (*Phragmites*), may maintain themselves in the marsh or fen and even spread within it by means of their strong horizontal rhizomes.

Marsh. Marsh commonly occurs on low river banks or on the shores of lakes and on the undrained flood plains of rivers, and its mineral soil often consists of alluvial silt; but marsh exists wherever mineral soil is waterlogged, irrespective of its origin. Where the relation of water level to soil level remains approximately stable, marsh (or fen) represents an edaphic climax. But if from any cause the soil surface is progressively built up above the water level (or the water table is lowered) so that the root systems of the plants are better aerated, the marsh (or fen) vegetation gives way to a more completely terrestrial type—grassland, scrub or forest—and ultimately to the climatic climax. This certainly happens quite normally as a continuation of the hydrosere wherever the soil is progressively and effectively raised, but the progress to climax forest is difficult to demonstrate in this country, because of the regulation of rivers checking the silting of alluvial plains, the draining of marshy ground disturbing the natural seres, the destruction of forest, and the resulting paucity of seed parents for the later stages of the succession, and especially for the climax dominants.

Marsh vegetation is commonly zoned round or along the edge of any permanent body of water, unless the bank is very steep so that the soil is not waterlogged at all. It is very various in detail, and the factors determining the different communities are not at all thoroughly understood.

Esthwaite marshes. Pearsall (1918) describes various zoned marsh communities on the shores of Esthwaite Lake with a substratum of rounded morainic stones, very little eroded or silted.

(1) On slightly eroded tracts of this morainic shore line are consocies of *Eleocharis palustris* and *Phalaris arundinacea*, accompanied by the purple loosestrife (*Lythrum salicaria*), extending below the summer water level and about 10–15 cm. above it. These are said to give way to *Phragmites* where peat accumulates or silting sets in; and all three communities may equally well be called reedswamp (cf. p. 585). The dominants and *Lythrum* all have tough rhizomes able to penetrate the hard substratum.

(2) Where pasture land abuts on the lake, and there is presumably softer soil, a mixed community occurs, extending about 10–30 cm. above the summer water-level, with no dominants and containing the following species:

Achillea ptarmica	f	Lysimachia nummularia	la
Caltha palustris	l	Lythrum salicaria	o
Carex diversicolor (flacca)	f	Myosotis caespitosa	f
C. flava	f	Polygonum hydropiper	o
C. goodenowii	o	Prunella vulgaris	f
Comarum palustre	l	Ranunculus flammula	f
Hydrocotyle vulgaris	f	Senecio aquaticus	l
Hypochaeris radicata	o		
Juncus articulatus	a	Climacium dendroides	l
J. effusus	f	Hypnum cuspidatum	r
J. conglomeratus	l	Rhacomitrium heterostichum	f
Littorella uniflora	o		

Apart from the bryophytes these are all genuine marsh plants except *Littorella*, which, as we have seen (Chapter xxx), is primarily a shallow-water aquatic, *Prunella*, which is a pasture plant, and *Hypochaeris*, a deep-rooted pasture or wayside species commonly found on drier ground.

(3) Where woods of *Quercus sessiliflora*, the climax forest of the hillsides of this region, approach the lake, the water's edge is fringed with a narrow

belt of alderwood with ash and grey sallow, and a ground vegetation which is partly marsh and partly woodland:

Alnus glutinosa	a	Juncus acutiflorus	ld
Fraxinus excelsior	f	J. articulatus	o
Salix atrocinerea	f	Lychnis flos-cuculi	f
		Lythrum salicaria	f
Achillea ptarmica	f	Meconopsis cambrica	r
Angelica sylvestris	la	Melandrium dioicum	f
Agrostis stolonifera	l	Molinia caerulea	ld
Brachypodium sylvaticum	la	Phalaris arundinacea	ld
Caltha palustris	f	Poterium officinale	o
Centaurea "nigra"	r	Ranunculus flammula	f
Circaea lutetiana	o	R. repens	o
Cirsium palustre	la	Rumex acetosa	l
Deschampsia caespitosa	la	R. obtusifolius	o
Epilobium montanum	o	Scrophularia nodosa	f
Filipendula ulmaria	f	Senecio aquaticus	o
Galium aparine	o	Succisa pratensis	a
Holcus lanatus	o	Valeriana officinalis	f
		V. sambucifolia	l

Phalaris and *Juncus acutiflorus* are locally dominant nearer the lake where the light intensity is 50 per cent, *Molinia*, *Filipendula* and *Deschampsia*, and finally *Circaea* and *Mercurialis perennis*, in the edge of the oakwood itself with a light intensity of 5 per cent.

(4) Behind the slope of unstable gravel, 1 m. wide, on the outward face of an open bank of morainic material which was formerly submerged and superficially eroded, but which now has a maximum elevation of 2·75 m. and a slope of 1 in 15, the following zoned communities occur:

(*a*) A line of *Carex hudsoni* (water level 7–14 cm. below the soil surface) which acts as a breakwater for the zones behind it:

(*b*)

Caltha palustris	Succisa pratensis
Lythrum salicaria	Valeriana sambucifolia

(*c*) A dense sward of

Carex panicea	a	Lythrum salicaria	f
C. stellulata	f	Molinia caerulea	f
Comarum palustre	f	Prunella vulgaris	o
Filipendula ulmaria	lf	Ranunculus flammula	a
Hydrocotyle vulgaris	f	R. repens	f
Juncus articulatus	la	Viola palustris	f
J. effusus	o		

(*d*) With accumulation of peat ("acid fen")

Molinia caerulea	Myrica gale

(*e*) A thicket of small trees and shrubs:

Alnus glutinosa	a	Fraxinus excelsior	l
Betula pubescens	a	Quercus sessiliflora	o
Corylus avellana	l	Salix atrocinerea	a
Frangula alnus	f		

The thicket is extending to the *Myrica-Molinia* zone, and as the water level falls (which it will do owing to the cutting back of the lake outlet) the successive zones will presumably move forward towards the lake. Here then we have acid fen succeeding marsh, but quickly overtaken by forest—Alnetum followed by Quercetum sessiliflorae.

Cornish marshes. In north Cornwall various types of alluvial marsh have been recognised by Magor (unpublished). The first two, situated near the sea over soil derived from Devonian slates, but not subject to flooding by salt or brackish water, may fairly be called reedswamp.

(1) The first is marked by the prevalence of tall rushes, grasses and dicotyledonous herbs, the dominants often forming almost pure local communities:

Epilobium hirsutum	co–d	Festuca elatior	f
Sparganium erectum	co–d	Filipendula ulmaria	f
Phalaris arundinacea	ld	Juncus articulatus	f
Juncus conglomeratus	va	Lotus uliginosus	f
Alopecurus geniculatus	a	Lycopus europaeus	f
Apium nodiflorum	a	Nasturtium officinale	f
Myosotis scorpioides (palustris)	a	Poa pratensis	f
Ranunculus acris	a	Potentilla anserina	f
Solanum dulcamara	a	Rumex conglomeratus	f
Caltha palustris	f	Veronica beccabunga	f
Calystegia sepium	f	Carex hirta	o
Carex remota	f	Juncus acutiflorus	o
C. vulpina	f	Oenanthe crocata	o
Equisetum fluviatile (limosum)	f	Orchis praetermissa	o
		Vicia cracca	o

(2) Another type, also dominated by similar tall plants but with moving water, has a distinctly different list of species, only about half being common to the two types: it is very local, occurring at Polzeath and Mennic Bay. Here the rare galingale (*Cyperus longus*) is co-dominant with *Sparganium erectum*:

Cyperus longus	co–d	Holcus lanatus	f
Sparganium erectum	co–d	Lotus uliginosus	f
Carex riparia	ld	Lychnis flos-cuculi	f
Galium palustre	va	Menyanthes trifoliata	f
Juncus inflexus (glaucus)	va	Oenanthe crocata	f
Mentha aquatica	va	Phalaris arundinacea	f
Polygonum persicaria	va	Poa pratensis	f
Epilobium hirsutum	a	Potentilla anserina	f
Equisetum fluviatile (limosum)	a	Pulicaria dysenterica	f
Myosotis scorpioides (palustris)	a	Ranunculus acris	f
		R. repens	f
Rumex conglomeratus	a	Rumex hydrolapathum	f
Iris pseudacorus	la	Senecio aquaticus	f
Apium nodiflorum	f	Solanum dulcamara	f
Bellis perennis	f	Plantago lanceolata	o
Caltha palustris var. minor	f	Scrophularia aquatica	o
Filipendula ulmaria	f	Plantago major	r
		Scirpus pauciflorus	r

(3) A third type is developed higher up the rivers and farther from the maritime influence. Here *Juncus articulatus* is dominant and woody plants (alders and sallows) have begun to colonise the marsh:

Juncus articulatus	d	Polygonum lapathifolium	f
Mentha aquatica	va	Potentilla procumbens	f
Cirsium palustre	a	Pulicaria dysenterica	f
Filipendula ulmaria	a	Rumex acetosa	f
Iris pseudacorus	a	Senecio aquaticus	f
Juncus effusus	a	Succisa pratensis	f
Lotus uliginosus	a	Veronica chamaedrys	f
Lychnis flos-cuculi	a	Athyrium filix-femina	o
Mentha aquatica × arvensis	a	Callitriche stagnalis	o
Poa pratensis	a	Galium palustre	o
Ranunculus acris	a	Hypericum acutum	
R. flammula	a	(tetrapterum)	o
Alnus glutinosa	la	Montia fontana	o
Sparganium erectum	la	Potentilla anserina	o
Epilobium hirsutum	la	Rubus sp.	o
Hydrocotyle vulgaris	f	Rumex conglomeratus	o
Juncus conglomeratus	f	Salix atrocinerea	o
Lythrum salicaria	f	S. aurita	o
Pedicularis sylvatica	f	Dryopteris spinulosa	o
Plantago lanceolata	f		

Of these thirty-eight species only fifteen occur in (2) and only nine in (1), eight species being common to all three types.

Magor comments on the different habitat requirements of the different species of rush: *Juncus conglomeratus*, a very abundant species in type (1), is wholly replaced by *J. inflexus* (*glaucus*) in the Cyperetum longi, and largely by *J. articulatus* with *J. effusus* in the riverside freshwater marsh farther from the sea, while yet other species of the genus occur in salt marsh and in peat bog.

Silting and peat formation. Although marsh has been defined as the wet soil-vegetation complex in which the soil is mainly mineral it must be understood, of course, that humus is constantly formed in marsh soil as a necessary result of the continual growth and decay of the vegetation. Under waterlogged conditions, indeed, this increase in humus will always tend to transform the soil from preponderantly mineral to preponderantly organic. We have seen (Chapter IV, p. 88) that "meadow soil", with impeded drainage, contains the largest proportion of humus among the terrestrial types. Increase in the proportion of humus can only be checked by an increased rate of disintegration following better aeration and concomitant increase in the population of soil organisms which carry out the disintegrative processes; or by the fresh accretion of mineral silt. On a river flood-plain periodically overspread by fresh silt we may thus have the soil continually raised, and a terrestrial succession initiated, leading to the climatic climax. In the earlier stages the soil surface is raised largely by the deposition of inorganic silt; in the later, when it is above the level of any but exceptional floods, mainly by the accumulation of plant debris, which, at such levels, is less waterlogged and consequently better aerated. In such cases marsh will form the section of the hydrosere between purely aquatic and purely terrestrial vegetation.

Fen. But where periodic mineral silting is a less important factor or is absent altogether the soil is mainly or entirely organic from the outset, and where the supply of basic ions is adequate the type of vegetation succeeding reedswamp is typical *fen*.[1] The organic soil is formed by the decay of the plant debris under relatively anaerobic conditions, and is therefore *peat*. It is however irrigated by water relatively rich in basic ions and often alkaline in reaction (see Chapter IV).

The East Anglian Fenland (Fig. 126, p. 148) lying between Cambridge and the Wash is the largest and best known area of fen in the country, and since its waters come mainly from calcareous rocks they are alkaline, and the fen vegetation (of which nothing more than fragments now remain in the East Anglian Fenland) is developed in a medium of relatively high pH value. The effect of this general alkaline "buffer" is mainly seen in the general absence of markedly oxyphilous species, which only occur locally where the peat is built up above the level of the alkaline ground water. The bulk of the fen vegetation consists of plants of waterlogged soil which are however in no way tied to alkaline conditions, but flourish equally well in marsh or fen with neutral or somewhat acid waters, though some "calcicolous" species are present. Smaller areas of fenland exist in many other parts of the country.

Esthwaite fens. The East Anglian fens are very little or not at all silted by river flooding, so that their soil is pure peat, but much vegetation which we cannot naturally separate from fens is silted to a greater or less degree. Thus Pearsall (1918) described fen at the N. end of Esthwaite Water (Figs. 124, 125), parts of which are heavily silted and whose waters are at the same time somewhat acid. Here we have waterlogged soils varying from mainly inorganic to mainly organic, i.e. transitions between marsh and fen according to our definition, but which, since their development and their vegetation are closely similar, are here treated with the fens.[2] The building up of such fens, while contributed to by inorganic silting, is largely brought about by the growth of peat. Only in the "reedgrass fen" does the in-

[1] Where the ground water is poor in basic ions and very acid in reaction, "bog" or "moss" is formed, characterised by quite a different type of vegetation and of peat (see Chapter XXXIV).

[2] Pearsall separates the apparently more or less static communities of waterlogged mineral soil surrounding the lake as "marsh" from those which are undergoing active development, which he calls "fens". It is usually better to make the distinction depend on the preponderance of mineral or inorganic material in the soil, irrespective of whether it is being raised by silting or not; but it must be recognised that no sharp line can be drawn between the two, which bear very similar vegetation so long as the supply of bases is adequate. The distinction is not in fact, so far as we know at present, of any great *vegetational* importance.

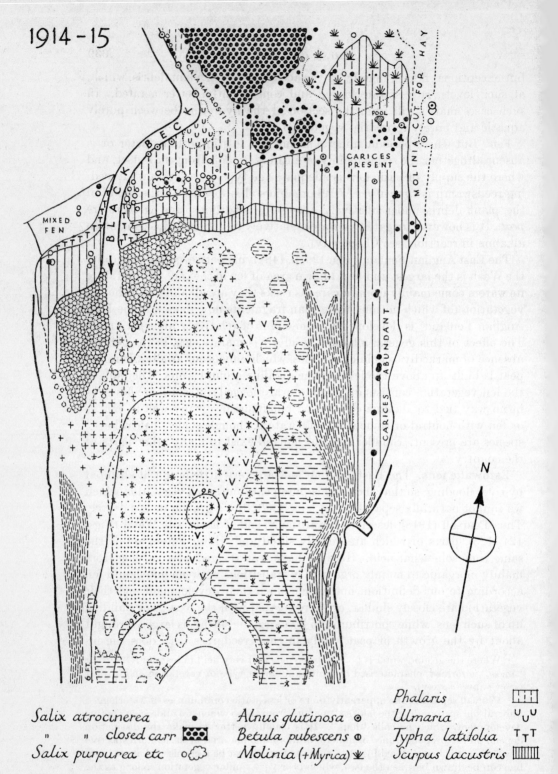

FIG. 124. NORTH FEN AT THE HEAD OF ESTHWAITE WATER, IN 1914–15

The aquatic communities occupying nearly three-quarters of the figure are described in connexion with Fig. 114 (p. 604) of which this is a reprint. The fen to the north is divisible into (1) a strip bordering the Black Beck, with rapid sedimentation, *Calamagrostis canescens*, *Phalaris arundinacea*, local *Filipendula ulmaria*, *Salix purpurea* and *S. decipiens*; (2) a central strip of moderate sedimentation with carr of *Salix atrocinerea*; and (3) an eastern strip of very slow sedimentation with *Molinia caerulea* and *Myrica gale*.

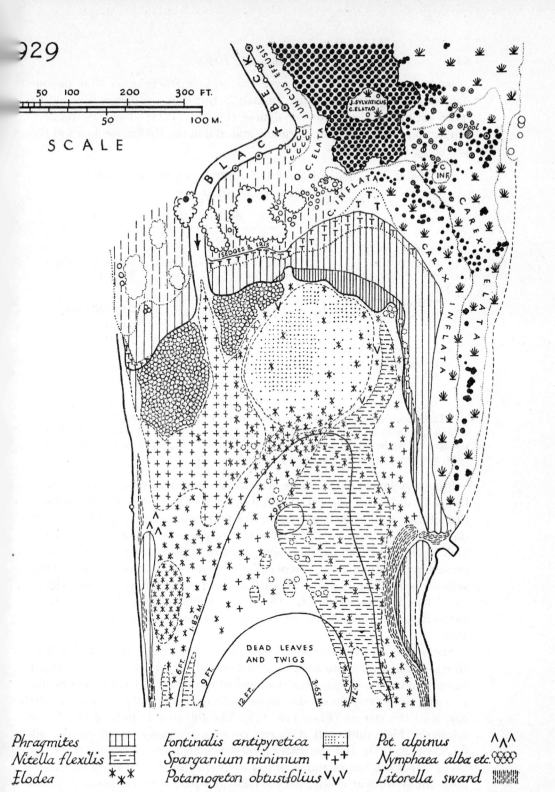

FIG. 125. NORTH FEN, ESTHWAITE WATER, IN 1929

Compare with Fig. 124. During the 15 years' interval between the two surveys the Carices have increased and invaded the central and western strips, the Salices have increased in all parts, and Molinietum has superseded Phragmitetum over a wide zone in the north-eastern and eastern parts. For symbols of fen species see Fig. 124.

organic largely preponderate over the organic fraction, so that on our definition it ought to be called "marsh". But such a separation would be quite artificial. The following figures (Pearsall 1918) show the relative proportions of organic and inorganic material in the Esthwaite fens and the preceding reedswamps:

Table XX. *Esthwaite fen soils: ratio of organic to inorganic constituents and pH values**

	Sedimentation		
	Rapid	Moderate	Slow
Reedswamp:			
Ratio O./I.	1·3 (*Typha*)	1·51 (*Phragmites*)	1·76 (*Phragmites*)
pH	6·4–6·0		5·5–5·4
Fen:			
"Mixed fen" Ratio ⎰1·06		?1·5–2·0	2·98 (*Molinia*)
"Reed grass" O./I. ⎱0·26			
pH	5·7–5·4		5·0–4·5
			(4·9–4·4 in 1936)
Carr (*Salix atrocinerea*):			
Ratio O./I.		2·2	
pH		Young 5·5–5·2	
		Middle 5·3–5·1	
		Old 5·0–4·8	

* The inorganic fraction necessarily includes "ash" derived from the plant debris, as well as inorganic silt. The pH values were taken in 1929. All the figures relate to the area shown in Figs. 124 and 125, sedimentation being rapid on the left near the beck, moderate in the middle, and slow on the right.

From the figures in this table, which are calculated from the averages of Pearsall's data, it is apparent that all the soils of the Esthwaite fens except that of the "reed grass" associes have a preponderance of organic material (in "mixed fen" the fractions are practically equal), that this is relatively high in the reedswamps, increasing with diminished sedimentation very greatly in the *Molinia* fen, where silting is at a minimum or absent and the organic is practically three times the inorganic fraction. In the richly silted fens it is much lower. All the pH values are on the acid side of neutrality, and there has been a general tendency for them to fall since 1929.

Succession on the Esthwaite fens. (1) *Area of rapid sedimentation*. The North Fen at Esthwaite is strongly influenced by the Black Beck, a rapid stream round whose exit into the head of the lake the fen has grown up, and which regularly floods and deposits silt upon the parts adjacent to the course of the stream (Figs. 124, 125). The soil in this part of the fen is relatively high and well drained, consisting largely of inorganic silt (O./I.=0·26, 1·06), and the water level depends on the rise and fall of the beck. When the bed of the beck has been cleared above the outfall much more silt is brought down into the lake, and part is also deposited on the adjacent fen during floods. A rapid development of the vegetation ensues.

Esthwaite Fens

(a) The reedswamp at the mouth of the beck is dominated by *Typha latifolia* (pH 6–6·4), this being the only part of the lake where the great reedmace occurs. *Phragmites* accompanies it, especially towards the open water, and the chief associates of the two dominants are *Caltha palustris*, *Menyanthes trifoliata*, *Ranunculus lingua* and *Scutellaria galericulata*.

(b) The "mixed fen" associes which follows this reedswamp in the succession has a black muddy soil with organic and inorganic fractions nearly equal and numerous abundant species, of which the following are typical.

Locally dominant or subdominant

Carex hudsonii (elata)	Menyanthes trifoliata
Galium palustre	Phalaris arundinacea
Iris pseudacorus	Typha latifolia

Abundant or locally abundant

Agrostis gigantea	Menyanthes trifoliata
Carex inflata	Myosotis scorpioides
Comarum palustre	Phragmites communis
Filipendula ulmaria	Ranunculus lingua
Lotus uliginosus	Salix atrocinerea subsp. aquatica
Lysimachia vulgaris	× S. decipiens
Lythrum salicaria	S. purpurea

(c) Adjoining the beck is a belt of what Pearsall calls the "reed grass associes" because relatively pure consocies of *Phalaris arundinacea* and *Calamagrostis canescens* are the conspicuous features, the former in the lower lying parts towards the lake and the latter in the older and drier parts of this zone of rapid silting. The soil is a grey-brown clayey silt with only about 20 per cent of humus. In summer the water level may be as much as a foot (30 cm.) below the surface. Besides the reed grasses, consocies of *Filipendula ulmaria* occur, as well as the following associates:

Caltha palustris	lf	× Salix decipiens	l
Galium palustre	lf	Scutellaria galericulata	o
Juncus effusus	o	Urtica dioica	o
Salix atrocinerea	l	Valeriana officinalis	l
S. purpurea	f		

Salis purpurea and ×*S. decipiens* are particularly characteristic of rapidly silted areas. A few alders fringe the stream. If developed extensively enough this type of habitat might ultimately produce woodland of alder and birch. It is probably only on well aerated gravelly soils that oakwood would eventually appear.

(2) *Area of moderate sedimentation.* This is the central vertical strip, containing "closed carr", in Figs. 124 and 125. (a) The reedswamp is here composed of *Scirpus lacustris* towards the open water and *Phragmites communis* inshore. The soil is peaty, and closer and tougher than in (1a). Herbaceous associates are sparse, the bog bean (*Menyanthes*) being the only common species.

(b) A zone in which *Carex hudsonii* (*elata*) or *C. inflata* is dominant succeeds the *Phragmites* reedswamp with but little change of soil, and the water level is close to the surface (5–10 cm.). The following species occur:

Carex hudsonii (elata)	d	Lythrum salicaria	f
C. inflata	l	Phalaris arundinacea	l
C. vesicaria	lsd	Phragmites communis	f
Comarum palustre	f	Scutellaria galericulata	f
Eriophorum angustifolium	r	Senecio aquaticus	l
Galium palustre	f	Typha latifolia	r
Hydrocotyle vulgaris	o		

The appearance of *Eriophorum angustifolium*, which is more commonly associated with acid bogs though not extremely oxyphilous, is of interest: it also occurs locally in the East Anglian fens.

(c) The next zone, called by Pearsall "open carr", is evidently a transitional zone in which woody plants are colonising the fen (Fig. 124). The soil is very variable in organic content and also in water level, which ranges from the surface to 15 cm. below it. Together with the woody plants there is also a great variety of herbaceous fen species, of which it is worth while to cite the full list given by Pearsall, for comparison with the lists from the East Anglian and Irish fens given on pp. 656, 664–6. It will be noted that a good many of the species, though by no means all, are the same:

Achillea ptarmica	Juncus articulatus
Agrostis stolonifera	J. effusus
A. tenuis	J. acutiflorus
Angelica sylvestris	Lathyrus pratensis
Anthoxanthum odoratum	Lotus uliginosus
Caltha palustris	Lychnis flos-cuculi
Cardamine pratensis	Lycopus europaeus
Carex canescens	Lysimachia vulgaris
C. goodenowii var. juncella	Lythrum salicaria
C. inflata	Mentha aquatica
C. hudsonii (elata)	M. arvensis
C. panicea	Menyanthes trifoliata
C. paniculata	Molinia caerulea
C. vesicaria	Orchis maculata
"Centaurea nigra"	Parnassia palustris
Cirsium palustre	Phalaris arundinacea
Comarum palustre	Phleum pratense
Crepis paludosa	Phragmites communis
Deschampsia caespitosa	Poterium officinale
Dryopteris spinulosa	Prunella vulgaris
Eleocharis palustris	Ranunculus repens
Epilobium parviflorum	Rumex acetosa
Equisetum fluviatile	R. crispus
Filipendula ulmaria	R. obtusifolius
Galeopsis tetrahit	Scrophularia nodosa
Galium palustre	Scutellaria galericulata
Holcus lanatus	Senecio aquaticus
Hydrocotyle vulgaris	Stachys palustris and var. canescens
Iris pseudacorus	Stellaria palustris

Succisa pratensis
Urtica dioica
Valeriana officinalis
V. sambucifolia
Veronica scutellata
Viola canina
V. palustris

Woody species

Alnus glutinosa
Betula pubescens
Rosa caesia (coriifolia)
Rubus idaeus
Salix atrocinerea
S. purpurea
Solanum dulcamara
Viburnum opulus

Of the herbaceous species *Filipendula ulmaria* and *Molinia* are the most abundant. Of the woody plants *Salix atrocinerea* is the most abundant species: drier, more inorganic soils bear *Salix purpurea*;[1] and *Betula pubescens* occurs where the organic content is very high.

(d) In the closed carr (Figs. 124, 125) the soil is a brown clayey mud, variable in the ratio of organic to inorganic constituents, but with a high average (2·2). The water level is also variable, frequently reaching the surface, and the drainage is very poor. *Salix atrocinerea* is dominant in close dense canopy, with *S. purpurea* and *S. aurita* rare, and *Alnus glutinosa*, *Betula pubescens* and *Frangula alnus* local. The light intensity beneath the sallows is low and obviously limits the ground vegetation, which is very sparse. In old carr the values range from 1 to 5 per cent, and there are practically no plants below the trees and shrubs. Where a scattered vegetation is present the light values are from 3 to 14 per cent. Where it rises to 70 per cent there is a dense vegetation of flowering *Filipendula ulmaria*, *Carex hudsonii* and *Lythrum salicaria*. Under the deeper shade of the carr, the Carices, *Phragmites* and *Iris* very seldom flower. The ground vegetation consists of the following species:

Agrostis gigantea	l	Iris pseudacorus	l
Caltha palustris	l	Lythrum salicaria	f
Carex hudsonii	la	Mentha arvensis	l
C. vesicaria	f	Molinia caerulea var.	
Comarum palustre	o	viridiflora	la
Filipendula ulmaria	la	Phragmites communis	la
Galium palustre	o	Valeriana sambucifolia	o

The mosses and liverworts are characteristic:

Amblystegium serpens	la	Plagiochila asplenioides	l
Fissidens taxifolius	f	Plagiothecium denticu-	
Hypnum patientiae	o	latum	f
Lophocolea bidentata	f	Pterygophyllum lucens	o
Mnium punctatum	f		

Between 1914 and 1929 the thickets of *Salix purpurea* and the closed carr of *S. atrocinerea* extended considerably (Figs. 124, 125), the latter filling up the interval between the main area of carr and the two outlying patches present in 1914.

[1] In 1929 *Salix atrocinerea* only occurred where the pH value was above 5 and *S. purpurea* where it was above 5·4.

No trace of regeneration is found in the carr, which appears to be transitory, the *Salix* bushes ultimately dying and disappearing. The transience of the grey sallow is in accord with what happens in the East Anglian fens. At Esthwaite, however, there is no evidence that any kind of woodland follows this stage. When the carr disappears *Molinia* is still present and it is probable that a bog with *Myrica gale*, *Erica tetralix*, etc., develops, as in (3) below. Here we encounter the influence of the cool wet climate which tends to produce bog as climax vegetation on all undrained soils with deficiency of oxygen.

(3) *Area of slow sedimentation.* This is the vertical strip on the right of Figs. 124 and 125.

(a) *Reedswamp.* As in (2a) this consists of an outer zone of *Scirpus lacustris* and an inner zone of *Phragmites communis*. The peaty soil has an organic/inorganic ratio of 1·76 and the peat is closer and less decayed. In 1914 *Phragmites* was still dominant for a distance of 45–50 m. beyond where the water level was at or just below the soil surface. In 1929, i.e. in the course of the 15 years since 1914, the outer edge of the *Phragmites* reedswamp advanced about 30 m. into the water and its inner portion was replaced by the Molinietum to a depth of about 100 metres (Figs. 124 and 125). Here then, this part of the succession is actually demonstrated as having occurred within a comparatively short term of years.

(b) *Molinietum.* As the *Phragmites* thins out on the inner edge of the reedswamp *Comarum palustre* becomes abundant, *Carex inflata* and *Lysimachia nummularia* frequent, and *Molinia caerulea* colonises the peat, gradually replacing *Phragmites* as the dominant when the water level is about 10 cm. below the peat surface. Here the organic/inorganic ratio is about 3. In striking contrast to the "mixed fen" and "open carr" of (2) the Molinietum is very poor in species, containing only the following:

Molinia caerulea	d	Myrica gale	lsd
Galium saxatile	f	Dryopteris dilatata	r
Potentilla erecta	l		
Succisa pratensis	l	Sphagnum acutifolium	l

Transition to bog. This Molinietum can barely be classed as fen, but is rather a stage in the development of bog. Later in the succession the *Molinia* dies out while *Myrica gale* persists. Thus under the conditions of this north-western climate, and with waters draining from mainly acidic rocks, the communities resembling those of "true fen" are only developed where there is abundant inorganic silting, and are very transitory.

The limits of fen. Though the alkaline fens of East Anglia have here been spoken of as "true fen", it is clear that no line based on soil reaction can be drawn between them and fens which are quite acid such as the "mixed fen" at Esthwaite (pH $c.$ 5·5). Many or most of the "true fen" species can and do flourish in peat of this degree of acidity. Probably more important is the base status of the peat, irrespective of soil reaction (cf. the exacting woodland flora of the acid clay-with-flints on the Chiltern plateau, Chapter XIX). What is wanted in order to understand the whole range of these fen communities is a close investigation of the conditions under which they can develop in different localities. As the ground water becomes poorer in bases and acidity increases a point seems to be reached at which an impoverished reedswamp of *Carex inflata* or *Equisetum fluviatile* (see Chapter XXX, p. 617) takes the place of *Phragmites* (itself able to tolerate a wide range of conditions in respect of acidity and base status). In poorer conditions still Sphagnetum probably follows directly on the acidic water vegetation. It is important to keep the terms *fen* and *moss* (or *bog*) for the two well-marked types of vegetation, whatever the particular combination of edaphic and climatic factors involved in producing them.

It is clear however that Molinietum is intermediate between luxuriant fen on the one hand, and bog or moss on the other (cf. Chapters XXXIII and XXXIV), and a re-investigation of the requirements of *Molinia*, its possible associates, and the range of conditions under which Molinietum develops is the immediate desideratum.

REFERENCES

MAGOR, E. W. Geographical distribution of vegetation in Cornwall: Camelford and Wadebridge district. (Unpublished.)

PEARSALL, W. H. The aquatic and marsh vegetation of Esthwaite Water. *J. Ecol.* 6, 53–74. 1918.

Supplementary information about the later development of the North Fen at Esthwaite has been kindly supplied by Dr Pearsall.

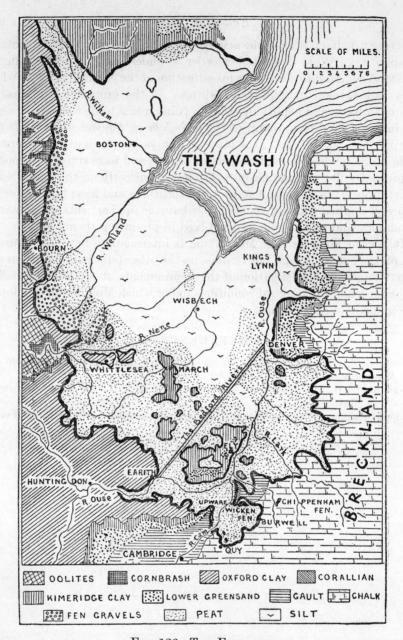

FIG. 126. THE FENLAND

The area of fenland is bounded by a thick black line, and consists of the "Marshland" (marine silt—unshaded) nearest the sea, and the "Fenland" in the narrower sense (peat—dotted) bordering the upland. The whole is drained and cultivated except a few small areas still bearing natural vegetation. Two of these, Wicken and Chippenham Fens, in the south are shown in black, the latter a "valley fen" outside the general peat area. The Cretaceous rocks (east and south-east) and the Jurassic rocks (south-west and west) bordering the Fenland and also forming "islands" in the southern part, are distinctively shaded, but the glacial deposits which cover much of their surface (Fig. 31, p. 108) are not shown. From Yapp, 1908.

Chapter XXXIII

THE EAST ANGLIAN FENS. NORTH IRISH FENS

THE EAST ANGLIAN FENS

The English "Fenland", *par excellence*, is the great tract of nearly flat peat and silt land about 1500 square miles in extent, scarcely raised above sea-level and occupying the northern portion of Cambridgeshire, with parts of the adjacent counties of Lincoln, Huntingdon, Norfolk and Suffolk (Fig. 126). It lies around the Wash, which is now the common estuary of the rivers Witham, Welland, Nen and Ouse. This great, very shallow basin, bounded to east and west respectively by Cretaceous and Jurassic rocks, was originally formed in pre-glacial times by an ancient river flowing northeast to join the confluent of the Rhine and the Thames in a plain which is now occupied by the North Sea. The whole region was invaded by icesheets during the Glacial Period, and since the withdrawal of the ice has suffered a depression of as much as 200 ft., broken by periods of stability or slight elevation, with corresponding transgressions and retreats of the sea. The result is that the floor of the Fenland consists of interdigitating wedges of marine silt and freshwater peat—the silt deposited by the invading tides, the peat built up by the fen plants which grew in marshes formed by the holding up of fresh water descending from the surrounding uplands. The surface soil is silt ("Marshland") towards the Wash, peat ("Fenland" in the narrower sense) towards the landward edge of the basin (see Fig. 126).

Practically the whole area is now cultivated. Drainage and reclamation began in Roman—perhaps in pre-Roman—times, but was not thoroughly carried out till the seventeenth century (see Chapter VII, pp. 186–7). The drained peat and silt (especially the latter) is very fertile, and is famous for its potatoes and locally for its fruit and its bulbs. Latterly, too, much sugar beet has been grown, since the relatively dry, sunny and somewhat continental climate is one of the most suitable in England for this crop. Before drainage the area must have been a waste of saltmarsh on the seaward, of fen and reedswamp on the landward side, the whole intersected by lagoons ("meres") and sluggish streams, tidal for considerable, though variable, distances from the sea. But when the land rose, or the peat grew up above the water level, the fen was extensively invaded by trees—oak, yew, pine and birch—the remains of which are still constantly being dug out of the peat. Recent evidence shows that raised bog also was formed in places towards the edges of the basin.

The few remaining fragments of peat fen bearing natural vegetation, which are all close to the edge of the basin, often owe their preservation to their use as "washes" or "catch-waters" into which flood water from the uplands could be turned when the cultivated areas were threatened with inundation. The surface of these surviving fens is at a higher level than that of the cultivated fenland, whose peat has wasted down, as the result of drying and ploughing, to a level markedly lower than that of the embanked

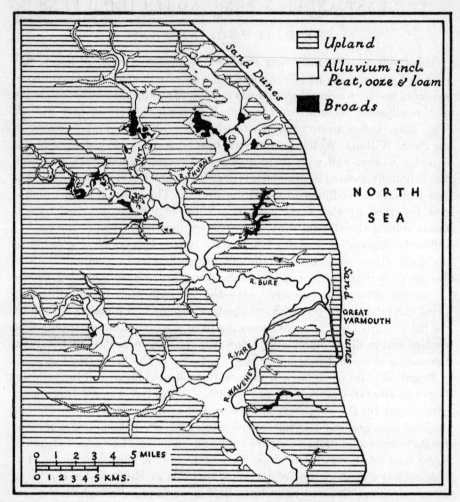

FIG. 127. THE FENLANDS OF EAST NORFOLK

The broads, shown in black, are confined to the upper parts of the old estuarine area where peat is formed. From Pallis, 1911.

rivers, into which the drainage water of the ditches ("lodes" or "dykes") is pumped—formerly by windmills, now by steam or oil (Diesel) engines. The whole drainage system is regulated by pumps and sluices. The present condition of the existing fragments of fen—of which Wicken, Chippenham, Woodwalton and Holme Fens are the largest and best known (the two first shown in Fig. 126)—is the result of human activity. The natural vegetation of Wicken Fen, for example, has been constantly cut, and since it has been occupied in strips by different owners, each of whom may have treated his strip according to a different regime varying from time to time, the fen shows a patchwork of different kinds of vegetation—dominated by *Cladium*, by *Molinia*, by mixed herbs, sedges and grasses, or by bushes. The disentangling of the detailed causes which have led to these varied results has not been an easy task, but prolonged observation and experiment, aided by comparison with fen regions elsewhere which have not been so closely under human control, have been successful in elucidating the

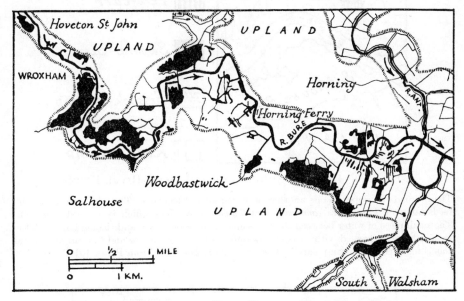

FIG. 128. PART OF THE BURE VALLEY (EAST NORFOLK)

The broads (black) lie just off the existing course of the river.

main outlines of the story together with many of the detailed mechanisms (Godwin *et al.*, 1929–1936).

East Norfolk fens. A fen region in which the vegetation has been in large part freer to develop on natural lines is found in East Norfolk. Here three rivers—the Bure, Yare and Waveney and their tributaries—flow through an alluvial plain, converging near Great Yarmouth, where their combined waters enter the North Sea (Fig. 127). This region of the Norfolk "Broads" —shallow lagoons which mostly lie surrounded by peat just off the courses of the rivers (Figs. 128–9)—though much smaller than the Fenland, is quite similar in its physical conditions, its soils and its natural vegetation; and there is the same contrast of peat in the upper parts of the river valleys with silt nearer the sea (Fig. 130). Before reclamation, indeed, the Fenland

[1] Fig. 129 and Pl. 88, phot. 213, show the situation of two broads to right and left of the River Bure from which they are separated by fen peat.

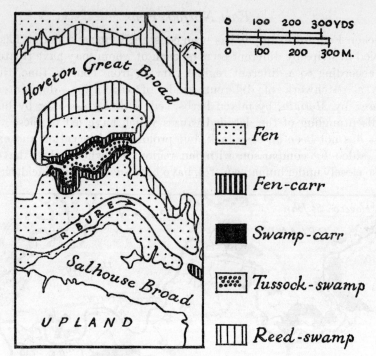

FIG. 129. HOVETON AND SALHOUSE BROADS (RIVER BURE)

These broads lie opposite one another on either side of the river. They are surrounded by reedswamp of *Phragmites communis* and *Typha angustifolia* which is succeeded by fen. The small piece of water between the fen island in Hoveton Broad and the fen bordering the river is however fringed with "tussock swamp" of *Carex paniculata* and *C. acutiformis*, and this is followed by "swamp carr" (cf. Pl. 88, phots. 214–15). From Pallis, 1911.

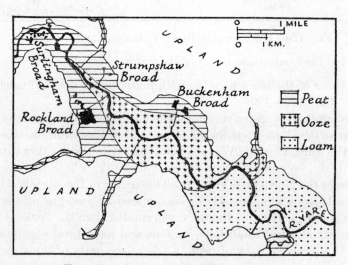

FIG. 130. PART OF THE YARE VALLEY

The peat, "ooze" and "loam" (silt) regions of "alluvium" are shown succeeding one another down the valley. The remains of the broads lie exclusively in the peat region (cf. Figs. 107, 108, pp. 593–4). From Pallis, 1911.

must have been much like parts of the Broads region to-day. The rivers of both areas drain calcareous soils so that their waters are alkaline, and in both the peat is correspondingly basic in reaction and bears an almost identical natural vegetation. But in East Norfolk the primary succession can be much more easily traced, and by its aid the different semi-natural plant communities seen at such a locality as Wicken Fen can be interpreted. The process of "Verlandung" has made considerable progress during the last century: several open broads of some size having "grown up" so that the open water is now restricted to quite small areas (Fig. 108, p. 594; Fig. 130).

Soil of Wicken Fen. The soil of Wicken Fen is a black, almost structureless peat. The water content of the surface peat (6 cm.) expressed as a percentage of the volume of the fresh sample is about 80, at greater depths often higher. The air content expressed in the same way is about 5 per cent, in one sample from "litter" (Molinietum) as high as 11 per cent.

The ratio of the weight of organic to inorganic substance in the surface peat (6 cm.) varies from 1·2 to 4·0: at greater depths similar values are generally obtained, but they may be higher, reaching as much as 11·5 in one case under carr, at a depth of 20 cm. Thus much higher ratios occur than in the silted Esthwaite fens (Chapter XXXII, p. 642).

The percentage of oxygen in the soil air at a depth of 20 cm. is about 18, of CO_2 about 4 at the end of the summer. These values are distinctly lower for oxygen and higher for CO_2 than at the same depth under grass turf growing on ordinary garden loam. The percentage of CO_2 dissolved in the fen water is much higher still, and of oxygen very low indeed, the highest values being less than 0·5 per cent of the atmospheric concentration. This suggests very clearly the lack of aeration that must be suffered by the root systems during flooding.

"Under former lacustrine conditions fresh-water shell marl was formed over much of the Cambridgeshire fenlands, and this bed, 5–10 cm. in thickness, is to be found about 30 or 40 cm. below the peat surface over large parts of Wicken Sedge Fen" (Godwin, Mobbs and Bharucha, 1932). The surface peat above this bed shows an alkaline reaction almost everywhere. The top centimetre or so is usually neutral or slightly acid: at somewhat greater depths the pH value varies from 7 to 8·3, with an average of 7·5.

Vegetation. In Chapter XXVIII a brief notice was given to the aquatic vegetation of the Broads, and the reedswamps were seen to consist generally of *Typha angustifolia* and *Phragmites communis*, with an open community of *Scirpus lacustris* on the side towards open water in Barton Broad on the Ant, and a closed swamp, passing to fen, of *Glyceria maxima* (*aquatica*) and *Phalaris arundinacea* in the Yare valley. These last dominants are probably associated with heavier silting and a higher pH value. The climax fen carr, dominated by alder, was considered in Chapter XXIII.

Cladietum. In the Bure valley and those of its tributaries the Ant and the Thurne the beginning of fen vegetation is marked by the incidence of the saw-sedge, *Cladium mariscus*, which colonises the very shallow water of closed reedswamp, and rapidly builds up the peat to the water surface. This also happens at Wicken Fen, where open water is absent except in the artificial and constantly cleared lodes and dykes. *Cladium* dominates the areas of very shallow water and maintains itself where it has formed peat just above the summer water level, i.e. in the lowest lying areas of fen. *Cladium* has long strap-shaped evergreen leaves, which may be as much as 3 m. long, with sharp serrated edges. These bend over at a height of about 1·5 m., so that the upper surface of the community is remarkably level. The leaves live for two or three years and when they die decay very slowly, remaining propped among the living leaves and forming a continuous elastic mass or "mattress". Owing to this habit *Cladium* excludes most other species, and at Wicken there are, within the Cladietum, only scattered individuals of tall yellow loosestrife (*Lysimachia vulgaris*), dwarf willow (*Salix repens* var. *fusca*), and *Phragmites*, which maintains itself for a long time in the fen by means of its vigorous and extensive underground rhizomes, pushing horizontally below the surface peat and sending up aerial shoots at intervals.

The Cladietum at Wicken Fen is invaded by bushes down to a level almost as low as the saw sedge itself extends,[1] the lower limit of bush growth being probably fixed by the depth of winter flooding (Godwin and Bharucha, 1932). Thus *Cladium* fen is almost immediately followed by carr (fen scrub or wood). All the other herbaceous fen communities at Wicken are the result of mowing or of removal of carr. But in East Norfolk, if the slope of the basin is slight and there is a deficiency of seed parents in the immediate neighbourhood the bush colonisation is slow and sparse, and the extent of Cladietum and of the other fen communities described by Pallis (1911) may be considerable and maintained for a long time before it is superseded by carr, though some of them are doubtless maintained by mowing as at Wicken. It may be presumed that the consocies of Glycerietum, Phalaretum *Glyceria maxima* and *Phalaris arundinacea* characteristic of Yare valley fen, where *Cladium* is said to be rare, are parts of a prisere corresponding with the Cladietum in the other East Norfolk valleys and at Wicken.

The following lists of typical species are given by Pallis (1911) for the East Norfolk fens:

Bure Valley Fen

Dominants:
  Cladium mariscus
  Juncus subnodulosus
  Phragmites communis

Local dominants:
  Molinia caerulea
  Carex lasiocarpa

Characteristic associated species:
  Liparis loeselii   r
  Pyrola rotundifolia   r
  Myrica gale   f

[1] Fig. 131.

Yare Valley Fen

Dominants:
  Glyceria maxima (forming consocies)
  Phalaris arundinacea (forming consocies)

Local dominant:
  Poa trivialis

Associated species:

Cladium mariscus	r	Myrica gale	o
Filipendula ulmaria	va	Thalictrum flavum	f
Galium palustre	va	Valeriana officinalis	va
Lychnis flos-cuculi	va		
Myosotis scorpioides (palustris)	va		

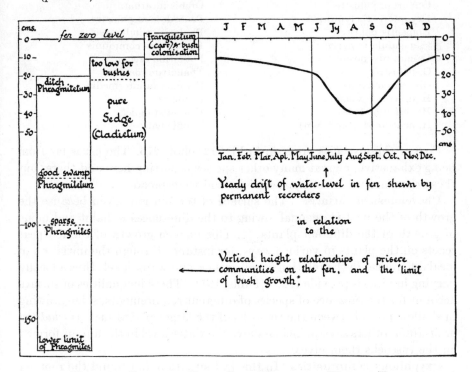

FIG. 131. LEVEL OF BUSH COLONISATION AT WICKEN FEN

The lowest level at which bushes can colonise the fen is about the winter water level, i.e. 30 cm. below the summer water level. From Godwin, 1936.

Certain oxyphilous plants have been omitted from these lists, because they are all quite local species which indicate the development of acidity and are referred to below, but the occurrence of *Pyrola* and the greater frequency of *Myrica* in the Bure valley fen already indicate a tendency to develop acidity. The woody species in Pallis's lists (other than the dwarf *Myrica*), which have also been omitted, are obviously the fore-runners of carr.

General list. The general list for all the East Norfolk fens includes:

Angelica sylvestris	Juncus subnodulosus
Caltha palustris	Lathyrus palustris
Calamagrostis canescens	Lychnis flos-cuculi
Carex disticha	Lysimachia vulgaris
C. hudsonii (elata)	Lythrum salicaria
C. flava	Menyanthes trifoliata
C. hornschuchiana (fulva)	Molinia caerulea
C. lasiocarpa	Myosotis scorpioides
C. panicea	Oenanthe fistulosa
C. paradoxa	O. lachenalii
Cladium mariscus	Ophioglossum vulgatum
Comarum palustre	Orchis incarnata
Dryopteris thelypteris	Peucedanum palustre
Epipactis palustris	Phalaris arundinacea
Filipendula ulmaria	Phragmites communis
Galium uliginosum	Potentilla erecta
G. palustre	Thalictrum flavum
Glyceria maxima	Utricularia intermedia
Hydrocotyle vulgaris	U. minor
Hypericum elodes	Valeriana dioica
H. acutum (tetrapterum)	V. officinalis

Five of these species are seen in Pl. 112, phot. 276. The list is far from being exhaustive, a great many other species occurring here and there, but those cited are the most characteristic and widespread.

Unevenness of surface. "The surface of the fen is uneven because the growth of the peat is unequal, owing to the differences in habit and mode of growth of the different plants.... This uneven growth of the fen peat reacts on the plants in various ways, for instance through the initiation of local differences in the relation of soil surface and water level; thus a locally varying habitat is provided..." (Pallis, 1911). These inequalities of surface account for the presence of species of different requirements, some growing in shallow pools between the tussocks of the larger plants such as *Cladium* or *Molinia*, others several inches above the water level in the humus formed on the tussocks themselves.

Oxyphilous communities. In this last situation and round the roots of colonising shrubs the natural acidity of the humus is not neutralised by the alkaline ground water, and here the Sphagna and other oxyphilous species such as sundew and cotton-grass occur locally (Pl. 112, phot. 278). The following are recorded by Pallis:

Drosera anglica	Sphagnum intermedium
Eriophorum angustifolium	(?=S. recurvum)
Sphagnum cymbifolium	S. squarrosum

Godwin and Turner (1933), from the immediate neighbourhood of Calthorpe Broad, whose water has a pH value between 7 and 8, recorded the following species growing on peat which showed pH values below 7, falling

PLATE 112

Phot. 276. Fen community: *Phragmites communis, Juncus subnodulosus, Dryopteris thelypteris, Epipactis palustris, Lysimachia vulgaris.* M. Pallis.

Phot. 277. "Mixed Fen." *Sium erectum, S. latifolium, Phragmites, Carex* sp., etc. C. G. P. Laidlaw.

Phot. 278. Oxyphilous society in fen: *Drosera anglica, Myrica gale,* with *Hydrocotyle, Carex,* etc. J. Massart.

FEN VEGETATION

PLATE 113

Phot. 279. Developing carr in the Norfolk Broads region: *Fraxinus excelsior*, *Betula pubescens*, *Salix atrocinerea*. The fen is dominated by *Phragmites*, with some *Juncus subnodulosus* and *Myrica gale*. M. Pallis.

Phot. 280. Young carr of *Frangula alnus* at Wicken Fen. Dead leaves of *Cladium* from the preceding Cladietum caught in the crotches. G. E. Briggs.

FEN CARR

to 5 (noted several times), while water squeezed from a *Sphagnum* tussock showed pH 4·2:

Polytrichum commune L.
Sphagnum fimbriatum Wils.
S. plumulosum Röll.
S. squarrosum Pers.
S. subsecundum Nees

These species, with the exception of *S. plumulosum*, are known from their habitats to have high mineral requirements. *Eriophorum angustifolium* also occurred (in one place a large area of nearly pure cotton-grass) in soil of pH values varying from about 7 down to as low as 4·3.

These local societies of Sphagna and associated species may well indicate the beginning of the tendency to develop raised bog on the basis of fen (see Chapter XXXIV, pp. 675–6), a tendency which in this region is overtaken and nullified by the colonisation of woody plants and the development of scrub and woodland.

On the edges of the fenland bordering the upland rather similar communities of oxyphilous plants may occur. One such, at Potter Heigham, mentioned by Pallis (1911) included the following:

Calluna vulgaris
Drosera rotundifolia
Erica tetralix
Eriophorum angustifolium
Potentilla erecta
Polytrichum commune
Sphagnum spp.
Aulacomnium palustre

These marginal oxyphilous communities may be due to the higher level of the peat built up over the edge of the upland, so that it is raised above the alkaline fen water, or alternatively to acid water drainage from neighbouring sandy upland soil; and they may correspond with the local tendency to form acidic bog on the edges of fenland, which seems to have found extensive expression in the past (p. 649).

Development of carr. Godwin and Bharucha (1932) have shown that the critical level for the successful establishment of woody plants is situated at about the height of the water table in winter (Fig. 131). Abundant shrub colonisation of the Cladietum at Wicken Fen can be seen at this level, and on old trenched ground where water stands in the furrows during summer this colonisation is confined to the intervening ridges.

Carr at Wicken Fen. The colonising woody species at Wicken are the alder buckthorn (*Frangula alnus*), which is far the most abundant, the grey sallow (*Salix atrocinerea*), the common buckthorn (*Rhamnus catharticus*) and the guelder rose (*Viburnum opulus*) (Figs. 132, 133). Privet (*Ligustrum vulgare*) is locally plentiful, while hawthorn (*Crataegus monogyna*) and blackthorn (*Prunus spinosa*) occur occasionally. Birch and alder have been planted here and there about the fen, and seedlings of birch are not uncommon. Curiously enough the alder, which is undoubtedly a natural dominant of carr in Norfolk, makes no headway at Wicken, though its pollen is abundant throughout the Wicken peat, and planted alders thrive and shed viable seeds. In a neighbouring fen there is some ash and *Populus canescens*, and oak seedlings are occasionally found.

Developing carr at Wicken is dominated by *Frangula alnus*, though why this shrub should be at first so much more abundant than *Rhamnus cath-*

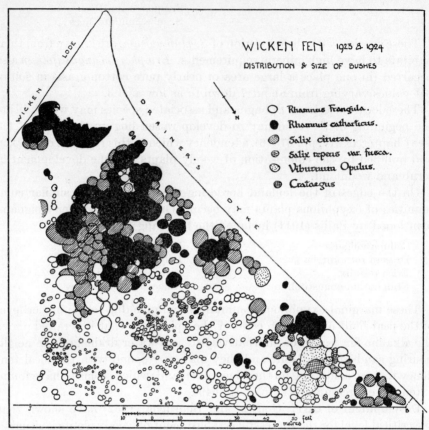

FIG. 132. BUSH COLONISATION IN A CORNER OF WICKEN FEN, 1923–24

An area of "mixed sedge" (*Cladium* and *Molinia*) colonised by bushes. To the south (upper part of the map) nearly closed carr of *Rhamnus catharticus* and *Salix atrocinerea* has become established. To the north (lower part of map) the sedge is being colonised by very numerous small bushes, mainly of *Frangula alnus*. All bushes more than 3 ft. high and 1 ft. in diameter are recorded. From Godwin, 1936. *Salix* "*cinerea*" = *S. atrocinerea*, "*Rhamnus Frangula*" = *Frangula alnus*.

articus is not clear. Abundance of *Salix atrocinerea* is characteristic of young carr, while older carr is commonly dominated by *Rhamnus catharticus* and *Viburnum opulus* (Figs. 132, 133). *Frangula* suffers from "die-back" caused by the fungi *Nectria cinnabarina* and *Fusarium* sp. and this is an important cause of its diminution and loss of dominance in adult carr (Godwin, 1936).

At first *Cladium* and other fen plants maintain themselves between the bushes, but as the canopy closes they are gradually suppressed. Dead

Development of Carr

Cladium may often be found below the shrubs of closed carr, the leaves sometimes caught up in the crotches of the branches (Pl. 113, phot. 280).

In time the ground becomes bare, and is then recolonised by a sparse but

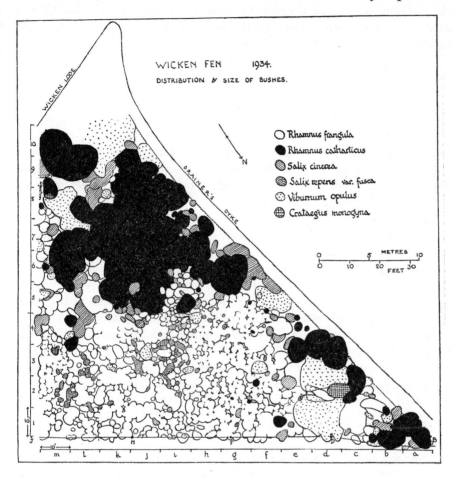

FIG. 133. PROGRESS OF BUSH COLONISATION IN 1934

During the 10 years' interval the clump of carr has extended and become quite closed, *Salix atrocinerea* giving way to *Rhamnus catharticus*. Small clumps of this dominant are established in the north corner. *Viburnum opulus* has also increased in this region and at the southern extremity of the plot. *Frangula alnus* has almost covered the main area of mixed sedge, though *Cladium* still survives between the bushes. From Godwin, 1936.

very characteristic vegetation consisting of the following species, none of which flowers under these conditions (Pl. 114, phot. 281):

Agrostis stolonifera	lva	Rubus caesius	va
Calystegia sepium	a	Symphytum officinale	f
Dryopteris thelypteris	va	Urtica dioica	f
Iris pseudacorus	a	Hypnum cuspidatum	a
Lysimachia vulgaris	a	Mnium affine	f

Carr in East Norfolk. The carr developed on the fens of East Norfolk is far more varied (Pl. 113, phot. 279), as is only to be expected from its much greater extent, and includes practically all the Wicken species, as well as alder. Some account of the alderwoods in the Broads district has already been given in Chapter XXIII, but the main facts may be repeated here to complete the story. Pallis (1911) gives the following list:

Developing carr

Alnus glutinosa	a	Quercus robur	o
Betula pubescens	a	Rhamnus catharticus	a
Frangula alnus	a	Ribes nigrum	o
Fraxinus excelsior	f	Salix atrocinerea	a
Ligustrum vulgare	o	Viburnum opulus	a

The undergrowth consists of the fen vegetation.

In the adult carr the same trees and shrubs occur, with alder typically dominant and a corresponding reduction of other species. *Betula pubescens* is sometimes dominant in fen carr.

The undergrowth characteristically includes the three shrubs *Ribes nigrum* (f), *R. rubrum* (f) and *R. grossularia* (o), the black and red currants and the gooseberry. It is sometimes said that these are bird-sown from cottage gardens, and such dispersal cannot of course be excluded, but from a comparison of fen and marsh woods as a whole there is every reason to suppose that these bushes are perfectly natural constituents of the carrs. Of scramblers and climbers the bittersweet (*Solanum dulcamara*) is characteristic, and the hop (*Humulus lupulus*) is fairly constant and may be very abundant.

The herb layer includes the following species:

Caltha palustris	f	Iris pseudacorus	a
Carex acutiformis	o	Osmunda regalis	r
C. paniculata	a	Urtica dioica	a
Dryopteris thelypteris	a	Mnium hornum	a
Filipendula ulmaria	f		

"Swamp carr." *"Swamp carr"*, also dominated by alder, is developed according to Pallis (1911) not on consolidated fen peat but on the tussocks or stools of reedswamp plants such as the large Carices (*C. paniculata, C. riparia, C. acutiformis*), on the edge of open water, and is a swampy wood often partially floating and with much open water between the peat islands (Pl. 88, phots. 214–15). It is commonly dominated by alder with *Salix atrocinerea* abundant and a ground vegetation of fen fern and the great Carices. The details of its origin and fate have not been sufficiently studied, but Godwin and Turner (1933) think that swamp carr may arise from young fen carr through depression of floating peat by the weight of the developing trees and shrubs (Chapter XXIII, p. 463) in the absence of consolidation by oak, etc.

There is some evidence (see Chapter XXIII) that alderwood may develop into oakwood and from the subfossil remains in the fen peat it seems that

PLATE 114

Phot. 281. Mature carr of *Rhamnus catharticus* at Wicken Fen. *Eupatorium cannabinum, Iris pseudacorus, Rubus caesius, Urtica dioica*: *Crataegus monogyna* on the extreme left. Cf. Pl. 90, phot. 220. *H. Godwin*.

Fen Carr

oakwoods have established themselves here in the past, possibly sometimes through an intermediate stage of birch and pinewood. But how far these various types of forest on peat are stages of the prisere and how far responses to fluctuations of climate is still uncertain. A drowned river basin so near the sea may never have allowed full expression to the development of later stages of the prisere.

Anthropogenic fen communities. Most of the existing fen, indeed all of it at Wicken and probably most in East Norfolk, is the result of human activity in mowing the sedge and clearing carr. At Wicken Fen Godwin and Tansley (1929) distinguished four such anthropogenic but semi-natural herbaceous fen communities, apart from the lodes and their artificial banks, and apart from the Cladietum which is a stage in the prisere: (1) "Mixed sedge" (Cladio-Molinietum), (2) "litter" (Molinietum), (3) the fen droves, and (4) "mixed fen" with an abundance of conspicuous flowers.

(1) *"Mixed sedge" (Cladio-Molinietum).* Cladietum, as we have seen, is in the first place a natural stage of the prisere following reedswamp and giving way in its turn to carr as the result of colonisation by bushes. But *Cladium* (the "sedge" *par excellence* of the fenmen) is useful for thatching or kindling and was at one time regularly mown for those purposes, as the *Phragmites* was for thatching. If Cladietum is mown once in three or four years the colonising bushes cannot make headway, but the *Cladium* survives and is maintained as a semi-natural community. Its vigour is however somewhat impaired, because *Cladium* is an evergreen plant and its vegetative capacity is reduced by the frequent removal of the living leaves. The plants consequently grow less luxuriantly, lose their exclusive dominance, and *Molinia* enters the community and comes to share dominance with the *Cladium*. *Phragmites*, less tall than in the reedswamp, is frequent, and a number of other species occur in local patches including:

Angelica sylvestris
Eupatorium cannabinum
Hydrocotyle vulgaris
Lysimachia vulgaris
Peucedanum palustre
Salix repens var. fusca

These are abundantly evident the first year after mowing, when they flower freely, but at the end of the 4-year interval they are hardly noticeable. The effect of the "coppicing cycle" on the field layer of coppiced oak-hazel wood (Chapter XIII, pp. 277–8) may be compared with this "mowing cycle". The mattress of dead *Cladium* leaves is still formed, and the peat of course continues to grow, raising the ground level of the mixed sedge several inches above the critical level at which bushes can first invade, and in fact above the soil level of many examples of mature carr. If cutting is stopped the community is quickly colonised by bushes.

(2) *"Litter" (Molinietum).* If the sedge or mixed sedge is cut every year the *Cladium* suffers so severely by the continuous removal of the living leaves that it cannot maintain itself, and the fen becomes dominated by *Molinia* which does not suffer from mowing, since only the dead leaves are

removed by the autumn cutting. In this Molinietum *Phragmites* is still frequent, while *Carex panicea* and *Juncus subnodulosus* become subdominant. Dicotyledonous plants are more abundant than in mixed sedge, and include:

    Angelica sylvestris   Succisa pratensis
    Cirsium anglicum   Thalictrum flavum
    Filipendula ulmaria   Valeriana dioica
    Hydrocotyle vulgaris

This community often occupies ground with the peat surface at precisely the same level as that of mixed sedge, and the difference is simply the result of the different treatment. On the cessation of mowing the ground is immediately colonised by bushes. The cut "litter" was used for cattle bedding and from this use the fenmen's name is taken.

(3) *The fen droves* (wide tracks kept clear to facilitate access to different parts of the fen) which are cut twice or thrice a year, are an extreme example of the effect of repeated cutting and traffic. Here there is quite a different flora, including many species of rush, sedge, grass and dicotyledonous plants, several of which are "dry land plants", not found elsewhere on the fen. Tall species are eliminated and many dwarf species are able to survive which do not exist in the communities of the fen proper.

(4) *"Mixed fen"* results from the clearance of carr. A host of plants, many of which have conspicuous flowers, is rapidly established with the access of full illumination. Such areas can be readily recognised by the presence of several relicts from the undergrowth of the carr. The following are among the species which occur:

   *Agrostis stolonifera   Lathyrus palustris
   Angelica sylvestris   Lysimachia vulgaris
   Calamagrostis epigejos   Lythrum salicaria
   C. canescens   Molinia caerulea
   Cirsium anglicum   Peucedanum palustre
   *Calystegia sepium   *Phalaris arundinacea
   *Dryopteris thelypteris   Phragmites communis
   Epilobium hirsutum   *Rubus caesius
   Eupatorium cannabinum   Sium erectum[1]
   *Filipendula ulmaria   S. latifolium[2]
   *Iris pseudacorus   Symphytum officinale
   Juncus subnodulosus   Thalictrum flavum

    * Species relict from the undergrowth of carr.

The "mixed fen", if left alone, rapidly reverts to carr, most rapidly of course if the stools of the cut carr shrubs have not been removed.

THE NORTH-EAST IRISH FENS

Around Lough Neagh in northern Ireland, but more especially round and to the south of its southern end lies an extensive area of fenland,

[1] Pl. 112, phot. 277 shows a small portion of "mixed fen" in which species of *Sium* are locally dominant.

probably the most extensive in the British Isles still remaining comparatively unspoiled. The water of Lough Neagh itself is approximately neutral in reaction, and in the drains cut through the fen peat it may have a pH value of 6–6.5, but is never more acid than that (Small, 1931). This relative alkalinity of the ground water seems to be the factor which determines the existence of fen in a region whose climate favours the production of acid bog and "moor". The rivers draining into Lough Neagh come partly from the basalts of Antrim, Derry, and Tyrone, and the basic material (largely calcium bicarbonate) which they bring, though much less than is dissolved from the chalk by the East Anglian streams, is sufficient to provide, as Small (1931) points out, an effective alkaline buffer, keeping the peat water from becoming at all markedly acid. This buffer action is the more efficient because of the deep winter flooding in these fens and the great reservoir of neutral water in Lough Neagh. The level of Lough Neagh itself normally rises a metre and sometimes nearly 2 m. when the rivers are in flood. The extreme range at Ballybay Bridge on the Ballybay River several miles south of Lough Neagh, recorded between November 1930 and October 1931, was about 120 cm. (White, 1932). About 50 cm. of water covered the communities of the lower fen during more than half the year 1930–1, though it was an abnormal year, since the usual winter level was never reached and unusual floods occurred during the summer. It seems probable that the zonation of the fen communities is largely determined in these fens by the winter water levels, as indeed it is at Wicken (Godwin, 1931), where the vertical range of both water level and plant communities is very much less, mainly because of the close regulation of the water level by the pumps and sluices.

The flora of the Lough Neagh fens is very much like that of the East Anglian fens, though a certain number of species, some of which are absent altogether from Ireland, are wanting: others, such as *Cladium* and *Rhamnus catharticus*, abundant in East Anglia, are local or rare.

Among biotic factors no mention is made of regular periodic mowing of the fens, as at Wicken, and there is no evidence that the cutting which is done has any decisive effect in altering the vegetation. Peat cutting is extensive, and the black fen peat (moulded into turves with mud) is of much better fuel quality than the yellow and brown bog and moor peats. Many of the "drains" which are met with are simply trenches dug for peat and have no effect on the general drainage.

Zonation. The zonation of the fen communities is very well marked and the different zones are dominated by a number of species forming local societies. The vertical range of most of these species is considerable, so that there is much overlapping, but their regions of dominance are rather sharply limited. The reedswamp, which occurs abundantly in "drains" showing open water throughout the year, as well as along the shores of the lakes, consists of the ordinary dominants *Scirpus lacustris*, *Phragmites* *communis* and *Typha latifolia*, as well as *Sparganium* and various other local dominants and associated species.

The vegetation of the fen itself appears to segregate into a great number of more or less zoned societies, and in this respect to differ from the East Anglian fens, which have fewer dominants. Miss White (1932) gives a list of the following fen societies, some of whose dominants are equally conspicuous in the reedswamps.

Lower fen. (1) Society of *Hippuris vulgaris*. This is often the initial society of the fen succession, growing as luxuriantly in the lower fen, where the water level is often an inch or two below the soil surface, as in the reedswamp. Associated may be *Bidens* spp., *Ranunculus* spp., *Alisma ranunculoides*, etc.

(2) Society of *Equisetum fluviatile* (*limosum*) and *E. palustre*. This also may be an initial society of the fen, or it may invade (1).

(3) The society of *Menyanthes trifoliata*, again, may initiate the fen succession, becoming established on the edge of a reedswamp (in which for example *Butomus* is dominant) or it may replace the *Equisetum* society.

(4) The *Lythrum salicaria* society occupies the next zone of dominance, and is frequently invaded by

(5) The society of *Lysimachia vulgaris*, with which *Lythrum* is also often co-dominant.

(6) A society of *Iris pseudacorus* may be initial, replacing reedswamp, or it may replace (1) and (2).

(7) *Phragmites* not only appears as a dominant of the reedswamp but also as a fen dominant, though it does not grow so tall in the latter situation. It ranges slightly higher than *Iris*.

(8) *Carex inflata*, too, with *C. hudsonii*, is often an important species in the reedswamp, occurring in the adjacent fen as a local dominant. It often colonises peat which has been cut to deep levels. *Cladium mariscus* and *Typha angustifolia* are confined to this zone.

(9) The society of *Comarum palustre* often invades and replaces the *Equisetum* society. It may be regarded as the uppermost society of the Lower Fen. The marsh cinquefoil ranges decidedly higher than the preceding species, extending as a subordinate species even into the Upper Fen.

Middle fen. None of the following societies is ever initial in the fen sere:

(10) The society of marsh marigold (*Caltha palustris*) frequently succeeds the marsh cinquefoil (9), or it may follow the iris (6) or the horsetails (2), but more often with the bog bean (3) intervening.

(11) The cuckoo-flower (*Cardamine pratensis*) society often follows (10), *Caltha* and *Cardamine* frequently maintaining co-dominance over the greater part of the zone they occupy. Again, *Cardamine* may succeed the marsh cinquefoil or the bog bean.

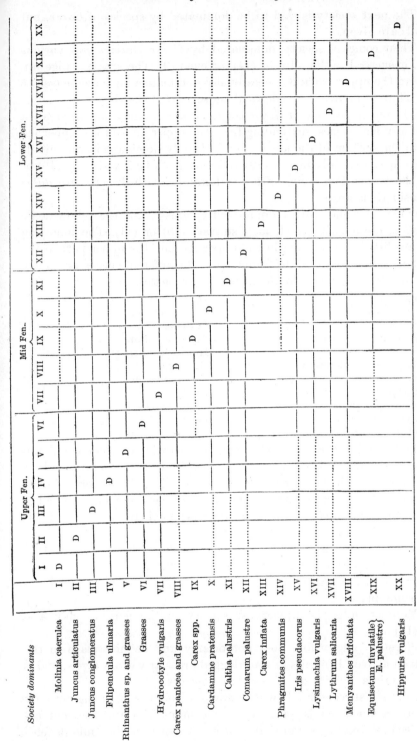

Fig. 134. Table of Fen Society Dominants in the Lough Neagh Fens (North-east Ireland)

Each of the dominant species is marked D opposite the number of the society it dominates. Its extension through other societies is shown by horizontal lines, continuous where the species is abundant, frequent or occasional, dotted where it is rare. From White, 1932.

(12) The next society is said to be dominated by species of *Carex*, but no details are given.

(13) Then comes a society dominated by *Carex panicea* and various grasses: it succeeds either (10) or (11), or even (9). The grasses range upwards into the upper fen and generally also dominate a higher zone (see 15 below). They form grassy meadows along several of the rivers.

(14) The marsh pennywort (*Hydrocotyle vulgaris*) society has the greatest range of any fen species—extending from the upper limit of the upper fen nearly to the bottom of the lower—but it is only dominant in the middle fen where it successfully invades (9) and may itself be invaded by (19) or (20) of the Upper Fen, being often co-dominant with *Juncus articulatus* over considerable areas.

Upper fen. (15) and (16) are dominated by grasses and the yellow rattle (*Rhinanthus crista-galli*) which may succeed (10) or (11). The following grasses are listed as occurring in the upper fen and the upper part of the middle fen, but no details of the frequency of the different species are given. Presumably several are co-dominant in these societies.

Agropyron repens	D. flexuosa
Agrostis canina	Festuca elatior
Aira praecox	F. ovina
Alopecurus geniculatus	Glyceria fluitans
A. pratensis	Holcus lanatus
Anthoxanthum odoratum	H. mollis
Briza media	Lolium perenne
Calamagrostis neglecta var. hookeri	Poa annua
Cynosurus cristatus	P. pratensis
Dactylis glomerata	P. trivialis
Deschampsia caespitosa	

Nothing is said of the effect of mowing or pasturing on these communities, nor of the relations of the different species in the vegetation. It is clear that the list includes moisture-loving species with others of drier situations, and some belonging to neutral with others of acid grassland.

(17) The meadowsweet (*Filipendula ulmaria*) dominates a society at this level and is said to succeed (12) and (14).

(18) The common rush (said to be *Juncus conglomeratus*) replaces (14) at several places, and in one instance is invading (9).

(19) *Juncus articulatus* may also invade (14) and the two are co-dominant in some places.

(20) Molinietum is the uppermost society of the fen. It may also invade (14).

Eriophoretum angustifolii does not occupy any definite place in the series given above. It occurs in waterlogged areas at various levels, and is said to contain a definite set of associated species.

Nature of the succession. It will be obvious from the foregoing not only that there is much overlapping between communities and frequent co-dominance in the ecotones, but that almost any society of the "middle fen" may invade and replace various societies (though usually the higher members) of the lower fen, and that the societies of the upper fen similarly succeed, not one another, but some society (very commonly the *Hydrocotyle* society) of the middle fen. In other words there is never anything like a complete linear series of the whole 20 societies on the ground, but very various shorter series consisting of perhaps half a dozen communities in any one local sere. Nevertheless Miss White insists that the dominant of each of the 20 societies has its own "niche" in the entire series. This is shown, together with the range of each dominant as an associated species, in Fig. 134 (p. 665), reproduced from her paper. The large number of these societies suggests that many differences of level have been brought about by peat cutting, and that this may account for the extreme fragmentation of the vegetation, in contrast to the greater extent of pure communities in the East Anglian fens.

Relation to "moor". The four uppermost societies (17–20) of the upper fen, taken together, "may be regarded as the climax fen community and as the pioneer moor community, because in each of these societies the substratum is more acidic than [in the] mid or lower fen societies. When *Sphagnum* becomes established in the fen there is soon a well-marked increase in the number of moorland species, and typical fen species become fewer in number." "This change in vegetation occurs when the level of the soil has been raised almost above the level of the winter floods. When the buffering of the flood waters ceases to control the acidity of the vegetation, the substratum changes to relative acidity" (White, 1932).

Vegetation of "ramparts". "Ramparts", i.e. ridges whose surface is above the general level, the peat between them having been cut away, are of frequent occurrence in many areas of the north Armagh fens. These are "really relicts of the moor formation, which, beyond doubt, was formerly much more extensive in this district than it is at present". They "serve as a place upon which to dry and stack the mud peat which is cut from the fen". "They are built up of yellowish brown peat composed of decayed *Sphagnum, Eriophorum* or *Calluna*...always acid in reaction." In the existing vegetation of the "ramparts" sometimes *Sphagnum*, sometimes *Calluna* is dominant, and accompanying these are sundews, rushes, *Erica tetralix, Narthecium ossifragum, Viola palustris, Molinia, Nardus*, and on well-drained summits *Blechnum spicant, Pteridium, Ulex europaeus* and *Galium saxatile*. Seedlings of fen plants are sometimes found on the "ramparts", having germinated in peat debris thrown up from neighbouring "drains", but they do not survive the summer. Inversely the seedlings of *Calluna, Erica tetralix*, etc., occasionally occur in the fen near the ramparts, but very rarely reach maturity. The small colonies of these species sometimes found in the fen are restricted to large sods of the "rampart" peat which have been isolated by cutting round and whose surface is well above the fen level.

Fen scrub or carr. The undershrubs *Myrica gale* and *Salix repens* occur here and there in the north Armagh fens, but their distribution is very limited and apparently sporadic. Where they are succeeded by taller bushes they do not persist, and are evidently unable to endure shading.

In carr proper *Salix atrocinerea* and *S. caprea* are most frequently the pioneers and continue as local dominants of the most advanced carr that is developed in the region. If we may judge from the analogy of the East Anglian fens this is probably evidence that the carr is young. The woody canopy never becomes dense enough to exclude the herbaceous fen vegetation, as it does in the East Anglian fens. It is not clear whether this is due to a difference of climate. "In very many places...it would appear that intensive grazing or the cutting of the fen vegetation is sufficient to keep the bush colonisation—the establishment of carr—at a standstill" (White, 1932, p. 278). But there seems no doubt that fen passes into an oxyphilous vegetation (*Sphagnum*, *Calluna*, etc.) far more frequently than in eastern England, and this difference is presumably correlated with the different climate.

The following woody species are recorded:

Salix atrocinerea and S. caprea	a, ld or co-d	
Alnus glutinosa la	Rhamnus catharticus	vr
Betula pubescens	Salix alba	
Crataegus monogyna	S. pentandra	
Fraxinus excelsior	S. purpurea	ld
Hedera helix	S. viminalis	ld
Ilex aquifolium	Ulex europaeus	la
Ligustrum vulgare r	Ulmus sp.	
Lonicera periclymenum	Viburnum opulus	r

The presence of these trees and shrubs is reported as without effect on the herbaceous fen vegetation, which is the same whether woody plants are present or not.

"Swamp carr" is formed in places on the shores of Lough Neagh by direct invasion of the reedswamp by the sallows: in other places it arises by extension of the adjacent fen carr over the reedswamp where this is passing into fen.

SUMMARY OF THE LATER HYDROSERE

By way of a link between the account of marsh and fen given in this chapter and the last and the following chapters dealing with bog or moss it will be convenient here to summarise the hydroseres from the point at which "land vegetation" begins, i.e. where the soil level is approximately the same as the summer water level. We have seen (p. 638) that it is theoretically possible for silting to continue (during floods) after Marsh this level has been attained, so that the hydrosere progresses through a stage of *marsh* (preponderantly mineral soil) to wet forest, and ultimately, by the accumulation of humus (which is then well aerated) to

climax forest. But no such successions have been described for the British Isles, few of whose alluvial plains are both ungrazed and also subject to silt-carrying floods; and thus there is little opportunity for this sequence to occur undisturbed by man. The commoner case is the development of *fen*
Fen at or above the summer water level, succeeded either by the invasion of woody plants (carr) or by the colonisation of *Sphagnum* and its associates.

A common fen pioneer and dominant in the calcareous basins of East Anglia is *Cladium mariscus*, and this is accompanied by a few other species of which the most abundant are *Carex panicea* and *Juncus subnodulosus*. Other fens are dominated by *Glyceria maxima*, *Phalaris arundinacea*, etc. In the small local calcareous fens of the south midlands, *J. subnodulosus* sometimes forms reedswamp and is succeeded by *Schoenus nigricans* as the first fen dominant. In other places *Phragmites communis* may dominate the fen as well as the reedswamp. In northern Ireland also the common reed may dominate the first stage of fen but the dominants of the "lower fen" are much more various, perhaps owing to the variety of water level and habitat introduced by extensive peat cutting.

It is only at Wicken Fen in East Anglia that exact observations have been made on the critical level for the colonisation of woody plants, which initiates the formation of *carr*. Godwin and Bharucha (1932)
Carr have shown that seedlings of *Frangula alnus* can establish themselves in Cladietum at about the average winter water level (i.e. about 30 cm. above the average summer water level), but not below this level, which suggests that the possible level of establishment is determined by winter flooding. Thus if we limit the term fen to the herbaceous peat communities whose soil surface is not below the average summer water level, it can develop here only over a vertical range of about 30 cm. (more at Lough Neagh). Most of the fen actually existing is above this level and is maintained by mowing, which prevents the development of woody vegetation.

At Wicken the most abundant woody colonist on fen is *Frangula alnus*, followed by *Salix atrocinerea* (which is relatively short lived), *Rhamnus catharticus* and *Viburnum opulus*—the two last usually dominating the tallest and oldest carr. In the more extensive natural fens and carrs of east Norfolk *Fraxinus excelsior*, *Betula pubescens* and *Alnus glutinosa* also enter into the structure of carr, both birch and alder, but especially alder, becoming dominant in the fen wood. At Calthorpe Broad two generations of *Quercus robur*, derived from planted parents, have established
Oakwood themselves freely in developing carr and can suppress *Alnus* (Godwin and Turner, 1933). It seems probable that oak forest would ultimately succeed fen wood, at least in East Anglia and probably over the greater part of England.

In the north-west, however, it is otherwise. Where a fen is heavily silted with inorganic silt, as on the North Fen at Esthwaite, "mixed fen" develops, often with *Calamagrostis* or *Phalaris arundinacea* dominant, and this is colonised by *Salix purpurea* and *S. fragilis*. *Alnus* follows, and probably *Betula pubescens* if the ground level is raised and sufficiently dry. Probably only on gravel would genuine woodland (Quercetum sessiliflorae) develop, and this would really be a marsh rather than a fen sere; but no examples have been described, though *zonations* of the kind are known (see Chapter XXXII, pp. 635–7). Where the silting is less and the proportion of organic material higher, *Molinia* enters the community (still containing many fen species), which is also colonised by *Salix atrocinerea*. The sallow is however transitory and the carr formed (Figs. 124, 125) does not regenerate. The soil is waterlogged, is freely colonised by *Sphagnum*, and passes into bog with *Molinia* and *Myrica gale*, *Erica tetralix* and some *Calluna*. Only when it is drained does *Betula* appear. Where sedimentation is very slight or absent the organic ratio is still higher, the reaction more acid and Molinietum immediately follows reedswamp. *Myrica gale* is characteristic of this Molinietum, which ultimately gives way to typical *raised bog*, with *Sphagnum* abundant or dominant, and the associated acid bog plants.

<small>Molinietum and transitory carr</small>

<small>Raised bog</small>

These are the conditions on flat undrained soil in the Lake District and apparently also in south-west Scotland. In north-east Ireland, also, woody vegetation does not follow the *Salix* carr, which never becomes closed, and there is a general tendency to the development of oxyphilous vegetation.

This is in contrast on the one hand to the East Anglian fens where it seems that oakwood would normally succeed carr in spite of the local tendency to develop *Sphagnum*, etc., and on the other to the blanket bogs of western Scotland and Ireland where fen and carr do not appear in the hydrosere at all, but blanket bog normally develops from the acidic aquatic vegetation of pools and on the general surface of undrained, flat or gently sloping ground. These conclusions are tentatively summarised in the accompanying diagram (Fig. 135).

<small>Blanket bog</small>

Correlation with climate. It is difficult to avoid the conclusion that these differences in the hydrosere are determined by the differences of climate—that the wetter and cooler the climate the greater the tendency to bog or moss development. In the East Anglian climate Sphagneta still appear locally and there is evidence that in some post-glacial periods raised bog was locally formed in this region, but oak forest would now seem to be the natural culmination of the sere. In the more oceanic climate of north-west England, south-west Scotland and north-east Ireland (which lie relatively close together) the climax on flat undrained ground is raised moss or bog, whether this develops from fen formed in a calcareous basin, as in the Armagh fens, or almost straight from aquatic vegetation (Lake District). Differences of inorganic silting or of the reaction of the ground waters suffice to modify the succession. Finally, in the extreme oceanic climate, moss or bog develops immediately following the aquatic stages of the hydrosere or directly on wet soil.

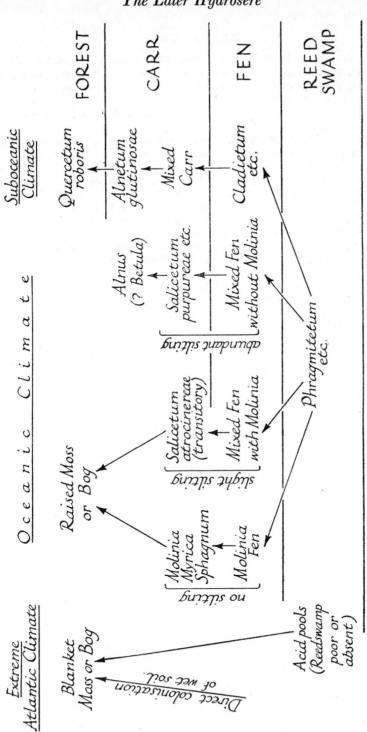

Fig. 135. Climatic Relations of Hydroseres

REFERENCES

DUFF, M. Ecology of the Moss Lane region. *Proc. Roy. Irish Acad.* **39**. 1930.
GODWIN, H. The "sedge" and "litter" of Wicken Fen. *J. Ecol.* **17**, 148–60. 1929.
GODWIN, H. Botany of Cambridgeshire in the *Victoria County History of Cambridgeshire*. Oxford, 1938.
GODWIN, H. Studies in the ecology of Wicken Fen. I. The ground water level of the fen. *J. Ecol.* **19**, 449–73. 1931.
GODWIN, H. and BHARUCHA, F. R. Studies in the ecology of Wicken Fen. II. The fen water-table and its control of plant communities. *J. Ecol.* **20**, 157–91. 1932.
GODWIN, H. and BHARUCHA, F. R. Studies in the ecology of Wicken Fen. III. The establishment and development of fen scrub. *J. Ecol.* **24**, 82–116. 1936.
GODWIN, H., MOBBS, R. H. and BHARUCHA, F. R. Soil factors in Wicken Sedge Fen. *The Natural History of Wicken Fen*, Part VI, 601–14. 1932.
GODWIN, H. and TANSLEY, A. G. The vegetation of Wicken Fen. *The Natural History of Wicken Fen*, Part V. 1929.
GODWIN, H., and TURNER, J. S. Soil acidity in relation to vegetational succession in Calthorpe Broad, Norfolk. *J. Ecol.* **21**, 235–62. 1933.
PALLIS, M. In *Types of British Vegetation*, Chapter x. 1911.
SMALL, J. The fenlands of Lough Neagh. *J. Ecol.* **19**, 383–8. 1931.
WHITE, J. M. The Fens of North Armagh. *Proc. Roy. Irish Acad.* **40**, 233–83. 1932.
YAPP, R. H. Wicken Fen [Sketches of vegetation at home and abroad, IV]. *New Phyt.* **7**, 61–81. 1908.

Chapter XXXIV

THE MOSS OR BOG FORMATION

TERMINOLOGY. VALLEY BOG, RAISED BOG, AND BLANKET BOG

BOG COMMUNITIES—(1) SPHAGNETUM

Terminology. The plant communities forming and continuing to grow upon constantly wet acid peat are almost everywhere called *mosses* in northern England and southern Scotland, and *bogs* in Ireland. In southern England, except on the high lands of the south-western peninsula, they are now almost confined to local depressions in sandy soil where the drainage is impeded, and there is no general common name for this type of vegetation.

The word *moor* (or *moorland*) applies in ordinary English speech primarily to (usually) high-lying country covered with heather and other Ericaceous dwarf shrubs, mainly Vaccinia; though it is often used more widely to refer to land bearing the whole series of oxyphilous communities and not regarded primarily as pastureland, from bog or moss to acidic grassland such as Nardetum or Molinietum. For example the elevated moorlands of the south-western peninsula, the largest areas of which are Dartmoor, Exmoor and Bodmin Moor, include both drier types on shallow sandy and peaty soil and wetter types on deep peat.

In German the word *Moor* applies to *any* area of *deep* peat, whether acid or alkaline, and the different types of peat were distinguished by Weber as *oligotrophic*, i.e. poor in nutritive (basic) salts, which we call in English "moor peat", "moss peat" or "bog peat", and *eutrophic*, i.e. rich in nutritive salts, which we call "fen peat". Under suitable climatic conditions oligotrophic peat may be built up on the top of eutrophic peat, i.e. fen (German *Niedermoor*) may be succeeded by moss or bog (German *Hochmoor*), often through a stage of *mesotrophic* peat, or as the Germans call it *Uebergangsmoor* ("transition moor") with vegetation intermediate between the two extreme types. The historical records of this change in the vegetation of a peat area may be preserved in the peat in the shape of remains of characteristic fen plants in the lower layers of peat, and of bog plants in the upper. Many examples of this succession have been described on the continent and some in England (see Fig. 47, p. 168 and Tables on pp. 692–5).

We cannot however use the English word "moor" as an ecological term in the German sense of an area covered with deep peat. According to the Oxford Dictionary moor means in modern English "a tract of unenclosed waste land" "usually covered with heather". The older meanings of "moor", which include waste marshland where peat might or might not be formed, are now obsolete. The root of the word is generally held by scholars to be connected with a word meaning "to die", and thus applied primarily to "dead" or barren, i.e. infertile, land. Its very wide original application is thus easily understood.

In fact the word is now generally used, as has been said, for *any* tract of unenclosed land (generally elevated) with acid peaty soil and not used *primarily* as pasture, so that it would not be applied to the bent-fescue hill pastures or to the limestone grasslands. Such "moorlands" may be wet or dry, and their soil may or may not be deep peat, but it is always acid and bears a related set of plant communities. These may be dominated by heather (*Calluna*), bilberry, etc. (*Vaccinium*), matgrass (*Nardus*), purple moorgrass (*Molinia*), cotton-grass (*Eriophorum*) or deer sedge (*Scirpus caespitosus*). In common speech the most typical "moors" are undoubtedly the "heather moors" or "grouse moors", dominated by *Calluna* and related vegetation and ordinarily used for the preservation and shooting of grouse (*Lagopus scoticus*). The soil may be of deep peat, but with a relatively dry surface, or the peat may be shallow and much mixed with mineral matter.

"Moss" or "bog". In the search for a suitable English word that can be used as a technical term for the natural group of *wet* peat-forming and peat-inhabiting communities we are thus limited to "moss" and "bog". Both have a strong claim founded on widespread common use for precisely the kind of soil-vegetation complex for which a designation is required. "Moss" has the additional advantage of linguistic correspondence with the Scandinavian words *Mosse* (Swedish), *Mose* (Danish and Norwegian) and the German *Moos*, which are applied to just the same vegetation. On the other hand it has the drawback—not perhaps very serious—of possible confusion with the taxonomic group of *Musci*, among which the Sphagna (the characteristic bog mosses) may be important, and are often primary, dominants or constituents of the formation in question. The disadvantage of the word "bog" is that it is sometimes loosely used in common language for *any* wet soil into which the foot sinks, but this again is not a very serious drawback.

In the present work, therefore, "moss" and "bog" are used synonymously for the wet acid peat vegetation; and "moor" and "moorland" are not used as technical terms, but only in the wide popular sense. The terms "low moor" and "high moor" (literal translations of the German *Niedermoor* and *Hochmoor*) are avoided altogether, both because "moor" cannot properly be used in English to include fen (*Niedermoor*) while it is always used for upland heath (which is neither *Niedermoor* nor *Hochmoor*), and because the adjectives inevitably carry misleading implications of altitude.

The plant communities which form and inhabit wet acid peat have often been divided into "lowland" and "upland", but they are more naturally classified as *valley bog*, *raised bog* and *blanket bog*—names which refer to real differences in habitat, structure and mode of development.[1]

[1] It is likely that with fuller knowledge additional types of bog will be recognised, but for the present the classification given will suffice.

Valley bog and Raised bog. Valley bog is developed where water, draining from relatively acidic rocks, stagnates in a flat bottomed valley or depression, so as to keep the soil constantly wet. In such situations species of *Sphagnum* and associated plants appear and produce a bog limited to the area of wet soil. Such bogs are common in the mountainous regions of Palaeogenic rocks in the north and west of the British Isles, but they have not been investigated ecologically, and very little is known of their structure and development. The "wet heath" communities (see pp. 734–41) occurring in depressions of the English lowland heaths, where the ground water is nearly level with the surface, are essentially of this type.

In a sufficiently moist climate the characteristic *raised bog* (German *Hochmoor*) may develop on the top of a valley bog. The valley bog itself, however, being fed by drainage water, is never so poor in soluble mineral constituents as a raised bog, and it contains plants, such as species of *Juncus* and *Carex*, which have no part in a typical raised bog. The stream running through a valley, however stagnant it may become in parts of its course and however acidic the rocks which it drains, brings a certain quantity of soluble salts from the upper reaches where erosion is taking place, and the water of a valley bog is typically less acid than that of a bog depending on precipitation alone. When a raised bog develops on the top of valley bog the stream is blocked or diverted by the growth of the bog vegetation and the peat which it forms. The original water course may thus be forced to one side, or split into two streams which find their way round the sides of the bog: part may run below the surface of the bog. The marginal watercourses form the *lagg* (a Swedish term) of raised bog, and the vegetation of the lagg is characteristically less extremely oxyphilous than that of the general surface of the raised bog.

Usually, however, raised bog is developed not on valley bog but on fen. As we saw in the last chapter, *Sphagnum* and associated plants, which require acid water in their habitat, locally colonise the surface of fen, especially where the large tussocks of certain fen plants have raised the level above the neutral or alkaline ground water, so that the natural acidity of the humus formed by the debris of the tussock is not neutralised. The small communities of oxyphilous plants arising in this way may remain very limited in extent, but there is abundant evidence in the British Isles, as well as on the continents of Europe and North America, that they have, in times past, spread over and superseded fen vegetation, replacing it by wide extents of the highly characteristic moss or bog formation. Where this formation has thus arisen on a fen localised in a basin it has itself remained restricted to the basin, though this may be of considerable extent, up to several miles in diameter. The surface of the moss is characteristically convex, sloping gently from the centre towards the periphery,[1] where it ends in a relatively steep bank bounded by a ditch or watercourse (lagg) repre-

---
[1] Pl. 115, phot. 283 a.

senting the original drainage channels of the fen basin and receiving the water draining from the bog. Beyond the lagg fen vegetation, not yet covered by the moss, may often be seen. Thus the moss as a whole is raised above the immediately surrounding fenland, and this is the origin of the German term *Hochmoor*. This is the type of bog so common in the great central limestone plain of Ireland,[1] where they are based on the local fen basins, and are known as "red bogs" from the red-brown colour of the dominant vegetation. Similar raised bogs occur in Scotland, northern England and Wales, and were formerly much commoner, many having been destroyed by draining and peat cuttings.

Blanket bog. It seems probable that raised bogs are formed in climates intermediate between that of the East Anglian fens, where the air seems too dry for the bog-moss vegetation to extend vigorously in dependence upon *atmospheric* moisture, and that of the west of Scotland and the west of Ireland, where the rainfall is high and the air so constantly moist that bog is the *climatic formation*, not necessarily arising in fen basins but covering the land continuously except on steep slopes and outcrops of rock. This is the third type of bog met with in the British Isles and may be called *blanket bog*, because it covers the whole land surface like a blanket.[2] The contrast can be well seen in the bogs developed east and west of Galway city in the west of Ireland. To the east is the great plain of Carboniferous Limestone with raised bogs developed in the numerous small fen basins. To the west is a region of acidic rocks under an extreme oceanic climate, and bearing almost continuous blanket bog. Blanket bog is independent of localised water supplies, depending on high rainfall and very high average atmospheric humidity. Raised bog, on the other hand, seems always to have an aquatic or semi-aquatic origin, being built upon fen, marsh or valley bog, so that the bog peat is often underlain by fen peat and lacustrine or estuarine silt. The climate in which raised bog is developed must, however, provide sufficient atmospheric humidity to make the upward growth of the bog possible.

Bog or moss communities. The dominants and associated species of blanket bog and raised bog are mostly, though not entirely, the same; but, as we have seen, the habitats (in the wide sense) are different, and the structure and development also differ in several respects. Blanket bog has never been studied by modern methods, but we know more about the development of raised bog, thanks to the labours of continental workers and to some quite recent work in the British Isles. The main plant communities of bog or moss are six—Sphagnetum, Rhynchosporetum, Schoenetum, Eriophoretum, Scirpetum and Molinietum—and of these Sphagnetum is the first and most fundamental.

(1) **Sphagnetum.** The species of the genus *Sphagnum*, as is well known,

[1] Pl. 115, phot. 283 and 283a, p. 686.　　　　[2] Pl. 121, p. 714.

have a highly specialised vegetative structure. The surface of the stem is covered with a layer of large empty cells, whose walls are strengthened by ribs and pierced by relatively wide holes; the leaves consist of a single layer of cells, the framework composed of cells similar to those covering the surface of the stem, and running between them are lines of narrow living cells containing chlorophyll. The network of fine capillary channels formed by the empty cells with their pierced walls results in the plants absorbing liquid water in contact with their shoots through the holes in the cell walls and holding it like a sponge, so that a considerable quantity can be easily squeezed out with the fingers from a living tuft of *Sphagnum*. Water is also held between the surfaces of the leaves and the stems. When this surface water has been lost by evaporation that contained in the cells is slowly evaporated into the air, the actual rate of evaporation depending of course primarily on the saturation deficit of water vapour in the air as well as on the structure of the plant.

Structure of Sphagnum

Variation of habitat and structure. Some species of *Sphagnum* are aquatic mosses, growing immersed in water: others live on constantly wet soil or more commonly on the wet bases of the aerial shoots of other plants. The habitat of these must be pretty constantly wet, either from soil water or abundant precipitation, and the "terrestrial" Sphagna are most abundant and flourish most luxuriantly under conditions of very high average atmospheric humidity. The species which live in the less constantly wet habitats, as Watson (1918) points out, possess xeromorphic characters in greater or less degree—compactness of habit, close imbrication of leaves on the branches, infolding of the leaf edges—characters which check loss of water by evaporation to the air. The more aquatic species are least xeromorphic and least able to resist desiccation.

Acidity and mineral requirements. The Sphagna are well known to flourish only where they are in contact with more or less acid water and to be killed by exposure to alkaline solutions. The acid substances held in their cell walls absorb the bases of nutrient salts, setting free the acid ions and thus maintaining an acid medium in contact with the moss. While this is true of all the species they vary a good deal in sensitiveness to an alkaline medium, and this is correlated with the degree of their own acidity (Skene, 1915). Broadly speaking the least acid species grow in the less acidic habitats, where they can usually obtain a greater supply of mineral salts, while the extremely acid species, most readily killed by alkaline solutions, have a very low mineral requirement and grow in the most acidic habitats. Thus Skene found that *Sphagnum contortum*, a species which, judged from its habitats, is of relatively high mineral requirements, had an average "primary acidity" (measured in terms of grams of acid hydrogen per hundred grams of *Sphagnum* material after thorough washing out of the absorbed bases) of 0·0715, while *S. rubellum*, a hummock-forming species of highly acidic raised bog, had an average primary acidity of 0·1092. This correlation is not however exact. Skene found that the acidity of different samples of the same species varied a good deal and that their ranges of acidity overlapped considerably; and there is evidence that what appears to be the same species may differ markedly in acidity and habitat in different regions. Further research is required on the relation of acidity and mineral requirements to habitat, and on the possible existence of different ecotypes of the same "species".

Oxidation-reduction potential. Pearsall (1938) has very recently found that the pH values in the Sphagnetum of raised bog, where the conditions were optimal for the growth of the bog moss (species not stated), varied between 4·17 and 4·62 at a depth of 10 cm., most of the determinations giving values about 4·2 or 4·3. These soils are "reducing". As the surface of the bog dries, as a result of drainage or otherwise, *Eriophorum* and ultimately *Calluna* becoming abundant, oxidation occurs and the acidity rises, pH values well below 4·0 being recorded. Thus while mor and acid peat have on the whole a low oxidation-reduction potential, *Sphagnum* peat shows an increase, on drying, with increase of acidity (cf. pp. 82, 84).

Habitats of various Sphagna.[1] Aquatic species of relatively high mineral requirements are *S. inundatum* and *S. platyphyllum*, and these grow in water which is less acid, e.g. in some fens. Typical aquatic species of low mineral requirements are *S. plumosum* and *S. cuspidatum*,[2] the latter the common dominant in wet hollows of raised bog in Ireland, Wales and Scotland as well as on the continent.

Species inhabiting situations of intermediate moisture are the following:

Relatively high mineral requirements:

S. amblyphyllum
S. contortum
S. squarrosum

S. subsecundum
S. teres
S. warnstorfii

Lower requirements:

S. angustifolium
S. cymbifolium (wide range of acidity)
S. papillosum[3]

Still lower:

S. apiculatum
S. balticum
S. tenellum (molluscum)

The following are species of the driest and most highly acid habitats, conspicuous in the formation of hummocks in raised bogs:

Less dry:

S. compactum
S. magellanicum (medium)
S. plumulosum (subnitens)

Drier:

S. acutifolium
S. fuscum
S. imbricatum
S. rubellum[3]

Role of Sphagna in vegetation. As has been said, certain species of the genus are the characteristic dominants and primary peat formers of the

---

[1] Kindly communicated by Prof. H. Osvald.   [2] Pl. 116, phot. 285, etc.

[3] *Sphagnum papillosum* and *S. rubellum* (Pl. 116, phot. 285) are the most important hummock formers in the British Isles: *S. tenellum* (*molluscum*) and *S. plumulosum* (*subnitens*) are also common species.

many mosses or bogs which cover wide areas of flat or gently sloping or undulating land in the cooler regions of the northern hemisphere, more especially the uneven morainic ground left by the Pleistocene ice-sheets. Starting with aquatic species in the innumerable pools or lakelets occupying the hollows, or with less hydrophytic species on valley bog, fen or forest vegetation in a sufficiently moist climate, the bog mosses may spread far and wide over surrounding vegetation, the more hydrophytic being succeeded by less hydrophytic species. In this way the moss bogs have destroyed and buried great areas of fen or forest more than once in post-glacial times, probably always as the result of change from a drier to a wetter climate. Owing to the peculiar structure of its tissues already described *Sphagnum* carries its own water with it as it grows upwards and outwards. Thus a pad or cushion of one of the less hydrophytic bog mosses forms an extending sheet saturated with water and with a convex upper surface, and the raised bog as a whole is an aggregate of such sheets, also with a slightly convex surface, since the centre of the bog represents its oldest and therefore its highest part.

The older, lower layers of the moss are cut off from light and air by the living surface layer, and progressively die. Compressed by the increasing superincumbent weight as the moss rises higher, and unable to decay completely owing to absence of free oxygen and of the normal action of soil bacteria, the lower layers of moss are converted into typical acid bog or moss peat, whose antiseptic properties are well known from the wonderfully complete preservation after many centuries of various objects, including the bodies of animals and men, that have been buried in the bog.

Such a raised bog is not however a simple mass of *Sphagnum*, but has a complex structure. In the first place it is composed of very numerous aggregated *Sphagnum* cushions or hummocks, in each of which progression from aquatic or subaquatic species at the base can be traced upwards to the more xeromorphic species at the summit of the cushion. And secondly it is inhabited by a number of species of oxyphilous vascular plants whose remains form part, sometimes the greater part, of the peat. The shape and growth of the individual cushions reflect *in petto* those of the bog as a whole. The structure of a raised bog will be described in more detail in connexion with the Irish raised bogs (pp. 686–96).

Sphagnetum in Great Britain. A great deal of *Sphagnum* peat has been formed in the British Isles during the wetter climatic epochs of the post-glacial period. But in the lowlands of England Sphagneta are now mainly met with in sandy heath areas where the drainage is impeded so that acid soil water accumulates. These "wet heath" Sphagneta with their associated species of flowering plants are really "valley bogs" of limited extent, and show a floristic composition quite similar to, though not identical with, those of raised and blanket bogs. Other species of Sphagna occur in certain wet woods and on certain fens under appropriate conditions. On the uplands Sphagnetum does not now cover any large areas of Great Britain. In many upland regions peat is now being actively eroded, though in the wettest climates, and locally where the necessary edaphic conditions are realised, peat is still being formed.

Sphagnum bogs of wet heath. The *Sphagnum* bog occupying the lowest portion of the "wet heath" described by Watson from Chard Common in Somerset (see Chapter XXXVI, pp. 740–1) is an example of the type of local lowland Sphagnetum (essentially a valley bog) referred to above, and similar bogs may be found in many parts of the country, though by far the greater number have been drained and destroyed. These of course are definitely local bogs determined by low lying ground with acid soil water overlying impermeable strata, or where the drainage is otherwise impeded. More extensive valley bogs have been described by Rankin (1911, pp. 259–64) from the New Forest in Hampshire. Here the low plateau of permeable Eocene beds has been dissected by small streams which have been cut down to the base level of erosion, so that their waters are extremely sluggish and in some places form chains of stagnant pools. Along the course of the stream reedswamps and alder thickets are developed, and outside these are Molinieta (see p. 523), but beyond the Molinietum, and abutting on the heath which covers the plateau, are Sphagneta. Besides these valley bogs or "valley moors", as Rankin calls them, he describes "spring moors" (better called "spring bogs"), also dominated by *Sphagnum*, which form round the springs of acid water that issue from the sides of the valleys along the lines of junction of permeable and impermeable strata. These valley and spring bogs tend to encroach on the adjacent heaths in the same way as typical raised bogs encroach on the fenland on which they are founded, and may in fact develop into raised bogs.

The following species are given by Rankin as composing the flora of the "spring" and "valley" bogs of the New Forest:

Sphagnum spp. (dominant, forming a matrix)

Eriophorum angustifolium a   Juncus acutiflorus   a
Eleocharis multicaulis a

The rhizomes and roots of these three species increase the firmness of the bog, and in the web so formed the following plants occur:

Drosera intermedia   Narthecium ossifragum
D. rotundifolia   Pinguicula lusitanica
Malaxis paludosa

A narrow edge zone overlapping the adjacent heath includes:

Drosera intermedia   Rhynchospora alba
Lycopodium inundatum   R. fusca

Calluna is said to colonise the drier bogs abundantly.

Raised bogs. Apart from these examples no typical raised bog has been

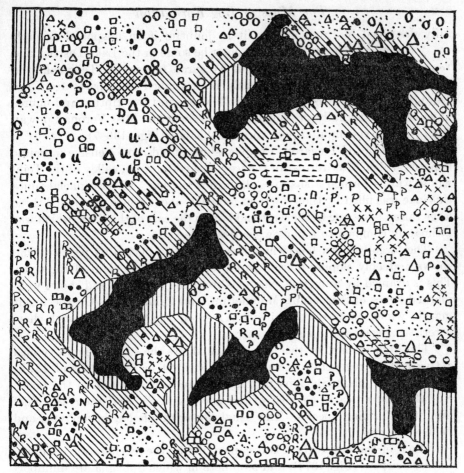

||||| *Sphagnum cuspidatum*
\\\\ *S. papillosum*
≡ *S. tenellum*
P *S. pulchrum*
✵ *Hypnum cupressiforme*

⋮⋮ *Cladonia sylvatica*
u *C. uncialis*
x *Rhacomitrium lanuginosum*
R *Rhynchospora alba*
N *Narthecium ossifragum*

D *Drosera rotundifolia*
• *Erica tetralix*
▲ *Eriophorum vaginatum*
△ *E. angustifolium*
□ *Scirpus caespitosus*
o *Calluna vulgaris.*

■ Open water or bare mud with or without *Zygogonium ericetorum*

FIG. 136. QUADRAT (5 m. square) OF "REGENERATION COMPLEX" (ACTIVE DEVELOPMENT) IN TREGARON BOG (CARDIGANSHIRE)

The hollows contain open water or are lined with *Sphagnum cuspidatum*. A little higher comes *S. pulchrum*. *S. papillosum* is the chief hummock-forming species, sometimes with *S. tenellum* capping the lower hummocks. *Hypnum cupressiforme*, *Rhacomitrium lanuginosum* and *Cladonia sylvatica* also occur on the hummocks. Of flowering plants *Rhynchospora alba* occupies the lowest and wettest positions, *Calluna* and *Scirpus caespitosus* the highest and driest. Godwin and Conway, unpublished.

fully described in England, and it is doubtful if an unspoiled example now exists, though there may be some in the Lake District and neighbouring regions, but Tregaron bog in the Teifi valley in central Wales has recently been studied by a party from the Cambridge Botany School, and the following short description and figures have been most kindly provided by Dr Godwin and Miss Verona Conway from their as yet unpublished work on this bog (Figs. 136–7).

Tregaron bog

"The village of Tregaron, Cardiganshire, stands on extensive morainic deposits which, after retreat of the ice, blocked the wide valley of the Teifi, whose waters formed a shallow lake below the bar. In post-glacial times this lake was filled by the growth of fen peat, above which developed three large raised bogs, one to the west and two to the east of the river, which was ultimately confined to a narrow central valley.

"The large western bog, which has been least affected by peat cutting, shows a steeply sloping margin (*rand*) along the flood plain of the river. This is drained by numerous channels from the bog itself and shows well-marked zones of Scirpetum, Molinietum, and Callunetum. On the uncut margin which abuts on the hillside is a small *lagg* (see p. 675), with characteristic species and a thin fen carr.

"The highest part of the bog surface shows the structure of a typical active bog or 'regeneration complex' illustrated in Figs. 136 and 137A. The most abundant Sphagna in this area are *S. cuspidatum* in the pools, *S. pulchrum*, an early colonist of the pools, and *S. papillosum*, the chief hummock-forming species. The usual flowering plants, bryophytes and lichens are present.

"Much of the bog surface is occupied, not by the regeneration complex, but by Scirpeta (Fig. 137B) and Molinieta in which Sphagna are unimportant and peat formation probably slow. The relation of these communities to the regeneration complex is not yet certain, nor is it known how far they reflect climatic or human influences."

Small bog at Loch Maree. The only available description of a Scottish raised bog is that of a small but very perfect example completely enclosed by native pinewood on the south-western side of Loch Maree in Ross-shire, at about 57° 38′ N. lat. and 5° 24′ W. long., not far above the shore of the lake. This is developed on a terrace of the hillside, and is only about 150 m. long by 70 m. wide and oval-oblong in shape (Fig. 138). It appears to be quite untouched and shows the typical features of such a bog very well indeed. The surface is slightly convex. (1) The lagg or drainage channel, initiated by a stream which descends the hill, is on one of the long sides of the bog against the steep hill which drains into it, and just above the lagg at the base of the steep slope are massive cushions of Sphagnum tailing out above into the vegetation of the pinewood. Within the lagg there is (2) a marginal zone only a few metres wide dominated by *Calluna*. This encloses the main area of the bog which contains many pools and consists of a more

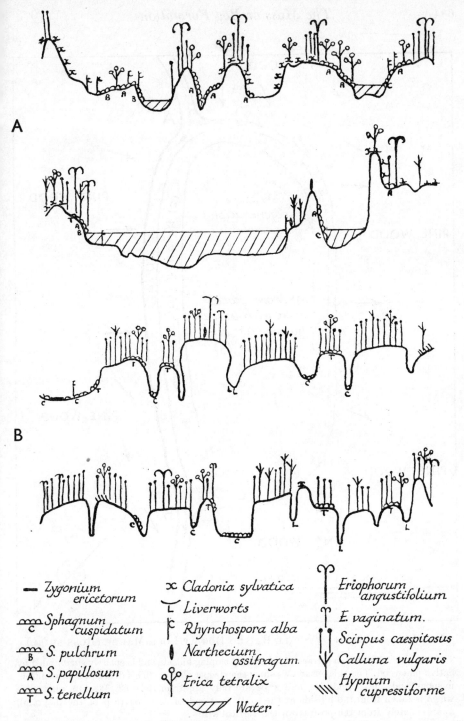

FIG. 137. PROFILES OF (A) REGENERATION COMPLEX, AND (B) SCIRPETUM CAESPITOSI IN TREGARON BOG

The hummocks and hollows are shown, with *Cladonia* usually on the highest hummocks in A. In B there is no open water and the Sphagna are much less prominent. Each profile is 5 m. long. Vertical scale 5 times the horizontal. Symbols not drawn to scale. Godwin and Conway, unpublished.

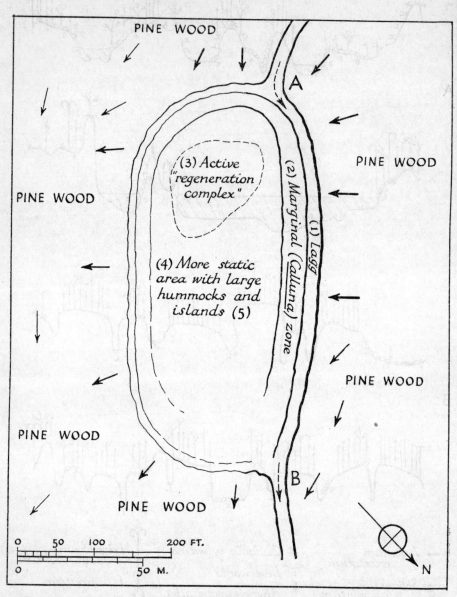

Fig. 138. Diagram of Small Raised Bog at Loch Maree (Ross-shire)

The bog is developed on a terrace of the steeply sloping hillside and is entirely surrounded by native pinewood. The thicker arrows show the steeper slopes. (This diagram is drawn from memory and the boundaries and scale are only approximate.) Cf. Pl. 115, phot. 282, which is taken from the hillside at the south-west end of the bog (top of the diagram), approximately from the position of the vertical arrow.

or less continuous carpet of *Sphagnum*. Part of this (3) appears to be actively growing (regeneration complex), the remainder (4) has reached a more static condition. It shows scattered raised "islands" (5), most of which are *Sphagnum* or *Rhacomitrium* hummocks, while the largest are rocky outcrops rising through the surface of the bog. Some of these bear dwarf pines.

The following species were recorded:

(1) *Lagg* (in and near the stream):

Molinia caerulea	d		
Sphagnum papillosum	cd	Calluna vulgaris*	f
S. recurvum	cd	Carex panicea	f
S. plumulosum	f	Potentilla erecta	f
Juncus effusus	o–la	Erica tetralix	o

* Not flowering freely

(2) *Marginal* Calluna *zone*:

Calluna vulgaris*	d	Potentilla erecta	f
Molinia caerulea	a	Drosera rotundifolia	o
		Sphagnum papillosum	f

* Flowering freely.

(3) *Regeneration complex*:

In pools and hollows		Forming hummocks	
Sphagnum cuspidatum var. submersum	a	Sphagnum papillosum	a
S. inundatum	r	S. plumulosum	f–a
Rhynchospora alba	a	S. magellanicum (medium)	o
Drosera anglica	a	S. rubellum (summits)	f
Eriophorum angustifolium	f–a		
Carex pulicaris	o		

Sphagnum tenellum (*molluscum*), *Rhacomitrium lanuginosum* and *Hypnum cupressiforme* were scattered on the hummocks.

(4) *More static area.* This contained the following additional species:

Eriophorum vaginatum	a–ld	Erica tetralix	f
Narthecium ossifragum	a	Pedicularis sylvatica	o–lf
Molinia caerulea	f–la	Juncus squarrosus	o
Potentilla erecta	f–la	Scirpus caespitosus	r–o
Drosera rotundifolia	f	Carex panicea	o
Calluna vulgaris	f	C. stellulata	r

Seedlings of *Pinus sylvestris* were abundant.

(5) "*Islands*" (rock outcrops):

Calluna vulgaris	a–d	Hypnum schreberi	f
Molinia caerulea	f	Dicranum scoparium	o
Pteridium aquilinum	f	Hylocomium loreum	o
Agrostis canina	o	H. splendens	o
Juncus effusus	o	Thuidium tamariscinum	o
Vaccinium myrtillus	o		
Pinus silvestris (dwarfed)	f	Cladonia sylvatica	o
Ilex aquifolium	o	C. uncialis	o

Raised mosses of Lonsdale. The mosses[1] of the Lonsdale lake and estuarine basins (north Lancashire, just south of the Lake District) were of raised bog type, but most of them have been partially or completely destroyed by draining and peat cutting and none is now dominated, according to Rankin's description (1911, pp. 247–59), by active Sphagnetum (regeneration complex). The best example extant in 1911 was Foulshaw Moss, which then occupied about two square miles (*c.* 5·2 sq. km.), though formerly, with adjacent mosses, it reached from the sea for seven miles inland. The surface sloped gently up from the edge where it abutted on the hillside, and was dominated by *Eriophorum vaginatum* mixed with *Scirpus caespitosus*. On the tussocks formed by these grew shrubby *Calluna*, *Erica tetralix* and *Andromeda*, while characteristic mosses, chiefly Sphagna, spread between them. "The occasional pools (writes Rankin) are filled with a matrix of *Sphagnum*, traversed by the rhizomes of *Eriophorum angustifolium* and *Rhynchospora alba*, and the lanky stems of the cranberry (*Oxycoccus quadripetalus*); here also the bog asphodel (*Narthecium ossifragum*) and the sundews (*Drosera rotundifolia*, *D. intermedia* and—most rarely—*D. anglica*) find a sufficiently firm substratum for their growth" (Rankin, 1911, p. 249). There was evidence however that Foulshaw was formerly (in the nineteenth century) much wetter, with a greater expanse of bog moss and it was then said to be quite impassable. The existence of several feet of *Sphagnum* peat below the surface *Eriophorum* peat (see Fig. 47, p. 168) shows that Foulshaw was at one time an actively growing raised moss.

Other raised mosses. Raised mosses are quite frequent in north-western England and south-western Scotland, and Pearsall (1938) has made observations on the acidity and oxidation-reduction potential of some of them, but they have not been systematically described.

Irish raised bogs. Raised bogs, in their actively growing condition dominated by species of *Sphagnum* and locally known as "red bogs", either because of the prevalence upon them of crimson-coloured bog mosses (mainly *Sphagnum rubellum*) or because of the warm reddish brown tint of the blended bog vegetation as seen from a distance, are a conspicuous feature of the Irish central limestone plain. In travelling through the plain one is constantly coming across these bogs as great brown uniform expanses of vegetation raised a few feet above the general level of the plain and with very slightly convex surfaces.[2] Thus between Athenry and Athlone, a distance of 42 miles, nine such bogs were seen to right or left of the road, and between Mullingar and Kinnegad (12 miles) at least three. South of Edenderry there is an area of about 20 square miles almost entirely covered with raised bog.

Distribution, climate and peat

The climate, though oceanic, is not so wet as in the far west, and the

[1] The distribution of some of these mosses is shown in Fig. 109 (p. 596).
[2] Pl. 115.

PLATE 115

Phot. 283. Raised bog near Shannon Bridge, Co. Roscommon, in the Irish central plain, from the top of a neighbouring esker. Callunetum (dark) occupies the cut bog margin across the road. Other raised bogs in the distance. *R. J. L.*

Phot. 282. Small raised bog on terrace at Loch Maree, Ross-shire, entirely surrounded by native pinewood. The dark marginal zone of Callunetum is seen in the foreground: behind is the regeneration complex with hummocks. Cf. Fig. 138.

Phot. 283*a*. Raised bog east-north-east of Athlone, West Meath, showing the convex surface in profile. The foreground, from which peat has been removed (turf stacks on the left) is occupied by Callunetum. The cut edge of the bog (vertical peat cliff) is seen as a dark band in the middle distance. *R. J. L.*

Raised Bog

PLATE 116

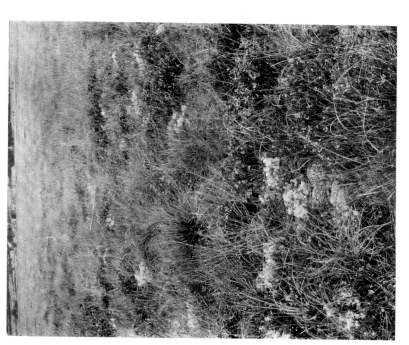

Phot. 284. Raised bog at Kilsallagh, south-east of Edgeworthstown, West Meath. Surface of "regeneration complex" showing hollows and hummocks. *Sphagnum cuspidatum* in hollow at centre of extreme front. Hummock with *S. papillosum* and *Cladonia sylvatica* (lightest) immediately above. The tufted plants are *Scirpus caespitosus* and *Eriophorum vaginatum*. *Calluna* is abundant. 2 ft. (60 cm.) rule in middle distance. R. J. L.

Phot. 285. Close view of hollow and adjacent hummock. *Sphagnum cuspidatum* with a plant of *Eriophorum angustifolium* in the hollow. *S. papillosum*, *S. rubellum* and *Cladonia sylvatica* (white) on the hummock. *Calluna vulgaris* on each side. R. J. L.

IRISH RAISED BOG—REGENERATION COMPLEX

bogs are topogenous, either built upon old lake areas which have passed through the stage of fen, sometimes of fen wood ("carr"), or directly on the site of forest which has been flooded or which has succumbed to the invasion of marsh and bog plants favoured by increasing wetness of climate. This is proved by the nature of the underlying peat (see below).

Some idea of the enormous mass of peat contained in these bogs can be gained from the fact that the population have been cutting the edges for fuel for many centuries, so that it is very rare to find an untouched bog edge, and the result is that the margins are almost everywhere "nibbled", so to speak; but the great mass of the bog remains intact. From only a few small bogs has the peat been completely removed.[1]

Structure and development. The convexity of the surface is everywhere evident, but the angle of slope is very slight, in one case which was measured a gradient of about 1 in 100, or 3°. The few untouched or apparently untouched edges slope much more steeply, and are occupied by luxuriant and dominant *Calluna*, *Pteridium* or *Ulex europaeus*, occasionally sparsely colonised by seedling birch (*Betula pubescens*) or subspontaneous pine. Such bog margins have apparently stopped growing, but the general surface of many of the bogs seems still quite active ("regeneration complex"). Below the steep marginal slope there can generally be traced the remains of the "lagg" or marginal ditch which is a constant feature of raised bog and receives the drainage water from the peat, but no example of an entirely natural lagg (such as is described on pp. 682, 685) was met with. Even where there has apparently been no peat removed from the edge of the bog the adjacent ground (originally fenland) has been modified by pasturing and drain cutting.

Raised bog vegetation is mixed, like that of blanket bog, and contains most of the same species, but there is a definite genetic sequence, or rather series of sequences, in the layers of peat which have built up the bog. When the bog has been raised on a substratum of fen the lower layers of the peat record the structure of the fen, and above this come successive *cycles* of different kinds of bog peat. The existence of these cycles depends upon the fact that the surface of a bog at any given time is not uniform, but consists of alternating *hummocks* and *hollows* inhabited by different species.[2] The peat at the bottom of each hollow is built up by the vegetation in the natural process of autogenic succession until it forms a new hummock, the surfaces of the adjacent pre-existing hummocks which have stopped growing thus coming to occupy a lower level than the new hummocks. The old hummocks thus become the sites of new hollows, the old hummock vegetation dies, and is replaced by species characteristic of hollows, these in their turn giving way to new

Hollow-hummock cycle

[1] The bogs are unlikely to remain much longer, since the systematic exploitation of the peat on a large scale is beginning.
[2] See Fig. 139; Pl. 116, phot. 284; and Pl. 117, phot. 286.

hummock formers. Thus the structure of the peat of a raised bog is lenticular,[1] each lenticle representing a complete cycle of "hollow-hummock" development, and all the phases of the cycle are represented at any given time on the surface of an actively growing bog ("regeneration complex"). The lenticular structure can be clearly seen in vertical section of the peat. The story of raised bog development was first worked out in detail by Osvald in his classical paper on the Swedish bog "Komosse",[2] and the Irish raised bogs are essentially similar, though some of the species are different.

Below are lists of the species occurring in different stages of the "hollow-hummock" development on a typical raised bog surface south of Athlone, in the middle of the central plain.

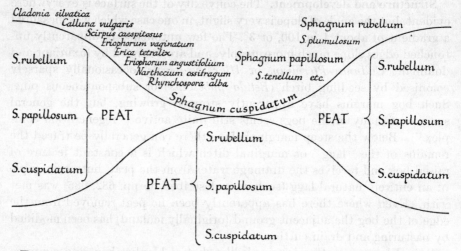

Fig. 139. Diagram of the succession of Species forming the Peat in a typical "Hollow-Hummock" cycle (Regeneration Complex)

Vegetation of seral stages. *Stage* 1. Semi-aquatic *Sphagnum* phase—wet hollow:

Sphagnum dusenii*	d	Rhynchospora alba	o
Eriophorum angustifolium	o		

* The dominant species of the hollows is generally *S. cuspidatum*.

Sphagnum cymbifolium and one plant of *Narthecium* occurred along the margin.

This community is increasingly colonised by vascular plants, species of semi-aquatic bog moss remaining dominant in the lower layer; and thus develops into

[1] Fig. 139.

[2] H. Osvald, Die Vegetation des Hochmoores Komosse, *Akad. Abhandlungen*, Uppsala, 1923. In the summer of 1935 the author had the privilege of examining some Irish raised bogs in the company of Prof. Osvald, who interpreted the details of their vegetation and peat structure in the field: the present account was thus made possible.

Phot. 286. Two pools, with hummock between, on raised bog near Edenderry. *Sphagnum cuspidatum* in pools, *S. papillosum*, etc. forming hummocks: *Scirpus caespitosus, Eriophorum vaginatum, Erica tetralix, Calluna.* R. J. L.

Phot. 287. Large hummock (height above water in pool 84 cm., diameter 137 cm.) on raised bog between Athlone and Ballinasloe. *Sphagnum cuspidatum* below; *S. magellanicum* and *S. rubellum* forming the mass of the hummock, which is capped by *Leucobryum glaucum, Cladonia sylvatica* and old *Calluna. Scirpus caespitosus* on the right. R. J. L.

IRISH RAISED BOGS—HUMMOCKS AND POOLS

PLATE 118

Phot. 289. Big tussock of *Scirpus caespitosus* on cut peat bank near Mullingar. *Calluna* to the right and below. *Molinia* behind. The peat is here drier than in untouched bog. *R. J. L.*

Phot. 288. Hummock and pool (in front) on raised bog between Athlone and Ballinasloe. *Sphagnum cuspidatum* and *Rhynchospora alba* in front. Most of the hummock is composed of *Sphagnum magellanicum*, darker (red) to right and *S. papillosum*, lighter (yellow) to left. *Cladonia sylvatica* and *Calluna* crown the hummock. *R. J. L.*

IRISH RAISED BOG

"Regeneration" Complex

Stage 2. *Rhynchospora* phase—hollow:

Rhynchospora alba	d	Oxycoccus quadripetalus	
Narthecium ossifragum	f	var. microcarpus	o
Erica tetralix	f		
Andromeda polifolia	o	Sphagnum cuspidatum	d
Calluna vulgaris	o	S. tenellum	f
Drosera anglica	o	Lepidozia sp.	o
D. rotundifolia	o	Cladonia uncialis	o
Eriophorum vaginatum	o		

Stage 3. *Narthecium* phase—hollow. *Narthecium* tends to assume dominance at the expense of *Rhynchospora*, while the semi-aquatic Sphagna and other species retrogress. In this stage the following list was made:

Narthecium ossifragum	d	Rhynchospora alba	o
Erica tetralix	f	Scirpus caespitosus	o
Andromeda polifolia	o		
Drosera anglica	o	Sphagnum papillosum	f
D. rotundifolia	o	S. rubellum	o
Eriophorum vaginatum	o	S. tenellum	o
Calluna vulgaris	o	Lepidozia sp.	o
		Kantia sp.	o

Stage 4. *Calluna—Sphagnum rubellum* stage—active phase of hummock formation. In this *Sphagnum rubellum* rapidly assumes dominance in the lower, and *Calluna* in the upper layer:

Calluna vulgaris	d	Oxycoccus quadripetalus	
Erica tetralix	f	var. microcarpus	o
Eriophorum vaginatum	f	Rhynchospora alba	o
Andromeda polifolia	o		
Drosera rotundifolia	o	Sphagnum rubellum	d
Eriophorum angustifolium	o	Hypnum schreberi	o
Narthecium ossifragum	o	Kantia sp.	o
		Mylia sp.	o

Stage 5. *Calluna-Cladonia* stage—last stage of hummock, in which the Sphagna have lost dominance and are dying:

Calluna vulgaris	d	Cladonia sylvatica	a
Erica tetralix	f	C. uncialis	o
Eriophorum vaginatum	f	Sphagnum rubellum	
E. angustifolium	o	S. plumulosum	f, but dying
Andromeda polifolia	o	S. tenellum	
Narthecium ossifragum	o	Kantia sp.	
Oxycoccus quadripetalus		Riccardia sp.	a
var. microcarpus	o	Lepidozia sp.	
Scirpus caespitosus	o		

Structure of hummocks. In the whole succession the Sphagna are of primary importance: first the species of high water requirement which dominate the wet hollows (*S. cuspidatum, S. dusenii, S. cymbifolium*), then the active hummock builders (*S. plumulosum, S. papillosum, S. rubellum*). *S. magellanicum*, better known in England as *S. medium*, is not so common in Ireland as the preceding species. *S. fuscum*, which is the characteristic species on the continent, appears to be rare in the Irish bogs. Some of the larger hummocks may be capped with *Leucobryum glaucum* (Pl. 117, phot. 287). Other mosses (e.g. *Hypnum schreberi*, *H. cuspidatum*) and liverworts may also settle among the Sphagna. *Calluna* may enter the less wet hollows, but it only becomes luxuriant and eventually dominant after active hummock formation has begun. As we shall see in the following chapters the ling has a very wide range of tolerance of soil water content, but it never becomes dominant unless there is good drainage for its roots and this it finds in the rising cushions of the second group of Sphagna. On the older hummocks, after the entrance of *Cladonia sylvatica*, the ling tends to become "leggy" and die. In view of the dryness of soil the ling can tolerate, it is difficult to attribute this to drought, and it may depend on insufficient nutrition or merely on the age of the plant. As we shall see later a similar phenomenon occurs with age on the English heaths and grouse moors.

The photographs on Pls. 116–118, taken by Dr Lythgoe on various raised bogs of the central Irish plain, show the positions in hollows and on hummocks of some of the more important species.

Peat profiles of raised bogs

Athlone bog. A boring was made by Prof. Osvald through this raised bog at a spot about 350 m. from its northern edge, in the middle of the typical "regeneration complex", as he calls the aggregation of active hummocks and hollows just described. With his permission a detailed record of this boring to a depth of 10 m., drawn to scale, is annexed (Table XXI, pp. 692–3).

It will be obvious, on consideration, that any given vertical section through a bog of the structure described will not always pass through the complete cycle of hollow-hummock development: in fact it will only do so if the bore strikes a hummock somewhere about its centre. Otherwise it will pass through the edges of the lenticular units, so that the series of phases represented in the bore will be incomplete. It will be noted that the basal layer of the cycle (the *S. cuspidatum* layer) and the layer immediately below it are generally the most "humified", i.e. decayed, and in these highly humified layers the *species* of *Sphagnum* present are rarely recognisable. The degree of humification is represented by numbers on a scale of 1 to 10. Thus 8 means highly humified, 2 very little.

The surface vegetation at the spot where the boring was made was that of a hollow partly grown up and dominated by *Narthecium* (see the list of Stage 3, p. 689).

Below the living vegetation cover came 20 cm. of moderately humified *Narthecium-Sphagnum* peat already formed in the hollow, and then 5 cm. of highly humified *Calluna-Sphagnum* peat marking the top of the preceding hummock, and underlain by alternating layers, much less humified, of peat consisting of *S. papillosum* and *S. rubellum*, with *Eriophorum vaginatum* and *Calluna*, to a total depth of 1·2 m. This represents the varying colonisation and growth of the typical hummock-forming bog-mosses, but the base of the hummock was not passed through.

Under this came a complete cycle, consisting, from above downwards, of 20 cm. of highly humified *Sphagnum-Calluna* peat, 45 cm. of *Sphagnum papillosum* peat, and finally 10 cm. of *S. cuspidatum* peat representing the vegetation of the hollow on which the hummock was built.

At 2 m. begins an incomplete cycle, followed by two which are fairly complete, each with typical *S. cuspidatum* peat at the base.

At 4 m., towards the bottom of the bog, we encountered half a metre of *Sphagnum* peat with *Eriophorum vaginatum* and *Calluna*, underlain again at 4·5 m. by a thin layer of *Sphagnum cuspidatum* and *Eriophorum angustifolium* peat, marking the wet conditions in which the raised bog began to develop.

Below this was the fen peat on which the bog was built, consisting of *Carex* and *Cladium* and passing down into reedswamp with *Phragmites*, *Menyanthes* and *Carex inflata*, the peat of the fen and reedswamp together having a thickness of more than 3 m. and becoming muddy towards the base.

At a depth of 7·77 m. from the present surface of the bog was a thin layer of yellow calcareous lake mud with remains of *Phragmites*, and below this 1·75 m. of cream-coloured lake marl with no remains of vegetation, representing the conditions of open water.

At 9·6 m. from the surface the bluish grey clay (glacial till) on which the lake was based was encountered, and at 10 m. the boring was stopped.

Edenderry bog. Another bore (Table XXII, pp. 694–5) was made in an extensive raised bog south of Edenderry. This was put down in a low hummock of *Calluna-Tetralix* (see Stage 4, p. 689) surrounded by *Sphagnum tenellum*. Immediately below came 25 cm. of *Sphagnum-Narthecium* peat, and then a fairly complete hollow-hummock cycle with *Sphagnum rubellum*, *S. papillosum* and *Eriophorum vaginatum* above and *Sphagnum cuspidatum* peat below. Then came about 2·5 m. of incomplete or doubtfully complete cycles, probably five in number, in which the basal *S. cuspidatum* peat could not be clearly identified, underlain by the first "regeneration cycle". Below this was peat composed of *Sphagnum, Calluna, Oxycoccus* and much *Eriophorum vaginatum*, and (at about 4·5 m. from the present surface of the bog) the topmost fen peat consisting of *Caricetum* colonised by *Sphagnum*. Under this was more than a meter of fen peat with alternating *Carex*, *Cladium* and *Equisetum*, and the moss *Amblystegium* towards the bottom. Below the fen peat we came to *Sphagnum* and *Calluna* again, then to more *Carex* peat with *Phragmites* and *Menyanthes*, and below also *Myrica*. Thus we have two fen horizons separated by a development of bog moss and heather. The upper fen was probably formed as the result of reflooding of

Table XXI. *Section of bog south-west of Athlone through moss (bog) peat and underlying fen and reedswamp peat to lake marl and basal glacial clay. H = degree of humification (scale of humification 1 to 10).*

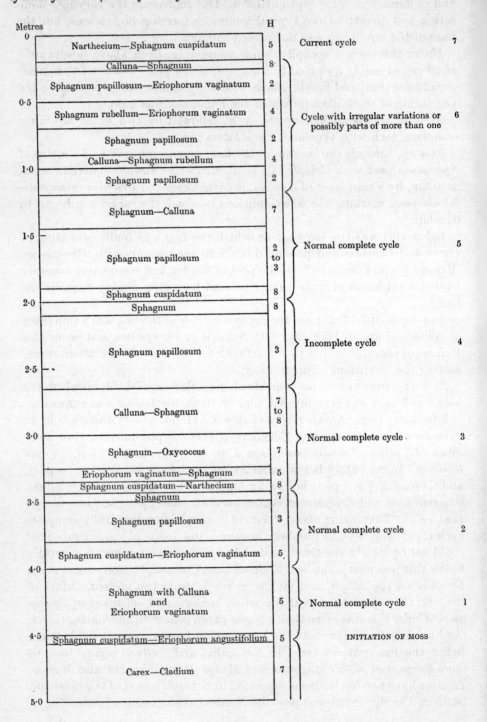

Table XXI *continued*

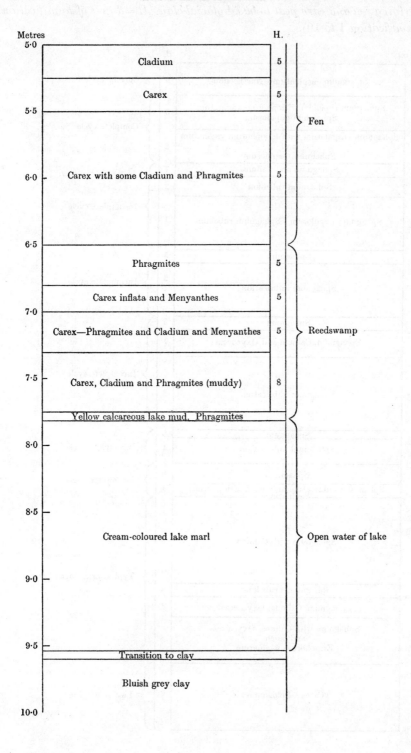

Table XXII. *Section of bog south of Edenderry through moss (bog) peat and underlying fen and carr peat to basal glacial clay. H = degree of humification (scale of humification 1 to 10).*

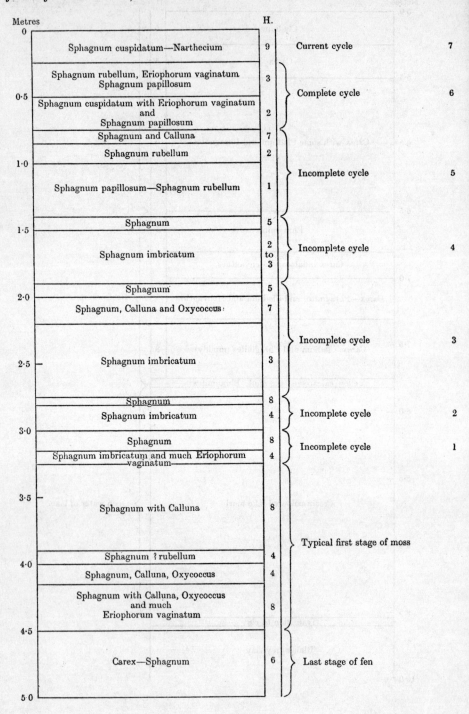

Table XXII *continued*

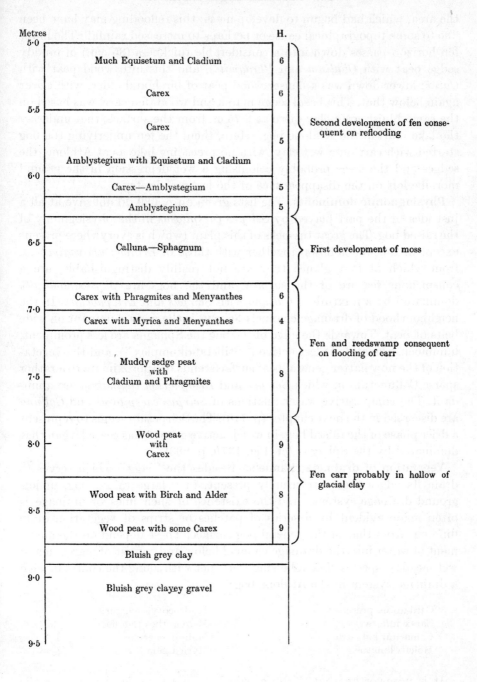

the area, which had begun to develop moss: this reflooding may have been due to some topographical cause or perhaps to increased rainfall. The lower fen horizon passes down into a considerable thickness (85 cm.) of muddy sedge peat with *Cladium* and *Phragmites*, and this into wood peat with *Carex*; lower down was a denser wood peat of birch and alder, with *Carex* again below that. This basal sedge marsh and wood (fen carr) was based on the same bluish grey clay (here at 8·75 m. from the surface) that underlay the lake marl in the Athlone bog. Here, then, the fen underlying the bog started with carr on a wet clay, with no preceding lake as at Athlone, the sedges and the trees probably colonising a wet depression in the ground moraine left on the disappearance of the last ice-sheets.

Physiognomic dominants. The lists given on p. 689 do not give at all a just idea of the part played by *Scirpus caespitosus* in the physiognomy of the raised bog. The great tussocks of this plant (which is everywhere present except in the wet hollows) together with those of *Eriophorum vaginatum*, from which at first glance they are not readily distinguishable, are a conspicuous feature of the surface, and the bogs are *physiognomically* dominated by a mixture of *Calluna* with *Scirpus* and *Eriophorum*. In the neighbourhood of drainage channels *Scirpus* is the main dominant on wide belts of peat. Towards the edge of the bog the Sphagna are less prominent, hummock formation is less active ("Stillstandkomplex"), and the vegetation of the now flatter general peat surface tends to approximate to a rather sparse Callunetum in which *Scirpus* and *Eriophorum vaginatum* are abundant. The comparative water relations of *Scirpus caespitosus* and *Calluna* are discussed in the next chapter (p. 710). The Scirpetum seems to represent a drier phase of the raised bog, in which active growth has ceased, than that dominated by the Sphagna (cf. Fig. 137b, p. 683).

Vegetation of drainage channels. Besides the "lagg" which serves to drain its edges there are usually present, in a large raised bog, underground drainage systems, and the existence of such lines of drainage is often made evident by a series of patches or strips of vegetation quite different from that of the general bog surface. These depend on the movement of water into the drainage channel below and consist of marsh, fen or wet meadow species. The following were met with along the course of such a drainage system in the Athlone bog:

Cardamine pratensis
Carex inflata
Comarum palustre
Holcus lanatus
Hydrocotyle vulgaris
Menyanthes trifoliata
Molinia caerulea
Myrica gale

It is noteworthy that *Molinia*, which occurs here, is absent from the regeneration complex of the raised bog. At intervals along the course of the underground water channel there were deep holes, with steeply sloping or precipitous sides, at the bottom of which flowing water could sometimes be seen. On the well-drained peat of the sides the following species occurred:

Crataegus monogyna
Orchis sp.
Osmunda regalis
Potentilla erecta
Pteridium aquilinum
Rubus fruticosus (agg.)
Ulex europaeus
Vaccinium myrtillus

Sphagnetum of blanket bog. In the blanket bogs of western Scotland and Ireland Sphagneta form quite local patches and hummocks, and in the elevated blanket bogs of the plateaux of western and northern England, Scotland and Ireland also they rarely dominate any considerable continuous area. The structure and development of these Sphagneta need detailed study by modern methods, and the following scattered data are all that can be furnished at present. Blanket bog is dealt with in the next chapter.

Sphagneta of the northern Pennines. Lewis (1904, p. 325) described extensive upland *Sphagnum* bogs in the northern Pennines, on the borders of Westmorland and Yorkshire, on either side of the Darlington and Tebay railway, about 6 to 10 miles south-west and south-south-west of Middleton-in-Teesdale. These mosses (e.g. Shacklesborough Moss, Red Gill Moss, Beldoo Moss) have an aggregate area of several square miles and are developed at an elevation of about 1300–1500 ft. ($=c.$ 400–450 m.) on the upland plateau in which arise the parent streams of the Balder, a tributary of the Tees, towards the head of its drainage basin on the eastern side of the main Pennine watershed. Elsewhere on the Pennines *Sphagnum* bogs are rare, and those which still exist are quite small and local, nor is *Sphagnum* peat at all common in the upland Pennine peat moors.[1]

The dominant species of *Sphagnum* in a similar north Lancashire Sphagnetum was said by Wheldon and Wilson (*Flora of Lancashire*, 1907) to be *S. recurvum* (one of the *cuspidatum* group), though the following species also occur (*Naturalist*, 1910):

S. rubellum
S. acutifolium
S. plumulosum
S. cuspidatum
S. undulatum
S. tenellum (molluscum)

Apart from the bog mosses the flora was reported by Lewis to be almost entirely composed of the following plants:

Erica tetralix	f	Rubus chamaemorus	f
Eriophorum angustifolium	f	Empetrum nigrum	o
Oxycoccus quadripetalus	f	Calluna vulgaris	r

Wheldon and Wilson add the following as "marginal or very subordinate constituents" in similar upland Sphagneta in north Lancashire:

Carex canescens
C. stellulata
Juncus effusus
Rhynchospora alba
Viola palustris

[1] These northern Pennine Sphagneta and those mentioned below require re-investigation to determine their relation to the types of bog described in this chapter and the next.

Sphagnetum in Cornwall. On Bodmin Moor in Cornwall, according to Magor (unpublished), Sphagnetum is a very local community and almost always restricted in area. Very wet spots form the so-called "piskie pits", usually of a vivid emerald green, in which *Sphagnum* peat may reach a depth of 12 ft. (3·5 m.) or more. Small quantities of the following plants may occur in the Sphagnetum:

Calluna vulgaris	Juncus effusus
Carex stellulata	Rhynchospora alba
Erica tetralix	Viola palustris
Eriophorum angustifolium	

Flora of Sphagnetum. According to Watson (1932, p. 293) Sphagnetum is well represented in Perthshire, North Wales, and on the higher parts of Exmoor, at 1100–1600 ft. (c. 335–480 m.), and the commonest species of bog moss "in the drier Sphagneta" are usually Cymbifolia and Acutifolia, especially *Sphagnum papillosum* and *S. quinquefarium*. According to Skene (1915, p. 79) these are species of medium to high acidity. The other species listed by Watson as abundant are the mosses *Polytrichum commune* and *Aulacomnium palustre*, while *Calluna* is occasional and the following are said to be frequent:

Anagallis tenella	Menyanthes trifoliata
Carex stellulata	Molinia caerulea
C. pulicaris	Myrica gale
Drosera rotundifolia	Narthecium ossifragum
Eriophorum angustifolium	Oxycoccus quadripetalus
E. latifolium	Pinguicula lusitanica (local)
Hydrocotyle vulgaris	Rhynchospora alba
Hypericum elodes	Viola palustris
Juncus acutiflorus	
Calypogeia fissa	Hypnum stramineum
Cephalozia connivens	H. cuspidatum
Gymnocolea inflata	Odontoschisma sphagni

Watson remarks that it is usual to find one particular sub-genus of *Sphagnum* dominating the whole or part of a bog, and one species may become the dominant or even the sole species over a given area. Thus over half an acre of bog on the Blackdown Hills in south Somerset *S. girgensohnii* was almost the only species present. *S. papillosum*, *S. cymbifolium* and *S. quinquefarium* often occur in almost pure patches. These are among the more acid species (Skene, 1915, p. 79), while *S. rubellum*, another common member of the Acutifolia, is the most acid of all. *S. cuspidatum* (one of the less acid) is the "most hydrophilous species and is frequently associated with *Drepanocladus fluitans*, *D. exannuletus*,...*Cephalozia fluitans*...and *Gymnocolea inflata*".

Many of the bryophytes accompanying *Sphagnum*, particularly those which normally live in drier communities, are characterised, in this habitat, by the increased length of their shoots (Watson, 1932).

REFERENCES

Lewis, F. J. Geographical distribution of vegetation of the basins of the rivers Eden, Tees, Wear and Tyne. Part I. Southern portion. *Geogr. J.* 1904.

Lewis, F. J. and Moss, C. E. In *Types of British Vegetation*, Chapter XII, "The Upland Moors of the Pennine Chain". Pp. 266–82. 1911.

Magor, E. W. Geographical distribution of vegetation in Cornwall: Camelford and Wadebridge district (unpublished).

The Naturalist, 1910, pp. 265, 313.

Osvald, H. Die Vegetation des Hochmoores Komosse. *Svenska Växtsoc. Sällsk. Handl.* I. *Akad. Abhandl.*, Uppsala. 1923.

Pearsall, W. H. The soil complex in relation to plant communities. III. Moorlands and bogs. *J. Ecol.* **26**, 298–315. 1938.

Rankin, W. Munn. In *Types of British Vegetation*, Chapter XI, pp. 249, 255, 259–64. 1911.

Skene, Macgregor. The acidity of *Sphagnum* and its relation to chalk and mineral salts. *Ann. Bot.* **29**, 65–87. 1915.

Watson, W. Sphagna, their habitats, adaptations and associates. *Ann. Bot.* **32**, 535–51. 1918.

Watson, W. The bryophytes and lichens of moorland. *J. Ecol.* **20**, 284–313. 1932.

Wheldon, J. A. and Wilson, A. *Flora of West Lancashire.* Eastbourne, 1907.

Chapter XXXV

THE MOSS OR BOG FORMATION (*continued*)

(2) RHYNCHOSPORETUM. (3) SCHOENETUM. (4) ERIOPHORETUM. (5) SCIRPETUM. (6) MOLINIETUM. IRISH BLANKET BOG. SUMMARY

(2) **Rhynchosporetum albae.** *Rhynchospora alba*, as we have seen in the last chapter, tends to dominate the hollows of raised bog in the phase immediately succeeding *Sphagnum cuspidatum*. It also dominates the wetter parts of the (blanket) bogs of Connemara (Co. Galway)[1] which lie but a few feet above sea level and are interspersed with rocky outcrops[2] and innumerable lakelets.[3] It seems to be a phase of succession preceded by a wetter stage in which the peat is colonised by *Zygogonium ericetorum*, often with a little *Carex limosa*, and followed by the dominance of other species, such as *Schoenus nigricans* or *Molinia*. The following species have been recorded·

Rhynchospora alba			d
Calluna vulgaris	l	E. latifolium	f
Cladium mariscus	l	Menyanthes trifoliata	l
Drosera anglica	f	Myrica gale	f
D. intermedia	f	Narthecium ossifragum	f
Erica tetralix	f	Phragmites vulgaris	
Carex panicea	f	Rhynchospora fusca	r
Potamogeton polygonifolius	l	Schoenus nigricans	va
Eriophorum vaginatum	f	Scirpus caespitosus	f
Campylopus atrovirens	la	Sphagnum magellanicum	l
Leucobryum glaucum	l	S. plumulosum	l
Rhacomitrium lanuginosum	l	S. rubellum	l
Eriophorum angustifolium	f	S. subsecundum	l

The *Phragmites* and *Cladium* form narrow reed swamps round parts of the lakelets, and where they occur in local patches elsewhere in the bog they probably represent, like *Menyanthes* and *Potamogeton*, relicts of this reed swamp community occurring on the sites of former pools, though in some cases they may indicate the sites of underground drainage water from neighbouring knolls.

(3) **Schoenetum nigricantis.** Considerable tracts of the west Irish blanket bogs are dominated by *Schoenus nigricans*,[4] which in England (except the extreme south-west) is mainly characteristic of calcareous swamps or fens. Such a Schoenetum may follow Rhynchosporetum in

[1] Pl. 122, phot. 298. [2] Pl. 121, phot. 296.
[3] Pl. 42, phot. 89.   [4] Pl. 122, phot. 298.

succession, but the list of species differs scarcely, if at all, from the list already given for Rhynchosporetum, or from that for mixed bog on p. 715. It has recently been suggested that the abundance or dominance of *Schoenus* in blanket bog near the western coasts may be favoured by the falling of sea spray, driven by inshore gales, on the surface of the bog, thus changing the soil reaction in the direction of its more normal habitat.

(4) **Eriophoretum.** Cotton-grass bogs, or mosses as they are always called, are characteristic of the whole Pennine chain, occupying very extensive areas on the flat or gently sloping plateaux at elevations between 1000, or more commonly 1200, and 2200 ft. (*c.* 370–670 m.) (Fig. 140). The actual altitude depends on the heights of the plateaux, the former representing the average lower limit in the southern, the latter the average upper limit in the northern Pennines.

Moss (1911) emphasises the dreary and monotonous character of the *Eriophorum* mosses. The cotton-grass is not only the dominant, but frequently the only vascular plant over wide areas. The cotton-grasses are active in the formation of peat, which may reach an extreme depth of about 30 ft. (9 m.) and is said to consist almost wholly of the remains of the dominant species.[1] It is usually saturated and frequently supersaturated with water, though the surface layer may become quite dry during a drought. Like the *Sphagnum* bogs these far more extensive Eriophoreta are always called "mosses" locally. "Moss" is much the commonest place name on the Pennine uplands. Of the two commoner species, *Eriophorum angustifolium* and *E. vaginatum*, the latter, which occupies drier peat than the former, is very much more commonly dominant, especially in the southern Pennines. Both species are relatively deep-rooted, and abundantly supplied with intercellular channels, but the former has creeping rhizomes and more rapidly colonises very wet peat (Pl. 120, phot. 293).

(*a*) **Eriophoretum vaginati.** Moss (1911, 1913), Adamson (1918) and Watson (1932) have studied this community, which is the characteristic bog or moss of the southern Pennines, almost everywhere covering the higher plateaux above 1000–1200 ft. (300–360 m.), though it is often in various stages of retrogression (Fig. 140 and Pls. 119, 120).

Soil. The community normally occurs on deep peat which it has itself formed. There is no evidence as to its beginnings: at present it is only extending locally, as in hollows where pools have existed. The peat, which varies in depth from 2 to 3 ft. up to 30 ft. (0·6–9 m.), is always relatively pure and free from admixture of sand or other mineral matter. The roots of all the plants of the community are entirely confined to the peat. Below this the mineral soil is podsolised: from 3 in. to 2 ft. (7–60 cm.) below the base of the peat there is nearly always a thin layer of "pan", which may be

---

[1] Doubt has recently been cast on this statement. The peat underlying the existing cotton-grass mosses requires re-investigation.

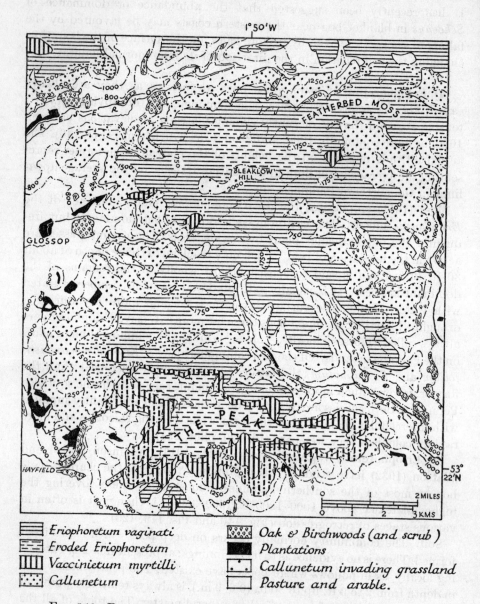

FIG. 140. PLATEAU (1500–2000 ft.) OF THE PEAK DISTRICT (NORTH DERBYSHIRE) OCCUPIED BY ERIOPHORETUM VAGINATI

Large areas of Eriophoretum are eroded and invaded by Vaccinietum myrtilli and Callunetum from the neighbouring "edges" and slopes. Contours in feet. After Moss, 1913.

quite hard and stone-like (Adamson, 1918). This "*B* horizon" is coloured red, presumably by ferric salts. Pearsall (1938) found that the *p*H values of pure undisturbed Eriophoretum vaginati varied from 2·98 to 3·40, and is thus markedly more acid than Sphagnetum. The oxidation-reduction potential varies round the critical point and in summer, when the surface dries out, the peat is feebly oxidising.

Floristic composition. Moss (1911) and Watson (1932) each gives fifteen species of vascular plants as occurring in this community: nine of these (marked with an asterisk *), including the dominant, are common to the two lists. Below is the composite list:

*Eriophorum vaginatum	d		
Agrostis stolonifera	o	Juncus squarrosus	o
A. tenuis	o	*Molinia caerulea	r
Andromeda polifolia	r	Narthecium ossifragum	r
Calluna vulgaris	lsd	Oxycoccus quadripetalus	r
*Carex canescens	o, la	Pinguicula vulgaris	r
*Empetrum nigrum	lsd	*Rubus chamaemorus	lsd
*Erica tetralix	r–la	*Scirpus caespitosus	r–la
*Eriophorum angustifolium	l	*Vaccinium myrtillus	r–la
Juncus effusus	o	V. vitis-idaea	o

Watson records thirty-five species of bryophytes and nine lichens from the community. The following bryophytes he marks as abundant or frequent:

Calypogeia trichomanis	f	Gymnocolea inflata	f
Campylopus flexuosus	a	Lepidozia repens	f
Cephaloziella starkii	a	Lophozia floerkii	f
Dicranella heteromalla	a	L. ventricosa	f
Diplophyllum albicans	f	Tetraphis pellucida	a
Drepanocladus fluitans var. atlanticus	f	Webera nutans	sd

Three species of *Sphagnum* (*S. recurvum, cymbifolium* and *papillosum*) which are of medium to high acidity, are recorded as occasional, and the following three lichens are frequent or abundant:

Biatora granulosa	f	Cladonia coccifera	a
Cetraria aculeata var. hispida	f		

Other areas. Lewis (1904) briefly describes Eriophoretum vaginati in the northern Pennines with a very small quantity of stunted *Calluna* only 2 or 3 in. high.

On the Cleveland Hills, on the eastern side of Yorkshire, pure Eriophoretum of any extent is unknown, but on the highest parts of the watershed (1450 ft.) on "peat of great thickness", the prevalent Callunetum may be mixed with an equal amount of *Eriophorum vaginatum,* and the latter species is locally dominant. Interspersed are "numerous pools full of floating *Sphagnum*, probably *S. cuspidatum*, which when drier also contain *Eriophorum angustifolium*" (Elgee, 1914, p. 10).

The Eriophoretum vaginati occupying the surface of Foulshaw Moss (a raised bog) and described by Rankin (1911) has already been mentioned (p. 686).

(b) **Eriophoretum angustifolii.** This community is very local in the southern Pennines, but more extensive on the northern part of the chain. According to Watson (1932) it is much more frequent on the Somerset moorlands than E. vaginati, and limited areas are described by Pethybridge and Praeger (1905) at high altitudes (about 500–700 m.) on the Wicklow mountains in eastern Ireland. The Wicklow Eriophoreta are small in extent and "below the uniform waving foliage of the cotton-grass is a continuous dense stunted growth of *Calluna* with several of the plants of the *Calluna* and *Scirpus* associations". The peat is "sopping wet" and spongy, unlike that of the Scirpetum caespitosi (see below), which is the characteristic community of most of the Wicklow mountain plateau. The species recorded from three sample areas were as follows:

Eriophorum angustifolium   d
Calluna vulgaris   sd

All three areas:

Eriophorum vaginatum	Cladonia sp. (two areas)
Empetrum nigrum	Erica tetralix (one area)
Scirpus caespitosus	Vaccinium vitis-idaea (one area)
Vaccinium myrtillus	Sphagnum spp. (one area)

The Eriophoretum angustifolii described by Lewis (1904) from the northern Pennines has *Eriophorum vaginatum* mixed in smaller quantity, together with a few individuals of *Rubus chamaemorus*, *Erica tetralix* and *Empetrum nigrum*.

Eriophorum angustifolium often occupies wet channels and depressions in the blanket bogs of *E. vaginatum*, and colonises cracks in the bare peat of such hollows by means of its creeping rhizomes (Pl. 120, phot. 293).

Bryophytes. Watson (1932) remarks that many Bryophytes present in the Eriophoretum vaginati are absent or rarer in the wetter E. angustifolii. He records the following species from the last-named community (of the Somerset moors) not listed from the E. vaginati:

Aulacomnium palustre	Lepidozia setacea
Campylopus brevipilus	Leptoscyphus anomalus
Cephalozia fluitans	Odontoschisma sphagni
Drepanocladus aduncus	Sphagnum acutifolium
D. fluitans (type)	S. plumulosum
D. revolvens	

Succession. There is no contemporary evidence as to any seral stages leading up to the formation of the upland Eriophoreta, but Rankin (1911, pp. 249, 254–5 and Fig. 16) showed that beneath 3 in. (7·5 cm.) or so of *Eriophorum* peat, on which the existing cotton-grass of Foulshaw Moss (see p. 686) is growing, there are several feet of very well preserved *Sphagnum* peat, the Sphagna of which it is composed belonging to the Cymbifolium

PLATE 119

Phot. 290. Eriophoretum at Wessenden Head, near Huddersfield, West Yorkshire (August).
Mrs Cowles.

Phot. 291. Eriophoretum near Huddersfield. *E. vaginatum* in fruit (June). *W. B. Crump.*

ERIOPHORETUM VAGINATI ON THE SOUTHERN PENNINE PLATEAU

PLATE 120

Phot. 292. The peat of Eriophoretum vaginati dissected by stream channels. *Vaccinium myrtillus* capping some of the peat haggs. Plateau of Fairsnape Fell, Lancashire. A. Wilson.

Phot. 293. South Pennine plateau. *Eriophorum angustifolium* in foreground, colonising peat cracks. *E. vaginatum* (tufted) fringing bare peat, *Empetum nigrum* beyond. Eriophoretum vaginati on the plateau behind.
J. Massart.

Phot. 294. Large tussock of *Eriophorum vaginatum* and carpet of *Empetrum nigrum* on side of peat channel. Eriophoretum on plateau beyond. J. Massart.

Phot. 295. *Rubus chamaemorus* on side of peat channel. J. Massart.

RETROGRESSIVE ERIOPHORETUM VAGINATI

group. Thus there can be no doubt that a Sphagnetum immediately preceded the existing Eriophoretum vaginati. And there is independent evidence that in the middle of the nineteenth century Foulshaw Moss was much wetter, quite impassable, and bore a greater expanse of bog moss than it did at the beginning of the twentieth century when Rankin described it. It must however remain an open question how far the change from Sphagnetum to Eriophoretum was a natural developmental change caused by the growing up of the bog moss above the level to which it could raise water. The drying of the surface was certainly greatly accelerated and may have been entirely determined by the trenching and removal of peat which had long been carried on.

The *Sphagnum* peat at Foulshaw is again underlain by other types of peat, showing that the existing and recently existing raised bogs were preceded by phases of forest and fen (see p. 168, Fig. 47).

Retrogression of Eriophoretum. Crowberry, ling and bilberry (*Empetrum nigrum, Calluna vulgaris* and *Vaccinium myrtillus*) occur repeatedly in the floristic lists of the Eriophoreta, and the first two also in some Sphagneta, though they do not grow with the luxuriance they show in drier habitats. These species are indicative of drier conditions than those which are optimal for the plants active in the formation of deep peat, and their presence in increasing quantity and vigour marks the beginning of conditions which when carried to their term will lead to the supersession of the typical "mosses" and their replacement by drier moorland types. Mixed communities of cotton-grass and ling, or at the higher levels above 2000 ft. (c. 600 m.), of cotton-grass, ling and bilberry, cover very great areas in the northern Pennines, and all of these may be interpreted as transitional from Eriophoretum to Callunetum or Vaccinio-Callunetum. The ling itself, as we shall see (Chapter XXXVI), has a very wide range of tolerance in respect of the total water content of the soil; and it seems that the purer the peat the more water the ling will tolerate, for Crump (1913) found that the W/H coefficient (i.e. water content divided by humus content) of *Calluna* soils varied within quite narrow limits.

The process of replacement of Eriophoretum by Callunetum or Vaccinietum is however often associated with active drainage and destruction of the cotton-grass peat. One way in which this can occur is by the cutting back towards their source of moorland streams rising in a peat-covered plateau. This process was carefully followed by Moss (1911, 1913) and Adamson (1918) on the plateau of the Derbyshire "Peak" (Fig. 140) and similar physiographic habitats in the southern Pennines, and it is a common phenomenon in other regions also. As the stream cuts back into the peat its banks become drier and the first sign of the change in vegetation is the appearance of a lining of *Empetrum* round its head. Lower down its course, i.e. nearer the edge of the peat plateau, where the drainage is freer and the peat of the stream banks becomes drier, *Empetrum* is followed by *Vaccinium myrtillus*. Meanwhile the bed of the stream is widened by erosion, and tributary streamlets cut back into its banks, increasing the general drainage of the surface peat, which then gradually becomes colonised by the bilberry.[1] When erosion is continuously active some of the tributary stream channels are cut back until they impinge upon others, so that a network of water courses is formed. The mounds (peat "haggs") left in the meshes of this network of channels are drained on all sides and the surface is occupied by dominant bilberry with crowberry (*Empetrum nigrum*),[2] cloudberry (*Rubus chamaemorus*),[3] and sometimes *Calluna*. The evidence that this succession has occurred on a large scale is quite convincing: the peat on which the bilberry grows is entirely composed of the remains of *Eriophorum vaginatum*, and the Ordnance Survey maps published between 1870 and 1880 show the stream channels about 1·2 km. longer than in the maps of 1830, while in the early years of this century they were about 0·4 km. longer still.

On the lower and more isolated hills of the southern Pennines Adamson (1918, pp. 98–9) describes a process of gradual drainage and drying of the peat without actual erosion of stream channels. Here *Eriophorum vaginatum* is gradually replaced by *Calluna*. "Passing from the pure Eriophoretum the first stage is a lack of luxuriance of the dominant plant, an almost complete absence of *Eriophorum angustifolium* and an admixture in increasing quantity of *Scirpus caespitosus* and *Calluna vulgaris*. With increasing dryness *Vaccinium myrtillus* and *Juncus squarrosus* appear, with some *Nardus stricta* and *Deschampsia flexuosa*, though the last seems seldom more than an occasional species. In certain cases this succession reaches the relatively stable condition of co-dominance of *Nardus* and *Juncus squarrosus*, with *Eriophorum* existing only on the wettest parts." The factor which leads to the establishment of this *Nardus* grassland instead of a *Calluna* climax is perhaps the binding of the old peat surface by the close tufts of the grass, which makes the invasion of *Calluna* difficult.

The action of wind may also result in the destruction of the Eriophoretum on isolated and projecting summits and on the crests of ridges. This may be so violent as to lead to the removal of the peat and the survival only of isolated tufts of *Eriophorum angustifolium*, which though primarily a more hydrophytic species, is much more resistant to such conditions than *E. vaginatum*. More commonly however the removal of the peat is less rapid and the crowberry and bilberry come in just as they do on the desiccating peat on the edges of stream courses.

All these forms of retrogression of Eriophoretum are assisted by burning, which is occasional on the cotton-grass moors, though not a deliberate and regular practice as on the grouse (heather and bilberry) moors. After destruction of the Eriophoretum by fire the peat is removed faster by the streams and there is more rapid immigration of bilberry, *Nardus* and *Juncus squarrosus*. The two last penetrate the moors along the lines of footpaths,

[1] Pl. 120, phot. 292.   [2] Pl. 120, phots. 293, 294.   [3] Pl. 120, phot. 295.

which can often be traced from a distance by the difference in colour of the vegetation.

(5) **Scirpetum caespitosi.** Scirpetum and Eriophoretum are in a sense alternative communities, for they occupy quite similar physiographical habitats and they are rarely both developed extensively in the same region. But the *Scirpus* peat ("pseudo-fibrous") is somewhat drier than *Eriophorum* peat, and is said to be markedly less fertile under cultivation. Like the Eriophoreta Scirpetum occurs mainly on the western side of the British Isles. According to Lewis (1908, p. 257) *Scirpus caespitosus* dominates many hundred square miles in the northwest Highlands (Sutherland, Ross and Inverness), the flat basaltic plateau of northern Skye, and the valley floors and gently sloping hillsides of North Uist and Lewis (Hebrides) and of the Shetlands. This Scirpetum also extends southwards through western Perthshire and Argyllshire.

Distribution

Scirpetum of Argyllshire. G. K. Fraser (1933) has recently given some account of the Scottish Scirpetum, particularly from the point of view of the peat which it forms and of its silvicultural possibilities; and he describes it as "the ultimate (climax) moorland vegetation in the west and north-west of Scotland". It is in fact typical "blanket bog". As the climatic climax it is the form of vegetation to which all other types tend to give place in this climatic region, and the process has been hastened and extended, according to Fraser, by the widespread grazing and burning which have taken place in the past. In the development of the Scirpetum the growth of *Sphagnum* cushions plays an important part.

Inverliver Forest. The "forest" of Inverliver on Loch Awe in Argyllshire, where Fraser's researches were largely conducted, has an annual rainfall varying from 74 to 88 in. (1850–2200 mm.), the monthly precipitation ranging from 7 to 12 in. from October to January, when rain falls on four days out of five, to not less than 4 or 5 in. in the drier months from March to July, when rain falls on two days out of three. Periods of ten days without rain are infrequent. Snow is of little account, as the small amount which falls usually melts within a few hours. Data as to atmospheric humidity are not available, but even after an exceptional 20 days of continuous bright weather with no rain and in the warmest season the soil is still thoroughly wet.

Climate

Days of bright sunshine average nine per month in winter and eleven in summer. The maximum summer temperature is about 21° C. but the mean monthly minima for July and August are less than 9° C., in January and February just under 0° C. The annual and daily ranges of temperature are slight. Frost may occur in nearly every week of the year, and it does occur frequently during the growing season except for about 7 weeks in July and August. The temperatures in the peat itself are considerably higher than in the air above it, the range is much less, and at a depth of 1 ft. (30 cm.) the thermometer remains stationary at about 7° C. throughout the year. But during bright sunshine very marked differences occur between the *Sphagnum* cover, the air above, and the peat below. Thus when the temperature 30 cm. above the bog moss was 9° C. and that at the surface of the peat below the *Sphagnum* 4° C., in the moss itself the thermometer recorded 22° C.

Vegetation. Under these climatic conditions the country is covered with a continuous sheet of blanket bog, except where special physiographic or edaphic factors are present. Since the topography varies widely there is considerable variation in vegetational detail, but the general character remains the same. The dominant *Scirpus caespitosus* forms an open, even or diffuse, tufted cover. *Calluna* is present in abundance as scattered open patches of low growth with short flowering shoots. *Narthecium ossifragum* is said to be always abundant, but of small size. *Potentilla erecta*, *Erica tetralix* and *Eriophorum angustifolium* constantly occur. Other species of vascular plants are considered by Fraser to be relict plants from previous vegetation, though this interpretation may be open to question. The moss vegetation, shows a fairly equal distribution of species of *Sphagnum* such as *S. magellanicum* (= *S. medium*), *S. papillosum* and *S. rubellum*, and of common heath mosses such as *Hypnum schreberi*, *H. cuspidatum* var. *ericetorum*, *Hylocomium splendens*, *Brachythecium purum*, with *Rhacomitrium lanuginosum* forming hoary pads on the tufts of heather. The liverwort *Pleurozia purpurea* forms compact masses in slight hollows.

Peat "hagging" by the formation of deep cracks and channels is due, according to Fraser, to the cutting away of the peat along drainage channels and subsequent erosion of their sides. Though this process leads to local and temporary change in the vegetation, such as increase of *Calluna* and *Rhacomitrium* along the edges of the haggs, it has no permanent effect on the bog as a whole.

Scirpetum in the northern Highlands. Superficial examination of several areas of blanket bog in the northern and north-western Highlands (September 1937) resulted in the following lists (see Table XXIII). The different areas, spread over a fairly wide region, are briefly characterised below. The peat in all except (8) appeared to be deep.

(1) Above Glen Garry (Inverness-shire), lat. 57° 4′ N., long. 4° 56′ W.:[1] undulating ground, alt. 600 ft.

(2) Above Guisichan Forest (Inverness-shire), lat. 57° 15′ N., long. 4° 50′ W.: nearly flat, alt. 1000 ft. Unusual species such as *Carex flava*, *C. pulicaris*, *Equisetum palustre* and *Selaginella selaginoides* were present, but the area had been somewhat disturbed by partial drainage in places.

(3) Dundonnell Forest (Sutherlandshire), lat. 57° 45′ N., long. 5° 7′ W.: undulating, alt. 1000 ft.

(4) "The Crask", 7 miles south of Altnaharra (Sutherlandshire): nearly flat, big pools and eroded haggs, alt. 800 ft.

[1] Latitudes and longitudes approximate only.

Table XXIII. *Composition of Scirpetum in different Highland areas*

								Rannoch Moor			
	1	2	3	4	5	6	7	8	9	10	11
Calluna vulgaris	f	cd	a	f, ld	f, la	lcd-ld	cd	va	f-a	f	f
Carex panicea	.	o-f	.	.	r	.	.	.	.	o	.
C. stellulata	.	o-f	.	f	r	.	.	.	la	.	.
Drosera anglica	o	o	.	.	.	.	.	.	.	.	.
D. rotundifolia	o	o	.	.	.	.	.	.	.	.	.
Erica cinerea	.	.	.	.	.	.	.	.	.	.	.
E. tetralix	a	f	a	f	f	f	o	.	f	d	f
Eriophorum angustifolium	lf	o	o	f (ld)	o–f	f–la	f–la	o, 1	a–ld	.	f
E. vaginatum	.	f	o	o	o	.	o	.	a	.	a
Juncus squarrosus	.	o	lf	.	.	.	.	.	.	l	.
Menyanthes trifoliata	f	.	f–d	.	f	lcd	.	edge, f	ld	f	.
Molinia caerulea	la	a–lf	f–la	a	a	f	.	a–va	f–a	o	f
Myrica gale	la	la	.	.	lf	.	o	o, 1	f	a	a
Narthecium ossifragum	f	.	.	.	.	.	.	f–a	f	o	.
Potentilla erecta	.	.	.	.	.	.	r	.	o–la	.	.
Scirpus caespitosus	d	cd	a–d	d	d	lcd	cd	d	.	.	f–a
Aulacomnium palustre	.	.						.	.	.	.
Campylopus flexuosus	.	.						1	.	l	.
Hylocomium loreum	.	.						.	.	.	.
H. splendens	o	+		Bryophytes and lichens not noted				.	.	.	.
Hypnum cupressiforme	o	+						1	o	.	.
H. schreberi	o	.						.	.	.	.
Rhacomitrium lanuginosum	.	.						.	.	.	.
Sphagnum compactum	.	+						.	.	.	.
S. cuspidatum var. submersum	a	.						.	+	.	.
S. cymbifolium	.	.						.	.	.	.
S. fallax	o	.						.	.	.	.
S. molluscum (tenellum)	a	.						.	+	.	.
S. papillosum	.	.						.	f	.	.
S. plumulosum	a	.						1	+	.	.
S. rubellum	.	.						.	.	.	.
S. subsecundum	.	.						.	.	.	.
Cladonia coccifera	f	f						o	.	.	f
C. floerkiana	.	.						o	.	.	f
C. sylvatica	.	.						f	.	.	.
C. uncialis	.	.						.	.	.	.

(5) Allt-a-Chraisg, 6 miles south of Altnaharra: slope of 10°. Wet, alt. 500 ft.

(6) Three miles south of Altnaharra; lat. 58° 16′ N., long. 4° 30′ W.; slope of 20° (more than 18 in. of peat); rather dry. Note the absence of *Sphagnum*, *Eriophorum*, *Narthecium*, *Myrica* and *Drosera*.

[(4)–(6) form a series of increasing dryness with increasingly steep slope.]

(7) Ord of Caithness (border of Sutherland and Caithness on the east coast); lat. 58° 8′ N., long. 3° 35′ W.; flat, not far from cliff edge, rather dry (more than 44 in. of peat), alt. 750 ft.

(8)–(11) Closely adjacent small areas on Rannoch Moor, head of Glencoe, Argyllshire, lat. 56° 39′ N., long. 4° 58′ W.; flat, slight differences of level, alt. 1000 ft.

  (8) Higher ground, peat shallow over rock (Eriophora absent, *Sphagnum* local).

  (9) Lower ground, flat peat (Eriophora abundant, Sphagna frequent).

  (10) Depression (*Eriophorum angustifolium* dominant, *Scirpus* absent).

  (11) Low haggs and pools with bare peat.

The characteristics of these lists are (*a*) the very restricted number of species of flowering plants, (*b*) the high constancy of nearly half the phanerogamic species present, (*c*) the presence of *Calluna* in every list. *Scirpus caespitosus* is present and usually dominant in every list but (10), which is from a local wet depression dominated by *Eriophorum angustifolium*. In the three driest areas, (2), (6) and (7), *Scirpus* and *Calluna* are co-dominant. This seems to indicate that the requirements of deer-sedge and heather are very nearly the same in this type of habitat, but nevertheless *Scirpus* is a true blanket bog *dominant*, while *Calluna* is not, in spite of its absolute constancy. *Scirpus* is typically dominant in the less wet areas of Highland blanket bog. When the dryness increases a very little further *Calluna* becomes co-dominant, but only attains its greatest luxuriance and overwhelming dominance on still drier peats which cannot be reckoned as bog.

Exmoor and Bodmin Moor. In England Scirpetum caespitosi is rare, but restricted areas occur in the south-west—Exmoor (Watson, 1932) and on Bodmin Moor (Magor). In north Devon *Scirpus caespitosus* is often more frequent on upland moors but is rarely dominant (Watson, 1932): in Cornwall it is rare (Magor). Occasionally, however, Scirpetum is formed in both counties and the following lists are given by Watson and by Magor from Exmoor and Bodmin Moor respectively. In the Bodmin Moor area the *Scirpus* is tufted, with bare intervals of black peat, as on the Wicklow Mountains, and one may infer from the lists that this area was decidedly wetter than that on Exmoor, which contains such species, unusual for Scirpetum caespitosi, as *Anthoxanthum odoratum* and *Deschampsia flexuosa*. Including the margin of the Exmoor area nine phanerogams are common to the two areas. Of these, four (*Calluna*, *Erica tetralix*, *Eriophorum angustifolium* and *Narthecium*) are the most constant in the eleven Highland

Composition of Scirpetum

areas listed above, quite constant in the five examples of Wicklow Scirpetum (p. 712), and are four out of the five given by Fraser as constant at Inverliver. *Potentilla erecta* comes next in constancy, and then *Drosera rotundifolia*.

Scirpus caespitosus d

	Exmoor	Bodmin Moor		Exmoor	Bodmin Moor
Anthoxanthum odoratum	o	.	Polygala serpyllifolia	o	r
Calluna vulgaris	o	f	Potentilla erecta	o	o
Carex binervis	.	o	Rhynchospora alba	.	a
C. caryophyllea	.	o	Scirpus fluitans	.	o
C. diversicolor (flacca)	.	r	Sieglingia decumbens	.	o
C. inflata	.	la	Vaccinium myrtillus	.	r
C. oederi	.	o			
Deschampsia flexuosa	a	.	Calypogeia fissa		
Drosera rotundifolia	m	va	f. aquatica	o	.
Empetrum nigrum	m	.	Campylopus flexuosus	a	.
Erica cinerea	r	.	Cephalozia bicuspidata	a	.
E. tetralix	r	a	Cephaloziella starkii	f	.
Eriophorum angusti-			Cladonia sylvatica	m	o
folium	m	va	Hylocomium squarrosum	o	.
E. vaginatum	.	f	Hypnum cupressiforme		
Juncus bulbosus	.	o	var. ericetorum	a	.
J. squarrosus	.	f	Polytrichum commune	.	a
Luzula multiflora			Rhacomitrium lanugin-		
f. congesta	o	o	osum	.	o
Molinia caerulea	.	va	Sphagnum cymbifolium		
Nardus stricta	a	o	f. congestum	o	.
Narthecium ossifragum	m	a	S. quinquefarium	o	.
			Sphagnum spp.	.	o

m = marginal.

Cleveland moors. On the eastern side of England Scirpetum is practically absent. Elgee (1914) describes it as "very rare" on badly drained clayey shale with thin peat on the Clevelands in north-eastern Yorkshire. *Erica tetralix* is said to be usually in about equal proportion to the *Scirpus*, and either may be dominant. "Other frequent though quite subordinate species are *Eriophorum vaginatum*, *Molinia caerulea* var. *depauperata*, *Juncus squarrosus* and *J. conglomeratus*."

Wicklow Mountains. In Ireland Scirpetum caespitosi is well developed on the Wicklow Mountains south of Dublin, where it was first described by Pethybridge and Praeger (1905). Here it occupies the flatter, less well-drained slopes between about 1250 and 2000 ft. (c. 380–610 m.).[1] It is "as a rule mixed with a considerable amount of stunted and apparently poorly thriving *Calluna*. The soil is thick peat and except in the hottest part of summer, when the soil may be comparatively dry, it is thoroughly saturated with water". In its "purest form" this Scirpetum "resembles a lawn",

[1] See Fig. 89, p. 505.

having a uniform height of about 6 in. (15 cm.) and contains the following species (five samples):

Andromeda polifolia	(2)	Eriophorum vaginatum	(2)
Calluna vulgaris	(5)	Narthecium ossifragum	(5)
Drosera rotundifolia	(2)		
Empetrum nigrum	(1)	Cladonia sp.	(2)
Erica cinerea	(1)	Rhacomitrium lanuginosum	(1)
E. tetralix	(5)	Sphagnum spp.	(4)
Eriophorum angustifolium	(5)		

More commonly, however, the *Scirpus* forms large separate tufts, the interspaces of "soppy peat" containing considerable amounts of *Calluna*,

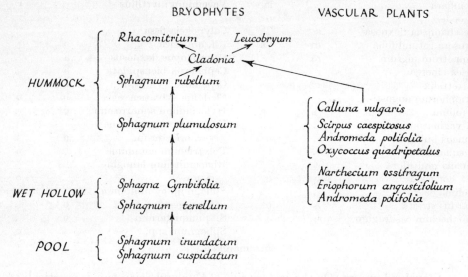

FIG. 141. SERAL STRUCTURE OF A HUMMOCK IN THE SCIRPETUM (PLATEAU BLANKET BOG) OF THE WICKLOW MOUNTAINS

The flowering plants associated with the Sphagna of succeeding developmental phases are shown in the right-hand column in approximate positions of greatest frequency.

Erica tetralix and *Eriophorum angustifolium*. *Narthecium ossifragum* is characteristic, and much more abundant than in other wet peat communities of the region. On the Wicklow Mountains *Andromeda polifolia* is confined to this community, and on its leaves the ascomycete *Rhytisma andromedae* Pers. occurs (Pethybridge and Praeger, 1905).

There is a good deal of hummock formation in the Wicklow Scirpetum, as in parts of the blanket bog of the west. The hummock succession as we[1] saw it in 1935 is quite similar, and the structure may be represented as in Fig. 141. The bryophytes are the actual builders of the hummocks, the vascular plants colonising at the levels indicated.

[1] Godwin, Osvald and Tansley.

Besides the species noted above we recorded in 1935 *Hylocomium splendens, Hypnum schreberi* and *Antitrichia curtipendula* below the *Calluna*. On flat ground at an altitude of about 1800 ft. (550 m.) at Sally Gap the Scirpetum had much less *Calluna*, with *Sphagnum subsecundum, S. papillosum* and *S. tenellum* in fairly extensive patches: also a little hummock building by *S. plumulosum* and *S. rubellum*.

This Scirpetum is still actively forming peat, and where it abuts on Callunetum the surface of the soil is seen to be several feet higher than in the latter. It is from the Scirpetum alone, according to Pethybridge and Praeger, that peat is cut for fuel in the Wicklow Mountains, and the increased drainage caused by the turf-cutting leads to invasion by the Callunetum.

In this area, as elsewhere in the British Isles, the peat is being extensively eroded in places and many of the Wicklow plateau bog communities are in process of degeneration. Remains of *Pinus sylvestris* and *Betula* (Sub-Boreal) have been found in the *Scirpus* peat at 1250 and 1700 ft. respectively, indicating drier climatic conditions anterior to the formation of the sub-Atlantic peat.

(6) **Molinietum caeruleae.** This community has already been described (Chapter XXVI, pp. 518–23) as a type of acid grassland. *Molinia* has a wide range of habitats, particularly in regard to pH value. We may recall that it resembles *Nardus* in rooting depth and in its ability to colonise mineral soil *or* peat, but requires both more water and a greater degree of root aeration, probably also a larger supply of mineral salts. On mineral soils it often forms, like *Nardus*, a peaty sod about 9 in. thick and easily separable from the mineral soil below. But like *Nardus* also it colonises peat formed by other species; and Molinietum, like Nardetum, as W. G. Smith remarked, may be described as "marginal", in a wide sense, to the great peat areas.

In the wet acid peat succession it is one of the "drier" communities, colonising in abundance, as it does, the drier parts of the mixed blanket bog of western Ireland (pp. 715 ff.), and also the sides of the older *Sphagnum* hummocks. It apparently plays a similar part in the Scottish Highlands. But the plant does not always become dominant in such situations, and Molinietum is probably always to be regarded as a transitional or marginal community. It is also said to cause the breakdown of peat already formed by other plants.

Callunetum vulgaris. When the surface of any bog or moss becomes dry the ling, which is nearly always already present, often in abundance, but is not luxuriant, tends to establish dominance; but this community is not to be regarded as part of the bog or moss formation. [See p. 710 and Chapter XXXVI, p. 724.]

IRISH BLANKET BOG

The plateau bog on the Wicklow Mountains in the east of Ireland has already been described.

Parts of the extreme west of Ireland are very largely covered by a thick layer of more or less wet acid peat, wherever the ground is flat or not too steeply sloping, supporting a carpet of moss or bog vegetation.[1] This is the same kind of bog which covers much of western and north-western Scotland and the Hebrides (there dominated mainly by *Scirpus caespitosus*), and belongs to the same general type as the Scirpetum caespitosi of the Wicklow Mountains and the cotton-grass bogs of the Pennine plateaux. The name "blanket bog" was suggested (in 1935) for this kind of vegetation, since it covers the country like a blanket in regions of high precipitation and very high atmospheric moisture; and there is no reason to suppose that it is built up (except locally) like the raised bogs of the central plain of Ireland, on fen or on the sites of old lakes through a stage of fen. Nevertheless the vegetation it supports is mainly composed of the same species as those of the raised bog.

Connemara and western Mayo. Western Galway and Mayo are typical regions in which by far the greater part of the surface is covered by blanket bog.[1] This is a country of acidic metamorphic rocks rising in bold rounded mountains between 2000 and 3000 ft. (600–900 m.) above the great stretches of flat or gently undulating land which lie but little above sea level. The bog covers the flat country and the greater portion of the lower mountain slopes up to an angle of at least 15°. Only where the rock stands out above the peat blanket[2] is the vegetation heathy, mainly dominated by *Calluna* and *Ulex gallii*, and in the more sheltered places by dwarf wood of *Quercus sessiliflora* with *Betula pubescens*, or by scrub composed of *Corylus*, *Crataegus*, *Salix aurita*, *Sorbus aucuparia* and *Ilex*. The blanket bog is the real climatic climax, the scrub and dwarf woodland are local communities belonging to the xerosere and conditioned by better drainage. At one time, however, probably in the Sub-Boreal period, this country was covered with pine forest, as is shown by the layer of pine stools below the bog peat which belongs to the more recent period (Sub-Atlantic), and has been formed during the last two and a half millennia.

The Connemara bog in western Galway is studded with innumerable lakelets of all sizes, from small pools in the peat to considerable areas of water, enclosed, or partly enclosed, by rocky rims supporting heath or a little scrub. In places there are low-lying basins, evidently the sites of old lakes which have become covered with bog vegetation, and some of these are marked by relict *Phragmites* and *Cladium*, the vestiges of old marginal reedswamps. Poorly developed reedswamp of the same species occurs on the margins of some of the existing lakes, with characteristic aquatic

---

[1] Pl. 121.   [2] Pl. 121, phot. 296.

PLATE 121

Phot. 296. The blanket bog of Connemara, near Roundstone. The expanse of bog is broken here and there by rocky outcrops. The "Twelve Beinns" behind. *Mrs Cowles.*

Phot. 297. The blanket bog of north-western Mayo with bog pools. Near Sheskin Lodge. *R. J. L.*

WEST IRISH BLANKET BOG

PLATE 122

Phot. 298. Rhynchosporetum albae (white heads) with *Myrica gale* (right and left centre) in foreground. Schoenetum nigricantis (black heads) behind. Large hummock capped with *Leucobryum glaucum* in background. R. J. L.

Phot. 299. Closer view of the same large hummock, eroded on the right by prevailing wind and covered with *Leucobryum glaucum* and *Molinia caerulea*. *Schoenus nigricans* and *Myrica gale* in foreground. R. J. L.

DETAIL OF CONNEMARA BLANKET BOG

plants such as *Lobelia dortmanna* and the American species *Eriocaulon septangulare*.

Peat. The pH values obtained from Connemara mixed blanket bog peat with *Schoenus* were 4·4–4·6. In bogs developed round small lakes with *Molinia* and *Myrica* present the pH was as high as 5·36. These permanently wet blanket bog peats appear to be all reducing (Pearsall, 1938).

Flora and vegetation. Over much of the blanket bog there is no general dominant. The vegetation consists mainly of a varying mixture of the following abundant species:

Eriophorum vaginatum	Schoenus nigricans
Molinia caerulea	Scirpus caespitosus
Rhynchospora alba	

Other more or less frequent species are:

Calluna vulgaris	Eriophorum angustifolium
Carex lasiocarpa	Menyanthes trifoliata†
C. panicea	Myrica gale
Drosera anglica	Narthecium ossifragum
D. intermedia	Pedicularis sylvatica
D. rotundifolia	Potamogeton polygonifolius†
Erica tetralix*	

* *Erica mackaiana*, which is found only here and in the Pyrenean region, occurs mainly on shallow peaty soil round rocks or by the wayside, where it hybridises freely with the closely allied *E. tetralix* to form the variable *E. praegeri*.

† These two species, which are of fairly frequent local occurrence in the general bog community, are probably, like *Cladium* and *Phragmites*, relicts from more open water conditions.

Of the five more abundant species mentioned, *Rhynchospora*, *Schoenus* and *Molinia*[1] are dominant, as we have seen, over considerable areas. The wetter parts of the bog are often dominated by *Rhynchospora alba*, and the drier by *Molinia caerulea* var. *depauperata*. In the shallow hollows bare peat is often covered by the purple-black conjugate alga *Zygogonium ericetorum*, and sparsely colonised by *Carex limosa*. *Rhynchospora alba* follows, and there may be a succession Zygogonietum →Rhynchosporetum → Molinietum as the general surface of the bog rises. *Schoenus nigricans* is a local dominant of parts of the bog, but not *Eriophorum*, and rarely *Scirpus caespitosus*.

The bryophytes, and especially species of *Sphagnum*, certainly play an important part in the formation of local hummocks,[2] though hummock formation does not constitute a universal genetic feature of the blanket bog, as it does of the raised bog. Pockets of *Sphagnum* are frequent on the surface of the peat among the higher plants, *S. subsecundum* on the lower ground, while *S. plumulosum*, *S. papillosum* and *S. rubellum* are the three most important hummock-building species, as on the raised bog. As the hummock increases in height, establishing a substratum better drained than

[1] Pl. 122.   [2] Pl. 122.

that of the general bog peat, it is colonised by *Molinia*, and by *Calluna*, which here grows and flowers much more luxuriantly than in the general bog. Finally *Cladonia sylvatica*, sometimes associated with *C. uncialis*, often joins *Calluna*, or the hummock is capped with a cushion of *Rhacomitrium*, more rarely of *Leucobryum*. Other bryophytes of the bog are *Campylopus atrovirens*, which forms conspicuous black patches, *Aulacomnium palustre* and *Dicranum* spp.

Blanket bog in north-western Mayo. The vegetation of the blanket bog on the hills near Glen Cullin, in the north-west of Mayo, was analysed. Parts of this were flat with a fairly even surface, and from a comparison of several neighbouring areas the following list may be taken as typical:

Flat bog between Carrefull and Glen Cullin (Co. Mayo)
(altitude 250 ft. (c. 75 m.))

Vascular plants:		Bryophytes and lichens:	
Schoenus nigricans	d	Campylopus atrovirens	f–a
Rhynchospora alba	ld	Dicranum sp.	o
Molinia caerulea	va	Leucobryum glaucum	o
Eriophorum vaginatum	a	Rhacomitrium lanuginosum	f
Scirpus caespitosus	a	Sphagnum cuspidatum	f
Narthecium ossifragum	f	S. cymbifolium	o
Calluna vulgaris	f	S. platyphyllum	o
Erica tetralix	f	S. plumulosum	f
E. cinerea	l	S. rubellum	f
Anagallis tenella	o	S. subsecundum	o
Drosera rotundifolia	o	S. tenellum	o
D. anglica	o	Blepharosia sp.	o
Eriophorum angustifolium	o	Cladonia sylvatica	f
Pinguicula vulgaris	o	C. uncialis	o
Potentilla erecta	o		
Polygala vulgaris	r		

Hummock formation, based mainly on the activity of *Sphagnum plumulosum* and *S. rubellum*, is local in these bogs. There is very often a capping of *Rhacomitrium lanuginosum*, which may form the bulk of the hummock. *Sphagnum plumulosum* usually starts the hummock by settling in a tussock of *Schoenus*, *Scirpus* or *Eriophorum vaginatum*, and still occurs round the base of the hummock when growth has been completed. It is followed by *Sphagnum rubellum* or directly by *Rhacomitrium*, and the sides of the mound so formed are colonised by *Molinia*, *Scirpus caespitosus* and *Eriophorum vaginatum*, the summit by *Calluna* and *Cladonia sylvatica*, sometimes by *Erica cinerea*.

Hummock structure and flora. The hummocks are usually quite small in size, rarely as much as a metre in diameter and 30–40 cm. in height; but occasionally a very large one is met with. One of these (Fig. 142) was 6 m. in diameter, the summit 75 cm. above the higher and 1·3 m. above the lower general surface of the sloping bog. Its base was formed as usual of *Sphagnum plumulosum*, and higher up its sides were *S. rubellum* and some

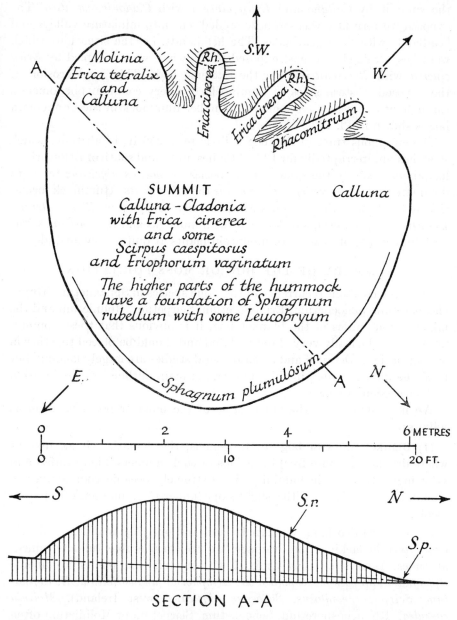

FIG. 142. DIAGRAM OF EXCEPTIONALLY LARGE HUMMOCK IN THE BLANKET BOG OF WESTERN MAYO. Drawn to true scale

Leucobryum glaucum. The slope of the exposed western side reached 25°, while the gentler eastern slope did not exceed 20°. The slopes were occupied by *Molinia*, *Eriophorum vaginatum* and *Scirpus*, with some *Erica tetralix*, the summit by *Calluna* and *Erica cinerea* with *Cladonia sylvatica*. The exposed western face was strongly eroded, cut into miniature valleys and headlands with bare peat sides. The headlands, i.e. the situations which were best drained and most exposed to the wind, were occupied by *Erica cinerea* with *Rhacomitrium* on the extreme ends. This wind erosion of the exposed western sides of hummocks is a very common phenomenon throughout the region. A somewhat smaller hummock in the Connemara bog is shown in Pl. 122.

The enormous tract of flat bog a little over 250 ft. in altitude, which stretches uninterruptedly for 12 or 14 miles north and south of Bellacorick,[1] has almost exactly the same list of species. *Schoenus nigricans* tends to dominate the wetter, *Scirpus caespitosus* the drier, parts. Hummock formation is local,[2] and the summits are generally occupied by *Rhacomitrium*. *Empetrum nigrum* occurs on some of the hummocks and *Lycopodium selago* on bare patches of peat—two mountain plants at quite a low altitude.

SUMMARY OF THE BOG OR MOSS FORMATION

From the constant recurrence of the same species in Sphagnetum, Rhynchosporetum, Schoenetum, Eriophoretum, Scirpetum, Molinietum and the mixed communities of the blanket bog, it is obvious that these communities are ecologically very closely related and should be placed together in one formation. Detailed and accurate seral studies are largely lacking, but from the data recorded in this chapter the main outlines of classification and succession are clear.

We may summarise the chief *forms* of the moss or bog formation as follows:

(1) **Blanket moss or bog**—ombrogenous, the climatic climax (except where drainage is quite free) in regions of cool summers, high rainfall and very high atmospheric humidity, i.e. extremely oceanic cool temperate climate. Surface flat or with a slight slope (under 15°): hummock formation local.

(*a*) Western Scotland, western Ireland and outlying islands, on almost every terrain lacking free drainage, up to considerable (undetermined) altitudes. Dominants usually mixed, with the following plants prominent: *Sphagnum* spp., *Rhynchospora alba*, *Eriophorum angustifolium*, *E. vaginatum*, *Scirpus caespitosus*, *Schoenus nigricans* (west Ireland), *Molinia caerulea*. Rhynchosporetum, Schoenetum, Scirpetum or Molinietum often developed over considerable areas.

(*b*) Plateaux and gentle slopes of mountain masses (Pennines, Wicklow

[1] Pl. 121, phot. 297.   [2] Pl. 123, phot. 300.

PLATE 123

Phot. 300. A local group of hummocks, each capped with *Rhacomitrium lanuginosum*. Pool with *Menyanthes trifoliata* in foreground. Near Sheskin Lodge, Co. Mayo. *R. J. L.*

Phot. 301. Hummock with *Sphagnum* spp., *Cladonia sylvatica*, *Molinia caerulea*, *Eriophorum angustifolium* and *Rhynchospora alba* (at base of hummock in front). *Potentilla erecta* on the right. Between Moylaw and Eskeragh, Co. Donegal. *R. J. L.*

HUMMOCKS IN BLANKET BOG

Mountains, Dartmoor, etc.) at elevations of 1500–2200 ft. (c. 450–650 m.) in less extremely maritime regional climates, where the elevation considerably lowers temperature and increases precipitation and atmospheric humidity compared with surrounding lowlands. Scirpetum, Eriophoretum, Calluno-Eriophoretum, Vaccinio-Eriophoretum.

(2) **Valley bog**—topogenous, formed in valleys and depressions where water, draining from acidic rocks, stagnates, and bog plants establish themselves. Surface flat or concave. In addition to the bog species proper other species of more exacting requirements are generally present. Insufficiently known.

(3) **Raised moss or bog**—topogenous, often based on the site of a former lake and built up above fen peat or valley bog peat, in less extremely humid climates. Surface more or less convex as a whole, with a marginal *lagg* or drainage channel which receives the drainage water. Surface composed in detail of hummocks and hollows determined by the constituent units of Sphagnetum, each built up from hollow to hummock by a succession of different species of *Sphagnum*, and colonised by various vascular plants, of which *Rhynchospora alba* and *Narthecium ossifragum* dominate the hollows, *Scirpus caespitosus*, *Eriophorum vaginatum*, *Calluna vulgaris* with *Cladonia sylvatica* the hummocks. *Scirpus*, *Eriophorum* and *Calluna* give the general tone to the bog.

Characteristic of the central plain of Ireland: a few in England and Wales and a good many in Scotland, especially the south-west, the Midland Valley and parts of the eastern coastal plain.

The blanket moss formation of the Pennine plateaux, the Wicklow Mountains, and certain other areas, is in one sense a relict of conditions introduced by the Pleistocene ice-age, for it occupies plateaux and gentle slopes which have been planed by ice and are often covered by a layer of impermeable boulder clay or "till", the ground moraine of the ancient ice-sheets, which holds up the drainage water. The peat made by the bog plants, and on which they still grow, began to be formed in wet post-glacial climates and often contains the remains of birch and pine forests developed during drier climatic phases. The latest (Sub-Atlantic) climate was cool and moist on these sub-oceanic mountains and has favoured moss development, but to-day many of the peat bogs are subject to severe erosion, probably due to the combined action of wind and water, so that the peat is extensively "hagged", large areas laid bare and sometimes removed down to the mineral soil below. Other mosses are progressively desiccated and occupied by other communities such as Vaccinietum, Callunetum and Nardetum, a process hastened of course by draining and burning.

The "lowland mosses" at quite slight elevations, are of the raised moss type, except in western Scotland and western Ireland, and were originally formed on the sites of old estuaries and lakes. In England most of them have been destroyed by drainage.

The accompanying scheme (Fig. 143) attempts to relate the main bog types to the hydroseres, to Callunetum, and to forest.

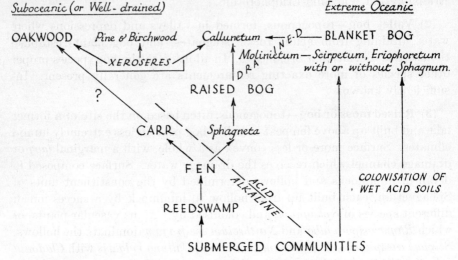

FIG. 143. SCHEME OF HYDROSERES IN RELATION TO FEN, BOG OR MOSS, CALLUNETUM AND FOREST

REFERENCES

ADAMSON, R. S. On the relationships of some associations of the Southern Pennines. *J. Ecol.* 6, 97–109. 1918.

CRUMP, W. B. The coefficient of humidity: a new method of expressing the soil moisture. *New Phytol.* 12, 125–47. 1913.

ELGEE, FRANK. The vegetation of the eastern moorlands of Yorkshire. *J. Ecol.* 2, 1–18. 1914.

FRASER, G. K. Studies of Scottish moorlands in relation to tree growth. *Bull. Forest. Comm.* No. 15. 1933.

LEWIS, F. J. Geographical distribution of vegetation in the basins of the rivers Eden, Tees, Wear and Tyne. *Geogr. J.* 1904.

LEWIS, F. J. In "The British Vegetation Committee's Excursion to the West of Ireland". *New Phytol.* 7, 253–60. 1908.

LEWIS, F. J. and MOSS, C. E. Chapter XII, "The Upland moors of the Pennine Chain", in *Types of British Vegetation*, pp. 266–82. Cambridge, 1911.

MAGOR, E. W. Geographical distribution of vegetation in Cornwall: Camelford and Wadebridge district (unpublished).

MOSS, C. E. *Vegetation of the Peak District.* Cambridge, 1913.

MOSS, C. E. See LEWIS and MOSS. 1911.

PEARSALL, W. H. The soil complex in relation to plant communities. III. Moorland bogs. *J. Ecol.* 26, 298–315. 1938.

PETHYBRIDGE, G. H. and PRAEGER, R. LLOYD. The vegetation of the district lying south of Dublin. *Proc. Roy. Irish Acad.* 1905.

RANKIN, W. M. "The Lowland Moors of Lonsdale" in *Types of British Vegetation*, pp. 247–59. 1911.

WATSON, W. The bryophytes and lichens of moorland. *J. Ecol.* 20, 284–313. 1932.

Part VII

HEATH AND MOOR

Chapter XXXVI

THE HEATH FORMATION

The heath formation of western Europe, of which the chief consociation is characteristically dominated by the common heather or "ling", *Calluna vulgaris* Salisb., occurs throughout the British Isles, and its presence is determined by a complex of factors. Of these the relatively moist air associated with an oceanic or suboceanic climate is one, while at least moderately free soil drainage and soil conditions (apparently mainly a low base status) correlated with high acidity, which permit the existence of the symbiotic fungus *Phoma* on which the ling seems to depend in nature (Rayner, 1915), are others.

For the dominance of heath some factor which prevents the successful gregarious establishment of trees must also be present, e.g. violent winds, recurrent fires, or a certain intensity of grazing, because in the absence of such factors trees will grow and may become dominant on most of the soils which are covered by heath when the factors hostile to trees are effective (see Chapter XVII). On the western coasts of Europe (apart from vegetation determined by salt spray) and on the exposed slopes of Irish and British oceanic and suboceanic hills where more or less violent winds are frequent, up to a level of about 2000 ft. (*c.* 600 m.), heath is commonly dominant. This is in the absence of *intensive* grazing, for while heath will support a certain number of animals, anything like heavy and continuous pasturing, either by sheep or rabbits, will quickly convert it into grassland.

On the western coasts and on more inland hills these conditions are satisfied on very various types of rock other than limestone, provided the slope is steep enough or the soil is sufficiently permeable to secure fair drainage, and heavy grazing is absent. Even on flat surfaces of limestone the strong leaching of the surface soil resulting from high rainfall, and the concurrent accumulation of acidic humus by colonising lichens and mosses, may produce acid conditions enabling *Calluna* and its symbiotic fungus to establish themselves, and once established the ling can even send its roots down into calcareous soil. Thus where the climate is exceptionally favourable to it we may have heath developed over limestone rock.

The lowland heaths so characteristic of acid sandy soils in Great Britain and north-western Europe generally depend for their maintenance on the anthropogenic factors of fire or felling.

Distribution. The general European distribution of heath follows the cool temperate oceanic and suboceanic climatic regions pretty closely. Thus it occurs throughout the British Isles, in southern Scandinavia, Denmark, north-west Germany, Holland, Belgium, and through most of

France. Northwards it is replaced by the arctic dwarf scrub vegetation, southwards by the Mediterranean *mâquis*. Eastwards heath becomes rarer and rarer till it disappears altogether as the climate becomes thoroughly continental in type, though *Calluna* itself occurs as far as the Ural Mountains and even beyond, but there it is said to be confined to woods.

Though forming luxuriant Callunetum only on soil with free drainage, the ling itself can nevertheless tolerate a considerable quantity of soil water and occurs mixed in various proportions with *Nardus*, *Molinia*, *Scirpus caespitosus*, and *Eriophorum*, over practically the whole of our northern and western peat moors and also, as we have seen, habitually colonises the drier parts of *Sphagnum* bog. If the surface dries out the ling commonly becomes dominant and forms genuine Callunetum. These "heather moors" lack certain species found on the southern English heaths with their drier climate and possess others which are characteristic of the wetter and colder climate and soil (see Chapter XXXVII).

LOWLAND HEATH (CALLUNETUM ARENICOLUM)

Southern heaths. On the Neogenic rocks of various age in the suboceanic climate of the midlands, eastern and south-eastern England heath is almost always developed on sandy or gravelly soil poor in bases and therefore acid in reaction and freely permeable to water—much more rarely on clays and loams; and the same is true of its occurrence on the continent of Europe. As we have already seen (Chapter XVII, p. 354) this lowland heath is in general a biotic or a fire climax and gives way to woodland in the absence of arresting factors, though there are apparently some heath soils which will not support tree growth.

Soil—the podsol profile. The usual subsoil of a southern English heath is a sandstone, sand or gravel, poor in basic salts and acid in reaction. Where a typical podsol profile is developed the horizons are as follows.

The soil surface is covered by a layer of raw humus (compacted dry peat or *mor*) varying from a centimetre or two to 4 or 5 in. (c. 10 or 12 cm.) in thickness ($A0$), usually derived largely from the mosses and especially the lichens—mainly species of *Cladonia*—which constitute the ground layer of the Callunetum. In extreme cases, as on the top of Hindhead (800–900 ft. $= c.$ 250–270 m.) in south-west Surrey the peat may reach a depth of 12 in. (30 cm.) (Haines, 1926). Below the dry peat comes a layer of humous sand ($A1$) commonly stained a grey-chocolate colour by the humous substances carried down from the surface peat, and containing white ("bleached") sand grains. This may pass into a pale whitish or ash-coloured layer of sand ($A2$) practically devoid of humus. The $A2$ horizon is however often absent, the "bleached" podsol layer being represented only by $A1$. Below this sand come the "pan" layers (B): first the "humus pan" ($B1$) which may be quite black but is usually dark brown, soft and friable in consistency, though often sharply bounded; and then the "iron pan" ($B2$)

containing ferric salts and typically a dark red-brown. $B1$ and $B2$ are, however, not always well separated. The pan is often underlain by somewhat light reddish sand ($B3$) into which some of the iron salts have been carried. This passes down into the unaltered sand (C) below. Fig. 144 is reproduced from Farrow's diagram of a typical soil profile from the Callunetum of Cavenham Heath in Breckland (north-west Suffolk).

Here we have all the features of a typical podsol, developed in a climate which is not a typical podsol climate, and probably owing its characters to the combination of the characteristic vegetation and the highly permeable soil initially very poor in basic salts. Great variation however occurs in the extent and definiteness of the podsol profile under Callunetum. Very often, as has been said, there is no distinct "bleached" layer, the whole of the A horizon being coloured by humous substances. Often also the "pan" is not sharply differentiated, though at least an ill-defined illuvial layer (B) can generally be made out. In gravelly soils, particularly, the differentiation of the horizons may be badly defined. In other cases there may be several B horizons with bleached layers between. The causes of these variations are still largely obscure. It has been suggested that many of the existing podsol profiles are "fossil", i.e. owing their origin to climatic conditions no longer present. Many have certainly been "truncated", i.e. the top layers eroded with subsequent fresh podsolisation.

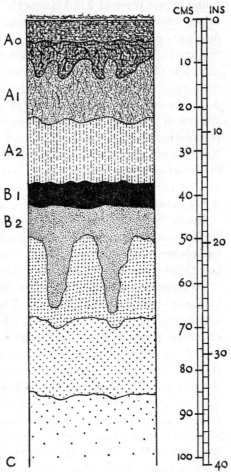

FIG. 144. A HEATH PODSOL PROFILE

For description see text. (Reprint of Fig. 28.) After Farrow, 1915.

Hindhead Common. The soil of the Callunetum on Hindhead Common in Surrey, developed over the Hythe Beds of the Lower Greensand and very thoroughly studied by Haines (1926), consists of a surface peat very dense and dark, especially in the lower part, slightly reddish brown, and 6–12 in. (15–30 cm.) thick, or even more—extremely deep for a southern English heath. This peat has a very low base status, is definitely poor in dissociable mineral matter, probably possessing no clay-forming material at all, and has a mean pH value of 3·42 at 2 in. (5 cm.) and of 3·9 at 9 in. (22·5 cm.), very few samples falling outside the range 3·0 to 3·6 at the former, and 3·5 to 4·3 at the latter depth. An extreme pH value of 2·3 was found in one place at 2 in.[1] Below the peat there was usually a thin layer of mixed humus and disintegrated sandstone not more than about a centimetre deep. Then came a layer of sand increasingly consolidated below, till the hard compact parent sandstone was reached. No trace of "pan", nor any clear illuvial layer, was found anywhere on the heath. There is probably little or no leaching, owing to the thickness and denseness of the peat. A rise in salt percentages found in some cases at 9 in., especially in the valleys, is probably due to disintegration of parent material and to lateral percolation from the higher ground in the layer of unconsolidated sand. No trace of calcium could be found in any of the samples, nor could nitrates be detected in the ordinary way; but from the appearance of a trace in an extract of unusual strength it was estimated that the nitrate content (as KNO_3) may have been of the order of 0·00027 per cent.

Heath fires occur at Hindhead every few years and consume most of the surface peat. The data given above refer mainly to an area in which 15 years had elapsed since the last fire and a considerable depth of peat had accumulated, but even here Haines regards the profile as still immature.

Structure and Flora. The structure of the Callunetum at Hindhead is typically two- or three-layered. The heather, which is generally overwhelmingly dominant, varies in height from a few inches to about 2 ft. 6 in. or 75 cm. (even more in old "leggy" growth) according to the soil and the age of the community, corresponding with the time that has elapsed since the last heath fire. A second ill-defined and interrupted layer developed beneath the dominant consists of flowering plants which can endure shade, such as the bilberry (*Vaccinium myrtillus*) and the dwarf gorse (*Ulex minor*). Both of these are local on the south English heaths but may be very abundant where they occur. Neither flowers under these deeply shaded conditions. The third or ground layer consists of mosses and lichens, *Polytrichum juniperinum*, *Hypnum cupressiforme* var. *ericetorum* and *H. schreberi* being the commonest mosses, while species of *Cladonia* are dominant among the lichens. It is largely this ground layer and especially *Cladonia*, typically forming by far its greater portion, that produces the raw material of the surface peat.

Erica cinerea[2] fairly often shares dominance with *Calluna*, and is sometimes dominant alone on rather dry sunny slopes. According to Fritsch and his co-workers (1913, 1915) the purple heath plays a prominent part in the development of the Callunetum after fire at Hindhead, and is therefore presumably somewhat more xerophytic than *Calluna*, at least when the

---

[1] This seems to be the highest acidity recorded for a British soil.
[2] Pl. 124, phot. 303; Pl. 125, phots. 307–8.

PLATE 124

Phot. 302. Young vigorous Callunetum with scattered Pteridium fronds of low growth. Toys Hill, Kent, on Hythe Beds. *S. Mangham*.

Phot. 303. Sandy heath on the raised beach of the Ayreland of Bride, Isle of Man, with *Erica cinerea* and *Ulex gallii*. From *E. J. Moore* (1931).

Phot. 304. Dense Pteridietum of low growth fringed with *Calluna* and *Vaccinium myrtillus* on edge of a chert quarry. Hythe Beds, Crockham Hill Common, Kent.

FACIES OF LOWLAND HEATH

PLATE 125

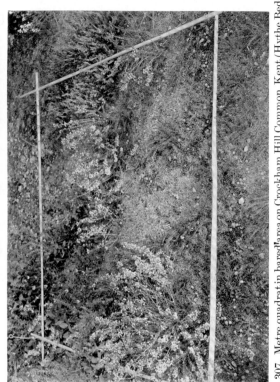

Phot. 307. Metre quadrat in bared area on Crockham Hill Common, Kent (Hythe Beds). *Erica cinerea*, *Calluna* and *Teucrium scorodonia* displacing *Polytrichum piliferum* and *Agrostis canina* turf (centre and right). Rabbit grazed; note abundant dung.

Phot. 308. Metre quadrat, later seral stage. *Erica cinerea, Ligustrum ...*

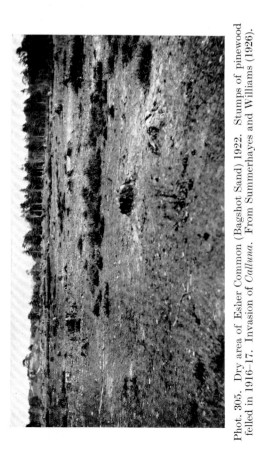

Phot. 305. Dry area of Esher Common (Bagshot Sand) 1922. Stumps of pinewood felled in 1916–17. Invasion of *Calluna*. From Summerhayes and Williams (1926).

Phot. 306. The same area in 1925 with nearly closed Callunetum. Marked growth in height of birches. From Summerhayes and Williams (1926).

plants are young. Fritsch in fact is of opinion that the purple heath would always give way eventually in competition with the ling. *Vaccinium myrtillus*[1] may also replace *Calluna* as the dominant, but what determines such a replacement is not clear. *Erica tetralix* often dominates areas of wetter soil, but on some heaths whose soil is not wet, as on the ridge at Hindhead, it is scattered through the general Callunetum (Fritsch and Parker, 1913). This may be due to high precipitation and atmospheric moisture at the relatively high altitude (270 m.) of this summit.

Ulex minor is very abundant on many heaths of the south-eastern and south central English counties. It occurs in two growth forms, a prostrate or semi-prostrate plant, which may be subdominant to *Calluna* or occupy gaps in the Callunetum, and an erect form 70–120 cm. high forming small clumps of dwarf scrub (Fritsch and Parker, 1913; Skipper, 1922). The former, typically prostrate, but sometimes with erect branches reaching a height of 45–75 cm., is the common form, and is often an important member of the Callunetum: the tips of the shoots dry off if they rise above the general level of the *Calluna*. At Hindhead the latter form is confined to the valleys and enters into the common gorse community, mainly dominated by *Ulex europaeus*. This last is not a member of the Callunetum except as a sporadic invader, mainly along paths where its seeds are probably carried by ants (Weiss, 1908; 1909). It is doubtful if it ever colonises undisturbed heathland. Juniper is not uncommon on some south-eastern heaths.

The bracken fern (*Pteridium aquilinum*), like the common gorse, is frequently found in and about heaths, but is hardly a true member of the Callunetum. This aggressive plant spreads actively by rhizome growth from centres of infection, and heath soils suit it very well. But it does not as a rule appear to invade undisturbed heath and its growth is severely restricted by the dominance of the heath undershrubs, the fronds in thick young heath being small, widely spaced and not more than a foot or so high.[2] With increasing growth of the heaths it seems to succumb altogether, perhaps because the highly compacted surface peat interferes with soil aeration. Furthermore *Pteridium* cannot stand wind nearly so well as the heaths, so that it is absent from exposed plateaux and slopes (cf. Chapter XXVI, p. 502).

On the other hand Farrow (1917) found *Pteridium* advancing along a broad front and invading Callunetum on Cavenham Heath in Breckland, apparently killing the *Calluna*.[3] As soon as the *Calluna* was thickly interspersed with bracken fronds the plants became unhealthy and soon died, seemingly as the result of the dead fronds falling on and smothering the *Calluna* bushes (Fig. 145). The surface peat is here very thin and the sand loose, conditions favouring the horizontal growth of bracken rhizomes.

The following list is taken from the adult Callunetum of Hindhead Common on the Hythe Beds of the Lower Greensand (Fritsch and Salisbury,

[1] Pl. 124, phot. 304.   [2] Pl. 124, phot. 302; Pl. 125, phot. 308.
[3] Pl. 126, phot. 311.

1915) in which the *Calluna* was 75–100 cm. high (Fritsch and Parker, 1913). The shrubs (other than *Ulex* and *Sarothamnus*) and trees are omitted (see below under *Succession*):

Calluna vulgaris	d	Ulex minor	sd
Vaccinium myrtillus	la	Erica tetralix	l
Ulex europaeus	l	Sarothamnus scoparius	r
Agrostis tenuis	r	Molinia caerulea	o
Blechnum spicant	lr	Polygala vulgaris	r
Carex pilulifera	f	Potentilla erecta	r
Deschampsia flexuosa	o	Pteridium aquilinum	l
Galium saxatile	r	Sieglingia decumbens	r

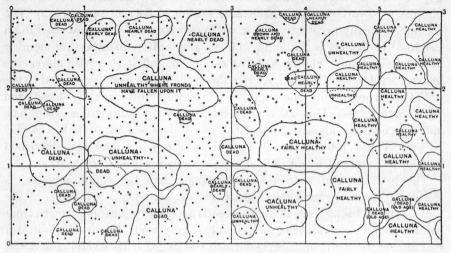

FIG. 145. INVASION OF CALLUNETUM BY *PTERIDIUM* ON CAVENHAM HEATH (BRECKLAND)

The invasion is from left to right. The bracken is just reaching the right hand edge of the chart, where its fronds are still sparse and the heather is thick and healthy. On the left where the bracken fronds are thick the heather is all dead. Each square represents a square metre. Cf. Pl. 126, phot. 311. From Farrow, 1917.

Mosses and lichens

Dicranum scoparium	f	Cladonia floerkiana	
Hypnum cupressiforme		f. trachypoda	l
var. ericetorum	f	C. pyxidata	lf
Leucobryum glaucum	f	C. sylvatica	f
Cladonia coccifera	r	Parmelia physodes	f

Algae and fungi

Gloeocystis vesiculosa	f	Mesotaenium violascens	f
Hormidium flaccidum	lr	Clitocybe sp.	f

Esher Common. An even poorer flora is recorded from the Callunetum of Esher Common and Oxshott Heath (Summerhayes, Cole and Williams, 1924) on Plateau Gravel and Bagshot Sand. This area is also in Surrey,

PLATE 126

Phot. 311. Callunetum (on the left) invaded and destroyed by Pteridietum from the right. Cavenham Heath, Breckland. From *E. P. Farrow* (1917).

Phot. 309. Displacement of Ammophiletum by Callunetum on the coast of South Haven Peninsula, Dorset. From *R. d'O. Good* (1935).

Phot. 310. Mature Callunetum invaded by young pines. Wych Cross, Ashdown Forest, Sussex, on Ashdown Sand. *S. Mangham.*

DEVELOPMENT AND DESTRUCTION OF CALLUNETUM

Flora

about 23 miles (c. 37 km.) north-east of Hindhead Common and at a much lower altitude. It is evidently drier than Hindhead Common. The dominant *Calluna* was about 45 cm. high.

Calluna vulgaris	d	Deschampsia flexuosa	
Erica cinerea	f	Potentilla erecta	
Ulex minor	lf	Cuscuta epithymum	
Dicranum scoparium	a	Polytrichum juniperinum	f
Hypnum cupressiforme		Leucobryum glaucum	o
var. ericetorum	a		
H. schreberi	f	Cladonia pyxidata	a

Cavenham Heath. From Cavenham Heath in Breckland—drier still— Farrow (1915) records *no* other angiosperms, except occasional seedlings of shrubs and trees, in *thick* Callunetum (though a good many grass heath species are found between the *Calluna* bushes where these are less crowded), but the following mosses and lichens occur in the ground layer:

Ceratodon purpureus	Cladonia alcicornis
Dicranum scoparium	C. cervicornis
Hypnum schreberi	C. coccifera
Leucobryum glaucum	C. furcata
Polytrichum piliferum	C. sylvatica
	C. uncialis

Sherwood Forest. From Sherwood Forest in Nottinghamshire, on the Bunter Sandstone, Hopkinson (1927) records a similar paucity of species in mature Callunetum. *Erica cinerea* accompanies the dominant where the latter is not completely closed. In other places the purple heath is dominant and the ling subdominant. This combination may perhaps represent a seral stage (see below under *Succession*). *Ulex minor* and *Genista anglica* are both also recorded, while *Ulex europaeus* is said to invade and displace the ling in some spots, "growing up from the centre of many well-grown shrubs of *Calluna*". Several of the mosses and lichens mentioned by Fritsch and by Farrow are also recorded by Hopkinson from the Sherwood Callunetum.

Thus it is clear that the mature closed Callunetum is exceedingly poor floristically. The lists of fairly numerous species given for the "heath association" in *Types of British Vegetation* and elsewhere include many which only occur in immature or modified Callunetum and properly belong to some seral stage or to the much richer "grass heath" flora described in Chapter XXVI (pp. 506 ff.).

SUCCESSION

(1) *The Burn Subsere*

This subsere is very commonly met with on heaths owing to the ease with which the vegetation is set alight and burns after prolonged dry weather. Fritsch and Salisbury (1915) give an excellent account of it as seen on Hindhead Common in Surrey. Hindhead Common, in fact, as they clearly show, is nothing but a mosaic of different seral stages, corresponding with the different periods which have elapsed since the last fire. Started sometimes from negligence or mischief, sometimes with deliberate intent to destroy the old "leggy" *Calluna* and make way for new and more vigorous growth, these fires constantly sweep across areas which have already been recently burned. Thus it is possible to recognise the effect of burning on different stages of the subseral development. When tall adult Callunetum is burned, there is a great mass of combustible material, the heat developed is greater, and more of the plants are killed outright by the fire. The proportion of ling and purple heath killed on a fire-swept area varies a good deal from this cause. A majority of the plants of *Ulex minor* always manage to survive. The sprouting of the surviving bases of the dwarf gorse after the fire aids materially in the rapid redevelopment of the formation, which, for the rest, is carried out by the establishment of sporelings of algae, lichens and mosses, and of seedlings of the undershrubs.

The pioneers—algae and lichens. According to Fritsch and Salisbury (1915, p. 129) the first colonists of the charred peat of burned Callunetum are "*Cystococcus humicola*, forming dark green granules in most of the countless small depressions of the soil", and the gelatinous unicellular algae, *Gloeocystis vesiculosa*, *Trochiscia aspera* and *Dactylococcus infusionum*, forming a thin, macroscopically invisible layer over the greater part of the surface. At a very early stage *Ascobolus atrofuscus* grows over the whole gelatinous layer in which its hyphae ramify profusely, enveloping groups of the *Cystococcus* in a dark pseudoparenchymatous investment and thus forming minute, rounded, olive-brown protuberances which give a velvety texture to the surface. Apothecia arise in quantity from the hyphae and are not specially developed from those associated with the *Cystococcus*. The *Ascobolus* persists on burned areas of all ages and also on parts of the mature heath. *Humaria melaloma* also occurs, but is rare.

The dark green *Cystococcus* granules also become pale yellowish green owing to investment by the colourless hyphae of another (unnamed) fungus, and thus lichens are formed. Of these "*Cladonia delicata* is at first the only one to be found in fruit, but other sterile forms are probably associated. The small scale-like lobes of their pale green thalli form numerous patches on the soil, and these are dominant for some years on the barer parts, until in fact the phanerogamic vegetation begins to close, when they become subordinated to other forms or completely disappear." Meanwhile the *Gloeocystis* and its associates, with *Mesotaenium violascens*, multiply over the general surface, making increasingly conspicuous dirty green jelly-like lumps which occupy practically the whole surface left bare by other forms. After about five years these lumps become visible to the naked eye on close observation. Finally, in the tall Callunetum the gelatinous algae form conspicuous masses a quarter to half an inch in diameter. Thus the algae and lichens form a closed layer from the first. The only other algae met with were *Hormidium flaccidum* and *Zygogonium ericetorum*, both of which were rare and local. Diatoms appear to be entirely absent.

Three or four years after a fire the following Cladoniae become associated with *Cladonia delicata*:

Cladonia furcata (first to appear)   C. squamosa
C. floerkiana f. trachypoda   C. sylvatica (rare at first)
C. pyxidata

It is suggested that the Cladoniae accumulate peat and thus allow of increasing luxuriance of *Calluna*. Measurements indicate that the average rate of growth of the ling is more rapid where the peat is deeper.

As the phanerogamic vegetation becomes more and more closed *Cladonia sylvatica* increases in amount, and with it are associated *C. pyxidata* and *C. coccifera*. The lichens reach their greatest development, both in number of species and individuals, in a phase of relatively poor development of *Calluna*, which on some heaths seems permanent, and in which *Cladonia sylvatica* is dominant in a richly developed and varied lichen flora (cf. Chapter XXVI, p. 511). In the tall Callunetum of Hindhead Common *C. sylvatica* is the only common lichen, apart from *Parmelia physodes* on the older stems and branches of the ling (see list on p. 728).

Mosses. Mosses occur in the earliest stages of the subsere, including *Ceratodon purpureus*, which is very abundant, and small patches of *Tortula subulata* and *Funaria hygrometrica*. In slightly later stages these are associated with *Campylopus brevipilus* and *Polytrichum piliferum*. The mosses form considerable sheets, but are not uniformly distributed over burnt areas like the algae and lichens. About two or three years after a fire *Ceratodon* tends to disappear and for a time the mosses are quite insignificant compared with the lichens. In the final stages a few species of moss are frequent (see list on p. 728), *Hypnum cupressiforme* var. *ericetorum* "attaining great development where light penetrates the *Calluna* canopy".

Vascular plants. Certain species which do not belong to the Callunetum (though they are often reckoned as "heath plants"), namely *Pteridium aquilinum*, *Molinia caerulea* and *Deschampsia flexuosa*, spread considerably after a fire, mainly from the valleys in which they are abundant, on to the peat bared of the ling, but they do not in the long run appear to extend their area at the expense of the redevelopment of the Callunetum (Fritsch, 1927), except perhaps when fires succeed one another at very short intervals.

Early herbaceous colonists are seedlings of *Carex pilulifera*, and *Polygala vulgaris*, *Rumex acetosella*, *Galium saxatile*, *Potentilla erecta*, *Epilobium angustifolium*. *Rumex acetosella*, which on many heaths is often locally dominant after burning, is never abundant over extensive areas on Hindhead. The only grasses whose seedlings occur in any quantity are *Agrostis tenuis*, *Deschampsia flexuosa*, *Festuca ovina* and *Poa annua*.

Cuscuta epithymum is also abundant in the early phases of the subsere, attacking a great variety of hosts, including:

Agrostis tenuis	Galium saxatile
Betula (sprouting)	Molinia caerulea
Calluna vulgaris	Potentilla erecta
Deschampsia flexuosa	Pteridium aquilinum
Erica cinerea	Sieglingia decumbens
E. tetralix	Ulex minor

In the adult Callunetum it appears to be absent.

Undershrubs. Of the undershrubs which enter into the structure of the climax *Ulex minor* is seldom killed by the fires, and sprouting at once from the base it secures a start in the recolonisation by vascular plants. Seedlings appear in great numbers but mostly die. A large number of the original *Calluna* plants are killed by the intense heat of the fires consuming the tall Callunetum, and here the ling regenerates mainly by seed. When the fire sweeps over the younger growth much of the *Calluna* survives and regeneration is largely by sprouting from the stocks. *Erica cinerea* more rarely survives burning but it produces very numerous seedlings. It seems that this species, at any rate in the young stages, can resist dry conditions better than *Calluna*, and hence may become dominant alone on slopes facing south and south-west. It is also normally prominent in the subfinal stage of the developing Callunetum ("CUE" phase of Fritsch), eventually yielding, except locally, to the dominance of the ling, with the dwarf gorse subdominant ("CU" phase of Fritsch).

Erica tetralix and *Vaccinium myrtillus* are much less abundant, but they both commonly survive the fires, the former because its tufts accumulate a protective covering of litter and humus, the latter because its rhizomes are 2 or 3 in. below the soil surface. These plants, then, regenerate *in situ* from the old stocks.

Summary of burn subsere. The following is a summary of the phases of succession in the burn subsere at Hindhead:

(1) Algal phase (*Cystococcus, Gloeocystis, Trochiscia*), with the fungi *Pyronema confluens* and *Ascobolus atrofuscus*, and seedlings and sprouting stools of *Ulex minor* which have survived the fire.

(2) Algae as in Phase 1, with soredial groups formed by *Ascobolus* and *Cystococcus*: mosses (*Ceratodon, Funaria*); sprouting stools as in Phase 1; seedlings of *Calluna* and *Erica cinerea*; a number of herbaceous species, especially *Deschampsia flexuosa, Carex pilulifera, Polygala vulgaris, Rumex acetosella* and *Cuscuta epithymum*.

(3) *Lichen phase* with *Cladonia* spp.: algae as before, with *Mesotaenium violascens*: increase in mass of phanerogamic vegetation, especially *Calluna* and *Erica cinerea*.

(4) Closing *Calluna, Ulex minor* and *Erica cinerea*, with numerous lichens: disappearance of herbaceous species ("CUE" phase).

(5) Adult Callunetum: *Calluna* dominant, *Ulex minor* subdominant, *Erica cinerea* suppressed; mosses, lichens and algae forming a ground layer ("CU" phase); considerable peat development. *Vaccinium myrtillus* co-dominant with *Calluna* in places with thickest peat.

(2) *Gravel or sand subsere*

At Hindhead The pioneer on this substratum, when it is bared at Hindhead by any means, is the terrestrial form of the conjugate alga *Zygogonium ericetorum*, which forms extensive dark purple sheets binding the surface layers of sand. These may be overgrown by the bright green threads of *Hormidium flaccidum*. The lichen *Baeomyces rufus* and the moss *Polytrichum piliferum* also play important parts, while scattered individuals of the fungus *Clavaria argillacea* occur in the autumn (Fritsch and Salisbury, 1915).

At Crockham Hill No information is available as to later stages of the gravel slide sere at Hindhead, but on horizontal gravel surfaces on the same rock (Hythe Beds) laid bare by chert diggings on Crockham Hill Common (700 ft.) in west Kent, the protonema of *Polytrichum piliferum* seemed to be the pioneer in recolonisation. It is possible however that the moss may have been preceded by algae. The protonema covered the bare soil in places with a close network and became studded with the dwarf leafy shoots of the moss. The thin sheet of incipient peat formed by the *Polytrichum* was then colonised by *Cladonia* spp., *Agrostis canina* (considerable patches) and numerous seedlings of *Calluna* and *Erica cinerea*, together with *Teucrium scorodonia*, and (in the square metre recorded) a few plants of *Galium saxatile*, *Veronica officinalis*, and one seedling each of *Rumex acetosella* and *Ulex europaeus*. Two years later the plants of *Erica* and *Calluna*, especially the former, had greatly increased in size and formed rapidly extending tufts, some of which had come into lateral contact, greatly restricting the extent of *Polytrichum* and *Agrostis*.[1] More heath seedlings had appeared and *Teucrium scorodonia* had also increased greatly in numbers, especially in the shelter of the heath tufts. Though the history of this quadrat could not be followed further, as it was destroyed by fresh chert digging, there is no doubt that the heaths would have closed up and eliminated practically all the *Polytrichum* and the other flowering plants. By following three separate quadrats, laid out in different places and representing different phases of succession, for 3 years, it was very roughly estimated that the whole sere, from bare ground to adult Callunetum, would have taken approximately 15–20 years to complete its development. Judging from the much faster progress during a wet than during a dry summer it is probable that the period occupied would be greatly affected by a run of wet or of dry years.

On an area in Breckland bared of vegetation by wind-blown sand and on

[1] Pl. 125, phot. 307. A later stage of the sere is seen in Pl. 125, phot. 308.

which the sandblast had ceased, Farrow (1919) records *Cladonia coccifera*, *C. cervicornis*, *C. uncialis*, *C. aculeata*, *Polytrichum piliferum* and *Campylopus flexuosus* as the agents of colonisation. Nothing is said of later stages in the succession (see Watt's account of the development of Agrostido-Festucetum, Chapter xxvi, pp. 512–14).

Thus the stages of succession appear to follow the same general course in the bare gravel and in the burn subseres, though there are minor differences between them.

(3) *Felling subsere*

At Oxshott Summerhayes and Williams (1926) briefly describe the colonisation by Callunetum of an area of pinewood on Oxshott Heath felled in 1917. They make no mention of pioneer algae, lichens and mosses, such as were found at Hindhead in the early stages of heath regeneration. *Calluna* itself comes slowly at first, since the seed is mainly derived from plants surviving in openings of the wood or under the original pines. It is not until these surviving plants of *Calluna* have recovered sufficiently to flower freely and produce abundant seed that colonisation is at all rapid. Also the undecayed pine needle litter is unfavourable to ecesis of seedlings of the ling. Thus after 5 years from felling (Pl. 125, phot. 305) the *Calluna* is seen to be quite sparse, but 3 years later (Pl. 125, phot. 306) the Callunetum is nearly closed. The frequency of *Erica cinerea* in the early development of the Callunetum is noted, in agreement with Fritsch. Of mosses, *Campylopus flexuosus*, *Webera nutans*, *Polytrichum juniperinum* and *P. formosum* occurred in very small quantity under the young heather canopy, while *Dicranum scoparium* and *Hypnum cupressiforme* var. *ericetorum*, abundant in the mature Callunetum, seemed to be spreading slowly in from what had been open heather areas in intervals of the original pinewood.

WET HEATH COMMUNITIES

In constantly wet places within an area of heath, for example where the ground water reaches the surface in a local hollow of the ground and forms a pond or bog, or where water oozes from the surface of a slope and trickles downhill, or in a low lying water-logged area where the nature of the subsoil determines acid conditions, characteristic wet-heath communities occur. These have a great deal in common with the wet peat moor, "bog" or "moss" communities developed in the extremely humid climates of western Scotland and Ireland, as well as on the mountain plateaux (see Chapter xxxiv).[1]

If the water level remains approximately constant in relation to the soil surface, the wet heath community may be permanent. It is not clear how

[1] On strictly vegetational lines these "wet-heath" communities are indeed to be considered as "valley bogs" belonging to the formation described in Chapters xxxiv and xxxv, but their occurrence in the lowlands as parts of the physiographic heath complex makes it convenient to describe them in this chapter.

far an actual autogenic hydroseral succession occurs through the gradual raising of the soil surface as a result of humus accumulation, though erosion and deepening of water channels (as for example between the tussocks of Molinietum on Chard Common described below) and the consequent increased drainage of the intervening areas undoubtedly leads to colonisation by a drier heath community. Where a dry heath slopes down to a lower-lying wet area there is of course a zonation of transitional vegetation between the two, but this does not necessarily represent an actual hydrosere.

Molinietum. Perhaps the most characteristic plant of damp or wet heath is *Molinia caerulea*, which is nearly as widespread in this habitat in southern England as it is on certain types of fen and on the margins of blanket bog in western Scotland and Ireland. Summerhayes and Williams describe the Molinietum of the low-lying damp parts of Esher Common: "...the community was strikingly uniform in appearance, consisting of large tussocks growing close together and producing inflorescences up to 3 ft. 6 in. (1 m.) high....The chief associates of the *Molinia* were *Calluna* and *Erica tetralix*, the latter being especially abundant locally, while *Calluna* was commoner nearer the upper edge of the community....It is probable...that permanent Molinieta do not exist on heaths in the low rainfall districts of England. Where the drainage is good the *Molinia* is replaced by other herbaceous or dwarf shrub species, while on damper soils woodland is soon established" (Summerhayes and Williams, 1926, p. 220). There were numerous associated species, but most of them were "far from common; indeed they rarely affect the uniform appearance of the *Molinia*" (p. 221).

Succession. During the long drought of the summer of 1921, there were numerous severe fires on Esher Common, some of which burned themselves out through having consumed all the combustible material. On the damper bare areas thus produced the early stages of recolonisation were traced from 1922 to 1925 and are represented in Fig. 146 *a–d*. The leading features of the subsere were the colonisation of the burned soil first by *Funaria hygometrica* and then by *Epilobium angustifolium*, which became completely dominant in the second year but was quite transient, and followed by the invasion of *Molinia*, *Betula* and *Calluna*, with *Polytrichum*, as soon as the alkalinity and intense nitrification of the soil resulting from the fire gave place to increasing acidity. The Molinietum would presumably give way to birch-heath and perhaps ultimately to pinewood.

Chard Common. The best existing account of wet-heath zonation is given by Watson (1915) from an area of Chard Common, Somerset. This wet heath occupies about 3 acres (1·2 hectares) in the middle of drained pasture land on non-calcareous Lias. The area drains to a stream on its edge by means of a complicated system of narrow water channels separating small polygonal areas or islands (Fig. 147), which vary from a few deci-

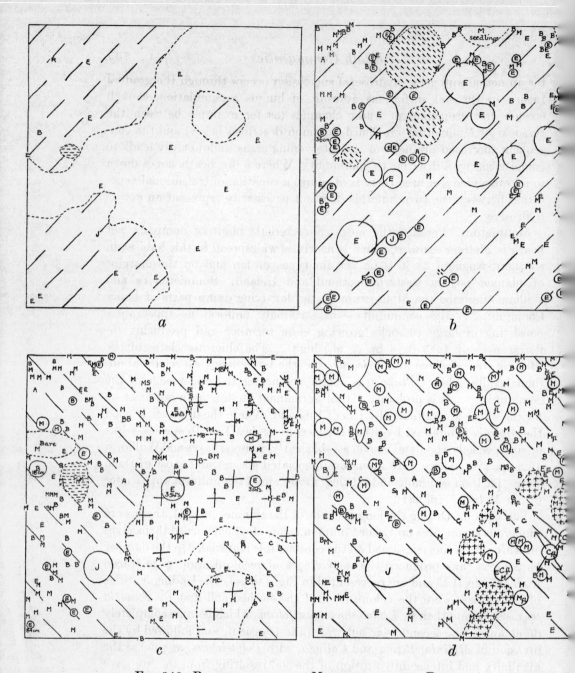

FIG. 146. REDEVELOPMENT OF MOLINIETUM AFTER BURNING

a, b, c, d. The same quadrat in four successive years, 1922–25. Esher Common, Surrey. From Summerhayes and Williams 1926.

BRYOPHYTES. //*Funaria hygrometrica*. = *Marchantia polymorpha*. *Polytrichum* spp. (mostly *P. juniperinum formosum*). + *Ceratodon purpureus*.

PHANEROGAMS. E, *Epilobium angustifolium*. J, *Juncus conglomeratus*. A, *Sorbus aucuparia*. B, *Betula* spp. M, *Molinia caerulea*. C, *Calluna vulgaris*. Sq., *Juncus squarrosus*.

(a) Bare area colonised by *Funaria hygrometrica* and *Epilobium angustifolium*.
(b) *Epilobium angustifolium* dominant. Invasion of *Molinia*, with *Betula* seedlings and a little *Calluna*: invasion *Polytrichum* spp. *Funaria* still dominant in the ground layer.
(c) Recession of *Epilobium*: *Molinia* and *Betula* increasing: *Polytrichum* and *Ceratodon* dominant in the ground layer.
(d) *Molinia* becoming dominant: increase of *Betula* and *Calluna*: *Polytrichum* dominant in the ground layer.

metres to nearly two metres in diameter, and are dominated by *Molinia caerulea*. In rainy weather the whole area is very wet and the channels all contain free water, but in dry summers the polygonal *Molinia* tussocks become dry and are separated by damp furrows. A certain amount of erosion of the sides of the tussocks occurs during wet weather when water is

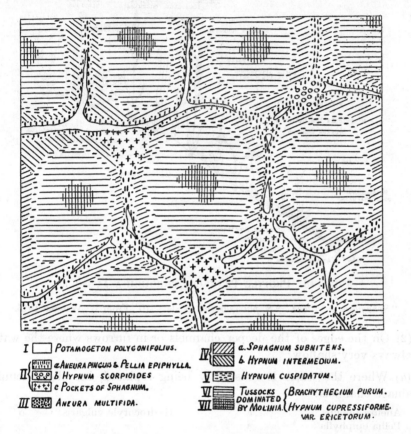

FIG. 147. WET HEATH COMPOSED OF TUSSOCKS SEPARATED BY CHANNELS

Chard Common, Somerset. Note the bryophyte zonation round each tussock. Scale 1:50. "*Sphagnum subnitens*" = *S. plumulosum*. From W. Watson, 1915.

actively draining away down the channels, and this tends to deepen the furrows and increase the height of the tussocks.

Zonation. The following zones were distinguished by Watson in passing from a permanent water channel to the summit of the adjacent tussock (Figs. 147, 148). Owing to the small size of the tussocks the whole of the zonation described only occupies about seven or eight decimetres, and some of the zones cannot in any case be considered as representing phases of succession, since they owe their origin to special conditions arising on the slightly eroded edges of the water channels.

(1) In the channels where water is constantly present *Potamogeton polygonifolius* occurs—alone where the depth of water exceeds 10 cm. This is the characteristic hydrophyte of the acid waters of wet heath. Desmids are very abundant and characteristic.

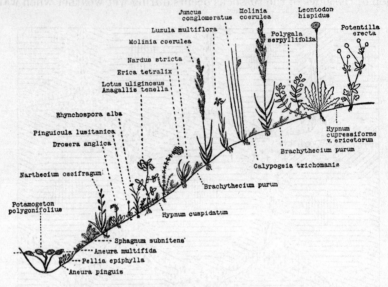

FIG. 148. PROFILE TRANSECT THROUGH A TUSSOCK

"*Sphagnum subnitens*" = *S. plumulosum*. Scale 1:10. From W. Watson, 1915.

(2) On the edges of the deeper channels or in furrows where the water is always very shallow:

(*a*) Where the water is constantly being replaced, i.e. along definite drainage channels:

Aneura pinguis	a	Hydrocotyle vulgaris	o
Pellia epiphylla	a		
Potamogeton polygonifolius	o		

The two liverworts typically have long narrow thalli.

(*b*) In shallow stagnant peaty pools, where the surface water may disappear:

Hypnum scorpioides (reddish)	d	Hypnum revolvens	o
		H. stellatum	o

(*c*) In wet pockets larger than those of (*b*):

Sphagnum cymbifolium (and varieties)	d	Sphagnum rufescens	o
S. recurvum	la	Aulacomnium palustre	a

(3) At the water level a line of *Aneura multifida* practically pure.

(4) Just above the water level:

(a) By the smaller (more stagnant) furrows: *Sphagnum plumulosum* is dominant, especially var. *violascens*, with *S. rufescens* (f) and other species of *Sphagnum*. Associated with the *Sphagnum* are *Cephalozia connivens* (a), *Calypogeia fissa* (a) and *Hypnum intermedium* (o), and the following flowering plants:

Drosera rotundifolia	a	Narthecium ossifragum	f
D. anglica	f	Pinguicula lusitanica	o
D. intermedia	f	Rhynchospora alba	o
Hydrocotyle vulgaris	f	Scutellaria minor	o
Mentha aquatica	o	Viola palustris	f

(b) By the sides of channels where there is a slow but constant current *Hypnum intermedium* is dominant, associated with *H. giganteum* (f), *H. revolvens* (f), *H. cuspidatum* (o) and *H. scorpioides* forma (o). Other mosses are *Sphagnum plumulosum* and *S. rufescens*, *Bryum pseudotriquetrum* and *Mnium affine* var. *elatum* (all o); and the following vascular plants:

Carex oederi	a	Equisetum fluviatile	
C. hornschuchiana	o	(limosum)	l
C. pulicaris	o	Galium uliginosum	a
C. inflata	l	G. palustre var. witheringii	l
C. paniculata	l	Glyceria plicata	l
C. stellulata	o	Hydrocotyle vulgaris	a
Eleocharis multicaulis	o	Juncus acutiflorus	a
Equisetum palustre	a	Mentha aquatica	a
var. polystachyum	o	Narthecium ossifragum	f
E. telmateia (maximum)	o	Pedicularis palustris	f
(in and near alder thickets	a)	Viola palustris	f

(5) At a slightly higher level *Hypnum cuspidatum* is sometimes dominant in a fairly definite zone, associated with *H. stellatum*, and occasionally with *H. intermedium* and *H. giganteum* belonging to 4 (b), from which this zone is not always separable in respect of its bryophytes. *Sphagnum subsecundum* is locally dominant in this zone, and *Calypogeia trichomanis* frequent.

The vascular plants are however characteristic; the cross-leaved heath being dominant, and the bog pimpernel very abundant:

Anagallis tenella	va	Erica tetralix	d
Carex helodes	l	Lotus uliginosus	f
Cirsium palustre	f	Lychnis flos-cuculi	f
C. anglicum	f	Lythrum salicaria	o
Epilobium palustre	a	Orchis ericetorum	o
Equisetum palustre	f	O. latifolia	o
E. telmateia (shaded places)	l		

(6) The next zone is dominated by *Molinia caerulea*, and the bryophytes are much less abundant owing to the competition of the grass. The characteristic moss is *Brachythecium purum*, one of the commonest grassland mosses in Britain. This is abundant, while *Sphagnum papillosum* and its

variety *confertum* are locally frequent. *S. plumulosum* and its variety *purpurascens*, *Hypnum cupressiforme* var. *ericetorum*, and the liverworts *Aneura multifida*, *Cephalozia connivens*, *Lophocolea bidentata* and *Calypogeia trichomanis* are occasional.

In addition to the dominant *Molinia* the following vascular plants occur in this zone:

Agrostis canina	f	Juncus effusus	a
Angelica sylvestris	f	Lotus uliginosus	o
Cirsium palustre	o	Luzula multiflora	a
C. anglicum	o	Lychnis flos-cuculi	o
Erica tetralix	o	Nardus stricta	f
Eupatorium cannabinum	o	Platanthera bifolia	l
Genista anglica	o	Rubus idaeus (thicket)	l
Juncus conglomeratus	a		

(7) In the driest parts of the *Molinia* tussocks, we have the following community:

Molinia caerulea d

Agrostis tenuis	f	Leontodon hispidus	f
Blechnum spicant	o	Linum catharticum	o
Dryopteris dilatata	o	Luzula multiflora var.	
D. spinulosa	o	congesta	f
Galeopsis tetrahit	f	Myrica gale	ld
Galium saxatile	o	Polygala serpyllifolia	f
Holcus lanatus	f	Potentilla erecta	a
Hypericum humifusum	o	Rubus spp. (fruticose)	f
H. pulchrum	o	Succisa pratensis	f
Juncus squarrosus	o	Senecio sylvaticus	o
Leontodon nudicaulis	o	Ulex gallii	a

The only mosses are *Hypnum cupressiforme* (o) and its variety *ericetorum* (f), and *Thuidium tamariscinum* (o). The common agaric *Laccaria laccata* also occurs.

Here we have a definitely "dry land" community, though the specific heath element is only partly represented, and *Calluna* has not succeeded in obtaining a footing in the *Molinia* tussocks.

A community with a drier soil than that of the tussocks occurs on the drier edges of the area and includes several species of fruticose Rubi, but the habitat of this has undoubtedly been much altered by human activity. An adjoining "dry heath" shows a more natural late stage of the heath succession, including such characteristic heath species as *Calluna vulgaris* and *Polytrichum juniperinum*, besides *Pteridium aquilinum*, *Ulex gallii* and *U. europaeus* and many of the heath species occurring on the driest parts of the *Molinia* tussocks, in addition to species of "grass heath" and others commonly found on relatively dry siliceous soil.

Sphagnetum. A small low-lying portion of the area bordering the stream into which the channels drain is occupied by a swamp which is not definitely marked out into *Molinia* tussocks and intervening channels and furrows, as is the sloping portion described above. This swamp is regarded by Watson as the most primitive stage of the succession (apart of course from the submerged community of *Potamogeton polygonifolius* occupying the deeper channels). But as has already been remarked, it is not clear how far actual succession has occurred except through the agency of erosion between the *Molinia* tussocks. The wetter pockets of the swamp are occupied by a community of *Sphagnum recurvum* (which is not dominant between the *Molinia* tussocks). *S. recurvum* (with other species of *Sphagnum*, *Hypnum cuspidatum*, *H. stellatum*, *H. intermedium* and *H. giganteum*) forms a matrix in which local societies of *Hypericum elodes*, *Menyanthes trifoliata*, *Mnium affine* and *Bryum pseudotriquetrum* occur, with an abundance of *Hydrocotyle vulgaris*, *Galium uliginosum* and *Anagallis tenella*.

The general zonation in this area may therefore be summarised as follows, omitting alternative phases and the liverwort communities confined to the edges of the water channels between the tussocks:

Potamogetonetum polygonifolii
|
Sphagnetum recurvi
|
Sphagnetum plumulosi
|
Molinietum caeruleae
|
Molinietum with abundant invasion of *Ulex gallii*, *Potentilla erecta*, etc.
|
? Callunetum

Other wet heath species. Besides the species occurring in the area described above, the following are recorded from various English wet heath communities:

Andromeda polifolia	vr	Malaxis paludosa	r
Carex leporina	f	Osmunda regalis	l
C. hudsonii	l	Oxycoccus quadripetalus	vr
Erica ciliaris (Dorset and Cornwall)	r	Peplis portula	o
		Pinguicula vulgaris	o–f
Eriophorum angustifolium	l	Ranunculus flammula	f
Gentiana pneumonanthe	l	Salix repens	la
Juncus acutiflorus	f	Schoenus nigricans	l
J. bulbosus (supinus)	la	Scirpus setaceus	r
Lycopodium inundatum	r	Wahlenbergia hederacea	o
Hypnum fluitans	f	Polytrichum commune	f

REFERENCES

FARROW, E. P. On the ecology of the Vegetation of Breckland. I. General description of Breckland and its vegetation. *J. Ecol.* **3**, 211–28. 1915. V. Observations relating to competition between plants. *J. Ecol.* **5**, 155–72. 1917. VII. General effect of blowing sand upon the vegetation. *J. Ecol.* **7**, 55–64. 1919.

FRITSCH, F. E. The heath association on Hindhead Common, 1910–26. *J. Ecol.* **15**, 344–72. 1927.

FRITSCH, F. E. and PARKER, WINIFRED M. The heath association on Hindhead Common. *New Phytol.* **12**, 148–63. 1913.

FRITSCH, F. E. and SALISBURY, E. J. Further observations on the heath association on Hindhead Common. *New Phytol.* **14**, 116–38. 1915.

HAINES, F. M. A soil survey of Hindhead Common. *J. Ecol.* **14**, 33–71. 1926.

HOPKINSON, J. W. Studies on the vegetation of Nottinghamshire. I. The ecology of the Bunter Sandstone. *J. Ecol.* **15**, 130–71. 1927.

RAYNER, M. C. Obligate symbiosis in *Calluna vulgaris*. *Ann. Bot., Lond.*, **29**. 1915.

SKIPPER, E. G. The ecology of the gorse (*Ulex*) with special reference to the growth-forms on Hindhead Common. *J. Ecol.* **10**, 24–52. 1922.

SUMMERHAYES, V. S., COLE, L. W. and WILLIAMS, P. H. Studies on the ecology of English heaths. I. *J. Ecol.* **12**, 287–306. 1924.

SUMMERHAYES, V. S. and WILLIAMS, P. H. Studies on the ecology of English heaths. II. *J. Ecol.* **14**, 203–43. 1926.

WATSON, W. A Somerset heath and its bryophytic zonation. *New Phytol.* **14**, 80–93. 1915.

WEISS, F. E. The dispersal of fruits and seeds by ants. *New Phytol.* **7**, 27. 1908.

WEISS, F. E. The dispersal of the seeds of the Gorse and the Broom by ants. *New Phytol.* **8**, 81–9. 1909.

Chapter XXXVII

THE HEATH FORMATION (*continued*)
UPLAND HEATHS AND "HEATHER MOORS"
BILBERRY MOORS
DISTRIBUTION OF BRITISH HEATHS
SUMMARY OF PEAT COMMUNITIES

"**Heath" and "Heather Moor".** It has been usual to separate "heaths" from "heather moors", corresponding with the German distinction of *Heide* and *Heidemoor*, the former described as developed on sandy or gravelly soil with a minimum of peaty humus, the latter on deep acid peat, and most commonly in Britain only at higher altitudes. This distinction was drawn by Robert Smith (1900), by W. G. Smith in *Types of British Vegetation* (1911), and by Elgee (1914); and it is maintained by Watson (1932) in his account of the bryophytes and lichens of moorland. W. G. Smith separated the "upland heaths" of Cleveland and of Scotland from the "heather moors" of the Pennines, and Watson similarly separates "upland heaths" in Somerset and north Devon from "heather moor" on Dartmoor and in other more northern regions.

Adamson, however (1918), contended that the so-called heather moors of the southern Pennines are really for the most part upland heaths, "very closely allied to the heaths of the lowlands and the south". The peat is generally quite thin and the roots penetrate to the sandy layer described by Crump (1913) as the "sub-peat" (the *A* horizon of modern pedology). Adamson held that the Calluneta (and Vaccinieta) of these moors (or upland heaths) are most closely related, not to the wet peat communities (Eriophoreta, etc.), but to the acid grasslands dominated by *Nardus* and *Deschampsia flexuosa*, reversible changes between heath and grassland occurring in one direction or the other in correspondence with the increase of grazing or of peat accumulation. *Vaccinium myrtillus*, however, as Moss showed, and under some circumstances *Calluna* also, may colonise drying *Eriophorum* peat and thus produce what some would call a "true" heather moor on deep peat.

Floristic differences. With regard to the floristic differences between "heath" and "heather moor", these are mainly functions of geographical position or marked difference of altitude, and therefore ultimately of climate, and they really affect a very small proportion of the common species, as can be readily seen by consulting the extensive comparative lists given by Watson (1932, pp. 288–92). The differences shown are mainly rather slight differences in frequency of the less common species. The "heather moors" however do possess, according to Watson, a number of rarer bryophytes and lichens not occurring on "upland heaths" (see p. 746).

Of the species confined to southern lowland heaths we have *Ulex minor* (south-eastern and south-central England), *E. ciliaris*[1] (wet heaths in Dorset and Cornwall) and *Erica vagans*[2] (Lizard peninsula). All three are "southern Atlantic" species (Matthews), i.e. species centred in the southern portion of the Atlantic coastline of Europe. *Pinguicula lusitanica*, another species of the same European distribution, which may be abundant on wet heaths in south-western England, is also widely distributed in Ireland (where it ascends to 1500 ft.—c. 450 m.—in the west) and reaches western Scotland. *Agrostis setacea* is another south-western plant, which ranges rather farther east than the western species previously mentioned, both in England and on the continent. The peculiar Breckland species are plants of Central European affinity, and have no connexion with the Callunetum as such, though this is one of the leading plant communities of Breckland. They are indeed almost confined to the "better" types of grassland and cannot properly be considered heath plants at all (see Chapter XXVI, p. 512).

The species associated with raw humus which are confined to the north or much commoner there are more numerous, and include the following:

Undershrubs

Arctostaphylos uva-ursi, south to Derbyshire only. *Arctous alpina*, Highlands only.
Betula nana, mainly in the Scottish Highlands, not south of Northumberland.
Empetrum nigrum, abundant in the north and west, rare in the south-west.
Rubus chamaemorus, not south of Derbyshire and Wales.
Vaccinium vitis-idaea, much commoner in the north, but extending to the south-west. *V. myrtillus* is much more abundant in the north and west but by no means rare (rather local) in the south-east. *V. uliginosum* not south of Durham.

Herbs

Pyrola media, rare in the midlands of England (woods), more frequent in Scotland.
P. minor, much commoner in the north: confined to woods in the south of England. Other species of *Pyrola* (p. 451) are also less rare in Scotland.
Trientalis europaea, not south of Yorkshire, frequent on Scottish heaths as well as in woods.
Listera cordata, very rare in the south.
Saxifraga hirculus, wet moorland, rare and local in northern England and central Scotland.
Cornus suecica, not south of Yorkshire, where it is very rare: practically Arctic-alpine in Scotland where it is said to be almost always associated with *Vaccinium myrtillus*.

Of these only the undershrubs (especially *Vaccinium myrtillus*) play any considerable part in the Callunetum.

When we come to consider the actual lists of species published from particular examples of different types of heath community, we find that the most useful comparative data are those of Watson (1932). He gives a list of sixty-four angiosperms, six pteridophytes and no less than 232 bryophytes and lichens, from five "upland heaths" in Somerset, seven "heather

[1] Pl. 127, phots. 312, 313. [2] Pl. 127, phot. 314.

moors" (mainly western) ranging from Perthshire to Dartmoor, and four lowland "wet heaths". Table XXIV shows the distribution of these 302 species in the three groups of communities.

Table XXIV

	"Upland heaths"	"Heather moors"	"Wet heaths"	Totals
Angiosperms	59	58	63	64
Pteridophytes	4	5	6	6
Mosses	70	83	72	83
Liverworts	37	46	37	48
Lichens	76	99	50	101
Totals	245	291	228	302

These figures make it clear that all three groups of communities contain the great majority of species of vascular plants recorded from any of them: of the mosses about seven-eighths occur on the "upland heaths" and the same proportion on the "wet heaths", all on the "heather moors": of the liverworts three-quarters of the whole number are recorded from "upland heaths", three-quarters from the "wet heaths" and all but two from the "heather moors": of the lichens three-quarters occur on the "upland heaths", only half on the "wet heaths", and all but two on the "heather moors".

Thus there is a very substantial community between the three types of vegetation, but the "heather moors" show a marked preponderance of species of non-vascular plants, which depend closely on damp air; and this is most conspicuous among the lichens.

If we consider the particular species which Watson found on one or more "upland heaths" but not on any "heather moor" and vice versa, we find that these are but few among the vascular plants, while the list of bryophytes and lichens occurring on "heather moors" but not on "upland heaths" is considerable. These however are almost all species occurring in one of the lower degrees of frequency ("occasional" or "rare"): the "abundant" and nearly all the "frequent" species are common to the two types.

The first of the following lists shows the very few species recorded from one or more of the upland heaths and from none of the heather moors, the second the numerous species of non-vascular plants recorded from heather moors and not from upland heaths.

Species recorded from "upland heaths", but not from "heather moors"

Vascular plants:
 *Agrostis setacea   a   * Frangula alnus   r–f
 *Pinguicula lusitanica   o   *Ulex minor   r

Bryophytes: none

Lichens:
 Biatorina littorella   o   Usnea florida   o

      * Also found on wet heaths.

Species recorded from "heather moors", but not from "upland heaths"

Vascular plants:
*Osmunda regalis	(r)†	*Comarum palustre	o
*Pinguicula vulgaris	(r–lf)		

Mosses:
*Campylopus atrovirens	o–f	Mnium serratum	r–o
*C. brevipilus	r	Oligotrichum hercynicum	f
*Dicranum spurium	r–o	Orthodontium heterocarpum	o
*Drepanocladus aduncus	(a)	Polytrichum alpinum	o
*D. sendtneri	o	*P. strictum	o
*D. uncinatus	o	Swartzia montana	o
*Hypnum imponens	r–o		

Liverworts:
*Cephalozia fluitans	r	Lophozia atlantica	o
*C. francisci	o	Marsupella funckii	r–o
C. leucantha	o	M. ustulata	o
*Leptoscyphus taylori	f	Odontoschisma denudatum	–
*L. anomalus	o	Sphenolobus minutus	o

Lichens:
Alectoria bicolor	o–f	Icmadophila ericetorum	o
A. chalyberiformis	o–f	Lecidea kochiana	r
*Biatora gelatinosa	r	Microglaena bredalbanensis	r
*Cetraria islandica	f	Nephromium parile	o
*C. var. tenuifolia	–	Peltidea apthosa	r–o
Cladonia bellidiflora	o	Polychidium muscicolum	o
C. degenerans	o	Psora demissa	o
Ephebia hispidula	o	*Pycnothelia papillaria	–
Gyrophora cylindrica	o–f	Sphaerophorus fragilis	r–o
G. polyphylla	o–f	S. melanocarpus	r–o
G. polyrhiza	o–f	Stenocybe bryophila	o
Haematomma coccineum	o	Stereocaulon condensatum	o
H. ventosum	o–f		

* Also found on wet heaths.
† The frequency letters enclosed in parentheses belong to species recorded from "non-typical habitats".

Furthermore, when we compare the frequencies with which species occurring both on "upland heaths" and "heather moors" are recorded for examples of the two groups we find that comparatively few show any markedly greater frequency on one or the other. Of the more widely distributed species *Polytrichum juniperinum* and *P. formosum* show greater abundance on the "upland heaths" while *P. commune* is more abundant on "moors". This is in accord with the preference of the last for wetter habitats.

Taking all these facts into consideration, we cannot but conclude that the *vegetational* distinction between upland heath and heather moor is very slight. What we have to deal with is a series of communities developed under a considerable range of conditions of climate and soil and thus showing minor differences in floristic composition of which the most conspicuous is the much greater number of species of bryophytes and lichens in the moister habitats, especially in the moister climates; but with all the communities essentially agreeing in the great majority of their species and in the frequencies with which they occur. We have no lists comparable with Watson's for the dry lowland south-eastern and eastern heaths, though there is no doubt that these are very much poorer in bryophytes and lichens, not only than the "heather moors" but also than the "upland heaths".

Depth of peat. There remains the criterion of depth of peat. Adamson would distinguish "upland heaths" on shallow peat in which the roots of the vascular plants penetrate into the "sub-peat" layer, from "heather moors" on deep peat in which they are confined to the peat itself. Heather moors so defined are based on peat not formed by *Calluna* but colonised by the ling as its surface layers dry out. In other words their successional development is different. We have no adequate study of this type of Callunetum, but there is no evidence that its flora and vegetation differ from those of Callunetum on shallow peat.

In what follows, therefore, no technical distinction will be drawn between heather moor and upland heath.

UPLAND HEATHS OR MOORS

(*Callunetum*)

"**Upland Heaths**" **of Somerset.** The following account is taken from Watson's paper (1932) already quoted. The chief vascular plants of the Callunetum of this south-western county are the following:

		Calluna vulgaris	d	
Agrostis setacea	a	Potentilla erecta	f
Deschampsia flexuosa	a	Pteridium aquilinum	la
Erica cinerea	f	Rumex acetosella	lf
E. tetralix	f*	Ulex gallii	a
Galium saxatile	f	Vaccinium myrtillus	f
Melampyrum pratense	f	Veronica serpyllifolia	f

* In slightly damper places, mixed with *E. cinerea*.

Cuscuta epithymum often attacks *Ulex gallii*, forming brown patches several yards square and visible from a considerable distance.

Where the *Calluna* is dense very little else grows except *Cladina* (*Cladonia*) *sylvatica*, its forma *implexa*, and *Hypogymnia* (*Parmelia*) *physodes*. The last-named is usually epiphytic, together with *Lecanora varia*, on the stems of the ling. Where the Callunetum is a little more open any of the following mosses may become abundant:

Brachythecium purum
Campylopus flexuosus
Dicranum scoparium
Hylocomium splendens

Hypnum cupressiforme var. ericetorum
Webera nutans

General flora. The following is a composite list (Watson, 1932) of species of the upland heaths (taken in the widest sense), from five regions of hill heathland in Somerset (Brendon, Blackdown, Mendip, Quantock and Exmoor) of an average altitude of 800–1000 ft. (*c.* 250–300 m.), the highest parts reaching 1300–1500 ft. (400–450 m.) and more (Watson, 1932). The underlying rocks are siliceous and belong to various geological systems. Near the summits of the higher plateaux the heath may give place to the moss (blanket bog) formation (see Chapter xxxv).

Agrostis canina	o–f	L. pilosa	f–a
A. setacea	a	Melampyrum pratense	r–f
A. tenuis	a	Molinia caerulea	la
Aira caryophyllea	o	Myrica gale	o
A. praecox	r–f	Nardus stricta	lf–la
Anthoxanthum odoratum	o–f	Pedicularis sylvatica	f
Betula pendula (alba)	o	Pinguicula lusitanica	o
B. pubescens	o–f	Polygala serpyllifolia	f
Calluna vulgaris	a–ld	Potentilla erecta	f
Carex binervis	o–f	Radiola linoides	r
Deschampsia flexuosa	f–a	Rumex acetosella	o–f
Empetrum nigrum	r–o	Sarothamnus scoparius	r
Erica cinerea	f	Scirpus caespitosus	o–la
E. tetralix	r–o	Sieglingia decumbens	o
Festuca ovina	a	Ulex europaeus	f
Frangula alnus	r–f	U. gallii	f–a
Galium saxatile	f	U. minor	r
Genista anglica	r	Vaccinium myrtillus	o–lf
Juncus squarrosus	o–f	V. vitis-idaea	r
Linum catharticum	o–f	Veronica serpyllifolia	f
Luzula multiflora	o–f		

Pteridophytes

Blechnum spicant	o	L. selago	r
Lycopodium clavatum	r–o	Pteridium aquilinum	lf–la

Mosses

Aulacomnium androgynum	o	Climacium dendroides	o
Bartramia pomiformis	o–f	Dicranella heteromalla	f–a
Brachythecium purum	f–a	D. cerviculata	o
B. rutabulum	f	Dicranum bonjeani	o
B. velutinum	o–f	D. scoparium	f–a
Bryum atropurpureum	o	Ditrichum homomallum	o
B. capillare	o	Eurhynchium myosuroides	f
B. erythrocarpum	o	E. praelongum	o–f
B. inclinatum	o	Funaria hygrometrica	o
B. pallens	o	Hylocomium loreum	o
B. pendulum	–	H. squarrosum	f–la
B. roseum	r–o	H. splendens	a
Campylopus flexuosus	a–f	H. triquetrum	o–f
C. fragilis	r–o	Hypnum cupressiforme	o
C. pyriformis	f–a	var. ericetorum	a–d
C. subulatus	o	H. schreberi	o–a
Catharinea undulata	f	Leptodontium flexifolium	r–o
Ceratodon purpureus	a	Mnium affine	o

Upland Heaths

M. cuspidatum	o	P. juniperinum	a
M. hornum	f	P. nanum	r
M. undulatum	f–a	P. piliferum	f–a
Plagiothecium denticu-		P. urnigerum	o
latum	o	Rhacomitrium canescens	o–f
P. elegans	o–f	R. fasciculare	r–o
P. sylvaticum	r–o	R. heterostichum	o
P. undulatum	o	R. lanuginosum	o
Pleuridium subulatum	o	Tetraphis pellucida	o
Polytrichum aloides	o–f	Thuidium tamariscinum	o–f
P. formosum	a	Webera nutans	f–a
P. gracile	o		

Liverworts

Alicularia geoscypha	r	L. setacea	r
Alicularia scalaris	o–a	Lophocolea bidentata	o
Aplozia crenulata	o	Lophozia bicrenata	r
Calypogeia arguta	o	L. attenuata	r
C. fissa	o	L. excisa	o
C. trichomanis	o–f	L. floerkii	r
Cephalozia bicuspidata	o–f	L. ventricosa	o
C. connivens	r	Plagiochila asplenioides	o
C. media	r–o	Ptilidium ciliare	o
Cephaloziella bifida	o	Scapania compacta	o
C. starkii	o–f	S. curta	o
Diplophyllum albicans	f	S. umbrosa	o
Eucalyx hyalinus	r–o	Sphenolobus exsectiformis	r
Frullania tamarisci	o–f		
Lepidozia reptans	o–f	Marsupella emarginata	r–f

Lichens

Acarospora fuscata	o	C. furcata	f
Bacidia umbrina	o	C. gracilis	r–o
Baeomyces roseus	o–f	C. foliacea	o
B. rufus	o–f	C. macilenta	o
Biatora coarctata	o–f	C. ochrochlora	o
B. granulosa	f–a	C. pityrea	o
B. uliginosa	f–a	C. pyxidata	f–a
Biatorina littorella	o	var. chlorophaea	f–a
Bilimbia lignaria	o	C. rangiformis	o–f
B. melaena	o	C. squamosa	o–f
B. sabuletorum	o	C. subcervicornis	f
Buellia myriocarpa	o	C. subsquamosa	r–o
Candelariella vitellina	o	Cetraria aculeata	o–f
Catillaria chalybeia	o	Coniocybe furfuracea	r
Cladina rangiferina	r	Coriscium viride	r
C. sylvatica	a	Crocynia lanuginosa	r–o
C. uncialis	f	Ephebe lanata	o
Cladonia bacillaris	r	Hypogymnia (Parmelia)	o–f
C. caespiticia	r	physodes	
C. coccifera	f	Lecanora polytropa	o–f
C. crispata	o	L. varia	o
C. fimbriata	f	Lecidea contigua	a
C. flabelliformis	o	L. crustulata	o
C. floerkeana	f–a	L. dicksonii	o

Lecidea expansa	o	P. polydactyla	o
L. lithophila	r	P. rufescens	f
L. lygea	r	Pertusaria dealbata	o–f
L. rivulosa	o–a	Physcia hispida	o
L. sorediza	f–a	Porina chlorotica	o
L. sylvicola	o	Rhizocarpon confervoides	o–f
Leptogium lacerum	o–f	R. geographicum	o
L. sinuatum	o–f	R. petraeum	o
L. microscopicum	o	Sphaerophorus globosus	o
Microglaena nuda	o	Stereocaulon coralloides	r
Parmelia fuliginosa	o	S. denudatum	o
P. omphalodes	o	Usnea florida	o
P. saxatilis	f	Verrucaria maculiformis	o
Peltigera canina	o	V. mutabilis	o

Alga

Botrydina vulgaris — f

The Cleveland heaths. Elgee (1914) has described the extensive Calluneta, always called "moors", of the elevated Cleveland district of northeast Yorkshire. These are developed on sandstones, coarse grits and sandy shales of Inferior Oolite age, rising to over 1400 ft. (*c.* 430 m.) on the highest parts of the plateaux, and almost completely destitute of overlying glacial drift except in the valleys. Callunetum "covers by far the widest areas on the high ridges and plateaux between the dales" but passes into moss with a mixture of *Calluna* and *Eriophorum* on the highest tracts of the watershed.

The soil of the Callunetum may be sandy and stony with a minimum of peaty humus ("thin moor"), or the peat may be as much as 2–4 ft., i.e. up to a metre or more in thickness ("fat moor"). The Callunetum attains its greatest luxuriance on peat at least 6 in. (15 cm.) deep, and frequently more, towards, but not on, the central watershed. This would be included as "fat moor", and here very few other species occur among the closely growing vigorous ling.

On the "thin moor", where *Calluna*, though dominant, is not so luxuriant, the following are the most characteristic associates:

Empetrum nigrum
Erica cinerea
Hypnum cupressiforme var. ericetorum

Potentilla erecta
Vaccinium myrtillus
Cladonia spp.

As the ling becomes old and straggly, leaving a space in the centre of the clump, this is colonised by *Hypnum* and *Cladonia*, by *Juncus squarrosus*, or by other species. On the wetter edges of "thin moor" *Nardus*, *Erica tetralix*, *Juncus squarrosus*, and *Cladonia* are interspersed with the *Calluna*, and on the slopes the following grasses become frequent:

Agrostis canina
Aira praecox
Deschampsia flexuosa

Festuca ovina
Sieglingia decumbens

Molinia also occurs freely among the heather where the soil is damp, both on deep peat and where there is a thin layer of glacial drift, usually sand and gravel.

Burn subsere. The Cleveland moors, like all grouse moors, are regularly fired at intervals of several years. The first plants to appear on the burned areas ("swiddens") are lichens, liverworts and mosses: *Cladonia* spp., *Lophozia inflata*, *Sphagnum papillosum* var. *confertum*, *Webera nutans*, and at a later stage *Ceratodon purpureus* and *Polytrichum commune*. *Calluna* regenerates by sprouting unless the fire has been severe enough to kill the whole plant, when the restocking is from seed. The other species prominent during the later stages of the burn subsere, before *Calluna* again becomes dominant, are *Agrostis canina*, *Aira praecox*, and *Festuca ovina*, which may form local swards; *Rumex acetosella* interspersed among *Empetrum*; *Erica cinerea*, which quickly becomes dominant on dry areas; *Vaccinium myrtillus* which often occupies recently burned areas, sometimes to the exclusion of *Calluna*; *Juncus squarrosus*, *Potentilla erecta*, etc.

The mixture of *Calluna* and *Eriophorum* on the deepest peat of the highest parts of the plateau, which is a transitional type between heath and blanket bog, is briefly described on p. 703.

Scottish heaths. No recent intensive work, either floristic or ecological, has been done on the heaths of Scotland, and the short account contributed by the late W. G. Smith to *Types of British Vegetation* (pp. 111–16) is still the best available.

W. G. Smith drew a distinction based on "habitat" between "*Calluna* heath" and "*Calluna* moor", but it is very doubtful, as we have seen, if any good vegetational distinction can be made. The former, he says, is developed over sandy or gravelly soil with a surface layer of dry peat, which may in extreme cases reach a depth of 12 in. (30 cm.), as on Hindhead in the south of England (p. 725), but is usually much shallower. These heaths are mainly met with on the eastern side of northern Britain and may occur at any altitude from sea level to about 2000 ft. (*c.* 600 m.). They are generally known as "moors" in common with all other open tracts of country (apart of course from broken rocky surfaces and definitely good grassland used primarily as hill pasture), having an acid peaty humous soil, whether or not deep peat is present and whatever the vegetation. The northern *Calluna* heaths of Smith are however typical "grouse moors" used for the preservation and shooting of the grouse (*Lagopus scoticus*) so far as they do not form part of "deer forests". For this purpose they are regularly burned over every 10 or 15 years to destroy the old "leggy" *Calluna* and make way for new thick growth which is best for feeding and sheltering the grouse. A few sheep may be pastured on some of them, but there is little other human interference. Typical areas occur, as we have seen, on the Cleveland Hills in north-east Yorkshire and there are many in the river basins of the Tay, Dee and Spey in the eastern Highlands. In the western

Highlands and in the southern Uplands of Scotland Callunetum, though developed locally, is much less extensive, and the same is true on the whole of the western sides of England, Wales and Ireland.

Flora. The following generalised list of the more characteristic species from four stations near Blair Atholl in the Tay valley, Perthshire (R. Smith), is taken from W. G. Smith's account in *Types of British Vegetation* (pp. 115–16):

Calluna vulgaris d

Locally subdominant:
    Arctostaphylos uva-ursi
    Empetrum nigrum
    Erica tetralix
    Vaccinium myrtillus
    V. vitis-idaea

Locally abundant:
    Erica cinerea
    Nardus stricta
    Hypnum spp.

Frequent:
    Agrostis tenuis
    Antennaria dioica
    Anthoxanthum odoratum
    Blechnum spicant
    Carex dioica
    C. goodenowii
    Deschampsia flexuosa
    Festuca ovina
    Galium saxatile
    Luzula multiflora
    Lycopodium clavatum
    Polygala vulgaris
    Potentilla erecta
    Cladonia spp.

Sparse or local:
    Genista anglica
    Melampyrum pratense var. montanum
    Juniperus communis
    Trientalis europaea

From this list it will be seen that several species are more or less commonly found which do not occur (or occur very rarely) on the southern English heaths, e.g. *Arctostaphylos uva-ursi, Empetrum nigrum, Vaccinium vitis-idaea, Trientalis europaea*; and to these may be added *Listera cordata, Cornus suecica* (rare), *Pyrola* spp. (cf. p. 744).

Burn subsere. No detailed studies of succession have been made on the northern heaths, but it has been noticed that during the first year or two after burning *Cladonia* spp. often form almost the only vegetation. *Erica cinerea* recovers quickly on dry ground, *Vaccinium myrtillus* and *V. vitis-idaea* on steep slopes, *Arctostaphylos uva-ursi* at altitudes towards 2000 ft. (600 m.), the grasses *Agrostis* spp., *Anthoxanthum, Deschampsia flexuosa* and *Festuca ovina* on sandy humus, *Nardus stricta* on moister humous soils. Though the ling typically becomes dominant after a few years, many of these species may maintain themselves in the Callunetum.

On some of the east Scottish coastal sand dunes, as elsewhere in Great Britain (Pl. 126, phot. 309), Callunetum develops, but the successional stages have not been followed in any detail (cf. Chapter XLI, pp. 860–1).

"**Heather Moors" of the Pennines.** These were originally described from west Yorkshire by Smith and Moss (1903), and by Smith and Rankin (1903), and from the borders of Westmorland and Yorkshire by Lewis (1904). Their accounts were summarised by Lewis and by Moss in *Types of British Vegetation* (1911), and Moss (1913) gave a fuller account from Derbyshire at the southern end of the Pennine range. Crump (1913) studied the water relation of the dominant *Calluna* to the soil on which it grows, and Watson (1932) gives extensive floristic lists. Reasons have already been given (pp. 741–7) for refusing to separate these "heather moors" from "upland heaths", but since they are among the best studied examples a short account of them is given here.

The Pennine range, extending some 140 miles (220 km.) from Northumberland southwards to central Derbyshire, and often called the backbone of northern England, bears some of the most extensive and best known of the English heather moors. In the southern Pennines heather moor may occur as low as 750 ft. (c. 250 m.) above the sea, but it usually begins about 1000 ft. (c. 300 m.), extending upwards to about 1500 ft. (c. 450 m.), or in the northern Pennines and the Wicklow Mountains of eastern Ireland, where the hills are higher, to about 2000 ft. (c. 600 m.). Above this level, everywhere in the British Isles, *Calluna* begins to lose its dominance under any conditions of topography and soil, and at about this level the extensive glaciated plateaux bearing the blanket bog or moss formation dealt with in Chapter XXXIV generally occur. Callunetum is the lowest "moorland" zone and occupies the lower, often steeper, slopes of the hill masses (see Figs. 89, 91, 140). In the southern Pennines and in Scotland too, its lower limit often coincides with the upper limit of enclosed grassland; and very often Callunetum adjoins and passes into the bent-fescue grassland dealt with in Chapter XXVI. As Adamson has shown (1918) Callunetum and this grassland have a reciprocal relation, the former passing into the latter under grazing, the latter into the former with cessation of grazing and accumulation of peat.

Water and humus content of soil. According to Crump (1913) and Adamson (1918) the peat of the typical Pennine heather moor "is as a rule so shallow that the heather roots regularly pass through it into the underlying coarse sandy soil that may be conveniently called the 'subpeat'" (Crump, 1913, p. 138). Adamson says that the peat is "rarely over a foot thick and usually only a few inches, and is always much mixed with sand and mineral matter" (1918, p. 100). The roots of the plants "penetrate freely beyond the peat proper into the sandy subpeat", which is "much darkened and discoloured by peat". Crump determined the "coefficient of soil-humidity", i.e. the ratio of water to humus ($W/H = m$), for a great number of peats and subpeats whose actual water and humus contents differed very widely, and found it to vary only between 1·7 and 2·8, with a mean of 2·3, for the peats and between 2·5 and 3·85 with a mean of 3·17 for the subpeats. By deducting the residual water held by the mineral particles of the subpeats he obtained the same mean coefficient ($m = 2\cdot32$) for both. The water content of the soil of the Callunetum is usually a function of its humus content, since clay and silt fractions are generally negligible.

Chemical nature of the peat. According to Pearsall (1938) all samples from the Yorkshire (Pennine) and Lake District "heather moors" "show remarkable similarity, with a pH value near to $3\cdot5$, and they are oxidising soils...which are markedly deficient in bases". The extreme range of ten samples from different localities was only $3\cdot41$ to $3\cdot63$. Nitrates were never present. In "heaths" with quite shallow peat over Triassic sands the pH value was rather more variable but still between $3\cdot2$ and $4\cdot0$. Surface samples were more acid and had a higher oxidation potential than deeper ones. On a burned and drained heath where *Calluna* had become completely dominant to the exclusion of *Eriophorum* and *Erica tetralix*, formerly present, the pH value was as low as $2\cdot98$.[1] These Callunetum soils are very similar to woodland soils on which *Deschampsia flexuosa* is dominant.

Floristic composition. The following list includes the principal species found in the Pennine heather moors (Lewis, 1904; Moss, 1913; Watson 1932):

Agrostis tenuis	f	Lycopodium clavatum	r
Anthoxanthum odoratum	o	L. selago	r
Blechnum spicant	o–a	Melampyrum pratense	o
Carex binervis	o	Molinia caerulea	lf
Deschampsia flexuosa	f	Nardus stricta	o
Empetrum nigrum	f	Polygala serpyllifolia	o
Erica cinerea	l	Potentilla erecta	o
E. tetralix	lf	Pteridium aquilinum	la
Eriophorum vaginatum	lf	Pyrola media	vr
Festuca ovina	f	Rumex acetosella	la
Galium saxatile	f	Salix repens	vr
Genista anglica	r, l	Scirpus caespitosus	o
Juncus bufonius	f	S. pauciflorus	o
J. inflexus	f	Trientalis europaea	vr
J. squarrosus	a	Ulex gallii	la
Lathyrus montanus	o	Vaccinium myrtillus	lsd
Linum catharticum	o	V. vitis-idaea	o
Listera cordata	vr	Veronica serpyllifolia	o
Luzula multiflora	o		

This list of vascular plants includes several species not forming part of the Callunetum, which, especially when typical and closed, is very poor floristically, as in other regions. Thus *Pteridium* is an invader of the edges of the moor from sheltered valleys, *Eriophorum*, *Scirpus* and *Erica tetralix* belong to bog and are confined to damp or wet places, and *Nardus* and *Molinia* belong to the acid grasslands whose relations to the Callunetum has been discussed on pp. 130 and 500.

[1] Cf. Haines' very low Hindhead values (p. 726).

Watson (1932) gives the following list of bryophytes and lichens from the heather moors of the southern Pennines:

Mosses

Aulacomnium androgynum	–	Hypnum stramineum	o
Bartramia pomiformis	–	Hylocomium loreum	o
Brachythecium purum	f	H. splendens	o
B. rutabulum	f	H. squarrosum	a
B. velutinum	o	H. triquetrum	o
Bryum atropurpureum	–	Leptodontium flexifolium	–
B. capillare	o	Leucobryum glaucum	o
B. erythrocarpum	o	Mnium cuspidatum	o
B. inclinatum	o	M. hornum	f
B. pallens	o	M. serratum	r
B. pendulum	o	M. undulatum	f
B. roseum	r	Oligotrichum hercynicum	f
Campylopus atrovirens	o	Orthodontium	
C. brevipilus	–	heterocarpum	o
C. flexuosus	a	Plagiothecium denticulatum	o
C. fragilis	o	P. elegans	f
C. pyriformis	a	P. sylvaticum	o
Catharinea undulata	f	P. undulatum	o
Ceratodon purpureus	a	Pleuridium subulatum	o
Climacium dendroides	o	Polytrichum aloides	o
Dicranella cerviculata	o	P. alpinum	o
D. heteromalla	o	P. formosum	f
Dicranum bonjeani	–	P. gracile	o
D. scoparium	f	P. juniperinum	f
D. spurium	–	P. piliferum	–
Ditrichum homomallum	–	P. strictum	o
Drepanocladus sendtneri	o	P. urnigerum	o
D. uncinatus	o	Rhacomitrium canescens	–
Eurhynchium myosuroides	o	R. fasciculare	–
E. praelongum	o	R. heterostichum	o
Funaria hygrometrica	o	R. lanuginosum	o
Hypnum cupressiforme	f	Swartzia montana	o
H. cuspidatum	o	Tetraphis pellucida	f
H. var. ericetorum	a	Thuidium tamariscinum	f
H. schreberi	o	Webera nutans	a

Liverworts

Alicularia scalaris	f	Lophocolea bidentata	f
Aplozia crenulata	o	Lophozia atlantica	o
Calypogeia arguta	o	L. attenuata	o
C. fissa	o	L. bicrenata	o
C. trichomanis	f	L. excisa	–
Cephalozia bicuspidata	f	L. floerkii	f
C. connivens	f	L. ventricosa	o
C. media	r	Odontoschisma denudatum	–
Cephaloziella bifida	a	O. sphagni	–
C. starkii	o	Plagiochila asplenioides	o
Diplophyllum albicans	f	Ptilidium ciliare	–
Eucalyx hyalinus	o	Scapania compacta	–
Frullania tamarisci	o	S. curta	o
Lepidozia reptans	f	S. umbrosa	o
L. setacea	o	Sphenolobus minutus	o
Leptoscyphus taylori	f		

Lichens

Acarospora fuscata	–	Coniocybe furfuracea	r
Bacidia umbrina	–	Coriscium viride	r
Baeomyces rufus	o	Gyrophora cylindrica	–
Biatora coarctata	–	G. polyphylla	–
B. granulosa	f	Hypogymnia physodes	o
B. uliginosa	–	Icmadophila ericetorum	–
Bilimbia lignaria	–	Lecanora polytropa	o
B. sabuletorum	o	Lecidia contigua	f
Botrydina vulgaris	o	L. dicksonii	–
Buellia myriocarpa	–	Leptogium lacerum	–
Candelariella vitellina	o	L. sinuatum	–
Cetraria islandica	o	Parmelia fuliginosa	–
Cladina sylvatica	f	P. omphalodes	o
C. uncialis	o	P. saxatilis	f
Cladonia bacillaris	o	Peltidea aphthosa	–
C. caespiticia	–	Peltigera canina	o
C. coccifera	f	P. polydactyla	–
C. degenerans	o	P. rufescens	o
C. fimbriata	o	Pertusaria dealbata	–
C. flabelliformis	o	Physcia hispida	o
C. floerkeana	f	Porina chlorotica	r
C. furcata	o	Rhizocarpon confervoides	–
C. gracilis	f	R. geographicum	o
C. macilenta	o	Sphaerophorus fragilis	r
C. pyxidata	f	S. globosus	r
var. chlorophaea	f	Stereocaulon condensatum	–
C. rangiformis	o	S. coralloides	–
C. squamosa	o	S. denudatum	–
C. subcervicornis	f		

Firing of heather moor. The heather moors of the Pennines, like those of the Cleveland district and of Scotland, are commonly used as grouse moors and are systematically fired every few years. This prevents the effective invasion of trees such as birch and rowan which would occur at the lower altitudes. But at the higher levels the moors are too windswept for good tree growth to occur and the heather moor here represents a climax community. The periodic burning of the heather has the effect of constantly rejuvenating it. According to Moss (1913, p. 178), when it is fired every four years *Calluna* does not grow more than ankle-high, but if eight or ten years elapse before the next burning it becomes knee-deep. In fifteen years or so the plant becomes definitely aged and "leggy" and its flowering capacity is greatly diminished. Twenty or twenty-five years is probably the span of its natural existence, but there are no available observations on the rejuvenation of the community when the bushes are left to complete their natural life.

Burn subsere. After burning, *Rumex acetosella, Deschampsia flexuosa* or *Nardus stricta* may first recolonise the burned peat surface. Where present in quantity in the burned community *Vaccinium myrtillus* often asserts itself by sprouting from its underground rhizomes before the heather returns. But the ling commonly sows itself abundantly from seed and soon reasserts its dominance, the bilberry remaining as a lower layer which does not flower below the *Calluna* canopy.

Heather moors of the Wicklow Mountains. From this small mountain complex lying to the south of Dublin, Pethybridge and Praeger (1905, pp. 158–62) describe the Callunetum as occupying the zone between the siliceous grassland of the lower slopes (here mainly dominated by *Ulex gallii*) and the Scirpetum of the summit plateaux, which covers most of the ground above 1500 ft. or 450 m. (see Fig. 89). They emphasise the fact that Callunetum depends on good drainage, and that the thick peat which it often occupies has been formed, not by the ling itself, but by some more hydrophytic plant such as *Sphagnum* or *Scirpus caespitosus*. Where drainage has been increased by turf-cutting *Calluna* has colonised the drying surfaces.

Where the drainage is best, particularly near the lower limits of the heather moor, *Calluna* makes a dense uniform growth 2–3 ft. (*c.* 60–90 cm.) in height, with *Listera cordata*, *Melampyrum pratense*, and a continuous undergrowth of mosses. On its upper limit there is a wide ecotone of *Calluna-Scirpus*, passing above into the pure Scirpetum (Chapter XXXV, pp. 711–13).

The Calluneta of the Wicklow Mountains, like those of Great Britain, are regularly fired for the sake of the grouse, and this burning, together with the strong winds, inhibits tree growth on the moors. The effect of wind may be seen in the few trees of *Sorbus aucuparia*, whose crowns are kept well below the edges of the gullies they inhabit.

Nine samples of Callunetum were examined by Pethybridge and Praeger, and the following species recorded. The number of samples in which a species occurred follows its name:

Calluna vulgaris   d (9)

Agrostis tenuis	4	Molinia caerulea	2
Anthoxanthum odoratum	1	Nardus stricta	8
Blechnum spicant	4	Narthecium ossifragum	1
Carex binervis	7	Oxalis acetosella	1
C. diversicolor (flacca)	3	Pedicularis sylvatica	1
C. pilulifera	2	Polygala serpyllifolia	1
Deschampsia flexuosa	2	Potentilla erecta	6
Dryopteris dilatata	2	Pteridium aquilinum	2
Empetrum nigrum	3	Scirpus caespitosus	5
Erica cinerea	6	Sieglingia decumbens	1
E. tetralix	1	Sorbus aucuparia	1
Eriophorum angustifolium	1	Succisa pratensis	1
E. vaginatum	2	Ulex gallii	3
Festuca ovina	4	Vaccinium myrtillus	8
Galium saxatile	6		
Juncus effusus	1	Polytrichum sp.	1
J. squarrosus	8	Rhacomitrium lanuginosum	2
Luzula campestris	1	Sphagnum spp.	5
L. sylvatica	5	Cladonia spp.	3
Melampyrum pratense	1		

No fuller list of the bryophytes and lichens is available.

758   *The Heath Formation*

The Mourne Mountains. Armstrong and his co-workers (1930, 1934) have briefly described the vegetation of the Mourne Mountains in north-eastern Ireland in which Callunetum occupies the most extensive areas (Fig. 149), and attains unusual altitudes. According to Armstrong's map (Fig. 149) it covers the summits of 2300 ft. though not those of 2400 ft. and over, which are occupied by alpine or "summit" grassland dominated by *Festuca ovina* and *Rhacomitrium lanuginosum*. But on the southern slope of the highest peak, Slieve Donard (2796 ft. or 850 m.), Callunetum is actually shown (Fig. 150) reaching in one place a height of 2710 ft. or 826 m. This highest Callunetum forms a low dense mat 5–8 cm. high and with very few other species.[1] The principal associates of *Calluna* at the higher levels are *Erica cinerea* and *Carex pilulifera*, and "on certain sections of the mountain side *Erica cinerea* is definitely the dominant plant". The basins and cols between the summits are occupied by extensive tracts of Molinietum and Scirpetum caespitosi (Fig. 149).

Bilberry moor (*Vaccinietum myrtilli*)

Vaccinium myrtillus (Bilberry, Whortleberry, Wimberry, or, in Lowland Scots, Blaeberry—compare the American "Blueberry") is a local associate of *Calluna* on the southern English heaths, but is there confined to the somewhat moister areas, being absent altogether from many of the drier heaths. In the west and north it is far more abundant, and becomes a constant constituent of the Callunetum, often preceding the ling in succession and frequently forming a subordinate layer of the vegetation. As we go north, in Scandinavia, and as we ascend the higher mountains in Scotland it replaces *Calluna* as a dominant species, and is evidently able to withstand more rigorous climatic conditions.

Highland Vaccinieta. Robert Smith (1900) was the first to describe the Vaccinietum in the Highlands of Perthshire. On the mountains composed of rocks which produce the "poorer" soils, and on which raw humus or acid peat tends to accumulate, *Calluna* is commonly dominant up to about 2000 ft. (c. 600 m.) and tends to be replaced by *Vaccinium* above that altitude, the Vaccinietum, either nearly pure or still accompanied by *Calluna*, occupying the mountain slopes up to about 3000 ft. (c. 900 m.). Robert Smith gives a list of thirty-five vascular plants found on the "heather moors" at lower altitudes, occurring also in this high-level Calluno-Vaccinietum or pure Vaccinietum, and ascending to altitudes varying from 2350 to 3600 ft. (c. 715–1100 m.). The following species are almost confined to this high-level community:

Betula nana	Lycopodium annotinum
Cornus suecica	Rubus chamaemorus
Loiseleuria procumbens	Vaccinium uliginosum

[1] Similar "alpine mats" of Callunetum have been described from high altitudes in other regions, e.g. by Crampton from Caithness and Sutherland in northern Scotland.

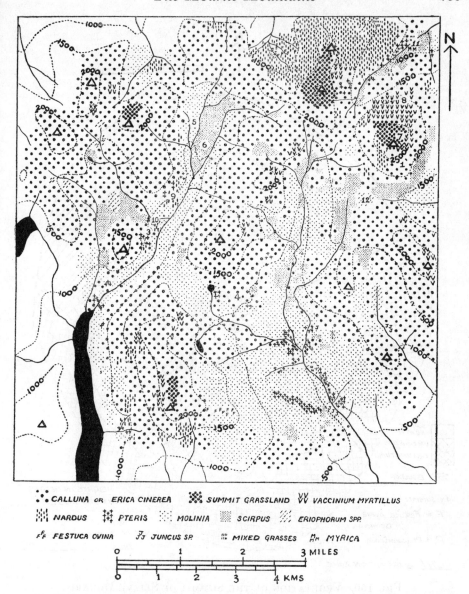

FIG. 149. VEGETATION OF THE MOURNE MOUNTAINS (N.E. IRELAND)

Callunetum occupies about half the area, covering the lower summits up to between 2300 and 2400 ft. and ascending above 2500 ft. on Slieve Donard in the north-west corner of the map. Vaccinietum occurs "on the steep faces of old boulder screes". The higher summits are covered with grassland dominated by *Rhacomitrium lanuginosum* and *Festuca ovina*. The stream valleys and flatter cols are occupied by Molinietum and Scirpetum caespitosi. Pteridietum is confined to sheltered valleys, and Eriophoretum and Nardetum are quite local. From Armstrong, Ingold and Vear, 1934.

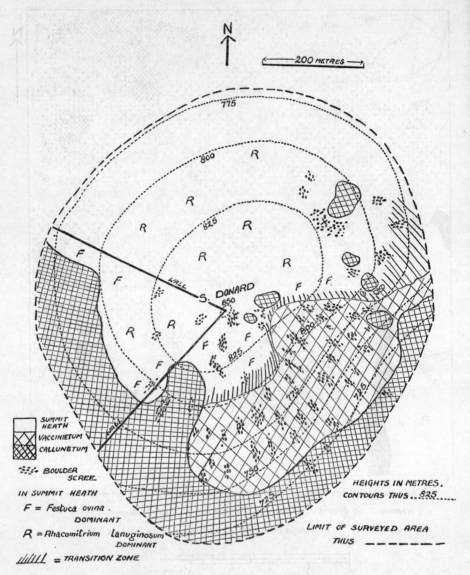

FIG. 150. VEGETATION OF THE SUMMIT OF SLIEVE DONARD,
MOURNE MOUNTAINS (IRELAND)

The actual summit (2800 ft.) and down to the 2600 ft. (800 m.) contour on the northern side is covered with "*Rhacomitrium* heath" (see pp. 787–8), with *Festuca ovina* very abundant, *Empetrum nigrum* abundant and *Salix herbacea* frequent. This is really an "arctic-alpine" community (see Chapter XXXVIII). On the southern side *Festuca* is dominant. The boulder scree to the south-east is occupied by Vaccinietum myrtilli, and below is Callunetum which covers most of the high ground of this mountain group. Contours in metres. From Armstrong, Calvert and Ingold, 1930.

Bilberry Moor (*Vaccinietum myrtilli*)

All of these have a claim to be considered "arctic-alpine" species, and the first three definitely belong to this category (compare Chapter XXXVIII, pp. 780–1).

Pennine Vaccinieta. Smith and Moss (1903) and Smith and Rankin (1903) recognised substantially the same displacement of *Calluna* by *Vaccinium myrtillus* at the higher levels in the southern Pennines of west Yorkshire, where Vaccinietum occupies the highest summits of the much lower hills (1400–1700 ft., c. 420–510 m.). These rocky boulder-strewn "Vaccinium summits" are composed of rocks with peaty humus between them, and the dominance of the bilberry is apparently determined by the better drainage and the greater exposure as compared with the peat-covered heather moor of the slopes below. The flora includes a small selection of the species of the Callunetum and lacks the characteristic arctic-alpine species of the high-level Scottish Vaccinietum.

"Vaccinium summits" and "edges"

Similar Vaccinietum occupies the very steep rocky escarpment slopes falling from the cotton-grass mosses of many of the Pennine plateaux (Fig. 140), but here the flora is somewhat richer, doubtless because of the greater shelter. Moss (1913, p. 182) gives the following list of vascular plants from such a "*Vaccinium* edge":

	Vaccinium myrtillus	d	
Arctostaphylos uva-ursi	la	Galium saxatile	la
Calluna vulgaris	lsd	Pteridium aquilinum	la
Deschampsia flexuosa	la	Rumex acetosella	o
Empetrum nigrum	la	Vaccinium vitis-idaea	lsd
Erica cinerea	la		

So good is the shelter that not only the bracken but even occasional individuals of hawthorn, rowan, birch and sessile oak have succeeded in establishing themselves.

Two rather sandy peats from "*Vaccinium* edges" at altitudes of 450 and 500 m. respectively showed pH values of 3·26 and 3·19. The soils are highly oxidising and nitrates are absent (Pearsall, 1938).

Northern Pennines. From the northern Pennines Lewis (1904) described, at about 2000 ft. (c. 600 m.), a "rocky slope with shallow peat in which *Calluna* tends to disappear, *Rubus chamaemorus* and *Eriophorum* are altogether absent, and the vegetation is made up of *Vaccinium myrtillus* and *V. vitis-idaea*, with *Juncus squarrosus* and a small quantity of *Nardus*". This seems to be essentially a "*Vaccinium* edge".

Mourne Vaccinieta. Armstrong, Ingold and Vear (1934) record Vaccinietum from the Mourne Mountains on steep boulder scree at 1500–2500 ft. also with *Blechnum*, *Deschampsia* and *Galium* and in other places *Calluna* and *Empetrum*, as well as *Anthoxanthum*, *Festuca ovina*, etc. (Figs. 149, 150).

Wicklow Vaccinieta. Pethybridge and Praeger (1905, p. 162) describe quite similar Vaccinieta "on well-drained rocky slopes" in the Wicklow Mountains, varying greatly in elevation "from a few hundred feet on Bray Head to over 2000 ft. on the hills". The vegetation of one of these "*Vaccinium* edges" (1600–1800 ft., c. 480–550 m.) contains a similar, though somewhat different, selection of species of the Callunetum (four species common to the two lists, excluding the dominant):

Vaccinium myrtillus   d
Blechnum spicant   Luzula sylvatica
Calluna vulgaris   Melampyrum pratense
Deschampsia flexuosa   Oxalis acetosella
Galium saxatile   Vaccinium vitis-idaea

Vaccinietum succeeding "Moss". On the plateau of "the Peak" in Derbyshire (southern Pennines), and on similar areas a little to the north (Fig. 140), Moss described the gradual conversion of an extensive moss, originally dominated by *Eriophorum vaginatum*, into a Vaccinietum, consequent on the gradual drainage and drying of the wet cotton-grass peat by the cutting back of streams into the edges of the Eriophoretum. This is described in Chapter XXXIV (pp. 705–6).

Mixture of *Vaccinium* with *Calluna* and *Eriophorum*. Very extensive mixed communities in which bilberry is associated with heather and cotton-grass are described by Lewis from the northern Pennines in a region where extensive peat erosion is proceeding. While it is not clear that these are exactly comparable with the Vaccinietum described by Moss on the Derbyshire Peak, it seems likely that here also *Calluna* and *Vaccinium* are colonising drying *Eriophorum* bogs, which still exist on these high-lying slopes and plateaux in immediate proximity to the mixed communities described. It is to be noted that while *Vaccinium myrtillus* is often mixed with *Calluna*, and both of these with *Eriophorum*, the bilberry does not seem in this region to appear as a co-dominant with the cotton-grass alone. This suggests that the relations of the three plants to drainage are in sequence—*Eriophorum*, *Calluna*, *Vaccinium*—a conclusion borne out by their general moorland habitats—so that a drying moss will tend to be colonised first by heather and then by bilberry. In the areas described by Moss it may be conjectured that drainage of the *Eriophorum* peat is so rapid that the bilberry is able to enter at once. The facts described by Adamson (1918) harmonise well with this conclusion. But there may well be other factors, not yet clear, connected with the autecology of the three species and helping to determine their distribution.

The peat bearing mixtures of *Calluna* and *Eriophorum vaginatum* shows somewhat higher pH values than that of pure Calluneta but they are below 4·0. The oxidation-reduction potential fluctuates about the critical range (Pearsall, 1938).

DISTRIBUTION OF THE HEATH FORMATION IN THE BRITISH ISLES

Though *Calluna vulgaris* is a ubiquitous species, being recorded from every vice-county in Great Britain and from every county in Ireland, Callunetum is not extensively developed on the western side of Ireland, Wales and Scotland, being confined to well-drained ground on poor soil which is not heavily pastured, where wind or frequent burning prevents the development of trees, and to acid peat which is drying as a result of natural or artificial drainage. This comparative scarcity of Callunetum in the west is probably due to the very wet climate, which favours the development of the bog or moss formation on all ground which is not exceptionally well drained.

Upland heaths. On the eastern side of Ireland and Wales, for example on the Wicklow Mountains, the Mourne Mountains, and on the Welsh Marches; and of Great Britain, as in the Eastern Highlands and on the Cleveland moors, upland heath dominated by *Calluna* is far more extensive, both on steep and on gentle slopes, and even on plateaux where the soil is permeable. These Calluneta are the typical grouse moors.

Lowland heaths. These are scattered throughout the English lowlands wherever there is sandy soil poor in bases and acid in reaction which has not been planted or cultivated—much more abundantly in the south and east where such soils are most prevalent. Many lowland heaths, except where they are exposed to violent winds, are always being colonised by birch, by pine (in the neighbourhood of plantations), less freely by oak, and locally by beech; but these are as constantly destroyed by felling, pasturing and burning. Many lowland heaths are still unfenced commonland and are pastured to a slight extent by the commoners, others are used as rabbit warrens, some are simply wasteland, a good deal of which is now being afforested. The following are some of the principal areas of English lowland heath:

(1) *East Anglian heaths*. These occupy considerable stretches of flat sandy country on the Pliocene crag of north-east Norfolk and south-east Suffolk, and especially on the overlying glacial sands and gravels there and in Essex. The Breckland heaths in south-west Norfolk and north-west Suffolk are developed on post-glacial sands and on leached sandy glacial till. Here however there are varying amounts of lime in the soil—the chalky boulder clay lying close below—and not only grass heath, but grassland rich in bases, sometimes approximating to chalk grassland, is often present.[1] Where rabbits are in excess Callunetum is converted into bent-fescue grassland. Colonisation by trees is not great except near plantations.

(2) *South-eastern (Lower Cretaceous and Wealden) heaths*. The heaths of

[1] See p. 511.

Kent, southern Surrey, and Sussex are extensive. They are developed on Lower Greensand (Cretaceous), and on the Ashdown Sand (Wealden) which forms the "Forest Ridge" in the centre of the Weald. Much of this land is occupied by "oak-birch heath", birchwood, and subspontaneous pinewood, some by oakwood, and locally there are small areas of beechwood.

(3) *Heaths of the London basin.* The Lower London Tertiaries (Reading Beds, etc.) on the edge of the London basin (together with overlying Plateau Gravels and Valley Gravels) and the Bagshot Sand, largely developed to the south-west of London, overlying the main formation of London Clay, also bear extensive heaths, those of the Bagshot Sands, especially, being much overgrown with subspontaneous pine.

(4) *Heaths of the Hampshire basin.* This, like the London basin, is a broad syncline of Tertiary rocks consisting of alternating sands and clays. The low plateau formed by these was covered in late Tertiary times by extensive sheets of river and estuarine gravels (Plateau Gravels) which have since been dissected by river systems forming shallow valleys. These sands and gravels bear wide stretches of heath, mainly to the west of Southampton Water in the New Forest area and in south Dorset. Much of this heathland is certainly very old: Domesday Book (eleventh century) and Leland (sixteenth century) mention the great "bruaria" of this region. The Callunetum alternates with dry oakwood, especially in the New Forest area. The soil of some of these Calluneta is very sterile—apparently toxic—and will not support tree growth, while the *Calluna* itself is poorly developed.

(5) *South-western heaths.* In west Dorset, south Somerset, Devon and Cornwall, heaths occur on non-calcareous rocks of very various age which form permeable soils. These are characterised by several south-western ("Atlantic southern") species. Thus large areas of the heath on the serpentine rock of the Lizard peninsula are dominated by *Erica vagans*,[1] a south-west European plant occurring nowhere else in the British Isles; and in the wet heaths of both Dorset and Cornwall *Erica ciliaris* occurs[2] and also *Pinguicula lusitanica*, both Atlantic southern species, the last of wider distribution in the west.

The heaths of the higher south-western hills—Exmoor, Dartmoor and Bodmin Moor on Devonian grits and on granite—are upland heaths and not extensively developed. Blanket bog and acid grassland are the typical communities of these high-lying moors.

(6) *Midland heaths.* On the Jurassic and Triassic rocks of the midlands heaths are not so common as on the Cretaceous, Tertiary and post-Tertiary sands and gravels of the south and east, but the Bunter Sandstone bears typical Callunetum in Sherwood Forest in Nottinghamshire. Delamere Forest in Cheshire, where the Bunter Sandstone is largely covered by recent deposits of alluvium and peat, has little Callunetum. The Lower Greensand

[1] Pl. 127, phot. 314.   [2] Pl. 127, phots. 312–13.

PLATE 127

Phot. 312. *Erica ciliaris* and *Molinia caerulea*. Wet heath near Perranwell, Cornwall. *Mrs Cowles.*

Phot. 313. *Erica ciliaris*, closer view. *J. Massart.*

Phot. 314. *Erica vagans* and *Ulex gallii* on the Lizard peninsula. *J. Massart.*

SPECIES OF SOUTH-WESTERN ENGLISH HEATHS

is well developed in Bedfordshire (south-east midlands) and bears Callunetum, but here it is very largely planted with conifers, especially Scots pine, which does exceedingly well.

STATUS OF THE HEATH FORMATION

The vegetational status of the Heath Formation requires some consideration. Primarily it is a west European formation, interdigitating so to speak with bog and forest, replacing the former on better drained soil, and the latter on the lighter and poorer soils, where pasturing or burning prevents the colonisation of trees. It can only be counted as a *climatic* formation on exposed coasts and on exposed mountain slopes between 300 and 600 m.,[1] that is, in situations where the wind factor is definitely inimical to tree growth, and where this is combined with good drainage and acid soil, either derived from acidic rocks or where the climate favours rapid leaching and surface peat formation. In these situations heath is certainly a climax, for, although the upland heaths of northern England and Scotland are practically always used as "grouse moors" and are fired periodically, there is no evidence that such heath, where it lies above the forest limit, would be superseded by any other type of vegetation if it were left entirely untouched.

The lowland heaths, so widespread in the south and east of England, are for the most part to be regarded as a stage in the succession to forest. In *Types of British Vegetation* (pp. 99–103) the heath formation was considered, following Graebner, as a result of the degeneration of oak forest on sandy soils which became podsolised, and some heaths have probably taken origin in this way, either through podsolisation of the soil, or more likely by felling, burning and pasturing. But many of them may well have been subject to burning or pasturing, or both, from the earliest times, and perhaps have never borne forest at all. Apart from some areas of Bagshot Sand, one of which (on Wareham Heath in Dorset) is now being investigated, there are few, if any, southern English lowland heath which will not support tree growth, at least of birch and pine, though the soil of some is probably too poor and dry for the good growth of oak. In general, however, they may be regarded as subclimax or "deflected climax" to dry oak or beech forest, belonging to the succession: heath → birch (→ pine) → oak → beech. It has been suggested that the development may really be cyclical, the beech, on these soils, creating soil conditions which render it incapable of regeneration, so that the forest ultimately dies and is replaced by heath, to begin the cycle anew (pp. 355, 415).

In regard to the development of heath itself, xeric subseres have been described (pp. 729–34, 751, 752, 756). Heath also develops on stabilised sand dunes (Pl. 126, phot. 309), but this prisere has not been ecologically investigated. It has been argued (pp. 734–5, 737) that the evidence for a hydrosere

[1] Higher than this in Scotland and north-east Ireland.

leading to heath is far from conclusive, though the possibility cannot be excluded. But the occupation of wet acid peat by Callunetum, as distinct from the abundant presence of *Calluna*, seems to depend, so far as the positive evidence goes, either on drainage or on change to a drier climate.

SUMMARY OF PEAT COMMUNITIES

It will be convenient to summarise here the seral relationships of the communities of peat and peaty soil described in the last six chapters.

Subaquatic peats. Peat may be formed under water by aquatic and reedswamp vegetation and Pearsall demonstrated that different conditions favour the colonisation of a lake floor by particular aquatic communities which form peat of different kinds; these seem to be analogous to the peats formed by the subaerial communities of fen and moss. But very little is known about British subaqueous peats, except that they have relatively high pH values (above 5·0) and are always reducing in character.

Ionic content of waters. We can draw a sharp distinction between the highly acidic waters of moorland pools, very poor in basic ions (or at least in calcium) and in nitrates, which are colonised by aquatic Sphagna and a few other water plants tolerant of such conditions and from which reedswamp is often absent, and the alkaline waters of a calcareous fen basin in which luxuriant and varied aquatic vegetation and reedswamp are present and typical fen is formed. Between the two come a series of waters, neutral or somewhat acid, which have never been given serious study. Most of them are fringed with some kind of reedswamp, and it is certain that the different reedswamp dominants show different degrees of toleration towards acidity and lack of calcium and nitrates. The "basic ratio", i.e. of potassium (and sodium) to calcium (and magnesium), may also be important. Until further study has been given to these matters there must remain a wide gap in our knowledge of the British hydroseres.

Reedswamp. Of the three commonest tall reedswamp dominants *Scirpus lacustris* and the two species of *Typha* (*T. latifolia* and *T. angustifolia*) are confined to reedswamp, *Scirpus* often occupying its outer zone in fairly deep water, while the commonest of all, *Phragmites communis*, maintains itself in fen. Of others, which may dominate the transition to fen, *Glyceria maxima* (*aquatica*) and *Phalaris arundinacea* seem to be characteristic of silting, *Cladium mariscus* is most luxuriant in alkaline, but not silting, waters, while some Carices and Equiseta, e.g. *Carex inflata* and *Equisetum fluviatile* (*limosum*), though by no means all or most, are tolerant of more acidic waters.

Phragmites communis is by far the most widespread and abundant of all the reedswamp-fen species, and is widely tolerant of different conditions. It forms reedswamp in various depths of water and is often abundant, or even dominant, in fen. By means of its strong, horizontal, deep-lying and wide-spreading rhizomes the common reed can freely invade wet soil (which may be relatively dry upon the surface), sending up its shoots among other vegetation. It is not always clear whether its presence in fen is due to survival from a reedswamp phase in which it has been dominant, or to secondary invasion. Nevertheless it is rather remarkably absent from many places which seem quite suitable for it.

Fen. *Cladium mariscus* is a very local plant in the British Isles, but it can tolerate very different conditions of climate and soil reaction, occurring alike in the East Anglian fens and in the west Irish bogs. It is a characteristic pioneer and dominant in the calcareous fen basins of East Anglia, forming a very pure community owing to its power of excluding other species.

In more mixed fens *Juncus subnodulosus* and *Carex panicea* are two of the most abundant species. In small calcareous basins in the south midlands *Juncus subnodulosus* sometimes forms reedswamp, maintaining itself as an abundant species in the fen, but followed as a dominant by *Schoenus nigricans*, a species characteristic of such habitats in most of England, though in the south-west and in Ireland it lives in and sometimes dominates acid bogs: it has recently been suggested that this depends upon salt spray. In northern Ireland the dominants of the "lower fen" are very various.

Succession in fen. It is only at Wicken Fen that exact observations have been made on the critical level for the colonisation of woody plants, and this is so close above the summer water level that there would seem little room for fen development between reedswamp and carr in the natural prisere. The decisive factors for the production of extensive fen may be (1) deep winter flooding as in north Ireland, or (2) lack of seed parents for the initiation of woody vegetation. Many existing fens are however certainly maintained by cutting, or in the drier fens, grazing. Under natural conditions the first shrubs of the fenwood or carr colonise Cladietum or other fen community while it is still very wet, and sometimes even reedswamp.

Cutting Cladietum not only prevents the development of carr but, when it is heavy and repeated at short intervals, handicaps and may eventually exterminate the dominant. At Wicken under these conditions *Molinia caerulea* enters the community and may become dominant, continuing to build up the peat level so as to produce a much drier type of fen than is formed by primitive carr. This is a good case of "deflected succession". If cutting is stopped the fen is at once colonised by bushes. *Molinia* is a species of great habitat range, occurring widely in fen, in wet heath, and in *relatively* dry bog, besides dominating extensive areas of wet upland grassland. It is a real peat former, though *Molinia* peat is seldom of any great depth.

Carr. Perhaps the commonest pioneer of the woody vegetation of fen is *Salix atrocinerea*. At Wicken this sallow is accompanied or preceded by *Frangula alnus*, which is the most abundant species in young carr, and followed by *Rhamnus catharticus* and *Viburnum opulus*, in Norfolk also by

Alnus glutinosa, and *Betula pubescens*. *Fraxinus excelsior* is often a frequent or abundant constituent of developing carr. *Crataegus monogyna* and *Ligustrum vulgare* also occur, and in some fens other constituents of calcareous woodland, such as *Viburnum lantana* and *Daphne laureola*, as well as calcicolous herbs like *Paris quadrifolia* and *Mercurialis perennis* and the climber *Tamus communis*.

Fully formed scrub carr at Wicken (where trees are practically absent) is dominated mainly by *Rhamnus catharticus*, locally by *Viburnum opulus* and *Salix atrocinerea*, or a mixture of species. Almost pure sallow carr is often met with on fens, but the sallow seems usually to give way to other species, sometimes dying in large numbers from a cause not yet ascertained. In Norfolk Alnetum is the typical ultimate carr, but Betuletum also occurs, and alder and birch are frequently mixed, together with ash, in other marsh and fen woods.

It seems probable that there would be a natural succession from fen carr to oakwood in this country under conditions which permitted of the gradual raising of the soil level by continued accumulation of humus, but it is difficult to demonstrate because of the almost universal interference with the soil and vegetation as soon as land conditions begin to be established.

Moss or bog. In sharp contrast with typical alkaline fen we have the moss or bog formation characteristic of highly acidic waters poor in basic ions, especially calcium.

The establishment of bog vegetation is a development of fen alternative to the progress to scrub and forest indicated above. It begins with the appearance of species of *Sphagnum* on the tussocks of the fen plants when they are raised somewhat above the alkaline ground water. The Sphagna are those of relatively high mineral requirements, such as *S. squarrosum* and *S. subsecundum*, which are more tolerant of alkaline solutions and do not themselves produce so much acid as other species that occur later in the succession. These are accompanied by a selection of such plants as *Eriophorum angustifolium*, *Pinguicula vulgaris*, *Polytrichum commune* and species of *Drosera*. The pH ranges of these have not been worked out—like the Sphagna themselves some are more, others less, oxyphilous. In eastern and central England, where the normal succession is from fen to scrub or woodland, the "acidic evolution" of fen usually seems to stop at the formation of small colonies of *Sphagnum* with their associated plants. This is probably correlated with the relatively dry climate, which may prevent the bog moss from growing far above the water table. In English fens *Sphagnum* colonisation and bush colonisation may be seen side by side, but the closing of the shrub canopy eventually kills out the bog moss, though some species may maintain themselves for a time in open carr.

Raised moss or bog. It is otherwise in Wales and western England, in parts of Scotland, and in central Ireland. In the moister climate of these regions, though tree colonisation of fen and carr formation may also occur,

Summary of Peat Communities

Sphagnum is able to build upon fen a *raised moss* or *bog*, entirely obliterating the fen vegetation and forming a lens-shaped cushion of acid peat several metres thick and many hundreds of metres in diameter. The development of a typical Irish raised bog, as recorded in the peat, is fully described on pp. 686–96.

There is a regular succession of species of *Sphagnum* from the semi-aquatic species, of relatively high mineral requirement, which originally colonised the fen at the base of the bog, to highly acid species of low mineral requirements in the bog itself. These last, independently in each of the elementary units of which the bog consists, show repeated cyclic successions, from species of high to species of low water requirements, as "hollows" give place to "hummocks"—the "regeneration cycle" of Osvald (pp. 687–8).

But though the development of raised bog depends essentially on the growth of different species of *Sphagnum*, its vegetation contains other and more conspicuous elements, flowering plants which determine the physiognomy, the tone, of the whole bog, and form a considerable fraction of the whole mass of plant material present. Of these *Scirpus caespitosus* and *Eriophorum vaginatum* are the most abundant and prominent, and *Calluna vulgaris* is nearly as conspicuous. The remaining species, except *Narthecium ossifragum* and *Rhynchospora alba*, which are the phanerogamic dominants of the "hollows", are of little vegetational importance.

There is no evidence in Ireland that trees ever colonise raised bog, except rather sporadically the well-drained marginal slope, where occasional isolated pines and birches may be seen. The slope normally bears Callunetum, with *Pteridium* and *Ulex* locally dominant. Drained bog peat also normally develops Callunetum.

Blanket bog. In the very wet climates of the extreme west of Scotland and Ireland another type of bog occurs, which may be called *blanket bog*. This is indeed the climatic climax formation of these regions, covering the general surface of the country, except on rocky outcrops and on the steeper slopes. Essentially the same general type of bog also covers the high plateaux, many about 2000 ft. (*c.* 600 m.), of the English, Scottish and Irish mountains, which also possess a wet climate.

Sphagnum rarely forms the essential basis of blanket bog as it does of raised bog (extensive Sphagneta occur only in one place in the Pennines). Hummocks of bog moss, of similar construction to those of raised bog,[1] may be formed locally (and sometimes abundantly), especially in the lowland blanket bogs of the west, but the dominants of blanket bog are vascular plants. The most widespread dominants are *Scirpus caespitosus* (west Ireland, west Scotland, Wicklow Mountains), *Eriophorum vaginatum* (Pennines), and *E. angustifolium* (northern Pennines), in that order, with *Schoenus nigricans* more local, and *Molinia caerulea* in drier places. *Rhynchospora alba*, *Narthecium ossifragum* or *Erica tetralix* may be locally

---

[1] Often capped by *Rhacomitrium lanuginosum* or *Leucobryum glaucum*.

dominant. Very often however blanket bogs are not dominated by single species.

There is no evidence and no likelihood that blanket bog is *generally* built up on fen, though doubtless it had this origin in certain calcareous basins in the west of Ireland. The plants of the blanket bog probably colonised the general surface of the country when the climate became sufficiently wet, whatever the previous vegetation. In many parts of the west of Ireland, it is clear from the continuous layers of pine stumps at the base of the peat that here bog succeeded pine forest in many places. The early stages of such successions are unknown, but from observations on current successions and from our knowledge of what happens in the north of Europe, we may conjecture that they have included the alga *Zygogonium* as well as *Sphagnum* itself. The soil on which blanket bog has developed, both in the lowlands and on the high plateaux, is very commonly glacial clay—the ground moraines left behind on the recession of the Pleistocene ice-sheets. Much of the blanket bog may have originated in the innumerable pools of acidic water that must have been present in local depressions of this ground moraine, colonised by aquatic species of *Sphagnum*, whose growth filled up the pools; but the details of such successions have not been studied.

Callunetum. The occurrence of *Calluna* in blanket bog is very widespread, practically universal, but it does not attain anything like luxuriant growth or abundant flowering except on drier knolls or the tops of local hummocks. A succession to Callunetum has not been observed except where the peat has been drained, and there it is a regular occurrence. It is probable that under natural conditions change to a drier climate is necessary for the supersession of the bog vegetation by Callunetum.

The Heath formation itself, as we have seen, generally occurs on a permeable mineral soil poor in bases, in drier climates developing only a shallow layer of surface raw humus. In damper climates it (or rather its accompanying lichens) forms a certain depth of acid peat, rarely exceeding 12 in. (30 cm.). Where Callunetum occurs on deeper peat this has been formed by some other community which has been superseded by the heather as a result of drainage or of the incidence of a drier climate.

The lowland Calluneta of the south of England show xeroseral succession, and it is not clear that wet heath can pass into Callunetum without the agency of erosion. Wet heath does not differ essentially from bog or moss except in the much lesser frequency of the Cyperaceae (*Scirpus*, *Eriophorum*, etc.), which are abundant or dominant in the latter. The upland Calluneta have been fully discussed, and from that discussion the conclusions emerge that while the *habitats* of *Calluna* are very varied, its *dominance* in the British climates depends upon relatively well-drained soils, and that it does not come to dominate the peat of moss or bog except through increased drainage or change of climate. Where erosion leads to

Summary of Peat Communities

rapid drying of the peat especially at relatively high altitudes on the Pennine plateaux, it is *Vaccinium myrtillus* which occupies the Vaccinietum ground. And at the higher altitudes on the high plateaux above 2000 ft. in the Scottish Highlands, the same species, in company with *V. vitis-idaea*, *V. uliginosum* and *Arctostaphylos*, replaces *Calluna* as the general dominant.

Synopsis of peat vegetation

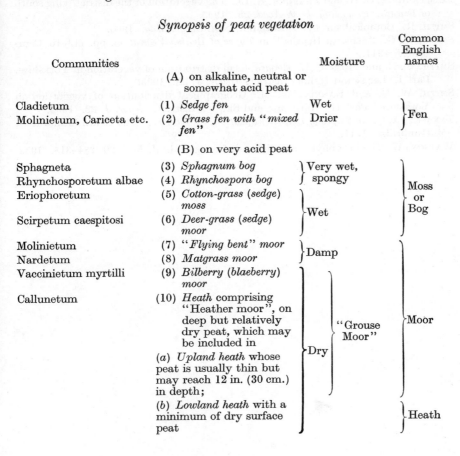

REFERENCES

ADAMSON, R. S. On the relationships of some associations of the Southern Pennines. *J. Ecol.* **6**, 97–109. 1918.

ARMSTRONG, J. I., CALVERT, J. and INGOLD, C. T. The ecology of the mountains of Mourne with special reference to Slieve Donard. *Proc. Roy. Irish Acad.* **39**, 440–52. 1930.

ARMSTRONG, J. I., INGOLD, C. T. and VEAR, K. C. Vegetation map of the Mourne Mountains, Co. Down, Ireland. *J. Ecol.* **22**, 439–44. 1934.

CRUMP, W. B. The coefficient of humidity: a new method of expressing the soil moisture. *New Phytol.* **12**, 125–47. 1913.

ELGEE, F. The vegetation of the eastern moorlands of Yorkshire. *J. Ecol.* **2**, 1–18. 1914.

LEWIS, F. J. Geographical distribution of the vegetation of the basins of the rivers Eden, Tees, Wear and Tyne. Parts I and II. *Geogr. J.* **23**. 1904.

LEWIS, F. J. and MOSS, C. E. "Heather Moor" in *Types of British Vegetation*, pp. 275–9. Cambridge, 1911.

MOSS, C. E. *Vegetation of the Peak district.* Cambridge, 1913.

PEARSALL, W. H. The soil complex in relation to plant communities. III. Moorlands and bogs. *J. Ecol.* **26**, 298–315. 1938.

PETHYBRIDGE, G. H. and PRAEGER, R. LL. The vegetation of the district lying south of Dublin. *Proc. Roy. Irish Acad.* **25**. 1905.

SMITH, R. Botanical survey of Scotland. *Scot. Geogr. Mag.* 1900.

SMITH, W. G. "Scottish Heaths", in *Types of British Vegetation*, pp. 113–16. Cambridge, 1911.

SMITH, W. G. and MOSS, C. E. Geographical distribution of vegetation in Yorkshire. Part I. Leeds and Halifax district. *Geogr. J.* **22**. 1903.

SMITH, W. G. and RANKIN, W. M. Geographical distribution of vegetation in Yorkshire. Part II. Harrogate and Skipton district. *Geogr. J.* **22**. 1903.

TANSLEY, A. G. In *Types of British Vegetation*, Chapter IV, "The Heath formation". Cambridge, 1911.

WATSON, W. The bryophytes and lichens of moorland. *J. Ecol.* **20**, 284–313. 1932.

Part VIII

MOUNTAIN VEGETATION

Chapter XXXVIII

THE UPLAND AND MOUNTAIN HABITATS

MONTANE AND ARCTIC-ALPINE VEGETATION. THE SCREE SUCCESSION. COMMUNITIES OF EXPOSED ARCTIC-ALPINE HABITATS

The highest hills in the south-east of England scarcely ever exceed 1000 ft. (304 m.) in height, and up to this altitude the vegetation shows no perceptible change. In the south-west there are considerable areas over 1000 ft. and some (Dartmoor) over 1500 ft. (457 m.) and up to 2000 ft. (610 m.) in altitude. These however are high-lying plateaux, and corresponding with the damp south-western climate they have a relatively high rainfall and often bear blanket bog (see Chapter XXXIV). Genuine mountain habitats are only to be found in parts of Ireland, Wales, the north of England, and especially in the Scottish Highlands.

Upper limit of forest. Except in the most exposed situations and on the wet waterlogged plateaux the former upper limit of forest must have commonly reached altitudes of 1500 and in places 2000 ft. where the mountain masses were of adequate height; and the existing vegetation of the hill slopes up to these levels has replaced forest, mainly as the result of grazing. The grassland and heath communities which now occupy these habitats have already been dealt with (Chapters XXVI and XXXVII). So far as flora is concerned there are a few species which may be called "montane" because while they are in no sense high mountain plants they are rarely found at altitudes of much less than 1000 ft.; but there are not many plant communities specially characteristic of these intermediate altitudes. Nardetum, upland Callunetum and Vaccinietum may however be considered as belonging to this category.

Montane vegetation

As we ascend to higher levels several important climatic factors show progressive change. Mean temperatures are lower, rainfall and humidity higher, snow lies much longer, and exposure to wind, except in sheltered places, is much greater. In this country very little attention has been given to the study of the factors which condition the characteristic mountain habitat in relation to vegetation, though their combined effect in determining the so-called arctic-alpine communities is sufficiently clear. In spite of the general severity of the mountain climate its total flora is extremely rich and the variety of small communities very great. This is because the mountain habitats are very varied and show many different combinations of factors, some of which are very favourable to plant life.

Talus and scree. There is one physiographic factor characteristic of craggy hills and mountains—namely the effect of gravity on the weathering debris of hard precipitous rocks (*crags*)—which produces a highly characteristic plant habitat, namely the accumulations of talus detached from crags as the result of alternate heating and cooling of the parent rock and of the freezing and thawing of water in rock fissures. The fragments so detached accumulate in a slope (*scree*) below the crag at the angle of repose, and as fresh rock fragments fall they disturb the equilibrium of the mass of talus so that there is a constant tendency for the stones to slide down the slope. Screes may be formed at any altitude, but they become more numerous and extensive at the higher levels where the parent crags are commoner, and are best considered as one type of mountain habitat, though when developed at low or moderate altitudes they may be entirely lacking in "arctic-alpine" species.

The scree succession. The vegetation of British screes has not received a great deal of attention, but Leach (1930) has published a good preliminary account of some non-calcareous screes in the Lake District and around Snowdon. In this he describes the leading factors of the scree habitat and the various successions that can be traced.

Screes differ much in their degree of stability as well as in the size of the fragments into which the parent rock weathers. Leach found that the most unstable type, most difficult of colonisation by plants, is that composed of tabular fragments of slate about 10–20 cm. in length and breadth by 1–4 cm. in thickness, while screes formed of more or less isodiametric fragments are much more stable. "Block screes" he does not consider.

Lithophytes such as *Andreaea petrophila*[1] and *Rhacomitrium fasciculare*, **Lithophilous bryophytes** which may colonise the actual surfaces of the rock fragments, play little or no active part in the sere, since they do not form a suitable nidus for the colonisation of higher plants. Mosses may however actually disintegrate the rock surface, which is apparently rendered porous by the action of the closely applied rhizoids of *Rhacomitrium heterostichum* and *Dicranella heteromalla*. The first really **Pioneers** effective stage in the sere is formed by chomophytes,[2] at first mosses and liverworts, followed by ferns, which settle in the spaces between the stones where they find small accumulations of fine detritus. The pioneer bryophytes in these situations are *Diplophyllum albicans*, *Rhacomitrium lanuginosum* and *Rh. heterostichum*, which occur on practically every non-calcareous scree examined between the altitudes of 800 and 2000 ft. (240 and 610 m.). In the cushions formed by these bryophytes, particularly *Diplophyllum*, the parsley fern (*Cryptogramma crispa*) finds enough water-retaining humus to enable its prothallus and young sporophyte to get a footing, and it is the principal pioneer vascular plant on all the more extensive siliceous screes examined. To enable the parsley fern to continue its growth a position of some stability is required; and this is generally provided by the

[1] Still conspicuous at 2000 ft. (610 m.). [2] See p. 787.

occurrence of a larger block which resists the flow of the surrounding talus, diverting the sliding fragments to each side. In the shelter of such a block, i.e. on its lower side, *Cryptogramma* can grow from year to year with the aid of the constantly increasing accumulation of humus supplied by the decay of its annually deciduous fronds. Below every large plant of this species there is a mass of root-bound humus composed entirely **Other ferns** of the debris of dead fronds (Pl. 128, phot. 317). Other ferns may occur in the same situations:

Athyrium filix-femina Dryopteris filix-mas
Blechnum spicant D. montana

but these are not so uniformly successful as *Cryptogramma crispa* in providing year by year for their own future requirements and thus acting as pioneer vascular plants.

Eventually *Cryptogramma* produces a mass of humus more than sufficient **Bryophytes** for its own needs, and then numerous bryophytes and some **and Flowering plants** flowering plants are able to settle down round the pioneer fern.

The following species occur in this situation:

Agrostis canina Festuca ovina and f. vivipara
A. tenuis Galium saxatile
Anthoxanthum odoratum Oxalis acetosella
Deschampsia flexuosa Viola riviniana

Campylopus flexuosus Hylocomium splendens
C. fragilis H. squarrosum
Catharinea undulata Plagiothecium elegans
Dicranella heteromalla Polytrichum alpinum
Dicranum fucescens P. commune
D. scoparium P. formosum
Eurhynchium pumilum P. gracile
*Hypnum cupressiforme var. P. juniperinum
   ericetorum P. piliferum
H. cuspidatum Webera nutans
H. schreberi
Hylocomium loreum Lophozia floerkii

* Still conspicuous at 2000 ft. (610 m.) on Lliwedd (Snowdon).

The amount of surface covered by bryophytes varies with the general humidity of the habitat. The number of species present decreases with altitude. Thus at 1000 ft. there were seventeen conspicuous species, at 1750 ft. eight, and at 2000 ft. six.

Formation of vertical strips. If the protecting block remains *in situ* the area of vegetation so arising tends to extend downwards from the site originally colonised, and the longer the vegetation remains the more permanent it is likely to become, because of the fixing and stabilising effect on the substratum of the continuous formation of root-bound soil between the rock fragments of the scree. The vertical strips of vegetation thus produced (Fig. 151, and Pl. 128, phots. 315, 316) are often dominated by *Calluna* and show the composition of a typical heath community. Besides those mentioned above as succeeding *Cryptogramma* the following species may be found in this heath community:

The heath community

Calluna vulgaris
Digitalis purpurea
Erica cinerea
E. tetralix
Euphrasia officinalis (agg.)
Lycopodium selago

Polygala serpyllifolia
Potentilla erecta
Solidago virgaurea
Teucrium scorodonia
Thymus serpyllum

Fig. 151. Tabular Scree, Semi-diagrammatic Sketch

Some of the vertical strips of vegetation are stabilised by protecting blocks at their upper ends, others (more permanent) are developed below areas at the top of the scree which are relatively quiescent. From Leach, 1925.

Many such patches and strips are however transient, being carried away by the general movement of the scree.

Besides the strips originating from this type of colonisation below large protecting blocks, others arise in places where the fragments of rock falling from the crags above are only accumulating slowly, so that there is comparative stability at the top of the scree. Such strips extend downwards from the foot of the crags to the bottom of the scree slope, though the lower ends may be overwhelmed and destroyed by the lateral spreading of fans of

PLATE 128

Phot. 315. Strip vegetation (heath) of scree on south slope of Causey Pike (Lake District) with the Birkrigg Oaks behind. Alt. 1100–1400 ft. From W. *Leach* (1930).

Phot. 316. Screes on south-west slope of Grasmoor above Crummock Water, showing vertical strips of Callunetum. Alt. 1500 ft. From W. *Leach* (1930).

Phot. 317. Large parsley fern (*Cryptogramma crispa*) with *Festuca ovina* on Hindscarth scree, alt. 1700 ft. From W. *Leach* (1930).

Phot. 318. *Vaccinium myrtillus* invading Grasmoor scree by rhizome growth. Alt. 1500 ft. From W. *Leach* (1930).

SCREE VEGETATION

unstable material from intervening strips of actively moving scree (Fig. 151). The vertical strips of vegetation are usually at a lower level than the scree on each side (Fig. 151, section *AB*), because they are protected from the piling up of fragments that occurs in a moving scree which is constantly being added to from above.

Destruction of seral stages. The complete succession from *Diplophyllum* and *Rhacomitrium* through *Cryptogramma* to Callunetum is rarely completed because the colonised areas are so often destroyed in one stage or another, either by movement of the scree as a whole or by the piling up and overflowing of the rock fragments, which obliterate the plant colonies. Thus an active scree shows fragments of seral communities, representing various stages of the succession, scattered upon its surface, and the more active the scree the sparser and younger the colonies.

Stable screes. The more stable screes, on the other hand, show more extensive and more advanced stages of progressive colonisation, and these connect with more gentle slopes and with areas of mountain-top detritus (see p. 787) where formation and disintegration of the rock fragments occur *in situ*. Here soil, often clayey, accumulates more rapidly between the fragments of rock, and *Agrostis tenuis*, *Alchemilla alpina* and *Festuca ovina*, accompanied by mosses such as *Polytrichum nanum* and *P. piliferum*, act as pioneers, in addition to *Diplophyllum*, *Rhacomitrium* and *Cryptogramma*. The roots of the grasses, and to some extent the moss rhizoids, bind the soil and prevent its being washed away by heavy rain.

Rhizomatous and creeping plants invade the edges of scree from neighbouring areas, either an adjacent hillside covered with vegetation, or an already fully stabilised scree. The most important of these marginal invaders are bracken (*Pteridium aquilinum*) and bilberry (*Vaccinium myrtillus*).[1] On the edge of the scree humus collects in the deeper and more stable layers of the substratum, and this is penetrated by the roots and rhizomes of the marginal plants which progressively colonise the margin of the scree. *Dryopteris linnaeana* and *Vaccinium vitis-idaea* are less frequent. The bilberry and crowberry (*Empetrum nigrum*) with their upright woody shoots here act as pioneers and have a markedly stabilising effect on the slowly moving stones. *Hedera helix* and *Silene maritima*, which creep over the surface, are occasionally met with.

Climax vegetation on screes. On completely stabilised scree up to an altitude of about 1500 ft. (450 m.) wood of *Quercus sessiliflora*, the climatic climax, may develop, and many of the hillside Querceta both in Wales and the Lake District undoubtedly occupy the sites of ancient scree (Pl. 128, phot. 315), just as Fraxinetum may be developed on calcareous scree (Chapter XXI, pp. 436–40). Woodland may develop as a result of the final weathering down of the crags causing cessation of the supply of falling stones.

[1] Pl. 128, phot. 318.

ARCTIC-ALPINE VEGETATION

Comparatively little work has been published in recent years on the higher mountain vegetation of the British Isles—the two most important contributions being Watson's account of the bryophytes and lichens of arctic-alpine vegetation (1925) and Price Evans' study of the vegetation of Cader Idris (1932). Both of these authors adopt the classification of the arctic-alpine plant communities proposed in 1911 (in *Types of British Vegetation*) by the late W. G. Smith, whose account is indeed so excellent as a preliminary treatment that the present chapter will be based upon it and considerable passages reproduced.

The arctic-alpine zone. "The recognition of an arctic-alpine zone of vegetation is based on the fact that in ascending the higher mountains one finds that at some altitude the general tone of the vegetation alters, and that new species characteristic of higher altitudes appear, forming plant communities distinct from the lowland ones. The species also present various growth-forms adapted to conditions ranging from the barrenness of windswept slopes and summits to the comparative shelter and generally favourable environment afforded by rock-crannies amongst precipitous crags and ravines. In the central Highlands of Scotland the contour line of 2000 ft. (*c.* 600 m.) indicates roughly the lower limit of this zone: above this the vegetation is characterised by an increase of such plants as saxifrages, dwarf willows and other low-growing perennials, generally associated with characteristic mosses and lichens. Such woodlands as still remain in the country cease either below or near this limit. On the subalpine ericaceous moorland the change is marked by decrease of *Calluna* and increase of *Vaccinium*, while the grasslands become poorer in species belonging to such families as Leguminosae and Umbelliferae, and assume a new physiognomy through increasing frequency of arctic-alpine plants, such as *Alchemilla alpina*, viviparous grasses, and arctic species or varieties of *Carex, Juncus, Luzula, Draba, Cerastium, Potentilla*, etc.

"On the Atlantic side of Scotland and Ireland a similar vegetation occurs at much lower altitudes, almost down to sea-level. These plant communities include many of the same species grouped in a manner comparable to that of the communities of the higher mountains. It is thus evident that in defining the arctic-alpine vegetation, too much stress should not be laid on zones of altitude expressed numerically; other factors must be sought which may explain the limited distribution of this vegetation.

"The arctic-alpine vegetation presents well-marked features both in its plant communities and in its characteristic growth forms. It can thus be characterised not only on a floristic, but also on a strictly ecological basis, especially as sub-alpine species occurring in the arctic-alpine zone assume the growth-forms of low-growing cushions, rosettes, or mats, more or less similar to those of the arctic-alpine species proper" (W. G. Smith, 1911, pp. 288–90).

Arctic-alpine species. The "arctic-alpine species proper" are those which are more or less confined in Britain to high altitudes and were called "Highland" by H. C. Watson (*Compendium* to the *Cybele Britannica*, 1868, Introduction, p. 28), and count between one and two hundred angiosperms, according to the number of arctic-alpine forms or ecotypes of lowland aggregates which we recognise as "species". These plants are characteristic of high northern latitudes in arctic and subarctic Europe, many of them occurring at sea-level in the high arctic and becoming restricted to higher and higher altitudes as we pass south from these regions. In central Scotland, as we have seen, they do not occur in any quantity below about 2000 ft. (*c.* 600 m.) and are much more abundant above 3000 ft. (*c.* 900 m.), while in the Swiss Alps many of the same species occur only in alpine communities above about 8000 ft. (*c.* 2400 m.). In western Scotland and western Ireland, however, several of them descend to sea-level.

Life forms of arctic-alpines. Both the arctic and the alpine angiosperms, as groups, have, as Raunkiaer showed (1908), a percentage of species of the chamaephytic life form[1] markedly exceeding that which obtains in the flora of most other climates, and the higher the latitude or altitude the greater the percentage of chamaephytes. Indeed the few species which inhabit the more remote arctic islands and those occurring above 4000 m. in the Alps are more than half chamaephytes, the rest being hemicryptophytes.[2] Of the eleven species of vascular plants occurring above 1000 m. (*c.* 3280 ft.) in the Clova region of the Scottish Grampians three are chamaephytes, seven are hemicryptophytes and one is a geophyte, i.e. 27 per cent are chamaephytes, as compared with 9 per cent in the flora of the world as a whole (Raunkiaer, 1908; Willis and Burkill, 1901–4). Between 800 and 900 m. (*c.* 2625–2950 ft.) while 60 per cent of the seventy-two species are hemicryptophytes there are already 22 per cent of chamaephytes, a very marked rise from lower altitudes.

Chamaephytes

The relatively large number of chamaephytes in the arctic-alpine vegetation is undoubtedly connected with the fact that the ground is covered with snow during most of the winter, in some places during the whole winter and the first half of summer,[3] so that the temperature just above the soil surface remains fairly constant round about 0° C., and the vegetation

---

[1] Chamaephytes are plants whose perennating buds are borne between the soil surface and a height of 25 cm. (10 in.) above it.

[2] Plants with perennating buds in or on the surface layer of soil.

[3] The summit of Ben Nevis in the western Scottish Highlands at an elevation of 4409 ft. (1344 m.)—the highest point in the British Isles—was only free from snow in August and September during 10 years' observation. The snow, which began to lie in October, steadily increased during the winter, reaching a mean maximum of 78 in. (198 cm.) at the beginning of May and not disappearing entirely till after the middle of July (Buchan and Omond, 1905).

is not subjected to the rapid alternations of frost and thaw which occur during winter in the lowlands of western Europe where the soil is exposed. The low-growing forms of the plants (the vegetative parts often forming rosettes or cushions) ensures that the vegetation is covered with snow during the cold period and may also be related to the violence of the winds that often blow at high altitudes during the growing period. The velocity of the wind is very much less close to the ground, and the close setting of the shoots and leaves gives a great deal of mutual protection. When the snow melts the surface of the soil warms up quickly under the rays of the sun, especially on southern slopes, and the rise in temperature of the soil surface and of the air immediately above it stimulates rapid development of the perennating buds into leaves and flowers. This is very necessary in a climate where the growing season is extremely short because the snow lies very late and fresh snowfalls occur in the early autumn.

Bryophytes and Lichens. Another marked feature of arctic-alpine vegetation is the prominence of mosses and especially of lichens. There are a large number of species of these groups which occur in, and many of which are peculiar to, the arctic-alpine zone, but even more striking is their massive conspicuousness in the vegetation. Everywhere they are locally dominant at the higher altitudes, and some dominate considerable communities, such as the "*Rhacomitrium* heath", especially on exposed summits and plateaux.

During the summer, extremes, and especially rapid fluctuations, of temperature and of humidity are much greater than in the lowlands, and mosses and lichens are particularly well adapted to endure such conditions of alternate drying and wetting, heating and cooling. Most lichens and many mosses are able to withstand almost complete desiccation, absorbing water again as soon as it is available and immediately resuming active life. The low-growing and tufted mat- or cushion-like habit of most mosses also gives them the maximum protection from wind. Furthermore, as W. Watson especially has shown, the leaves of many species of moss have structural characters which tend to check evaporation from their leaves. Thus most of these non-vascular plants are extremely well protected against killing by drought; but for the reproductive processes of bryophytes, unlike those of the higher plants, external water is essential in the sexual stage and dry air in the stage of spore distribution. In fact some species very rarely fruit, reproducing themselves almost entirely by vegetative means.

Though the habitats of the exposed upper slopes and the summits of mountains are frequently very dry, and the vegetation is correspondingly xerophytic, it must not be forgotten that there are, on many mountains, both very wet and also very well-protected habitats with abundant soil water. It is in these last that, side by side with arctic-alpine species, various lowland plants (or arctic-alpine ecotypes of lowland plants) may flourish very luxuriantly, though they often do not flower.

Effect of rock and soil. The effect of the nature of the rock (which conditions the type of erosion) and of the soil which it produces on the kind of vegetation developed on the higher Scottish mountains is well marked. On the acidic rocks of which the majority of the mountains are composed, subalpine moorland with regular acid peat formation extends up to a high level, e.g. 3000 ft. on the Clova-Canlochan plateau. This type of moorland, dominated by *Vaccinium myrtillus*, though often with *Calluna* associated (Calluno-Vaccinietum) has been described in Chapter XXXVII (pp. 761–2). The flora is poor and typical of sub-alpine moorland, including very few arctic-alpine and those mostly arctic species, such as *Loiseleuria procumbens*, *Betula nana* and *Cornus suecica*. But on exposed knolls, where the drainage is good and peat cannot form, a few other arctic-alpines occur, such as *Alchemilla alpina*, *Carex rigida*, *Gnaphalium supinum* and *Salix herbacea*. Even on the summits which rise from such moorlands to a much higher level (in the neighbourhood of 4000 ft. or 1200 m.) the arctic-alpine flora is relatively poor.

Richer flora of basic rocks. Sharply contrasted with these floristically poor mountains are those composed of easily weathering rocks rich in mineral salts. Actual limestone is very rare in the higher Scottish mountains but some of the schists and other metamorphic and volcanic rocks are relatively rich in bases (calcium, magnesium and potassium). It is on the slopes, particularly the southern slopes, especially if they are irrigated by numerous springs, in the corries (steep-sided valleys), and on the rock ledges of such mountains that the arctic-alpine flora is seen in its greatest profusion. The two typical Scottish areas are the Breadalbane Mountains in the central Highlands of Perthshire, of which Ben Lawers is the best known example, and the Clova region on the borders of Forfarshire and Aberdeenshire in the eastern Highlands; though here, as we have seen, the elevated plateau bears peat moor and most of the numerous arctic-alpine species are confined to the slopes and corries. The Cairngorm group, composed mainly of granitic rocks, on the borders of Aberdeen, Banff and Inverness, though it has by far the most extensive area of high mountain land, including some of the highest summits, is not so rich as these two, though considerably richer than the western Highlands.

The same contrast between the flora and vegetation of acidic rocks and of those rich in basic salts is seen in the higher English and Welsh mountains, for example in the Lake District, in Snowdonia and on Cader Idris (see pp. 786, 808).

ARCTIC-ALPINE GRASSLAND

On the old, stabilised, scree-covered slopes, especially the southern slopes, of those mountains whose rocks produce a more basic soil, generally "where the individual summits begin to be differentiated from the more continuous undulating slopes of the valleys", at some altitude between 2000 and

3000 ft. (about 600–900 m.) there is developed a characteristic arctic-alpine grassland, or at least a vegetation in which grasses are prominent and not dominated by dwarf shrubs.

By its fresh bright green colour this differs markedly in appearance from the dingy, brownish green, peaty grassland often dominated by *Nardus*, and developed on the flatter morainic terraces below. The soil of arctic-alpine grassland is derived by direct weathering of the basic rocks of the mountain, and occurs in continuous sheets between and over the overgrown boulders and outcropping rocks (Pl. 130, phot. 322). Raw humus formation is at a minimum and peat is absent, except locally. "The habitat is kept moist throughout the greater part of the year by numerous springs emerging at various levels, many of them marked by swelling cushions of bright green mosses. This water is mainly derived [on Ben Lawers] from schistose rocks comparatively rich in calcium, magnesium and potassium, and owing to the steepness of the slopes it drains away and does not become stagnant and acid. An essential condition of the habitat is that the soil is stable and little disturbed by surface erosion" (W. G. Smith). In these respects this grassland resembles the limestone grasslands at lower altitudes and contains many of the same species, though it is characterised by an assemblage of arctic-alpine plants. It may certainly be considered as a climax community, developed above the natural forest zone and not grazed to any considerable extent nor interfered with by artificial drainage.

Floristic composition. The floristic composition of a rich arctic-alpine grassland, like that of Ben Lawers, is very varied. There is no general dominant, but the two commonest local dominants are *Festuca ovina*, forma *vivipara* and *Alchemilla alpina* (both arctic-alpine plants), which often cover considerable areas.[1] The community has not been analysed in detail, and we can only give a general list of species mostly recorded by Robert Smith (1900) and taken from W. G. Smith's account (1911, pp. 300–1). Species limited to wet stream banks are omitted.

Abundant and locally dominant lowland species

Agrostis tenuis	
Anthoxanthum odoratum	
Deschampsia flexuosa[2]	general and abundant
Festuca ovina (agg.)	
Carex goodenowii (forma)	
Molinia caerulea	locally dominant[2]
Nardus stricta	

"Highland" (arctic-alpine) species

Alchemilla alpina	
Festuca ovina f. vivipara	abundant, locally dominant

[1] Pl. 130, phot. 322.

[2] *Nardus, Molinia, Carex goodenowii* (forma) and *Deschampsia flexuosa* are evidently species of peaty soil and must be regarded as local intrusions determined by local conditions and not belonging to the community described.

Carex capillaris
C. rigida
Cerastium alpinum
Luzula spicata
Lycopodium alpinum
L. selago

Phleum alpinum
Poa alpina
Potentilla crantzii
Sagina saginoides
Selaginella selaginoides
Sibbaldia procumbeus

Lowland species

Achillea millefolium
Alchemilla vulgaris (agg.)
Anemone nemorosa
Antennaria dioica
Avena pratensis
Bellis perennis
Blechnum spicant
Botrychium lunaria
Campanula rotundifolia
Carex binervis
C. stellulata
C. flava (forma)
C. pilulifera
Cerastium vulgatum
Deschampsia caespitosa
Euphrasia officinalis (agg.)
Galium saxatile
Heracleum sphondylium
Hypericum pulchrum
Juncus squarrosus
Lathyrus montanus
Leontodon autumnalis
Linum catharticum

Lotus corniculatus
Luzula multiflora var. congesta
Lycopodium annotinum
Melampyrum pratense var. montanum
Orchis maculata
Oxalis acetosella
Plantago lanceolata
Polygala serpyllifolia
Potentilla erecta
Ranunculus acris
Rumex acetosa
R. acetosella
Sagina procumbens
Scirpus caespitosus
Succisa pratensis
Taraxacum officinale
Thymus serpyllum
Trifolium repens
Vaccinium myrtillus (sparse and dwarf)
Veronica serpyllifolia
Viola lutea f. amoena

Similar types of high altitude grassland have been described from the slopes of other mountains, but over rocks producing soil poorer in mineral salts they are much less rich floristically, and contain but few arctic-alpine species. Thus Price Evans (1932) gives lists from five different sites of "dry grassland on a porous substratum" on Cader Idris in Wales. This grassland occurs on "parts of the summit plateau, and on drained ledges of the great escarpment, but it reaches its greatest development on the lower slopes where the high summits begin to be differentiated from the moorland". These sites, which vary from 1600 ft. (488 m.) to 2700 ft. (823 m.) in altitude, are dominated or partly dominated by *Festuca ovina*. They are mostly situated at somewhat lower absolute altitudes than the corresponding grassland described from Ben Lawers (the summit of Cader Idris is only 2927 ft. while Ben Lawers reaches 3984 ft.) and the conditions of the habitat are clearly not so favourable. The list of species is therefore much shorter and there are only one or two which can be considered arctic-alpines, but there are very few which are not recorded as occurring in the Ben Lawers grassland. It is clearly the same type of community. The following is a composite list from the five sites:

Festuca ovina		d or co-d	
Agrostis spp.	o	Luzula campestris	r-o
Anthoxanthum odoratum	o	*Lycopodium alpinum	f-a
Blechnum spicant	r	L. clavatum	r-la
Campanula rotundifolia	r-o	*L. selago	o-f
Carex pilulifera	r-f	Nardus stricta	r-f
Cryptogramma crispa	l	Polygala vulgaris	o
Deschampsia flexuosa	r	Potentilla erecta	o-f
Empetrum nigrum	o-a	Vaccinium myrtillus	f-co-d
Galium saxatile	a	V. vitis-idaea	o-a
Juncus squarrosus	r-o	Viola sp.	o

Species of *Cladonia* are locally abundant or sub-dominant on some of the sites, and *Rhacomitrium lanuginosum* varies from "frequent" to "co-dominant".[1] Four other species of lichen were noted:

| *Cerania vermicularis | Cetraria glauca |
| *Cetraria aculeata forma hispida | C. islandica |

* "Highland" or arctic-alpine species.

Lower Festucetum. Moist facies
On the lower slopes of the northern face of Cader Idris, but still at a relatively high altitude (1500–1800 ft., or 457–550 m.), in the neighbourhood of springs and flushes, under a band of calciferous rocks, there are small detached patches of mixed grassland locally dominated by *Festuca ovina*, *Agrostis tenuis* or *Poa annua* accompanied by a number of lowland species more than half of which are also recorded from Ben Lawers arctic-alpine grassland, and some of which are present as arctic-alpine varieties or forms. Among these are the following, besides the local dominants just mentioned:

Achillea millefolium	f	Prunella vulgaris	f
Bellis perennis	la	Ranunculus acris	o
Campanula rotundifolia	l	R. repens	o
Cerastium vulgatum	o	Rumex acetosa	o-f
Cirsium palustre	la	R. acetosella	o-f
Dactylis glomerata	o	Sagina procumbens	o
Euphrasia officinalis	la	Sieglingia decumbens	o
Ficaria verna	o	Stellaria media	o
Leontodon autumnalis	l	Thymus serpyllum	o
Oxalis acetosella	o-f	Trifolium repens	la
Plantago lanceolata	f	Veronica officinalis	o
P. major		Viola sp.	o

Out of a total of fifty-nine species of vascular plants recorded from the Ben Lawers grassland and forty-four from Cader Idris, twenty-nine are common to both.

[1] Compare also the Festucetum ovinae with *Rhacomitrium* on the summit of Slieve Donard (p. 758 and Fig. 150).

These two communities on Cader Idris, taken together, may fairly be regarded then as the equivalent of the Ben Lawers grassland, the dry and the moist facies being separated on Cader Idris, which also has the arctic-alpine element very feebly represented. The lower is transitional, as Price Evans says, to the *Agrostis-Festuca* grassland of lower altitudes (Chapter XXVI).

UPPER ARCTIC-ALPINE ZONE

Above the level of the continuous high-level moorland or arctic-alpine grassland, i.e. at an altitude of about 3000 ft. (900 m.) or more on the higher mountains, we come to the most typical arctic-alpine communities, consisting very largely of "highland" species. The habitats may be broadly distinguished as "exposed" and "sheltered". The exposed higher slopes, for example the northern slopes of Ben Lawers, and the summits, are covered with so-called "mountain-top detritus", a waste of loose rocks and rock debris which has accumulated *in situ* from the continuous action of frost and wind on the solid rock of the higher mountain slopes and summits, and is often of great depth. This is the most extreme habitat and supports only an open community of mosses and lichens with a few associated phanerogams, which, on the higher mountains, are almost exclusively arctic-alpine species; or a closed Rhacomitrietum (often called a "*Rhacomitrium* heath") with fewer "highland" species of flowering plants and more heath plants which depend on the accumulation of raw acid humus by the dominant moss.

<small>Exposed habitats</small>
<small>Mountain-top detritus</small>

In strong contrast to the exposed habitats are the *corries* with their sheltered hollows, fissures and ledges containing pockets of mineral debris and soil, with little humus, but moist and well aerated. These are much the most favourable habitats at the high altitudes and support by far the greater number of the total list of arctic-alpine species. W. G. Smith groups the plants of these habitats together as *chomophytes*—a term due to Oettli, from χῶμα a heap or pile of earth. Along with the numerous arctic-alpine species the fissures and ledges of the corries also hold a number of lowland plants. This vegetation is dealt with in the next chapter.

<small>Protected habitats</small>
<small>Chomophytes</small>

FORMATION OF MOUNTAIN-TOP DETRITUS

The succession. "In the earlier phases [of succession] lichens form crusts and xerophytic mosses gradually cover the rock detritus and bridge the gaps, till, by their growth and decay, cushions of humus become available for the growth of flowering plants. At the same time fine soil collects in the pockets amongst the stones.

"The formation includes two extreme phases. There is an open plant community on a substratum where stones and boulders are still visible, but are covered with crusts of lichens and a partially discontinuous carpet of moss and lichen (moss-lichen open community). This leads by gradations to the *Rhacomitrium* heath in which the substratum is often completely hidden by masses of woolly fringe-moss (*Rhacomitrium*) which frequently form a thick layer of humus. This latter is the final or closed stage of the formation, but at the lower altitudes it may advance a stage further and become a community of ericaceous shrubs. The final stage has however a precarious existence on the summit ridges, as the moss carpet may frequently be observed completely torn up by wind so that the rocky floor is exposed, and the succession begins again. No hard and fast line can be drawn between the extreme communities, and all transitional stages between them may frequently be seen within a limited area (Pl. 129).

<small>Moss-lichen associes</small>

<small>Rhacomitrium heath</small>

"The formation coincides topographically with the 'alpine plateau' of Robert Smith. The two most marked communities are described in the Faröes by Ostenfeld as the 'alpine formation on the rocky flat' and the '*Grimmia* (*Rhacomitrium*) heath'; the species given by Ostenfeld correspond closely with those recorded on the Scottish mountains. The *Rhacomitrium* community was recognised by R. Smith (1900):—'above 3000 ft. (*c.* 910 m.), and even on bare exposed places at lower elevations, such as the summits of hills, the ground is usually stony and sparsely covered with vegetation. Mosses and lichens dominate: in particular, the woolly fringe-moss (*Rhacomitrium lanuginosum*) which on many of the steep mountain rubbles forms the peat on which the alpine humus plants develop.'[1]

Status of the community. "The arctic-alpine formation of mountain-top detritus is primarily climatic and represents the extreme limit of plant-life on an extremely porous substratum in places greatly exposed to wind, subject to extremes of temperature for the greater part of the year and probably under snow for a considerable period. On the one hand, it may be regarded as an outpost of vegetation represented in the extreme case by the first lithophytes on bare rock, and progressing towards a condition of permanent covering. On the other hand, passing upwards from below it appears as the remnant of the grass slope and the ericaceous moorland of the lower arctic-alpine zone, with all species eliminated which cannot adapt themselves successfully to the more extreme ecological conditions. At the same time the vegetation is recruited as regards 'highland' species, e.g. *Silene acaulis*, *Minuartia* (*Arenaria*) *sedoides*, *Saxifraga oppositifolia* on

---

[1] Pethybridge and Praeger (1904, p. 168) record *Rhacomitrium* "associations" from Killakee and other summits in Ireland below 2000 ft.; two forms of this community are recognised by them, one in which this moss forms high bosses or cushions on wet boggy moorland and another in which *Rhacomitrium* forms a flatter carpet over granite debris. The former seems to be identical with the *Rhacomitrium* hummocks of blanket bog (pp. 716, 718), but the latter is analogous to that described in the text, though arctic-alpine species play no part.

PLATE 129

Phot. 319. Mountain top detritus on summit of Cross Fell (northern Pennines), alt. 2900 ft. (c. 880 m.). *Rhacomitrium lanuginosum* between the boulders. J. Massart.

Phot. 320. Near the summit of Cross Fell. *Carex rigida*, *Rhacomitrium lanuginosum*. J. Massart.

Phot. 321. "*Rhacomitrium* heath", with open associes of mosses and lichens in foreground and distance. Summit of Mynydd Moel, 2700 ft. (c. 820 m.), Cader Idris. From E. Price Evans (1932).

VEGETATION OF MOUNTAIN TOP DETRITUS

Lawers, from the more sheltered corries where these species find their most favourable habitat.

Habitat conditions. "The surface is dry and well-drained, either because of its porous character or from steepness of slope, but the frequency of rain, snow or mist makes it periodically wet. There is little or no soil layer, so that the competition of the closed moorland communities is excluded. The plant covering is distinctly xerophilous in response to frequent dry periods as a result of wind action, less frequently to insolation and lack of precipitation. The growth forms are those indicated by Warming as characteristic of the arctic mat-vegetation and 'Felsenflur', the moss tundra or heath, and the lichen tundra or heath. Most of the species of flowering plants are deeply rooted amongst the stones and boulders in pockets of soil which are deeper than a superficial glance might suggest. Advantage is taken of any shelter afforded by boulders, so that depressions are more rapidly occupied than the more exposed ridges. An observation (supplied by Mr W. E. Evans) illustrates this; while ascending the Cairngorms, the stony waste appeared devoid of green plants, but on looking down the slope from above there was a distinct green tint, due mainly to *Juncus trifidus* growing in the shelter afforded by the slightly raised edges of a series of terraces of rock debris.

Life forms. "The more characteristic growth [life] forms on the drier substrata are as follows: cushion plants (*Silene acaulis, Minuartia (Arenaria) sedoides*), low mats of decumbent intertwined branches (*Loiseleuria procumbens, Sibbaldia procumbens, Empetrum nigrum, Saxifraga oppositifolia* which also has a cushion form), mats formed by rhizomes and shoots intertwined just below the surface (*Luzula spicata, Carex rigida*), mats with rhizomes deeply buried (dwarf *Vaccinium myrtillus, Salix herbacea*), rosettes (dwarf *Ranunculus acris, Festuca ovina* f. *vivipara*). These growth forms give much protection, and this is further increased by hairs and other protective adaptations. Most of the species are well adapted for vegetative propagation and several are viviparous.

Recruitment. "Where the moss-lichen associes is adjacent to exposed rock-faces and corries with a rich flora (as on Ben Lawers) the list of species is considerably increased. This indicates a great wastage of plant life under the extreme conditions of this formation, a loss which can only be replaced from habitats more favourably situated. While the closed *Rhacomitrium* association would appear to increase the shelter for other species, it is noteworthy that the proportion of 'highland' species is generally less in it than on the more open stony waste" (W. G. Smith, 1911, pp. 310–13).

Flora of moss-lichen associes. This is the pioneer open associes on mountain-top detritus developed on summits exposed to extreme variations of temperature and humidity. While the general physiognomy is extremely uniform the micro-habitats differ very much according to whether they are moist or dry. The associes consists of a great assemblage of species and varieties of bryophytes and lichens, which include not only a large number of strictly arctic-alpine forms, but also many species that are common at low altitudes, as for example the following very widely distributed species: *Polytrichum piliferum, Dicranum scoparium, Dicranella heteromalla, Eurhynchium myosuroides, Hypnum cupressiforme, Hylocomium splendens, H. triquetrum, Cephalozia bicuspidata, Peltigera canina, Cladonia sylvatica.* These are accompanied by a much smaller number of scattered vascular plants, nearly all arctic-alpine species.

The lists of mosses (70), liverworts (39) and lichens (172) given by Watson (1925) in his comprehensive study of this vegetation are reproduced in full below. Asterisks are attached to the names of species or varieties which are strictly arctic-alpine:

Mosses

*Andreaea alpina (moister places)	o	Hylocomium splendens	a
A. crassinervia (usu. on rock)	f	H. squarrosum	o
		H. triquetrum	o
A. nivalis (moister places)	l	Hypnum callichroum	la
A. petrophila	a	*H. bambergeri	o
*var. alpestris (rocks)	o	H. cupressiforme	f
Blindia acuta	f	var. ericetorum	f
*Brachythecium glaciale	r and l	*H. hamulosum	f
*B. plicatum	r and l	H. incurvatum	r
B. rutabulum	o	H. molluscum	lf
Bryum pseudotriquetrum (moist)	o	*H. procerrimum	l and r
		*H. revolutum	r
Ceratodon purpureus	o	H. schreberi	la
*Conostomum boreale	f	Oedipodium griffithianum (usu. rock nooks)	l
Dichodontium pellucidum (moist)	o	Oligotrichum hercynicum	a
		Orthothecium rufescens	o
*var. fagimontanum (moist)	o	Plagiothecium denticulatum	f
		var. obtusifolium	la
Dicranella heteromalla	o	P. elegans	o
var. interrupta	o	*P. muhlenbeckii	l
Dicranoweisia crispula	lf	*Polytrichum aloides	o
Dicranum falcatum	lf	*P. alpinum	a
D. fuscescens	a	P. commune	o
*var. congestum	f	P. juniperinum	o
*D. molle (usu. on ledges)	r	P. piliferum	a
D. schisti	lf	*P. sexangulare	la
D. scoparium	o	Pterigynandrium filiforme	o
var. spadiceum	o	Rhacomitrium canescens	o
D. starkei (often on rock)	o	R. heterostichum	o
*Ditrichum zonatum	f	var. gracilescens	a
*var. scabrifolium		R. lanuginosum	a
*Eurhynchium cirrosum	r and l	Swartzia montana	f
E. myosuroides	o	*var. compacta	f
E. swartzii		Webera albicans (moist)	
Hylocomium loreum	a	W. elongata	o
*H. pyrenaicum	l	*W. ludwigii (moist)	o
H. rugosum	l	W. nutans	a

Liverworts and Lichens of Mountain-top Detritus

Liverworts

*Alicularia breidleri	lf	Gymnomitrium crenulatum	o
A. scalaris	a	*G. obtusum	o
*Anthelia julacea	a	*G. varians	la
*A. juratzkana	a	Leptoscyphus taylori	o
Bazzania triangularis	o	*Lophozia alpestris	f
B. tricrenata	o	L. floerkii	o
*Cephalozia ambigua	l	L. incisa (damp pl.)	o
C. bicuspidata	f	L. quinquedentata	o
Cephaloziella byssacea	f	L. ventricosa	o
Diplophyllum albicans	a	*Marsupella condensata	la
*D. taxifolium	r	*M. ustulata	o
Eucalyx subellipticus	o	Plagiochila asplenioides	f
Gymnocolea inflata	o	var. minor	lf
f. compacta (drier pl.)	o	*Pleuroclada albescens	lf
f. laxa (moist gr.)	o	Ptilidium ciliare	f
*Gymnomitrium adustum	o	Scapania dentata (moist pl.)	o
*G. alpinum	o	S. irrigua (small form)	o
*G. concinnatum	la	S. undulata (moist pl.)	o
*G. corallioides	o	*Sphenolobus politus	o
*G. crassifolium	la		

Lichens

Alectoria bicolor	lf	*C. nivalis	r
*A. divergens	r	*Cladonia amaurocraea	r
*A. ochroleuca	o	*C. bellidiflora	la
*A. sarmentosa	r	C. cervicornis	a
*var. cincinnata	r	var. subcervicornis	a
*Arthopyrenia bryospila	r	C. coccifera	o
Baeomyces roseus	o	C. crispata	lf
B. rufus	o	C. deformis	o
Bacidia flavovirescens	o	C. degenerans	o
*Biatorella fossarum	l	var. phyllophora	la
*Biatorina contristans	la	*C. destricta	lf
B. cumulata	l	C. flabelliformis	o
Bilimbia lignaria	o	C. floerkeana f. trachypoda	la
B. melaena	la	C. furcata	a
B. sabulosa		var. pinnata	a
*var. montana		C. gracilescens	lf
B. sabuletorum	o	C. gracilis	la
var. simplicior	r	C. pyxidata	o
*B. rhexoblephara	l	C. rangiferina	
*Buellia alpicola (on stones)	l	C. squamosa	o
*B. badioatra (on stones)	l	C. sylvatica	f
var. atrobadia (on stones)	l	C. uncialis	a
B. myriocarpa (on stones)	o	f. adunca	a
*B. pulchella (rock nooks)		f. turgescens	f
*Cerania vermicularis	la	C. verticillata	o
Cetraria aculeata	f	*Collema ceraniscum	r and l
var. alpina	o	Collema tenax	o
C. crispa	o	*Dacampia hookeri	la
*C. cucullata	r	Dermatocarpon cinereum	la
*C. hiascens	r	*var. cartilagineum	lf
C. islandica	a	D. hepaticum	o
*forma platyna	o	D. lachneum	o

*Gyalecta foveolaris	o	L. fuscocinerea (rocks and stones)	l
Gyrophora cylindrica	a	L. granulosa	o
*G. erosa (rock)	r	*var. escharoides	la
G. polyphylla	o	L. griseoatra (rocks and stones)	o
*G. torrefracta (rock)	r	L. kochiana (rocks and stones)	o
Haematomma ventosum (rock)	la	L. lapicida (rocks and stones)	o
Icmadophila ericetorum	o	*L. limosa	la
*Lecania curvescens	vr	L. lithophila (rocks and stones)	o
Lecanora badia (rocks)	f	*L. nigroglomerata	vr
*L. epibryon	l	*L. pycnocarpa (rocks and stones)	l
*L. geminipara	vr	*L. rhizobola	r
L. polytropa	a	L. sanguinaria	o
L. tartarea	f	*var. affinis	
*var. frigida	la	L. sorediza (rocks and stones)	o
*var. gonatodes		L. sublatypea (rock)	o
L. subtartarea	o	*L. tabidula (rocks and stones)	r
*L. upsaliensis	r	*L. tesselata (rocks and stones)	o
Lecidea aglaea	o	L. uliginosa	a
*L. alpestris	l and r	*L. vernalis	
*L. arctica	la	Leptogium lacerum	o
L. atrofusca (on moss)	o	var. pulvinatum	o
L. auriculata (rocks and stones)	lf	Pannaria brunnea	f
var. diducens (on rocks and stones)		*P. hookeri	l
*L. berengeriana	l	*Parmelia alpicola (often on quartz)	r
*var. lecanodes		*P. pubescens	a
*L. breadalbanensis	r	*var. reticulata	
L. coarctata	o	P. saxatilis	o
var. glebulosa	o	P. tristis (usu. on rock)	la
var. elacista	o	*P. vittata	r
L. confluens (rocks and stones)	f	*Parmeliella lepidota	r
*L. consentiens (rocks and stones)	r	Peltidea aphthosa	o
L. contigua (rocks and stones)	f	Peltigera canina (small forms)	o
L. contigua v. flavicunda (on rocks and stones)	f	P. polydactyla	o
		P. rufescens	o
var. platycarpa (on rocks and stones)	f	*Pertusaria bryontha	l
L. crustulata (rocks and stones)	f	*P. dactylina	l
*L. cuprea		*P. glomerata	l
L. demissa	f	*Pertusaria oculata	la
*L. deparcula (rocks and stones)	r	*P. xanthostoma	l
L. dicksonii (rocks and stones)	lf	*Placynthium delicatulum	r
L. epiphorbia	r	*Platysma fahlunense	o
L. fuscoatra (rocks and stones)	o	*P. hepatizon	la
		*P. polyschizum	o

*Polyblastia gelatinosa	r	*Solorina bispora	r
P. nigritella	r	*S. crocea	la
*P. sendtneri	r	S. spongiosa	r
*Porina furvescens	r	Sphaerophorus fragilis	o
Rhizocarpon confervoides (on stones)	la	S. corallioides f. congestus (drier)	a a
R. geographicum (on stones)	a	*Stereocaulon alpinum S. corallioides	f f
var. atrovirens (on stones)	o	S. evolutum S. denudatum	a f
R. oederi (on stones)	lf	*S. tomentosum	r
*R. postumum	r	*Thelopsis melathelia	r
*Schizoma lichinodeum	la	*Varicellaria microsticta	l and r

Adaptations to habitat. Watson notes that the species are preponderantly xerophytic, but that many which are common to the lowland and the arctic-alpine habitats show variations in structure which he thinks indicate that they are living in moister conditions on the mountain tops. Thus the hair points of the leaves of *Polytrichum piliferum* and *Rhacomitrium canescens*, well developed in the typical lowland habitat, are often less conspicuous at high levels, and in *Rhacomitrium* may be almost absent. Similar variations are found in the lowlands when the plant grows in ditches. Again *Dichodontium pellucidum, Bryum pseudotriquetrum, Sphenolobus politus, Scapania dentata* and *S. undulata*, usually found in more constantly wet situations at lower altitudes, are able to exist in this exposed community because they can absorb water through the whole of their surfaces.

Plagiothecium denticulatum and *Hypnum schreberi*, on the other hand, usually flourishing best in damp and somewhat shaded situations, have intensified xeromorphic characters in this high-lying habitat.

Vascular flora. The following vascular plants are listed by W. G. Smith (1911) as members of this associes:

*Alchemilla alpina	*Lycopodium selago
*Arabis petraea	*Minuartia (arenaria) sedoides
Deschampsia flexuosa (dwarf)	Nardus stricta
*Draba incana	*Polygonum viviparum
Empetrum nigrum	Ranunculus acris (dwarf)
*Festuca ovina f. vivipara	*Salix herbacea
*Gnaphalium supinum	*Saxifraga oppositifolia
*Juncus trifidus (dwarf)	*S. stellaris (dwarf)
*Loiseleuria procumbens	*Silene acaulis
*Luzula arcuata	Solidago virgaurea
*L. spicata (dwarf)	*Sibbaldia procumbens
*Lychnis alpina (Clova)	

* Arctic-alpine species.

Watson (1925) records *Rhytisma salicinum* on *Salix herbacea* on Ben Doran at 3000 ft., and *Ticothecium erraticum* on the crustaceous lichens.

The moss-lichen associes on Cader Idris. On the summit plateau of Cader Idris, which lies roughly between 2500 and 2900 ft. (762 and 883 m.),

Price Evans (1932, pp. 32–3) describes this associes as "a meagre vegetation consisting chiefly of mosses and lichens" with a few flowering plants. He ascribes its poverty not only to the severe climatic conditions but also to the infertility of the granophyre of which the detritus is composed.

Rhacomitrium lanuginosum is the most abundant and characteristic plant, and associated with it are:

Dicranella heteromalla		Polytrichum piliferum	a
var. interrupta	—	Scapania gracilis	—
Lophozia floerkii	—		
Baeomyces roseus	—	Cladonia furcata	f
*Cerania vermicularis	—	C. gracilis	f
*Cetraria aculeata f. hispida	—	C. rangiformis	f
C. glauca f.	f	C. subcervicornis	a
C. islandica	f	C. uncialis	f
Cladonia cervicornis	f		

Of rock mosses and lichens on the stones of the debris there are:

Andreaea petrophila	Rhizocarpon geographicum
Gyrophora cylindrica	R. oederi
Lecidea contigua f. flavicunda	Rhacomitrium gracilescens
L. dicksonii	*Parmelia pubescens

and of vascular plants only the following:

Campanula rotundifolia	r	*Lycopodium alpinum	a
Empetrum nigrum		*L. selago	r–o
(prostrate)	o	Potentilla erecta	a
Festuca ovina	a	Vaccinium myrtillus	f
Galium saxatile	f	V. vitis-idaea	o

* "Highland" or arctic-alpine species.

The flora of the detritus on the summit of this Welsh mountain is very meagre and the flowering plants are nearly all "lowland" forms, arctic-alpine species being represented only by two species of *Lycopodium*. Nevertheless, the vegetation is certainly a fragmentary example of the same community as that which occurs on the tops of the floristically rich Scottish mountains.

Flora of *Rhacomitrium* heath. As W. G. Smith pointed out (see pp. 787–8) this community arises where a species of *Rhacomitrium* (usually *R. lanuginosum*), already abundant in the moss-lichen associes, has succeeded in establishing continuous dominance, at the same time forming a nidus of raw humus open to the colonisation of such ericaceous plants as can tolerate the extreme conditions of exposure. The bryophytes and lichens are naturally greatly reduced from the enormous numbers occurring in the open associes, owing to the general dominance of *Rhacomitrium*, but, as Watson says, many species of the open associes may still occur in the closed "heath". Though a few species and varieties recorded by Watson for the latter are not included in the lists for the former community, he informs me (*in litt.*) that it is doubtful if there are any bryophytes or lichens in the *Rhacomitrium* heath which may not occur in the open community. Liverworts are rare or absent in the Rhacomitrietum.

The following are recorded:

Bryophytes

Rhacomitrium lanuginosum   d

Dicranum fucescens	o	Rhacomitrium canescens	
D. uncinatum	r	var. ericoides	o
Hylocomium loreum	o	R. heterostichum	
H. splendens	o	var. gracilescens	la
Hypnum cupressiforme	o	Polytrichum alpinum	o
H. schreberi	o		
Leptoscyphus taylori	o	Ptilidium ciliare	o

Lichens

Alectoria nigricans	la	Cladonia flabelliformis	o
*A. ochroleuca	o	C. furcata	la
Bilimbia melaena	r	C. gracilis	o
*Cerania vermicularis	o	C. rangiferina	a
Cetraria aculeata	o	C. sylvatica	a
f. subnigrescens	o	C. squamosa	o
*var. alpina	o	C. uncialis	a
C. crispa	la	f. obtusata	a
C. islandica	la	f. turgescens	a
*Cladonia bellidiflora	o	Galera hypnorum	r
C. deformis	la	Icmadophila aeruginosa	r
C. degenerans		Lecanora tartarea	o
f. pleolepidea	o	Peltigera canina	o
*C. destricta	l		

The following is an amended list of the vascular plants given by W. G. Smith (1911) for the *Rhacomitrium* heath:

Arctostaphylos uva-ursi	Empetrum nigrum
*Arctous (Arctostaphylos) alpina	Euphrasia officinalis
*Astragalus alpinus	Galium saxatile
Calluna vulgaris	*Lycopodium alpinum
Campanula rotundifolia	Potentilla erecta
*Carex rigida	Rumex acetosa
*Cerastium alpinum	*Salix herbacea
*C. arcticum	Vaccinium myrtillus
*Cornus suecica	*V. uliginosum
Deschampsia flexuosa	V. vitis-idaea

For the *Rhacomitrium* heath on the summit plateau of Cader Idris on thin black peat with a pH value of 5·6 (Pl. 129, phot. 321), Price Evans gives:

Rhacomitrium lanuginosum   d

*Cetraria aculeata f. hispida	o	Cladonia uncialis	f
C. islandica	la		
Cladonia furcata	f	Hylocomium loreum	o
C. gracilis	f	Hypnum schreberi	o
C. rangiferina	f	Polytrichum piliferum	o–f
C. sylvatica	f		

Campanula rotundifolia	r–o	*Lycopodium alpinum	la
Carex pilulifera	o	*L. selago	o
Empetrum nigrum (dwarf)	r–o	Potentilla erecta	o
Festuca ovina	o–f	Vaccinium myrtillus (dwarf)	o–f
Galium saxatile	a	V. vitis-idaea	o

The species of *Cladonia* and *Cetraria* are sometimes co-dominant with *Rhacomitrium*. On exposed parts of the summit plateau "*Rhacomitrium* moor" occurs. This is a community dominated by the same moss, but moister and with more peat formation. It has several of the same species as *Rhacomitrium* heath but without the Cladoniae, and with *Juncus squarrosus*, *Luzula maxima*, *Nardus stricta*, *Eriophorum vaginatum*, *Plagiothecium* sp., *Sphagnum* sp. and *Polytrichum commune* (f–ld).

The summit plateaux of many British mountains have the same type of vegetation but an extremely poor flora. *Rhacomitrium lanuginosum* and *Carex rigida* are fairly constant species (Pl. 129, phots. 319, 320).

REFERENCES

BUCHAN, A. and OMOND, R. T. The meteorology of the Ben Nevis observatories. Part III. *Trans. Roy. Soc. Edinb.* **43**. 1905.

LEACH, W. A preliminary account of the vegetation of some non-calcareous British screes (Gerölle). *J. Ecol.* **18**, 321–32. 1930.

PRICE EVANS, E. Cader Idris: a study of certain plant communities in south-west Merionethshire. *J. Ecol.* **20**, 1–52. 1932.

RAUNKIAER, C. Livsformernes Statistik som Grundlag for biologisk Plantegeografi. *Bot. Tidskr.* **29**, 1908. Translated as Chapter IV of *The Life forms of Plants and Statistical Plant Geography*. Oxford, 1934.

SMITH, R. Botanical Survey of Scotland. Part II. North Perthshire. *Scot. Geogr. Mag.* 1900.

SMITH, W. G. "Arctic-alpine Vegetation", Chapter XIII of *Types of British Vegetation*, 1911, pp. 288–329.

WATSON, W. The Bryophytes and Lichens of Arctic-alpine Vegetation. *J. Ecol.* **13**, 1–26. 1925.

WILLIS, J. C. and BURKILL, I. H. The Phanerogamic Flora of the Clova Mountains in special relation to Flower Biology. *Trans. Bot. Soc. Edinb.* **22**, 109–25. 1901–4.

Chapter XXXIX

ARCTIC-ALPINE VEGETATION (*continued*)
"SNOW-PATCH" COMMUNITIES. VEGETATION OF PROTECTED HABITATS. LITHOPHYTES AND CHOMOPHYTES

"SNOW-PATCH" COMMUNITIES

Effect of snow lie. The length of time for which snow lies during the year has an important effect in differentiating arctic-alpine plant communities. Snow, as we have seen, is very effective in protecting the low-growing vegetation during winter. This can be well observed in places where the snow can get no permanent lodgment because it is constantly removed by the wind. Such places are either quite barren or support a scanty population of the most resistant species. On the other hand, where snow lies very late, so that the soil is exposed for a short time only during the growing season, vegetation is necessarily much restricted, and in extreme cases, where the soil is uncovered on the average for a few weeks or even for a few days only, plants may not have time to develop at all. This can be very well seen in late August in the Scandinavian mountains. Some patches of soil where the last snow is only just melting are obviously quite barren and will necessarily remain so during the two or three weeks before snow falls again in September. Other places, where the last snow disappeared in July, support a few scattered plants.

In situations where the snow lies late, but the soil is exposed long enough for a well-developed vegetation to flourish, these peculiar conditions may determine a special vegetation. "The chief factors are patches of fine silty soil derived from dust accumulated in the snow, and an abundant supply of cold snow-water by which the soil is saturated: then as drying proceeds, there results a slippery slime, and finally, in summer, a dry cracked surface" (W. G. Smith, 1911). The Swiss botanists have long recognised the peculiar vegetation of such areas ("Schneeflecken" and "Schneetälchen") in the Alps. Our knowledge of the Scottish representatives is mainly due to W. G. Smith (1912) and Watson (1925).

Like the arctic-alpine communities previously described the snow-patch vegetation is at first mainly composed of bryophytes and lichens, and a liverwort, *Anthelia juratzkana*, is the dominant pioneer. According to Watson (1925) three of Rübel's five Swiss snow-patch communities are represented in the British Isles—Anthelietum, Polytrichetum sexangularis and Salicetum herbaceae. These apparently stand in successional relationship, the pioneer *Anthelia* forming a crust on the soil surface, which may then be colonised by *Polytrichum* (*P. sexangulare* and *P. alpinum*), and later by vascular plants, of which the most constant and sometimes the dominant is *Salix herbacea*.

The following is a collective list of species recorded from these communities, mostly taken from Watson (1925):

Lichens

Alectoria bicolor	o	Cetraria crispa	la
*Cerania vermicularis	la	C. islandica	a
Cetraria aculeata	—	*Lecidea consentiens	la
*var. alpina	—	*Solorina bispora	la

Liverworts

Alicularia breidleri	—	Gymnomitrium crassifolium	la
A. scalaris	a	G. varians	la
*Anthelia julacea	a	Lophozia alpestris	—
*A. juratzkana	ld	L. floerkii	—
Cephalozia bicuspidata	o	L. ventricosa	o
Diplophyllum albicans	a	*Marsupella condensata	la
D. taxifolium	—	*M. ustulata	o
*Gymnomitrium adustum	o	*Moerckia blytti	la
*G. alpinum	o	*Pleuroclada albescens	ld
G. concinnatum	—	Ptilidium ciliare	—
G. coralloides	—	Scapania undulata	o

Mosses

*Brachythecium glaciale	o	Oligotrichum hercynicum	la
*B. plicatum	o	*Polytrichum alpinum	o
*Bryum arcticum	o	*P. sexangulare	ld
*B. muhlenbeckii	—	Rhacomitrium fasciculare	o
Dicranum falcatum	—	Webera commutata	f
D. schisti	—	W. cucullata	o
D. starkei	—	W. nutans	a

* Arctic-alpine species.

Watson remarks that the whitish appearance of *Pleuroclada albescens*, from which its name is derived, is due to the moribund condition of the plant after the snow has melted. Underneath the snow-patch, if sufficient light and air have access, this liverwort is quite green and active.

The flowering plants recorded from Scottish snow-patch vegetation are few, all but one arctic-alpines:

- *Alchemilla alpina
- *Cochlearia micacea
- Galium saxatile
- *Gnaphalium supinum
- *Juncus biglumis
- *Juncus triglumis
- *Loiseleuria procumbens
- *Phleum alpinum
- *Salix herbacea
- *Saxifraga stellaris

VEGETATION OF RELATIVELY PROTECTED HABITATS

In habitats protected from the more violent winds conditions are naturally much more favourable for vegetation in general and though the plant habitats based on mountain-top detritus are far more extensive they possess many fewer species of flowering plants than the sheltered corries, ledges and fissures, which are by far the richest collecting grounds for the rarer species of all classes. Here the conditions are very varied, owing to extreme fragmentation of the topography; and in place of extensive and more or less uniform communities we have small groups of plants constantly varying from place to place. The dynamic factors of the habitat—constant erosion by frost, wind and water—bring about continual change, so that the vegetation furnishes typical examples of "migratory" communities. From another point of view these represent early stages of succession which never develop into climax vegetation.

Lithophytes, chomophytes and chasmophytes. The stations available for plant growth are of two kinds: exposed rock surfaces inhabited by *lithophytes*, and the piles and pockets of debris resulting from erosion and supporting the so-called *chomophytes*. On the Scottish mountains the arctic-alpine chomophyte communities reach their greatest development in the corries of the mica-schists. Plants inhabiting rock fissures are often called *chasmophytes*, but here they will not be separated from chomophytes, since the two groups pass into one another. These habitats with several of the species they bear are shown in Pls. 130–132.

Succession. "The succession begins on the bare rock faces with lithophytes, coatings of algae and lichens, followed by typical rock-mosses (*Andreaea*, etc.); flowering plants rarely occur in this phase. As weathering proceeds, and soil accumulates in fissures, the chomophytes become established. Xerophilous species (*Minuartia* (*Arenaria*) *sedoides, Silene acaulis, Saxifraga oppositifolia*) here precede the mesophytes, which will ultimately occupy the deeper deposits of fine soil and humus wherever they are accumulated. Later still, as the ledges become more or less obliterated by soil accumulations, and as the crevices become filled up, there is invasion by surface-rooting and mat-forming plants ('exochomophytes' of Oettli) from the closed plant communities of the grassland or heath of the lower arctic-alpine or sub-alpine zone. This last phase is well seen on many Scottish hills (e.g. Ben Lawers), where the lower terraces of the corrie area are covered with closed grassy swards, although they are clearly parts of the same system of rock exposures as the ledges and ravines bearing open communities at a higher level. There is reason to believe that if erosion of the rocks ceased, the whole system of rock-ledges and corries would ultimately become closed plant communities, and, as is the case now on the closed terraces, the characteristic arctic-alpine chomophytes would be largely exterminated" (W. G. Smith, 1911, pp. 319–20).

Screes. "The same phases may be traced on screes, beginning with lithophytes and ending with a closed vegetation. Mosses play a larger part in the intermediate phases, they become established on the blocks, and extend over the narrower cavities, and by their growth and decay provide a humus covering, but frequency of drought prevents any but xerophilous flowering plants from becoming established. Within the larger interstices, vegetation is restricted to liverworts, mosses and such species as can live in deficient light, but as soil and humus accumulate the conditions become not unlike those of the chasmophytes or shade chomophytes of fissures in the rock face."

Humidity. "Other factors, besides the progressive disintegration of rock, obviously play an important part in modifying the vegetation, especially the supply of moisture, and its conservation for plants through shelter from wind. Atmospheric humidity is probably quite as essential for many species as soil moisture. Lithophytes and other plants on the rock faces depend mainly on aerial sources for moisture, and are therefore subject to recurrent drought. On the ledges and in crevices there is frequently telluric water available, more constant in supply and richer in food materials. Sometimes, as on Ben Lawers, this supply is so ample that throughout the summer water continues to run over extensive ledges and forms the sources of numerous streamlets; this is the habitat for the community of hydrophilous chomophytes referred to later. Another aspect of water-supply is the temporary supply of cold water from melting snow which in early summer can be seen trickling down the rocks in streamlets or spreading over ledges that later become comparatively dry. Here we have a 'snow-valley' (Schneetälchen) and with it the occurrence of certain species of plants, e.g. *Saxifraga rivularis, Veronica alpina, Arabis petraea*" (W. G. Smith, 1911, pp. 320–1).

Lithophytes. "The first plants to find a footing on the bare rock are algae and lichens, the former requiring a damper substratum than the latter in order to become noticeable constituents of the vegetation, though *Trentepohlia* gives reddish-orange hue to many fairly dry rocks. The structure of some lichens renders them specially able to act as lithophytes. Collemoid lichens seldom occur except when the rocks are almost constantly damp and are rarely very dry or very wet. Crustaceous lichens are very abundant on dry rocks, and those which are common are widely distributed species frequent at lower elevations, e.g. *Lecanora parella, L. tartarea, L. atra, L. polytropa, L. contigua, L. sorediza, L. crustulata, Rhizocarpon confervoides* and *R. geographicum*....

"A slight amount of erosion enables such lichens as *Dermatocarpon miniatum*, and species of *Gyrophora, Umbilicaria, Stereocaulon*[1] and *Sphaerophorus* to establish themselves on exposed rock faces, and these are accompanied or followed by mosses, species of *Andreaea* and *Grimmia* being the earliest representatives, different species of the same genus varying in their powers of establishment. For example the very rare moss *Blindia caespiticia* occurs in some of the small crevices, its near relative, *B. acuta*, being scattered over the face of the same rock. Some pleurocarpous mosses appear later, and then phanerogamic chomophytes become well established in fissures or on rock ledges. In this succession hepatics are rare at first—gradually

---

[1] Pl. 132, phot. 332.

becoming more abundant till on ledges they often form the chief constituents. The opposite is true of lichens" (Watson, 1925, p. 13).

Lichens.

On rocks in earlier stages of erosion

*Acarospora admissa	—	L. picea	r
A. discreta	—	L. polytropa	a
*A. peliocypha	—	*var. alpigena	—
*Aspicilia alpina	r	L. subtartarea	a
*A. chrysophana	r	L. tartarea	a
A. cinerea	o	Lecidea albocoerulescens	—
*A. cinereorufescens		*L. armeniaca	—
*f. diamarta (moist)	r	L. auriculata	—
*A. depressa	o	var. diducens	—
A. gibbosa	o	L. cinerascens	—
*A. leucophyma	r	*L. commaculens	—
*A. pelobotrya	r	L. confluens	—
*A. superiuscula	r	L. contigua	a
Bacidia flavovirescens var.	—	var. flavicunda	a
alpina	o	var. platycarpa	—
Biatorella simplex	o	*L. contiguella	—
Biatorina biformigera	—	L. crustulata	—
B. candida	—	L. dealbatula	—
*B. confusior	—	*L. deparcula	r
*B. contristans	—	L. dicksonii	la
*B. cumulata	—	L. fuscoatra	—
*B. rhypodiza	r	L. goniophila	—
*Buellia alpicola	r	L. griseoatra	—
*B. badioatra	r	L. kochiana	—
*var. atrobadia	r	L. lapicida	—
B. colludens	—	L. leucophea	—
*B. deludens	r	L. lithophila	—
*Callopisma siebenhaariana	r	L. mesotropoides	r
Coenogonium ebeneum		*L. nigroglomerata	—
(damp)	—	L. plana	—
Dermatocarpon miniatum		L. polycarpa	—
var. complicatum (moister)	—	*L. pycnocarpa	—
D. hepaticum	—	L. rivulosa	la
*Gyrophora arctica	—	L. sanguinaria	—
G. cylindrica	a	L. sorediza	—
*var. delisei	—	*L. subgyratula	—
*G. erosa	—	L. subkochiana	—
*G. hyperborea	—	L. sublatypea	—
*G. leiocarpa	—	*L. tabidula	—
G. polyphylla	—	*L. tessellata	—
G. polyrrhiza	—	*L. umbonella	—
G. proboscoidea	—	Microglaena bread-	
*G. torrefracta	—	albanensis	—
Haematomma ventosum	a	M. corrosa	—
*Lecanora austera	r	*var. nericensis	—
L. badia	f	Microthelia exerrans	—
*L. frustulosa	—	*Parmelia alpicola	—
L. intricata	f	P. conspersa	f
*var. leptacina	—	var. stenophylla	la
L. parella	a	*P. encausta	la

* Arctic-alpine species and varieties (and so throughout the lists).

Parmelia omphalodes	f	P. theleodes	—
*P. pubescens	a	Porina chlorotica	o
Pertusaria ceuthocarpa	—	Racodium rupestre (damp)	—
P. concreta	—	Rhizocarpon confervoides	a
P. dealbata	—	*R. geminatum	r
P. lactea	—	R. geographicum	a
P. monogyna	—	R. obscuratum	o
*Placodium elegans	o	R. oederi	la
*Platysma fahlunense	—	R. petreum	la
*P. hepatizon	f	*R. plicatilis	r
*P. polyschizum	—	*Squamaria chrysoleuca	r
Polyblastia fuscoargillacea	—	S. gelida	f
P. intercedens	—	Thelidium papulare	o
*P. scotinospora	—	Umbilicaria pustulata	—

On rocks in later stages of erosion

Cetraria aculeata (usually on ground)	a	Ephebe lanata	o
C. islandica (usually on ground)	a	*Europsis granatina	—
		*Leptogium glebulentum (moist)	—
C. nivalis	lf	L. rhyparodes (wet rocks)	lf
C. odontella	—	Peltidea aphthosa (usually on ground)	—
*Cladonia bellidiflora	—		
C. cervicornis	—	Peltigera canina (usually on ground)	—
f. stipata	la		
var. subcervicornis	a	P. polydactyla (usually on ground)	—
C. furcata	—		
C. pyxidata	—	P. rufescens (usually on ground)	—
C. rangiferina	—		
C. squamosa	—	*Psorotichia furfurella	r
C. sylvatica	—	*Pterygium pannariellum	—
C. uncialis	—	*Pyrenopsis furfurea (damp)	r
C. verticillata	—	P. fuscatula	—
*Collema ceraniscum	r	*P. haematopsis	—
C. granuliferum	—	*P. homoeopsis	—
C. multifidum	—	*Schizoma lichinodeum	lf
f. marginale	—	*Solorina crocea[1]	
C. tenax	o	Sphaerophorus compressus	—
Coriscium viride (usually on ledges)	—	S. corallioides	a
		f. congestus	a
Ephebe hispidula	—	Thermutis velutina (damp)	r

Liverworts. These plants are rarely present on rocks unless they are wet or shaded or with a soil-cap, but they are often abundant on rocky ledges and may there form conspicuous members of the community.

A. *On rocks in the early stages of erosion*

*Chandonanthus setiformis	*Radula lindbergii var. germana
Frullania tamarisci	*Sphenolobus saxicolus
*Gymnomitrium obtusum	

[1] Pl. 132, phot. 333.

B. *On rocks which are moist, shaded or partly covered with a thin soil cap*

Alicularia scalaris
*Anthelia julacea
*A. juratzana
Diplophyllum albicans
*D. taxifolium
*Gymnomitrium alpinum
*G. concinnatum
 *var. intermedium

*Gymnomitrium coralloides
G. crenulatum
Herberta adunca
Lophozia floerkii
*L. quadrifolia
*Marsupella sparsifolia
Plagiochila asplenioides var. minor
Ptilidium ciliare

Mosses.

A. *On rocks or boulders in the earlier stages of erosion*

*Andreaea alpina
A. crassinervia
*A. nivalis
A. petrophila
 *var. alpestris
 var. acuminata
A. rothii
Blindia acuta
*B. caespiticia
*Campylopus schwarzii
Dichodontium pellucidum (moist)
 *var. fagimontanum
*Dicranum elongatum
*D. falcatum

D. fulvellum (clefts)
D. longifolium
D. schisti
Grimmia apocarpa
 *var. alpicola (wet)
 var. pumila
 var. rivularis
*G. atrata (wet)
G. conferta
G. doniana
G. funalis
G. torquata
G. trichopylla

B. *In cracks, on soil caps or on wet rock faces, etc.*

Anoectangium compactum
Bartramia oederi
*Brachythecium glaciale
B. glareosum
*B. plicatum
B. rutabulum
*Bryum arcticum
B. pseudotriquetrum
Ceratodon purpureus
*Conostomum boreale
Cynodontium gracilescens
*C. virens
*C. wahlenbergii
Dicranella heteromalla
Dicranoweisia crispula
Dicranum fuscescens
 *var. congestum
D. scoparium
D. starkei
Encalypta ciliata
*E. commutata
 *var. imberbis
E. rhabdocarpa
*Eurhynchium cirrhosum
E. myosuroides
Fissidens osmundoides

*Grimmia atrata (wet)
*G. alpestris
*G. unicolor
*Hylocomium pyrenaicum
Hypnum bambergeri
H. callichroum
H. cupressiforme
*Hypnum halleri
*H. hamulosum
*H. sulcatum
*Mnium lycopodioides
M. orthorrhynchum
M. serratum
*M. spinosum
Myurella apiculata
M. julacea
Orthochecium rufescens
*Plagiobryum demissum
Plagiothecium denticulatum
 var. obtusifolium
*P. muhlenbeckii
P. pulchellum
*Polytrichum alpinum
*Pseudoleskea atrovirens
*P. catenulata
*P. patens

*Pseudoleskea striata var. saxicola
Pterigynandrium filiforme
Rhacomitrium fasciculare
R. heterostichum
  var. gracilescens
R. lanuginosum
R. ramulosum

Swartzia montana
Trichostomum tortuosum
  *var. fragilifolium
Webera commutata
W. nutans
Zygodon lapponicus

Vascular plants. Terrestrial vascular plants are never strict lithophytes, at least in this country, since they must have *some* soil or humus in which to root. In the foregoing lists of bryophytes and lichens, however, many forms are included which are equally unable to colonise bare rock surfaces, but require at least a thin layer of disintegrated mineral or organic material. A distinction is drawn in Watson's lists between the species living on rocks "in the earlier stages of erosion", corresponding roughly to lithophytes proper, and those which settle in cracks, on thin soil over the tops of rocks, or on wet rocks where a thin film of organic substance is provided by algae. It is obvious that no hard and fast line can be drawn between such habitats and those providing enough soil to support the typical chomophytes of arctic-alpine vegetation. W. G. Smith (1911, p. 322) gives a list of thirty flowering plants and ferns forming "open communities on exposed rock faces", very nearly all arctic-alpine species, the habitat being an "open ledge without much soil" (the habitat of *Erigeron alpinus* can be seen from Pl. 130, phot. 324), but these, he says, generally have "a single rootstock extending into a fissure, within which a branching root system anchors the plant firmly"; and this is obviously a character which renders the plant independent of *surface* soil. The following species are included:

on open ledges

*Alchemilla alpina[1]
*Arabis petraea
*Asplenium viride
*Astragalus alpinus (rare[2])
*Bartsia alpina (rare[2])
Campanula rotundifolia
*Cerastium alpinum
Cystopteris fragilis
*Draba incana
*D. rupestris
*Dryas octopetala
*Erigeron alpinus (rare[2])[3]
*Juncus trifidus
*Luzula spicata
*Minuartia rubella

*Minuartia sedoides[4]
*Oxytropis uralensis (rare[2])
*Potentilla crantzii
*Sagina nivalis
*Salix reticulata[5]
*Saxifraga cernua[6]
*S. hypnoides
*S. nivalis
*S. oppositifolia
*Sedum rosea[7]
*Sibbaldia procumbens[8]
*Silene acaulis
Thymus serpyllum
*Woodsia alpina

[1] Pl. 131, phot. 328.
[2] "Rare" means rare within the habitat: the majority of the species are "rare" in the ordinary sense because their habitats are very restricted.
[3] Pl. 130, phot. 324.   [4] Pl. 131, phot. 328.   [5] Pl. 130, phot. 323.
[6] Pl. 132, phot. 330.   [7] Pl. 130, phot. 325.   [8] Pl. 132, phot. 331.

PLATE 130

Phot. 322. Arctic-alpine grassland on Ben Lawers, alt. 2700 ft. (c. 820 m.). *Alchemilla alpina*, *Festuca ovina* f. *vivipara*, etc. Abundance of mosses and lichens on boulders.

Phot. 323. *Salix reticulata* and *Rhacomitrium lanuginosum* on a dry ledge near the summit of Ben Lawers, alt. c. 3500 ft. (1060 m.). *F. F. Laidlaw*.

Phot. 324. *Erigeron alpinus* and *Festuca ovina* f. *vivipara* on an open ledge, alt. c. 3500 ft. *F. F. Laidlaw*.

Phot. 325. *Sedum rosea* (*rhodiola*) and *Alchemilla alpina* on steep rocks, alt. c. 3600 ft. *J. Massart*.

Phot. 326. Hydrophilous chomophytes—*Webera albicans* var. *glacialis*, *Chrysosplenium alternifolium*. *F. F. Laidlaw*.

ARCTIC-ALPINE VEGETATION OF BEN LAWERS

PLATE 131

Phot. 327. *Athyrium alpestre, Cystopteris fragilis, Polystichum lonchitis, Oxyria digyna, Saxifraga stellaris, Oxalis acetosella*, etc. At top of cleft, more exposed, *Alchemilla alpina, Festuca ovina* f. *vivipara, Cerastium alpinum*, etc. *N. F. G. Cruttwell.*

Phot. 328. *Minuartia sedoides, Alchemilla alpina. J. Massart.*

Phot. 329. *Oxyria digyna, Saxifraga aizoides, Nardus stricta. J. Massart.*

ARCTIC ALPINES (BEN LAWERS)

CHOMOPHYTES

Rock ledges offer very varied habitats, according to shelter, exposure and insolation, the slope and width of the ledge, and whether the surface is relatively dry or constantly irrigated. More or less mineral soil and humus will accumulate upon the ledge according to the conditions. Watson gives the following lists of bryophytes and lichens from this type of habitat. Though numerous in species they do not so greatly outnumber the vascular plants as in the more exposed habitats dealt with in the last chapter. Watson remarks that the bryophytic constituents are often very varied, though sometimes pure masses of one species occur. Foliose liverworts are generally abundant, but lichens and Sphagna are rare.

Mosses

Andreaea petrophila	—	Mnium hornum	o
Bartramia ithyphylla	o	*M. lycopodioides	—
B. oederi	o	M. orthorrhynchum	—
Bryum pseudotriquetrum (moist)	f	M. punctatum (moist)	lf
		var. elatum	lf
*Campylopus schimperi	—	*M. spinosum	o
Dicranum fulvellum	—	*Myurella apiculata	r
D. fuscescens	o	*M. julacea	o
*D. molle	lf	Oedipodium griffithianum	f
D. scoparium	o	Oligotrichum hercynicum	—
*Eurhynchium cirrhosum	lf	Plagiothecium denticulatum	—
E. myosuroides	o	var. obtusifolium	—
E. swartzii	o	P. elegans	—
Grimmia doniana	—	*P. muhlenbeckii	—
Heterocladium squarrosulum	—	P. pulchellum	—
Hylocomium loreum	—	*Polytrichum alpinum	o
H. splendens	o	P. commune	—
H. squarrosum	o	P. piliferum	—
H. rugosum	—	*Pseudoleskea atrovirens	—
H. triquetrum	o	Pterigynandrium filiforme	o
Hypnum callichroum	—	Rhacomitrium fasciculare	—
H. cupressiforme	—	R. heterostichum	—
*H. halleri	—	R. lanuginosum	a–d
*H. hamulosum	—	Sphagnum acutifolium	r
H. molluscum	—	S. molluscum	o
H. schreberi	—	Webera elongata	—
		W. nutans	—

Liverworts

Alicularia scalaris	lf	*C. pleniceps	—
Anastrepta orcadensis	lf	Diplophyllum albicans	a
Anastrophyllum donianum	—	*D. taxifolium	l and r
Bazzania triangularis	—	*Gymnomitrium concinnatum	—
B. tricrenata	f		
B. trilobata	—	Herberta adunca	f
Blepharostoma trichophyllum (moist)	o	H. hutschinsiae	f
		Jamesoniella carringtonii	lf
Cephalozia bicuspidata (moist)	o	Leptoscyphus taylori	la
		Lophozia floerkii	o

Lophozia hatcheri	—	Pleurozia purpurea	lf
*L. heterocolpa	r	Preissia quadrata	o
*L. incisa	f	Ptilidium ciliare	o
*L. lycopodioides	—	Reboulia hemispherica	lf
*L. obtusa	r	Scapania aequiloba	o
L. quadriloba	r	S. aspera	o
L. quinquedentata	f	S. irrigua	o
L. ventricosa	o	*var. alpina	r
Marsupella emarginata	—	*S. nimbosa	r
Metzgeria furcata	o	*S. ornithopodioides	o
Pellia neesiana	o	*Sphenolobus politus	
Plagiochila asplenioides	f	*var. medelpadicus	r

Lichens

Biatorina candida	o	Lecidea decipiens	o
*Buellia pulchella	r	L. demissa	o
Cetraria islandica	lf	L. lurida	o
Cladonia cervicornis	o	*L. rubiformis	r and l
var. stipata	lf	*Schizoma lichinodeum	r
Coriscium viride	o	Solorina saccata	
Dermatocarpon lachneum	o	Sphaerophorus coralloides	o
*Lecidea cupreiformis	r		

Vascular plants

Smith (1911, p. 323) gives the following eight species from dry, well-drained rock-ledges, of southern exposure, with abundance of soil and relative freedom from competition:

Botrychium lunaria†
*Gentiana nivalis†
Linum catharticum
*Myosotis alpestris
*Potentilla crantzii
*Rhinanthus borealis
*Veronica saxatilis†
Viola lutea

† Three species recorded together in various stations.

And the following eighteen from ledges with shelter and deep soil, the community nearly closed:

*Alchemilla alpestris
Angelica sylvestris
*Carex atrata
*Deschampsia alpina
Dryopteris dilatata
Galium boreale
Geranium sylvaticum
*Hieracium alpinum, etc.
Luzula sylvatica
Melandrium dioicum (rubrum)
Orchis maculata
*Poa alpina, etc.
Pyrola minor
P. rotundifolia
P. secunda
Rubus saxatilis
Rumex acetosa
Trollius europaeus

Here under the better conditions we have a much larger proportion of "lowland" species.

Shade chomophytes. Where light is deficient, as in wide and deep fissures in the rock face, or in hollows among block screes, there is also very good shelter, exceptionally high atmospheric humidity, and soil moisture. Among the mica schists such cavities have earthy or rocky walls, moist and bare except where covered by mats of mosses and liverworts. Here and

PLATE 132

Phot. 330. *Cochlearia micacea, Sagina linnaei, Saxifraga cernua.* J. Massart.

Phot. 331. *Sibbaldia procumbens, Gnaphalium supinum, Euphrasia scotica, Galium saxatile.* J. Massart.

Phot. 332. *Stereocaulon alpinum.* J. Massart.

Phot. 333. *Solorina crocea.* J. Massart.

Arctic-Alpines (Ben Lawers)

there grow a few flowering plants, more or less pale and etiolated. On shaded rocks, where water oozes down through cushions of algae and mosses, flowering plants are more abundant. The deeper fissures are evidently buried in snow early in the winter, but the plants obtain water trickling under the snow before it melts altogether (W. G. Smith).

The following characteristic species are listed by Smith (1911, pp. 324–5) and W. Watson (1925):

Vascular plants

Adoxa moschatellina	Galium saxatile
*Agropyron donianum	Geranium robertianum
Anemone nemorosa	*Lactuca alpina
*Athyrium alpestre[1]	*Luzula spicata
*A. flexile	Oxalis acetosella[1]
Cardamine flexuosa	*Oxyria digyna[1]
C. hirsuta	*Polygonum viviparum
*Carex atrata	*Polystichum lonchitis[1]
Cystopteris fragilis[1]	Potentilla erecta (f.)
*C. montana	*Salix lapponum
Chrysosplenium alternifolium	*Saussurea alpina
C. oppositifolium	*Saxifraga stellaris[1]
Dryopteris linnaeana	Taraxacum officinale (f.)

In this habitat more woodland species are mixed with the arctic-alpines. Watson gives the following list of bryophytes mainly from the hollows of block screes above 3000 ft. on the east side of Ben Lawers:

Mosses

Anoectangium compactum	o	M. punctatum var. elatum	
Brachythecium glareosum		(well-shaded or moist	
*B. plicatum		places among boulders)	
B. rutabulum		*M. spinosum	f
Dicranum fuscescens	f	Plagiobryum zierii	
Eurhynchium confertum		Plagiothecium denticulatum	f
(among boulders)	f	*P. muhlenbeckii	
E. myosuroides		P. pulchellum	
E. praelongum		*Polytrichum alpinum	f
E. swartzii		*Pseudoleskea atrovirens	f
Fissidens osmundioides	o	Rhacomitrium canescens	
Grimmia patens	f	var. ericoides	
Hylocomium loreum		R. heterostichum	
*H. pyrenaicum	lf	R. lanuginosum	
H. squarrosum		*R. sudeticum	
H. umbratum	lf	*Thuidium philiberti var.	
Hypnum callichroum	f	pseudotamarisci	
H. cupressiforme		T. tamariscinium	o
H. falcatum (well-shaded or		*Timmia norvegica (moist	
moist places among		places among rocks)	
boulders)		Trichostomum tortuosum	o
H. molluscum	a	Webera annotina	
*H. procerrimum		W. elongata (also mossy	
*Mnium lycopodioides		boulders)	
M. orthorrhynchum		W. nutans	

[1] Pl. 131, phot. 327.

The following are more usually found in small nooks of the rocks

Bartramia ithyphylla
Dicranum fulvellum
Ditrichum zonatum
*Grimmia elongata
Myurella julacea
Oedipodium griffithianum

Polytrichum aloides
Rhabdoweisia crenulata
R. denticulata
R. fugax
*Saelania caesia
Zygodon lapponicus

Hepatics

Alicularia scalaris
Aneura pinguis
Anastrepta orcadensis
Bazzania tricrenata
Cephalozia bicuspidata
Diplophyllum albicans
*D. taxifolium
*Gymnomitrium concinnatum
  *var. intermedium
Lophozia floerkii
L. hatcheri
*L. lycopodioides
L. obtusa
Lophozia quinquedentata f

L. ventricosa
*Marchantia polymorpha var.
  alpestris f
Pellia neesiana
Plagiochila asolenioides
  var. minor, f. laxa (in
  damper places)
Preissia quadrata
Ptilidium ciliare
Scapania curta
S. dentata
  var. ambigua (damper
  places)

Lichens

Lichens are naturally not characteristic but the following may be found occasionally:

Cladonia furcata
C. sylvatica
Collema multifidum, etc.
Pannaria brunnea

Peltidea aphthosa
Peltigera canina, etc.
Racodium rupestre
*Solorina bispora

Cader Idris. On the calciferous "pillow lavas" (upper and lower basic volcanic series) of Cader Idris, Price Evans (1932) records a rich

Basic rocks

flora including a fair number of arctic-alpine species. His lists give "representative" species and do not profess to be exhaustive, but they are interesting to compare with the general lists given above which are mainly taken from the Scottish Highlands. The lists below are composite, including the upper and lower bands of "pillow lava" (1500–1700 ft. and 1800–2400 ft., c. 450–520 and 550–730 m.) and do not distinguish between the types of habitat:

Lichens

si	Lecanora parella	f	si	Pertusaria dealbata	o
	L. subtartarea	a		Rhizocarpon petraeum	f
	L. tartarea	a	si	Sphaerophorus globosus	o
	Lecidea contigua f. calcarea	o–f		Stictina fuliginosa	la
	Peltigera canina	o	ca	Solorina saccata	f

Liverworts

	Aneura pinguis	o		Frullania tamarisci	f
	Anthelia julacea	a	ca	Reboulia hemisphaerica	f
	Fissidens adiantoides	f–a	si	Scapania dentata	o

Cader Idris

Mosses

	Species	Freq		Species	Freq
	Bartramia pomiformis	f	ca	Neckera crispa	f–a
ca	Hypnum commutatum	f		Plagiobryum zierii	f
	Mnium undulatum (shade)	f		Polytrichum urnigerum	o

Vascular plants

	Species	Freq		Species	Freq
ca	Adiantum capillus-veneris	vr		H. pulchrum	o
	Adoxa moschatellina	o		Hymenophyllum peltatum	f
	Alchemilla alpestris	f		Lathyrus montanus	o
	Anemone nemorosa	o		Linum catharticum	o
	Antennaria dioica	o		Luzula maxima	o
ca	Arabis hirsuta	o	*	Lycopodium selago	o
	Asplendum ruta-muraria	r–o		L. clavatum	o
	A. trichomanes	f		Meconopsis cambrica	o–f
*ca	A. viride	f		Molinia caerulea	o–f
si	Blechnum spicant	o		Oxalis acetosella	o
si	Calluna vulgaris (lower alt.)	f	*ca	Oxyria digyna	r–la
	Campanula rotundifolia	o	ca	Pimpinella saxifraga	o, l
	Cardamine hirsuta	o		Pinguicula vulgaris	o
	Carex flava	o		Polygala vulgaris	o
	C. pulicaris	o		Polypodium vulgare	o
	Chrysanthemum leucanthemum			Primula vulgaris	o
	Chrysanthemum leucanthemum	o		Ranunculus repens	o
	Chrysosplenium oppositifolium			*Rhinanthus cristagalli, forma	r
	Chrysosplenium oppositifolium	la		*Saxifraga hypnoides	la
	Conium maculatum	o		*S. stellaris	o–f
si	Cryptogramma crispa	o		Sedum anglicum	o
	Cystopteris fragilis			*S. rosea	r–la
	*ca. f. dentata	f		Scirpus caespitosus	o
si	Deschampsia flexuosa	o		Selaginella selaginoides	o–f
si	Digitalis purpurea	o		Solidago cambrica	o–f
	Dryopteris filix-mas	o		S. virgaurea	o
	D. linnaeana	o		Succisa pratensis	o–f
si	D. phegopteris	o–f		Taraxacum officinale	o
	Epilobium angustifolium	la	ca	Thalictrum minus (agg.)	lf
	E. palustre	o		Thymus serpyllum	f
	Euphrasia officinalis	o		Urtica dioica	o
	*Festuca ovina f. vivipara	o		Valeriana dioica	o
	Ficaria verna	o		Veronica officinalis	o–f
	Filipendula ulmaria	o		Viola sp.	
	Geum rivale	o	si	Vaccinium myrtillus (higher alt.)	f
	Geranium robertianum	o			
	Hieracium sylvaticum	o			
	Hypericum humifusum	o			

* Arctic-alpine species, ca, calcicolous, si, silicicolous (calcifuge).

Price Evans points out that a few species (ca) in these lists are calcicolous, but the majority are "indifferent". The rich flora must be attributed to the generally favourable conditions for vegetation on these lava soils, conditions of which the high base status is one. There are however a certain number of "calcifuge" species (si) and the presence of these he attributes to the formation of acid humus on some of the rock ledges and also to the soil produced by acidic beds interstratified with the calciferous lavas.

Acid rocks. On the granophyre escarpment (the great northern cliff) of Cader Idris the available calcium is low. Much of the surface is bare rock and the variety of species is much less than on the basic lavas, but the high altitude (up to 2900 ft. = 884 m.) determines the presence of a few arctic-alpine forms, notably the mountain forms of the sea pink, sea campion and sea plantain.

The following are some characteristic species:

Agrostis stolonifera
Alchemilla alpina
*Armeria maritima, forma
Athyrium filix-femina
Blechnum spicant
Campanula rotundifolia
Carex flava
Cryptogramma crispa
Deschampsia flexuosa
Dryopteris phegopteris
Galium saxatile
*Lycopodium selago
Luzula maxima
Molinia caerulea
Narthecium ossifragum

Oxalis acetosella
Pinguicula vulgaris
*Plantago maritima, forma
Potentilla erecta
*Saxifraga stellaris
Scirpus caespitosus
*Sedum rosea (rhodiola)
*Silene maritima, forma
Solidago cambrica
Sorbus aucuparia (1·2 m. high, up to 700 m.)
Succisa pratensis
Thymus serpyllum
Viola riviniana

On the ledges dwarf *Calluna* is frequent at lower levels and *Vaccinium myrtillus* at higher. *Festuca ovina* forma *vivipara* and *Rhacomitrium lanuginosum* are generally abundant, with *Polytrichum commune* and *Sphagnum* in the wetter places.

The following are abundant on the rock faces:

Andreaea petrophila
Lecidea contigua

Lecidea dicksonii
Rhizocarpon confervoides

Gyrophora cylindrica and *Parmelia pubescens*, which are frequent on the summit plateau, are occasional on the rocks at the base of the escarpment, i.e. so low as 550 m., and cushions of *Anthelia julacea, Scapania* sp. and *Tetraplodon mnioides* are abundant on wet surfaces in the gullies, the last named in association with *Peziza rutilans* saprophytic on the decayed bones of sheep.

Hydrophilous chomophytes. Wet habitats are common and varied on many of the high mountains. Springs emerge at quite high altitudes and form numerous small shallow runlets before they converge to form the "burns" or mountain streams at lower levels. In the areas of schist, as on Ben Lawers, series of terraces are formed, and excess of water from the numerous springs, not yet confined to definite channels, flows gently over these. Such ground is covered with a bright green, wet carpet of mosses, liverworts and algae, with a certain number of flowering plants. On Ben Lawers the lower corries show the main development of this vegetation, but on a smaller scale it occurs also at higher levels. The substratum of the vegetation is fairly uniform and is formed mainly of flakes of mica schist and other rocks caught among larger stones and boulders. Thus the habitat is constantly wet and at the same time well aerated and supplied with abundant nutrient salts from the moving spring water. Along the lower stream flats, on the other hand, where drainage is impeded, more stagnant conditions prevail, and dull green bog or marsh is formed, dominated by *Sphagnum* and by rushes and sedges.

Societies of bryophytes. Watson (1925, p. 19) distinguishes six societies dominated by bryophytes:

(1) *Philonotis fontana* society (described by Ostenfeld in *Botany of the Faroes*, 1908, and mentioned by W. G. Smith, 1911). The dominant is often pure, but may be mixed with species of *Hypnum, Bryum, Mnium, Sphagnum* and *Scapania*. This forms the typical bright green wet carpet irrigated by slowly moving spring water and referred to above.

(2) *Sphenolobus politus* society in places similar to (1):

*Harpanthus flotowianus
Hypnum sarmentosum
Pellia neesiana

*Scapania obliqua
Sphagnum inundatum

(3) *Scapania obliqua* society in a spring at 3000 ft. on Ben Lui, associated with:

Aulacomnium palustre
Bryum pseudotriquetrum
Fissidens adiantoides
Hypnum vernicosum

Scapania dentata
S. uliginosa
S. undulata

(4) *Hypnum trifarium* society on wet boggy slopes, sometimes associated abundantly with *H. exannulatum, revolvens, sarmentosum* and *stramineum*.

(5) *Marsupella* society (*M. aquatica* and *M. pearsoni*) in water flowing over stony ground.

(6) *Aplozia* society in a boggy spring on Ben Lawers over 3000 ft.:

Aplozia cordifolia	a	Hypnum falcatum	a
A. atrovirens		H. revolvens	a
A. riparia		Scapania dentata	o
Alicularia scalaris	f	S. irrigua	
Aneura pinguis	o	S. undulata	
Bryum pseudotriquetrum	a	Sphagnum cymbifolium	a
Fissidens osmundoides	o		

Such bryophytic societies require further study before they can be properly defined and their relations to habitat established.

Flora. Watson (1925) records the following hydrophilous bryophytes from the arctic-alpine region. For notes on their habitats and associates his paper should be consulted:

Mosses

Aulacomnium palustre f
Blindia acuta
Bryum alpinum

B. pallens
Cinclidium stygium
*Cynodontium virens

*Cynodontium wahlenbergii		H. uncinatum	
Dichodontium pellucidum		H. undulatum	
Dicranella squarrosa		H. vernicosum	o
Fissidens osmundoides		Meesia trichodes	
Grimmia apocarpa var. rivularis		Mnium cinclidoides	r
		M. punctatum var. elatum	a
*H. arcticum		M. subgobusum	
H. commutatum	a	Orthothecium rufescens	
H. cuspidatum		Philonotis adpressa	
*H. decipiens		P. calcarea	
H. exannulatum var. orthophyllum	a	P. fontana	
		*Polytrichum alpinum	
H. falcatum	a	P. commune	
H. fluitans	o	*P. sexangulare	
H. lycopodioides	o	Splachnum vasculosum	
*H. molle		*Tayloria lingulata	
H. ochraceum		*Thuidium philiberti var. pseudotamarisci	
H. revolvens	a		
H. sarmentosum	f–a	*Timmia norvegica	
H. scorpioides		Webera albicans	
H. stellatum		*var. glacialis (Pl. 130, phot. 326)	
H. stramineum	a	W. elongata	
H. trifarium	la	W. ludwigii	

Liverworts

Alicularia scalaris		*L. kunzeana	r
Aneura pinguis		L. lycopodioides	f
*Anthelia julacea var. gracilis		L. muelleri	r
		*L. wenzelii	r
*Aplozia atrovirens var. sphaerocarpoidea		Marsupella aquatica	a
		M. emarginata	
A. cordifolia	a	M. pearsoni	a
A. riparia	f–a	Pellia epiphylla	o
*A. sphaerocarpa var. nana		P. neesiana	a
Blepharostoma trichophyllum	o	Ptilidium ciliare	o
		*Scapania crassiretis	
Cephalozia bicuspidata	a	S. dentata var. ambigua	a
Chiloscyphus polyanthus var. fragilis			
		S. irrigua	
*Eremonotus myriocarpus		*S. obliqua	lf
Eucalyx obovatus	o	*S. paludosa	r
Gymnocolea inflata forma nigricans	o	S. subalpina	
	a	S. uliginosa	a
*Harpanthus flotowianus	lf	S. undulata	a
*Hygrobiella laxifolia		Sphenolobus politus	a
Lophozia bantriensis	la		

Lichens on wet rocks. The following lichens may occur on stones in streams or on wet rocks by stream sides (Watson, 1925):

Aspicilia lacustris	E. pubescens
Collemodium fluviatile	Lecidea contigua forma hydrophila
Dermatocarpon miniatum var. complicatum	
	Polyblastia theleodes
Ephebe hispidula	*P. inumbrata

*Psorotrichia furfurella
*Pyrenopsis phylliscella
Staurothele clopima
Verrucaria aethiobola
V. laevata
V. margacea
V. submersa

Representative vascular hydrophytes. W. G. Smith gives the following flowering plants as "representative species" occurring in the hydrophilous moss communities referred to above, or in other hydrophytic communities: one or two are added from Watson's paper:

Highland species
Carex saxatilis
C. vaginata
Cerastium alpinum
C. cerastioides
Epilobium alsinifolium
E. anagallidifolium
Juncus biglumis
J. triglumis
Oxyria digyna
Polygonum viviparum
Sagina saginoides
Saxifraga aizoides
S. stellaris
Thalictrum alpinum
Tofieldia palustris
Veronica alpina

Lowland species
Alchemilla vulgaris (agg.)
Caltha palustris var. minor
Carex curta
C. dioica
C. stellulata
Chrysosplenium oppositifolium
Cochlearia officinalis (agg.)
Equisetum hyemale
Montia fontana
Narthecium ossifragum
Pinguicula vulgaris
Sedum villosum
Stellaria uliginosa
Veronica humifusum

REFERENCES

Price Evans, E. Cader Idris: a study of certain plant communities in south-west Merionethshire. *J. Ecol.* **20**, 1–52. 1932.

Smith, W. G. Arctic-alpine vegetation. Chapter XIII in *Types of British Vegetation.* 1911.

Smith, W. G. Anthelia: an arctic-alpine plant association. *Scott. Bot. Rev.* pp. 81–9. 1912.

Watson, W. The bryophytes and lichens of arctic-alpine vegetation. *J. Ecol.* **13**, 1–26. 1925.

Part IX

MARITIME AND SUBMARITIME VEGETATION

Chapter XL

INTRODUCTORY. THE SALT-MARSH FORMATION

Maritime vegetation. It is convenient to distinguish *maritime vegetation*, that is land vegetation affected more or less profoundly by the neighbourhood of the sea, from *marine vegetation*, living in the sea. This of course cannot be an absolute distinction, because vegetation which is covered with sea water at high tide, and exposed to the air at low tide is clearly intermediate between the two. Of this "tidal" vegetation the seaweed communities of the intertidal zone of rocky coasts are reckoned as "marine" and are not dealt with in this book, while the "salt-marsh" communities of more or less protected mud or sand flats, dominated mainly by flowering plants, though marine algae are present and are often important constituents dominating some of the communities, are included in maritime vegetation.

The habitats. Maritime vegetation is here taken to include the communities which occupy the following habitats:

(1) The sand and mud of the intertidal zone where these are protected from the more violent wave action (*salt-marsh formation*) (Chap. XL).

(2) The uppermost zone of the sea shore on exposed coasts, barely reached by the spring tides (*seashore community*) (Chap. XLI).

(3) Coastal sand dunes (*sand-dune formation*) (Chap. XLI).

(4) Coastal shingle beaches (Chap. XLII).

(5) Brackish marshes (Chap. XLIII, p. 896).

(6) Sea cliffs and coastal rocks out of reach of the tides, but exposed to spray (Chap. XLIII, p. 897).

(7) Maritime and submaritime grasslands, above the limit of high tide but affected by salt spray (Chap. XLIII, p. 899).

The characteristic plant formation (1) inhabiting the sand and mud of the intertidal zone is the nearest to marine vegetation proper because it is exposed to immersion in salt water at some high tides. It is essentially composed of halophytes and includes many salt-water algae. This formation is based on marine alluvium and occurs only where it is protected from the erosive action of waves and the scour of swift currents and tide races, particularly where the flat shores of an estuary are protected by headlands, shingle banks or sand dunes. The intertidal zone on coasts formed of hard rock and exposed to the waves is inhabited by communities of marine algae, zoned according to their degree of tolerance of exposure to the air and also of wave action. These seaweeds are firmly fixed to their rock substratum, and depend for their mineral nutrition entirely on what they can absorb from the sea water while they are covered by the tide. Salt-marsh plants, on the other hand, are rooted in mud or sand and, like rooted fresh-water aquatics and land plants, get their water and mineral nutrients through their root-systems.

Sandy or muddy sea shores between tide marks on *exposed* coasts are destitute of higher plants because of their instability under wave action:[1] it is the uppermost zone alone, reached only by the highest spring tides, that often bears a scattered population of rooted phanerogams—*littoral* plants proper—and often known as the "beach" or "sea shore" community (2).

Other maritime plant communities, less exposed to the direct effect of the tides but showing more or less maritime influence, are (3) coastal sand-dune vegetation, which is not composed of halophytes and passes over into typical land vegetation; (4) shingle-beach vegetation, inhabiting a local and special habitat, and having much in common with that of sand dunes, but including halophytic species; (5) brackish marshes in the upper reaches of estuaries or wherever salt water, diluted with fresh water, has access; (6) the vegetation of sea cliffs and coastal rocks out of reach of the tide, but exposed to drenching by spray and therefore largely halophytic; and finally (7) grasslands close to the sea and subject to the effect of sea spray driven on to the land by the wind, which show a mixture of halophytic and non-halophytic species.

The types of maritime community just enumerated are not only constantly found in proximity because they all depend on maritime influence, they are also frequently interdependent. Thus the measure of protection required for the development of salt marsh is often provided by shingle spits (cf. pp. 569–73) or dune complexes, while sand dunes themselves are often formed on a basis of shingle beach. These relations are illustrated on Plates 133 and 134 as well as in several of the photographs reproduced on other plates in this and the following chapters.[2]

Plate 133 shows the parallel dune ridges of the Headland at Blakeney Point which are largely built on pre-existing parallel shingle banks. The continuation of one of these banks as a naked shingle spit is seen in phot. 335 curving round and protecting a large salt marsh which is traversed by smaller and older spits. This marsh was doubtless formed in successive compartments, each developed after the formation of the spit which established the condition of protection. All the compartments of the marsh now drain into a common channel shown on the right of the photograph. Plate 135, phot. 338, is a view of this marsh from a north-western bay and shows three old worn-down spits, two of which nearly shut off the marsh compartment seen in the foreground.

Plate 134, phot. 336, shows a series of much older spits with their enclosed mature marshes farther up the estuary, and collectively known as "The

[1] Pl. 133, phots. 334–5.
[2] E.g. Pl. 135; Pl. 136, phot. 341; Pl. 147, phot. 373; Pl. 158, phot. 405.

PLATE 133

Phot. 334. The Headland at Blakeney Point showing the parallel lines of old shingle beach overlaid by sand dunes, with lows between. The outermost dunes are young and unconsolidated. The white lines crossing the main dune ridges are human foot tracks. Mottled and ribbed sand and mud of the estuary occupy the bottom of the picture. The continuation of the main shingle bank leading to the Far Point is seen in the top left-hand corner *Major J. C. Griffiths* (alt. 7000 ft.), 1921.

Phot. 335. Part of the *Salicornia-Pelvetia* salt marsh behind the Headland. The marsh is enclosed and protected by an old lateral shingle spit running across the centre of the picture. Two minor spits ending freely in the marsh are seen (centre and right) at the top of the picture. The spits are fringed with bushes of *Suaeda fruticosa* (black dots). Tidal access to the marsh is by the channel on the right. Ribbed sandy mud of the estuary at bottom. *Major J. C. Griffiths* (alt. 7000 ft.), 1921.

BLAKENEY POINT FROM THE AIR

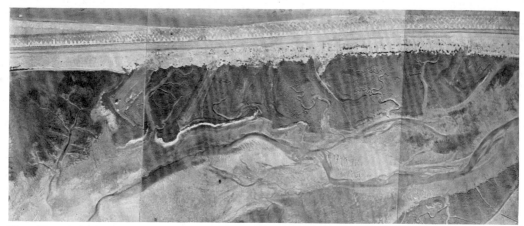

Phot. 336. At the top of the picture is the edge of the sea with mottled sand and lines of drift marking different high tide levels; then the main shingle bank with dark *Suaeda fruticosa* (dots and irregular lines) and other vegetation (fainter). Below is the series of salt marshes in compartments, bounded and protected by the lateral spits or "hooks" and drained by sinuous branching channels up which the tide gains access. Below is the channel of the River Glaven bordered by expanses of estuarine mud. The length of the photograph is about 1¼ miles (2 km.). *Major J. C. Griffiths* (alt. 7000 ft.), 1921.

Phot. 337. Central portion of the Marams on a larger scale. The three irregular zones of *Suaeda fruticosa* on the main bank, the deltas of shingle invading the marshes and the *Suaeda* bushes on the laterals can be distinguished. *Major J. C. Griffiths* (1921).

"The Marams" at Blakeney Point from the Air

PLATE 135

Phot. 338. Salt marshes protected and partly enclosed by lateral shingle spits, which are attached to the dune-covered main spit (left). Two run across the centre of the picture, and the proximal end of a third is in the foreground. All three are fringed with *Suaeda fruticosa* and end freely in the marsh, which is dark coloured and dominated by *Salicornia* with unattached *Pelvetia* as an understorey. Waters of the estuary to the right behind. The photograph was taken from a north-western bay of the marsh shown in Pl. 133, phot. 335, just out of the top of the air photograph. F. W. Oliver.

Phot. 339. The other side of the marsh shown in phot. 338. On the right is an old lateral spit covered with dunes much worn down and abundant *Senecio jacobaea*. The marsh is fringed with *Suaeda fruticosa* on marsh soil covered with shingle, severely rabbit-eaten on the side towards the spit. Dune-covered main spit (separating the marsh from the open sea) in the distance. F. W. Oliver.

MARITIME VEGETATION AT BLAKENEY POINT (SAND DUNE, SHINGLE AND SALT MARSH)

PLATE 136

Phot. 340. Centre of the marsh shown on Pl. 135, with drainage channels and pans. Dune-covered main shingle bar in the distance. *F. W. Oliver.*

Phot. 341. Old lateral shingle spit at "the Marams", curving round a salt marsh. Zone of *Suaeda fruticosa* on slope of the spit: above (right foreground) is the zone of *Festuca rubra* (see p. 889). Other *Suaeda*-covered spits enclosing salt marshes beyond. *Mrs Cowles* (1911).

MARITIME VEGETATION AT BLAKENEY POINT (SALT MARSH AND SHINGLE SPITS)

Marams". Here the spits are seen to be "laterals" arising from the main shingle bank shown at the top of the picture (see Chapter XLII, pp. 870–1). Pl. 136, phot. 341, shows one of these lateral spits curving round a salt-marsh compartment.

(1) THE SALT-MARSH FORMATION

In tidal estuaries into which sand and mud are carried by the tide and laid down on the flat shores, where wave action is at a minimum, a luxuriant phanerogamic vegetation, with associated algae, is developed. This vegetation is generally known as *salt marsh*, its flora, like the factors of its habitat, is extremely distinct, and it constitutes a well-defined plant formation. The master factor which differentiates this from other formations is the salt water which bathes the whole plant body during the periodic immersions and forms the soil solution (though varying in concentration) at all times. Hence the plants are said to be *halophytes* or *halophilous*, many of them growing well only when supplied with salt water, though others can do without it. Halophytes occur not only in coastal salt-marsh formations but also inland, wherever there is a high concentration of salt in the soil, for example round existing salt lakes or on the sites of old lake basins in arid regions, or again in the neighbourhood of salt springs.

Relation to tides. The salt-marsh formation is periodically immersed by the tide. It extends upwards from levels which are just about reached by high water of neap tides (tides of least range) and are covered by all the tides of intermediate range, to the highest levels of the marsh, which are only reached by high water of the highest spring tides (i.e. the tides of greatest range). At low water of spring tides the tide recedes far below the level of the lowest salt-marsh community and uncovers stretches of bare sand or mud and also the fields of "sea wrack", "grass wrack" or "sea grass" (*Zostera*), a marine monocotyledon with flexible, band-shaped leaves, which sometimes occupy this position. The Zosteretum must be reckoned as belonging to marine vegetation for *Zostera* is one of the small group of marine angiosperms, and has a different habit and economy from the salt-marsh plants; but Zosteretum, where present, may be considered as the starting point of the halosere since it exists on similar substrata and shows transitions to the Salicornietum.

Different levels of the marsh are thus covered by the sea water for very different lengths of time, the lowest zones for long periods, the higher for much shorter ones, while the highest of all are visited only by a few tides, usually twice a year, around the equinoxes.

Salt content of ground water. Besides undergoing periodic immersions in sea water the plants of the salt marsh have a ground water which is normally salt, though the amount of salt it contains may vary at different periods from a percentage higher than that of sea water (about 3·3) to quite a low percentage approximating to that of fresh water. The variation in the saltness of the ground water is due to the interaction of various factors—immersion by the tide, the rate of evaporation when the tide is down, and the washing out of salt by rain during the periods of emersion. One of the characteristic features of most salt-marsh plants is their succulence, caused by the presence of abundant "water tissue" whose cells are typically large, thin walled, and swollen with cell sap. The halophytes absorb large amounts of salt, which remains in their tissues, and the concentration of osmotically active solutes in their cell sap is necessarily brought more or less into equilibrium with the medium surrounding their roots, the salt ground water, and, during the periods of immersion, with the sea water of the tide which covers the plants. It has been shown that the osmotic pressure of halophytic cells is not only much higher than that of mesophytes, a character which is shared by the cells of xerophytes, but that it changes in response to the concentration of the surrounding medium.

Zonation and succession. As always where some decisive ecological factor undergoes a regular spatial change in intensity ("gradient") the salt-marsh formation is very definitely zoned. These zones undoubtedly correspond, in a general way, with the salt-marsh succession (*halosere*) passing from pioneer communities, on wetter, more mobile mud or sand exposed to the air for shorter periods, to those on drier more stable substrata which are exposed for longer periods, the highest being very rarely immersed by the tide. The cause of the succession is the rise of level brought about by continuous tidal silting. As soon as the substratum becomes sufficiently stable for colonisation by plants the silt is trapped between them and the general level raised, with increasing rapidity at first, as the density of the vegetation increases (F. J. Richards, 1934), then more slowly, as the surface is less and less often covered by the silt-bringing tide. At the highest levels of the marsh silt deposition is almost negligible and any further rise in level depends mainly on humus accumulation.

Ecological factors. The most important factor differentiating the zones of the marsh is probably the relative lengths of time during which they are submerged or exposed to the air. Other factors are the nature of the soil and its water content, determining aeration and depending on drainage. Besides these physical factors there is of course the competition between the plants of the later, closed, communities.

Submersion and exposure. It is not generally realised for how long a time even the lower salt-marsh communities are exposed to the air. This is well brought out in V. J. Chapman's very thorough studies (1934, 1938) of the times of exposure of the salt marshes at Scolt Head island on the north coast of Norfolk. He shows that at the bottom of the Zosteretum, i.e. well below the level of the salt marsh proper, the ratio of hours of submergence to hours of emergence per month (calculated for the period 31 July to 9 October) is 0·628, representing a submersion of 282·4 out of 732 hr., while at the top of the marsh this S/E ratio is 0·004, representing a total submersion of only 3 hr. per month. The whole extent of the salt-marsh vegetation is thus submerged by the tide for much less than half the time, and the upper levels for very small fractions only.

Effects on different zones. Further, the duration of submergence varies a good deal at different levels according to the season. The upper communities of the marsh, which Chapman classes together as Group X (see Fig. 152), are most frequently submerged at the spring equinox, that is during the period of seedling growth when a good supply of water is essential, and at the autumn equinox, the period of fruiting, when the tides can distribute the seeds. During the summer there is a long uninterrupted period of exposure, when, in the absence of rain, the surface soil may become very dry, so that the plants of the X communities have to endure drought, besides having to tolerate brief periods of immersion in sea water around the equinoxes. The most frequent submergences of the lower communities, which Chapman separates as Group Z, occur in midwinter and at midsummer, i.e. during the period of maximum growth, so that the plants are unlikely to suffer from water deficiency at a time when those of the X communities may be wilting or dying from drought. Thus the Z communities are probably limited upwards by the inability of the species of which they are composed to withstand drying, while those of the X communities are limited downwards by their inability to tolerate frequent immersions in sea water during their period of maximum growth. Most of the communities of the salt marsh belong either to Group X or to Group Z, very few crossing the line between them. These few intermediate communities Chapman classes as Group Y.

SALT-MARSH COMMUNITIES

Variation in detail. The general zonation and succession of communities in salt marsh is very uniform, but different marshes show very considerable variation in detail, owing to differences in the substratum (sand and mud) and also in physiography and topography. A salt marsh rarely shows a uniform gentle slope with a series of zones of uniform width: it is nearly always dissected by creeks of varying depth and is very often developed in bays as individual marshes of varying size and varying relation to other physiographic features such as sand dunes and shingle beaches; and all these differences involve variations in detailed development.

A very usual arrangement on the east coast of England, where shingle "spits" are common (see Fig. 168, p. 872), is the development of separate marshes on the estuarine mud or sand of the bays enclosed by lateral shingle banks attached to the inner side of a main shingle spit which runs between the open sea and the estuary.[1] These laterals extend inwards at various angles from the main bank and end freely in the estuary, and the

[1] Cf. Pl. 134.

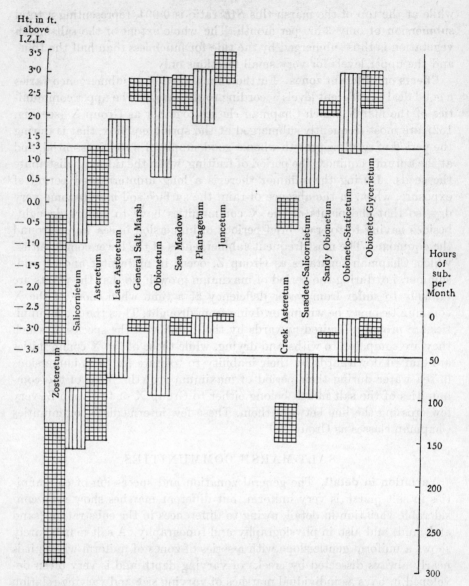

FIG. 152. RANGES OF SALT MARSH COMMUNITIES AT SCOLT HEAD, NORFOLK

The upper part of the figure shows the ranges in vertical heights. The critical height, separating the X communities (above the horizontal line 1·3) from the Z communities below, is 8·3 ft. above Ordnance Datum (I.Z.L.=7 ft. above O.D.). The lower part of the figure shows the range in hours of submergence per month. 50 hours is the critical number separating the X communities (above) from the Z communities (below). The communities in the right hand group do not belong to the primary series, but are partly determined by local soil conditions. From Chapman, 1934.

marshes they enclose commonly form a progressive series, the most recently arisen nearest the extremity of the main spit. Such marshes also differ according to the nature of the substratum and according to whether the bay has a wide or a narrow opening.

The most various phases of retrogression and regeneration of salt marsh are to be found, especially on the edges of creeks and also on the seaward face of an old marsh which has reached a relatively high level and has then been eaten away by the tides owing to a change in currents. Here there is often a vertical crumbling cliff or escarpment, several feet high, at the top of which the old salt marsh persists, with soft mud below formed from the debris of the cliff, on which much younger communities are developed.

Individual salt-marsh species sometimes appear at different stages of the succession owing to causes not fully understood. For example at Scolt Head on the north coast of Norfolk, *Glyceria maritima*, one of the most widespread and ubiquitous salt-marsh plants, and generally entering the succession very early, does not appear till a later stage than in most marshes; and at the neighbouring Blakeney Point it is only prominent in upper marsh communities.

General zonation. The following is the general zonation, though one or more of the zones may be absent:

(1)* Algal communities beginning below the level reached by high water of neap tides. This zone may be occupied by a Zosteretum, associated with the same algae.

(1) Salicornietum herbaceae.

(1a) Spartinetum townsendii.

(2a) Glycerietum maritimae.

(2b) Asteretum tripolii.

(3) Limonietum vulgaris.

(4) Armerietum vulgaris.

(4a) Obionetum portulacoidis.

(4b) Suaedetum fruticosae.

(5) Festucetum rubrae (sandy marshes).

(6) Juncetum maritimi.

Fig. 153 shows the general zonation in the sandy salt marsh at Ynyslas on the shore of the Dovey estuary in Cardiganshire. Here (3), (4a) and (4b) are (exceptionally) absent.

(1*) *Algal communities and Zosteretum.* The first colonists of the still mobile sand or mud are algae, most commonly species of *Rhizoclonium, Ulothrix, Chaetomorpha, Vaucheria, Monostroma* and *Enteromorpha* among the green algae (of which the first and last genera are the most constant and abundant), and in some muddy estuaries a rich flora of diatoms and blue-green algae (Fig. 154) as described by Carter (1932). In sandy estuaries the algal flora is much poorer (Fig. 155). The filamentous forms undoubtedly play a considerable part in binding the loose mud, and probably serve as a good nidus for the colonisation of *Salicornia*. The brown seaweed *Fucus vesiculosus* var. *evesiculosus* has also been recorded from this zone (Chapman, 1934).

It is at this level too that a Zosteretum may be developed. This is dominated in some localities by *Zostera marina* (Philip, 1936), in others by

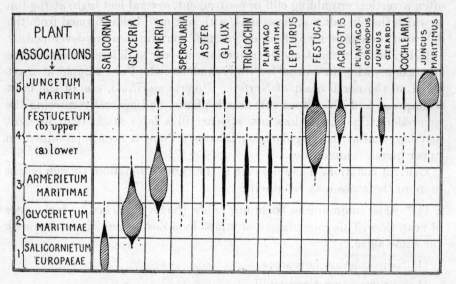

Fig. 153. Zonation of Angiospermic Communities at Ynyslas Salt Marsh, Dovey Estuary (Cardiganshire)

The ranges of the principal species in the zoned communities are shown. Dominance or subdominance is indicated by diagonal shading. From Yapp, 1917.

Z. nana (Chapman, 1934); *Ruppia rostellata* may also be present (Philip, 1936).

(1) **Salicornietum herbaceae** (Samphire or Glasswort Marsh) (Pls. 137-8, phots. 344-6). In the great majority of salt marshes that have been studied this is the first (earliest) community of the salt marsh proper. It is dominated by and (apart from algae) often exclusively composed of herbaceous annual species of *Salicornia*, which may be included in the aggregate *S. herbacea*: segregate or closely allied species such as *S. dolichostachya*, *S. stricta*, *S. ramosissima*, may also be present in some quantity. The colonisation is at first by isolated plants, which are very liable to be uprooted by wave action on the mobile mud (Wiehe, 1935) especially within range of daily tides (Fig. 156). At a higher level the numbers increase, but the Salicornietum remains open with bare mud between the plants[1] still

[1] Pl. 137, phot. 344-5.

PLATE 137

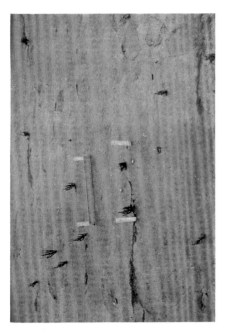

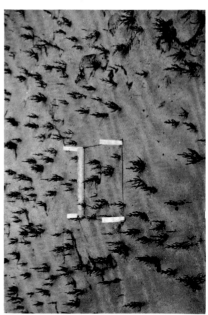

Phot. 344. Very sparse Salicornietum averaging one plant per sq. ft. subject to daily covering by the tide. The pegs enclose a square foot. Dovey estuary. *P. O. Wiehe* (1935).

Phot. 345. Denser though still sparse Salicornietum (av. 24 plants per sq. ft.) just above the range of daily tides. *P. O. Wiehe* (1935).

Phot. 342. Bare cracked mud of the Severn estuary with severely eroded cliff whose level top bears old salt marsh grassland. One plant of *Aster tripolium* has colonised the mud. *C. G. P. Laidlaw.*

Phot. 343. Looking down the Severn estuary. Deeply cut channel with slopes of bare unstable mud; eroded cliff on the right above. *C. G. P. Laidlaw.*

BARE MUD AND SALICORNIETUM

PLATE 138

Phot. 346. Salicornietum with primary depression pan. Dovey salt marshes. *R. H. Yapp* (1917).

Phot. 347. *Spartina townsendii* invading sparse Salicornietum. *Fucus* sp. in background and extreme foreground. Hayling Island, Hants. *S. Mangham*.

SALICORNIETUM AND *SPARTINA*

occupied by the algae of the lower zone. The first phanerogam to become associated with *Salicornia* is usually *Glyceria maritima*, the most characteristic of the salt-marsh grasses and commonly occurring through a considerable range of level. In some marshes *Glyceria* may colonise the mud *pari passu* with *Salicornia* (Marsh, 1915), or even act as a pioneer (Heslop

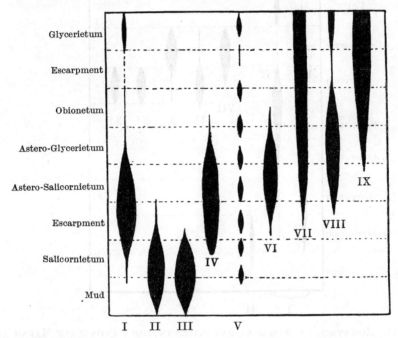

FIG. 154. ZONATION OF ALGAL COMMUNITIES IN THE MUDDY SALT MARSH AT CANVEY ISLAND IN RELATION TO THE COMMUNITIES OF ANGIOSPERMS

I. General community of Green Algae, of great range but most developed in the lower zones. II. Marginal community of Diatoms. III. Marginal community of Blue-green Algae. These two have a maximum development in the unstable mud of the channel and in the adjacent Salicornietum. IV. *Ulothrix flacca* community—transitory, pre-vernal, covering at that time all the lower zones of the marsh. V. *Enteromorpha minima—Rhizoclonium* community—epiphytic, throughout the marsh, only observed at Canvey. VI. *Anabaena torulosa* community—middle zones of the marsh. VII. Filamentous Diatom community on mud, middle and upper zones (not seasonal). VIII. Autumn community of Blue-green Algae. IX. *Phormidium autumnale* community, middle and upper zones. Most of the algal communities are markedly seasonal. From Carter, 1933, p. 390.

Harrison, 1918; Morss, 1927), but usually there is a pure Salicornietum preceding the advent of the grass. Other species which may settle in the pioneer Salicornietum are *Aster tripolium* and *Suaeda maritima*, the latter especially but not exclusively on sandy substrata. The dwarf *Fucus caespitosus*, embedded in the soil, is characteristic of the Salicornietum at Scolt Head. The other algae are broadly the same as those of the lower zone. Fungi are few and far between (see p. 833).

At Scolt Head the lower fringe of the Salicornietum is exposed to the air only at low tide except for a very short period in March, but the upper extremity is exposed for long periods, amounting in all to about seven times the periods of submergence (Chapman, 1934).

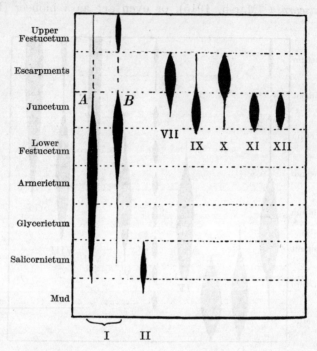

FIG. 155. ZONATION OF ALGAL COMMUNITIES IN THE SANDY SALT MARSH AT YNYSLAS (DOVEY ESTUARY) IN RELATION TO THE COMMUNITIES OF ANGIOSPERMS

Compare Figs. 153, 154.

I. General community of Green Algae, of greatest range. The green components (A) are developed most strongly in the lower and middle, the blue-green components (B) in the middle and upper zones. II. Marginal community of Diatoms, only in unstable mud and the Salicornietum. (Communities III–VI inclusive of the Canvey Marsh are absent.) VII. Filamentous Diatom community, confined to escarpments and Juncetum, owing to the closed vegetation of most of the other zones. (Community VIII of Canvey Marsh is absent.) IX. *Phormidium autumnale* community, Juncetum (mainly) and escarpments only. X. *Rivularia-Phaeococcus* community, almost confined to escarpments. XI, *Pelvetia canaliculata* (ecad *muscoides*) community, and XII, *Catenella opuntia-Bostrychia scorpioides* community, are confined to the Juncetum. The last three communities are absent at Canvey, except X, which exists on the escarpments to a slight extent. From Carter, 1933, p. 394.

Wiehe (1935) has published an interesting study of the effects of the tides on Salicornietum. He finds that a "threshold" time of 2 or 3 days undisturbed by tides is necessary for the establishment of a dense population of *Salicornia*. Daily tides drag the seedlings from their anchorage in the mud (Fig. 156).

(1a) **Spartinetum townsendii** (Pl. 138, phot. 347; Pl. 139, phots. 348–9).

On deep mobile mud, often too mobile for the successful establishment of the Salicornietum, though approximately on the same level, the tall strongly growing perennial grass *Spartina townsendii* (Fig. 157), a hybrid of *S. alterniflora* (an American species) with the European *S. stricta*, forms almost pure communities on the south coast of England, especially in Southampton

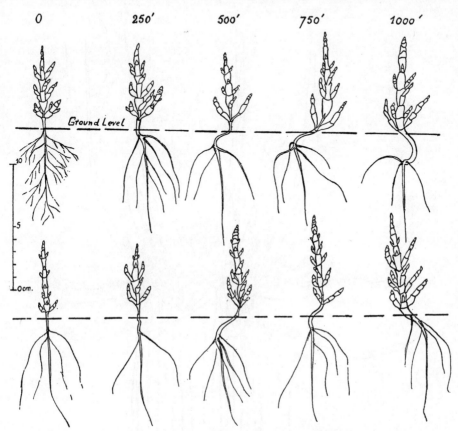

Fig. 156. *SALICORNIA HERBACEA* IN THE SALICORNIETUM OF THE YNYSLAS SALT MARSH (DOVEY ESTUARY)

The plants show increasing distortion of the axis in passing from left to right of the figure. Those at 500 ft. and more from the zero point are subject to the effect of daily neap tides, which drag the seedlings from the erect position and frequently dislodge them altogether. From Wiehe, 1935.

Water and Poole Harbour, as also on the opposite coast of France. This hybrid was first reported from Southampton Water in 1870, and since then has extended its area very rapidly. It has also been planted in several other places on our coasts, and in some of these is well established and spreading. Pl. 139 illustrates the remarkable transformation that can occur on stretches of tidal mud as the result of the activity of the rice-grass.

S. townsendii is a much larger plant than the native *S. stricta* which plays a subordinate part in some of our salt marshes. The stout stem bears stiff erect leaves: from the bases of the stems stolons radiate in all directions, binding the soft mud; and feeding roots, mostly horizontal in direction,

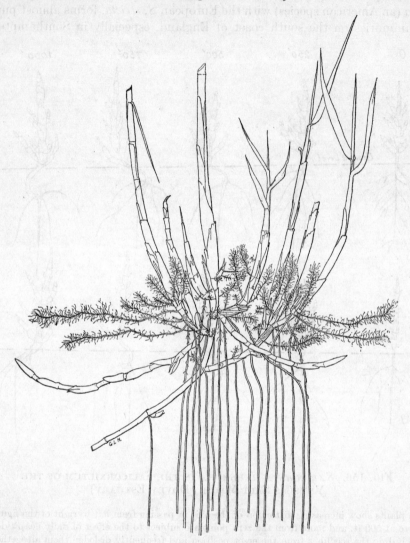

FIG. 157. *SPARTINA TOWNSENDII*

Showing bases of stout ascending aerial shoots, horizontal stolons and feeding roots, and deeply penetrating vertical anchoring roots. From Oliver, 1926.

ramify through the surface layers of the mud, while stouter anchoring roots extend vertically downwards (Fig. 157). The leaves offer broad surfaces to the silt-bearing tidal water, and their points catch and hold fragments of seaweed and other flotsam. The thick forest of stems and leaves breaks up the tidal eddies, thus preventing the removal of mud which has once

PLATE 139

Phot. 348. View across Holes Bay, Poole Harbour, Dorset, showing *Spartina townsendii* colonising the soft mud. June 1911. *R. V. Sherring.*

Phot. 349. View from approximately the same spot showing dense Spartinetum developed in 13 years. June 1924. *F. W. Oliver.*

SPARTINETUM TOWNSENDII

PLATE 140

Phot. 350. *Salicornia herbacea* (foreground), *Aster tripolium* and Glycerietum maritimae, near Gedney Drove End, The Wash. *R. H. Yapp.*

Phot. 351. *Salicornia dolichostachya*, on bare mud, and dense Glycerietum developed between 1911 and 1921. Berrow mud flats, Somerset. *H. S. Thompson* (1921).

Phot. 352. Another view of the Glycerietum, same locality. Luxuriant flowering tuft of *Aster tripolium* (centre) on edge of channel and *Triglochin maritimum* (left). *H. S. Thompson* (1921).

GLYCERIETUM MARITIMAE

settled on the marsh. In this way large areas of mobile mud are fixed, the surface level is raised, and "reclamation" eventually made possible. At the same time the tidal currents and channels are profoundly altered and the whole aspect of the intertidal zone completely changed. No other species of salt-marsh plant, in north-western Europe at least, has anything like so rapid and so great an influence in gaining land from the sea.

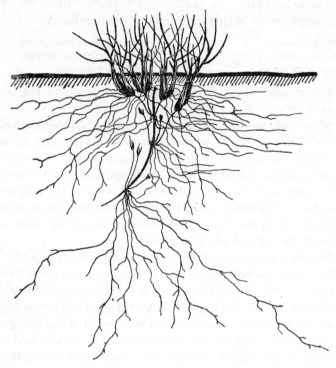

FIG. 158. *GLYCERIA MARITIMA*

Tiller formation giving tufted habit under grazing (tillers separated for drawing). ¾ natural size. From Yapp, 1917.

The Spartinetum townsendii is a remarkably pure community, containing only occasional examples of species flourishing at about the same level, such as *Salicornia herbacea* and *Aster tripolium*. The Salicornietum is thus very much restricted in areas in which the rice grass has successfully established itself.

(2*a*) **Glycerietum maritimae.**[1] The upper edge of the Salicornietum is invaded, and *Salicornia* eventually replaced as a dominant, either by *Glyceria maritima* or *Aster tripolium*. *Glyceria maritima* (Fig. 158) is a vigorous perennial grass which quickly extends horizontally and begins to form a turf. After it is once established *Glyceria* rapidly collects silt and often forms low flat hummocks, thus restricting the space available for the

[1] Pl. 140.

single plants of *Salicornia*. The plant is generally shallow rooted (2–4 in. = 10 cm.), but under exceptionally favourable conditions (good aeration) roots may penetrate to a much greater depth, reaching 2 ft. (60 cm.) or even more. Sometimes a community is formed in which *Glyceria* and *Salicornia herbacea* are co-dominant, with a little *Suaeda maritima* and other species of herbaceous *Salicornia*, such as *S. disarticulata* and *S. ramosissima* (Heslop Harrison, 1918). In the Severn estuary (Priestley, 1918) the following species occur in this zone besides the dominant:

Salicornia herbacea	Spergularia marginata
Suaeda maritima	Triglochin maritimum
Plantago maritima	Glaux maritima

The last four species usually occur at higher levels of the marsh, and *Glaux* flowers very little in this habitat.

In the estuary of the River Nith (Dumfriesshire) on the north side of the Solway Firth (Morss, 1927), where *Salicornia* plays quite a minor role, *Glyceria maritima* is the pioneer on all the areas studied, and is predominant on most parts of the marshes. *Cochlearia officinalis* occurs freely in the middle Glycerietum of the Nith estuary and also on the soft mud of the sides of creeks. It affects comparatively open soil and may also occur (at a much higher level) in the Juncetum maritimi, where the shade of the tall sea-rush keeps the surface of the soil bare.

In the Bouche d'Erquy, a sandy Breton salt marsh, studied by Professor F. W. Oliver and his colleagues and students in the early years of the century, most of the area of the marsh is occupied by Glycerietum, composed of alternating low flat hummocks or tracts of somewhat higher lying flat ground, and shallow channels or lower lying flats, the grass being regularly associated with *Suaeda maritima* on the former and with *Salicornia herbacea* in the latter. All this area, uncovered during the whole of the neap tide cycle, is used as sheep pasture.

(2*b*) **Asteretum tripolii.** In some marshes, in the lower zones of which Glycerietum is absent or plays a relatively insignificant part, the Salicornietum is invaded instead by *Aster tripolium*. Thus at Scolt Head (Chapman, 1934) the new dominant displaces *Salicornia*, which remains in the barer areas between. The following species are present:

Aster tripolium var.		Pelvetia canaliculata	
discoideus	d	forma libera	
Salicornia herbacea	ld	Bostrychia scorpioides	
Spartina stricta	ld		
Suaeda maritima var.			
flexilis	la		

Along the banks of the creeks, farther up in the marsh but at about the same horizontal level, an Asteretum is developed in which *Suaeda maritima* var. *flexilis* is replaced by *Salicornia perennis*, and *Pelvetia* by *Fucus caespitosus* and *F. volubilis*. The soft mud deposited on the banks of creeks cut by the tide in consolidated marshes is a favourite habitat of *Aster*.

At Canvey Island in the Thames estuary (Carter, 1933) *Aster* also plays a prominent part at this level, invading the Salicornietum along with *Glyceria* and remaining abundant after *Glyceria* has established dominance, but it is not itself dominant. A number of green and blue-green algae are prominent in this zone (see Fig. 154). *Ulothrix flacca* is dominant in February, giving place in March to *Enteromorpha prolifera*, which covers the *Ulothrix* like a blanket during the spring and summer and decreases in August, when a mixture of dark blue, olive green or brown Cyanophyceae appear, including several species of *Oscillatoria* and *Phormidium*. *Microleus chthonoplastes* forms local streaks on the mud, and circular colonies of *Anabaena* and *Nodularia* occur near the moist margins of the zone. Diatoms are frequently present among the Cyanophyceae. This phase lasts till October when the mud becomes increasingly bare.

The marshes up to this point are included in Chapman's Group Z (see Fig. 152), belonging to the lower marsh.

(3) **Limonietum vulgaris** (Sea-lavender marsh). The common sea lavender, *Limonium vulgare* (formerly known as *Statice limonium*) invades the upper edge of the Glycerietum in many salt marshes, and becomes co-dominant. Flowering throughout the later summer its lavender-purple blossoms make magnificent sheets of colour. This is a very typical community of what may be called the middle marsh and often occupies wide areas. It forms a transition to the communities of the upper marsh (Group X).

In the Tees marshes the following species occur (Heslop Harrison, 1918):

*Limonium vulgare and *Glyceria maritima   co-d

*Aster tripolium	a	*Suaeda maritima	o
Plantago maritima	a	*Salicornia herbacea (agg.)	r
Spergularia salina	f	Limonium humile	vr
*S. marginata	f	*Triglochin maritimum	vr
*Armeria maritima	f		

The eight species marked with an asterisk also compose the flora of the Limonietum at Holme in Norfolk (Marsh, 1915).

In some salt marshes the sea lavender appears to be quite absent, e.g. in the Dovey and Nith marshes. Yapp (1917) and Morss (1927) suggest that this may be due to heavy grazing.

(4) **Armerietum maritimae** (Thrift marsh). The thrift or sea pink, *Armeria maritima* (Fig. 159), often dominates a community at a slightly higher level than the sea lavender. Here we have typically a close turf composed of a number of species in which the compact rosettes of *Armeria*, together with the tufts of *Glyceria*, form the largest part. *Armeria* cannot tolerate such high salt concentrations as *Salicornia* and *Glyceria*. It roots more deeply (to 23 in. in the Nith estuary) and grows more vigorously with

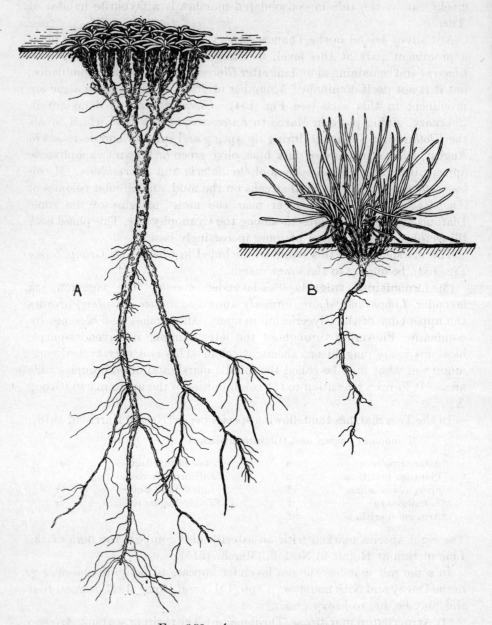

FIG. 159. *ARMERIA MARITIMA*
A. From the Armerietum—rosette habit under grazing.
B. From the Juncetum—more diffuse habit under partial shade and protection from grazing. About ¾ natural size. From Yapp, 1917.

decreasing salt, only giving way before the still less halophytic competitors of the upper marsh. The thrift stands grazing well and its compact rosettes (Fig. 159 A) often form a considerable part of the turf of sea meadows. The thrift has a much longer flowering season than the sea lavender, beginning in April and being at its best in May and June, when the massed flowers show sheets of rose-pink, comparable in beauty with the purple stretches of sea lavender in the following months.

In the Tees marshes a rather open Armerietum has the following species:

Armeria maritima and Glyceria maritima		co-d	
Plantago maritima var. latifolia	f	Suaeda maritima	o
Spergularia spp.	f	Aster tripolium	o
Salicornia herbacea	lf	Limonium vulgare	o
		Triglochin maritimum	r

In another part of these marshes the dark green turf of the Armerietum is much closer and the annuals are practically absent. Here *Festuca rubra*, *Glaux maritima* and *Artemisia maritima* occur, though none is more than rare or occasional.

In the Dovey marshes on the coast of Wales (Yapp, 1917) the Armerietum (Pl. 141, phot. 354; Pl. 143, phot. 359; Pl. 144, phot. 364) contains the following species:

Armeria maritima	d	
Glyceria maritima	co-dominant in the lower part of the zone	
Triglochin maritimum		Aster tripolium
Plantago maritima		Spergularia sp.
Glaux maritima		Pholiurus (Lepturus) filiformis

Green algae. Three green algae compete for space in the Dovey Armerietum—*Rhizoclonium*, *Vaucheria* and *Enteromorpha percursa*, the first-named dominant during winter and spring, *Vaucheria* and Cyanophyceae prominent in the late summer when the marsh is exceedingly dry and *Rhizoclonium* and *Enteromorpha* are not visible. *Vaucheria* probably tolerates drought better than other green forms because its filaments penetrate the soil to some depth (Carter, 1933, p. 388).

Soil fungi. The soil fungi of the Dovey salt marshes were investigated by Elliott (1930). This is one of the very few natural plant formations whose soil fungi have been studied.

In the Salicornietum fungi were few and far between. The species found there all occurred also in the Glycerietum and Armerietum (from which most of the samples examined were taken) except *Chaetomium spirale* and *Macrosporium commune*. The soil is a badly aerated stiff tenacious clay (pH 8). Generally speaking the same species were common to the two, and those found at a depth of $3\frac{1}{2}$ in. (c. 8·75 cm.) were for the most part the same as at $1\frac{1}{2}$ in. (c. 3·75 cm.). Direct microscopic examination of the soil yielded little information. Conidia and fragments of hyphae were rarely met with and then only in association with organic matter such as dead or dying roots, stems and leaves. In pure culture, however, forty-eight species were isolated, but the fungi were not abundant, or at least not very active, since many fewer colonies were developed in culture from a suspension of a gram of waterlogged salt-marsh soil than from a gram of cultivated soil. Samples taken in June (soil temperature at $3\frac{1}{2}$ in. 19·5° C.) gave the largest crops.

Of the whole number of species recorded, most of which are known above ground as saprophytes, twenty-seven were Fungi Imperfecti, fourteen Ascomycetes and seven Mucorales. Besides these several septate sterile mycelia were found, and one, showing clamp connexions, was presumably a Basidiomycete. No evidence of active parasitism was found.

The commonest species were *Torula allii*, *Penicillium hyphomycetis* and *Fusarium oxysporium* var. *resupinatum*. The first two were not previously recorded as soil fungi. Almost equally common were *Trichoderma lignorum*, *T. königi*, *Hormodendron cladosporoides*, *Mucor circinelloides*, and *Periconia felina*. Several of the species isolated grew in culture by sending out a few long branches, fruiting as they grew, followed by others taking the same course and coiling round the pioneer branches so that rope-like strands were formed, from which similar strands branched out later. In this way the fungus quickly covered a large area of substratum.

Both *Glyceria maritima* and *Armeria maritima* are mycorrhizal plants (though *Glyceria* at least can be cultivated in the absence of mycorrhiza), the endophytic fungus extending throughout the plant body of the former and probably also of the latter. It is likely that some of the mycorrhizal fungi are the same as those recorded from the soil, and one was seen in the root of *Glyceria* which appeared identical with *Monilia pruinosa* occurring in the soil. *Stachylidium cyclosporum* and *Cladosporium herbarum* appeared in cultures from the rootlets of *Glyceria* and these may also perhaps be mycorrhizal. *Glyceria* has a compact and densely lignified root stele which resists the decay that destroys the cortex, but the roots of *Armeria* have little lignified tissue and possess large air spaces. At a depth of a foot the taproots of *Armeria* are little more than hollow tubes of cork, the whole of the interior having decayed from the attacks of fungi and bacteria. This superior resistance of *Glyceria* to decay helps to explain its much greater efficiency in binding the tidal silt. Where tidal erosion has taken place the soil of the Glycerietum shows a tangle of roots and rhizomes which is absent in the Armerietum.

It is to be noted that the salt marsh dominated by these middle communities has long been used as sheep pasture and that the excrement is no doubt an important factor in determining the fungal vegetation. It should also be noted that the culture media employed (bread-, potato- and raisin-agar) may have had a selective effect on the fungi.

"**General salt-marsh community.**" In *Types of British Vegetation*,

PLATE 141

Phot. 353. Early stage of primary salt marsh formation. Thin Glycerietum with hummocks of *Armeria* (which traps silt more quickly). Winding channels beginning to be established. Dovey estuary. *R. H. Yapp* (1917).

Phot. 354. Later stage: Glycerietum denser, *Armeria* hummocks more numerous and larger. Primary depression pan in foreground, narrow channel crossing the picture from left to right. *R. H. Yapp* (1917).

GLYCERIETUM AND ARMERIETUM

PLATE 142

Phot. 355. General salt marsh community: *Glyceria maritima*, *Obione portulacoides* and *Limonium vulgare* (left), *Limonium reticulatum* (right), *Armeria maritima* (flower heads on extreme left), *Plantago maritima*, *Spergularia marginata* (centre). Holme-next-the-sea, Norfolk.
R. H. Compton.

Phot. 356. *Obione* fringing raised edge of deeply cut channel S. Hastings.

Phot. 357. Obionetum succeeding Glycerietum. Near Gedney Drove End, The Wash. R. H. Yapp.

Phot. 358. Old salt marsh, largely covered with Obionetum, and with deeply cut channels.

General Salt-marsh Community

written when but few British salt marshes had been carefully surveyed, a "general salt-marsh association",[1] in which no species was generally dominant, was described as occupying the middle zones of the formation; but most subsequent workers have recognised specific dominants of all the zones. Chapman (1934, p. 111), however, describes a "general salt-marsh association" at Scolt Head island with the six species listed below "more or less equally important":

Limonium vulgare
Armeria maritima
Glyceria maritima

Spergularia salina (media)
Triglochin maritimum
Obione portulacoides

He thinks that the co-dominance of these plants in the general salt marsh is due to the separation in space of their root systems, which reach different depths (Figs. 160, 161), and to the difference in time of their flowering seasons. It is clear that the "general salt-marsh" community corresponds in a general way with the Limonietum and the Armerietum, but that in many if not most marshes the sea lavender and thrift (in conjunction with *Glyceria*) are definitely dominant in zones of slightly different level. Pl. 142, phot. 355, shows this community at Holme, Norfolk.

Salisbury (in Oliver and Salisbury, 1913) gives the following representative list for the "saltings" (general salt-marsh community) on the south side of "Blakeney Harbour" (estuary of the River Glaven):

Armeria maritima	a	Spergularia salina	f
Limonium vulgare	a	Salicornia disarticulata	o
Glyceria maritima	a	Suaeda maritima	r
Salicornia ramosissima	a	Cochlearia officinalis	r
Glaux maritima	f		

(4a) **Obionetum.** *Obione* (*Atriplex*) *portulacoides* is a mealy greyish white undershrub which plays an important part in many salt marshes, especially on the eastern coast of England.[1] The common large bushy form of the plant (*O. portulacoides* (L.) Gaertn. var. *latifolia* (Gussone) Chapman) has a wide vertical range on the salt marshes of north Norfolk (6·66–10·17 ft. above O.D. at Scolt Head, Chapman, 1937); but at the lower levels it is especially dominant on the banks of the tidal channels which become established quite early in the development of the marsh. Here the *Obione* occupies a relatively well-drained soil, and it is probably this factor which mainly determines its habitat. The water channels deepen considerably as the marsh develops and eventually form deep creeks up which the high tides run to gain access to the upper levels of the marsh. When the creek is full the silt-bearing water pours over the sides on to the surrounding flats. A portion of the silt is strained off by the fringing vegetation, consisting largely of *Obione*, and thus raises the level of the bank above that of the adjacent marsh flats.[2] In this way the bank becomes exceptionally well drained and *Obione* increases in luxuriance, forming a conspicuous greyish

[1] Pl. 142.   [2] Phot. 356.

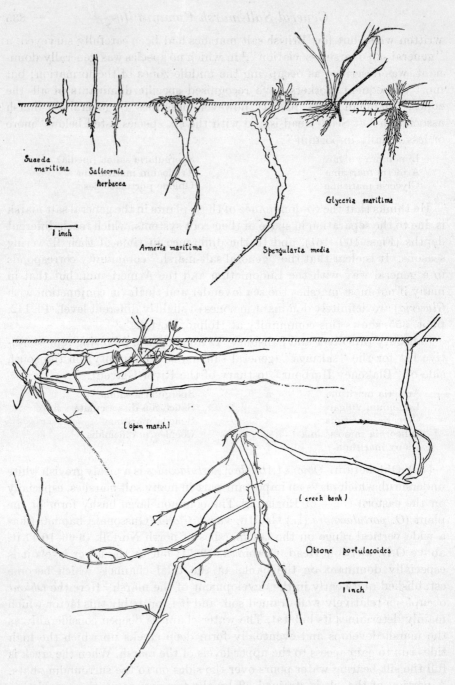

FIG. 160. ROOT SYSTEMS OF SALT MARSH PLANTS

Suaeda maritima and *Salicornia herbacea* (shallow rooted): *Glyceria maritima*, *Armeria maritima*, *Spergularia media* (medium depths), *Obione portulacoides* (different depths of penetration). *Obione* shows by far the deepest roots in the well aerated habitat of the creek bank. Plover Marsh, Scolt Head (Norfolk). From Chapman, 1934.

white band by which the courses of the creeks can be traced from a distance.[1] Where the plant is growing with maximum luxuriance there is little or no room for other species, and from the creek banks it spreads over the drier marshes, forming a dense carpet one to two feet in depth and almost completely obliterating their former vegetation.[2] The almost ubiquitous

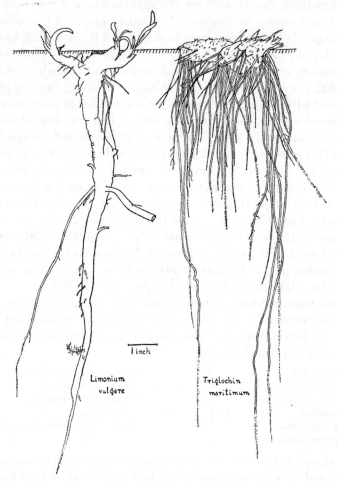

FIG. 161. ROOT SYSTEMS OF *LIMONIUM VULGARE* AND *TRIGLOCHIN MARITIMUM*

Plover Marsh, Scolt Head (Norfolk). From Chapman, 1934.

Glyceria maritima is said by Salisbury (Oliver and Salisbury, 1913) to be the only species which grows successfully in this thick Obionetum.

Obione also becomes dominant on areas where blown sand has covered the salt marsh ("sandy Obionetum" of Chapman, 1934). Here it grows in slightly raised tussocks owing to its action in arresting and binding the

[1] Pl. 142, phot. 356,   [2] Pl. 142, phot. 357–8.

sand. The small prostrate form (*O. portulacoides* var. *parvifolia* Rouy) with very much smaller leaves is locally abundant on sandy salt marshes in Norfolk (Chapman, 1937). The two varieties maintain their character when transplanted into each other's habitat.

(4*b*) **Suaedetum fruticosae.** *Suaeda fruticosa*, a Mediterranean species, which occurs (only locally but very abundantly where it does occur) on the south and east coasts of England, is another plant with a considerable vertical range. It forms dark green bushes 2 or 3 ft. high, and, like *Obione*, seems to depend on the drainage factor. It is specially associated with shingle beaches, where these overlie salt marsh, and the belts of dark green *Suaeda* mark the boundary between shingle and marsh (see Chapter XLII).

Associated with *Suaeda fruticosa* on the north Norfolk coast are two other Mediterranean species, *Limonium reticulatum* and *Frankenia laevis*.[1] They are also abundant in the "lows" or hollows of the sand-dune system (often floored on this coast with mixed sand and shingle brought in by exceptionally high tides) and on flat tracts of redistributed sandy shingle lying between the dunes and the salt marshes. They occur too in some of the upper marsh communities such as Chapman's "sandy Obionetum" at Scolt Head. *Limonium binervosum* (*bellidifolium*) is also found on the shingly sand of the lows and flats, but it is more abundant at a somewhat higher level on the sides of the old lateral shingle spits or "hooks" separating the marshes of the Marams at Blakeney, in a zone even more rarely visited by the high spring tides (see Chapter XLII, p. 889).

(5) **Festucetum rubrae.** Where the soil is very sandy (Dovey, Severn, Holme) this community occupies a zone above the "general salt marsh". At Holme *Festuca* invades the upper edge of the Armerietum (Marsh, 1915). The following is a list of species:

Festuca rubra	d	Spergularia marginata	o–f
Juncus gerardi	ld	Armeria maritima	o
Plantago coronopus	f	Pholiurus (Lepturus)	
P. maritima	f	filiformis	o
Glaux maritima	f–a	Triglochin maritimum	o
Agropyron pungens	o		

The first three of these species and *Agropyron pungens* are confined to the upper zones of the salt marsh and *Glaux* is characteristic of the higher levels, though it occurs in some marshes lower down. *Agrostis stolonifera* is locally dominant or co-dominant at the higher levels of the Festucetum in the Dovey marshes, and *Hordeum nodosum* occurs in this community in the Severn estuary. In the close turf of the upper Festucetum of the Dovey

Mosses and Algae — marshes two mosses are recorded—*Pottia heimii* and *Amblystegium serpens* var. *salinum*.

In the lower part of the Festucetum of the Dovey estuary *Rhizoclonium* is more successful in the summer, *Vaucheria*, accompanied by a number of

[1] Pl. 143, phot. 360.

blue-green algae and diatoms, in the wetter winter months. In the late summer, below the green algae, there is often a tough stratum of Cyanophyceae, and the drier the conditions the more prominent these become in relation to the green forms. The algal flora of the soil of the higher Festucetum most nearly approaches that of non-saline soils. In the summer Cyanophyceae predominate and are more abundant than in the lower zones, *Rivularia* and *Nostoc* being the most important, the latter in the wetter months (Carter, 1933).

(6) **Juncetum maritimi.** Where the upper edge of salt marsh is relatively undisturbed it is commonly occupied by a community in which the sea-rush (*Juncus maritimus*) is dominant.[1] This grows much taller than any of the accompanying species, and its dominance tends to destroy the close turf of the sea meadow, providing more open soil and maintaining moist air between its shoots. This often leads to the reappearance of more moisture-loving plants from the lower zones, such, for example, as *Cochlearia officinalis*, *Plantago maritima*, *Aster tripolium* and even *Salicornia* spp., and to an increase in the number and variety of the algae.

Moister conditions

At Blakeney the Juncetum maritimi contains:

Armeria maritima	a	Limonium vulgare	f
Glyceria maritima	a	Plantago maritima	f
Glaux maritima	f	Spergularia salina	f
Salicornia spp.	f		

In the Dovey marshes, where it follows the Festucetum:

Festuca rubra	a	Glaux maritima	f
Agrostis stolonifera	a	Plantago maritima	f
Armeria maritima	f	Spergularia salina	o
Cochlearia officinalis	f	Aster tripolium	o

The algal vegetation of the Dovey Juncetum is more stable than in the lower zones of the marsh (Carter, 1932) and includes *Vaucheria* and *Rhizoclonium* among the green algae, *Lyngbya aestuarii*, *Lyngbya* spp. and *Microcoleus chthonoplastes* among the blue-green: of brown seaweeds there are *Sphacelaria* and *Pelvetia canaliculata*; and of red *Catenella opuntia* and *Bostrychia scorpioides*. The last three large algae are very local in occurrence and prefer stretches of bare mud between the *Juncus* plants.

Algae

Transition to land vegetation. The upper edge of the salt marsh, only touched by the highest spring tides, is invaded by non-halophytic species of considerable variety—pasture plants on the drier soils, fresh-water marsh plants where the soil is wet but the water supply is mostly fresh. Among the latter *Oenanthe lachenalii* and *Samolus valerandi* are characteristic. *Phragmites communis* may also occur at this level when the salt content is low, as well as other species of *Juncus*, *Phalaris arundinacea*, *Ranunculus*

[1] Pl. 143, phot. 359.

flammula, etc. and submaritimes such as *Scirpus maritimus* and *S. tabernaemontani* (cf. Chapter XLIII).

Salt-marsh pasture. The middle and upper levels of salt marsh are very generally used as sheep pasture and are sometimes grazed by cattle. Where the sandy Glycerietum ((2a), p. 829), associated with *Suaeda maritima* on the drier higher lying areas, and with *Salicornia herbacea* on the wetter slightly lower lying flats, is extensive and is not covered by the neap tides, so that it is exposed for nearly a fortnight at a time as in the Bouche d'Erquy (p. 830), this is the main area on which the sheep depend. In other less sandy marshes, as at Blakeney and Scolt Head, *Glyceria maritima* becomes important and often co-dominant in later stages of the succession, as in the Limonietum and the Armerietum; and here the so-called "sea meadow" is developed, interrupted only by the Obionetum, on the upper levels of the marsh. In very sandy marshes *Glyceria*, as we have seen, is replaced by *Festuca rubra* in this upper zone. The short close turf of the sea meadow, whether Glycerietum, Armerietum or Festucetum, is determined largely by the grazing.

Drainage channels. The surface of a primary salt marsh is never quite level, because of the unequal growth and unequal powers of trapping silt of the different marsh-building species. Thus there is a local formation of hummocks by the plants which hold silt in greatest quantity (Pl. 141, phot. 353) and the flow of the tide tends to take winding courses round the largest hummocks. On the one hand the hummocks extend horizontally as well as vertically and often coalesce into ridges and continuous raised areas, while their growth in height slows down: this process tends to the development of flat uniform stretches of marsh carpet. On the other hand the channels followed by the tide between the hummocks are at first shallow and shifting, but they are gradually deepened by the flowing water. A high spring tide rushes up the channels with considerable force, especially if backed by a strong wind, but it is mainly at the ebb of such a tide that erosion of the channels occurs. The growth of the marginal plants increases their power of trapping silt and thus the banks of a channel rise while its bed is deepened by the scour of the water. In this way regular drainage systems showing both erosion and deposition are established in the older marsh, and these are very similar to river systems with main streams and tributaries, though the action of the water is intermittent and both up and down the channels. The greater the tidal range and the vertical height of the marsh the stronger the scour and the deeper the channels are cut. The edges of the deep channels in mature marsh are raised above the general level owing to the trapping of silt by the larger marsh plants, such as *Obione*, which line the crests of the banks (Pl. 142, phot. 356).

Owing to various causes the flow of water up and down a particular channel may become insufficient to keep its bed scoured and clear of

PLATE 143

Phot. 359. Juncetum maritimi abutting on sward of *Festuca rubra* with *Armeria maritima* in flower. Dovey salt marshes. *R. H. Yapp* (1917).

Phot. 360. Society of *Frankenia laevis* in mature salt marsh. Blakeney Point. *J. Massart.*

UPPER SALT MARSH

PLATE 144

Phot. 362. Formation of pans from channel. *R. H. Yapp* (1917).

Phot. 363. Old pan drained into channel and invaded by *Glyceria* at upper end. *R. H. Yapp* (1917).

Phot. 365. Formation of subterranean channel by bridging. Drained and revegetated pan behind. *R. H. Yapp* (1917).

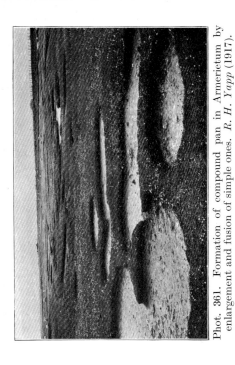

Phot. 361. Formation of compound pan in Armerietum by enlargement and fusion of simple ones. *R. H. Yapp* (1917).

Phot. 364. General view of channels and pans in salt marsh turf. Dovey estuary. *R. H. Yapp* (1917).

vegetation. The channel is then invaded by the marsh plants, and sometimes bridged across the top, for example by stolons of *Glyceria maritima*. Thus parts of such a dormant or semi-dormant channel may "grow up" or be "roofed in" by silt-trapping marsh vegetation. The roofing in of a channel which still conducts a certain amount of water may result in its becoming "subterranean" for parts of its course, opening to the surface at intervals (Pl. 144, phot. 365; cf. the similar phenomena in raised bog, Chapter xxxv).

Formation of "pans". If a group of hummocks formed in a primary marsh coalesce so as to surround an area which then forms a relative depression (Pl. 141, phot. 354) the water is hindered from draining away at the ebb, and a "primary depression pan" is formed in which water remains after the ebb. This water may slowly percolate away if the substratum is sufficiently permeable, or it may remain for a long time, from one spring tide cycle to the next, and slowly evaporate, the water level rising of course if there is heavy rainfall. Such pans are normally bare of the ordinary salt marsh vegetation, probably because of their stagnancy and the great variations of salt content due to alternating evaporation and precipitation. The sequence of conditions in stagnant pans and their algal vegetation have not, however, been adequately investigated.

The pans formed in the earlier phases of salt marsh development, e.g. in the Salicornietum (Pl. 138, phot. 346 and Pl. 141, phot. 353) are often transitory, but some of them persist into the later stages, and these are generally remarkably stable and permanent, e.g. in the *Armeria* sward. Pans may also be formed by the blocking up of a channel at intervals, when the scour has become insufficient to keep the channel open, and vegetation has extended across it. "Channel pans" may be recognised by the fact that they occur in series, marking the course of the former channel (Pl. 144, phots. 362, 364).

If a pan is secondarily drained into a drainage channel, vegetation may invade its surface (Pl. 144, phot. 363), and ultimately it may be obliterated. Very often however this invasion is partial and smaller "residual pans" are formed on the floor of the original one (Pl. 144, phot. 365). Erosion of the sides of a pan may occur as a result of swirling water entering the pan at a high spring tide, and in this way the turf intervening between neighbouring pans may be cut through and compound pans formed (Pl. 144, phot. 361).

For a detailed description and discussion of drainage channels and pans in the salt marshes of the Dovey estuary the paper by Yapp, Johns and Jones (1917) should be consulted. On some salt marshes pans may be formed in other ways than those described.

Algal communities of the salt marsh. Chapman (1938) recognises a number of algal communities of the salt marsh distinct from those dominated by flowering plants. These, he says, are not in general correlated with marsh level, and the factors which determine their distribution are still largely unknown. In view of the important part played by algae in the salt-marsh formation it seems worth while to reproduce Chapman's list here:

I. General "association" of Chlorophyceae:
  (a) Low sandy community.
  (b) Sandy community.
  (c) Muddy community.
II. Marginal Diatom community.
III. Marginal community of Cyanophyceae.
IV. Vernal *Ulothrix* community.
V. *Enteromorpha minima* community.
VI. Community of gelatinous Cyanophyceae.
VII. Community of filamentous Diatoms.
VIII. Autumn community of Cyanophyceae.
IX. *Phormidium autumnale* community.
X. *Rivularia-Phaeococcus* community.
XI. Dwarf *Pelvetia* community.
XII. *Catenella-Bostrychia* community.
XIII. *Pelvetia limicola* community.
XIV. *Enteromorpha clathrata* community.
XV. *Fucus limicola* community.
XVI. Pan community.

Some of these were previously described by Carter (1933). Cf. Figs. 154, 155. For details the reader is referred to Chapman's paper.

REFERENCES

CARTER, NELLIE. A comparative study of the alga flora of two salt marshes. Part I. *J. Ecol.* **20**, 341–70. 1932. Part II. **21**, 128–208. Part III. 385–403. 1933.

CHAPMAN, V. J. "Ecology" in Steers, *Scolt Head Island*. Cambridge, 1934.

CHAPMAN, V. J. A note upon *Obione portulacoides* (L.) Gaertn. *Ann. Bot., Lond.*, N.S. **1**. 1937.

CHAPMAN, V. J. Studies in Salt Marsh Ecology. *J. Ecol.* **26**, 144–79. 1938.

ELLIOTT, JESSIE S. B. The soil fungi of the Dovey salt marshes. *Ann. Appl. Biol.* **17**. 1930.

HESLOP HARRISON, J. W. A survey of the lower Tees marshes and of the reclaimed areas adjoining them. *Trans. Nat. Hist. Soc. Northumb.*, N.S. **5**, pt. I. 1918.

MARSH, A. S. The maritime ecology of Holme-next-the-Sea, Norfolk. *J. Ecol.* **3**, 65–93. 1915.

MORSS, W. L. The plant colonisation of Merse lands in the estuary of the River Nith. *J. Ecol.* **15**, 310–43. 1927.

OLIVER, F. W. Blakeney Point Reports, 1913–29. *Trans. Norfolk Norw. Nat. Soc.*

OLIVER, F. W. *Spartina* problems. *Ann. Appl. Biol.* **7**. 1920.

OLIVER, F. W. *Spartina townsendii*, its mode of establishment, economic uses and taxonomic status. *J. Ecol.* **13**, 74–91. 1925.

OLIVER, F. W. and SALISBURY, E. J. The Topography and Vegetation of the National Trust Reserve known as Blakeney Point, Norfolk. *Trans. Norfolk Norw. Nat. Soc.* **9**, pt. 4. 1913.

PHILIP, G. An enalid plant association in the Humber Estuary. *J. Ecol.* **24**, 205–19. 1936.

PRIESTLEY, J. H. The pelophilous formation of the left bank of the Severn Estuary. *Bristol Nat. Proc.*, 4th Series, **3**, pt. 1. 1911.

RICHARDS, F. J. The salt marshes of the Dovey Estuary. IV. The rates of vertical accretion, horizontal extension, and scarp erosion. *Ann. Bot., Lond.*, **48**, 1934.

WIEHE, P. O. A quantitative study of the influence of tide upon populations of *Salicornia europea* [*herbacea*]. *J. Ecol.* **23**, 323–32. 1935.

YAPP, R. H., JOHNS, D. and JONES, O. T. The salt marshes of the Dovey Estuary. Part II. The salt marshes (Yapp and Johns). *J. Ecol.* **5**, 65–103. 1917.

Chapter XLI
THE FORESHORE COMMUNITIES. COASTAL SAND-DUNE VEGETATION
(2) THE FORESHORE COMMUNITIES

On open sea shores undisturbed by man, and where the substratum is not composed of rock, a scattered vegetation of flowering plants is often met with along the zone barely reached by the highest spring tides. The existence of these littoral communities depends on the maintenance for a time of conditions that are not too constantly and violently disturbed. The substratum must be reasonably stable, not subject to erosion by wind or waves and not overwhelmed by sand or shingle. Such conditions are realised on the upper limit of a fringing shingle beach (p. 868) where shingle is sparsely scattered over a clayey or loamy substratum; and again at the seaward foot of sand dunes which are not growing very actively nor being eroded by wind or the sea. These habitats are of course *liable* to considerable disturbance, e.g. by specially high tides, and correspondingly the foreshore communities are essentially "migratory".

Tidal drift. The soil of the foreshore communities is fed by sea drift (organic debris largely composed of dead seaweeds) thrown up by exceptionally high tides, and this humus supply is important for the luxuriant growth of the plants. The soil also contains a considerable amount of sea salt, and the plants may fairly be reckoned as halophytes, some being characteristically succulent.

Characteristics. The foreshore communities are very open, consisting of more or less widely and irregularly spaced individuals, because the conditions are not stable enough for long periods to permit of the development of closed vegetation, and for the same reason they are inconstant. They are composed of scattered annuals, nearly all belonging to the families Chenopodiaceae, Cruciferae or Polygonaceae, with a few perennials (*Beta, Crambe, Mertensia*), which settle in this zone when and where they can, some favouring sandy and some muddy shores, disappearing and reappearing as the conditions fluctuate, and marking a true "tension line" between tidal and non-tidal habitats. The characteristic species are few, but they are sometimes reinforced, especially on shingly and muddy coasts, by annual weeds which can tolerate a certain amount of salt in the soil; and where the conditions are sufficiently stable, by perennials like *Beta maritima* and *Crambe maritima*.

Flora. The following are the most characteristic species of this community:

Atriplex glabriuscula	f	A. patula	o
A. hastata	f	Beta maritima	f
A. littoralis	f	Cochlearia spp.	o–f

Sand-dune Formation

		Sandy shores	
Mertensia maritima	r		
Polygonum littorale	o	Cakile maritima[1]	f
P. raii	o	Crambe maritima	r
Raphanus maritimus	o	Salsola kali[2]	f

This foreshore vegetation has never been given serious study, so that it is impossible to say more about the exact conditions of its existence, nor about the particular requirements of the individual species. On sandy shores where dunes are formed it immediately precedes, and is sometimes reckoned as the first stage of the dune succession (psammosere).

(3) THE SAND-DUNE FORMATION

Conditions of existence. The accumulations of blown sand known as sand dunes are not confined to sea coasts. On the contrary, by far the greatest dune areas of the world are the so-called "continental" sand dunes of desert and arid or semi-arid regions generally, while inland sandy regions in a relatively though not excessively dry climate may show smaller local dune areas. Coastal dunes are also formed on the shores of large bodies of fresh water such as the Great Lakes of North America. Wherever plants can establish themselves in mobile dune areas their life forms have a general resemblance, for in order to survive they must be able to cope with the sand which is continually blowing over them.

In this country sand dunes are almost confined to the sea coast where collectively they cover a considerable area, second only to that occupied by the salt marshes; and it is mainly because of this that their plant covering is included in maritime vegetation, for the plants which grow on sand dunes are not halophytes and will not, for the most part, endure immersion in salt water.

The supply of sand for the formation of coastal dunes comes from shoals formed offshore on a flat coast and exposed at low tide, and from gently sloping sandy beaches over which the tide advances and recedes for a long distance. At low tide great areas of sand are thus laid bare, and as it dries the grains are driven landwards by onshore winds. At any small obstacle the stream of sand grains is checked and accumulated round it, both on the windward and still more on the leeward side,[3] for there the air is relatively quiet, and the lighter grains are carried by eddies up to the top of the miniature dune, and then slide down the longer and gentler leeward slope. On a fixed obstacle, such as a large stone or a lump of stranded seaweed, the sandhill can grow no higher than the top of the obstacle. Thus a brushwood fence planted across a sand-laden windway will accumulate sand up to the height of the fence. But a growing plant can push its shoots upwards and sideways as it becomes covered with sand (though different species vary greatly in this power), and the embryo dune thus grows in height and width until it is arrested from other causes.

Effect of growing plants

[1] Pl. 145, phots. 366–8. [2] Pl. 146, phot. 371.
[3] Pl. 148, phot. 379; Pl. 149, phot. 381.

It is in this way that large and lofty systems of dunes are formed by plants which have exceptional powers of pushing up their shoots and of continually forming fresh roots in the moist sand just below the surface. Species which have only limited powers of such adjustment form dunes of

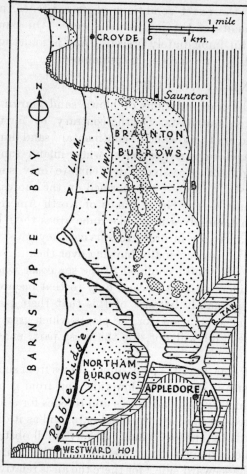

Fig. 162. Sketch Map and Section of Braunton Burrows (North Devon)

Sand-covered area dotted, height of dunes indicated by increasing closeness of dots. Horizontal lines are added to show the mud (mixed with sand) bordering the estuaries and bearing saltmarsh in places. Cultivated land represented by vertical ruling.

1. Flat sand covered at high tide. 2. Foredunes. 3. Very mobile dunes rising to 50 ft. 4. Brackish "slacks". 5. Mobile dunes rising to 100 ft. 6. Second line of slacks. 7 and 8. Low stable sandhills.

Horizontal scale of section twice that of map: vertical scale five times horizontal scale. After Watson, 1918.

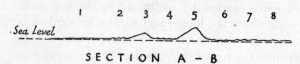

little height. Sand-dune complexes are formed in bays facing prevailing onshore winds (Fig. 162) and in other situations where blowing sand is trapped.

Parallel dune ranges. If a new series of dunes arises on the seaward side of a range already formed the latter is more or less protected from the wind, the supply of blown sand is cut off so that it ceases to grow in height,

PLATE 145

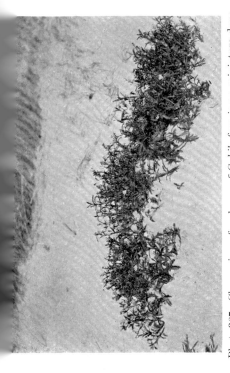

Phot. 367. Closer view of a clump of *Cakile* forming a miniature dune.

Phot. 369. *Honckenya peploides* (with isolated *Ammophila* shoots forming miniature dunes in front of the main ridge). Holme-next-the-Sea. *R. H. Compton.*

Phot. 366. *Cakile maritima* in flower in front of low *Ammophila* dunes. Near Rye, Sussex. November 1904.

Phot. 368. Zonation on seaward face of dunes near Hunstanton, Norfolk. From front to back: (1) sand mixed with shingle, highest limit of ordinary spring tides; (2) foreshore community of *Cakile maritima*, *Salsola kali*, *Honckenya peploides*; (3) foredune, Agropyretum juncei; (4) Ammophiletum with *Eryngium maritimum*. *S. Mangham.*

FORESHORE AND FOREDUNES

PLATE 146

Phot. 370. Tidal drift and low dunes dominated by *Agropyron junceum*. Wind-eroded *Ammophila* dunes behind. Southport, Lancs. *W. Ball.*

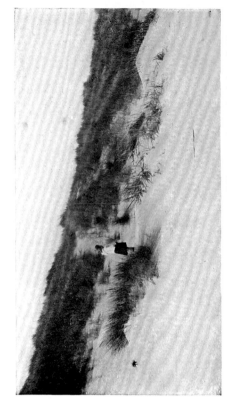

Phot. 371. Foreshore community of *Salsola kali*. Sparse Agropyretum in front. *Ammophila* dunes behind. Paul Graebner and F. W. Oliver. Near Southport, Lancs. *Mrs Cowles* (1911).

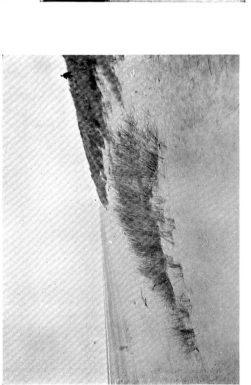

Phot. 372. Another part of the shore near Southport. *Ammophila* is here the first colonist. *Mrs Cowles.*

Phot. 373. Foredunes formed by *Ammophila arenaria* on the left and *Elymus arenarius* on the right. Hemsby, Norfolk.

FORESHORE AND FOREDUNES

and thus a series of parallel dune ranges is often formed.¹ The protected older dunes gradually become stabilised and covered with a continuous carpet of vegetation, interrupted only where strong winds gain access and form the so-called "blow-outs" (Pl. 153).

Sand-dune vegetation shows a regular successional series of communities (psammosere) beginning with those forming and inhabiting the embryo dunes on the sea shore and ending with the stabilised vegetation on the landward side.

MOBILE DUNE COMMUNITIES

(a) **Sea couchgrass consocies** (Agropyretum juncei). This community, which is not represented in all dune systems, is dominated by *Agropyron junceum*, a plant of somewhat similar habit to that of the common couch-grass or "twitch" (*Agropyron repens*) which is such a pestilent weed of clayey arable land. It can withstand short immersions in seawater and is thus able to grow within reach of the high spring tides; and its extensively creeping rhizomes ramify through the sand and send up new aerial shoots through the fresh sand blown over the plant.

Foredunes. The sea couchgrass is peculiarly well fitted to act as a pioneer in accumulating sand. Its underground runners penetrate deeply into the substratum, even shingle, and are thus protected from exposure by erosion, while the leaves are often prostrate and thus tend to prevent the removal by wind of accumulated sand. In this way *Agropyron* hummocks rapidly spread and in a few years a single seedling may form a dune 15–20 ft. across and 3 or 4 ft. high (Oliver, Blakeney Point Report, 1929). *Agropyron* cannot however grow indefinitely in height, and the dunes it forms, often called *foredunes*, are thus comparatively low (Pl. 145, phot. 368).

Associated species. Very commonly associated with the sea couchgrass is the succulent sea sandwort *Honckenya* (*Arenaria*) *peploides*, which itself may produce miniature dunes (Pl. 145, phot. 369); and since the low dunes are formed by these two species on the sandy foreshore just along the limit of high spring tides, representatives of the sandy facies of the foreshore community, such as *Cakile maritima* and *Salsola kali* (Pl. 145, phot. 366–8), as well as the littoral species of *Atriplex*, frequently occur between the miniature sand hills. *Cakile* and *Salsola*, although annuals, may form tiny dunes during their season of growth, and their persistent dead bodies hold the sand throughout the winter, thus assisting the establishment of *Ammophila* (see below) at the slightly raised level.

The first five species to appear on the foreshore of the "Far Point" at Blakeney, when the shingle bank had attained the level of the highest spring tides were *Salsola*, *Cakile*, *Agropyron*, *Honckenya* and *Ammophila*. Occasional plants of species more characteristic of the Ammophiletum, such as the horned poppy (*Glaucium flavum*), the sea holly (*Eryngium mariti-*

¹ Cf. Pl. 133, phot. 334.

mum) and the common sea spurge (*Euphorbia paralias*) may also be found on the larger and older *Agropyron* dunes. A rayed variety of the common groundsel (*Senecio vulgaris* var. *radiatus*) is frequently met with in the Agropyretum of the Lancashire dunes.

(*b*) **Marram grass consocies** (Ammophiletum arenariae).[1] This is the principal consocies of mobile sand dunes, and is almost the sole agent in the building of the main dune ranges. Seedlings of *Ammophila* establish themselves on sand or sandy shingle just above the reach of high tides.[2] Unlike *Agropyron junceum* the marram grass cannot endure immersion in salt water and is therefore unable to colonise sand reached by the sea. Furthermore, its underground runners spread by preference in pure sand rather than shingle, and in the young stage its leaves stand up like a shaving brush, so that the plant is more liable than *Agropyron* to destruction by erosion (Oliver, Blakeney Point Report, 1929).[3] On the other hand, once established in deep sand, it has far more vigorous powers of vertical and lateral extension and is able to grow up through many metres of sand provided this is not deposited too quickly. When a high dune has been partly removed by wind, as often happens, the exposed face of sand down to the base level, representing a vertical section of the dune, is seen to be completely penetrated by the rhizomes and roots of *Ammophila*, mostly dead, but those nearer the surface still living. The capacity to form fresh adventitious roots as the shoots grow up, at higher and higher levels in the moist layer of sand which is constantly maintained a little below the dried surface, is the power which enables any plant to dominate and increase the height of a dune. On the great sand dunes of the southern shore of Lake Michigan a certain number of trees and shrubs possess this power, while others do not, and it is the former alone whose living tops maintain themselves on the surface of dunes more than 100 ft. (*c*. 30 m.) high. Those which cannot root in this way are killed when they are covered by sand.

Limitations and powers of Ammophila

On many sandy shores the Agropyretum is absent, and here *Ammophila* forms the foredunes as well as the main ranges. Another grass which plays a similar role, but by no means so universally, is the sea lyme grass (*Elymus arenarius*). The two are seen forming foredunes side by side in Pl. 146, phot. 373. In some sandy bays of the west coast of Ireland, as at Dog's Bay in Connemara, *Ammophila* is absent, and here the dunes are all low and are entirely dominated by *Agropyron* (Pl. 154, phot. 397).

The speed with which new dunes are formed by *Agropyron* and *Ammophila* under favourable conditions is well illustrated by the development of the "Far Point" beyond Blakeney headland (cf. Fig. 170, p. 875). Based on the continuation of the main shingle spit westwards and the subsequent formation of numerous "hooks", sand dunes have developed and become

---

[1] Pls. 146–9.   [2] Pl. 148, phot. 377.
[3] The different habit of the two grasses is well seen in Pl. 147, phot. 375.

PLATE 147

Phot. 374. The Far Point in 1927, looking west. Early stage of development. Low *Agropyron* dunes in foreground accumulating sand on the shingle. *Ammophila* and *Agropyron* complexes in the distance and middle distance. *F. W. Oliver*.

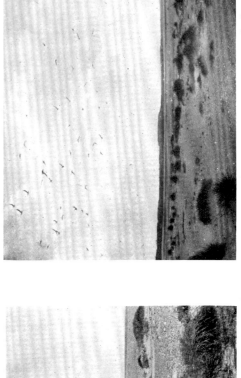

Phot. 376. Terns rising from their nests in foredunes. *F. W. Oliver* (1927).

Phot. 375. *Agropyron* and *Ammophila* recently covered by tide in foreground. Single *Ammophila* tufts in distance to the left. *F. W. Oliver* (1927).

FOREDUNES ON BLAKENEY FAR POINT

PLATE 148

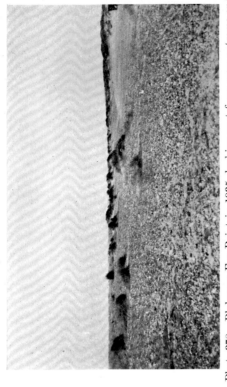

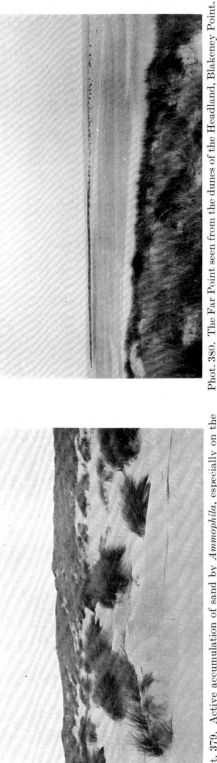

Phot. 377. Colonisation of sandy shingle by *Ammophila* seedlings. Blakeney, Norfolk. *F. W. Oliver* (1910).

Phot. 378. Blakeney Far Point in 1935, looking east from near extreme end. New scattered dunes on the sandy shingle in front. Consolidated dunes behind on the right.

Phot. 379. Active accumulation of sand by *Ammophila*, especially on the lee (right hand) side of each tuft, main dune ridge behind. Blakeney Point. *R. H. Compton* (1910).

Phot. 380. The Far Point seen from the dunes of the Headland, Blakeney Point. Consolidated Ammophiletum to the left, scattered foredunes to the right. 1935.

ACTIVE AMMOPHILETUM

consolidated on the sandy shingle of the spit and hooks in a very few years. In 1928 there were only a few scattered *Agropyron* and *Ammophila* dunes, whose development had been slow during the preceding years.[1] In 1928 (Oliver *in litt.*) *Ammophila* seeded very heavily and the seeds germinated everywhere, forming numerous "lawns" of seedlings in the following year. Since then development and consolidation have been extremely rapid, so that now (1937) there are many acres of continuous consolidated dunes[2] whose crests are 2·5 m. above the beach, in place of the few scattered pioneer dunes of 1927–8. New hook formation and new colonisation by the dune pioneers is still proceeding (Fig. 170, 1937).

Features of Ammophiletum. The marram grass community is never closed on the surface, the dominant forming separate tufts or clumps with dry loose sand between.[3] These interspaces are colonised by scattered individuals of a few other non-maritime species, such as ragwort (*Senecio jacobaea*),[4] the hawkweed *Hieracium umbellatum*, and the two thistles *Cirsium lanceolatum* and *C. arvense*. The sand fescue (*Festuca rubra* var. *arenaria*) is the first grass to associate itself with *Ammophila*, and other maritime species which are frequent on the edges of the Ammophiletum, though sometimes found on older dunes, are the two sea spurges, *Euphorbia paralias* (of fairly general occurrence in England except on the north-east coast) with the rarer *E. portlandica* (only on south-western and western coasts), the sea holly (*Eryngium maritimum*) and the sea convolvulus (*Calystegia soldanella*). The rare *Brassica monensis* occurs in this community in the Isle of Man (Moore, 1931).

While the marram grass is practically the sole agent in forming and holding together the structure of the main dunes it does little to consolidate the surface sand (Salisbury, 1913). Where it is present alone, or almost alone, the wind still freely removes the loose sand from between the tufts of the grass, so that unless the rate of supply of sand is at least equal to the rate of removal the dune will be eroded. Consolidation of the surface is often effected by the sand fescue (*Festuca rubra* var. *arenaria*). In relatively sheltered places the sand sedge (*Carex arenaria*) is also effective in binding the superficial sand with its long rhizomes, which grow just below the surface in remarkably straight lines whose course is marked by a series of aerial shoots bearing rosettes of leaves above and bunches of roots below. The sea spurges and the sea convolvulus also possess creeping stems below the surface of the sand and carry out a similar function, but on a much less extensive scale.

The physiognomy of the main mobile dune complex, with its steeply rolling, highly irregular crests and valleys, is very characteristic. Stretches of perfectly bare sand on the slopes and in the hollows alternate with clumps and tufts of marram, which occupy most of the crests. This is due

---

[1] Cf. Fig. 170, 1921. [2] Seen in the distance as a dark band in Pl. 148, phot. 380.
[3] Foreground of phot. 380, Pl. 148. [4] Pl. 154, phot. 394.

to the complex distribution of wind currents and eddies, as the streams of quickly moving air are deflected and broken up by the slopes and crests. Throughout this maze of elevations and depressions the same types of surface form are constantly repeated, following the laws of motion of the loose wind-driven sand, partially fixed here and there by the marram grass.

Mobile dunes are often called "white" or "yellow" dunes according to the prevailing colour of the fresh sand, in contrast to fixed "grey" dunes dominated by the lichens *Cladonia* and *Cetraria*.

Flora. The following list of species is given by Watson (1918) for the very mobile Ammophiletum of the west coasts of England:

	Ammophila arenaria	d	
Anagallis arvensis	o	Bryum argenteum	r
Cynoglossum officinale	r	Potentilla anserina	r
Erodium cicutarium	r	Sedum acre	o
Euphorbia paralias	o	Senecio jacobaea	o
Leontodon hispidus	r	Teucrium scorodonia	r
Nepeta hederacea	r	Viola canina	r

For the Ammophiletum of the headland at Blakeney Point, Norfolk, the following species are recorded by Salisbury (Oliver and Salisbury, 1913).

Ammophila arenaria d
Festuca rubra var. arenaria sd

Abundant:
  Cerastium semidecandrum
  C. tetrandrum
  Cirsium arvense
  Erophila verna (agg.)
  Myosotis collina
  Phleum arenarium
  Sedum acre
  Senecio jacobaea
  Stellaria pallida (boraeana)
  Tortula ruraliformis

Frequent:
  Anagallis arvensis
  Cirsium lanceolatum
  Erodium cicutarium
  Galium verum

Rare:
  Agropyron junceum
  Silene maritima

Very rare:
  Corynephorus canescens
  Elymus arenarius
  Hieracium pilosella
  Solanum dulcamara

From the north coast of the Isle of Man, where the supply of sand is very meagre, so that the dunes do not exceed a few feet in height, Moore (1931, p. 121) records the following species from the Ammophiletum:

	Ammophila arenaria	d	
Atriplex sp.		Hypochaeris radicata	o
Brassica monensis	r	Leontodon nudicaulis	o
Cakile maritima	o	Matricaria inodora	r
Calystegia soldanella	o	Ononis repens	f
Eryngium maritimum	la	Senecio jacobaea	f
Festuca rubra var. arenaria	f		

PLATE 149

Phot. 381. Wind-blown *Ammophila* accumulating sand, Blakeney Far Point. *F. W. Oliver* (1927).

Phot. 382. Great Sandy Low, Blakeney Point, holding water left by the last high spring tide. In the foreground and between the low and the sea are young *Ammophila* dune complexes. These have since formed consolidated dune ridges. Old consolidated dune ridge on the left behind. *F. W. Oliver* (1910).

ACTIVE AMMOPHILETUM

PLATE 150

Phot. 383. Moss carpet of fixed calcareous dune. *Tortula ruraliformis, Brachythecium albicans, Camptothecium lutescens, Hypnum cupressiforme, H. chrysophyllum: Ammophila arenaria, Eryngium maritimum, Leontodon nudicaulis, Centaurium littorale, Asperula cynanchica*, etc. Castle Gregory, Co. Kerry. R. J. L.

Phot. 384. Moss carpet of fixed calcareous dune near Rosapenna, Co. Donegal. *Camptothecium lutescens* and *Cylindrothecium concinnum* the most abundant species; also a rich phanerogamic flora (not shown) with *Ammophila* relicts. R. J. L.

FIXED DUNE WITH MOSS CARPET

Fixation of Dunes

Variability of flora. A comparison of these three lists illustrates the considerable variation of the accessory flora of "white" or "yellow" dunes in different localities. This variability is doubtless largely due to accidental seeding (at Blakeney partly from the arable mainland), and most of the species are extremely inconstant since they have little or no means of coping with superincumbent sand. Their successful germination and establishment depend no doubt on periods of damp weather when the surface of the sand is temporarily stable. From some areas of Ammophiletum other vascular plants are entirely absent.

In the very wet climate of the west of Ireland it would seem that almost any creeping plant which roots at the nodes can fix sand on a limited scale. *Ranunculus repens*, *Cirsium arvense* and *Potentilla anserina* especially have been found playing this role on the smaller dunes of Connemara, Mayo and Donegal.

Role of mosses in fixation.[1] Owing to their very limited rooting systems neither mosses nor lichens can establish themselves on very mobile soils such as loose incoherent sand. They are therefore absent from the typical Ammophiletum in the earlier stages of its development. But when a ridge of dunes bearing Ammophiletum finds some shelter from wind and from blowing sand by the growth of a new range on its seaward side, certain species of these groups appear. Thus Richards (1929), at Blakeney Point, found four species in this situation: *Tortula ruraliformis*, confined to sand dunes, and characteristic of places on the dunes where a moderate amount of fresh sand is being received; *Brachythecium albicans*, also chiefly on these comparatively young dunes, but persisting through the whole succession in fair abundance; *Ceratodon purpureus*; and *Bryum* sp., perhaps *B. capillare*. All four species form separate patches growing between the dense tufts of *Ammophila*. No lichens had yet appeared.

On the west coasts of England (Watson, 1918), besides the four species of moss occurring at Blakeney, the sand dune form of *Camptothecium lutescens* is abundant on young dunes as well as three species of *Bryum* (o). These indeed, are recorded by Watson from the Agropyretum juncei, which more typically has no bryophytes at all; but if the development of the foredunes is considerable and sheltered areas are produced the conditions would be much the same as in similar situations on the main dunes. The occurrence of the calcicolous *Camptothecium* and the greater variety of the moss flora are doubtless due to the fact that in some of the areas described the sand is largely composed of comminuted seashells. On the highly calcareous dunes of the west of Ireland, also, *Camptothecium* is very abundant and *Hypnum cupressiforme* and *H. chrysophyllum* are frequent (Pl. 150).

Nature of dune sand. Owing to the varying nature of the sand which composes them dunes differ very much in their original content of calcium

[1] Pl. 150; Pl. 153, phot. 393; Pl. 154, phot. 394.

carbonate. The two extreme cases are those composed almost entirely of minute shells or comminuted fragments of larger shells, nullipores and corallines, and those which are composed almost entirely of quartz grains, more or less coloured by iron salts. Typically these form "white" and "yellow" dunes respectively. Every grade of mixture of quartz grains and calcareous shells or fragments may occur. The effect is seen particularly in the flora of the fixed dunes—where the sand contains much lime the vegetation is rich in calcicolous species.

<small>White and yellow dunes</small>

FIXED DUNE COMMUNITIES

As the surface of sand becomes gradually consolidated and less mobile by the action of the various plants which settle down in the relative protection afforded by the marram grass, and on the landward side by the dune crests themselves, many other species colonise the soil and eventually establish a continuous carpet of vegetation. Thus the mobile dunes gradually pass, with protection from wind and fresh blowing sand, into fixed dunes.

Variability of vegetation. The plants which actually make up fixed dune vegetation are extremely varied, and form, according to geographical situation, climatic region, and various physiographic, edaphic, and particularly biotic factors, a considerable series of different communities. Thus sand dunes in most forest climates eventually develop forest; in western Europe, including parts of the British Isles, often *Calluna*-heath; where they are used for grazing or as rabbit warrens, grassland, often bearing local scrub.

In the British Isles natural forest communities do not occur on fixed dunes, though plantations of various trees in suitable situations, especially pines, and particularly the Austrian pine, *P. nigra* var. *austriaca*, are often quite successful. The failure to develop forest may be partly due to violent coastal winds, but is mainly because of the paucity of suitable seed parents in the neighbourhood, alluded to in earlier chapters in connexion with the widespread lack of spontaneous forest establishment, and this is especially marked near the coast. The development of Callunetum on dunes has been recorded in eastern Scotland (W. G. Smith, 1903), on Walney Island off the coast of north Lancashire (Pearsall, 1934), and on the south coast of England, in Dorset (Good, 1935),[1] but the stages have not been closely studied.

Most of the British fixed dunes are occupied by some form of grass or mixed grass and herb vegetation, with scrub frequently well developed here and there. Such dunes have either been left as "wasteland", almost entirely deserted, or are more commonly used as rabbit warren or for golf links.

The consolidated sand of the low dunes of the Ayreland of Bride at the

[1] See Pl. 126, phot. 309.

northern end of the Isle of Man bears a turf in which the marram grass is still abundant, and the majority of species of the Ammophiletum are still present, but have been joined by numerous others (Moore, 1931, p. 123):

Agrostis spp.	f	Hypochaeris radicata	o
Ammophila arenaria	a	Jasione montana	o
Brassica monensis	r	Lotus corniculatus	f
Cakile maritima	r	Ononis repens	f
Carduus tenuiflorus	o	Orchis maculata	r
Carex arenaria	la	Rosa spinosissima	l
Cerastium semidedandrum	o	Sedum anglicum	a
Erodium cicutarium	o	S. acre	o
E. maritimum	o	Senecio jacobaea	o
Eryngium maritimum	o	Silene maritima	o
Festuca rubra var. arenaria	a	Thymus serpyllum	f
Galium verum var. maritimum	f	Viola canina var. ericetorum	o
Hieracium pilosella	o		

The consolidation is mostly brought about by *Carex, Festuca, Thymus* and *Lotus*.

From the dunes at the north end of Walney Island, Pearsall (1934) records the following species from the fixed dune community:

Ammophila arenaria	a	Lotus corniculatus	f, la
Calystegia soldanella	f	Ononis repens	la
Festuca rubra var. arenaria	a	Rosa spinosissima	la
Galium verum	f, la	Salix repens (hollows)	la
Geranium sanguineum	f, la	Thalictrum dunense	la

The *p*H of the soil is between 6 and 6·5.

Mosses and lichens. In the formation of closed communities on the sand surfaces, lichens and mosses usually play the most important part, but not before the loose sand surface has been to some extent stabilised by the higher plants or protected by the formation of new dunes on the seaward side. Among lichens, species of *Peltigera* and *Cladonia* are prominent, and it is the abundance of the latter genus which has given the name of

Grey dunes   grey dunes to certain fixed dunes, as contrasted with the mobile white or yellow dunes. Of mosses *Tortula ruraliformis* is the most important pioneer, often forming luxuriant carpets on protected patches of bare sand in the Ammophiletum itself (Salisbury in Oliver and Salisbury, 1913).

Watson (1918) gives the following list of mosses and lichens, for the less mobile dunes, still dominated by *Ammophila*, of the west coasts of England:

Mosses

Tortula ruraliformis	d	Bryum argenteum	f
T. ruralis	o	B. caespiticium	o
Camptothecium lutescens (sand dune form)	a	B. inclinatum	f
		B. pendulum	f
Barbula convoluta	f	Trichostomum crispulum	f
Brachythecium albicans	a	T. flavovirens	f
Ceratodon purpureus	f		

Lichens

Peltigera canina	a	P. spuria	o
P. rufescens	a	Collema crispum	o

For "somewhat stable" and "almost fixed" sandhills he gives the following lists:

Mosses

Tortula ruraliformis	d	Camptothecium lutescens	a–d
Acaulon muticum	o	Dicranella heteromalla	o
Amblystegium serpens	o	Dicranum scoparium	o
Barbula convoluta	a	Ditrichum flexicaule	o
B. fallax	a	Encalypta rhabdocarpa	r
var. brevifolia	a	E. streptocarpa	r
B. gracilis	o	Eurhynchium confertum	o
B. hornschuchiana	o	E. megapolitanum	r
B. rubella	a	Hylocomium splendens	o
B. unguiculata	o	var. gracilius	o
B. vincalis	o	H. squarrosum	f
Brachythecium albicans	a	Hypnum cupressiforme	a
B. glareosum	o	var. ericetorum	a
B. rutabulum	f	var. tectorum	o
Bryum argenteum	f	Pleuridium alternifolium	o
B. atropurpureum	r	Swartzia inclinata	r
B. caespiticium	o	Thuidium abietinum	r
B. capillare	o	T. filiberti	r
B. donianum	r	Tortula ruralis	r
B. inclinatum	f	T. subulata	o
B. pendulum	a	Trichostomum fragile	o
B. roseum	o	T. mutabile	o
Ceratodon conicus	o	var. cophocarpum	o
C. purpureus	a	var. littorale	o
Climacium dendroides			
f. depauperata	o		

Liverworts

Frullania dilatata	r	Scapania aspera var.	
Lophocolea bidentata	o	inermis	o
L. cuspidata	f	S. aequiloba	–
Ptilidium ciliare	r		

Lichens

Bacidia muscorum	o	L. scotinum var. sinuatum	a
Biatorina caeruleonigricans	o	L. tenuissimum	r
Bilimbia sabuletorum	o	Peltigera canina	a
Cladonia furcata	a–d	P. horizontalis	r
C. fimbriata	o	P. rufescens	a
C. pungens	f	P. spuria	o
f. foliosa	f	Ramalina farinacea	o
Collema ceranoides	o	Rhizocarpon petraeum	
C. cheileum	r	(stones)	r
C. crispum	f	Squamaria crassa	r
C. pulposum	o	Urceolaria bryophila	f
Evernia prunastri	o	f. lichenicola	f
Leptogium lacerum	a	U. scruposa	f
pulvinatum	o	Usnea hirta	o

Phases of succession. The succession from yellow to grey dunes at Blakeney Point is marked, according to Richards (1929) by the enlargement and fusion of the separate patches of the four mosses mentioned (p. 851) as colonising protected areas of the Ammophiletum (*Tortula ruraliformis, Brachythecium albicans, Ceratodon purpureus* and *Bryum* spp.) into wide carpets, and by the appearance of tufts of *Hypnum cupressiforme* var. *tectorum*. At the same time the following lichens appear:

Cladonia fimbriata	Parmelia caperata
C. furcata	P. physodes
C. pyxidata	Peltigera polydactyla
Evernia prunastri	P. rufescens
var. stictoceras	

Nearly all these lichens are represented by young thalli growing in moss tufts, and "it is probably safe to say that no significant colonisation of the dunes by lichens could take place without the aid of mosses".

On the typical grey dunes the mosses are surpassed both in number of species and in abundance by the lichens. The most abundant moss is *Hypnum cupressiforme* var. *tectorum*, forming wide, golden-brown patches which tend to break up in the middle after reaching a certain size, probably due to the scratching of rabbits. *Brachythecium albicans* and especially *Tortula ruraliformis* are still abundant, and *Ceratodon* remains fairly frequent. Some of the lichens have now become very abundant, especially *Cladonia furcata*; two new species, *Cladonia foliacea* and *Cetraria aculeata*, appear and become conspicuous, and later *Cladonia sylvatica* joins them. In some places, indeed, mosses are almost entirely ousted by lichens, among which the large circular thalli of *Parmelia physodes* and *P. saxatilis* are conspicuous.

Flora of typical grey dune. The following list (Richards, 1929) illustrates the cryptogamic flora of a particular well-established grey dune at Blakeney Point:

Bryophytes

Brachythecium albicans	r	Dicranum scoparium var.	
Bryum sp.	o	orthophyllum	la
Ceratodon purpureus	lf	Hypnum cupressiforme var.	
Cephaloziella starkii	vr	tectorum	o–f
(round the roots of an		Tortula ruraliformis	la
Ammophila tuft)			

Lichens

Cladonia fimbriata	vr	Lecidea uliginosa	r
C. floerkeana	la	Parmelia physodes	lf
C. foliacea	la	P. saxatilis	r
C. furcata	a–va	Peltigera canina	+
C. pyxidata	o–a	P. polydactyla	+
C. sylvatica	la–o	P. rufescens	o
Evernia prunastri var.		Usnea florida var. hirta	vr
stictoceras	r		

Vascular plants. Unfortunately there is no record of the phanerogams occurring on the particular areas studied by Richards, but one dune which bore nearly all the above-mentioned bryophytes and lichens was occupied by Caricetum arenariae.

When the surface of the soil has been stabilised and enriched with humus by the growth of the bryophytes and lichens the tufts of *Ammophila* become sparser and less luxuriant, with many dead leaves, they rarely or never flower, and are sometimes obviously moribund. The contrast between the vigorous green marram grass of the mobile dunes and the impoverished, sparsely scattered, dingy tufts of the fixed dunes at once strikes the observer.[1] Though it may persist for a long time in this condition the plant has obviously become a dying relict. Various explanations of this loss of dominance and vigour have been proposed, of which the two most plausible are increasingly severe competition for available water, and the effect of lack of oxygen, or of increased carbon dioxide or some other toxic product of the new vegetation carpet. But the problem has not received serious experimental investigation.

Degeneration of Ammophila

The vascular plant covering of fixed dunes is very varied, and shows a mixture of maritime with non-maritime species. On dunes whose soil is not very calcareous the latter are preponderantly "grass-heath" species, but mixed with these are a good many others of various categories, including weeds of light arable soil which are able to find a foothold in the still open communities.

The following general list is taken from Salisbury's lists (Oliver and Salisbury, 1913) for two areas of fixed dune, "the Long Hills" and "the Hood", at Blakeney Point:

*Agrostis maritima	l	Melandrium album	vr
Aira praecox	f–a	Myosotis collina	f
*Ammophila arenaria	a	M. versicolor	r
Anagallis arvensis	f	Phleum arenarium	f
*Armeria maritima	lf	*Plantago coronopus	lf
Bromus hordeaceus (mollis)	r	†Polypodium vulgare	l
*Carex arenaria	ld	Rumex acetosella	f
*Catapodium (Desmazeria)		*Sagina maritima	o
loliaceum	vr	Sedum acre	f
Cerastium semidecandrum	f–a	Senecio jacobaea	a
C. tetrandrum	f–a	S. sylvaticus	r
C. vulgatum	vr	Sherardia arvensis	vr
Cirsium arvense	f	*Silene maritima	la
C. lanceolatum	r	Solanum nigrum	o
*Corynephorus canescens	ld	Stellaria pallida	f–a
Erodium cicutarium	f	Taraxacum erythro-	
Filago minima	lr	spermum	o–f
Galium verum	la	Urtica dioica	vr
Geranium molle	f	Valerianella olitoria	f
Hypochaeris glabra	f	Veronica arvensis	o–f
H. radicata	r	V. officinalis	lf
Luzula campestris	la		

* Maritime species. † On humus at the base of *Ammophila* tufts.

One plant of *Athyrium filix-femina* was found in a disused rabbit hole.

[1] Pl. 150.

The Blakeney Point dunes are of rather special interest. Though of small extent and not particularly rich in species, they have been comparatively well studied, the relative age of the different ridges is more or less accurately known, and the whole area is separated from the nearest land vegetation by more than a mile (about 2 km.) in a direct line across the estuary of the River Glaven.

"The Hood" at Blakeney Point. The oldest dune extant (estimated at about 300 years) is the small area known as "the Hood" separated from the rest of the system by about a quarter of a mile (400 m.). It is a curved patch of old dune sand, not more than 200 m. across in greatest diameter, considerably worn down by wind, and almost completely surrounded by shingle. The sand bears the majority of the species given in the list on p. 850 (Salisbury, 1913), but is specially notable for the local dominance of *Corynephorus canescens*, a rare and local species of our east coast dunes. This occurs abundantly in sheltered places on the edge of the old dune ridge. *Carex arenaria* is found in similar situations. *Ammophila* is still present in scattered tufts—the typical impoverished state on fixed dunes. In 1927 (Richards, 1929) most of the vegetation consisted of a very close sward of tall *Carex arenaria*, and the cryptogamic plants were completely different from those of the typical grey dunes described on pp. 853–5, being much more like those of an inland grass heath. The lichens had completely lost their dominance and were relatively scarce, while the moss vegetation had become rich and abundant (14 species in all), a number occurring which are not found elsewhere on the Point, including tall "hypnoid" mosses such as *Brachythecium purum*, *Hylocomium triquetrum* and *Hypnum cupressiforme* var. *ericetorum*, well adapted for living in closed communities of higher plants. The Hood is also almost the only place where liverworts occur on the Point.

The Hood, therefore, illustrates very well the evolution of fixed dune vegetation towards grass heath, though the dominant phanerogams are still entirely "maritime" species.

Calcareous dunes. The Blakeney dunes contain very little lime, but the sand of many dune systems, as already mentioned (p. 852), is largely composed of minute shells and comminuted fragments of larger shells and calcareous algae, and a large number of calcicolous species occur. Thus at Dog's Bay, on the south coast of Connemara (Co. Galway)[1] the sand consists almost entirely of minute shells. *Ammophila* is entirely absent and the low dunes of the centre of the spit fully exposed to the wind are dominated by *Agropyron junceum* and heavily eroded. The soil of the fixed dune community has 75 per cent of calcium carbonate, pH 8·1. The vegetation contained the following species:

*Festuca rubra var. arenaria		d	
Achillea millefolium	o f, la	Luzula campestris	o
Agrostis sp.	o	Plantago lanceolata	f
Anthyllis vulneraria	o–f	Polygala vulgaris	o
Asperula cynanchica	a	Prunella vulgaris	o
Bellis perennis	f	Ranunculus bulbosus	o
*Carex arenaria (relict)	la	Rumex acetosa	o
Cerastium vulgatum	o	Sedum acre	
Chrysanthemum		Trifolium dubium	o
leucanthemum	o	T. pratense	o
Hypochaeris radicata	f	T. repens	f–a
Lotus corniculatus	f		

[1] Pl. 154, phot. 397.

In an older fixed and pastured community on dune sand accumulated between rocks there were in addition:

Campanula rotundifolia
Carex diversicolor (flacca)
Euphrasia sp.
Galium verum
Hieracium pilosella
Koeleria cristata

Linum catharticum
Leontodon nudicaulis
*Plantago coronopus
*P. maritima
Sagina nodosa
Spiranthes spiralis

* Maritime species.

From the same locality Praeger records further:

Anacamptis (Orchis) pyramidalis
Arabis brownii (ciliata)
Blackstonia perfoliata
Carlina vulgaris

Centaurea scabiosa
Euphrasia salisburgensis
Sesleria caerulea

Out of approximately forty species in each of these lists, from Blakeney Point and Dog's Bay respectively, it will be observed that there are only seven species common to the two. The Dog's Bay list contains several markedly calcicolous plants, and if we except the very few maritime species, might have been taken from a non-maritime highly calcareous grassland.

Rosapenna   On the extensive fixed dunes at Rosapenna in Co. Donegal, several hundred metres from the sea, the soil of which showed a pH of 8·3 and a carbonate content of 48·6 per cent, the mosses are on the whole dominant over much of the area, especially in the hollows.[1] The following species occur in this mixed moss community:

Brachythecium albicans	lf	Ditrichum flexicaule	o
Camptothecium lutescens	a, ld	Hypnum chrysophyllum	r
Ceratodon purpureus	o	H. cupressiforme	f
Cylindrothecium concinnum	va	Tortula ruraliformis	f, ld

with the lichens *Peltigera canina* and *P. rufescens* (lf).

Among the cryptogamic vegetation there are scattered flowering plants which sometimes close to form a continuous carpet:

Ammophila arenaria (scattered, relict)
Arenaria serpyllifolia
Bellis perennis
Cardamine hirsuta
Centaurium umbellatum
Cerastium vulgatum
Cirsium arvense
C. lanceolatum
Crepis capillaris
Daucus carota
Erodium cicutarium
Euphrasia sp.
Festuca rubra var. arenaria

Fragaria vesca
Galium verum
Gentiana campestris
Geranium molle
Hieracium pilosella
Holcus lanatus
Leontodon nudicaulis
Linum catharticum
Lotus corniculatus
Luzula campestris
Poa compressa
Potentilla anserina
Prunella vulgaris
Ranunculus repens

[1] Pl. 150, phot. 384.

Dune Grassland

Senecio jacobaea
Sedum anglicum
Sonchus asper
Taraxacum officinale
Thymus serpyllum
Trifolium repens
Veronica chamaedrys
Viola curtisii

This list again has only seven or eight species in common with each of the other two, and there is no species which is characteristically maritime except *Ammophila*, though *Daucus* and *Erodium* may be considered submaritime.

Dune grassland. When rabbits are abundant on a fixed dune area, or when it is used as sheep or cattle pasture, woody plants have little chance of establishing themselves except sporadically.

Under pasturage by sheep or cattle a continuous turf develops, made up of a mixed assemblage of grasses, herbs and mosses, but usually reduced in numbers and modified in composition as a result of systematic grazing. At Castle Gregory on the north coast of the Dingle peninsula in Co. Kerry the following species occurred in a moist flat low-lying cattle pasture, evidently derived from a "slack", on calcareous dune sand:

Festuca rubra, thinly dominant

Agrostis stolonifera	a	Leontodon nudicaulis	f
Bellis perennis	f	Linum catharticum	a
Carex arenaria	a	Lotus corniculatus	f–la
C. diversicolor (flacca)	a	Potentilla anserina	o
Euphrasia sp.	a	Prunella vulgaris	a
Hydrocotyle vulgaris	la	Ranunculus repens	o
Hypochaeris radicata	a	Salix repens	f–la
Juncus gerardi	f	*Samolus valerandi	o
J. maritimus	ld	*Trifolium repens	f
Camptothecium lutescens	a		

* In Juncetum.

Here submaritime, marsh and calcicolous elements are all in evidence.

At Tramore in Co. Donegal a sloping cattle pasture on dune sand showed practically no maritime elements:

Festuca rubra   d

Achillea millefolium	f–a	Prunella vulgaris	f–a
Bellis perennis	a	Ranunculus repens	o–f
Carex arenaria	f–a	Senecio jacobaea	r–o
C. diversicolor (flacca)	la	Taraxacum erythro-	
Cerastium vulgatum	f, la	spermum	o
Galium verum	va	Trifolium repens	a
Lotus corniculatus	f	Veronica chamaedrys	la
Tortula ruraliformis	a	Camptothecium lutescens	f
Plantago lanceolata	va		

Pteridium aquilinum often invades non-calcareous dune grassland, just as it does other grassland on light soils, and closed Pteridieta may become established.

Acidic dunes. Walney Island. Pearsall (1934) describes a grassland community occupying the more rounded and lower dunes on Walney Island behind the first fixed dune community mentioned on p. 853. This community contains the following species:

Most constant species:
- Aira praecox
- Anthoxanthum odoratum
- Carex arenaria
- Festuca rubra (most abundant)
- Luzula campestris
- Polygala serpyllifolia
- Potentilla erecta
- Salix repens (la, in hollows)

Widely distributed:
- Campanula rotundifolia
- Galium saxatile
- Holcus lanatus
- Viola canina

- Brachythecium purum
- Dicranum scoparium
- Polytrichum juniperinum

The pH of the soil is between 4·8 and 5·5.

At points farthest from the sea this dune grassland is gradually passing into heath, though the process is not complete. The constant species are now:

- Agrostis tenuis
- Aira praecox
- Calluna vulgaris
- Carex arenaria
- Erica cinerea
- Rumex acetosella
- Polytrichum juniperinum

Frequent, but more local:
- Sedum anglicum
- Sieglingia decumbens
- Dicranum scoparium

And in the hollows:
- Nardus stricta
- Juncus squarrosus

The pH is 4·2 at the surface and 4·6–4·8 at a depth of 3 in. Leaching has evidently been considerable, judging from the very marked reaction with alcoholic ammonium thiocyanate.

Dune heath. Ayreland of Bride. Heath forms a zone within the low dunes on the Ayreland of Bride in the Isle of Man (Moore, 1931), and though it is difficult to demonstrate actual primary succession from one to the other this has probably occurred, judging from the occasional *Ammophila* present, which can scarcely be interpreted otherwise than as a persistent relict from a dune stage. The heath has the following composition, the numbers indicating the individuals on a line transect about 190 yards long:

Species	No.	Species	No.
Agrostis spp.	2	Festuca ovina	19
Ammophila arenaria	8	F. rubra	2
Calluna vulgaris	10	Hieracium pilosella	1
Carex arenaria	12		
Deschampsia flexuosa	1	Dicranum scoparium	1
Erica cinerea	20	Polytrichum juniperinum	3
Erodium cicutarium	1	P. strictum	1

Dune Heath. "Slacks"

Jasione montana	1	Ulex gallii	10
Lotus corniculatus	3	Viola canina var.	
Polygala vulgaris	1	ericetorum	1
Rosa spinosissima	3		
Sedum anglicum	6	Cladonia cervicornis	1
Silene maritima	1	Peltigera sp.	1
Thymus serpyllum	1	Lycoperdon caelatum	1

The sample of heath on which this transect was taken is obviously incompletely developed, since the sand dune elements are still considerable and the heaths have no overwhelming dominance over the grasses and herbs. Pl. 124, phot. 303 shows a later stage of development.

Subseres Subseres initiated by rabbit scratching show a stage of colonisation by *Cladonia cervicornis* and *Parmelia physodes* (which are said to furnish humus by scorching during summer), then by *Polytrichum juniperinum*, followed by *P. strictum*, *Dicranum scoparium* var. *spadiceum* and *Hypnum cupressiforme*. Soil in which sand is predominant (the substratum of the whole area is largely composed of pebbles) is rapidly colonised by the suckers of *Rosa spinosissima*. More pebbly soil is colonised by *Sedum anglicum*, which increases rapidly both by seed and vegetatively, and then by *Thymus serpyllum* and *Jasione montana*. Other species follow, including the grasses *Festuca* spp., *Deschampsia* and *Agrostis*, which form a turf containing such herbs as *Viola canina*, *Lotus corniculatus*, *Erodium cicutarium*, *Ononis repens*, and *Galium verum*. Finally *Erica cinerea* and *Calluna vulgaris* enter and become dominant. Where the acidity is marked (pH 5·8) the heaths may, however, directly colonise a lichen and moss carpet.

South Haven peninsula (Dorset). Here Good (1935, pp. 387–9) has traced a complete psammosere passing from the Ammophiletum of the dunes to typical Callunetum (Pl. 126, phot. 309). After the appearance of a number of grasses and herbs among the *Ammophila* the vegetation is invaded by *Erica cinerea* and later by *Calluna* which ultimately becomes completely dominant, the marram grass progressively disappearing. In some places *Ulex europaeus* replaces the heaths.

Vegetation of "slacks"

On the west coasts the damp or wet hollows left between the dune ridges, where the ground water reaches or approaches the surface of the sand, are known as "slacks". The following general account is summarised from Watson (1918):

Flora of permanent or semi-permanent pools. These may contain:

Amblystegium filicinum var. whiteheadii	H. intermedium
	H. revolvens
Hypnum aduncum, with numerous varieties	H. scorpioides
	H. wilsoni and var. hamatum
H. cuspidatum	
H. elodes	Potamogeton perfoliatus
H. giganteum	Zannichellia palustris

Around the margin the vegetation does not differ from that of an inland pool:

Eleocharis palustris
Epipactis palustris
Glyceria maxima (aquatica)
Holcus lanatus
Hypnum aduncum and varieties

Hydrocotyle vulgaris
Iris pseudacorus
Juncus effusus
Phalaris arundinacea
Pellia fabbroniana

Dune marshes.[1] These again show a vegetation which contains only a slight representation of maritime or submaritime species:

Anagallis tenella	sd	Juncus maritimus	l
Carex spp.	a	Littorella uniflora	ld
Galium palustre	a	Lycopus europaeus	a
Glaux maritima	o	Mentha aquatica	a
Hydrocotyle vulgaris	sd	Orchis incarnata	o
Iris pseudacorus	o	Parnassia palustris	lf
Juncus acutus	l	Sagina nodosa	f
J. bufonius	o	Samolus valerandi	f
J. effusus	f	Scirpus holoschoenus	l
J. inflexus	o	Teucrium scordium	l

Together with a great variety of Hypna (Harpidia):

Hypnum aduncum, and several varieties	a	Hypnum polygamum var. stagnatum	f
H. cuspidatum	f	H. scorpioides	o
H. giganteum	o	H. sendtneri	–
H. intermedium	lf	H. wilsoni	o
H. lycopodioides	f	var. hamatum	lf

And other bryophytes and algae:

Amblystegium filicinum var. whiteheadii	f la	Pellia fabbroniana Riccia crystallina	f o
Bryum neodamense	r		
B. pseudotriquetrum	f	Mougeotia sp.	a
		Tribonema bombycina	a
Cephalozia bicuspidata	o	Vaucheria sessilis	a
Moerckia flotowiana	f	Various Cyanophyceae	a

In moist places of the "slacks" of Braunton Burrows on the north coast of Devon (4 and 6 in Fig. 162, p. 846) occur patches of algae and liverworts:

Arthopyrenia areniseda
Collema glaucescens
C. pulposum

Riccia crystallina
Vaucheria dichotoma
V. sessilis

Good (1935, pp. 390 ff.) describes the vegetation of various damp places and marshes on the South Haven peninsula.

***Salix repens* dunes.** On many of our western dunes, as on those of the north coasts of France and Belgium, the damp soil of the slacks is colonised by the creeping willow (*Salix repens*).[2] If quantities of sand are not constantly blown on to the Salicetum, the latter maintains itself at or near the

[1] Pl. 151, phot. 387.   [2] Pl. 151, phot. 385.

PLATE 151

Phot. 385. Salicetum repentis occupying a depression ("slack") in the dune complex at Southport, Lancs. *W. Ball.*

Phot. 386. Edge of "slack" with low dune dominated by *Salix repens*. Eroded high dunes behind. *Mrs Clements, I.P.E.* 1911.

Phot. 387. Vegetation of a wet "slack"— *Parnassia palustris, Anagallis tenella, Centaur.um littorale,* etc. *J. Massart.*

"SLACKS" OR DUNE VALLEYS

Phot. 388. Scrub of *Hippophaë rhamnoides* (sea buckthorn) on back of main dune ridge at Hemsby, Norfolk. *Rubus rusticanus* is abundantly associated and *Ammophila* still conspicuous in the gaps.

Phot. 389. Detail of dune scrub: *Hippophaë* (right), *Rubus rusticanus* (left), *Polypodium vulgare* (centre), *Lonicera periclymenum*, *Festuca rubra*, *Holcus lanatus* and *Ammophila* (relict). Hemsby.

Phot. 390. Detail of dune scrub. *Hippophaë, Ammophila, Calystegia soldanella*. Hemsby. J. Massart.

Phot. 391. Detail of dune scrub. *Ulex europaeus, Pteridium, Polypodium vulgare*. Hemsby. J. Massart.

Dune Scrub (Hippophaëtum)

original level, the surface soil remaining more or less wet and then supporting a variety of marsh and damp grassland species. These include *Selaginella selaginoides* (Anglesey), and a long list of bryophytes (see Watson, 1918, p. 142) with eleven species of *Bryum*, five of *Hypnum*, and five of *Aneura*, as well as other liverworts, such as *Petalophyllum ralfsii* and *Moerckia flotowiana*, characteristic of this habitat.

When enough sand is supplied to form low dunes or hummocks through which the shoots of *Salix* grow up, carrying the surface well above the ground water, a drier community is established,[1] including, on the Lancashire dunes, such plants as *Pyrola rotundifolia* and *Monotropa hypopitys* which live in the humus accumulated from the decay of the willow leaves.

If more sand is constantly supplied from the neighbouring *Ammophila* dunes the shoots of the willow continue to grow up through it, producing new roots in the damp layer just below the surface, and thus form mobile dunes of considerable height capped by the living shoots of the willow. These larger *Salix* dunes, in all stages of development and wind erosion, are a conspicuous feature of some of our west coast sandhills, as they are of the Channel coasts of Belgium and north-eastern France. Their flora hardly differs from that of the ordinary mobile *Ammophila* dunes.

At Sandscale opposite Walney Island (Pearsall, 1934) the older dunes become almost completely covered by *Salix repens* developed from the carpet which has established itself in the slacks. Here the ground water level is high, held up by an underlying stratum of clay. The larger slacks may show all stages from bare sand to a continuous carpet of *Salix repens* or occasionally *Potentilla anserina*. The following species occur in the earlier stages of colonisation:

Carex arenaria	Listera ovata
C. oederi	Orchis maculata
Centaurium umbellatum	O. purpurella
Epipactis palustris	Parnassia palustris
Equisetum variegatum	Potentilla anserina
Galium verum	Salix repens
Gentiana amarella (agg.)	Samolus valerandi
Juncus articulatus var. litoralis	

At a later stage *Salix repens* becomes dominant and begins to collect sand, forming low dunes on which a drier and somewhat acid facies develops, with:

Carlina vulgaris	Pyrola minor
Carex arenaria	Rhacomitrium lanuginosum
Festuca rubra var. arenaria	Tortula sp.

These dunes resemble the Southport dunes in a general way, and are much less acid than those on Walney Island. The pH value of the soil of the advanced *Salix* dunes is about 6 (lowest recorded 5·4). This may per-

[1] Pl. 151, phot. 386.

haps be accounted for by the height of the water table or the protection from leaching afforded by the covering of *Salix*. Pearsall says there are indications that woodland might develop here, starting from the damper places.

Effects of disturbance. The surface and vegetation of fixed dune areas are generally subject to various kinds of disturbance. Sand is often blown on to it from the mobile dunes nearer the sea, and in this way smaller secondary dunes arise, sometimes colonised by *Ammophila* or by *Salix repens*; or, if the sand is comparatively slight in quantity, by some of the pioneer mosses, such as *Tortula ruraliformis*, previously mentioned. A very important factor is disturbance by rabbits. Most dune areas are infested by these animals, which find the easily excavated sand peculiarly suitable for making burrows, and some such areas are preserved as rabbit warrens. Because they do not eat the bryophytes their abundance may lead to the dominance of these plants. (Cf. Chapter XXVII, p. 543 and Pl. 102, phot. 248.) The rabbits not only browse on the plants, but also, by their continual scratching in particular spots, destroy the carpet of vegetation, break up the stabilised surface of the sand and give the wind a purchase. In this way "blow-outs"[1] are started, and these, when at all extensive, throw back the course of succession to an earlier stage. "Blow-outs" are also (and more generally) initiated by specially violent winds, particularly from an unusual direction. Thus, in addition to the normal course of the main successional sequence, every stage of slow sand deposition, of erosion and rejuvenation, may be met with on a fixed dune area.

Ultimate vegetation. The ultimate vegetation of fixed dunes is various, as we have already seen. Three main types may be distinguished, Callunetum, scrub, and grassland. Where Callunetum develops the heather colonises the stabilised surface of the sand and heath conditions are established.

Dune scrub may consist of any of the common spinous shrubs which can colonise light sandy soils, e.g. gorse (*Ulex europaeus* and in the west *U. gallii*), bramble (*Rubus fruticosus* agg.), blackthorn (*Prunus spinosa*), or species of rose. The burnet rose (*Rosa spinosissima*) is often abundant in highly calcareous dune areas. Elder (*Sambucus nigra*), that ubiquitous "weed shrub", which is markedly resistant to rabbit attack, may also be present in quantity. The most characteristic dune shrub, confined in Britain to this habitat, is the sea buckthorn (*Hippophaë rhamnoides*), but this is very local in occurrence, and confined, as a native plant, to some of the east coast dunes,[2] with *Polypodium vulgare* conspicuous beneath it (Pl. 152, phots. 389, 391). Elsewhere it is frequently planted. The species does not in fact occur in maritime habitats over most of its range. In central Europe, though not in the British Isles, it inhabits shingle banks by the sides of rivers and similar situations. The establishment, history and fate of dune scrub has not been closely studied.

Hippophaëtum

[1] Pl. 153.   [2] Pl. 152.

Phot. 392. An old blow-out in half fixed *Ammophila* dune near Southport, Lancs, partly revegetated. *Mrs Cowles, I.P.E.* 1911.

Phot. 393. An old blow-out in old fixed dune near Carrigart, Co. Donegal, partly revegetated. Moss carpet in foreground and on the left. *R. J. L.*

"Blow-outs"

PLATE 154

Phot. 394. Back of half fixed dunes at Blakeney Point. *Ammophila* and *Senecio jacobaea*. Complete fixation of the sand is effected by the mosses (*Tortula ruraliformis* dominant) in the foreground. *R. H. Compton.*

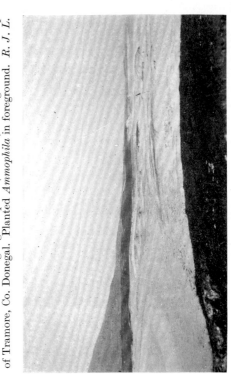

Phot. 395. View including highest point (c. 150 ft.) in the great dune complex of Tramore, Co. Donegal. Planted *Ammophila* in foreground. *R. J. L.*

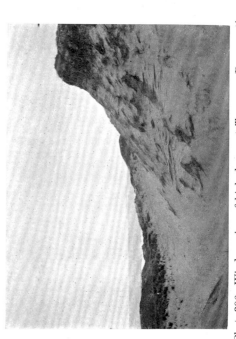

Phot. 396. Wind erosion of high dunes at Tramore. Regeneration on the left. Tory Island on the horizon.

Phot. 397. Spit of mobile, highly calcareous dunes, worn down by wind erosion, between two bays. *Agropyron junceum* dominant, *Ammophila* absent. Fixed dune grassland in foreground and distance. Dog's Bay, near Roundstone, Connemara.

Dune soils

Dune soils. The various composition of dune sand has been mentioned on pp. 851–2. The pH values fall with the development and progress of fixed dunes owing to progressive leaching. The initial alkalinity is due primarily to sea salt, and the sand of the foredunes is always on the alkaline side of neutrality. But dune sand largely composed of shells shows much higher pH values and contrasts with the relatively neutral dunes and even more strongly with those which develop high acidities.

Salisbury (1922) investigated some of the chemical characters of the dune sand at Blakeney Point in the different stages of succession. The youngest dunes had an appreciable but low carbonate content, generally under 1·0 per cent. (The highest value in the area, 4·2 per cent, was obtained from the drift zone of a lateral shingle hook.) This gradually decreased in passing to the older sandhills, many samples from which yielded no carbonates at all. The average pH values and humus contents (loss on ignition) obtained by Salisbury are shown in Table XXVI:

Table XXVI

	pH	Humus (loss on ignition) per cent
Youngest dunes	7·1 (max. 7·4)	0·36
Main ridge	7·03	0·5
Older ridges	6·9	0·52 / 0·86
Oldest ridges	6·8 (min. 5·5) / 6·24	1·15 (max. 6·34) / 2·7

These figures clearly show both progressive leaching and progressive humus accumulation with time. The drift of the two processes is graphically shown for the Blakeney Point and Southport dunes in Figs. 163, 164 (Salisbury, 1925).

Quite comparable data were obtained by Moore (1931) from the sand of the very pebbly soil in different zones of the Ayreland of Bride in the Isle of Man. "Coarse sand" enormously preponderated over the other fractions separated by mechanical analysis. Table XXVII shows some selected figures:

Table XXVII

	pH	Humus (loss on ignition) per cent	Carbonates per cent
Embryo dune	7·6	0·86	2·73
Seaward face of dune	6·9	Nil	3·32
Consolidated dune	7·0	1·24	0·75
Grass heath	6·4	5·54	Nil
Heath	5·8	10·86	Nil

The sand of dunes consisting mainly or almost entirely of minute shells and fragments of larger shells, or of nullipores and coralline algae, show very much higher carbonate contents and pH values. Table XXVIII gives

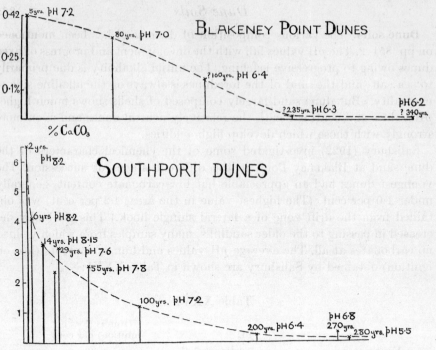

FIG. 163. RATES OF LEACHING AND DECREASE OF pH VALUES WITH AGE OF SAND DUNE SOILS

Note the much greater initial lime content and higher pH values of the Southport dunes compared with those of Blakeney Point. In both cases the pH value is reduced to the neighbourhood of 6 in about 200 to 300 years. From Salisbury, 1925.

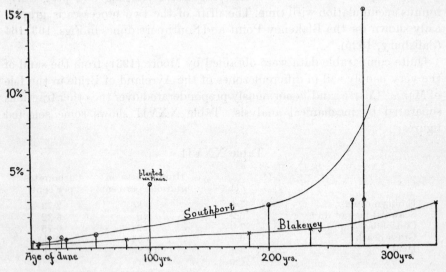

FIG. 164. INCREASE IN ORGANIC CONTENT WITH INCREASE OF AGE OF DUNE SOILS

Note the higher content and greater rate of increase at Southport. From Salisbury, 1925.

data from single samples of partly fixed dunes on the west coast of Ireland.

Table XXVIII

	pH	Carbonates per cent
Castle Gregory (Co. Kerry)	7·3	17·9
Tramore (Dunfanaghy, Co. Donegal)	8·3	25·4
Rosapenna (Co. Donegal)	8·3	48·6
Dog's Bay (Roundstone, Co. Galway)	8·1	75·0

The highly calcareous nature of these dunes is reflected in the vegetation (see pp. 857–9).

The foredunes and mobile dunes at Walney Island (Pearsall, 1934) show pH values of 7·5 and 7, falling to 6–6·5 on the fixed dunes (p. 853), 4·8–5·5 in the dune grassland, and 4·2–4·8 as heath develops (p. 860).

REFERENCES

CAREY, A. E. and OLIVER, F. W. *Tidal Lands: a study of shore problems.* London, Blackie and Sons, 1918.

GOOD, RONALD. Contributions towards a survey of the plants and animals of the South Haven peninsula, Studland Heath, Dorset. II. General ecology of the flowering plants and ferns. *J. Ecol.* **23**, 361–405. 1935.

McLEAN, R. C. The ecology of the maritime lichens at Blakeney Point, Norfolk. *J. Ecol.* **3**, 129–48. 1915.

MOORE, E. J. The ecology of the Ayreland of Bride, Isle of Man. *J. Ecol.* **19**, 115–36. 1931.

OLIVER, F. W. Blakeney Point Reports, *passim*, 1913–29. *Trans. Norfolk Norw. Nat. Soc.*

OLIVER, F. W. and SALISBURY, E. J. Topography and Vegetation of Blakeney Point, Norfolk. *Trans. Norfolk Norw. Nat. Soc.* **9**, 1913 (also issued separately).

PEARSALL, W. H. North Lancashire sand dunes. *Naturalist*, pp. 201–5. 1934.

RICHARDS, P. W. Notes on the ecology of the bryophytes and lichens at Blakeney Point, Norfolk. *J. Ecol.* **17**, 127–40. 1929.

SALISBURY, E. J. The soils of Blakeney Point: a study of soil reaction and succession in relation to plant-covering. *Ann. Bot.* **36**, 391–431. 1922.

SALISBURY, E. J. Note on the edaphic succession in some dune soils with special reference to the time factor. *J. Ecol.* **13**, 322–8. 1925.

Types of British Vegetation, pp. 339–52. Cambridge, 1911.

WATSON, W. Cryptogamic vegetation of the sand dunes of the west coast of England. *J. Ecol.* **6**, 126–43. 1918.

Chapter XLII

(4) SHINGLE BEACHES AND THEIR VEGETATION

Nature of shingle. Shingle beaches, which fringe many miles of our southern and eastern coasts, form another well-marked maritime habitat, though their total area in the British Isles is not nearly so great nor is their distribution so wide as that of the sand dunes. They are formed of water-worn and more or less rounded pebbles of very various size, derived by wave erosion from hard rocks or from flints originally embedded in chalk, driven along the coast by shore currents and eventually accumulated in banks or flat expanses on low-lying shores.

Below the limits reached by ordinary high tides they are destitute of plant life, for the pounding of the shingle in a rough sea, apart from the force of the waves themselves, makes existence impossible within the zone of surf. But above this zone the substratum becomes accessible to invasion.

Sand and shingle. On most beaches sand and shingle are mixed in various proportions, and where sand forms any large part of the substratum the vegetation above high water of spring tides is essentially arenicolous. For example, the low dunes of the Ayreland of Bride on the north coast of the Isle of Man, described in the last chapter, are formed from scanty blowing sand, the substratum of beach, dunes and heath—based on old "raised beach"—being composed of pebbles. At Blakeney Point and Scolt Head, on the Norfolk coast, more considerable dunes are formed above a substratum of shingle, which is often in evidence in the "lows" or valleys between the dune ridges. The presence of shingle tends of course to stabilise the sand, but the vegetation depends on the latter.

Beaches composed almost entirely of pebbles show, however, quite distinctive features as a plant habitat, and these have been investigated mainly by Oliver and his fellow-workers (1911–29, see References, pp. 894–5).

Types of shingle beach. Oliver (1912) classifies coastal shingle beaches as follows:

(1) *Fringing shingle beaches* are the simplest and commonest type (Pl. 156, phot. 399; Pl. 158, phot. 406). The shingle, driven by a shore current parallel with the coast, forms a strip in contact with the land along the top of the beach. *Examples*: most of the shingle beaches along the Sussex coastal plain.

When the current leaves the shore other types are produced.

(2) *The shingle spit* (Fig. 165). When the coast line changes its direction while the shore current continues in a straight line the latter necessarily leaves the shore, and the shingle it carries forms a bank or causeway diverging from the land and often extending for several miles. The apex is

liable to sudden deflection landward as a "hook". *Examples*: Hurst Castle and Calshot spits, Orfordness, Blakeney bank (Pl. 133, phot. 335; Pl. 134; Pl. 136, phot. 341 and Figs. 168–70).

(3) *The shingle bar* (Fig. 166) is formed if a shingle spit once more reaches the land, cutting off a closed lagoon between spit and land. *Examples*: Chesil Bank (Pl. 155, phot. 398), which may however have originated as a fringing beach, the lagoon behind (Fleet) being later excavated by erosion: Looe Bar (Cornwall).

(4) The *apposition shingle beach* (Fig. 167) is formed when new shingle is deposited on the flank of an earlier beach, where it accumulates till lifted

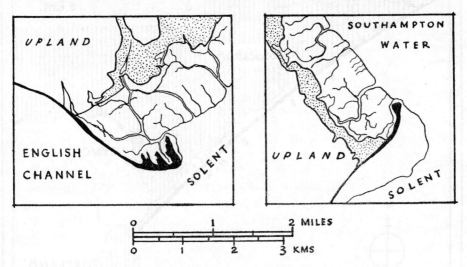

FIG. 165. TWO SHINGLE SPITS (HURST CASTLE AND CALSHOT SPITS), ON THE SOLENT, HAMPSHIRE

Note the continuation into the curved spit of the line of fringing shingle beach (black). The spits protect areas of soft mud, exposed at low tide and traversed by drainage channels. On this mud salt marsh is developed. The dotted areas on the landward side are mature saltings and partly reclaimed.

above tidal limits by an exceptionally high tide caused by a gale. If the process is repeated a succession of closely approximated, more or less parallel banks are formed, producing a very extensive area of shingle (Pl. 161). *Example*: Dungeness.

The fringing and apposition types of shingle beach have not been investigated in detail. The former is always relatively narrow, and the strip colonisable by vegetation usually extremely narrow. On the latter, owing to their great extent, the most advanced (non-maritime) vegetation occurs.

Most of the British work on shingle vegetation has been done on the spits and on the famous "bar" of Chesil Bank.

Morphology and development of the shingle spit. A longshore current carrying shingle tends to continue in a straight line when the coast line, along which the shingle has been deposited as a fringing beach up to that point, turns landward. The continuation of the fringing beach then forms a spit, diverging at an angle from the inward curve of the coast and protecting the waters of the bay or estuary in this inward curve. The shingle

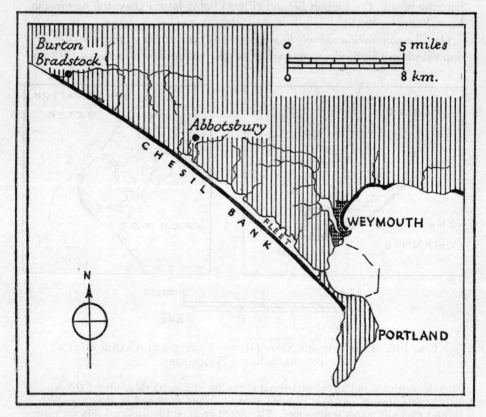

FIG. 166. THE CHESIL BANK SHINGLE BAR

The western end of the Chesil Bank and the beaches east of Weymouth are of the fringing type. Compare Phot. 398 (Pl. 155) which is taken from the air about two miles southeast of Abbotsbury looking south-eastward along the Bank towards Portland.

spit itself is usually gently curved inwards, its convex side facing the open sea (Figs. 165, 168), and the apex generally shows a marked landward deflection known as a *hook*. Besides the terminal hook there are commonly several others ("laterals") running landwards from the sub-terminal portion of the main spit, and these may form quite complex systems, as at Blakeney Point and Scolt Head (Fig. 168). Each of these sub-terminal or "lateral" hooks was at one time the apical hook, deflected landwards when it was the growing apex of the main spit. Thus the spit grows in a manner

PLATE 155

Phot. 398. The south-eastern half of the Chesil Beach shingle bar seen from the air. The lagoon called the Fleet separates the bar from the mainland to the left; Portland Harbour, with shipping, beyond. The distance from the front of the picture to the harbour is more than 6 miles (nearly 10 km.). The high ground of Portland (left centre distance) is another 2 miles along the beach.

The "camm", with its "ravines" and "buttresses" is plainly seen in the foreground along the inner side of the beach. "Deltas" project into the Fleet, forming a sinuous line, and the dark edging is the belt of *Suaeda fruticosa*. Photograph from *The Times*.

CHESIL BEACH FROM THE AIR

PLATE 156

Phot. 399. Sheep feeding on *Silene maritima* at the end of a long drought (September 1911). Near Burton Bradstock. The beach is here a fringing beach. *F. W. Oliver.*

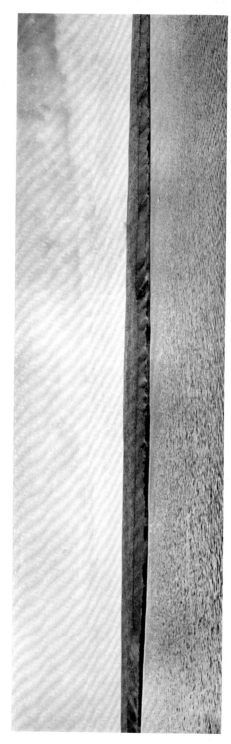

Phot. 400. Chesil Beach seen across the Fleet, showing the "camm" with its ravines and buttresses, the fringing belt of *Suaeda fruticosa* on the edge of the water, and the "back" of the beach above. About 2 miles east of Abbotsbury. *F. W. Oliver*, 1911.

CHESIL BEACH

analogous to that of a sympodial rhizome, the lateral hooks corresponding with successive apices.

The actual cause of hook development, i.e. of the landward deflection of the apex of the spit, appears to be the rapidly increasing force of the incoming tidal current, greatly augmented by the waves formed by a gale blowing in the same direction, which fills the bay or estuary inside the spit, as this lengthens and encloses a greater area of tidal ground. The volume of water required to fill this space in a given time obviously increases as the square of the length of the spit if this makes an angle of 45° with the coast line. Spits rarely in fact make such a wide angle with the shore line, and the space to be filled is progressively diminished by the silting up of the bay or estuary, but the scour of the tide is certainly greatly increased by the widening of the mouth as the spit diverges farther and farther from the

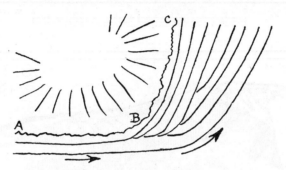

FIG. 167. DIAGRAM OF APPOSITION SHINGLE BEACH

The current here followed the shore line round the corner at B so that successive strips of shingle were added outside the stretch B–C. From Oliver, 1912.

mainland. To this increased force of the tidal current, backed by wave action on the flood when a strong wind is blowing, is also due the frequent deflection of the hooks backwards (i.e. towards the proximal end of the main spit) seen especially in the Blakeney Point system (Figs. 168, 169, 170). Oliver actually observed the formation of an L-shaped end to a subterminal hook during a gale in 1911, the hook having recently been exposed to the tidal scour by the wasting of the terminal dune-covered head of the main bank, which had formerly protected the sub-terminals (Oliver, 1913).

The concentration of hooks in the sub-terminal region of the spit is probably due to the increased proportional strength of the tidal current entering the bay or estuary with growth in length of the main spit, as compared with the shore current responsible for lengthening the main spit. Thus the whole spit system passes through a period of youth, when the shore current is alone active, resulting in the continuous lengthening of the main spit, to a period of maturity or hook formation when the incoming tidal current becomes strong enough, backed by a gale, to deflect the tip. Since this deflecting action is intermittent, depending on the occurrence of

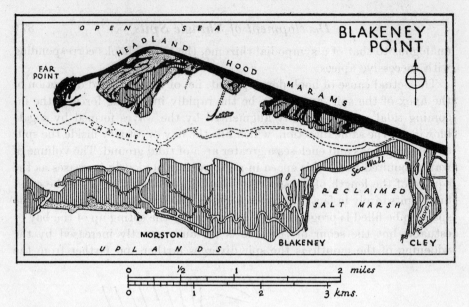

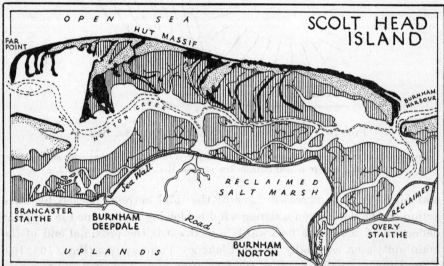

FIG. 168. BLAKENEY POINT (1923–24) AND SCOLT HEAD ISLAND (1925)

Two shingle spit systems on the north coast of Norfolk with associated sand dunes and salt marshes. Shingle *black*, sand dunes *dotted*, salt marshes *vertically ruled*. Sea, coastal and estuarine sand and mud, reclaimed marshes and uplands *white*. The general physiographic correspondence of the two systems is obvious at a glance.

The shore current runs from east to west, and the growing tips of the banks are consequently at their western extremities. Note that the banks are curved inwards towards their distal ends. The "hooks" or lateral banks connected with the inner side of the main banks form complex irregular groups, each group corresponding to a period of active hook formation (prevalence of storms?) with intervening periods of continuous growth of the main bank. Most of the hooks are more or less curved, with the convexity facing west or north-west. Each hook was originally an apex of the main bank, though its shape has often been much modified subsequently. The "Far Points" are the most recent additions, each forming a main group of hooks.

The sand dunes overlie the broader expanses of shingle: the salt marshes are formed mainly in the concavities of the hooks, where protection is greatest.

gales, there is now an alternation of hook formation and growth of the main spit.

It is necessary to understand the developmental morphology of shingle spits because the different parts of the system bear different vegetation.

Mobility of shingle. Except in the apposition type, which will be referred to later, and the narrow landward edge of the fringing type, the shingle of a sea beach never reaches a state of complete repose. The mobility of the stones is however very different in the different regions of a beach, and on decrease in mobility depends the possibility of the existence of vegetation.

There are three factors responsible for the movement of shingle. The first is *wave impact* on the seaward face of the beach, which keeps the shingle constantly on the move with every incoming tide, most violently of course when an onshore gale is blowing. This throws up shingle on to the "storm shelf" (Figs. 172, 174) and even over the crest of the beach when there is a high spring tide combined with an onshore gale. The crest and back of the Blakeney bank are awash under such conditions, so that fans of shingle spread over the salt marsh behind (Pl. 157, phot. 401) and the whole bank gradually travels landwards. Where different stretches of a shingle bank are subject to unequal incidence of wave action or are unequally resistant to it a "sagging" of the bank landwards can be observed. The landward travel of the main bank will necessarily encroach on the basal parts of the lateral hooks and thus abbreviate their length (Fig. 171, p. 876). This may affect the vegetation of that part of the main bank which has incorporated the base of the hook.

The second factor is *percolation* of water through the beach. As a result of its open structure shingle is readily traversed by water. At Blakeney at high tide, especially if a heavy sea is running and its level is well above that of the salt marshes ("saltings") on the inner side of the bank, sea water normally traverses the bank, breaking out in numerous springs just above the level of the saltings. A certain amount of displacement of shingle then occurs, but it is relatively insignificant because of the small height of the bank—only 6 or 7 ft. (*c.* 2 m.) above ordinary high spring tide mark.

"Camm" formation. On the Chesil Bank, however, where the crest rises 25 ft. (*c.* 8 m.) above high water mark and the slope often approaches the critical angle of rest of the shingle, the results of this landward percolation are important. Seen from across the Fleet (the lagoon separating the shingle beach from the land) the bank rises steeply above the low flat terrace which fringes the Fleet, and is seamed by a series of ravines ("cans") 10 or 12 ft. deep, which with the intervening buttresses form a conspicuous "camm" (Pl. 156, phot. 400 and Fig. 173). The back and sides of a "can" are inclined at the angle of repose (about 34°), and the floor is generally scored by a shallow gully connecting with a detrital fan of shingle projecting into the Fleet some 8 or 10 ft. beyond the edge of the terrace (Pl. 158, phot. 404).

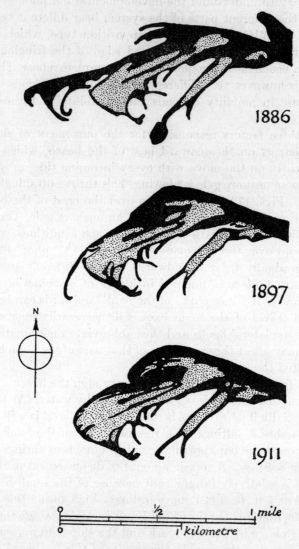

FIG. 169. DEVELOPMENT OF THE HEADLAND OF BLAKENEY POINT, (NORFOLK), 1886–1911

Shingle *black*, sand dunes *dotted* (salt marshes not shown).

The hooks forming a "Far Point" complex in 1886 were obliterated by 1897 and a new single hook formed which had itself disappeared by 1911. Note also the progressive development of new dunes on the northern side of the Headland.

From Oliver and Salisbury, 1913. (The first two maps are from 6 in. O.S. maps, the third is based on a survey by the Botanical Department. University College, London.)

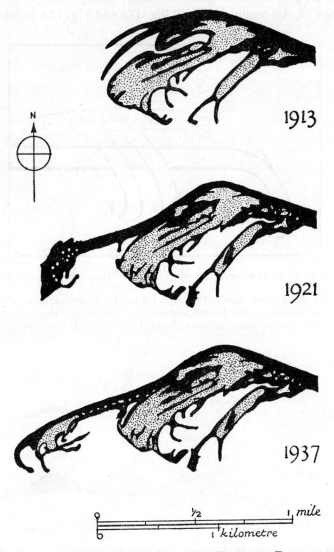

FIG. 170. DEVELOPMENT OF THE FAR POINT AT BLAKENEY 1913–1937 (continuation of Fig. 169)

By 1913 the growth of the main spit was resumed, and by 1921 it had extended half a mile, the apex broadening into an expanse of shingle (the Far Point) bearing scattered embryo dunes. Between 1927 and 1935 these consolidated into considerable dune areas and several new hooks were formed, the latest (westernmost) in 1936–37. Note the modifications of several old hooks. The charts of 1913 and 1921 (based on air photographs) are from *Blakeney Point Reports* (1917–19 and 1920–23); the Far Point in the 1937 chart is from a plane table survey by the Botanical Dept., University College, London, by kind permission.

The slopes of the buttresses are inclined at a lower angle than those of the "cans" and merge without material change of angle into the gentle slope at the back of the beach crest (see Fig. 173 and Fig. 179, p. 886).

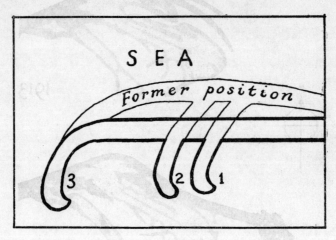

FIG. 171. DIAGRAM OF PART OF A MAIN SHINGLE SPIT WITH HOOKS

1 and 2 are "hooks or former apices of the main bank, 3 is the present apex—all deflected inwards. The thin lines indicate the former position of the main bank, which has been driven inwards by cumulative wave impact and has embedded the bases of the hooks. From Oliver, 1912.

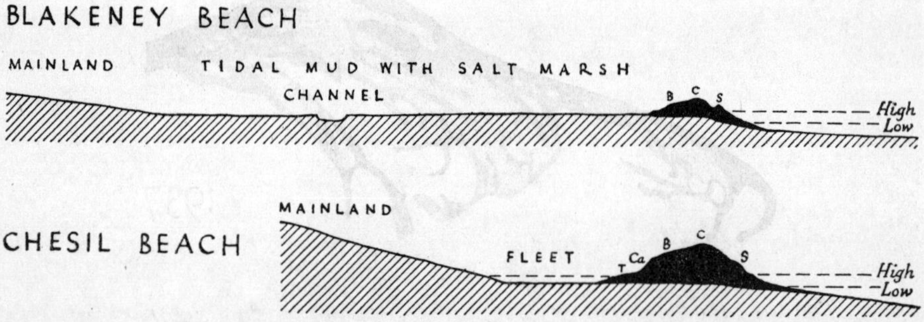

FIG. 172. BLAKENEY AND CHESIL SHINGLE BANKS

Profiles showing the relation of the shingle (*black*) to the marine terrace. S = Storm shelf, C = crest, B = back, Ca = camm, T = terrace.

Vertical scale about four times the horizontal, but drawing not accurately to scale. After Oliver, 1912.

These ravines ("cans") are formed by the sea water running through from the front of the beach and gushing out at the back. At the points of exit shingle is displaced, dislodging the stones above, which slide down their "slope of repose". Thus the "cans" are cut deeper into the bank. The amount of shingle shifted in ravine formation is large, and the ravine-buttress or camm structure of the bank above the terrace is important in determining the vegetation. The general result of the process is clearly the transfer of shingle from the higher parts of the bank to the terrace below, to which it is added, while the stones which are carried farthest form the fans which project into the Fleet. Active "can" formation and cutting back seems to be an intermittent phenomenon, some "cans" apparently remaining quiescent for long periods. The whole process is a minor feature

FIG. 173. PART OF THE "CAMM" ON THE INNER FACE OF CHESIL BANK

View taken from the terrace about 20 ft. from the edge of the Fleet looking up a gully or "can" with bare floor. The terrace bears scattered plants and a bush of *Suaeda fruticosa*. On the flat top of the nearest "buttress" and on the "back" of the bank (at the top of the picture) are plants of *Silene maritima*. Reproduction (by kind permission) of a sketch by Prof. T. G. Hill. From Oliver, 1912.

of the landward creep of the whole bank, primarily caused by wave impact, and made up of the piling of fresh shingle from the sea on the storm shelf and crest and its gradual transference down the landward face.

A third factor affecting the mobility of a shingle bank is *undercutting* by tidal or other currents, e.g. the current of a river impinging on the inner side of a bank. This of course results in wasting the bank and may retard its general inward creep.

Structure of a shingle spit or bar. The primary division of a shingle spit is into *main bank* and *laterals* (*hooks*), of which the former is, broadly speaking, mobile, while the latter are immobile.

On the main bank of a spit like that at Blakeney or a bar like Chesil Beach we can distinguish the following topographical features (Fig. 172, 174). On the seaward face, which is normally the steeper, there is a series of "steps" corresponding with different high tide marks, culminating in the *storm shelf* (S), not reached by ordinary spring tides. Above this is the *crest* (C) or summit of the bank, built up only by quite exceptional tides when these coincide with onshore gales. The lower steps, where the mobility of the shingle is at a maximum, are destitute of vegetation. On the storm

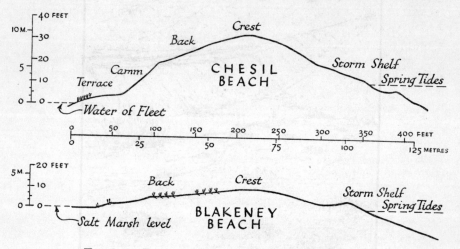

FIG. 174. BLAKENEY AND CHESIL SHINGLE BANKS

Profiles drawn to the same scale. Note the three zones of *Suaeda fruticosa* on the back of the Blakeney beach. After Oliver and Salisbury, 1913.

shelf the vegetation, when present at all, is very sparse indeed, consisting mainly of prostrate species of *Atriplex* and sometimes *Beta maritima* (cf. the foreshore community p. 844). The crest of the bank, like the storm shelf, bears very little vegetation, though there may be a few poor specimens of plants derived from the back of the bank. Both crest and storm shelf are obviously very much exposed to wind and spray, making plant establishment difficult, and the species which do establish themselves are occasionally drenched with sea water. On a low bank like that at Blakeney the *back* (B) of the bank slopes gently to the marshes. On the lofty Chesil Bank it slopes more steeply, up to 10–15° on the general face (B), while on the sides of the ravines (Ca) the angle is the critical angle of rest of the shingle (about 34°). Below is the flat terrace (T) receiving shingle from the bank, most actively from the "cans", and encroaching on the Fleet within. There is no terrace on the landward side of the Blakeney bank, the back slope passing straight on to the marshes, over which spread fans of shingle. The back of the bank is by far the most important plant habitat and bears the vegetation characteristic of mobile shingle (Pl. 157, phots. 401–2; Pl. 158, phot. 405; Pls. 159–60).

The laterals of a shingle spit show later stages of succession and possess a much more numerous and less characteristic flora.

Shingle and sand. The shingle of our coastal beaches is sometimes almost pure but very often mixed with sand. Sand and pebbles are both constantly thrown up by the sea and both are carried up on to shingle beaches, to which wind-borne sand may also be contributed; but much of this is again washed down by retreating waves and on the ebb of the tide. Because of the very different carrying power required to move sand and stones the tides and waves tend to sort them out one from the other, so that comparatively pure strata of each are often met with. Thus the storm shelf normally consists of pure shingle. On the other hand, in boring through an old shingle beach, strata representing very various degrees of mixture of the two constituents are met with (Hill and Hanley, 1914).

Water content. The water content of shingle beaches is considerable. "In the hottest summers and on the hottest days water may be found within a few inches of the surface of a shingle bank." This water is commonly "fresh", wells sunk in shingle yielding potable water, and on the inner slope of the main bank at Blakeney Point a series of samples at a depth of 5 cm. yielded from 0·07 to 0·44 per cent of chlorides (sea-water 3·19 per cent). At a depth of from 5–9 ft. from the surface of the shingle, on a level with the standing water table at the surface of the adjacent salt marsh, the percentage of chlorides varied quite similarly, from 0·08 to 0·45, much higher values being obtained only near the edge of the marsh, from which there was probably some lateral percolation of salt water. This fresh water is no doubt derived largely from rain, but there is evidence that dew is an important contributor to it, and "internal condensation", i.e. dew formation on the surfaces of the stones in the interior of the beach, may also be a considerable factor. "Diffusion of salt from the sea to this fresh water must be so slow that the additions of fresh water are sufficient to prevent a relative increase in the salinity" (Hill and Hanley, 1914). Shingle plants do not apparently ever suffer from drought.

Humus. The humus of young shingle beaches, like that of the youngest dunes, is entirely formed from the disintegration of tidal drift, and the supply is always sufficient to provide adequate food for the shingle plants which establish themselves. The considerable amounts present in such beaches can easily be tested by delving among the pebbles with the hands, which quickly become soiled by the dark humous material in the interstices between the stones. Salisbury (1922) found the loss on ignition of the fine material from the youngest lateral at Blakeney Point to be 0·486 per cent with a pH of 7·6. On the old stabilised laterals which are covered with vegetation, the humus content was 3·19 per cent and the pH 6·9. The high "elbows" of some intermediate laterals show more extreme figures (humus 22·45 per cent, pH 6·38) and bear a much larger population of non-maritime plants.

FLORA AND VEGETATION

There are no species dominants of shingle beaches in the sense in which *Ammophila arenaria* and *Agropyron junceum* are dominant on sand dunes, and the vegetation is very poor floristically, except on certain old stabilised beaches where the flora is mainly composed of inland species, just as old stabilised dunes support inland heath vegetation. A few species are, however, very characteristic of shingle, and of these several occur also on sand dunes.

Suaeda fruticosa. One species, the shrubby sea-blite, *Suaeda fruticosa*, occupies a special position.[1] It is a perennial sometimes reaching 4 ft. (1·2 m.) in height with a woody stem sometimes 2 in. (5 cm.) in diameter and numerous more or less upright branches arising close together near the ground. The sub-cylindrical, dark green, purplish or crimson leaves are evergreen in the south but fall in late autumn in the more exposed situations on the Norfolk coast. In habitat, over most of its range, in anatomy, and presumably in physiology, the species is a halophyte, but in Britain, where it is extremely local in occurrence, it grows abundantly and luxuriantly only in connexion with shingle beaches, and mainly with the shingle spits of the north coast of Norfolk and the Chesil beach in Dorset. According to Salisbury (1922) *Suaeda fruticosa* grows in ordinary garden soil with extreme luxuriance.

Distribution at Blakeney Point. At Blakeney Point *Suaeda fruticosa* is almost restricted to the main shingle beach and to the numerous laterals, though it occurs also near the edges of neighbouring salt marsh. On old laterals it forms a dense zone on either flank[2] and is continuous over the crests of the lower laterals, which are almost always reached or covered by the higher spring tides. It also occurs on the landward slope of the main beach,[3] but only where this is protected from wave action by the lateral beaches, which act as groynes. From the more exposed stretches of the inner edge of the main beach it is absent (Pl. 157, phot. 403).

Along the inner margin of the main beach, where this abuts on well developed salt marsh the shingle advances intermittently on the marsh, driven down by the highest tides flowing over the bank when backed by an onshore gale, and forms local fans of stones spreading over the marsh (Pl. 158, phot. 404 and Fig. 175). The spring tides flow from the estuary over the marshes (i.e. in the opposite direction), and may rise several feet up the landward side of the bank, leaving a drift line of debris, but the onset of the tide here is always quiet and there is no wave action.

[1] See also Chapter XL, p. 838.   [2] Pl. 146, phot. 341.   [3] Pl. 157, phot. 401.

Dissemination and establishment. In a good year this drift is laden with ripe seeds of *Suaeda*, a handful containing several hundred seeds, and it is in this zone that germination occurs in great numbers, the young plants rapidly developing long tap-roots which reach down to the moister layers of the shingle. Their density often amounts to 20–30 per sq. ft. (*c.* 900 sq.

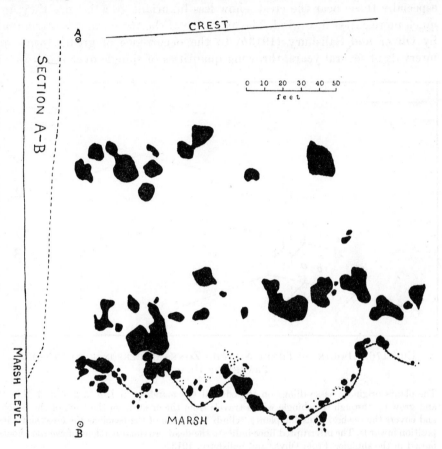

FIG. 175. DISTRIBUTION OF *SUAEDA FRUTICOSA* IN THREE ZONES ON THE BACK OF BLAKENEY MAIN SHINGLE BANK

Fans of shingle projecting on to the marsh. Dotted areas crowded with *Suaeda* seedlings. Vertical scale of section is twice the horizontal scale. From Oliver and Salisbury, 1913.

cm.), the maximum density being reached in the bays between the shingle fans where the thickest drift is deposited (Fig. 175). The conditions for germination and growth are extremely favourable since the aspect is full south and there is complete protection from north winds as well as from wave action. The shoots of *Suaeda* attain a height of 2–3 in. in the first year, 6 in. in the second, while in five years a bushy plant is developed 18 in. (45 cm.) in height and an inch in stem diameter.

882   *Shingle Vegetation*

Behind this littoral zone there are two other zones of *Suaeda* bushes higher up the bank, one just above the shingle fans and the third near the crest of the beach (Figs. 175, 178),[1] and here also seedlings occur and establish themselves, though in much smaller numbers, probably because of the sparsity of tidal drift at the higher levels. The higher lying seedlings, especially those near the crest, show less luxuriant growth since they are much more exposed to wind. The existence of the three zones is attributed by Oliver and Salisbury (1913*b*) to the occurrence of great storms at intervals of several years, throwing quantities of shingle over the crest of

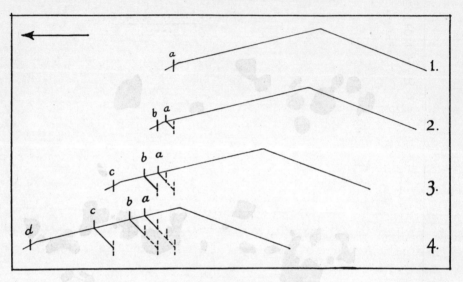

Fig. 176. Origin of Three *Suaeda* Zones in response to Inward Travel of Beach

The plants originate as seedlings on the edge of the marsh (*a* in 1, *b* in 2, *c* in 3, *d* in 4) and grow up through the shingle which travels from the crest down the back of the beach, and covers the bushes. Thus the plants "climb" the back of the beach as the crest shifts its position inwards. The interrupted lines indicate the dead proximal portions of *Suaeda* shoots buried in the shingle. From Oliver and Salisbury, 1913*b*.

the beach and down on to the marsh behind. During the intervals between these the shingle fans fringing the marsh are quiescent and for a number of years opportunities occur for the establishment of fresh *Suaeda* seedlings. The two upper zones of *Suaeda* bushes are thus interpreted as the records of former periods of quiescence of the inner fringe of the beach (Fig. 176). The bushes of these zones have grown up through the shingle with which they have been partially covered during storms that have **Effects of shingling over** forced the whole bank landwards, and this is shown by their habit. When shingle has flowed over a plant the lower part of the shoot system is bedded in the shingle and tilted forwards, the buried base

[1] Pls. 134 and 157, phot. 401.

showing repeated branching and tufts of roots arising at irregular intervals. In the middle and upper zones, where each individual bush covers a large area and is more exposed to shingle flow the original main axis of the plant is rarely recognisable, its place being taken by an underground system of prostrate shoots thickly beset with roots, and mainly directed more or less horizontally towards the downward slope of the bank. From these arise crowded leafy shoots which together form the existing sub-aerial portion

FIG. 177. BRANCH OF *SUAEDA FRUTICOSA*

The plant has been beaten down and covered with shingle, and has sent up new vertical shoots from its distal portion. One-third natural size. From Oliver and Salisbury, 1913*b* (drawing by Sarah M. Baker).

of the bush (Fig. 177). The horizontally running underground shoots with their numerous roots form a dense mat in the surface layers of shingle, checking the travel of stones down the landward slope of the beach. The older parts of these buried shoots die off and then rapidly disintegrate. Three or four feet from the insertion of the green aerial shoots the "rhizome" is usually dead and cannot be traced for more than another foot or so to the mouldering end. This rapid disintegration is probably important as a contribution to the food supply of the shingle vegetation.

The prostrate habit of the underground stems of *Suaeda fruticosa* seems

884   *Shingle Vegetation*

to be imposed by its conditions of life: they do not *grow* horizontally. On the dormant shingle of the laterals the habit of the plant is erect.

According to the observations of Oliver and Salisbury (1913*b*) the more *Suaeda fruticosa* is "shingled over" the more vigorous its vegetative response, and this applies also to other characteristic and abundant shingle plants, such as *Silene maritima* and *Honckenya* (*Arenaria*) *peploides*.

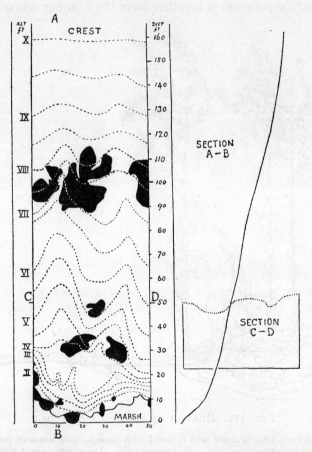

Fig. 178. Contoured strip of Blakeney Main Bank from Crest to Marsh

Two gullies are shown with *Suaeda* bushes of the three zones mainly on the ridges between. Vertical scale of section five times the horizontal scale. From Oliver and Salisbury, 1913*b*.

After a winter in which the shingle has been much disturbed there is not only an unusual vegetative display but also an extraordinary profusion of flowers on these two species.

Shallow gullies run in the shingle from near the crest of the beach to the shingle fans on the edge of the marsh. These represent the lines of travel of the shingle, and the clumps of *Suaeda* in the two upper zones lie, on the whole, between them (Fig. 178). This suggests that the plants effect a real

PLATE 157

Phot. 401. The back of the beach at the Marams from one of the mature salt marshes, showing the three zones of *Suaeda fruticosa*, the lowest (youngest) on the edge of the marsh, the highest (oldest) just short of the crest. Cf. Figs. 175, 178. *F. W. Oliver.*

Phot. 402. The crest of the beach looking west. Saltings to the left, open sea to the right. Clumps of *Suaeda fruticosa*. The tall upright plants are *Rumex crispus* var. *trigranulatus* and *Glaucium flavum*. *F. W. Oliver*, 1908.

Phot. 403. Main beach from salt marsh just west of the Marams. Here the inner side of the beach is devoid of *Suaeda fruticosa* owing to tidal scour from the west (left). On the right is the last (youngest) lateral of the Marams, lined with a dark belt of *Suaeda* which also occurs at its junction with the main bank. *F. W. Oliver*, 1927.

BLAKENEY BEACH

Phot. 404. Looking north-west along the Fleet: "deltas" of shingle project from the terrace into the Fleet, with the belt of *Suaeda fruticosa* behind and the "back" of the beach above. *F. W. Oliver* (1911).

Phot. 405. The back of the beach with patches of *Silene maritima*. Belt of *Suaeda fruticosa* on the edge of the Fleet. *F. W. Oliver* (1911).

Phot. 406. The Chesil Bank west of Abbotsbury; here a comparatively narrow fringing beach, poorly vegetated. Belt of *Tamarix anglica* on the right (inner edge of beach). *F. W. Oliver* (1911).

CHESIL BEACH

arrest of shingle travel, a suggestion confirmed by the heaping up of stones behind individual bushes or clumps. The gullies tend to be permanent when once formed, carrying fresh shingle driven down from the crest by a new series of storms.

Distribution on Chesil beach. Besides the north Norfolk beaches the other British locality where *Suaeda fruticosa* occurs abundantly and grows luxuriantly is the great Chesil shingle bank on the coast of Dorset, an account of whose structure has been already given (pp. 873, 876–8). Here *Suaeda* lines the low terrace, which abuts on the Fleet for a great part of its length (Pl. 156, phot. 400, and Pl. 158). It will be recalled that the shingle of this terrace is fed from the main mass of the bank through a series of steep-sided ravines caused by the active percolation of water through the beach from the sea at high spring tides, not, like the shingle fans spreading over the salt marsh at Blakeney, by the travel of shingle over the general surface of the bank. Fig. 179 is a chart of one of these ravines, locally known as "cans".

Correspondingly the Chesil *Suaeda* shows no power of climbing the bank, but merely sends finger-like extensions of its zone on to the floors of some of the gullies, which lie scarcely higher than the surface of the terrace itself. This incapacity to spread up the bank may probably be accounted for by the excessive mobility of the sides of the "cans" and the stability of the intervening buttresses. The stability of the Chesil shingle, except in the gullies, is evidenced by the abundance and variety of the lichens which cover its stones. The *Suaeda* bushes abutting on the stable shingle are the least healthy and some are moribund. Furthermore, the arrival of tidal drift bringing humus and seeds is much scantier than at Blakeney, for most of the Fleet is little affected by the tides owing to its very narrow and protected entrance and its great length. Altogether the conditions for active invasion and spread of the *Suaeda* are much less favourable than at Blakeney.

Suaeda fruticosa, then, though primarily a halophyte, finds, in this **Preference for shingle habitats** country, its best conditions for growth on shingle, whose soil may show quite a low chloride content; and it is actually benefited by partial covering with travelling stones. It seems probable that the freer aeration of its root system is one factor that leads to its preference for this habitat, and the parallelism with the behaviour of *Ammophila* (see p. 856) is noteworthy. *Obione* (*Atriplex*) *portulacoides* seems to share this preference for well aerated soil, though to a less marked degree.

Other flowering plants. Salisbury (in Oliver and Salisbury, 1913 a, p. 25) enumerates 60 species of flowering plants as occurring on the main shingle bank at Blakeney. Of these 20 are marked as "very rare" and 16 more as "rare". Six at least of the rare or very rare species are clearly casuals on shingle, while others are occasional colonists from the adjoining salt marsh or sand dune formations.

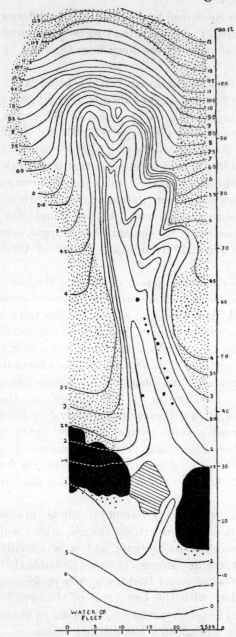

FIG. 179. CONTOURED CHART OF RAVINE ("CAN") IN THE CAMM OF THE CHESIL BANK

Stable shingle dotted, mobile shingle white. *Suaeda fruticosa* black. *Obione portulacoides* diagonally ruled. From Oliver and Salisbury, 1913*b*.

[1] Pl. 159, phot. 408.
[2] Pl. 156, phot. 399; Pl. 158, phot. 405; Pl. 160, phot. 409; Pl. 161, phot. 413.

The following twelve species are distinguished by Salisbury as "the more important":

Agropyron junceum
A. pungens
Festuca rubra
Glaucium flavum
Honckenya peploides
Rumex crispus var. trigranulatus
Sedum acre
Senecio vulgaris (forma)
Silene maritima
Sonchus arvensis var. angustifolius
S. oleraceus
Suaeda fruticosa

Suaeda fruticosa has already been dealt with at some length. Next in order of importance come *Honckenya peploides*[1] and *Silene maritima*,[2] which are abundant on all the less barren portions of the bank. The former is a characteristic sand plant of the foredunes (p. 847). It often occurs also on secondarily bared areas where the loose sand is not too deep, and is notable for its power of burrowing into underlying shingle. Its presence on shingle is almost certainly determined by the existence of considerable sand among the stones. It approaches the sea more closely than the other shingle plants (compare its position on foredunes), sometimes occurring on the seaward side of the crest of the bank. Though somewhat succulent, and easily enduring occasional immersion by spring tides, it is not quite a complete halophyte, since prolonged immersion in salt water turns its aerial shoots yellow. *Silene maritima*, while not infrequent on partly fixed dunes, is

PLATE 159

Phot. 407. *Suaeda fruticosa* (rabbit eaten) in a "low" of sandy shingle which is covered at the highest spring tides. The bushes accumulate a little blown sand. *Mrs Cowles* (1911).

Phot. 408. *Honckenya peploides*, with one plant of *Rumex crispus* var. *trigranulatus*, on the back of Blakeney main bank. The crest of the bank, with the sea beyond, is seen in the distance. *Mrs Cowles* (1911).

SHINGLE VEGETATION AT BLAKENEY POINT

even more characteristic of shingle and these two species often form extensive carpets on the banks. Their creeping stems are able to endure covering by stones as by sand, and with their deep roots and abundant aerial shoots they certainly contribute considerably to the stabilisation of the beach.

The horned poppy (*Glaucium flavum*) and a variety of the curly dock (*Rumex crispus* var. *trigranulatus*)[1] are the next most abundant species. Individuals of both are conspicuous by their erect habit and are numerous on all the less sterile portions of the bank, i.e. along the stretches where there is abundant tidal drift which brings the seeds and also the humus materials. *Glaucium* is a perennial in more protected situations on the landward side, but on the crest a biennial, not surviving its second (flowering) year. *Sedum acre*, the yellow stonecrop, is also a common plant on the same drift-laden stretches of the beach and it also behaves as a biennial, the flowering plants of one summer being almost entirely replaced by seedlings in the next, so that the greatest abundance of the species shifts from place to place.

Agropyron pungens is a characteristic plant of the edge of the zone above the marsh just touched by the high spring tides, while *A. junceum* is a pioneer psammophyte which can extend its runners in shingle (see pp. 847–8).

Other common species are the red fescue, the two sow-thistles and the common groundsel. *Festuca rubra* usually occurs between the ranks of the *Suaeda* bushes, on the sheltered side, and thus (with *Sonchus oleraceus* and *S. arvensis* var. *angustifolius*) is characteristic of the most stable shingle. (Cf. p. 889 and Pl. 136, phot. 341.) The ephemeral *Senecio vulgaris* is fairly common throughout the length of the bank.

In troughs and shallow depressions ("lows") on the landward side of the beach, of greater surface stability and with finer shingle, the grasses *Poa annua*, *Aira praecox*, *Catapodium* (*Desmazeria*) *loliaceum* and *Pholiurus* (*Lepturus*) *filiformis* occur, the first named common and characteristic, *Catapodium* local and *Pholiurus* rather rare.

Halophytes. Apart from *Suaeda fruticosa* the following halophytes occur on the main shingle bank at Blakeney Point:

Artemisia maritima	Limonium vulgare
Aster tripolium	Obione portulacoides
Glyceria maritima	Plantago maritima

Glyceria maritima and some *Obione* occur away from the salt marshes, but in regions where drift has accumulated, carried up by storms. The remaining four species and *Obione* are found in the shingle opposite the fringing marshes, mostly on the edge of the latter where the marsh soil has

---

[1] Pl. 157, phot. 402; Pl. 160, phot. 409; Pl. 161, phot. 413.

recently been covered with shingle, but isolated plants occur well up on the bank. *Artemisia maritima* was found in more than a dozen separate stations, sometimes in large patches, and mainly on the landward edge of and quite close to the lowest line of *Suaeda* bushes. It occupies a corresponding position on the lateral spits, and must thus be regarded, together with most of the other shingle halophytes, as surviving from the uppermost edge of salt marsh now covered by shingle. The plants of *Aster, Limonium* and *Plantago* close to the inner edge of the beach still have their roots in the marsh soil below.

Psammophytes. Besides *Agropyron junceum, Honckenya peploides* and *Silene maritima*, referred to above, the following sand dune plants occur as rarities on the shingle:

Carex arenaria	Eryngium maritimum
Calystegia soldanella	Senecio jacobaea
Elymus arenarius	Stellaria pallida (boraeana)

Ammophila arenaria is a not infrequent component of the shingle vegetation, and may be a relict of old dunes, now vanished, on the older part of the bank.

The sea-holly (*Eryngium maritimum*), typically a sand dune plant, occurs at Blakeney only on the shingle.

Sandy or gravelly heath plants are rare on the shingle, but *Lotus corniculatus, Festuca ovina, Holcus lanatus* and perhaps *Arrhenatherum elatius* may be reckoned as very sporadic colonists belonging to this category.

Habit of shingle plants. All the shingle plants, apart from *Suaeda*, are very low growing, except *Glaucium, Sonchus* and *Rumex*. The two former pass most of their life in the state of rosettes, only sending up their flowering shoots in the milder season. *Rumex* protects its young leaves with the old withered foliage, and all the flowering spikes may be killed by the salt-laden spray driven before a violent wind.

Vegetation of lateral banks

While the main bank as a whole shows the earlier and characteristic stages of shingle vegetation, which is here "open", the plants for the most part being scattered without very definite arrangement, the stabilised laterals exhibit the later stages of succession leading to non-maritime vegetation. The stages of succession can on the whole be very accurately traced from the younger to the older banks, on which the same zoning is constantly repeated.

Each bank presents three main topographical habitats: the sloping *flanks*, steeper on the side turned towards the apex of the main bank, i.e. originally the seaward side; the flattened or slightly convex *crest*; and the *high elbow*, or crest of the sharp bend of the L-shaped termination. The zonation from below upwards is as follows, the first three zones occupying the flank of the bank.

(1) The *Suaeda fruticosa* zone with dense bushes about 2 ft. (60 cm.) high and of varying width.[1] On its outer (lower) side it abuts on the zone of *Artemisia maritima*, here taken as the uppermost salt marsh zone. *Obione portulacoides* is common in the *Suaeda* zone, especially where the bank abuts on marshes in which it is dominant. *Aster tripolium* (scattered) and *Glyceria maritima* (sometimes frequent) are also present. *Cochlearia anglica* is restricted to this zone, but is rather rare and on the older laterals only.

Suaeda bushes in decreasing number occur in the higher zones up to the crest, but they are very scattered and of poor growth above the limit of drift at the top of zone 2.

(2) The *Festuca rubra* zone, of variable width and sometimes broken, but consisting of an almost pure sward of the grass.[1] At the upper edge of this there is often a subsidiary zone of *Agropyron pungens* (also var. *aristatum*), with which is associated *Cochlearia danica*, and more rarely *Atriplex littoralis*. Here is the upper limit of the drift.

(3) The *Limonium binervosum* zone, with much less abundant *Frankenia laevis*. The substratum is sandy mud covered with bare shingle. Numerous scattered plants of *Obione*, prostrate in habit and only 2 or 3 in. high, *Armeria maritima*, and stunted *Glyceria maritima* and *Plantago coronopus* are also present. This zone occupies the whole top of the fifth lateral, which is broad and flattened and devoid of any crest.

(4) The *crest*, which is of greater width than any of the other zones, is occupied by a sward dominated by *Armeria maritima*, *Silene maritima* and *Agrostis stolonifera*. The following associated species occur:

Arenaria serpyllifolia	a	*Pholiurus filiformis	o
Artemisia maritima	vr	Plantago coronopus	a
Catapodium (Desmazeria) loliaceum	l	*Sagina maritima	o
		Sedum acre	a
*Cerastium tetrandrum	f	*Trifolium arvense	f
Honckenya peploides	r	T. procumbens	o
Leontodon autumnalis	vr	*T. striatum	l
Limonium binervosum (relict)	f		

* Mainly on the tracks along the crests, and probably determined by the relative bareness of the banks and the low growth of the vegetation.

(5) The *high elbow* bears several "crest" species, and many others which do not occur elsewhere on the shingle banks. The following is a complete list:

Aira praecox	f	Bromus hordeaceus (mollis)	f
Arenaria serpyllifolia	o	Cirsium lanceolatum	vr
Armeria maritima	a	Erodium cicutarium	o
Bellis perennis	vr	Festuca ovina	o

[1] Pl. 136, phot. 341.

"Festuca maritima"	vr	Rumex acetosella	a
Galium verum	f	R. crispus var. trigranu-	
Geranium molle	r	latus	r
Hordeum murinum	r	Sedum anglicum	r
Koeleria cristata	vr	Senecio jacobaea	r
Lotus corniculatus	a	S. sylvaticus	vr
Plantago lanceolata	a	Silene maritima	f
Poa pratensis	a	Vicia angustifolia	o

It will be noted that the most abundant plants, except the thrift, are non-maritime species occurring on gravelly heaths, while *Galium*, *Aira*, *Festuca ovina*, *Koeleria*, *Sedum* and *Senecio sylvaticus* also belong to light inland soils. *Bellis*, *Cirsium* and *Hordeum* are only found close to the watch house and their introduction is doubtless due to human agency.

This "high elbow" vegetation clearly represents the most advanced stage of succession in the whole Blakeney Point system.

The foregoing lists are taken from the more or less stable zonation of the mature banks. The younger banks are mostly dune-covered, but one shows a younger and still open phase of the crest community, with *Honckenya peploides* and *Silene maritima*, recalling the open vegetation of the main bank, but with *Glaucium* absent, *Rumex trigranulatus* only occasional, and the addition of *Armeria maritima* and such species of the crest or high elbow of the more stable banks as *Lotus corniculatus*, *Plantago coronopus* and *Rumex acetosella*, as well as *Filago minima*.

Vegetation of Chesil beach. This has not been investigated in detail, but owing to the absence of laterals and their more varied flora as seen at Blakeney, and probably also owing to the poverty of drift on the inner margin bordering the Fleet, it is not so rich in flowering plants. The occurrence of *Suaeda fruticosa* on the Chesil terrace has already been described (p. 885). The main vegetation of the crest is dominated by *Lathyrus maritimus*, a species which does not occur at Blakeney, but which according to Oliver (1912) is at its best on mobile shingle. The stable buttresses between the percolation gullies are occupied by *Silene maritima*, which also dominates the stable shingle of the flat crest of the beach at its western end. Remarkable evidence of the good and permanent water content of shingle beaches is cited by Oliver (1912) à *propos* of this plant. At the end of the long drought of 1911, when the pasture on the neighbouring downs was so parched that sheep could find no food there, they wandered down on to the shingle bank to feed on the shoots of the sea campion, which were perfectly green and fresh (Pl. 156, phot. 399). Near the same spot *Polygonum amphibium* was found growing luxuriantly on the shingle. None of the shingle beach vegetation observed in 1911 suffered from the drought.

Other characteristic shingle species. Another species plentiful on the stable buttresses of the Chesil Bank is *Geranium purpureum* (Pl. 160, phot. 410), a very characteristic plant of stable shingle which is common on

PLATE 160

Phot. 409. *Rumex crispus* var. *trigranulatus* and *Silene maritima* (in flower) on the back of Blakeney main bank.

Phot. 410. *Geranium purpureum* Vill. in flower and fruit on shingle at the back of the present tidal beach, near Rye. This is a characteristic species of the south-eastern shingle beaches. November 1904.

Phot. 411. Freshwater marsh with *Triglochin palustre* in a "low" of an apposition beach near Rye, Sussex. Zone of *Glaucium flavum*, etc. on the slope of the "full" to the left.

SHINGLE BEACH VEGETATION

PLATE 161

Phot. 413. Shingle "fulls" with *Rumex crispus* var. *trigranulatus* and *Silene maritima*. The "lows" between the fulls are here uncolonised by plants.

Phot. 412. General view of the great apposition shingle beach west of the River Rother (Dungeness lies beyond the picture to the right). Shingle "fulls" (ridges) covered with grass used as sheep pasture in foreground. Beyond is largely bare shingle with patches of scrub.

Phot. 415. Flat expanse of shingle with patches of grass. The darker patches are prostrate blackthorns (*Prunus spinosa*).

Phot. 414. Old shingle fulls with vegetation. A tree of elder (*Sambucus nigra*) and a stack of cut gorse (*Ulex europaeus*). Scrub in the distance.

APPOSITION SHINGLE BEACHES

many south coast fringing beaches but is not recorded from Blakeney Point. *Solanum dulcamara* var. *marinum* is frequent in similar situations on the Sussex fringing beaches.

Dungeness. The great areas of apposition shingle at and near Dungeness on the borders of Kent and Sussex show much later stages of succession, for which there is no opportunity of development on the beaches hitherto described. This vegetation has never been studied at all closely, but the shingle relatively remote from the sea is known to support not only a variety of ruderal plants and grasses, but also scrub of bramble, gorse blackthorn, hawthorn and elder. This varied land vegetation is certainly correlated with greatly increased accumulation of soil, as well as progressive removal from maritime conditions, and there seems no reason why such old shingle areas should not develop climatic climax vegetation if seed parents were available. (Pl. 161 illustrates phases of apposition shingle beach vegetation.)

SHINGLE LICHENS AND MOSSES

The lichen vegetation of shingle is naturally almost confined to stable areas where the stones remain undisturbed for long periods. At Blakeney Point, where the lichens were studied by McLean (1915), the species are comparatively few,[1] while on the Chesil Bank, a provisional list from which was drawn up by Watson (1922), the number of species is larger and the lichen vegetation much more abundant. This difference is probably partly due to the much greater areas of stable shingle and partly to the milder and moister climate of the Dorset coast.

Blakeney Point. All the lichens over the whole Blakeney area, according to McLean, "with the possible exception of those upon the main bank, are subject to submersion at some time or other during the year". *Verrucaria maura* is the species most constantly submerged. It is only found on the edge of shingle where this abuts on the mud on the inner side of the bank, and here "it is submerged for several hours during every high tide, even the lowest neaps". The thallus is intensely black and slow-growing. With *Verrucaria*, Watson (Oliver, 1929) found *Placodium lobulatum* very abundant. At Burnham Overy, a few miles to the west, though not at Blakeney, *Verrucaria* is followed by an overgrowth of *Xanthoria parietina*. To the mud below the high water mark of ordinary spring tides *Collema ceranoides* and *Lecanora badia* are restricted, probably because they depend on a mud substratum: above this level the shingle is comparatively clean.

Among the colonising *Suaeda fruticosa*, on the inner edge of the bank where tidal immersion is regular and frequent, humus-bearing mud is deposited around and over the stones, and stability is high. This shingle

---

[1] The number recorded was augmented by Watson and by Richards (Oliver, 1929).

bears the following lichens, which cover almost every available spot on the firmly embedded pebbles:

†Lecanora atra	‡Rhizocarpon confervoides
*L. citrina[1]	*Verrucaria maura
*L. badia	†Xanthoria parietina

* Primary colonisers.   † Saprophytic.   ‡ Ubiquitous.

The mud slowly overwhelms the stones and reduces the lichens, so that the area will ultimately become a mud flat devoid of lichens.

On the main bank itself the seaward slope is of course destitute of lichens, and so is most of the crest. Here mobility is considerable and where the pebbles are small and there is much sand in the shingle the scour of the wind-driven sand also keeps the stones free from lichens, except in minute depressions where very small thalli may occur. Towards the landward edge of the bank patches of crustose lichens occur opposite the proximal ends of the dormant lateral banks. This is probably because the stable shingle of the lateral resists the inward thrust of mobile shingle from the sea, reducing it to a condition of comparative stability. Since however the whole bank is moving shorewards and actually embedding the proximal ends of the laterals (Fig. 171), this stability cannot be permanent. Probably it is destroyed at intervals of some years, during violent onshore gales. The intervening periods of relative quiescence during which the lichens can develop are therefore limited. The lichens of these patches are young and vigorous—evidently recent colonisers—in contrast with the mature growths on the old laterals. The thalli of these young lichens are remarkably large, as much as 2–3 in. (5–7½ cm.) in diameter, indicating quite rapid growth. The following species occur in such situations:

Lecanora citrina[1]	Rhizocarpon confervoides
*Parmelia fuliginosa	*Squamaria saxicola
P. saxatilis	Xanthoria parietina
Physcia tenella	

The two species marked with an asterisk are confined, at Blakeney, to this community. *Rhizocarpon* is almost ubiquitous above tide marks and *Lecanora citrina*[1] occurs elsewhere only on low shingle among *Suaeda*.

On loose shingle mixed with sand there are:

Lecanora atra	Lecanora galactina
L. citrina[1]	Rhizocarpon confervoides

L. galactina is a decidedly arenicolous species and colonises some of the loosest and sandiest shingle areas.

On the "bound" shingle of the lateral banks there is, according to McLean, a marked distinction between the lichens of the open communi-

---

[1] Probably *Placodium lobulatum* according to Watson (Oliver, 1929).

ties, in which crustose forms occur on the stones, and those of the closed communities dominated by grasses (*Festuca rubra* and *Agropyron pungens*) and characterised by foliose forms. The former has the following species:

Aspicilia gibbosa
Biatorina chalybeia
Buellia colludens
Lecanora atra
Lecanora atroflava
Physcia tenella
Verrucaria microspora
Xanthoria parietina

The latter community is composed of Cladonias and *Cetraria*:

Cetraria aculeata
Cladonia furcata
C. pungens

The dominant *Cetraria* and *Cladonia furcata* were present in astonishing quantity, though quite overshadowed by the grasses. *Peltigera rufescens*, a characteristic species of the grey dunes (see p. 855) occurs also on the "high elbow" of one of the banks.

Lichens of Chesil Bank. With the comparatively poor lichen flora of Blakeney Point on which McLean noted 31 species,[1] most of which have been mentioned in the foregoing account, we may contrast the richer lichen vegetation of the Chesil Bank (Watson, 1922). There is a probably larger number of species, but the more favourable conditions are mainly evidenced by the greater luxuriance of the lichen vegetation, especially on the terrace bordering the Fleet.

Biatorina chalybeia
B. lenticularis
Buellia colludens
B. myriocarpa
B. stellulata
Callopisma cerinum
C. citrinum
Cladonia rangiformis
Evernia prunastri
Lecania erysibe
Lecania porella
L. prosechoides
Lecanora expallens
L. galactina
L. hageni
L. polytropa
L. sophodes
L. umbrina
Lecidea contigua
L. nigroclavata
L. protensa
Parmelia caperata
P. conspersa
P. dubia
P. perlata
P. physodes
P. saxatilis
P. sulcata
Physcia tenella
Porina chlorotica
Ramalina cuspidata
R. farinacea
R. polymorpha
R. scopulorum
R. subfarinacea
Rhizocarpon confervoides
Rinodina demissa
R. exigua
Verrucaria maura
V. nigrescens
Xanthoria parietina

Mosses. Mosses are not common on loose shingle beaches, and they do not colonise shingle until it has been more or less bound and stabilised.

[1] Twelve additional species were however recorded by Watson (Oliver, 1929).

On the main bank at Blakeney mosses were found (Richards, 1929) only on the landward face, where species of *Bryum* were prominent:

- *Amblystegium serpens
- *Brachythecium rutabulum
- Bryum pallens
- B. pendulum
- B. spp.
- Ceratodon purpureus
- *Eurhynchium praelongum
- Tortula ruraliformis
- Trichostomum flavovirens

* In the shelter of *Silene maritima*.

On the stable lateral banks the following occurred:

- Brachythecium albicans
- Bryum spp.
- Ceratodon purpureus
- Dicranum scoparium var. orthophyllum
- Eurhynchium praelongum
- Hypnum cuspidatum var. tectorum
- Tortula ruraliformis
- Trichostomum flavovirens

Polytrichum piliferum was found on the Yankee bank. It will be noted that this list of species is much like that occurring on the grey dunes (p. 855).

From the Chesil Bank *Bryum capillare, Ceratodon pupureus* and *Hypnum cupressiforme* have been recorded (Watson, 1922).

REFERENCES

HILL, T. G. and HANLEY, J. A. The structure and water content of shingle beaches. *J. Ecol.* **2**, 21–38. 1914.

McLEAN, R. C. The ecology of the maritime lichens at Blakeney Point, Norfolk. *J. Ecol.* **3**, 129–48. 1915.

OLIVER, F. W. The maritime formations of Blakeney Point in *Types of British Vegetation*, 1911, pp. 354–66.

OLIVER, F. W. The shingle beach as a plant habitat. *New Phytol.* **11**, 73–99. 1912.

OLIVER, F. W. Some remarks on Blakeney Point, Norfolk. *J. Ecol.* **1**, 4–15. 1913.

OLIVER, F. W. Blakeney Point Report. *Trans. Norf. and Norw. Nat. Soc.* **12**, 1929.

OLIVER, F. W. and SALISBURY, E. J. Topography and vegetation of Blakeney Point. *Trans. Norfolk and Norw. Nat. Soc.* **9**, 1913a. (Also issued separately.)

OLIVER, F. W. and SALISBURY, E. J. Vegetation and mobile ground as illustrated by *Suaeda fruticosa* on shingle. *J. Ecol.* **1**, 249–72. 1913b.

RICHARDS, P. W. Notes on the ecology of the bryophytes and lichens at Blakeney Point, Norfolk. *J. Ecol.* **17**, 127–40. 1929.

SALISBURY, E. J. The soils of Blakeney Point: a study of soil reaction and succession in relation to the plant covering. *Ann. Bot.* **36**, 391–431. 1922.

WATSON, W. List of lichens, etc. from Chesil Beach. *J. Ecol.* **10**, 255–6. 1922.

Chapter XLIII

SUBMARITIME VEGETATION

Besides the well-marked maritime vegetation described in the last three chapters there are various plant communities developed near the sea but less directly exposed to maritime influences, and these we may consider together as *submaritime vegetation*.

Because of the gradual dying away of the effects of proximity to the sea as we pass away from the zones directly affected by the tides or by constant sea spray it is impossible to draw a sharp line between maritime and submaritime vegetation. Spray may be carried many miles inland by violent storms, and many non-maritime species are clearly tolerant of its incidence, while some plants which are normally halophytes can equally clearly flourish perfectly well in certain inland habitats. Again, on certain flat coasts salt water may penetrate far up the rivers and drains, rendering the soil water brackish at a considerable distance from the sea. Thus many communities developed near the coast show a mixture of maritime or submaritime and non-maritime plants. It is impossible to analyse these satisfactorily until they have been properly studied and their ecological requirements determined, including the degree of toleration of sea salt.

Tolerance of sea salt

Submaritime species. The *species* which may be called submaritime are more abundant near the sea though not absent from inland stations. Some of these are evidently able to endure considerable amounts of salt, and this capacity may well facilitate their existence in habitats free from the competition of less tolerant species to which the submaritimes might succumb in many inland habitats. Apart from the flora of brackish marshes a number of species occur more commonly in various habitats as the sea is approached, and of these the following are a small selection:

Cerastium tetrandum	Erodium maritimum
Carduus tenuiflorus	Ononis reclinata
Daucus carota	Plantago coronopus
Erodium cicutarium	Urtica pilulifera

(5) BRACKISH WATER AND MARSHES

These are developed especially on flat coasts behind the zone of salt marsh proper where the tides have no direct access and the salt water which penetrates is much diluted by fresh water coming from the land. Brackish marshes may occur along the courses of tidal estuaries at some little distance from the sea, as well as on flat land behind fringing shingle beaches and narrow belts of sand dunes where the water table is brackish. Such land has nearly always been originally reclaimed from salt marsh by "inning" and drainage and is thus intersected by ditches, and where the drainage is effective may be made into good pasture land. The drainage ditches themselves, where the water is brackish, contain aquatic species characteristic of such situations. And where the drainage is ineffective, so that the ground is still largely waterlogged, a number of reedswamp and marsh species occur. The original habitat of such plants is the brackish water of slow tidal rivers and their adjoining creeks and marshes, where the fresh water flowing from the land is mixed with salt water brought by the flood tides.

Besides the characteristic brackish water species a number of fresh water and marsh plants can tolerate small quantities of salt, and on the other hand some brackish species can grow in quite fresh water. The vegetation of the habitats described thus consists of various mixtures of all three categories. No systematic work has been done on this vegetation and very little is known about the actual requirements and degree of toleration of the species concerned.

The following are some of the species met with in these habitats, but it must be understood that the list is far from being exhaustive, and that a number of different, mainly unexplored, communities are represented.

Brackish water	*Brackish reedswamps and marshes*
Potamogeton pectinatus	Althaea officinalis
P. filiformis	Blysmus rufus
Ranunculus baudotii	Carex distans
Ruppia maritima	C. divisa
R. rostellata	C. extensa
Zannichellia palustris, etc.	C. punctata
Enteromorpha spp.[1]	C. vulpina
Vaucheria spp.	Eleocharis uniglumis
and many other algae	Oenanthe lachenalii
	Rumex maritimus
	Samolus valerandi
	Scirpus carinatus
	S. maritimus[1]
	S. tabernaemontani
	S. triqueter

Where water from fresh springs enters the upper edge of a salt marsh there is a mixture of halophytes and freshwater plants, with or without some of these semihalophytes or "brackish" species (see pp. 839–40). The salt content of the soil water often varies greatly within a very short distance.

(6) SPRAY-WASHED ROCKS AND CLIFFS

Well-drained rocks and cliffs above the reach of high tides but constantly exposed to spray, support a characteristic vegetation of halophytes. Most of these (e.g. *Limonium vulgare, Armeria maritima, Plantago maritima*) are

[1] Pl. 162, phot. 416.

Brackish Water and Marshes

abundant also in salt marshes, but the samphire (*Crithmum maritimum*)[1] is quite characteristic of the rocky spray-washed habitats. Other less frequent species of such habitats are beet (*Beta maritima*),[2] fennel (*Foeniculum vulgare*), wild cabbage (*Brassica oleracea*), queen stock (*Matthiola incana*) and sea spleenwort (*Asplenium marinum*).

As examples of definite small halophytic sea cliff communities the following are recorded by Tansley and Adamson (1926, p. 11) from ledges of a fully exposed chalk cliff on the Sussex coast (Cliff End at Cuckmere Haven). All but *Crithmum* and *Beta*, it will be noted, are salt marsh plants.

10 ft. above the beach	6 ft. above the beach
Glyceria maritima	Beta maritima
Limonium vulgare	Crithmum maritimum
Spergularia marginata	Obione portulacoides

The general vegetation of sea cliffs has never been fully described. It is actually very mixed according to the nature of the rock, the slope, the exposure and the height above the sea. The halophytic species are accompanied by chomophytes in the rock clefts and constituents of grassland and heath communities where the slope is not too precipitous and soil can accumulate; and on the mountainous west coasts of Scotland, and especially of Ireland, many alpine plants descend almost or quite to sea level. On the Irish cliffs, as Praeger has described (many publications summarised in 1934), we have the most extraordinary mixture of species of very various ecological as well as geographical status.

Petch (1933) describes the vegetation of the "high cliffs" up to 1300 ft. (395 m.) of St Kilda, a small isolated island in the Atlantic about 50 miles (80 km.) west of the Outer Hebrides, as consisting of "isolated patches of moorland growing in such cracks and ledges as afford a foothold", but with, in addition, the following curious mixture of species, several of them woodland plants, confined to this habitat:

Angelica sylvestris	f	Rumex acetosa	f
Athyrium filix-femina	f	*Salix herbacea	vr
Dryopteris dilatata	f	*Sedum roseum	a
Lonicera periclymenum	r	Silene maritima	o
Polypodium vulgare	f	Taraxacum palustre	r
Primula vulgaris	o		

* Arctic-alpine species.

The "low cliffs", within reach of the sea spray, similarly bear fragments of surrounding communities, with the addition of the following, all maritime or submaritime species:

Armeria maritima	f	Ligusticum scoticum	lf
Atriplex babingtonii	o	Matricaria maritima	a
Cerastium tetrandrum	a	Silene maritima	f
Cochlearia anglica	a		

Finally the "moorland" which descends to within a few hundred feet of the sea and is therefore affected by the salt spray consists of a very well-

[1] Pl. 162, phot. 417.
[2] Pl. 162, phot. 418.

marked community quite comparable with Praeger's *Plantago*-sward (see below) though poorer floristically. All the species are presumably tolerant of salt.

Agrostis tenuis	o	Festuca ovina	a
Armeria maritima	a	Juncus articulatus	o
Carex goodenowii	f	Leontodon autumnalis	f
C. oederi	o	Plantago coronopus	a
Cerastium vulgatum	r	P. lanceolata	f
Euphrasia officinalis	r	Sagina procumbens	f

(7) MARITIME AND SUBMARITIME GRASSLANDS

Grassland close to the sea, though quite out of reach of the tides, is always likely to be more or less affected by spray and usually contains maritime or submaritime species, besides the definite halophytes mentioned in the last section; and this whether it is situated close to sea level or on the tops of comparatively lofty cliffs. Very few ecological observations are available, and determinations of the actual amounts of salt deposited on the sward, together with a study of the autecology of the species present are especially wanted. Nearly all such grassland is more or less grazed, and this of course affects the composition of the turf; but a great deal of such land would certainly be grassland even in the absence of grazing.

Worm's Head. The most elegant observations bearing on the point are contained in McLean's paper (1935) on the maritime grassland of Worm's Head at the extremity of the Gower Peninsula in South Wales. This is developed, like that of the immediately adjoining mainland, on a heavy loam derived from the underlying Carboniferous Limestone by weathering *in situ*, and 2–6 ft. (0·6–1·8 m.) deep. This loam is ferruginous but practically non-calcareous, the calcium carbonate having been leached out during and after the process of weathering. The surface pockets of the limestone contain a dark alkaline rendzina soil (pH 7·2–7·8).

The grassland grazed by sheep contains the following species:

Agrostis tenuis
*‡Armeria maritima
†‡Bellis perennis
Calluna vulgaris
†Cerastium vulgatum
Cirsium lanceolatum (occ.)
Cynosurus cristatus
†Dactylis glomerata
Erica cinerea
Festuca ovina
‡Galium verum
Holcus lanatus
Leontodon nudicaulis
†‡Lotus corniculatus
Medicago lupulina (rare)
Picris hieracioides
*†‡Plantago coronopus

P. lanceolata
*‡P. maritima
†Poa annua
P. pratensis
Potentilla erecta
‡Poterium sanguisorba
Rubus ulmifolius (rusticanus)
Rumex acetosa
†‡Sedum acre
Spiranthes spiralis
‡Thymus serpyllum
Trifolium repens
Ulex gallii
Viola riviniana

Marasmius oreades (fairy rings)

* Maritime and submaritime species. † "Lair flora." ‡ Shallow soil over rock.

PLATE 162

Phot. 416. Brackish channel behind the present tidal beach, near Rye. *Scirpus maritimus* in the water, *Enteromorpha* sp. floating and on the bank in the foreground.

Phot. 417. *Crithmum maritimum* on maritime rocks. Kynance Cove, Cornwall. *J. Massart.*

Phot. 418. *Beta maritima* and *Silene maritima* on maritime rocks. Kynance Cove, Cornwall. *J. Massart.*

MARITIME AND SUB-MARITIME VEGETATION

The "lair flora" consists of species specially concentrated in places where sheep habitually lie and thus forms a community dependent on abundant supply of organic nitrogen.

Worm's Head itself is a rocky promontory extending due west into the sea and consisting of three separate eminences. Of these the innermost, on which sheep are grazed and which bears the species enumerated above, is separated from the middle one by a low rocky neck which the sheep do not cross. Thus the last two eminences are entirely free from the effect of the grazing factor. The outer head is too precipitous and rocky for the development of grassland, but the vegetation of the middle head, rising not more than 50 ft. (15 m.) above high water mark, contains the following species, arranged in order of frequency:

Festuca rubra subsp. eu-rubra, var. genuina, sub-var. pruinosa Hack., dominant and largely pure.

Dactylis glomerata
Holcus lanatus
*Beta maritima
*Silene maritima
Rumex acetosa
Trifolium repens
*Armeria maritima
Sonchus oleraceus
Rumex crispus
Plantago lanceolata
*Limonium binervosum
Cirsium lanceolatum
Spergularia rupicola
Leontodon nudicaulis
Galium verum

* Maritime species.

The soil is manured by sea birds (the inner headland both by sheep and by sea birds) whose droppings are full of mussel shells, which also abound in the soil, and is rich in humus (15·1 per cent). Its content in calcium carbonate averages 4·25 per cent, and the pH 7·8. It is colloidal in texture and full of ants.

The dominant fescue forms a thick mattress, through which, in places, a stick may be thrust down 2 ft. (60 cm.) before reaching the soil. The leaves are over a foot (30 cm.) long and the plant flowers freely. The associated species occur as large isolated plants half buried among the dominant grass.

This extreme dominance of the red fescue under conditions of extreme exposure to wind and spray but in the absence of grazing is a very interesting phenomenon which has not been recorded elsewhere (but see below). It will be noted that the number of species present is only half that of the equally exposed but grazed community, largely owing no doubt to the overwhelming dominance of the fescue whose competitive powers are known to be diminished by grazing. The disappearance of the maritime and sub-maritime plantains (*Plantago maritima* and *P. coronopus*) which normally form rosettes is probably due to this cause. On the other hand four maritime species occur (*Beta maritima, Silene maritima, Limonium binervosum* and *Spergularia rupicola*) which are absent from the grazed areas. The other species of this Festucetum rubrae absent from the grazed areas are the tall *Sonchus oleraceus* and *Rumex crispus*, which are doubtless intolerant of grazing.

Festucetum on sea stacks. Praeger (1911, p. 40) records "dense deep masses of *Festuca ovina*[1] and other plants" on "some small sea stacks[2] inaccessible to sheep" at Clare Island (Co. Mayo) contrasting "strongly with the closely nibbled grass of the adjoining slopes"; and this seems comparable with McLean's Festucetum on Worm's Head.

Effect of grazing. The great majority of cliff top and cliff slope grasslands are more or less grazed. Sheep reach the most unlikely looking spots and it is hard to say how far the vegetation of such places has been modified, or even in some areas determined as grassland in contrast to heath, by this biotic factor. "Even on the great scarp of Croaghmore" (in Clare Island), writes Praeger, "the sheep have left their mark behind them in the little colonies of *Poa annua, Stellaria media, Cerastium vulgatum*, etc." But it is probable, as already said, that many such sub-maritime habitats would be dominated by grasses even in the absence of sheep.

Plantago sward. In extremely exposed situations on the west coast of Ireland "where in winter gales the soil becomes soaked with spray" Praeger (1911, 1934) describes constantly recurring communities dominated by *Plantago maritima*, usually in company with *P. coronopus*, and often containing several species of grass. "A mild form" from the south coast of Clare Island, Co. Mayo, at an altitude of 50–100 ft. (15–30 m.) "forms a dense sward $\frac{1}{2}$ in. in height with flower-stems rising to about 2 in.", and has the following composition:

Plantago maritima, P. coronopus   d

Aira praecox	Hypochaeris radicata
Anagallis tenella	Koeleria cristata
Bellis perennis	Lotus corniculatus
Calluna vulgaris	Ophioglossum vulgatum
Carex diversicolor (flacca)	Plantago lanceolata
C. oederi	Polygala depressa
Centaurium umbellatum	Potentilla erecta
Cerastium tetrandrum	Prunella vulgaris
C. vulgatum	Radiola linoides
Cynosurus cristatus	Sagina procumbens
Euphrasia officinalis (agg.)	Succisa pratensis
Festuca ovina	Sieglingia decumbens
Hieracium pilosella	Thymus serpyllum
Hydrocotyle vulgaris	Trifolium repens
Holcus lanatus	Viola riviniana

"An extreme example" consisted of "a smooth shining sheet of *P. maritima* (at least 80 per cent) with mere scraps of other plants. The rosettes

---

[1] Dr Praeger (*in litt.*) is not quite certain that the species was not *F. rubra*.
[2] Precipitous rocky islets.

of *P. coronopus* and *P. maritima* measured ½–¾ in. across." The composition was as follows:

Plantago maritima	d
Aira praecox	P. lanceolata
Anagallis tenella	Potentilla erecta
Festuca ovina	Radiola linoides
Galium saxatile	Sedum anglicum
Jasione montana	
Luzula campestris	Mnium hornum
Plantago coronopus	

The same community occurs on a number of other exposed western islands (Praeger, 1934). Thus on the Great Blasket, a steep and lofty island off the western extremity of the Dingle peninsula in Co. Kerry, entirely exposed to the Atlantic weather, "the south slope is occupied mainly by short grass full of *Plantago maritima* and *P. coronopus* with patches of *Pteridium* in the hollows". On Loop Head (Co. Clare) "*Plantago* sward is well developed, giving way on the low hill tops (about 300 ft.) to dwarf *Calluna*". "Above, over a considerable area, an almost pure *Armeria* sward occupies the ground." On the island of Inishturk the community over a large area is "often composed entirely of *P. maritima* and *P. coronopus* without any other ingredient...as close and smooth as if shaved with a razor". On Achill Island (Co. Mayo) it is well developed in maximum exposure up to 300 or 400 ft. (90–120 m.), as at the extreme point of Achill Head, and forms a dense smooth mat as follows:

Plantago maritima, P. coronopus	d
Agrostis tenuis	Radiola linoides
Aira praecox	Sagina maritima
Armeria maritima	S. procumbens
Cerastium tetrandrum	Spergularia rupicola
Festuca ovina	

Here, it will be seen, other maritime species are present. On Inishkea, Co. Mayo, it is also well developed.

These *Plantago* swards are closely grazed by sheep and rabbits, and their existence, seems to be determined by a combination of extreme exposure to wind, high salt content and close grazing.

Grassland of the Sussex cliff tops. On the flat top of Beachy Head (alt. 500 ft. = c. 150 m.) the turf is very short and is quite a typical sample of chalk grassland without maritime or sub-maritime species. Farther west, where the cliffs are not so high (100–300 ft. = 30–90 m.) and may be supposed to receive more spray during strong onshore gales, the following appear in the grassland of the cliff tops:

Agropyron pungens	Erodium cicutarium
Armeria maritima	Glaucium flavum
Carduus tenuiflorus	Plantago coronopus

and *Daucus carota* becomes locally very abundant. But the maritimes do not form a quantitatively important part of the vegetation. The turf is uniformly short ($\frac{1}{2}$ in.) except in local shelter, and though grazing is widespread the extreme exposure must be an important factor. In sunny summers deficient in rainfall the drought must be very severe, and the exposure to wind maximal at all times (Tansley and Adamson, 1926, pp. 10–11).

REFERENCES

McLean, R. C. An ungrazed grassland on limestone in Wales, with a note on plant "dominions". *J. Ecol.* **23**, 1935.

Petch, C. P. The vegetation of St Kilda. *J. Ecol.* **21**, 1933.

Praeger, R. Lloyd in Clare Island Survey. Part X. Phanerogamia and Pteridophyta. *Proc. Roy. Irish Acad.* **31**, 1911.

Praeger, R. Lloyd. *The Botanist in Ireland*. Dublin, 1934.

Tansley, A. G. and Adamson, R. S. A preliminary survey of the chalk grasslands of the Sussex downs. *J. Ecol.* **14**, 1926.

INDEX

The names of plants in the lists of the various communities are not, as a rule, indexed. Exceptions are made, however, for the more important dominants, for a few large genera whose species occupy markedly different habitats, and for some other plants of vegetational importance; but the selection is necessarily arbitrary.

Aberdeen (climate), 56 (Fig. 23), 62
Aberdeenshire (beech and beechwood), 249, 406, 416, 419, 421, 424, 425
Abies alba, 188, 248
Acer campestre, 257, 272, 276, 278, 289, 294, 306, 314, 368, 373, 383, 391, 429, 469
  pseudo-platanus, 188, **257**, 294, 325, 388, 391, 392, 433, 436, 457, 461, 469
Acheulian period, 152
Achill Island, 901
Achillea millefolium, 436, 529 (Fig. 93), 530, 539, 554, **566**, 567, 568, 572, 785, 786, 857, 859
  ptarmica, 573
Achnashellach Forest, 445
acid (acidic) bog, 644, 657, 663
"acid fen", 637
acid peat, 84, 92, 100, 104, 105, 125, 673, 674, 678, 679
acid soils, 81–3, 118
acidic dunes, 860
  intrusions, 123
  peats, *see* acid peat
  rocks, 124
  schists, 123
acorns, 140–1, 291–2
Adamson, R. S., on light values in oak-hazel coppice, 274
  on societies of the oakwood field layer, 283
  on succession to pedunculate oakwood, 295
    to sessile oakwood, 311–13
  on invasion of other communities by *Pteridium*, 503
  on Nardetum, 514–16
  on changes in Eriophoretum, 706
  on "upland heath" and "heather moor", 743, 753
Adoxa moschatellina, 282, 287, 308, 430, 807, 809
adsorption, 81, 84
aeration of Cumbrian lakes, 68
  of soil, 97
  of water, *see* oxygen supply
afforestation (placing under "Forest Law"), 176, 177, 178
agricultural area, 187
  land, 87
  soils, 106
agriculture, 101–2, 178, 183, 191
  Belgic, 171
  climate and, 74–7
  mediaeval, 180
  Neolithic, 165
  Romano-British, 171

Agrostis alba, see *A. stolonifera*
  canina, 322, 344, 346, 494, 499, 500, 507, 516, 522, 666, 685, 733, 740, 748, 750, 751, 777
  gigantea, 643, 645
  "*maritima*", 856
  setacea, 501, 507, 522, 745, 747, 748
  spp., 207, 452, 492, 506, 514, 515, 554, 575, 786, 853, 860
  stolonifera, 311, 317, 464, 466, 467, 469, 494, 497, 499, 507, 510, 513, 516, 520, 534, 538, 546, 549, 550, **563**, 567, 569, 570, 572, 573, 644, 659, 662, 703, 810, 838, 839, 859, 889
  tenuis, 286, 297, 298 (Fig. 54), 311, 313, 325, 331, 335, 340, 344, 346, 353, 354, 389, 406, 415, 438, 490, 494, 496, 497, 499, 500, 501, 507, 510, 513, 516, 518, 521, 534, 538, 551, 552, 557, **563**, 567, 572, 573, 575, 644, 703, 728, 732, 740, 748, 752, 754, 757, 777, 779, 784, 860, 898, 901
Agrostis-Festuca grassland, 132, 136, 207, 313, 490, 494, **499–514**, 787
  development in Breckland, 512–14 (Fig. 90)
  flora, 501–2, 507–9, 510–11
  invasion by heath, 500
  soil, 500
Agropyretum juncei, 847–8, 851
Agropyron junceum, 847, 848, 849, 850, 857, 880, 886
  pungens, 838, 886, 889, 901
  repens, 497, 847
Ailsa Craig (bird cliffs), 144
Ajuga pyramidalis, 452, 453
Alchemilla alpestris, 317, 806, 809
  alpina, 491, 779, 780, 783, 784, 793, 798, 804, 810
  vulgaris (agg.), 785, 813
alder, 154, 162, 166, 198, 203, 243, **255**, 316, 342, 445, 446, 460, 643, 657, 660, and *see Alnus glutinosa*
alder buckthorn (*Frangula alnus*, q.v.), 265
alder carr, 461, 463, 468–70
alder wood, 273 (Fig. 51), **460–71**, 636, 660
alkaline buffer, 94, 527, 639, 663
  soils, 82, 83
  waters, 639, 663
Allium ursinum society, 430, 441
alluvial grassland, 208, 568
  soils, 97, 519, 559, 566
alluvium, 20 (Fig. 6), 26, 27 (Fig. 9), 28, 79, 88, 103
Alnetum glutinosae, 464–7, 637, *and see* alder carr, alder wood

Alnus glutinosa, 162, 203, **255**, 278, 305, 314, 316, 325, 342, 343, 344, 433, 452, 457, 460, 461, 463, 466, 467, 468, 469, 636, 637, 638, 640 (Fig. 124), 643, 645, 660, 668, 669, 670, *and see* alder
Alopecurus pratensis, 492, 496, 497, **562**, 567, 569, 570, 572, 666
Alpine earth movements, 16, 23
alumino-silicates, 81, 97, 98
Ammophila arenaria, **847–51**, 856, 857, 859, 860, 861, 888
Ammophiletum arenariae, 848–51, 853, 855, 861
 flora, 850
 physiognomy, 849–50
Anderida (Andred), Forest of, 113, 115, 176 n.
Anderson, M. L., on "alder-birch", "birch-rowan" and "birch-aspen" woods, 454
 on ash-alder woods, 470
 on rowan-birchwood, 258
Anderson, V. L., on rooting depths in chalk soil, 528, 529 (Fig. 93)
 on water content of chalk soil, 527–8
Andromeda polifolia, 686, 689, 712, 741
Anemone nemorosa, 275 (Fig. 52), 279, 281, 282, 307, 310, 316, 317, 344, 346, 347, 370, 435, 454, 500, 807, 809
 society, 281–2
animal ecology, 127
animals, effect on vegetation, 127–45
Ant, River, 592, 593, 653
Anthelia juratzkana, 797
Anthelietum, 797
Anthoxanthum odoratum, 287, 301, 311, 315, 317, 322, 323, 324, 325, 326, 340, 344, 346, 347, 392, 438, 452, 492, 496, 497, 501, 507, 515, 516, 518, 533, 537, 554, **563**, 569, 570, 572, 573, 784, 786
anthropogenic climax, 129, 225
 communities, 661–2
 factors, 129, 145, 223
anticyclone, continental, 45
anticyclones, 45
anticyclonic conditions, 44
ants, 262, 899
aquatic formations, 208
 habitats, 579
 plants, 579
arable and grassland, 205–6
 cultivation, 8, 14
 land, 75, 106, 117, 179, 182, 185, 186
arbutus, 122
Arbutus unedo, 328, 336, 338–9
Archaean rocks, 8, 10, 11, 12, 123–4, 319
arctic-alpine bryophytes, *see under* bryophytes
 climate, 43, 63–6, 781–2
 communities, 204
 edaphic factors, 783
 formations, 202, 204
 grassland, 783–7
  Ben Lawers, 784–5
  Cader Idris, 785–7
 lichens, *see under* lichens
 life forms, 781–2
 vegetation, 73, 780–813
  of mountain-top detritus, 787–96

arctic-alpine vegetation (*cont.*):
 of protected habitats, 798–813
 zone, 499, 780–1
 upper, 787–813
Arden, Forest of, 176 n.
Armeria maritima, 810, 831, 832 (Fig. 159), 833, 834, 835, 836 (Fig. 160), 841, 901
Armerietum maritimae, 831–3, 834, 840
Armorican folds, 9 (Fig. 4), 13, 18, 23
Armstrong, J. L., on Callunetum of the Mourne Mountains, 759
 on Vaccinietum of the Mourne Mountains, 762
Arrhenatherum elatius, 293, 297, 298 (Fig. 54), 317, 433, 438, 534, **536**, 547, 554, **564**, 569, 888
Artemisia maritima, 887, 888, 889
Arum maculatum, 282, 293, 308, 316, 370, 397, 435, 479
ash (*Fraxinus excelsior, q.v.*), 115, 117, 163, 188, 198, 243, **252–3**, 272, 288, 305, 316, 317, 318, 319, 320, 342, 344, 353, 374, 377, 379, 381, 382, 384, 388, 391, 397, 400, 401, 404, 405, 410, 427, 442, 443, 445, 658
ash-alder wood, 470
ash-birch wood, 442–3
ash charcoal, 173
"ash copse", 432
ash-hazel scrub, 432
ash-oak wood, 272, 288, 361, 390, 391, 394, 396–8, 442
 field layer societies, 289
 shrubs, 289
 species, 289
(ash-)oak-hazel wood, 319, 397
ash saplings, 481
 scrub, 382
ashwood, 252, 358, **427–43**, 470, 474–5
 on limestone pavement, 434–5
 on scree, 779
 and see Fraxinetum
aspect societies, 230
aspen (*Populus tremula, q.v.*), 203, **259**, 391, 429, 446, 452
association, 230, 232
associes, 232
Aster tripolium, 139, 825, 829, 830, 839, 887
Asteretum tripolii, 830–1
Atlantic period, 150 (Fig. 38), **159**, **161–3** (Fig. 44), 164, 177, 255
 times, 198, 206, 246
"Atlantis", 6
Atriplex spp., 844, 847
Atropa belladonna, 136, 375, 479
Auenwald, 208, 252
augite, 81
Avena pratensis, 527, 530, 534, 535, 536, 537, 549, 554, 563, 785
 pubescens, 433, 527, 530, 534, 535, 536, 537, 547, 554, 563
"Azores high", 44 (Fig. 15), 45 (Fig. 16), 46

back (of shingle bank), 873, 876, **878** (Fig. 174), 879
Bagshot Sand, 23, 110 (Fig. 33), 111, 175, 351, 411

Baker, H., on grazed and mown Thames-side meadows, 568–70
"balance of nature", 142, 244
Ballochbuie Forest, 253, 445, **446–7**
bare stage (in plateau beechwood), 402, 403 (Fig. 78)
barley, 76
barometric pressures (distribution of), 43–5
Barrett-Hamilton, G. E. H., on history of the rabbit, 133–4
on food of the bank vole, 141
bases, 82, 84
exchangeable, 82
basic ions, 81, 83, 84, 86, 639
rocks, 82, 554, 783
grassland, 495, 525–58
soil, 525, 783
basifuge plants, 82
Bates, G. H., on the vegetation of footpaths, 496–8
Beachy Head, 36, 901
"Beaker settlements", 165
beech (*Fagus sylvatica*, q.v.), 157 (Fig. 41), 158 (Fig. 42), 164, 188, 189, 200, 243, 245, **248–9**, 357, 374
and oak, 249–50, 361
consociation, 230
height growth, 421, 422 (Fig. 84)
immigration, 358
status, 358, 424
beechwood, 250, 265, **358–426**, *and see* Fagetum
distribution, 359 (Fig. 65)
on loam, 386–407
on rendzina, 366–72
on sand and gravel, 408–16
seres, 423 (Fig. 85)
soils, 363–6
types, 366
beechwoods, summary of, 421–6 (Figs. 84–5)
Belgic settlements, 171, 174, 175
Ben Lawers (vegetation), 783, 784, 785, 789, 799, 800
Ben Nevis, climate of, 37, 40, 43, 54, **64–6**, 781 n. 3
Benglog ashwood, 436–40
bent, common, *see Agrostis tenuis*
bent-fescue grassland, *see Agrostis-Festuca* grassland
Beta maritima, 844, 897, 899
Betula alba, *see B. pendula*
nana, 154, 744, 758, 783
pendula, 154, **250–1**, 278, 305, 311, 313, 319, 344, 347, 415, 433, 435, 444, 453, 456, 467, 468, 469
pubescens, 154, 155, **250–1**, 278, 305, 311, 313, 314, 317, 319, 325, 326, 331, 334 (Fig. 61), 335 (Fig. 62), 336 (Fig. 63), 339 (Fig. 64), 340, 342, 343, 344, 347, 352, 353, 415, 433, 437, 442, 443, 444, 446, 450, 451, 452, 453, 456, 461, 463, 465, 466, 467, 468, 469, 471, 475, 636, 640 (Fig. 124), 645, 657, 660, 669, 670, 687
spp., 312 (Fig. 57), 313, 348
tortuosa, 155, 444

Bilham, E. G., on air humidity, 34–5
on coastal and inland stations, 55
on fog frequencies, 36
on influence of western position on climate, 31
on lowest precipitations, 54
on thermal abnormality, 45–6
on thunderstorms and shallow depressions, 47
biotic climax, 129, 225, 238
factor, the, **127–146**
factors, 128, 129, 223
subclimax, 129
zonation, 132
birch, 145, 149, 154, 174, 189, 243, 351, 356, 444, 657, 756, *and see Betula*
and alder, 471
charcoal, 173
forest, 251
birch-juniper wood, 453–4
birches, the two, 154, 156, **250–1**, *and see Betula* spp.
birchwood, 119, 217, 414 (Fig. 82), 423 (Fig. 85), 444, **451–6**
Caithness, 451–3
seral, 455–6
bird cherry (*Prunus padus*, q.v.), 203, 258
"bird cliffs", 144
birds, 129, 141, 144
birdsfoot trefoil, *see Lotus corniculatus*
Birkrigg Oaks (Cumberland), 320–4
bittersweet (*Solanum dulcamara*, q.v.), 266, 660
black currant, *see Ribes nigrum*
Black Italian poplar (*Populus serotina*), 259, 467 n.
black medick, *see Medicago lupulina*
black poplar (*Populus nigra*), 259, 467
Black Wood of Rannoch, 253, 445
Blackdown Hills (Sphagnetum), 698, 748
blackthorn (*Prunus spinosa*, q.v.), 203, 261, 263, 373, 382
Blakeney Point (Norfolk), 818
salt marshes, 823, 839
sand dunes, 847, 848–9, 850, 851, 855, 856–7, 865, 866 (Figs. 163–6)
shingle beaches, 869, 870, 871, 872 (Fig. 168), 874–5 (Figs. 169–70), 876 (Fig. 172), 878 (Fig. 174), 879, 880–6 (Figs. 175–9), 885–90, 891–3, 894
blanket bog, 33, 34, 201, 674, **676**, 718, 768
climate, 35
Irish, 714–18
peat, 85, 92
Blechnum spicant, 301, 315, 322, 325, 326, 331, 340, 342, 343, 345, 346, 438, 447, 449, 450, 453, 516, 728, 740, 752, 754, 757, 785, 786, 809, 810
"bloomeries", 172
"blow outs", 847, 864
blown sand, 28, 97, 845
Bodmin Moor, 13, 17 (Fig. 5), 121, 673
Festuca-Agrostis grassland, 501–2
Molinietum, 521–2
Scirpetum caespitosi, 710–11
Sphagnetum, 698

bog ("moss"), 634, 673, **674**, 768–9
  blanket, *see* blanket bog
  communities, 676–713
  mosses (Sphagna), 82, 92–3, 99, 201, 674, **676–8**, *and see Sphagnum*
  peat, 92, 673
  raised, *see* raised bog
Boreal period, 150 (Fig. 38), **157–9**, 160 (Fig. 43), 162, 163, 167 (Figs. 45, 46), 260
Boswell, P. G. H., on four British glaciations, 152
boulder clay, 25, 26, 105, 108 (Fig. 31)
  calcareous, 272
  chalky, *see* Chalky Boulder Clay
box (*Buxus sempervirens*), 265, 368
Box Hill, 265, 368
Boxley, 265
Boxwell, 265
Brachypodium pinnatum, 146, 495, 533, 534, **536**, 537, 552, 554
  sylvaticum, 287, 289, 293, 342, 370, 375, 393, 438, 469, 470, 474, 479, 482, **536**, 554, 636
Brachythecium albicans, 511, 851, 853, 855, 858, 894
  purum, 288, 289, 318, 322, 326, 347, 397, 419, 462, 481, 530, 540, 547, 549, 550, 747, 748, 755, 857, 860
  rutabulum, 288, 289, 296, 322, 371, 382, 397, 462, 465, 469, 540, 549, 550, 748, 755, 807, 854, 894
bracken fern (*Pteridium aquilinum*, *q.v.*), **131**, 135, 136, 276, 280, 281, 307, 353, 356, 390, 502, 503, 727, 779
brackish marshes, 818, 895–6
  water, 895–6
Braid, K. W., on the spread of *Pteridium*, 503
bramble ("*Rubus fruticosus*", *Rubus* sect. *Eubatus*, *q.v.*), **263**, 276, 281, 308 (Fig. 56), 311, 382, 383, 384, 403, 404, 407, 414, 423 (Fig. 85), 476
Bramhope ponds (Leeds), 585, 587 (Fig. 104)
Brassica monensis, 849
Braunton Burrows, 846 (Fig. 162)
Breadalbane Mountains, 783
Breckland, 17 (Fig. 5), 27 (Fig. 9)
  characteristic species, 511–12
  climate, 70, 509
  grass heath, 507, **509–14**
  rabbit effects, 136–9 (Figs. 36–7)
  soils, 107–8 (Fig. 31), 507, 509
Brenchley, W., on "Broadbalk Wilderness", 293, 297
  on "Geescroft Wilderness", 294
brick-earth, 20 (Fig. 6), 125
Briza media, 433, 530, 534, 537, 547, 549, 554, **564**, 666
"Broadbalk Wilderness", 293–4, 297
Broads, *see* Norfolk Broads
Brockmann-Jerosch, H., on climatic limits of forest, 455
Bromus erectus, 495, 529 (Fig. 93), 533, 534, **535**, 536, 537, 541, 552, 554
  ramosus, 287, 316, 370, 384, 389, 393, 430
Bronze Age, 149, 150 (Fig. 38), 151, 166, 169, 172, 358

brown earth, **85–8** (Fig. 27), 92, 96, 97, 107, 273, 363
  climate, 86, 90
Bruchwälder, 252, 255
bryophyte communities at Killarney, 331–4
bryophytes of alderwood, 462–3, 465, 468, 469
  arctic-alpine, 782, 788, 790–1, 793–5, 797–8, 802–4, 805–6, 807–9, 811–12
  ashwood, 431, 439–40
  beechwood, calcicolous, 371, 384
    on loam, 389, 407
    on podsols, 411, 413, 415, 417, 419
  bent-fescue community, 501, 502, 509, 511
  birchwood, 453, 454
  chalk grassland, 540–1, 549
  Cumbrian lakes, 611
  Esthwaite fens, 645
    marshes, 635
  hazel scrub, 474
  hawthorn scrub, 481
  heath, 726, 728, 729, 731, 733, 734, 736, 737–41, 746, 747, 748–9, 755
  Molinietum, 520
  moss (bog), 676–98, 703, 704, 705, 709, 711, 712, 715, 716
  Nardetum, 516
  oakwood, 286, 288, 289, 301, 309–10, 318, 322, 326, 331–4, 341, 342, 345, 347, 348
  pinewood, 447, 449, 450, 457
  river, 624, 625
  salt marsh, 838
  sand dune, 851, 853–4, 855, 857–62, 864
  scree, 776–7
  shingle, 894
Bryum spp., 851, 894
buckthorn, alder (*Frangula alnus*), 265, 412, 657, 658, 669
  common (*Rhamnus catharticus*), 265, 373, 429, 432, 657–8
  sea (*Hippophaë rhamnoides*), 864
Bunter sandstone, 16, 116–17, 125, 352
Bure, River, 104, 592, 593 (Fig. 107), 595, 651 (Fig. 128), 652 (Fig. 129), 653
  valley, 653
burn subsere (Callunetum), 729–33, 735, 736 (Fig. 146), 751, 752, 756
burnet saxifrage, *see Poterium sanguisorba*
Burnham Beeches, 411–15
  composition and structure, 412–14
  situation and soil, 411–12
  succession, 414–15
Burren, the (Co. Clare), 473
  hazel scrub, 474
  limestone heath, 473
    pavement, 473
    succession, 473
bush colonisation of fen, 654, 655 (Fig. 131), 657, 658 (Fig. 132), 659 (Fig. 133), 668
bushland, 472–84
Butcher, R. W., on the vegetation of rivers, 622–33 (Figs. 122–3)
Butser Hill (Hampshire), 374, 378 (Fig. 71)
buttresses (of shingle bank), 873, 876–7 (Fig. 173)
Buxus sempervirens, **265**, 368

Cader Idris, 518, 783, 785–7, 808–10
Cairngorm Mountains, 40, 447, 783
Caithness, 165, 203, 522, 710
  birchwoods, 451–3
Cakile maritima, 139, 845, 847, 853
Calamagrostis canescens consocies, 643, 670
  neglecta, 666
calcareous boulder clay, 11, 26, 125
  clay, 125
  dune sand, 851–2
  river water, vegetation of, 624
  soils, 94, **98–9**
  type of lake, 609
calcicolous coppice, 382
  plants, 94, 99
  shrubs, 264–5, 272, 373, 390, 429, 432
Calciferous Sandstone, 120
calciferous pillow lavas of Cader Idris, 808–9
calcifuge plants, 82, 94, 98
calcium, 81, 82, 86, 98, 364, 527, 784
  carbonate, 80, 82, 94, 98, 389, 509, 525
  salts, 609
Caledonian folds, 9 (Fig. 4), 11, 12, 13, 14
"Caledonian Forest", 183
Calluna heath, *see* Callunetum
Calluna vulgaris, 130–1, 135, 145, 199, 203, 311, 315, 321–4, 334–6, 343, 347–8, 354, 356–7, 364, 390, 415, 417, 421, 438, 446, 449, 450, 455, 457, 500, 509, 510, 521, 551, 657, 667, 670, 682, 687, 689, 691, 704, 705, 706, 708, 710, 716, 718, 723, 726–8, 729–34, 735, 740, 758, 762, 763, 769, 770
Callunetum (*Calluna* heath and moor), 131, 136, 137 (Fig. 36), 138, 139 (Fig. 37), 207, 262, 456, 503, 504, 505 (Fig.89), 506, 519, 705, 713, **723–34**, 743, 744, **747–58**, 763, 764, 770
  arenicolum, **724–34**
    Cavenham Heath, 729
    Esher Common, 728–9
    flora of, 728–9
    Hindhead Common, 725–6
    podsol profile, 724–6 (Fig. 144)
    Sherwood Forest, 729
    soil of, 724
    structure of, 726
    succession, 729–34
      burn subsere, 729–33
      felling subsere, 734
      gravel subsere, 733–4
  on dunes, 852, 861, 864
  upland, 747–58
Calluno-Vaccinietum, 783
Calthorpe Broad, Norfolk, 463–4 (Fig. 87)
Calystegia sepium, 461, 464, 470, 659, 662
  soldanella, 139, 849, 850, 888
Cam, River, submerged vegetation, 630
Cambrian rocks, 8, 10, 11, 12, 123
Cameron, A. G., on food and shelter of red deer, 143
"camm" (shingle), 873, 876–7 (Fig. 173)
Camptothecium lutescens, 540, 541, 851, 853, 854, 858, 859
  sericeum, 371
Campylopus atrovirens, 700, 716, 755
  brevipilus, 704, 731, 755

Campylopus flexuosus, 509, 703, 709, 711, 734, 747, 748, 755, 777
"can" (shingle gulley), 873, 876–7 (Fig. 173)
Carboniferous (Mountain) Limestone, 6, 10, 11, 13, 14, 101, 119, 122, 126, 252, 253, 260, 338, 358, 366, 427, 473
  ashwood, 427–36
  grassland, 553–7
  sessile oakwood, 320
  swamps, 557
Carex associes, 603, 617
  peat, 691, 692–5, 696
Carex acuta, 557
  acutiformis, 462, 466, 468, 569, 588, 595, 652 (Fig. 130), 660
  arenaria, 136, 138 (Fig. 37), 139, 507, 509, 849, 857, 860, 863, 888
  atrata, 806, 807
  binervis, 438, 502, 507, 516, 518, 522 711, 754, 757, 785
  canescens, 520, 644, 697, 703
  capillaris, 785
  caryophyllea, 507, 510, 533, 551, 555, 568, 572, 711
  curta, 813
  digitata, 431
  dioica, 520, 752, 813
  distans, 447, 863, 896
  disticha, 467, 557, 569
  diversicolor, 530, 538, 542 (Figs. 97–8), 546, 547, 555, 557, 570, 573, 635, 711, 757, 900
  divisa, 896
  divulsa, 507
  elata, *see C. hudsonii*
  ericetorum, 512
  extensa, 896
  filiformis, 589 (Fig. 105), 591 (Fig. 106)
  flacca, *see C. diversicolor*
  flava, 520, 521, 590, 635, 785, 809, 810
  goodenowii, 516, 520, 521, 522, 569, 586, 587 (Fig. 104), 635, 644, 752, 784, 898
  helodes, 287, 316, 739
  hirta, 467, 569, 637
  hudsonii, 636, 643, 644, 645, 664, 739, 741
  hornschuchiana, 739
  inflata, 589 (Fig. 105), 590, 591 (Fig. 106), 592, 603, 617, 620, 621, 643, 644, 646, 647, 664, 665 (Fig. 134), 691, 696, 711, 739, 766
  laevigata, *see C. helodes*
  lasiocarpa, 603, 617, 654, 715
  limosa, 326, 715
  leporina, 739, 741
  muricata, 341
  oederi, 711, 739, 863, 898, 900
  ornithopoda, 555
  pallescens, 309, 317
  paludosa, 572
  panicea, 434, 453, 469, 518, 520, 521, 569, 636, 644, 662, 665 (Fig. 134), 669, 685, 700, 709, 715, 767
  paniculata, 256, 288, 461, 466, 467, 468, 469, 470, 471, 595, 644, 652 (Fig. 130), 660, 739
  pendula, 288, 289, 317, 557
  pilulifera, 309, 315, 411, 413, 507, 555, 728, 731, 732, 757, 758, 785, 786, 796

Carex pseudocyperus, 466
 pulicaris, 453, 685, 698, 739, 809
 remota, 287, 316, 326, 637
 rigida, 785, 789, 795
 riparia, 462, 467, 637, 660
 saxatilis, 813
 stellulata, 453, 520, 521, 522, 636, 685, 697, 698, 709, 739, 785, 813
 strigosa, 309, 316, 557
 sylvatica, 287, 288, 316, 317, 340, 342, 370, 430, 474, 480, 557
 vaginata, 813
 vesicaria, 588, 603, 617, 644, 645
 vulpina, 585, 637, 896
Caricetum, 691
 arenariae, 856
 (associes), 603, 617
carnivorous birds, 142, 244
 mammals, 142, 244
Carpinus betulus, **256–7**, 278, 305, 311, 313, 374
 continental range, 257
carr (fen wood), 251, 252, 255, **460–4**, 468–70, 669, 767–8
 development of, 657–60
 East Norfolk, 660
 Esthwaite, 643–6
 north Armagh, 668
 Wicken Fen, 657–9 (Figs. 132–3)
Carter, N., on salt-marsh algae, 824, 825 (Fig. 154), 826 (Fig. 155), 831, 833, 838–9
cattle, 74, 99, 130, 171, 182, 185
 fodder, 261
"Celtic fields", 172, 174, 177
Celtic cultivation, 270
 tribes, 175
Census of woodlands, 197, 202 n.
Central Plain, Irish, 9 (Fig. 4), 10, 11, 101, 125, 126, 686
Ceratodon purpureus, 509, 511, 729, 731, 736 (Fig. 146), 748, 751, 755, 790, 808, 851, 855
Cetraria, 746, 850
 aculeata, 509, 511, 512–13, 791, 798, 802
 var. *hispida*, 703, 786, 794
Chair-making industry, 359
Chalk, the, 9 (Fig. 4), 15, 17 (Fig. 5), 19, 20 (Fig. 6), **21**, 22 (Fig. 7), **23**, 24 (Fig. 8), 26, 93, **112**, 113, 115, 126, 134, 135, 164, 358, 361
 escarpment, 19, 21, 359, 360 (Fig. 66), 362 (Fig. 67), 363 (Fig. 68), 366, 372, 373, 525, 526
chalk grassland, **525–51**
 floristic composition, 528
 bryophytes, 540–1
 characteristic species, 534
  grasses, 534–7
  herbs, 538–9
 constant species, 530, 533
 exclusive species, 533–4
 lichens, 541
 rooting depths, 528, 529 (Figs. 93–6)
 soil, 524–8
  water content, 527–8
 stabilisation by grazing, 550–1
 structure, 541–5 (Figs. 97–103)
 succession, 545–52
chalk heath, 551–2

chalk plateaux, 23, 107, **172**, 361, **386**
 scrub, 372–6
 soils, 93–5, 113, 164
 uplands, 107, 163–4, 165
Chalky Boulder Clay, 26, 27 (Fig. 9), 106–7, 108–9 (Figs. 31, 32), 112, 125, 152, 174, 252
chamaephytes, 235, 407, 781–2
Chapman, V. J., on salt marsh, 820–1, 822 (Fig. 152), 824, 830, 836 (Fig. 160), 837 (Fig. 161)
 on *Obione portulacoides*, 835, 837–8
 on salt marsh algal communities, 841–2
charcoal burning, 173
chasmophytes, 799
cherry, wild, see *Prunus avium*, *P. cerasus*
Chesil Beach, 103, 869, 873, 876–8 (Figs. 172–4)
 vegetation of, 890–1, 893–4
Chiltern commons, 390–2
 beechwoods, 359, 361, 362–3 (Figs. 67–8), 366–72, 383–4, 387–93, 409–11
 escarpment, 366, 373, 374, 383
 Hills, 15, 21, 249, 359, 361, 372, 386
 plateau, 175, 177, 359, 361, 387, 390
 soils, 408–10 (Fig. 81)
chomophytes, 787, 799, 805
 shade, 806–7
 hydrophilous, 810–13
Christy, Miller, on hornbeam, 257
 on distribution of *Primula elatior*, 289
Christy, Miller, and Worth, Hansford, on Wistman's Wood, 300–2
Cirsium acaule, 529 (Fig. 93), 530, 538, 542 (Fig. 98), 544 (Fig 101)
 arvense, 849, 851
 lanceolatum, 135, 849
 palustre, 135, 544 Fig (101)
Cladietum, 653–4, 657, 661, 671 (Fig. 135), 767
Cladio-Molinietum, 661
Cladium mariscus, 186, 225, **654**, 656, 658–9, 661, 663, 669, 700, 766, 767
 peat, 691, 692, 693, 695, 696
Cladonia cervicornis, 861
 coccifera, 511, 514, 703, 709, 728, 729, 749, 756, 791
 floerkeana, 511, 514, 709, 728, 729, 731, 749, 756, 791
 foliacea, 855
 furcata, 511, 514, 731, 756, 791, 794, 795, 808, 855
 spp., 507, 509, 511, 704, 712, 724, 726, 733, 750, 757, 786, 791, 802
 sylvatica, 511, 512–13, 518, 681 (Fig. 136), 683 (Fig. 137), 685, 689, 690, 709, 711, 716, 718, 728, 729, 731, 747, 749, 756, 791, 795, 802, 808, 855
 uncialis, 511, 514, 518, 681 (Fig. 136), 685, 689, 709, 716, 729, 749, 756, 791, 794, 795, 802
Cladonia-Cetraria stage (of grass heath), 513–14
Clapham, A. R., on the alder hydrosere at Sweat Mere, 464–7
 on alderwood in the Cothill basin, 468–70
Clapham, A. R. and Baker, H., on Broadbalk Wilderness, 293–4
Clare Island, 900
Clark, J. G. D., on the Mesolithic period, 163
clay, 80, 86, **97–8**, 125

Index

clay loam, 87, 98
Clay-with-flints, 20 (Fig. 6), 22 (Fig. 7), 23, 82, 107, 361, 525
Clematis vitalba, 266, 368, 373, 382, 397, 429, 479, 547
Clements, F. E., on vegetational terminology, 231, 232 n.
climate, **29–77**, 78, 79
  and agriculture, 74–7
  and the hydrosere, 670, 671 (Fig. 135)
  and vegetation, 29–30, 68–74
  damp, 43
  general characters of British, 43
  mildness of, 55
  temperate, 43
climate-soil-vegetation complex, 95
climates, arctic-alpine, 43
  Breckland, 70
  coastal and inland, 55
  east coast, 61–3
  English midland, 66–7
  extreme Atlantic, 57–9
  Irish inland, 67
  maritime, 55
  northern inland, 63
  regional, 55–77
  Scottish mountain, 63–6
  south coast, 60–1
  west coast, 59–60
climatic climax, 129, 222, 223 (Fig. 49), 444
  formations, 202
  regions, 68–74, 85
  soil types, 79, 85, 90
climax, anthropogenic, 129
  biotic, 129
  climatic, 129, 222, 223 (Fig. 49), 444
  community, 217, 218
  dominants, 235
  fire, 145
  forest, 444
climax vegetation, 85, 218–19
closed communities, 213–14
cloth-weaving, 182
Clova-Canlochan plateau, 783
clover, red, see *Trifolium pratense*
  suckling, see *T. dubium*
  white, see *T. repens*
Coal Measures, 15, 119, 173
coastal dunes, see dunes, sand dunes
cocksfoot, see *Dactylis glomerata*
Cole, Grenville, on the "Huronian continent", 6, 11
colloidal clay, 81, 84, 86
  humus, 81, 84, 86
colonisation of gaps (regeneration) in plateau beechwoods, 403–6 (Figs. 79, 80)
colony, 232
Colt Park wood (W. Yorks.), 434–6
common fields, 178, 179 n., 180, 186
"commons", 390–1, 560, 561
Commons, Open Spaces and Footpaths Preservation Society, 192
competition, 217
  between members of a community, 217, 227
  between species, 136

compound (soil) particles, 81
Connemara and Mayo, woodland in, 343
  blanket bog of, 714–16
conifers, 197, 199
  planting of exotic, 185, 192
coniferous forest, see Northern Coniferous Forest
consociation, 230, 232
consocies, 232
continental shelf, 2 (Fig. 1), 3, 5 (Fig. 2)
Convallaria majalis, 286, 431
convectional rainfall, 47
coppice, 184, 197, 198, 270, 305, 310
coppice-with-standards, 181, 245, 267, 271–2
  origin of, 181, 270–1
  structure of, 277–8
coppicing, effect of, 310
Corn Laws, repeal of, 205
Cornish marshes, 637–8
  winters, 49, 73
Cornus sanguinea, **265**, 272, 279, 289, 294, 296, 368, 373, 382, 384, 391, 397, 429, 432, 481
  suecica, 744, 752, 783, 795
Cornwall, mild winters of, 49
corries, 783, 787, 789, 799
Corylus avellana, 155 (Fig. 39), 157 (Fig. 41), 158 (Fig. 42), 160–1 (Figs. 43–4), 203, **259–60**, 279, 289, 294, 296, 311, 313, 314, 317, 325, 326, 329 n., 340, 342, 343, 344, 368, 373, 382, 390, 391, 392, 393, 433, 435, 436, 437, 469, 473, 474, 475, and see hazel, hazel coppice, hazel scrub
Corynephorus canescens, 857
Cothill basin (Berks), 468
Cotswold Hills, 17 (Fig. 5), 18, 115, 126, 171, 180 n., 206, 358, 366, 372
  beechwoods, 116, 358, 372
  oolite escarpment, 177
cotton grass, 701, and see *Eriophorum*
  moss, 701, and see Eriophoretum
country planning, 192
cowberry (*Vaccinium vitis-idaea*), 203
crab apple (*Pyrus malus*), 258
crack willow (*Salix fragilis*), 259
Crampton, C. B., on Caithness birchwoods, 451–3
  on destruction by red deer, 143
  on Highland Molinieta, 522
  on stable and migratory vegetation, 238–9
cranberry (*Oxycoccus quadripetalus*), 686, 712 (Fig. 141)
Crataegus, 260, 306, 311, 313, 434
  monogyna, 260, 306, 313, 480, 483 (Fig. 88), 657, and see hawthorn, hawthorn sere, hawthorn scrub
  oxyacantha, see *C. oxyacanthoides*
  oxyacanthoides, 260, 306, 480, 483 (Fig. 88)
Crawford, O. G. S., on "Celtic fields", 172
crest (of shingle bank), 878 (Fig. 174)
Cretaceous rocks, 16, 18–23, 112–15
  sea, 4, 15
crowberry, see *Empetrum nigrum*
crumb structure (of soil), 81
Crump, W. B., on heather moor peat, 753
Cryptogramma crispa, 776–7, 778, 779

cryptophytes, 235
Culm measures, 119
Cumbrian lakes, vegetation of, 597–618
  deep-water communities, 609–15
  floating leaf communities, 616
  reed swamp, 616–17
  shallow-water communities, 615–16
  succession, 614–15, 617, 618 (Fig. 120)
  and see under Esthwaite Water
current, river, 580, 622–3, 625–8, 630–1
  medium to slow, 628
  moderate, 627–8
  moderately swift, 626
  very rapid, 624–5
  very slow or negligible, 631
curved timber, 270
cyclical alternation of forest and heath, 355
cyclones, 46
cyclonic precipitation, 47
Cynosurus cristatus, 496, 497, 537, 550, 562, 569, 572, 573, 575, 666
Cyperus longus, 637

Dactylis glomerata, 297, 327, 417, 433, 436, 438, 452, 496, 497, 533, 534, 537, 549, 550, **562**, 567, 569, 570, 572, 573, 575, 666, 786
Damp oakwood, 271, 304
  field layer societies, 282–3
  species, 287
Danes, invasions of, 176
Darby, H. C., on Royal Forests, 176 n., 177, 177 n.
  on drainage of the Fens, 186–7
  on enclosure of forest, 178
  on open field cultivation, 178
Dartmoor, 7 (Fig. 3), 10, 13, 17 (Fig. 5), 34, 36, 121, 125, 298–9, 673
Davies, A. M., on the Palaeozoic region, 8
Davies, Wm., on grass verges, 497–8
  on percentage of Pteridietum in Wales, 502
  on percentage of Ulicetum in Wales, 504
  on percentage of Nardetum and Molinietum in Wales, 516
deciduous summer forest, 85, 95, 162, 199, 202, 444
  woodland, 197
De Geer, G., on geochronology, 149 n.
deer, fallow, 144
  ravages of, 448, 451
  red, 143, 446
  roe, 144, 448
deer forest, 124, 446, 751
deflected seres, 225
  succession, 297, 767
Denmark, 4
  percentage of woodland, 196
depletion of forest, 181
depressions (barometric), 46
  secondary, 46
Deschampsia alpina, 806
  caespitosa, 284 (Fig. 53), 288, 316, 345, 346, 353, 389, 392, 430, 438, 447, 470, 474, 493, 520, **538**, 541, **564**, 569, 572, 573, 636, 644, 666, 785

Deschampsia flexuosa, 207, 300–1, 315, 322, 325, 326, 337, 347, 353, 354, 411, 412, 413, 415, 417, 418–19, 421, 437, 438, 447, 449, 450, 452, 454, 494, 501, 506–7, 515, 516, 521, 522, 554, 710, 711, 728, 731, 732, 743, 747, 748, 750, 752, 754, 756, 757, 759, 762, 777, 784, 793, 795, 809, 810, 860
destruction of forest, 178, 179, 183, 184, 189, 192, 194
  of seeds and seedlings, 245
Devonian age, 10, 121
  conglomerates, 13
  limestones, 13, 126
  rocks, 10, 14, 121
Devonshire, 13, 61, 67
  east, 19, 21, 62, 441
dewberry (*Rubus caesius*, q.v.), 263, 429, 470, 480
Digitalis purpurea society, 285
Diplophyllum albicans, 310, 318, 322, 333, 345, 411, 413, 703, 749, 755, 776, 779, 791, 798, 803, 805, 808
disafforestation (removal from Forest Law), 176, 177
distribution of vegetation, 194–210
Dixon, H. N., on endemic mosses, 4
dog rose, *see Rosa canina*
dogstail, crested, *see Cynosurus cristatus*
dogwood (*Cornus sanguinea*, q.v.), 265, 272, 373, 396, 429, 432
Domesday survey, 175, 176, 270, 568
dominance, 214
dominant trees, 246–56
Donegal woodland and scrub, 342–3
Dove, River, 624, 626
downs, 172, 180 n., *see* North Downs, South Downs
draining of the Fens, 186–7
Dry oakwood, 271, 304
  field layer societies of, 280–2
  species, 286
Dryas, 150 (Fig. 38), 154
Dryopteris thelypteris, 256, 461, 462, 468, 656, 659, 660, 662
Du Rietz, G. E., on vegetational terminology, 232 n.
duckweeds (*Lemna*, q.v.), 595
dune areas:
  Ayreland of Bride, 852–3, 865
  Blakeney Point, 847–8, 848–9, 850, 851, 855, 856, 857, 865
  Castle Gregory, 859, 867
  Dog's Bay, 848, 857–8, 867
  Rosapenna, 858, 867
  Sandscale, 863
  South Haven peninsula, 861, 862
  Southport, Lancs, 863, 865
  Tramore, Donegal, 857, 867
  Walney Island, 852, 867
dune grassland, 859
  heath, 860–1
  marsh, 862
  scrub, 864
  slacks, 861–3
  soils, 851–2, 865–7

dune soils (*cont.*):
    Ayreland of Bride, 865
    Blakeney Point, 865, 866 (Figs. 163–4)
    Southport, 865, 866 (Figs. 163–4)
  subseres, 861
  succession, 847, 855
dunes, fixed, 852–61
  grey, 850, 853, 855
  mobile, 847–52
  white, 850, 852
  yellow, 850, 852, 855
Dungeness, 56 (Fig. 23), 869
  climate, 60, 61
  vegetation, 891
durmast oak, *see Quercus sessiliflora*
dwarf birch (*Betula nana q.v.*), 154
dwarf gorse (*Ulex minor, q.v.*), 261–2
dykes, 650

earthworms, 83, 364
East, G., on increase of tillage in the eighteenth century, 187
East Anglia, 26, 27 (Fig. 9), 125, 151, 173, 179, 251
East Anglian climate, 54, 670
  Fenland, 104, 125, 648 (Fig. 126), 649–51
    draining, 186–7
  fens, 252, 254, 265, 639, **649–62**, 670
  Heights, 15, 17 (Fig. 5)
  soils, 107, 108 (Fig. 31), 109 (Fig. 32)
  Tertiary deposits, 107
East Norfolk fens, 651–3 (Figs. 127–30)
ecosystem, 127, 128, 142, **228**
ecotone, 215
ecotope, 228
edaphic climax, 224
  factors, 222
  formation, 236
edge colonisation, 261
elder (*Sambucus nigra, q.v.*), **264**, 306, 368, 373, 375, 382, 429, 891
Eleocharis palustris, 585, 586, 587 (Fig. 104), 635
Elgee, F., on Cleveland Scirpetum caespitosi, 711
  on upland heath and heather moor, 743, 750–1
Elliott, J. S. B., on salt-marsh fungi, 833–4
Ellis, E. A., on carr at Wheatfen Broad, 461–3
elm, 154, 162, 163, 188, 198
elms, 258, *and see Ulmus*
Elodea canadensis, 584, 588, 599, 600, 612, 613, 614, 627, 628, 629 (Fig. 122), 630, 631
Elodetum, 588, 589 (Fig. 105), 591 (Fig. 106)
eluvial horizon, 87, 89
eluviation, 79
Elymus arenarius, 848, 850, 888
  europaeus, 389, 430
embryo dunes, 847
Empetrum nigrum, 167 (Fig. 45), 448, 450, 518, 697, 703, 704, 705, 706, 711, 712, 718, 744, 748, 750, 752, 754, 757, 761, 762, 779, 786, 789, 793, 796
enclosure, 179, 180, 182, 185, 186, 188, 191
endemic species, 4
Eocene beds, 23, 107

Epilobium angustifolium, 285, 735, 736 (Fig. 146)
  society, 285
epiphytes, at Killarney, 328
  in Wistman's Wood, 301
epiphytic bryophytes, 301, 331–3
Epping Forest, 177
  vegetation of, 415–16
equilibrium (of vegetation), 128, 142, 143, 215, 217, 221, 225, 226, 227, 228, 234
Equisetum fluviatile (limosum), 585, 588, 589 (Fig. 105), 590, 603, 621, 637
  consocies, 617, 620 (Fig. 121), 647
Erdtman, G., on Scottish pollen diagrams, 444
Erica ciliaris, 741, 744, 764
  cinerea, 323, 324, 356, 390, 415, 508, 522, 551, 711, 712, 716, 717 (Fig. 142), 718, 726, 729, 732, 733, 734, 747, 748, 750, 751, 752, 754, 757, 758, 761, 861
    in Highland oakwoods, 345, 346, 347
    in Irish oakwoods, 337, 340
    refused by animals, 130, 135
  mackaiana (mackaii), 715
    × *praegeri*, 715
  tetralix, 135, 415, 449, 502, 518, 520, 521, 522, 621, 646, 667, 670, 681 (Fig. 136), 683 (Fig. 137), 685, 686, 688 (Fig. 139), 689, 697, 698, 700, 703, 704, 708, 709, 710, 711, 712, 715, 716, 718, 727, 728, 732, 735, 740, 747, 748, 752, 754, 757, 769
  vagans, 744, 764
Eriophoretum, 676, **701**–7, 771
  invasion by other species, 705–6
  retrogression, 705–7
Eriophoretum angustifolii, 666, **704**
Eriophoretum vaginati, **701**–4 (Fig. 140)
Eriophorum, 667, 674, 710, 715, 750, 751, 761, 762, 770
  peat, 704, 707, 743, 762
  angustifolium, 471, 644, 656, 657, 680, 681, (Fig. 136), 683 (Fig. 137), 685, 686, 688 (Fig. 139), 689, 691, 697, 698, 700, 701, 703, 704, 706, 708, 709, 711, 712, 716, 718, 741, 757, 768, 769
    peat, 692
  gracile, 471
  latifolium, 698, 700
  vaginatum, 681 (Fig. 136), 683 (Fig. 137), 685, 686, 688 (Fig. 139), 689, 691, 696, 700, 701, 703, 704, 706, 709, 711, 712, 715, 716, 718, 719, 754, 757, 769
    peat, 692, 694
Eryngium maritimum, 847–8, 849, 850, 853, 888
escarpment beechwoods, 359, 366, *and see Fagetum calcicolum*
Esthwaite fens, 639–46 (Figs. 124–5), 653
  fen soils, 639, 642
  fen succession, 642–6
  marshes, 635–7
Esthwaite Water, 597, 598 (Fig. 110)
  vegetation of, 597–606 (Figs. 112–15)
    deep-water communities, 597, 599–600
    floating leaf communities, 603
    reedswamp communities, 603
    shallow-water communities, 600

Euonymus europaeus, **265**, 272, 279, 289, 368, 373, 382, 391, 397, 429, 437, 469, 473, 474 n., 475
Euphorbia amygdaloides, 283, 287, 370
   society, 283
  paralias, 848, 849, 850
  portlandica, 849
eutrophic peat, 673
evaporation, 31–4, 79, 80
Evelyn, John, influence of, 184, 185
even age of trees, causes of, 401–2
Exmoor, 17 (Fig. 5), 143, 191, 673, 748
exploitation of woods, methods of, 245

Fagetum calcicolum, 366–72
  biological spectrum, 371
  bryophytes, 371
  development, 383–4
  distribution, 366
  larger fungi, 371–2
  mycorrhiza, 368
  structure and composition, 367–72
  succession, 372–84
Fagetum ericetosum, 408–21
  composition and structure, 410
  larger fungi, 413, 416
  mycorrhiza, 410
Fagetum rubosum, 386–407
  bryophytes, 389
  field layer, 388–9
  larger fungi, 389
  mycorrhiza, 387
  role of *Rubus*, 388
  slope and soil, 386
  structure and composition, 387–9
  succession, 390–403
Fagus sylvatica, **248**–9, 325, 475
  continental range, 248
  distribution in Britain, 248–9
  soil preferences, 249
  status, 358, 424
  and see beech
Farrow, E. P., on *Agrostis tenuis* favoured by increased water supply, 500–1
  on Breckland lichens, 734
  on Breckland podsols, 725 (Fig. 144)
  on *Pteridium* invading Callunetum, 727, 728 (Fig. 145)
  on rabbits and vegetation, 129, 136, 139, 500
  on rabbit-resistant plants, 135
  on a subsere initiated by blown sand, 733–4
  on valley fen woods, 467–8
fallow deer (*Dama vulgaris*), 144
felling subsere to Callunetum, 734
felspars, 81
fen, 208, 583, 634, **639–68**, 669
  and moss (bog), 647
  carr, *see* carr
  droves, 662
  limits of, 646–7
  peat, 84, 85, **88**, 99, 100, 104, 124–5, 639, 649, 653, 673, 676, 682, 691, 693, 695
Fenland, the, 179, 592, 648 (Fig. 126), 649, 651
  draining of, 186–7

fens, East Norfolk, 650 (Fig. 127), 651
  Esthwaite, 639–46, 653
  North Armagh, 662–8
  Salix repens var. *fusca* on, 264
Fenton, E. Wyllie, on sheep grazing, 131–2
fescue, sheep's, *see Festuca ovina*, 136
  meadow, *see Festuca pratensis*
  red, *see Festuca rubra*
fescues, the fine-leaved, 499, 500–1, 534–5, 562–3
Festuca elatior, 497, 527, 538, 541, 554, 557, **562**, 569, 572, 666
  gigantea, 340, 389, 393
  loliacea, 569, 572
  ovina, 225, 490, 494, 497, 499, 500, 501, 507, 510, 513, 516, 518, 521, 522, 530, **534–5**, 536, 537, 541, 542 (Figs. 97–9), 546, 549, 550, 551, 552, 553, 554, 557, **563**, 567, 572, 573, 575, 666, 731, 777, 779, 784, 785, 786, 794, 796, 860, 888, 890, 898, 900
    forma *vivipara*, 777, 784, 789, 793, 809, 810
  pratensis, 496, 534, 541, **562**, 569, 570, 572, 573
  rubra, 208, 225, 311, 490, 491, 494, 495, 499, 506, 507, 530, **534–5**, 538, 541, 546, 552, 554, 562, 567, 569, 570, 572, 575, 833, 838, 840, 859, 860, 886, 887, 899
    var. *arenaria*, 849, 850, 853, 857, 858, 863
Festucetum rubrae, 823, 824 (Fig. 153), 838–9, 899–900
Festuco-Agrostidetum, development, 512–14 (Fig. 90)
Ficaria verna, 275 (Fig. 52), 279, 282, 308, 310, 316, 438
field layer, 214, 276
field maple, *see Acer campestre*
Filipendula ulmaria society, 283
Finland, percentage of woodland, 196
fire, 145–6
fixed dunes, 852–6, 857, 858, 859
  effect of disturbance on, 864
"floating-leaf", 582–3
floods, fenland, 187
  river, 623
fog, 35–6, 38
Fontinalis antipyretica consocies, 599, 606, 614
footpath vegetation, 496–8
foredunes, 847, 848
foreshore communities, 844–5
forest climate, 236
  law, 176
  regions, 172–3
  resources, 181
  upper limit of, 173, 775
Forest of Anderida (Andred), 113, 115
  of Arden, 176 n.
  Ballochbuie, 253, 446–7
  "Caledonian", 183
  of Dean, 176 n., 177, 319
  Epping, 177, 415–16
  Glenmore, 449
  Rothiemurchus, 253, 447–51 (Fig. 86)
  Sherwood, 182, 352–4
  Windsor, 176 n., 177

Forestry Commission, 191–2, 344, 346, 445, 448
  census of woodlands, 197
formation (plant), 229, 232
formation-type, 229, 232
Fort Augustus, 63, 64
Fort William, 64, 65
fossil podsol, 90 n.
foul waters, 584
Foulshaw Moss (N. Lancs), 686, 704, 705
Fox, Cyril, on Neolithic and Beaker settlements, 164, 165
foxes, 142, 244
fox coverts, 245, 270
  hunting, 244–5
foxtail, *see Alopecurus pratensis*
France, percentage of woodland, 196
Frangula alnus, 255, **265**, 306, 357, 412, 461, 463, 470, 645, 657, 658 (Fig. 132), 659 (Fig. 133), 669
Fraser, G. K., on Inverliver Forest, 707–8
  on the Meyer ratio, 33–4
  on Molinietum, 519, 520, 522
  on Scirpetum caespitosi at Inverliver, 708
Fraxinetum calcicolum, **427–43**
  distribution, 427–8
  composition and structure, 428–31
  situation and soil, 428
  succession, 432–3
Fraxinetum on scree, 779
Fraxinus excelsior, 163, **252**–3, 278, 289, 296, 313, 316, 326, 368, 461, 467, 469, 474, 475, 482, 660, *and see* ash
Fream, W., on flora of water meadow, 571–2
free-floating plants, 582–3
freezing point, 39
Fritsch, F. E., on endemic algae, 4
Fritsch, F. E. and Parker, W. M., on Hindhead Callunetum, 727–8
  on *Ulex minor* in Callunetum, 727
Fritsch, F. E. and Salisbury, E. J., on the burn subsere at Hindhead Common, 729–33
  on the gravel subsere, 733
frost, **39–40**, 78, 707
Fruticetum, 472–84
fungi, soil, 78, 83
  salt marsh, 833–4
  the larger, of Fagetum calcicolum, 371–2
  of Fagetum ericetosum, 413, 416
    rubosum, 389
  of pedunculate oakwood, 286–7, 288
  of subspontaneous pinewood, 458

Galeobdolon luteum society, 283
gales, 42, 43
  on Ben Nevis, 65
galingale (*Cyperus longus*), 637
Gault, 19, 20 (Fig. 6), 22 (Fig. 7), 113
gean (*Prunus avium, q.v.*), 258, 305, 374
"Geescroft Wilderness", 294
geochronology, 149
geodynamic factors, 239
geographical elements, 234
geophytes, 235
Geranium purpureum on shingle, 891
Germany, percentage of woodland, 196

Gilbert, E. W., on Roman wool export, 172
  on Roman iron smelting, 173
glacial deposits, 25–6
  drift, 27 (Fig. 9), 101, 105–7
  gravels, 26, 125
  sands, 26, 125
  succession on, 311
  till, 105
glaciations, 149, 151–3, 154
Glasspoole, J., on highest precipitations, 54
Glaucium flavum, 847, 886, 887, 888, 890
Glaux maritima, 139, 830, 833, 835, 838, 839, 862
Glengariff oakwoods, 340
Glenmore, 7 (Fig. 3), 8, 63, 123, 344, 445
Glenmore Forest, 447, 449
gley horizon, 86
Glyceria maritima, 139, 208, 491, 823, 825, 829 (Fig. 158), 831, 833, 834, 835, 836 (Fig. 160), 837, 840, 887
  maxima (*aquatica*), 186, 595, 653, 654, 669
    fen, 654
    reedswamp, 630–1
Glycerietum maritimae, **829–30**, 833, 834, 840
Glycerietum maximae, 630, 653–5
goat willow, *see Salix caprea*
Godwin, H., on beech pollen in Cambridgeshire, 358
  on deflected succession, 225
  on developing carr at Wicken Fen, 658–9 (Figs. 132–3)
  on post-glacial tree pollen, 155 (Fig. 39), 157 (Fig. 41), 158 (Fig. 42), 160 (Fig. 43), 161 (Fig. 44)
  on Tregaron Bog, 681–3 (Figs. 136–7)
  on Trent valley ponds, 584–5
Godwin, H. and Bharucha, F. R., on Cladietum at Wicken Fen, 654
  on critical water level at Wicken Fen, 657
Godwin, H., Godwin, M. E. and Clifford, M. H., on Sub-Boreal advance of pine, 165
Godwin, H., Mobbs, R. H. and Bharucha, F. R., on peat of Wicken Fen, 653
Godwin, H. and Tansley, A. G., on anthropogenic communities at Wicken Fen, 661–2
Godwin, H. and Turner, J. S., on carr succession at Calthorpe Broad, 463, 660, 669
Good, R. d'O., on sand dune Callunetum, 852, 861
Goodwood area, 360 (Fig. 66)
gooseberry, *see Ribes grossularia*
gorse (*Ulex, q.v.*), 146, 261–2, 263, 313, 391, 475, 891
  scrub, 313, 354, 504
Gower Peninsula (S. Wales), 898
grass heath, 490, 499, **506–14**
  flora, 507
  of Breckland, 509–14
  succession in Breckland, 512–14
grass minimum temperatures, 39
grassland, 74, 129–33, 178–9, 204–8, **487–576**
  acidic, 494, 499–524
  alluvial, 491, 559, 568–70, 571–2
  arctic-alpine, 491, **783–7**
  basic, 495, **525–58**

grassland (*cont.*):
  chalk, 206–7, 491, **525–51**
  limestone, 206–7, 490–1, 552–7
  maritime, 491, 898–902
  neutral, 495, **559–76**
  siliceous, 207–8, 490, 499
  sub-maritime, 898–902
grassland climate, 235
gravel, 80, **97**
  glacial, 26, 125
gravel subsere (Callunetum), 733–4
grazing, 129, 205, 375, 454, 475, 488–9, 496, 499, 560–1, 723, 753
  aftermath, 567
  by cattle, 129–31
  by rabbits, 133–40
  by sheep, 129–31
  factor, 132–3
  pressure, 261
  regime, 496, 560
"Great Ice Age", 149
Great Ridge Wood (Wiltshire), 174
"greens", 560, 561
grey dunes, 850, 853, 855
grey poplar (*Populus canescens, q.v.*), 259
ground layer, 214, 276
grouse (*Lagopus scoticus*), 4, 751, 757
grouse moor, 124, 128, 751, 765
Grovely Wood (Wiltshire), 174, 176 n., 177
guelder rose (*Viburnum opulus, q.v.*), 265

habitat, **215–16**
Hagley Pool (Oxford), 586, 588
Haines, F. M., on Hindhead Common soil, 724–6
Hall, A. D. and Russell, E. J., on Kent and Surrey sands, 352
halophytes, 819, 845
halophytic vegetation, 99
halosere, 820, 825, 826
Hampshire basin, 107, 111, 125
Hampshire uplands, 107, 171, 180 n.
hares, 142
Harley, J. L., on beech mycorrhiza, 368, 387
  on two upland ashwoods, 433–6
Harris, G. T., on species from Wistman's Wood, 301
Hastings beds, 19, 113
hawthorn (*Crataegus monogyna*), **260–1**, 262, 306, 353, 372, 373, 377, 379, 383, 384, 391, 396, 397, 429, 432, 475, 476, 480, 657
  large fruited (*C. oxyacanthoides*), 260, 306, 480
hawthorn scrub, 295, 372–5, 376, 377, 380, 382, 383, 441, 480–4
  sere, 372, 373, 375, 382, 383
hawthorn-mercury sere, 384
hay crop, 179, 205
Hay Meads (Oxford), 568–70
hazel (*Corylus avellana, q.v.*), 154, 181, 203, **259–60**, 267, 306, 311, 382, 383, 429, 432, 442, 445, 474
  coppice, 245, 250, 259, 267
  scrub, 260, 474
heat, 30, 78, 79

heath, 199–200, 202, 203, 351, 353, 354, 356, 357, *and see* Callunetum
  and heather moor, 743–7
  formation, **723–72**
    distribution, 723–4, 763–5
    status, 765–6
  grasses, 353
  invasion by, 490, 500
  "pasture", 489
  plants, 490
heather, *see Calluna*
"heather moor", 128, 674, 743–7, 753–7
heathland, 176, 224
Hedera helix, **266**, 279, 293, 298, 300, 315, 331, 336, 341, 343, 344, 368, 369, 382, 384, 392, 429, 435, 437, 461, 473, 474
helophytes, 235
hemicryptophyta rosulata, 371
  scaposa, 371, 407
hemicryptophytes, 235, 369, 371, 487
Hercynian continent, 15, 16
Herefordshire plain, 14, 17 (Fig. 5), 121, 175
Hertfordshire sessile oakwoods, 304–11, 350–1
Heslop Harrison, J. W., on salt marsh, 825, 830, 831
"high moor", 674
"Highland Line", 2 (Fig. 1), 9 (Fig. 4), 17 (Fig. 5)
Highland oakwoods, 343–9
"Highland zone", 194
Highlands, *see* Scottish Highlands
Hill, T. G. and Hanley, J. A., on the water content of shingle beach, 879
"hill pasture", 489, 499 n.
Hippophaë rhamnoides, 266, 864
Hippophaëtum, 864
historical factors, 216
Historical Period, the, 150 (Fig. 38), **171–93**
Hochmoor, 201, 673, 674
Holcus lanatus, 316, 318, 345, 496, 521, **538**, 547, 550, 554, 564, 570, 572, 575, 666, 696, 888
  mollis, 281, 285, 286, 301, 307, 308, 315, 318, 325, 327, 345, 346, 353, 389, 392, 406, 412, 413, 417–19, 452, 506, 507, 562
Holland, percentage of woodland, 196
holly (*Ilex aquifolium, q.v.*), **257**, 276, 306, 328, 330, 336, 374, 391, 410, 412, 442
Honckenya peploides, 847, 884, 886, 889
honeysuckle (*Lonicera periclymenum, q.v.*), **266**, 279, 353, 429
hooks (of shingle beaches), 869, 870–3 (Fig. 168), 874 (Fig. 169), 875 (Fig. 170), 876 (Fig. 171), 878
hop, *see Humulus lupulus*
Hope Simpson, J. F., on chalk grasses, 535
  on chalk bryophytes, 540–1
  on chalk spoil heaps, 546–8
Hopkinson, J. W., on Sherwood Forest, 352–4
  on Sherwood Forest Callunetum, 729
Horwood, A. R., on relict *Nardus*, 518
hornbeam (*Carpinus betulus*), **256–7**, 305, 311, 374
  coppice, 305–7
hornblende, 81

humidity (air), 30, 33–5, 65
Humulus lupulus, 461, 660
humus, 81, **83–5**
   podsol, 90
hybridism of oaks, 348–9, 351, 352
hydrogen, 82
   ions, 81, 82
hydrolysis, 81, 84
hydrophytes, 235
hydrosere, 220, 464, 491, 582–3
hydroseres, the, **579–740**
   the later, 668–71
hydroxyl ions, 81
Hylocomium, 203, 375, 447, 502
Hylocomium brevirostre, 333, 341, 474
   loreum, 289, 301, 318, 322, 326, 333, 341, 347, 397, 440, 447, 450, 453, 454, 709, 748, 755, 777, 790, 795, 805, 807
   splendens, 288, 318, 322, 333, 335, 341, 345, 347, 348, 417, 431, 439, 440, 447, 449, 450, 453, 454, 540, 708, 709, 748, 755, 777, 790, 795, 805, 814
   squarrosum, 288, 318, 347, 447, 474, 481, 501, 533, 540, 711, 748, 755, 777, 790, 805, 807, 854
   triquetrum, 288, 289, 341, 342, 345, 347, 348, 382, 397, 419, 439, 440, 447, 449, 454, 474, 481, 533, 540, 748, 755, 790, 805, 857
Hymenophyllum peltatum, 809
   tunbridgense, 331, 332
   unilaterale, 332, 334, 335
Hypericum hirsutum, 289, 375, 397, 430
   pulchrum, 285, 286, 318, 322, 325, 337, 340, 341, 345, 346, 411, 473, 785, 809
Hypnum chrysophyllum, 540, 851, 858
   crista-castrensis, 447, 450
   cupressiforme, 288, 318, 322, 333, 341, 345, 348, 371, 415, 439, 462, 465, 501, 511, 709, 748, 755, 790, 795, 803, 805, 807, 851, 854, 858, 861
      var. *ericetorum*, 322, 333, 708, 711, 726, 728, 729, 731, 734, 740, 747, 748, 755, 777, 790, 857
      var. *elatum*, 540
      var. *filiforme*, 322, 333, 342, 348, 462
      var. *tectorum*, 854, 855
   cuspidatum, 462, 465, 469, 540, 549, 611, 635, 698, 737, 739, 741, 755, 777, 812, 861, 862
   molluscum, 318, 371, 440, 540, 549, 805, 807
   schreberi, 318, 322, 326, 341, 348, 411, 417, 431, 447, 449, 450, 453, 457, 501, 509, 511, 685, 689, 708, 709, 713, 726, 729, 748, 755, 777, 790, 793, 795, 805

Ice age, 25, 149, 151
ice sheets, 25, 105, 149
Iceland, 1, 3, 46 n.
"Icelandic low", 44 (Fig. 15), 45 (Fig. 16), 46, 47
"ideal river", 633
igneous rocks, 6, 8, 10, 11, 12, 14, 80, 81
Ilex aquifolium, **257**, 276, 279, 294, 296, 305, 314, 328, 336, 343, 344, 368, 374, 392, 393, 410, 412, 466, 474 n., *and see* holly
illuvial layer, 87, 89
immature soil, 94, 95

impeded drainage, 88, 95
interglacial periods, 25, 152, 153
intertidal zone, 817
Inverliver Forest (Argyllshire), 707–8
invertebrates, 144
   soil, 83
ions, 81
Ireland, 1, 3, 4, 5 (Fig. 2), 6, **10–11**, 12, 13, 14, 15, 16, 21, 25, 30, 42, 49, 57, 58, 71, 74, 76, 133, 427, *and see under* Irish
Irish blanket bog, 714–18 (Fig. 142)
   central plain, 9 (Fig. 4), 10, 11, 101, 125, 126, 686
   hare, 4
   inland climate, 67
   raised bogs, 686–96
      development, 687–90
      distribution, 686
      drainage channels, 696
      "hollow-hummock" cycle
      hummock structure, 689–90 (Fig. 139)
      peat profiles, 690–6
         Athlone bog, 690–1, 692–3
         Edenderry bog, 691, 694–5, 696
      physiognomic dominants, 696
      seral stages, 688–89
   Sea, 10
   stoat, 4
   woodlands, 327–43
Iris pseudacorus, 255, 461, 462, 466, 468, 588, 628, 637, 643, 644, 645, 659, 660, 664, 862
Iron age, 149, 150 (Fig. 38), 151, 169, 171, 173, 175
iron podsol, 90
Isoetes lacustris, 600, 609, 611, 612, 613, 614, 619
   consocies, 600, 611
isohyets, 34
isopleths, 34
isotherms, 17 (Fig. 48), 18 (Fig. 49), 48 (Fig. 17), 49 (Fig. 18), 50 (Fig. 19), 51 (Fig. 20), 52 (Fig. 21)
Itchen, River, 624, 628, 632 (Fig. 123)
ivy (*Hedera helix, q.v.*), 266, 279, 298, 368, 369, 429

Jefferies, T. A., on Molinietum, 519–21
Jessen, K., on peat in the North Sea area, 156 (Fig. 40)
Juncetum effusi, 573–4, 586, 587 (Fig. 104)
   maritimi, 839
Juncus acutiflorus, 468, 572, 573, 636, 637, 644, 698, 739, 741
   acutus, 862
   articulatus, 317, 586, 632, 635, 636, 637, 638, 644, 665 (Fig. 134), 666, 863, 898
   biglumis, 798, 813
   bufonius, 288, 309, 573, 754, 862
   bulbosus, 453, 586, 611, 711, 741
   compressus, 557
   conglomeratus, 288, 468, 573, 635, 637, 638, 665 (Fig. 134), 666, 711, 736 (Fig. 146), 740
   effusus, 288, 317, 466, 573, 586, 587 (Fig. 104), 635, 636, 638, 643, 644, 685, 697, 698, 703, 740, 757, 862

Juncus fluitans, 609, 611, 613, 614, 619, 620
    consocies, 611
    socies, 616
  gerardi, 838, 859
  inflexus (glaucus), 288, 557, 572, 573, 637, 638, 754, 862
  lamprocarpus, see *J. articulatus*
  maritimus, 839, 859
  obtusiflorus, see *J. subnodulosus*
  squarrosus, 516, 703, 706, 709, 711, 736 (Fig. 146), 748, 750, 751, 754, 757, 761, 785, 786, 860
  subnodulosus, 468, 469, 654, 656, 662, 669, 767
  trifidus, 789, 793, 804
  triglumis, 798, 813
juniper (*Juniperus communis*), 203, **263**, 343, 373, 377, 448, 454, 727, 752
  scrub, 372–3, 375, 376, 383
  sere, 372, 377, 383, 384
juniper-sanicle sere, 384
Juniperus communis (*see* juniper)
  var. *montana*, 263
Jurassic claylands, 173
  limestones, 18, 26, 93, 125, 126
  scarps, 17 (Fig. 5)
  sea, 15
  system, 16, 18
  rocks, 16–18, 115, 764

Keskadale Oaks (Cumberland), 320–4
Keuper marls, 16, 116–17, 125
Killarney climate, 67, 327
  oakwoods, 257, **327–40**
Kingley Vale, 374, 377 n., 379 (Fig. 72)
"knee-pieces", 270
Knollentypus (of mycorrhiza), 410
Koeleria cristata, 473, 508, 510, 530, 534, 537, 547, 551, 554, 568, 890

lagg, 460, 470, 675, 687
Lagopus scoticus, 4, 751, 757
lair flora, 901
  type of grassland, 493
lakes, 579, *and see* pp. 588–621
Lake District, 174, 191, 252, 596 (Fig. 109), 597, 783
  oakwoods, 173
"lammas shoots", 302
land utilisation, 178, 185
Larix decidua, 188, 433
Lark, River (Suffolk), 467, 628, 629 (Fig. 122), 631
Lathyrus maritimus, 890
  palustris, 656, 662
  pratensis, 297, 436, 555, **565**, 567, 569, 572, 575
Lawes, J. B., Gilbert, J. H. and Masters, M. T., on flora of park grassland, 571–2
layer (of a community), 214
  society, 230
Leach, W., on Birkrigg and Keskadale oaks, 320–4
  on succession on scree, 776–9
leaching, 79, 86, 88

Leland, John, on sixteenth-century woodland, 181
  on "fyrres" in Co. Durham, 444
Lemna gibba, 582
  minor, 582, 584, 588
  trisulca, 582, 584, 628
Leucobryum glaucum, 309, 318, 347, 357, 411, 412, 413, 415, 457, 500, 690, 700, 716, 718, 728, 729, 755
Lewis, F. J., on ash-birch woods, 442–3
  on Eriophoretum angustifolii, 704
  on "grass heath", 489–90
  on limestone grassland, 553–6
  on post-glacial forests, 159, 167
  on Scirpetum caespitosi, 707
  on Sphagnetum in the northern Pennines, 697
  on Vaccinieta of the northern Pennines, 761
Lias, 18, 116, 735
Liassic marls, 125
lichens, arctic-alpine, 782, 786, 788, 791–3, 794, 795, 798, 800–2, 806, 808, 812
  as pioneers, 219
  Breckland grass heath, 511, 513–14
  chalk grassland, 541
  heath, 728, 729, 730–1, 745, 746, 749, 756
  Killarney, 337–8
  Norfolk carr, 463
  river, 625
  sand dune, 853, 854, 855, 857, 858, 861
  shingle beach, 891–4
life form, 213, 234, 235, 236
  classes (Raunkiaer's), 234–5
life forms, 487
  of aquatic plants, 581–3
light in beechwood, 369
  in lakes, 607–8
  in pedunculate oakwood, 274–6
  in sessile oakwood, 305
light phase (of woodland), 274, 369
Ligustrum vulgare, 265, 272, 279, 289, 294, 296, 373, 391, 392, 429, 657
lime (*Tilia*, *q.v.*), 162, 163, 198, **258**
limestone, 80, 84, 85, 86, 93, 94, 98, 99
  copse, 432
  grassland, 206–7, 490, 552–7
  heath, 432, 473, 557
  pavement, 120
  swamp, 557
Limonietum vulgaris, 831, 835
Limonium binervosum, 140, 838, 889
  reticulatum, 838
  vulgare, 140, 831, 833, 835, 837 (Fig. 161), 839, 888
  zone, 889
linear-leaved associes (Esthwaite Water), 599
ling, *see Calluna*
Ling Ghyll wood (W. Yorks), 433–4
Lipman, T., on "one-layered associations", 214 n. 2
Listera cordata, 447, 451, 452, 453, 744, 752, 754, 757
lithophytes, 219, 799, 800
lithosere, 219
litter (leaf), **83**, 100

Index

"litter" (Molinietum, in the Fens), 653, 661–2
"littoral" river communities, 631, 632
Littorella uniflora, 590, 592, 600, 602, 603, 615, 617, 618, 619, 621
Littorella-Lobelia associes, 600, 601 (Fig. 112), 602, 606, 615
liverworts, endemic, 4
  and see bryophytes
loam, 80, 86, **98**
  overlying chalk, 525
Lobelia dortmanna, 600, 602, 611, 615, 617, 621
Loch Maree pine forest, 445, 450–1
  small bog at, 682, 684 (Fig. 138), 685
Locheil Old Forest (Inverness-shire), 253
lodes, 650
Lolium perenne, 496, 497, 498, 534, 537, 541, 550, **561**, 564, 565, 567, 569, 570, 572, 573, 575, 666
London basin, 23, 24 (Fig. 8), 25, 107, 110 (Fig. 33), 111, 125
London Clay, 22 (Fig. 7), 23, 24 (Fig. 8), 173, 247
"long leys", 128, 496
Lonicera periclymenum, **266**, **279**, 281, 300, 317, 325, 331, 341, 343, 344, 346, 390, 392, 393, 412, 429, 435, 436, 437, 452, 474
Lonsdale raised mosses, 686
Lotus corniculatus, 494, 501, 508, 530, 538, 542–3 (Figs. 99–100), 546, 547, 548, 549, 550, 555, **565**, 570, 572, 573, 575, 785, 888, 890
Lough Neagh, 663
  fens, 104, 662–8
    carr, 668
    lower fen, 664
    middle fen, 664–5
    "ramparts", 667
    relation to "moor", 667
    succession, 666–7
    upper fen, 666
    zonation, 663–6
"low moor", 674
lowland ponds, 584
"Lower Forestian" horizon, 159, 167 (Fig. 45)
Lower Greensand, 19, 20 (Fig. 6), 22 (Fig. 7), 24 (Fig. 8), 110 (Fig. 33), 113, 114 (Fig. 34), 125
Lycopodium alpinum, 785, 786, 794, 795, 796
  annotinum, 758, 785
  clavatum, 748, 752, 754, 786, 809
  inundatum, 741
  selago, 718, 748, 754, 778, 785, 786, 793, 794, 796, 809, 810
lynchets, 172, 174, 361

Mackinder, H. J., on "Proto-Britain", 6
McLean, R. C., on Blakeney Point lichens, 891–3
  on Festucetum rubrae at Worm's Head (Glam.), 898–900
Magdalen Meadow, 570
Magnesian limestone, 16, 94, 117, 126
magnesium, 81, 527, 784
  salts, 609
Magor, E. W., on composition of Cornish bent-fescue grassland, 501

Magor, E. W., on Cornish Molinieta, 521–2
  on Cornish marshes, 637–8
  on Scirpetum, 710–11
  on Sphagnetum, 698
Malvern oakwoods, 318–19
maples (*Acer, q.v.*), 257
maquis, 199, 339
marginal (wood edge) society, 309
  species, 477–9
maritime formations, 209–10
  grassland, 208
  habitats, 817–19
  soils, 102–3
  vegetation, 817–902
    foreshore, 844–5
    salt marsh, 819–43
    sand dune, 845–67
    shingle beach, 880–94
    submaritime, 895–902
market gardens, 103
markets, 101–2
marl, 80, 84, 125
Marsh, A. S., on salt marsh, 825, 831
marsh, 208, 583, **634–9**, 668
  in calcicolous ashwood, 430
marshes, 579
  Cornish, 637–8
  Esthwaite, 635–7
"Marshland", the, 649
mat-grass (*Nardus stricta, q.v.*), 131, 490
Matthews, J. R., on the White Moss Loch, 588–92 (Figs. 105–6)
"mattress" (of dead herbage), 130, 489, 661
mature soil, 85, 95
Mayo blanket bog, 716–18 (Fig. 142)
meadow, 179
  grasses, 488, 562, 567–9
  soils, 85, **88**, 559
  vetchling, *see Lathyrus pratensis*
meadows, 566–70
mechanical analysis (of soil), 80
Medicago lupulina, 297, 508, 539, **565**, 572
Megalithic monuments, 165
megaphanerophytes, 235
Melandrium dioicum (*rubrum*), 283, 287, 316, 325, 430, 434, 470, 479, 806
  society, 283
Mendip Hills, 13, 119, 126, 427, 557
Mercia, 175
Mercurialis perennis, 274, 280, 282, 288, 289, 293, 294, 296, 308, 316, 318, 345, 362 (Fig. 67), 369, 371, 372, 376, 377, 382, 383, 384, 388, 392, 397, 406, 430, 432, 434, 435, 438, 441, 460, 470, 480, 482, 636, 768
mercury, *see Mercurialis*
  society, 282, 308, 430
  woods, 363, 369, 371
Mesolithic (period), 149, 150 (Fig. 38)
mesophanerophytes, 235
mesotrophic peat, 673
metamorphic rocks, 6, 8, 10, 11, 12, 86
Meyer ratio, 33, 34
mica, 81, 129, 140–1, 267, 271, 291–3, 402
microclimates, 29
microphanerophytes, 234, 472

Middle Ages, 74, 115, 172, 175, 177, 181, 250, 267, 270
Midland plain (English), 9 (Fig. 4), 14, 16, 17 (Fig. 5), 106, 116, 117, 166, 172, 175, 318
Midland valley (Scottish), 8, 9 (Fig. 4), 11, 63, 105
migratory communities, 844
　habitats, 239
　vegetation, 238–9
mild humus, 83, 85, 98
mild winter climate, 45, 49, 74
milfoil, *see Achillea millefolium*
Milium effusum, 287, 316, 389
mineral soil, 80
Minnikin, R. C., on velocity of river currents, 623
mist, 35
"mixed fen", 662
mixed oak forest (Quercetum mixtum), 162 n. 2, 267, 350 n.
　oakwood, 350–4
"mixed sedge" at Wicken Fen, 661–2
mobile dune communities, 847
　dunes, 845–50
Moerckia flotowiana, 863
moisture (as climatic factor), 30
Molinia bog, 519
　fen, 642, 671 (Fig. 135)
　grassland, 494, 518–23
　meadow, 519
　peat, 520
Molinia caerulea, 207, 225, 315, 339 (Fig. 64), 345, 346, 414, 447, 449, 450, 453, 457, 490, 492, 493, 494, 502, 515, 519 (Fig. 92), 520, 521, 522, 636, 640–1 (Figs. 124–5), 644, 645, 646, 647, 651, 654, 656, 661, 662, 663 (Fig. 134), 667, 670, 685, 696, 698, 709, 711, 713, 715, 716, 718, 724, 728, 732, 735, 736 (Fig. 146), 737, 738 (Fig. 148), 739, 748, 751, 754, 757, 767, 769, 784, 809, 810
　var. *depauperata*, 519, 715
Molinietum, 490, **518–23**, 646–7, 653, 661, 666, 670, 673, 676, 682, **713**, 735, 741, 758, 759 (Fig. 149), 771
　habitat factors, 519
　Cornish, 521–2
　Highland, 522
　Irish, 522
　Pennine, 520–1
　southern Scottish, 520
Moneses uniflora, 451
Monotropa hypopitys, 371, 393, 416, 863
montane vegetation, 775
moor, 673
moor grass, purple, *see Molinia caerulea*
moor peat, 673
Moore, Barrington, on distribution of acorns, 141
Moore, E. J., on the Ayreland of Bride (Isle of Man), 849, 850, 852–3, 860
　on dune soil, 865–6
moorland, 673
moorpan, 89
mor, **83**, 84, 100, 364
moraines, 26
Morss, W. L., on salt marsh, 825, 830, 831

moss (bog), 200–2, 203, 670, 673, **674**, 720, 768–9
　formation, 200–1, 673–720
　peat, 84, 85, 93, 99, 104, 673
Moss, C. E., on Eriophoretum vaginati, 701–3
　on Fraxinetum calcicolum, 428–32
　on limestone grassland, 553–6
　　heath, 557
　　swamp, 557
　on "oak-hazel wood", 272
　on Pennine Callunetum, 754–6
　　Vaccinietum, 761
　on progressive and retrogressive scrub, 475
　on Quercetum sessiliflorae, 313–17
　on retrogression of Eriophoretum, 705–6
　　on limestone, 440–1
　on vegetational concepts and terms, 236
mosses, endemic, 4, *and see* bryophytes
mountain climate, 43, 63–6, 202, 775, 781–2
　habitat, 782–3
Mountain Limestone, *see* Carboniferous Limestone
mountain-top detritus, 787
　formation of, 787–96
　adaptations, 793
　bryophytes, 790–1
　growth forms, 789
　habitat conditions, 789
　lichens, 791–3
　moss-lichen associes, 788, 789–94
　succession, 787–8
　vascular flora, 793–4
Mourne Mountains, 6, 7 (Fig. 3), 10, 12, 763
　Callunetum, 758, 759 (Fig. 149), 760 (Fig.150)
　Vaccinietum, 760 (Fig 150), 761
　vegetation, 758, 759–60 (Figs. 149–50)
mull, **83**, 84, 85
Murray, J. M., on Scottish planting, 185, 188, 190
　on sheep *v.* plantations, 190
mycorrhiza, 227, 368, 387, 410
Myrica gale, 646, 654, 655, 668, 670, 696, 698, 700, 709, 710, 715, 740, 748
Myrica-Molinia zone, 637, 646
Myriophyllum alterniflorum, 590, 600, 612, 616
　verticillatum, 612, 613, 616
　spicatum, 586, 588, 590, 593, 612, 613, 616, 618, 619, 624, 627, 633

Naias flexilis, 599, 600, 615
　marina, 595
nanophanerophytes, 235, 472
Nardetum, 490, 724, **514–18** (Fig. 91), 713, 719, 771, 775
Nardus grassland, 494, **514–18** (Fig. 91), *and see* Nardetum
Nardus stricta, 131, 207, 490, 492, 493, 494, 500, 501, 514, **515**, 516, 518, 519, 522, 706, 711, 713, 724, 748, 752, 754, 757, 761, 784, 786, 793, 796, 860
National Parks, 192
National Trust, 192
native forest, 243
　woods, 245
"natural pasture", 432, 490

"natural" vegetation, 194
Neogenic region, 2 (Fig. 1), 6, 15
  rocks, 6, 9 (Fig. 4), 15, 101
  soils, 101
Neolithic period, 149, 150 (Fig. 38), 163–6, 172, 173, 550
Neottia nidus avis, 371, 393, 416
neutral grassland, 495, **559–76**
  grasses of, 561–4
  non-gramineous species of, 564–6
New Forest, 176 n., 177, 471, 680, 764
Nicholson, W. E., on endemic liverworts, 4
nightshade, woody (*Solanum dulcamara*, q.v.), 266
  deadly (*Atropa belladonna*), 136, 375, 479
Nitella associes, 612
Nitella flexilis, 597, 599, 606, 611, 612, 614
nitrates, 82, 609
non-silted river communities, 626, 632
Norfolk Broads, 255, 460, **592–5** (Figs. 107–8), 597, 651 (Fig. 128), 652 (Figs. 129–30)
Norman conquest, 176, 178
  devastation, 176
Norsemen (invasions of), 176
North Downs, 15, 17 (Fig. 5), 21, 126, 171, 172, 256, 361, 368, 526
North Sea, 3, 5 (Fig. 2), 16, 25, 62, 151, 154, 156 (Fig. 40), 162, 267, 651
Northern Coniferous Forest, 91, 95, 199, 202, 444
Norway, 247, 248
Norway Deep, 3
Nuphar intermedium, 616
  luteum, 585, 603, 613, 627, 630, 631
    consocies, 616
  pumilum, 620, 621
Nymphaea alba, 595, 599, 602 (Fig. 113), 603, 605 (Fig. 115), 606, 617
  occidentalis, 603, 620, 621
    consocies, 603, 616
Nymphaeetum albae, 602 (Fig. 113), 603, 605 (Fig. 115)

oak, 154, 162, 164, 173, 182, 188, 189, 190, 198, 200, 243, 246, *and see Quercus robur, Q. sessiliflora*
  and beech, 249–50, 361–2
  bark, 181, 184
  charcoal, 173, 182, 184
  forest, 163, 172, 181, 195, 246, 267, 670
    former extent of, 270
    mixed (Quercetum mixtum), 160 (Fig. 43), 161 (Fig. 44), 163, 350 n.
  saplings, 182, 291, 296, 302, 353
  seedlings, 247 n., 269 n. 2, 291, 292, 295, 302, 353
  standards, 245, 267
  timber, 181
oak-birch heath, 354–7, 456–7, 764
oak-hazelwood, 177, 272
oak-hornbeam forest, 270, 281
oaks on acid soils, 269
  on shallow soils, 269
  the two British, 246, 268–9, 303–4
oakwood, 173, 217, 230, 236, 244, 245, 250, 256, 258, 259, 263, 267, 669

oakwood (*cont.*):
  culture form, 248
  disappearance in the far north, 348
  facies of, 273 (Fig. 51)
  Highland, 445, 455
  native, 270
  on peat, 661
  soils, 273–4
  under extreme conditions, 298–302, 320–4
oakwood, pedunculate, 269–302
  light, 274–6 (Fig. 52)
  societies of field layer, 279–85 (Fig. 53)
  soils, 273–4
  species, 285–9
  structure, 276–80
  succession, 293–8
oakwood, sessile, 303–43
  at highest altitudes, 320–4
  Hertfordshire, 304–11
  Irish, 327–43
  Killarney, 327–40
  Malvern, 318–19
  on Mountain Limestone, 320
  Pennine, 313–17
  succession, 311–12
  Welsh, 325–7
oakwood, mixed (Quercetum ericetosum), 350–4
  Quercetum mixtum, 162 n. 2, 267, 350 n.
oat grass, *see Arrhenatherum, Avena, Trisetum*
oats, 76
Obione portulacoides, 140, 835, 836 (Fig. 160), 837, 838, 840, 885, 887
Obionetum, 835, 837–8, 840
oceanic climate, 52, 670
  winds, 50
oceanicity of climate, 60
Oenanthe fistulosa, 586
  fluviatilis, 626, 627, 628, 630
Old Red Sandstone, 10, 121–2
  planted beechwoods on, 406
oligotrophic peat, 673
Oliver, F. W., on *Agropyron junceum*, 847–8
  on *Ammophila arenaria*, 848–9
  on shingle beaches, 868–80
  on the vegetation of shingle beaches, 880
Oliver, F. W. and Salisbury, E. J., on *Suaeda fruticosa*, 880–5
Olsen, C., on *Urtica dioica* and nitrates, 283
Oolite, 18, 115–16
  beechwood on, 358, 372
  grassland, 552
oolitic limestone, 126
open canopy, 271
  communities, 213–14
open field cultivation, 178, 179 n., 188
Ordovician rocks, 8, 9, 10, 12, 122–3
organic material in soil, 83
  soils, 92, **99–100**, 124–5
orographical precipitation, 47
orography, 7 (Fig. 3)
osiers, 264
Osvald, H., on raised moss or bog, 688
outcrops and soils, distribution of, 101–24
overgrazing, 131
  effects of, 489

Oxalis acetosella, 286, 287, 307, 309, 316, 318, 322, 326, 331, 340, 341, 342, 345, 346, 347, 388, 392, 402, 403, 406, 419, 434, 435, 438, 447, 452, 454, 466, 500, 555, 757, 762, 785, 786, 807, 809, 810
Oxalis stage in Fagetum rubosum, 402–3 (Fig. 78)
oxbows, 584, 585, 622
oxidation-reduction potential, 82
  of Callunetum, 754
  of Eriophoretum, 703
  of Sphagnetum, 678
oxlip (*Primula elatior*), 283, 289
Oxycoccus quadripetalus, 520, 686, 689, 691, 692, 694, 697, 698, 703, 712 (Fig. 141), 741
oxygen supply in water, 579–80
  in waterlogged soils, 85
oxyphilous communities (in fen), 656–7
  plants, 82, 92, 675, 679
  species (in fen), 655, 657
oxyphobes, 82
oxyphytes, 98, 124

Palaeogenic complex, 16
  limestones, 14
  region, 2 (Fig. 1), 6, 8, 17 (Fig. 5), 28
  rocks, 6, 10, 14, 15, 101, 187
  soils, 14, 101, 117
  systems, 8
Palaeolithic (period), 149, 150 (Fig. 38)
Palaeozoic, 6, 8, 10, 13, 117, 118, 122
Pallis, M., on alder carr in east Norfolk, 461, 463, 660
  on aquatic vegetation of the Norfolk Broads, 592–5
  on "fen carr" and "swamp carr", 463, 660
  on oxyphilous communities in fen, 656–7
  on typical species of east Norfolk fen, 654–5
  on unevenness of fen surface, 656
  on varieties of reedswamp, 595
pannage, 270
pans (in salt marsh), 841
parsley fern, see *Cryptogramma crispa*
pasturage in forest, 178
pasture, 179, 223, 487, 488–90 (*and see* grazing)
  grasses, 567–9
pastures, 566–70, 574–5
  dairy, 575
  fatting, 575
  sheep, 575
  untended, 575
"path society", 308–9
Paulson, R., on Epping Forest, 415–16
Peak, the (Derbyshire), 176 n., 702 (Fig. 140), 705
pear, wild (*Pyrus communis*), 258
Pearsall, W. H., on aquatic vegetation in the Cumbrian lakes, 597–621
  on *Calluna* peat, 754
  on Esthwaite fens, 639–46
    marshes, 635–7
    Molinietum, 523–6, 646
    Water, 597
  on high-level grazing, 174
  on mull and mor, 83–4
  on oxidation-reduction potential, 82, 678

Pearsall, W. H., on pH of sand-dune soil, 867
  on sand-dune Callunetum, 852
  on the silting factor, 581
  on *Sphagnum* peat, 678
  on upper limit of forest, 173
  on *Vaccinium* peat, 761
  on vegetation of "slacks", 863
  on Walney Island sand dunes, 853, 860
peat, 28, 84, 85, 92–3, 99–100, 104–5, 634, 639, 646, 649, 656, 667, 673, 674, 687, 700, 701, 703, 714, 719, 724–6, 743, 747, 750, 753, 754, 771
  acid, 84, 92, 104, 674, 679
  communities (summary), 766–71
  fen, see fen peat
  haggs, 708
  moor, 734
  moss (bog), 734
  pollen preserved in, 149
  profiles, 690–6
  Sphagnum, 678–9, 687–8 (Fig. 139), 690–6
  subaqueous, 609, 766
pedology, 78
pedunculate oak (see *Quercus robur*)
Pelham, R. A., on mediaeval agriculture, 180
  on Sussex wool, 181
  on export of timber and bark, 181
Peltigera, 541, 853
  canina, 858
  rufescens, 858
Pennine anticline, 119
  ashwoods, 427–33
  axis, 10, 16
  bilberry moors, 761
  cotton grass mosses, 701–3, 705–7
  heather moors, 753–7
  Hills, 8, 9 (Fig. 4), 10, 14, 15
  oakwoods, 313–17
  region, 119
  ridge, 14, 151
Pennines, 126, 152, 174, 175, 769
  northern, 166, 251, 252, 254, 427, 557, 701, 703, 761, 769
  southern, 166, 313, 428, 701, 753, 755, 761
Pepys, Samuel, on dependence of the Navy on imported timber, 184
perennating buds, 234
"permanent grass", 194, 195–6, 206, 496, 560
  pasture, 74–5, 129
Permian, 16, 117
  marl, 352–3
Petalophyllum ralfsii, 863
Pethybridge, G. H. and Praeger, R. Ll., on Calluneta of the Wicklow Mountains, 757
  on *Rhacomitrium* "associations", 788 n.
  on Scirpetum caespitosi of the Wicklow Mountains, 711–12
  on Vaccinieta of the Wicklow Mountains, 761
  on woodland plants in grassland, 502–3
pH range, 82, 83, 269
  value, 81, 82, 83
Phalaris arundinacea, 572, 595, 643, 644, 653, 656, 669, 670
  consocies, 643
  fen, 654

phanerophytes, 234, 235
pheasant preserves, 245
pheasants, 270, 291
Philip, G., on *Zostera* and *Ruppia*, 824
Phleum pratense, 537, **562**, 569, 572
  var. nodosum, 548, 549
Phragmites communis, 470, 590, 592, 595, 603, 606, 617, 653, 654, 664, 669, 700
  reed swamp, 643, 644, 646, 653
Phragmites-Scirpus associes, 603, 616
Phragmitetum, 590
Phyllitis scolopendrium, 431, 434, 435, 441
physiographic climax, 224
  factors, 223
  regions, 17 (Fig. 5)
phytoclimates, 235
Picea abies, 199, 254, 444
Pimpinella saxifraga, 530, 538, 547, 551, 555, 572
pine, common, *see Pinus sylvestris*
  forest, 157, 159, 165, 166, 167, 253, 444–5
Pinner Wood, Middlesex, 416
Pinus nigra var. *austriaca*, 852
Pinus sylvestris (common pine, "Scots pine", "Scotch fir"), 145, 154, 156, 158–9, 160–1 (Figs. 43–4), 165–7, 189, 199, 203, 243, **253–5**, 356–7, 374, 444, 457
  continental range, 253
  size of big, 446
  var. *scotica*, 444
pinewood, 199, 203, 444, **446–51**
  characteristic species, 451
  field layer, 446, 449
  subspontaneous, 457–8
pioneer community, 218 (Fig. 48), 219
pioneers, 217, 220
plagioclimax, 225, 226 (Fig. 50), 231, 487
plagiosere, 225, 226 (Fig. 50)
plant communities, 213–14
  naming of, 227–8
  nature of, 228–31
Plantago maritima, 810, 830, 831, 833, 838, 839, 887, 899–901
Plantago sward, 900–1
plantations, 188–91, 195, 197
plateau beechwoods, 359 (*and see* Fagetum rubosum)
  gravels, 125
Pleistocene glaciations, 95, 149, 151–3, 236
  ice age, 25, 149, 151
Pliocene deposits, 25, 151
  times, 236
ploughland, 178, *and see* arable
plum (*Prunus domestica*), 258
Poa, 492
Poa annua, 546, 569, 572, 666
  on disturbed soil, 498
  nemoralis, 287, 316, 370, 474
  pratensis, 318, 537, 554, **562**, 569, 570, 572, 573, 666
    on trampled soil, 497–8
  trivialis, 318, 496, 554, **562**, 569, 570, 572, 573, 575, 666
podsol, 82, 85, **88–92**
  climate, 88–9
  heath, 724–6 (Fig. 144)

podsolisation, 92, 97, 124, 354
"polar front", 46
pollen analysis, 149, 177, 444
Polygonum amphibium on shingle beach, 891
Polytrichetum sexangularis, 797
Polytrichum alpinum, 746, 755, 777, 790, 795, 797, 798, 803, 805, 807, 812
  commune, 341, 347, 447, 450, 516, 657, 698, 711, 741, 746, 749, 751, 768, 777, 790, 810
  formosum, 286, 309, 318, 322, 326, 331–3, 341, 345, 347, 407, 411, 413, 440, 734, 736 (Fig. 146), 746, 749, 755, 777
  juniperinum, 309, 322, 440, 509, 511, 514, 729, 734, 736 (Fig. 146), 740, 746, 749, 755, 777, 790
  nanum, 749, 779
  piliferum, 322, 509, 511, 512, 729, 731, 733, 734, 749, 755, 777, 779, 790, 794, 795
  sexangulare, 790, 797, 798, 812
  spp. 364, 419, 557, 757
Polypodium vulgare, 301, 450, 864
ponds, vegetation of, 584–8
Pontoon (Co. Mayo) oakwoods, 340–2
poplars, 259
Populus canescens, 259, 278, 466, 467, 469
  italica, 467 n.
  nigra, 259, 467
  serotina, 259, 467 n.
  tremula, 203, 259, 278, 317, 347, 429, 446, 452, 453, 461, 468
Port Meadow, Oxford, 568–70
post-climax, 217, 218 (Fig. 48)
post-glacial history, 149–51
Potamogeton acutifolius, 585
  alpinus, 586, 600, 601 (Fig. 112), 613, 614, 615
    consocies, 613
  angustifolius, 615
    associes, 590, 615
  compressus, 631
  crispus, 599, 612, 625, 627, 628, 630, 631
    var. *serratus*, 599, 600, 612, 630
  densus, 626, 627, 628, 629 (Fig. 122), 632 (Fig. 123)
  filiformis, 588
  gramineus, 590, 615
  heterophyllus, 600
  lucens, 588, 613, 627, 630, 631
  natans, 585, 586, 587 (Fig. 104), 588 (Fig. 105), 590, 591 (Fig. 106), 610 (Fig. 118), 611, 614, 615, 621, 627, 631
    consocies, 585, 586, 616, 617
  nitens, 614 (Fig. 119), 615
  obtusifolius, 590, 599, 612, 613, 614
    consocies, 612–13
  panormitanus, 599, 612, 613
  pectinatus, 593, 626, 627, 629 (Fig. 122), 631
    subsp. *lacustris*, 599, 611, 612, 613, 619
  perfoliatus, 588, 590, 611, 612, 613, 614, 615, 627, 628, 630
    consocies, 613
  polygonifolius, 620, 700, 715, 737 (Fig. 147), 738
    var. *pseudofluitans*, 611, 619
  praelongus, 612, 613, 628, 630
    consocies, 613
  pusillus, 599, 609, 612, 613, 625, 627, 628

Potamogeton pusillus, subsp. *lacustris*, 599, 611, 612, 613, 619
    community, 612
  spp. 610 (Fig. 118)
  trichoides, 590
potash, 527, 609, 614, 618
potassium, 81, 609, 784
Potentilla anserina fixing dune sand, 851
Poterium sanguisorba, 527, 528, 529 (Fig. 93), 530, 532 (Fig. 96), 533, 538, 541, 542 (Fig. 97), 545 (Fig. 103), 547, 551, 556, **566**, 567, 572
Praeger, R. Ll., on submaritime *Plantago* sward, 900–1
Pre-Boreal period, 150 (Fig. 38), **154–7** (Figs. 39, 40)
precipitation, 30–4, 79, 80
  highest, 53–4
  lowest, 54
precipitation-evaporation ratio, 79, 86
preclimax, 217, 252, 254
pre-history, 149–70
prevernal aspect, 279
Price Evans, E., on arctic-alpine vegetation of Cader Idris, 785–7, 793–4, 795–6, 808–10
  on Cader Idris Nardetum, 518
  on Craig-y-Benglog ashwood, 436–40
Priestley, J. H., on salt marsh, 830
"primitive" lakes, 581, **618–21**
Primula elatior, 283, 289
  veris, 482, 533, 540, 550, 556, 567, 572
  vulgaris, 280, 281, 282, 289, 296, 316, 318, 341, 345, 406, 434, 438, 452, 474, 897
    society, 282
privet (*Ligustrum vulgare*), 265, 272, 289, 294, 296, 373, 382, 383, 391, 392, 429, 432, 657
prisere, 218 (Fig. 48), 219
profile of peat in raised bog, 690–6
  soil, 78
"progressive succession", 236–7
"Proto-Britain", 6
Prunus avium, **258**, 278, 296, 368, 374, 388, 391
  cerasus, 258
  lauro-cerasus, 328
  lusitanica, 328
  padus, 203, **258**, 316, 317, 429, 433, 436, 452, 453
  spinosa, 203, **261**, 279, 294, 313, 314, 317, 325, 373, 390, 397 n., 433, 434, 436, 437, 461, 467, 473, 474, 475, 476, 481, 657
psammosere, 847, 861
Pteridietum, 280, 284 (Fig. 53), 307, 354, 356, 502–4
Pteridium aquilinum, 131, 135, 276, **280**, 281, 285, 286, 296, 299, 300, 301, 307, 311, 313, 315, 318, 320, 321, 322, 323, 326, 331, 335, 336, 341, 342, 343, 345, 346, 353, 390, 413, 438, 449, 450, 453, 456, 473, **502**, 503, 507, 526, 727, 728, 731, 732, 740, 747, 748, 754, 779
Pteridium and Callunetum, 727, 728 (Fig. 145)
Pteridium-Holcus-Scilla society, 281
Pyrola media, 451
  minor, 451
  rotundifolia, 451, 654–5, 863

Pyrola secunda, 451
Pyrus malus, 258, 278, 305, 314, 317, 352

quaking grass, see *Briza media*
Quantock Hills, 748
quartz, 81
quartzite, 85
Quaternary deposits, 101, **102–7**
Quercetum ericetosum, 351–4
  on scree, 779
  mixtum, 162 n. 2, 267, 350 n.
  roboris, 259, **269–302**, 314, 315
    coppice-with-standards, 270–1
    damp, 271–2
    distribution, 269–70
    dry, 271–2
    field layer societies, 279–85
      species, 285–9
    light, 274–6 (Fig. 52)
    regeneration, 291–3
    shrub layer, 279
    soils, 273–4
    structure, 276–8
    succession, 293–8
    trees, 278–9
    under extreme conditions, 298–302
    woody climbers, 279
  roboris et sessiliflorae, 350–4, 355
  sessiliflorae, 259, 260, 281, **303–43**, 635, 670
    at highest altitudes, 320–4
    Hertfordshire (oak-hornbeam), 304–11
    Irish, 327–43
    Killarney, 327–40
    Malvern, 318–19
    Pennine, 313–17
    succession, 311–13
    Welsh, 325–7
    west of England, 319–20
Quercus petraea, see *Q. sessiliflora*
  robur, 162–3, **246–7**, 268, 269, 272, 278, 294, 300, 303, 304, 305, 311, 344, 346, 347, 349, 350, 351, 374, 388, 392, 415, 441, 442, 463, 559, 669
  sessiliflora, 162, **247–8**, 251, 268, 269, 272, 303, 304, 305, 311, 313, 314, 319, 320, 323, 325, 326, 327, 330 (Fig. 60), 340, 341, 342, 343, 344, 346, 347, 349, 350, 351, 415, 437, 442

rabbit (*Oryctolagus cuniculus*), in Britain, 133–4
  grazing, 133, 134–5, 445
  pressure, 375, 381
  warrens, 134
rabbit-resistant plants, 135–6, 381, 508
rabbits, 129, **133–40**, 141, 142, 185, 245, 264, 267, 402
  and maritime vegetation, 138, 140
  effect on Callunetum, 136, 137 (Fig. 36)
    on chalk grassland, 134–6, 542–5 (Figs. 98–103)
    on scrubland, 375, 381, 480, 481
rainfall, 30, 31, 32 (Fig. 11)
  seasonal distribution of, 67–8
  and see precipitation
rain shadow, 54

rainwash, 79
raised bog (moss), 201, 619, 670, 674, **675**, 679, **680–96**, 719, 768
  peat, 85, 92
Raistrick, A., on nunataks, 152
Rankin, W. M., on alderwood marginal to peat bog, 470–1
  on New Forest Molinietum, 523–6
    Sphagnetum, 680
  on peat of Lonsdale mosses, 168
  on raised mosses of Lonsdale, 686
Ranunculus fluitans, 625
    community, 626–8
  repens, fixing dune sand, 851
    society in oakwood, 283
Raunkiaer, C., system of life forms, 234–5, 487, 781
ravines (in shingle bank), 873, 876–7 (Fig. 173)
raw humus, 83, 100, *and see* mor
  species associated with, 744
Rayner, M. C., on *Phoma* symbiotic with *Calluna*, 723
"red bogs", 675
red currant, *see Ribes rubrum*
red deer (*Cervus elaphus*), 143, 446
reed, common (*Phragmites communis, q.v.*), 470
reed grass associes, 643
reedswamp, 582–3, 585, 586, 590, 603, 616–17, 646
  of slow rivers, 630, 631
  open, 595
  relict, 700
regeneration of oakwood, 291–3
  of Fagetum rubosum, 403–6 (Figs. 79, 80)
Reid, E. M. R. and Chandler, M. E. J., on interglacial floras, 152
relative humidity, 33
rendzina, 85, **93–5**, 363, 525, 526
retrogression of Eriophoretum, 705
"retrogressive succession", 236–7
  on limestone, 440–1
Rhacomitrium canescens, 755, 790, 793, 807
  fasciculare, 755, 776, 798, 804, 805
  heterostichum, 755, 776, 790, 795, 804, 805, 807
  lanuginosum, 700, 708, 709, 711, 712 (Fig. 141), 716, 718, 749, 755, 757, 776, 786, 790, 794, 795, 804, 805, 807, 810, 863
  ramulosum, 804
Rhacomitrium heath, 760 (Fig. 150), 782, 787, **788**, 794–6
  hummocks, 685
  moor, 796
Rhamnus catharticus, 265, 373, 429, 463, 657, 658 (Fig. 132), 659 (Fig. 133), 660, 663, 669
Rhamnus frangula, see Frangula alnus
Rhododendron ponticum at Killarney, 328, 339–40
Rhynchospora alba, 521, 680, 681 (Fig. 136), 683 (Fig. 137), 685, 686, 688, 689, 697, 698, 700, 711, 715, 716, 718, 719, 738 (Fig. 148), 739, 769
  fusca, 700
Rhynchosporetum albae, 676, **700**
Ribes alpinum, 429
  grossularia, 429, 461, 660

Ribes nigrum, 429, 461, 660
  rubrum, 429, 461, 470, 660
Riccia fluitans, 584
Ricciocarpus natans, 584
Richards, F. J., on silting of salt marsh, 820
Richards, P. W., on Blakeney Point mosses, 851, 855
  on Killarney bryophytes, 331–4
  on sand-dune mosses and lichens, 851, 855, 857
  on shingle-beach mosses, 894, 895
Richardson, H. G., on destruction of woodland for iron and glass manufacture, 183
  on fourteenth-century timber importation, 181
  on use of sea coal, 182
  on Wealden fuel, 181–2
  on West Saxon fines for destroying trees, 175 n.
rift valley, 8
riparian communities, 631–2
river alluvium, 103
rivers, 579, 580, 622–4
  vegetation of, 624–33
Roberts, Alun, on *Ulex europaeus* in Wales, 504
rodents, 140, 245
roe deer (*Capreolus capreolus*), 144, 448
Roman Britain (Ordnance map of), 173
  occupation, 171
  times, 177, 267
Romano-British agriculture, 171
Romell, L. G., on mull and mor, 83
Romney Marsh, 575
Rosa arvensis, 263, 314, 383, 384, 390, 391, 392, 429, 470, 474
  canina, 263, 314, 317, 373, 383, 390, 429, 437, 461, 469, 474 n., 481, 482, 483 (Fig. 88)
  micrantha, 263, 373, 383, 429, 481, 482, 483 (Fig. 88)
  rubiginosa, 263, 373, 383
  spinosissima, 264, 473, 861, 864
  spp., 263–4, 294, 296, 311, 313, 383, 429, 476
  villosa, 317, 433, 435
roses, *see Rosa* spp.
Ross, R., on thicket scrub on Chalky Boulder Clay, 480–4
Rothamsted park grassland, 560, 564, 571–2
Rothiemurchus Forest, 253, 445, **447–9** (Fig. 86)
rough grazings, 129, 131, 194, 221, 489, 494, 499, 525
rowan (*Sorbus aucuparia, q.v.*), 203, 258, 276, 374, 445, 446, 451, 756
Rowan, W., on rabbits and maritime vegetation, 139–40
Royal Forests, 174, **176–7**, 179, 183, 352
Rubus caesius, 263, 311, 374, 429, 470, 480, 481, 659
  chamaemorus, 703, 704, 706, 758, 761
  corylifolius, 263
  dasyphyllus, 314 n.
  dumetorum, 263
  "*fruticosus*" (*Rubus* sect. *Eubatus*), 263, 276, 286, 294, 300, 301, 311, 313, 314, 322, 325, 326, 341, 343, 345, 388, 389, 390, 391, 392, 393, 403, 405, 422 (Fig. 84), 425, 429, 438, 461, 469, 470, 474, 480, 481, 482, 548, 864

Rubus idaeus, 314, 317, 325, 383, 429, 435, 436, 438, 450, 466, 469
  köhleri, 307
  lindleianus, 314 n.
  leucostachys, 296, 307
  macrophyllus, 296
  rhamnifolius, 307
  rudis, 388 n.
  saxatilis, 431
  selmeri, 314 n.
  suberectus, 263
  ulmifolius (rusticanus), 263
  vestitus, 388 n.
Rubus layer (in Fagetum rubosum), 388, 393, 403
  society, 307
  stage in Fagetum rubosum, 402–3 (Fig. 78)
Rubus-Pteridium society, 281, 307
Rumex crispus var. *trigranulatus*, 886, 887, 888, 890
Ruppia maritima, 896
  rostellata, 824, 896
rye grass, perennial, *see Lolium perenne*

Sagittaria sagittifolia, 583 n. 1, 615, 627, 628, 630, 631
  forma *vallisneriifolia*, 627, 630
Sagittaria community, 627, 630
Salicornia disarticulata, 830
  dolichostachya, 824
  herbacea, 139, 824, 825, 826, 827 (Fig. 156), 829, 830, 831, 836 (Fig. 160), 840
  perennis, 830
  ramosissima, 824, 830
  stricta, 824
Salicornietum herbaceae, 819, 824–9, 831, 833, 841
saline soils, 99
Salisbury, E. J., on *Ammophila arenaria*, 849
  on Breckland climate, 70
  on climate and distribution of species, 29
  on coppice flora, 278, 478
  on dune soils, 865, 866 (Figs. 163–4)
  on fixed dune flora, 856
  on flora of Ammophiletum, 850, 853
  on flora and vegetation of shingle beaches, 886–89
  on floral elements in Hertfordshire scrub, 476–7
  on hedgerow flora, 478
  on Hertfordshire sessile oakwoods, 303–11, 350
  on light values in oak-hornbeam coppice, 276
  on marginal (wood-edge) flora, 477–8
  on salt marsh, 835, 837
  on societies in oak-hornbeam woods, 281, 282, 283
  on succession on arable and heathland, 294–5 to sessile oakwood, 311
  on thicket scrub, 132, 475
  on woodland scrub, 475–6
Salisbury, E. J. and Tansley, A. G., on Malvern oakwoods, 318–19
Salisbury Plain, 17 (Fig. 5), 21, 112, 126, 166, 172, 180 n.

Salix alba, 259, 467 n., 668
  atrocinerea, 255, 264, 279, 296, 300, 311, 313, 342, 343, 412, 429, 463, 636, 638, 643, 645, 646, 657, 658, 660, 668, 669
  aurita, 264, 317, 325, 343, 348, 410
  caprea, 264, 317, 342, 353, 391, 429, 433, 668
  cinerea, *see atrocinerea*
  decipiens, 643
  fragilis, 259, 314, 316, 467 n.
  herbacea, 154, 783, 789, 793, 798
  lapponum, 154, 264, 807
  myrsinites, 264
  nigricans, 264
  purpurea, 264, 643, 645, 668, 670
  repens, 264, 862–4
    var. *fusca*, 264, 654
  reticulata, 154, 804
  viminalis, 264, 668
sallows, 264, 442, *and see Salix atrocinerea, S. aurita, S. caprea*
Salsola kali, 847
salt marsh, 99, 103, 817, 818, 819–43
  algae, 823–4, 825 (Fig. 154), 826 (Fig. 155), 833, 839, 841–2
  communities, 821–3 (Fig. 152), 823–40
    algal, 823–4 (Figs. 154–5), 841–2
    Armerietum maritimae, 831–3 (Fig. 159)
    Asteretum tripolii, 830–1
    Festucetum rubrae, 838–9
    "General salt marsh", 834–5
    Glycerietum maritimae, 829–30 (Fig. 158)
    Juncetum maritimi, 839
    Limonietum vulgaris, 831
    Obionetum portulacoidis, 835, 837–8
    Salicornietum herbaceae, 824–9 (Fig. 156)
    Spartinetum townsendii, 826–9 (Fig. 157)
    Suaedetum fruticosae, 838
    Zosteretum, 824
  drainage channels, 840–1
  formation, 819–43
    ecological factors, 820
    relation to tides, 819
    salt content of water, 819–20
    submersion and exposure, 820–1
    succession, 820
    zonation, 820
  pans, 841
  pasture, 830, 834, 840
  soil fungi, 833–4
Salter, M. de C., on cyclonic and convectional rainfall, 47
  on deviations from mean rainfalls, 68
Sambucus nigra, 136, **264**, 279, 289, 296, 314, 368, 373, 381, 384, 429, 469, 470, 864
sand, 80, **97**, 125
  blown, 97, 99, 125, 837, 845, 846
  coarse, 80
  fine, 80
sand-dune communities, 847–64
  mobile dunes, 847–52
    Agropyretum juncei, 847–8
    Ammophiletum arenariae, 848–51
  fixed dunes, 852–61
    Tortuletum ruraliformis, 853–4
    Cladonietum, 855

sand-dune, fixed (*cont.*):
    Festucetum arenariae, 857, 859
    Hippophaëtum, 864
    slacks, 861–2
    Salicetum repentis, 862–4
sand-dune formation, 845–7
sand-dunes, 102–3, 845–64
Sandscale dunes, 863
sanicle, *see Sanicula europaea*
    woods, 362 (Fig. 67), 363 (Fig. 68), 369 n. 2, 371, 384
Sanicula europaea, 280, 282, 316, 327, 345, 369, 388, 397, 406, 438, 474
    society, 282
    woods, 363, 369, 371
saturation deficit, **33–5**
saw sedge, *see Cladium mariscus*
Saxon settlements, 174–5
    times, 177
Scandinavia, 25, 48, 154, 159, 199, 251, 256
Scandinavian highlands, 151
    ice sheet, 151
    mountains, 797
    relict species, 154
scar woods, 432
Schimper, A. F. W., on climatic and edaphic formations, 236
Schoenetum nigricantis, 676, **700–1**
Schoenichen, W., on yews in Germany, 256
Schoenus nigricans, 469, 700, 701, 715, 716, 718, 741
Scilla non-scripta society, 282, 308
Scilly Islands, 56 (Fig. 23), 57, 58, 59, 60, 73, 121
Scirpetum caespitosi, 676, **707–13**
    Argyllshire, 707
    Bodmin Moor, 710–11
    Cleveland moors, 711
    Exmoor, 710–11
    Inverliver Forest, 707–8
    Northern Highlands, 708–10
    Wicklow Mountains, 711
Scirpetum lacustris, 643, 646
Scirpus caespitosus, 681 (Fig. 136), 683 (Fig. 137), 685, 686, 689, 696, 700, 703, 704, 706, 707, 708, 709, 710, 711, 712 (Fig. 141), 714, 715, 716, 717 (Fig. 142), 718, 719, 724, 754, 757
    and *Calluna vulgaris*, 710
    peat, 707
    lacustris, 585, 595, 601 (Fig. 112), 602 (Fig. 113), 603, 604–5 (Figs. 114–15), 606, 617, 628, 630, 631, 643, 653, 663
    reedswamp, 643, 646
sclerophyll, 213
Scolt Head island, 820–1, 822 (Fig. 152), 823, 826, 835, 838, 870, 872 (Fig. 168)
"Scoto-Icelandic Rise", 3
Scottish beechwoods, 406–7, 416–21
    on fertile loam, 406–7
    on acid and podsolised soils, 416–21 (Fig. 83)
Scottish Highlands, 6, 8, 9 (Fig. 4), 10, 11, 12, 13, 14, 25
    arctic-alpine vegetation, 775, 780–5, 787–93, 794–808, 810–13

Scottish Highlands (*cont.*):
    birch forest, 199, **451–5**
    Callunetum, 751–2, 763
    climate, 63–6, 73
    deer, effects of, 143–4
    destruction of forest, 189
    former limit of forest, 173
    juniper, 263
    Molinietum, 522
    Nardus, spread of, 131
    native pine, 253
    oak forest, 267, **343–9**, 455
    pine forest, 254, **444–51**
    planting of conifers, 190, 192
    precipitation, 53–4
    Scirpetum caespitosi, 707–10
    sheep farming, 188
    shrubby willows, 264
    soils and vegetation, 118, 123, 124, 125
    Sub-Boreal pine, 166, 167 (Figs. 45–6)
    Vaccinietum, 758, 771
    waste land, 195
scree, arctic-alpine, 799–800
    stable, 779
    succession on, 776–9 (Fig. 151)
        bryophyte pioneers, 776
        bryophytes and vascular plants, 777
        Cryptogramma crispa, 776–7
        vertical strips of vegetation, 777–9 (Fig. 151)
scrub, 132–3, **472–84**
    climatic, 472
    climax, 472–3
    gorse, 313, 354, 504
    hawthorn, 295, 372–5, 376, 377, 380, 382, 383, 441, 480–4 (Fig. 88)
    hazel, 260, 474
    seral, 472
    subseral, 472, 475
    thicket, **475–6**, 480–4
    woodland, **475, 476**
sea breeze, 55, 57
sea buckthorn (*Hippophaë rhamnoides*), 266, 864
seasonal aspects, 231, 275 (Fig. 52), 279–80
    changes in the River Lark, 629 (Fig. 122)
sedimentary rocks, 80–1
Sedum acre, 135, 508, 556, 850, 887
    anglicum, 301, 437, 439, 502, 508, 809, 860, 861, 890
seed parents, 244, 383, 852
seeds and seedlings, destruction of, 140–1, 245, 291–2
Selaginella selaginoides, 708, 785, 809
    in dune grassland, 863
selective felling, 245
seminatural communities, 195, 198, 243
    grassland, 194, 195–6, 487, 559
    oakwood, 198, 267
    vegetation, 194
    woods, 195, 198, 243–4, 267
Senecio integrifolius (*campestris*), 533, 539
    jacobaea, 135, 139, 435, 546, 573, 849, 850, 859, 888
    vulgaris on shingle, 887
        var. *radiatus* on sand dune, 848

seral ashwood, 380, 382
  ash-oak wood, 391, 396
  communities, 227
sere, 218, 239
Sernander, R., on post-glacial climates, 159
Sesleria caerulea, 435, 473, 554, 556, 557
sessile oak, *see Quercus sessiliflora*
  oakwood, *see* Quercetum sessiliflorae
shade phase, 274, 275 (Fig. 52)
shading of water plants, 632 (Fig. 123)
sheep, 74, 99, 102, 113, 119, 129–32, 171, 180, 182, 184, 185, 262
  farming in the Highlands, 188
  grazing, 119, 124, 130–2, 172, 445, 536, 543
  pasture, 206, 208, 494, 495, 500–1, 830, 834, 840, 859
sheep's fescue, *see Festuca ovina*
Sherwood Forest, 16, 176 n., 182, 456
  heath, 729
  oakwood, 352–4
shingle, 868, 873–80
  and sand, 868, 879
  humus, 879
  lichens, 891–4
  mosses, 893–4
  mobility, 873
  vegetation, 880–94
  water content, 879–80
shingle banks, lateral ("hooks"), 870, 872 (Fig. 168), 873, 879, 888–90
  crest, 889
  "high elbow", 889
  vegetation, 888–90
shingle beaches, 103, 264, 868
  apposition, 869, 871 (Fig. 167)
  bar, 869, 870 (Fig. 166)
  fringing, 868
  spit, 868–9 (Fig. 165), 870–5 (Figs. 168–70)
  hooks, 869, 870, 871, 872 (Fig. 168), 873, 874–5 (Figs. 169–70), 876 (Fig. 171)
shingle spit, 868–9
  morphology and development, 870–3
  structure, 878–9
shingle spits, 821, 823, 869 (Fig. 165), 872 (Fig. 168), 874–5 (Figs. 169–70), 876 (Fig. 171)
shipbuilding, 270
shrub layer, 214, 276
  of ashwood, 429
  of oakwood, 279, 304
shrubs, the more important British, 259–66
  calcicolous, 264–5, 272, 373, 390, 429, 432
Sieglingia decumbens, 309, 326, 341, 345, 501, 502, 518, 522, 538, 551, 552, 554, 711, 728, 732, 748, 750, 757, 786
Silene maritima, 137, 139, 779, 810
  on sand dune, 850, 853, 856
  on shingle beach, 886, 889, 890
silica, 81, 364, 490, 609
siliceous grassland, 207, 490, 499
  rocks, 499
silt (in ponds, lakes and rivers), 579, 580, 581, 649
silt (soil), 80, 97, 125
silted river communities, 627–31
  waters, 581

silting, 581, 608–9, 622–4, 668
  and peat formation, 638–9
  changes in, 618
Silurian limestones, 126, 252, 319, 320
  rocks, 8, 9, 10, 12, 122–3
Simpson, G. C., on theory of glacial periods, 152
Simpson, J., on enemies of rabbits, 142
  on rabbit warrens, 134
Skene, Macgregor, on acidity of *Sphagnum*, 677–8
"slacks" (dune), 861–2, 863
smelting (iron), 172, 173, 182
Smith, Robert, on alpine plateau, 788
  on birch and pinewood soils, 455
  on "hill pasture", 489
  on flora of limestone exposure, 553
  on "upland heath" and "heather moor", 743
Smith, W. G., on arctic-alpine vegetation, 780–1, 784–5, 787–9, 794, 795, 799–800, 804, 806, 813
  on snow-patch communities, 797–8
  on basic grassland, 553–6
  on Nardetum, 514–16, 517 (Fig. 91)
  on sand-dune Callunetum, 852
  on "upland heath" and "heather moor", 743, 751–2
Smith, W. G. and Crampton, C. B., on grassland habitats, 492–3, 495
  on habit types of grassland, 493
Smith, W. G. and Moss, C. E., on "heath pasture", 489
  on limestone grassland, 553–6
  on Pennine Vaccinieta, 761
Smith, W. G. and Rankin, W. M., on limestone heath, 557
  on Pennine Vaccinieta, 761
  on scarwoods, 432
snow, 40–1
  lie, 40, 65, 797
snow-patch communities, 797–8
  Anthelietum, 797
  Polytrichetum sexangularis, 797
  Salicetum herbaceae, 797
Snowdonia, 7 (Fig. 3), 12, 53, 783
sociation, 232 n.
socies, 232
society, 230, 232
sodium, 81, 609
soil, 78–100
  bacteria, 83, 84
  fungi, 83, 833–4
  horizon, 78
  profile, 78
  stratification, 78
Solanum dulcamara, 266, 294, 390, 461, 464, 467, 469, 479, 548, 637, 645, 660
  var. *marinum*, 891
  nigrum, 135
"sole" of grassland, 488
Solms-Laubach, Graf zu, on box at Boxhill, 265
Solway Firth, 16, 830
Sonchus arvensis var. *angustifolius*, on shingle, 886, 887
  oleraceus, on shingle, 886, 887

Index

Sorbus aria, **258**, 368, 374, 379, 384, 412, 429, 548
   aucuparia, 156, 203, **258**, 276, 278, 300–1, 305, 314, 321, 330 (Fig. 60), 331, 334–6 (Figs. 61–3), 337, 339 (Fig. 64), 340, 342, 343, 344, 347, 352, 357, 374, 412, 419, 421, 433, 434, 437, 443, 446, 450, 451, 452, 453, 757
   torminalis, 258, 278, 305
South Downs, 15, 17 (Fig. 5), 21, 22 (Fig. 7), 112, 126, 130, 164, 171, 172, 177, 180–1, 206
   ash on, 252, 380
   beechwood, 249, 359, 361, 372, 374, 386–90, 393–406
   chalk grassland, 206, 526–52
   chalk scrub, 372, 374
   yew wood, 256, 376, 377–80
South Haven peninsula (Dorset), 861, 862
Southern Uplands (Scotland), 8, 9 (Fig. 4), 12, 14, 15, 17 (Fig. 5), 25, 26, 106, 118, 122, 123, 166, 174, 183, 517 (Fig. 91)
Sparganium erectum, 585, 586, 587 (Fig. 104), 588, 628, 629 (Fig. 122), 631, 637
   reedswamp, 631
   minimum, 599, 606, 611, 613, 614, 616, 620, 621
     consocies, 599, 601 (Fig. 112), 602 (Fig. 113), 606, 613–14
   simplex, 627, 628, 629 (Fig. 122), 630, 631
     community, 628
Spartina townsendii, 827, 828 (Fig. 157)
Spartinetum townsendii, 826–9
Sphagnetum, 647, **676–90**, 697–8, 705, 719, 740, 769
Sphagnum, 201, 413, 667, 670, 675, **676–7**, 679, 680, 686, 699, 708, 710, 757, 768, 769, 770, 810
   acidity, 677–8
   bog, 163, 471, 680, 697, 701
   cushions, 679, 682
   habitats, 677, 678
   hummocks, 679
   mineral requirements, 677
   peat, 168 (Fig. 47), 678, 679, 686, 690, 691, 704, 705
     oxidation-reduction potential, 678
   role in vegetation, 678–9
   structure, 677
   acutifolium, 646, 678, 697, 704
     var. *subnitens*, 657
   amblyphyllum, 678
   angustifolium, 678
   apiculatum, 678
   balticum, 678
   compactum, 678, 709
   contortum, 677, 678
   cuspidatum, 678, 681 (Fig. 136), 682, 683 (Fig. 137), 688 (Fig. 134), 689, 697, 698, 703, 712 (Fig. 141), 716
     peat, 691, 692
     var. *submersum*, 685, 709
   cymbifolium, 447, 656, 678, 688, 689, 698, 703, 709, 716, 738
     var. *congestum*, 711
   dusenii, 688, 689

Sphagnum fallax, 709
   fimbriatum, 657
   fuscum, 678, 689
   girgensohnii, 447, 698
   imbricatum, 678
   intermedium, 656
   inundatum, 678, 685, 712 (Fig. 141)
   magellanicum (*medium*), 678, 685, 689, 700, 708
   medium, see *magellanicum*
   molluscum, see *tenellum*
   obesum, 447
   papillosum, 447, 678, 681 (Fig. 136), 682, 683 (Fig. 137), 685, 688 (Fig. 139), 689, 695, 703, 708, 709, 713, 715
     peat, 691
     var. *confertum*, 739–40, 751
   platyphyllum, 678, 716
   plumosum, 678
   plumulosum (*subnitens*), 447, 469, 678, 685, 688 (Fig. 139), 689, 697, 700, 704, 709, 712 (Fig. 141), 713, 715, 716, 737 (Fig. 147), 739
   pulchrum, 681 (Fig. 136), 682, 683 (Fig. 137)
   quinquefarium, 698, 711
   recurvum, 471, 656, 685, 697, 703, 738, 741
   rubellum, 677, 678, 685, 686, 688 (Fig. 139), 689, 697, 698, 700, 708, 709, 712 (Fig. 141), 713, 715, 716
   rufescens, 738, 739
   spp., 450, 453, 621, 657, 704, 711, 712, 715, 718, 719, 741, 757, 769
   squarrosum, 447, 656, 657, 678, 768
   subnitens, see *plumulosum*
   subsecundum, 471, 657, 678, 700, 709, 713, 715, 716, 739, 768
   tenellum (*molluscum*), 678, 681 (Fig. 136), 683 (Fig. 137), 685, 688 (Fig. 139), 689, 697, 709, 712 (Fig. 141), 713, 716
   teres, 678
   warnstorfii, 678
Sphagnum-Calluna peat, 690, 691
Sphagnum-Narthecium peat, 690
spindle tree (*Euonymus europaeus, q.v.*), 265, 272, 373, 429
spray (sea), 701, 895, 896, 898
spray-washed cliffs, 896–8
   rocks, 896
spruce (*Picea abies*), 199, 254, 444
squirrels, 142
stable vegetation, 238–9
Stapledon, R. G., on composition of sheepwalk turf, 500
   on grades of pasture, 574–5
   on *Molinia* bog and *Molinia* meadow, 519
   on National Parks, 192
   on species of meadow and pasture, 567–8
storm shelf (of shingle bank), 878 (Fig. 174)
Strathoykell (Sutherlandshire), 445
stratification (of a community), 214
stratum (of a community), 214
structure (of a community), 213
Suaeda fruticosa, 140, 838, **880–5** (Figs. 175–8), 886, 887, 889, 890
   maritima, 825, 830, 836 (Fig. 160), 840
     var. *flexilis*, 830

Suaedetum fruticosae, 838
Subarctic period, 150 (Fig. 38), **154**
Sub-Atlantic climate, 194
  period, 150 (Fig. 38), 151, **166–9**, 173, 255
Sub-Boreal period, 150 (Fig. 38), 151, **163–6**, 167 (Fig. 46), 177, 358
  forest, 445
subclimax, 222
submaritime vegetation, 895–902
  brackish water and marshes, 895–6
  grassland, 208, 898–902
  spray-washed rocks and cliffs, 896–8
submerged leaves, 583
  plants, 582–3
  vegetation in the Cam, 630
subseral scrub, 475
subseres, 220–1, 238
subsoil, 78
succession, theory of, 217–18
sugar beet, 76
Summerhayes, W., on voles and Molinietum, 141
Summerhayes, W., Cole, L. W. and Williams, P. H., on Esher Callunetum, 728–9
  on subspontaneous pinewood, 457–8
Summerhayes, W. and Williams, P. H., on development of Callunetum, 734
  of Molinietum, 735, 736 (Fig. 146)
sunshine, 36, 37 (Fig. 12), 38 (Fig. 13)
Sussex Downs, *see* South Downs
swamp, 634
"swamp carr", 463, 660
Sweat Mere, alderwood at, 464–7
Sweden, 248, 254
  percentage of woodland, 196
sweetbriar (*Rosa rubiginosa*), 263, 373, 383
sweet vernal grass, *see Anthoxanthum*
sycamore (*Acer pseudo-platanus*, q.v.), 188, 257, 294, 388, 391
symbiosis, 227

talus and scree, 775–6
Tansley, A. G., on autogenic and allogenic factors, 230 n.
  on neutral grassland, 495–6
  on rabbit-eaten grassland, 135, 136, 139
Tansley, A. G. and Adamson, R. S., on chalk grassland, 487, 541–5 (Figs. 97–103), 546–8, 548–50
  on submaritime vegetation of cliffs, 901–2
tarns, mountain, 581, **618–21**
Taxus baccata, **256**, 278, 336, 337, 343, 357, 368, 374, 393, 429, *and see* yew
  continental range, 256
Tees, River, 624, 625
temperature, 30, 31 (Fig. 10), 33, 34, 35, 38, 39, 40, 43, 49, 51, 52
  range, 52, 53 (Fig. 22), 55
temperatures, mean minimal, 51 (Fig. 20)
  mean maximal, 52 (Fig. 21)
  summer maxima, 51, 52 (Fig. 21)
Tern, River, 624, 631
Tertiary basalts, 10
  deposits, 107
  granites, 10, 121

Tertiary lavas, 12
  times, 151
textile industry, 197
texture (of soil), 80, **96**
Thames, River, 624, 626, 627, 630, 631
thicket scrub, 261, 263, 277, 356, **475–6**
thunderstorms, 47
Thurne, River, 592, 593, 653
tidal drift, 844
timber, 181
  exportation of, 181
  dearth of, 181, 182, 183
  importation of, 181
Tilia cordata, 163, **258–9**, 437
  platyphyllos, **258–9**
timothy, *see Phleum pratense*
torrential river communities, 624–5, 632
Tortula ruraliformis, 851, 853–5
traveller's joy (*Clematis vitalba*, q.v.), 266
treatment of grassland, 560
tree layer, 214, 276
  planting, English, 182, 185, 188, 191
    Scottish, 183, 184–5, 189, 190
Tregaron Bog (Cardiganshire), 125, 158 (Fig. 42), 681 (Fig. 136), 682, 683 (Fig. 137)
Trent valley ponds, 584–5
Trias, 16, 116, 764
Triassic claylands, 173
Trientalis europaea, 346, 347, 417, 451, 452, 453, 454, 744, 752, 754
Trifolium dubium, 556, **565**, 567, 568, 572
  pratense, 530, 539, 547, 549, 550, 556, **565**, 567, 570, 572, 573, 575
  repens, 496, 497, 540, 547, 549, 550, 551, 556, 561, 564, **565**, 570, 572, 573, 575, 785, 786, 859
Triglochin maritimum, 830, 831, 833, 835, 837 (Fig. 161), 838
Trisetum flavescens, 533, 534, **537**, 549, 554, 564
turbid rivers, 624
  vegetation of, 631
Turner, J. S., on Killarney woods, 327, 329–31, 339
Tüxen, W., on Mesobrometum, 535 n. 3
Typha angustifolia, 585, 595, 653, 664, 766
  latifolia, 466, 585, 606, 617, 628, 631, 643, 644, 664, 766
Typhetum angustifoliae, 595, 653
  latifoliae, 465, 603, 604–5 (Figs. 114–15), 643

Ulex, 146, 261–2, *and see* gorse
  europaeus, 146, 261, **262**, 295, 311, 313, 342, 356, 390, 397, 476, 502, 504–6, 507, 551, 687, 697, 727, 728, 729, 740, 748
  gallii, 146, 261, **262**, 299, 319, 338, 339 (Fig. 64), 340, 342, 343, 502, 504–6, 522, 714, 740, 747, 748, 754, 757
  minor, 261, **262**, 415, 551, 726, 727, 728, 729, 730, 732, 733, 745, 748
Ulicetum, 504
  europaeae, 504–5 (Fig. 89)
  gallii, 504–5 (Fig. 89)
Ulmus glabra (*montana*), 162, 163, **258**, 326, 429, 435, 475, *see also* wych elm
  minor, 162 n., 480, 481

Ulmus nitens, 162 n.
  procera, 162 n.
undergrazing, 130
  effects of, 489
Upland and mountain habitats, 775–6
  ashwoods, 433–40
  heaths (incl. "heather moors"), 743–7
    Callunetta, 747–58
      Cleveland, 750–1
      Mourne Mountains, 758, 759 (Fig. 149)
      Pennine, 753–7
      Scottish, 751–2
      Somerset, 747–50
      Wicklow, 757
    Vaccinieta, 758–62
      Highland, 758
      Mourne Mountains, 760 (Fig. 150), 762
      Pennine, 761
      Wicklow, 761–2
  heathy oakwoods, 314–15, 321–4
Upper Greensand, 19, 113, 252, 441
"Upper Forestian" horizon, 167 (Fig. 45)
Urtica dioica, 135, 144, 255, 283, 293, 382, 434, 438, 461, 466, 468, 470, 482, 508, 645, 659, 660, 809
  society, 283
  pilulifera, 895
  urens, 135, 508

Vaccinietum myrtilli, 705, 719, 758, 759 (Fig. 149), 760 (Fig. 150), 762, 771
Vaccinio-Callunetum, 705
Vaccinio-Eriophoretum, 719
Vaccinium, Calluna and *Eriophorum*, 762
Vaccinium peat, 761
Vaccinium myrtillus, 203, 301, 315, 318, 320, 321, 322, 323, 324, 326, 327, 331, 335, 336, 337, 413, 417, 418, 419, 438, 446, 449, 450, 452, 454, 457, 557, 674, 703, 704, 706, 711, 726, 728, 733, 744, 747, 748, 750, 752, 754, 756, 757, **758**, 761, 762, 771, 779, 783, 785, 786, 794, 795, 796, 809
  uliginosum, 758, 771, 795
  oxycoccus, see *Oxycoccus quadripetalus*
  vitis-idaea, 203, 315, 323, 446, 449, 450, 454, 703, 704, 744, 748, 752, 754, 761, 762, 771, 794, 795, 796
Valentia Island, 57, 58, 59
valley bog, 674, **675**, 679, 680, 719
  fen wood, 467–8
  gravels, 125
"varves", 149
verges, grass road, 132, 491, 496–8, 561
vernal aspect, 279
Verrucaria maura, 891, 893
Vevers, H. G., on "bird cliffs", 144
Viburnum lantana, **265**, 272, 289, 368, 373, 382, 397, 429, 470, 481, 768
  opulus, 255, **265**, 279, 306, 314, 317, 325, 391, 392, 393, 461, 463, 469, 470, 474, 657, 658 (Fig. 132), 659 (Fig. 133), 660, 668, 669
voles, 140–1, 267, 271, 291–3, 402

Walker, N., on small ponds, 585–6, 587 (Fig. 104)

Walney Island sand dunes, 853, 860, 863, 867
Warburg, E. F., on northern birches, 251, 444 n.
"warplands", 103
Wash, the, 54, 103, 186, 592, 648 (Fig. 126), 649
"waste", 179 n.
waste grassland, 132
waste land, 195, 221
water, aeration of, 608
  conditions of life in, 579–80
  light penetrating, 598, 600, 607 (Fig. 116)
  silting in, 608
  temperature of, 608 (Fig. 117)
  wave action in, 598
waterlilies, 595, 614, 615, *and see Nuphar, Nymphaea*
water meadows, 566–72
  vapour, 30, 33, 35
waterlogged soil, 84–5
Watson, H. C., on "Highland species", 781
Watson, W., on arctic-alpine bryophytes and lichens, 790–3, 794–5, 800–4, 805–6, 807–8, 811–13
  on Blakeney Point lichens, 891
  on Chesil Bank lichens, 893–4
    mosses, 893–4
  on Exmoor Scirpetum caespitosi, 710–11
  on flora of Eriophoretum angustifolii, 704
    vaginati, 703
    of Sphagnetum, 698
  on mountain torrent bryophytes, 624–5
  on submerged mosses, 624
  on sand dune flora, 850
    lichens, 854
    mosses, 851, 853–4
  on snow-patch communities, 797–8
  on "upland heath" and "heather moor", 743–50, 754–6
  on vegetation of "slacks" 861–2
  on wet heath zonation, 735, 737–41 (Figs. 147–8)
Watt, A. S., on Agrostido-Festucetum, 509–11, 512–14
  on beechwood soils, 359, 363–6, 386–7, 408–10
  on Breckland climate, 70
  on Breckland grass heath, 509–11, 512–14
  on bryophytes in the ash-oak associes, 289
  on causes of even age, 401
  on colonisation of gaps in beechwood, 403–6
  on destruction of seedlings by mice and birds, 129
    of acorns by mice and voles, 140
  on deterioration of soil under beech, 424
  on Fagetum calcicolum, 366–71
    rubosum, 386–9
  on failure of oakwood to regenerate, 291–3
  on Killarney soils, 329
  on Scottish beechwoods, 406–7, 416–21
  on succession on chalk soil, 372–6, 381–4
    on loam overlying chalk, 390–401
  on "the balance of nature", 142
  on yew woods, 376–80
Watt, H. B., on Gaelic names indicating woodland, 183
  on Scottish 16th-century planting, 183
  on Scottish 18th-century planting, 185, 188

Waveney, River, 104, 651
wayfaring tree (*Viburnum lantana*, q.v.), 265, 383, 432
Weald, the, 9 (Fig. 4), 19, 20 (Fig. 6), 22 (Fig. 7), 113, 114 (Fig. 34), 115, 125, 126, 166, 169, 172, 173, 175
    Clay, 19, 113, 125
    forest, 113, 115, 172–3, 175, 178, 181
Wealden anticline, 19
    beds, 19, 22 (Fig. 7), 113, 114 (Fig. 34)
    dome, 19, 21
    sands, 19, 113, 115
weathering, 78, 80
    chemical, 81
    secondary, 80
weathering complex, 81, 85
Welsh Marches, 67, 171, 180 n., 318
    grassland, 499, 500, 501, 504, 516, 518
    oakwoods, 325–7
Wenlock Limestone, 319
wet heath, 675, 679, 680, **734–41**
    succession, 735–6 (Fig. 146)
    zonation, 735–41 (Figs. 147–8)
wet oakwood, field layer societies of, 283, 285
    species, 288
Wharfe, River, 626
wheatland, 75
Wheldon, J. A. and Wilson, A., on upland Sphagneta in north Lancashire, 697
White, J. M., on fenland at Lough Neagh, 663–8
whitebeam (*Sorbus aria*, q.v.), 258, 374, 419, 429
white clover, see *Trifolium repens*
white dunes, 850, 852
White Moss Loch, 588–92 (Figs. 105–6)
white willow (*Salix alba*, q.v.), 259
Wicken Fen, 104, 650, 651, **653–5**, **657–60**, **661–2**, 663, 669
    anthropogenic communities, 661–2
    carr, 657–9 (Figs. 132–3)
    Cladietum, 654
    Cladio-Molinietum, 661–2
    "litter", 661–2
    "mixed fen", 662
    "mixed sedge", 661
    Molinietum, 661
    "sedge", see Cladietum
    soil, 653
Wicklow Mountains, 6, 7 (Fig. 3), 10, 12, 125, 714
    Eriophoretum angustifolii, 704
    Scirpetum caespitosi, 711–13 (Fig. 141)
    Ulicetum, 505 (Fig. 89)
Wiehe, P. O., on Salicornietum, 824, 826, 827 (Fig. 156)
willows, 259
Wilmott, A. J., on relict species, 153
Wilson, M., on *Quercus sessiliflora* on London Clay, 247, 268, 351
wind, 41, 42 (Fig. 14), 43, 78
winds, moisture-laden, 47, 53
    prevailing, 44 (Fig. 15), 45 (Fig. 16), 47
    westerly, 45, 47–51, 55

Windsor Forest, 176 n., 177
"winter-green" grasses, 204
winter minima (of temperature), 51 (Fig. 20), 57
Wistman's Wood, 246, 268, **298–302**
Woodhead, T. W., on nunataks, 152
    on former forest in the southern Pennines, 500
woodland scrub, 133, 475–6
Woodruffe-Peacock, E. A., on assimilation of oak plantations to natural woods, 195, 268
Woodwalton Fen, 104
wool, 172, 180, 181
Wooldridge, S. W., on Belgic and Saxon colonisation of loams, 174–5
woollen industry, 172
"world groups" (soil), 79, 85, 90
Worm's Head, Festucetum rubrae on, 898–900
Worth, Hansford, on Wistman's Wood, 299, 300
Wright, W. B., on limits of glaciation, 152
wych elm (*Ulmus glabra* = *U. montana*, q.v.), 163, **258**, 305, 319, 429, 442

xerophytes, 219
xeroseral stages, 240
xerosere, 219–20

Yapp, R. H., on the Dovey salt marshes, 824 (Fig. 153), 829 (Fig. 158), 831, 832 (Fig. 159), 833
    on drainage channels and pans in salt marsh, 840–1
Yare, River, 104, 593, 594 (Fig. 108), 595, 651
    valley, 652 (Fig. 130), 653
yarrow, see *Achillea millefolium*
yellow dunes, 850, 852
yew (*Taxus baccata*, q.v.), **256**, 278, 337, 343, 357, 368, 374, 376, 377, 380, 388, 396, 429
yew-ashwood, 377
yew wood, 256, 338, **376–80** (Figs. 71–3)
    distribution, 376
    localisation, 381
    origin and development, 376–7
    structure, 377
"Yorkshire fog", see *Holcus lanatus*
Yorkshire chalk wolds, 166

Zannichellia palustris, 631, 861, 896
"Zechstein Sea", 16
zonation of aquatic vegetation, 582, 585–92 (Figs. 104–6)
    biotic, on common land, 132–3
    of vegetation on the border of the Upper Lake, Killarney, 334 (Fig. 64)
    round an isolated rabbit burrow, 138 (Fig. 37)
    on the Wicklow Mountains, 505 (Fig. 89)
Zostera, 819
    marina, 824
    nana, 824
Zosteretum, 819, 824

Date Due